# Lecture Notes in Computer Science 5550

# Lecture Notes in Computer Science

Commenced Publication in 1973

Founding and Former Series Editors:

Gerhard Goos, Juris Hartmanis, and Jan van Leeuwen

Editorial Board

Gerhard Weikum

Max-Planck Institute of Computer Science, Saarbruecken, Germany

Luigi Fratta Henning Schulzrinne
Yutaka Takahashi Otto Spaniol (Eds.)

# NETWORKING 2009

8th International IFIP-TC 6 Networking Conference
Aachen, Germany, May 11-15, 2009
Proceedings

Springer

Volume Editors

Luigi Fratta
Politecnico di Milano
Dipartimento di Elettronica e Informazione
via Leonardo da Vinci 32, 20133 Milano, Italy
E-mail: fratta@elet.polimi.it

Henning Schulzrinne
Columbia University
Department of Computer Science, M/S 0401
1214 Amsterdam Avenue, New York, NY 10027-7003, USA
E-mail: hgs@cs.columbia.edu

Yutaka Takahashi
Kyoto University
Graduate School of Informatics, Department of Systems Science
Kyoto 606-8501, Japan
E-mail: takahashi@i.kyoto-u.ac.jp

Otto Spaniol
RWTH Aachen, Lehrstuhl für Informatik 4
Informatikzentrum
Ahornstr. 55, 52074 Aachen, Germany
E-mail: spaniol@nets.rwth-aachen.de

Library of Congress Control Number: Applied for

CR Subject Classification (1998): C.2, C.4, H.4, D.2, J.2, J.1, K.6, K.4

LNCS Sublibrary: SL 5 – Computer Communication Networks
and Telecommunications

ISSN 0302-9743
ISBN-10 3-642-01398-8 Springer Berlin Heidelberg New York
ISBN-13 978-3-642-01398-0 Springer Berlin Heidelberg New York



Typesetting: Camera-ready by author, data conversion by Scientific Publishing Services, Chennai, India
Printed on acid-free paper SPIN: 12666836 06/3180 5 4 3 2 1 0

# Preface

**"What a difference a year makes – 52 little weeks"**

This variation of the first line from Dinah Washington's famous song, which originally reads, "What a difference a day makes - 24 little hours," brings it to the point:

According to all experts, the press, and most people's impression we are today in a serious economic recession. Less than one year ago, we practically lived on the "island of the blessed" (namely, at Networking 2008 that was held on the island of Singapore), or in the famous country where "milk and honey flow" (or "where wine and liquor flow"). This convenient situation has changed abruptly within less than 52 weeks. It looks like the same kind of problems has emerged in all areas – and the "Networking" area has, of course, been affected, too.

Looking into the 2009 proceedings, however, you will immediately notice that the manuscripts are largely unaffected by any aspect of the economic crisis (which should be a bit of a consolation). Apparently, research directions are dictated by a process that is all too sluggish in order to be quickly and radically changed by a "tsunami." Likewise, the conference itself was prepared in spite of such a crisis.

Some comments about the conference venue and some statistics appear to be in order:

A. Aachen is the westernmost major city in Germany, located on the borders with The Netherlands and Belgium. Other names of the town include Aix-la-Chapelle, Aken, Aquisgran, Aquisgrana, Cachy and many more – indicating its very international atmosphere. Despite its 'borderline' location, Aachen is situated in the very center of Western Europe. And more than 1200 years ago, it was the capital of Charlemagne's empire that comprised Germany, the Benelux, France, Switzerland and a large part of Italy.
B. Perhaps influenced by this international location, the conference received 232 submissions from 47 countries from all continents (except for Antarctica). Only 48 submissions survived the review process (i.e., just 20%) and were accepted as full papers. Another 28 more manuscripts (i.e., 12%) were selected as work-in-progress papers. This rather low acceptance rate should guarantee timely and high-quality papers. I would once again like to highlight the fact that six out of seven continents are represented in the conference program.

Some words of appreciation:

- To IFIP TC6 (Communication Systems) who entrusted us with the organisation of Networking 2009 despite three very strong competitors.
- To the local organisers in Aachen who did an excellent, efficient and time-consuming job despite sometimes having to suffer from the strange and unrealistic ideas that came spontaneously to the mind of the General Chairman of the conference.
- To the Chairpersons of the Technical Program Committee, namely (in alphabetical order): Luigi Fratta, Henning Schulzrinne, and Yutaka Takahashi, who did a fantastic and extremely efficient and constructive job.
- To the keynote speakers (once again in alphabetical order) Paul Francis, Mario Gerla, and Paul Kühn, for presenting up-to-date information concerning today's most important networking questions.
- To the invaluable work of the TPC members and the other reviewers who did a perfect serious and (in most cases) timely job under very heavy time pressure.
- To the IFIP TC6 delegates and other experts who were prepared to serve as Session Chairpersons.
- To all the (yet largely unknown) attendees who made the conference possible in difficult financial times.
- To all those who may have been forgotten in this list but who also deserve a "thank you."

May 2009 Otto Spaniol

# Organisation

## Executive Committee

| | |
|---|---|
| **General Chair** | Otto Spaniol,<br>RWTH Aachen University, Germany |
| **Publicity Co-chair** | Arnd Hannemann, Martin Krebs,<br>Alexander Zimmermann,<br>RWTH Aachen University, Germany |
| **Publication Chair** | Ulrich Meis, Benjamin Schleinzer,<br>RWTH Aachen University, Germany |
| **Finance Chair** | Jan Kritzner,<br>RWTH Aachen University, Germany |

## Local Organisation

| | |
|---|---|
| Juan Miguel Espinosa Carlín | RWTH Aachen University, Germany |
| Markus Fidler | Leibniz-Universität Hannover, Germany |
| Arnd Hannemann | RWTH Aachen University, Germany |
| Mesut Güneş | Freie Universität Berlin, Germany |
| Kai Jakobs | RWTH Aachen University, Germany |
| Martin Krebs | RWTH Aachen University, Germany |
| Jan Kritzner | RWTH Aachen University, Germany |
| Ulrich Meis | RWTH Aachen University, Germany |
| Andriy Panchenko | RWTH Aachen University, Germany |
| Jürgen Rapp | RWTH Aachen University, Germany |
| Benjamin Schleinzer | RWTH Aachen University, Germany |
| Jens Schmitt | University of Kaiserslautern, Germany |
| Dirk Thißen | RWTH Aachen University, Germany |
| Alexander Zimmermann | RWTH Aachen University, Germany |

## Steering Committee

| | |
|---|---|
| Georg Carle | TU Munich, Germany |
| Pedro Cuenca | University Castilla-la-Mancha, Spain |
| Guy Leduc | Université de Liège, Belgium |
| Ioannis Stavrakakis | University of Athens, Greece |

## Workshop Organisers

| | |
|---|---|
| Hakima Chaouchi | TELECOM SudParis, France |
| Udo Krieger | Otto-Friedrich University Bamberg, Germany |
| Maryline Maknavicius | TELECOM SudParis, France |
| Peter Reichl | Telecommunications Research Center Vienna (ftw.), Austria |
| Burkhard Stiller | University Zürich, Switzerland |
| Bruno Tuffin | INRIA Rennes, France |

## Technical Programme Committee Co-chairs

| | |
|---|---|
| Luigi Fratta | Politecnico di Milano, Italy |
| Henning Schulzrinne | Columbia University, USA |
| Yutaka Takahashi | Kyoto University, Japan |

## Technical Programme Committee Members

| | |
|---|---|
| Finn Arve Aagesen | NTNU Trondheim, Norway |
| Khaldoun Al Agha | Université de Paris-Sud 11, France |
| Fan Bai | General Motors, USA |
| Ernst Biersack | Eurecom, France |
| Chris Blondia | University Antwerpen, Belgium |
| Olivier Bonaventure | Université catholique de Louvain, Belgium |
| Raouf Boutaba | University of Waterloo, Canada |
| Torsten Braun | University of Bern, Switzerland |
| Antonio Capone | Politecnico di Milano, Italy |
| Georg Carle | TU Munich, Germany |
| Augusto Casaca | INESC, Portugal |
| Matteo Cesana | Politecnico di Milano, Italy |
| Piotr Cholda | Krakow University, Poland |
| Marco Conti | IIT-CNR, Italy |
| Mark Crovella | University of Pittsburgh, USA |
| Pedro Cuenca | University of Castilla-la-Mancha, Spain |
| Amitabha Das | Nanyang Technological University, Singapore |
| Wolfgang Effelsberg | University of Mannheim, Germany |
| Lars Eggert | Nokia Research, Finland |
| Laura Feeney | SICS, Sweden |
| Andreas Fasbender | Ericsson Research, Germany |
| Anja Feldmann | TU Berlin, Germany |
| Markus Fiedler | Blekinge TH, Sweden |
| Eric Fleury | ENS Lyon, France |
| Andrei Gurtov | HIIT Helsinki, Finland |
| Günter Haring | University of Vienna, Austria |

| | |
|---|---|
| Boudewijn Haverkort | University of Twente, The Netherlands |
| Oliver Heckmann | Google Labs Zürich, Switzerland |
| Markus Hofmann | Bell Labs, USA |
| David Hutchison | University of Lancaster, UK |
| Anthony Joseph | UC Berkeley, USA |
| Holger Karl | University of Paderborn, Germany |
| Martin Karsten | University of Waterloo, Canada |
| Andreas Kassler | University of Karlstad, Sweden |
| Ulf Körner | Lund University, Sweden |
| Gabriele Kotsis | J. Kepler University, Austria |
| Udo Krieger | University of Bamberg, Germany |
| Pieter Kritzinger | University of Cape Town, South Africa |
| Guy Leduc | Université de Liège, Belgium |
| Chang-Gun Lee | Rice University, USA |
| Francis Lee | Nanyang Technological University, Singapore |
| Jörg Liebeherr | University of Toronto, Canada |
| Christoph Lindemann | University of Leipzig, Germany |
| Claudia Linnhoff-Popien | LMU Munich, Germany |
| Gerald Maguire | KTH Stockholm, Sweden |
| Petri Mähönen | RWTH Aachen University, Germany |
| Olli Martikainen | ETLA, Finland |
| Peter Martini | University of Bonn, Germany |
| Rudolf Mathar | RWTH Aachen University, Germany |
| Friedemann Mattern | ETH Zürich, Switzerland |
| Martin Mauve | University of Düsseldorf, Germany |
| Martin May | ETH Zürich, Switzerland |
| Tommaso Melodia | SUNY Buffalo, USA |
| Paulo Mendes | INESC, Portugal |
| Michael Menth | University of Würzburg, Germany |
| Wojciech Molisz | Gdansk University of Technology, Poland |
| Edmundo Monteiro | University of Coimbra, Portugal |
| Josep Pareta | Polytechnic Catalunya, Spain |
| Craig Partridge | BBN, USA |
| Ioannis Psaras | University of Surrey, UK |
| Thomas Plagemann | University of Oslo, Norway |
| Dario Pompili | Rutgers University, USA |
| Ana Pont-Sanjuán | University of Valencia, Spain |
| Ramon Puigjaner | University of Illes Balears, Spain |
| Guy Pujolle | University of Paris 6, France |
| Serugudi V. Raghavan | IIT Madras, India |
| Peter Reichl | Telecommunications Research Center Vienna, Austria |
| James Roberts | Orange, France |
| Harsha Sirisena | Canterbury University, New Zealand |

| | |
|---|---|
| Robin Sommer | ICSI/LBNL, USA |
| Ioannis Stavrakakis | University of Athens, Greece |
| Ralf Steinmetz | TU Darmstadt, Germany |
| Burkhard Stiller | University of Zürich, Switzerland |
| Phuoc Tran-Gia | University of Würzburg, Germany |
| Piet Van Mieghem | TU Delft, The Netherlands |
| Ning Wang | University of Surrey, Guildford, UK |
| Klaus Wehrle | RWTH Aachen University, Germany |
| Lars Wolf | TU Braunschweig, Germany |
| Adam Wolisz | TU Berlin, Germany |
| Krzysztof Walkowiak | Wroclaw University of Technology, Poland |
| Martina Zitterbart | University of Karlsruhe, Germany |

## Additional Reviewers

Issam Aib
Emilio Ancillotti
Steinar Hidle Andresen
Markus Anwander
Pere Barlet-Ros
Lee Begg
Bastian Blywis
Eleonora Borgia
Fatma Bouabdallah
Nizar Bouabdallah
Raffaele Bruno
Albert Cabellos-Aparicio
Davide Careglio
Marcel Castro
Baozhi Chen
Chao Chen
Florin Ciucu
Lucia Cloth
Peter De Cleyn
Franca Delmastro
Nicolas Diez
Lei Ding
Christian Doerr
Jeroen Doumen
Michael Duelli
Norman Dziengel
Juan M. Espinosa C.
Jabed Faruque
Markus Fidler
Stefan Götz
Jorge Granjal
Mesut Güneş
Carlos Guerrero
Aun Haider
Arnd Hannemann
Tobias Harks
Pieter Hartel
Matthias Hartmann
Masum Hasan
Mahbub Hassan
Bernhard Heep
Gavin Holland
Llorenç Huguet
Philipp Hurni
Brent Ishibashi
Loránd Jakab
ShanShan Jiang
Felix Juraschek
Nour Kadi
Emmanouil Kafetzakis
Georgios Karagiannis
Mehdi Khouja
Tom Kleiberg
Alexander Klein
Olivier Klopfenstein
Martin Krebs
Jan Kritzner
Fernando Kuipers
Indraneel Kulkarni
Pardeep Kumar
Georg Kunz
Li-Chung Kuo
Olaf Landsiedel
Rami Langar
Eun Kyung Lee
Franck Legendre
Frank Lehrieder
Paulo Loureiro
Sadeq Makram
Lefteris Mamatas
Steven Martin
Leonardo Martucci
Fabien Mathieu
Mickael Meulle
Pietro Michiardi
Luca Muscariello
Qasim Mushtaq
Sofia Nikitaki
Dusit Niyato
Simon Oechsner
Konstantinos Oikonomou
Beatriz Otero
Sara Oueslati
Sergio Palazzo
Antonis Panagakis
Andriy Panchenko
P. Papadimitriou
Giovanni Pau
George Pavlou

Juan Perez
Fabio Picconi
Paolo Pileggi
Tatiana Polishchuk
Aiko Pras
Rastin Pries
Scott Pudlewski
Joseph Rahmé
Francisco Rente
Kay Roemer
Sylwia Romaszko
Ramin Sadre
David Salyers
Fahad Samad
Germán Santos-Boada
Raimondas Sasnauskas
Daniel Schlosser
Michele Sciuto
Tim Seipold
Bartomeu Serra
Kurt Smolderen
Rute Sofia
Kathleen Spaey
Barbara Staehle
Dirk Staehle
Thomas Staub
Moritz Steiner
Kirsten Terfloth
Despina Triantafyllidou
Hsin-Mu Tsai
Leonidas Tzevelekas
Athanasios Vaios
Erwin Van de Velde
Wil van der Aalst
Samu Varjonen
John Paul Varkey
Constantinos Vassilakis
Michael Voorhaen
Markus Waelchli
Gerald Wagenknecht
Sonia Waharte
Elias Weingärtner
Maarten Weyn
Joachim Wilke
N. Wisitpongphan
Christos Xenakis
Jin Xiao
Ou Yang
Bo Yu
Fan Yu
Alexander Zimmermann
Thomas Zinner

## Sponsoring Institutions (in alphabetical order)

CISCO Germany
Ericsson Eurolab, Herzogenrath
Euro-NF
Gesellschaft für Informatik (GI)
NEC Europe, Heidelberg
REGINA e.V.
RWTH Aachen University
Software AG, Darmstadt
T-Mobile, Bonn
UMIC Excellence Cluster

# Table of Contents

## Quality of Service: New Protocols



## Wireless Networks: Planning and Performance

## Applications and Services: System Evaluation



## Peer to Peer: Topology



## Next Generation Internet: Transport Protocols

## Wireless Networks: Protocols



## Next Generation Internet: Network and Transport



## Modelling and Performance Analysis: Infrastructure

## Applications and Services: Streaming and Multimedia



## Wireless Networks: Availability



## Modelling and Performance Evaluation: Network Architectures

## Peer to Peer: Frameworks and Architectures



## All-IP Networking: Frameworks

## Next Generation Internet



## Performance and Wireless

# Calibrating Wireless Sensor Network Simulation Models with Real-World Experiments

Philipp Hurni and Torsten Braun

Institute of Computer Science and Applied Mathematics
University of Bern
{hurni,braun}@iam.unibe.ch

**Abstract.** This paper studies the energy-efficiency and service characteristics of a recently developed energy-efficient MAC protocol for wireless sensor networks in simulation and on a real sensor hardware testbed. This opportunity is seized to illustrate how simulation models can be verified by cross-comparing simulation results with real-world experiment results. The paper demonstrates that by careful calibration of simulation model parameters, the inevitable gap between simulation models and real-world conditions can be reduced. It concludes with guidelines for a methodology for model calibration and validation of sensor network simulation models.

**Keywords:** Wireless Sensor Networks, Energy Efficient Medium Access Control, Model Validation, Model Calibration, Networking 2009.

## 1 Introduction

In the past years, many energy efficient medium access and routing protocols for wireless sensor and actor networks have been proposed. Although few protocols have been implemented and evaluated on real sensor hardware testbeds, countless papers rely on network simulation results. Simulation tools are a valuable, manageable and yet cheap test environment for evaluating wireless sensor network mechanisms. Investigations on scalability issues, e.g studies on the behavior of network characteristics in large-scale ad hoc or wireless sensor networks, would simply be unfeasible without simulation tools. Simulations provide essential insights when developing and evaluating protocol mechanisms. Regrettably, evaluations on real-world systems are often neglected. Simulating performance improvements in protocols for wireless ad-hoc or sensor networks undoubtedly is easier and more convenient than realizing and proving them on real-world prototypes.

Simulation studies inevitably take assumptions and apply simplified models, e.g. 2-dimensional topologies and perfectly circular transmission ranges. The properties of the wireless channel, such as signal dispersion, environmental noise, multipath propagation, scattering and fading effects are often not incorporated at all. Model verification using real-world implementations is often omitted. The credibility of simulation studies has therefore frequently been questioned. Several

L. Fratta et al. (Eds.): NETWORKING 2009, LNCS 5550, pp. 1–13, 2009.

recent studies underlined the lack of rigor in the application of simulation tools. Inadequate simulation models and improper data analysis have been shown to produce inconsistent or misleading results. *Kurkowski et al.* [1] lately surveyed papers published in the MobiHoc conference [2] between 2000 and 2005 and found severe flaws and inconsistencies in the simulation methodology. The survey concludes that the large majority of the research papers are not independently repeatable because of lack of documentation, omitted simulation input parameters, lacked statistical validity, inappropriate radio models and unrealistic traffic and/or mobility models. They emphasize that in any case of communication protocol study, researchers must validate the simulation model as a baseline to start any experimentation. *Andel et al.* similarly criticise the lack of realism of simulation studies in [3]. They emphasize that "properly validating simulation models against the intended or real-world implementation and environment can mitigate many of the problems of simulation", such as incorrect and unrealistic parameter settings and improper level of detail.

In this paper we analyze the energy-efficiency and service characteristics of a recently developed energy-efficient MAC protocol. By conducting the same experiments in simulation and on real sensor hardware and reasoning over anomalies in the results, we outline a methodology to calibrate and validate wireless sensor network simulation models. We show that model validation and calibration is feasible with reasonable effort. By cross-comparing and simulation and real-world implementation of the energy-efficient MAC protocol, we show that careful investigations on parameter settings play a major role.

The paper first portrays the mechanism of the power saving protocol WiseMAC [4] in section 2.1. Section 2.2 describe the simulation model of WiseMAC, and 2.3 the WiseMAC prototype on real-world sensor hardware. Section 3 evaluates the protocol in a simple test scenario, and illustrates how we carried out our proposed model validation and parameter calibration process on the WiseMAC simulation and real-world implementation. Section 4 concludes the paper.

## 2 WiseMAC

### 2.1 WiseMAC Protocol Operation

WiseMAC [4] is an unscheduled, contention-based sensor MAC protocol. It is very energy-efficient in scenarios with low or variable traffic. WiseMAC's wake-up scheme consists of periodic duty cycles of only a few percents in order to sense the carrier for a preamble signal, as depicted in Figure 1. All nodes in the network sample the medium with a common basic cycle interval T, but their wake-up patterns are independent and left unsynchronized. A preamble of variable length is prepended to each frame in order to alert the receiving node. When the receiver's wake-up pattern is yet unknown, the duration of the preamble equals the full basic cycle interval duration T, as illustrated in Figure 1 in the first transmission. The own schedule offset is then piggybacked to the frame and transmitted to the receiver. After successful frame reception, the receiver

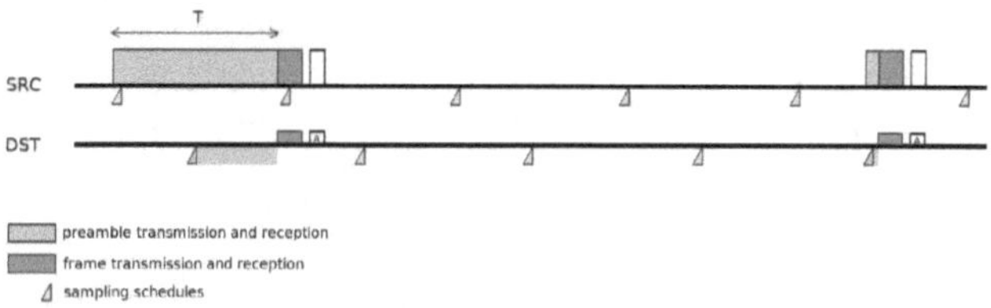


**Fig. 1.** WiseMAC

node piggybacks its own schedule to the respective frame acknowledgment. Received schedule offsets of all neighbor nodes are subsequently kept in a table and are periodically updated. Based on this table, a node can determine the wake-up patterns of all its neighbors, which in turn allows minimizing the preamble length for the upcoming transmissions. As the sender node is aware of the receiver's wake-up pattern, it only prepends a preamble that compensates for the maximum clock drift that the two involved node's clocks may have developed during the time since the last schedule exchange, as illustrated in Figure 1 in the second transmission.

### 2.2 WiseMAC in the OMNeT++ Simulator

We implemented the WiseMAC protocol [4] in the OMNeT++ Network Simulator [5] using the Mobility Framework [6], which supports simulations of wireless ad hoc and mobile networks on top of OMNeT++. It calculates SNR (Signal-to-Noise) ratios according to a free space propagation model.

The energy consumption model calculates and sums up the amount of energy that is used by the transceiver unit. [7] models the energy consumption of a IEEE 802.11 wireless device with a transceiver state model with the three states sleep, idle, receive and transmit. Experimental results in [7] confirm the adequateness of the state model with three different energy consumption levels. As many low-power and low-bandwidth transceivers used in sensor networks are of very low complexity, the energy consumption in idle and receive mode is often almost equal and do not need to be treated differently. Pursuing the same methodology as [7], we modelled the energy consumption of the sensor nodes with a state transition model with respect to the time spent in three operation

**Table 1.** OMNeT++ parameters

| | |
|---|---|
| path loss coefficient $\alpha$ | 3.5 |
| carrier frequency | 868 $MHz$ |
| transmitter power | 0.1 $mW$ |
| SNR threshold | 4 $dB$ |
| sensitivity | -101.2 $dBm$ |
| carrier sense sensitivity | -112 $dBm$ |
| communication range | 50 $m$ |
| carrier sensing range | 100 $m$ |

**Table 2.** Transceiver parameters

| | |
|---|---|
| supply voltage | 3 $V$ |
| transmit current | 12 $mA$ |
| recv current | 4.5 $mA$ |
| sleep current | 5 $\mu A$ |
| recv to transmit | 12 $\mu s$ |
| transmit to recv | 12 $\mu s$ |
| recv to transmit | 518 $\mu s$ |
| recv to sleep | 10 $\mu s$ |

**Table 3.** WiseMAC parameters

| | |
|---|---|
| basic interval duration T | 500 *ms* |
| duty cycle | 1% |
| bit rate | 19'200 *bps* |
| minimum preamble | 1 *ms* |
| medium reservation preamble | u(0,2) *ms* |
| MAC header | 104 bit |
| payload | 96 bit |

modes sleep, receive and transmit, weighted with the respective energetic costs. In a first step we applied the transceiver parameters of the TR1001 low-power radio transceiver module [8] (transmission rate, state transition delays, power consumption in sleep/recv/transmit states). The TR1001 is the radio chip of quite a few sensor nodes, including our own sensor hardware testbed. The parameters of the simulation environment and the energy consumption model are listed in Tables 1 and 2. WiseMAC-specific parameters are listed in Table 3.

## 2.3 WiseMAC on Embedded Sensor Boards

The simulation environment described in section 2.2 is an attempt to model a wireless sensor network, with respect to effects of signal dispersion, environmental noise, bandwidth limitation, energy constraints and clock drifts. Yet many other aspects that may play a role for wireless sensor networks are still left aside. Only measurements on real hardware make sure that all influences and side effects are taken into account. We therefore ported the original WiseMAC mechanism to the Embedded Sensor Boards (ESB) [9], a sensor node platform along with its own operating system, ScatterWeb OS [10]. WiseMAC requires a very accurate timing in order to keep track of and reach nodes in their particular wake-ups. In the presence of several sources of imprecisenesses (e.g. clocks drifts, software-based timers), this proved to be a challenging task. Yet the main features of WiseMAC outlined in [4] could be realized. We chose the same essential parameters of the WiseMAC simulation model as listed in Table 3.

**Frame Transmission.** When a packet has to be sent, the network handler determines whether the frame receiver is already known and its schedule offset is already stored in the WiseMAC neighbor table. If the medium is free, the node calculates the necessary preamble duration, contends for medium access, switches to the transmit state and transmits preamble and frame subsequently. If the medium is not free, the node turns to the sleep state again and schedules the next transmission attempt for the next wake-up of the receiver. In case the receiver is unknown yet, the preamble duration is set to the basic interval duration $T$ and the transmission attempt is initiated immediately.

**Preamble Sampling and Frame Reception.** As the transceiver switches need a certain turnaround time, and carrier detection is bound to the recognition of a sequence of predefined start bytes, nodes need a certain minimum duty cycle to actually recognize whether a preamble is being sent. The duration of the wake duty cycle calculates as $\triangle t = T \cdot dutycycle = 5ms$. In fact, $5ms$ is only the time between the instants when the transceiver switches are initiated. The transition delay for changing from sleep state to the wake state has to be subtracted from the duty cycle. The net duty cycle therefore is in the range of only $3 - 4ms$ in each interval $T$. If a node does not recognize the start byte sequence within its duty cycle $\triangle t$, it immediately returns to the sleep state until the next wake interval. If it recognizes the start byte sequence, it stays in the receive state until preamble and frame are correctly received. After reception, the node checks the type of the frame. In case of a broadcast frame, the node immediately returns to the sleep state. In case of a unicast frame, it returns a 10-byte acknowledgement and goes back to the sleep state.

# 3 Comparing Simulation and Real-World Experiments

## 3.1 Measurement Methodology on the ESB

We investigated the energy consumption of ESB nodes via measurements on the node lifetime. This methodology is widely accepted and has been used in [11], [12]. It consists in charging so-called GoldCap capacitors and measuring the time a node can live on this given charge. When two nodes are charged with the same initial amount of energy, the node with a lower overall energy consumption can live longer. This allows evaluating the energy consumption of sensor MAC protocols in small-scale test scenarios.

Applying the GoldCap methodology, we obtained robust results with low variance. The following results form an entry point to the lifetime measurements on the ESB. In a first step, we compared the node's energy consumption in the three different states of the transceiver (sleep, receive, transmit). Figure 2 depicts the lifetimes of nodes in the particular states. The first bar illustrates the lifetime of an ESB node with a permanently turned-off transceiver. The second bar illustrates the lifetime of an ESB node running ScatterWeb CSMA, which keeps the transceiver permanently in the receive state. The third bar corresponds

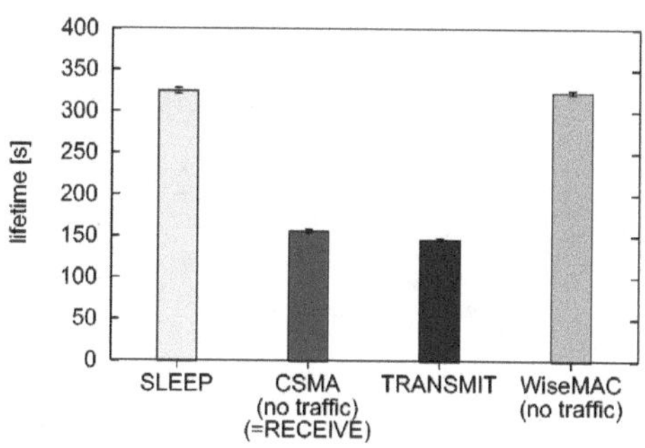


**Fig. 2.** Lifetimes of ESB nodes in different states

to a node in the most costly transmit state. As ESB nodes apply on-off keyed (OOK) modulation, the signal is simply turned on and off for bits '1' and '0', respectively. We therefore measured the transmit state (third bar) when sending a strictly alternating sequence of '1' and '0'. In regard of Figure 2, we can conclude that the energy consumption of the entire ESB node is highest in the transmit state. Receiving is almost equally expensive, and approximately twice as costly as the sleep state.

The fourth bar in Figure 2 illustrates the lifetime of WiseMAC nodes in the absence of traffic. It becomes obvious that the WiseMAC implementation on the ESB with only 1% duty cycle leads to a very low idle energy consumption. Its lifetime is almost equal to the lifetime of a node with the permanently turned-off transceiver. We measured a mean lifetime reduction of 0.47% with a standard deviation of 1.19% in respect to the average lifetime in the sleep state.

Comparing the lifetime of the WiseMAC node to the lifetime of simple ScatterWeb CSMA, the lifetime could be increased by approximately 120%. To the best of our knowledge, this WiseMAC prototype is the most energy-efficient implementation of a sensor MAC protocol implemented on the ESB nodes research platform.

### 3.2 Evaluation Scenario

We evaluated the energy consumption and basic properties of the ESB WiseMAC prototype with increasing traffic load. We have chosen a linear chain topology with six nodes forwarding traffic from one source to one sink, as depicted in Figure 3. The nodes are all in the transmission range of each other. They build up a full mesh topology, but only the bold links are used. In the measurements on the simulator and the real-world implementation on the ESB nodes, we measured the one-way delay and the time until intermediate node 5 depleted.

An external node synchronizes the nodes by emitting a SYNC packet, upon reception all the nodes set back their internal clocks. Node 1 starts generating traffic and addresses all frames to node 2. The application layer in node 1 generates a packet and logs the exact time. It then passes the packet to the MAC layer, which buffers it and sends it at the next appropriate instant. Node 2 receives the frame and subsequently forwards all frames to node 3, until the packets reach node 6. When node 6 receives the frame, it passes it to the application layer, where it is decapsulated and the transmission time of node 1 is extracted. Like this, the one-way delay is calculated as the time between the instant when the application layer in node 1 passes the packet to the MAC and the instant when the application layer

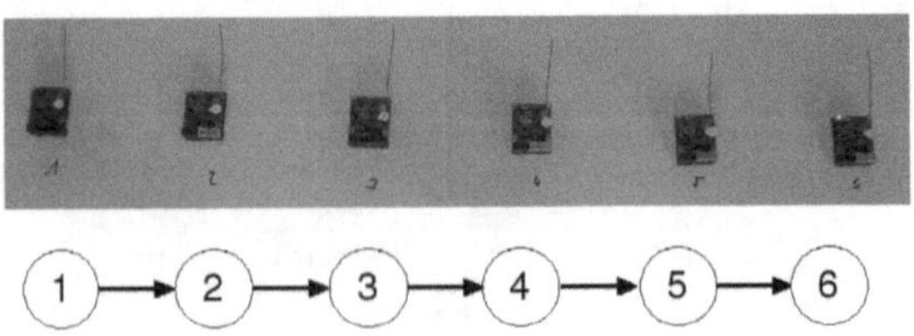


**Fig. 3.** 6 nodes chain topology

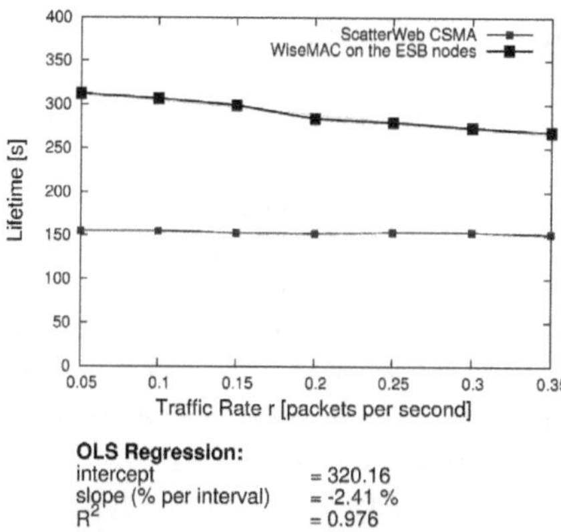


**Fig. 4.** WiseMAC & CSMA on ESB

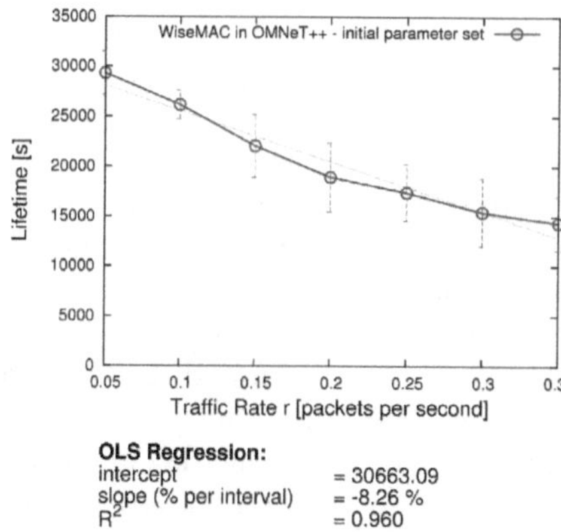


**Fig. 5.** OMNeT - initial parameters

in node 6 decapsulates the frame. The resulting inaccuracies of the one-way delay due to synchronization and packet processing are, given a 16 MHz processing unit on the ESB, in the range of some $\mu s$ and can be considered negligible.

### 3.3 Node Lifetime in Simulation and Real-World Experiment

Many wireless sensor network simulation studies investigate on node lifetimes in the context of energy-balanced routing protocols, in most cases by assuming a linear battery model without self-discharge (e.g. [13]). We accordingly emulated the GoldCap lifetime methodology in the simulation model in OMNeT++, and measured the time until the intermediate node 5 depleted of an initial energy endowment of 20 Joules.

Figure 4 depicts the lifetime of node 5 applying WiseMAC on the ESB measured with the GoldCap capacitors. Figure 5 depict the lifetime of the same node in the same simulation experiment in OMNeT++. The y-axis corresponds to the traffic rate $r$ being generated by the source node 1. In Figure 4 the lifetime is measured as the time the intermediate node 5 can live of its initial GoldCap charge ($t_{charge} = 120s$). We focus on the slope of the lifetime curves in Figures 4 and 5. With traffic increasing linearly, a linear decrease of the lifetime could be observed in both the real-world and the simulation experiment. We applied OLS regression analysis to measure intercept, slope (measured as the relative change per measurement interval in respect to the intercept) and the correlation coefficient $R^2$ to assess the *goodness of fit* of the linear model. Notice that the absolute values of the lifetimes are of no particular importance. As we could not assess the absolute value of the charge of the GoldCap capacitor, the absolute values of the lifetime curves can not be put into relation.

### 3.4 Impact of Inappropriate Simulation Parameters

Figure 5 displays the lifetime curve when applying the initially chosen parameters (c.f. Table 2) to the WiseMAC simulation in OMNeT++. As one can clearly see by comparing Figure 5 to Figure 4, the impact of the traffic applied to the chain is much stronger in the simulation model than in the real-world implementation of WiseMAC. The WiseMAC ESB lifetime curve decreases with −2.41% of the intercept per measurement interval ($\Delta r = 0.05$). The high correlation coefficient $R^2$ approves the appropriateness of the linear model. Applying the same linear OLS regression model, we measured a slope of −8.26% with the curve of the simulation model and a similarly high correlation of the linear model.

We investigated on the reason for the much higher negative slope in the simulation experiment and found that some of the initial assumptions of the simulation model had been too simplistic: as we had relied the choice of the parameters only on the technical specification of the transceiver module (see Table 2), as done in many other simulation studies of wireless sensor MAC protocols, we had omitted the energy consumption of the CPU and board circuitry. With the initial parameter set listed on Table 2 in Section 2.2, receive and transmit states are approximately 1000 times more costly than the sleep state. These parameters only account for the energy spent by the transceiver unit, and neglect the energy consumed by CPU and board circuitry.

The sad truth is that the major portion of nowadays wireless simulation studies base on such inappropriate simulation parameters and assumptions. Relying on the scarce information of the wireless transceiver manufacturer's datasheets and feeding those parameters into simulators has become the de-facto standard procedure of most wireless sensor MAC protocol studies. When comparing the slopes of lifetime curves of the real world experiment in Figure 4 with Figure 5, one must admit that the simulation model basing solely on the datasheet parameters of the transceiver chip does not yet deliver a reasonable energy model for sensor network simulations.

### 3.5 Simulation Model Calibration

As we measured the ratio between sleep and receive and transmit on the ESB nodes with the GoldCap capacitor methodology in Section 2 to be merely in the range of 1:2.25:2.5, we went on to calibrate the energy parameters of the simulation model accordingly. We find that the cross-comparisons between real-world

**Table 4.** Calibrated OMNeT++ simulation parameter set

| | | | |
|---|---|---|---|
| supply voltage | 3 $V$ | recv to transmit | 4 $ms$ |
| transmit current | 5.0 $mA$ | transmit to recv | 2 $ms$ |
| recv current | 4.5 $mA$ | sleep to recv | 1 $ms$ |
| sleep current | 2.0 $mA$ | recv to sleep | 1 $ms$ |
| | | | |
| WiseMAC parameters: | | | |
| basic interval duration T | 500 $ms$ | | |
| duty cycle | 1% | | |
| minimum preamble | 5 ms | | |
| medium reservation preamble | u(0,6) ms | | |

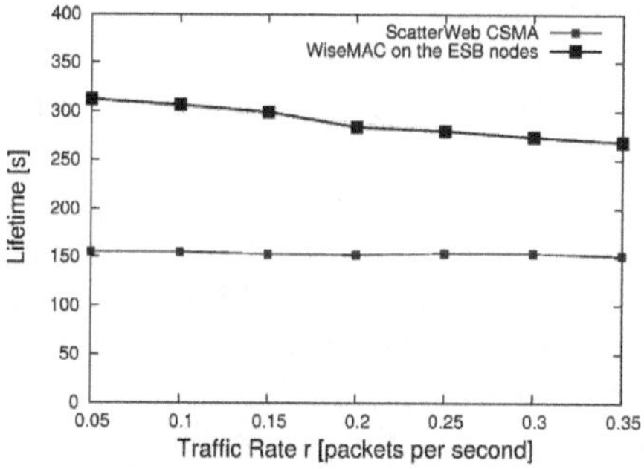


**Fig. 6.** WiseMAC & CSMA on ESB

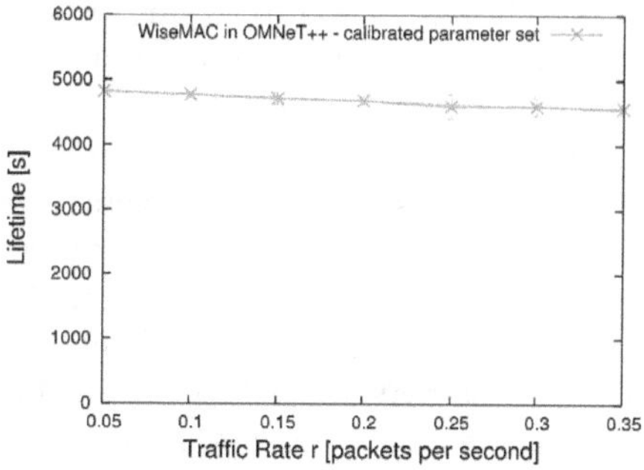


**Fig. 7.** OMNeT - calibrated parameters

implementation results of the WiseMAC ESB prototype and the OMNeT++ simulation results of Section 3.3 were of enormous value and helped to find suitable and realistic simulation parameters. We adjusted the energy consumption parameters of Table 2 to reflect the measurement results of the experimental evaluation of the nodes lifetimes in the different operation modes with the GoldCap methodology. Measuring the currents in receive and transmit modes with a customary multimeter is almost impossible because of high variation during signal (de)modulation. With the GoldCap methodology, we obtained stable average values for the energy consumption of the entire node in each state of the transceiver. We measured the node's energy consumption in sleep state to roughly 2 $mA$, and accordingly estimated transmit and receive currents with 4.5 $mA$ and 5 $mA$.

In a next step, we calibrated the state transition delays to realistically reflect the properties of the medium access mechanism of the ESB nodes. Switches generally require more time than indicated by the technical datasheet of the transceiver manufacturer [8]. To switch from receive to transmit, the network interface driver of the ESB nodes needs to go through different implementation-specific steps (e.g. disable certain interrupts, initialize and tune the radio interface). This procedure requires roughly 4 $ms$, whereas the datasheet of the transceiver only accounts for 12 $\mu s$. Similarly, switches from transmit to receive and from sleep to receive need more time. The calibrated parameter set containing all adjusted values is listed in Table 4.

### 3.6 Simulation Model Validation

For the ease of illustration, Figure 6 again depicts the measurements on the ESB nodes. One can clearly see the astonishing impact of the adaptation of the simulation parameters in Figure 7. The figure depicts the simulation results when applying the *calibrated* parameter set. The lifetime decreases only slowly with a slope of $-0.91\%$ per measurement interval, and reflects the real-world measurements of Figure 6 far better than the curve obtained with the initial datasheet-based parameter set in Figure 5. Both Figures 6 and 7 illustrate a similarly low lifetime decrease with increasing traffic along the chain, as well as a similarly high correlation coefficient $R^2$. As transmitting and receiving is only twice as expensive as the sleep state, the increasing traffic has a lower impact on the curve, unlike with the initial datasheet parameters of Table 2. The difference between the slopes of the real-world curve of $-0.91\%$ and the calibrated parameter set in Figure 7 of $-2.41\%$ can be explained by the retransmissions which occur in the real-world experiment due to the inherently unreliable channel, and the absence of retransmissions in the small-scale simulation, as well as the slight self-discharge of the GoldCap capacitors in the real-world experiment.

We underline that the investigations on the energy-consumption and transition delays of the real-world testbed noticeably paid off, as calibration and validation of the simulation model led to simulation results which come very close to the real-world experiment results. The ratio between the energy consumption of the transceiver states proved to be the decisive parameters for the slope of the lifetime curve. With the calibrated parameter set, we definitely obtained a more realistic energy consumption model for WiseMAC on the ESB nodes than with the initial datasheet-based parameter set.

### 3.7 One-Way Delay

Figure 8 depicts the one-way delay of the packets sent along the six nodes chain, measured both on the ESB prototype and in the OMNeT++ simulation. As expected, the delay proved to be independent from the examined traffic rates, as no congestion effects yet occur. Obviously, the results of simulation and the ESB implementation fit quite well. The per-hop delay of roughly $300ms$ could be obtained both in simulation and on the ESB nodes, and can be explained as follows: If a packet has to be sent, the sender node first waits for the next wake-up of the receiver. The expected time to wait for the next instant of the next

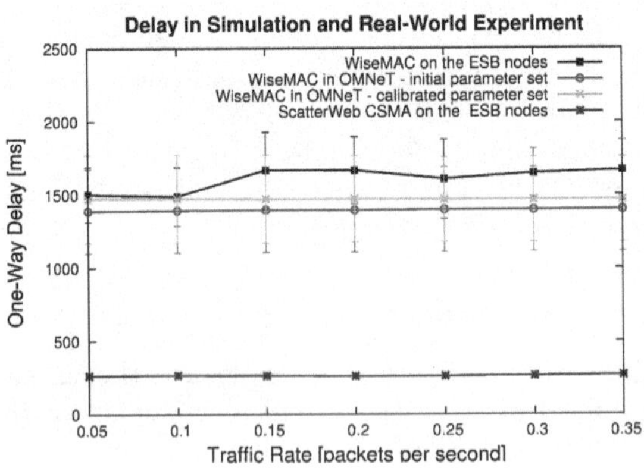


**Fig. 8.** One-way delays of WiseMAC in simulation and the ESB prototype

node is $E(t_{wait}) = \frac{T}{2} = 250ms$. The expected delay per hop $E(d_{hop})$ therefore consists of $t_{wait}$ and the delays caused by the actual frame transmission and acknowledgement, i.e. the time for the medium reservation preamble $t_{MRP}$, the minimum preamble $t_{MP}$, the transmission delay of the frame $t_f$, the transceiver switches $t_{rxtx}$ and $t_{txrx}$ and transmission delay of the acknowledgement $t_{ack}$. With the examined traffic rates and a clock drift of $\theta = 30ppm$, the preamble does not yet exceed the minimal preamble $t_{MP}$ of $5ms$.

We analytically obtain an expected per-hop delay $E(d_{hop})$ of

$$\begin{aligned} E(d_{hop}) &\cong E(t_{wait}) + E(t_{MRP}) + t_{MP} + t_f + t_{rxtx} + t_{txrx} + t_{ack} \\ &\cong 250ms + 3ms + 5ms + 20ms + 4ms + 2ms + 10ms \cong 294ms \end{aligned}$$

This results in a 5-hop delay of $1470ms$, which is the latency that was actually measured in simulation using the calibrated parameter set. The gap between the delays of simulation and real-world experiment has been slightly reduced by applying the calibrated parameter set in Figure 8.

The measurements of the one-way delays on the ESB are between $1501ms$ and $1665ms$, and differ from the simulation results by 2-13%. The reasons for the slightly higher values in the one-way delay of the ESB WiseMAC prototype are manifold. First, transmissions on the ESB still take longer as the calibrated parameters model it. There is an additional delay between frame transmission and acknowledgement reception, as ESB nodes first prepare and buffer the acknowledgement frame and then pass it to the network interface driver. Other implementation-specific issues may also play a role, i.e. the packet scheduling was implemented to include a safety margin of some milliseconds, such that the sender can carefully check the carrier before accessing it. In addition, retransmissions in the real-world implementation increase the average one-way delay. As nodes have to wait for $T = 500ms$ for the next duty cycle after each transmission attempt, retransmissions inevitably increase the one-way delay. We intend to integrate an error and packet loss model in future investigations, which however will require additional investigations on channel behavior and parameters (e.g. error model, packet error rate).

## 4 Conclusions

This paper illustrates by case study how cross-comparisons of real-world experimental results gained in small-scale experiments can to help to calibrate and validate wireless sensor network simulation models. We propose the following five steps as a baseline for simulation experiments with sensor network protocols:

1. Development and implementation of a simulation model with respect to the most important physical real-world and hardware constraints (e.g. signal dispersion, transceiver transition delays, energy consumption, etc).
2. Implementation of a prototype on real sensor hardware with the basic functionality of the proposed protocol mechanism.

3. Definition of small-scale experiment scenarios. Measurement and estimation of the prototype parameters (e.g. energy consumption, transition delays, bandwidth, packet error rate, etc) by analyzing the protocol behavior in small scale.
4. Calibration of the simulation model by reintegration of the measured or estimated real-world parameters.
5. Validation of the calibrated simulation model by cross-comparison with the real-world results of the small-scale experiments.

Careful investigations on realistic models for wireless ad-hoc and sensor network simulations are valuable and inevitable when confronting the legitimate scepticism against simulation studies. In order to achieve and provide confidence in own simulation studies' results, resarchers should exert themselves to cross-compare their proposed ideas and mechanisms with at least small-scale real-world experiments. Although this might be more costly, time-consuming and exhausting than relying on pure simulation results, it leads to more valuable and more solid research results.

Mechanisms that only pay off and that can only be reproduced in simulation are of no particular value. Without any validation, simulations of wireless sensor networks mechanisms only produce unverifiable and possibly even misleading results. Anchoring the simulation model to real-world experiments undoubtedly increases trust and confidence into simulation results, although scalability effects might still be unaccounted for. The methodology applied in this case study shall therefore be a guideline for model calibration and validation of sensor network simulation models.

## References

[1] Kurkowski, S., Camp, T., Colagrosso, M.: MANET simulation studies: the incredibles. SIGMOBILE Mob. Comput. Commun. Rev. 9(4), 50–61 (2005)
[2] MobiHoc: The ACM International Symposium on Mobile Ad Hoc Networking and Computing, `http://www.sigmobile.org/mobihoc`
[3] Andel, T.R., Yasinsac, A.: On the Credibility of Manet Simulations. IEEE Computer Magazine (2006)
[4] El-Hoiydi, A., Decotignie, J.D.: WiseMAC: An Ultra Low Power MAC Protocol for Multihop Wireless Sensor Networks. In: Nikoletseas, S.E., Rolim, J.D.P. (eds.) ALGOSENSORS 2004. LNCS, vol. 3121, pp. 18–31. Springer, Heidelberg (2004)
[5] Varga, A.: The OMNeT++ Discrete Event Simulation System. In: European Simulation Multiconference (2001), `http://www.omnetpp.org`
[6] Drytkiewicz, W., Sroka, S., Handziski, V., Koepke, A., Karl, H.: A Mobility Framework for OMNeT++. In: 3rd Intl. OMNeT++ Workshop (2003)
[7] Feeney, L.M., Nilsson, M.: Investigating the Energy Consumption of a Wireless Network Interface in an Ad Hoc Networking Environment. IEEE INFOCOM (2001)
[8] RF Monolithics: Datasheet for the TR1001 868.35 MHz hybrid transceiver, `http://www.rfm.com/products/data/TR1001.pdf`

[9] Schiller, J., Liers, A., Ritter, H., Winter, R., Voigt, T.: ScatterWeb - Low Power Sensor Nodes and Energy Aware Routing. In: 38th Annual Hawaii International Conference on System Sciences (2005)
[10] Schiller, J.H., Liers, A., Ritter, H.: ScatterWeb: A wireless sensornet platform for Research and Teaching. Elsevier Computer Communications (2005)
[11] Ritter, H., Schiller, J., Voigt, T., Dunkels, A., Alonso, J.: Experimental evaluation of lifetime bounds for wireless sensor networks. In: European Workshop on Wireless Sensor Networks (EWSN) (2005)
[12] Staub, T., Bernoulli, T., Anwander, M., Waelchli, M., Braun, T.: Experimental Lifetime Evaluation for MAC Protocols on Real Sensor Hardware. In: ACM REALWSN (2006)
[13] Chang, J.H., Tassiulas, L.: Maximum lifetime routing in wireless sensor networks. IEEE/ACM Transactions on Networking (2004)

# Experimental Study: Link Quality and Deployment Issues in Wireless Sensor Networks

Monique Becker, Andre-Luc Beylot, Riadh Dhaou,
Ashish Gupta, Rahim Kacimi, and Michel Marot*

Institut TELECOM, CNRS-SAMOVAR, 9 Rue Charles Fourier, Evry, France
{Monique.Becker,Ashish.Gupta,Michel.Marot}@it-sudparis.eu
University of Toulouse, IRIT-ENSEEIHT, 2 Rue Charles Camichel, Toulouse, France
{Andre-Luc.Beylot,Riadh.Dhaou,Rahim.Kacimi}@enseeiht.fr

**Abstract.** In this paper, we highlight the extent of the effects of topological specificities on the deployed solutions, which can be useful to refine already proposed models as well as to carry out protocol tuning or adjustments. We present, an intensive experimental study on wireless Link Quality Indicator (LQI). Using Moteiv's Tmote Sky sensors, we deployed multiHopLQI algorithm of TinyOS in various network configurations: homogeneous and heterogeneous; straight-line and grid topologies with various transmission power levels and distances.

Initially, we study LQI time-varying and try to understand the relationship between transmission power level, distance and link quality and present how some random disturbances due to external (physical changes) or internal phenomena (node movement,power variation) may affect the dynamics of the network. Later, we address impacts and side effects of position and power transmission level of some important nodes in the network like the Base Station in such LQI based algorithms.

**Keywords:** Wireless Sensor Networks, CC2420 Radio, Link Quality Indicator, Transmission power.

## 1 Introduction

Most of the sensor applications are designed to use simple, cheap, tiny devices with limited battery power. Furthermore, when real sensors are deployed, usually they do not have access to GPS (which is high energy consuming as well as expensive). Therefore, they flood the network with messages like ROUTE_REQUEST and then wait for the replies from the neighboring sensors to identify their exact location in the network. Based on the replies, they also construct their neighbor table or routing table to build the network as well as topology.

The sensors do not know the exact physical locations of the other sensor nodes, the decision which sensor is near or far is dependent upon the received signal quality or in case of our experiments Link Quality Indicator (LQI). For

* The authors name appear in alphabetical order and the experimental work has been jointly conducted by Rahim Kacimi and Ashish Gupta.

L. Fratta et al. (Eds.): NETWORKING 2009, LNCS 5550, pp. 14–25, 2009.

each received packet, this value is obtained through Chipcon CC2420 [1] radio module provided in the Moteiv's Tmote Sky sensor [2]. As per the matrix of Chipcon, the higher the LQI value is, the better the link quality between the two nodes is. Therefore, in these sensors, if the LQI between two sensors is above a given threshold, they can communicate directly, taking into account the overall network topology. In this paper, we consider scenarios (presented in the following section) where the sensors calculate the LQI between them and the top 3 neighbors and construct the neighbor table accordingly, with number of hops (distance) from the Base Station (BS). The choice of the next hop is based on the link cost estimation, in order to connect with the BS. The link cost depends jointly on the LQI and on the minimum number of hops algorithm. The most interesting aspect in any sensor network is the transmission power of the sensor, a major component of energy consumption in any sensor. Higher transmission power leads to better signal quality over a large area, nonetheless resulting in higher energy consumption and vice-versa.

The remainder of the paper is organized as follows. Section 2 introduces prior works and problems. Then, we discuss our experimental methodology in Section 3. In Section 4, we present and analyze the experimental results. And finally, Section 5 concludes the paper.

## 2 Problem and Background

Last few years have witnessed the tremendous leap in sensor network domain. Indeed, researchers try to exploit all the parameters that this domain provides to improve the performance criteria of the proposed solutions, protocols, and algorithms. In [3], authors present a resource-aware and link quality based routing metric for wireless sensor and actor networks in order to adapt to variable wireless channel conditions in such heterogeneous networks. In the field of localization, Blumenthal *et al.* use the LQI to estimate a distance from a node to some reference points [4]. More currently, the experimental/deployment analysis become one of the forefront subject in WSN field. Recent experimental studies [5], [6], [7], [8] and [9] have shown that in real sensor network deployments, wireless link quality varies over space and time. In [6], authors investigated performance issues related to node placement, packet rate and distance. In [5], Wahba *et al.* used two motes and evaluated link quality over distance and various power levels. Polastre *et al.* [10] presented preliminary evaluation results for Telos motes (based on CC2420) and suggested that the average LQI was a better indicator of packet reception rate (PRR). In all the work, authors have taken into account the homogeneous nature of the network, where all the nodes have equal transmission power. Higher energy emission leads to better signal over a large area, resulting in higher energy consumption or vice-versa.

The most interesting aspect in any sensor network is the transmission power of the sensor, a major component of energy consumption in any sensor. This paper compares the various homogeneous and heterogeneous scenarios (described in next section) and their effect on Link Quality and hence on the connectivity of the network.

Thus, a sufficient reason for our interest in the link quality is to answer the following questions. Is this parameter time-varying? What are the factors of this variation? How LQI depends on transmission power and distances between the nodes? And finally, what is its impacts on routing and network topology?

## 3 Experimental Set-Up

In order to experience and understand how few fundamental aspects of deployment can influence the sensor network as a whole, let us analyze some real time deployment issues. We have conducted 40 different scenarios and have recorded observations for more than 800 minutes (grand total of all scenarios) per sensor. All these scenarios are different either in terms of number of nodes, distance between the nodes, transmission power level of nodes, transmission power level of Base Station (BS) or finally, in terms of topology i.e. straight-line/grid (Fig. 1). All the scenarios are conducted in indoor conditions. The experiments are performed at several power levels.

In fact, all these scenarios helped us to compare several as well as relevant configurations for a given sensor network. We started with simple straight line topology, observed the network with time, node displacement, positioning, connectivity, etc. Then applied those observations by adding node redundancy (grid-topology) to the network.

Tmote Sky is a small platform including a microcontroller operating at 8MHz, 48K of ROM, 10K of RAM, a 2.4GHz ZigBee wireless transceiver, and a USB interface for device programming and logging. Each device operates on 2 AA batteries. Tmote Sky node provides an interface to parameterize its transmission power. The parameter varies from 1 (-25 dBm, minimum Transmission Power Level (TPL)) to 31 (0 dBm, the maximum TPL). Therefore, just by varying the TPL parameter transmission power can be increased or decreased. Additionally, in all the scenarios only printed antenna on the sensor has been used (no additional external antenna). Furthermore, all the sensors are placed on the floor.

All scenarios, as described in Table 1 later are based on following assumptions:

- Sensors usually have low quality of radio antenna.
- Deployment in an area with steady environment is not possible.

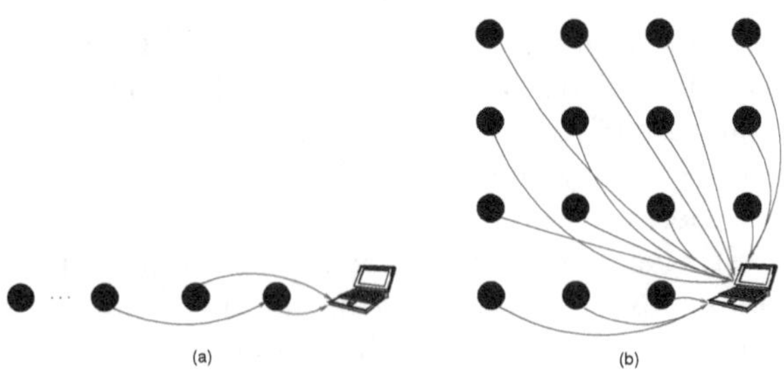


**Fig. 1.** Straight-line (a) and Grid (b) deployment

**Table 1.** Scenario Description

| Scenario | Nodes count | BS-TPL | Node-TPL | Distance |
|---|---|---|---|---|
| 1 | 12 | 31 | 25 | 3 |
| 2 | 7 | 31 | 25 | 6 |
| 3 | 7 | 25 | 25 | 6 |
| 4 | 5 | 25 | 25 | 9 |
| 5 | 5 | 31 | 25 | 9 |
| 6 | 12 | 31 | 20 | 3 |
| 7 | 12 | 20 | 20 | 3 |
| 8 | 7 | 31 | 20 | 6 |
| 9 | 7 | 20 | 20 | 6 |
| 10 | 5 | 20 | 20 | 9 |
| 11 | 5 | 31 | 20 | 9 |
| 12 | 12 | 31 | 15 | 3 |
| 13 | 12 | 15 | 15 | 3 |
| 14 | 7 | 15 | 15 | 6 |
| 15 | 5 | 15 | 15 | 9 |
| 16 | 5 | 31 | 15 | 9 |
| 17 | 7 | 31 | 15 | 6 |
| 18 | 12 | 31 | 10 | 3 |
| 19 | 12 | 10 | 10 | 3 |
| 20 | 7 | 31 | 10 | 6 |
| 21 | 7 | 10 | 10 | 6 |
| 22 | 5 | 31 | 10 | 9 |
| 23 | 5 | 10 | 10 | 9 |
| 24 | 12 | 31 | 5 | 3 |
| 25 | 12 | 5 | 5 | 3 |
| 26 | 7 | 31 | 5 | 6 |
| 27 | 7 | 5 | 5 | 6 |
| 28 | 5 | 31 | 5 | 9 |
| 29 | 5 | 5 | 5 | 9 |

# 4 Analysis and Observation

In Scenarios 1 to 29 (Tab. 1), nodes are placed in 2m (approx.) wide indoor corridor (in straight line, direct visibility) along the wall. Further, the area is open to public and have experienced frequent movements of people during the measurements.

In these scenarios, we varied the number of nodes (respectively, 12, 7 then 5, including BS), separated by 3, 6, and 9 meters respectively. For each set of above parameters, we have used two different sets of Transmission Power Level (TPL), namely SetMax{31} and SetLow{25,20,15,10,5}.

## 4.1 Real Time Evolution

The channel quality of a given sensor network is dynamic i.e., not only it is being affected by the limited battery of sensors but also by the periodic/random

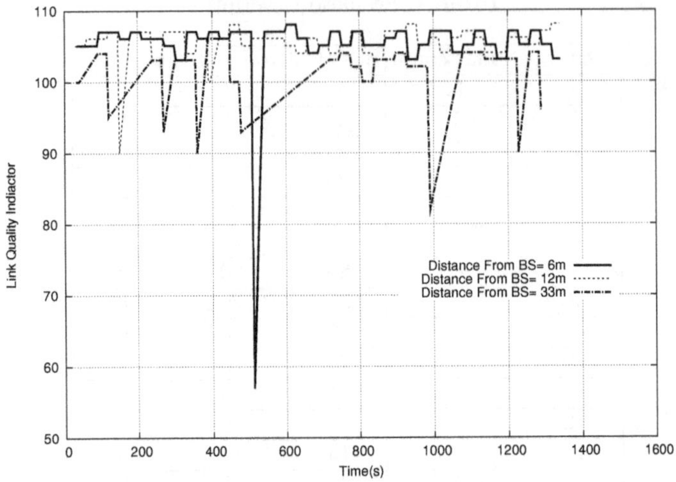


**Fig. 2.** Real time evolution of LQI

change in the physical properties of the channel, e.g. a group of people passing around the sensors can easily change the dynamics of the network. In Fig. 2, we plan to summarize this effect and will discuss the *Scenario 1*. Whenever, there has been a movement of group of people, in and around the network, we have experienced connectivity problems. The troughs which are being presented in Fig. 2, represent the deterioration of communication channel. Furthermore, the sharper curves leads to change in the connectivity and topology in the network. Let us remember, only the LQI readings between the sensor nodes and the BS are being discussed. In fact, it shows network instability and its vulnerability to physical medium, even as in *Scenario 1*, considering connectivity range, nodes are very powerful and more are or less are very near.

### 4.2 Impact of the Position of the Base Station

In most of the sensor networks, the role of a BS is to collect data and send it to a remote server or end-user. The BS can be selected statically or dynamically. The LQI usually determines the connectivity between the various nodes. Here, we will discuss Scenario 24. In this scenario, we have 12 nodes including BS. Each node is separated by 3 meters and all the nodes are in straight line (direct visibility). Fig. 3 presents the LQI values between various sensors and the BS. We have observed that troughs deal with the discontinuity in the network and shaper troughs in LQI reading lead to disruption of communication channel/link. Furthermore, positioning of the BS can have a subtle effect on the performance of BS. Also, all the nodes are placed on the floor next to wall. We have run this scenario for over 1200 seconds. Even though, there are another 6 nodes (excluding BS) in the network. For Clarity reasons, we present the relevant results only for few nodes. Initially, we have observed that, the node which is 33 meters away from the BS, is not connected directly with BS. Thereafter, (from time 50-500

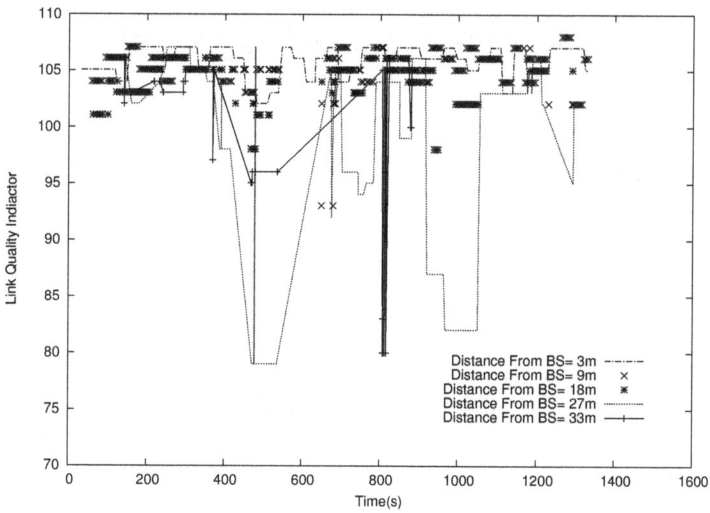


**Fig. 3.** Impact of position of Base Station

seconds) we raise the position of the BS by about 0.5 meter from the floor. Again, we can see from Fig. 3, merely, by raising the position of BS with respect to other sensors, we observe the major shift in LQI values. Later on, we play with BS with intermittingly raising and lowering the position of BS and finally, at around time = 800 seconds we end this procedure. Between these periods, we can easily distinguish the various LQI troughs being made repeatedly. And once, we are over with this process, farthest node is connected via multiple hops with BS i.e. no more direct connectivity with BS. The fluctuation in LQI values due to these random movements is obvious in Fig. 2.

### 4.3 Impact of the High Power of the Base Station

Another important aspect of any sensor network is the transmission power of its nodes. Transmission power limits the range of any given sensor. Sensor network relies upon neighbor discovery and route discovery mechanism to communicate with BS. Therefore, it is interesting to see, how different level of BS energy may affect sensor network. Scenarios 26 and 27 are different only in terms of TPL level of BS. In both the scenarios, we have 7 sensor nodes, separated by 6 meters in the straight line. Fig. 4 presents the LQI readings of each sensor with BS, in a two different networks (for ease of clarity, again only few nodes are depicted). As, we compare LQI values, we can observe, that just by increasing the TPL of the BS, the LQI between the nodes and the BS improves tremendously. Also, the lower the TPL of BS the lower is the LQI (apart from sensor which is nearest to BS). Further, we can clearly observe the difference of LQI readings of sensor which is being placed at a distance of 18 meters from the BS. Due to difference in LQI values of these sensors, sensor with BS (31 points, scenario

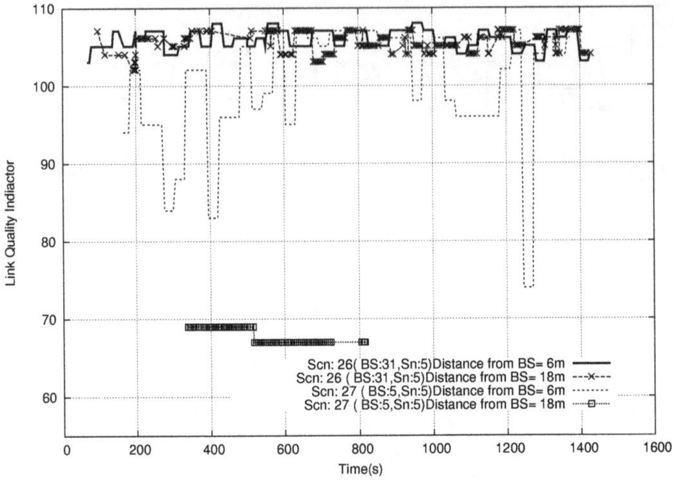


**Fig. 4.** Impact of high power of Base Station

26) remains connected continuously with BS, the other sensor in Scenario 27 is rather connected via its neighbours.

Higher power level for a given node leads to a natural single hop cluster, since each node sees the BS as being close (even if it is far) and consumes lot of energy because of its high transmission power level. And then, each node tries to communicate directly with the BS instead of communicating to BS via a set of hops. This raises some more issues for example in terms of traffic where a traffic can be captured by a single high power node. In fact, in the next subsection, we will magnify this effect in the grid topology and the ramification of this phenomenon.

### 4.4 Impact of Different Power Level i.e Heterogeneous Sensor Network

Table 2 presents another set of tests (Scenarios 30-40). These tests are executed in an indoor room but in an area cut-off from the public. We have used two different grids of size 4x4 and 3x6. In both cases, sensors are separated by 3 meters. In these scenarios, two TPL sets are defined as SetMax{31} and SetLow{10,5,3}.

#### Link Quality with Distance

Here, the link quality variations are not completely due to the change in the physical properties of the channel because of the closed environment (a classroom) without any presence of people. Generally (Fig. 5), all the collected values for every combination of distance and transmission power vary between 103 to 108. Furthermore, if we refer to other kind of experiments [5],[6],[11] these values remain interesting because the packet received rate for such LQI values is high.

**Table 2.** Grid Scenario Description

| Scenario | Nodes count | BS-TPL | Node-TPL | Distance |
|---|---|---|---|---|
| 30 | 4x4 | 31 | 10 | 3 |
| 31 | 4x4 | 10 | 10 | 3 |
| 32 | 4x4 | 3 | 10 | 3 |
| 33 | 4x4 | 31 | 5 | 3 |
| 34 | 4x4 | 5 | 5 | 3 |
| 35 | 3x6 | 31 | 10 | 3 |
| 36 | 3x6 | 10 | 10 | 3 |
| 37 | 3x6 | 3 | 10 | 3 |
| 38 | 3x6 | 31 | 5 | 3 |
| 39 | 3x6 | 5 | 5 | 3 |
| 40 | 3x6 | 3 | 5 | 3 |

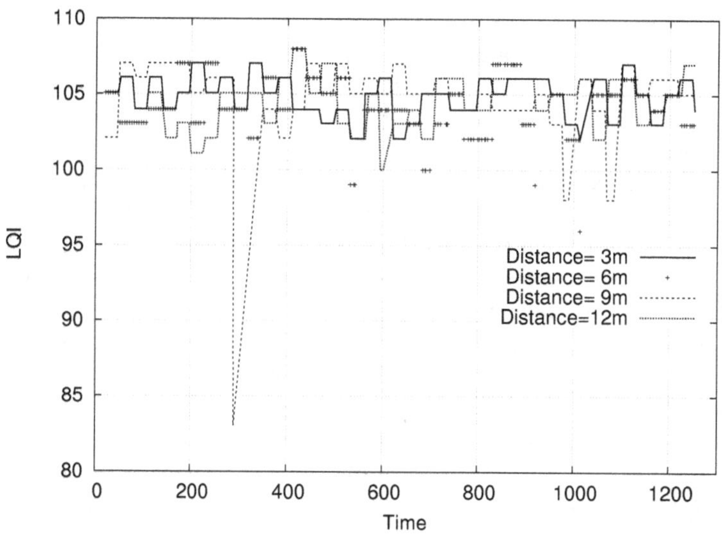


**Fig. 5.** LQI variation with time, scenario 33

For a given transmission power level, the LQI values are slightly different (i.e they decrease when distance from the Base Station increases); and for a given distance, these values decrease slightly when we reduce the transmission power. When the distance from BS is higher than 3 or 6 meters, we notice some dramatical decreases in the LQI variations. We also observe that the variations of LQI are more frequent with the nodes placed along the wall, than when they are placed in diagonally.

## Influence of BS Transmission Power on Topology

To conduct our experiments, we have used multiHop LQI routing algorithm [12] in TinyOS, because the code for the Tmote Sky platform was available.

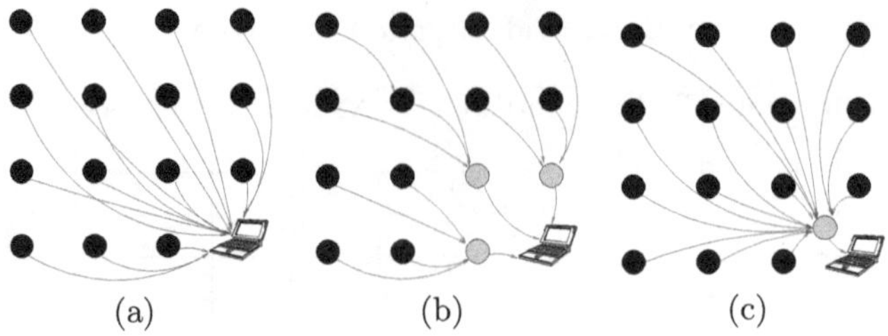

Fig. 6. BS transmission power effects

According to this algorithm, we noticed that the transmission power of the Base Station is a crucial parameter. Moreover, the BS has an important role in the network topology and the route changing. Indeed, in order to allow the nodes to choose their routes to reach the Base Station, the Base Station required to send beacon packets regularly.

As shown in Fig. 4.4, we analyze the results of these tests according to three distinct cases. The first case, when the Base Station transmits with a higher power than the power of the nodes. In this case, all the nodes note that the link quality with the BS is sufficiently high to choose direct connections (Fig. 6(a)). The second case, when the Base Station transmits with the same power than the nodes, we observed some multi-hop routes especially for the furthest nodes. The third case when the Base Station transmits with a lower power than the power of the nodes, several multi-hop connections appear with an important traffic overload on the nodes closer to the BS. (Fig. 6(b)). Indeed, the routing algorithm issues that getting through these nodes constitutes the most optimal way (number of hops) and the most effective (link quality). We proved that by adding another node with a high transmission power beside the Base Station and all the traffics are transmitted via this node (Fig. 6(c)).

### Influence of Nodes Transmission Power on LQI and Multi-hopping

While the routing algorithm is mainly based on the link quality, thus, varying the nodes TPL implies certainly changes in the network topology. Here, we consider the scenarios 30 to 40 to analyze these changes according to the nodes- and BS-TPL.

Fig. 7 plots the average number of hops as a function of the BS- and node-TPL, observed in a grid of 3x6 nodes. We can note that the number of hops increases with the reduction in BS-TPL. This result endorses the observations of the preceding paragraph on BS-TPL impact. The number of hops remains reasonable (3) even with the lowest BS-TPL because the area is relatively small (6x15m).

On the other hand, Fig. 8 illustrates the difference between two sets of scenarios Set{35,36,37} and set{38,39,40}. In these two sets we used two distinct Node-TPL (respectively 10 and 5). In this figure we can clearly note that the number of multi-hop paths is always higher when the node-TPL is lower.

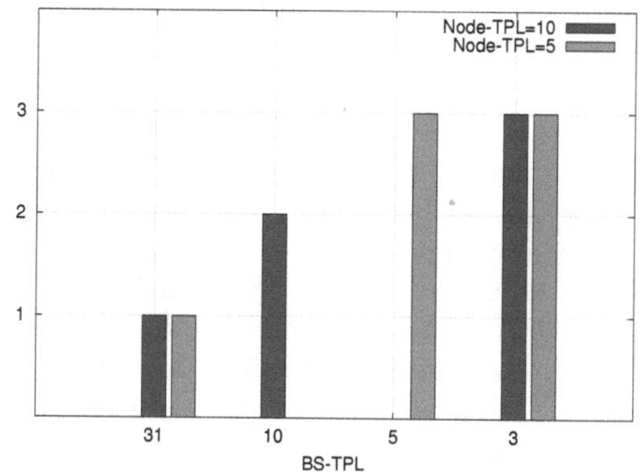


**Fig. 7.** Average number of hops

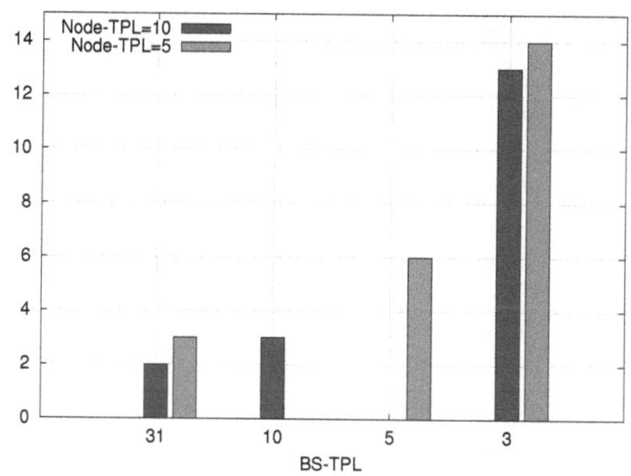


**Fig. 8.** Number of multi-hop routes

## 5 Conclusion

Focusing on using a commercial hardware platform in sensor systems, we carried out in this work, an experimental study on the link quality in wireless sensor networks. In the first set of experiments, we studied the LQI evolution over time and observed the dynamics of transmission channel. Very briefly, we discussed the significance of positioning of the Base Station in any given sensor network. We saw, how network is sensitive to small node displacements. With these experiences, we presented LQI time-varying and some random disturbances due to external phenomena and physical changes. It is very important to study these issues, as sensors may not be subjected to steady state deployment. Finally, we studied the impact of transmission power of BS and observed, how sensors in networks with high TPL of BS can miss-construct network topology and the effects on the connectivity between nodes and BS. We saw, how a high power node creates a natural single hop cluster. We used these observations and experiences to conduct further experiments.

In the second set of experiments, we have also investigated the impact of nodes transmission power on the LQI which affects consequently the network topology. Indeed, with high BS-TPL and Node-TPL, often we observed only one cluster (BS as a cluster-head). When we varied the TPL between nodes (heterogeneous nodes), several clusters appeared (cluster-head with high TPL). So, it may be a possible solution to organize the network on clusters. However, such heterogeneity may affect the lifespan of these nodes and the network connectivity.

We also proved in this study that the Base Station TPL may have a misleading effect for the furthest nodes. Indeed, these nodes notice that the link quality with the BS is sufficiently high to choose direct connections. But the distance is large and the risk of packet loss might increase. Indeed, the link quality on the another direction (node to BS) is not necessarily the same because of the distance or the weak Node-TPL. As a concluding remark, routing protocols should not be entirely based on LQI.

Considering our measurement results, it seems that it may be interesting to reduce the transmission power in order to save energy, or to deploy heterogeneous nodes for topology issues. As future work, we plan to endorse these conclusions by an evaluation of the network performance such as energy-efficiency, fairness, transmission delay, etc.

## Acknowledgments

The authors gratefully acknowledge material and technical support from the CNRS-SAMOVAR (Evry) , IRIT-ENSEEIHT (Toulouse), National Polytechnic Institute of Toulouse (INPT) and TELECOM-SudParis. This work is also supported by the "CAPTEURS" grant, a National Telecommunication Research Network (RNRT) project.

## References

1. CC2420 Radio, `http://www.chipcon.com`
2. Tmote Sky datasheet, `http://www.moteiv.com/products/docs/tmote-sky-datasheet.pdf`
3. Gungor, V.C., an Sastry, C., Song, Z., Integlia, R.: Resource-aware and link quality based routing metric for wireless sensor and actor networks. In: IEEE International Conference on Communications, 2007. ICC 2007, pp. 3364–3369 (June 2007)
4. Blumenthal, J., Grossmann, R., Golatowski, F., Timmermann, D.: Weighted centroid localization in zigbee-based sensor networks. In: IEEE International Symposium on Intelligent Signal Processing, 2007. WISP 2007, pp. 1–6 (October 2007)
5. Wahba, S.K., LaForce, K.D., Fisher, J.L., Hallstrom, J.O.: An empirical evaluation of embedded link quality. In: International Conference on Sensor Technologies and Applications, 2007. SensorComm 2007, pp. 430–435 (October 2007)
6. Lal, D., Manjeshwar, A., Herrmann, F., Uysal-Biyikoglu, E., Keshavarzian, A.: Measurement and characterization of link quality metrics in energy constrained wireless sensor networks. In: IEEE Global Telecommunications Conference, 2003. GLOBECOM 2003, vol. 1, pp. 446–452 (December 2003)

7. Son, D., Krishnamachari, B., Heidemann, J.: Experimental analysis of concurrent packet transmissions in wireless sensor networks. In: Proceedings of the Fourth ACM SenSys Conference, Boulder, Colorado, USA, pp. 237–249. ACM, New York (2006)
8. Srinivasan, K., Dutta, P., Tavakoli, A., Levis, P.: Understanding the causes of packet delivery success and failure in dense wireless sensor networks. In: ACM SenSys. (2006)
9. Zhao, J., Govindan, R.: Understanding packet delivery performance in dense wireless sensor networks. In: SenSys 2003: Proceedings of the 1st international conference on Embedded networked sensor systems, pp. 1–13. ACM, New York (2003)
10. Polastre, J., Szewczyk, R., Culler, D.: Telos: enabling ultra-low power wireless research. In: IPSN: Information Processing in Sensor Networks, pp. 364–369 (2005)
11. Holland, M., Aures, R., Heinzelman, W.: Experimental investigation of radio performance in wireless sensor networks. In: 2nd IEEE Workshop on Wireless Mesh Networks, 2006, pp. 140–150 (2006)
12. TinyOs MultiHopLQI routing algorithm, `http://www.tinyos.net/tinyos-1.x/tos/lib/MultiHopLQI`

# Aggregation Protocols for High Rate, Low Delay Data Collection in Sensor Networks*

Jie Feng, Derek L. Eager, and Dwight Makaroff

Department of Computer Science, University of Saskatchewan,
Saskatoon, SK S7N 5C9, Canada
{jif226,eager,makaroff}@cs.usask.ca

**Abstract.** Sensor network applications commonly require sensor data to be periodically collected. Aggregation protocols can make this process considerably more efficient. This paper considers the problem of devising aggregation protocols for applications that must achieve as "real-time" a view of the monitored area as possible, entailing a high sampling rate and a low data collection delay, at the possible cost of some modest amount of data loss. We examine in particular broadcast-based protocols that minimize the number of packet transmissions, relying on multipath delivery rather than ARQ for reliability, and consider the question of whether such protocols can achieve lower collection delays and support higher sampling rates than conventional aggregation protocols. Our results suggest that broadcast-based protocols can yield significantly improved performance in some scenarios, when sensor data can be aggregated into packets of size that is independent (or largely independent) of the number of values being aggregated.

**Keywords:** sensor networks, aggregation protocols, performance evaluation.

## 1 Introduction

Sensor networks consist of a potentially large number of sensor nodes capable of capturing measurements of their immediate environment, together with one or more "sink" nodes at which sensor data can be collected. Data collection may be either periodic [1][2][3][4], as in environmental monitoring applications for example, or as needed (aperiodic) [5][6][7] in response to exceptional events. We consider here the case of monitoring applications that must achieve as "real-time" a view of the monitored area, at a single sink node, as is possible. For this purpose, periodic data collection is required, with a high sampling/collection rate and a low data collection delay.

The efficiency of data collection can be vastly improved by aggregating the sensor data received at intermediate nodes into fewer numbers of packets, as this data

* This work was supported by the Natural Sciences and Engineering Research Council of Canada.

L. Fratta et al. (Eds.): NETWORKING 2009, LNCS 5550, pp. 26–39, 2009.

is being forwarded to the sink [1] [2][3][4][5][6]. By reducing the number of transmitted packets, aggregation can reduce energy usage, increase the achievable data collection rate, and (owing to reduced network contention) decrease the data collection delay. Aggregation may be achieved in an application-independent manner by simply concatenating (possibly in compressed form) the sensor values received from multiple sensors in one packet [8]. Alternatively, aggregation may make use of application semantics, such as when only the maximum sensor reading is required; in this case, only the largest value need be forwarded [1].

Existing protocols for data collection using aggregation ("aggregation protocols") may be classified as either asynchronous or synchronous and according to whether they use unicast or broadcast communication [1][2][3][9][10][11]. The former distinction concerns how a node determines when to wait for additional data from upstream nodes, and when to send downstream an aggregate packet containing whatever data it has received up until then. In TAG, for example, a synchronous approach is used, in which all nodes $i$ network hops from the sink forward their (aggregated) data to their parent nodes in the aggregation tree during interval $d - i$ of each "round" of communication, where one set of sensor readings is collected each round, and where $d$ is the maximum number of hops from the sink [1]. In contrast, with asynchronous protocols, each node adaptively determines when to send versus when to wait based on its local history of past packet receptions from upstream nodes.

Most prior aggregation protocols use unicast transmission, with reliability achieved using an acknowledgement/retransmission (ARQ) facility, as provided by the link layer for example. The prior work concerning aggregation protocols using broadcast transmission has focused on protocol mechanisms for "duplicate sensitive" aggregation [9][10], in which the sink must never receive multiple aggregates including the same sensor value (or the same share of a sensor value, if "value splitting" [1] is employed).

We focus on protocol mechanisms that broadcast-based protocols can employ to maximize the achievable sampling/collection rate and minimize the collection delay with some modest amount of data loss, and the question of whether such protocols can achieve better performance in these respects than unicast-based protocols. It is assumed acceptable for the sink (and intermediate nodes) to receive multiple aggregates including the same sensor value, either because aggregation is duplicate insensitive (e.g., only the maximum sensor value is needed), or duplicates can be filtered (each aggregate is a concatenation of sensor values).

Both synchronous and asynchronous broadcast-based aggregation protocols are developed. They take advantage of the fact that multiple downstream nodes may be potential receivers of a single broadcast transmission, and of the ability of a node to listen to transmissions from neighbouring nodes so as to determine which (if any) have included that node's broadcast data in their own transmissions. The latter ability is used as a substitute for acknowledgements: in addition to the reliability improvement that arises from potential multipath delivery to the sink, we also achieve improved reliability through use of two-phase protocols in which a node may repeat (once) its broadcast.

We compare unicast-based protocols and the new broadcast-based protocols mostly using simulation. Rather than assuming some particular aggregation function, two extreme cases are considered. In one of these, it is assumed that sensor data can be aggregated into packets of size that is independent of the number of values being aggregated. In the other, required packet size is assumed to increase linearly with the number of values included in the aggregate. The first case applies with duplicate-insensitive aggregation functions such as "maximum". The first case would also clearly apply with duplicate-sensitive aggregation functions such as "average", were it not for our assumption that it is possible to recognize and filter out duplicates that have been included in multiple aggregates. Which of these two cases more closely reflects reality may depend on the extent to which sensor values and identifiers can be encoded in highly-compressed form in the aggregates.

We find that for a fixed required packet size, broadcast-based protocols are able to support higher sampling/collection rates with lower collection delays, than unicast-based protocols. The performance improvements are particularly pronounced for synchronous aggregation. When the required packet size is assumed to increase linearly with the number of values included, however, we find no significant performance advantage with broadcast-based protocols.

The remainder of the paper is organized as follows. Sections 2 and 3 review the synchronous and asynchronous unicast-based protocols, respectively, on which our new broadcast-based protocols are based, and then describe the new protocols. Section 4 presents simulation results evaluating the performance of the new protocols in comparison to that of the unicast-based protocols. Section 5 provides initial experimental results. Section 6 concludes the paper.

## 2 Synchronous Aggregation

With synchronous aggregation protocols, all sensor nodes at the same distance (number of hops) from the sink are given the same interval of time in which to transmit, within each round of communication. Nodes farther away from the sink are given earlier transmission intervals, so as to allow their data to be aggregated with that of nodes closer to the sink before these latter nodes make their own transmissions. We first briefly describe the unicast-based synchronous protocol on which our new broadcast-based synchronous protocol is based, in Section 2.1, and then describe the design of the new protocol in Section 2.2.

### 2.1 Unicast-Based

The unicast-based synchronous protocol is a variant proposed in previous work[3] of the original synchronous aggregation protocol as used in TAG [1]. Aggregation is performed over a tree rooted at the sink. Sensor readings are made periodically with period duration $t$. Nodes at different tree levels are assigned to different transmission intervals within each round of communication (i.e., collection of one set of sensor values) based on their distances to the sink. It is assumed that each node $i$ knows its hop count $h_i$ to the sink and the maximum hop count $H$ in the tree, and accordingly chooses its transmission interval within each round.

Let $I$ denote the interval duration. In each round $j$, node $i$ picks a random value $r_i^j$ between 0 and $\lambda I, 0 \leq \lambda \leq 1$, aggregates the data it has received for this round, and sends out its packet at time $T_0 + t(j-1) + (H - h_i)I + r_i^j$. Here it has been assumed that all nodes agree on the same base time $T_0$ defining the beginning of the first round. It has been found that randomizing transmissions over $\lambda I$ yields better performance than when all nodes at the same tree level attempt to transmit at approximately the same time, and that setting $\lambda$ to 0.8 yields good performance over a wide variety of network configurations [3].

As described previously, we make no assumptions regarding the type of aggregation that is performed (application-independent, duplicate-sensitive, duplicate-insensitive, etc.), but rather attempt to model a range of scenarios in our performance experiments through consideration of two extreme cases with respect to how packet size grows with the number of aggregated values.

### 2.2 Broadcast-Based

In our broadcast-based synchronous aggregation protocol, nodes are organized into a ring topology [11], as shown in Fig. 1. The sink is the only node that is located in ring 0, nodes one hop away from the sink are in ring 1, etc. As in the unicast-based protocol, nodes in different rings are allotted different time intervals within each round of communication for their transmissions. As before, the duration of the period between sensor readings is denoted by $t$, and the interval duration by $I$.

Unlike the unicast-based protocol, in our broadcast-based protocol each interval is divided into a first and second phase with durations $\alpha I$ and $(1-\alpha)I$, respectively. Each node (except for the sink) broadcasts a packet containing (possibly aggregated) sensor data during the first phase of its interval in each

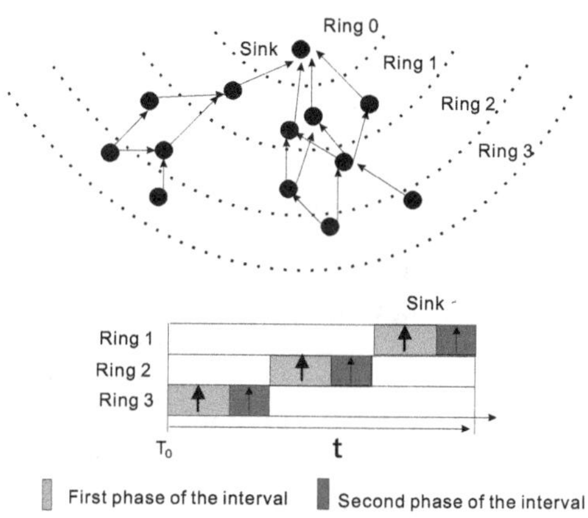


**Fig. 1.** Synchronous Broadcast-based Aggregation

round. Nodes may also make a second broadcast during the second phase, as described below. Each broadcast packet includes a bit vector indicating the nodes whose data is aggregated in the packet. Nodes aggregate all of the data they have received from broadcasts for the current round (including broadcasts from neighbouring nodes in the same ring), for their own broadcasts.

Specifically, in each round $j$, node $i$ picks a random value $r1_i^j$ between 0 and $(\alpha - 0.1)I$, aggregates the data from the broadcasts it has received for this round, and makes its own first broadcast at time $T_0 + t(j-1) + (H - h_i)I + r1_i^j$. Node $i$ then picks another random value $r2_i^j$ between 0 and $(1-\alpha-0.1)I$. A broadcast is made in the second phase of the round, at time $T_0+t(j-1)+(H-h_i)I+\alpha I+r2_i^j$, if, by this time, node $i$ has not heard a broadcast transmission from some other node in ring $i$ that has included node $i$'s data (owing to the other node having heard node $i$'s first broadcast, prior to its own first broadcast). Similarly to the unicast-based protocol, randomizing the transmissions within each phase yields better performance than when all of the nodes in the same ring attempt to transmit at approximately the same time.

The second broadcasts are important to improve reliability. For nodes with few neighbours, or nodes that are the last, among the nodes in the same ring, to make their first phase broadcast, a second broadcast increases the likelihood that their data is received by at least one node in the next ring. We have found that this two-phase strategy can reduce the overall end-to-end loss rate significantly, at minimal additional cost in terms of numbers of transmissions.

## 3 Asynchronous Aggregation

With synchronous aggregation, the time interval during which each node transmits is statically determined. In contrast, in the asynchronous aggregation protocols considered here each node adaptively determines when to transmit based on its history of past packet receptions from its children in the aggregation tree. We use the unicast-based "adaptive asynchronous" aggregation protocol proposed in previous work [3] as a basis for our new broadcast-based asynchronous aggregation protocol. The former protocol is reviewed in Section 3.1, while Section 3.2 describes the design of the new broadcast-based protocol.

### 3.1 Unicast-Based

In each round of the unicast-based asynchronous protocol, a timeout is set at each non-leaf node establishing the maximum time the node will wait to receive packets from its children in the aggregation tree. The timeout value is determined adaptively, based on the timings of packet receptions from the child nodes in the previous rounds. The node transmits its packet (aggregating its own data with whatever it has received from its children) either when packets have been received from all children, or when the timeout expires. Note that the choice of the timeout is critical: a long wait until timeout may cause excessive delay, while substantial data loss may be incurred if the timeout occurs too soon, since in this case packets may arrive too late to be aggregated (and will be dropped).

As before, we assume all nodes agree on the same base time $T_0$ defining the beginning of the first round. Each node $i$ picks a random value $r_i$ between 0 and $R$, where $R$ is a protocol parameter, and at time $T_0 + r_i$ unicasts a packet containing its sensor data for the first round to its parent.

In each subsequent round $j$, each node $i$ that is a leaf in the aggregation tree sends its sensor data at time $T_0 + r_i + (j-1)t$, where $t$ denotes, as before, the duration of the period between sensor readings. Each non-leaf node transmits a packet containing aggregated sensor data when it has received packets from all of its children for round $j$, or when its timeout for that round has expired. Timeouts are established as follows. Let $L_i^j$ denote the time at which a non-leaf node $i$ receives the last packet for round $j$ from its children in the aggregation tree. (Note that "last packet" means the last received packet; some packets sent by the children may not be received, owing to communication errors.) Let $TO_i^j$ denote the timeout for round $j$ at node $i$ (i.e., the time at which node $i$ will transmit if packets have not yet been received for round $j$ from all of node $i$'s children). After the first round, the timeout at node $i$ for round two is set to $TO_i^2 = L_i^1 + t$. After each subsequent round $j, j > 1$, the timeout for round $j+1$ at node $i$ is set according to the following rules:

1. If node $i$ received packets for round $j$ from all its children prior to time $TO_i^j$, its timeout for round $j+1$ is set to $TO_i^{j+1} = (1-\delta)(TO_i^j+t)+\delta(L_i^j+t+e)$. The parameter $\delta$ is used to implement an exponential weighted moving average, so as to control how quickly a node reacts to changes in network conditions. Parameter $e$ allows for transmission variance.
2. If the timeout for round $j$ expires before node $i$ receives packets from all of its children, its timeout for round $j+1$ is tentatively set to $TO_i^{j+1} = TO_i^j + t$. If node $i$ receives one or more packets for round $j$ from its children subsequent to the expiry of its timeout for that round, it updates $TO_i^{j+1}$ to $L_i^j + t$. The packets that arrive too late to be aggregated are simply dropped because only up-to-date values are of interest in real-time monitoring.

It is important to randomize the transmission times of leaf nodes to avoid congestion at the beginning of each round. Parameter $R$ controls the duration of the randomization interval. The adaptive protocol achieves improved performance by adjusting the value of $R$ so that it is kept within a certain range, as measured relative to the *average data collection delay* $D$. Intuitively, if $R$ is "large" relative to $D$, a substantial portion of the delay $D$ may be due to the randomization delay at the leaf nodes. In this case, it may be advantageous to reduce $R$. If, on the other hand, $R$ is "small" relative to $D$, $R$ might be fruitfully increased, so as to better spread out the transmissions of the leaf nodes.

Specifically, the data collection delay for each round $j$ is measured at the sink as the time duration from the start of the round $(T_0 + (j-1)t)$, until the sink has received the last packet for round $j$. The average data collection delay is measured as $D = \alpha D + (1-\alpha)D^*$, where $D^*$ is the latest measurement of the data collection delay, and $\alpha$ is a smoothing factor that determines the weight given to the old value. The adaptive protocol attempts to keep the value of $R/D$ in the interval $[\beta - \Delta, \beta + \Delta]$, where $\beta$ and $\Delta$ are protocol parameters. When $R/D$

moves out of this range, the value of $R$ is updated (and sent by the sink to all of the sensor nodes) as follows. If $R/D < \beta - \Delta$, $R$ is updated to $R = D(\beta + \Delta)$. If $R/D > \beta + \Delta$, $R$ is updated to $R = D(\beta - \Delta)$.

### 3.2 Broadcast-Based

In the unicast-based synchronous aggregation protocol used in TAG, aggregation is performed over a tree structure, with one parent for each node except the sink [1]. The authors observe that a node could potentially transmit to multiple parents, each one hop closer to the sink, depending on the density of the network, and propose a "value splitting" aggregation protocol in which each node may have two parents. In the broadcast-based protocol proposed in this section, nodes may similarly have two parents[1], but the same sensor value (rather than only distinct shares of a sensor value) may be aggregated at each parent and forwarded on up the tree. Such multipath routing can yield improved reliability, but requires that nodes be able to receive multiple aggregates including the same sensor value, either because aggregation is duplicate insensitive, or because aggregation is performed in such a way that it is possible to recognize and filter out duplicates.

As with our broadcast-based synchronous protocol, each node (excepting the sink) broadcasts either once or twice during each round. Each broadcast may include not only data from its children, but also data overheard from broadcasts from other neighbours in the tree. Each broadcast packet includes a bit vector indicating the nodes whose data is aggregated in the packet. A non-leaf node makes its first broadcast during a round either when it has received the data from each of its children (either directly from a broadcast by that child, or indirectly via a broadcast from some other child), or upon timeout. Timeouts are established in a similar manner as in the unicast-based asynchronous protocol described in Section 3.1. A second broadcast is made if the node does not hear a broadcast transmission from any other node that includes its data, or if it receives additional data from its children, before a second timeout occurs.

As in the unicast-based protocol, each node $i$ (other than the sink node) picks a random delay $r_i$ between 0 and $R$, where $R$ is a protocol parameter that is adapted so as to keep the ratio of $R$ to the average data collection delay $D$ within a certain range, as defined by protocol parameters $\beta$ and $\Delta$. At time $T_0 + r_i$, node $i$ broadcasts a packet containing its own data for the first round.

In each subsequent round $j$, each leaf node $i$ makes its first broadcast at time $T_0 + r_i + (j-1)t$, aggregating its own data and any other data it has overheard from other broadcasts. Let $A_i^j$ denote the time at which a non-leaf node $i$ receives all of the data from its children for round $j$ that it will receive during this round. Let $L_i^j$ denote the time at which non-leaf node $i$ receives the last of the first broadcasts from its children for round $j$ that it will receive during this round. Note, $A_i^j$ may not equal $L_i^j$. Let $TO_i^j$ denote the timeout for the first transmission of round $j$ at node $i$. We set $TO_i^2 = L_i^1 + t$.

[1] When a node has only one neighbour closer to the sink it can choose a sibling that has multiple neighbours closer to the sink as its second parent.

After each round $j$, $j > 1$, $TO_i^{j+1}$ is set as follows:

1. If node $i$ received the data for round $j$ from all of its children prior to time $TO_i^j$, its timeout for round $j+1$ is set to $TO_i^{j+1} = (1-\delta)(TO_i^j + t) + \delta(A_i^j + t + e)$, similarly as in the unicast-based asynchronous protocol.
2. If the first transmission timeout for round $j$ expires before node $i$ receives the data from all of its children, its timeout for round $j+1$ is tentatively set to $TO_i^{j+1} = TO_i^j + t$. If node $i$ receives one or more first-time broadcasts for round $j$ from its children subsequent to the expiry of its timeout for that round, it updates $TO_i^{j+1}$ to $L_i^j + t$. Data from these packets has been received too late to be aggregated in node $i$'s first broadcast for round $j$, but can be included in node $i$'s second broadcast.

As noted above, following the first broadcast, a second broadcast is made if the node does not hear a broadcast transmission from any other node that includes its data, or if it receives additional data from its children, prior to a second timeout. The second timeout is set to a time duration $e$ following the time of the first broadcast.

# 4 Simulation-Based Performance Evaluation

## 4.1 Goals, Metrics, and Methodology

The performance of the unicast-based and the broadcast-based protocols is evaluated through ns2 simulation. The primary performance metrics are: (1) the end-to-end loss rate (the ratio of the number of sensor readings not included in the aggregates arriving at the sink to the total number of readings the nodes generate), (2) the maximum data age ($t$ plus the average data collection delay $D$), a measure of how "stale" the data received at the sink from one round can become, before that for the next round is received, (3) the average number of MAC layer packet transmissions per round (for unicast, including both ACK and data packet transmissions), and (4) the average number of bytes transmitted per round. The last metric yields insight into relative energy usage.

The sensor fields are generated by randomly scattering nodes in square areas. The sink is located in the center of the network. The physical layer packet loss rate is specified as a simulation input parameter. The uniform random error model is used for all experiments. An 802.11 MAC layer is simulated for unicast-based protocols, without RTS/CTS [12], with a transmission range of 40 meters and rate of 2Mbps. Different maximum numbers of MAC layer retransmissions are simulated for when the sender fails to receive an ACK. The same transmission range and data rate are used for the broadcast-based protocols in the simulations. The protocols are evaluated with both fixed and increasing packet sizes. Fixed-sized packets are 52 bytes. Otherwise the packet size grows linearly with the number of values aggregated in the packet at a rate of 4 bytes per value.

### 4.2 Principal Performance Comparison

Fig. 2 and Fig. 3 show the performance of the synchronous and asynchronous aggregation protocols, respectively, with our default system protocol parameter settings. The different points on each curve are generated by varying the sampling period duration $t$, and measuring the resulting maximum data age and end-to-end loss rate. For synchronous aggregation, the interval duration is set to $t$ divided by the maximum hop count to the sink. The initial value for $R$ is 0 for the asynchronous protocols.

Previous work shows that for asynchronous unicast-based aggregation, all parameters except $R$ can be fixed to certain values that yield good performance for a broad range of network configurations [3]. Through experimentation, we find that the same conclusion holds for asynchronous broadcast-based aggregation as well. In all the simulations whose results are reported here, $e$ is fixed at 0.03 seconds and 0.02 seconds for asynchronous broadcast-based aggregation with increasing packet size and fixed packet size respectively. For asynchronous unicast-based aggregation, $e$ is set to 0.15 seconds for increasing packet size and 0.1 seconds for fixed packet size. Parameter $\beta$ is fixed at 0.7, 0.6, 0.7 and 0.5 for asynchronous unicast-based aggregation with fixed and increasing packet size, and asynchronous broadcast-based aggregation with fixed and increasing packet size, respectively. Other common parameters for both asynchronous unicast-based and broadcast-based aggregation, $\delta$, $\alpha$, and $\Delta$ are fixed at 0.05, 0.875, and 0.15 respectively. Figures showing the impact of the parameters are omitted due to space limitations.

For each specific $t$, we measure the end-to-end loss rate and the maximum data age of the protocols and plot the results in the figures. As $t$ gets smaller, the end-to-end loss rate starts to grow as more packets are lost because of collisions. For synchronous aggregation, the interval duration gets shorter as $t$ decreases. When the interval is too small, nodes in the same ring cannot make their transmissions within their own transmission interval. Packets that arrive after the receiver sent out its partial aggregate are not aggregated, and the end-to-end loss rate

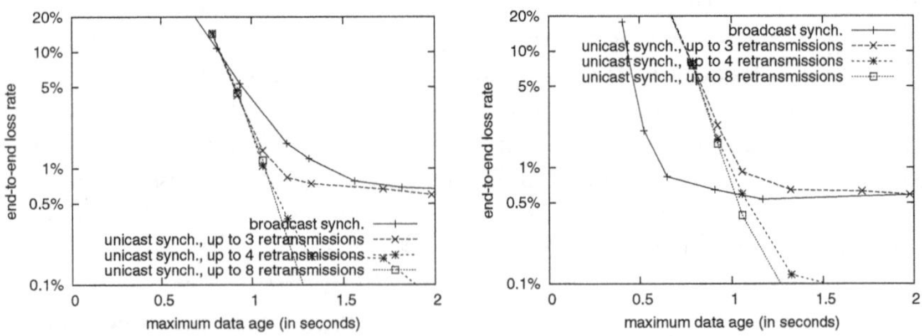


(a) Increasing Packet Size (b) Fixed Packet Size

**Fig. 2.** Synchronous Unicast-based and Broadcast-based Aggregation (160 nodes, 250m×250m, 20% physical layer loss rate)

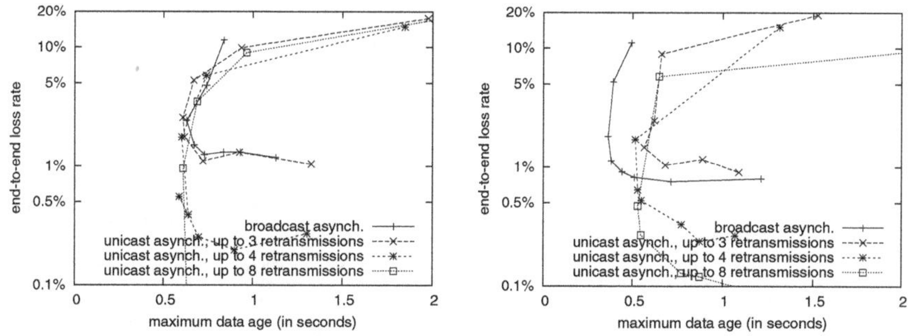


(a) Increasing Packet Size

(b) Fixed Packet Size

**Fig. 3.** Asynchronous Unicast-based and Broadcast-based Aggregation (160 nodes, 250m×250m, 20% physical layer loss rate)

increases sharply. With asynchronous aggregation, not only does the loss rate increase sharply when the sampling rate is too high, but the maximum data age also increases, since nodes increase their timeout values (and therefore their delays before sending their aggregates) to reflect the late arrival times of packets that have been retransmitted. This explains why the curves turn sharply and move to the upper right corner in Fig. 3.

The figures show that both synchronous and asynchronous broadcast-based aggregation are outperformed by their unicast-based counterparts for increasing packet size. For fixed packet size, broadcast-based aggregation is able to achieve lower maximum data age than unicast-based aggregation. Comparing Fig. 2 and Fig. 3, it appears that a broadcast-based approach may be most promising for synchronous (rather than asynchronous) aggregation. For this reason, and owing to space limitations, in the remainder of the paper only results for synchronous aggregation are presented.

### 4.3 Loss Rate, Density, and Traffic Volume

Aggregating the same data at different nodes improves reliability but also increases the packet size for broadcast-based aggregation with increasing packet size. When the physical layer loss rate is 20% as in Fig. 2, the cost of the retransmissions in unicast-based aggregation is relatively modest, and, as seen in Fig. 2(a), unicast-based aggregation outperforms broadcast-based aggregation. As the physical layer loss rate increases, however, the cost of the retransmissions in unicast-based aggregation becomes more substantial.

Fig. 4 shows that for both increasing and fixed packet size, synchronous broadcast-based aggregation is able to achieve lower maximum data age than the unicast-based protocol at 40% physical layer loss rate. The increase of the physical layer loss rate has a similar impact on the asynchronous protocols.

As the network density increases, the broadcast-based protocols are expected to achieve higher reliability as a broadcast is likely to be received by more nodes. However, for broadcast-based aggregation with increasing packet size,

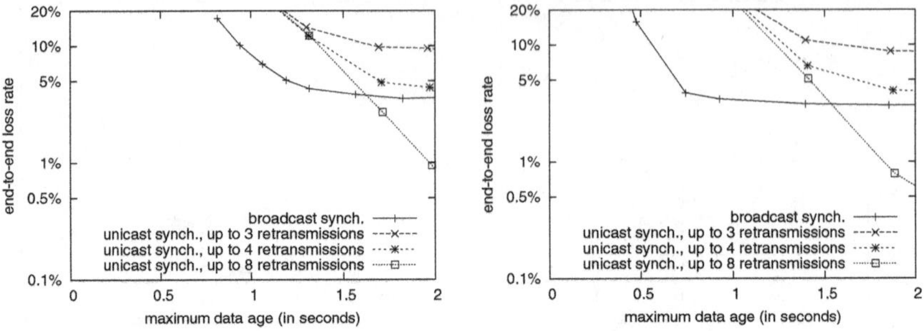


(a) Increasing Packet Size (b) Fixed Packet Size

**Fig. 4.** Impact of Physical Layer Loss Rate on Synchronous Aggregation (160 nodes, 250m×250m, 40% physical layer loss rate)

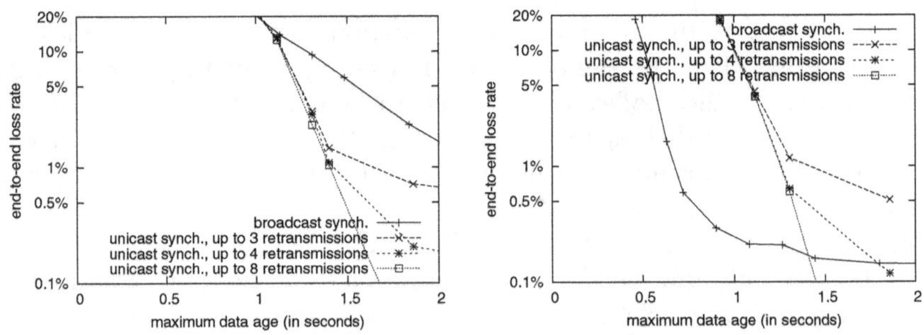


(a) Increasing Packet Size (b) Fixed Packet Size

**Fig. 5.** Impact of Density on Synchronous Aggregation (240 nodes, 250m×250m, 20% physical layer loss rate)

larger packets are produced as the same data is aggregated by more nodes, which means a longer packet transmission time and greater network congestion. Meanwhile, packet loss recovery is quite feasible for unicast-based protocols with the packet loss rate of 20% that is used for these figures. As the result of these two effects, Fig. 5 shows that synchronous broadcast-based aggregation is outperformed even more by unicast-based aggregation, for increasing packet size, when the number of nodes (and density) is increased. For fixed packet size, however, performance with the broadcast-based protocol improves but degrades with unicast-based aggregation. Similar results are seen with the asynchronous protocols.

Fig. 6 shows the average number of bytes that are transmitted per round with the synchronous aggregation protocols, for a 20% physical layer packet loss rate. While broadcast-based aggregation sends fewer packets per round than unicast-based aggregation, as shown in Fig. 7, a larger volume of data is produced by broadcast-based aggregation in the case of increasing packet size. This helps to explain why broadcast-based aggregation yields poorer performance in this case.

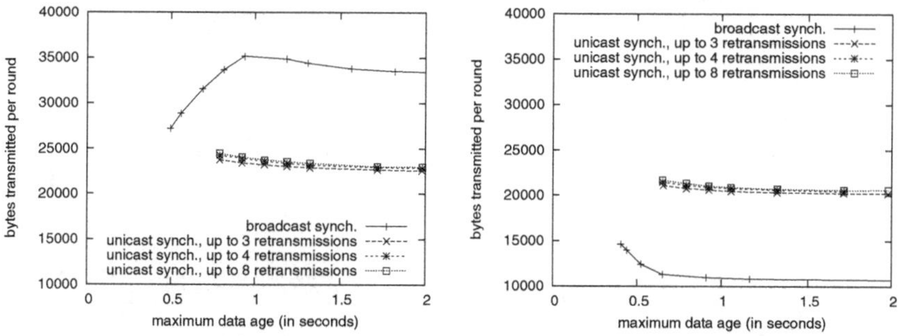


(a) Increasing Packet Size (b) Fixed Packet Size

**Fig. 6.** Average Number of Bytes Transmitted per Round of Synchronous Aggregation (160 nodes, 250m×250m, 20% physical layer loss rate)

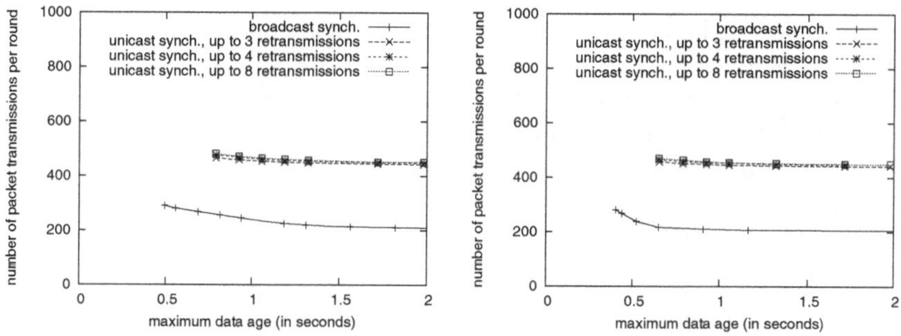


(a) Increasing Packet Size (b) Fixed Packet Size

**Fig. 7.** Average Number of Packet Transmissions per Round of Synchronous Aggregation (160 nodes, 250m×250m, 20% physical layer loss rate)

## 5 Preliminary Experimental Results

We report here on results from some preliminary experiments using implementations of the synchronous protocols on Crossbow MICAz motes[2]. For the experiments whose results are reported here, 24 MICAz motes are deployed in an area approximately 9m × 9m, with one mote at the corner as the sink. The transmission power level of the motes is set to 3, yielding an approximate transmission range of 2.5m to 3m. (However, the actual connectivity region around each node is highly irregular, as has been observed in previous studies.) In all experiments with unicast-based aggregation, the same aggregation tree shown in Fig. 8 is used, as built using a simple flooding algorithm. For synchronous broadcast-based aggregation, a ring topology is formed with nodes $i$ hops away in ring $i$. By default, the MICAz uses CSMA/CA at the MAC layer, without packet retransmission. For unicast communication, we activate the automatic retransmission functionality at

[2] Crossbow: http://www.xbow.com

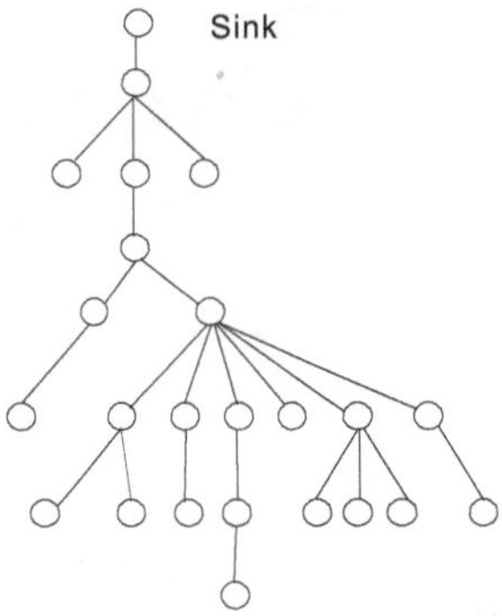


**Fig. 8.** Aggregation Tree

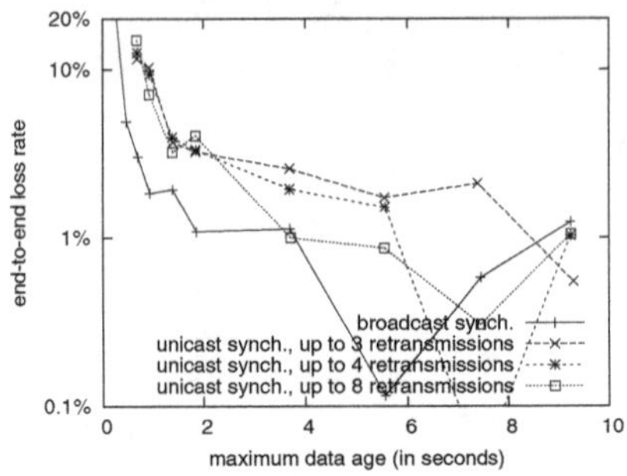


**Fig. 9.** Experimental Results for Synchronous Aggregation (24 MICAz nodes, 9m×9m, fixed packet size)

the motes and set the maximum number of retransmissions to 3, 4 and 8 in the experiments. A fixed payload size of 28 bytes is used.

We carry out multiple sets of experiments, with the experiments in each set using a range of values of the sampling period duration $t$. Each single experiment runs for 200 sampling rounds. Fig. 9 shows the measured performance of the synchronous protocols from one set of these experiments. Data from the other sets of experiments shows similar results.

Unlike the results from simulations, the experimental results show substantial variability, especially at low sampling rates. Packet loss in the experiments comes from two main sources, contention and physical layer link error. Packet loss due to contention plays a key role in the end-to-end loss at high sampling rates. As $t$ increases, the amount of packet loss from contention decreases consistently and physical layer link error becomes the main reason for packet losses. In our experimental environment, the physical layer link loss is highly bursty, and communication may fail even with large numbers of retransmission attempts.

Where a packet loss happens in the tree also has significant impact on the end-to-end packet loss rate. As can be seen in Fig. 8, there are several key nodes in the aggregation tree, the loss of whose packets results in the loss of the data of most nodes. Since each experiment only lasts for 200 sampling rounds, packet losses at those key nodes, even in just one or two rounds, can have a substantial impact on the measured end-to-end packet loss rate.

Despite the performance variability seen in Fig. 9, taking into account the differing characteristics of the experimental and simulated networks, the experimental results appear to be quite consistent with the simulation results shown in Fig. 2(b), 4(b), and 5(b).

## 6 Conclusions

In this paper, new synchronous and asynchronous broadcast-based aggregation protocols are proposed and compared to their unicast-based counterparts in the context of real-time monitoring systems. The results show that for aggregation with fixed packet size, broadcast-based aggregation is able to achieve lower maximum data age than unicast-based aggregation, particularly in the case of synchronous protocols. However, for increasing packet size, the results show no significant performance advantage (and sometimes poorer performance) with broadcast-based protocols.

## References

1. Madden, S., Franklin, M.J., Hellerstein, J.M., Hong, W.: TAG: a tiny aggregation service for ad-hoc sensor networks. SIGOPS Operating System Review 36(SI), 131–146 (2002)
2. Solis, I., Obraczka, K.: The impact of timing in data aggregation for sensor networks. In: ICC 2004, Paris, France, pp. 3640–3645 (June 2004)
3. Feng, J., Eager, D., Makaroff, D.: Asynchronous data aggregation for real-time monitoring in sensor networks. In: Akyildiz, I.F., Sivakumar, R., Ekici, E., de Oliveira, J.C., McNair, J. (eds.) NETWORKING 2007. LNCS, vol. 4479, pp. 73–84. Springer, Heidelberg (2007)
4. Heinzelman, W.R., Chandrakasan, A., Balakrishnan, H.: Energy-efficient communication protocol for wireless microsensor networks. In: HICSS 2000, Maui, HI, pp. 8020–8029 (January 2000)
5. Fan, K.W., Liu, S., Sinha, P.: Scalable data aggregation for dynamic events in sensor networks. In: SenSys 2006, Boulder, CO, pp. 181–194 (November 2006)
6. Zhang, W., Cao, G.: DCTC: dynamic convoy tree-based collaboration for target tracking in sensor networks. IEEE Trans. Wireless Commun. 3(5), 1689–1701 (2004)
7. Zhang, W., Cao, G.: Optimizing tree reconfiguration for mobile target tracking in sensor networks. In: INFOCOM 2004, Hongkong, China, pp. 2434–2445 (March 2004)
8. He, T., Blum, B.M., Stankovic, J.A., Abdelzaher, T.: AIDA: Adaptive application-independent data aggregation in wireless sensor networks. ACM Trans. on Embedded Computing Systems 3(2), 426–457 (2004)
9. Gobriel, S., Khattab, S., Mosse, D., Brustoloni, J., Melhem, R.: Ridesharing: Fault tolerant aggregation in sensor networks using corrective actions. In: SECON 2006, Reston, VA, pp. 595–604 (September 2006)
10. Motegi, S., Yoshihara, K., Horiuchi, H.: DAG based in-network aggregation for sensor network monitoring. In: SAINT 2006, Phoenix, AZ, pp. 292–299 (January 2006)
11. Nath, S., Gibbons, P.B., Seshan, S., Anderson, Z.R.: Synopsis diffusion for robust aggregation in sensor networks. In: SenSys 2004, Baltimore, MD, pp. 250–262 (November 2004)
12. Xu, K., Gerla, M., Bae, S.: Effectiveness of RTS/CTS handshake in IEEE 802.11 based ad hoc networks. Ad Hoc Netw. 1(1), 107–123 (2003)

# Event Based Fairness for Video Surveillance Sensor Networks*
## (Work in Progress)

Yunus Durmus, Bahri Atay Ozgovde, and Cem Ersoy

Computer Networks Research Laboratory
Department of Computer Engineering
Boğaziçi University
Bebek 34342 Istanbul, Turkey
{yunus.durmus1,ozgovde,ersoy}@boun.edu.tr

**Abstract.** With their ease of installation, infrastructureless mode of operation and flexible deployment style, *Video Surveillance Sensor Networks* (VSSNs) provide more opportunities than legacy surveillance methods for applications such as habitat monitoring and border surveillance. We argue that events created in the coverage area of a VSSN are the application level messaging units and propose to employ *Event Based Fairness* (EBF) which aims at a fair distribution of nodes' resources according to the event flows. We carried out simulation experiments to compare the application level performances of two different EBF implementations with that of FCFS based queueing. We observe that EBF enhances the VSSN performance in two ways: Firstly, when the video traffic due to the events created exceed the total capacity of the network, EBF increases the overall number of events properly reported to the sink. Secondly, EBF reduces the initial event reporting delay, thus decreases the response time to the occurring events within the network.

**Keywords:** Video Surveillance Sensor Networks, Fairness, Queue Management.

## 1 Introduction

When compared with legacy Wireless Sensor Networks (WSNs) that operate on scalar data such as humidity and temperature, the visual information provided by Multimedia Wireless Sensor Networks (MWSNs) in general and Video Surveillance Sensor Networks (VSSNs) in particular, increase the accuracy of event identification and decrease the false alarm rate considerably. However, this enhanced identification capability comes with the additional complexity of increased traffic volume that needs to be processed according to the realtime QoS requirements. This is very challenging since, in spite of the increased application and networking level complexity, VSSNs are typically implemented on similar hardware designed for scalar WSNs [1,2].

* This work is supported by TUBITAK under the grant number 108E207.

L. Fratta et al. (Eds.): NETWORKING 2009, LNCS 5550, pp. 40–51, 2009.

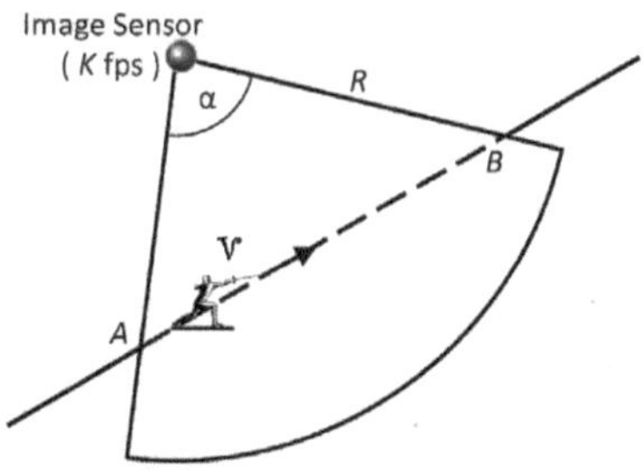


**Fig. 1.** Number of frames $F$ produced by a single sensor node upon detection is $F = K\frac{D_{AB}}{V}$ where $K$, $V$ and $D_{AB}$ stands for the camera frame rate, target speed and pathlength respectively. The pathlength $D_{AB}$, in turn, depends on the sensing radius $R$ and the FoV, $\alpha$.

A VSSN operates in event-triggered mode where nodes start pumping video frames as soon as they detect an event and continue to do so as long as the target is within the sensing radius and the Field of View (FoV). The number of frames triggered by an event is variable and as shown in Fig. 1, it is a function of the duration of the event and the camera frame rate. Event duration is actually the target residence time inside the coverage area, which in turn depends on the target speed, $V$, and the path length, $D_{AB}$, covered inside the FoV.

The data traffic created by an individual node upon detection will be termed as an *event-flow*. An event-flow is identified as the sequence of image frames produced by the same source node triggered by the detection of a target. As opposed to the time-triggered (periodic) traffic pattern, for event-triggered traffic, the number of events created per unit time may easily reach high values depending on the number and mobility of the target(s). When combined with the large video frame sizes, this leads to instantaneous traffic volumes that exceeds the capacity of the network, which in turn results in packet drops due to buffer overflow. Our aim in this work is twofold. Firstly, we want to maximize the overall visual information content conveyed to the sink in the presence of packet drops. Secondly, we aim at reducing the delay experienced by the initial frames of an event. To achieve these, we introduce an application level fairness scheme called *Event Based Fairness* (EBF). Here, we identify event flows as the application level entities and we seek their fair treatment in the network. By distributing network resources equally among event-flows, we aim at lowering the impact of packet drops on the visual information carried. We provide two implementations of EBF, namely, Round Robin Fair Queueing based EBF (EBF-RR), and Least Attained Service (LAS) scheduling based EBF (EBF-LAS). In the simulation experiments, both EBF-RR and EBF-LAS are shown to perform better than the FCFS queueing where frames are serviced in their order of arrival. EBF-RR operates on the snapshot of the queue, hence, provides fairness among flows whose frames currently exist in the queue. Event-flows in a VSSN show an intermittent behavior, therefore, steady state flow rates may not always be attained. In this respect, EBF-LAS, which considers not only the current queue content but also the service history of the event-flows, is more successful in dealing with short-lived event-flows, hence provides better event level fairness. Operationally, when event-flows have to be handled simultaneously by a sensor node, EBF-LAS gives priority to the flow which has sent the least number of frames so far. This has two implications: (i) when a packet is dropped due to overflow, it is guaranteed to belong to a flow that has sent the maximum number of frames so far, i.e. the drop will decrease the information content of a flow that already transmitted the maximum visual information, (ii) in cases when no overflow is experienced, the frames of an event which has the smallest sequence number will have priority over frames of other event flows. Therefore, the delay of the initial frames will be decreased.

The rest of this paper is organized as follows: Section 2 gives an overview of the existing fairness approaches in both wired and wireless context. The motivation behind the EBF scheme is presented in Section 3. Internals of Event Based Fairness is discussed in Section 4. Experiment setup and results obtained are presented in Section 5. Finally, Section 6 concludes the paper.

## 2 Related Work

Fairness is well-studied in the context of wired networks [3,4,5]. For wireless communication, fairness is generally discussed according to the OSI level that fairness is supported. For instance, authors of [6] advocates that MAC level fairness alone cannot ensure the fairness of the whole wireless network, although MAC support can increase the efficiency for the fairness provided at the network layer. An option to make the network independent of the fairness issues at lower layers is to achieve fairness at the transport flow layer. A centralized max-min fairness approach for wireless mesh networks (WMN) which strives to achieve end-to-end fairness at the transport layer is presented in [7]. The centralized solution discussed is justified for WMNs but it is not applicable for the WSN case whether it be a scalar or a video based WSN. There are studies that specifically address fairness in WSNs [8,9,10], among which Rangwala et al. proposes IFRC that combines fair bandwidth allocation with rate control [8]. In [11], feedback based congestion control mechanisms to enhance data delivery such as ESRT, CODA and SPEED [12,13,14] are classified as reactive and the authors come up with a collision-free scheduling that provides max-min fairness in a proactive and distributed manner. In that sense, our work also can be characterized as proactive. A similar work by Tassiulas et al. proposes a scheduling scheme which achieves max-min fairness without giving the implementation level details of the MAC protocol [15].

As a transport protocol, ESRT [12] tries to carry the optimum number of packets from an event with a feedback mechanism from the sink to the nodes. However, the effectiveness of ESRT depends on the length of decision intervals ($\approx$ 10 sec) and the feedback latency. If the duration of the event is short as in surveillance applications and the feedback latency is high (e.g., the network diameter is high), the notification may arrive to the source after the end of the event and cannot avoid congestion. Moreover, ESRT is not designed to decrease the reporting delay of the events.

As mentioned in a recent survey on *Least-Attained Service* (LAS) [16], the *Shortest Remaining Processing Time* (SRPT) is optimal for minimizing the mean response time. SRPT gives precedence to the jobs with the shortest remaining time left by assuming that the queue dispatcher is aware of the residual size of the job that has not arrived yet. However in blind systems as in WSNs, although the job size may not be known, a job's age is always known, therefore instead of SRPT, a more practical policy, LAS scheduling is a better choice. In the literature, LAS scheduling exists in different names, such as *Foreground-Background* and *Shortest-Elapsed-Time* [16]. Among them, the performance of LAS with respect to the variability of the job size is analyzed in [17]. Authors indicate that while 99% of the jobs encounter a reduced conditional mean slowdown under LAS, less than 1% of the largest jobs experience a negligible increase of their conditional mean slow down. In [18], Wierman et al. showed that LAS outperforms *Processor-Sharing* with respect to the mean response time and the mean slowdown when the job size distribution has a decreasing failure rate. Furthermore Wierman et al. presents a classification of scheduling policies considering the unfairness in [19]. Furthermore, the effect of LAS on heavy-tailed traffic in wireless networks is presented in [20]. The authors compare LAS with *Round-Robin* (RR) based scheduling and show that LAS outperforms RR in a single bottleneck link and also in a one hop wireless shared link.

## 3 Motivation

When we focus on the contents of an event-flow, we observe that there is spatio-temporal redundancy among consecutive frames. This is mainly because the camera module of the sensor node takes continuous snapshots of the scene with a certain frame rate. It is not possible to generically define the number of frames to be received at the sink for healthy reception of the event. This depends on the type of detection method run on the back end. This could range from simple event detection in which only the existence of the event is notified to the classification or the identification of the target. Also the frames received could be an input to an image recognition engine or to an human operator. Another factor is the specific positioning and movement of the target within the visual sensing range. A target closer to the camera module takes a bigger portion of the picture, however assuming that the target is mobile, proximity to the sensor also implies shorter residence time inside the sensing range, hence a shorter event-flow. Therefore, we can crudely conclude that event-flows as they become longer, they contain more frames of the scene and likely to have more redundancy among frames. The information contribution of the individual frames of a generic event-flow is depicted in Fig. 2.

With this observation in mind, we propose that irrespective of the duration of the events, initial frames of an event deserve special care. That is because they contain much of the visual information and also the delay experienced by them

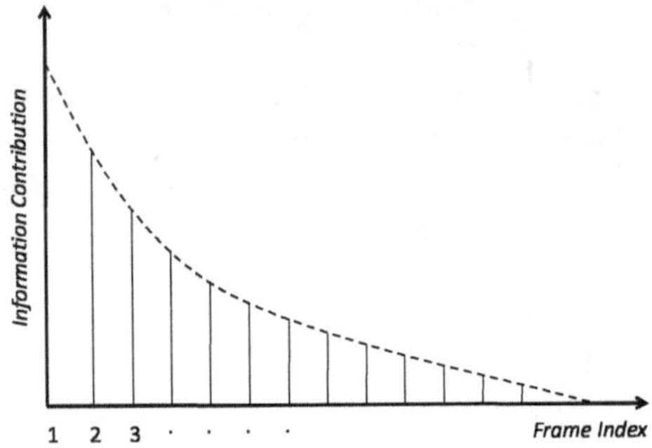


**Fig. 2.** Information contribution of individual frames of a surveillance video event-flow

directly affects the reporting delay. In this work, we give priority to the initial frames via an application level fair queue management scheme, namely *Event Based Fairness.*

## 4 Providing Application Level Fairness with EBF

WSNs, including VSSNs, have the unique characteristic that the network is designed to achieve a common specific task, in which all the nodes operate collaboratively. Opposingly, in the previous wired and wireless networking paradigms, we can see the clear distinction between the applications running on the nodes and the communication service provided by the network. In this picture, the nodes care about maximizing their own utility and not necessarily respecting the network wide resource scarcity. Therefore, previous studies on the fairness mainly focused on the per-node and the per-flow based fairness. For a WSN, on the other hand, the individual nodes may not need to obtain fair service at all times. However, fairness becomes a crucial issue when considered in terms of the application performance.

The performance of a VSSN application depends on how well the events are reported, i.e. the video quality of the events conveyed and the initial reporting delay which are related to how events are handled and processed in the network. In the standard FCFS queueing approach, the frames of the events are queued in the nodes according to the their sequence of arrival. A burst event with many frames fills the head of the queue in the relay node and other events can only utilize the space left from the former event. The rest of the events have to wait till the frames of the previous events are served. Moreover, in the case of buffer overflow, only a few of the packets of the forthcoming events can be relayed. As a result, when FCFS is preferred, while some events are reported with high quality and low latency, others are received in low quality and high latencies. In order to guarantee a certain Quality of Service (QoS) in video quality and acceptable reporting latency, we propose an event-aware fair queue management scheme called EBF that is streamlined for VSSN applications.

### 4.1 Round Robin Based Implementation of EBF

Round Robin based EBF implementation (EBF-RR) strives to give fair service to all events that are currently enqueued in a VSSN node. The queue is composed of frames received from the network for relaying purposes and the frames received from the application layer, i.e. the video frames produced by the node itself. EBF-RR operates by servicing frames of events in a round robin manner, one frame from each event at a time. Internally, EBF-RR dynamically forms logical queues for each event and gives service to each queue in a time-shared manner. The duration in which all event queues are served once is called an *epoch*. During an epoch, the available bandwidth is equally divided among each event. The overall service rate an event gets from EBF-RR depends on the length of the event (in terms of frames) occupied in the main queue, the total number of events in the queue, the length of each event in the queue and the congestion level experienced at the MAC level (available effective bandwidth). When the incoming frames are more than the capacity of the node, buffer overflow occurs. In that case, EBF-RR drops the frame from the longest event queue. With this scheme a received frame that arrives at the full main buffer need not be dropped unless it belongs to the event that currently has the longest logical queue. When compared with the FCFS behavior, EBF-RR provides fair bandwidth allocation to events and also gives priority to events with fewer frames. This latter property is especially more pronounced in the case of buffer overflows in which frames of events with longer queues are dropped. In that sense, EBF-RR tries to homogenize the service rate among events according to the snapshot of the queue.

One point to note in the above discussion is that the queue manipulation is done in terms of frames and not packets. Therefore in our VSSN implementation, *SMAC* [21] with *Message Passing* feature is used as the MAC layer. Since Message Passing allows frames to be passed among nodes intact which makes our assumption about frame based queue manipulation possible.

### 4.2 LAS Based Implementation of EBF

The main idea behind LAS Based EBF (EBF-LAS) is that an event is a sequence of frames flowing in the network and at a specific time instance, only a portion of it may be contained in the buffer of a VSSN node. This is due to the buffer size limitations and earlier frame drops that an event may experience. EBF-RR operates on the instantaneous snapshot of the buffer and provide fairness among events according to what is currently present. In this respect, a way to provide better fairness among events is to consider not only the current buffer composition but also to take into account the frames of an event that has been relayed previously. EBF-LAS, like EBF-RR, forms logical queues of frames per event and service one frame from each queue in an epoch in a round robin fashion. However, unlike the RR implementation, EBF-LAS keeps track of the sent frames and inserts a virtual frame to the event queues as place holders for each frame of an event that is relayed. Therefore, a logical queue for an event contains both real frames that are waiting to be send and virtual frames that

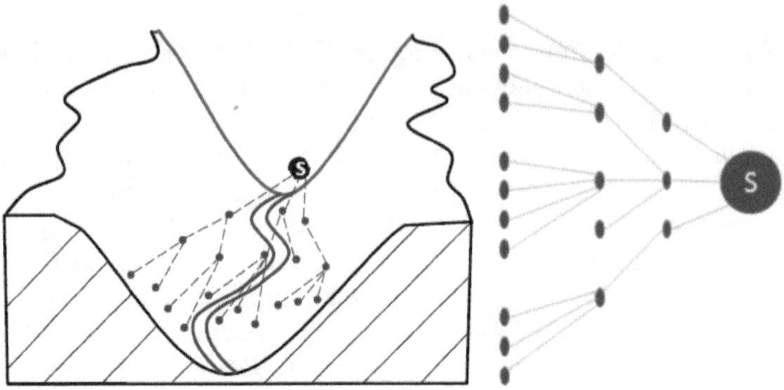


(a) Trespassers' Favorite Path (TFP) (b) Detailed view of the TFP multi-hop simulation scenario

**Fig. 3.** Deployment Scenario. Intruders follow the favorite path in which the sensors are deployed more densely.

are already sent. In every logical event queue, virtual frames are placed in the front of the queue, therefore, when deciding on the next frame to get relayed, EBF-LAS gives explicit priority to the events that have fewer frames sent.

# 5 Comparative Evaluation of EBF-RR, EBF-LAS and FCFS

## 5.1 Experimental Setup

We examine the effect of EBF-RR and EBF-LAS using the OPNET simulation environment [22]. In order to observe the improvements on the reporting latency and the video quality of events in detail, a surveillance scenario is examined. In a geographical area that is under surveillance not every path is equally likely to be used. The paths which intruders have higher tendency to follow are called *Trespassers' Favorite Path* (TFP) [23]. TFPs are preferred over other alternatives due to the reasons such as easy geographical conditions and remoteness to checkpoint locations. For effective operation, existence of TFPs should be considered when designing a surveillance network, as large portion of the total traffic is likely to be originated from sensor nodes located within the TFPs. In our surveillance scenario, we have simulated the traffic created in a valley-type TFP which contains 19 VSSN nodes, as depicted in Fig. 3.

In video surveillance applications the volume of the data traffic is related to the dwell time of the target, the camera frame rate and the compression algorithm. The duration of an event and the number of frames created during the event varies according to these parameters. In our tests, we model the traffic creation using event size, $\Psi$, frame interarrival time, $\Delta_F$, and event interarrival time, $\Delta_E$. $\Psi$ denotes the number of frames contained in an event, whereas $\Delta_F$ is the frame

**Table 1.** Simulation Parameters

| **Parameter** | **Value** |
|---|---|
| Event Size ($\Psi$) | Normal distributed with ($\mu = 15$,$\sigma^2 = 7$) Frames |
| Video Frame Size | 10 Kbits |
| Packet Size | 1 KBits |
| Frame Interarrival Time ($\Delta_F$) | Uniformly distributed with $\mu = 1/3$ sec |
| Event Interarrival Time ($\Delta_E$) | Exponentially distributed with $\mu = 25$ sec |
| Duty Cycle | $10\%, 20\%^*, 30\%, 40\%$ |
| Bandwidth | 250 Kbps |
| Buffer size | 100 Kbits |
| MAC layer | SMAC [21] with Message Passing feature |
| Number of Repetitions | 20 |
| Confidence Interval | 95% |

(*)Unless otherwise specified, 20% duty cycle is the default in the experiments.

generation rate of the camera modules of the VSSN nodes. $\Delta_E$ models the time between two consecutive events a sensor node detects. Values for $\Psi$, $\Delta_F$, $\Delta_E$ and other related simulation parameters are presented in Table 1.

Preliminary experiments are run to fix the buffer capacity that is allocated on the individual nodes. Enlarging the buffer enhances the throughput at the expense of increased delay. After a threshold value, larger buffer sizes result in intolerable delay levels [24]. The chosen buffer size (100 Kbits) is within an operationally feasible region in which the delay-throughput balance is observed. Please refer to [25] for a detailed study in which the effects of the factors like buffer size, camera frame rate and MAC duty cycle on the VSN performance are systematically examined.

### 5.2 Results

A histogram summarizing the events according to the number of successfully received frames at the sink is presented in Fig. 4(a). Out of the 2220 events generated, for the FCFS case, for instance, around 130 events are reported only with a single frame whereas around 90 events are reported with 14 frames. The number of events reported for all queueing mechanisms are close to each other 2122, 2215 and 2211 for FCFS, EBF-RR and EBF-LAS respectively. However, it is observed that the variance in the frequency of the frames per event is decreased by EBF-RR and EBF-LAS. In the $1-3$ fps interval and $10-20$ fps interval, the number of events are less in EBF-RR and EBF-LAS cases than FCFS since EBF-LAS and EBF-RR decreases the number of over reported events and share the available excess bandwidth among the under reported events. Thus, most of the events are reported similarly which is due to the fair treatment of frames according to the related events.

The total number of frames required to be received at the sink in order a triggered event to be considered as *detected* depends on the application. However,

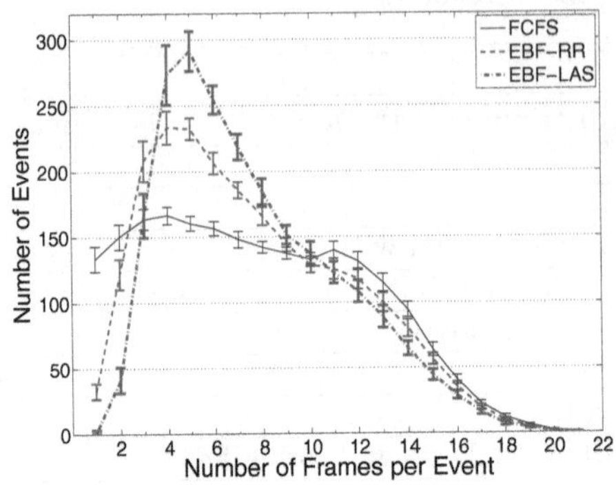


(a) Histogram of the number of frames received at the sink for each event.

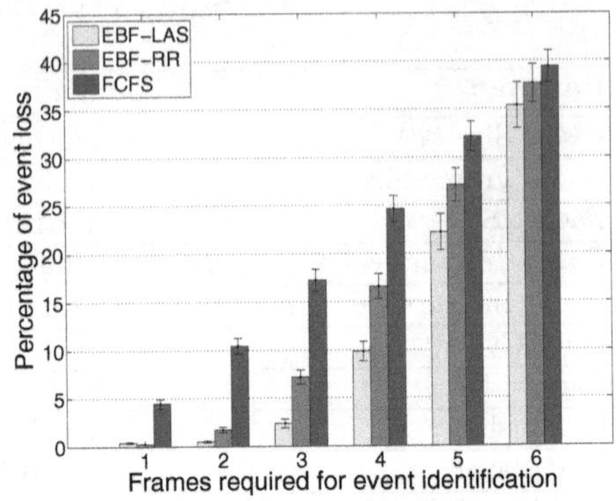


(b) Missed event ratios with different number of frames required for the identification.

**Fig. 4.** Histogram and miss ratio for events which are composed of variable number of frames

as previously pointed out, it is the initial frames of an event that contribute more to the visual information received at sink. In order to observe the effect of EBF-LAS and EBF-RR on the reliable event reporting, we plot the ratio of missed events in Fig. 4(b). As expected, when the required frames per event is increased, missed event ratio also increases in all queueing techniques. However, the ratio of missed events are clearly less in EBF-RR and EBF-LAS cases compared to that of the FCFS case. For instance, when the required frames for event identification is set to 4, while FCFS misses 24% of the events, EBF-RR misses 17% and EBF-LAS misses less than 10%. Additionally, the difference between the FCFS, EBF-RR and EBF-LAS systems decreases as the number of required frames increases. The reason is that EBF-LAS punishes the large events while giving precedence to smaller ones especially when the network load becomes high.

Besides the video quality, the reporting latency of the events are also important. Especially the frame of an event which arrives first to the sink has the most significant contribution for the event reporting since it makes the sink aware of that event. Fig. 5(a) presents the average of the first frame latency of the events with various duty cycles and indicates that EBF-RR and EBF-LAS improve the event reporting delay significantly compared to the FCFS case.

To have a more general understanding of the latency behavior of the events, Fig. 5(b) depicts the average delay a certain frame of an event experiences, e.g., the average latency of the $8^{th}$ frame of the events. It is observed that EBF-LAS decreases the delay for all frames of the events, whereas EBF-RR performs better than FCFS up until the $4^{th}$ frames of the events. In other words EBF-LAS decreases the mean response time of the events.

On the other hand, when the duty cycle is increased, most of the frames arrive to the sink. Therefore as in Fig. 6(a), the difference between the intelligent queueing techniques and the FCFS decreases. However, as observed in 6(b),

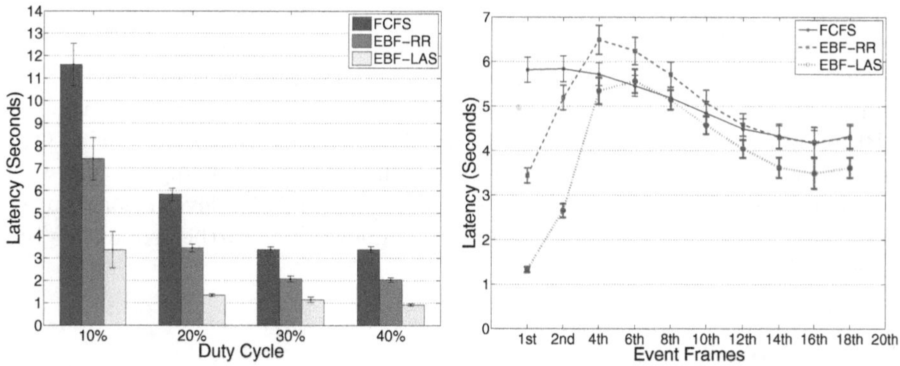


(a) Average delay for the first arriving frame of an event. (using various duty cycles)

(b) Average latency of frames.

**Fig. 5.** Frame latencies of the events.(20% default duty cycle)

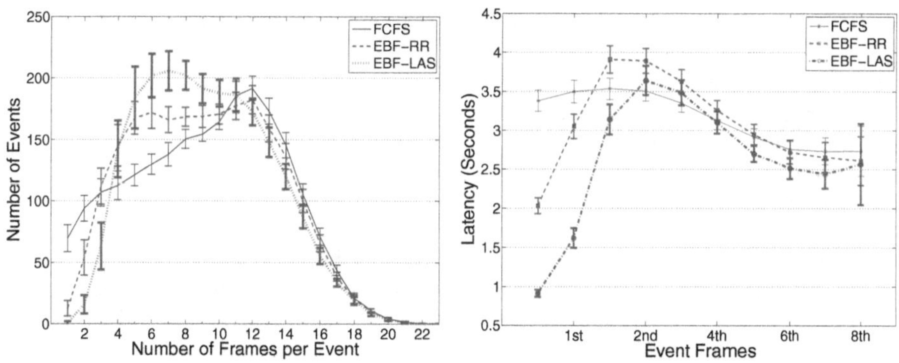


(a) Histogram of the number of frames reached to sink for each event

(b) Average latency of frames

**Fig. 6.** Latency and Histogram of the events. (40% duty cycle).

compared with FCFS, EBF-RR and EBF-LAS can relay all frames without a significant sacrifice in terms of latency.

## 6 Conclusions and Future Work

In this work, *Event Based Fairness* (EBF) for Video Surveillance Sensor Networks (VSSNs) is introduced. Main goals of EBF are to increase the event reception ratio at the sink node and to decrease the initial reporting delay of the events. There is inherently high volume of traffic in a VSSN when an intruder(s) is present. Most of the time, the traffic produced is more than that can effectively be carried by the network. To be able to achieve its design goals, EBF defines an *event*, as opposed to a video *frame* or a network *packet*, to be the logical messaging unit in a VSSN and introduces the fair treatment of events which are application level entities. Two different EBF implementations, namely EBF-RR and EBF-LAS are compared with the legacy FCFS style queue management. EBF is shown to not only increase the number of events that are reported properly but also to lower the initial reporting delay considerably. As a future work, we plan to implement cross-layer solutions in which MAC level priority mechanisms are co-implemented with the EBF framework to further enhance VSSN functionality.

## References

1. Downes, I., Rad, L., Aghajan, H.: Development of a mote for wireless image sensor networks. In: Proc. of COGnitive systems with Interactive Sensors (COGIS), Paris, France (March 2006)
2. Rahimi, M., Baer, R., Iroezi, O.I., Garcia, J.C., Warrior, J., Estrin, D., Srivastava, M.: Cyclops: in situ image sensing and interpretation in wireless sensor networks. In: SenSys 2005: Proceedings of the 3rd international conference on Embedded networked sensor systems, pp. 192–204. ACM Press, New York (2005)
3. Demers, A., Keshav, S., Shenker, S.: Analysis and simulation of a fair queueing algorithm. In: SIGCOMM 1989: Symposium proceedings on Communications architectures & protocols, pp. 1–12. ACM, New York (1989)
4. Shreedhar, M., Varghese, G.: Efficient fair queueing using deficit round robin. SIGCOMM Comput. Commun. Rev. 25(4), 231–242 (1995)
5. Bertsekas, D., Gallager, R.: Data networks, 2nd edn. Prentice-Hall, Inc., NJ (1992)
6. Jun, J., Sichitiu, M.: Fairness and qos in multihop wireless networks. In: IEEE 58th Vehicular Technology Conference, VTC, vol. 5, pp. 2936–2940 (2003)
7. Raniwala, A., De, P., Sharma, S., Krishnan, R., cker Chiueh, T.: End-to-end flow fairness over ieee 802.11-based wireless mesh networks. In: IEEE INFOCOM, pp. 2361–2365 (2007)
8. Rangwala, S., Gummadi, R., Govindan, R., Psounis, K.: Interference-aware fair rate control in wireless sensor networks. In: Rizzo, L., Anderson, T.E., McKeown, N. (eds.) SIGCOMM, pp. 63–74. ACM, New York (2006)
9. Fan, K.-W., Zheng, Z., Sinha, P.: Steady and fair rate allocation for rechargeable sensors in perpetual sensor networks. In: Proc. of ACM SENSYS, Raleigh, NC (November 2008)
10. Chen, S., Fang, Y., Xia, Y.: Lexicographic Maxmin Fairness for Data Collection in Wireless Sensor Networks. IEEE Transactions On Mobile Computing, 762–776 (2007)
11. Sridharan, A., Krishnamachari, B.: Max-min fair collision-free scheduling for wireless sensor networks. In: Workshop on Multihop Wireless Networks (MWN 2004), IPCCC (2004)
12. Akan, O.B., Akyildiz, I.F.: Event-to-sink reliable transport in wireless sensor networks. IEEE/ACM Trans. Netw. 13(5), 1003–1016 (2005)
13. Wan, C.-Y., Eisenman, S.B., Campbell, A.T.: Coda: congestion detection and avoidance in sensor networks. In: SenSys 2003: Proceedings of the 1st international conference on Embedded networked sensor systems, pp. 266–279. ACM, New York (2003)

14. He, T., Stankovic, J.A., Lu, C., Abdelzaher, T.F.: Speed: A stateless protocol for real-time communication in sensor networks. In: ICDCS, pp. 46–55 (2003)
15. Tassiulas, L., Sarkar, S.: Maxmin fair scheduling in wireless networks. In: INFOCOM (2002)
16. Nuyens, M., Wierman, A.: The Foreground-Background queue: A survey. Performance Evaluation 65(3-4), 286–307 (2008)
17. Rai, I., Urvoy-Keller, G., Biersack, E.: Analysis of LAS scheduling for job size distributions with high variance. In: Proceedings of the 2003 ACM SIGMETRICS international conference on Measurement and modeling of computer systems, pp. 218–228. ACM Press, New York (2003)
18. Wierman, A., Bansal, N., Harchol-Balter, M.: A note on comparing response times in the M/GI/1/FB and M/GI/1/PS queues. Operations Research Letters 32(1), 73–76 (2004)
19. Wierman, A., Harchol-Balter, M.: Classifying scheduling policies with respect to unfairness in an M/GI/1. ACM SIGMETRICS Performance Evaluation Review 31(1), 238–249 (2003)
20. Shao, Z., Madhow, U.: Scheduling heavy-tailed data traffic over the wireless Internet. In: Proceedings of IEEE 56th Vehicular Technology Conference, 2002. VTC 2002-Fall, vol. 2 (2002)
21. Ye, W., Heidemann, J., Estrin, D.: Medium access control with coordinated adaptive sleeping for wireless sensor networks. IEEE/ACM Trans. Netw. 12(3), 493–506 (2004)
22. Inc, O.: OPNET Modeler
23. Komar, C., Ersoy, C.: Optimization of power consumption using trespassers favorite path and variable sensing range integrated sleep schedule in surveillance wireless sensor networks. In: 23rd International Symposium on Computer and Information Sciences, 2008. ISCIS 2008, pp. 1–5 (2008)
24. Ozgovde, A., Demirkol, I., Ersoy, C.: Effect of sleep schedule and frame rate on the capabilities of Video Sensor Networks. In: 3rd International Symposium on Wireless Pervasive Computing, 2008. ISWPC 2008, pp. 156–159 (2008)
25. Durmus, Y., Ozgovde, A., Ersoy, C.: Exploring the effect of the network parameters of video sensor networks. In: Proceedings of ISCN 2008 8th International Symposium on Computer Networks, pp. 188–192 (2008)

# Humpty Dumpty: Putting iBGP Back Together Again

Ashley Flavel[1], Jeremy McMahon[2], Aman Shaikh[3], Matthew Roughan[1], and Nigel Bean[1]

[1] School of Mathematical Sciences, University of Adelaide
{ashley.flavel,matthew.roughan,nigel.bean}@adelaide.edu.au
[2] TRC Mathematical Modelling, University of Adelaide
jeremy.mcmahon@adelaide.edu.au
[3] AT&T Labs - Research
ashaikh@research.att.com

**Abstract.** Humpty Dumpty is the anthropomorphic nursery-rhyme egg broken into many pieces. Similarly, we have many pieces of measurement data to represent the current iBGP state. However, unlike the nursery-rhyme where the King's men couldn't put Humpty together again, we present a systematic approach to putting all the pieces of measured iBGP data together to obtain a more complete picture of a network's routing state.

Our technique determines the decisions made by all routers in a network. It is efficient, has no assumptions about router configuration and is accurate. We present a case-study of a large Tier-2 ISP, finding for those routers with adequate measurement infrastructure, we consistently find the egress location for 99.9999% of $(router, prefix)$ pairs. Further, for the 85% of routers without measurement infrastructure we predict their decisions. This technique has been successfully applied in a 'what-if' scenario and has future applications in the real-time analysis of routing decisions.

**Keywords:** iBGP, Route prediction.

## 1 Introduction

Measurement plays a crucial role in management of IP networks since it allows operators to determine how the network is currently operating. The measurement data can be used for tasks such as deriving traffic demands in operational networks [1], finding traffic matrices [2] and their dynamics [3], and oscillation detection [4]. A majority of such tasks require some knowledge of the path traffic takes through a network — hence the need for routing measurements. However, due to high storage requirements, operational setup costs and dependency between routes selected across routers, BGP monitors collecting BGP routing information are often only connected to a subset of routers. In this paper we provide a methodology to make use of the high dependency between router decisions to systematically "fill in the gaps" left by partial measurements.

L. Fratta et al. (Eds.): NETWORKING 2009, LNCS 5550, pp. 52–65, 2009.

An Autonomous System's (AS) BGP routing decisions are not atomic. When multiple routes are available to a destination, individual routers within the AS can make different decisions as to their selected route based on their own perspective of the 'best' route. The *network solution*, that is, the decision of all routers in the network for a particular destination, is dependent on the *subset* of AS-wide routes which are learned at each individual router. Hence, it is not valid to assume *all* AS-wide routes are learned at each router for selection [5]. The iBGP configuration employed, such as full mesh or route-reflection [6], determines whether all routes or a subset of routes are available at every router. In this paper we focus on the route reflector iBGP configuration since it is used widely in large enterprise and service provider networks.

In [4], we introduced a model to analyze the oscillatory properties of a two-tier route-reflector (RR) iBGP topology. We now extend this model to determine the network solution of a general RR iBGP topology. The model captures the reliance of a router on other routers for choosing its best route. This model underpins a methodology for determining routes selected by *all* routers based on the knowledge of routes selected by a subset of routers and the iBGP configuration. Another benefit of our methodology is that it can also be used for 'what-if' analysis. Compared to the methodology proposed by Feamster and Rexford [7], which provides similar functionality, our methodology is applicable to *any* RR iBGP configuration, not just configurations satisfying recommendations of Griffin [8]. Further, our approach is topology independent making it extensible to topologies other than a RR iBGP configuration.

We applied our methodology to the topology of a large Tier-2 AS and using measurements collected from 15% routers (mostly RRs), we could determine routes for all the routers in the network. Of over 12.7 million routing decisions, we predicted a decision consistent with the observed data for all but seven routers. In the process, we also detected several configuration and data collection issues on routers when routes predicted by our methodology were inconsistent with the measurement data — highlighting an additional benefit of our analysis.

## 2 Background and Related Work

The Internet is comprised of a collection of Autonomous Systems (ASes) which co-operate to ensure connectivity between any two hosts. BGP is the protocol used to disseminate reachability information between ASes. A given AS has several routers at its border that connect with other ASes. These routers use BGP to learn routes to external prefixes from other ASes. These routes are then propagated to other routers inside the AS using iBGP (internal BGP). Every router selects a 'best' route from all the routes learned for every prefix. The selected route is then sent to other routers. For route dissemination (via iBGP), routers inside an AS need to form a full mesh of sessions. However, in large networks, a RR hierarchy is used to mitigate scalability issues of a full mesh. One consequence of RR hierarchy is that the set of routes available at each router is usually a subset of all routes learned across AS [5].

Prior work has recommended guidelines for designing iBGP configurations [9, 10, 11], proposed alterations to iBGP to disseminate more information AS-wide [12,13] and proposed centralized AS-wide route selection [14]. Our approach is orthogonal to these. Our aim is very pragmatic — to understand the current operation of a network — irrespective of whether it satisfies certain guidelines or not.

One approach to discovering the network solution is to simulate BGP by propagating information in an arbitrary order between routers until no router alters its decision (for example, C-BGP [15]). However using this approach, a significant number of intermediate router states are evaluated prior to converging to an arbitrary final solution (there may be multiple feasible solutions [4]). Consequently, determining if a network is in the convergence process or if it is persistently oscillating is difficult. In contrast, we avoid many intermediate states to quickly find a valid solution and more importantly, converge to a solution consistent with observed data. This enables our approach to predict the route traffic actually takes in the network. In addition, if the configuration has an oscillatory state, we can quickly identify it and pinpoint the responsible routers.

The most closely related work to ours is by Feamster and Rexford [7]. Their motivation was to predict the network solution *as designed*. That is, they assume recommended guidelines for network configuration are satisfied resulting in a unique network solution. We make no such assumptions allowing the network to be analyzed as it is currently operating — whether it satisfies guidelines or not. In addition, they assume complete visibility of input routes. In contrast, our technique works with even limited knowledge of input routes from the network. Further, our technique is designed to use observed data to influence which of multiple network solutions is actually chosen by the network. Finally, it can efficiently analyze the impact of small changes to the network without a significant re-analysis.

The primary assumption of Feamster and Rexford requires all RRs to prefer a client (a directly connected router in a lower level of the RR hierarchy) learned route over any other. This constraint is a *sufficient* condition to prevent persistent oscillation and guarantees a unique solution [8]. However as the condition is not *necessary*, it can be overly restrictive and not satisfied in practice [10, 4]. Removing this assumption removes the benefit of always converging to a unique solution — the timing of BGP updates can determine which of multiple solutions is settled upon. In addition, the tie-breaking option employed in operational networks, such as the one we examined, can be non-deterministic, resulting in an even greater number of feasible network solutions. Our technique always converges to a valid solution and in almost all instances, converges to a solution consistent with the observed data.

## 3 Two-Level Route-Reflector Reliance Graph

We first introduced, in [4], the concept of *reliance* between router decisions to determine if a network configuration was oscillatory. In this paper, we use reliances to efficiently and accurately determine the *actual* routes selected by any router. We say a router $u$ is reliant on another router $v$ if it can learn of its best route for a particular prefix (after convergence) from $v$. We denote this reliance as $u \leftsquigarrow v$. Reliances are represented by a directed edge in the direction of information flow in the reliance graph. Routing information can only flow over iBGP sessions between routers. Consequently, the reliance graph is a sub-graph of the iBGP signaling graph. The rules governing route-propagation in an iBGP topology determine which links are pruned from the signaling graph to form the reliance graph. Note that a reliance graph is a directed graph and we term strongly connected components *co-reliance groups*[1].

By the construction of the reliance graph, the only location a router can learn of its best route is from an inbound edge from a neighboring router in the reliance graph. Consequently, if all neighbor's decisions are static, then the router's decision is also static (as the set of available routes remains constant). Hence, if there are only singleton co-reliance groups in the reliance graph, then a topological sort of all routers will result in an ordering where all routers' decisions must only be evaluated once. However, if there are non-singleton co-reliance groups, then we cannot topologically sort the routers (as there is at least one loop in the reliance graph). We can still topologically sort the co-reliance groups. By evaluating each co-reliance group in a topological ordering, we can ensure we do not need to re-visit a co-reliance group, but we may need to visit routers *within* a non-singleton co-reliance group multiple times.

Consider Fig. 1(a) as an example. In this example we have not shown the full-mesh iBGP sessions between RRs. When a RR is closer to a non-client egress router[2] the IGP distances are shown. Routers $e$, $f$ and $g$ learn routes from outside the AS. We create reliances when a router can learn of its best route from another (see Fig. 1(b)). For example, $a$ is reliant on $e$ as $e$ propagates its externally learned route to $a$. However $a$ is also reliant on $b$ as $b$ can inform $a$ of the route which it learns from $f$. Similarly, $b$ is reliant on $a$. As $c$ is closer to its own client ($g$) than any non-client, it will never select any route learned via another RR. Router $d$ will select the best route from routes learned from other RRs. Clients $h$ and $i$ will only ever select a route which they learn from their parent RRs. In Fig. 1(c) we highlight all strongly connected components in the reliance graph. Evaluating router decisions in any topological ordering of co-reliance groups (for example the numerical ordering $D_1, D_2, ...$) will result in a valid solution. Notice there are two valid solutions to this example based on the order of evaluation of routers in $D_5$. If we evaluate $a$'s decision first, both routers in $D_5$ will select the egress $e$. Conversely if we evaluate $b$'s decision first, both routers will select egress $f$.

The reliance rules in the three-level hierarchy are somewhat more complicated than the two-level case examined in [4]. We first present a brief recap of the notation used in [4] before outlining generalized reliance rules and showing an example of how they apply. Similar techniques can be applied to any iBGP topology if the topology can be abstracted to a reliance graph.

[1] For every $u$ and $v$ in a strongly connected component, there is a path from $u$ to $v$.

[2] An egress router is a router which learns a route directly from a neighboring AS.

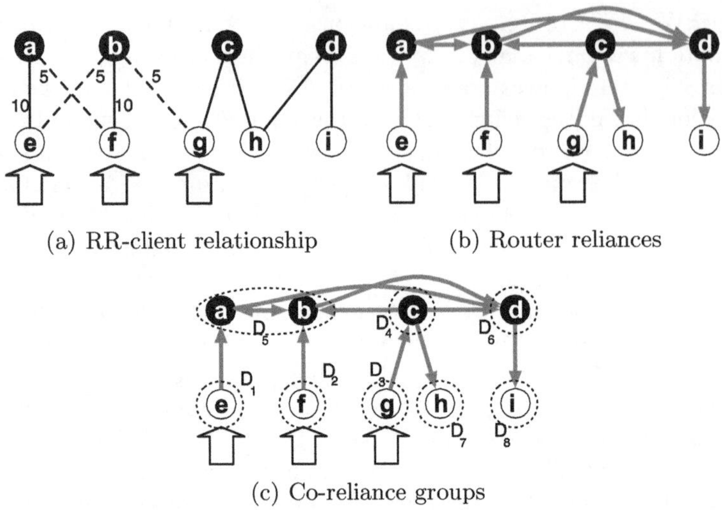


**Fig. 1.** Example two level RR topology. Solid nodes are RRs and transparent nodes are clients. IGP distances shown when non-client router is closer than client router, thus for example $a$ prefers $f$ over $e$ when possible. Large arrows indicate a route is learned from a neighboring AS.

## 4 General Route-Reflector Reliance Graph

The rules we examined in [4] were only applicable to the two-level RR hierarchy. We now generalize these rules to a multi-level hierarchy. The rules governing reliance in the multi-level hierarchy are significantly more complex than the two-level case as RRs can hide information propagated to their parents (see [16] for details). Consequently, we use an example topology shown in Fig. 2(a) as we describe the rules to assist the reader in following the concepts. In this figure, routers $1, 2, 3$ and $4$ form the central core mesh of RRs, while $c, d, g, h, i, l$ and $n$ form the middle level RRs. The solid lines represent iBGP sessions and the dashed lines represent a router's preference for a non-downstream egress. Where no dashed line exists, a downstream egress is preferred. We explicitly define several vital preferences in the caption of the figure [3]. Routes learned directly from neighbor ASes are denoted by large arrows, i.e., routers $b, e, f, i$ and $m$ are the egress routers.

### 4.1 Notation Recap

We now present a brief recap of the important notation used for the remainder of this chapter. For a thorough description, we refer the reader to [4].

An iBGP configuration $C$ is a pair $C = (\mathcal{G}_\mathcal{P}, \mathcal{G}_\mathcal{S})$ where $\mathcal{G}_\mathcal{P}$ is the physical graph on which the IGP is run to determine the shortest path between two

[3] The ranking function $\lambda$ is defined later in Section 4.1.

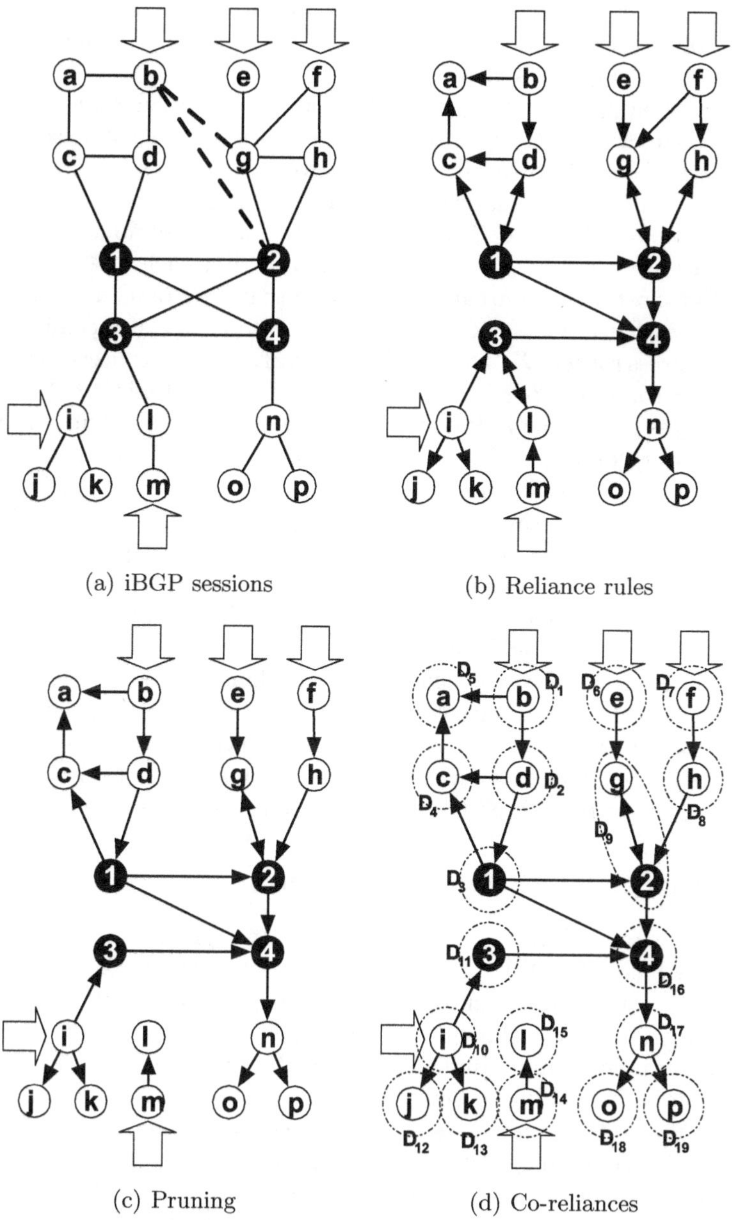


**Fig. 2.** An example 3-level RR topology. Black nodes represent RRs at the top level. Arrowed lines represent a reliance. We explicitly define the following preferences: $\lambda_g(b) > \lambda_g(e) > \lambda_g(f), \lambda_2(b) > \lambda_2(e) > \lambda_2(f), \lambda_3(i) > \lambda_3(m), \lambda_4(m) > \lambda_4(i) > \lambda_4(b) > \lambda_4(e) > \lambda_4(f)$.

routers. The iBGP signaling graph $\mathcal{G}_S = (V, A_S)$ is overlaid on top of the physical graph with routers $V$ connected by directed arcs in $A_S$.

Three types of arcs exists in $A_S$. An arc $(u, v) \in$ **down** represents an arc from a RR $u$ to one of its clients $v$. An arc $(u, v) \in$ **up** if and only if $(v, u) \in$ **down**. Arcs in **up** are acyclic — consistent with a hierarchy rather than an arbitrary network design. An arc $(u, v) \in$ **over** represents a vanilla iBGP session from router $u$ to $v$. If $(u, v) \in$ **over** then $(v, u) \in$ **over**.

A valid signaling path $S$ satisfies the following property. The path $S$ can be split into sub paths $S = PQR$ where $P = p_1p_2...p_a$ for some $a \geq 0$ such that each $p_i \in$ **up**, $R = r_1r_2...r_b$ for some $b \geq 0$ such that each $r_i \in$ **down** and $Q$ consists of at most a single arc $q \in$ **over**. Note that $P, Q$ or $R$ may be empty.

An *egress instance* [8] $I = (C, X)$ corresponds to a pair of a configuration $C$ and a set of egress routers $X$. The set $X$ consists of all egress routers that learn an external BGP route to a particular prefix which are not eliminated by the BGP decision process (up-to the IGP distance step) when compared with all AS-wide routes [7]. In our example of Fig. 2(a), $b, e, f, i$ and $m$ form $X$. An *egress ancestor set* $E$ can be recursively defined as the set of egress routers $X$ and all parents of routers in $E$. In our example, $E = \{b, d, e, f, g, h, i, m, k, l, 1, 2, 3\}$. Note that although an egress router may learn multiple routes (to a prefix) it will only advertise its best route to neighbors. Hence, there is a one-to-one mapping from egress routers to available routes. Therefore, we will refer to an egress router and its available route interchangeably.

The BGP decision process is denoted by a ranking function $\lambda_u$ for a router $u$ such that if a route $a_k$ is preferred over a route $a_j$ at router $u$, then $\lambda_u(a_k) > \lambda_u(a_j)$. If two routes $a_k$ and $a_j$ are equivalent up-to the tie-break option and the actual route chosen is dependent on message timing, then $\lambda_u(a_k) = \lambda_u(a_j)$. For convenience, we denote the preference of the null route $\phi$ as $\lambda_u(\phi) = -\infty$.

### 4.2 Reliance Rules for Route Reflection

In this section we generalize the reliance rules we previously defined for the two-level RR hierarchy [4] to an arbitrary hierarchy. Although there is a strict set of reliances which are a subset of arcs (of type **up, down** and **over**) in the signaling graph $A_S$, defining where a router can learn of its best route in an $n$-level hierarchy is more difficult than in the two-level case. An important consideration is that failing to define reliances can result in *incorrect* decisions, while defining additional reliances simply increases the computational complexity of predicting selected routes (as it may create larger co-reliance groups than really exist). Consequently, we start with a relatively conservative definition of reliances before pruning many of those which cannot exist. We assume the MED attribute is filtered or compared AS-wide in this section.

**Downstream Egress Set.** Let us generalize the best downstream egress function defined in [4] to return a *set* of *downstream* egresses $\Lambda(u)$ for a router $u$. If $u$ has no downstream egresses, $\Lambda(u) = \phi$. Unlike the two-level hierarchy, in an arbitrary hierarchy, it is no longer guaranteed that a router will learn of all

**Table 1.** Downstream egress sets for routers in example topology of Fig. 2

| Router($u$) | 1 | 2 | 3 | $d$ | $g$ | $h$ | $l$ | all other routers |
|---|---|---|---|---|---|---|---|---|
| $\Lambda_1(u)$ | $\phi$ | $\phi$ | $i$ | $b$ | $e$ | $f$ | $m$ | $\phi$ |
| $\Lambda_n(u)$ | $b$ | $e, f$ | $\phi$ | $\phi$ | $\phi$ | $\phi$ | $\phi$ | $\phi$ |

downstream egresses since the set of available routes is restricted by the selection of intermediate routers.

We first define $\Lambda_1(u)$ as the set of best downstream egresses which are one downstream iBGP hop away from $u$. Router $u$ is guaranteed to learn of these routes due to the direct iBGP session and so these routes will always be available. Formally, for $u \notin X$,

$$\Lambda_1(u) = \{v \in X : (u,v) \in \textbf{down} \text{ and } \lambda_u(v) = \max_{w \in X:(u,w) \in \textbf{down}} \lambda_u(w)\}$$

We show for our example in Fig. 2 the sets $\Lambda_1(u)$ for all routers in Table 1. Notice that as $\lambda_g(e) > \lambda_g(f)$, $\Lambda_1(g) = e$.

Now, let us consider other egresses $u$ could learn (and select as best) from clients which are not direct egresses, i.e., those more than one hop away. We denote this set by $\Lambda_n(u)$ and define it for $u \notin X$ as,

$$\Lambda_n(u) = \bigcup_{w \in E \backslash X:(u,w) \in \textbf{down}} \{v \in \Lambda(w) \backslash \Lambda_1(u) : \lambda_u(v) \geq \max_{r \in \Lambda_1(u)} \lambda_u(r)\}.$$

Note that egresses not preferred over an "always available egress" are not in $\Lambda_n(u)$. We also show in Table 1 for our example in Fig. 2 the sets of $\Lambda_n(u)$ for all routers. As router 2 can learn of both $e$ and $f$ from its children ($g$ and $h$), both egresses are in $\Lambda_n(2)$. Also notice that as $i$ is an always available downstream egress of 3 and $i$ is preferred over $m$, $m$ is *not* in $\Lambda_n(3)$. Finally, we define $\Lambda(u) = \Lambda_1(u) \cup \Lambda_n(u)$. Note, $\Lambda(u)$ is well defined as we define $\Lambda(u)$ recursively up the hierarchy.

**Rules for Reliance.** Reliance rules are adapted from the route propagation rules [6] and indicate where a router can learn of its best route. Arcs in the reliance graph are a subset of the arcs in the signaling graph $A_S$. There are three types of arcs in $A_S$ which may be part of the reliance graph. Consider the arc $(u,v) \in A_S$:

1. $(u,v) \in$ **down:** a RR $u$ is reliant on its child $v$ iff $u \notin X$ and $v \in E$.
2. $(u,v) \in$ **up:** a client $u$ is reliant on its parent $v$ iff $u \notin X$.
3. $(u,v) \in$ **over:** a router $u \notin X$ is reliant
   (a) on another router $v \in E \backslash X$ iff

$$\min_{r \in \Lambda(u)} \lambda_u(r) \leq \max_{s \in \Lambda(v) \backslash \Lambda_1(u)} \lambda_u(s)$$

(b) on another router $v \in X$ iff

$$\min_{r \in \Lambda(u)} \lambda_u(r) \leq \lambda_u(v)$$

We demonstrate the above rules for our example in Fig. 2(b). Applying rule 1, we see any router in $E$ which does not have a direct egress is reliant on its children, for example, 2 is reliant on $h$. Applying rule 2, all client routers are reliant on their parents (unless they are a direct egress), for example $l$ is reliant on 3. Rule 3 applies when a router can learn of a better route via an **over** edge than any client-learned route. For instance, rule 3(a) applies to the reliance of 2 on 1, as 2 will select the route from $b$ if it ever learns of it, whereas rule 3(b) applies for the reliance of $a$ on $b$, as $a$ can learn the egress directly from $b$, via an **over** edge.

**Pruning Reliances.** Our technique for determining router decisions would work on the reliance graph defined by the rules above. However, ideally we would like to have the smallest possible co-reliance groups in the reliance graph to minimize computation. The reliance rules are essentially a pruning of the signaling graph. We can continue in this vein by pruning even more reliances:

1) $u \not\leftrightsquigarrow v$ if $(u, v) \in$ **down** and $v \in X \backslash \Lambda_1(u)$.

That is, a RR is reliant only on its best client with a direct egress. In our example, as $g$ prefers $e$ over $f$, and as $e$ is always available, $f$ will never be chosen and is pruned in Fig. 2(c).

2) $u \not\leftrightsquigarrow v$ if $(u, v) \in$ **down**, $v \in E \backslash X$ and

$$\min_{r \in \Lambda_1(u)} \lambda_u(r) > \max_{s \in \Lambda(v)} \lambda_u(s)$$

That is, if a RR $u$ has a client $r$ with a direct egress and no possible egress which it can learn from another client $v$ is better than $r$, then $u$ cannot be reliant on $v$. In our example, as 3 prefers $i$ over $m$, 3 is not reliant on $l$ and this edge is also pruned in Fig. 2(c).

Our next rule is for **up** edges. Before specifying the rule for a $(u, v) \in$ **up**, we define $L(u, v)$ as the egresses that can be learned by a router $u$ from the parent router $v$ which are not available from a direct client. For a general hierarchy, the exact form of $L(u, v)$ can be quite complicated. In practice, RR hierarchies do not tend to be larger than three-levels, and for three levels, we can formally define

$$L(u, v) = \bigcup_{w \in E : v \leftrightsquigarrow w} \Lambda(w) \backslash \Lambda_1(u).$$

3) $u \not\leftrightsquigarrow v$ if $(u, v) \in$ **up** and

$$\min_{r \in \Lambda_1(u)} \lambda_u(r) > \max_{s \in L(u,v)} \lambda_u(s)$$

That is, a RR in the second level of the hierarchy is only reliant on its parent if the parent can learn a better route than any of the always available egress at the RR. In our example, $l$ prefers $m$ over any egress it can learn from 3 (as $\lambda_l(m) > \lambda_l(i)$). Hence the reliance of $l$ on 3 is pruned. Other reliances which are pruned using this technique are $h \leftsquigarrow 2$ and $d \leftsquigarrow 1$. Notice $g \leftsquigarrow 2$ is not prunable as $g$ may select an egress learned from 2.

**Finding a Valid Solution.** As the reliance graph precisely identifies which routers' decisions a particular router is dependent upon, if there are no cycles in the reliance graph structure, we can topologically sort the routers and evaluate them in-order (visiting them exactly once). However, as shown in our example, it is likely there are cycles in the reliance graph. Hence, we partition the reliance graph into co-reliance groups before undertaking a topological sort on co-reliance groups. We show these co-reliance groups in Fig. 2(d). One topological ordering (any topological order will result in the same network solution) is the numerical ordering of $D_1 - D_{19}$ in Fig. 2(d). If multiple routers are present in a co-reliance group (such as $D_9$) the decisions of the routers may be dependent on message timing and we evaluate their decisions until a *valid* solution is found. The ordering in which we evaluate the routers within a co-reliance group determines which of possibly multiple valid solutions we converge upon [4]. Our desire is to converge to the *actual* solution selected by the network. We describe how we achieve this in the next section.

## 5 Finding the Actual Solution

A walk of the reliance graph can have multiple valid solutions when either

1. co-reliance groups have multiple routers; or
2. the non-deterministic oldest-route tie-breaker is used.

When such conditions exist, we want to find the *actual* solution chosen by routers. If decisions of some routers in the network are known (through measurement), we can use them as constraints while determining the solution[4]. Two feasible approaches to using these constraints are: (i) find all possible solutions and select one which satisfies the constraints; or (ii) gravitate towards a solution satisfying all constraints by ensuring that when we visit each co-reliance group, we select a solution consistent with the constraints. We take the latter approach as it reduces unnecessary computation of infeasible solutions. However, it can result in discrepancies if we reach a co-reliance group and there are no solutions satisfying the constraints. We could backtrack along the reliance graph to resolve discrepancies, however in our examined network, we found only seven in over 12.7 million decisions had such discrepancies and so we did not implement a backtracking algorithm.

The ordering of router evaluation within a co-reliance group determines which of multiple valid solutions we converge to. Due to space constraints, we refer the reader to [16] for full details of ordering of routers within a co-reliance group.

[4] There may be multiple feasible solutions which match the known route selections.

Once we have an ordering for co-reliance groups and an ordering for routers within a co-reliance group we can calculate the decisions of all routers by simply walking the reliance graph. We visit each co-reliance group in-order, and visit each router within the co-reliance group in-order. If the co-reliance group is non-singleton, we continue evaluating the router decisions until no routers alter their decision. Our algorithm does not rely on the underlying topology, only its description in terms of a reliance graph. Consequently, any topology describable by a reliance graph can be analyzed using this algorithm.

It is possible that a co-reliance group never converges to a solution [4]. However, if this is the case, then the actual network also has oscillatory properties. The non-convergent co-reliance group isolates the routers responsible for oscillatory modes so administrators can take corrective action.

When the oldest-route tie-breaker is used, there may be multiple routes available at a router with equal IGP distances to an egress. Any route with this equally good IGP distance may be chosen by the router. We again use any available constraints to assist our decision (see [16] for details).

## 6 Evaluation

We have implemented our techniques and tested them on a large Tier-2 AS. The AS has a three-level RR hierarchy and uses the oldest-route tie-break option. The MED attribute is reset by the AS. A BGP monitoring infrastructure collects BGP updates from 15% of routers, majority of which are RRs. We use the known routes from these routers as the set of input routes to the network. Each such route contains a "next-hop" attribute which corresponds to the egress router for the route. IGP distances are determined based on data collected by an OSPF monitor [17].

Our algorithm discovers the decisions made by routers once the network has converged to a solution. Consequently, we only examine *stable* prefixes – those prefixes with no updates witnessed from any router under observation in the 6 hours prior and 6 hours past the examined time. Our evaluation is based on data collected on the 26th May 2008, although we found similar results for several other examined intervals. During the analysis process, our model discovered several minor configuration errors. In this case, our model predicted the "correct" outcome, although the network selected an "incorrect" outcome due to a configuration error on several egress routers. We exclude the prefixes affected by these configuration errors from our analysis.

To speed up our analysis, we group all prefixes with the same set of egress routers and the same egress selection by all routers that have session to the BGP monitor. This allowed us to discover the egress router selected by all routers (including the remaining 85% of routers without BGP monitors) for 224, 870 stable prefixes with only 827 reliance graphs and 1154 "walks" of the reliance graph.

As our technique is based on the rules of route propagation, it will *always* find a valid solution given any configuration. With the addition of monitor information (or any other constraints available), we can converge to a solution satisfying such constraints. In practice, we found our technique always found valid solutions and only seven inconsistencies with BGP monitor data in over 12.7 million known $(prefix, router)$ pairs. This discrepancy was resulting from a tie-break decision at a router without a BGP monitor session. The predicted egresses were in the same PoPs as the actual selected egress. Backtracking to alter the random tie-break decision would correct this.

We found 99.99% of co-reliance groups were singleton and the maximum size of a co-reliance group was five routers which occurred only four times ensuring our technique very rarely required the re-evaluation of router decisions.

The execution time of this case study lasted several hours. However, the evaluation time was dominated by the conversion of the enormous amounts of compressed binary BGP data on disk to an ASCII readable format in memory. The construction and walk of the reliance graph did not significantly contribute to the execution time. Consequently, the incorporation of incremental changes to BGP and IGP may allow near real-time analysis of router decisions. We leave the incorporation of such routing dynamics to future work.

## 7 General Router Reliance Graph

In previous sections, we focused on the reliance rules for an iBGP route-reflector topology. However, our technique is applicable to *any topology* describable by a series of reliances. That is, if a set of rules defining where a router can possibly learn of its best route can be defined, our technique will find a solution. This is in contrast to [7] where different iBGP topologies required separate algorithms. Further, for the iBGP topology examined in this paper the conditions required by Feamster-Rexford technique are not satisfied. Our technique, in the best case can precisely order the execution of router decisions (when all co-reliance groups are singular) and in the worst case (all routers in a general topology form a single co-reliance group) reverts to an arbitrary ordering of router decisions (such as in [15]). The addition of reliance rules increases the efficiency of our technique.

An example of a different iBGP topology is an RR topology with the MED attribute respected. The authors of [7] recommended a simulator as the best technique to evaluate the network solution to this topology. However, in [16] we detail reliance rules for such a topology, finding that although the number of edges in the reliance graph can increase (in comparison to when the MED attribute is not respected), routers outside the egress ancestor set do not form non-singleton co-reliance groups. Hence, the maximum co-reliance group size is bounded by the size of the egress ancestor set which is commonly an order of magnitude smaller than the total number of routers — making our technique significantly more efficient than a pure simulator. Consequently, the ordering of router decisions outlined in this paper may also be used to improve the convergence times of existing BGP simulators such as C-BGP [15].

The separation of the topology and the rules for route propagation from the algorithm used to evaluate the network solution has possible applications not only in iBGP described in this paper, but also in the eBGP context. A topology inferred by a technique such as [18] could form a starting point to predict the (Internet-wide) solution for a particular prefix and may help to answer Internet-wide 'what-if' questions.

## 8 Conclusions

In this paper, we presented a reliance graph model to capture the dependence amongst routers for route selection. The input to the model is the iBGP topology and IGP distances. The model allows one to efficiently calculate the network solution (set of routes selected by all routers) with no assumptions on the iBGP configuration. The model also works when only partial information about routes is available. We demonstrated the efficacy of the model by applying it to a Tier-2 containing over 220,000 prefixes. Our methodology was able to find a valid solution and a solution consistent with observed routes for all but seven $(prefix, router)$ pairs even when routes from only about 15% routers were known.

One significant benefit of using a reliance graph model is that dynamics of iBGP topology or IGP distance that do not affect the reliance graph does not have any effect on the actual routing choices. Furthermore, BGP route dynamics only require the re-evaluation of routers in the portion of the network so affected. We believe that these two features should allow our methodology to work in real-time for filling gaps to BGP monitors as well as for 'what-if' analyses. In fact, we have applied the methodology successfully to determine the current router decisions and predict changes under modified route availability [16].

BGP monitors are used to determine the route selected by routers in the network *i.e.*, the selected network solution. Our reliance graph analysis identifies where routes can be learned from. Consequently, a new direction of research could be to identify the optimal placement of BGP monitors to minimize the number of random tie-break decisions while maximizing the information about the available egresses.

## References

1. Feldmann, A., Greenberg, A., Lund, C., Reingold, N., Rexford, J., True, F.: Deriving Traffic Demands for Operational IP Networks: Methodology and Experience. IEEE/ACM Transactions on Networking (2001)
2. Zhang, Y., Roughan, M., Lund, C., Donoho, D.: Estimating Point-to-Point and Point-to-Multipoint Traffic Matrices: An Information-Theoretic Approach. IEEE/ACM Transactions on Networking 13(5), 947–960 (2005)
3. Teixeira, R., Shaikh, A., Griffin, T.G., Voelker, G.M.: Network Sensitivity to Hot-Potato Disruptions. In: ACM SIGCOMM (2004)
4. Flavel, A., Roughan, M., Bean, N., Shaikh, A.: Where's Waldo? Practical Searches for Stability in iBGP. In: IEEE International Conference on Network Protocols (2008)
5. Buob, M., Meulle, M., Uhlig, S.: Checking for Optimal Egress Points in iBGP Routing. In: International Workshop on the Design of Reliable Communications Networks (2007)

6. Bates, T., Chandra, R., Chen, E.: BGP Route Reflection - An Alternative to Full Mesh IBGP, RFC 2796 (2000)
7. Feamster, N., Rexford, J.: Network-Wide Prediction of BGP Routes. IEEE/ACM Transactions on Networking 15(2), 253–266 (2007)
8. Griffin, T., Wilfong, G.: On the Correctness of IBGP Configuration. In: ACM SIGCOMM (2002)
9. Feamster, N., Balakrishnan, H.: Correctness Properties for Internet Routing. In: Forty-third Allerton Conference on Communication, Control, and Computing (2005)
10. Vutukuru, M., Valiant, P., Kopparty, S., Balakrishnan, H.: How to Construct a Correct and Scalable iBGP Configuration. In: IEEE INFOCOM, Barcelona, Spain (April 2006)
11. Buob, M., Uhlig, S., Meulle, M.: Designing Optimal iBGP Route-Reflection Topologies. In: IFIP Networking (2008)
12. Bonaventure, O., Uhlig, S., Quoitin, B.: The Case for More Versatile BGP Route Reflectors, Work in progress, draft-bonaventure-bgp-route-reflectors-00.txt (2004)
13. Poduri, K., Alaettinoglu, C., Jacobson, V.: BST - BGP Scalable Transport. In: NANOG 27 (2003)
14. Caesar, M., Caldwell, D., Feamster, N., Rexford, J., Shaikh, A., van der Merwe, J.: Design and Implementation of a Routing Control Platform. In: Symposium on Networked Systems Design and Implementation (2005)
15. Quoitin, B., Uhlig, S.: Modeling the Routing of an Autonomous System with CBGP. IEEE Network Magazine, Special Issue on Interdomain Routing (2005)
16. Flavel, A.: BGP, Not As Easy As 1-2-3. Ph.D thesis, University of Adelaide (2009)
17. Shaikh, A., Greenberg, A.: OSPF Monitoring: Architecture, Design and Deployment Experience. In: Symposium on Networked Systems Design and Implementation (2004)
18. Mühlbauer, W., Maennel, O., Uhlig, S., Feldmann, A., Roughan, M.: Building an AS-Topology Model that Captures Route Diversity. In: ACM SIGCOMM (2006)

# Performance Evaluation of Weighted Fair Queuing System Using Matrix Geometric Method

Amina Al-Sawaai, Irfan Awan, and Rod Fretwell

Mobile Computing, Networks and Security Research Group
School of Informatics, University of Bradford,
Bradford, BD7 1DP, U.K.
{a.s.m.al-sawaai,i.u.awan,r.j.fretwell}@bradford.ac.uk

**Abstract.** This paper analyses a multiple class single server M/M/1/K queue with finite capacity under weighted fair queuing (WFQ) discipline. The Poisson process has been used to model the multiple classes of arrival streams. The service times have exponential distribution. We assume each class is assigned a virtual queue and incoming jobs enter the virtual queue related to their class and served in FIFO order.We model our system as a two dimensional Markov chain and use the matrix-geometric method to solve its stationary probabilities. This paper presents a matrix geometric solution to the M/M/1/K queue with finite buffer under (WFQ) service. In addition, the paper shows the state transition diagram of the Markov chain and presents the state balance equations, from which the stationary queue length distribution and other measures of interest can be obtained. Numerical experiments corroborating the theoretical results are also offered.

**Keywords:** Weighted Fair Queuing (WFQ), First Inter First Out (FIFO), Markov chain.

## 1 Introduction

Queue based on weighted fair queuing is a service policy in multiclass system. Consider a weighted service link that provides service for customers belonging to different classes. In WFQ, traffic classes are served on the fixed weight assigned to the related queue. The weight is determined according to the QoS parameters, such as service rate or delay. The WFQ [1] is a scheduling discipline usually applied to QoS enabled routers.

In this work, a two class single server M/M/1/K queue with a finite capacity under WFQ scheduling discipline is analyzed. The Poisson process is used to model two classes of arrival streams. The service times have exponential distribution. We assume each class is assigned a virtual queue and incoming jobs enter the virtual queue related to their class and served in FIFO order. The queue $i$ is served at rate $w_i$ for some $w_i > 0$ when queue $1-i$ is not empty and at rate unity when that queue is empty. We model our system as a two dimensional Markov chain and use matrix geometric method to solve for its stationary probabilities.

L. Fratta et al. (Eds.): NETWORKING 2009, LNCS 5550, pp. 66–78, 2009.

There were many solutions proposed to offer the solution for the WFQ system with two classes of customers on infinite buffer in [4] [5] and [6]. However, the class based with finite buffer is applied in a lot of computer and communication system and more realistic. The main aim of our work is to provide a solution to the WFQ system with a finite buffer. To the best of our knowledge to analytical and numerical solutions for such a system have not been in the literature.

The rest of this paper is organized as follows: An overview of related work is shown in Section 2. The queueing model of M/M/1/K queue with WFQ is described in Section 3. Section 4, offers a matrix geometric solution for M/M/1/K queue under WFQ discipline, while section 5 presents and explains the numerical results of the model, followed by the conclusions and future work in Section 6.

## 2 Related Work

The WFQ scheduling discipline is an important method for providing bounded delay, bounded throughput and fairness among traffic flows [2], [3]. The subject of WFQ has been investigated by many authors. Models similar to our WFQ system with two classes of customers have been analyzed for Poisson arrivals and exponential service times but with infinite buffers [4] and [5]. In [4], a two class system with two queues is considered; however the WFQ system is approximated with a two server, two queue systems and a numerical solution is provided. In [5], the same problem in [4] is considered; however an analytical solution for the system is given.

In [6], the authors provide an analytical solution for a WFQ system with a long range dependent traffic input with an infinite buffer and prove that analytical results provide an accurate estimation of queue length distribution and can be helpful in choices of WFQ weights. The work described in [7], defines an analytical solution of a WFQ system with two classes of customers with exponential service times in an unsteady state with an infinite buffer. The authors introduce an analytical model for the system and derive the exact expression of the tail of the probability distribution of the numbers of customers.

In [8], an analytical solution for the WFQ system with more than two queues and time connected variable service rates with an infinite buffer is provided. In addition a different design proposal for a dynamic WFQ scheduler is analyzed to provide quality of service guarantees.

This paper, we analyses the WFQ system with two classes of customers with a finite buffer in steady state and studies the effect of weights on the system mean a queue length, throughput and mean response time. To the best of our knowledge, the analysis of a WFQ system with finite buffer has not been proposed in the literature.

## 3 The WFQ System Model

Throughout this paper, we consider the same model in [4]; however instead of an infinite buffer, we use a finite buffer with size K. The maximum number of customers who can be in the system at any time is K and any additional arriving customers will be refused entry to the system and will depart immediately without service.

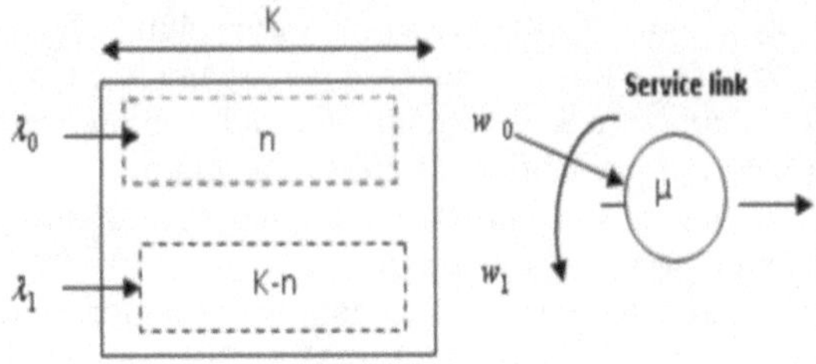


**Fig. 1.** The Weighted Fair Queuing System

For our WFQ system, we assume two classes of jobs; Jobs of class 1 and class 2 arrive according to a Poisson process with rate $\lambda_i$ , $i = 0,1$ and require exponential service times with mean $1/\mu i$ , $i = 0,1$ . Each class is assigned a virtual queue and arriving jobs enter the virtual queue related to their class and are served in FIFO order.

The queue $i$ is served at rate $w_i$ for some $w_i > 0$ when queue $1-i$ is not empty and at rate unity when that queue is empty. The coefficients $w_i$ are such that $w_0 + w_1 = 1$. The server is work conserving, i.e., it serves jobs only if at least one queue is not empty. Fig. 1 depicts the WFQ system.

Fig. 2, illustrates the state-transition-rate diagram of the WFQ system where each state denotes the number of customers in the system. A generalized Markov model can be described by a two dimensional Markov chain with state $(i, j)$, where $i$ and $j$ are the number of customers in class 1 and class 2 at each state, respectively. When the process is in state $(i, j)$ , it can transfer to the state $(i+1, j)$, $(i, j+1)$, $(i-1, j)$ if $i > 0$, $(i, j-1)$ . The transition rate from state $(i, j)$ to $(i+1, j)$ where $(0 \le i \le K-1)$ is the arrival rate of class 1, i.e. $\lambda_0$, of the Poisson process. A transition out of state $(i, j)$ to $(i, j+1)$ where $(0 \le j \le K-1)$ is the arrival rate of class 2, i.e. $\lambda_1$ , of the Poisson process. When no customers of class 1 are in the system, the transition rate from $(0, j)$ to $(0, j-1)$ is the service rate of class 2, i.e. $\mu_1$ . The change from state $(i,0)$ to $(i-1,0)$ is the service rate of class 1, i.e. $\mu_0$ . However, the transition rate from state $(i, j)$ to $(i-1, j)$ is the service rate of class 1 multiplied by the weight of class 1, i.e. $w_0\mu_0$ . A transition from $(i, j)$ to $(i, j-1)$ is the service rate of class 2 multiplied by the weight of class 2, i.e. $w_1\mu_1$ .

The infinitesimal generator of this process is given by:

$$
\begin{aligned}
&Q_{(i,0),(i-1,0)} = \mu_0, i \ge 1 \\
&Q_{(i,j),(i-1,j)} = w_0\mu_0 \\
&Q_{(i,j),(i,j-1)} = w_1\mu_1, i \ge 1, j \ge 1 \\
&Q_{(i,j),(i+1,j)} = \lambda_0
\end{aligned}
\quad (1)
$$

and

$$Q_{(i,j),(i,j+1)} = \lambda_1, i \geq 0, j \geq 0 \tag{2}$$

The generator matrix $Q$ of the Markov chain is given by

$$Q = \begin{pmatrix} B_0 & A_1 & & & & \\ C_0 & B_1 & A_2 & & & \\ & C_1 & B_2 & A_3 & & \\ & & \ddots & \ddots & \ddots & \\ & & & C_{n-2} & B_{n-1} & A_n \\ & & & & C_{n-1} & B_n \end{pmatrix} \tag{3}$$

Where

$$C_n = \begin{pmatrix} \mu_1 & & & \\ w_0\mu_0 & w_1\mu_1 & & \\ & w_0\mu_0 & w_1\mu_1 & \\ & & & \\ & & & \mu_0 \end{pmatrix}, \quad A_n = \begin{pmatrix} \lambda_1 & \lambda_0 & & & \\ & \lambda_1 & \lambda_0 & & \\ & & & & \\ & & & \lambda_1 & \lambda_0 \end{pmatrix},$$

$$B_{n-1} = \begin{pmatrix} -(\lambda+\mu_1) & & & \\ & -(\lambda+w\mu) & & \\ & & -(\lambda+w\mu) & \\ & & & \\ & & & -(\lambda+\mu_0) \end{pmatrix}, \quad B_n = \begin{pmatrix} -\mu_1 & & & \\ & -w\mu & & \\ & & -w\mu & \\ & & & \\ & & & -\mu_0 \end{pmatrix}$$

In addition

$$\lambda = \lambda_0 + \lambda_1 \tag{4}$$

and

$$w\mu = w_0\mu_0 + w_1\mu_1 \tag{5}$$

The block entries $B_0, B_1, \ldots, B_n$ are square matrices, while the block entries $A_1, A_2, \ldots, A_n$ and $C_1, C_2, \ldots, C_n$ are rectangular matrices. The Markov chain with generator matrix $Q$ is irreducible; the matrices along the diagonal are non-singular. Below is the state equilibrium equations for all of the states of the Markov model of Fig. 2.

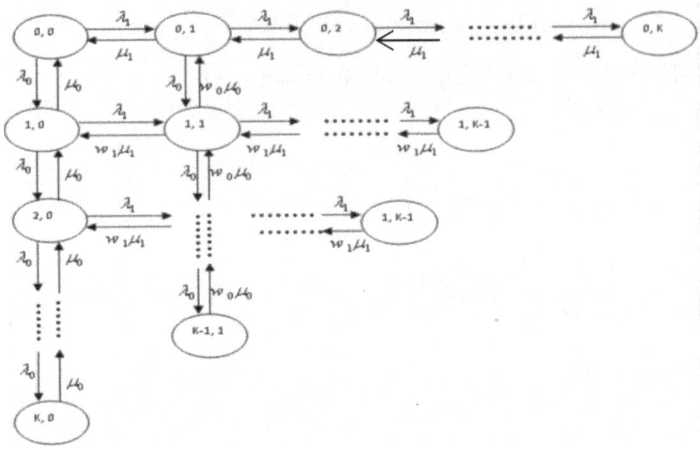


**Fig. 2.** The state transition diagram for an M/M/1/K queue under WFQ scheduling discipline

# 4 Matrix Geometric Solution:

## 4.1 State Equilibrium Equations

The stationary probability vector $\underline{\pi}$ for $Q$ is generally partitioned as $\underline{\pi} = [\pi_0, \pi_1, \pi_2, \ldots \pi_N]$. Solving $\underline{\pi} Q = 0$ along with normalizing equation $\underline{\pi} e = 1$, yields the following set of equations in matrix form:

$$\pi_0 B_0 + \pi_1 C_0 = 0 \tag{6}$$

$$\pi_0 A_1 + \pi_1 B_1 + \pi_2 C_1 = 0 \tag{7}$$

$$\pi_{n-1} A_n + \pi_n B_n + \pi_{n+1} C_n = 0, 1 < n < N-1 \tag{8}$$

$$\pi_{N-2} A_{N-1} + \pi_{N-1} B_{N-1} + \pi_N C_{N-1} = 0 \tag{9}$$

$$\pi_{N-1} A_N + \pi_N B_N = 0 \tag{10}$$

Considering the above state equilibrium equations, it can be assumed that between any two states there is ''flow in, flow out'' equilibrium without any effect on the other remaining neighboring states.

## 4.2 Matrix Geometric Method

Suppose there exists a matrix $R$ as

$$\pi_n = \pi_{n-1} R \quad \forall n \geq 1. \tag{11}$$

then, we get by successive substitutions into the state equilibrium equations that

$$\pi_n = \pi_0 R^n \quad \forall n \geq 0. \tag{12}$$

A solution of the (12) is called a *matrix geometric* solution [9]. The explanation for solving a matrix geometric system is to state the matrix $R$ , the rate matrix, which is discussed below.

### 4.3 Computation of the Rate Matrices

By a simple algebraic stage management of the state equilibrium equations $R_n$ '$s$ is formed as follows:

From Equation (6) and we know that $B_0$ is non-singular and we obtain,

$$\pi_0 B_0 = -\pi_1 C_0 \rightarrow \pi_0 = -\pi_1 C_0 B_0^{-1} \rightarrow \pi_0 = \pi_1 R_0 \tag{13}$$

Equation (10) leads to the following expression for $\pi_N$ and $R_N$ where $B_N$ is required to be non-singular,

$$\pi_N B_N = -\pi_{N-1} A_N \rightarrow \pi_N = -\pi_{N-1} A_N B_N^{-1} \tag{14}$$

Equation (9) leads to the following expression of $\pi_{N-1}$ and $R_{N-1}$

$$\begin{aligned}
&\pi_{N-1} B_{N-1} + \pi_{N-1} R_N C_{N-1} = -\pi_{N-2} A_{N-1} \\
&\pi_{N-1}(B_{N-1} + R_N C_{N-1}) = -\pi_{N-2} A_{N-1} \\
&\pi_{N-1} = -\pi_{N-2} A_{N-1} (B_{N-1} + R_N C_{N-1})^{-1} \\
&\pi_{N-1} = \pi_{N-2} R_{N-1}
\end{aligned} \tag{15}$$

Finally from equation (8) we get a general relation between $\pi_{n-1}, \pi_n$ and $R_n$,

$$\begin{aligned}
&\pi_n = -\pi_{n-1} A_n (B_n + R_{n+1} C_n)^{-1} \\
&\pi_n = \pi_{n-1} R_n \quad 2 \le n < N - 1
\end{aligned} \tag{16}$$

$R_n$ can be calculated from Algorithm 1.

**Algorithm 1.** Calculate $R_n$

```
R_N ← −A_N B_N^{-1}
if n ≥ 1 then
   for j = N − 1 → 1 do
       R_j ← −A_j (B_j + R_{j+1} C_j)^{-1}
   end for
   return R_n ← −A_n (B_n + R_{n+1} C_n)^{-1}
end if
if n = 0 then
   return R_0 ← −C_0 B_0^{-1}
end if
```

### 4.4 Stationary Probabilities

**Theorem 1.** For any QBD process with a finite state space, having an infinitesimal generator matrix given by Equation $Q$ , the stationary probabilities are given in matrix-geometric form by

$$\underline{\pi}_n = \underline{\pi}_1 R_n^* \tag{17}$$

Where $R_n^* = \prod_{j=2}^{N} R_j$ and $R_j$ is computed using Algorithm 1.

**Proof.** The system of linear equations is solved for $\underline{\pi}_1$, and from equation (14), $\underline{\pi}_n$ is obtained as:

$$\begin{aligned} \underline{\pi}_n &= \underline{\pi}_{n-1} R_n \\ &= \underline{\pi}_{n-2} R_{n-1} R_n \\ &\vdots \\ &= \underline{\pi}_1 R_2 \ldots R_{n-2} R_{n-1} R_n \\ &= \underline{\pi}_1 \prod_{j=2}^{n} R_j = \underline{\pi}_1 R_n^* \end{aligned} \tag{18}$$

Solving Equation (6) and (7) for $\underline{\pi}_1$ and use $\underline{\pi}_2 = \underline{\pi}_1 R_2$ leads to,

$$\underline{\pi}_1 (R_0 A_1 + B_1 + R_2 C_1) = 0 \tag{19}$$

Thus, after an exchange and mathematical manipulation, Equation (18) follows from normalizing condition $\sum_{n=0}^{N} \pi_n e = 1$ and equation (6).

$$\underline{\pi}_1 (R_0 e + \sum_{n=1}^{N} R_n^* e) = 1 \tag{20}$$

It is worth noting that Theorem1 giving the structure of the vector $\pi$ and that of the vector $\pi_1$ still needs to be determined. The vector $\pi_1$ could be computed from either one of the equations

$$\begin{aligned} &\underline{\pi}_1 (R_0 A_1 + B_1 + R_2 C_1) = 0 \\ &\underline{\pi}_1 (R_{n-1}^* A_n + R_n^* B_n + R_{n+1}^* C_n = 0, \quad 1 < n < N-1 \\ &\underline{\pi}_1 (R_{N-1}^* A_N + R_N^* B_N) = 0 \end{aligned} \tag{21}$$

and

$$\underline{\pi}_1 (R_0 e + \sum_{n=1}^{N} R_n^* e) = 1$$

| **Algorithm 2.** Calculate $\underline{\pi}$ |
| --- |
| 1: **for** $j = 2 \rightarrow n$ **do** |
| 2: $\underline{\pi}_j \leftarrow \underline{\pi}_1 \prod_{j=2}^{n} R_j$ |
| 3: **end for** |

Algorithm 2 is used to compute the stationary probabilities $\underline{\pi}$.

## 5 Performance Analyses

In this section, we present the results of the numerical calculations for an M/M/1/K queue under WFQ using (15) with different values for $\lambda_0, \lambda_{1,} \mu_0$ and $\mu_1$. The maximum buffer size for the system is 50; hence this model has 1326 states. The following is the numerical evaluation of the analytical model results for some performance measurements based on and derived from the Markov chain described in the previous section.

### 5.1 Mean Queue Length

The mean queue length $L$ is calculated from the model as follows:

$$L = \sum_{i}^{K} i\pi(i) \tag{22}$$

Where $\pi(i) = \pi(0), \pi(1), \ldots, \pi(K)$.

The following equations are derived from (22) to find the mean queue length for class 1, $L_0$ and class 2, $L_1$, respectively:

$$L_0 = \sum_{i=0}^{K} \sum_{j=0}^{K-i} i\pi(i, j) \tag{23}$$

$$L_1 = \sum_{j=0}^{K} \sum_{i=0}^{K-j} j\pi(i, j) \tag{24}$$

where $\pi(i, j)$ is the steady state probability at state $(i, j)$.

Fig. 3 depicts the mean queue length for class1 and class2 for different arrival rates and with $w_0 = w_1 = 0.5$. We choose $\mu_0 = \mu_1 = 1$ to only examine the effect of the queue weights. Using these parameters, the WFQ system reduces to two M/M/1/K systems. It can be seen from Fig. 3 that the mean queue length for both classes increases with the increase of the arriving traffic until λ=1 then $L_0 = L_1 \approx 24.5$. The total of $L_0$ and $L_1$ equals the buffer size: K=50. $L_0$ and $L_1$ have identical values for all arrival rates.

In Fig. 4, we choose $w_0 = 0.6$ and $w_1 = 0.4$ for different values of $\lambda$ when $\mu_0 = 2, \mu_1 = 1$. It indicates that the mean queue length for class1 is lower than the mean queue length for class2, as expected. Class1 is served faster than class2 at rate $w_0 \mu_0 = 1.2$. We find the same results for $L_1$, when $w_0 = 0.4$, $w_1 = 0.6$ and $\mu_0 = 1, \mu_1 = 2$.

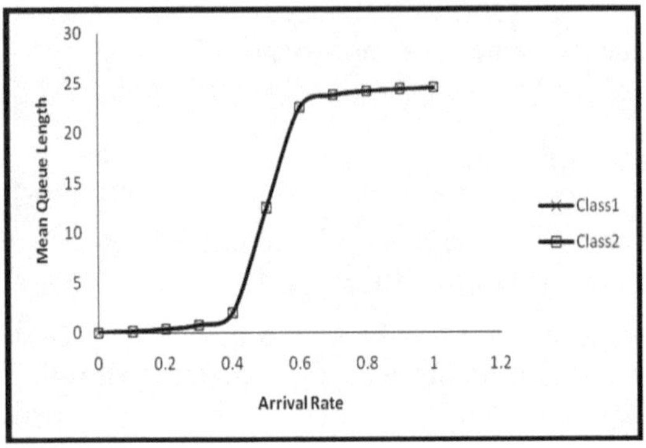


**Fig. 3.** Mean queue length, $w_0 = w_1 = 0.5$ and $\mu_0 = \mu_1 = 1$, $\lambda_0 = \lambda_1$

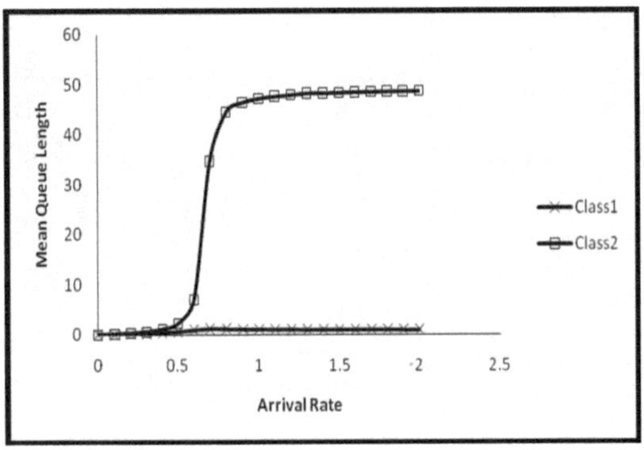


**Fig. 4.** Mean queue length, $w_0 = 0.6, w_1 = 0.4$ and $\mu_0 = 2, = \mu_1 = 1$, $\lambda_0 > 0, \lambda_1 > 0$

Fig. 5 depicts the mean queue length for different values of class1 with an arrival rate $\lambda_0 > 0$ and fixed arrival rate $\lambda_1 = 0.2$ with $w_0 = 0.6$, $w_1 = 0.4$ . We make a note of that as $\lambda_0$ increases, and then $L_0$ increases until the mean queue length approaches to 50 and then it is stable. However, the mean queue length of class 2 increases and then decreases to zero. The $L_1$ is lower in this case compared to $L_0$.The reason why $L_0$ is higher than $L_1$ is the fact that class1 is getting a higher service rate, $w_0 \mu_0 = 1.2$ and its increasing arriving traffic causes class2 to be refused entry to the system. This can be observed in Fig. 5.

### 5.2 System Throughput

The following equation is used to find the system throughput ($T$) for the M/M/1/K queue from the analytical model:

$$T = \mu(1 - \pi(0)) \quad (25)$$

- $\mu$ is the output rate, when the server is busy .
- $\pi$ (server is busy) = (1- $\pi$ (server is idle) = $(1 - \pi(0))$ .
- when the server is idle, the output rate equals zero.

Equation (25) is used to derive the throughput ($T$) for class1, $T_0$ , and class2, $T_1$ , respectively:

$$T_0 = w_0\mu_0 (1 - \pi(0)) \quad (26)$$

$$T_1 = w_1\mu_1 (1 - \pi(0)) \quad (27)$$

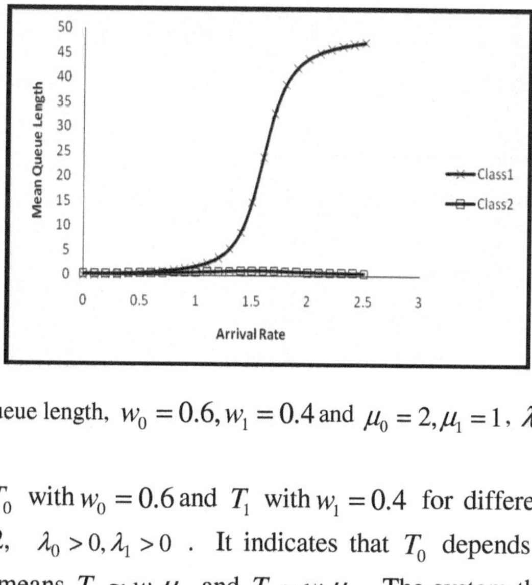


**Fig. 5.** Mean queue length, $w_0 = 0.6, w_1 = 0.4$ and $\mu_0 = 2, \mu_1 = 1$, $\lambda_0 > 0, \lambda_1 = 0.2$

Fig. 6 shows $T_0$ with $w_0 = 0.6$ and $T_1$ with $w_1 = 0.4$ for different arrival rates of class1 and class 2, $\lambda_0 > 0, \lambda_1 > 0$ . It indicates that $T_0$ depends on the weight of class1 and that it means $T_0 \approx w_0\mu_0$ and $T_1 \approx w_1\mu_1$. The system throughput will not exceed the services rate (in this example: $w_0\mu_0 + w_1\mu_1 = 1.6$).

In Fig. 7, we obtain something like the results for $T_0$ with $w_0 = 0.6$ and $T_1$ with $w_1 = 0.4$ for different arrival rates of class1 and class 2, $\lambda_0 > 0, \lambda_1 > 0$ . However, the graph will be stable with arrival rate $\lambda_0 = \lambda_1 = 2.3$ for both classes.

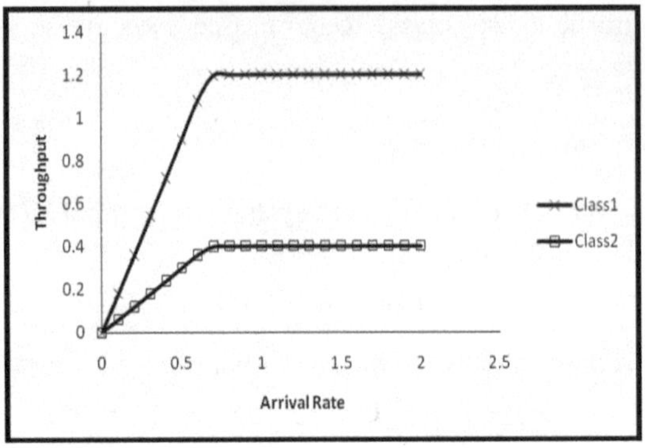


**Fig. 6.** Throughput, $w_0 = 0.6, w_1 = 0.4$ and $\mu_0 = 2, \mu_1 = 1$, $\lambda_0 > 0, \lambda_1 > 0$

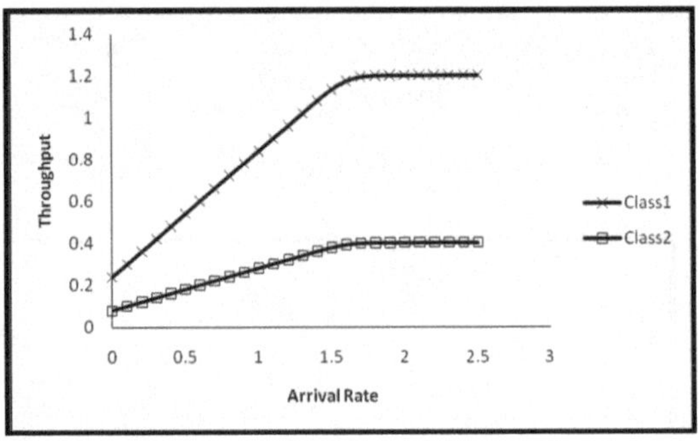


**Fig. 7.** Throughput, $w_0 = 0.6, w_1 = 0.4$ and $\mu_0 = 2$, $\mu_1 = 1$, $\lambda_0 > 0, \lambda_1 = 0.2$

### 2.3 The Mean Response Time

Using Little's law [10]:

$$L = T\,S \tag{28}$$

We can derive the mean response time for class 1, $S_0$, and class 2, $S_1$, from (28) respectively as:

$$S_0 = L_0 / T_0 \tag{29}$$

$$S_1 = L_1 / T_1 \tag{30}$$

where $T_0$ and $T_1$ are measured system throughput when packet loss is considered ($T_{Effective}$).

Fig. 7 shows the mean response time of class 1 and class 2 when $\mu_0 = 2$, $\mu_1 = 1$ and $w_0 = 0.6, w_1 = 0.4$. We can observe that $S_1$ is higher than $S_0$ and class 2 stays longer in the system. The reason why the mean response time has turned out to be higher for class 2 is the fact that it has a low weight.

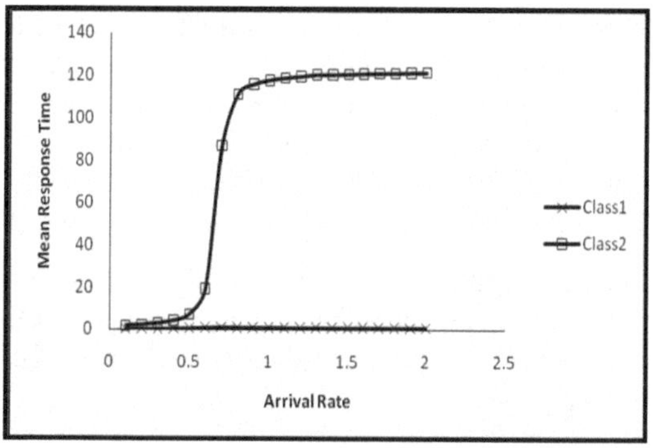


**Fig. 8.** Mean response time, $w_0 = 0.6, w_1 = 0.4$ and $\mu_0 = 2,\ \mu_1 = 1,\ \lambda_0 > 0, \lambda_1 > 0$

Fig. 8 depicts the mean response time for different values of class1 with the arrival rate $\lambda_0 > 0$ and the fixed arrival rate $\lambda_1 = 0.2$ with $w_0 = 0.6,\ w_1 = 0.4$ . We can calculate that as $\lambda_0$ increases, $S_0$ increases until the mean response time approximately equals 40 and then it is stable. However, the mean response time of class2 increases until 2.4 and then decreases to zero. The reason why $S_0$ is a higher than $S_1$ is the fact class1 is receiving different arrival rates and a higher service rate, $w_0\,\mu_0 = 1.2$.

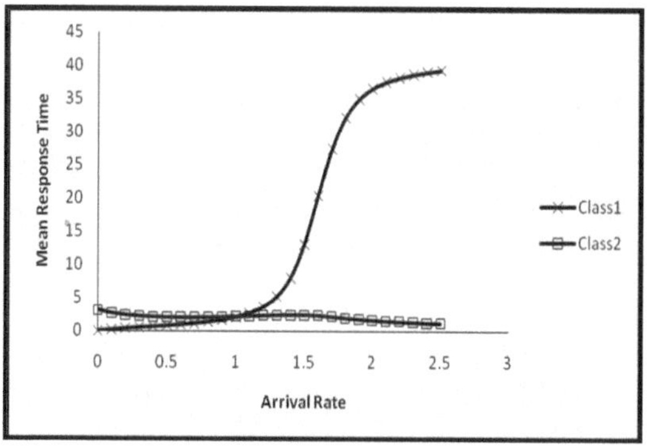


**Fig. 9.** Mean response time, $w_0 = 0.6, w_1 = 0.4$ and $\mu_0 = 2,\ \mu_1 = 1,\ \lambda_0 > 0, \lambda_1 = 0.2$

For all obtained performance measurements, the WFQ system behaved as expected. These numerical results confirm the validity of the analytical model.

## 6 Conclusions and Future Work

In this paper, we provide an analysis of the two class single server M/M/1/K queue with a finite capacity under a weighted fair queuing scheduling discipline. The

Poisson process has been used to model the multiple classes of arrival streams. The service times have exponential distribution. An analytical expression for the flow balanced equations has been derived using a Markov chain. Queue length distribution has been derived by solving these expressions. We derived a general expression for the steady state probabilities for any finite buffer with size K. In addition, we found the steady state probabilities $\pi_{i,j}$ for M/M/1/50 queue with WFQ as an example and we presented the numerical results. Future work will focus on deriving general equations for the performance measure for an M/M/1/K queue with a finite capacity under a weighted fair queuing (WFQ) with more than two classes. We will also extend it with more realistic traffic models.

## References


1. Li, C., Tsao, S., Chen, M., Sun, Y., Huang, Y.: Proportional Delay Differentiation Service Based on Weighted Fair Queuing. In: Proceedings Ninth International Conference on Computer Communications and Networks, 2000, pp. 418–423 (2000)
2. Parkekh, A.K., Gallager, R.G.: A generalized processor sharing approach to flow control in integrated services networks: The single-node case. IEEE Trans. On Networking 1(3), 344–357 (1993)
3. Parekh, A.K., Gallager, R.G.: A generalized processor sharing approach to flow control in integrated services networks: The multiple node case. IEEE Trans. On Networking 2(2), 137–150 (1994)
4. Horváth, G., Telek, M.: Approximate analysis of two class WFQ systems. In: Workshop on Preformability Modeling of Computer and Communication Systems - PMCCS 2003, Arlington, IL, USA, September 2003, pp. 43–46 (2003)
5. Guillemin, F., Pinchon, D.: Analysis of the Weighted Fair Queuing System with Two Classes of Customers with Exponential Service Times. Journal of Applied Probability (2004)
6. Ashour, M., Le-Ngoc, T.: Performance of Weighted Fair Queuing System with Long Range Dependent Traffic Inputs. In: Electrical and Computer Engineering, pp. 1002–1005 (2005)
7. Davis, P.F., Rabinowitz, P.: Methods of numerical intergration, 2nd edn. Academic Press, London (1984)
8. Ashour, M., Le-Ngoc, T.: Performance of Weighted Fair Queuing System with Variable Service Rates. In: International Conference, pp. 51–57 (2006)
9. Daigle, J.: Queueing Theory for Telecommunications, 1st edn., pp. 138–147. Addison-Wesley Longman (1991)
10. Adan, I., Resing, J.: Queueing Theory, Eindhoven University of Technology, Department of Mathematics and Computing Science, February 14 (2001)

# Counting Flows over Sliding Windows in High Speed Networks

Josep Sanjuàs-Cuxart, Pere Barlet-Ros, and Josep Solé-Pareta

Universitat Politècnica de Catalunya (UPC), Computer Architecture Dept.
Jordi Girona, 1-3 (Campus Nord D6), Barcelona 08034, Spain
{jsanjuas,pbarlet,pareta}@ac.upc.edu

**Abstract.** Counting the number of flows present in network traffic is not trivial, given that the naive approach of using a hash table to track the active flows is too slow for the current backbone network speeds. Several algorithms have been proposed in the recent literature that can calculate an approximate count using small amount of memory and few memory accesses per packet. Fewer works have addressed the more complex problem of counting flows over sliding windows, where the main challenge is to continuously expire old information. One of the existing proposals is a straightforward adaptation of the direct bitmaps technique to the sliding window model. We present an algorithm called Countdown Vector that also builds upon the direct bitmaps technique. Our algorithm, however, obtains significant cost reductions both in terms of memory and CPU, by introducing an extra approximation in the mechanism in charge of the expiration of old information.

**Keywords:** traffic measurement, counting active flows, sliding windows.

## 1 Introduction and Related Work

The design of efficient algorithms to count the number of active flows in high-speed networks has recently attracted the interest of the network measurement community [1,2,3]. Counting the number of flows in real-time is particularly relevant to network operators and administrators for network management and security tasks. For example, this metric is the basis of most network intrusion detection systems to detect port scans and DoS attacks.

However, the naive solution of tracking flows using a hash table is becoming unfeasible in high-speed links. First, it requires several memory accesses per packet with the overhead of creating new flow entries and handling collisions. Second, this solution uses large amounts of memory, since it requires storing all flow identifiers. The number of concurrent flows present in high speed networks is very large, and can be well over one million in current backbone links [4]. Therefore, hash tables must be stored in DRAM, which has an access time greater than current packet interarrival times. For example, access times of standard DRAM are in the order of tens of nanoseconds, while packet interarrival times can be up to 32 ns and 8 ns in OC-192 (10 Gb/s) and OC-768 (40 Gb/s) links

L. Fratta et al. (Eds.): NETWORKING 2009, LNCS 5550, pp. 79–91, 2009.

respectively. Thus, flow counting algorithms must be able to process each packet in very few nanoseconds to be suitable for high-speed links.

In order to reduce the large amount of memory required to store flow tables, most routers (e.g., Cisco NetFlow [5]) and network monitoring systems [6] resort to packet sampling. However, it has been shown that packet sampling is biased towards large flows and tends to underestimate the total number of flows [7].

Recently, several probabilistic algorithms have been proposed to efficiently estimate the number of flows in high-speed networks [1,2,3]. These algorithms share the common approach of using specialized data structures to approximately count the number of flows, which need a very small amount of memory, as compared to the traditional approach of keeping per-flow state, while requiring one or very few memory accesses per packet. This drastic reduction in the memory requirements allows the storage of these data structures in fast SRAM, with access times below 10 ns. These techniques can therefore be implemented in router line cards or, in general-purpose systems, the data structures can reside in cache memory.

The direct bitmaps technique is the basis of most probabilistic algorithms to estimate the number of flows. This technique was first proposed by Whang et al. [8] in the database community and popularized in the networking community by Estan et al. in [1], which also presents several variants of the direct bitmaps that require less memory. One of the variants, called multiresolution bitmaps, obtains similar accuracy as Loglog counting [9]. Giroire's proposal [10] is to estimate the count from the minimum of the hashed values. These and other techniques are compared in [11]. A remarkable conclusion from this study is that, from a practical standpoint, direct bitmaps offer the best tradeoff between complexity and accuracy.

The basic idea behind direct bitmaps is to use a small vector of bits (i.e., bitmap). For each packet, a hash of the flow identifier is computed and the corresponding bit is set in the bitmap. At the end of a measurement interval, the number of flows can be simply estimated according to the number of non-set bits and the collision probability [8]. The main problem of these algorithms is that they can only operate over fixed, non-overlapping measurement intervals and, therefore, cannot obtain continuous estimates of the number of flows.

The *sliding window model* is increasingly gaining interest in the networking and database communities, given the streaming nature of many current data sources (e.g., network traffic, sensor networks or financial data). For example, a new class of systems is emerging in the database [12,13] and network monitoring [14,15,16] communities in order to support continuous queries over sliding windows. Under this new paradigm, queries process streaming data, instead of static databases, and compute metrics over time windows that advance continuously (e.g., the number of active flows during the last 5 minutes). An interesting introduction to the field of data stream research can be found in [17], while a more informal discussion that motivates this research area is [18].

Datar et al. [19] analyze the basic problem of counting the number of ones within the last N elements of an arbitrary stream composed of zeros and ones.

They present an algorithm based on a technique called Exponential Histograms that can provide an approximate solution to this problem. Using this basic algorithm as a building block, they approach related problems such as calculating sums, averages, maximum and minimum values. For the distinct count problem, they propose adapting a bitmap-based counting technique to the sliding window model by storing timestamps instead of bits in each bitmap position.

Kim and O'Hallaron [3] propose a technique that addresses the flow counting problem using this approach, called Timestamp Vector (TSV). TSV is based on the original direct bitmap algorithm and consists of replacing the bitmap for a vector of timestamps. However, the main limitations of TSV are that (*i*) it requires a significantly larger amount of memory compared to the original direct bitmap (a 64-bit timestamp for each vector position) and (*ii*) the cost of the queries is linear with the size of the vector, which renders this solution impractical in scenarios where a continuous estimate of the number of flows is needed.

In this paper, we propose a new algorithm called Countdown Vector (CDV) to estimate the number of flows over sliding windows. The basic idea behind our method is the use of a vector of small timeout counters, instead of full timestamps, that are decremented independently of the per-packet update and query processes. Our algorithm requires less resources (i.e., CPU and memory) than existing solutions, and has $O(1)$ query cost. This way, a network monitoring system can implement our method using less memory, and can react faster to changes in the number of flows (e.g., network anomalies or attacks), since queries can be issued more frequently than in previous proposals. Another interesting advantage of our technique over other alternatives is that it is possible to degrade the accuracy of the estimates according to a given CPU and memory budget.

We implement our algorithm in an existing network monitoring system and present experimental evidence of the accuracy and cost of our technique using real-world packet traces. Our results show that our technique is able to estimate the number of flows over a wide range of sliding windows with a similar accuracy to the TSV, while reducing the memory requirements to up to 1/20th and the number of memory accesses to up to 1/30th.

The rest of the paper is organized as follows. Section 2 describes our algorithm to estimate the number of flows over sliding windows, while Section 3 presents a performance evaluation of our technique using real packet traces collected at the access link of a large university network. Finally, Section 4 concludes the paper and discusses future work.

## 2 Countdown Vector Algorithm

This section describes our method to count flows over sliding windows. A flow is defined as a sequence of packets that share equal values for a subset of the TCP/IP header fields. Typically, a flow is identified by the 5-tuple that consists of the source and destination IP addresses and ports, and protocol field. However, our method is agnostic to the particular definition of a flow.

Since our algorithm is based on direct bitmaps, we review this technique in greater detail. We also describe the Timestamp Vector algorithm (TSV), which

**Table 1.** Notation

| |
|---|
| $b$: Number of positions of the vector or bitmap. |
| $c$: Value to which counters are initialized. |
| $f$: Time between queries in a jumping window model. |
| $n$: Number of flows in the traffic during a time window. |
| $\hat{n}$: Estimation of the number of flows $n$. |
| $s$: Time between counter updates. |
| $w$: Length of the time window. |
| $z$: Number of positions with value 0 in the bitmap or the vector or bitmap. |

we use to compare the accuracy and overhead of our method. Additionally, we propose a simple improvement of the TSV that significantly reduces its memory requirements under certain conditions.

## 2.1 Direct Bitmaps

Like the naive algorithm of using a hash table to track flows, direct bitmaps are based on hashing, but do not create one table entry per flow. Instead, they just record whether a position in the hash table would be used or not (hence the name bitmap). The algorithm uses a pseudo-random hash function (e.g., [20]) to evenly distribute the positions corresponding to each flow identifier.

One could think of extrapolating the number of ones in the bitmap as the number of flows in the original dataset. However, this would not be correct, since there is a non-negligible probability that several flow identifiers collide (i.e., hash to the same bitmap position), given the reduced size of the bitmap. While the bitmap cannot be used to extract the exact count of flows, instead, an estimate can be obtained from the bitmap size ($b$) and the number of zeros in the bitmap ($z$) as shown in [8]:

$$\hat{n} = b \, ln \left( \frac{b}{z} \right) \tag{1}$$

We refer to the process of obtaining an estimate of the number of flows as the process of *querying* the bitmap. Table 2.1 summarizes the notation used throughout this section.

The principal advantage of this technique is that, with a small amount of memory (especially when compared to the naive algorithm), the number of flows can be estimated with high accuracy. A second important characteristic of this approach is that the accuracy can be arbitrarily increased or reduced by the bitmap size. To correctly dimension the bitmap, the appropriate size must be chosen so that the expected amount of unique elements can be estimated within the desired error bounds, as explained in [8].

To give the reader an idea of how little memory this algorithm requires, it suffices to state that this technique can count $10^6$ elements by using around 20KB with errors below 1% [8].

### 2.2 Timestamp Vector

As discussed in the previous section, direct bitmaps are a very efficient technique to count the amount of flows in the network traffic. However, in order to provide meaningful values when monitoring a network traffic stream, measurement statistics must be bounded in time, i.e., must correspond to a particular *time window*. Direct bitmaps do not incorporate a sense of time and thus must be periodically reset to avoid lifetime flow counting.

Periodically querying and resetting a direct bitmap would provide flow counts over consecutive, non-overlapping windows. While there is value to this application of the technique, it imposes the restriction that queries must be aligned with bitmap resets. In contrast, in the *sliding window* model, queries can arrive at any time. This requirement implies that old information must be removed as newer arrives, in order to continuously maintain the data structures to be able to provide an estimate at any time.

A straightforward solution to adapt the direct bitmaps to the sliding window model is the Timestamp Vector algorithm [3]. Instead of a bitmap, a vector of timestamps is now used. When a packet hashes to a particular position, its timestamp is set to the timestamp of that packet. Using this vector, when a query for a time window of $w$ time units arrives at time $t$, the number of flows can be estimated by using Equation 1, where in this case $z$ corresponds to the number of positions with timestamp less than $t - w$, and $b$ to the number of positions in the vector. Note that this requires a full traversal of the vector for each query.

### 2.3 Extension of the Timestamp Vector

The principal limitation of the Timestamp Vector technique is the increase in the amount of memory that it requires. Timestamps in network monitoring are typically 64 bits long, e.g. as provided by libpcap [21] or Endace DAG cards [22]. Hence, the final size of the vector is increased by a factor of 64 compared to the original bitmap.

A relaxation of the sliding window model is the *jumping window* model [23], where the window does not advance continuously, but discretely in fractions of the measurement window. When operating under this model, we propose the following improvement over the Timestamp Vector algorithm that significantly reduces its memory requirements. Instead of full timestamps, only the fraction of the window where the packet arrived is stored. This idea can be implemented using $log_2(w/f + 1)$ bits per position when measuring a window of $w$ time units with query periods of $f$. For example, with a window of 30 s and queries every 1 s, the original Timestamp Vector will require 64 bits per position, while this approach would require only 5 bits per position, using below 10% of the memory. Note however that, using this extension, the Timestamp Vector can only be queried every $f$ time units.

One limitation of both variants of the Timestamp Vector algorithm is, thus, their additional memory requirements. The jumping window variant reduces memory usage by limiting the time where queries may be performed. The second disadvantage of this scheme is that, for each query, one full traversal of the array is required to calculate the number of positions whose timestamp is older than $t - w$.

### 2.4 Countdown Vector

In this section we present our technique for flow counting over sliding windows. Our scheme is, like the TSV algorithm, an adaptation of the direct bitmaps to the sliding window model. We start by outlining the basic intuition behind the technique we propose.

The main difficulty when adapting the direct bitmap to the sliding window model is to remove old information from the data structure as time advances. Let us start by defining an ideal algorithm which would precisely calculate $z$ in a sliding window of $w$ time units. Recall that $b$ (vector size) and $z$ (number of zeros in it) suffice to estimate the number of flows in the traffic using Equation 1. A vector of $b$ positions could be used, with the values set to $w$ time units every time a packet hashed to the corresponding position. Every time unit, all the positions of the vector with non-zero value would be decremented by one. The count of positions with counter value zero would correspond to $z$ in order to estimate the number of flows seen in the time window.

In order to obtain a perfect resolution, this scheme would require defining the time unit to the maximum resolution of the system clock. This ideal scheme would therefore be very costly in terms of both memory and CPU. First, a high resolution counter would have be stored in each vector position, thus increasing the overall memory required to store the vector. Second, all of the counters would need to be updated for each time unit. These additional costs make this technique infeasible as described, especially when it would achieve results equivalent to the TSV.

The technique that we propose is very similar to the ideal algorithm just described but, instead, using small integer counters in each position. Using small values requires less memory and calls for a low counter decrement frequency for counters to reach zero after $w$ time units, in exchange of introducing small inaccuracies. In the following paragraphs we describe our technique with detail. The key to the effectiveness of our algorithm is that surprisingly low values suffice to achieve good accuracy to estimate network flow counts. In practice, then, we require less memory and introduce lower overhead compared to the original TSV algorithm and to its extension presented in Section 2.3.

**Algorithm.** Our algorithm starts by allocating a vector of counters, all of which are initialized to zero. Our algorithm can then be seen as divided in two concurrent processes: the first updates the vector for each packet, while the second is in charge of decreasing the counters at a fixed rate.

The first process is described in Algorithm 1 and runs synchronously with the packet stream. When a packet hashes to a position, we store a maximum value $c$ in the corresponding position. This process remains the same independently of the time window that is being measured.

In contrast, the second process, which is described in Algorithm 2, performs a continuous maintenance of the vector. It decreases one counter every $s$ time units,

**Algorithm 1.** Initialization and packet-synchronous operations

```
Data: b: vector size, c: counter initialization value
z ← b;
vector ← {0,...,0};
start continuous maintenance procedure;
foreach packet p do
    key ← hash(p.flow_identifier);
    if vector[key mod b] = 0 then                /* update count of zeros */
        z ← z − 1;
    end
    vector[key mod b] ← c;
end
```

advancing one position in the vector at every step. The desired time window $w$ plays a role in this data structure maintenance process, where it conditions the speed at which counters are decreased.

To determine $s$ we proceed as follows. Since the packet arrivals and the maintenance are independent processes, and each flow hashes to a random position, on average, the first decrement of a counter (after it is set to $w$ by the first process) will happen after $b/2$ counter updates. Afterwards, the counter will be decremented after $b$ additional counter updates. Therefore, on average, counters reach zero after $b/2 + b\ (c-1) = b\ (c - 1/2)$ updates. Since this time must correspond to the time window $w$, we calculate $s$ the following way:

$$s = \frac{w}{b\ \left(c - \frac{1}{2}\right)} \tag{2}$$

Both the first and the second processes maintain the count of positions of the vector with value zero, updating the value of $z$ as values of the vector are modified. This has the advantage that query operations run in constant time $O(1)$, by simply applying Equation 1, as described in Algorithm 3.

Our algorithm is then governed by the following configuration parameters: $(i)$ the desired measurement window $w$, $(ii)$ the size of the vector $b$, and $(iii)$ the maximum values to which counters are set $c$.

The precision of our algorithm increases with larger $b$ and $c$ values. Larger values of $b$ (vector size) make the estimation error of Equation 1 decrease, as explained in further detail in [8]. On the other hand, increasing $c$ also has a positive impact on the accuracy of the method, since, the larger $c$ is, the more our algorithm approaches the ideal algorithm explained in the previous subsection.

However, larger values of $c$ increase both the memory and CPU requirements of our algorithm, since, the higher $c$ is, the more space is required to store the counters, and the higher the counter decrease frequency, i.e., $s$ decreases, according to Equation 2.

Our algorithm has two sources of error. First, the approximation introduced by the original technique upon which ours builds, the direct bitmaps. The second source of error is introduced by the fact that old information is expired after $w$

**Algorithm 2.** Continuous maintenance procedure

**Data**: $b$: vector size, $c$: counter initialization value, *vector*: vector of counters, $z$: count of positions with value 0

```
s ← w / (b × (c − 1/2));
i ← 0;
while True do
    sleep for s time units;
    if vector[i] = 1 then                    /* update count of zeros */
        z ← z + 1;
    end
    vector[i] ← max(0, vector[i] − 1);
    i ← (i + 1) mod b;
end
```

**Algorithm 3.** Query procedure

**Data**: $b$: vector size, $z$: number of zeros in *vector*

```
return b × ln(b/z);
```

time units only *on average*. Inevitably, some counters will, on the worst case, be set to $c$ right after the maintenance process has decremented the corresponding position, and will thus will reach zero after $c * b * s = \frac{c}{c-1/2} w$ time units. Conversely, others will reach zero after $\frac{c-1}{c-1/2} w$ time units. In the worst case a counter will be inaccurate only during this small period of time. This explains why the accuracy increases with larger values of $c$, which tighten these bounds around $w$.

In order to choose a value for $b$, the tables provided by [8] can be looked up to determine the an appropriate vector size for the expected number of flows in the traffic. In the next section we analyze the impact of $c$ over the accuracy of the method and show the overhead reduction of our method compared to both variants of the Timestamp Vector.

## 3 Evaluation

In order to obtain sensible results, we have tested our technique using real traffic. We collected a 30-minute packet-level traces in November 2007 at the access link of the Technical University of Catalonia (UPC), which connects 10 campuses, 25 faculties and 40 departments to the Internet through the Spanish Research and Education network (RedIRIS). The trace accounts for 106M packets with an average data rate of 271.6 Mbps. The average number of flows is around 50000 in a ten seconds window, and 1.8 million in a 10 minute window. Our technical report version of this paper includes the results obtained with other traces [24], which are similar to the ones presented in this section.

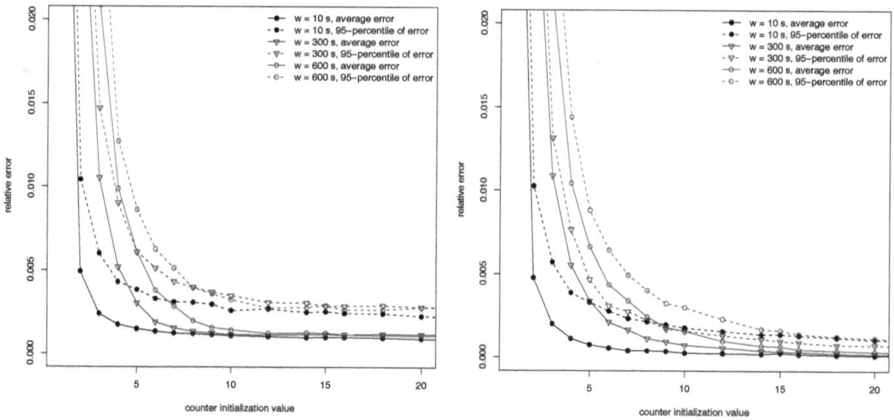


**Fig. 1.** Error of the Countdown Vector algorithm (left) and error of the Countdown Vector algorithm compared to the estimates of the Timestamp Vector algorithm (right)

Figure 1 (left) shows the results of running our algorithm with window sizes of 10, 300 and 600 seconds with different counter initialization values, using a fixed vector size. The size of the vector has been chosen according to [8] so that the average number of flows for the largest window can be counted with errors below 1%. Each point in the figure corresponds to a full pass on aforementioned trace, querying the algorithm every second, and shows either the average relative error or the 95th percentile of the relative error compared to a precise calculation of the number of flows.

As expected, for every window size, the relative error decreases as $c$ increases. It is interesting to observe that, while the error is intolerable for very small values of $c$ (5 and below), when it reaches values as small as 10 the error stabilizes, and does not decrease significantly beyond that point, even for a window as large as 600 s. This observation explains the great overhead savings that our technique shows in comparison to the Timestamp Vector, as we show in the next paragraphs. We have observed that the error for these values of $c$ is almost equal to that of the Timestamp Vector algorithm. This source of error can be imputed to the underlying method of estimation of the direct bitmaps (see Equation 1). Figure 1 (right) shows the error of our method relative to the values obtained using the Timestamp Vector algorithm and confirms this observation.

We now examine the cost of the Countdown Vector algorithm and compare it to that of the Timestamp Vector. Figures 2 and 3 summarize a different set of executions of the algorithms using the same trace. To obtain realistic overhead calculations, in this case we dimension the vectors for both TSV and CDV appropriate for the observed number of flows in each time window, using [8], bounding the error introduced by the estimation formula to 1%. We dimension our variant of the TSV for a 1 second query frequency. For performace reasons, our implementation restricts the vector sizes to powers of two; we choose the smallest suitable sizes.

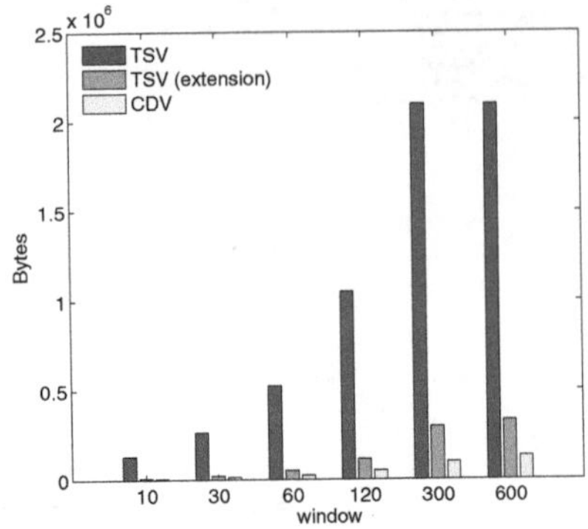


**Fig. 2.** Memory consumption for the three algorithms with varying window sizes

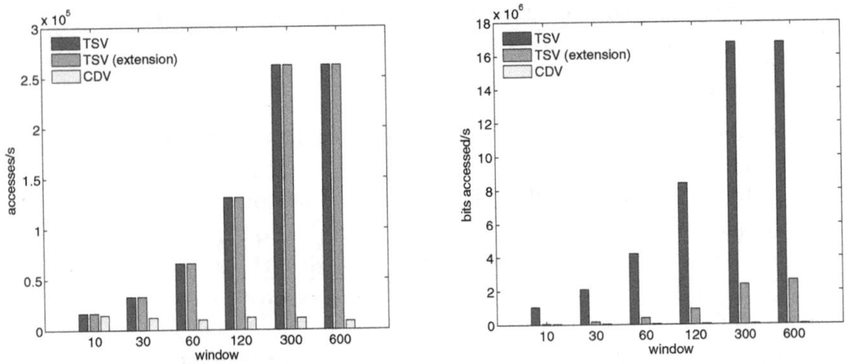


**Fig. 3.** Number of memory accesses per second (left) and number of bits accessed per second (right) for the three algorithms with varying window sizes

Our algorithm has, besides the size of the vector, an additional parameter: the counter initialization value ($c$). We have run the experiments with various $c$ values, and have chosen, for each time window, the smallest that obtains at most 0.1% more relative error than the Timestamp Vector algorithm.

Figure 2 shows the high memory savings that our variant of the TSV introduces, at the expense of reducing the query frequency to only one second. In contrast, the CDV we propose further reduces the memory to roughly one half compared to our TSV variant, without introducing such a restriction.

We compare the cost of the algorithms using the number of memory accesses per second, assuming one query every second, and including the maintenance cost in the case of the CDV. However, we omit the cost of updating the vectors for every packet, since this cost is common to all of the algorithms. It suffices to state that it is only in the order of 50000 accesses per second, which roughly corresponds to the average packet rate in our trace. In Figure 3 (left), it can be observed that the cost of the Timestamp Vector grows with the size of the vector, since it has to be traversed for every query. In contrast, the cost of the Countdown Vector remains small.

This figure is unfair to our variant of the TSV since, while the number of memory accesses are equal to those of the original TSV, these are accesses to smaller chunks of memory. Depending on the architecture, then, specific optimizations could be employed to improve its performance (e.g., read several positions with a single memory access). To correct this, we also present the cost in terms of bits accessed per second in Figure 3 (right).

The cost of both variants of the TSV grows proportionally to the vector size, since full vector traversals per query are required. Vectors grow with window sizes, since higher flow counts have to be obtained. In contrast, the CDV's query cost is constant; in our algorithm, the bulk of the cost is in the maintenance phase. However, since very low counter values can obtain an accuracy very similar to the TSV variants, counter decrements are performed at a low frequency, therefore incurring significantly lower costs.

## 4 Summary and Future Work

We have presented an algorithm called Countdown Vector that efficiently calculates the number of flows present in the network traffic over sliding windows. Our scheme introduces a continuous maintenance cost but, unlike previous proposals, can be queried in constant time. We have performed an evaluation using real traffic, and compared the cost in terms of memory and CPU to the state of the art Timestamp Vector algorithm. The Countdown Vector shows comparable accuracy with significatly lower costs.

The basic idea behind our scheme is to use a vector of timeout counters that expires old information in an approximate fashion. While in this work we implement this idea on the direct bitmaps technique, we plan on adapting our scheme to other more efficient flow counting algorithms. In particular, this scheme can be easily applied to the algorithms proposed by Estan et al. in [1].

Another important piece of future work is to perform a theoretical analysis of the accuracy of the algorithm we propose. While we have performed an empirical analysis using network traffic traces, it is important to be able to provide error bounds that apply to other scenarios.

## Acknowledgments

This work was partially funded by the Spanish Ministry of Science and Innovation (MICINN) under contract TSI2005-07520-C03-02 (CEPOS). The authors thank the Supercomputing Center of Catalonia and UPCnet for allowing them to collect the packet traces used in this work. Authors are also grateful to Manel Martinez Torres and Gianluca Iannaccone for useful discussion in early stages of this work.

## References

1. Estan, C., Varghese, G., Fisk, M.: Bitmap algorithms for counting active flows on high speed links. In: Proc. of ACM SIGCOMM Internet Measurement Conf. (October 2003)
2. Fusy, E., Giroire, F.: Estimating the number of Active Flows in a Data Stream over a Sliding Window. In: Proc. of SIAM Workshop on Analytic Algorithmics and Combinatorics (January 2007)
3. Kim, H., O'Hallaron, D.: Counting network flows in real time. In: Proc. of IEEE GLOBECOM (December 2003)
4. Fang, W., Peterson, L.: Inter-AS traffic patterns and their implications. In: Proc. of IEEE GLOBECOM (December 1999)
5. Cisco Systems: NetFlow services and applications. White Paper (2000)
6. Barlet-Ros, P., Iannaccone, G., Sanjuàs-Cuxart, J., Amores-López, D., Solé-Pareta, J.: Load shedding in network monitoring applications. In: Proc. of USENIX Annual Technical Conf. (June 2007)
7. Duffield, N., Lund, C., Thorup, M.: Properties and prediction of flow statistics from sampled packet streams. In: Proc. of ACM SIGCOMM Internet Measurement Workshop (November 2002)
8. Whang, K.Y., Vander-Zanden, B.T., Taylor, H.M.: A linear-time probabilistic counting algorithm for database applications. ACM Trans. Database Syst. 15(2) (June 1990)
9. Durand, M., Flajolet, P.: Loglog Counting of Large Cardinalities. In: Proc. of Annual European Symposium on Algorithms (September 2003)
10. Giroire, F.: Order statistics and estimating cardinalities of massive data sets. In: Proc. of Intl. Conf. on Analysis of Algorithms (June 2005)
11. Metwally, A., Agrawal, D., El Abbadi, A.: Why go logarithmic if we can go linear?: Towards effective distinct counting of search traffic. In: Proc. of Intl. Conf. on Extending Database Technology: Advances in Database Technology (March 2008)
12. Babcock, B., Babu, S., Datar, M., Motwani, R., Widom, J.: Models and issues in data stream systems. In: Proc. of ACM SIGMOD-SIGACT-SIGART Symp. on Principles of Database Systems (June 2002)
13. Golab, L., Özsu, T.M.: Issues in data stream management. SIGMOD Record 32 (June 2003)
14. Cranor, C., Johnson, T., Spataschek, O., Shkapenyuk, V.: Gigascope: A stream database for network applications. In: Proc. of ACM SIGMOD (June 2003)
15. Iannaccone, G.: Fast prototyping of network data mining applications. In: Proc. of Passive and Active Measurement Conf. (March 2006)
16. Reiss, F., Hellerstein, J.M.: Declarative network monitoring with an underprovisioned query processor. In: Proc. of IEEE Intl. Conf. on Data Engineering (April 2006)
17. Muthukrishnan, S.: Data Streams: Algorithms And Applications. Now Publishers Inc. (2005)
18. Hayes, B.: The Britney Spears Problem, `http://www.americanscientist.org/issues/pub/the-britney-spears-problem`
19. Datar, M., Gionis, A., Indyk, P., Motwani, R.: Maintaining stream statistics over sliding windows. In: Proc. of ACM-SIAM Symp. on Discrete Algorithms (January 2002)
20. Carter, J.L., Wegman, M.N.: Universal classes of hash functions. J. Comput. Syst. Sci. 18 (April 1979)

21. Jacobson, V., Leres, C., McCanne, S. (libpcap) Lawrence Berkeley Laboratory, Berkeley, CA. Initial public release (June 1994), http://www.tcpdump.org
22. Endace: DAG network monitoring cards, http://www.endace.com
23. Golab, L., DeHaan, D., Demaine, E., Lopez-Ortiz, A., Munro, J.: Identifying frequent items in sliding windows over on-line packet streams. In: Proc. of ACM SIGCOMM Internet Measurement Conf. (October 2003)
24. Sanjuàs-Cuxart, J., Barlet-Ros, P., Solé-Pareta, J.: Counting network flows over sliding windows in high-speed networks. UPC Technical Report, http://loadshedding.ccaba.upc.edu/papers/counting-network-flows.techrep2008.pdf

# Heterogeneous Protection in Regular and Complete Bi-partite Networks (Work in Progress)

Jasmina Omić, Robert E. Kooij, and Piet Van Mieghem

Delft University of Technology,
Faculty of Electrical Engineering, Mathematics and Computer Science,
P.O Box 5031, 2600 GA Delft, The Netherlands
{P.VanMieghem,J.S.Omic,R.E.Kooij}@ewi.tudelft.nl
Dr. ir. Kooij is also with TNO Information Communication Technology,
P.O Box 5050, 2600 GB Delft

**Abstract.** We examine the influence of heterogeneous curing rates for a $SIS$ model, used for malware spreading on the Internet, information dissemination in unreliable networks, and propagation of failures in networks. The topology structures considered are the regular graph which represents the homogenous network structures and the complete bi-partite graph which represents the hierarchical network structures. We find the threshold in a regular graph with $m$ different curing rates.

Further, we consider a complete bi-partite graph with 2 curing rates and find the threshold for any distribution of curing rates among nodes. In addition, we consider the optimization problem and show that the minimum sum of the curing rates that satisfies the threshold equation is equal to the number of links in the graph. The optimization problem is simplified by assuming fixed curing rates $\delta_1, \delta_2$ and optimization of the distribution of curing rates among different sets of nodes.

**Keywords:** Virus spread, epidemic threshold, heterogeneous networks.

## 1 Introduction

The Susceptible-Infected-Susceptible ($SIS$) infection model, which arose in mathematical biology, is often used to model the spread of computer viruses [1], [2], [3], epidemic algorithms for information dissemination in unreliable distributed systems like P2P and ad-hoc networks [4], [5], and propagation of faults and failures in networks like $BGP$ [6].

The $SIS$ model assumes that a node in the network is in one of two states: *infected* and therefore infectious, or *healthy* and therefore susceptible to infection. The $SIS$ model usually assumes instantaneous state transitions. Thus, as soon as a node becomes infected, it becomes infectious and likewise, as soon as a node is cured it is susceptible to re-infection. There are many models that consider more aspects like incubation periods, variable infection rate, a curing process that takes a certain amount of time and so on [7], [1], [8]. In epidemiological

L. Fratta et al. (Eds.): NETWORKING 2009, LNCS 5550, pp. 92–103, 2009.

theory, many authors refer to an epidemic threshold $\tau_c$, see for instance [7], [9], [1] and [3]. If it is assumed that the infection rate along each link is $\beta$ while the curing rate for each node is $\delta$ then the effective spreading rate of the virus can be defined as $\tau = \beta/\delta$. The epidemic threshold can be defined as follows: for effective spreading rates below $\tau_c$ the virus contamination in the network dies out - the mean epidemic lifetime is of order $\log n$, while for effective spreading rates above $\tau_c$ the virus is prevalent, i.e. a persisting fraction of nodes remains infected with the mean epidemic lifetime [2] of the order $e^{n^\alpha}$. In the case of persistence we will refer to the prevailing state as a metastable state or steady state. It was shown in [10] and [2] that $\tau_c = 1/\rho(A)$ where $\rho(A)$ denotes the spectral radius of the adjacency matrix $A$ of the graph. Recently, the epidemic threshold formula has also been verified by using the $N$-intertwined model [11], which consists of a pair of interacting continuous Markov chains.

It is the aim of this paper to derive results for the epidemic threshold in the case of heterogeneous curing rates for regular and complete bi-partite graphs. A regular graph is an approximation of the random graph for large $N$ and it represents a significant set of networks used in telecommunications. Further, the complete bi-partite graph represents a hierarchal type of topology, also frequently used in telecommunications. Notice that (core) telecommunication networks often can be modeled as a complete bi-partite topology. For instance, the so-called double-star topology (i.e. $K_{M;N}$ with $M = 2$) is quite commonly used because it offers a high level of robustness against link failures. For example, the Amsterdam Internet Exchange,[1] one of the largest public Internet exchanges in the world, uses this topology to connect its four locations in Amsterdam to two high-throughput Ethernet switches. Sensor networks are also often designed as complete bi-partite graphs.

The rest of the paper is organized as follows. In Section 2, we present the classical model by Kephart and White which describes the homogenous spread of a virus on regular graphs and the $SIS$ model for the complete bi-partite graph also analyzed in [12]. In Section 2.1, we derive and analyze the spread of viruses in regular graphs in case of $m$ curing rates. In Section 2.2, we discuss a specific case of regular graphs with 2 curing rates. In Section 2.3, we consider the spread of viruses on the complete bi-partite graphs with two curing rates. In the following section, we give solution for the optimization problem on complete bi-partite graph in the heterogenous case. We summarize our results in Section 4.

## 2 Virus Spread on Regular and Bi-partite Graphs

In order to explain our model of spread for computer viruses with heterogeneous curing rates, it is useful to first discuss the spread of viruses with homogeneous curing rate.

The homogenous model for regular graph is based on a classical result by Kephart and White [1] for $SIS$ models. We consider a connected graph with $N$ nodes, where every node has degree $k$. We denote the number of infected

[1] see www.ams-ix.net

nodes in the population at time $t$ by $X(t)$. The probability that a randomly chosen node is infected is $v(t) \equiv X(t)/N$. Now, the rate at which the infection probability changes is due to two processes: susceptible nodes becoming infected and infected nodes being cured. The change in probability $v(t)$ due to the curing of infected nodes is $\delta v(t)$. The rate at which the infection probability $v(t)$ grows is proportional to the probability of a node being susceptible, i.e. $1 - v(t)$. For every susceptible node, the rate of infection is the product of the infection rate per node ($\beta$), the degree of the node ($k$) and the probability that on a given link the susceptible node connects to an infected node ($v(t)$). Therefore, we obtain the following differential equation describing the time evolution of $v(t)$:

$$\frac{dv(t)}{dt} = \beta k v(t)(1 - v(t)) - \delta v(t). \tag{1}$$

The solution to Eq. (1) is

$$v(t) = \frac{v_0(1-\rho)}{v_0 + (1 - \rho - v_0)e^{-(\beta k - \delta)t}}, \tag{2}$$

where $v_0$ is the initial probability of infected nodes. The steady state solution is

$$v_\infty = \frac{\beta k - \delta}{\beta k} \tag{3}$$

An epidemic steady state only exist for $v_\infty > 0$, therefore, the epidemic threshold equals to $\tau_c = \frac{1}{k}$. For $k$-regular graphs, the spectral radius of the adjacency matrix [13] is equal to $k$, therefore $\tau_c = \frac{1}{k}$ is in line with the result in [10].

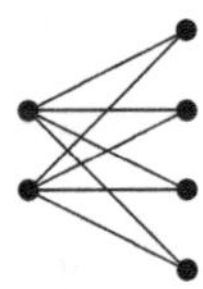

**Fig. 1.** Complete bi-partite graph $K_{2,4}$

Further, we will consider the complete bi-partite graphs with one curing rate $\delta$. The $SIS$ model for the complete bi-partite graph is presented in [12]. A complete bi-partite graph $K_{M,N}$ consists of two disjoint sets $S_1$ and $S_2$ containing respectively $M$ and $N$ nodes, such that all nodes in $S_1$ are connected to all nodes in $S_2$, while within each set no connections occur. Figure 1 gives an example of a complete bi-partite graph with 6 nodes. Since there are two sets of nodes with different degrees, equation (1) does not hold. A node from the set $S_1$ is connected to $N$ nodes from the set $S_2$. The probability that a randomly chosen node is infected in set $S_1$ is $v_1(t) \equiv X_1(t)/M$. The rate at which probability $v_1(t)$ grows is proportional to the probability that a node in the set $S_1$ is susceptible multiplied by the degree of the node $N$ and the probability that a node connects to an infected node from set $S_2$, which is $v_2(t) \equiv X_2(t)/N$. For the set $S_1$ and $S_2$, we can write differential equations

$$\frac{dv_1(t)}{dt} = \beta N v_2(t)(1 - v_1(t)) - \delta v_1(t),$$
$$\frac{dv_2(t)}{dt} = \beta M v_1(t)(1 - v_2(t)) - \delta v_2(t)$$

For $\frac{dv_1(t)}{dt} = 0$ and $\frac{dv_2(t)}{dt} = 0$, we have the steady state solution

$$v_{1\infty} = \frac{\beta^2 MN - \delta^2}{N\beta(\delta + \beta M)}, \; v_{2\infty} = \frac{\beta^2 MN - \delta^2}{M\beta(\delta + \beta N)}$$

Now, the epidemic threshold equals $\tau_c = \left.\frac{\beta}{\delta}\right|_{v_\infty = 0} = \frac{1}{\sqrt{MN}}$, which is the reciprocal of the spectral radius of the adjacency matrix for the complete bi-partite graph [13].

## 2.1 Virus Spread on Regular Graphs with $m$ Curing Rates

In this section, we derive the threshold for the spread of viruses on regular graphs with $m$ curing rates.

Assume that $n_1, n_2, .., n_m$ denotes the fraction of nodes with curing rate $\delta_1, \delta_2, .., \delta_m$ ($\sum_{i=1}^{m} n_i = 1$). It is important to note that one of the assumptions is complete symmetry of the problem. For every node $i$, a fraction $n_1$ of neighbors has the curing rate $\delta_1$, a fraction $n_2$ has curing rate $\delta_2$ and so on.

Denote the number of infected nodes of type $i$ in the population at time $t$ by $X_i(t)$. The probability that a randomly chosen node of type $i$ is infected is $v_i(t) \equiv \frac{X_i(t)}{Nn_i}$. Now, the rate at which the probability of infection for nodes of type $i$ changes is due to two processes: susceptible nodes becoming infected and infected nodes being cured. The curing rate for an infection probability $v_i$ is $\delta_i v_i$. The rate at which the probability $v_i$ grows is proportional to the probability of a node of type $i$ being susceptible, i.e. $1 - v_i$. For every susceptible node the rate of infection is the product of the infection rate per node ($\beta$) and the probability that on a given link the susceptible node connects to an infected node ($\sum_{j=1}^{m}(n_j k)v_j$).

Therefore, we obtain the following differential equation describing the time evolution of $v_i(t)$:

$$\frac{dv_i}{dt} = \beta k (\sum_{j=1}^{m} n_j v_j)(1 - v_i) - \delta_i v_i, \; i = 1, .., m \tag{4}$$

Note that for $\delta_1 = \delta_2 = .. = \delta_m$, the system of equations (4) reduces to Eq. (1) with $v = \sum_{j=1}^{m} n_j v_j$.

For the general case with different curing rates, it is impossible to obtain an explicit solution for the system of equations (4). The standard approach for this type of system of nonlinear differential equations, is to study the qualitative behavior in the phase space.

**Theorem 1.** *Consider connected regular graphs where each node has exactly $k$ neighbors. Assume that the infection rate along each link is $\beta$ while the curing* *rate for each node is $\delta_i$ for a fraction $n_i$ of the nodes, with $i = 1,..,m \leqslant k$ and $\sum_{i=1}^{m} n_i = 1$. Complete symmetry is assumed, where each node sees the same fraction of different curing rates. If we define the effective spreading rate as $\tau = \frac{\beta}{\delta^*}$, where $\delta^*$ is defined as the weighted harmonic mean of $\delta_1, ..., \delta_m$, i.e. $\delta^* = \left(\sum_{i=1}^{m} \frac{n_i}{\delta_i}\right)^{-1}$, then the epidemic threshold satisfies $\tau_c = \frac{1}{k}$.*

*Proof.* We denote the fraction of infected nodes of type $i$ ($1 \leqslant i \leqslant m$) at time $t$ as $v_i(t)$. This leads to a system of $m$ differential equations (4).

We will use a Lyapunov function [14] to show that, under the condition $\beta\sum_{t=1}^{m} \frac{n_t}{\delta_t} - \frac{1}{k} \leq 0$, the origin is a global attractor for $\{v_1 \geq 0, v_2 \geq 0, .., v_m \geq 0\}$, hence, that the virus dies out. Let $V(v_1, v_2, .., v_m) = \prod_{j=1}^{m} \delta_j \sum_{s=1}^{m} \frac{v_s}{\delta_s}$. Then, we have

$$\frac{dV}{dt} = -\left(\sum_{s=1}^{m} v_s\right)\left(\beta k V - \beta k \prod_{j=1}^{m} \delta_j \sum_{t=1}^{m} \frac{n_t}{\delta_t} + \prod_{j=1}^{m} \delta_j\right)$$

$$= -\left(\sum_{s=1}^{m} v_s\right)\left(\beta k V - k \prod_{j=1}^{m} \delta_j \left(\beta \sum_{t=1}^{m} \frac{n_t}{\delta_t} - \frac{1}{k}\right)\right).$$

The claim follows directly by applying Lyapunov's stability theorem. Next we consider the case $\beta\sum_{t=1}^{m} \frac{n_t}{\delta_t} - \frac{1}{k} > 0$. We first note that any trajectory of the system (4) can never leave the box $B = \{(v_1, ..., v_m) | 0 \leq v_1 \leq 1, ..., 0 \leq v_m \leq 1\}$. This follows from $\frac{dv_1}{dt}|_{v_1=0} = \beta k(\sum_{j=1}^{m} n_j v_j) \geq 0$, and similar inequalities at the borders of the box $B$.

From the construction of the above Lyapunov function $V$, we can see that for $\beta\sum_{t=1}^{m} \frac{n_t}{\delta_t} - \frac{1}{k} > 0$, and for $(v_1, ..., v_m) \in B$ and sufficiently close to the origin, $\frac{dV}{dt} > 0$. This implies that the origin has an unstable manifold in $B$. Therefore, since any trajectory of system (4) can never leave the box $B$, system (4) has an attractor as the $\omega$-limit set and, hence, the virus does survive. This finishes the proof of the theorem. □

## 2.2 Virus Spread on Regular Graphs with Two Curing Rates

The two dimensional case ($m = 2$) of virus spread on a regular graph can be analyzed in more details. Applying Theorem 1, the spreading process has a threshold at $\tau = \frac{\beta}{\delta^*} = \frac{1}{k}$, where $\delta^* = \frac{\delta_1 \delta_2}{n_1 \delta_2 + n_2 \delta_1}$.

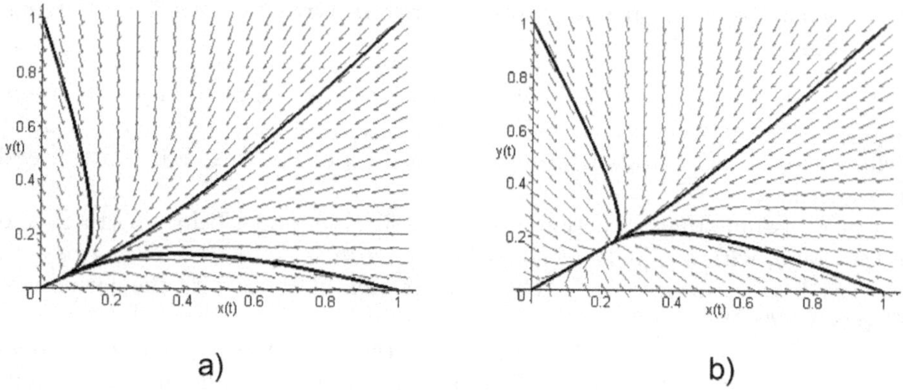


**Fig. 2.** Phase portrait for a regular graph with the two curing rates where a) virus dies out $\beta = 0.2$, $\delta_1 = 0.8$, $\delta_2 = 1.2$, $k = 4$, $n_1 = n_2 = 0.5$. b) virus survives $\beta = 0.4$, $\delta_1 = 0.8$, $\delta_2 = 1.2$, $k = 4$, $n_1 = n_2 = 0.5$.

The phase portrait of two examples are depicted in Figure 2. The attractor for the case where virus survives is given by $(v_1, v_2) = (0.22, 0.17)$.

For system (4) where $m = 2$, it can be proven that the attractor is an equilibrium point of a nodal type, situated on a straight line $L$. It can also be shown that the system does not contain other equilibrium points in $A$ or closed orbits. Therefore, in the case $m = 2$, this equilibrium point is a global attractor of system (4) in $A$.

**Lemma 1.** *The set of differential equations given by (4) for* $m = 2$, *has a straight line solution of the form* $v_2 = \lambda v_1$.

*Proof.* We have that

$$\left(\frac{dv_2}{dt} = \lambda\frac{dv_1}{dt}\right)_{v_2=\lambda v_1}$$

$$-v_1(\beta k n_1 \lambda^2 + (\beta k(n_1 - n_2) - \delta_1 + \delta_2)\lambda - \beta k n_2) \equiv -v_1 f(\lambda)$$

$f(\lambda)$ has got exactly one negative root and one positive root. The positive root $\lambda_1$ satisfies

$$\lambda_1 = \frac{\beta k(n_2 - n_1) + \delta_1 - \delta_2 + \sqrt{\Delta}}{2\beta k n_1},$$

where $\Delta = \beta^2 k^2 + 2\beta k(n_1 - n_2)(\delta_2 - \delta_1) + (\delta_1 - \delta_2)^2$. Therefore the straight line $L : v_2 = \lambda_1 v_1$ is a solution of system (4) for $m = 2$, which for $0 \leq v_1 \leq 1$ is situated in $A$.

By application of the Poincaré-Bendixson theorem [14] on $A$, the $\omega$-limit set for the system (4) for $m = 2$, can be either an equilibrium point or an isolated periodic orbit. From the fact that there is a line solution through the equilibrium point, it follows that the $\omega$-limit set is the equilibrium point. □

### 2.3 Virus Spread on Complete Bi-partite Graphs with Two Curing Rates

We will now derive a model for virus spread on the complete bi-partite graph $K_{M,N}$ with two different spreading rates. The result is general and it can be reduced to the case with all nodes in one set ($S_1$) having one curing rate $\delta_1$ and in the other ($S_2$) $\delta_2$.

Let us assume that a fraction $p$, with $p \in [0,1]$, of nodes belonging to $S_1$ and a fraction $q$, with $q \in [0,1]$, of nodes belonging to set $S_2$ have a curing rate $\delta_1$, the rest have a curing rate $\delta_2$. The total fraction of nodes with the curing rate $\delta_1$ is $s = \frac{Mp+Nq}{M+N}$.

Denote the number of infected nodes of type 1 in the population of nodes from set $S_1$ at time $t$ by $X_{i1}(t)$. The probability that a randomly chosen node of type 1 from set $S_1$ is infected is $v_{i1}(t) \equiv \frac{X_{i_1}(t)}{Mp}$. Similarly, let $v_{i2}$ denote the infection probability for nodes of type 2 from set $S_1$, ($v_{j1}$ denotes type 1, set $S_2$; and $v_{j2}$ denotes type 2, set $S_2$). Now, the rate at which the probability of infection for nodes of type 1, set $S_1$ changes is due to two processes: susceptible nodes becoming infected and infected nodes being cured. The curing rate for an infection probability $v_{i1}$ for nodes of type 1, set $S_1$ is $\delta_1 v_{i1}$. The rate at which the probability $v_{i1}$ grows is proportional to the probability of a node of type 1, set $S_1$ being susceptible, i.e. $1 - v_{i1}$. For every susceptible node the rate of infection is the product of the infection rate per node ($\beta$), the degree of the node ($N$) and the probability that on a given link the susceptible node connects to an infected node $(qv_{j1} + (1-q)v_{j2})$.

Similarly, we obtain the differential equations for the other probabilities $(v_{i2}, v_{j1}, v_{j2})$:

$$\begin{cases} \frac{dv_{i1}}{dt} = \beta N(qv_{j1} + (1-q)v_{j2})(1 - v_{i1}) - \delta_1 v_{i1}, \\ \frac{dv_{i2}}{dt} = \beta N(qv_{j1} + (1-q)v_{j2})(1 - v_{i2}) - \delta_2 v_{i2}, \\ \frac{dv_{j1}}{dt} = \beta M(pv_{i1} + (1-p)v_{i2})(1 - v_{j1}) - \delta_1 v_{j1}, \\ \frac{dv_{j2}}{dt} = \beta N(pv_{i1} + (1-p)v_{i2})(1 - v_{j2}) - \delta_2 v_{j2}, \end{cases} \tag{5}$$

The same set of equations can be obtained by the $N$-intertwined model [11].

In order to simplify the system of equations, we will substitute

$$i_1 = pv_{i1},\ i_2 = (1-p)v_{i2},\ j_1 = qv_{j1},\ j_2 = (1-q)v_{j2}$$

and

$$i = i_1 + i_2,\ j = j_1 + j_2$$

Therefore, we obtain the following differential equations for $i_1(t)$, $i_2(t)$, $j_1(t)$, $j_2(t)$:

$$\begin{cases} \frac{di_1}{dt} = p\beta Nj - \beta Nji_1 - \delta_1 i_1, \\ \frac{di_2}{dt} = (1-p)\beta Nj - \beta Nji_2 - \delta_2 i_2, \\ \frac{dj_1}{dt} = q\beta Mi - \beta Mij_1 - \delta_1 j_1, \\ \frac{dj_2}{dt} = (1-q)\beta Ni - \beta Nij_2 - \delta_2 j_2, \end{cases} \tag{6}$$

By solving the system of equations 6 for the steady state ($\frac{di_1}{dt} = \frac{di_2}{dt} = \frac{dj_1}{dt} = \frac{dj_2}{dt} = 0$) we can calculate the threshold:

$$\frac{\beta}{\delta^*} = \tau_c = \frac{1}{\sqrt{MN}} \tag{7}$$

$$\delta^* = \frac{\delta_1\delta_2}{\sqrt{\delta_1^2(1-p)(1-q)+\delta_2^2 pq+\delta_1\delta_2(p(1-q)+q(1-p))}}$$

**Theorem 2.** *Consider complete bi-partite graphs $K_{M,N}$ consisting of two disjoint sets $S_1$ and $S_2$ containing respectively $M$ and $N$ nodes. Assume that the infection rate along each link is $\beta$. For the nodes in $S_1$ a fraction $p$ has curing rate $\delta_1$ and in $S_2$ a fraction $q$ of the nodes has curing rate $\delta_1$, while the curing rate for a fraction $(1-p)((1-q))$ of the nodes is $\delta_2$. If we define the effective spreading rate as $\tau = \frac{\beta}{\delta^*}$, where $\delta^*$ is defined as $\delta^* = \frac{\delta_1\delta_2}{\sqrt{(1-p)(1-q)\delta_1^2+pq\delta_2^2+\delta_1\delta_2(p(1-q)+q(1-p))}}$, then the epidemic threshold satisfies $\tau_c = \frac{1}{\sqrt{MN}}$.*

*Proof.* First, we will show that if $\frac{\beta}{\delta^*} \leqslant \frac{1}{\sqrt{MN}}$, the virus dies out. $(0,0,0,0)$ is an equilibrium point for system (5). We will use a Lyapunov function to show that, under the condition $\frac{\beta}{\delta^*} \leqslant \frac{1}{\sqrt{MN}}$, the origin is a global attractor for $i_1 \geq 0, i_2 \geq 0, j_1 \geq 0, j_2 \geq 0$.

Let $V(i_1,i_2,j_1,j_2) = \delta_1\delta_2^2 i_1 + \delta_1^2\delta_2 i_2 + \beta N(p\delta_2+(1-p)\delta_1)(\delta_2 j_1+\delta_1 j_2)$. Then,

$$\begin{aligned}\frac{dV}{dt} &= (\beta^2 MN((1-p)(1-q)\delta_1^2+pq\delta_2^2+\\ &+\delta_1\delta_2((1-p)q+(1-q)p)) - \delta_1^2\delta_2^2)(i_1+i_2)\\ &-\beta N\delta_2(\beta M(p\delta_2+(1-p)\delta_1)+\delta_1\delta_2)i_1 j_1\\ &-\beta N\delta_1(\beta M(p\delta_2+(1-p)\delta_1)+\delta_2^2)i_1 j_2\\ &-\beta N\delta_2(\beta M(p\delta_2+(1-p)\delta_1)+\delta_1^2)i_2 j_1\\ &-\beta N\delta_1(\beta M(p\delta_2+(1-p)\delta_1)+\delta_1\delta_2)i_2 j_2.\end{aligned}$$

The extinction of the virus follows directly by applying Lyapunov's stability theorem. Next we will show that if $\frac{\beta}{\delta^*} > \frac{1}{\sqrt{MN}}$, the virus survives. We first note that any trajectory of the system (5) can never leave the box $B = \{(i_1,i_2,j_1,j_2)|0 \leq i_1 \leq 1, 0 \leq i_2 \leq 1, 0 \leq j_1 \leq 1, 0 \leq j_2 \leq 1\}$. This follows from $\frac{di_1}{dt}|_{i_1=0} = p\beta N(j_1+j_2) \geq 0$, and similar inequalities at the borders of the box $B$.

From the construction of the Lyapunov function, we can observe that for $\beta^2 MN((1-p)(1-q)\delta_1^2+pq\delta_2^2+\delta_1\delta_2((1-p)q+(1-q)p)) - \delta_1^2\delta_2^2) - \delta_1^2\delta_2^2 > 0$ and for $(i_1,i_2,j_1,j_2) \in B$ and sufficiently close to the origin, $\frac{dV}{dt} > 0$. This implies that the origin has an unstable manifold in $B$. Therefore, because any trajectory of system (5) can never leave the box $B$, system (5) has an attractor as the $\omega$-limit set and hence the virus does survive. □

The result from Theorem 2 holds for non-symmetric cases: a node from set $S_1$ sees different portion of nodes with curing rate $\delta_1$ than a node from set $S_2$

$(p \neq q)$. In the symmetric case $(p = q)$, a more general result with $m$ different curing rates can be derived, as in the case of the regular graph, described in Theorem 1.

## 3 Optimal Heterogenous Protection of Complete Bi-partite Graphs

We will not consider the simple case of optimization for a regular graph.

For any bi-partite graph, the threshold for the heterogenous case is fixed and equal to $\delta^* = \beta\sqrt{MN}$. The threshold can be reached for different values of $\delta_1, \delta_2, p$ and $q$. For example, for $(\delta_1 = \beta M, \delta_2 = \beta N, p = 1, q = 0)$ the threshold is reached with $\delta_1$ applied on nodes from set $S_1$, while for $(\delta_1 = \beta M, \delta_2 = \beta N, p = 1, q = 0)$ the threshold is also reached and the curing rate $\delta_1$ is now applied on the nodes from the other set. The question is how can we decide which solution is better. One of the options is to minimize the total protection strategy applied on the network, while reaching the threshold. The total protection strategy can be defined as a sum of all protection strategies and we will denote it by $D$

$$D = \sum_{l=1}^{M+N} \delta_l = Mp\delta_1 + M(1-p)\delta_2 + Nq\delta_1 + N(1-q)\delta_2 \tag{8}$$

For the previous two cases, the total protection strategy is different. In case $(\delta_1 = \beta M, \delta_2 = \beta N, p = 1, q = 0)$, the total protection strategy is $D = \beta(M^2 + N^2)$, and in the other case, $D = 2\beta MN$, which is always smaller than or equal to the first case.

Let us formulate the optimization problem as follows:

*Problem 1. Minimize*

$$D = Mp\delta_1 + M(1-p)\delta_2 + Nq\delta_1 + N(1-q)\delta_2 \tag{9}$$

*subject to the conditions*

$$\beta\sqrt{MN} = \frac{\delta_1\delta_2}{\sqrt{(1-p)(1-q)\delta_1^2 + pq\delta_2^2 + \delta_1\delta_2(p(1-q) + q(1-p))}} \tag{10}$$
$$0 \le p, q \le 1$$
$$0 < \delta_1, \delta_2$$

The optimization problem is non-linear with non-linear conditions. However, from [15], we know that the minimum of the function $D$ for any graph and any set of curing rates is equal to the number of links $L$ in the network, multiplied by $2\beta$,

$$D_{\min} = 2\beta L.$$

In the case of the complete bi-partite graph, the minimum is $D_{\min} = 2\beta MN$ and it is reached for $(\delta_1 = \beta M, \delta_2 = \beta N, p = 1, q = 0)$ or $(\delta_1 = \beta N, \delta_2 = \beta M, p = 0, q = 1)$. This means that for $N > M$, the larger curing rate proportional to the number of links in set $S_1$ will be assigned to the nodes from that set. The larger curing rate is assigned to the more connected nodes.

Further, we can have a situation, where curing rates $(\delta_1, \delta_2)$ are fixed and we will optimize the parameters $(p, q)$. This optimization problem can be formulated as follows.

*Problem 2. For two fixed curing rates $\delta_1, \delta_2$, minimize function (8), subject to the conditions (10).*

From the threshold condition we can determine one of the variables $p$ or $q$. We will derive equations for variable $q$ (the case with $p$ is analogue),

$$q = \frac{\delta_1(MN\delta_1(1-p) + MN\delta_2 p - \delta_1\delta_2^2)}{MN(\delta_1^2(1-p) + \delta_1\delta_2(2p-1) + \delta_1^2)} \tag{11}$$

By substituting $q$ in $D$, the total sum of curing rates becomes a function of parameter $p$ only and optimization is simplified. The function is of the form $D(p) = \frac{P_2(p)}{P_1(p)}$ where $P_1(p)$ is a polynomial of the first order in $p$ and $P_2(p)$ is a polynomial in the second order in $p$.

**Lemma 2.** *For any fixed $\delta_1, \delta_2$, the optimal solution of minimization problem 2 is on the boundary of the region ($p = 0$ or $p = 1$ or $q = 0$ or $q = 1$).*

*Proof.* The function $D(p)$ is not defined for $P_1(p) = 0$, which holds for $p = \frac{\delta_1}{\delta_1 - \delta_2}$. The value $\frac{\delta_1}{\delta_1-\delta_2}$ does not belong to the interval $[0, 1]$. The second derivative of $D(p)$ is strictly negative in the interval $q \in [0, 1]$.

$$\frac{d^2D(p)}{dp^2} = -\frac{2\delta_1^2\delta_2^2(\delta_1 - \delta_2)^2}{(\delta_1(1-q) + \delta_2 q)} < 0, q \in [0, 1]$$

Therefore, $D(p)$ is concave in the interval of interest and minimum is on the boundaries of the interval. □

For given $\delta_1, \delta_2$, it is not always possible to reach the threshold. In the case $\delta_1, \delta_2 < \beta\sqrt{MN}$, the threshold cannot be reached and the network is in the state of permanent infection. For example, if $\delta_1, \delta_2 < \beta\sqrt{MN}$ and $\delta_1 > \delta_2$, if we take only the larger curing rate for the whole network, we have $\frac{\beta}{\delta_1} < \beta\frac{1}{\sqrt{MN}}$. In the case $\delta_1, \delta_2 > \beta\sqrt{MN}$, the network is cured, however, the network is above the threshold and a higher curing rate than necessary is applied.

If the threshold can be reached, Lemma 2 shows that either set $S_1$ or set $S_2$ is completely protected with only one curing rate. In order to minimize the sum of curing rates we are interested how many times we can apply smaller curing rates. Without loss of generality, let $\delta_1 < \beta\sqrt{MN} < \delta_2$ and $N > M$. Firstly, we will assign $\delta_2$ to all the nodes from larger set with $N$ nodes and $\delta_1$ to the smaller set. If the effective spreading rate obeys $\frac{\beta}{\delta^*} > \frac{1}{\sqrt{MN}}$, than $p = 1$, and $q$ can be calculated from equation (11). In the case $\frac{\beta}{\delta^*} < \frac{1}{\sqrt{MN}}$, the network is cured and below the threshold. Then $q = 0$ and $p$ can be calculated from the condition for the threshold.

## 4 Conclusion

The epidemic theory is widely applied on many networking problems. The *SIS* model, on which we have focused here, is applied in malware modeling in the Internet [1], [2], [3], information dissemination in P2P and ad-hoc networks [4], [5] and propagation of faults and failures [6]. The two types of topologies that we considered, namely the regular graph and the complete bi-partite graph, arise as subnet structures in telecommunication networks. We have studied the influence of heterogenous protection in regular and complete bi-partite graphs.

Using Lyapunov's stability theorem, we have shown that for regular graphs, the epidemic threshold satisfies $\frac{\beta}{\delta^*} = \frac{1}{k}$, where $\delta^*$ is defined as the weighted harmonic mean of $\delta_1, ..., \delta_m$. This result holds under the assumption of complete symmetry, where each node sees the same fraction of different curing rates. Without this assumption, the problem becomes significantly complex [16].

Further, we have considered the heterogenous case with 2 curing rates for the complete bi-partite graph. The threshold, given by Eq. (7) becomes the geometric mean of curing rates $\delta_1, \delta_2$ for $p = 1, q = 0$ and the weighted harmonic mean if $p = q$. For other values of $p$ and $q$, total curing rate $\delta^*$ belongs to the interval $[\delta_1, \delta_2]$.

Many different pairs of curing rate can satisfy the threshold equation, therefore the question which solution is more optimal rises. We consider the optimality of heterogeneous protections for complete bi-partite graph with the respect to sum of all applied curing rates and concluded that global optimum in this respect is equal to the number of links in the complete bi-partite graph. For the case of fixed $\delta_1$ and $\delta_2$, the optimal solution is on the boundaries of $(p, q) \in [0, 1] \times [0, 1]$.

## References

1. Kephart, J.O., White, S.R.: Measuring and modeling computer virus prevalence. In: IEEE Symposium on Security and Privacy (1993)
2. Ganesh, A., Massoulié, L., Towsley, D.: The effect of network topology on the spread of epidemics. In: Proceedings of IEEE INFOCOM, pp. 1455–1466 (2005)
3. Pastor-Satorras, R., Vespignani, A.: Epidemic spreading in scale-free networks. Phys. Rev. Lett. 86(14), 3200–3203 (2001)
4. Chakrabarti, D., Leskovec, J., Faloutsos, C., Madden, S., Guestrin, C., Faloutsos, M.: Information survival threshold in sensor and p2p networks. In: Proceedings of IEEE INFOCOM (2007)
5. Eugster, P., Guerraoui, R., Kermarrec, A.M., Massoulie, L.: From epidemics to distributed computing. IEEE Computer 37, 60–67 (2004)
6. Coffman, F.G.J., Ge, Z., Misra, V., Towsley, D.: Network resilience: exploring cascading failures within bgp. In: Proc. 40th Annual Allerton Conference on Communications, Computing and Control (2002)
7. Daley, D.J., Gani, J.: Epidemic modelling: An Introduction. Cambridge University Press, Cambridge (1999)
8. Wang, Y., Wang, C.: Modeling the effects of timing parameters on virus propagation. In: WORM 2003: Proceedings of the 2003 ACM workshop on Rapid malcode, pp. 61–66. ACM, New York (2003)

9. Bailey, N.T.J.: The Mathematical Theory of Infectious Diseases and its Applications. Charlin Griffin & Company (1975)
10. Wang, Y., Chakrabarti, D., Wang, C., Faloutsos, C.: Epidemic spreading in real networks: An eigenvalue viewpoint. In: IEEE Symposium on Reliable Distributed Systems. IEEE Computer Society, Los Alamitos (2003)
11. Van Mieghem, P., Omic, J., Kooij, R.: Epidemic spreading in scale-free networks. IEEE Transactions on Networking, 1–14 (2009)
12. Omic, J., Kooij, R., Van Mieghem, P.: Virus spread in complete bi-partite graphs. In: Proceedings of BIONETICS (2007)
13. Cvetkovic, D.M., Doob, M., Sachs, H.: Spectra of graphs, Theory and Applications, 3rd Revised and Enlarged Edition. Johan Ambrosius Barth Verlag (1998)
14. Guckenheimer, J., Holmes, P.: Nonlinear oscillations, dynamical systems, and bifurcations of vector fields. Springer, New York (1983)
15. Omic, J., Orda, A., Van Mieghem, P.: Protecting against network infections: A game theoretic perspective. In: Proceedings of IEEE INFOCOM (2009)
16. Van Mieghem, P., Omic, J.: In-homogeneous virus spread in networks. TUDelft Technical report 20080801 (2008)

# Conducting and Optimizing Eclipse Attacks in the Kad Peer-to-Peer Network

Michael Kohnen, Mike Leske, and Erwin P. Rathgeb

University of Duisburg-Essen, Institute for Experimental Mathematics,
Ellernstr. 29, 45326 Essen
Michael.Kohnen@iem.uni-due.de, Mike.Leske@stud.uni-due.de,
Erwin.Rathgeb@iem.uni-due.de

**Abstract.** The Kad network is a structured P2P network used for file sharing. Research has proved that Sybil and Eclipse attacks have been possible in it until recently. However, the past attacks are prohibited by newly implemented security measures in the client applications. We present a new attack concept which overcomes the countermeasures and prove its practicability. Furthermore, we analyze the efficiency of our concept and identify the minimally required resources.

**Keywords:** P2P security, Sybil attack, Eclipse attack, Kad.

## 1 Introduction and Related Work

P2P networks form an overlay on top of the internet infrastructure. Nodes in a P2P network interact directly with each other, i.e., no central entity is required (at least in case of structured P2P networks). P2P networks have become increasingly popular mainly because file sharing networks use P2P technology.

Several studies have shown that P2P traffic is responsible for a large share of the total internet traffic [1, 2]. While file sharing probably accounts for the largest part of the P2P traffic share, also other P2P applications exist which are widely used, e.g., Skype [3] for VoIP or Joost [4] for IPTV. The P2P paradigm is becoming more and more accepted also for professional and commercial applications (e.g., Microsoft Groove [5]), and therefore, P2P technology is one of the key components of the next generation internet.

P2P networks can be categorized into two main types, the unstructured and the structured ones. The structured networks have gained popularity recently in science and are becoming increasingly popular also in practice. One of the most popular structured P2P networks is the file sharing network Kad which is used, e.g., by eMule [9] and aMule [10].

Structured networks introduce identifiers for both content and nodes and a notion of closeness between those identifiers (see below). Thereby, they define which nodes are responsible for a certain content item, e.g., a file provided for sharing in the network. The logical closeness metric is completely independent from physical closeness.

L. Fratta et al. (Eds.): NETWORKING 2009, LNCS 5550, pp. 104–116, 2009.

The structured approach offers more precise and comprehensive lookup compared to unstructured networks as it ensures that all content that exists will actually be found. The well-defined responsibility of nodes for specific content, however, also introduces specific weaknesses. Users can create multiple identifiers [13] and select them such that they control all nodes that are responsible for a certain content item. If they succeed, they can control access to it and can also limit its availability, which is called Eclipse attack.

In [7], Steiner et al. state that an Eclipse attack is easily possible in the Kad network, as the nodes can freely choose their node identifiers, and present results of successfully performed Eclipse attacks. However, since the publication of [7], new versions of eMule and aMule have been released which include security measures against these kinds of attacks. Steiner himself states on [14] that the changes affect the practicability of his attacks.

Motivated by Steiner's work, we analyzed the security measures built into the state-of-the-art Kad clients and developed a way of circumventing them in an efficient way. In this paper, we describe the basic approach for the improved attack as well as ways to make it highly efficient. The experiments described provide a proof of concept and also give an indication of the effort required for the attack in a realistic scenario.

## 2 How the Kad Network Works

Kad is based on the Kademlia algorithm which was presented by Maymounkov and Mazières [8]. It is only defined by its implementations in several file sharing applications such as eMule [9], aMule [10] and MLDonkey [11] – no formalized protocol specification exists. Details of the aMule Kad implementation can be found in [12]. This paper concentrates on eMule, as it is the most widespread application.

Every file that is available in the Kad network has a *file ID* that is derived by calculating an MD4 hash value of the contents (not the name) of the file. Every Kad node has a *node ID* that is – normally – randomly generated the first time the node enters the Kad network and retained afterwards. Both IDs are 128 bits long. The logical distance between two identifiers is calculated using the XOR operation. Hence, the more bits match at the beginning of the IDs, the smaller is the distance between them.

The nodes with the smallest distance from a certain file ID are responsible for that file. The set of responsible nodes is not statically defined, but varies as nodes join and leave the network. Therefore, the set of responsible nodes needs to be determined again during every lookup process.

Nodes that offer a file store their contact information (IP address, port numbers, node ID) on the responsible nodes. Nodes that look for a file ask the responsible nodes for source nodes for that file. This concept is called indirect storage, as the responsible nodes do not store the file itself, but only pointers to nodes that store it.

To download a file from the network, its ID must be known. The default approach to retrieve the ID of the desired file is to use the keyword search mechanism of Kad, the other way is to use specific websites which list file IDs of popular files, thus bypassing the keyword search function of Kad.

When a node offers a new file, it stores a pointer to itself (*source information*) at the nodes responsible for the file ID, which allows other nodes to retrieve the file.

Additionally, it generates keywords by dividing the filename into parts (e.g., single words). It calculates hash values for every keyword (*keyword IDs*) and stores pointers at the nodes that are responsible for each keyword ID which contain the whole filename and the ID of that file. Each process of storing information for a file or keyword ID is called *publication process*.

If another user performs a *keyword search*, the client software calculates the keyword ID of the first keyword provided by the user and looks for the nodes that are responsible for that ID. In the subsequent request to those nodes, it includes the other keywords in clear text. The polled nodes use the provided keywords to filter their pointer lists and answer with matching files including their respective file IDs. The user can select the file(s) he wants to download from that list.

After the file ID of the desired file has been determined, the node performs a *source search* by asking the nodes that are responsible for the file ID for source nodes for that file and starts to download the file (parts) directly from those nodes.

Attacking the source search mechanism is our main target, as it is possible to eclipse files from the network this way regardless of how the user determined the file ID. During our analyses, we decided to attack the publication process as well, because only this way it can be ensured that only our nodes know about the content (see below). The attack against the publication process can be directed against both keyword and file IDs. The keyword search mechanism behaves the same way the source search process does, so no specific attack is necessary.

## 2.1 Kad Message Types

Two versions of Kad messages exist; we only use version 1 message types. All current versions of the client applications understand both versions of the messages. By using the older version, we circumvent the necessity to implement the obfuscation and advanced handshake features of newer versions in our tools. The usage of version 1 messages only has no impact on our tests.

Kad messages are transmitted via UDP. Every Kad message begins with a one-byte identifier indicating that the message is a Kad message, which is followed by a one-byte opcode. Message parameters are appended, if applicable.

Information about other peers is transmitted in a *peer information* data structure which can be found in many message types. This data structure contains contact information about the node, such as its ID, its IP address and the UDP port used for Kad communication, and various other pieces of information.

*Hello Requests* are used to check the availability of other nodes. We use the fact that Hello Request messages can also be used to insert a node into another node's routing table for our attack. The queried node is supposed to reply with a *Hello Response* message. Both messages only contain the peer information data structure of the respective sending peer.

*Request* messages are used to retrieve information about other nodes during a lookup process – explained in detail in section 2.2. They contain a Type field identifying the type of lookup (e.g., source search, keyword search, publication, etc.), the target ID and the ID of the queried node. A *Response* message contains the queried target ID and a peer information data structure for each matching node the queried node knows.

*Publish Requests* are used during a publication process to store information about offered files or keyword information on the responsible nodes. They contain the target ID and information about the published objects such as the filename or the bit rate of audio files. A node acknowledges a successful publication with a *Publish Response*.

During a source or keyword search, *Search Requests* are used to ask other nodes for sources or files matching a keyword. This message type contains the target ID and a field indicating whether the ID is a file or a keyword ID. In case of a keyword search with a search term consisting of more than one part, the second to last part of it are appended in clear text to the message. The queried node replies with a *Search Response* message containing the target ID and a result list.

## 2.2 The Lookup Process

A node performs a lookup process every time it requires (more) sources for a file, when it publishes a file or when the user performs a keyword search in order to find the currently responsible nodes for that file or keyword ID. When these nodes are identified, the specific action is performed, e.g., a query for sources.

As several lookup processes may run concurrently, eMule uses independent search processes for each lookup which are controlled by the so-called SearchManager. SearchManager invokes search processes and terminates them when their maximum run time is exceeded (default: 45 seconds) or their maximum number of results has been achieved (default: 300 sources during a source search, 10 nodes during a publication).

Each search process is split into two phases: In phase 1, the responsible nodes are identified. Up to three nodes are queried in parallel in order to accelerate the lookup process and to cope with non-responding nodes. The number n of required nodes depends on the action: During a publication process, a node contacts ten nodes to store the source information on them. A source search queries as many nodes as necessary until either the maximum number of sources or a timeout is reached. During phase 2, the specific action is performed on the identified nodes.

**Phase 1: Locating the responsible nodes.** A search process manages four lists: POSSIBLE, BEST, TRIED and RESPONDED. All lists are sorted by distance to the target with the closest node at the top.

At the beginning of a lookup, POSSIBLE will be filled with the 50 currently known nodes that are closest to the target ID. BEST stores the three closest nodes that have been found so far, TRIED stores the nodes that have already been queried, and RESPONDED stores all nodes that have responded to queries.

After initializing, the three BEST nodes are queried for closer nodes and copied to TRIED. The nodes answer with a list of nodes that are closer to the target. Those nodes are inserted into POSSIBLE and also into BEST, if applicable. Nodes in BEST are queried immediately for closer nodes.

Eventually, this process starves, as no closer nodes will be found. BEST then contains the three currently closest nodes to the target. Fig. 1 illustrates this process.

A search process cannot transition to phase 2 on its own. SearchManager periodically evaluates the status and transitions the search process to phase 2 if no response has been received for more than three seconds.

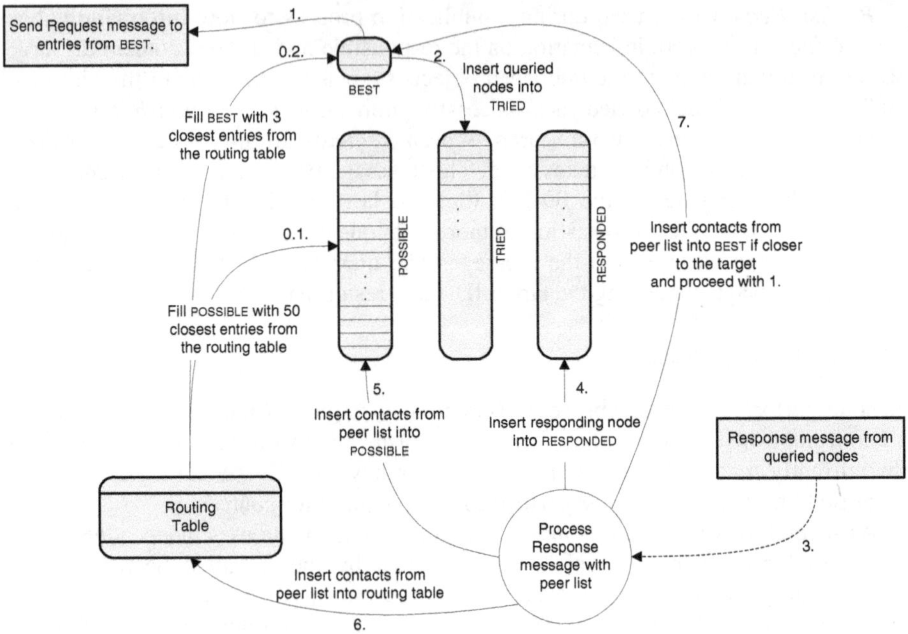


**Fig. 1.** Phase 1 of the lookup process. The initialization procedures are displayed as "0.x". All lists are sorted by distance to the target.

**Phase 2: Performing the specific action.** In phase 2, the search process evaluates the nodes contained in POSSIBLE, starting with the first node. If the node is also contained in TRIED and RESPONDED, the specific action is performed and the node is removed from POSSIBLE; if it is not in TRIED and RESPONDED, the search process falls back into phase 1 and queries this node for closer nodes.

### 2.3 Security Issues: Sybil and Eclipse Attacks and Countermeasures

Kad is susceptible to both the Sybil and the Eclipse attacks. A Kad node selects its own ID autonomously. Therefore, a malicious user can select specific IDs instead of randomly generated ones. Additionally, he can use multiple IDs concurrently, which is called a Sybil attack [13]. If an attacker thereby provides or controls the n closest nodes to the victim ID, all queries for this ID will converge on his nodes, as the nodes whose IDs are closest to a queried ID are responsible for managing content identified by it. This way, the attacker performs an Eclipse attack, as he is able to control the victim ID and can decide whether and how to respond to incoming requests.

As a consequence, the eMule developer community has designed security measures against this attack that have been implemented in eMule 0.49a [9]. Prior versions do not include any countermeasures. The new countermeasures constrain the routing table as follows:

- One IP address must not have more than one node ID assigned.
- A Routing Bin[1] must not contain more than two nodes from the same /24 IP subnet.
- The whole routing table must not contain more than ten nodes from the same /24 IP subnet.

These constraints only apply to the routing table, so they are only checked when nodes are added to it. They are not applied during lookup processes, so lookups are as susceptible to Sybil and Eclipse attacks as before.

Without these restrictions, an attacker could easily create an arbitrary number of Sybil nodes using one IP address and therefore perform an Eclipse attack using only a single machine. The new restrictions force the attacker to use several machines located in different subnets, so the attack effort rises significantly.

Modifications ("mods") of eMule are developed besides the official eMule application. SafeKad (included in the modification MorphXT [15], e.g.) implements a similar countermeasure that is also applied during the lookup process: It only allows one node per /20 IP subnet in a single response message. This mechanism leads to a more secure lookup process, as an attacker now needs nodes that reside in different /20 IP subnets. However, we will show a way to also circumvent this security measure. SafeKad is not developed further, though, as its developers regard the security measures of the official client as superior [15].

## 3 An Improved Eclipse Attack on the Kad Network

Based on the attack described in [7], we developed an improved attack procedure that circumvents the new security measures of eMule 0.49a and SafeKad. To demonstrate its feasibility, we developed a tool suite consisting of specialized Kad clients conducting the attacks. We attack both the publication process and the source search process, as they differ in the amount of queried nodes (ten vs. as many as necessary to obtain the maximum amount of sources or until the timeout is reached). Both processes perform a lookup process to determine the nodes they need to query.

### 3.1 Basic Attack Concept

We define the *Target ID* as the Kad ID which we want to eclipse. For the source search process, this ID represents a file. The publication process attack can either be directed against a keyword ID or a file ID, depending on what the attacker desires to eclipse (e.g., keyword "ubuntu" vs. file "ubuntu-8.04.2-desktop-i386.iso").

The Target ID resides in what we call a *CrawlArea* – the part of the Kad ID space surrounding the Target ID that we analyze for nodes that need to be manipulated. The size of the CrawlArea is defined by the length of the common prefix of the CrawlArea nodes and the Target ID; e.g., a "/20 CrawlArea" means that the first 20 bits of the CrawlArea nodes' IDs are the same as the first 20 bits of the Target ID.

---

[1] The routing table of a Kad node is divided into Routing Bins which contain up to 10 nodes and represent up to 6.25% of the Kad address space, depending on their position in the node's routing table. For details refer to [12].

In order to perform the attack, we use a chain of *Sybil nodes* (Sybil 0 to Sybil n). The Sybil nodes have different, specifically chosen Kad IDs where Sybil 0 has the largest distance to the Target ID and Sybil n the smallest. We call this order of nodes "decreasing order". The chosen node IDs and the Target ID have the first 125 bits in common, so it can be assumed that our nodes have smaller distances to the Target ID than any other node in the network. As a result, our nodes – and only our nodes – are responsible for the Target ID.

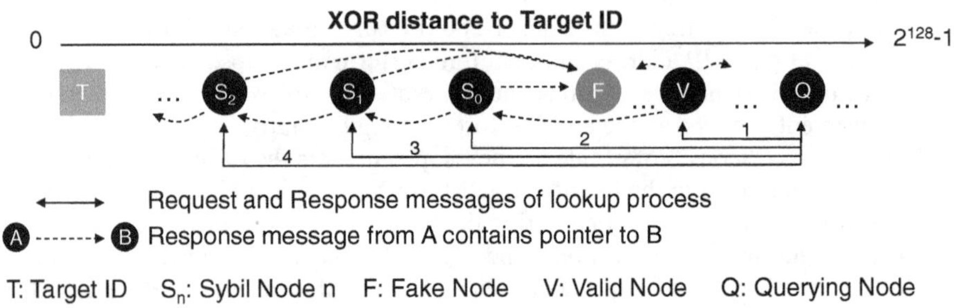


**Fig. 2.** Illustration of how a querying node is lured into the Sybil node chain. The Sybil nodes are arranged in decreasing order. All IDs are arranged according to their distance to the Target ID.

The new security measures of the official eMule client and the SafeKad modification require new attack techniques: To circumvent eMule 0.49a's restriction that one Routing Bin may only contain up to two nodes from the same /24 IP subnet, only Sybil 0 is actively inserted into other nodes' routing tables. As this check is not applied during lookup processes, we use Sybil 0 to successfully direct the querying node to our other Sybil nodes in the same /24 subnet.

SafeKad's security measures prevent the usage of a single Response message containing all Sybil nodes, because they would discard all except one of them as they reside in the same /20 IP subnet. Hence, we include only one Sybil node and another randomly generated faked node in a Response message[2]. The random node's ID is chosen to be farther away from the Target ID than the Sybil's one.

When a node performs a lookup process, it will eventually be directed to Sybil 0 by other valid nodes that have been manipulated. Sybil 0 will respond with a Response message containing Sybil 1 and a faked node, which is illustrated in Fig. 2. When receiving this message, the node will immediately put Sybil 1 into its POSSIBLE and BEST lists, as it is closer than all previously found nodes, and query it for closer nodes. The Sybil node in turn answers with the next Sybil node and a faked node. This process is repeated.

Attacking a publication process requires ten Sybil nodes to eclipse a file or a keyword by deceiving the publishing nodes. This number is fixed, as a node stores the information always on ten nodes.

During a source search, more Sybil nodes can be required: When a Kad node tries to find sources for a file, it queries as many nodes as possible until it has found the

[2] The faked node is required because Kad expects a Response message to contain at least two nodes.

maximum amount of sources or reaches a timeout. Using the first condition, in theory, only one Sybil node is required: The search terminates when 300 sources have been found (default value). If the first queried node replies with 300 randomly generated, non-existent sources, no further Sybil nodes are required. The drawback of this approach is that it is visible to the user, as the client application will first report 300 sources which will then drop to 0 after all result nodes have been queried. We chose a more stealthy approach using the second termination criterion, the timeout. We try to keep the querying node busy long enough so that the search is terminated due to a timeout. To perform this attack, we introduced a delay after the reception of a Request message before sending the Response message. The number of required Sybil nodes is variable and depends on the delay interval (see 4.1).

Summarizing, our attack bases on a chain of Sybil nodes of which only the first node (Sybil 0) is actively proclaimed. Sybil 0 points querying nodes to the next node in the chain which then points to the next Sybil node and so forth. The Sybil nodes answer after a configurable delay with only the next Sybil node and an arbitrarily chosen fake node.

### 3.2 Optimizations

During our test phases, we discovered optimizations for the attacks on the publication and the source search processes.

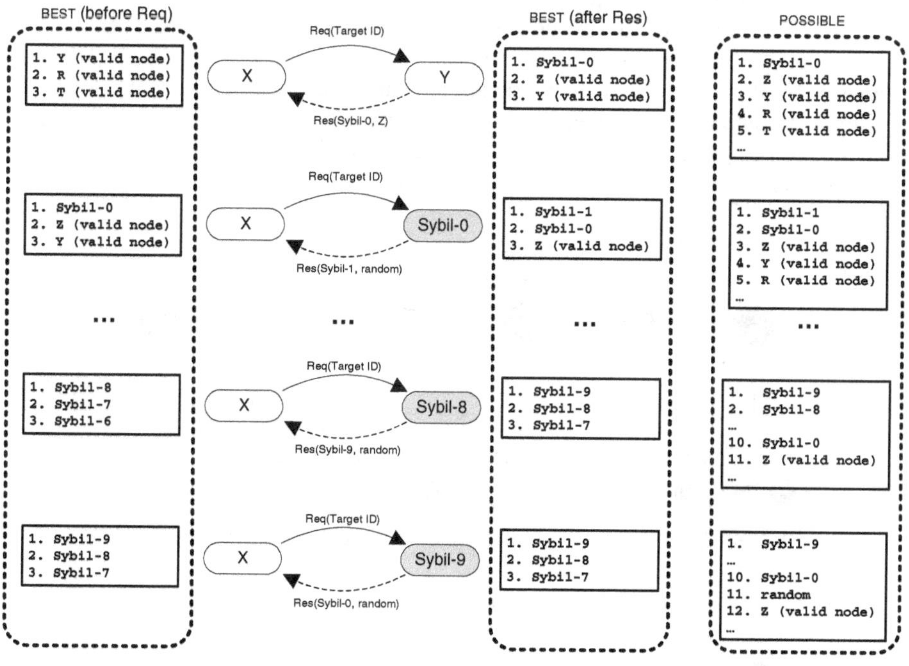


**Fig. 3.** The BEST and POSSIBLE lists before and after each message of the Sybil node chain

**Accelerating the publication attack.** To have a node enter phase 2 of the lookup process more quickly when we attack the publication process, we have Sybil 9 send Sybil 0's contact information instead of Sybil 10's ones. This way, the node notices that there are no new closer nodes and that it has found the ten currently closest nodes to the Target ID – which are our Sybil nodes 0 to 9 – and will start to send the Publish Requests to them immediately. Additionally, we chose the node ID of the fake, non-existent node included in Sybil 9's response message to be only a little farther away than our Sybil nodes to place it between the last Sybil and the first valid node in the list, in case one of the Sybils' answers does not arrive in time. If this happened, the attacked node would send a Publish Request to the first valid node on its list. This way, the request is sent to the non-existent node. The sending node does not receive an acknowledgement, so it tries the closest non-queried node on its POSSIBLE list, which should contain all ten Sybil nodes by then. Fig. 3 shows the status of the BEST and POSSIBLE lists before and after each message exchange when a node is lured into our Sybil node chain.

**Minimizing the number of required Sybil nodes for the source search attack.** During our test runs, we discovered that the number of required Sybil nodes for the attack on the source search can be minimized by reversing the order of the Sybil nodes so that Sybil 0 is the closest node to the target and Sybil n is the farthest one ("increasing order") – Fig. 4 illustrates this order of Sybil nodes. This change causes a node to repeatedly transition between phases 1 and 2 during the lookup process: The node tries to find closer nodes by only querying the currently known closest nodes. As Sybil 0 is the closest node, the process starves when Sybil 0 is queried and returns Sybil 1 which is farther away. The starvation causes a transition to phase 2 for the first time. The node queries Sybil 0 for sources, does not obtain any and discovers that it has not found the maximum amount of sources yet. Therefore, it continues its lookup by also querying the next node in POSSIBLE – which is Sybil 1 – for closer nodes and thereby transitions back to phase 1. This process repeats until the timeout is reached.

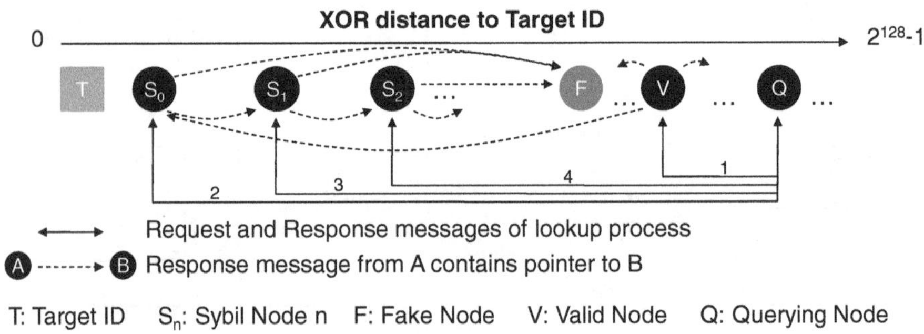


**Fig. 4.** Illustration of the increasing order of the Sybil nodes

### 3.3 Tool Suite

To perform our analyses, we developed a tool suite consisting of three parts: *KadCrawler*, *Capo* and *SybilNode*. KadCrawler is responsible for finding all nodes in the CrawlArea and storing them in a database. The area of the address space containing the nodes that are poisoned is a subset of the CrawlArea which we call *PoisonArea*.

The routing tables of the nodes in the PoisonArea are then poisoned by the second tool, Capo: It selects the PoisonArea nodes from the database and sends a Hello Request message to them containing peer information about itself, so Capo takes the role of Sybil 0. Using the Hello Requests, Capo tries to insert itself into the routing table of the receiving node. Unfortunately, the reception of a Hello Response message cannot be interpreted as a successful insertion into the routing table, as a node replies before checking if the node should be inserted, so there is no way of determining whether the insertion was successful.

Capo performs the poisoning process every ten minutes to ensure that the currently closest nodes are manipulated. It also responds to incoming Request messages querying for closer nodes for a given ID. If the queried ID is the Target ID, Capo responds by sending information about Sybil 1 and a faked node to lure the querying node into our chain of Sybil nodes. Hello and Publish Requests for arbitrary IDs are answered positively and logged; Search Requests are logged, but not answered.

The third tool, SybilNode, is a functionally reduced Capo as it does not actively poison any nodes. It only reacts to incoming messages the same way Capo does. The Sybil nodes 1 to n are instances of this tool.

## 4 Results

Using our tool suite, we performed two test series. First, we performed short tests using one file with random content and name and ten test runs for each test scenario. The scenarios vary by the order of the Sybil nodes (decreasing vs. increasing) and the answering delay.

For the second test, we created 12 such files and tested for 90 hours to conduct the efficiency analysis. We refrained from eclipsing file or keyword IDs of content offered by third parties due to legal reasons.

### 4.1 Proof of Concept

For the test with one file, we selected a 12-bit PoisonArea. The test runs were performed using eMule 0.48a with SafeKad and eMule 0.49b as client applications. After each run, all configuration files were reset to defaults.

All attacks against the publication process were successful, as all ten Publish Requests of each test run were sent to our Sybil nodes. Both eMule 0.48a with SafeKad and eMule 0.49b were susceptible to our attacks.

The attacks against the source search process were conducted in both decreasing and increasing order of the Sybil nodes. We varied the answering delay of the Sybil nodes using values of 0.0, 0.5, 1.0, 1.5 and 2.0 seconds. Longer delays are not reasonable, as a lookup process is transitioned to phase 2 after a timeout of 3 seconds. If this would happen before the node's lookup timeout had been reached, the node would query valid nodes as well which would avert the attack. Table 1 presents the average number of Sybil nodes required for the attack and shows that our optimization of the order of the Sybil nodes remarkably reduces the amount of necessary Sybil nodes.

For delays < 1.0s and an decreasing order of Sybil nodes, more than 30 Sybil nodes are required for the attack to be successful. The number of required Sybil nodes decreases with increasing delay. The optimal configuration is to arrange the Sybil nodes in increasing order and to use a delay of 2.0 seconds.

**Table 1.** The average number of required Sybil nodes to let the source search process reach the timeout

| | | Answering Delay [s] | | | | |
|---|---|---|---|---|---|---|
| | | **0.0** | **0.5** | **1.0** | **1.5** | **2.0** |
| **Decreasing Order** | **eMule 0.48a (SafeKad)** | >> 30 | > 30 | 24.7 | 16.9 | 13.0 |
| | **eMule 0.49b** | >> 30 | > 30 | 24.8 | 14.9 | 11.0 |
| **Increasing Order** | **eMule 0.48a (SafeKad)** | 8.0 | 7.0 | 5.5 | 5.5 | 5.1 |
| | **eMule 0.49b** | 7.2 | 6.5 | 5.8 | 5.1 | 4.6 |

### 4.2 Efficiency Analysis

After proving that our tools work, we analyzed the required amount of nodes that need to be poisoned in order for the publication attack to be successful. We chose to attack the publication process although it requires more Sybil nodes and messages and therefore resources, because only by attacking the publication, an attacker can ensure that he controls all nodes that store information about the content item. If only the search process would be attacked, there would exist valid nodes that store the information and which could be found by chance, e.g., due to network failures or misbehaving client applications.

Each of the 12 test files was published every 20 minutes. We only used an unmodified eMule 0.49b as a client application as we had proved before that also eMule 0.48a with SafeKad does not avert our attack.

We used 13- to 24-bit masks as PoisonArea sizes. The results can be divided into three groups: Attacks using 13- to 20-bit masks were immediately successful. The ones with 21- and 22-bit masks were successful after 60 and 5 hours respectively. The attacks using 23 and 24 matching bits failed.

**Immediately successful attacks.** The attacks using 13- to 19-bit masks were immediately successful. Every publication process performed by one of our eMule clients succeeded. By contrast, 2 out of 270 attacks using 20-bit masks failed. The reason was that the lookup process of the publishing eMule client starved before it had found a poisoned node and therefore transitioned to phase 2 before our Sybil nodes had been found. This led to our Sybil nodes receiving only 8 of 10 Publish Requests.

**Delayedly successful attacks.** As we limited the amount of nodes chosen to be poisoned by demanding that the first 21 resp. 22 bits of the Target ID and the node ID match, the attacks on these two identifiers were not successful at first, because no matching nodes were found. This is not surprising, as [7] assumes that the Kad network contains up to 4 million nodes: If the node IDs were uniformly distributed, 0.95 to 1.91 nodes should match our criterion on an average (e.g., $4 \cdot 10^6 / 2^{22} = 0.95$).

After 57 hours, the poisoning process for the 21-bit mask area was successful for the first time; the last success occurred after 67 hours. In total, only six different Kad nodes were poisoned during 15 out of 490 poisoning processes. However, after the first node was poisoned, all subsequent publication processes of our eMule clients until the end of the measurement were successfully directed to our Sybil nodes, although no nodes were poisoned during the last 23 hours. This proves that the information of Capo's presence was spread in the target area by the poisoned nodes. This information persists in the routing tables as long as Capo responds to Hello Requests. So the attack is effective without requiring a re-poisoning as long as the poisoned nodes are available.

The poisoning process using the 22-bit mask was successful after 5 hours of measurement. For the same reason the 20-bit mask attack failed two times, 3 of 256 attacks failed.

**Unsuccessful attacks.** The attacks on the 23- and 24-bit mask areas never succeeded. For the 24-bit mask area, no matching node was found. Information about a single node having the first 23 bits in common with the Target ID was indeed found, but the node did not respond to Capo's Hello Request.

### 4.3 Scalability

Due to legal reasons, we only attacked self-generated content that was demanded only by our clients. However, by chance, one of our generated IDs lay close to the ID of a keyword that was accessed by other nodes. We received information about 75,000 content items in more than 125,000 Publish Requests during the 90 hour measurement. Only 14,714 Search Requests were sent to our Sybil nodes (and left unanswered), so the keyword was published more frequently than demanded.

The queries caused by the keyword did not have any impact on our tools or our other attacks. We therefore suppose that the only factor limiting the success of the attack is bandwidth: If requests of publishing nodes are not answered due to packet loss or timeouts, the nodes will publish on other nodes and the attack fails. Assuming an average Kad packet size of 100 bytes, 20 messages per direction and a duration of 5 seconds per publication process results in a network load of 3.2 kbit/s on average. Adding 50% of overhead for the poisoning and Hello and Search Request messages results in a load of 4.8 kbit/s. Using a symmetrical 2 MBit/s connection, more than 400 publications per second would be necessary for the attack to fail. The attack would fail then because the querying nodes would ask other than the Sybil nodes, as their query packets to the Sybil nodes would be (partially) lost due to congestion. As eMule's default re-publish interval is 5 hours, this would require 7.5 million users to publish the same file or keyword, which is 87.5% more users than the number of 4 million found by [7].

## 5 Conclusion and Outlook

We demonstrated that the Kad network is still susceptible to eclipse attacks despite the inclusion of security measures into the current official eMule client. The tool suite we presented is able to eclipse arbitrary popular content items using little resources.

We showed that it suffices to poison one single – namely the currently closest – valid Kad node to be able to perform the attack.

In order to protect against our attack, we recommend to further harden the lookup process by having the IP subnet checks ignore nodes even though they are closer. However, if the attacker controls nodes in different IP subnets, these checks would not help. Hence, we will continue our research by implementing and evaluating the concept of disjoint routing paths. Research in this direction might also help to identify means of disabling "botnets" such as the Storm Worm network, as they also employ P2P technology for communication and receiving orders [6].

## References


[1] ipoque GmbH: ipoque Internet Study (2007), http://www.ipoque.com/resources/internet-studies/internet-study-2007/

[2] Haßlinger, G.: ISP Platforms Under a Heavy Peer-to-Peer Workload. In: Steinmetz, R., Wehrle, K. (eds.) Peer-to-Peer Systems and Applications. LNCS, vol. 3485, pp. 369–381. Springer, Heidelberg (2005)

[3] Skype website: http://www.skype.com/

[4] Joost website: http://www.joost.com/

[5] Microsoft Groove website: http://www.microsoft.com/groove/

[6] Holz, T., Steiner, M., Dahl, F., Biersack, E., Freiling, F.: Measurements and Mitigation of Peer-to-Peer-based Botnets: A Case Study on Storm Worm. In: Proceedings of the 1st Usenix Workshop on Large-Scale Exploits and Emergent Threats, pp. 1–9 (2008)

[7] Steiner, M., En-Najjary, T., Biersack, E.W.: Exploiting KAD: possible Uses and Misuses. SIGCOMM Computer Communication Review 37(5) (2007)

[8] Maymounkov, P., Mazières, D.: Kademlia: A Peer-to-peer information system based on the XOR metric. In: Druschel, P., Kaashoek, M.F., Rowstron, A. (eds.) IPTPS 2002. LNCS, vol. 2429, pp. 53–65. Springer, Heidelberg (2002)

[9] eMule website: http://www.emule-project.net/

[10] aMule website: http://www.amule.org/

[11] MLDonkey website: http://www.mldonkey.org/

[12] Brunner, R.: A performance evaluation of the Kad-protocol. Master's thesis, University of Mannheim and Institut Eurécom, Sophia-Antipolis, France (2006)

[13] Douceur, J.R.: The Sybil attack. In: Druschel, P., Kaashoek, M.F., Rowstron, A. (eds.) IPTPS 2002. LNCS, vol. 2429, pp. 251–260. Springer, Heidelberg (2002)

[14] Computer Communication Review Online, Editorial Zone of [7]: http://ccr.sigcomm.org/online/?q=node/288

[15] MorphXT website: http://www.emule-mods.de/?mods=morphxt/

# On the Optimal Scheduling of Streaming Applications in Unstructured Meshes

Luca Abeni, Csaba Kiraly, and Renato Lo Cigno*

DISI – University of Trento, Italy
{abeni,kiraly,locigno}@disi.unitn.it

**Abstract.** Unstructured, chunk-based P2P streaming (TV and Video) systems are becoming popular and are subject of intense research. Chunk and peer selection strategies (or scheduling) are among the main driver of performance. This work presents the formal proof that there exist a *distributed* scheduling strategy which is able to distribute every chunk to all $N$ peers in exactly $\lceil \log_2(N) \rceil + 1$ steps. Since this is the minimum number of steps needed to distribute a chunk, the proposed strategy is optimal. Such a strategy is implementable and an entire class of deadline-based schedulers realize it. We show that at least one of the deadline-based schedulers is resilient to the reduction of the neighborhood size down to values as small as $\log_2(N)$. Selected simulation results highlighting the properties of the algorithms in realistic scenarios complete the paper.

**Keywords:** P2P, Streaming, Optimality.

## 1 Introduction

P2P streaming and in particular P2P support for IP-TV are becoming not only hot research topics, but also available systems and services like [1,2,3,4,5].

Fundamental to support live streaming is the guarantee of a low distribution delay of the information to all peers. This is strictly related to the overlay characteristics and the scheduling that distribute chunks to peers.

The community has been divided on whether structured systems, i.e., an overlay with known and controlled topological properties like a tree or a hypercube, or unstructured systems based on general meshes are better for this scope. The advantage of structured systems lies in the possibility of finding deterministic scheduling that achieve optimal performance, but they are normally fragile in face of churn (coming and leaving of nodes), require signaling for the overlay maintenance, and can be complex to manage. Unstructured systems, instead, are robust and easy to manage. Overlay maintenance only requires connectivity: each node autonomously search and contact its own neighbors. Their disadvantage has been so far the impossibility of finding a *distributed* scheduling algorithm that is optimal and robust under normal operating conditions.

* This work is supported by the European Commission through the NAPA-WINE Project (Network-Aware P2P-TV Application over Wise Network – www.napa-wine.eu), ICT Call 1 FP7-ICT-2007-1, 1.5 Networked Media, grant No. 214412.

L. Fratta et al. (Eds.): NETWORKING 2009, LNCS 5550, pp. 117–130, 2009.

This paper tackles this problem, demonstrating the existence of an entire class of optimal schedulers under the assumption that the overlay is fully connected, and showing that at least one of these schedulers is robust against the reduction of the neighborhood down to $\log_2(N)$, where $N$ is the number of peers.

## 2 Problem Statement

We study the scheduling (chunk and peer selection) for dissemination at each peer in non structured overlay networks. It is well known that the lower bound on the dissemination delay of any piece of information, given that nodes have exactly the bandwidth necessary for the streaming itself, is $\delta_{lb} = (\lceil \log_2(N) \rceil + 1)T$ where $T$ is the transmission time[1]. It is also known [1] that centralized schedulers can distribute every chunk of a stream in exactly $\delta_{lb}$. Also, in [6] it was proved that a bound holds for several *distributed* schedulers if $N \to \infty$ and $M_c \to \infty$ ($M_c$ is the number of chunks). However, when real-time distribution systems are considered such an asymptotic bound is not equivalent to $\delta_{lb}$.

This paper focuses on formally proving the existence of a distributed optimal algorithm, and in finding robust, feasible schedulers that with restricted neighborhoods perform within a reasonable bound of the optimal one. This is the starting point (a reference optimum) for further research on heterogeneous systems, on the interaction of the overlay with the underlying IP network, and on all those 'impairments' that forbid finding closed-form formal solutions to problems in real networking scenarios.

### 2.1 System Description

We consider an overlay of peers connected with a general mesh topology. The total number of peers is $N$. Each peer is connected to $NN$ other peers[2] which constitute its *neighborhood*. A special case is $NN = N - 1$, which define a fully connected mesh. We consider the presence of one more "special peer" that is the source of the video. The source never receives chunks, so its links are logically unidirectional and it is not part of any neighborhood, i.e., its unidirectional links are additional to the others. Fig. 1 reports two sample topologies.

The source distributes a (possibly live) video or TV program. The video is divided in $M_c$ *chunks* of equal duration emitted periodically. All peers have unit bandwidth (i.e., they can transmit a chunk in exactly the inter-chunk generation time) on the uplink and no limitations on the downlink. We do not consider churn and we focus, as main performance parameter, on the diffusion delay of chunks, which is the delay with which chunks are received by all peers. Formally, if $r_i$ is the emission time of chunk $C_i$, then its diffusion delay is $f_i = t - r_i$ such that

[1] The bound comes from the fact that each node can transmit the information only after receiving it, and the number of nodes owning the chunk at most doubles every $T$.

[2] For the sake of simplicity we restrict discussion to $n$-regular topologies: random graphs with symmetric connectivity and $n$ links per node.

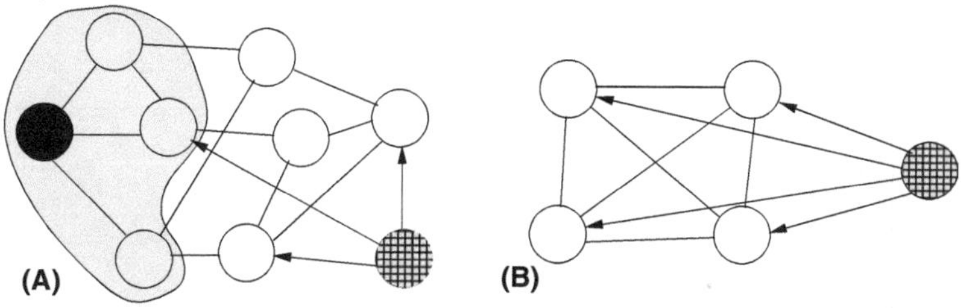


**Fig. 1.** (A) – General mesh topology with $N = 8$ and $NN = 3$; the shaded (pink) area is the neighborhood of the black node; the source is the checkered (yellow) node; (B) – Full mesh with $N = 4$

all $N$ peers have received $C_i$. Each peer has a perfect knowledge of the status of its neighbors.

The assumptions above means that: i) no global ordering of peers is required; ii) the system is not structured; iii) schedulers' decisions are independent one another; vi) peers know exactly the subset of chunks already received or being received by all neighbors; and v) signaling delay is negligible.

The first scheduling decision is whether a peer *pushes* information to other peers or if it *pulls* it from other peers . . . or a mix of the two policies. Sometimes in the literature it is stated that pushing information is a behavior typical of structured systems, and pull methods are more adapt for non-structured overlays. Recent papers like [6,7] instead use push schedulers on non-structured meshes. Indeed, the choice of whether it is better to push or pull information is not related to the structure (or the lack of it) of the system, but to the bandwidth bottleneck, which can create conflicts in scheduling decisions.

Push-based systems are suitable for systems where the bottleneck is the uplink, because this guarantees a priori that only one chunk will be scheduled for transmission on the uplink, and that scheduling conflicts arising from the distributed nature of the scheduling will insist on the downlink of other peers.

If the situation were reversed (uncommon in networks dominated by ADSL access, but technically possible), then pull-based schedulers would solve a priori the conflict on the downlink, and more bandwidth-endowed uplinks would accommodate scheduling conflicts. Interestingly, a scenario with symmetric up- and downlink capacities does not offer an easy logical choice on whether pushing or pulling information is the best choice.

We consider push-based schedulers, but we claim that reversing the bottleneck hypothesis, pull-based schedulers which are dual to those we prove optimal in the sequel can be easily derived.

### 2.2 Formal Notation and Definitions

A system is composed by a set $\mathcal{S} = \{P_1, \ldots P_N\}$ of $N$ *peers* $P_i$, plus a special node called *source*. Each peer $P_i$ receives chunks $C_j$ from other peers, and send them out to other peers at a rate $s(P_i)$. The source sends chunks with rate

**Table 1.** Definitions and symbols used in the paper

| Symbol | Definition |
|---|---|
| $\mathcal{S}$ | The set of all the peers |
| $N$ | The number of peers in the system |
| $M_c$ | The total number of chunks |
| $P_i$ | The $i^{th}$ peer |
| $C_h$ | The $h^{th}$ chunk |
| $r_h$ | The time when the source generates $C_h$ |
| $NN$ | The neighbourhood size |
| $f_h$ | The diffusion delay of chunk $C_h$ (the time needed by $C_h$ to reach all the peers) |
| $\mathcal{C}(P_i, t)$ | The set of chunks owned by peer $P_i$ at time $t$ |
| $\mathcal{C}'(P_i, t)$ | The set of chunks owned by $P_i$ at time $t$ which are needed by some of $P_i$'s neighbours |
| $\mathcal{N}_i$ | The neighborhood of peer $P_i$ |
| $s(P_i)$ | The upload bandwidth of peer $P_i$ |

$s(source)$. The set of chunks already received by $P_i$ at time $t$ is indicated as $\mathcal{C}(P_i, t)$.

The source, not included in $\mathcal{S}$, generates chunks in order, at a fixed rate $\lambda$ ($C_j$ is generated by the source at time $r_j = \frac{1}{\lambda} j$). We normalize the system w.r.t. $\lambda$, so that $r_j = j$. Also, we set $\forall i, s(P_i) = s(source) = \lambda = 1$, which is the limit case to sustain streaming.

If $\mathcal{D}_j(t - r_j)$ is the set of nodes owning chunk $C_j$ at time $t$, the worst case diffusion delay $f_j$ of chunk $C_j$ is defined as the time needed by $C_j$ to be distributed to every peer: $f_j = \min\{\delta : \mathcal{D}_j(\delta) = \mathcal{S}\}$. According to this definition, a generic peer $P_i$ will receive chunk $C_j$ at time $t$ with $r_j + 1 \le t \le r_j + f_j$. Considering an unstructured overlay $t$ will be randomly distributed inside such interval. Hence, in an unstructured system $P_i$ is *guaranteed* to receive $C_j$ at most at time $r_j + f_j$. To correctly reproduce the whole media stream, a peer must buffer chunks for a time of at least $F = \max_{1 \le j \le M_c}(f_j)$ before starting to play. For this reason, the worst case diffusion delay $F$ is a fundamental performance metric for P2P streaming systems, and this paper will focus on it.

When $\forall i, s(P_i) = \lambda = 1$, at time $t$ the source sends a chunk $C_j$ (with $r_j = t$) to a peer and every peer $P_i$ sends a chunk $C_h \in \mathcal{C}(P_i, t)$ to a peer $P_k$. All these chunks will be received at time $t + 1$.

As discussed earlier, the minimum possible diffusion delay $f_j$ for chunk $C_j$ is $\lceil log_2(N) \rceil + 1$. Chunk diffusion is said to be optimal if $\forall j, f_j = \lceil log_2(N) \rceil + 1 = F$.

The most important symbols used in this paper are recalled in Table 1.

## 3 Scheduling Peers and Chunks

In a push-based P2P system, when a peer $P_i$ sends a chunk, it is responsible for selecting the chunk to be sent and the destination peer. The chunk $C_j$ to be sent is selected by a *chunk scheduler*, and the destination peer $P_k$ is selected by a *peer scheduler*. This paper focuses on algorithms which first select the chunk $C_j$, and then select a target peer $P_k$ which needs $C_j$, but the definition of optimality presented in this paper is valid for any chunk-based P2P streaming system.

Some well known chunk scheduling algorithms are *Latest Blind Chunk*, *Latest Useful Chunk*, and *Random Chunk* (again, blind or useful). The Latest Blind Chunk algorithm schedules at time $t$ the latest chunk: $C_j \in \mathcal{C}(P_i, t) : \forall C_h \in \mathcal{C}(P_i, t), r_j \geq r_h$ ($C_j$ is scheduled even if all the other peers already have it). The Latest Useful Chunk (LUc) algorithm selects a chunk that is needed by at least one peer: $C_j \in \mathcal{C}'(P_i, t) : \forall C_h \in \mathcal{C}'(P_i, t), r_j \geq r_h$ where $\mathcal{C}'(P_i, t)$ is a subset of $\mathcal{C}(P_i, t)$ containing only chunks that have not already been received (or are not currently being received) by some other peers. The Random Chunk algorithms select a random chunk in $\mathcal{C}(P_i, t)$ (Random Blind Chunk) or in $\mathcal{C}'(P_i, t)$ (Random Useful Chunk – RUc).

Once the chunk $C_j$ to be sent has been selected, the peer scheduling algorithms selects a peer $P_k$ which needs $C_j$. The most commonly used peer scheduling algorithm is Random Useful Peer, which randomly selects a peer which needs $C_j$. In theory, the chunk scheduling algorithm can select $P_k \in \mathcal{S}$, but in practice peer $P_i$ will only know a subset of all the other peers, and will select $P_k$ from a subset of $\mathcal{S}$ called *neighborhood*. The neighborhood of $P_i$ is indicated as $\mathcal{N}_i$. The case in which $\forall i, \mathcal{N}_i = \mathcal{S} - P_i$ is special, and corresponds to a fully connected graph.

### 3.1 Optimal Peer Scheduling

Random peer selection prevents achieving optimality, because the selected peer might be unable to further distribute the chunk. The rationale behind optimal peer selection should be the following: the selected destination peer should be able to immediately take on the role of redistributing the chunk.

We define the "Earliest-Latest" peer scheduler (ELp) as follows: ELp selects as target a peer $P_l$ that needs $C_h$ and owns the latest chunk $C_k$ with the earliest generation time $r_k$:

$$C_h \notin \mathcal{C}(P_l, t) \wedge \forall P_j \in \mathcal{N}_l, L(P_l, t) \leq L(P_j, t) \tag{1}$$

where $L(P_i, t) = \max_k\{r_k : C_k \in \mathcal{C}(P_i, t)\}$ is the latest chunk owned by or in arrival to $P_i$ at time $t$. If at time $t$ $P_i$ has not received any chunk yet, $L(P_i, t) = 0$. If more peers exist that satisfy (1) one is chosen at random.

### 3.2 Optimal Chunk Scheduling

We show in Theorems 1 and 2 that a LUc/ELp scheduler is optimal in the full mesh case; however, LUc/ELp provides large worst-case diffusion delays when the neighbourhood size is reduced (as will be shown in Section 5). Such a bad behaviour is common to all the LUc schedulers, and is caused by the fact that such schedulers always select the latest useful chunk. Hence, if for some reason (such as a restricted neighbourhood size or a limited knowledge of the neighbourhood) a chunk $C_k$ with $r_k > r_h$ arrives to a peer before $C_h$ is completely diffused, then the peer is not able to diffuse $C_h$ anymore and its diffusion delay is increased by a large amount. In other words, every time that limited knowledge of the neighborhood makes a later chunk arrive to a peer before an earlier one, the diffusion of this latter might be stopped.

For this reason, a new scheduling algorithm has been developed to be equivalent to LUc/ELp in the full mesh case, and to perform reasonably well when the graph is not fully connected. The new algorithm is based on a deadline-based chunk scheduling algorithm, named Dl. The Dl scheduling algorithm works based on *scheduling deadlines* $d_k$ associated to every chunk instance. The scheduling deadline is initialized to $d_k = r_k + 2$ when the source sends $C_k$ at time $r_k$. The chunk scheduler then works by selecting the chunk $C_k$ with the minimum scheduling deadline:

$$C_k : \forall C_h \in \mathcal{C}'(P_i, t), d_k \leq d_h; \tag{2}$$

Before sending $C_k$ its scheduling deadline is postponed by 2 time units: $d_k = d_k + 2$ (both $P_i$ and the destination peer will see $C_k$ with its new scheduling deadline, while chunk instances present in other peers are obviously not affected).

The scheduling strategy based on selecting the chunk with a minimum deadline is known in literature as Earliest Deadline First (EDF), and is mentioned as "Deadline Driven Scheduling" in a seminal paper by Liu and Layland [8], but to the best of our knowledge, it has never been applied with dynamic deadlines in distributed systems.

**Observation 1.** *The scheduling deadline $d_k$ of a chunk instance $C_k$ at peer $P_i$ is equal to $r_k + 2d$, where $d$ is the number of times that $C_k$ has been selected by the Dl schedulers along the path taken by the chunk till $P_i$.*

## 4 Analysis with Full Meshes

In this section, some important properties of the LUc/ELp and Dl/ELp scheduling algorithms are proved for the case of a fully connected overlay. In Theorems 1 and 2, it is proved that LUc/ELp achieves optimality, while in Theorem 3 the optimality of Dl/ELp is shown.

**Lemma 1.** *When using ELp, $\forall i, t \leq \lfloor \log_2(N+1) \rfloor \Rightarrow ||\mathcal{C}(P_i, t)|| \leq 1$.*

*Proof.* During an initial transient, at time $t$ the system contains $2^t - 1$ chunk instances (because at every time instant the source emits a new chunk and all the peers having at least one chunk send a chunk); hence, there are $N - (2^t - 1)$ peers having no chunks. By definition, the ELp scheduler selects such peers as targets, hence a peer $P_i$ can have more than 1 chunk only if $2^t - 1 > N \Rightarrow 2^t > N + 1 \Rightarrow t > log_2(N+1)$.

**Lemma 2.** *If $\forall i, s(P_i) = \lambda = 1 \wedge \mathcal{N}_i = \mathcal{S} - P_i$, if a LUc/ELp scheduling algorithm is used, then*

$$\forall \delta,\ 0 < \delta \leq \lfloor log_2(N) \rfloor \Rightarrow ||\mathcal{L}_j(\delta)|| = 2^{\delta - 1}$$

*where* $\mathcal{L}_j(\delta) = \{P_i : \max_k\{r_k : C_k \in \mathcal{C}(P_i, r_j + \delta)\} = r_j\}$ *is the set of peers having* $C_j$ *as their latest chunk at time* $r_j + \delta$.

*Proof.* The lemma is proved by induction on $\delta = t - r_j$, and by considering the latest chunk owned by the peers at time $t = r_j + \delta$, so that $\mathcal{S}$ is partitioned into three subsets:

- $\mathcal{X}(\delta) = \bigcup\{\mathcal{L}_j(i) : i > \delta\}$ is the set of peers with latest chunk later than $C_j$;
- $\mathcal{Y}(\delta) = \mathcal{L}_j(\delta)$ is the set of peers having $C_j$ as their latest chunk;
- $\mathcal{Z}(\delta) = \bigcup\{\mathcal{L}_j(i) : i < \delta\}$ the set of peers with latest chunk earlier than $C_j$.

The above is a partitioning into disjoint subsets, therefore $||\mathcal{X}(\delta)|| + ||\mathcal{Y}(\delta)|| + ||\mathcal{Z}(\delta)|| = ||\mathcal{S}|| = N$. The lemma can be now proved by induction on $\delta$.

**Induction base:** After chunk $C_j$ is generated by the source at time $r_j$, it is sent out to a peer $P_i$, which will receive it at time $t = r_j + 1 \Rightarrow \delta = 1$. Hence,

$$\mathcal{D}_j(1) = \{P_i\} \Rightarrow ||\mathcal{D}_j(1)|| = 1$$

As $C_j$ is the newest chunk in the system, $\mathcal{X}(\delta)$ is empty and $C_j$ becomes the latest chunk on $P_i$:

$$\forall C_k \in \mathcal{C}(P_i, r_j + 1), r_j > r_k$$

Thus, $\delta = 1 \Rightarrow ||\mathcal{L}_j(\delta)|| = ||\mathcal{D}_j(\delta)|| = 1 = 2^{\delta-1}, ||\mathcal{X}(\delta)|| = 0 = 2^{\delta-1} - 1$. Also note that $||\mathcal{Z}(\delta)|| = N - 1 > ||\mathcal{X}(\delta)|| + ||\mathcal{Y}(\delta)||$.

**Inductive step:** First of all, it is easy to notice that $||\mathcal{X}(\delta - 1)|| \leq 2^{\delta-2} - 1$: in fact, at every time unit a new chunk $C_k : r_k > r_j$ is generated, and all the peers $P_i \in \mathcal{X}(k-1)$ can send their latest chunk to another peer. As a result, $||\mathcal{X}(\delta-1)||$ will be at most equal to $2||\mathcal{X}(\delta-2)||+1$. But $||\mathcal{X}(\delta-2)|| \leq 2^{\delta-3}-1$ (by induction), so

$$||\mathcal{X}(\delta - 1)|| \leq 2(2^{\delta-3} - 1) + 1 = 2^{\delta-2} - 1$$

Now, if $\delta \leq \lfloor log_2(N) \rfloor$, then

$$\delta \leq \lfloor log_2(N) \rfloor \Rightarrow 2^{\delta} \leq N \Rightarrow 2(2^{\delta-2} + 2^{\delta-2}) \leq N$$

and since $||\mathcal{L}_j(\delta - 1)|| = 2^{\delta-2}$, $||\mathcal{X}(\delta - 1)|| \leq 2^{\delta-2} - 1$ and $||\mathcal{Z}(\delta - 1)|| = N - ||\mathcal{X}(\delta - 1)|| - ||\mathcal{Y}(\delta - 1)||$, the above equation can be rewritten as

$$2(||\mathcal{X}(\delta-1)||+1+||\mathcal{Y}(\delta-1)||) \leq N \Rightarrow ||\mathcal{X}(\delta-1)||+||\mathcal{Y}(\delta-1)|| \leq ||\mathcal{Z}(\delta-1)||-2$$

As a result, at $\delta - 1$, $||\mathcal{Z}(\delta - 1)||$ is more than half of $N$, therefore there are enough peers with latest chunk older than $C_j$ to receive chunks from both $\mathcal{X}(\delta - 1)$ and $\mathcal{Y}(\delta - 1)$, so $||\mathcal{L}_j(\delta)|| = ||\mathcal{D}_j(\delta)|| = 2^{\delta-1}$, hence the claim.

**Theorem 1.** *If* $N = 2^i$, *algorithm LUc/ELp is optimal.*

*Proof.* By definition, an algorithm is optimal iff $\forall j, f_j = \lceil log_2(N) \rceil + 1$. In this case, this means $\forall j, f_j = i + 1$. By Lemma 2,

$$\forall j, \delta \leq \lfloor log_2(N) \rfloor \Rightarrow ||\mathcal{L}_j(\delta)|| = 2^{\delta-1}$$

hence, $\forall j, ||\mathcal{L}_j(i)|| = 2^{i-1}$. As a result, $||\mathcal{D}_j(i+1)|| = 2||\mathcal{L}_j(i)|| = 2^i = N$, and $f_j = i + 1$.

**Theorem 2.** *Algorithm LUc/ELp is optimal also if* $N \neq 2^i$.

*Proof.* If $N = 2^i + n$, with $n < 2^i$, by Lemma 2 it comes $\forall j, ||\mathcal{L}_j(i)|| = 2^{i-1}$. Hence, for $\delta = i$ chunk $C_j$ is sent $2^{i-1}$ times and chunks with $r_k > r_j$ are sent $2^{i-1} - 1$ times. As a result, $||\mathcal{D}_j(i+1)|| = 2^i$, $||\mathcal{X}(i+1)|| = 2^i - 1$, $||\mathcal{Z}(i+1)|| = 0$, and $||\mathcal{L}_j(i+1)|| < ||\mathcal{D}_j(i+1)||$. To compute the exact value of $||\mathcal{L}_j(i+1)||$, let $x$ be the number of chunks sent by peers in $\mathcal{X}(i)$ to peers in $\mathcal{Z}(i)$ and let $y$ be the number of chunks sent by peers in $\mathcal{Y}(i)$ to peers in $\mathcal{Z}(i)$. According to the peer scheduling rules, $x + y = ||\mathcal{Z}(i)||$ (because chunks are sent to peers having the earliest latest chunk). Moreover, $||\mathcal{L}_j(i+1)|| = y + ||\mathcal{L}_j(i)|| - (||\mathcal{X}(i)|| - x)$. Hence,

$$||\mathcal{L}_j(i+1)|| = ||\mathcal{Z}(i)|| - x + 2^{i-1} - (2^{i-1} - 1 - x) =$$

$$= (N - 2^{i-1} - (2^{i-1} - 1)) - x + 2^{i-1} - 2^{i-1} + 1 + x = N - 2^i + 1 + 1 = N - 2^i + 2$$

Finally,

$$||\mathcal{D}_j(i+2)|| = \min\{N, ||\mathcal{D}_j(i+1)|| + ||\mathcal{L}_j(i+1)|| = 2^i + N - 2^i + 2\} = N$$

Hence, $f_j = i + 2 = \lceil log_2(N) \rceil + 1$.

**Observation 2.** *If an optimal chunk scheduling is used, all the copies of every chunk* $C_k$ *are forwarded from time* $r_k$ *to time* $r_k + f_k - 2$.

Based on the optimality of LUc/ELp, it is now possible to prove that Dl/ELp is an optimal algorithm too. This is done by showing that on a full mesh it generates the same schedule as LUc/ELp.

**Theorem 3.** *If* $\forall i, s(P_i) = \lambda = 1$, $\forall i, \mathcal{N}_i = \mathcal{S} - P_i$, *then the chunk distribution produced by Dl/ELp is identical to the chunk distribution produced by LUc/ELp.*

*Proof.* By contradiction: assume that at any time $t_0$ the chunk distribution produced by Dl/ELp starts to differ from the one produced by LUc/ELp, i.e., assume that Dl at peer $P_i$ at time $t_0$ selects chunk $C_j$ while LUc would select chunk $C_k$ (so, $r_k > r_j$). However, it will be shown that choosing $C_j$ with Dl implies $r_j \geq r_k$ contradicting the hypothesis $r_k > r_j$.

If $t_0 < \lfloor \log_2(N+1) \rfloor$, then Lemma 1 guarantees that all chunk schedulers are identical under ELp peer scheduling.

If $t_0 \geq \lfloor \log_2(N+1) \rfloor$, we have from the hypotheses that $\forall t < t_0$ the schedules produced by Dl/ELp and LUc/ELp are identical. By definition at time $t_0$ in $P_i$ LUc/ELp choses

$$C_k \in \mathcal{C}'(P_i, t_0) : \forall C_h \in \mathcal{C}'(P_i, t_0) r_k \geq r_h.$$

Since the source only produces a single chunk at every time unit, $r_k$ and $r_h$ cannot have the same value, hence $r_k > r_h$. To obtain a different schedule Dl/ELp must choose $C_j \neq C_k$.

Since for $t < t_0$ Dl/ELp produced the same schedule as LUc/ELp, $C_j$ and $C_k$ have been transmitted $t_0 - r_j$ and $t_0 - r_k$ times respectively (see Observation 2); hence, $d_i = r_i + 2(t_0 - r_i)$ for both $C_j$ and $C_k$.

Since Dl/ELp chooses $C_j \in C'(P_i, t_0) : \forall C_h \in C'(P_i, t_0) d_j \leq d_h$ we have $d_j \leq d_k \Rightarrow r_j + 2(t_0 - r_j) \leq r_k + 2(t_0 - r_k) \Rightarrow -r_j \leq -r_k \Rightarrow r_j \geq r_k$ which contradicts the hypothesis $r_k > r_j$.

**Observation 3.** *Note that the Dl scheduler postpones the scheduling deadline by two time units per transmission as* $d_k = d_k + 2$. *If a generic constant* $q$ *was used instead of* 2 *and the scheduling deadline was postponed as* $d_k = d_k + q$, *then the last equation in the proof of Theorem 3 would have become*

$$r_j + q(t_0 - r_j) \leq r_k + q(t_0 - r_k) \Rightarrow (q-1)r_j \geq (q-1)r_k$$

*which contradicts* $r_k > r_j$ *if* $q > 1$. *Hence, if a generic constant* $q > 1$ *is used to postpone the scheduling deadline, then Dl/ELp is still equivalent to LUc/ELp. In this sense, Dl can be seen as a whole class of deadline-based algorithms.*

## 5 Neighborhood Restriction and Selected Results

Although both LUc/ELp and Dl/ELp have been proved to provide optimal performance in the case of a fully connected graph their performance in more realistic situations is still unclear. Besides these two algorithms we consider various combinations with LUc, RUc and RUp algorithms for comparison.

### 5.1 Simulating P2P Streaming and Measuring Performance

The behavior of the scheduling algorithms introduced in Section 3 is analyzed using the SSSim simulator [9], by setting up an overlay of $N$ peers with unit upload and infinite download bandwidth. The source distributes $M_c$ chunks.

As explained in Section 2 the performance metric considered in this paper is the worst case diffusion time $F$, and (as stated in Section 3), a scheduling algorithm is optimal iff $F = \lceil log_2(N) \rceil + 1$.

First of all, the algorithms have been simulated on a fully connected graph, as shown in Figure 2. In accordance with Theorems 2 and 3, LUc/ELp and Dl/ELp achieve optimal performance, outperforming the other algorithms (in particular, RUc/ELp achieves a value of $f_i$ near the double of the optimal, and all the other algorithms achieved even worse performance).

### 5.2 Restricting the Overlay

In realistic situations a restricted overlay is used instead of a fully connected graph. Such a restricted overlay is modeled assuming bidirectional relations and

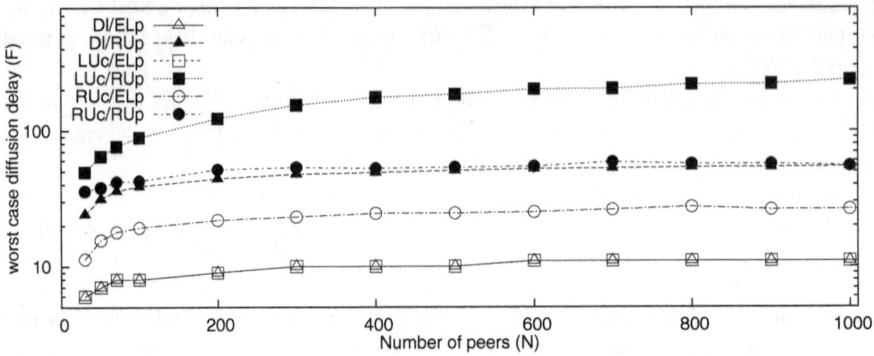


**Fig. 2.** Full mesh overlay; maximum diffusion delay as a function of N; 500 chunks

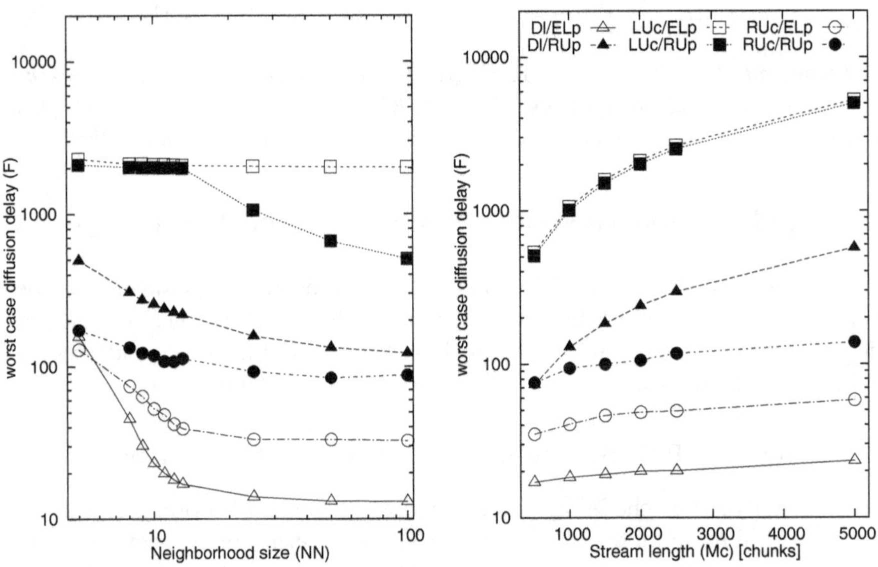


**Fig. 3.** Worst case chunk diffusion delay of algorithms, with 1000 peers, as a function of: (left) neighborhood size, with 2000 chunks; (right) number of chunks, with $\mathcal{NN} = 11$

a pre-defined number ($\mathcal{NN} = ||\mathcal{N}_i||$) of neighbor nodes. The resulting graph is a random $\mathcal{NN}$-regular graph. In the following simulations, the algorithms are evaluated on 10 instances of the random $\mathcal{NN}$-regular graph. We have verified that confidence intervals were always within 5% of the reported mean values with a confidence level of 90%.

The left hand side of Figure 3 shows performance of different streaming algorithms as a function of $\mathcal{NN}$ and shows how the LUc/ELp algorithm (which is optimal on a full mesh) is highly sensitive to neighborhood restrictions and performs badly when $\mathcal{NN} < N - 1$. Dl/ELp, on the other hand, works better than all the

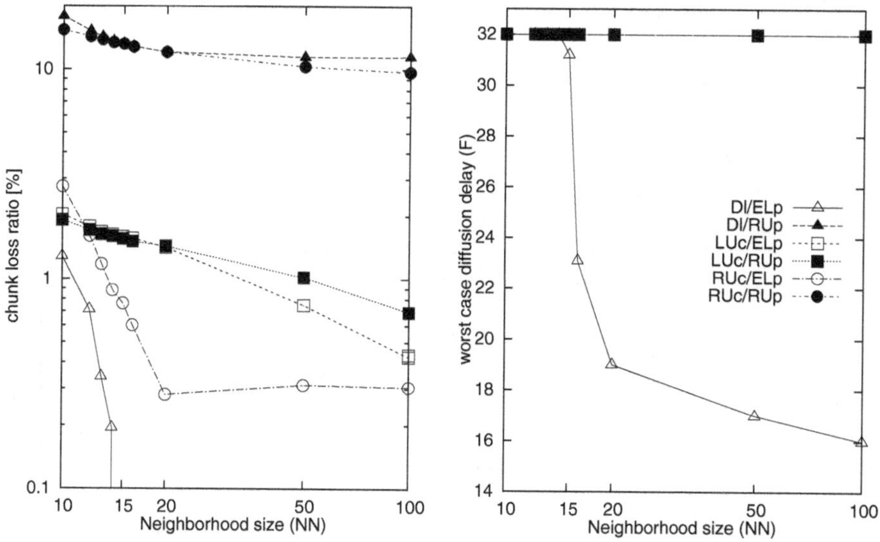


**Fig. 4.** Chunk loss and F as a function of the neighborhood size ($N = 10000, D = 32$)

other algorithms and is able to achieve values of $F$ near the optimum (which in this case is 11). The right side of Figure 3 shows how the number of chunks affects $F$ for $NN = 11$ (note that $\log_2(N) = 9.9658$). Dl/ELp keeps good performance even for long streams, while for several other algorithms $f_i$ increases with $i$ (the performance of the algorithm depends on the stream length), hence these distribution mechanism results to be unstable in a streaming context.

### 5.3 Limiting the Chunk Buffer Size

The only solution to the instability problem is to define a playout delay $D$, and to discard chunks $C_j$ at time $r_j + D$. This causes some chunk loss (for chunks $C_j$ that would have $f_i > D$), but can make the distribution system stable again. Moreover, the playout delay $D$ can be used to dimension the chunk buffers in the peers (in particular, each peer needs to buffer at most $D$ chunks)[3].

Since some chunks can be lost, the performance should be evaluated based on both chunk loss ratio and the maximum delay. Figure 4 plots the chunk loss ratio (left) for the various algorithms as a function of the neighborhood size with $D = 32$. Note that for $NN > 14$ the chunk loss ratio for Dl/ELp is 0, showing that it is possible to dimension the chunk buffer size so that it does not affect the algorithm's performance (to the authors' best knowledge, this is not possible for the other algorithms). The worst case diffusion time $F$ (right) fastly approaches the optimum with Dl/ELp, while it is obviously 32 for all other algorithms.

[3] Implementing the *chunks buffer size* in the simulator can enable optimizations which allow to simulate larger task sets, hence we move to $N = 10000$.

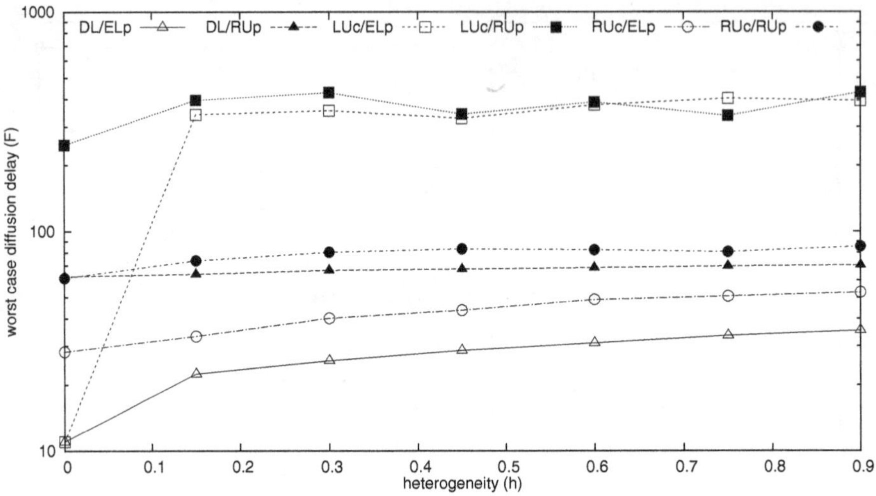


**Fig. 5.** F as a function of bandwidth heterogeneity ($N = 600$, $M_c = 600$)

### 5.4 Heterogeneous Upload Bandwidth

Finally, we evaluate the performance of Dl/ELp in heterogeneous networks. We use a scenario similar to that of [6]. The system is composed of $N$=600 nodes, devided in 3 classes based on their upload bandwidth: bandwidth of 2 for $(h/3)N$ nodes[4], bandwidth of 0.5 for $(2h/3)N$ nodes, and unit bandwidth for $(1 - h)N$ nodes, thus keeping the mean bandwidth at 1. We vary $h$ from 0 (homogeneous case) to 1.

Figure 5, plotting the diffusion delays for $M_c = 600$ chunks and an infinite buffer size, shows that Dl/ELp performs better in this specific setting than the other algorithms studied for the whole range of $h$. These initial studies indicate that Dl/ELp could be a strong contender also in heterogeneous settings. We leave more detailed studies, including studies of the effect of Dl's increment parameter on performance (see observation 3), for future work.

## 6 Related Work and Contributions

Optimality of schedulers has been extensively studied in the literature. For the case of full mesh overlay and unit upload bandwidth limits, the generic (i.e., valid for any scheduler) lower bound of $(\lceil \log_2(N) \rceil + 1)T$ is well known. [1] proves that this bound is strict in a streaming scenario by showing the existence of a *centralized* scheduler that achieves such bound. A similar proof (although for the case of file dissemination) can be found in [11]. Our work improves on these results by proving

[4] In order to validate our results with heterogeneous bandwidth, we implemented our algorithms also in the P2PTVSim [10] simulator. For this reason, we had to use a smaller number of peers and chunks.

the existence of *distributed* schedulers (LUc/ELp and Dl/ELp) that achieve the same strict bound.

Generic upper bounds as well as upper bounds on the distribution times achieved by different distributed schedulers can also be found in literature. The fundamental work of [12] studies asymptotic properties of distributed gossiping algorithms in a similar setting, showing an upper bound for any pull based algorithm of all the messages in $\mathcal{O}(M_c + log(N))$ time with high probability even for blind algorithms. Generic asymptotic bounds are also shown for blind push based algorithms, although in this case full dissemination cannot be guaranteed. A blind algorithm that distributes chunks with a high probability in $(9*M_c+9*\log_2(N))T$ is also shown. Note that this suggest a distribution delay for the individual chunk that grows with $M_c$.

Authors of [11] also evaluate blind distributed strategies in the case of file distribution, showing distribution delays dependent on the number of chunks.

[6] studies upper bounds for specific well known algorithms, showing that the combination of random peer selection and LUc achieves asymptotically good delays, however this demonstration is provided in the case of upload bandwidth higher than 1.

The distributed LUc/ELp and Dl/ELp schedulers presented in our paper perform significantly better than the generic upper bounds shown in [12] and [11] in that it achieves full diffusion of all chunks in $(M_c + \lceil log_2(N) \rceil)T$, i.e. a chunk diffusion delay independent of $M_c$.

It also differs from the streaming algorithms studied in [6], since for LUc/ELp and Dl/ELp this strict delay bound holds for any $N$(not just asymptotically), and it is valid even in the boundary case of unit upload bandwidth, without relying on redundant source coding.

[6] uses ER graphs to model the restricted neighborhood. With $N = 600$ and $\mathcal{NN} = 10$, authors find that the studied algorithms suffer significant losses. These chunk losses are confirmed by our results (even if our random graph model is slightly different) for the algorithms considered therein. However, we also show (through simulations) that the new Dl/ELp algorithm performs near the optimum with any $M_c$ and any $N$, even with significant overlay restrictions. Namely, reducing the neighborhood size to any $\mathcal{NN} >= \lceil log_2(N) \rceil$, our algorithm keeps distributing all chunks with a delay only slightly above the lower bound and always (on all simulated $\mathcal{NN}$-regular random graphs) below $2 * (\lceil log_2(N) \rceil + 1)T$. Note that the neighborhood of $\lceil log_2(N) \rceil$ practically means less than 30 in any reasonable setting.

## 7 Conclusions and Future Work

This paper presented the formal proof that *distributed* algorithms can achieve optimal diffusion for streaming applications in unstructured meshes. The paper introduced a class of deadline based algorithms Dl/ELp which are optimal in full meshes and maintain very good properties also in realistic scenarios with small neighborhoods.

Future work includes on the one hand extending the theoretical results to scenarios with different constraints, including large bandwidth and heterogeneous scenarios, and, on the other hand, exploiting these algorithms to implement real P2P streaming systems.

## References

1. Liu, Y.: On the minimum delay peer-to-peer video streaming: how realtime can it be? In: MULTIMEDIA 2007: Proceedings of the 15th international conference on Multimedia, Augsburg, Germany, pp. 127–136. ACM Press, New York (2007)
2. Hefeeda, M., Habib, A., Xu, D., Bhargava, B., Botev, B.: Collectcast: A peer-to-peer service for media streaming. In: ACM Multimedia 2003, vol. 11, pp. 68–81 (2003)
3. Hei, X., Liang, C., Liang, J., Liu, Y., Ross, K.W.: Insights into pplive: A measurement study of a large-scale p2p iptv system. In: Proceedings of the Workshop on Internet Protocol TV (IPTV) services over World Wide Web in conjunction with WWW 2006 (2006)
4. Chu, Y., Ganjam, A., Ng, T.S.E., Rao, S.G., Sripanidkulchai, K., Zhan, J., Zhang, H.: Early experience with an internet broadcast system based on overlay multicast. In: ATEC 2004: Proceedings of the annual conference on USENIX Annual Technical Conference, Boston, MA, June 2004, USENIX Association (2004)
5. Pianese, F., Keller, J., Biersack, E.W.: Pulse, a flexible p2p live streaming system. In: IEEE INFOCOM (2006)
6. Bonald, T., Massoulié, L., Mathieu, F., Perino, D., Twigg, A.: Epidemic live streaming: optimal performance trade-offs. In: Liu, Z., Misra, V., Shenoy, P.J. (eds.) SIGMETRICS, Annapolis, Maryland, USA, pp. 325–336. ACM, New York (2008)
7. Couto da Silva, A., Leonardi, E., Mellia, M., Meo, M.: A bandwidth-aware scheduling strategy for p2p-tv systems. In: Proceedings of the 8th International Conference on Peer-to-Peer Computing 2008 (P2P 2008), Aachen (September 2008)
8. Liu, C.L., Layland, J.: Scheduling alghorithms for multiprogramming in a hard realtime environment. Journal of the ACM 20(1) (1973)
9. Abeni, L., Kiraly, C., Cigno, R.L.: TR-DISI-08-074: SSSim: Simple and Scalable Simulator for P2P streaming systems. Technical report, University of Trento (2008), `http://disi.unitn.it/locigno/preprints/TR-DISI-08-074.pdf`
10. The NAPA-WINE Project: P2PTVSim home page, `http://www.napa-wine.eu/cgi-bin/twiki/view/Public/P2PTVSim`
11. Mundinger, J., Weber, R., Weiss, G.: Optimal scheduling of peer-to-peer file dissemination. J. of Scheduling 11(2), 105–120 (2008)
12. Sanghavi, S., Hajek, B., Massoulié, L.: Gossiping with multiple messages. In: Proceedings of IEEE INFOCOM 2007, Anchorage, Alaska, USA, May 2007, pp. 2135–2143 (2007)

# Identify P2P Traffic by Inspecting Data Transfer Behaviour*

Mingjiang Ye[1], Jianping Wu[1], Ke Xu[1], and Dah Ming Chiu[2]

[1] Tsinghua National Laboratory for Information Science and Technology, Department of Computer Science, Tsinghua University, Beijing, 100084, P.R. China
yemingjiang@csnet1.cs.tsinghua.edu.cn, jianping@cernet.edu.cn, xuke@csnet1.cs.tsinghua.edu.cn

[2] Dept. of Information Engineering, The Chinese University of Hong Kong, Hong Kong, P.R. China
dmchiu@ie.cuhk.edu.hk

**Abstract.** Classifying network traffic according to its applications is important to a broad range of network areas. Since new applications, especially P2P applications, no longer use well-known fixed port numbers, the native port based traffic classification technique has become much less effective. In this paper, we propose a novel approach to identify P2P traffic by leveraging on the data transfer behaviour of P2P applications. The behaviour investigated in the paper is that downloaded data from a P2P host will be uploaded to other hosts later. To find the shared data of downloading flows and uploading flows online, the content based partitioning schema is used to partition the flows into data blocks. Flows sharing the same data blocks are identified as P2P flows. The effectiveness of this method is demonstrated by experiments on various P2P applications. The results show that the algorithm can identify P2P applications very accurately while only keeping a small set of data blocks. The method is generic and can be applied to most P2P applications.

**Keywords:** traffic management, P2P traffic identification, data transfer behaviour, content based partitioning, Rabin fingerprint.

## 1 Introduction

Classifying network traffic according to its applications is important to a broad range of network areas including network monitoring, network management and optimization, network security, traffic accounting etc. Different from the traditional applications (http, email, ftp), new applications, especially P2P applications, usually use dynamic port numbers. The traffic classification technique based on native port has become less effective. However, the payload signature

* This research is supported by NSFC-RGC Joint Research Project (20731160014), 973 Project of China (2009CB320501), 863 Project of China (2008AA01A326, 2006AA01Z205, 2006AA01Z209), and Program for New Century Excellent Talents in University.

L. Fratta et al. (Eds.): NETWORKING 2009, LNCS 5550, pp. 131–144, 2009.

based traffic classification technique [1,2] can achieve a high accuracy. But the technique also has its limitations. It is ineffective in classifying encrypted traffic. Besides, a lot of P2P applications use proprietary protocols. Lacking open protocol specifications makes analysing signatures and maintaining up-to-date signatures very difficult.

Recently, machine learning algorithms which classify network traffic using flow statistical information [5,6,7,8,9] have been proposed. There are several challenges in classifying network traffic by using flow properties. First, the statistical characteristics used in classification are unstable since the delays and/or the packet loss ratios of the networks are dynamic. Second, flows belonging to different applications can have similar per-flow statistical characteristics. It is hard to distinguish these similar flows by using flow properties.

This paper proposes a novel approach to identify P2P traffic based on its data transfer behavior. The idea of the approach is based on the observation that a P2P peer uploads data to other peers after downloading it. The idea is first proposed by Xing Lu [10]. In their method, the first k bytes of each packet in downloading flows are stored for each host. When the same content are found in uploading flows of the host, the flows associated with the content are classified as P2P flows. The partitioning schema in their method is named the head packet partitioning schema in this paper. As shown in our experiments, the performance of the head packet partitioning schema is very poor in identifying some P2P applications.

The content based partitioning schema is proposed in the paper to solve the problem. The schema divides payloads of flows into data blocks (a data block is a contiguous content block of payload). The shared files and videos in P2P applications are usually divided into small data pieces during exchanges. For flows sharing the same data piece, the schema can synchronize the boundary of the data piece, and extract the same part of the data piece as a data block. The schema is generic and can be applied to most P2P applications.

Our contributions are as follows. First, we proposed the content based partitioning schema in identifying data transfer behaviour. As shown in experiments, The schema performs much better than the head packet partitioning schema. Besides, Head tail partitioning schema, a simple enhancement schema of the head packet partitioning schema, also can improve performance greatly, though not as good as the content based partitioning schema. Second, we have studied the performance impactions of some important issues. They are the size of the data block, the number of the data blocks to be stored, the data block replacement algorithms and the ratio of unobservable communications.

From the studies, we have drawn several conclusions. First, 256 bytes is a suitable size for data block. Second, the random replacement algorithm is suggested in replacing the old data blocks. Third, with the random replacement algorithm, the method performs quit well while only keeping a small data block set (3 minutes). Last, even though the communications of a large fraction of peers (about 30%) are not observed, the performance degradation is rather small.

## 2 Related Work

Payload signatures are useful in classifying network traffic [1,2]. Application signatures are the common strings in the P2P protocols to identify P2P traffic whereas our method focuses on the data being shared in P2P applications.

In addition to signature based methods, other payload based methods are also proposed. ACAS [3] uses the first N bytes payload as the input to train a machine learning model and uses it to classify flows. Levchenko et al. build several probabilistic models on payload, including the statistical model treating each n-byte flow distribution as a product of N independent byte distributions and the Markov process model which relies on introducing independence between bytes [4]. The models still employ frequently appearing bytes in application protocols to classify traffic since the random bytes in application data are meaningless in statistics.

The machine learning algorithms classify network traffic by using traffic characteristics [5,6,7,8,9], such as average length of packets, arrival interval of packets, etc. The algorithms can be further summarized into supervised and unsupervised methods. Zander et al. compare several supervised algorithms, including Naive Bayes, C4.5 decision tree, Bayesian Network and Naive Bayes Tree [9]. They find that the classification precision of the algorithms is similar, but computational performance can be significantly different. McGregor et al. use the unsupervised expectation maximum algorithm to cluster the flows [6]. The experiment finds that the average precision of classification is very high, but some applications are very difficult to distinguish.

The special communication patterns of applications are used in traffic classification. Karagiannis et al. study the communication pattern of P2P traffic in transport layer to identify P2P traffic [11]. BLINC attempts to capture the communication pattern of a host at three levels: the social level, the functional level and the application level [12]. Graphlet is used to describe the communication pattern of each application and classify network traffic.

It is hard to find a general communication pattern to fit all P2P applications. Y Hu et al. propose a method to build behavioural profiles of the target application which then describes the dominant communication patterns of the application [13]. The profiles of each application are built by data mining algorithms in the training phase. The data transfer behaviour used in our method is general for all the P2P applications, so our method does not need any training phases.

Using the data transfer behaviour to identify P2P traffic was first proposed by Xing Lu et al. [10]. But they only use the first k bytes of packet to find the shared data. As shown in the experiments, our schemas performances much better. Our method is general for different P2P applications while their schema has some limitation to identify some P2P applications. Besides, we have studies some important issues in identifying the data transfer behaviour which are not studied before.

# 3 Method

## 3.1 Payload Partitioning Schemas

The basic idea of the method is simple. The method inspects the P2P data transfer behaviour on the host level. From the view of a host, flows can be classified as downloading flows and uploading flows. If a host downloads a data piece in a downloading flow, and the uploads it to other hosts in some uploading flows. These flows are classified as P2P flows.

The biggest obstacle in the method is how to find the same data pieces in different flows. One way is comparing the payloads of flows directly. But it is time consuming. Besides, complete payloads of flows have to be saved in the memory before comparing. It is quit difficult - and therefore, unrealistic - as an online process.

In our method, payloads of flows are divided into data blocks and then the signatures of the data blocks are compared. A payload partitioning schema decides how to divide the payload into data blocks. The ideal result is that each data block is an exact data piece in a P2P application. For example, for two distinct flows in Fig. 1, the ideal case is that the generated data blocks in the first flow equal to data pieces 1, 5 and 7 and the generated data blocks in the second flow equal to data pieces 9, 4 and 1. Thus, the shared data (pieces 1) of the two flows can be found.

However, a prerequisite to generating such ideal results is the detailed protocol information. As shown in Fig. 1, there are several challenges in locating the boundaries of data pieces. First, a flow can contain protocol fields with variable sizes. Second, the sizes of data pieces are variable in some P2P live streaming applications. In these applications, a data piece contains 1 second of video content. The size of the data piece is variable in most video coding schemas.

Several partitioning schemas are considered. The first one is the method used in [10]. We called it as the head packet partitioning schema. If the payload size of a packet exceeds the threshold S, the first S bytes of the packet is extracted as a data block.

The second one is the head tail packet partitioning schema. If the payload size of a packet exceeds the threshold value S, the first S bytes of payload and the last S bytes of payload are extracted as two data blocks. This schema is an

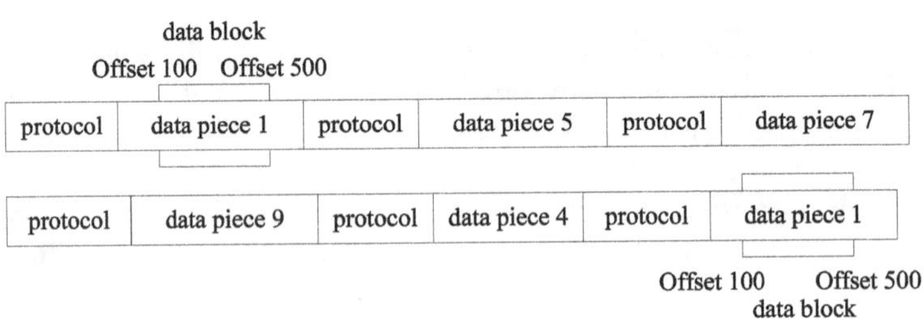


**Fig. 1.** Shared data piece

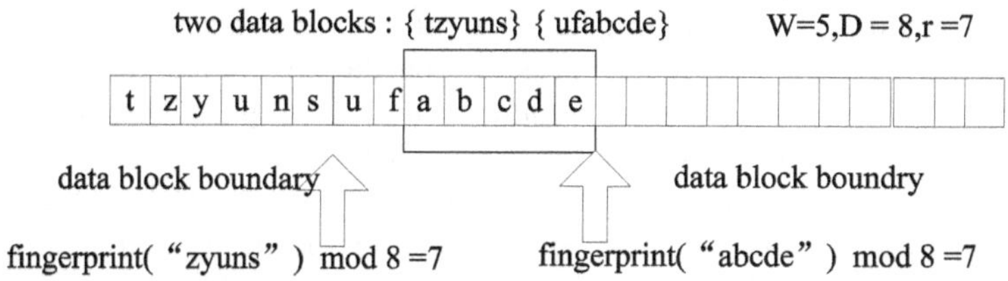


**Fig. 2.** Content based partitioning schema

enhancement of the previous one. If the target packets contain some protocol fields at the beginning or at the end, the schema can skip them.

The last one is the content based partitioning schema. It has been used in saving bandwidth in network file systems [14] and automatically generating signatures of worms in security fields [10]. The schema divides a flow into variable-size, non-overlapping data blocks. The schema works as follows. First, a pair of pre-determined integers $D$ and $r$ ($r < D$) are set. Then a sliding window of fixed width $W$ moves across the byte sequence. The window begins from the first $W$ bytes in the sequence, and slides one byte at a time toward the end. At every position of the byte sequences, a fingerprint $F$ is calculated according to the content in the current window. If $F$ mod $D$ equals $r$, the end of the window is a data block boundary.

For example, in Fig. 2, the windows size is 5, $D$ is 8 and $r$ is 7. A window of width 5 moves across the byte sequence and fingerprints are calculated in every position. When the window reaches abcde, two positions have satisfied the condition. Two data blocks tzynuns and ufabcde are extracted. The fingerprint is calculated by using Rabin fingerprint [16] which has a low collision rate. Using the pre-computed table, it is vey efficient to calculate the Rabin fingerprint [17].

The principle behind the content based partitioning schema is that, because the schema determines the boundaries of data blocks based on the local content of payload in a small window, the boundaries can synchronize in the same data pieces. For example, in Fig. 3, there are two flows containing the same data piece (the black part) in different positions. In the first flow, there are two positions satisfying the condition. They locate at offset 100 and 500 from the beginning of the data piece. In the second flow, the positions at the same offsets from the beginning of the data piece should also satisfy the condition, as the contents in the windows are the same. The same data block is extracted in both flows.

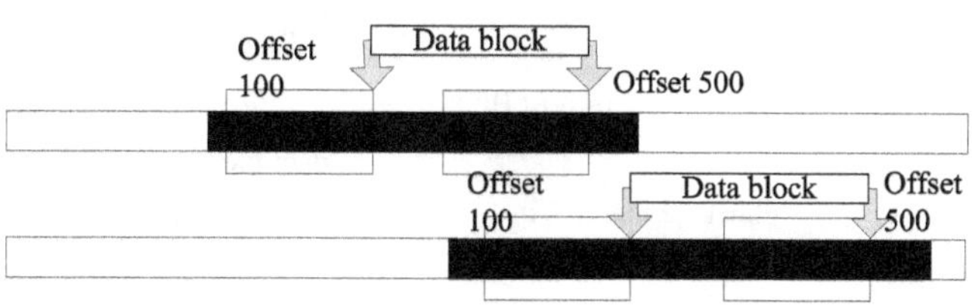


**Fig. 3.** Principle of content based partitioning

The content based partitioning schema can create blocks with various sizes. Although the average size of blocks is D, the data blocks can be as short as several bytes. Short data blocks introduce many false positives, so the results are further filtered to only keep data blocks whose sizes exceed threshold $S$.

The value of $D$ decides the average size of data blocks. If $D$ is much smaller than $S$, many data blocks which are smaller than $S$ will be generated and filtered. So $D$ is equal to $S$ in the algorithm. Since the performance is not sensitive to the value of $r$, we set $r$ to $D-1$ in the algorithm.

The boundaries of data blocks are determined by the local content in the window. The window size should be much smaller than the size of data pieces to generate data blocks inside a data piece. On the other side, a small window is sensitive for random content in flows. In the experiments, the window size is 32 bytes.

### 3.2 Algorithm

The details of the algorithm are as follows. For each host, the algorithm keeps a hash set, which is called the download set, to save data blocks. Payloads of flows are divided into data blocks. Data blocks of each downloading flow and flow identifiers (a flow identifier is the 5-tuple: source IP address, source port, destination IP address, destination port, and protocol) are saved in the corresponding download set. To save the memory capacity, signatures of data blocks (Ranbin fingerprint) are calculated and saved instead of the original data blocks. If the size of the download set exceeds the limitation, then data block replacement is applied. Data blocks of each uploading flow are checked whether they have been saved in the corresponding download set already. If a data block of the uploading flow has been saved in the download set, the uploading flow, the downloading flow and their reverse flows are identified as P2P flows.

A flow has two roles. One is the downloading flow of the target host. The other is the uploading flow of the source host. The data blocks of the flow are saved in the download set of the target host and checked in the download set of the source host.

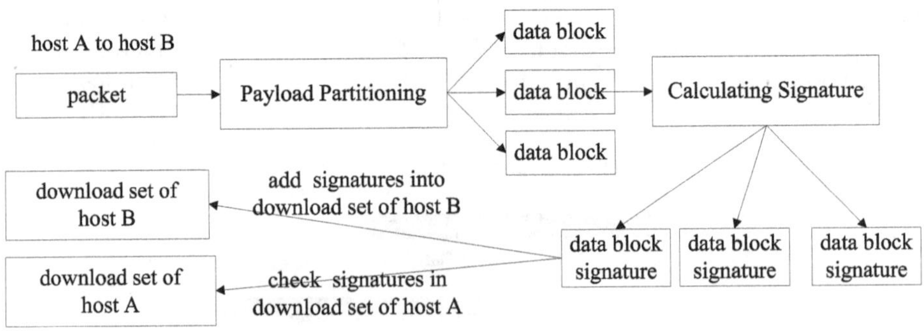


**Fig. 4.** Online process of the algorithm

The online process of the algorithm is illustrated in Fig. 4. When a packet arrives, it is first partitioned into several data blocks. And then, the signatures of the data blocks are calculated. Finally the signatures are saved into the download set of the destination host and checked in the download set of the source host.

### 3.3 Practical Problems

Our method is payload based, which focuses on the same data being shared in P2P applications, so it is ineffective in classifying encrypted traffic. Besides, the P2P applications using network coding [18] are also undetectable under our method. We argue that encryption and network coding pose a burden on the P2P application developers, so the mainstreaming P2P applications are still transferring data without any transformation. To identify encrypted traffic, non-payload based methods are more efficient. There are some studies using machine learning algorithms to identify encrypted traffic [19].

The computation and memory overhead are important in identifying P2P traffic online by using our method. The head packet partitioning and head tail packet partitioning have little computation overhead, while the content based partitioning required calculating the Rabin fingerprint which is also very fast. The hash set can be used to check whether a data block have been saved in the download set. In average, there are only several look-up operations for each packet. So the computation overhead is affordable.

Memory is used mainly for two purposes. First, in the payload partitioning, the byte level partitioning schemas have to keep some intermediate information for each flow. For example, to update the fingerprint in the content based partitioning schema, the fingerprint and the content of the last window are saved for each flow. The intermediate information is quite small and 40 bytes are enough to keep the intermediate status. Suppose there are 1M concurrent flows, only 40MB memory is needed.

Second, the signatures and flow identifiers of the data blocks are saved. Suppose the size of a signature is 16 bytes and the size of the flow identifier is 4 bytes (an index in the flow table is used instead of original 5-tuple), 20 bytes are needed to record a data block. If a data block is as large as 512 bytes, a bidirectional 1Gbps link with 50% link utilization can generate at most $1Gb/8/512*20 = 5MB$ records per second. It costs about 300MB of memory for saving 1 minute records and 1.46GB of memory for saving 5 minutes records.

The experiments show that keeping data blocks for several minutes is enough. So the memory consumption is affordable in current hardware capabilities. Besides, because a lot of small data blocks are not saved, the simple estimation causes the results to be upper bounds. Other methods can further reduce memory consumption. For example, flows which are too small or too short are unnecessary to be saved since they are unlikely to be P2P downloading flows.

# 4 Evaluation

## 4.1 Experiments Setup

Two metrics are used in the evaluation [20]. The first one is the True Positive (TP). It is used to measure the traffic fraction that can be recognized by the algorithm out of the traffic belonging to the given application.

$$TP = \frac{\text{application traffic classified as the applciation}}{\text{Total applciation traffic}} \tag{1}$$

The second one is the False Positive (FP). It is used to measure the traffic fraction which is not really produced by a given application out of the traffic classified as the application.

$$FP = \frac{\text{Non-applicationtraffic classified as the applciation}}{\text{Total traffic classified as the applciation}} \tag{2}$$

They are very important metrics especially when flow type identification is used in traffic management. For example, if network operators differentiate the service qualities according to the flow types, high false positive or low true positive rates can make high priority traffic suffer from performance degradation. A good algorithm should have low false positive and high true positive rates.

As shown in table 1, a lot of popular P2P applications are evaluated in the experiments. They are classified into three types: P2P file sharing, P2P live streaming and P2P video on demand (VOD).

**Table 1.** P2P applications

| type | applications |
|---|---|
| P2P file sharing | BitTorrent, Emule |
| P2P live streaming | PPLive,PPStream,TVAnt,FeiDian,PPMate,SinaLive,TVULive |
| P2P video on demand | PPLive VOD, XL VOD, PPStream VOD |

To evaluate the method, traffic traces are captured and replayed. Two kinds of traffic traces are used in the experiments. The first ones are the pure application traces. The traces are captured from a host which only has the given application running on it. Each P2P file sharing applications trace contains the traffic generated by downloading several files. Each P2P live streaming or P2P VOD applications trace contains the traffic generated by watching a video. The time slot of each trace file varies from half an hour to 1 hour. The traces are only used to evaluate the True Positive.

The second ones are the mixed traces, which are captured from a host running various applications. Each trace contains the traffic generated by a P2P application and some non-P2P applications, such as HTTP, FTP, and POP3 etc. P2P applications in them are labelled by their payload signatures. In each mixed trace, the P2P traffic accounts for 30%-40% of the total traffic. The traces are used to evaluate both the True Positive and the False Positive. we have generated mixed traces with two P2P applications. They are BitTorrent and PPLive.

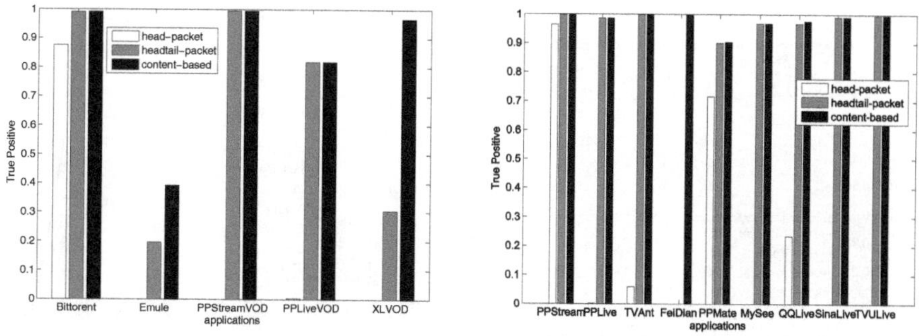


**Fig. 5.** True Positives

### 4.2 Comparing Partitioning Schema

First, the True Positive of different partitioning schemas are studied by applying them on the pure application traces. In the experiments, the threshold $S$ is 256 bytes. To study the best True Positive that each schema can achieve, the size of the download set is unlimited.

True Positives are shown in Fig. 5. Among the three partitioning schemas, performance of the head packet partitioning schema is the worst. Quit a few P2P applications can't be recognized by the schema. Performance of the head tail packet partitioning schema is almost as effective as the content based partitioning schema, except for the FeiDian live streaming application. If the applications have protocol fields in both the head and the tail of the packets, the packet level based partitioning schemas don't work. The content based partitioning schema is more generic.

For most applications, the True positives in bytes are more than 90%, but it is only 40% for the Emule application. There are two reasons. First, Emule will upload the files which have been downloaded are older while data uploading in BitTorrent are fresher. Second, Emule transfers a part of a file with 1300 bytes in a separate packet each time to avoid fragmentation. The start position of the part in the file is specified in a request message. The data pieces exchanged in BitTorrent protocol are globally divided, but the data pieces exchanged in the Emule protocol are not. It is more difficult to find the shared data pieces in flows of the Emule application.

### 4.3 Data Block Size

The threshold $S$ in the content based partitioning schema is important.The effect of threshold $S$ is evaluated in Fig. 6. The size of the download set is unlimited.

The consequences of a smaller threshold $S$ are: the generation of more data blocks; higher memory consumption and increased False Positives. On the other side, the probability of finding the shared data pieces decreases as the threshold $S$ increases.

True Positives of most applications do not decrease significantly when S is smaller than 1024 bytes, except the XLVOD application. It implies that the data

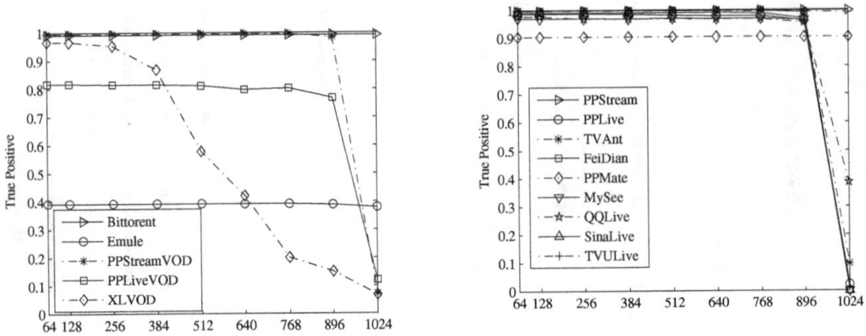


**Fig. 6.** True Positives of different threshold S

pieces of most P2P applications are not less than 1024 bytes. The threshold $S$ of 256 bytes is suggested in the algorithm and used in the following experiments.

## 4.4 Download Set Size

The algorithm keeps recent data blocks in a download set for each host. Saving data blocks for an extended period can improve the True Positive, but it requires a lot of memory.

The effect of the download set size is shown in Fig. 7. The threshold $S$ is 256. The size of the download set is measured in time windows. The size of 1 minute means that the download set will keep data blocks in last 1 minute.

Because peers exchange the video content in a small time window in p2p live streaming applications, the True positives of P2P live streaming applications are much better than P2P file sharing and P2P VOD applications for small download sets. For P2P file sharing and P2P VOD applications, downloaded data pieces can be uploaded after a long time. Keeping the most recent data blocks requires a large time window for the applications.

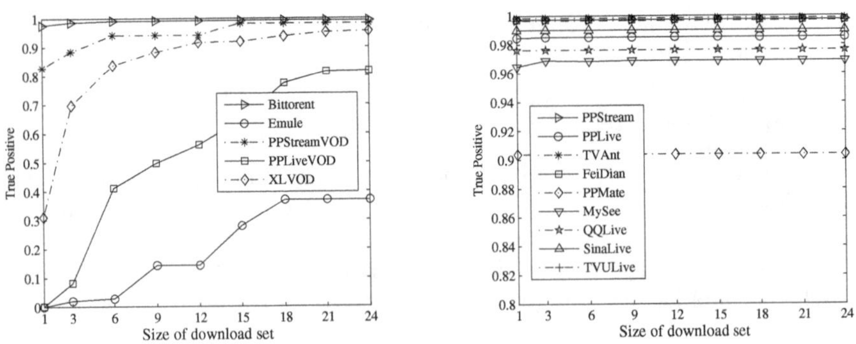


**Fig. 7.** True Positives of different download set sizes

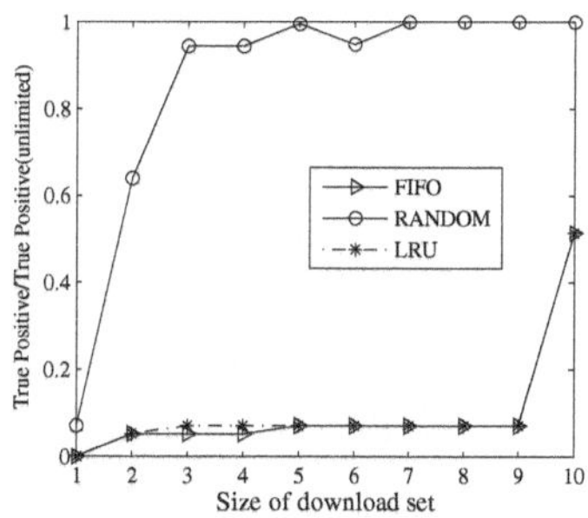


**Fig. 8.** True Positives for Emule

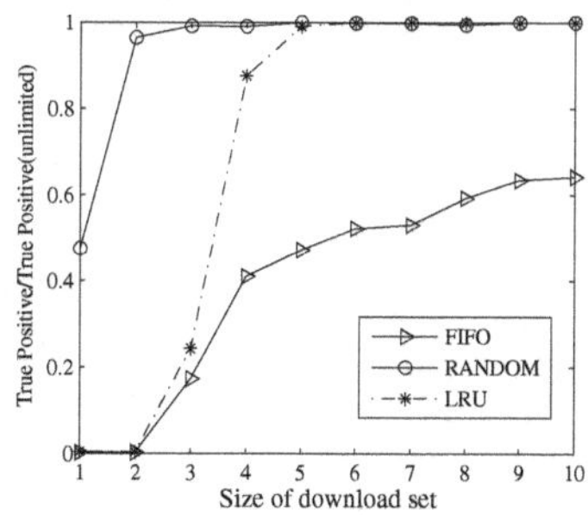


**Fig. 9.** True Positives for PPLive VOD

The performance can be further improved using other replacement policies instead of the current first in and first out(FIFO) policy. There are two other data block replacement policies to consider. They are least recently used (LRU) and random replacement (RANDOM).

We have evaluate the performance of all the P2P applications. Limited by the space, only the result of Emule and PPLive VOD are shown in Fig. 8 and Fig. 9. The x-axis is the size of the download set and the y-axis is the normalized True Positive. It is calculated by dividing the current True Positive by the True Positive of the unlimited download set. If the value is 1, it implies that the performance of a small download set is as good as the one using the unlimited download set.

The experiment results indicate that 1 minute time window is large enough for P2P live streaming applications, each of the three policies work almost the same for them. But for P2P file sharing and P2P VOD applications, the random replacement algorithm works much better than the other algorithms. A download set of a 3 minutes time window is enough for all P2P applications using random replacement policy.

The idea behind the random replacement policy is that, old data blocks are more likely to be kept by the algorithm than other algorithms. Keeping the old data blocks can improve True Positives for P2P file sharing and P2P VOD applications. Besides, the algorithm has a positive bias to big flows, which can also improve True Positives in bytes. A flow transferring more traffic has a higher probability to be recorded and identified in random replacement.

### 4.5 False Positive

The mixed traces are used to evaluate False Positives for BitTorrent and PPLive. The size of the download set is unlimited in the experiment. The results are shown in Fig. 10. The x-axis is the threshold S while the left y-axis is the False Positive and the right y-axis is the True Positive.

As the result, a threshold of 256 bytes can help to guarantee a low False Positive while non-P2P applications are transferring random data. But some behaviors of non-P2P applications may lead to false positives. For example,

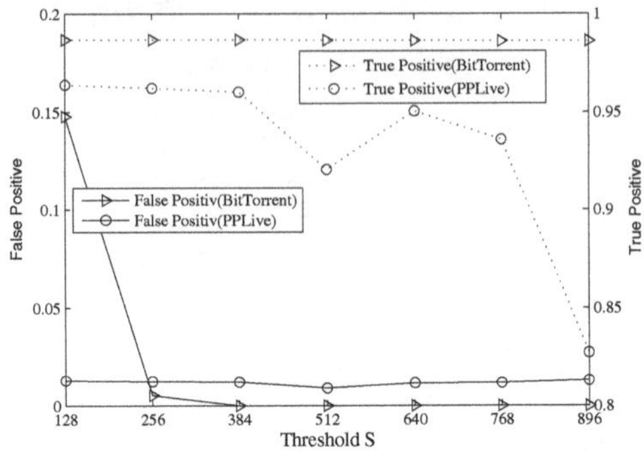


**Fig. 10.** False Positives and True Positives for BitTorrent and PPLive

people forward email they received. Some methods can be used to eliminate these false positives [11]. Since a P2P host always have a service port, we can further identify the P2P {IP, port} pairs which associates with many identified flows. The flows which are not associated with a P2P {IP, port} pairs are filtered to eliminate false positives.

### 4.6 Deploying Place

In the previous experiments, all the communications of the host can be observed and analysed. When deploying the algorithm in the gateway of an institute or an edge router, the communications inside the intra network are unobservable. The performance of absent communications is studied in the experiment. The threshold S is 256 byte, the size of the download set is 3 minutes and the random replacement policy is used in the experiment.

The results are shown in Fig. 11. The x-axis is the fraction of hosts which are unobservable. For example, 0.2 means that flows between the monitored host and 20% of the other hosts are filtered. The missed hosts are selected randomly and the flows are removed from the original trace. The y-axis is the True Positive on the filtered trace.

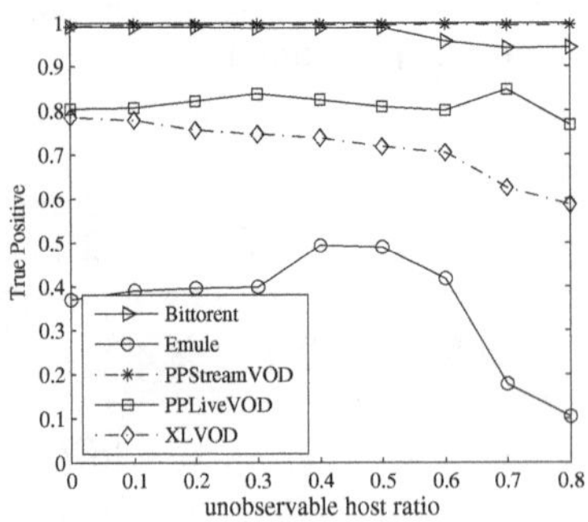


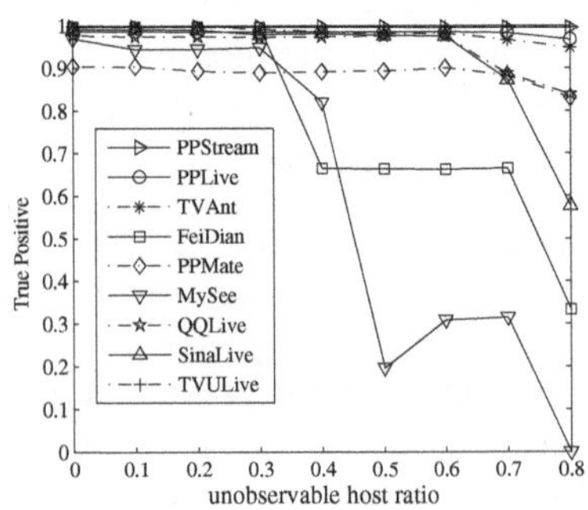


**Fig. 11.** True Positives of different unobservable host ratios

The results indicate that even though a large fraction of hosts (about 30%) are not observed, the algorithm can still achieve a high True Positive. It also implies that the algorithm can work well even when the deploying place is not close to the hosts being inspected, such as the gateways of large institutes and edge routers.

## 5 Conclusion

The paper proposed the content based partitioning schema to identify the P2P data transfer behavior. The schema is generic in identifying P2P applications. Besides, some important issues are studied by experiments. The experiments show that the method can achieve a high True positive and a low False Positive while only keeping a rather small data block set with random replacement policy.

Our future work is to extend our approach to distinguish different P2P application flows by using their relationships in data exchange and flow properties.

## References

1. Moore, A., Papagiannaki, K.: Toward the Accurate Identification of Network Applications. In: Passive and Active Measurements Workshop, Boston, MA, USA, March 31- April 1 (2005)
2. Sen, S., Spatscheck, O., Wang, D.: Accurate, Scalable In-Network Identification of P2P Traffic Using Application Signatures. In: Proc. of ACM WWW 2004 (2004)
3. Haffner, P., Sen, S., Spatscheck, O., Wang, D.: ACAS: automated construction of application signatures. In: Proceedings of the 2005 ACM SIGCOMM Workshop on Mining Network Data 2005, pp. 197–202. ACM, New York (2005)
4. Ma, J., Levchenko, K., Kreibich, C., Savage, S., Voelker, G.M.: Unexpected means of protocol inference. In: Proceedings of the 6th ACM SIGCOMM Conference on internet Measurement 2006, pp. 313–326. ACM, New York (2006)
5. Erman, J., Mahanti, A., Arlitt, M., Cohen, I., Williamson, C.: Semi-supervised network traffic classification. In: Proceedings of the 2007 ACM SIGMETRICS (2007)
6. McGregor, A., Hall, M., Lorier, P., Brunskill, J.: Flow Clustering Using Machine Learning Techniques. In: Barakat, C., Pratt, I. (eds.) PAM 2004. LNCS, vol. 3015, pp. 205–214. Springer, Heidelberg (2004)
7. Zander, S., Nguyen, T., Armitage, G.: Self-Learning IP Traffic Classification Based on Statistical Flow Characteristics. In: Dovrolis, C. (ed.) PAM 2005. LNCS, vol. 3431, pp. 325–328. Springer, Heidelberg (2005)
8. Moore, A.W., Zuev, D.: Internet Traffic Classification Using Bayesian Analysis Techniques. In: ACM SIGMETRICS (2005)
9. Williams, N., Zander, S., Armitages, G.: A Preliminary Performance Comparison of Five Machine Learning Algorithms for Practical IP Traffic Flow Classification. SIGCOMM Computer Communications Review, 36(5), 5–16 (2006)
10. Lu, X., Duan, H., Li, X.: Identification of P2P traffic based on the content redistribution characteristic. In: Communications and Information Technologies, 2007. ISCIT 2007, pp. 596–601 (2007)
11. Karagiannis, T., Broido, A., Faloutsos, M.: Transport Layer Identification of P2P Traffic IMC (2004)

12. Karagiannis, T., Papagiannaki, K., Faloutsos, M.: BLINC: Multilevel Traffic Classification in the Dark. In: ACM SIGCOMM, Philadelphia, PA (August 2005)
13. Hu, Y., Chiu, D.M., Lui, J.C.S.: Application Identification based on Network Behavioral Profiles. In: IEEE IWQoS (2008)
14. Muthitacharoen, A., Chen, B., Mazieres, D.: A low-bandwidth network file system. In: Proceedings ofthe 18th ACM Symposium on Operating SystemsPrinciples (SOSP 2001), Banff, Canada, pp. 174–187 (2001)
15. Kim, H., Karp, B.: Autograph: Toward automatic distributed worm signature detection. In: Proc. of the USENIX Security Symp. Diego, pp. 271–286 (2004)
16. Rabin, M.O.: Fingerprinting by Random Polynomials. Tech. Rep.TR-15-81, Center for Research in Computing Technology, Harvard University (1981)
17. Broder, A.Z.: Some applications of Rabin's fingerprinting method. In: Capocelli, R., De Santis, A., Vaccaro, U. (eds.) Sequences II: Methods in Communications, Security, and Computer Science, pp. 143–152. Springer, New York (1993)
18. Gkantsidis, C., Miller, J., Rodriguez, P.: Comprehensive view of a live network coding P2P system. In: Proceedings of the 6th ACM SIGCOMM Conference on internet Measurement, IMC 2006, pp. 177–188. ACM, New York (2006)
19. Bonfiglio, D., Mellia, M., Meo, M., Rossi, D., Tofanelli, P.: Revealing skype traffic: when randomness plays with you. SIGCOMM Comput. Commun. Rev. 37(4), 37–48 (2007)
20. Salgarelli, L., Gringoli, F., Karagiannis, T.: Comparing traffic classifiers. SIGCOMM Comput. Commun. Rev. 37(3), 65–68 (2007)

# Where Is My Peer? Evaluation of the Vivaldi Network Coordinate System in Azureus
## (Work in Progress)

Moritz Steiner[1,*] and Ernst W. Biersack[2]

[1] Bell-Labs / Alcatel Lucent
791 Holmdel-Keyport Rd
Holmdel, NJ 07733, USA
moritz@bell-labs.com
[2] Eurecom
2229, route des Crêtes
06904 Sophia–Antipolis, France
erbi@eurecom.fr

**Abstract.** Network coordinates allow to estimate the latency among a large number of hosts in a scalable way. Recently, Azureus, a popular implementation of BitTorrent, has implemented network coordinates. We have developed a crawler that allows us to obtain from the network coordinates over one hundred thousand peers running Azureus and to measure the network and application level round trip times to these peers.

Our measurements confirm that network coordinates allow to correctly estimate the round trip time between two peers. Our measurements also show that the round trip times from our crawling host to a set of peers located in the same country can vary between a few tens of milliseconds to more than one second. This high variance is due to the large buffers in the ADSL access links, which can increase the round trip time by hundreds of milliseconds. As a consequence, network coordinates and round trip estimations in general cannot be used to select peers that are "nearby", such as peers connected to the same ISP or located in the same country.

**Keywords:** peer-to-peer, measurement, network coordinate system.

## 1 Introduction

Internet Coordinate Systems [1,2,3,4] are very popular today, since selecting nodes based on their location in the network is a basic building block for many distributed systems. The Euclidean distance between the coordinates of two hosts is used as an estimator of the round trip time between these hosts; the actual measurement needs not to be done. In this paper, we consider Azureus whose peers use the network coordinate system Vivaldi [1] to compute their

* This work was done while the first author was working towards his Ph.D. at Eurecom.

L. Fratta et al. (Eds.): NETWORKING 2009, LNCS 5550, pp. 145–156, 2009.

network coordinates. With the help of these coordinates, they try to find other peers physically close to them in order to reduce download times and to keep the traffic local, e.g. inside an ISP or a country.

Our study aims at evaluating the Vivaldi coordinate system in "action". By crawling Azureus, we collect round trip times at the application layer and at the network layer of several hundred thousand of clients around the world. We also collect their Vivaldi coordinates and analyze their accuracy. Previously, only the work of Ledlie et al. [5] evaluated the Vivaldi coordinate system in the "wild" and introduced what is known as the *second version* of the Vivaldi coordinate system. However, all the measurements in [5] are performed from PlanetLab nodes. Since most users are connected to the Internet with ADSL, we feel that these results are biased (cf. Sect. 4.3).

As we will discuss in detail, our results indicate that the Vivaldi network coordinates of Azureus are well suited to predict the round trip times between two Azureus hosts. However, Vivaldi network coordinates often do not allow to infer the geographical distance between hosts. Extremely long round trip times of several seconds are a strong indication for heavily loaded ADSL links and not for a large distance between those peers. Due to this fact, groups of peers in geographical proximity, e.g. countries or ISPs, are not reflected in the coordinate space and it is not possible to find close by peers based on their Vivaldi coordinates in Azureus.

In Sect. 2, we give a brief background on the use of Kademlia [6] based peer-to-peer systems and on the Vivaldi network coordinates system. In Sect. 3, we detail our measurement methodology to learn about the peers present in the Azureus network, to get their Vivaldi coordinates, and to measure the application layer and the network layer round trip times to those peers. In Sect. 4, we examine how well the application layer round trip times used by Azureus to compute the Vivaldi coordinates are correlated to the network layer round trip times. Moreover we analyze the accuracy of the network coordinates. Finally, in Sect. 5, we conclude and recall an alternative approach to find close by peers.

## 2 Background

### 2.1 Azureus

Distributed Hash Tables (DHTs) map a large identifier space onto the nodes that participate in the system in a deterministic and distributed way. The DHT Kademlia [6] is implemented by several peer-to-peer applications such as Azureus [7], Overnet [8], eMule [9], aMule [10], and lately the Storm worm [11]. The two open–source projects eMule and aMule share the same implementation of Kademlia and they do have the largest number of simultaneously connected users, 3 – 4.5 million users [12], followed by Azureus with about 1 million users.

### 2.2 Vivaldi Network Coordinates

Internet coordinate systems allow a host to predict the round trip times to other hosts without actually measuring them. Explicit measurements are often unattractive, because the cost of measurement can outweigh the benefits of exploiting proximity information. Coordinates are assigned to hosts such that the distance between their coordinates predicts the RTT between these hosts. Simulation–based systems map nodes and latencies into a physical system whose minimum energy state determines the node coordinates. Vivaldi [1] calculates the coordinates as the solution to a spring relaxation problem.

The system envisions a spring between each pair of nodes with the resting position of the spring equaling the network latency between the pair. Each node updates and refines its own position successively by taking into account newly reported RTT measurement toward its communication partners. Since this information is piggybacked on other network messages, e.g. route requests, no additional messages are sent through the network. In other words, nodes allow themselves to be pulled or pushed by a connected spring. Vivaldi attempts to minimize the potential energies over all springs.

Vivaldi uses Euclidean coordinates of $d$ dimensions augmented with a height value: $x = x_1, \dots, x_d, x_h$. The coordinates without the height vector can be seen as reflecting the distance across the high–speed Internet core to which the end users are attached. The last mile that may suffer from queuing delays due to large buffers, as it is the case for ADSL, is represented by the height value. Without using the height, the coordinate space would be erroneous since it is possible to measure latencies in the order of seconds between peers in the same country, which is more than the propagation time needed to make the tour of the globe (≈500 ms). To calculate the distance between two nodes $x$ and $y$, first the distance of their Euclidean coordinates is calculated and then the heights of both nodes are added:

$$Vivaldi\ distance(x, y) = \sqrt{\sum_{i=1}^{d} (x_i - y_i)^2} + x_h + y_h$$

For further details on Vivaldi we refer the reader to [1].

# 3 Measurement Methodology

## 3.1 Crawlers

The first step in order to do any latency measurements is to learn about the hosts we want to measure to. Therefore, we used our crawler Blizzard for KAD [12] and adapted it to work for Azureus as well. For each peer $P$, our crawler logs the time of the crawl, the IP address of $P$, the port number used for the control traffic, the DHT ID of $P$, and the Vivaldi network coordinates of $P$.

The implementation of Blizzard is straightforward: it runs on a local machine and starts by contacting a seed which is run by us. The crawler asks the seed peer for a set of peers to start with and uses a simple breadth first search and iterative queries. It queries the peers it already knows in order to discover new peers.

At the beginning of each crawl, the number of newly discovered peers grows exponentially before it asymptotically approaches the total number of peers in the system. The crawl is done when no new peers are discovered and all the discovered peers have been contacted.

The crawler implements two asynchronous threads: one thread sends the `REQUEST_ FIND_NODE(id)` messages and the other thread receives and parses the `REPLY_FIND_NODE` messages returned. A list containing all the peers discovered so far is used and maintained by both threads: the receiving thread adds the peers extracted from the `REPLY_FIND_NODE` messages to the list, whereas the sending thread iterates over the list and sends $n$ (8 to 16) `REQUEST_FIND_NODE(id)` messages to every peer in the list. The value of the DHT ID `id` is different in each message. This is done in order to minimize the overlap between the sets of peers returned.

Not all the peers discovered can be contacted directly by the crawler. Approximately half of the peers queried do not respond to the crawler. There are two main reasons why a peer does not respond to our queries: either the peer has left the system or the peer is behind a NAT that blocks our query. For the crawler, it is not possible to distinguish between these two cases.

To crawl the DHT implemented by a BitTorrent client is a new approach to learn about BitTorrent peers, since to this day the classical approach was to learn about torrents on portal web sites and to contact the corresponding trackers afterwards. Since in BitTorrent every torrent forms a swarm of peers, all existing torrents need to be tracked in order to get a full view on all clients participating. Our approach is not based on torrents, but we make use of the DHT in which all peers participate and in which information about all torrents is published. For the moment, this DHT is used as a fall-back mechanism in case the tracker goes down. Not all DHTs implemented by the different BitTorrent clients are compatible with each other. Therefore, with our method of crawling a DHT, we do not learn about all peers using BitTorrent, but only about those using a DHT compatible to the DHT of Azureus.

### 3.2 Application Layer Round Trip Time

The crawler exploits the control messages of the DHT and uses the messages intended for routing in order to learn about other peers. The *application layer*

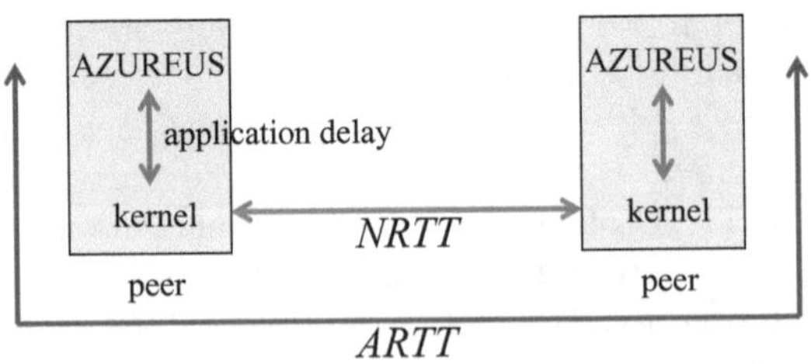


**Fig. 1.** The ARTT is composed of the NRTT and the additional delay induced by the application

*round trip time* (ARTT) of those messages is composed by two parts: the network layer round trip time (NRTT), which will be explained in the next paragraph, and the additional delay induced by the computation of the application that depends on the load of the end-system (Figure 1). Therefore, we always expect the application layer round trip time to be larger than the network layer round trip time. In the remaining of this chapter, the round trip time is always expressed in milliseconds.

### 3.3 Network Layer Round Trip Time

The REPLY_FIND_NODE message contains the IP address and the port of a peer. After the discovery of a peer, a TCP packet with the ACK flag set is sent on this very port. The peer is expected to reply with another TCP packet having the RST flag set [13]. We call the delay between the emission of the TCP ACK and the reception of the TCP RST the *network layer round trip time* (NRTT). The TCP ACK packet we send is directly processed by the kernel of the operating system of the queried peer. Therefore, we expect the network layer round trip time to be equal to the round trip time of an ICMP ping. We send a TCP ACK instead of an ICMP ping since some network providers filter ICMP packets and most (Wlan-) routers and personal firewalls do not reply to ICMP pings. Moreover, by default, all ports are closed, which is why it is so important to learn about an open port by first crawling the peer-to-peer network.

### 3.4 Network Coordinates

The reply messages received during a crawl also contain the Vivaldi [1] network coordinates of the queried peer. Depending on the version of the Azureus client, different versions of the Vivaldi network coordinates are communicated: none, version 1 (made of 2 dimensions plus the height), or version 1 and version 2.4 (made of 4 dimensions plus the height). The difference between the two implementations of the Vivaldi network coordinate system is not limited to the number of dimensions, also the age of the coordinate is transmitted. Moreover, the ways new measurements are used to update the coordinates have become more sophisticated. In total, 16 additional bytes are transmitted for version 2.4. See [5] for details about the different versions of the network coordinates. 97.2% of the Azureus clients use the latest version of the DHT protocol which transmits both version 1 and version 2.4 of the network coordinates. Therefore, we considered only these peers.

We run an Azureus client on the machine performing the crawl in order to learn about the network coordinates of the crawler itself. Using those coordinates, we are able to compute the Vivaldi distance between the crawler and the queried peers. It is expected to approximate the round trip time of the packets sent to this peer. Since the Azureus application is not aware of the NRTT, but only of the ARTT, we expect the Vivaldi distance to be more tightly correlated with the ARTT than with the NRTT.

## 4 Results

For the analysis presented in this paper, we collected the following datasets.

- **Mannheim:** The crawler is at the University of Mannheim, which is attached to the German research network. One single crawl performed on March 31, 2008, at 08:00 CET. 1,044,155 peers have been discovered, 291,850 of them responded to the crawler (Azureus ARTT and network coordinates are available). 157,205 peers replied to the TCP ACK. For those peers, the full data is available. The crawl duration was 12 minutes. This dataset is available online:
`http://www.eurecom.fr/~btroup/vivaldiazureus/`.
- **Eurécom:** The crawler is at Eurécom, which is attached to the French research network. Starting on 15th of February, 2008, we performed 3 full crawls of the Azureus network a day (05:00, 13:00, and 21:00 CET). On each crawl, 1 – 1.4 million were discovered. About 300,000 – 400,000 of them responded to the crawler, thus, of those peers, Azureus ARTT and network coordinates are available. For about 50% of the responding peers, the NRTT is also available, the other peers did not respond to the TCP ACK packet sent. Each crawl has a duration of about 20 minutes.
- **ADSL:** The crawler is connected via an ADSL line. One single crawl was performed on March 26, 2008. This crawl took 12 hours and, out of 1,267,822 discovered peers, 118,548 peers responded. Whereas the NRTTs are only available for 37,346 of those peers.

Table 1 gives an overview of the results obtained on Azureus in the Mannheim dataset. The results of the two other vantage points are omitted for space constraints since they are qualitatively very similar. We mapped the IP addresses

**Table 1.** Overview of the Azureus results: ARTTs, NRTTs, and coordinate distances in milliseconds. Measurement host is located in Mannheim.

| 1 | 2 | 3 | 4 | 5 | 6 | 7 | 8 | 9 | 10 |
|---|---|---|---|---|---|---|---|---|---|
| Country | # Clients | ARTT | NRTT | | | | | Coordinates | |
| | | mean | mean | st. dev. | 5th perc. | median | 95th perc. | v2 | v1 |
| France | 13,775 | 359 | 306 | 626 | *44* | 92 | 1,235 | 542 | 344 |
| Germany | 11,439 | 435 | 236 | 598 | *21* | 64 | 1,143 | 736 | 415 |
| Spain | 8,281 | 641 | 566 | 984 | *55* | 125 | 2,604 | 1,043 | 581 |
| Italy | 3,464 | 389 | 325 | 671 | *57* | 119 | 1,286 | 560 | 368 |
| Canada | 12,349 | 360 | 298 | 512 | 120 | 169 | 948 | 454 | 259 |
| US | 32,528 | 394 | 319 | 543 | 111 | 176 | 1,052 | 488 | 269 |
| Venezuela | 530 | 851 | 765 | 1,175 | 169 | 258 | 3,299 | 1,197 | 657 |
| Brazil | 2,364 | 776 | 718 | 1,067 | 233 | 312 | 2,828 | 1,053 | 598 |
| China | 315 | 563 | 513 | 302 | 324 | 413 | 1,101 | 656 | 381 |
| Korea | 362 | 413 | 377 | 134 | 308 | 346 | *563* | 372 | 231 |
| Japan | 1,283 | 443 | 370 | 358 | 281 | 300 | *586* | 466 | 246 |
| Australia | 3,733 | 934 | 872 | 1,326 | 340 | 392 | 3,729 | 1,162 | 634 |
| All countries | 157,205 | 454 | 375 | 739 | 37 | 151 | 1,541 | 647 | 376 |

of the peers to their countries using Maxmind [14], a database containing geolocation information. In the table, we list several countries from different parts of the world that are representative for their continents: countries close to our crawling site, countries far away, countries with good and with poor Internet connectivity. The second column shows the number of unique Azureus clients measured, the third column shows the average ARTT followed by 5 columns for the NRTT. Columns 9 and 10 show the average Vivaldi distance from our crawling site to the Azureus peers.

The mean NRTT value does not allow to distinguish between continents, compare 566 ms for Spain with 319 ms for the US or 377 ms for Korea. In fact, the *mean* NRTT is strongly biased by outliers, as can be seen from the 95th percentile (Column 8 and Fig. 2). A much better indicator for geographic proximity is the 5th percentile: 21 to 57 ms for European countries, 111 to 120 ms for North America, 169 to 233 ms for South America, and 281 to 340 ms for Asia and Australia. Even if it is possible to make a difference between the NRTT distributions for different continents, this does not imply that, from the NRTT to a single peer, one can deduce its continent of origin.

We do explain the very low variance for peers in Korea with the widely deployed fiber to the home in these countries. The high variance of the NRTTs, e.g. in Europe, is introduced by the last mile to these users that are often connected with ADSL [15]. The ADSL access links own buffers that can add an additional delay ranging from tens of milliseconds to more than a second.

Compared to the mean NRTTs (Column 4), the mean ARTTs (Column 3) are slightly higher and they show a higher variance, which reflects the additional delay introduced by the Azureus application. However, the overall shape of the corresponding cumulative distribution function per country remains the same. In Fig. 2, the measured CDF of the NRTTs are plotted.

## 4.1 Network Coordinates

We compared the calculated Vivaldi distances for both implementations of the coordinate system to the application round trip time measurements and to the network layer round trip time measurements we performed.

The ARTTs and the NRTTs do have a positive linear correlation of 0.88 (Table 2, column 2). For some countries, such as Germany and Japan, the correlation

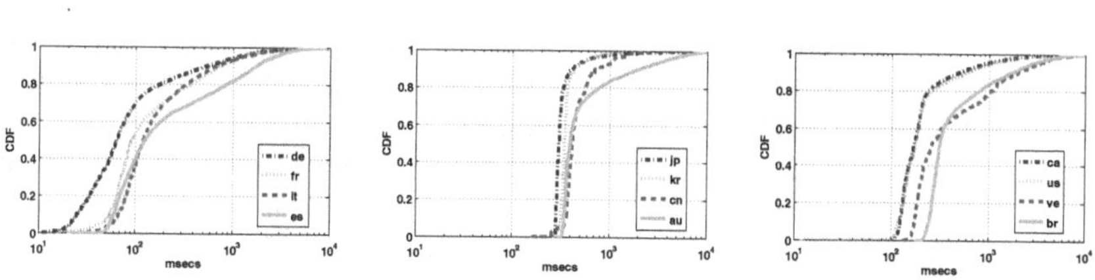


(a) Europe (b) Asia (c) America

**Fig. 2.** NRTT for different countries. Origin of the measurements is in Mannheim / Germany.

**Table 2.** Correlations of the results for RTTs and Vivaldi distances shown in table 1

| 1 | 2 | 3 | 4 | 5 | 6 | 7 |
|---|---|---|---|---|---|---|
| | ARTT NRTT | v.1 v.2 | v.2 ARTT | v.2 NRTT | v.1 ARTT | v.1 NRTT |
| France | 0.94 | 0.89 | 0.91 | 0.89 | 0.85 | 0.84 |
| Germany | *0.65* | 0.91 | 0.92 | 0.63 | 0.87 | 0.61 |
| Spain | 0.93 | 0.94 | 0.94 | 0.89 | 0.91 | 0.87 |
| Italy | 0.90 | 0.92 | 0.91 | 0.89 | 0.85 | 0.84 |
| Canada | 0.85 | 0.81 | 0.83 | 0.81 | 0.77 | 0.74 |
| US | 0.81 | 0.86 | 0.85 | 0.82 | 0.79 | 0.74 |
| Venezuela | 0.97 | 0.84 | 0.93 | 0.94 | 0.79 | 0.80 |
| Brazil | 0.94 | 0.90 | 0.92 | 0.93 | 0.87 | 0.88 |
| China | 0.75 | 0.80 | 0.93 | *0.54* | 0.77 | 0.63 |
| Korea | 0.94 | 0.74 | 0.83 | 0.80 | 0.60 | 0.59 |
| Japan | *0.48* | 0.90 | 0.93 | 0.47 | 0.93 | 0.47 |
| Australia | 0.98 | 0.87 | 0.91 | 0.91 | 0.85 | 0.84 |
| All countries | 0.88 | 0.89 | 0.90 | 0.84 | 0.84 | 0.80 |

between the ARTTs and the NRTTs is much lower than compared to other countries. Therefore, the correlation between the ARTTs and the distances computed with the coordinates are lower, too. The weak correlation for the German peers is not due to the measurement origin being in Germany, the other two datasets also confirmed these results.

The correlation between the two versions of the network coordinates is 0.89 (Column 3). This strong correlation indicates that the two additional dimensions introduced in version 2 do not have a big impact. The correlation between the coordinates version 1 and the ARTTs is 0.84 (Column 6), for version 2 this increases to 0.90 (Column 4). The correlation between the network coordinates version 1 and the NRTT is 0.80 (Column 7), for version 2 this value increases to 0.84 (Column 5). We can conclude that the two additional dimensions, the introduced age of the coordinates, and the resulting additional overhead in version 2 do not result in a significant improvement of the network coordinates' accuracy.

Our direct measurements of the NRTT are all taken from hosts based in Europe, thus the CDFs for the different countries do all have Europe as a point of origin (Figure 2). To get a different point of origin, we need to make use of the *network coordinates*. We chose an Azureus client in the US and computed its Vivaldi distance to all the other peers. In Fig. 3, the Vivaldi distances of this US peer to peers in Japan, Canada, and Germany are plotted. Surprisingly, in most of the cases, it is not the peers in Canada that appear closer to peers in the US, but the peers in Japan. Thus, based on Vivaldi distance, a peer located in the US would often prefer a Japanese peer over a Canadian peer.

The CDFs of the coordinate distances of a peer based in Germany have a shape similar to the CDFs of the NRTTs shown in Fig. 2.

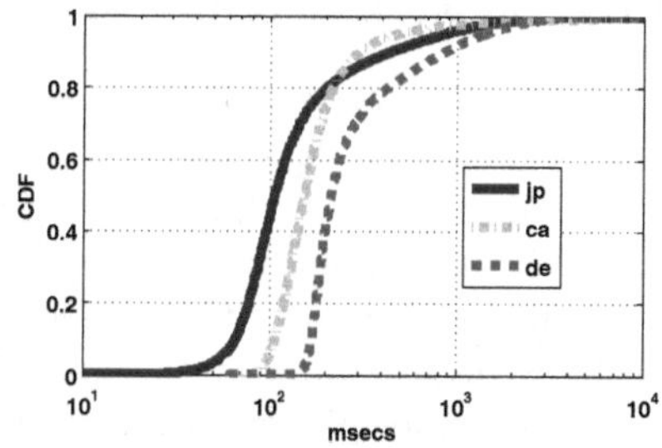


**Fig. 3.** Vivaldi distances for different countries. Origin peer is in the US.

## 4.2 Height

The height value in the network coordinates should reflect the latency introduced by buffers on the last mile toward the end-user. The usage of the height is necessary in order to not distort the coordinate system by large latencies introduced on ADSL lines. We extracted all 112 peers from our dataset that are customers of the provider France Télécom and that are located in Nice / France. The propagation delay between any pair of those clients is in the order of a few milliseconds.

The Vivaldi distance between pairs of those peers is very short if the height value is discarded given the small geographic distances between the hosts. Considering the height, however, they can be far away from one to another. In Fig. 4(a), the CDF of the pairwise distances is plotted. We see that the latency introduced by the ADSL links is completely reflected in the height value and not in the coordinates. For version 2, the results are even better than for version 1, the distance ignoring the height is only of 38 milliseconds in median.

Figure 4(b) shows the pairwise Vivaldi distances of all 12,438 Azureus peers in France. Again, the distances without the height do reflect the geographic distances, whereas the distances including the height are of one order of magnitude larger due to the queuing delay on the last mile.

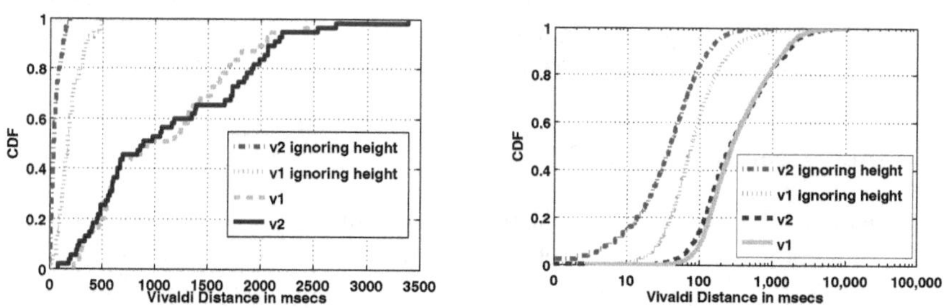


(a) Clients of France Télécom in Nice (b) All clients in France

**Fig. 4.** CDF of the pair wise distances of Azureus clients, with and without considering the height value of the coordinates

### 4.3 Visual Check of the Network Coordinates

Since a network coordinate system assigns coordinates based on the measured latency between the hosts, one should expect to see clusters of peers when plotting the coordinates that correspond to peers on different continents. Theses clusters should be separated by gaps, the oceans. In Fig. 5(a), we plotted the network coordinates version 1 (without the height) of the German and the Australian peers. There is a strong overlap between them, no clear separation and no gap. This is the same for other pairs of countries plotted together. In Fig. 5(b), the peers' coordinates of all countries are plotted. It is not possible to distinguish between countries or even continents. Geographical distances are not at all represented.

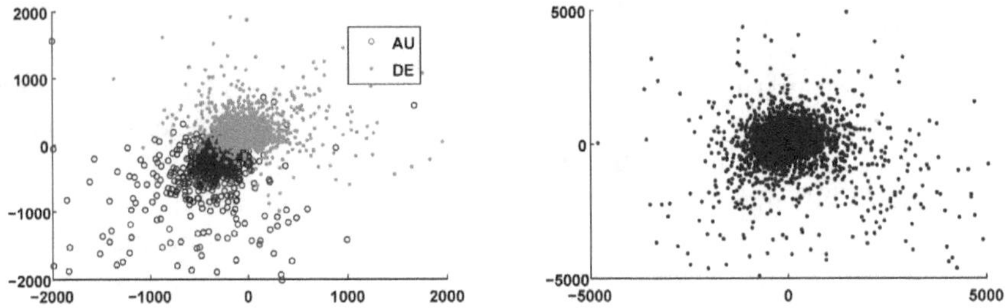


(a) Germany and Australia: overlap, no separation, no gap.

(b) All countries: no distinction possible.

**Fig. 5.** Azureus network coordinates version 1

These findings are in contradiction to Ledlie et al. [5] who claim that embedding the Internet, which interconnects peers across the globe (the Earth), into an Euclidean space works well due to the fact that traffic between Europe and Asia is routed via the US. The same paper also states that the peers of different continents (Asia, Europe, and North America) cluster together in the network coordinate space, which is in clear disaccord to our findings. In a technical report [16] by the same group, snapshots are presented (Figure 11) of the coordinates of the peers they run on PlanetLab where three clusters representing the continents are distinguishable. We believe that such results can be obtained on PlanetLab but not with peers that are connected via ADSL.

### 4.4 Which Peer to Choose?

The classical use of an Internet coordinate system is to choose the peers from which to download from. To check if the coordinate system implemented in Azureus fulfills that request, we set up a very simple experiment. We dumped the Vivaldi coordinates of 2 peers (running on 2 machines in the same LAN) at Eurécom, 2 peers (running on 2 machines in the same LAN) at the University of Mannheim, and one connected via a France Télécom ADSL line every 5 minutes for 9 days, starting on February 25, 2008. Using the coordinates, the peers located

**Table 3.** Vivaldi distances between two peers. The first column indicates the location of the two peers, the second column indicates the NRTT between them which is stable.

| peers | NRTT | version 1 | | | version 2 | | |
|---|---|---|---|---|---|---|---|
| | ping | avg. | st.dev. | max | avg. | st.dev. | max |
| ma 1 / ma 2 | 0 | 133 | 117 | 1,208 | 121 | 56 | 336 |
| eur 1 / eur 2 | 0 | 182 | 204 | 2,606 | 140 | 72 | 414 |
| ma 1 / eur 2 | 39 | 158 | 120 | 786 | 95 | 45 | 302 |
| ma 2 / eur 1 | 39 | 144 | 171 | 2,563 | 167 | 71 | 390 |
| eur 1 / adsl | 70 | 177 | 126 | 1,066 | 108 | 37 | 241 |
| ma 1 / adsl | 70 | 179 | 121 | 981 | 106 | 37 | 269 |

in Mannheim, respectively Eurécom, should be able to choose the other peer in the same LAN.

In the following, we computed the Vivaldi distances between those peers (Table 3). When using network coordinates, the distance values for the different pairs of peers make these pairs practically indistinguishable, or, as in the case of version 2, make two peers in Eurécom / Mannheim look closer to each other than peers that are adjacent.

## 5 Conclusion and Outlook

We have studied the Vivaldi network coordinate system currently implemented in Azureus and evaluated its possibilities and limitations. We saw that the latencies estimated using Vivaldi network coordinates exhibit a high correlation with the round trip times at application layer (ARTT) and to a lesser degree with the round trip times at network layer (NRTT).

In general, the Vivaldi coordinates are not suitable for selecting close-by (geographically or within same ISP) peers: The round trip time is composed of three elements: propagation delay, transmission delay, and queuing delay. As many peers are connected to the Internet via ADSL, the queuing delay of the ADSL access link can dominate the round trip time and hide the contribution of the geographical distance completely, which is reflected in the propagation delay. Extremely long round trip times of several seconds are a strong indication for heavily loaded ADSL links and not for a very large distance between those peers. Due to this fact, groups of peers in geographical proximity, e.g. same country or ISP, cannot be determined using Vivaldi coordinates.

For an interesting approach to locate "close-by" peers, we refer the reader to recent work of Choffnes and Bustamente [17] who developed an Azureus plug-in called Ono. The plug-in builds a coordinate system based on the measurements performed to several landmarks which are edge servers of the Akamai and Limelight CDN networks.

Given this limitation, Vivaldi coordinates are still very useful for peer selection whenever the round trip time has an impact on the performance, as is the case in query routing and when downloading content via TCP connections.

Some of our traces, the dataset called "Mannheim" (cf. Sect. 4), are available under `http://www.eurecom.fr/~btroup/vivaldiazureus/`.

## References

1. Cox, R., Dabek, F., Kaashoek, F., Li, J., Morris, R.: Vivaldi: A Decentralized Network Coordinate System. In: Proceedings of SIGCOMM (2004)
2. Ng, E., Zhang, H.: Predicting Internet network distance with Coordinates-Based Approaches. In: Proceedings of INFOCOM (2002)
3. Tang, L., Crovella, M.: Virtual Landmarks for the Internet. In: Proceedings of the Internet Measurement Conference, IMC (2003)
4. Shavitt, Y., Tankel, T.: Big-Bang Simulation for Embedding Network Distances in Euclidean Space. In: Proceedings of INFOCOM (2003)
5. Ledlie, J., Gardner, P., Seltzer, M.: Network Coordinates in the Wild. In: 4th USENIX Symposium on Networked Systems Design & Implementation, pp. 299–311 (2007)
6. Maymounkov, P., Mazieres, D.: Kademlia: A Peer-to-Peer Information System Based on the XOR Metric. In: Druschel, P., Kaashoek, M.F., Rowstron, A. (eds.) IPTPS 2002. LNCS, vol. 2429, pp. 53–65. Springer, Heidelberg (2002)
7. Azureus: `http://azureus.sourceforge.net/`
8. Overnet: `http://www.overnet.org/`
9. E-Mule: `http://www.emule-project.net/`
10. A-Mule: `http://www.amule.org/`
11. Holz, T., Steiner, M., Dahl, F., Biersack, E.W., Freiling, F.: Measurements and Mitigation of Peer-to-Peer-based Botnets: A Case Study on Storm Worm. In: First Usenix Workshop on Large-scale Exploits and Emergent Threats, LEET (2008)
12. Steiner, M., En-Najjary, T., Biersack, E.W.: A Global View of KAD. In: Proceedings of the Internet Measurement Conference, IMC (2007)
13. Postel, J.: Transmission Control Protocol – Protocol Specification. Request for Comments (Standard) RFC 793, Information Sciences Institute, USC (1981)
14. Maxmind: `http://www.maxmind.com/`
15. Dischinger, M., Haeberlen, A., Gummadi, K.P., Saroiu, S.: Characterizing Residential Broadband Networks. In: Proceedings of the Internet Measurement Conference, IMC (2007)
16. Ledlie, J., Gardner, P., Seltzer, M.: Network Coordinates in the Wild. Technical report, University of Harvard, Cambridge, MA, US (2007)
17. Choffnes, D.R., Bustamante, F.E.: Taming the Torrent: A practical approach to reducing cross-ISP traffic in P2P systems. In: Proceedings of ACM SIGCOMM (2008)

# IP Fast ReRoute: Lightweight Not-Via

Gábor Enyedi[1], Gábor Rétvári[1,*], Péter Szilágyi[1], and András Császár[2]

[1] Department of Telecommunications and Media Informatics
Budapest University of Technology and Economics
Magyar tudósok körútja 2., Budapest, Hungary, H-1117
{enyedi,retvari,szilagyi}@tmit.bme.hu
[2] TrafficLab, Ericsson Research
Laborc utca 1., Budapest, Hungary, H-1037
Andras.Csaszar@ericsson.com

**Abstract.** In order for IP to become a full-fledged carrier-grade transport technology, a native IP failure-recovery scheme is necessary that can correct failures in the order of milliseconds. IP Fast ReRoute (IPFRR) intends to fill this gap, providing fast, local and proactive handling of failures right in the IP layer. Building on experiences and extensive measurement results collected with a prototype implementation of the prevailing IPFRR technique, Not-via, in this paper we identify high address management burden and computational complexity as the major causes of why commercial IPFRR deployment still lags behind, and we present a lightweight Not-via scheme, which, according to our measurements, improves these issues.

**Keywords:** QoS, resilience, IP, fast reroute, redundant trees.

## 1 Introduction

IP has come a long way to become a cost-effective bearing platform for commercial services, providing scalable QoS, point-and-click management, secure VPN services, unpaired scalability, etc. There is, however, an important piece still missing in the puzzle: a resilience scheme capable to treat transient and persistent failures in some tens of milliseconds. Nowadays, IP networks rely on the somewhat outdated resilience scheme built into routing protocols, like Open Shortest Path First (OSPF), hardly fast enough for multimedia applications. Hence, operators resort to working around the limitations of IP, deploying for instance MultiProtocol Label Switching (MPLS) Fast ReRoute, and tolerate the implied boost in capital and operational expenditures.

In response to these challenges, the Internet Engineering Task Force has initiated the IP Fast Reroute framework [1] to introduce fast, local and proactive failure recovery in IP networks. "Local", in this context, means that only routers in the vicinity of the failed component repair, and "proactive" means that detours are precomputed well in advance. To our days, many IPFRR proposals have

* G. Rétvári was supported by the János Bolyai Fellowship of the Hungarian Academy of Sciences.

L. Fratta et al. (Eds.): NETWORKING 2009, LNCS 5550, pp. 157–168, 2009.

come to existence, yet the largest industrial backing is undoubtedly behind the technique based on the notion of "Not-via addresses" [2]. In order to distinguish detoured packets from ordinary packets, so that special routing can be applied to them, Not-via introduces an alternative address space: packets tunneled around the failed component are destined to certain not-via addresses, clearly separable from normal addresses and unambiguously communicating which component the sender believes to be the cause of the failure. This way, detours simplify into shortest paths in a topology with the failed component deleted, and rerouting boils down to pushing the packet to a pre-established IP-in-IP tunnel, a cheap operation now commonly implemented in the fast-path of IP routers. This makes Not-via a practical and easy-to-deploy IPFRR technique.

This paper came into being in reaction to the vast operational experience we gathered on a Not-via-enabled IP testbed deployed at BME-TMIT [3]. Thanks to our prototype system, we are now in a position to be able to thoroughly judge on Not-via's pros and cons. We found that Not-via raises serious address management issues, originating from the need to hand out many not-via addresses, and it poses substantial additional CPU-load on IP routers. This is objectionable, as contemporary IP infrastructure, even without IPFRR, is actually struggling to keep up with the ever-increasing routing tables. The significant additional management and computational cost makes operators reluctant to adopt IPFRR, despite of its potential benefits.

To improve the manageability of Not-via, we present a lightweight Not-via scheme. The main idea is, on the traces of [4], to adopt the concept of node-redundant trees (simply redundant trees in the sequel) for IPFRR and apply them directly to Not-via. As shall be shown, this modification reduces the number of not-via addresses, cuts the computational complexity down to the level of plain shortest path routing, and it removes many corner cases that plague the original Not-via proposal.

The rest of the paper is organized as follows. In Section 2, we discuss Not-via and we summarize our operational experiences. In Section 3, we recast Not-via over redundant trees, we discuss the issue of additional addresses and we report on a related theoretical result: a distributed algorithm which finds next-hops in redundant trees corresponding to all nodes in linear time. In contrast, this was only possible in quadratic time or worse previously. We implemented the modified Not-via in our prototype and in Section 4 we present observations and measurement results we gathered on our testbed. Finally, in Section 5 we conclude the paper.

## 2 IPFRR Using Not-Via Addresses

IP Fast ReRoute attains fast response time by handling failures locally, with only the routers in the vicinity of the failure participating in the repair but other, distant routers not being informed of the failure in any ways. Therefore, IPFRR applies special routing to packets being forwarded along a detour. Otherwise, loops might emerge as a distant router not aware of the failure might blindly loop the detoured packet back along the default forwarding path. Not-via uses the destination address in IP packets to mark whether the packet is being forwarded on the default path or in an IP-to-IP tunnel along a detour. The starting node of the detour is the router whose next-hop has become unreachable, and the tunnel is terminated at the next-next-hop (NNH), the second closest node along the shortest path tree. This facilitates common handling of node and link failures.

Perhaps an example is in order here. Consider the network depicted in Fig. 1, and suppose that a packet entering the network at node $A$ is forwarded to the egress node $C$. Furthermore, assume that the shortest path (marked by bold arrows) goes through node $B$, but $A$ suddenly loses contact with $B$. Now, $B$ encapsulates the packet in a new IP header with a special not-via address set as destination address, which has the semantics "forward me to $C$ (the NNH) not via $B$", and sends it to node $D$. Thanks to the special destination address, $D$ will not send the packet back to $C$ on the shortest path (as it would be the case if default routing applied), but instead sends it to $E$ through LAN $L$. Node $E$ forwards the packet to node $C$ where it is decapsulated and passed further along the default forwarding path as if no failure happened.

Unfortunately, Not-via is a bit more difficult than that, and anyone trying to implement it faces painful exceptions and complex corner cases. Consider, for instance, the so called *LAN problem* that arises when $D$ tries to send a packet to node $E$ using LAN $L$ but LAN $L$ fails. On one hand, node $D$ could assume that all the nodes connected to this LAN failed, in which case it would lose all connectivity to node $E$. On the other hand, if selective fault detection was available on the LAN, then $D$ could distinguish between a LAN failure (when more than one router attached to the LAN becomes unavailable) and multiple single router failures. This would provide more efficient recovery, at the cost of quadratic number of additional not-via addresses to cover all the possible fault scenarios. Similar corner cases arise at the decapsulation point of the detours (the so called *last-hop problem*) and at bridge nodes [2].

Despite these issues, Not-via is still a practical and rather straight-to-the-point solution, therefore, we chose Not-via to base our IPFRR testbed onto. After dealing with all the intricacies of implementing the standard and experimenting with it in operation, we are now feeling confident enough to judge on Not-via's merits and identify some of its pressing limitations.

*Burdening address management:* The first question an implementor inevitably faces is how to assign and distribute not-via addresses. As of this writing, there is no official protocol support for advertising not-via addresses into the routing domain. The situation is worsened by the fact that a not-via address has a compound meaning, as it encodes both a destination node and a component to be avoided, and there is currently no way to communicate this rich semantics between routers. As a work-around, network operators may assign not-via addresses statically, but this is inflexible, subject to human configuration errors and breaks down rapidly as the network increases. Just the sheer number of not-via addresses can pose problems: the simple network of Fig. 1 would require a total of 17 not-via addresses.

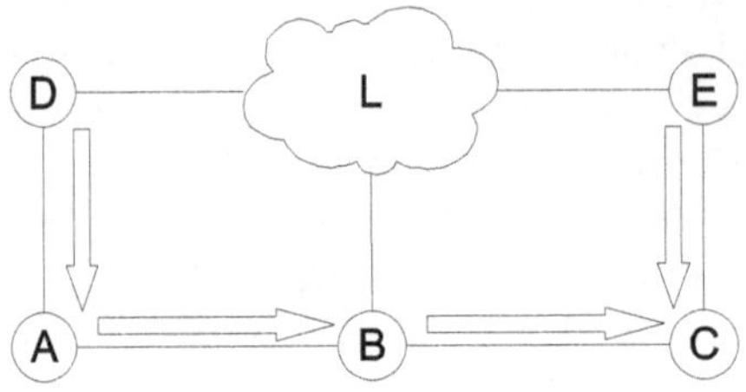


**Fig. 1.** Sample network with IP routers $A$, $B$, $C$, $D$ and $E$ and LAN $L$. Bold arrows mark the shortest path to $C$.

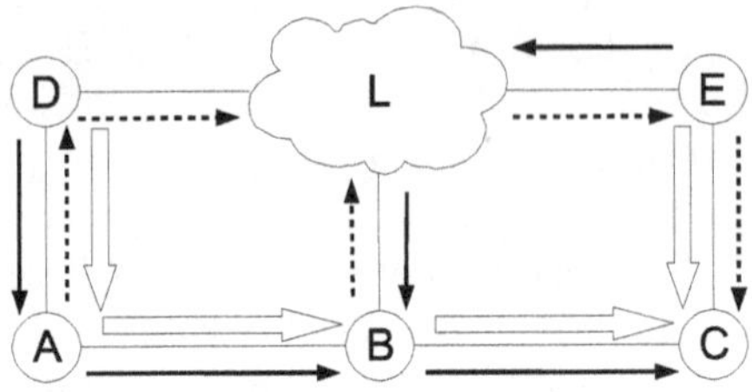


**Fig. 2.** Sample network. Bold arrows mark the shortest path to $C$, dashed arrows mark the primary and solid arrows mark the secondary redundant tree rooted at $C$

*Considerable computational overhead:* In an ordinary IP network, the next-hops towards all destinations are obtained by a single shortest path tree (SPT) calculation. With Not-via, a router must execute as many SPT instances as there are components that can fail, with the failed component removed from the topology. Using some simple heuristics one can go down to some few dozen additional SPT calculations [5], which is still significant. Note that substantial additional costs show up due to having to deal with an increased number of entries in the routing tables, establish, maintain and tear down tunnels, etc.

*Complexity and special cases:* As mentioned above, Not-via brings in subtle intricacies into routing and in many cases it overrides well-known IP routing mechanisms. The corner cases mentioned above make implementations convoluted and operation of the protocol hardly tractable by operators.

In Section 4, we shall support the above claims with measurement results obtained on an operational IP testbed. In addition, we note that similar observations were reported elsewhere [5]. In the next section, we propose deliberate modifications to Not-via in order to remove, or at least mitigate, these compelling issues.

## 3 An Improved Lightweight Not-Via

Our modified Not-via technique uses the concept of redundant trees [6]. Redundant trees are basically a pair of directed spanning trees, which have the appealing property that a single node or link failure destroys connectivity through only one of the trees, leaving the path along the other tree intact. The concept was first applied to IP Fast ReRoute in [4]. In contrast, in this paper we apply redundant trees directly to the prevailing IPFRR technique, Not-via. As shall be shown below, organizing the detours over redundant trees gives rise to an easily implementable and deployable "lightweight Not-via" scheme: it significantly decreases the number of Not-via addresses, with clever modifications it reduces computational complexity to linear, and it eliminates most of Not-via's corner cases without introducing new ones.

### 3.1 Redefining the Semantics of Not-Via Addresses

Our lightweight Not-via uses ordinary shortest paths for default forwarding, and a pair of redundant trees (the *primary* and the *secondary* backup tree) for resilience. Correspondingly, a node $v$ has three IP addresses: a default ($\mathcal{D}_v$), a primary ($\mathcal{P}_v$) and a secondary ($\mathcal{S}_v$). If there is no failure, packets are forwarded along the shortest paths as usual. On the other hand, if a failure shows up, packets are tunneled along either the primary or the secondary tree. This is achieved by encapsulating the packets into a new IP header with the primary (respectively, secondary) address of the Next-next-hop set as the outer destination address. Since, by definition of redundant trees, a single failure leaves at least one of the trees intact, it is guaranteed that packets avoid the failed component.

Consider Fig. 2, depicting the same sample network as before, but now not only the shortest path but also the primary and the secondary backup trees directed towards node $C$ are given (observe that the paths in these trees are node-disjoint). Suppose $A$ has a packet to send to node $C$. As long as its default next-hop, $B$, is alive, $A$ simply passes the packet to $B$. If, however, $B$ goes down, $A$ must find a backup path, or at least a next-hop that can push the packet further, towards $C$. So it encapsulates the packet, sets the outer destination address to the primary backup address of the NNH (node $C$) and passes it to the next-hop along the primary tree, node $D$. Assuming that $D$ computed the exact same redundant tree to $C$ (which is not hard to ensure), $D$ will pass the packet through LAN $L$ and node $E$ to $C$, where it gets decapsulated and sent further. If, instead, it is now node $E$ that has to get a packet to $C$ and it finds that connectivity to $C$ went away, both its shortest path and its primary backup path are affected by the failure. In this case, the packet is encapsulated to the secondary backup path and sent through $L$ to $B$. Note that the secondary backup path can not be impacted by the failure in this case, as it is node disjoint from the primary path. Finally, a packet forwarded along the primary path gets rerouted to the secondary path should it encounter a failure on its path (this might be the very same failure that pushed the packet to the backup in the first place) but not *vice versa*.

The forwarding process of the lightweight Not-via scheme is given in Algorithm 1. Note that the operation `push` $\mathcal{X}$ in routing terminology means "encapsulate the packet into an IP-in-IP tunnel and set its outer destination address to $\mathcal{X}$". The operation $\mathcal{X} \leftarrow$ `pop` does the reverse: decapsulates the packet and puts the address of the innermost IP header to $\mathcal{X}$. $\mathcal{D}$, $\mathcal{P}$ and $\mathcal{S}$ are the address spaces of the default, primary and secondary backup addresses.

It is easy to see intuitively that this forwarding rule is correct. First, in the absence of failures, packets get to their destination along the shortest path as usual. In case of a single failure, a packet first gets to the NNH along either the primary or the secondary backup path, provided that such paths exist, which is always true as long as the network is 2-connected (see more on this matter later). Both backups can not be affected by the failure at the same time, as they are node disjoint. So single node or link failures are handled correctly. Finally, packets can not get into loops in the presence of multiple simultaneous failures,

**Algorithm 1.** Forwarding process at node $u$ for a packet destined to address $\mathcal{A}$, given the set of unavailable neighbors $F$. The next hop of $\mathcal{X}$ is given by $\mathrm{nh}(\mathcal{X})$.

**if** $\mathcal{A} = \mathcal{P}_u$ or $\mathcal{A} = \mathcal{S}_u$ **then** # This is the end of the tunnel
    $\mathcal{A} \leftarrow$ pop;
**end if**
**if** $\mathcal{A} = \mathcal{D}_u$ **then** # This is the destination
    consume the packet; **return** ;
**end if**
**if** $\mathrm{nh}(\mathcal{A}) \notin F$ **then** # Next hop is operational
    forward packet to $\mathrm{nh}(\mathcal{A})$; **return** ;
**end if**
**if** $\mathcal{A} \in \mathcal{D}$ **then** # Default path failed
    let $v$ be the NNH to $\mathcal{A}$;
    **if** $\mathrm{nh}(\mathcal{P}_v) \notin F$ **then** # Forward to primary next hop
        push $\mathcal{P}_v$ and forward packet to $\mathrm{nh}(\mathcal{P}_v)$; **return** ;
    **else if** $\mathrm{nh}(\mathcal{S}_v) \notin F$ **then** # Forward to secondary next hop
        push $\mathcal{S}_v$ and forward packet to $\mathrm{nh}(\mathcal{S}_v)$; **return** ;
    **end if**
**else if** $\mathcal{A} \in \mathcal{P}$ **then** # Primary backup path failed
    $\mathcal{X} \leftarrow$ pop;
    $\mathcal{S}_X \leftarrow$ the secondary backup address for $\mathcal{X}$;
    **if** $\mathrm{nh}(\mathcal{S}_X) \notin F$ **then** # Forward to secondary next hop
        push $\mathcal{S}_X$ and forward packet to $\mathrm{nh}(\mathcal{S}_X)$; **return** ;
    **end if**
**end if**
drop the packet; # Secondary backup failed

as a packet is unconditionally dropped should it encounter a failure along the secondary path.

With this modification, a not-via address protects multiple failures; the primary address protects the default path and the secondary address protects the primary backup. In this way, the number of addresses is decreased to 3 per node, the absolute minimum realizable by the original Not-via only in special topologies (point-to-point rings). What is more, in certain cases it is possible to completely avoiding using extra addresses. In traditional IP networks, routers have a loopback address and a unique address for each interface. Hence, one can use any two interface addresses as the primary and the secondary address. Since these addresses are always disseminated by the IGP, other routers can easily learn them. Naturally, in this case routers should be addressed via their loopback, otherwise traffic destined directly to routers would not be protected. While in a conventional IP network this technique removes the need to maintain additional not-via addresses, it must be emphasized that it is not applicable to any arbitrary IP network. Namely, IP backbones running over unnumbered point-to-point links (e.g., MPLS LSPs) still need to maintain at least two additional addresses per router, since interfaces usually don't have unique IP addresses in such cases.

We have seen previously that the original Not-via proposal has some subtle details, making it difficult to implement it correctly and understand it in operation. Though, redefining Not-via in terms of redundant trees removes most of the corner cases. For instance, LANs no more need special treatment: a LAN is handled like any ordinary node except that it does not get not-via addresses. Additionally, in [7] we show an easy way to tackle the problem of bridge nodes that show up in non-2-connected networks, another corner case in Not-via. Finally, the last-hop problem is treated by simply repairing to the next-hop, similarly to Not-via (as a matter of fact, we have already seen this case when we examined the case of $E$ sending a packet to $C$ and losing connectivity to it).

### 3.2 Reducing the Computational Complexity of Not-Via

The computational cost of Not-via is dominated by the large number of SPT calculations, since an SPT with respect to all potentially failing components needs to be obtained. Suppose there are $N$ nodes and $E$ point-to-point links in a network and there are no LANs. In this case, Not-via's complexity is $O(N(N \log N+E))$ ($N$ times the complexity of Dijkstra's SPT algorithm), which is worse than quadratic in the number of nodes. Unfortunately, without careful modifications the lightweight Not-via would have essentially the same complexity: although a pair of redundant trees comes in linear time, $O(E)$ [8], we need redundant trees with respect to all destination nodes yielding $O(NE)$ steps in general. In this section, we show how to reduce this complexity to $O(E)$ using a simple distributed algorithm.

The idea is that for our lightweight Not-via to work correctly, we do not need the entire redundant tree instances to all destinations, we just need the corresponding next-hops. Thus, we compute a single pair of redundant trees, the primary $P$ and the secondary $S$, rooted at some designated node $r$. Then, for any $d \neq r$ we rewire these trees with $d$ set as root, and we take the corresponding next-hops along the rebased trees. Since computing the initial redundant tree takes $O(E)$ steps and, as shall be shown below, we can decide on the next-hops for a particular node in $O(1)$, the overall complexity is $O(N + E) = O(E)$.

**Proposition 1.** *Let $P$ and $S$ be a pair of redundant trees, rooted at some $r$, and perform the following steps to obtain a graph $D$:*

1. *reverse the edges in $S$*
2. *take the union of the edges of the resultant trees*
3. *split $r$ into two nodes, $r^+$ and $r^-$, so that edges only enter $r^+$ and only leave $r^-$*

*We assume that $P$ and $S$ were so that the graph $D$ yielded by the above steps is:*

*(i) a directed acyclic graph (DAG) and*
*(ii) there is only one edge entering $r^+$.*

While these requirements seem somewhat strong, in reality the majority of the redundant tree algorithms in the literature easily satisfy Proposition 1. We used the linear time algorithm in [8]. An alternative is to modify a redundant tree algorithm so that it immediately produces the DAG (a good candidate would be the algorithm in [9]). Henceforward, we shall assume that the DAG $D$ is at our disposal, it satisfies *(i)* and *(ii)* and it can be computed in linear time.

**Definition 1.** *Let the node set of $D$ be $V$ and define a relation $(\prec)$ on $V$ as follows: $u \prec v : u, v \in V$ if and only if there is a directed path from $u$ to $v$ in $D$.*

It is easy to see that $(V, (\prec))$ makes up a bounded partially ordered set (poset). Because $D$ is a DAG, $(\prec)$ is unambiguous. Additionally, since edges only leave $r^-$, the minimal element is exactly $r^-$. Similarly, $r^+$ is the maximal element.

**Definition 2.** *For some node $u$, let $V_u^+$ be the set of the nodes larger than $u$. Similarly, let $V_u^-$ be the set of nodes smaller than $u$:*

$$V_u^+ = \{v \in V | u \prec v\} , \qquad V_u^- = \{v \in V | v \prec u\} .$$

*Additionally, let $f_u^+(d)$ denote the first hop along some path from $u$ to any $d \in V_u^+$, and $f_u^-(d)$ be the same for any $d \in V_u^-$. For the root node $r$, we define $f_r^+(d) = f_{r^-}^+(d)$ and $f_r^-(d) = f_{r^+}^-(d)$ for all $d \in V \setminus \{r^+, r^-\}$.*

$V^+$ and $f^+(\cdot)$ can be computed by a Breadth-First-Search (BFS) traversal of $D$. Similarly, $V^-$ and $f^-(\cdot)$ come from a reverse BFS. This way, $f^+(\cdot)$ and $f^-(\cdot)$ encode the next-hop along the minimum-hop path, which makes our detours shorter. Note that in general $V_u^+ \cap V_u^- = \emptyset$ but $V_u^+ \cup V_u^- \neq V \setminus \{u\}$, because some nodes might not be ordered with respect to $u$.

**Theorem 1.** *Given nodes $u$ and $d$, $u \neq d$, choose the primary next-hop $h_u^P(d)$ and the secondary next-hop $h_u^S(d)$ from $u$ to $d$ as follows:*

1. *If $d \in V_u^+$: $h_u^P(d) = f_u^+(d)$ and $h_u^S(d) = f_u^-(r^-)$*
2. *If $d \in V_u^-$: $h_u^P(d) = f_u^+(r^+)$ and $h_u^S(d) = f_u^-(d)$*
3. *Else: $h_u^P(d) = f_u^-(r^-)$ and $h_u^S(d) = f_u^+(r^+)$*
4. *Special rules apply at the root node (if $u = r$):*
   *$h_r^P(d) = f_r^+(d)$ and $h_r^S(d) = f_r^-(d)$*

*Then, interleaving the primary next-hops $h^P(d)$ and the secondary next-hops $h^S(d)$ makes up a pair of redundant trees rooted at $d$.*

*Proof.* To prove the theorem, it is enough to show that following the primary and the secondary next-hops comprises two loop-free, node-disjoint paths. The rules encode the intuitive idea: following the next-hops $h^P(d)$ we move in increasing direction in the poset, along $h^S(d)$ in decreasing direction, and if $u$ and $d$ are not mutually ordered, we move downwards in the poset until we can move upwards (and *vice versa*).

First, we show that for two nodes $v, w : v \prec w$, what we obtain by following the primary next-hops $h^P(w)$ is a loop-free $v \to w$ path. Observe that either $w = f_v^+(w)$ and we arrive to $w$ in the next step, or $w \in V_x^+ : x = f_v^+(w)$ and we can step to $h_x^P(w)$ and repeat the same reasoning to eventually arrive to

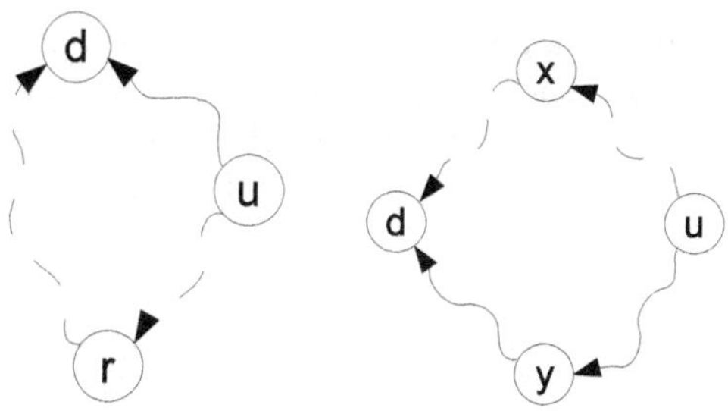

(a) ordered case (b) unordered case

**Fig. 3.** Illustration for Theorem 1

$w$. Along the similar lines, following $h^S(w)$ yields a loop-free $v \to w$ path for $v, w : v \succ w$.

If $d = r$, the claim is trivial. Suppose $d \neq r$ and there is an ordering between $u$ and $d$, say $u \prec d$. Now, following $h^P(d)$ yields an $u \to d$ path $p_p$ (the path marked by solid arrow in Fig. 3a), and following $h^S(d)$ yields first a $u \to r^-$ path $p_s^1$ and then an $r^+ \to d$ path $p_s^2$ (dashed arrow in Fig. 3a). Based on the observation above, these subpaths are indeed paths and they are loop-free. The concatenation of $p_s^1$ and $p_s^2$ gives the secondary path $p_s$. Finally, $p_p$ and $p_s$ are node-disjoint: nodes along $p_p$ belong to the interval $[u, d]$, $p_s^1$ to $[r^-, u]$ and $p_s^2$ to $[d, r^+]$, and these intervals are disjunct except the endpoints.

If there is no ordering between $u$ and $d$, the situation is slightly more difficult: following $h^P(d)$ first yields an $u \to y$ path $p_p^1$ and then a $y \to d$ path $p_p^2$, where $y$ is the first node for which $u \succ y$ and $y \prec d$ holds (see the solid arrows in 3b). Similarly, $h^S(d)$ yields first a $u \to x$ path $p_s^1$ and then an $x \to d$ path $p_s^2$ for the first $x : u \prec x$ and $x \succ d$ (dashed arrows in 3b). Again, concatenation of the corresponding subpaths yields two node-disjoint paths: first, $p_p^1$ and $p_s^1$ are node-disjoint because $p_p^1 \in V_u^-$, $p_s^1 \in V_u^+$ and $V_u^- \cap V_u^+ = \emptyset$; second, $p_p^1$ and $p_s^2$ are also node-disjoint because the nodes of $p_p^1$ are not ordered with respect to $d$ but those of $p_s^2$ are; third, $p_P$ and $p_S$ can not both traverse $r$, because $x \prec r^+$ (due to Proposition 1, condition *(ii)*, we have a node $m$ for which $v \prec m : v \in V \setminus \{r^+, m\}$, so the secondary path turns back in $m$ at the very latest). Similar reasoning applies to see that the rest of the subpaths are mutually node-disjoint too. □

Hence, computing the primary and the secondary next-hops with respect to each node in the network involves first obtaining a pair of redundant trees, then converting them to a DAG using Proposition 1, two BFS traversals to compute $V^+$ and $V^-$ and finally cycling through all nodes to compute the corresponding next-hops using Theorem 1. All these steps can be performed in linear time, therefore, the overall complexity of our method is $O(E)$. Thanks to these modifications, now it takes $O(N \log N + E)$ steps to compute the default next-hops and an extra $O(E)$ steps specific to IPFRR. Therefore, the computational complexity of the lightweight Not-via is dominated by the cost of standard shortest path routing and the additional penalty of IPFRR simply disappears in the long run.

Finally, we point out that basically any protection and restoration scheme relying on redundant trees faces with the problem of finding redundant trees to all destinations at the same time. Therefore, the result that this can be done in linear time might have generic interest beyond IPFRR.

## 4 Performance Evaluation

In this paper, we argue that it is not some deep theoretical limitation or trade-off that hampers the wide-scale deployment of IPFRR the most, but rather a couple of very technical and very concrete practical issues. In order to confirm this claim, we implemented and tested both the prevailing IPFRR proposal, Not-via, and also our lightweight Not-via in an operational IP testbed. Our test system is a full-fledged IPFRR prototype, deployed on 9 PC routers running a stock Debian GNU/Linux distribution, the Open Shortest Path First routing protocol (OSPF) from the `Quagga` suite of routing daemons [10] and `kbfd`, a kernel-based implementation of the Bidirectional Forwarding Detection [11] protocol[1]. Below, we briefly report on some of our most important observations. For a complete coverage on the measurement results, the reader is referred to [3].

Our experiences indicate that IP Fast ReRoute is just what it promises to be: fast. Configuring BFD so that any failure is detected in at most 9 ms (BFD interval = 3 ms, BFD multiplier = 3), Not-via repairs single failures in 16.65 ms on average and 18.5 ms at maximum. With conventional OSPF, on the other hand, one can measure anything between 120 ms and several seconds depending on the actual topology, the nature and the location of the failure, etc.

Our measurements were primarily aimed at identifying the management cost of Not-via. We found that considerable management complexity arises from the need to hand out and maintain vast numbers of not-via addresses. Fig. 4 gives this number for both the original Not-via and our lightweight Not-via, as computed by our prototype system for some commonplace ISP topologies. To simulate the effect of LANs, we treated 20% of the routers as if they were LANs. Observe that with the lightweight Not-via, the number of additional addresses remains modest even in very large topologies. We found, in addition, that the second most important cost of Not-via comes from its considerable computational complexity. Fig. 5 shows the CPU time needed to compute the default and the backup next-hops and downloading them into the forwarding engine.

These measurement results cast Not-via in a completely different light: although the computational complexity of Not-via is substantial, yet it is the extra management burden caused by the extension of the address pool that dominates its complexity. Our measurements reproduce this burden spectacularly even in small and middle-sized topologies, and we expect it to become prohibitive in larger networks. On the other hand, it is exactly this burden where the advantages of the

[1] Our modifications to `Quagga` and `kbfd` are maintained separately at `http://opt.tmit.bme.hu/~kbfd` until all of our upstream patches go into the respective production releases.

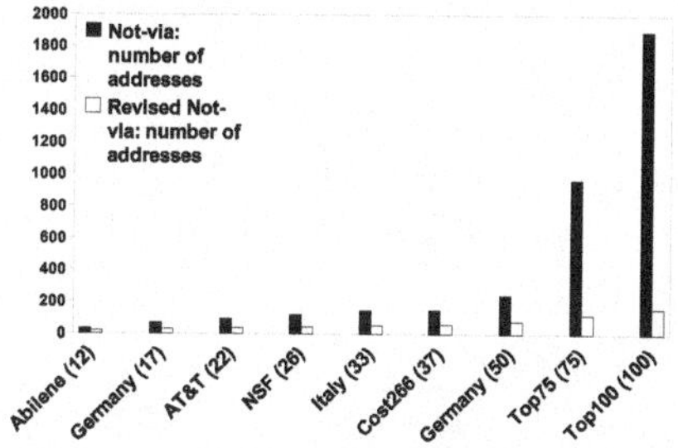


**Fig. 4.** Number of additional addresses for the original and lightweight Not-via in commonplace ISP topologies (number of nodes is given in parentheses), with every fifth node substituted by a LAN

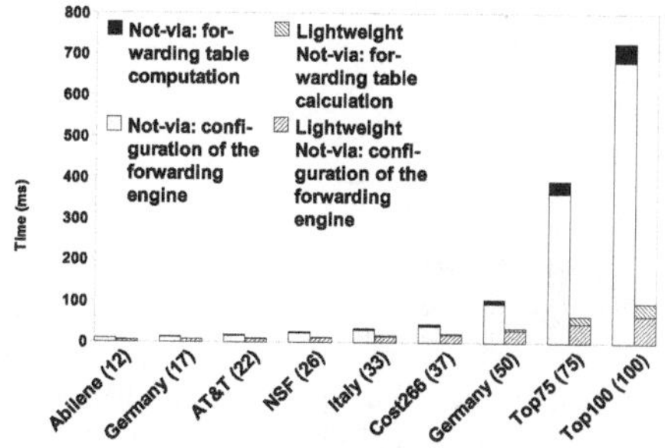


**Fig. 5.** Execution time of computing the default and backup next-hops and configuring the forwarding engine for the original and the lightweight Not-via

lightweight Not-via really manifest themselves: the time of computing the next-hops and configuring the forwarding engine decreases by an order of magnitude into the range of some few hundred milliseconds, which falls well within the time range contemporary IP routers perform ordinary shortest path routing [12].

## 5 Conclusion

IP Fast ReRoute is one of the last missing technological components from the IP protocol suite on its way to become a mature carrier-grade transport technology. In this paper, we argued that, despite of the strong incentives, wide-spread adoption of IPFRR will not occur until the additional cost of IPFRR is reduced to a level tolerable to network operators. To support our claims, we presented a formal performance evaluation of IPFRR obtained on a full-fledged prototype. As far as we are aware of, this is the first time that such an evaluation is published in the literature.

Our measurements showed that the immense number of not-via addresses imposes considerable load on both IP routers and network management. However, by reformulating Not-via in terms of redundant trees we could decrease the number of additional addresses substantially. We also improved the complexity of computing the detours to strict linear time from the worse than quadratic complexity of Not-via. Hence, in the lightweight Not-via the extra computational complexity of fast reroute amortizes as compared to even shortest path routing. We discovered, however, that a more significant improvement comes from redefining the semantics of Not-via addresses, so that one address covers not just one but many failure scenarios, since fewer additional addresses caused a spectacular drop in the associated management cost. This demonstrates that, with clever modifications to Not-via, the extra load of IPFRR can be brought down to a tolerable level. We believe that this will further incentivize network operators to seriously consider deploying IPFRR in the future.

## References


1. Shand, M., Bryant, S.: IP Fast Reroute framework. Internet Draft (February 2008), http://tools.ietf.org/html/draft-ietf-rtgwg-ipfrr-framework-08
2. Bryant, S., Shand, M., Previdi, S.: IP fast reroute using Not-via addresses. Internet Draft (February 2008), http://www.ietf.org/internet-drafts/draft-ietf-rtgwg-ipfrr-notvia-addresses-00.txt
3. Szilágyi, P., Tóth, Z.: Design, implementation and evaluation of an IP Fast ReRoute prototype. Technical report, BME (2008); First prize at Scientific Student Conference (2008), http://opti.tmit.bme.hu/~enyedi/ipfrr/
4. Cicic, T., Hansen, A.F., Apeland, O.K.: Redundant trees for fast IP recovery. In: Broadnets, pp. 152–159 (2007)
5. Li, A., François, P., Yang, X.: On improving the efficiency and manageability of NotVia. In: Proc. of ACM CoNEXT, pp. 1–12 (2007)
6. Médard, M., Barry, R.A., Finn, S.G., Galler, R.G.: Redundant trees for preplanned recovery in arbitary vertex-redundant or edge-redundant graphs. IEEE/ACM Transactions on Networking 7(5), 641–652 (1999)
7. Enyedi, G., Rétvári, G., Császár, A.: On finding maximally redundant trees in strictly linear time. In: IEEE ICC (2008) (submitted), http://opti.tmit.bme.hu/~enyedi/ipfrr/
8. Zhang, W., Xue, G., Tang, J., Thulasiraman, K.: Linear time construction of redundant trees for recovery schemes enhancing QoP and QoS. In: INFOCOM 2005 (March 2005)
9. Enyedi, G., Rétvári, G.: Finding redundant trees in linear time. IEEE Communications Letters (2008) (submitted), http://opti.tmit.bme.hu/~enyedi/ipfrr/
10. GNU Quagga routing software, http://www.quagga.net
11. Katz, D., Ward, D.: Bidirectional forwarding detection. Internet Draft (2008), http://tools.ietf.org/html/draft-ietf-bfd-base-08
12. Shaikh, A., Greenberg, A.: Experience in black-box OSPF measurement. In: IMW 2001: Proceedings of the 1st ACM SIGCOMM Workshop on Internet Measurement, pp. 113–125 (2001)

# QoS Support for Mobile Users Using NSIS

Roland Bless and Martin Röhricht*

Institute of Telematics
Universität Karlsruhe (TH)
Zirkel 2, D–76128 Karlsruhe, Germany
{bless,roehricht}@tm.uka.de

**Abstract.** Resource reservations in the Internet become more and more important with the advent of real-time multimedia services like Voice-over-IP and IPTV. At the same time we see an increasing interest in accessing Internet services by using mobile devices. In this paper we describe how Quality-of-Service guarantees can be achieved in mobile environments across different domains using the *Next Steps in Signaling* (NSIS) framework. We provide an analysis of mobility scenarios in combination with QoS signaling and propose to use an additional node local *Flow Information Service* element that supplies the necessary mobility support within NSIS capable mobility-aware nodes. We show that reservations can be setup quickly along the new path after a handover happened. Even the tear down of the reservation of the old path after a successful handover is performed quickly.

**Keywords:** QoS, NSIS, Signaling, MobileIP.

## 1 Introduction

Controlling resources in the Internet requires manipulation of state in network elements along the path of a given data flow. In order to install, maintain, or tear down state on nodes on a given path signaling must be performed accordingly. The *Resource reSerVation Protocol* (RSVP) was once designed as a signaling protocol for Quality-of-Service establishment in IP networks. In response to some deficiencies of RSVP the IETF working group *Next Steps in Signaling* (NSIS) [1] was formed to design a framework for generic signaling on the IP layer.

With the advent of bandwidth demanding Internet applications and multimedia streams such as video broadcasts, voice-over-IP, or IPTV a continuously growing need for Quality-of-Service (QoS) arises. NSIS elaborated a QoS signaling protocol as its first use case which enables applications to reserve resources along a given path.

As mobile devices are becoming increasingly powerful, mobility and mobile computing become more and more attractive. That rises the desire to have continuous network connectivity. Since the Internet Protocol was not designed to

* The authors would like to thank Max Laier for his implementation efforts and conceptual input. Part of this work was supported by Deutsche Telekom Laboratories and T-Systems within the ScaleNet project.

L. Fratta et al. (Eds.): NETWORKING 2009, LNCS 5550, pp. 169–181, 2009.

cope with mobility, MobileIP [2] was developed by the IETF in the mid 1990's as an extension in order to provide transparent mobility support of applications.

However, QoS resource reservations are established for one particular data path and are thus not aware of mobility. MobileIP on the other hand does not cover QoS mechanisms and is primarily concerned with the correct routing of data packets towards the mobile endpoint. Even though the NSIS framework was designed from the beginning with support for mobility scenarios in mind, only very basic protocol mechanisms—e. g. a session identifier—were specified to accomplish this goal.

This paper provides an analysis of mobility scenarios in combination with QoS signaling. We propose an additional node local *Flow Information Service* element that supplies the necessary mobility support within NSIS capable mobility-aware nodes. We show that this support enables the NSIS QoS signaling protocol to work well in mobile scenarios and that the implied overhead in terms of additional reservation setup latency is small compared to non-mobile scenarios.

Though an existing Internet-Draft [3] discusses some mobility aspects of the NSIS protocol suite, a careful and detailed analysis of mobility supported is, however, still required. For instance, there are only a few mobility scenarios elaborated and considerations as well as more practical guidance for implementers is missing.

We will briefly introduce the NSIS framework and the basic operation of MobileIP in the following section. The use of QoS NSLP is of particular interest and will be discussed in more detail along different mobility scenarios in Section 3. As a further step we will discuss and propose solutions to these problems before we outline the necessary design decisions. The applicability of the NSIS protocols in mobile scenarios will be proven by evaluation in Section 5.

## 2 NSIS-Based Signaling and Mobility

In order to develop an understanding for the main problems of NSIS-based signaling in mobile environments, we introduce the Next Steps in Signaling protocol suite and the basic operation of MobileIP that is relevant for the findings and solutions discussed in the remainder of this paper.

### 2.1 The Next Steps in Signaling Framework

Reliable QoS guarantees along a path and across different networks can only be accomplished by means of (possibly aggregated) resource reservations at routers residing on this path. In order to negotiate, install, and maintain on-demand resource reservations, signaling and admission control must be performed. The NSIS framework [4] was designed to perform signaling in IP-based next generation networks and employs a two-layer approach by separating the transport of signaling messages from the signaling application logic.

The protocol stack is conceptually divided into two layers. The lower layer is called the *NSIS Transport Layer Protocol* (NTLP). It is responsible to discover the next NSIS capable node on the path which also supports the specific signaling application, to setup state between both nodes, and to transport the higher layer signaling messages. Instead of deploying a new set of transport and security protocols it makes use of independent and wide-spread protocols, such as UDP, TCP, or TCP with TLS. The *General Internet Signaling Transport* (GIST) [5] is a protocol that fulfills the requirements of an NTLP.

The NSIS framework basically distinguishes between the *initiator* of a signaling message exchange (e.g., the entity initiating a reservation), the *responder* to such a request, and the *sender* or *receiver* of the corresponding data flow. Thus, if a *receiver-initiated* reservation is performed, the data flow receiver initiates the resource reservation. We use this important distinction during the further discussion, especially the one in section 3.4.

The *NSIS Signaling Layer Protocol* (NSLP) builds the higher layer of the protocol stack. Unlike GIST, which operates only between two hops, the NSLP provides end-to-end signaling functionality. Currently there are two signaling applications defined, namely a QoS NSLP [6] and a NAT/FW NSLP. Whereas the former establishes state among nodes in order to fulfill resource reservation requests, the latter is concerned with configuration of NAT-Gateways and Firewalls.

### 2.2 Mobility Management by Mobile IP

Mobility in an IP-based network can be achieved by using MobileIPv6 [7] which was proposed by the IETF as an extension to the base IPv6 protocol. Figure 1 gives a conceptual overview of MobileIPv6's basic operations, where each cloud denotes a different IP network (could be also different provider domains or Autonomous Systems). A mobile node is assigned two IP addresses, one is the *Home Address* (HoA) that is assigned at the mobile node's home network and is used to identify the communication endpoints, i. e. all connections to and from the mobile node. In addition one or more *Care-of-Addresses* (CoA) are assigned to the mobile node that represent its current location. MobileIPv6 was designed to perform a transparent mapping between both address types, i. e. it hides the actual location and therefore the CoA of a mobile node from its communication peer, the *Correspondent Node* (CN). In order to fulfill this request two operational modes were specified: the *Tunnel mode*, which is used initially and whenever the CN is not MobileIPv6-aware, and the *Route-optimization mode*, which can be used whenever the CN is MobileIPv6-aware. In this case the MN and the CN establish a binding between CoA and HoA, which can then be used to send traffic directly from one endpoint to another, instead of redirecting it through the MN's home network. In order to differentiate route-optimized packets from normal packets, special IPv6 options in extension headers are used.

Tunnel mode is used initially and whenever the CN is not MobileIPv6-aware. The traffic from the MN is then tunneled to a specific service node within the MN's home network, called the *Home Agent* (HA). Once the HA retrieves a data packet destined to the CN, it forwards the traffic on behalf of the MN. If the CN sends data towards the MN, the traffic is intercepted by the HA and forwarded using the established tunnel.

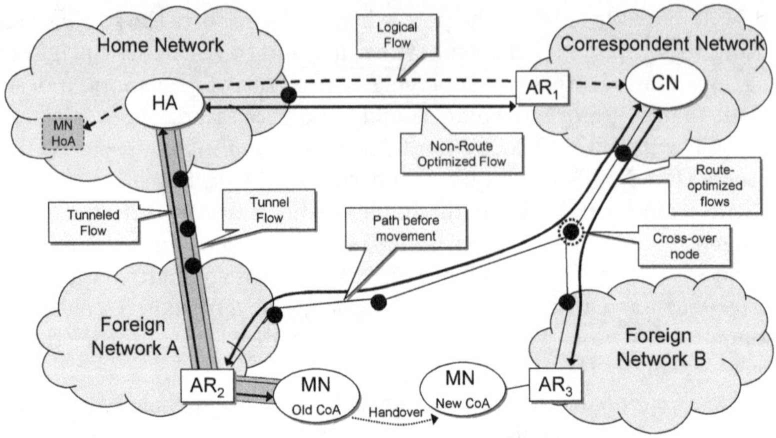


**Fig. 1.** Basic operations of MobileIPv6

It is important to note, that the route-optimized flow is not related to the *logical flow*. Especially in case of signaling under these circumstances the signaling application must be aware of the *actual flow*.

## 3 Mobility Analysis for QoS NSLP

Now we briefly describe some issues that we found while trying to implement mobility support for QoS NSLP. Due to limited space we can present only some issues very briefly, whereas more details can be found in a technical report [8]. Solutions are discussed in Section 4.

### 3.1 Mobility Awareness

The NSIS mobility applicability draft [3] assumes that QoS NSLP or the application are mobility aware, e. g., that they know the current *Care-of-Address* (CoA). It proposes that the reservation refers to the actual current flow, so that a CoA is contained in the corresponding *Path-Coupled Message Routing Information* (PC-MRI), which describes all relevant addressing information of the concerned data flow. On the one hand such a scheme requires knowledge of the current CoA within GIST, QoS NSLP, and the application. On the other hand this approach contradicts the use of Mobile IP that hides mobility from applications and transport protocols. However, since Mobile IP adds overhead to IP layer packets, either in form of tunneling packets or by adding IPv6 extension headers, the QoS NSLP *must* be aware of a node's mobility to take this overhead into account[1] when requesting resources. More specifically, it must adjust the TMOD parameter of the QSPEC object [9] that conveys the necessary resources

[1] Even if this overhead is considered to be small in total, it may be relatively large when small packets such as Voice-over-IP packets are transmitted.

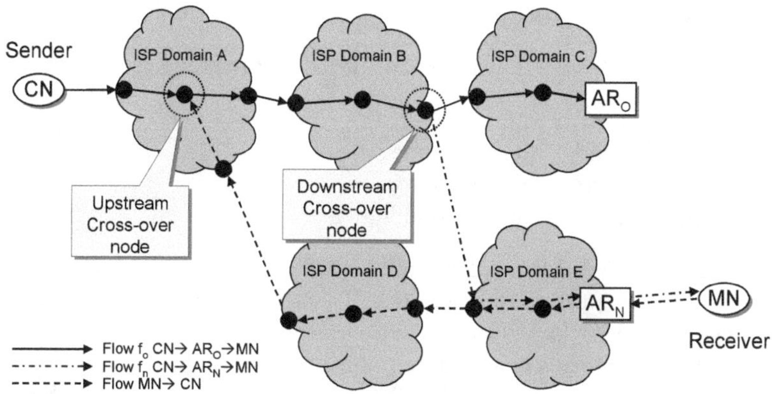


**Fig. 2.** Example for signaling and data flows, upstream and downstream paths may be different

information, depending on whether the *Mobile Node* (MN) is in its home domain, in a foreign domain and/or using route optimization. In this respect it does not make sense trying to hide mobility from QoS NSLP.

### 3.2 Upstream Signaling

Path-coupled signaling, which is the default message routing method for GIST, makes it difficult to signal for a new flow $f_n$ in upstream direction (cf. Figure 2). This could be required in order to trigger a reservation request from the other end or cross-over node. Upstream forwarding of a message is a problem if no message routing state exists already in the next GIST hop. The upstream message would require the new flow's MRI and the resulting GIST Query must be sent in upstream direction. In this case there is the problem of choosing the correct encapsulation for the GIST Query: only upstream Q-mode encapsulation would be an option, but this is not appropriate to use in this case due to its limited applicability to certain environments (e. g., restricted topologies with only one default router). Even if the MN could send the NOTIFY to its new access router, there is no routing state installed yet in upstream direction in this node, i. e., it does not know any next peer in upstream direction.

### 3.3 Resource Release

After performing a handover resources along the inactive path should be released as soon as possible in order to reduce reservation blocking for new reservation requests. Thus, there must be a possibility to release resources as soon as they are not used anymore. Though NSIS protocols use a soft-state approach that automatically removes unused state, it may be of advantage to release resources explicitly.

### 3.4 Mobility Scenarios

At least four different cases must be considered in the context of QoS NSLP:

1. Mobile node is sender and initiator
2. Mobile node is sender and responder
3. Mobile node is receiver and initiator
4. Mobile node is receiver and responder

The case of the MN being sender is not so difficult to handle, because mobility events may trigger appropriate QoS NSLP actions in order to adapt existing reservations. In case 1 the *MN* can initiate a new RESERVE message for the new flow $f_n$ directly after it has changed its point of attachment and got a new CoA. In case of a vertical handover the QSPEC may be adapted, but the Session ID will stay the same and a new PC-MRI corresponding to $f_n$ is used, so that the message can be sent downstream towards the *CN*. Even in case of an unchanged QSPEC, the signaling message must be forwarded up to the *CN*, because the PC-MRI has been changed and the related states must be updated along the whole path. In case 2 the *MN* could simply emit a new QUERY QoS NSLP message with a RESERVE-INIT flag set.

The case of the MN being receiver (scenarios 3 and 4) is more difficult, because it is especially hard to notify the sender (*CN*) of any change from the *MN*'s side using QoS NSLP as described in Section 3.2. Since the path from *CN* to *MN* is the downstream direction, the signaling path upstream from $AR_N$ towards the *CN* as source is not known in advance (cf. Figure 2). It is even difficult to determine the correct cross-over node, because the data path downstream (from $CN \rightarrow AR_N$) may be asymmetric to the upstream signaling path ($AR_N \rightarrow CN$). Furthermore, it may be often the case that the message must go back up along the old path to the sender in the worst case: If the flow address has changed, the profile in the first-hop router must be updated at sender side. In some cases the available QoS at the new point of attachment ($AR_N$) may be different from the one before at $AR_O$. In this case the changed resources require a re-negotiation along the whole path in most cases. Probably, in addition to that an application level signaling (e. g., SIP negotiation) is required in order to re-negotiate the content that should be sent or its coding, respectively.

### 3.5 Messaging Associations

In case an MN uses reliable or secure message transport, GIST establishes a Messaging Association to the next signaling peer, i. e., the next QoS NSLP aware node. In most cases this would be the access router. After changing the access router, the MN must establish a new MA to the new access router. Therefore, the new CoA must be used as source address, in order to avoid that the signaling connection is established via the home agent, which would be a long detour, resulting in long additional signaling delays in most cases.

## 4 Solutions to Mobility Problems

Since the solution space varies with the underlying assumptions about the environment, we have to distinguish between the following cases:

A. The application is mobility-aware and tries to manage mobility itself, i.e., no use of MobileIP, but SIP mobility support for instance.
B. Mobile Node (MN) and Correspondent Node (CN) are using MobileIPv6 and route optimization.
C. MN uses MobileIP, but the CN is not using MobileIP or does not support Route Optimization.

Case A does not impose any problems, because RESERVE and QUERY messages can be generated as soon as the MN moves. The CN can be notified about flow changes by some application level signaling protocol if it must send the RESERVE or QUERY.

In case B mobility events should be reported to GIST which in turn should notify the affected NSLPs. They can initiate new RESERVE/QUERY messages upon such a NetworkNotification. This works for all four cases of mobility scenarios that were described above in Section 3.4 and avoids the upstream signaling problem described in Section 3.2.

In case C the home agent must split the reservations and could act upon receiving binding updates from the MN by re-initiating reservations for the tunnel to the MN's current CoA as in the previous case. This HA-based solution also includes the case when MobileIPv4 is going to be used that does not support route optimization.

In general, we must deal with the case that nodes have probably QoS NSLP and MobileIPv6 support, but the QoS NSLP is not mobility-aware and consequently gets not notified about the corresponding events. In most cases, it is sufficient if the MN has such mobility support in the NSLP, because it can detect for instance that the HA is not initiating a reservation for the tunnel although a reservation request for a flow was received. In such a case the MN may request a reservation for the tunnel on its own. In some other cases a tunnel reservation will not be possible or reservations for the route optimized flow paths cannot be made.

With respect to the release of resources mentioned in Section 3.3, there are several options possible: the MN could resend a RESERVE with the "Replace" flag set so that the cross-over router would automatically initiate a tear down of the reservation along the old branch. Since a teardown can only be initiated in direction from the *QoS NSLP Initiator* (QNI) to the *QoS NSLP Responder* (QNR), the cross-over router can send a RESERVE with a Tear flag set only if the branch is downstream. Another option would be a NOTIFY message that is sent in order to initiate a tear down of the old branch from the right direction.

As sketched in Section 3.4, it is important for a QoS NSLP to be mobility aware: it must reserve the right amount of resources that depends on the current location and mode (tunnel or route optimization) due to the involved MobileIP

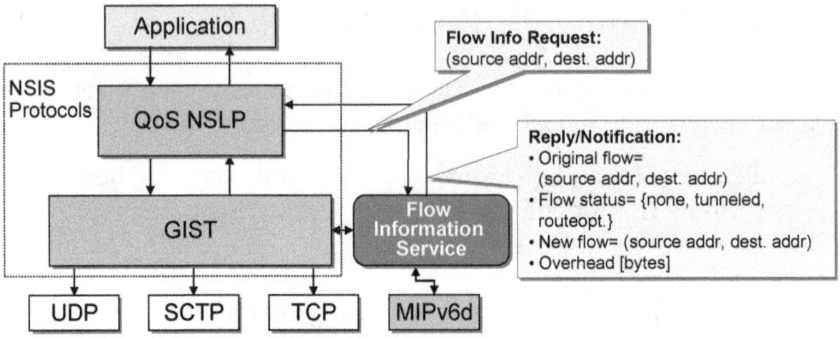


**Fig. 3.** Flow Information Service

packet overhead and it must know the current care-of-addresses in order to change the flow identifier for route optimized flows. Very important is also that mobility events, e. g., binding updates, trigger corresponding RESERVE or QUERY messages immediately in order to update the reservation accordingly. Therefore, an NSLP must be able to query the current state from the MobileIP mobility management and the MobileIP implementation must send notifications to the NSLP for significant events, such as handovers, new CoAs and a change in the binding cache and binding update list.

Although [10] specifies a MobileIPv6 Management Information Base (MIB) that would allow implementation independent access to the required information, this is not implemented in any of the MobileIPv6 implementations available to us at the time. Therefore, we created an additional component that provides the necessary additional information about actual flow addresses. For the QoS NSLP it is crucial to use the right CoA and the corresponding overhead values. Figure 3 shows the interaction of the *Flow Information Service* element with GIST and other NSLPs. The Flow Information Service is able to access state information directly from the MobileIPv6 implementation (MIPv6d). Basically, the same information could be supplied by using an SNMP agent that provides access to the MobileIPv6 MIB and SNMP traps could be used to realize mobility triggers.

The interface to the Flow Information Service is mainly based on a simple request/response interface. An NSLP entity sends a request indicating that it wishes to retrieve information about a certain flow. The Flow Information Service replies with the current state of that flow. In addition the Flow Information Service sends notifications whenever the state of an active flow changes. This way the consumer (the NSLP) is able to cache the results provided by the Flow Info Service, but does not need to bootstrap and mirror the complete state of the MobileIPv6 implementation. State information can also be polled every time the NSLP needs current flow addressing information.

As depicted in Figure 3 the request needs to contain a flow address as argument by specifying two IP-addresses (flow source and destination). This could be, for example, the address pair of a logical flow where one of the addresses is the home address. The reply needs to describe three possible flow states:

1. *No MobileIP flow*—because the source is not a HoA of the node and there is currently no active MobileIP state for the destination.[2]
2. *Tunnel mode*—the flow will enter or exit a tunnel at the current node, on the MN or the HA respectively. This happens when the flow source is a HoA and the peer is either MobileIP unaware or state is not established yet. The response must include the tunnel source and destination in order to enable the NSLP to establish a bound session for the tunnel section or update an existing session accordingly.
3. *Route optimization mode*—there is an active state for this flow and the flow source and/or flow destination will be rewritten. The information required in the reply consists of the new flow addresses.

Notifications reuse the reply message format and simply inform the requester (NSLP entity) about the new state of a flow. The requester must then internally identify the affected flow state information.

Furthermore, GIST is notified in case of mobility events and NSLPs are notified via the internal GIST API [5, Appendix B] primitive *NetworkNotification*. We defined a new *Network-Notification-Type* of *Mobility Event* for this purpose. Mobility-aware NSLPs may use this indication to request new information from the flow information service, for which further details can be found in [8].

## 5 Evaluation

The proposed Flow Information Service was implemented in the freely available NSIS-ka protocol suite [11] that is based on C++ and Linux. In order to evaluate the design we set up a testing environment consisting of six virtual hosts residing on one physical machine as shown in Figure 4. On each of them runs a slightly modified Linux kernel (2.6.26) to be MobileIPv6-aware. All virtual hosts are connected to a *smart switch* on another dedicated physical machine where the topology is set up by bridging VLAN interfaces accordingly.

Given this setup we could easily obtain packet captures of all signaling messages exchanged between the hosts. Furthermore it allows for accurate and easy measurement of the delays obtained, resulting from a handover event and it was not necessary to take care of clock synchronization between each single host. As the smart switch runs on real hardware and the signaling messages travel over a real physical wire, virtualization should not affect the measurements beneficially.

### 5.1 Signaling Performance Benchmarks

The signaling performance was evaluated based on the time between receiving the *Binding Update/Binding Acknowledgment* on the MN/CN respectively and the time the final RESPONSE is received for the new reservation. This time was sampled for 48 consecutive movements of the MN from $AR_3$ to $AR_2$ to $AR_1$ and

[2] It might turn out later on that the peer is indeed a mobile node, but for the moment the flow is sent unmodified.

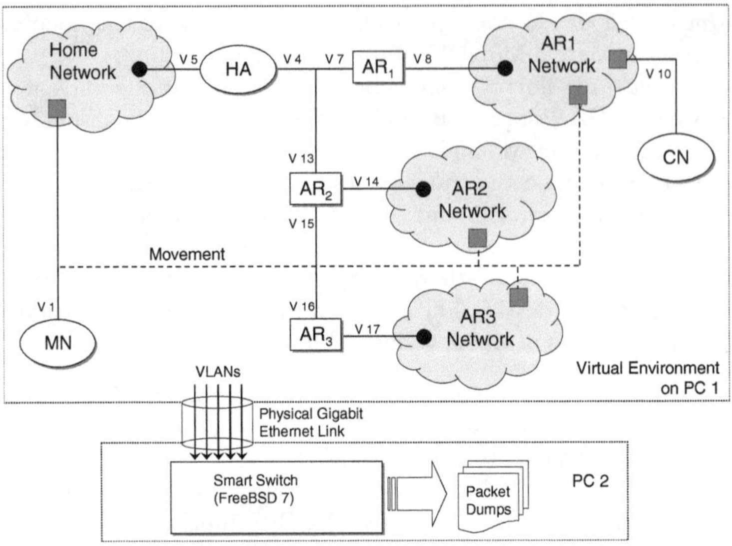


**Fig. 4.** Layout of testing environment and assignment of VLAN interfaces

back. Further we measured the time needed to tear down state on the old path between $AR_2$ and $AR_3$ in order to release resources not needed any longer. The results are shown in Table 1 representing the median of all runs. Note that no artificial delay was introduced, which results in round trip times of under 1 ms between the virtual hosts.

In order to setup reservation between the CN and the first hop $AR_1$ we need less than 14 ms on average. In case of a reservation setup for $AR_2$ and $AR_3$ belonging to foreign networks, state is established in less than 34 ms and 44 ms respectively, no matter whether the MN is sender or receiver and whether sender- or receiver-initiated reservations are performed. The time needed to tear down state on the old path differs only—but significantly—for the case of the MN being a sender and by using sender-initiated reservations where approximately 21 seconds are necessary. The time depends on the lifetime of the routing state between the MN at the old CoA and the $AR_3$ which times out eventually. This process could be sped up by actively acknowledging the NOTIFY on QoS NSLP level.

**Table 1.** Median of the measurement results of reservation setup delays and old path tear down delays after movement

| **Testcase** | **$AR_1$ setup** | **$AR_2$ setup** | **$AR_3$ setup** | **Tear** |
|---|---|---|---|---|
| MN sender, sender-initiated | 11.8 ms | 26.9 ms | 37.5 ms | 20600 ms |
| CN sender, sender-initiated | 13.3 ms | 27.4 ms | 40.3 ms | 26.8 ms |
| MN sender, receiver-initiated | 11.3 ms | 29.0 ms | 43.1 ms | 28.0 ms |
| CN sender, receiver-initiated | 12.5 ms | 33.4 ms | 42.2 ms | 31.9 ms |

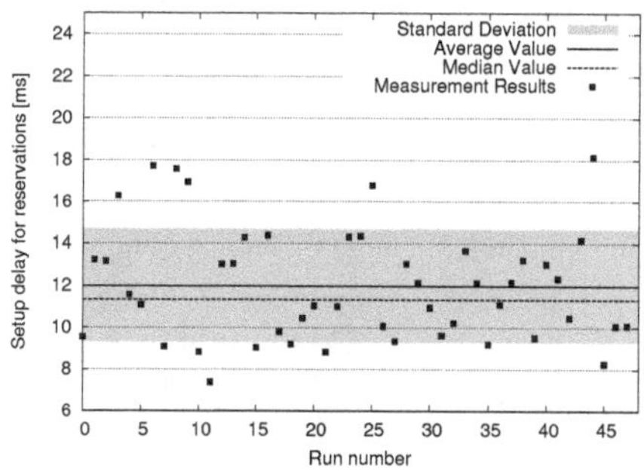


**(a)** Access Router 1

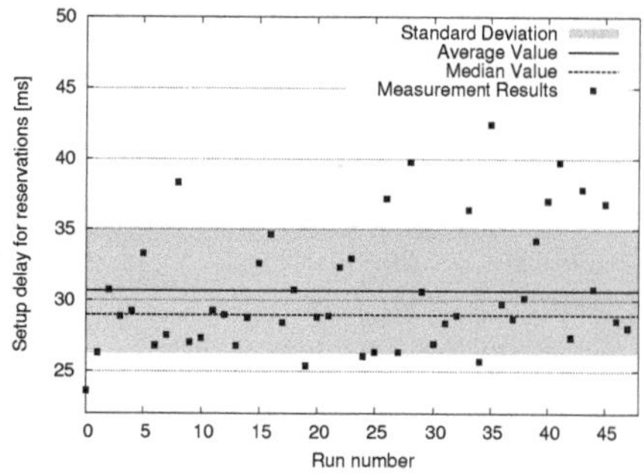


**(b)** Access Router 2

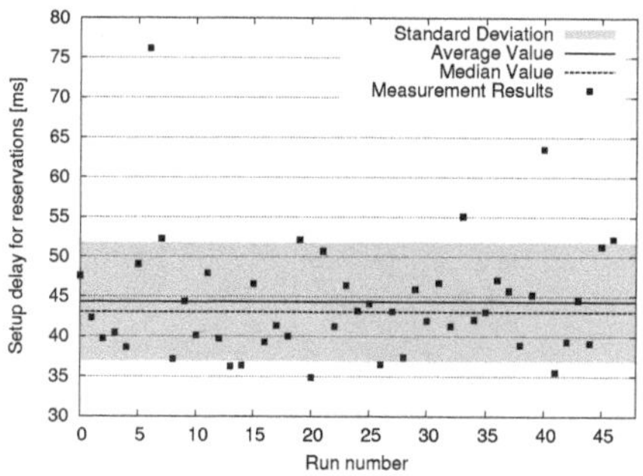


**(c)** Access Router 3

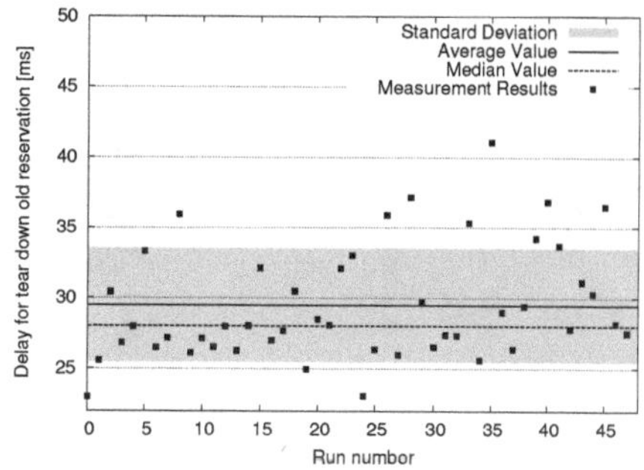


**(d)** Tear down of old path

**Fig. 5.** Measurement results for reservation setup and tear down delay if MN is sender and receiver-initiated reservations are used

Figure 5 shows the measurement results for one of our four possible setups (MN is sender and uses receiver-initiated reservations). The standard deviation of our measurements shows, that the dispersion is relatively small, ranging from 2.7 ms for $AR_1$ to 7.4 ms for $AR_3$.

We set up a further benchmark by introducing artificial delays on the smart switch in order to simulate a more realistic Internet scenario. A 50 ms delay was configured between $AR_1$ and $AR_2$ and another 25 ms delay between $AR_2$ and $AR_3$. Given these settings, we obtained an RTT of 104 ms between the MN being located at $AR_2$ and the CN and 160 ms while the MN is located at $AR_3$ respectively.[3]

Results of measurements with additional artificial delays are shown in Table 2. The "optimal" delay that can be obtained by using the RTT values (2 RTT in case of sender-initiated mode, 2.5 RTT in case of receiver-initiated mode) are printed in parentheses after each measured value. Figure 6 illustrates the difference from the retrieved results compared to the theoretical possible

[3] The difference compared to theoretical 100 ms and 150 ms is due to scheduling granularity and queuing overhead on the smart switch.

**Table 2.** Median of the measurement results of reservation setup delays with an artificial delay between different hosts

| Testcase | $AR_2$ setup delay (optimal value) | $AR_3$ setup delay (optimal value) |
|---|---|---|
| MN sender, sender-initiated | 228.8 ms (208 ms) | 348.9 ms (320 ms) |
| CN sender, sender-initiated | 231.5 ms (208 ms) | 353.1 ms (320 ms) |
| MN sender, receiver-initiated | 285.5 ms (260 ms) | 429.3 ms (400 ms) |
| CN sender, receiver-initiated | 287.8 ms (260 ms) | 429.1 ms (400 ms) |

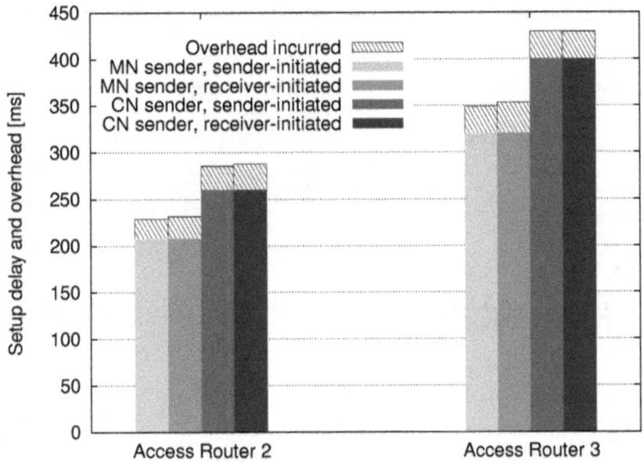


**Fig. 6.** Setup delay and overhead of reservations for movement from $AR_2$ to $AR_3$

performance for $AR_2$ and $AR_3$. The overhead incurred by our mobility-aware reservations ranges between 6.78% and 10.15% only.

## 6 Conclusion

QoS NSLP works basically well in mobility scenarios if it has enough information about the actual data flow and gets the necessary mobility triggers. In this paper we introduced a node local component, the *Flow Information Service* element, that allows for getting required mappings from *logical* flows to *actual* flows (using current CoAs) as well as any mobility-related per packet overhead. In case a binding for a data flow changes the NTLP will notify any NSLPs of the change and mobility-aware NSLPs can request more information from the Flow Information Service if required. This makes it possible to re-reserve resources as soon as a MobileIP handover occurred.

Currently, we are working towards a seamless handover solution for NSIS using an anticipated handover concept. This requires, however, larger protocol changes of QoS NSLP and also additional support in GIST for providing path-decoupled signaling as proposed in [12].

## References


1. Next Steps in Signaling (nsis) Working Group: NSIS Charter (February 2009), http://www.ietf.org/html.charters/nsis-charter.html
2. Perkins, C.E.: Mobile IP. IEEE Communications Magazine 40(5), 66–82 (2002)
3. Sanda, T., Fu, X., Jeong, S.H., Manner, J., Tschofenig, H.: Applicability Statement of NSIS Protocols in Mobile Environments. Internet-Draft draft-ietf-nsis-applicability-mobility-signaling-11 (November 2008), http://tools.ietf.org/html/draft-ietf-nsis-applicability-mobility-signaling
4. Fu, X., Schulzrinne, H., Bader, A., Hogrefe, D., Kappler, C., Karagiannis, G., Tschofenig, H., den Bosch, S.V.: NSIS: A New Extensible IP Signaling Protocol Suite. IEEE Communications Magazine 43(10), 133–141 (2005)
5. Schulzrinne, H., Hancock, R.: GIST: General Internet Signalling Transport. Internet Draft draft-ietf-nsis-ntlp-17 (October 2008), http://tools.ietf.org/id/draft-ietf-nsis-ntlp
6. Manner, J., Karagiannis, G., McDonald, A.: NSLP for Quality-of-Service Signaling. Internet Draft draft-ietf-nsis-qos-nslp-16 (February 2008), http://tools.ietf.org/id/draft-ietf-nsis-qos-nslp
7. Johnson, D., Perkins, C., Arkko, J.: Mobility Support in IPv6. RFC 3775 (Proposed Standard) (June 2004)
8. Laier, M.: Analysis and Design of Mobility Support for QoS NSLP. Technical Report TM-2009-1, Institute of Telematics, Universität Karlsruhe, Telematics Technical Reports ISSN 1613-849X (February 2009), http://doc.tm.uka.de/tr/TM-2009-1.pdf
9. Ash, G., Bader, A., Kappler, C., Oran, D.: QoS NSLP QSPEC Template. Internet-Draft draft-ietf-nsis-qspec-20.txt, Work in progress (April 2008), http://tools.ietf.org/html/draft-ietf-nsis-qspec-20
10. Keeni, G., Koide, K., Nagami, K., Gundavelli, S.: Mobile IPv6 Management Information Base. RFC 4295 (Proposed Standard) (April 2006)
11. Institute of Telematics: NSIS-ka – A free C++ implementation of NSIS protocols (April 2009), https://projekte.tm.uka.de/trac/NSIS/
12. Bless, R.: An Explicit Signaling Target Message Routing Method (EST-MRM) for the General Internet Signaling Transport (GIST) Protocol. Internet-Draft draft-bless-nsis-est-mrm-01.txt, Work in progress (July 2008), http://tools.ietf.org/html/draft-bless-nsis-est-mrm-01

# Real Time Identification of SSH Encrypted Application Flows by Using Cluster Analysis Techniques

Gianluca Maiolini[1], Andrea Baiocchi[2], Alfonso Iacovazzi[2], and Antonello Rizzi[2]

[1] Elsag Datamat, Automation, security and transportation division,
V. Laurentina 760, 00143 Rome, Italy
gianluca.maiolini@elsagdatamat.com
[2] INFOCOM Dept., University of Roma "Sapienza"
Via Eudossiana 18 - 00184 Rome, Italy
{name.surname}@uniroma1.it

**Abstract.** The identification of application flows is a critical task in order to manage bandwidth requirements of different kind of services (i.e. VOIP, Video, ERP). As network security functions spread, an increasing amount of traffic is natively encrypted due to privacy issues (e.g. VPN). This makes ineffective current traffic classification systems based on ports and payload inspection, e.g. even powerful Deep Packet Inspection is useless to classify application flow carried inside SSH sessions. We have developed a real time traffic classification method based on cluster analysis to identify SSH flows from statistical behavior of IP traffic parameters, such as length, arrival times and direction of packets. In this paper we describe our approach and relevant obtained results. We achieve detection rate up to 99.5 % in classifying SSH flows and accuracy up to 99.88 % for application flows carried within those flows, such as SCP, SFTP and HTTP over SSH.

**Keywords:** Traffic analysis, statistical traffic classification, SSH, cluster analysis, k-means.

## 1 Introduction

The control of QoS becomes a very important stake, even more with the arrival of new applications with very different profiles, such as real time multimedia. These applications have very different behavior according to packet sizes, offered traffic profile, transport level protocol (UDP/TCP) and different requirements on throughput, transit delay, jitter, packet loss rate.

Traffic shaping and scheduling techniques for edge routers aim at managing bandwidth resources according to QoS policy model. Before applying QoS policy it is important to classify correctly application flows. Most routers use port based classification of protocols but new applications are forwarded using well-known ports, i.e 80. Deep packet inspection (DPI) systems can be quite effective in recognizing applications, thanks to IP payload analysis, even if they fail in classifying ciphered flows. From provider side, it will be very important to recognize applications encoded within ciphered flows in order to apply QoS policies; in fact encoded traffic could contain

L. Fratta et al. (Eds.): NETWORKING 2009, LNCS 5550, pp. 182–194, 2009.

real time applications, transactional applications (ERP, finance,..), comfort application (web, mail,..) and so on. In this situation, future requirements could consist in optimizing performance in terms of QoS for application natively encoded.

Secure Shell or SSH is a network protocol that allows data to be exchanged using a secure channel between two networked devices. It is often used to login to a remote computer but it is also applied for tunneling, file transfer and forwarding arbitrary TCP ports over a secure channel between a local and a remote computer. What makes the detection of this protocol interesting is that its traffic is encrypted. Thus any payload analysis based classification method is irrelevant since the payload is encrypted. Actually DPI technology cannot recognize application delivered within SSH flows.

The objective of our work is to develop a real time system to recognize and classify SSH flows by analyzing statistical features of first IP packets belonging to a SSH connection, such as arrival times, directions and lengths. By recognition we mean identifying which flows belong to SSH protocol as opposed to other application level protocols. By classification we mean to identify the kind of service carried within each SSH connection, such as SCP, SFTP and HTTP over SSH. Experiments show that our approach permits us to achieve great recognition accuracy up to 99.2% for SSH identification and, once SSH has been identified, applications in those SSH tunnels are classified with accuracy up to 99.8%.

The paper is organized as follows. Section II establishes the position of this work, relative to earlier research. Problem statement has been discussed in section III. In section IV we have described dataset creation, in particular data collection and preprocessing. The machine learned based approaches for real time SSH classification are presented in section V. Experimental results are presented in section VI, and conclusions are drawn in section VII.

## 2 Related Works

Different approaches to traffic classification have been developed, using information available at IP layer such as inter-arrival times, bytes transferred, packet size. Some proposals [4][5] need also semantically complete TCP flows as input.

In [1], Karagiannis et al. developed a heuristic that uses social, functional and application level behaviours of a host to identify all traffic flows originating from it. This approach, although really innovative, is tailored onto a specific source host.

Salgarelli et al. [2] used only size and inter-arrival time of first *n* packets to create a statistical descriptor (a Fingerprint) of an application layer protocol: this fingerprint is then used to measure the similarity of a certain flow to the corresponding protocol.

The Hidden Markov Models (HMM) theory is used in [3]: packets size and inter-arrival time are used to build a model describing a certain protocol. The results of the training phase is a HMM model describing the behaviour of each protocol. Even though this approach can classify distinct encrypted applications, its performance on SSH is (76% detection rate and 8% false negative) is not as good as well known application traffic such as WWW and instant messaging.

Moore et al. [4] used a supervised machine learning algorithm called Naive Bayes (and its generalization, Kernel Estimation) on a wide set of characteristics (tens or hundreds), as flow duration, packets inter-arrival time and payload size and their statistics (mean, variance...). Moreover, they use a filtering technique to identify the best characteristics to be used with the mentioned methods.

A number of works [5][6][7] rely on unsupervised learning techniques. McGregor et al. [5] explore the possibility to use cluster analysis to group flows using transport layer attributes, but they do not evaluate the accuracy of the classification. Zander et al. [6] extend this work using another Expectation Maximization (EM) algorithm named Autoclass. They also analyze the best set of attributes to use. Both these works only test Bayesian clustering technique trained by an EM algorithm, which has a slow learning time.

Bernaille et al. [7] use faster clustering algorithms representing data in different spaces: K-means and Gaussian Mixture Models (GMM) for euclidean space and Spectral clustering in HMM based space. The only features they use are packet size and packet direction: they demonstrate the effectiveness of these algorithms even using a small number of packets (e.g. the first four of a TCP connection).

Alshammari et Al [8], work attempted to classify/identify applications services running over SSH. They have shown the utility of two supervised learning algorithms AdaBoost and RIPPER for classifying SSH traffic without using features such as payload, IP addresses and source/destination ports. Results indicate that a detection rate of 99% and a false positive rate of 0.7% can be achieved using RIPPER. Moreover, promising preliminary results were obtained when RIPPER was employed to identify which service was running over SSH. They can recognize applications inside SSH flows such SCP and SFTP with accuracy up to 99.8% but they have performed off-line analysis on complete traces. We aim at classifying applications inside SSH flows in real time mode just analyzing the firsts 4 packets after SSH negotiation. We rely on K-means cluster analysis machines algorithm.

## 3 Problem Statement

In this paper, we focus on the classification of IP flows generated from network applications communicating through TCP protocols. Our objective is to recognize SSH flows out of other applications such as HTTP, FTP, POP3, etc. and, once that is accomplished, to identify which service is actually carried within the encrypted SSH tunnel. Then, we first need to define exactly what we mean for TCP flow.

**Definition:** A flow $F$ is the bi-directional, ordered sequence of IP packets exchanged during a TCP connection.

Within a TCP connection, application level data are delivered as well as control packets, such as those related to three way-handshake (RFC-793) and TCP ACK packets. So, TCP flow will be composed by packets from SYN ($PK_0$) to FIN ($PK_{N-1}$). Each flow could be seen as a sequence of ($PK_0$, …., $PK_{N-1}$), where $PK_j$ represents the $j$-th IP packet exchanged during TCP connection. Since we aim at classifying application flows relying on statistical features of IP packets, such as length, direction and absolute-arrival times, we will characterize each TCP flow $F$ as an ordered sequence of $N$-tuples ($d_j$, $l_j$. $t_j$), with $0 \leq j \leq N-1$, where:

- $d_j \in [0,1]$ where 1 encodes the direction detected for SYN packet and 0 the opposite direction;
- $l_j$ length of IP $PK_j$, in bytes;

- $t_j = T_j - T_0$, where $T_j$ is the timestamp of $PK_j$ at capture point and $T_0$ is the timestamp of the $PK_0$ at the capture point.

The packet length ranges between a minimum and a maximum. The latter is the MTU (Maximum Transmission Unit) of the interfaces crossed by TCP connections packets. In all experiments we found out MTU=1500 bytes has never been exceeded, which is just the largest allowed MTU of most Ethernet LANs and hence most of the Internet [10]. As for the minimum length, it corresponds to those carrying a TCP ACK and is denoted as $l_{ACK}$ in the following. It is the smallest length detectable for a TCP packet as we tested during our experiments and as RFC 793 refers, typical values ranging between 40 and 56 bytes, depending on options in the TCP and IP headers.

## 4 Data Set Creation

Given our aim as stated in the introduction, we assume a trained machine learning approach, exploiting cluster analysis. To that end, we need both a test and a train data set. A data set for our purposes is composed of a collection of flows in the sense defined in Section III along with metadata per flow, reporting the *known* application layer protocol the flow belongs to, the absolute timestamp of its first packet ($T_0$), the capture date and location.

Knowing the application protocol each flow belongs to is needed to reliably train our algorithm. Since publicly available traces have payloads stripped off (for obvious privacy reason, e.g. CAIDA traces) and classification results cannot be checked reliably, we resorted to artificial traffic carefully generated by exploiting network premises at the University campus, the Elsag Datamat site and a private home. This way we encompass three major kinds of Internet access points: institutional, business and domestic. The controlled traffic generation is a must specifically for collecting SSH traces whose service content is known, i.e. to further label each SSH flow with a metadata reading which service it is carrying among SCP, SFTP and HTTP.

### 4.1 Data Collection

**IP traces generation of TCP based application:** We focus our attention on four different plain application layer protocols, namely HTTP, FTP-control, POP3 and of course SSH, which seem to be the ones accounting for the majority of traffic flows in the INTERNET (except of peer-to peer traffic). As for HTTP and FTP-Control (FTP-C in the following), we collected traffic traces coming from the Networking Lab at our Department. By means of automated tools mounted on machines within the Lab, thousands of web pages have been downloaded in a random order, over thousands of web sites distributed in various geographical areas (Italy, Europe, North America, Asia). FTP sites have been addressed as well and control FTP session established with thousands remote servers, again distributed in a wide area. The generated traffic has been captured on our LAN switch configuring a mirroring port; we verified that the TCP connections bottleneck was never the link connecting our LAN to the big Internet. This experimental set up, while allowing the capture of artificial traffic that (realistically) emulates user activity, gives us traces with reliable application layer protocol classification. In order to complete our traces repository, in the next section we describe how we collected SSH traces.

**SSH IP traces generation:** Our data collection approach is to simulate possible network scenarios using one or more computers to capture the resulting traffic. In order to have realistic traces and technology independent implementations of SSH (version 2) protocol, we used computers with heterogeneous operative systems, namely Linux and Windows. We simulate SSH connections by connecting three client computers deployed in three different LAN to one server. As shown in figure 1, client LANs and SSH server have been connected to the Internet by using different geographic links. We run the following SSH services: SCP, SFTP and HTTP over SSH. SCP and SFTP are transfer file services natively available on OpenSSH [9]. In particular we downloaded/uploaded files from clients to server using both SCP and SFTP protocols collecting eight thousands flows. HTTP over SSH traces have been collected downloading web pages through SSH tunnels (one SSH tunnel for each HTTP session). We get four thousands of flows.

SSH connections can tunnel several TCP flows at the same time: we are working in the case where each flow is assigned by SSH a separated channel, each with specific SSH identifier. Finally we will consider flows without SSH compression feature.

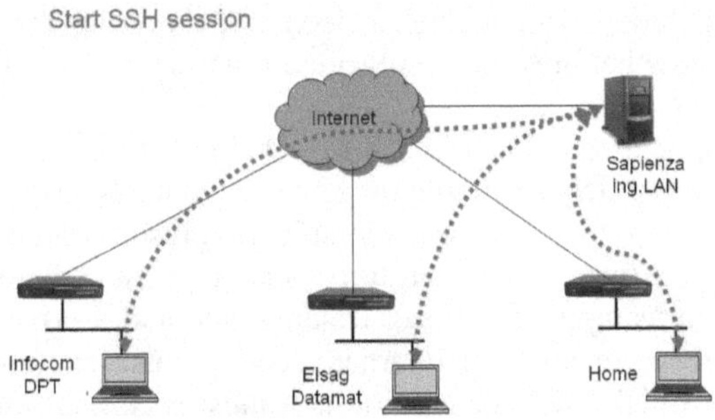


**Fig. 1.** Platform used to generate SSH traffic: SSH server is inside the University campus network; clients are at University, Elsag Datamat and a private home premise, respectively

## 4.2 Data Set Creation: Pre-processing of Traces

In order to create data sets we pre-processed collected traffic traces. In particular we think that removing packets related to TCP control messages from each flow *F* can help us highlighting the differences among various applications. Therefore we remove from each flow *F* packets related to:

- Three-way handshake of TCP: $PK_{0=SYN}$, $PK_{1=SYN\text{-}ACK}$, $PK_{2=ACK}$;
- TCP ACK packets, i.e. those packets carrying only a TCP level ACK and no payload data;
- Retransmitted packets.

According to TCP protocol (RFC 793) the third packet ($PK_{2=ACK}$) of each TCP connection flow *F* carries an ACK. In order to remove ACK packets and TCP header length at the same time, we detect $PK_{ACK} = <d_{ACK}, l_{ACK}, t_{ACK}>$ of each session, where:

- $d_{ACK}$ is 1, because ACK direction in three way handshake is always consistent with that of SYN packet;
- $l_{ACK}$, is the length of packet containing TCP ACK;

- $t_{ACK}$, is the relative capture time of that packet.

So, for each flow $n$ packets will be pre-processed as:

$$\underline{PK_j}^* = <\delta_j = d_j,\ \lambda_j = l_j - l_{ACK},\ \tau_j = t_j - t_{ACK}>, \quad j \in [3,\ldots, N - N_{ACK} - 1]$$

where $N_{ACK}$ is the number of ACK packets detected in the pre-processed flow. Packets with $\lambda_j$=0 are removed and packets shown in $\underline{PK_j}^*$ will contain bytes concerning just application contribution. In particular, $\lambda_j \in [0, l_{MTU} - l_{ACK}]$.

In order to make our analysis in real time, we need to run our classification method exploiting the very firsts packets of each flow. After tests and analysis of results we set how many packets will be required to strike a convenient trade-off between high classification accuracy and an acceptable classification delay. So our dataset will be composed as shown in Figure 2. In our dataset we collected traces belonging to the following application layer protocols: HTTP, FTP-C, POP3 and SSH.

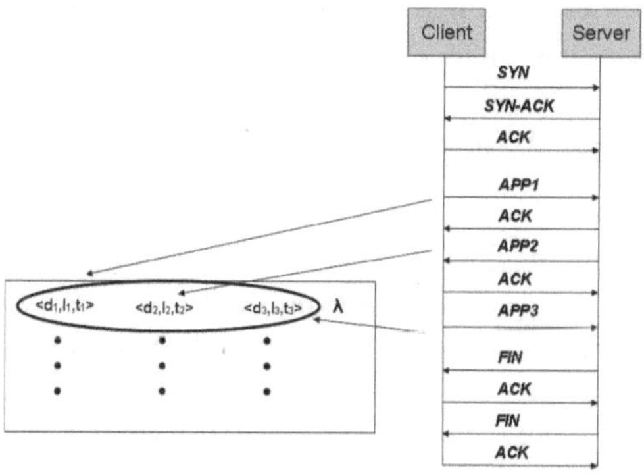


**Fig. 2.** Pre-processing TCP flows

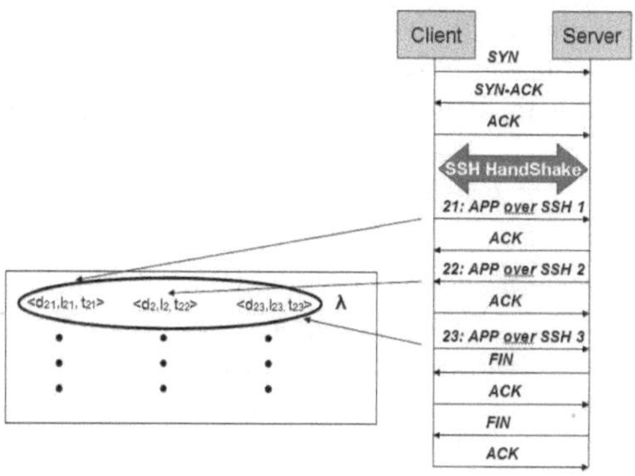


**Fig. 3.** Pre-processing of SSH flows

Once SSH application flows has been detected, we aim at identifying application within SSH tunnels. Then, we further process SSH flows by removing packets related to the SSH initial handshake (see Figure 3). We consider the following services inside encrypted SSH tunnels: SCP, SFTP and HTTP over SSH.

# 5 Classification Method

## 5.1 A K-Means Based Approach

In this section some details about the adopted classification system are exploited. Basically a classification problem can be defined as follows. Let $P : X \rightarrow L$ be an unknown oriented process to be modeled, where $X$ is the domain set and the codomain $L$ is a label set, i.e. a set in which it is not possible (or misleading) to define an ordering function and hence any dissimilarity measure between its elements.

If $P$ is a single value function, we will call it *classification function*. Let $S_{tr}$ and $S_{ts}$ be two sets of input-output pairs, namely the training set and the test set. We will call *instance* of a classification problem a given pair ($S_{tr}$ , $S_{ts}$) with the constrain $S_{tr} \cap S_{ts} = \emptyset$ . A classification system is a pair ($M$ , $TA_i$), where $TA$ is the training algorithm, i.e. the set of instructions responsible for generating, exclusively on the basis of $S_{tr}$, a particular instance $\overline{\text{M}}$ of the classification model family M, such that the classification error of $\overline{\text{M}}$ computed on $S_{ts}$ will be minimized. The *generalization capability*, i.e. the capability to correctly classify any pattern belonging to the input space of the oriented process domain to be modeled, is for sure the most important desired feature of a classification system. From this point of view, the mean classification error on $S_{ts}$ can be considered as an estimate of the expected behavior of the classifier over all the possible inputs. In the following, we describe a classification system trained by an unsupervised (clustering) procedure.

When dealing with patterns belonging to the $R^n$ vectorial space we can adopt a distance measure, such as the Euclidean distance; moreover, in this case we can define the prototype of the cluster as the centroid (the mean vector) of all the patterns in the cluster, thanks to the algebraic structure defined in $R^n$. Consequently, the distance between a given pattern $x_i$ and a cluster $C_k$ can be easily defined as the Euclidean distance $d(x_i\,; \mu_k)$ where $\mu_k$ is the centroid of the pattern belonging to $C_k$:

$$\mu_k = \frac{1}{\mu_k} \sum_{x_i \in C_k} x_i$$

A direct way to synthesize a classification model on the basis of a training set $S_{tr}$ consists in partitioning the patterns in the input space (discarding the class label information) by a clustering algorithm (in our case, by the K-means).

Successively, each cluster is labeled by the most frequent class among its patterns. Thus, a classification model is a set of labeled clusters (centroids); note that more than one cluster can be associated with the same label, i.e. a class can be represented by more than one cluster. Assuming to represent a floating point number with four bytes, the amount of memory needed to store a classification model is $K \cdot (4 \cdot n + 1)$ bytes, where n is the input space dimension and assuming to code class labels with one byte. An unlabeled pattern $x$ is classified by determining the closest centroid $\mu_i$ (and thus the closest cluster $C_i$) and by labeling $x$ with the same class label associated with $C_i$.

It is important to underline that, since the initialization step of the K-Means is not deterministic, in order to compute a precise estimation of the performance of the classification model on the test set $S_{ts}$, the whole algorithm must be run several times, averaging the classification errors on $S_{ts}$ yielded by the different classification models obtained in each run.

### Normalization

This paragraph describes how data sets have been normalized. The available data is split in training set and test set. The training and test data sets are normalized in order to guarantee that each feature will contribute to distance computations with equal weights; for each feature we adopted the "affine normalization", consisting in:

$$y = \frac{x - \min}{\max - \min}$$

where $x$ is the value we want to normalize, *max/[min]* is the maximum/[minimum] value of x in a reasonable range of possible values of x and $y$ is the normalized value.

The direction takes the values in the domain (0,1), and doesn't need to be normalized. As concerns the lengths, min and max values are respectively 0 byte and $l_{MTU}$–$l_{ACK}$ bytes. Absolute arrival time was intended between 0 and 200 sec. During normalization, timestamps over two hundreds seconds has been normalized to 1.

### Cross-validation

In order to choose the optimal number of clusters to be used in the K-means clustering procedure, we have performed cross validation. It help us to estimate "a priori" the performance of the classification system using only the training set and give us an estimate of the homogeneity of training set. The cross-validation is a function which receives as inputs the training set (containing $a \times N$ patterns, in which $a$ is the number of applications and $N$ is the number of patterns for each application), the length range of clusters' number to use and the number of parts to split the training set ($n$-fold cross-validation), and it returns a vector of length $l$ containing the estimated accuracy for each classifier synthesized during the cross-validation procedure. In more detail, the function divides in $n$ parts (simply called sub-datasets) the train dataset, assigning $n$ patterns for each application in a random way to each part. All the new created datasets will have he same number of patterns for each application and will be balanced in the same way of the original dataset. Next, for each value of $k$ belonging to a considered range of reasonable values, the function calculates the classifier performances taking in turn as test set one of the n sub-dataset and as training set the remaining $n$-$1$. Therefore the training procedure will be run $n$ times, as long as each of the sub-dataset $n$ has been test set once. The results are inserted into a matrix $\prod$ of $k \times n$ size:

$$\Pi = \begin{pmatrix} p_{1,1} & p_{1,2} & .. & p_{1,n} \\ p_{2,1} & .. & .. & .. \\ .. & .. & .. & .. \\ p_{k,1} & p_{1,2} & .. & p_{k,n} \end{pmatrix}$$

where the $p_{i,j}$ represents the mean accuracy of the classifier that has been obtained by the i[th] dataset as test set and by clustering data fixing $k$ equal to $j$. The mean accuracy is the average of the accuracies on the single application $acc_h^{(i)}(k)$ divided by the application number.

$$p_{k,i} = \frac{1}{a}\sum_{h=1}^{a} acc_h^{(i)}(k)$$

Finally the function computes a further average of the accuracies in the matrix $\prod$:

$$\pi(k) = \frac{1}{n}\sum_{i=1}^{n} p_{k,i}$$

obtaining in this way the $\pi(k)$ performance vector of the classifier, as a function of the number of clusters fixed in the clustering procedure. Finally, the optimal number of clusters to be used in the classification model is determined as follows:

$$k = \arg\max_{1 \le k \le l}\{\pi(k)\}$$

Actually, in this way it is very likely that the highest $k$ value in the range will be chosen because the performances of the classifier tends to improve by increasing of the number of clusters. However, the cross-validation purpose is to find the number k of clusters representing a trade-off between performances and complexity of clustering algorithm. We have chosen the minimal value of k after which the rate of the performance indicated from vector $\pi$ increases slower (or decreases).

## 5.2 On-Line Procedure for Capturing and Classifying Application Flows

In this section, we will describe how our algorithm works. Our method is divided into four phases.

First phase: traffic retrieving. This phase consists in connecting a network protocol analyzer, such Wireshark, to the network we want to investigate. In the case of LAN traffic classification, Wireshark workstation will be connected to a mirroring port of the edge switch in order to sniff all traffic traces flowing through the geographical link connecting LAN to the Internet.

Second phase: detecting TCP flows. Wireshark allows our tool to manage traffic in real time. As SYN packet is detected, related IP source and destination are identified as peers involved in a new TCP flow. Datagram exchanged by those IP addresses are retrieved and stored until the number of packets required for classifying application has been reached. Number of packets analyzed for each flow is settable (for instance: six after three-way handshake). This method enables us identifying TCP flows through IP addresses and TCP flags, as well as retrieving plain information for classifying applications such as lengths, direction and arrival time of each packet.

Third phase: preprocessing and normalization phase. Each TCP flow previously detected includes TCP control information as well as retransmitted packets, so lists of TCP flows are preprocessed according to specifications described in Section 4.2 and normalized as depicted in Section 5.1, in "normalization" sub section.

Fourth phase: classification. Output of the previous phase is a list of application flows processed and normalized. Our k-means based classification machine is configured off-line through process shown in Figure 4 and described in Section 5.1. It is trained by using training set composed by traces collected as described in Section 4.1. As depicted in Figure 5, decision phase consists in classifying flow (that could be seen as a point in $R^n$, $p^{TS}$) to the application of the nearest centroid in $R^n$. Each centroid has been assigned to a well known application during training phase ($c^{new}$), according to section 5.1.

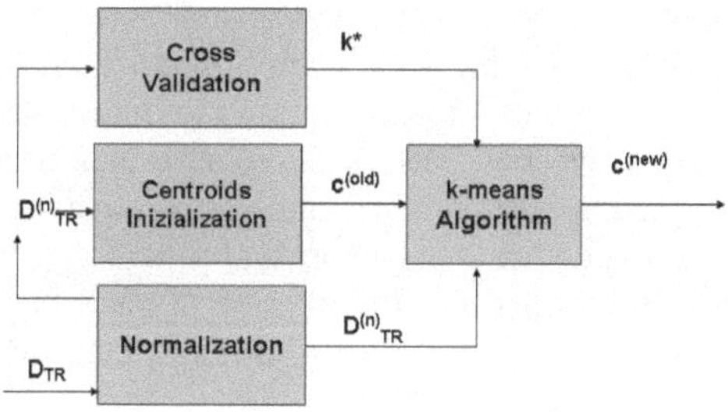


**Fig. 4.** A k-means based approach: off-line training phase

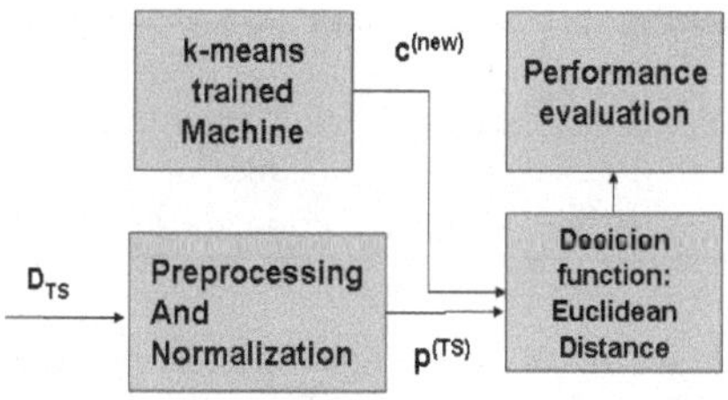


**Fig. 5.** A k-means based approach: on-line traffic classification

## 6 Experimental Results

By classifying plain flows with our approach, we obtained results shown in table 1.

**Table 1.** Plain application flows classification results

| | | Accuracy (%) | | | | |
|---|---|---|---|---|---|---|
| Pkt <l,d,t> | K* | HTTP | FTP-C | SSH | POP3 | Average |
| 3 | 32 | 99.0 | 34.1 | 99.0 | 95.7 | 81.95 |
| 4 | 35 | 98.9 | 82.9 | 98.9 | 92.3 | 93.25 |
| 5 | 26 | 98.6 | 32.0 | 99.3 | 90.6 | 80.13 |
| 6 | 23 | 99.2 | 92.2 | 98.4 | 88.3 | 94.53 |
| 7 | 28 | 99.5 | 92.9 | 99.1 | 90.2 | 95.43 |

We can notice that by analyzing the firsts six packets of each flow we achieve average accuracy up to 94.53%, with 23 clusters as the optimal value found by the cross validation technique. We can demonstrate that accuracy in recognizing SSH application flows is up to 98.4% relying on the statistical features of lengths, directions and absolute times. We emphasize that increasing the number of analysed packets improves accuracy, even if it delays the classification intended to be real time. Results demonstrate that six packets are necessary to have a trade-off good enough to guarantee high classification accuracy as well as acceptable classification delay.

Once SSH has been classified, we have applied our classification method to identify application flows forwarded in SSH tunnels.

In classifying SSH tunnel content, we have not used arrival times. In fact times related to HTTP over SSH are greater than others due to delay introduced by navigation trough the main server. This could affect our classification results. To make our analysis more realistic for classifying SSH tunnels we have represented each packet using only its size and direction, discharging arrival times.

We tried out processing all possible combination of packets up to ten packets after end of SSH negotiation (i.e. the initial common handshake phase, same in all SSH flows).

**Table 2.** Encoded SSH applications flows

| 1° | 2° | 3° | 4° | 5° | HTTP over SSH | scp | sftp |
|---|---|---|---|---|---|---|---|
| 0 | 0 | 1 | 1 | 1 | 99.80% | 98.93% | 99.75% |
| 0 | 0 | 1 | 1 | 0 | 99.88% | 99.30% | 99.05% |

As shown in Table 2, we tested different patterns representations, increasing the considered number of packets for each flow in order to identify which one contains more information to emphasize difference among applications. As shown in Table 2, the K-means based algorithm yields very interesting results in terms of identification of encoded applications. We can detect different applications with accuracy up to 99.8 for HTTP over SSH protocol, just analyzing third and fourth packets after SSH negotiation. We can notice that analyzing also the fifth packet does not improve significantly accuracy. Moreover, increasing the considered number of packets means introducing delay for real time recognition.

**Table 3.** Encoded SSH application flows (with scp and sftp upload and download)

| 3° | 4° | 5° | 6° | 7° | 8° | 9° | HTTP over SSH | scp up | scp down | sftp up | sftp down |
|---|---|---|---|---|---|---|---|---|---|---|---|
| 1 | 1 | 0 | 0 | 0 | 1 | 0 | 90.00% | 96.78% | 95.65% | 96.74% | 4.61% |
| 1 | 1 | 1 | 0 | 0 | 1 | 0 | 89.78% | 96.83% | 95.87% | 96.70% | 4.61% |
| 1 | 1 | 1 | 0 | 0 | 0 | 1 | 92.44% | 94.65% | 88.57% | 96.17% | 5.61% |

We have also tried to classify uploads and downloads of SCP-SSH and SFTP-SSH flows. Results are less encouraging: in fact accuracy decreases for every application. At a glance, SFTP downloads get confused with other applications. By inspecting data set we concluded that this method mistakes this kind of classification due to the fact that SFTP and SCP are characterized by similar patterns in terms of directions and packet's lengths. Further tests have been performed by analyzing a greater number of packets up to the ninth without achieving significant improvements.

## 7 Conclusion and Future Works

In this paper we describe a cluster analysis based method to classify in real time encoded traffic flows, overcoming actual limits of deep packet inspection. We are able to identify SSH flows out of other plain TCP based applications with accuracy up to 99.2% after collecting and analysing up to six application packets. Other protocols have been correctly classified with accuracy up to 94.53%. Once SSH flows have been detected, we can classify the nature of each SSH tunnel continuing in gathering session packets. In doing so, we gain accuracy up to 99.88% in classifying HTTP over SSH just analyzing the third and fourth packet after the end of the SSH negotiation phase. The same encouraging results have been obtained by classifying SCP (up to 99.3) and SFTP (up to 99.05) applications. Further works should be performed in order to improve results for classification of download and upload flows for SCP and SFTP. Moreover, it will be necessary to investigate the applicability of the approach on wider application dataset.

Recovering large quantity of well-known traffic is a critical point in traffic analysis due to privacy issues in collecting traces from geographic links. We also tried to classify CAIDA traces [11], but direction of packets is unavailable making our pre-processing assumption useless. We have obtained these good results in our own dataset; yet more traces would help assessing the robustness of our methodology.

Currently on-going work includes extension of the classification tool to more powerful classification algorithms, well beyond k-means; in this respect, k-means shall be regarded as a first use attempt, to verify the soundness of our approach, before proceeding to more complex yet reliable classification algorithms. We are also working on preprocessing and normalization in order to improve the extraction of statistical behavior of application level flows.

## References


1. Karagiannis, T., Papagiannaki, D., Faloutsos, M.: BLINC: Multilevel traffic classification in the dark. In: Proc. of ACM SIGCOMM 2005, Philadelphia, PA, USA (August 2005)
2. Crotti, M., Dusi, M., Gringoli, F., Salgarelli, L.: Traffic Classification through Simple Statistical Fingerprinting. ACM SIGCOMM Computer Communication Review 37(1), 5–16 (2007)
3. Wright, C., Monrose, F., Masson, G.: On Inferring Application Protocol Behaviors in Encrypted Network Traffic. Journal of Machine Learning Research (JMLR): Special issue on Machine Learning for Computer Security 7, 2745–2769 (2006)
4. Moore, A.W., Zuev, D.: Internet traffic classification using Bayesian analysis techniques. In: ACM SIGMETRICS 2005, Banff, Alberta, Canada (June 2005)

5. McGregor, A., Hall, M., Lorier, P., Brunskill, J.: Flow clustering using machine learning techniques. In: Barakat, C., Pratt, I. (eds.) PAM 2004. LNCS, vol. 3015, pp. 205–214. Springer, Heidelberg (2004)
6. Zander, S., Nguyen, T., Armitage, G.: Automated traffic classification and application identification using machine learning. In: LCN 2005, Sydney, Australia (November 2005)
7. Bernaille, L., Teixeira, R., Salamatian, K.: Early Application Identification. In: Proceedings of CoNEXT (December 2006)
8. Alshammari, R., Nur Zincir-Heywood, A.: A Flow Based Approach For Ssh Traffic Detection. In: IEEE International Conference on Systems, Man and Cybernetics, 2007. ISIC (2007)
9. http://www.openssh.com/
10. MTU: RFC 879
11. http://www.caida.org

# Evaluation of a Multiobjective Alternative Routing Method in Carrier IP/MPLS Networks
## (Work in Progress)

Lúcia Martins[1,4,*], Catarina Francisco[2], João Redol[2], José Craveirinha[1,4], João Clímaco[3,4], and Paulo Monteiro[2]

[1] Department of Electrical and Computers Engineering, University of Coimbra 3030-290 Coimbra, Portugal
{lucia,jcrav}@deec.uc.pt
[2] Nokia Siemens Networks S.A., Rua Irmãos Siemens,1, Ed. Alfravip 3, Piso 3, Alfragide, 2720-093 Amadora, Portugal
{catarina.francisco,joao.redol,paulo.1.monteiro}@nsn.com
[3] Faculty of Economics, University of Coimbra 3030-290 Coimbra, Portugal
jclimaco@fe.uc.pt
[4] INESC-Coimbra, Rua Antero de Quental 199, 3000-033 COIMBRA, Portugal

**Abstract.** We present a first simplified version of the MultiObjective Dynamic Routing (MODR) method, more suitable for a realistic network environment as the computational effort is very much reduced while good results can still be reached. The simplified version presented herein is based on the results obtained from a discrete event simulation study which shows that, in case of overload, more important than the alternative routing algorithm itself is to control the excess of alternative routing traffic. Moreover, in a multiservice network in the case of lightly loaded traffic conditions, when alternative routing starts to be effective, network performance can still be improved if we can avoid alternative routing for specific traffic flows. Classical dynamic alternative routing methods for traditional ISDN networks have a trunk reservation mechanism with a similar purpose but apparently without the same performance. Our method applies to MPLS strongly meshed networks which are typical of core networks.

**Keywords:** QoS, MPLS Networks, alternative routing, multiobjective optimization.

## 1 Introduction and Motivation

The rapid transformation of the Internet into a commercial infrastructure supporting many types of services which can integrate not only best effort traffic but also IP-telephony, IP-multimedia as well as other types of services, gives rise to new routing protocols based on QoS (Quality of Service) parameters. These

* Corresponding author.

L. Fratta et al. (Eds.): NETWORKING 2009, LNCS 5550, pp. 195–206, 2009.

new services have network performance requirements like end-to-end delay, delay jitter, required bandwidth and packet loss probability, that must be fulfilled. This evolution seems to lead to the absorption of traditional ISDN networks by these new IP-based networks (e.g., British Telecom [1]).

Any Transport over MPLS (AToM) is a solution for transporting Layer 2 packets like ATM, Frame Relay, Ethernet, PPT or HDLC, over a single, integrated, packet-based MPLS backbone network, instead of separate networks with different network management environments. In a nutshell, AToM inserts a label in the packets at the provider-edge router, based on the Forwarding Equivalence Class (FEC), and then transports them over the backbone. A FEC is a group of IP packets which are forwarded in the same manner and for that reason they belong to the same label-switched path (LSP).

Regarding MPLS, each explicit LSP is treated as a point-to-point path that, for a given time duration, has a constant bandwidth. In MPLS, if an explicit path specified with a 'non-mandatory' preference rule attribute value is not feasible, an alternative route (or path) may be chosen [2]. Hence, if a flow request does not find available resources it needs in the first choice path, a second chance may be given to that flow as it will be possible to try a pre-computed alternative path.

On the other hand, DiffServ [3] is a coarse-grained, class-based mechanism for traffic management. DiffServ networks operate on the principle of traffic classification, where each data packet is placed into one of a limited number of traffic classes, rather than differentiating network traffic based on the requirements of an individual flow (like in IntServ networks). DiffServ has two important advantages over IntServ: all of the processing takes place before the flows enter the network, at the boundaries, and the flows are aggregated so that there is no need for routers to analyze the requirements of each individual flow, eliminating the scalability issues. However, DiffServ does not solve the problem of call admission control (CAC), which is essential for QoS guarantees. This implies that QoS, with Diffserv, is usually guaranteed by overprovision, which is not always possible.

To implement CAC there is the pre-congestion notification (PCN) architecture suggested by IETF [4] which enforces QoS by marking packets based on the utilization of links and gives early warnings before congestion occurs. To cope with the issue of exceeding bandwidth allocation, a per flow admission control is suggested for a DiffServ network, in particular a measurement-based admission control (new flow requests are blocked dynamically in response to actual (incipient) congestion on a router within the DiffServ network). In this context, instead of a lost connection, a second chance may be given to these flow requests by allowing alternative routing.

The method developed in this paper applies to DiffServ-aware-MPLS meshed networks with PCN. In our multiservice model, traffic with different bandwidth requirements is classified into the same FEC and because of that is carried in the same LSP between adjacent nodes.

In the multiservice model it is considered that, for the time duration of each flow, it requires constant bandwidth on each LSP corresponding to the effective bandwidth that is characteristic of that type of flow. Effective bandwidth can encapsulate traffic behaviour and QoS issues at the cell and packet levels [5,6]. In addition, we can forget the bursts and bandwidth variations because of the PCN-threshold-rate which allows PCN-boundary-nodes to convert measurements of PCN-markings into decisions about flow admission. At this point, blocked requests may be rerouted to an alternative path.

Our approach treats each explicit LSP as a multiservice point-to-point path with a constant bandwidth shared by all services, were each flow is admitted in the LSP if the effective bandwidth necessary for that flow is available, otherwise the flow is rejected. This behaviour together with the proper adjustment of the the PCN thresholds, allows the consideration of a quasi circuit switching capability superimposed on the current Internet routing model [2].

The necessity of dealing with multiple and multifaceted QoS requirements in the new network technological platforms makes that there are potential advantages in formulating many routing optimisation problems as multicriteria models. In the particular multiobjective formulations enable the trade-offs among different objective functions (QoS metrics or cost functions) to be treated mathematically in a fully consistent manner. In this type of formulation instead of the concept of optimal solution the concept of non-dominated solution should be used that is a solution such that it is not possible to improve one of the objective functions unless at least one of the others is worsened. A state of art review on applications of multicriteria analysis in telecommunication network design is in [7]. A recent review on multicriteria routing models with an application study, is in [8]. In references [9] (for single-service networks) and [10] (for multiservice networks) a multiobjective dynamic routing model designated as MODR was formulated and solved through a heuristic approach. This model may be considered as a particular case of the network-wide optimisation meta-model for multiobjective routing in MPLS networks proposed in [11].

The main contribution of this paper is to present a first simplification of the former MODR method, which was developed in order to obtain a more suitable version for application to a realistic IP/MPLS network environment. In particular this version aims at a significant reduction in the computational effort required by the method while maintaining good results in terms of network performance measures. The routing algorithm presented herein is based on a procedure that selectively eliminates each alternative path and is a much simplified version of the one proposed in the MODR method. Also the use of Howard costs (much easier to compute) instead of implied costs is analysed in this context.

The paper is structured as follows. In section 2, general features of the MODR method are reviewed and the new simplified version aimed at a reduction on path computational effort, is described. In section 3, simulations regarding the decrease of computational effort and some procedures will be presented. In section 4 a comparative study between two metrics (implied costs, which is the metric used by MODR method, and Howard costs, suggested in the Separable Routing scheme in [12]) is presented in order to decide which one is more effective in our routing algorithm. In section 5 conclusions are presented and discussed.

# 2 The Multiobjective Dynamic Routing Method

## 2.1 Review of the MODR Method

The MODR method applies to strongly meshed networks, in which it has been extensively documented in the literature that the first choice route should always be the direct one if it exists. This article describes a simplification/adaptation of MODR, a hierarchical Multiple Objective Dynamic Routing model for telecommunication networks, presented in [9,10]. The general purpose of MODR is to find, in a strongly meshed network, in each route updating interval, the set of alternative paths for all flows that adapt the best to the offered traffic conditions, in order to fulfill the objectives at network and service levels. In the present context we consider as strongly meshed networks, those with topological density close to a complete graph and such that for each origin-destination pair there are at least two 2-link paths. We begin by reviewing the hierarchical multiobjective alternative routing model that MODR addresses.

Notation:

- $G = (V, L)$ - undirected graph representing the network topology where V is the node set and L the arc set;
- $f_s \equiv (v_o, v_t, \gamma)$ where $v_o, v_t \in V$ and $v_o \neq v_t$ - is a traffic flow from node $v_o$ to node $v_t$ of service type $s$ where $\gamma$ represents a traffic descriptor which enables a complete definition of the associated stochastic process (e.g. mean service $s$ time $h_s$, number $n_s$ of links required by each connection of traffic flow $f_s$ in every arc of each attempted path);
- $F$ - set of all traffic flows in the network;
- $A_t(f_s) = I_t(f_s)\, h_{fs}$ where $I_t(f_s)$ represents the average arrival intensity during time period $t = nT\,(n = 1, 2, ...)$ - traffic offered (in Erlangs) for traffic flow $f_s = (v_i, v_j, \gamma) \in F$ at time $t$ and $h_{f_s}$ is the mean occupation time of $f_s$ flow calls;
- $B(f_s)$ - point-to-point blocking probability for traffic flow $f_s \in F$;
- $R_t(f_s) = \{r^1(f_s), r^2(f_s) : r^1(f_s), r^2(f_s)\}$ - where $r^1(f_s)$ and $r^2(f_s)$ are loopless paths. $R_t(f_s)$ is the ordered set of paths which may be used by flow $f_s$ at time $t$;
- $\bar{R}_t = \{R_t(f_1), \ldots, R_t(f_{|F|})\}$ - routing plan for the network at time $t$;
- $B_{ks}$ - blocking probability experienced by a service $s$ call on link $l_k = (v_i, v_j) \in L$;
- $C_k$ - capacity of link $l_k = (v_i, v_j) \in L$;
- $\rho_{ks}$ - service $s$ total offered traffic to link $l_k$ (the mean of the total number of calls of type $s$ offered to $l_k$ during calls mean service time);
- $L_{r^i(f_s)}$ - mean blocking probability on route $r^i(f_s)$, experienced by a call of $f_s$;
- $\underline{d_k} = [d_{k1}, \ldots, d_{k|S|}]$ - required bandwidth on link $l_k$ by a call of service $s \in \{1, 2, \ldots, |S|\}$, which may be interpreted as its effective bandwidth;
- $\mathcal{D}(f_s)$ - routing domain for traffic flow $f_s$ which encompasses the set of all possible paths from origin node $v_o$ to destination node $v_t$.

As stated in [9], it is assumed the following: all traffic flows are homogeneous Poissonian and independent, service times are negative exponentially distributed, there is statistical independence in the occupations of the links and routes $r^1(f_s), r^2(f_s)$ are node disjoint.

The blocking probability of a connection of type $s$ in arc $l_k$ is given by $B_{ks} = \mathcal{L}_s\left(\overline{d_k}, \overline{\rho_k}, C_k\right)$. As explained in [10], functions $\mathcal{L}_s$ represent the traffic calculation model that enables the marginal blocking probabilities on the links to be computed namely according to the methods in [13,14].

The MODR method relies on a heuristic for route calculation and selection based on two mechanisms: first, a biobjective shortest path algorithm (MMRA) to obtain the subset of candidate non-dominated alternative path solutions $R_t(f_s)$ for each flow, and second, a procedure to decide which alternative paths should be updated in each time interval. The problem formulation for MMRA is as follows:

$$(\text{Problem } \mathcal{P}^2) \qquad \min_{r_s \in \mathcal{D}(f_s)} m^n(r_s) = \sum_{l_k \in r_s} m^n_{ks}, n = 1, 2 \tag{1}$$

where $m^i_{ks}$ is the value of metric $i$ associated with link $l_k$ and service $s$, $m^i(r_s)$ is the value of objective function $i$ for path $r_s$. These metrics are the implied costs $m^1_{ks} = c_{ks}$, as defined in [15,10,11], and the blocking probability $m^2_{ks} = -\log(1 - B_{ks})$. The log is used to transform blocking probability into an additive metric.

Let's consider the following simplifications:

$d_{ks} = d_s\left(\forall l_k \in r^i(f_s) \wedge \forall s \in S\right)$ which will also be made equal to the revenue associated with a call of all traffic flows $f_s$. $A^o_s$ and $A^c_s$ are the service $s$ total offered and total carried traffic, respectively. A heuristic was developed to discover, in each time interval and among the set of non-dominated solutions discovered by MMRA, the set of alternative paths to update in order to guarantee a compromise solution in terms of the network level objective functions (o. fs.), (aiming at maximizing network expected revenue $W_T$ and minimizing the maximal service mean blocking probability $B_{Mm}$) and service level o. fs. (in order to minimize the service mean blocking probabilities $B_{ms}$ and the maximal point-to-point blocking probability, $B_{Ms}$, for each service s). The formalization of the hierarchical multiple objective dynamic alternative routing problem for multiservice networks is (Problem $\mathcal{P}^{G_S}$):

$$NL: \qquad \min_{\overline{R}_t} -W_T = -\textstyle\sum_{s \in S} d_s A^c_s = -\sum_{s \in S} d_s A^o_s (1 - B_{ms}) \tag{2}$$

$$\min_{\overline{R}_t} B_{Mm} = \max_{s \in S}\{B_{ms}\} \tag{3}$$

$$SL: \qquad \min_{\overline{R}_t(s)} B_{ms} = (A^o_s)^{-1} \textstyle\sum_{f_s \in F_s} A_t(f_s) B(f_s), \; s = 1, \ldots, |S| \tag{4}$$

$$\min_{\overline{R}_t(s)} B_{Ms} = \max_{f_s \in F_s}\{B(f_s)\}, \; s = 1, \ldots, |S| \tag{5}$$

$s.t.$ Equations of the teletraffic model to calculate$\{B(f_s)\}$ in terms of $\{A_t(f_s)\}$ and $\overline{R}_t$

Important to note that this is a hierarchical optimization problem where the first level objective functions (NL) have priority over the second level objective functions (SL).

Finally, an additional mechanism (APR - Alternative Path Removal) was introduced in the original heuristic as a service protection scheme the objective of which consisted of preventing blocking degradation in overload network situations due to excessive use of alternative routing. In this scheme, elimination of alternative routes occurs whenever the following condition stands:

$$m^1(r_s) > d_s \wedge m^2(r_s) > -z_{APR} \times \log(1-0.3) \tag{6}$$

where $z_{APR}$ is just an empirical parameter used by the heuristic, which varies dynamically between 0 and 1 in the inner cycle of the procedure.

It was already proved that, assuming quasi-stationary conditions, such that the offered traffic stochastic features remain stationary during periods which are relatively long compared to the solution time, the single objective alternative routing problem is NP-complete in the strong sense. Since the problem $\mathcal{P}^{\mathcal{G}s}$ is a multiobjective one and having in mind the interdependencies between network mean blocking and maximal marginal blocking probabilities and their dependencies on the routing plan, it is expected great intractability for this problem. The foundation of the heuristic procedure is the search for a subset of the alternative path set for all flows, the elements of which should possibly be modified in a given route update period. This leads to a heuristic with two internal cycles of solution improvement that is very heavy in terms of computational cost. Details are given in [9,10]. This heuristic is now replaced by a simplified version, more suitable to be applied in real networks as described in the next sub-section.

### 2.2 Proposal of a Simplified Method

Our main objective, in this paper, is to propose another simpler heuristic in order to fulfil as far as possible the original objectives for the alternative routing problem. Our approach consists of seeking to update sequentially, in each time interval, only a subset of the available pairs of routes, instead of all route pairs (complete routing plan) as in the original heuristic. The number of paths to update in each time interval is directly related to the speed at which the network evolves due to changes in the offered traffic. However, as explained in [9], neither the update of all pairs nor the update of only one origin-destiny pair in each time interval is a good policy. In addition, experience has shown that at least as important as the routing algorithm itself, is the way in which direct traffic is protected in overloaded networks, as suggested in [16]. These two different but related aspects of the problem will be explained next.

Concerning the first aspect of the problem discussed above, and after a number of experiments a first simplified strategy was considered which consists of updating, in each period, the alternative routes for $\alpha$ pairs of nodes alone for every service. In our case study networks the recommended value was $\alpha = N/2$ where $N = |V|$ (number of network nodes). Note that this implies that all alternative routes for all services can be updated every $\frac{|F|}{|S|\alpha}$ (where $|F|$ is the total number of node to node flows) route updating periods. In other network structures different values of $\alpha$ might have to be considered after an experimental study with the routing method.

Let $t = n\mathcal{T}\,(n = 1, 2, \cdots)$ where $\mathcal{T}$ is the path update time interval, $\overline{R}_t^{(n)}$ the routing plan for the $\mathrm{n^{th}}$ update interval. In addition, let's considerer $\overline{R}_t^* = \{r^2(f_s) : r^2(f_s) \text{ is updated by MMRA at } t = n\mathcal{T}\}$. Consider also that the initial origin-destiny pair value in the pseudocode below is 1-1.

1. $\overline{R}_t^{(n)} \leftarrow \overline{R}_t^{(n-1)}$
2. Calculate $\overline{B}$, $\overline{c}$, $\{B_{ms}\}$ and $B_{Mm}$, for $\overline{R}_t^{(n)}$ and a given $\overline{A}_t$ estimate using the fixed point iterators. Consider $\overline{R}^*_{t_{old}} = \{\}$, $\overline{R}^*_{t_{new}} = \{\}$ and counter $\leftarrow 0$
3. while (counter $<\ \alpha$) do
   (a) destiny $\leftarrow$ destiny + 1
   (b) if (destiny = N+1 ) origin $\leftarrow$ origin + 1 and destiny $\leftarrow$ 1
   (c) if (origin = N+1) origin $\leftarrow$ 1 and destiny $\leftarrow$ 2
   (d) if (origin = destiny $\wedge$ destiny $\neq$ N) destiny $\leftarrow$ destiny + 1
   (e) if (origin = destiny $\wedge$ destiny = N ) origin $\leftarrow$ 1 and destiny $\leftarrow$ 2
   (f) for (s=1 until s=S) do
      i. $\overline{R}^*_{t_{old}} \leftarrow \overline{R}^*_{t_{old}} \cup \{r^2(f_s) : f_s \equiv (v_o, v_t, \gamma) \wedge\ v_o \equiv \text{origin} \wedge v_t \equiv \text{destiny}\}$
      ii. Use MMRA to determine the new $r^2(f_s)$
      iii. $\overline{R}^*_{t_{new}} \leftarrow \overline{R}^*_{t_{new}} \cup \{r^2(f_s)\}$
      iv. Selective elimination of $r^2(f_s)$ (according to criterion (8) later explained)
   (g) counter $\leftarrow$ counter +1
4. $\overline{R}_t^{(n)} \leftarrow \overline{R}_t^{(n)} \setminus \overline{R}^*_{t_{old}} \cup \overline{R}^*_{t_{new}}$

The experimentation showed that the original MODR heuristic achieves better results in terms of global performance than the presented approach because it recalculates the routing plan for all the network flows in each update period. This was already expected, nevertheless, the gain achieved with the speed and simplicity of this new method was an incentive for the continuation of our study and lead us to second aspect of our problem.

We can define a numerical complexity value for MODR in terms of the upper bound of the number of alternative routing solutions that may be analysed in the heuristic. This complexity is of the order of $|S||\bar{F}|^2$ where $|\bar{F}|$ is the average number of traffic flows per service. For the 6 node network in the experimental study in section 2.3 this gives 2700 while the simplified heuristic only analyses $\alpha|S| = 9$ solutions, hence leading to a quite significant complexity reduction. The CPU time for the original heuristic is 21.844 seconds in a 2.8 GHz Pentium 4 while the new heuristic takes 94.3 milliseconds.

In this experimental study, extensively explained in [17], the direct traffic protection mechanism in case of overloads is based on alternative path elimination because from our experience (and [6]), it gives better global performance than trunk reservation schemes. So, with $z_{APR} = 1$ (the initial value of the parameter, which varies in the original heuristic, but which disappeared with this new approach), the path implied cost ($m^2(r_s)$) and the path blocking probability ($m^1(r_s)$) are calculated so that the alternative path is eliminated whenever condition (7) is verified:

$$m^2(r_s) > -\log(1-0.3) \wedge m^1(r_s) > d_s \tag{7}$$

### 2.3 Performance Evaluation

A discrete-event simulator was used for the comparative study of the MODR performance, considering the reformulation of the heuristic. Two fully meshed networks with six nodes ('A' and 'M') presented in [17] were used for this study to allow a comparison with previous work and also because simulation time for the original heuristic is very high. These networks were engineered with three services: telephone, data and video, with the required bandwidth $\overline{d} = [1, 6, 10]$ for each service and call durations of 1, 5 and 10 minutes, respectively.

Simulations were carried out with different path elimination criteria for both test networks.

Note that the constant 0.3 in equation (7) corresponds to a threshold of 30% for the blocking probability which in practice tends to protect (from excessive alternative routing) the more demanding services (since these tend to have higher blocking) leaving the less demanding services with potentially excessive alternative routes. To overcome this limitation a new factor $0.1\frac{B_{ms}}{B_{Mm}}$ was introduced in that condition so that the smaller is the mean blocking of the services relative to maximal mean of all services $B_{Mm}$, the lower is the blocking threshold above which the alternative route is eliminated. Our next modification consists of the substitution of the AND by the OR operator, which allows us to take more advantage of implied costs in the sense that it seems advisable to eliminate an alternative route when the corresponding implied cost is greater than the expected revenue per connection of the current traffic flow, independently of the condition on the blocking probability. This leads us to the following condition:

$$m^2(r_s) > -\log\left(1 - 0.1\frac{B_{Mm}}{B_{ms_d}}\right) \vee m^1(r_s) > d_s \tag{8}$$

We can state what we had already concluded from extensive simulation in both networks: that the original criterion (7) allows the highest blocking probabilities to be obtained for the less demanding services, and is slightly better for low load situations. On the contrary, condition (8) which implements fairness in the removal process of alternative paths for the different services, is the best suited for nominal and overload situations. In order to clearly evaluate the advantages of alternative routing and justify equation (8), we also present the results from a simulation with the direct routing scheme, where no alternative routes exist.

Another topic of importance is the estimation of the average traffic offered to the network by a given flow.In the simulator, the estimated offered traffic $\tilde{x}$ in the $n^{th}$ time interval for traffic flow $f_s$ is obtained from an estimate $\tilde{X}(n-1)$ of the offered traffic in the previous interval calculated from on-line measurements, for the same traffic flow, by using a first order moving average iteration: $\tilde{x}_{f_s}(n) = (1-b)\tilde{x}_{f_s}(n-1) + b\tilde{X}_{f_s}(n-1)$ (as suggested in [15]) with $b = 0.1$ (which is the value proposed in [18]) because while relying in traffic history still allows a slow adaptation in case of changes in the network, which is better suited for overload situations.

Another evaluated topic was the influence of the network load in path update intervals. In this respect, a comparison was made regarding a 10 seconds (a typical value in circuit-switching networks) and a 1 minute (previously used) update interval. A smaller route update interval achieves better results in underloaded situations as it allows traffic flows to be better accommodated with the frequent changes in path allocations, while a 1 minute interval has a better performance for overloaded situations because sudden changes in the offered traffic do not result in a "bad" set of paths in the following interval.

Regarding the possibility of a service dependent path update interval, different values were used depending on the service average duration at stake. None of the simulations with different service update intervals achieved good results, the 10 seconds choice (instead of the 1 minute in use in the original heuristic) being the one with the most appealing performance. This smaller update interval complements the reduced number of paths being updated in each time interval in the new heuristic, when compared with the original one.

In conclusion, the decisive factors in this approach are the load situation, the arrival rate and the alternative path selective elimination.

The comparative analysis between the original MODR and other reference dynamic routing methods is out of the scope of this work and can be consulted in [10]. A comparative study of different method variants for the test network presented in [10] is in figure 1. Note that the assumed 'nominal load' considered in these experiments is 20% less than in [10]. The simulations results presented, obtained by the method of the independent replications, are the mid points of a 95% confidence interval. Regarding the two global network performance metrics we can conclude that the original heuristic behaves better than our simpler heuristic, if we consider the same alternative path elimination mechanism in both methods. However, if we make use of criteria (8) for the path elimination, we can obtain results that are comparable with the ones of the original heuristic, and even improved in terms of expected revenue in overload situations, achieving other non-dominated solutions in terms of global network performance. In fact, from extensive experimentation, it is possible to confirm an interesting conclusion: in a meshed network, in case of overloads, it is much more important to control the excess of alternative traffic than the alternative routing algorithm itself. Details related to service performance analysis are in [17].

Implied costs have already demonstrated to behave well in the proposed routing model. However, we decided to make a comparison with a different and much lighter metric in terms of computational effort, namely Howard costs.

In [19] a scheme is presented called Forward-Looking Routing (FLR) based on Howard costs. These costs, $\Delta(k, j)$, can be interpreted as the expected increase in the number of future blocked calls on a link $l_k$ due to the acceptance of a call

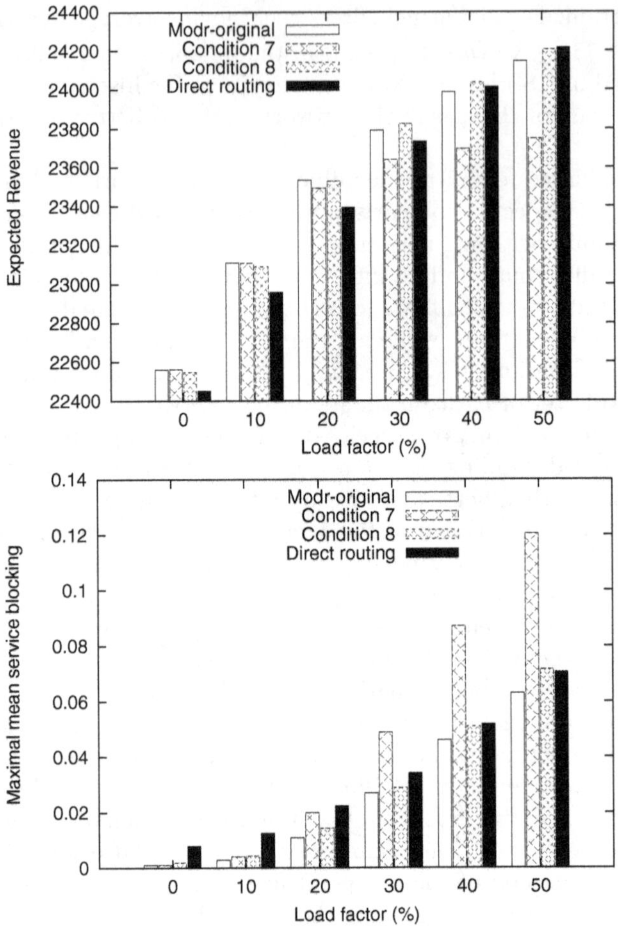


**Fig. 1.** Comparison with respect to the original version of MODR and direct routing – Global Performance (expected revenue and maximal mean service blocking)

when $j$ calls are already in progress. Howard costs were adapted in a simplistic way to a multiservice environment as follows: $\Delta(k,j) = \frac{B_{ks}}{B_{k_j s}}, 0 \leq j \leq C_k$, where $B_{k_j s} = \mathcal{L}_s\left(\overline{d_k}, \overline{\rho_k}, j\right)$ are the blocking probabilities calculated by the algorithms mentioned in the previous section.

Paths with the minimal Howard cost tend to contribute to the maximization of throughput and to an adequate load balancing, as routes with less calls in progress are the ones which tend to be chosen. As Howard costs are additive, the path cost is given by: $m^1(r_s) = \sum_{l_k \in r_s} \Delta(k_i, j_i)$. These costs replace the implied costs in the bi-objective shortest path sub-algorithm MMRA, in our revised simpler heuristic. The results for network global performance are presented in the report [17] and for this test network are very similar to those obtained with implied costs. However in other test networks Howard costs lead to worse results then implied costs using the same simplified heuristic. Therefore the introduction of Howard costs requires a careful pre-evaluation.

## 3 Conclusions and Further Work

Best-effort architecture does not meet the requirements of the current integrated services network Internet carrying heterogeneous data traffic. For this reason, high-speed wide area networks are likely to be connection-oriented for real-time traffic. Traffic engineering with Multiprotocol Label Switching (MPLS) is an attempt to take the best out of connection-oriented traffic engineering techniques.

The approach described in this paper attempted to implement alternative routing in IP/MPLS networks. This type of networks based on shortest-path routing have frequently localized congestion which may be smoothed by alternative routing. To achieve this, MODR formalized the routing problem as a multiobjective hierarchical routing problem in order to promote global fairness in terms of the QoS of the multiple services. Our starting point for solving this difficult problem was an 'heavy' heuristic which is here replaced by a new one, with slightly worse but similar results. Nevertheless this simplified heuristic is more suited to a realistic environment as it is a few hundred times lighter in terms of computational effort.

An interesting conclusion which confirms, in the context of MODR, the remarks in [16] is that in a meshed network, in case of overloads, it is more important to control the excess of alternative traffic than the alternative routing algorithm itself.

The work presented above is the starting point to a QoS future modelling approach to routing optimisation aiming to be applied to DiffServ-aware-MPLS meshed networks. Future work will also include MPLS Fast Reroute because MPLS was designed to meet the needs of real-time applications and, for that reason, rapid route restoration upon failure becomes crucial.

## References


1. Cherry, S.: Nothing but net. IEEE SPECTRUM (January 2007)
2. Group, I.N.W., 2702, Awduche, D., Malcolm, J., Agogbua, J., O'Dell, M., McManus, J.: RFC2702 - Requirements for Traffic Engineering Over MPLS (September 1999)
3. Blake, S., Black, D., Carlson, M., Davies, E., Wang, Z., Weiss, W.: RFC2475 - An Architecture for Differentiated Service (December 1998)
4. Chan, K., Charny, A., Eardley, P.: Pre-congestion notification problem statement: Internet-draft, work in progress (2006)
5. Mitra, D., Morrison, J.A., Ramakrishnan, K.G.: Optimization and design of network routing using refined asymptotic approximations. Performance Evaluation An International Journal 36-37, 267–288 (1999)
6. Conte, M.: Dynamic Routing in Broadband Networks. Kluwer Academic, Dordrecht (2003)

7. Clímaco, J., Craveirinha, J.: 22, Multicriteria Analysis in Telecommunication Planning and Design - problems and issues. In: Multiple Criteria Decision Analysis: State of the Art Surveys. International Series in Operations Research & Management Science, vol. 78, pp. 899–951. Springer, Heidelberg (2005)
8. Clímaco, J., Craveirinha, J., Pascoal, M.M.: Multicriteria Routing Models in Telecommunication Networks - Overview and a Case Study. In: Advances in Multiple Criteria Decision Making and Human Systems Management: Knowledge and Wisdom, pp. 17–46. IOS Press, Amsterdam (2007)
9. Martins, L., Craveirinha, J., Clímaco, J., Gomes, T.: On a bi-dimensional dynamic alternative routing method. European Journal of Operational Research - Special Issue on Advances in Complex Systems Modeling 166(3), 828–842 (2005)
10. Martins, L., Craveirinha, J., Clímaco, J.: A new multiobjective dynamic routing method for multiservice networks: Modelling and performance. Computational Management Science 3(3), 225–244 (2006)
11. Craveirinha, J., Girão, R., Clímaco, J.: A meta-model for multiobjective routing in MPLS networks. Central European Journal of Operations Research 16(1), 79–105 (2008)
12. Krishnan, K., Ott, T.: Forward-Looking Routing: A New State-Dependent Routing Scheme. In: Bonatti, M. (ed.) TELETRAFFIC SCIENCE for New Cost-Effective Systems, Networks and Services, ITC-12, pp. 1026–1032. Elsevier Science B.V, North-Holland (1989)
13. Kaufman, J.S.: Blocking in a shared resource environment. IEEE Transactions on Communications 29(10), 1474–1481 (1981)
14. Mitra, D., Morrison, A.J.: Erlang capacity and uniform approximations for shared unbuffered resources. IEEE/ACM Transactions on Networking 2(6), 558–570 (1994)
15. Kelly, F.P.: Routing in circuit-switched networks: Optimization, shadow prices and decentralization. Advances in Applied Probability 20, 112–144 (1988)
16. Medhi, D.: QoS Routing Computation with Path Caching: A Framework and Network Performance. IEEE Communications Magazine 40(12), 106–113 (2002)
17. Martins, L., Francisco, C., Redol, J., Craveirinha, J., Clímaco, J., Monteiro, P.: A first evaluation of multiobjective alternative routing in strongly meshed MPLS networks. Technical Report ISSN: 1645-2631, INESC-Coimbra (November 2008)
18. Girard, A.: Routing and Dimensioning in Circuit-Switched Networks. Addison-Wesley Publishing Company, Reading (1990)
19. Ott, T., Krishnan, K.: State dependent routing of telephone traffic and the use of separable routing schemes. In: Akiyama, M. (ed.) International Teletraffic Congress, ITC-11. Elsevier Science B.V, North-Holland (1989)

# Dimensioning and Location Planning for Wireless Networks under Multi-level Cooperative Relaying

Bin Lin[1,2] and Pin-Han Ho[1]

[1] Department of Electrical and Computer Engineering, University of Waterloo, Waterloo, ON N2L 3G1, Canada
[2] Department of Information Technology, Dalian Maritime University, Dalian, China, 116026
{b2lin, pinhan}@bbcr.uwaterloo.ca

**Abstract.** This paper studies the problem of network dimensioning and location planning in multi-hop wireless networks. A key technology of wireless multi-level cooperative relaying (CR) is incorporated, which has been recognized as an effective design paradigm for achieving throughput enhancement. A mathematical formulation is presented to capture the nature of the problem, and characterize the behavior of multi-level CR. The tasks of dimensioning, relay placement, relay allocation, and relay sequence design have been considered in a unified optimization framework. The formulation is a nonlinear integer program. To avoid the intractable computation complexity in solution, an efficient two-phase algorithm is developed. We conduct a series of case studies to verify the proposed algorithm, in which the results demonstrate the efficiency of our proposed algorithm and the significant benefits in terms of deployment cost reduction under multi-level CR.

**Keywords:** Networking Dimensioning and Location Planning, Multi-level Cooperative Relaying, Optimization.

## 1 Introduction

Network planning is a never-ending process with the wireless network's evolution and growth. A new round of work is forced when it is needed to meet new requirements of boosting system throughput and/or changing of network environments [1]. Typically, new shopping centers and communities in a cell or at a cell's edge may lead to new hotspots or demand for new (micro-) cell sites, and newly constructed buildings may change the multipath environments, possibly resulting in "black holes" in the coverage area. To address these newly arisen problems in existing wireless networks, placing relay stations (RSs) between a base station (BS) and subscriber stations (SSs) provides a viable approach, instead of the conventional setup of a new (micro-) cell by deploying BS. The quality of wireless channels can be improved by multi-hop relaying. And RSs can be placed close to black holes, as they can establish extra transmission

L. Fratta et al. (Eds.): NETWORKING 2009, LNCS 5550, pp. 207–219, 2009.

links to circumvent obstacles. Moreover, compared with BSs, RSs can be deployed more easily, faster, and cheaper, which are rather appealing to network operator.

In this paper, we study the problem of **D**imensioning and **L**ocation **P**lanning (DLP) of RSs in multi-hop wireless networks. A key technology of *multi-level cooperative relaying (CR)* is incorporated since cooperative communication has been recognized as an effective paradigm for achieving throughput enhancement. Thus, incorporating promising technical breakthrough will lead to new problem, new model and new solution so as to achieve a cost-effective design.

The *multi-level CR* in this paper refers to a multi-hop transmission that both the source and internal relays cooperatively transmit to a destination. It is different from the traditional *multi-hop non-cooperative transmission (NCT)*, in which the received data are simply forwarded at each internal nodes to the next-hop node. In other words, with *multi-level CR*, the destination receives multiple copies of data packets (possibly in different codes) from the source and all the upstream relays; while, with traditional *multi-hop NCT*, the destination only receives one copy of data packet from its previous-hop node. Therefore, multi-level CR transmission can achieve even higher throughput. Moreover, multiple simultaneous active links with *multi-level CR* can provide better robustness and fault tolerance than a single link.

However, the relay locations have a critical effect on the destination's achievable rate [2] [3], even in the simplest relay model with only 3 nodes (source-relay-destination) [4]. In wireless networks with multi-level CR, deciding the location of each RS and estimating the amount of RSs needed are much more challenging. In literatures, numerous research efforts have been undertaken to address the placement problems. So *et. al* studied minimum cost configuration of relay and channel infrastructure in heterogeneous wireless mesh networks (WMNs) in [5], where Bender's decomposition-based heuristic was proposed to solve it efficiently. The placement of tetherless relay points (TRPs) to improve the throughput of a WLAN was investigated in [6], in which a Lagrangian relaxation iterative algorithm was developed. In [7] and [8], for planning WMNs, Amaldi *et. al* presented a novel optimization model and a relaxation-based heuristic. The issues of traffic routing, interference, rate adaptation, and channel assignment were also considered.

In this paper, we develop an optimization framework for the DLP problem. A mathematical model is provided to capture the nature of DLP problem, and characterize the behaviors and constraints under multi-level CR. We have jointly considered RS placement, relay allocation, and relay sequence design in a unified framework. The formulation is an integer nonlinear program, which is nonetheless subject to intractable computation complexity in solution. Thus, a two-phase algorithm is developed for solving the problem effectively and efficiently. Simulation and case studies are conducted to verify the optimization framework and demonstrate the economic and performance benefits of the multi-level CR against traditional multi-hop NCT.

## 2 Modeling of Multi-level Cooperative Relaying

In the fundamental multi-level CR model, a source (node 0) communicates with a destination (node $G$) through multiple relays (node $1, 2, \ldots, G-1$). The achievable rate for a fixed destination can be expressed as [9]

$$R = \min_{1 \leq j \leq G} [C(\frac{1}{N_0} \sum_{k=1}^{j} (\sum_{i=0}^{k-1} \sqrt{P_{ik} d_{ij}^{-\alpha}})^2)] \tag{1}$$

where $\alpha$ is the path loss exponent, $P_{ik}$ is the transmit power of node $i$ to node $k$, $N_0$ is the power of the background noise, $C(.)$ is the *Shannon function* such that $C(x) = \frac{1}{2}\log(1+x)$ for $x \geq 0$, with $d_{ij}$ as the distance between node $i$ and $j$. In this paper, we assume that all the nodes use the same power density for transmission, i.e., $P_{ik} = P_0$ for $\forall i, k$. The more complex issue of power control will be deferred for future research.

## 3 Network Model and Problem Formulation

### 3.1 Network Model

A three-tier network architecture is considered, which consists of three network entities: a BS, RSs, and fixed SSs. The BS serves as a central controller in the cell. The RSs are responsible for relaying data between the BS and the associated SSs under multi-level CR [9]. The RSs have no direct connections to the core network. Within a cell, there exists a certain number of new SSs, which could be a residential or business subscriber. Each SS submits a certain data rate requirement. These information must be known prior to the network planning, and they can be acquired by statistical data analysis of traffic measurement as well as anticipation of future growth. Since an RS cannot be placed anywhere due to some geographic limitations, certain physical locations in the cell are taken as candidate positions (CPs) for placing RSs. The COST231-Hata model is adopted as the channel model which is applicable to the transmissions inside an urban environment [10]. Small scale fading is not explicitly included in this study since a long-term planning and design is targeted. Additionally, in consideration of transmission delay, the maximum number of allowable intermediate RSs between the BS and an SS in a cell is assumed to be limited.

### 3.2 Problem Formulation

We consider a practical deployment scenario in a broadband wireless access network. The goals of the DLP problem are

(i) to obtain the minimal number of RSs to be deployed;
(ii) to obtain the optimal locations of the RSs,
(iii) to allocate serving RS(s) to each SS,
(iv) to determine the relay sequence of the serving RS(s) for each SS,

**Table 1.** Definitions of important symbols

| Symbol | Definition |
|---|---|
| $N_{CP}$ | The set of CPs for RSs within the cell, $\|N_{CP}\| = M$. |
| $N_{SS}$ | The set of SS within the cell, $\|N_{SS}\| = N$. |
| $\Omega$ | The set of all the nodes within the cell, $\Omega = \{BS\} \cup N_{CP} \cup N_{SS}$, $\|\Omega\| = 1 + M + N$. |
| $\rho_n$ | The minimal rate requirement for $SS_n$. |
| $P_0$ | The transmit power of an RS. |
| $N_0$ | The average thermal noise power in cell. |
| $\alpha$ | The path loss exponent. |
| $ML$ | The maximal allowable number of internal RSs in a cell. |
| $\overrightarrow{\mathscr{F}^n}$ | The virtue directed flow for $SS_n$. |
| $\mathcal{C}$ | The objective value of the problem DLP. |
| $\mathbb{Z}$ | The CP-RS location incidence vector. |
| $\mathbb{U}$ | The CP-SS association incidence matrix. |
| $\mathbb{F}$ | The virtue flow matrix. |
| $\mathbb{Q}$ | The relaying node level (stage) matrix. |
| $\mathbb{H}$ | The CP-relay-level incidence matrix. |
| $\mathscr{F}_j^{n,t}$ | The $j$th column of the virtue flow (relay sequence) for $SS_n$ with $t$ relays in FRS table. |
| $r_j^{n,t}$ | The $j$th column of the achievable rate for $SS_n$ with $t$ relays in FRS table . |
| $\xi(n, m)$ | The contributive index of $CP_m (\in N_{CP})$ to $SS_n (\in N_{SS})$. |
| $\Xi(m)$ | The set contributive index of $CP_m (\in N_{CP})$ to $N_{SS}$. |
| $\mathcal{C}^{LB}$ | The lower bound of the objective value. |

such that each SS's minimal rate requirement is satisfied. The notations used in the problem are listed in Table 1.

Let $\overrightarrow{G} = (\Omega, \overrightarrow{E})$ represent the directed graph, which consists of a node set $\Omega = \{BS \bigcup N_{CP} \bigcup N_{SS}\}$ and a directed edge set $\overrightarrow{E}$. $N_{CP}$ and $N_{SS}$ are the set of CPs and SSs, respectively. To formulate the problem, we start by defining the decision variables for RS location ($\mathbb{Z}$) and allocation ($\mathbb{U}$), respectively. $\mathbb{Z} = (z_m)_{1 \times M}$ is a CP-RS incidence vector (*location variable*) where $z_m = 1$ if $CP_m$ is selected for placing an RS; otherwise, $z_m = 0$. $\mathbb{U} = (u_{mn})_{M \times N}$ is a CP-SS incidence matrix (*CP allocation variable*) where $u_{mn} = 1$ if $CP_m$ is allocated to serving $SS_n$; otherwise, $u_{mn} = 0$.

Since the DLP problem targets at minimizing the total number of RSs, denoted by $\mathcal{C}$, the objective function can be expressed as,

$$\text{minimize} \quad \mathcal{C} = \sum_{m \in N_{CP}} z_m \tag{2}$$

The RS location and allocation variables (i.e., $\mathbb{Z}$ and $\mathbb{U}$) are subject to the following constraints.

$$u_{mn} \leq z_m, \quad \forall m \in N_{CP}, n \in N_{SS} \tag{3}$$

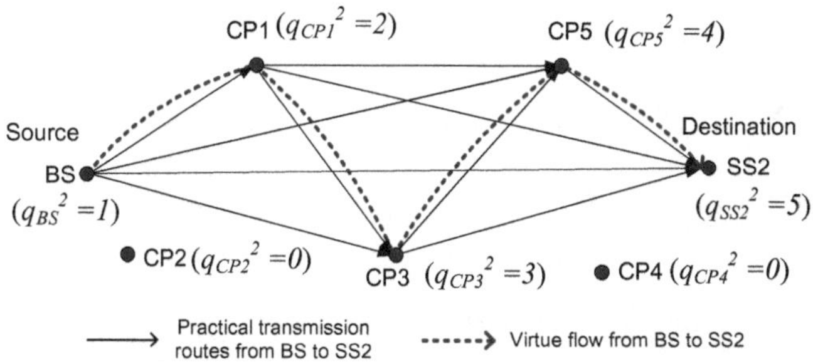


**Fig. 1.** An illustration of practical relay routes and virtue flow (BS→ $SS_2$)

$$\sum_{m \in N_{CP}} u_{mn} \leq ML, \quad \forall m \in N_{CP}, n \in N_{SS} \in N_{SS} \tag{4}$$

Constraint (3) ensures that if $CP_m$ is associated with $SS_n$, an RS must be placed at $CP_m$. Constraint (4) makes a limit for the maximal number of internal relays (i.e., associated CPs) for an SS.

The complexity of the formulation arises at representing the achievable rate for each SS in practical deployment scenario. It is because each CP is still undetermined whether to place an RS and whether to be allocated to an SS. Moreover, even the allocation of RSs to an SS has been determined, their exact sequence in a relay route are still undecided, since there are totally $n!$ permutations of possible sequences for $n$ RSs. By observing (1), we find that the destination's achievable rate heavily depends on the locations and relay sequence of the allocated RSs. To represent the relay sequence for each SS, we adopt the concept of *virtue flow*, which is defined as the route with the maximal number of hops among all the practical transmission routes from BS to $SS_n$, denoted by $\overrightarrow{\mathscr{F}^n}$. As an example shown in Figure 1, suppose $CP_1$, $CP_3$ and $CP_5$ are allocated to provision transmissions to $SS_2$, while $CP_2$ and $CP_4$ are not. The virtue flow (relay sequence) from BS to $SS_2$ ($\overrightarrow{\mathscr{F}^2}$) is illustrated with a blue dotted line. Namely, $\overrightarrow{\mathscr{F}^2}$ : {BS, $CP_1$, $CP_3$, $CP_5$, $SS_2$}, for simplicity, $\overrightarrow{\mathscr{F}^2}$ : {$CP_1$, $CP_3$, $CP_5$}.

Then, we define a virtue flow matrix $\mathbb{F} = (f_{ij}^n)_{(M+1)\times(M+1)\times N}$, where $f_{ij}^n = 1$ if both node $i$ and node $j$ are in $\overrightarrow{\mathscr{F}^n}$, and node $i$ is on the upstream of node $j$ (for $i \in BS \bigcup N_{CP}$, $j \in N_{CP} \bigcup SS_n$, $n \in N_{SS}$); otherwise, $f_{ij}^n = 0$. For each SS, the virtue flow matrix is subjected to :

$$f_{ij}^n + f_{ji}^n \leq 1, \quad \forall i, j \in N_{CP}, n \in N_{SS} \tag{5}$$

$$\sum_{m \in N_{CP}} f_{BS,m}^n = 1, \quad \forall n \in N_{SS} \tag{6}$$

$$\sum_{i \in \Omega} f_{i,SS_n}^n = 1, \quad \forall n \in N_{SS} \tag{7}$$

$$\sum_{i \in N_{CP} \cup BS} f_{mi}^{n} = u_{mn}, \quad \forall m \in N_{CP}, n \in N_{SS} \tag{8}$$

$$\sum_{i \in N_{CP} \cup BS} f_{im}^{n} = u_{mn}, \quad \forall m \in N_{CP}, n \in N_{SS} \tag{9}$$

$$\sum_{j_1 \in N_{CP}} f_{j_1 m}^{n} + \sum_{j_2 \in N_{CP}} f_{m j_2}^{n} = 2u_{mn}, \quad \forall m \in N_{CP}, n \in N_{SS} \tag{10}$$

Constraint (5) stipulates the virtue flow is directed where $f_{ij}^{n} = 1$ if node $i$ is on the upstream of $j$ in $\overrightarrow{\mathscr{F}^n}$. Constraints (6) and (7) ensure that the BS and each terminal node ($SS_n$) in $\overrightarrow{\mathscr{F}^n}$ has only one downstream node and one upstream node, respectively. Constraints (8) and (9) ensure that each CP in $\overrightarrow{\mathscr{F}^n}$ also has exactly one upstream node as well as one downstream node, if the CP is associated with $SS_n$ (i.e., the CP is in $\overrightarrow{\mathscr{F}^n}$). Constraint (10) formulates the flow conservation property; i.e., the number of upstream and downstream nodes of a CP equals two if the CP is in $\overrightarrow{\mathscr{F}^n}$.

To represent the position of a node in a relay sequence, we now define the *relay level* of a node as below.

**Definition 1. Relay level.** The *relay level* of a node $A$ for a destination $D$ (denoted by $q_A^D$) is defined as $q_A^D = \sum_{A \in \Omega} \deg^-(A) + 1$,where $\sum_{A \in \Omega} \deg^-(A)$ represents the *indegree* of node $A$ in graph $G$. Note that $q_A^D = 0$ indicates that node $A$ is not associated with destination $D$.

In Figure 1, the relay levels of BS, $CP_1$-$CP_5$ are 1, 2, 0, 3, 0 and 4, respectively.

We define a node relay level matrix $\mathbb{Q} = (q_i^n)_{|\Omega| \times N}$, where $q_i^n$ is the relay level of node $i \in \Omega$. According to the constructed directed virtue flow, the relay level of the BS should always be 1. The relay level of node $j$ increases by one, if both $i$ and $j$ are in $\overrightarrow{\mathscr{F}^n}$ and the upstream node of $j$ is $i$ ; otherwise, the relay level is set to 0. The following constraints (11)-(13) stipulate the above definition of relay level.

$$q_{BS}^{n} = 1, \quad n \in N_{SS} \tag{11}$$

$$q_j^n - q_i^n \geq f_{ij}^n - \beta(1 - f_{ij}^n), \forall i \in N_{CP} \cup BS, j \in N_{CP} \cup SS_n, n \in N_{SS} \tag{12}$$

$$q_i^n \geq 0, i \in \Omega, n \in N_{SS} \tag{13}$$

where $\beta = |\Omega|$. To establish a connection between the relay level of a CP and Eq. (1), a new set of variable $\mathbb{H}$ is needed. $\mathbb{H} = (H_{hi}^n)_{ML \times M \times N}$ is a CP-relay-level incidence matrix such that $H_{hi}^n = 1$ if node $i \in N_{CP}$ is in $\overrightarrow{\mathscr{F}^n}$ and $i$'s relay level is $h \in \{2, \ldots, ML+1\}$; otherwise, $H_{hi}^n = 0$. The following constraints (14) and (15) stipulate the definition of $\mathbb{H}$. Constraint (16) ensures that there is at most one node at each level in $\overrightarrow{\mathscr{F}^n}$.

$$H_{hi}^n \geq \gamma + \gamma(q_i^n - h), \forall i \in N_{CP}, n \in N_{SS}, h \in \{2, \ldots, ML+1\} \tag{14}$$

$$1 - H_{hi}^n \geq \gamma + \gamma q_i^n, \quad \forall i \in N_{CP}, n \in N_{SS}, h \in \{2, \ldots, ML+1\} \tag{15}$$

$$\sum_{i \in N_{CP}} H_{hi}^n \leq 1, \qquad \forall n \in N_{SS}, h \in \{2, \ldots, ML+1\} \tag{16}$$

where $\gamma = \frac{1}{|\Omega|^2}$. To ensure the throughput of each SS is larger than the minimal rate requirement, the constraints can be expressed as follows.

$$\log[1 + (\frac{1}{N_0} \sum_{k=2}^{h_2} (\sum_{h_1=1}^{k} \sqrt{P_0} (D_{h_1 h_2}^n)^{-\alpha})^2)] \geq \rho_n, h_2 = 2 \ldots ML+1, \forall n \in N_{SS} \tag{17}$$

$$D_{h_1 h_2}^n = \sqrt{(\sum_{l \in \Omega} H_{h_1 l}^n x_l - \sum_{l \in \Omega} H_{h_2 l}^n x_l)^2 + (\sum_{l \in \Omega} H_{h_1 l}^n y_l - \sum_{l \in \Omega} H_{h_2 l}^n y_l)^2}, \tag{18}$$

$$\forall 2 \leq h_1 < h_2 \leq ML+1, \forall n \in N_{SS}$$

$$D_{1,h_1}^n \leq \frac{D_{1,h_2}^n}{\varepsilon + \sum_{l \in \Omega} H_{h_2 l}}, \forall 2 \leq h_1 < h_2 \leq ML+1, n \in N_{SS} \tag{19}$$

where $\varepsilon$ is a very small number. Constraint (19) ensures that the relay sequence of the internal relays (the selected CPs) is the one with maximal achievable rate among all the permutations of the nodes.

In summary, the objective function (2) and the constraints (3)-(19) form the mathematical formulation for problem DLP. Due to the nonlinearity of constraints (17), (18) and (19), problem DLP is a nonlinear integer program, which cannot be solved by any systematic method [11]. In the next section, an effective algorithm will be presented by exploiting the nature of the problem.

# 4 An Algorithm to Solve Problem DLP

## 4.1 Algorithm Description

We propose a two-phase algorithm to solve problem DLP, which is shown in Algorithm 1. The core idea is as follows. First, we reduce the search space such that only feasible candidate relay sequences are left. Then, in Phase II, the best relay sequence for each SS can quickly be found by iteratively removing the least contributive CP.

- **Phase I: Feasible-Relay-Sequence (FRS) Table Setup Phase**

The search space is defined in the FRS table, which is the basis for deciding the RSs' final locations in the next phase. As an example, assume $ML = 3$, the FRS table is shown in Figure 2. Each SS has its own sub-table. The sub-table consists of two rows: a row of achievable rates, which are sorted in non-increasing order, and a row of relay sequences, which one-to-one map the rates in the same column in the first row. $r_j^{n,t}$, $\mathscr{F}_j^{n,t}$ represent the achievable rate of $SS_n$ via $t$ relays, and the corresponding relay sequence (i.e., virtue flow) for $SS_n \in N_{SS}$, respectively. $j$ is the column index.

**Algorithm 1.** A Proposed Two-Phase Algorithm for Problem DLP

**Input**: $N_{CP}, N_{SS}, (\rho_n)_{1\times N}, M, N, c, P_0, \alpha, ML$;
**Output**: $\mathcal{C}, \mathbb{Z}, \mathbb{U}, \mathbb{Q}$;
Initialization: Input the given cell layout and system parameters ;
**PHASE I: Feasible-Relay-Sequence (FRS) Table Setup**
**for** $\forall\ SS_n \in N_{SS}$ **do**
  Calculate $r_1^{n,1}$, the achievable rate of $\mathrm{SS}_n$ with direct transmission;
  **for** $t = 2$ *to* $ML$ **do**
    **for** $j = 1$ *to* ${}^{M}C_t$ **do**
      Calculate $r_j^{n,t}$, the maximal achievable rate of $\mathrm{SS}_n$ for the $j^{th}$ combination of $t$ selected CPs in $N_{CP}$;
      Store the corresponding relay sequence in $\mathscr{F}_j^{n,t}$ to achieve the rate $r_j^{n,t}$;
    **end**
    Sort $r_j^{n,t}$s in non-increasing order;
    Map the corresponding $\mathscr{F}_j^{n,t}$ and write relay sequence entries in $\mathrm{SS}_n$'s sub-table;
  **end**
  **if** $\rho_n > r_1^{n,ML}$ **then**
    Print{"No solution exists with current network configuration!"};
    **return**;
  **else**
    Determine $t_n$, the required number of relays, for $\mathrm{SS}_n$ by Eq. (20);
    Truncate the $\mathrm{SS}_n$'s sub-table such that only the columns satisfying $r_j^{n,t} \geq \rho_n$ and $t = t_n$ are remaining;
    Calculate the lower bound of the objective value $C^{LB} = \max_{\forall n \in N_{SS}} J_n$;
  **end**
**end**
**PHASE II: Minimum Cost Relay Planning**
Set the initial set of $N_{RS}$ as the union of all the CPs in the FRS table, i.e., $N_{RS} = \bigcup_{n\in N_{SS}} \bigcup_{j\in\{1\ldots J_n\}} \mathscr{F}_j^{n,t_n}$ ;
Calculate contributive index $\xi(n,m)$ and set contributive index $\Xi(m)$ for $\forall m \in N_{RS}, \forall n \in N_{SS}$;
/* FRS_Filter procedure:*/
**repeat**
  Compute $\Delta = \max_{\forall m\in N_{RS}}(\Xi(m)) - \min_{\forall m \in N_{RS}}(\Xi(m))$;
  Find $\mathrm{CP}_m$ with minimal set contributive index, i.e., $m = \arg\min_{\forall m \in N_{RS}}(\Xi(m))$;
  **if** $m \in \bigcap_{j\in\{1\ldots J_n\}} \mathscr{F}_j^{n,t_n}$ **then**
    Set $\mathrm{CP}_m$ as optimal, i.e., $\mathbb{Z} \leftarrow m$;
  **else**
    Update $N_{RS}$ by removing $\mathrm{CP}_m$, i.e., $N_{RS} \leftarrow N_{RS}/\{m\}$;
    Update $\mathrm{SS}_n$'s sub-table for $\forall n \in N_{SS}$ by removing the column that contains $\mathrm{CP}_m$ in $\mathrm{SS}_n$'s sub-table; $J_n \leftarrow J_n - \xi(n,m)$;
    Re-calculate $\xi(n,m)$ and $\Xi(m)$ for $\forall m \in N_{RS}$, $\forall n \in N_{SS}$;
  **end**
**until** $(\Delta == 0)$ *or* $(|\mathscr{F}^n| == 1$ *for* $\forall n \in N_{SS})$ *or* $(|N_{RS}| == C^{LB})$
Set the optimal relay sequences of RSs ($\mathbb{Q}$) as the first column ($\mathscr{F}_1^{n,t_n}$) in each SS's updated sub-table;
Determine the final locations ($\mathbb{Z}$), the allocation ($\mathbb{U}$) to each SS, and total number of RSs ($\mathcal{C}$) according to $\mathbb{Q}$;
**return**;

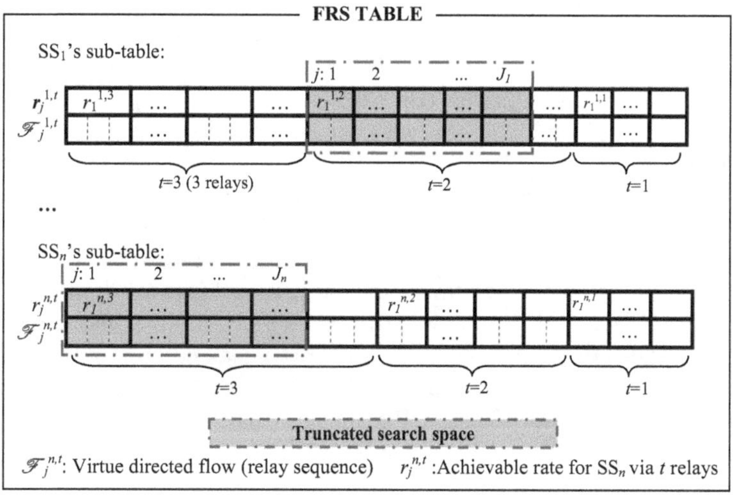


**Fig. 2.** An illustration of FRS table ($ML = 3$)

Phase I focuses on setting up such a FRS table. To construct it, for each SS, $r_j^{n,t}$ is calculated with Eq. (1) by enumerating all the possible combinations of $t$ relays (CPs) in $N_{CP}$, where $t = \{1 \ldots ML\}$. Note that there are $t!$ number of permutations for a specific set of $t$ relays, which lead to $t!$ achievable rates. Thus, the maximal one is set as $r_j^{n,t}$ and the corresponding relay sequence is stored in $\mathscr{F}_j^{n,t}$. Then, in Line 10, we sort all the $r_j^{n,t}$s in non-increasing order, and map the corresponding $\mathscr{F}_j^{n,t}$. The entries of $r_j^{n,t}$s and $\mathscr{F}_j^{n,t}$s are then written in the FRS table. In Line 13, we check if there exists a feasible solution with current network configuration (the set of CPs). More specifically, we compare $\rho_n$ and $r_1^{n,ML}$, which is the maximum possible rate for $\text{SS}_n$. If $\rho_n > r_1^{n,ML}$, it indicates more better CPs are needed in $N_{CP}$. If $\rho_n \leq r_1^{n,ML}$ for $\forall n$, there must exist a feasible solution. By exploiting the monotonicity of achievable rate row in $\text{SS}_n$'s sub-table, we can use the following criteria to determine the required number of relays (denote as $t_n$) for $\text{SS}_n$ to achieve its minimum rate requirement.

$$t_n = i + 1, \text{ if } r_1^{n,i+1} \geq \rho_n > r_1^{n,i}, \forall n \in N_{SS} (i = 0, \ldots, ML - 1) \qquad (20)$$

Afterwards, we truncate the search space in each SS's sub-table such that only the columns satisfying $r_j^{n,t} \geq \rho_n$ and $t = t_n$ are remaining. $J_n$ is the total number of columns in $\text{SS}_n$'s sub-table after truncation.

- **Phase II: Minimum Cost Relay Planning Phase**

In this phase, we propose a FRS_Filter method to find the minimum cost placement, which is a procedure of iteratively removing the least contributive CP to $N_{SS}$.

**Definition 2. Contributive Index.** The *contributive index* of $\text{CP}_m (\in N_{CP})$ to $\text{SS}_n (\in N_{SS})$ (denoted as $\xi(n, m)$) is defined as the total number of $\text{CP}_m$ in the relay sequence row of $\text{SS}_n$'s sub-table .

**Definition 3. Set Contributive Index.** The *set contributive index* of $\mathrm{CP}_m(\in N_{CP})$ to the set $N_{SS}$ (denoted as $\Xi(m)$) is defined as the total number of $\mathrm{CP}_m$ in all the relay sequence rows in the FRS table, i.e., $\Xi(m) = \sum_{n \in N_{SS}} \xi(n,m)$.

*Remark.* $\Xi(m)$ represents the contributive index of a specific $\mathrm{CP}_m$ to the whole set of $N_{SS}$. The smaller the value of $\Xi(m)$, the less chance that $\mathrm{CP}_m$ is selected to the final optimal set of RSs. Based on this concept, in Phase II, the minimal cost placement will be realized through the process of FRS_Filter.

In Line 22, we first get the initial set of RSs ($N_{RS}$), which is the union of all CPs in the FRS table. Then, the $\xi(n,m)$s and $\Xi(m)$s for $\forall m \in N_{RS}, \forall n \in N_{SS}$ are calculated according to DEFINITION 2 and 3. In Lines 24-35, the procedure of FRS_Filter is executed. Suppose $\mathrm{CP}_m(\in N_{RS})$ is the one with minimal set contributive index. We then check if it appears in each column of relay sequence in the $\mathrm{SS}_n$'s sub-table. If yes, it indicates that $\mathrm{CP}_m$ must be selected to place an RS and it can not be removed, i.e., $\mathrm{CP}_m$ is an optimal location for RS; otherwise, we can update $N_{RS}$ by removing $\mathrm{CP}_m$. Then all the SSs' sub-tables should also be updated by removing the whole column that contains $\mathrm{CP}_m$. Afterwards, all the $\xi(n,m)$s and $\Xi(m)$s are re-calculated so as to conduct next loop of FRS_Filter operations. The loop terminates when any of the following is true: (1) $\Delta$ equals zero, where $\Delta = \max_{\forall m \in N_{RS}}(\Xi(m)) - \min_{\forall m \in N_{RS}}(\Xi(m))$. It indicates that all the CPs left in the FRS table have an equal chance to be finally selected, namely, none is worse than the others. Thus, the current $N_{RS}$ is the optimal, and $\mathbb{Z}$ can be determined accordingly. (2) $|\mathscr{F}^n|$ equals one for $\forall n \in N_{SS}$. It indicates that there is only one column remaining in each SS's sub-table, in which the left relay sequence is definitely optimal. (3) $|N_{RS}|$ equals $C^{LB}$. It indicates that the lower bound of the objective value has been achieved, so the remaining relay sequences are definitely optimal.

## 5 Numerical Results

We consider a practical network scenario where an IEEE 802.16 WirelessMAN air interface [12] is assumed. $ML$ is set to 2. Each node transmits with a unit power density. The path loss exponent $\alpha$ is set to 3. To simulate the presence of buildings, trees and other obstructions in practical network environments, the shadowing effects typically result in 5-10 dB losses are also taken into account in the simulations. Without loss of generality, the BS is located at the origin point (0,0). We conduct a series of case studies. Figure 3 shows one of the network layouts, in which the normalized coordinate of each SS and CP is illustrated and the amount of minimal rate requirement for each SS is proportional to its radius.

To verify the optimality of our proposed two-phase algorithm, we compare the results with the optimal one (obtained by an exhaustive search). We find that the final objective value of our proposed algorithm is the same with the optimum for all the cases we studied. Figure 3 also shows the resulting configurations of the RS placement and the optimal relay sequences (virtue flows), which are represented with the dotted line from BS to each SS.

The proposed algorithm demonstrates excellent computation efficiency as shown in Figure 4. The computation time of our proposed algorithm increases

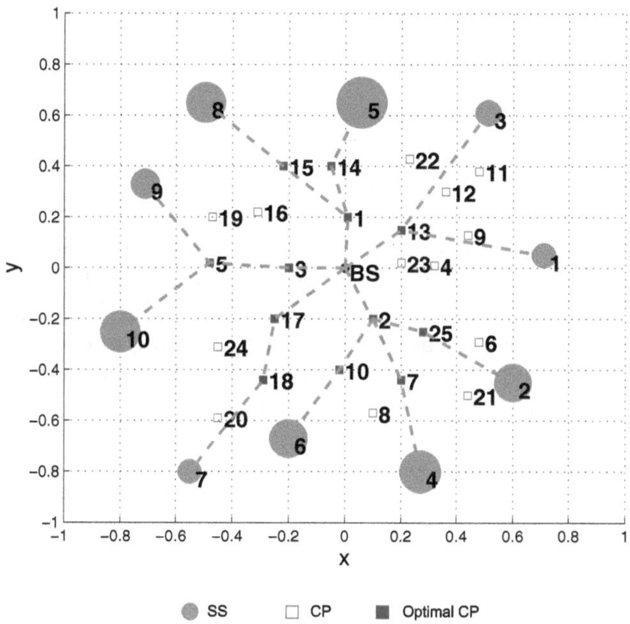


**Fig. 3.** The optimal results obtained by proposed 2-phase algorithm

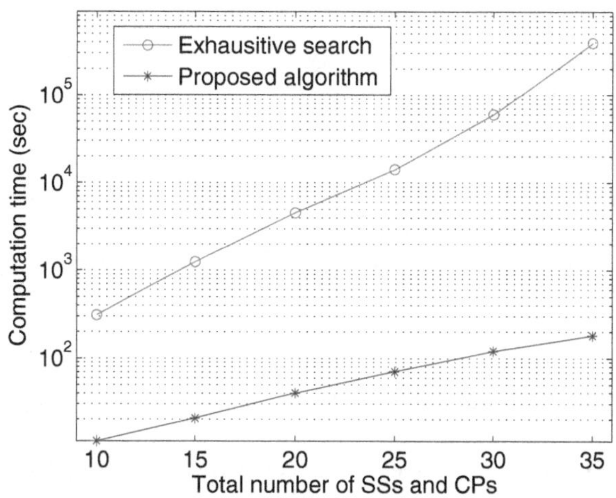


**Fig. 4.** The comparison of computation time of proposed algorithm against exhaustive search

linearly with the problem size (i.e., the total number of SSs and CPs), while the computation time shows an exponential growth with the exhaustive search. Therefore, our proposed algorithm shows a good scalability in practical large-scale wireless network design.

**Table 2.** Objective Value Comparison

| | Number of SSs | 4 | 5 | 6 | 7 | 8 | 9 | 10 |
|---|---|---|---|---|---|---|---|---|
| Objective value | multi-level CR | 5 | 7 | 8 | 9 | 9 | 10 | 12 |
| | multi-hop NCT | 8 | 12 | 15 | 17 | 19 | 22 | 24 |
| | Saved Cost | 37.5 % | 41.7 % | 46.7 % | 47.1 % | 52.6 % | 54.6 % | 50.0 % |

Table 2 compares the objective value (i.e., the number of RSs) with multi-level CR against traditional multi-hop NCT. It is observed that the number of RSs to be deployed can be substantially reduced (37.5% $\sim$ 54.6%) when multi-level CR is employed. Therefore, by incorporating advanced cooperative transmission technology, a salient economic benefit in terms of RS deployment cost reduction can be achieved. On the other hand, given the same RS placement, a significant improvement on end-user and system throughput can be achieved with cooperative transmission, which just meets the ever-increasing demands of future residential and business premises.

## 6 Conclusions

In this paper, we have conducted a comprehensive study on the issue of dimensioning and location planning in wireless networks under multi-level CR. The tasks of RS placement, relay allocation, and relay sequence design have been jointly considered and formulated in a unified optimization framework. To avoid intractable computation complexity in solving the nonlinear formulation, a two-phase algorithm has been developed to solve it efficiently and effectively. Simulation and case studies have been conducted, in which the results have demonstrated the RS deployment cost can be reduce by 37.5% $\sim$ 54.6% with multi-level CR compared with that with traditional multi-hop NCT. The proposed framework should provide a guideline for the operators in the effort of RS placement and evaluation of network expenditure in practice.

## References


1. Korhonen, J.: Introduction to 3G Mobile Communications, 2nd edn., p. 251. Artech House (2002)
2. Lin, B., Ho, P.H., Xie, L.L., Shen, X.: Optimal Relay Station Placement in IEEE 802.16j Networks. In: IWCMC, Hawaii, USA (August 2007)
3. Lin, B., Ho, P.H., Xie, L.L., Shen, X.: Relay Station Placement in IEEE 802.16j Dual-Relay MMR Networks. In: IEEE ICC, Beijing, China (May 2008)
4. Cover, T.M., Gamal, A.A.E.: Capacity Theorems For The Relay Channel. IEEE Trans. on Inf. Theory 25(5), 572–584 (1979)
5. So, A., Liang, B.: Optimal Placement And Channel Assignment Of Relay Stations In Heterogeneous Wireless Mesh Networks By Modified Bender'S Decomposition. The Elsevier Ad Hoc Networks journal (invited from IFIP Networking 2007) 7(1), 118–135 (2009)

6. So, A., Liang, B.: Enhancing WLAN Capacity by Strategic Placement of Tetherless Relay Points. IEEE Trans. on Mobile Computing 6(5), 522–535 (2007) (Extended version of paper in IFIP Networking 2006)
7. Amaldi, E., Capone, A., Cesana, M., Malucelli, F.: Optimization Models for the Radio Planning of Wireless Mesh Networks. In: Akyildiz, I.F., Sivakumar, R., Ekici, E., de Oliveira, J.C., McNair, J. (eds.) NETWORKING 2007. LNCS, vol. 4479, pp. 287–298. Springer, Heidelberg (2007)
8. Amaldi, E., Capone, A., Cesana, M., Filippini, I., Malucelli, F.: Optimization Models and Methods for Planning Wireless Mesh Networks. Computer Networks 52(11), 2159–2171 (2008)
9. Xie, L.L., Kumar, P.R.: A Network Information Theory for Wireless Communication: Scaling Laws and Optimal Operation. IEEE Trans. on Inf. Theory 50, 748–767 (2004)
10. Mark, J.W., Zhuang, W.: Wireless Communications and Networking. Prentice Hall, Englewood Cliffs (2003)
11. Garey, M.R., Johnson, D.S.: Computers and Intractability: A Guide to the Theory of NP-completeness, pp. 245–248. W. H. Freeman and Company, New York (1979)
12. IEEE Standard for Local and Metropolitan Area Networks Part 16: Air Interface for Fixed Broadband Wireless Access Systems, IEEE Std. 802.16-2004 (2004)

# Multi-user OFDMA Frame Aggregation for Future Wireless Local Area Networking*

James Gross[1], Oscar Puñal[1], and Marc Emmelmann[2]

[1] Mobile Network Performance Group, UMIC Research Centre,
RWTH Aachen University,
Templergraben 55, 52065 Aachen, Germany
{gross,punal}@umic.rwth-aachen.de
[2] Telecommunication Networks Group,
Einsteinufer 25, 10587 Berlin, Germany
emmelmann@ieee.org

**Abstract.** State-of-the-art wireless local area networking enables frame aggregation as approach to increase MAC efficiency. However, frame aggregation is limited to the aggregation of packets destined for the same station. In order to serve different stations, the access point still has to contend for the channel multiple times. In this paper we propose and evaluate a novel approach that enables multi-user frame aggregation. We combine this concept with channel-dependent OFDMA resource assignments, yielding a higher PHY efficiency (by exploiting multi-user diversity and instantaneous channel state information) as well as a higher MAC efficiency. The downside to this approach is the increase in protocol overhead to enable such multi-user OFDMA frame aggregation. However, we show that the proposed approach outperforms state-of-the-art 802.11n for different packet sizes and stations to be served.

**Keywords:** WLAN, OFDMA, adaptive modulation, frame aggregation.

## 1 Introduction

IEEE 802.11 wireless local area networks (referred on to as WLANs) are almost omnipresent today. Nevertheless, the research and standardization activities in this field are still very intense, addressing a wide range of improvements. However, the issue of increasing the network capacity has always drawn much attention. Recently, the 802.11 working group has started discussion on future WLAN advancement based on two different approaches: utilizing the 60 GHz frequency band [1], and increasing the aggregated throughput by exploiting multi-user diversity [2]. For both projects there is little doubt that orthogonal frequency division multiplexing (OFDM) will remain the basic prevailing transmission scheme.

The main feature of OFDM is that it splits the bandwidth into many sub-channels, also referred to as sub-carriers. Instead of transmitting symbols sequentially through one (broadband) channel, multiple symbols are transmitted

* Part of the work has been done while all authors were with the Telecommunication Networks Group, TU Berlin.

L. Fratta et al. (Eds.): NETWORKING 2009, LNCS 5550, pp. 220–233, 2009.

in parallel. This mitigates the impact of intersymbol interference leading to a better system performance in broadband wireless channels. If channel state information is available at the transmitter, system performance can be further improved. Due to the frequency diversity of broadband channels, adaptation of transmission parameters (transmit power, modulation) on a per sub-carrier basis are more efficient than applying these parameters uniformly to the whole set of sub-carriers [3]. These adaptations for single-user links (one transmitter, one receiver) are referred to as *loading algorithms*. In contrast, a significant improvement can be achieved for multi-user settings: multiple data streams are transmitted in parallel, e.g. from the access point to several stations. In this case, multi-user diversity is present in the system (different stations experience different channel states for the same sub-carrier) which can be exploited by so called dynamic OFDMA assignment algorithms [4].

As multi-user diversity is to be exploited by future WLAN systems, OFDMA is one candidate technology. This approach provides the advantage that a higher MAC efficiency can be combined with a higher PHY efficiency. Multi-user frame aggregation is in contrast to the frame aggregation implemented in state-of-the-art 802.11n systems, which enables the transmission of several packets destined for the same station within one channel access. Hence, multi-user frame aggregation is expected to improve MAC efficiency. By combining multi-user frame aggregation with dynamic OFDMA schemes, higher MAC and PHY efficiencies can be obtained. On the other hand, significant overhead has to be added to the protocol to enable dynamic OFDMA (acquiring the channel states and signaling the resource assignments). Hence, it is open if (and when) this additional overhead pays off due to the higher efficiency in the PHY and MAC when comparing such a system proposal with IEEE 802.11n. Note that [5] presents a related protocol enhancement for 802.11a. However, the approach does not support backward compatibility to legacy 802.11a devices and is not evaluated in comparison to frame aggregation for single stations in WLAN.

In this paper we study these performance effects based on a new protocol concept enabling multi-user frame aggregation via dynamic OFDMA schemes. Our focus is less on the physical layer aspects of dynamic OFDMA. Instead, we are interested in evaluating our protocol design in comparison to the performance of 802.11n. Specifically, our evaluation focuses on the frame aggregation performance of the two different approaches taking a fixed MIMO-OFDM physical layer as comparison basis. Our protocol design is novel as it allows multi-user frame aggregation by dynamic OFDMA resource assignment while still being fully backward compatible to legacy (802.11n and 802.11a) devices. The second contribution of this paper is the discussion of the performance difference between 802.11n (as state-of-the-art technology) and our proposed extension. Note that our work has been already presented to the IEEE [6].

The paper is structured as follows. In Section 2 we first summarize the amendment IEEE 802.11n. Next, we present our protocol proposal in Section 3. Next, the performance evaluation is discussed in Section 4 before we conclude the paper in Section 5.

## 2 Overview of IEEE 802.11n WLAN

IEEE 802.11 chose an OFDM physical layer for its operation in the 5 GHz band [7] as well as for its *extended rate PHY (ERP)* amendment for 2.4 GHz operation in order to provide data rates up to 54 Mbit/s. The available bandwidth is divided into 52 sub-carriers from which four are exclusively used as pilots [8]. IEEE 802.11n, aiming at providing even higher throughput up to 600 Mbit/s, introduces enhancements at both the physical and MAC layer. Regarding the first, it increases the number of sub-carriers from 52 to 56 and incorporates new error correction coding schemes (5/6 convolutional codes as well as Low-Density-Parity-Check codes). It also defines optional features such as a shorter guard period between symbols and channel-bonding, a technique that practically doubles the capacity by simply doubling the available transmission bandwidth. All OFDM-based PHYs utilize link adaptation: prior the transmission the payload data is first convolutionally encoded. The resulting data block is transmitted via *all* available sub-carriers employing the *same* modulation type on each sub-carrier [8, 9]. The choice of the employed modulation and coding scheme (referred further on to as PHY mode) is crucial for the performance but not standardized. For that purpose, the MAC may make usage of, e.g., the radio signal strength indicator (RSSI) level gained during reception of previous packets.

Nevertheless, the most significant enhancement of the PHY is the introduction of multiple antenna capabilities at the transmitter and receiver side (MIMO). Specifically, these can be distinguished into transmit beamforming, space-time-coding and spatial multiplexing. In this work we will only consider the employment of the latter. Spatial multiplexing is a technique that enables the transmission of several different data flows over a set of multiple antennas (without requiring more radio spectrum).

With respect to the MAC layer, the major improvement is the introduction of frame aggregation. This technique allows the transmission of several payload packets within one channel access. Obviously, this improves the efficiency as the overhead for framing and channel access is only spent once. On the other hand frame aggregation is more sensitive to interference as the medium is blocked for a longer time. The IEEE 802.11n draft suggests two different ways of performing frame aggregation: aggregated MSDU (A-MSDU) and aggregated MPDU (A-MPDU). The first performs the aggregation of packets without the specific 802.11 framing, while A-MPDU aggregates payload packets each of them with its own MAC header. Clearly, A-MSDU reduces the overhead at the cost of an increased packet error probability. In contrast, A-MPDU enables to check each single packet for an error while featuring a higher overhead. In addition, the latter permits the usage of the 'block acknowledgment' technique which allows the differentiated retransmission of the incorrect packets out of the set of all aggregated ones. Both frame aggregation types have a maximal data aggregation size: 8 kByte in case of the A-MSDU and 64 kByte in case of the A-MPDU type. However, the major constraint (of both schemes) is that the destination address of all aggregated packets has to be the same.

## 3 Multi-user Frame Aggregation Based on Dynamic OFDMA: 802.11 DYN

The presented approach applies dynamic OFDMA to the payload part, i.e. the Data field of packet transmissions in IEEE 802.11 WLANs based on the HT-mixed frame format [9]. Packets with this format can be decoded by 802.11n *high-throughput* stations (HT-stations) as well as by 802.11a/g/b ones. We refer to our new proposal as 802.11 DYN. It consists of two different modes, a *single-user* mode and a *multi-user* mode[1]. In this paper we solely focus on multi-user frame aggregation implemented through dynamic OFDMA - we refer to this as the multi-user mode of 802.11 DYN. We propose the usage of the multi-user mode for down-link transmissions. As a result, multiple packets can be transmitted in parallel to different stations while the medium is acquired only once. The multi-user mode employs adaptive modulation, where the modulation type per sub-carrier is chosen individually according to the channel conditions. By contrast, the link adaptation technique, as used in legacy 802.11a/g/n stations, does not consider a sub-carrier granularity. Finally, all presented protocol modifications are compatible with existing equipment such that operating a mixture of enhanced stations and "legacy" stations in one cell is feasible.

For supporting the multi-user mode we propose modifications regarding the frame format and the frame exchange sequence. We first describe the new frame format and afterwards discuss the modifications to the frame exchange sequence. Any 802.11 DYN payload frame uses HT-mixed format PPDU with a slightly modified *Signal* field (cf. Figure 1). The modified frame starts with the usual preambles [9]. Afterwards, the first 24 bits of the signal field are in total compliance to legacy IEEE 802.11a/g/n, with the exception that in the Rate field a different bit sequence is inserted, which is not specified in the standard (causing legacy stations to discard the frame). For instance, the bit sequence 1100 could be used to identify the 802.11 DYN frame. After the Tail field a new element of the signal field is introduced, the *Signaling field*. This field contains all the information to decode the payload transmission within the Data Field according to 802.11 DYN. The precise layout of the Signaling field is discussed in detail below. Then dynamic OFDMA is applied to the transmission of the payload in the Data Field. For the multi-user case this consists of several pairs of a Signaling field and a corresponding PSDU. These pairs are transmitted on the sub-carrier sets assigned to the respective station. Finally, for each PSDU also a Tail field and potentially some padding bits are transmitted to complete the data frame. The modified L-SIG field of the PLCP frame (including the Signaling field) is transmitted applying BPSK with rate 1/2 convolutional coding. Compared to legacy IEEE 802.11a/g/n systems, the main overhead stems from the Signaling field. We suggest the following format for the Signaling field (cf. Figure 1). Initially, an ID field is transmitted with 2 bit in length, indicating either a multi-user (using a sequence of 11) or a single-user transmission (using 00). Next, a Length

[1] Both modes have been presented to the IEEE standardization group recently [6, 10]. Furthermore, the single-user mode was presented in [11].

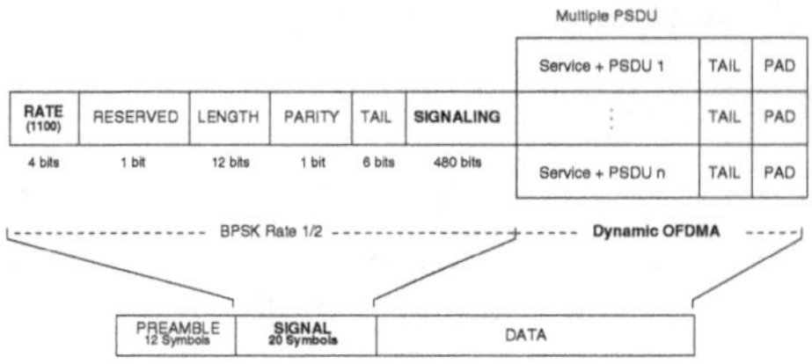


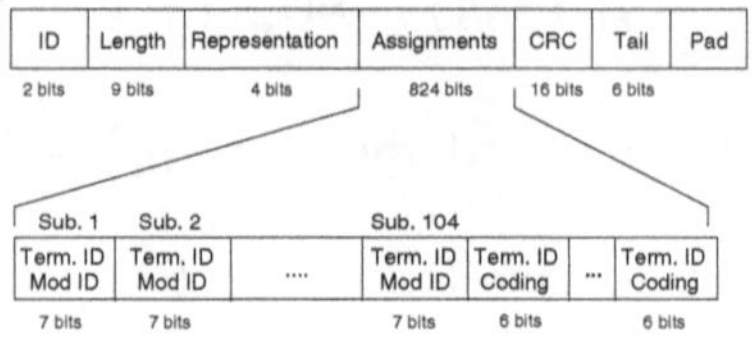


**Fig. 1.** Multi-user mode 802.11 DYN framing – **Left figure** shows the overall structure of the new PLCP frame – **Right figure** shows details of the required Signaling field of the new PLCP frame

field of 9 bit is inserted, which indicates the entire size of the Signaling field. The third field is the Representation field. It is 4 bit long and indicates primarily different types of representing the signaling information (for example, compressed signaling information). Then, the information about the assignments per sub-carrier follows. In case of the multi-user mode, an assignment per sub-carrier is characterized by a station identifier and a modulation identifier. One possible, straight-forward representation of the signaling information could be a sort of fixed signaling size field: Every assignment tuple <Terminal ID|Modulation ID> for each sub-carrier is signaled. Hence, the position of the tuple indicates the sub-carrier this tuple refers to. In 802.11 DYN a station is represented by a 4 bit sequence while a modulation type is represented by 3 bit (leaving some bit combinations for future use). This yields to 7 bit per assignment. Depending on the PLCP frame format, the Signaling field has to contain the assignments for 48 sub-carriers in case of 802.11a or 52 sub-carriers per spatial stream in the HT-mixed format of 802.11n. Focusing in the following only on the latter with 2 spatial stream, the total length of the assignments is 728 bits. However, each payload packet is also protected by FEC which has to be indicated to the stations as well. Hence, after the end of the assignments, further tuples are appended consisting of <Terminal ID|Coding ID>, requiring 6 bit in total. We propose to limit the number of stations included in one multi-user burst to 16 (which is not a limit of the total amount of stations that can be served in the cell). Hence, the coding assignment field has a maximum length of 96. This leads to a total length of 861 uncoded bits for the Signaling field.

Next, let us discuss the modification of the frame exchange sequence. The new sequence proposed for the multi-user mode is shown in Figure 2. Initially, the access point holds packets for several stations in its cell. Hence, the access point first has to acquire the medium by the standard rules of DCF. After it acquires the medium, it first transmits a CTS-to-self frame (for reasons related to the NAV, as discussed below). Next, the access point has to acquire the channel knowledge. A modified RTS frame is introduced, which carries a polling list in it. According to this polling list stations answer with a CTS frame which enables the access point to estimate the sub-carrier states using the received frame preamble. The polling order is indicated by transmitting the RTS frame based on the new frame format (as discussed above). The Signaling field

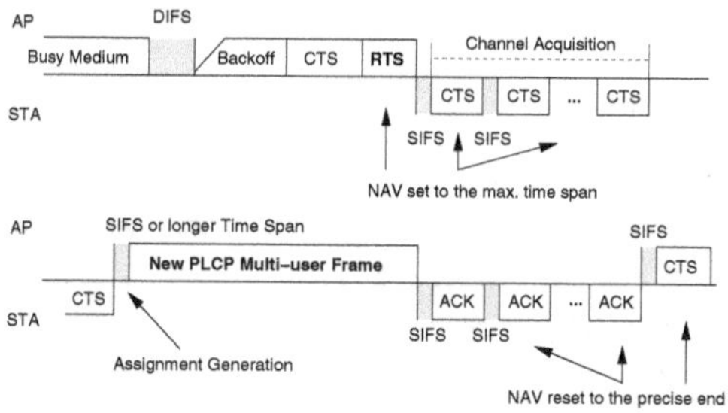


**Fig. 2.** Transmission sequence of the new 802.11 DYN multi-user mode. In order to set the NAV correctly, a slightly modified transmission sequence is required.

of the new frame carries pairs of 48-bit addresses and 4 bit IDs. The sequence of the pairs indicates the sequence with which stations transmit a CTS frame. Furthermore, the pairs also assign 4 bit IDs to each station such that during the following payload transmission stations do not have to be addressed by 48 bits each but by 4 bits. After the last station replied with a CTS frame, the access point starts computing the dynamic OFDMA assignments. Then follows the multi-user payload transmission (we also refer to this as multi-user burst) which employs the new frame format mentioned above. After the multi-user burst transmission, the stations confirm correctly received packets with a legacy ACK frame in the same order already used for channel acquisition. At the end, the access point finishes the multi-user mode frame exchange with a CTS-to-self frame.

From the description above, several open issues arise. First of all, the management of the NAV is more complicated for the dynamic scheme: After winning the channel during the contention phase, the access point does not know how long the packet transmission will take due to the still unknown sub-carrier states. Hence, up to the payload transmission, it has to announce a pessimistic estimate of the NAV setting, e.g., reserving the channel for the time span needed to convey all scheduled packets if for all stations only BPSK with rate 1/2 encoding can be used. Once the sub-carrier assignments are computed, the exact NAV can be distributed in the cell. However, as the initial RTS frame and the payload frame are transmitted with the new PLCP frame format, legacy stations have to be informed by a different way of the pessimistic NAV estimate and the updated NAV. This is the reason for starting the whole sequence with a CTS-to-self frame. The CTS frames, coming back from the stations, announce this pessimistic NAV value within the entire transmission range. After the access point has acquired the channel knowledge, it can announce the correct NAV value in its multi-user burst frame. However, this is not decoded by legacy stations due to its new frame format. Hence, after the ACK frames reset the NAV value within their transmission range, the access point has to ensure by a final CTS-to-self frame (carrying the correct NAV setting) that all stations reset their NAV.

# 4 Performance Evaluation

## 4.1 Simulation Model

The following system model is assumed for our simulations. Located within a single WLAN cell there are $J$ stations and one access point. Packets arrive at the latter for down-link transmission. We assume the access point queue is never empty, thus we consider the saturation mode. Per simulation run, all payload packets have a fixed size of $\varsigma$ bits.

**PHY model.** The maximum transmit power equals $P_{\max} = 10$ mW. The bandwidth, the number of sub-carriers, the symbol duration and the guard interval are all chosen in accordance to IEEE 802.11n (see Section 2). We assume in the following the application of 2 by 2 MIMO employing spatial multiplexing with MMSE reception to separate the spatial streams. In order to generate the MIMO channel matrix the 802.11n task group published a MATLAB module to generate traces of MIMO channel states [12, 13]. We use this tool to generate the channel matrix. We consider channel type 'E' representing a large office environment with a certain path loss model and a fading characterized by a delay spread of 100ns [13]. In general, the sub-carrier gains are assumed to be stable during the transmission of a PLCP payload frame – either in the legacy mode or in the 802.11 DYN case. The noise power $\sigma^2$ is computed at an average temperature of 20° C over the bandwidth of one sub-carrier (312.5 kHz).

**Packet Error Rate Model.** Clearly, we require a detailed packet error model for the link layer, which takes the fading and modulation setting of individual OFDM sub-carriers into account. Packet error rate investigations for OFDM transmission over a frequency-selective channel can be found for example in [14]. We follow a similar approach relying on an upper bound for the packet error probability, which takes the average bit error probability (of the modulation types per sub-carrier) as input. Note that in case of 802.11 DYN the system can control the bit error probability $\theta_j$ by setting the respective switching levels of the adaptive modulation. These switching levels refer to the SNR points at which the modulation employed should be changed to a higher or lower one. For a detailed presentation of the error model we refer the interested reader to [11].

**Simulation Methodology.** All results are generated with OPNET Modeler Version 14.0.A PL2. Modifications of standard models required to support 802.11 DYN are with regard to the OPNET model library as of September 2007. For the simulation of the 802.11 system, we generally follow the standard as close as possible. We only consider long preambles. All non-payload frames of 802.11 DYN are transmitted in base mode (BPSK with rate 1/2 encoder) and are assumed to be always received correctly. For all our simulations we vary the distance between access point and stations as well as the number of stations present in the cell. For a single simulation run we do not consider mobility. Furthermore, for a single simulation run all stations have the same distance to the access point and therefore the same average SNR due to path loss. The fading components of the OFDM sub-carrier channel gains are randomly regenerated for each payload packet transmission and therefore the error behavior for two sequentially transmitted packets can be assumed to be statistically independent. However, for all packet transmissions correlation of the fading in frequency is fully taken into account. The 99% confidence levels of all our results are very high and are not shown in the following graphs due to their small size.

**Comparison Schemes and Metrics.** We are interested in the saturation mode goodput in bit/s of 802.11 DYN versus legacy 802.11n. For this specific investigation, we assume in addition to the above mentioned model that the access point always holds a packet for each station in the cell. Stations do not have any data to send - we are only interested in the down-link performance. Hence, no collisions occur. We compare two different schemes:

- 802.11n : The access point serves one station after the other one using state-of-the-art 802.11n. Optionally, frame aggregation can be activated. In that case the AP transmits multiple packets within a single channel access to the corresponding station (note that the AP always has enough packets queued in order to fill any depth of frame aggregation). The depth of the frame aggregation is key to the observed performance. We explicitly comment the chosen depths below. Furthermore, we consider the performance of each transmission PHY mode separately over the full SNR range.
- 802.11 DYN Multi-user mode: The multi-user mode enables the transmission of several packets simultaneously to different stations by applying multi-user frame aggregation based on dynamic OFDMA. The PHY applies now adaptive modulation with respect to a pre-specified target bit error probability $\theta_j$ per station. Again, the setting of this error probability has a significant impact on the system performance, as demonstrated for the 802.11 DYN single-user mode in [11]. In this section, we only provide results for the optimum setting of $\theta_j$. We consider up to eight stations in the cell for the saturation mode investigations. This keeps transmission times reasonably short even if large packet sizes are assumed. Notice that in addition to multi-user frame aggregation, the access point can also apply frame aggregation per station.

One particular important issue in case of the multi-user mode is how sub-carriers are assigned to stations (in general, how the scheduling of packets to sub-carriers works). The focus of our work is not on optimal scheduling schemes and resource assignment algorithms. Hence, we pick a straightforward approach for assigning sub-carriers to packets: Each station receives the same amount of sub-carriers. Given this fixed sub-carrier allocation, a simple algorithm is employed to assign the specific sub-carriers to each station [15]. Basically, it considers one station after the other and assigns the preallocated number of best sub-carriers to the corresponding station from the set of remaining sub-carriers. For fairness reasons, the order of picking sub-carriers for stations is shifted for each frame.

### 4.2 Results

We consider four different parameters for our down-link saturation mode evaluation: average SNR, packet size $\varsigma$, frame aggregation depth and number of stations present in the cell. In the following we first discuss the results related to a small packet size (equaling roughly VoIP packets). Afterwards, we consider the results for large packet sizes. Before starting this discussion, recall that it is generally known that saturation mode performance in 802.11 systems depends heavily on the considered packet size (the larger the considered packet is the less does the MAC overhead influence the performance results).

**Small Packets of 234 Byte.** For the following discussion notice that we assume a maximum (single station) frame aggregation of 4 packets. This is motivated by the fact that such small packets are assumed to stem from a VoIP flow. As such flows are well known to be delay sensitive, we set the maximum frame aggregation depth to four (considering a G.711 voice encoder at a rate of 64 kbps, one such VoIP packet is generated every 20 ms.

In the first scenario we consider four stations to be present in the cell. In case of 802.11n always four packets are aggregated (for the same station) while in case of 802.11 DYN four packets, each for a different station, are multiplexed. Hence, per station 802.11 DYN does not apply frame aggregation initially. In this first scenario 802.11n also employs the RTS/CTS handshake prior to transmission of the aggregated frame due to the large payload size. In Figure 3, the corresponding average goodput per station is presented for an increasing SNR. Notice that the figure shows eight different curves for 802.11n as we show initially the performance results of all eight PHY modes (in all other graphs we will only show the upper envelope of the PHY modes). We observe a significant performance gain in the case of 802.11 DYN at low SNR values, however from 10 dB on the legacy modes outperform 802.11 DYN clearly. At very high SNR values, the legacy scheme achieves a goodput which is roughly 150% higher than the one of 802.11 DYN. What is the reason for this performance behavior? First notice that for small packets the raw PHY performance plays a smaller role as a lot of time is spent for resolving medium contention at the MAC. In fact, the faster the PHY transmits the payload (at high SNR values, for example) the more important gets the overhead due to RTS/CTS handshake, ACK frames and framing. From the above sections it is clear that 802.11 DYN is related to a large additional overhead in order to implement dynamic OFDMA (channel acquisition, signaling of assignments, NAV management). From Figure 4 we see that 802.11 DYN features a much lower PER rate while providing a higher spectral efficiency. Especially, the low PER leads to the observed performance improvement for small SNR values. However, as packet error rates improve in case of 802.11n, the protocol overhead becomes much more an issue and so 802.11n outperforms 802.11 DYN. Notice in general the superior performance of 802.11 DYN regarding the PHY efficiency in the left graph of Figure 4. The performance of 802.11 DYN increases continuously along the whole SNR range delivering significant gains (100-300%) compared to IEEE 802.11n. Apart from

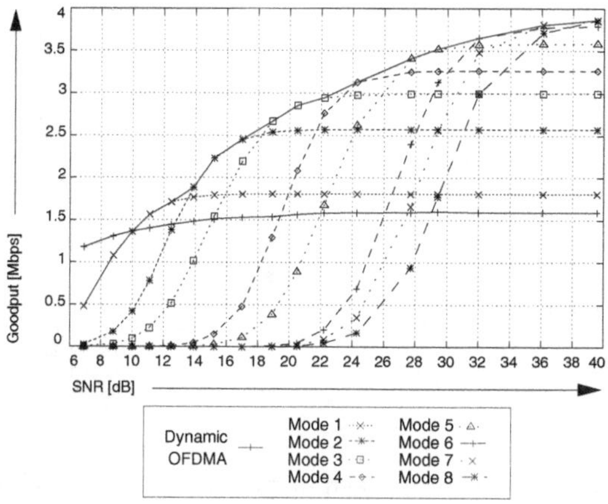


**Fig. 3.** Comparison of the average goodput per terminal between 802.11 DYN and selected PHY modes of 802.11n (showing also as the envelope of the maximum performance) for various different SNR levels. The packet size is set to 234 Byte. $J = 4$ stations are present in the cell. The legacy scheme performs an aggregation of 4 MPDUs, while 802.11 DYN does not make usage of this technique.

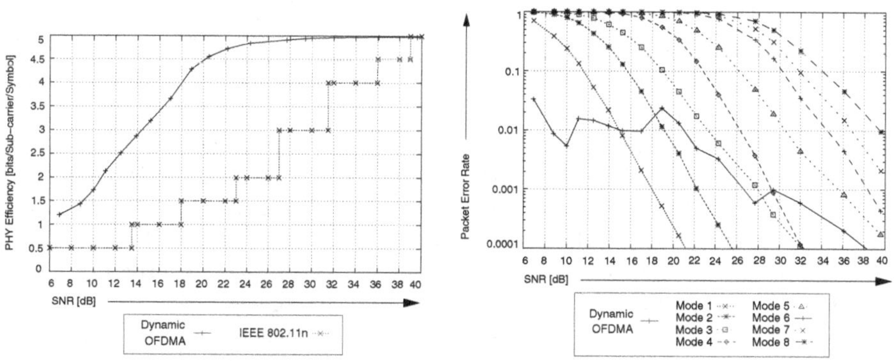


**Fig. 4.** Performance comparison of 802.11n and 802.11 DYN for small packets and four stations in the cell. **Left figure** Average PHY efficiency per sub-carrier and symbol. **Right figure** Average packet error probability per station.

the more precise modulation and coding selection, 802.11 DYN also benefits from multi-user diversity, which helps to further increase the physical layer efficiency. Notice that these performance improvements are coupled with a much lower packet error rate (right graph in Figure 4). Recall that the packet error probability is controlled in 802.11 DYN due to exploiting channel state information and adapting modulation types with respect to the target bit error probability.

Even though 802.11 DYN has a very efficient behavior at the physical layer, the overhead caused by the protocol may strongly reduce the achieved gain, as

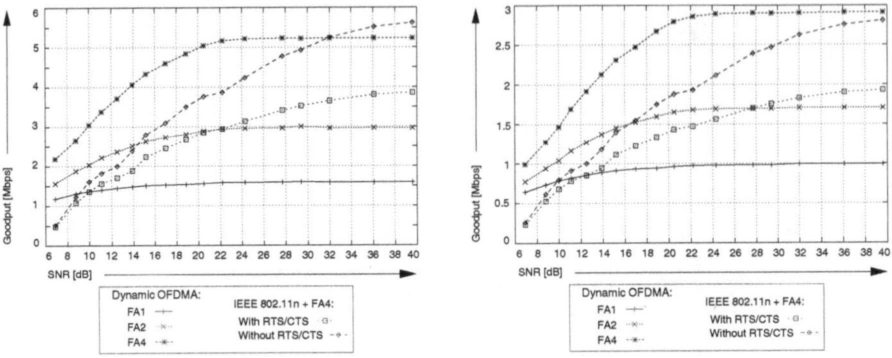


**Fig. 5.** Comparison of the average goodput per station of 802.11 DYN and 802.11n. The packet size is set to 234 Byte. **Left figure** $J = 4$ stations are present in the cell. 802.11 DYN applies a frame aggregation of 1, 2 and 4 packets, while 802.11n aggregates always 4 packets. Also, 802.11n results are given depending on the usage of RTS/CTS frame exchange. **Right figure** Same comparison with $J = 8$ stations in the cell.

seen in Figure 3. Hence, in Figure 5 (left graph) we activate single station frame aggregation in case of 802.11 DYN, allowing a maximum aggregation depth of 4 MPDUs. In this case, 802.11 DYN transmits 16 packets per (successfull) medium access. This leads to a better MAC efficiency of 802.11 DYN and reduces the impact of the protocol's overhead. When frame aggregation is deactivated, a maximal goodput per user of about 1.5 Mbps can be obtained by 802.11 DYN. When 2 MPDUs are aggregated, the maximal goodput reaches 3 Mbps and it rises slightly above 5 Mbps when the aggregation of 4 MPDUs is performed. Again, IEEE 802.11n aggregates 4 MPDUs, but in contrast to Figure 3 only the envelope of the best performing legacy mode at any SNR is plotted. In this case, the presented results consider both, the activation and deactivation of the RTS/CTS frame exchange. Notice that the results corresponding to the deactivation of RTS/CTS handshake have to be seen rather as upper bound as we assume that the access point always chooses the optimal PHY mode. Without an initial RTS/CTS handshake this can not be always assumed. In Figure 3 it can be observed that the higher the aggregation depth used for 802.11 DYN, the larger the range where it outperforms the legacy scheme. In the comparison there is always a cross-over point, from which one IEEE 802.11n outperforms 802.11 DYN. Next, consider the right graph of Figure 5. Here we consider the same scenario but increase the number of stations in the cell to $J = 8$. Since the total number of aggregated packets is considerably higher in the case of 802.11 DYN, its MAC efficiency is increased as the overhead's impact is reduced. Taking IEEE 802.11n without RTS/CTS as reference, 802.11 DYN without frame aggregation is outperformed by the legacy scheme from 10 dB on. When 802.11 DYN aggregates 2 MPDUs the crossing-point is shifted to 17 dB and when 4 MPDUs are aggregated 802.11 DYN performs better over the whole SNR range plotted. Notice that the increase in PHY efficiency due to multi-user diversity is modest

- the major performance improvement stems from a better MAC efficiency in case of a higher frame aggregation depth.

**Large Packets of Size 1536 Byte.** Next, we increase the packet size to 1536 Byte, corresponding to large IP packets. Assuming the packets to belong to web traffic or data transfers, delay is less an constraint (compared to VoIP) allowing a larger aggregation depth. However, aggregation depth is still limited by the time-selective behavior of the channel. This time-selectivity is measured for example by the coherence time. Coherence time is defined as the amount of time during which the autocorrelation function of the channels random gain is greater than a certain value (for example 0.9). Hence, coherence time can be assumed to be the period during which the channel is roughly constant. In the following we will limit the aggregation depth to the coherence time of the channel. Regarding a system operating at the 5.2 GHz band and considering that all the objects in the environment do not move faster than 1 m/s (pedestrian velocity), the coherence time is about 12.5 ms. When analyzing if the coherence time is exceeded, for simplification purposes, the time needed to transmit packet headers and control frames is not considered. Since the data size is considerably large, overhead is negligible and this simplification does not introduce much error in the calculations. Furthermore, we only do the calculation for the worst considered SNR (of about 6 dB). When 4 stations are present in the cell, 802.11 DYN can aggregate a maximum of 7 large MPDUs per station (leading to 28 packets transmitted in total), while IEEE 802.11n is able to aggregate 12 such packets. In case of $J = 8$ stations present in the cell, 802.11 DYN can aggregate a maximum of 3 packets per station while 802.11n aggregates again at most 12 packets per station.

The goodput results that correspond to these aggregation depths are shown in Figure 6. The left figure shows the results for $J = 4$ stations, while the right figure shows the results for $J = 8$ stations in the cell. All 802.11n curves are based on RTS/CTS handshake. Comparing the best performing curves of each transmission scheme, the range where 802.11 DYN outperforms IEEE 802.11n goes up to 36 dB for J = 4 stations in the cell. For all SNR points below 36 dB a huge gain is observed, sometimes about 200-300%. Frame aggregation increases significantly the performance of both systems, especially at medium and high SNR values. Notice that in IEEE 802.11n the use of frame aggregation starts paying off from an SNR of 18 dBs on. Prior to that point the system's performance does not vary significantly for different aggregation depths. In the low SNR range, where the PHY efficiency is considerably small, the transmission of the payload packet takes much more time than needed to transmit protocol overhead and control frames. If control information is negligible, increasing the data packet size by a factor $\alpha$ approximately increases the whole transmission time also by a factor $\alpha$. Frame aggregation only increases the MAC efficiency of the protocol significantly if the time required for payload transmission is roughly in the same order as the duration for transmitting the control information. Since 802.11 DYN introduces much more overhead than the legacy scheme, the gain that the system can obtain by using this technique is correspondingly higher.

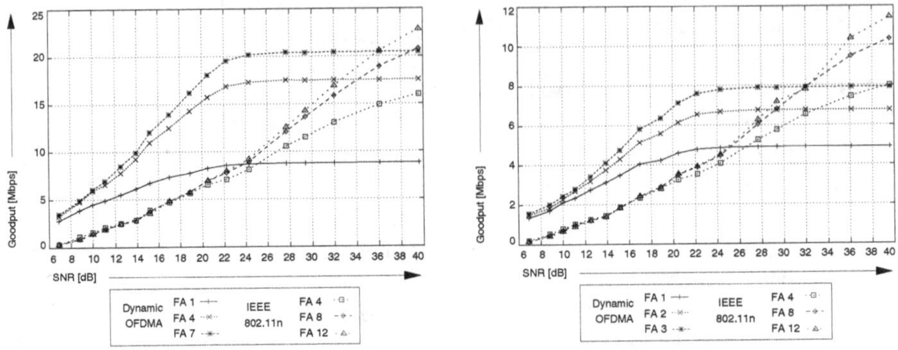


**Fig. 6.** Comparison of the average goodput per station for 802.11 DYN and 802.11n (with RTS/CTS handshake) for various different SNR levels. The packet size is set to 1570 Byte. **Left figure** $J = 4$ stations are present in the cell. 802.11 DYN applies a frame aggregation of 1, 4 and 7 packets, while 802.11n aggregates 4, 8 and 12 packets. **Right figure** $J = 8$ stations are present in the cell. 802.11 DYN applies a frame aggregation of 1, 2 and 3 packets, while 802.11n aggregates 4, 8 and 12 packets.

Similar observations regarding the performance difference of 802.11 DYN and 802.11n can be found for $J = 8$ stations in the cell (right graph of Figure 6). Notice that the limitations on the frame aggregation depth do not hold for larger SNR values. Under better channel conditions higher aggregation depths could be considered, thus increasing the gain.

## 5 Conclusions

In this paper we have presented a novel protocol which allows multi-user frame aggregation in 802.11 WLANs. The approach is coupled with channel-dependent (i.e. dynamic) OFDMA resource assignments which exploits multi-user diversity. This combination promises an increase in PHY efficiency as well as in MAC efficiency. On the other hand, significant additional overhead is required to guarantee backward compatibility and enable dynamic OFDMA resource assignments.

In comparison to 802.11n we show that our novel approach outperforms state-of-the-art WLAN technology whenever payload packet sizes or frame aggregation depths are large. Under such conditions, the proposed protocol enhancement benefits from the improvement in PHY efficiency since the required overhead is less important. However, if only few bytes are to be transmitted, the considered multi-user approach only outperforms 802.11n for low SNR ranges. We believe that additional research should focus on ways to reduce the associated overhead with multi-user frame aggregation via dynamic OFDMA to improve its performance in case of smaller payload packet sizes even for larger SNR ranges.

## References


[1] Smith, M., et al.: Proposal for PAR and 5 Criteria for Very High Throughput (VHT) SG for 60 GHz. doc. 08/223, IEEE 802.11 VHT SG Very High Throughput Study Group, Jacksonville, FL, USA (May 2008)
[2] de Courville, M., et al.: Proposal for PAR and 5 Criteria for Very High Throughput (VHT) SG for below 6 GHz operation. doc. 08/609, IEEE 802.11 VHT SG Very High Throughput Study Group, Jacksonville, FL, USA (May 2008)
[3] Czylwik, A.: Adaptive OFDM for wideband radio channels. In: Proc. of the Global Telecommunications Conference (1996)
[4] Bohge, M., Gross, J., Meyer, M., Wolisz, A.: Dynamic Resource Allocation in OFDM Systems: An Overview of Cross-Layer Optimization Principles and Techniques. IEEE Network 21(1), 53–59 (2007)
[5] Valentin, S., Freitag, T., Karl, H.: Integrating multiuser dynamic OFDMA into IEEE 802.11 WLANs - LLC/MAC extensions and system performance. In: Proc. of IEEE International Conference on Communications, ICC (May 2008)
[6] Gross, J., Emmelmann, M., Puñal, O., Wolisz, A.: Dynamic Point-to-Point OFDM Adaptation for IEEE 802.11a/g Systems. doc. 11-07/720, IEEE 802.11 WNG SC Wireless Next Generation Standing Committee, Montreal, Canada (May 14-18, 2007)
[7] IEEE: 802.11a – High-speed Physical Layer in the 5 GHz Band, Amandment 1 to 802.11 (2000)
[8] IEEE: 802.11 – Wireless LAN Medium Access Control (MAC) and Physical Layer (PHY) Specifications (June 12, 2007)
[9] IEEE: 802.11n – Enhancements for Higher Throughput, Amendment 4 to IEEE 802.11 Part 11: Wireless LAN Medium Access Control (MAC) and Physical Layer (PHY) specifications (December 11, 2007)
[10] Gross, J., Emmelmann, M., Puñal, O., Wolisz, A.: Dynamic Multi-User OFDM for 802.11 Systems. doc. 07/2062, IEEE 802.11 VHT SG Very High Throughput Study Group, San Francisco, CA, USA (July 2007)
[11] Gross, J., Emmelmann, M., Puñal, O., Wolisz, A.: Dynamic Single-User OFDM Adaptation for IEEE 802.11 Systems. In: Proc. ACM/IEEE International Symposium on Modeling, Analysis and Simulation of Wireless and Mobile Systems (MSWIM), Chania, Crete Island, pp. 124–132 (October 2007)
[12] Schumacher, L.: WLAN MIMO Channel Matlab Program (Technical report)
[13] Erceg, V., Schumacher, L., Kyritsi, P., Molisch, A., Baum, D.S., Gorokhov, A.Y., Oestges, C., Li, Q., Yu, K., Tal, N., Bas Dijkstra, N., Jagannatham, A., Lanzl, C., Rhodes, V.J., Medbo, J., Michelson, D., Webster, M.: Tgn channel models (May 2004)
[14] Awoniyi, O., Tobagi, F.: Packet Error Rate in OFDM-based Wireless LANs Operating in Frequency Selective Channels. In: Proc. IEEE INFOCOM (April 2006)
[15] Gross, J., Karl, H., Fitzek, F., Wolisz, A.: Comparison of Heuristic and Optimal Subcarrier Assignment Algorithms. In: Proc. Int. Conference on Wireless Networks, ICWN (June 2003)

# WIFE: Wireless Indoor Positioning Based on Fingerprint Evaluation*

Apostolia Papapostolou and Hakima Chaouchi

Telecom-Sudparis, CNRS SAMOVAR, UMR 5157, LOR department
{apostolia.papapostolou,hakima.chaouchi}@it-sudparis.eu

**Abstract.** Location awareness of a user or a device has always been considered as a key element for enhancing network performance and improving user experience by enabling efficient mobility and innovative location-based services. In this paper, we present our proposal named *WIFE* (Wireless Indoor positioning based on Fingerprint Evaluation), a user-based location determination system which utilizes the information of the Signal Strength (SS) received from the surrounding Access Points (APs) inside a building. We focus on a WiFi environment for its low cost and ease of deployment and study fingerprint-based deterministic techniques for their simplicity and reduced processing time and resource requirements. We first address the inherent impairments of an indoor environment which prevent a positioning system from being accurate and then describe our proposed methodology for mitigating them.

**Keywords:** Wireless networks, Location, Positioning System.

## 1 Introduction

Advances in mobile computing, wireless networking and portable devices have fostered mobility of the users while connected. Information regarding the location of the user at any time is important not only for facilitating network management tasks, like resource management, network planning, load balancing etc. but also for enabling the so called location-based services or more broadly speaking context-aware applications, with location being a key attribute of the term context. The user can interact with his surrounding environment based on this location information. The goal of a position determination scheme is to provide an accurate, real-time and reliable estimation of the user or device current location.

Positioning systems can be classified regarding various parameters: indoor or outdoor, hardware dependent or independent, deterministic or probabilistic based on the technique, WiFi, Bluetooth, infrared, ultrasound, ultra-wideband depending on the technology. In this paper, we focus on WiFi-based indoor positioning systems and study deterministic methods for their simplicity and low

* This paper is conducted in the French national research project ANR 2006 SUN: Situated and Ubiquitous Networks.

L. Fratta et al. (Eds.): NETWORKING 2009, LNCS 5550, pp. 234–247, 2009.

complexity. The strength of the signal received at an unknown position from an AP is related to its distance from the AP's position and thus this information can be utilized for estimating the unknown position. The main benefits of employing WiFi technology for positioning are that WLAN infrastructure is already available and the number of WiFi-enabled devices is rapidly increasing in the markets.

On the other hand, using WiFi for localization has shortcomings as well. Since the main target of a WLAN is the communication between its components, the placement and deployment of the APs are such that minimum overlapping is achieved, undesirable for localization purposes. Moreover, the inherent characteristics of the wireless medium, the so called propagation losses, and uncontrollable environmental changes cause undesirable variations of the SS deteriorating the positioning process. These hindrances become more intense in an indoor environment due to the presence of walls, objects and people movements.

In this paper, we first identify how these inherent impairments of the wireless and indoor characteristics cause SS variations and then describe how our proposed system, WIFE, tries to mitigate them. More specifically, we propose orientation-specific SS calibration of the area before the real-time localization process. Also, a simple pre-processing of both the calibrated and real-time SS samples is proposed for testing their quality and reliability before applying them to the position estimation algorithm. Finally, a modified version of the basic position estimation technique is described for improving the performance of the system regarding both the accuracy and time response factors.

The remainder of this paper is organized as follows: in section 2 we refer to work related to localization issues. In section 3 the main hindrances while applying localization methods in an indoor environment are addressed. In section 4 we describe and model the architecture and the positioning methodology followed by our system and in section 5 we some discuss further enhancements that can be applied. Section 6 gives the simulation results for the experimental data we applied to our system. Finally, in section 7 we conclude our work and identify future directions.

## 2 Related Work

The significance of the location information for many applications but also the difficulty in obtaining an accurate and fast location estimation without expensive hardware-dependent solutions have attracted the interest of the research and industry communities for many years in this topic. For indoor positioning the proposed algorithms can be classified in three major classes: *triangulation*, *proximity* and *scene analysis*, [9].

*Triangulation* can be done via *lateration*, which uses multiple distance measurements between known points, or via *angulation*, which measures angle or bearing relative to points with known separation. Lateration-based techniques assume a radio propagation model which is easy to derive for an outdoor environment but almost impossible for a complicated indoor scene. Angulation-based approaches, on the other hand, rely on complex and expensive hardware. Another family of localization techniques is known as *proximity* which measures the nearness to a known set of points. Thus, the estimated location is not very accurate and can be utilized for specific only applications. Finally, *scene analysis* or *fingerprint-based* examines a view from particular vantage points and tries to infer the given unknown location based on the similarity between these views. Depending on the selected format for quantifying this view we can further classify this technique in *deterministic* or *probabilistic*. *Scene analysis* schemes are mostly preferred for indoor positioning.

Many proposals can be found in the literature. Some of them, such as [6], [5], [7], are based on specific hardware. Some others, like [8], use sensors to sense environmental changes and adapt to them. WLAN-based systems, more related to our work, include RADAR, [1], HORUS, [3] and COMPASS, [4]. In RADAR a deterministic based approach is followed and a propagation model is also proposed for predicting the SS variations. HORUS is a probabilistic-based approach which tries to exploit the correlation between successive samples for increasing the accuracy of the system. Finally, COMPASS, also based on a probabilistic model, incorporates orientation information for dealing with SS variations due to the human body blocking effects.

## 3 Problem Description

The main idea of a WiFi-based positioning system is that the strength of the signal received from an AP is indicative of the distance from this AP because of its attenuation with the increase in distance. Thus, combining SS levels from multiple APs the location of the receiver can be inferred. The stability of the SS values at a specific point and the uniqueness of these SS values among different points are two desirable factors for inferring accurately a location based on the available SS information. Unfortunately, unpredictable and uncontrollable SS variations due to inherent characteristics of the wireless medium and the indoor environment make positioning inside a building, compared to an outdoor environment, more complicated and fusing. In the following these main impairments are addressed.

### 3.1 Orientation Issues

It has already been identified that the power level of the signal received from an AP at a fixed location depends on the orientation of the user due to the blocking effect of the human body. For instance, consider the case when a user's orientation is such that his WiFi-receiver has direct line of sight with an AP but for the opposite orientation his body fully prevents a signal from being received from the same AP and the same physical position. Thus, the relation between a position and SS level is not unique (1-1), but, in contrast, different SS values correspond to the same position.

In RADAR, the authors try to deal with this by sampling SS for 4 different orientations and taking the maximum value as representative for this position.

In COMPASS, the authors assume the existence of digital compasses during the calibration and the online phases.

### 3.2 Environmental Changes

The existence of walls, objects, people in an indoor environment and the often rearrangement of these cause propagation losses (reflection, scattering, diffraction) and thus prevent the signal strength from remaining stable. In RADAR, the authors propose a radio propagation model depending on the number of walls and the construction material in order to model and predict the SS variations. However, we believe that it is difficult to identify the appropriate model for each indoor environment and furthermore some environmental changes are completely random and cannot be incorporated in any model. For example, consider the case when one or more persons enter between the user and an AP while the SS sampling task is performed. It is obvious that the strength of the signal will change even though both location and orientation are fixed. In the Horus system, the authors observe high autocorrelation between successive samples and propose a first order autoregressive model in order to account for this high autocorrelation.

### 3.3 Aliasing

Aliasing is a term used to describe the phenomenon when two points even physically far apart are very similar in signal space. This is mainly because of the complex indoor propagation environment. Apparently, since a WiFi-based location system is based on the similarity in SS to infer the location, aliasing should be eliminated. In [2], considering the location history of the tracked user is proposed. Thus, if a point close in signal space is far away from the previous location of the user, it is not considered as a candidate for inferring the current position. Hardware-based solutions can provide the location system with a location approximate and thus points not within this range are excluded.

## 4 WIFE: Architecture and Methodology

This section presents the general architecture and components of the positioning process followed by the WIFE system.

### 4.1 Positioning Architecture

WLAN-based location determination systems can be categorized into two classes from architectural point of view: switch-based and user-based. In the former, a specific positioning component is required which measures the SS of the mobile user devices within its range and estimates their location based on a specific positioning algorithm. In the user-based approach, the user device is responsible for measuring the SS from the visible APs and in the sequence uses this SS information and a positioning algorithm to infer its location. We assume *WIFE* follows a user-based architecture so no special component is required and therefore it can easily work with the available WLAN system. However, the limited resources of user devices impose the requirement for a positioning process with low processing overheads.

### 4.2 Positioning Methodology Overview

Figure 1 illustrates an overview of the positioning methodology followed by the WIFE system. It includes two main phases, namely the *offline* and *online*. During the offline phase, orientation-specific calibration of the area under study is conducted. These SS sample measurements are processed by the SS sample filter module, transformed to an appropriate format and finally stored in a database. The final format of the SS samples is called *SS fingerprint* and the database *Radio Map* (*RM*). Note that the offline phase is conducted only once and repeated only in the case of big environmental changes which may affect the SS characteristics of this area. The *online* phase is actually the main phase of the positioning process. Initially, as in the offline phase, run-time SS sampling and filtering of these samples are performed. The resulting online *SS fingerprint* is compared with the *RM fingerprints* and a subset of RM entries is selected according to a mathematical *positioning algorithm*. Finally, during the *location estimation* step, the unknown location of the user device is estimated based on the known location of the selected RM entries.

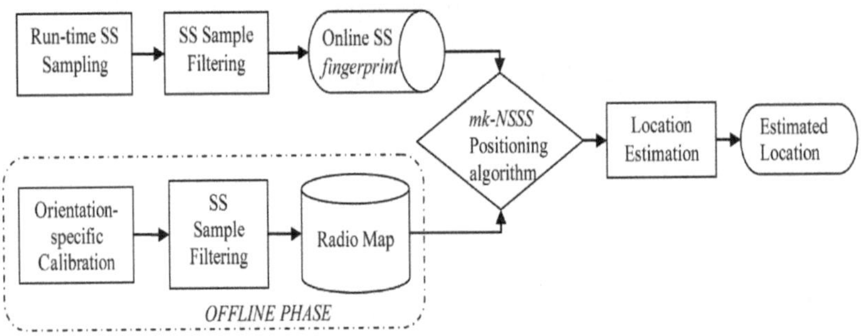


**Fig. 1.** Positioning Process Overview

In the following the system components are described with more details.

### 4.3 Orientation-Specific Calibration and Radio Map

Since orientation is responsible for SS variations in large extent (see subsection 3.1), we claim that the orientation information should certainly be considered by the positioning system. During the offline phase, we perform an orientation-specific calibration of the area under study. Assume $r$ physically distinct points within the building are selected for training the system and $i : i \in \{1, \ldots, r\}$ one of these with known location $(x_i, y_i)$. Instead of ignoring the orientation at this point and simply sampling SS measurements from all APs, we propose repeating this SS sampling process for eight possible orientations, i.e. $\forall \omega_j$ with $j = 1 \ldots 8$. Let $\mathbf{S}_i(\omega_j) = [SS_{i1}(\omega_j), SS_{i2}(\omega_j), \ldots, SS_{in}(\omega_j)]$ the SS sample vector at position $(x_i, y_i)$ and orientation $\omega_j$, where $SS_{i1}(\omega_j)$ the SS from $AP_1$. This has already been proposed in the COMPASS system where digital compasses are employed for that purpose.

However, we follow a different approach during the Radio Map construction. Let $(x_i, y_i, \mathbf{S}_i)$ a RM entry, where $\mathbf{S}_i$ the SS fingerprint at point $(x_i, y_i)$, which should have a format such that the orientation factor is accounted. In COMPASS, the format of $\mathbf{S}_i$ depends on the current user orientation, $\omega$, which implies that users need to be equipped with a compass. More precisely, the SS samples of the similar orientations of each calibrated point $(x_i, y_i)$ are merged to finally describe the *SS fingerprint* of the corresponding RM entry $i$, i.e. $\mathbf{S}_i = \bigcup_{j:|\omega_j - \omega| < T} \mathbf{S}_i(\omega_j)$, where $\bigcup$ the merging operation and $T$ an orientation threshold. In contrast, in WIFE, we do not require from users to know their orientation and construct the RM according to two possible formats,

$$\mathbf{S}_i^A = \bigcup_{j=1,\ldots,8} \mathbf{S}_i(\omega_j), \forall i \in \{1, \ldots, r\} \tag{1}$$

$$\mathbf{S}_i^B = [\mathbf{S}_i(\omega_1), \mathbf{S}_i(\omega_2), \ldots, \mathbf{S}_i(\omega_8)]^T, \forall i \in \{1, \ldots, r\}. \tag{2}$$

In the first case the *SS fingerprint* $\mathbf{S}_i^A$ actually merges the *SS fingerprints* from all 8 orientations, whereas in the second case, $\mathbf{S}_i^B$, we differentiate among the *SS fingerprints* from the 8 different orientations and consider them as 8 individual *RM* entries, even though their physical location is identical.

## 4.4 Signal Strength Filtering

Uncontrollable environmental changes distort the SS characteristics and prevent the RF-based location systems from being accurate, (see section 3.2). We try to cope with this problem with a simple, not hardware-dependent solution. We believe that even though these environmental changes are random, their impact in the sample space can be detected and alleviated. Consider a set of SS samples (for a given point and a specific AP) most of which have absolute value higher than zero and some of them with value very close to zero. It is obvious, that since signal from this AP is received, this point is within the AP's range and the almost zero-valued samples are most probably due to the instantaneous blocking effect of obstacles temporally placed during sampling. However, considering these zero values in the average [1] reduces the correspondence with the real strength level and the higher the strength of the signal the worse this distortion is, which is even more undesirable since high SS values are more indicative and significant for the location estimation process.

[1] The average for summarizing the SS samples is chosen as explained in section 4.5.

Based on this observation, we propose the incorporation of a *Sample Filtering* module after the SS sampling step. This additional module initially removes the almost zero-valued samples in the case of existence of non-zero valued samples. More complicated filtering approaches, taking into account the frequency of the samples added complexity without an accuracy improvement, at least for our experimental data.

### 4.5 Fingerprint-Based Positioning Algorithm

During the online phase the *SS fingerprint* of a user $u$, denoted as $\mathbf{S}_u$ is compared with the *RM SS fingerprints*, $\mathbf{S}_i, i \in \{1, \ldots, r\}$ and the entries corresponding to the most similar ones are retrieved. The way of defining the format of an *SS fingerprint*, the similarity between them and the way of selecting the *RM* entries are defined by the positioning algorithm. In general, the positioning techniques are categorized into two main types: namely, deterministic and probabilistic. Since WIFE is a user-based positioning system we focus on deterministic schemes, which are simpler and have less processing requirements than the probabilistic.

In the deterministic approaches, a single scalar, usually the mean value of the measured SS samples is the format selected for representing the fingerprints. Accordingly, the merging operation in equation (1) is the average. The metric for quantifying the similarity between them is their distance in signal space, i.e.

$$d_{ui} = |\mathbf{S}_u - \mathbf{S}_i| = \sum_{l=1}^{n} |SS_{ul} - SS_{il}| \tag{3}$$

where $n$ the total number of APs. The basic idea of the positioning algorithm is the *Nearest Neighbor in Signal Space* (*NNSS*) concept, since the closeness in signal information is considered. According to this, the RM entry $i$ with minimum $d_{ui}$ is selected as the NN of the user $u$. If more than one closest matches need to be determined we have the *k-NNSS* positioning algorithm, where $k$ is a parameter defining the number of these *NN*s.

We propose a modified version of the *k-NNSS* positioning algorithm which actually defines a different method for searching the NNs in the *RM*. Before giving the details of the proposed algorithm, we present the motivation behind it.

After the orientation-specific calibration, an increased number of calibrated points is available, i.e. if $r$ is the number of physically distinct points, a total number of $8 \times r$ *SS fingerprints* is available. Including them individually in the RM, see (2), and considering all of them during the online searching phase would on the one hand increase the probability of finding a closest match in SS but on the other hand, a large search space would degrade the time response and the resource utilization performance. Furthermore, even though orientation is an important factor for SS variations, the actual location still remains the main indicative one and thus, it should be prioritized during the SS similarity testing. Finally, the phenomenon of aliasing, as addressed in section 3.3, would become more possible since the probability of two physically far located points having the same SS characteristics would increase. In other words, both complexity and accuracy related reasons triggered us in proposing a two-level algorithm, called *mk-NNSS*,

1. Initially, we consider (1) as the RM representation and the $m$ NNs are determined.
2. In the second level, we consider (2) as the RM representation but only for the selected $m$ $NN$s, leading to a search space of $8 \times m$ points, and the $k$ NNs are finally determined.

### 4.6 Location Estimation

After having selected the $k$ NNs of user $u$, their locations are utilized for estimating the unknown location of the user. The most common method is to simply average their coordinates and take the corresponding result as the location estimate. Thus,

$$(\widehat{x}, \widehat{y}) = (\frac{\sum_{i \in \mathcal{N}_u} x_i}{k}, \frac{\sum_{i \in \mathcal{N}_u} y_i}{k}) \tag{4}$$

where $\mathcal{N}_u$ the set of the $k$ NNs of users $u$. Different approaches can also be used, like a weighted sum of the NNs' coordinates but we prefer to maintain the positioning system as simple as possible. Furthermore, for our experiments the simple average gave the best accuracy.

## 5 Enhancements

In this section further enhancements are discussed for further improving the performance of the WIFE system.

### 5.1 Number of Access Points

An important parameter is the number of APs from which the received SS is sampled and considered in the final format of the SS fingerprint. In general, it holds that the more the available information, the more accurate the final decision. However, this statement is more rational for triangulation-based positioning techniques, according to which the position of a device is estimated as the overlapping region of the surrounding APs' ranges. In our fingerprint-based approach the similarity between fingerprints is a single scalar which contains SS information from all APs without discriminating among them (see (3)). Thus, if many APs are considered it is more possible two different fingerprints have the same distance from the user fingerprint, and consequently fusing the system.

### 5.2 Number of Nearest Neighbors

In this section we explore the influence of the parameters $m$ and $k$, of the *mk-NNSS* algorithm, on the system performance.

**Parameter $m$.** Obviously, higher values of $m$ result in larger search space at the second level of the *mk-NNSS*. Thus, the processing time and the consumed resources for performing more comparisons are increased. Its effect on the accuracy is not so obvious. A bigger search space increases the probability of finding a very similar point but also increases the probability of aliasing.

**Parameter $k$.** Parameter $k$ defines how many NNs are finally selected for estimating the unknown location. Thus, the computational complexity is not affected by its value. However, it does affect the accuracy.

Intuitively, considering more points increases the accuracy but adding locations which are far from the real position may distort the final result. Thus, finding the optimal value is an issue. In most works which follow the $k$-NNSS method the value of $k$ is fixed for all user-cases and is more related to the grid-geometry employed during the calibration phase. However, the implicating factors of an indoor environment make each case different from the others. Therefore, making the value of $k$ adaptive for each user, i.e. $k_u$ instead of a fixed $k$ $\forall u$, appears to be promising for increasing the accuracy.

The distance in SS, $d_{ui}$, would maybe give a hint for the optimal value of $k_u$. Assuming a fixed value for $k$, we observed that for some users, $k$ or more NNs had relatively small $d_{ui}$ but for some other users only a subset of these $k$ *NN*s had relatively small $d_{ui}$. Thus, adding or excluding *NN*s based on the relativity of their $d_{ui}$ value can lead to the optimal value of $k_u$.

The following algorithm, `AdaptiveK(Du,rlt)`, gives the details for estimating the optimal value of $k_u$ for user $u$. The input `Du` is a vector of SS distances between $u$ and the RM entries, sorted in ascending order and `rlt` is a parameter which defines the number of these distances that should be considered for defining a *relatively small* distance in SS. The operations `mean` and `std` give the mean value and the standard deviation of the first `rlt` smallest distances, respectively, and their purpose is to define the *relatively small distance* (`RSdist`) and the *relatively small deviation* (`RSdev`). The main idea of the algorithm is: if the absolute difference between the distance in SS of a NN $i$ and the `RSdist` is smaller than the `RSdev`, then this NN should be considered as candidate for estimating the unknown location of user $u$.

```
function ku = AdaptiveK(Du, rlt)

RSdist := mean(Du(1:rlt))
RSDev:= std(Du(1:rlt))
ku := 1
for i=1:length(Du)  do
    if  | d(u,i) -  RSdist | < RSdev  then
        ku = ku + 1;
    end if
end for
```

## 6 Performance Evaluation

For evaluating the performance of the WIFE system we chose to use a real experimental scenario instead of simulating one. Also, instead of performing a new experiment we preferred to utilize the already available measurement results from another one performed for the same purpose, [10], so that we can fairly compare our schema with other systems. In this section, we first describe the details of the experimental environment and the process for collecting the data appropriate for the positioning algorithm. After defining the performance metrics, we depict the performance improvement achieved by our proposed schema over two other positioning systems for the same experimental scenario.

### 6.1 Experiment Specifications

**Test Environment.** The test environment where the experiment was performed corresponds to the hallway of an office building on the campus of the University of Mannheim. The floor plan of the operation area is nearly $15 \times 36\ m^2$.

**Hardware and Software setup.** As described in [4] and [10], the test environment is equipped with five Linksys/Cisco WRT54GS and four Lancom L-54g APs. All APs support 802.11b and 802.11g. One Lancom and all Linksys APs are located on the same floor with the testing area whereas three Lancom APs are located in other places inside the building.

As a client, a Lucent Orinoco Silver PCMCIA network card supporting 802.11b was used. This card was plugged into an IBM Thinkpad R51 running Linux kernel 2.6.13 and Wireless Tools 28pre. To collect signal strength samples, the implemented framework contained two parts: a library cooperating with the network card driver to perform scans and capture internal driver information, and an easy-to-use application that stores this information in a file together with additional data such as the physical position and a timestamp. Further, the application configures the library to select a scan frequency and scan technique for the signal strength measurements. For these experiments active scanning was used. Active scanning is defined in the 802.11 standard1 and it is a technique to find a suitable gateway to the Internet by measuring the SS of APs within communication range. Therefore, the available data we had and used for testing WIFE was multiple scans with the following information,

- t: timestamp in milliseconds
- id: MAC address of the scanning device
- pos: the physical coordinate of the scanning device
- degree: orientation of the user carrying the scanning device in degrees
- $MAC_i$: MAC address of a responding peer $i$ (e.g. an AP or a device in adhoc mode) with the corresponding values for signal strength in dBm, the channel frequency and its mode (access point = 3, device in adhoc mode = 1).

For keeping consistency throughout the paper, we did not discriminate between the cases of whether the responding peer is an access point or a device in adhoc mode, i.e. mode 1 or mode 3. Thus, we refer to each responding peers as an access point, leading to an operational area covered in total by 17 APs (the total number of unique MAC addresses).

To obtain the orientation of the user the Silicon Laboratories C8051F350 Digital Compass Reference Design Board was used. This device provides a USB-to-Serial bridge to access the data and is powered by the USB electricity supply. The compass in the middle of the operation area was calibrated. In a closer area around the calibration point a variation of $1^o$ was measured. However, variations up to $23^o$ were rarely detected at a few points of the testing area. These measurement errors occurred always close to electro magnetic objects such as high-voltage power lines and electronic devices.

**Data Collection.** A grid of reference points was applied to the operation area including 166 points with a spacing of 1 meter. During the offline phase, the signal strength was measured at these reference points for different orientations. At each reference point and for each orientation 110 signal strength measurements. This led to 146,080 measurements for the offline phase. Almost 10 hours were spent to collect all the data. For the online phase 60 coordinates and orientations were randomly selected and at each one of them 110 signal strength measurements were sampled, leading to 6,600 measurements in total.

### 6.2 Performance Evaluation Metrics

A user-based positioning system is successful when it can estimate the user current location accurately, fast and without wasting the limited user-device resources. As accuracy metric we choose the *Mean Location Error (MLE)* between the real and estimated locations, denoted as $(x_u, y_u)$ and $(\widehat{x}_u, \widehat{y}_u)$, respectively,

$$MLE = \sum_{u=1}^{N} \sqrt{(x_u - \widehat{x}_u)^2 + (y_u - \widehat{y}_u)^2} \quad (5)$$

where $N$ the total number of users. For our data $N = 60$.

For evaluating the computational complexity, we measure the number of comparisons that need to be made during the online phase. For the *mk-NNSS* this is $(166 + m \times 8)$, thus it is directly related to the parameter $m$.

### 6.3 Simulation Results

In this section we evaluate the performance of WIFE for the real experiment we already described. To facilitate the comprehension of the following figures, we employ the following naming for the different cases examined,

- *MergedRM* is the case when orientation is not considered, i.e. only (1) is used during the Radio Map construction.
- *FullRM* is the case when only (2) is used.
- *HierarchicalRM* when both (1) and (2) are used.

In fig. 2, the superiority of the *FullRM* cases over the *MergedRM* cases illustrates the improvement in *MLE* after orientation-specific calibration. *MLE* is further decreased for both cases after applying the *SS Sample Filtering* module.

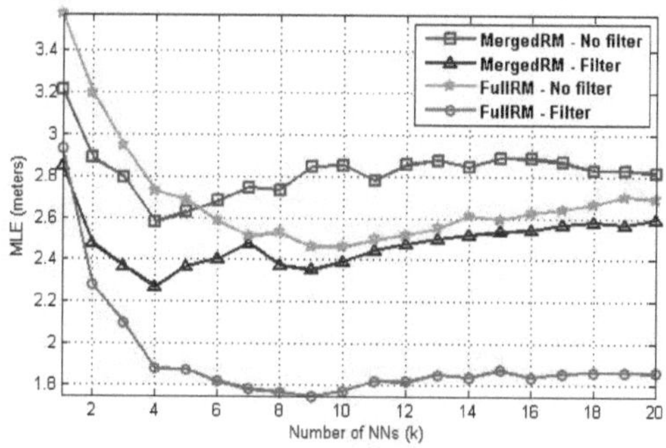


**Fig. 2.** Accuracy improvement after orientation-specific calibration and filtering

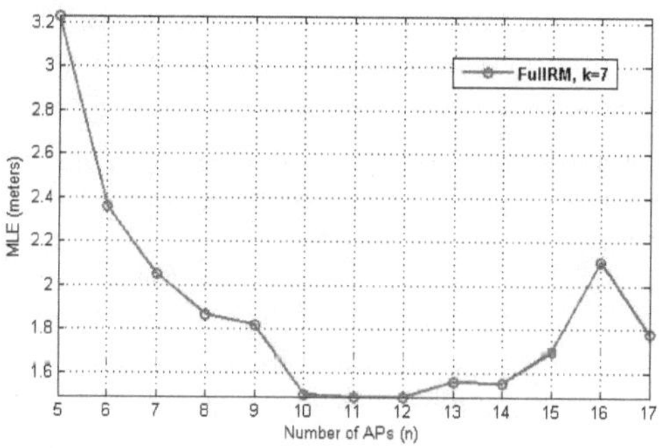


**Fig. 3.** MLE versus number of Access Points

The $x$ axis corresponds to the number of NNs, $k$. The best accuracy is achieved when we perform both enhancements and $k = 7 - 10$.

In fig. 3 we examine if we can further improve this best achieved accuracy by changing the number of APs, $n$ , that should be considered in (3). The $x$ axis corresponds to the number of APs. Note that it is important not only how many APs but also which APs are considered. However, drawing such conclusions was not trivial since the large number of the available APs would lead to a big number of possible combinations. Thus, we decided to first consider only the mode-1 APs (real APs) and then include the mode-3 APs (peers in adhoc mode), since the former were more frequently scanned. We observe that the *MLE* is improving as $n$ increases until it exceeds a certain value. This justifies our claim that more information improves the accuracy but at the same time the phenomenon of aliasing becomes more possible. The minimum *MLE* is achieved when $n = 11$ APs, which is also the average number of the visible APs from each position. Thus, a general conclusion that could be made after these observations is that, for each online user case $u$ an AP $l$ should be included in (3) if it is visible by this user $u$, i.e. $SS_{ul} \neq 0$.

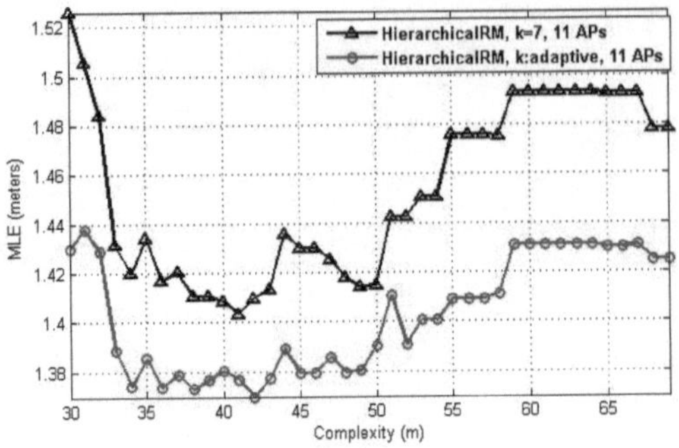


**Fig. 4.** MLE versus $m$ for $k = 7$ or $k$: adaptive

**Table 1.** Comparison with other systems

| System | Average Error (meters) | Worst Error (meters) | Complexity (♯ comparisons) | Type |
|---|---|---|---|---|
| WIFE-$k = 7$ | 1.41 | 4.10 | $166 + 8 \times 42$ | Deterministic |
| WIFE-$k$ adaptive | 1.37 | 3.93 | $166 + 8 \times 34$ | Deterministic |
| RADAR | 2.26 | 15 | 166 | Deterministic |
| COMPASS | 1.65 | 11 | $2 \times 166$ | Probabilistic |

The main drawback of the *FullRM* scheme is the large size of the resulting Radio Map, which degrades the time response and resource utilization. The *hierarchicalRM* scheme was proposed for reducing the time for searching the possible locations. In fig 4 the *MLE* is shown for different values of $m$ and for the cases when $k$ is either 7 or adaptive, as described in section 5.2. We observe that the *HierarchicalRM* can be more accurate than the *FullRM* scheme and this justifies our motivation behind the *mk-KNSS* algorithm. Making $k$ adaptive improves further the system performance. Regarding the parameter $m$, we see that for $m = 34 - 50$ the accuracy is optimal.

Finally Table 1 compares the performance of WIFE with RADAR and COMPASS for the same experiment.

## 7 Conclusion - Future Venues

In this paper we have presented WIFE, a WLAN-based system for inferring user location accurately and with low processing requirements. The main innovations proposed are: including the orientation information while training the system, performing a simple processing on the SS sample set and a two-level positioning algorithm for selecting the most probable locations. We also examined the impact on the accuracy if we consider the SS from less APs and if, for each user case, we adapt the number of calibrated points selected as possible locations. As future work, we plan to experiment the WIFE system in our own testbed.

**Acknowledgments.** We would like to thank the *CRAWDAD - Community Resource for Archiving Wireless Data At Dartmouth* for the provision of the experimental data we used for evaluating our system.

## References

1. Bahl, P., Padmanbhan, V.N.: RADAR: An In-Building RF-based User Location and Tracking System. In: IEEE Infocom 2000, vol. 2, pp. 775–784 (2000)
2. Bahl, P., Padmanbhan, V.N.: Enhancements to the RADAR User Location and Tracking System, Technical Report MSR-TR-2000-12, Microsoft Research, Microsoft Corporation One Microsoft Way Redmond, WA 98052 (February 2000)
3. Youssef, M., Agrawala, A.: The Horus Location Determination System. In: Mobisys 2005, pp. 205–219 (2005)
4. King, T., Kopf, S., Haenselmann, T., Lubberger, C., Effelsberg, W.: COMPASS: a Probabilistic Indoor Positioning System Based on 802.11 and Digital Compasses. In: WinTeck 2006, pp. 24–40 (2006)
5. Want, R., Hopper, A., Falcao, V., Gibbons, J.: The Active Badge Location System. ACM Transactions on Information Systems 1992 40(1), 91–102 (1992)
6. Priyyantha, N.B., Chakraborty, A., Balakrishnan, H.: The Cricket Location-Support System. In: 6th International ACM MOBICOM 2000 (2000)
7. Ni, L.M., Liu, Y.: LANDMARK: Indoor Location Sensing Using Active RFID. In: Wireless Networks 2004, pp. 701–710 (2004)
8. Chen, Y.C., Chiang, J.R., Chua, H.H., Huang, P., Tsui, A.W.: Sensor Assisted Wi-Fi Indoor Location System for Adapting to Environmental Dynamics. In: MSWiM 2005, pp. 118–125 (2005)
9. Hightower, J., Borrielo, G.: Location Systems for Ubiquitous Computing. In: IEEE 2001 (2001)
10. `http://crawdad.cs.dartmouth.edu/meta.php?name=mannheim/compass`

# Performance Analysis of Packet Aggregation in WLANs with Simultaneous Multi-user Access
## (Work in Progress)

Andreas Könsgen, Md. Shahidul Islam, Andreas Timm-Giel, and Carmelita Görg

Communication Networks
Center for Computing Technologies (TZI)
University of Bremen
Otto-Hahn-Allee 1, 28359 Bremen, Germany
ajk@comnets.uni-bremen.de

**Abstract.** A cross-layer wireless LAN system is considered which transmits packets for a number of users simultaneously using OFDMA or SDMA transmission. Due to varying packet lengths and physical bitrates for each user, this results in different transmission times so that users with faster transmissions have to wait until the transmission of the slower users is complete. Packet aggregation can reduce the loss of airtime and enhance the throughput without affecting the delay by filling in the gaps with additional packets which are ready for transmission. In this paper, the throughput is determined analytically and by simulation with and without packet aggregation for some given distribution functions of the packet size and the transmission time. For a more realistic channel model, the throughput enhancement due to packet aggregation is determined by simulations. It is shown that dependent on the distribution of the different flows, a significant throughput enhancement can be achieved.

**Keywords:** wireless LANs, multi-user access, packet aggregation.

## 1 Introduction

This paper reports about work in progress where a cross-layer based WLAN transmission system is designed which considers user quality-of-service demands from the application layer, queue filling states on the MAC layer and the radio channel conditions on the PHY layer. The general concept of the system was introduced in [1] and extended by QoS support in [2]. In each turn of the scheduling process, the MAC scheduler assigns a priority to the data packet at the top of each queue according to throughput and delay requirements of the application. An OFDMA/SDMA platform was introduced in [3] which provides simultaneous transmission to multiple users by adaptively allocating channel resources to each of them acording to the determined priorities. The decisions of the channel allocation algorithm is then fed back to the queues which have to keep track how much data was transferred for each user.

L. Fratta et al. (Eds.): NETWORKING 2009, LNCS 5550, pp. 248–259, 2009.

The cross-layer transmission system is located in the access point which has full control of the channel access, similar to the Hybrid Coordination Function Controlled Channel Access (HCCA) specified in IEEE 802.11e. In this paper, only the downlink from the access point to the mobile stations is considered.

In the previous investigations, it was assumed that the packet sizes for all users were constant and equal. In this paper, the scenario is extended to variable packet size which, when the users face variable transmission rates because of fading effects of the radio channel, results in widely varying times that the individual users need to transmit a packet. The cross-layer scheduler adaptively allocates power to the different users and the OFDM subcarriers, which requires that the transmission of all users start at the same time. Hence they have to wait until the slowest user has completed his transmission before the next set of packets can be sent so that airtime is wasted unused.

This paper analyzes the gain which can be achieved if the above-mentioned unused airtime gaps are filled by aggregation of multiple packets. The basic idea is that each user usually has packets waiting in the queue which may fit into the gap. In the analysis discussed here, first the mean gap size which the users experience is calculated; after that, the upper boundary for the increase of the throughput is analytically determined. Finally, simulation results are given which illustrate the effect of packet aggregation.

The rest of the paper is organized as follows. An overview about related work is given in sect. 2. An analytical model for the packet aggregation is developed and elaborated in sect. 3 for simplified distribution functions which can be closely expressed. In sect. 4, the analytic calculations are compared with simulation results. Section 5 concludes the paper.

## 2 Related Work

Quality-of-service based scheduling in wireless LANs has become an important research topic in recent years. For example, in [4], a QoS scheduler is shown which is derived from an optimization problem and then simplified to a heuristic algorithm. QoS is also considered in the scheduler described in [5] where 802.11e transmission parameters are modified; [6] proposes a scheduler which takes a scheduling decision on a packet-level basis and considers the multiple PHY rates which a WLAN offers. In [7], demands by the application are considered by abstracting them to a single numerical value. An analysis of a channel state aware scheduler is given in [8] where the channel is modeled as a Markovian process. An enhanced channel model is used in [9] where the channel characteristics are included into a cross-layer approach for the scheduler. The algorithm discussed here is cross-layer based as well, considering a MIMO-OFDMA transmission; in this paper, the focus is on a theoretical analysis of the scheduler performance.

The introduced transmission system extends the proposal of the IEEE 802.11n standards draft, where centralized channel access is specified as Hybrid Coordinated Channel Access (HCCA). On the physical layer, MIMO is used, however only one user is served at a time using TDMA. The scheduler discussed here introduces the parallelized transmission using OFDMA or SDMA.

# 3 Calculation of the Capacity

A scenario is considered where an access point simultaneously transmits packets to $n$ users. Each of the packets has a different transmission duration, which can result from varying channel conditions or packet lengths.

Each of the simultaneously transmitted packets take a different time for transmission. For all data flows which transmit their packet faster than the packet with the slowest transmission time, there is a time gap until the next transmission can be started. The algorithm which assigns the channel capacities for the data flows needs to know the priorities of the packets which are determined at the MAC layer so that a transmission start at the same time is required.

Because of the analytical calculation, simplified distribution functions for the packet size and the channel capacity are used for which closed expressions are available. For comparison, the distribution function for a more realistic simulated channel is shown afterwards along with the resulting throughput figures.

In the calculation, first the mean length of the packet with the longest transmission time interval and the mean lengths of the other (shorter) packets are calculated. From this, the mean length of the gap $\delta$ between the end of the transmission and the start of the next one is determined.

## 3.1 Calculation of the Mean Gap Size

The sample of the probability distribution function (PDF) for the transmission duration for flow $i$ at time $t$ is written as $f_i(t)$. The CDF is then

$$F_i(t) = \int_0^t f_i(x)dx. \tag{1}$$

The maximum transmission time within a turn of the scheduler is written as $x_{\max} = \max\{x_1, x_2, \ldots x_n\}$. The probability that $x_{\max}$ is shorter than a time $t$ is

$$\begin{aligned} P(x_{\max} \leq t) &= P(\max\{x_1, x_2, \ldots\} \leq t) \\ &= F_1(t)F_2(t)\ldots F_n(t). \end{aligned} \tag{2}$$

Assuming the same distribution functions for all flows, there is $f_i(t) = f(t)$ and $F_i(t) = F(t)$ for all $i$, so that $P(x_{\max} \leq t)$ can be written as $(F(t))^n$.

The PDF for the transmission time of the longest packet $f_{\max}(t)$ is then

$$\begin{aligned} f_{\max}(t) &= \tfrac{\mathrm{d}}{\mathrm{d}t}(F(t))^n \\ &= nF(t)^{n-1} \cdot f(t). \end{aligned} \tag{3}$$

To get the mean gap size $\delta$, the difference between the mean value of the longest packet and the mean value of an individual route is calculated:

$$\delta = \int_0^\infty t \cdot f_{\max}(t)\mathrm{d}t - \int_0^\infty tf(t)\mathrm{d}t. \tag{4}$$

In case of $n$ data flows with $f(t)$ being a uniform distribution, the minimum transmission time $T_{\text{min}}$ and the maximum transmission time $T_{\text{max}}$, one gets as a solution for (4):

$$\delta = \frac{nT_{\text{max}} + T_{\text{min}}}{n+1} - \frac{T_{\text{max}} + T_{\text{min}}}{2}. \tag{5}$$

For the exponential distribution with mean transmission time $\mu$, a close expression (4) as a function of the route number $n$ is not possible. When solving for a fixed number of $n = 8$ which is used as the number of data flows in the example scenario which is discussed here, one gets as the solution

$$\delta = \frac{481}{280}\mu. \tag{6}$$

### 3.2 Calculation of the Throughput

After determining the time gaps, the throughput can be calculated. The calculations are done for uniform and exponential distribution of the transmission duration. When the time gap is filled for a certain data flow by packet aggregation and interruptions of the transmission due to interframe spacings and acknowledgement packets are neglected, a continuous transmission is possible. The available throughput is then equal to the physical bitrate. With the equations given for the mean gap size, the increase of the available data rate can then be calculated. In the following calculations, $S_{\text{max}}$ is the throughput with packet aggregation, $S_0$ is the throughput without, $\Delta S = S_{\text{max}} - S_0$ and $T_{\text{meanmax}}$ is the mean value of the TX time needed for the longest packet.

The delay of the packets is not increased by packet aggregation because only existing gaps are filled in by additional packets which otherwise would be transmitted later. No transmission is delayed in order to aggregate the packets.

The following calculations are presented for three cases, where either the packet size or the transmission rate or both are variable. For each of these cases, first the expression for a general distribution is given, after that the equations for uniform and negative exponential distributions are deduced.

**Constant Transmission Rate, Variable Packet Size.** When the TX rate $R$ is assumed to be constant and the packet size is subject to a general distribution, there is

$$S_0 = \frac{\text{MeanPacketSize}}{\text{MeanLongestTxTime}}. \tag{7}$$

with

$$T_{\text{meanmax}} = \int_0^{\infty} t \cdot f_{\text{max}}(t)\text{d}t. \tag{8}$$

$f_{\text{max}}$ can be calculated according to (1) to (4), where $x_i = P_i/R$, $P_i$ being the size of the packet of flow $i$ which is currently transmitted.

*Uniform distribution.* For the increment of the throughput it can be calculated:

$$\Delta S = R \cdot \left( \frac{n \cdot T_{\max} + T_{\min}}{n+1} - \frac{T_{\max} + T_{\min}}{2} \right). \tag{9}$$

*Negative exponential distribution.* There is no close expression which expresses $S$ as a function of $n$. In case of $n = 8$ flows, there is

$$\Delta S = \frac{481}{761} R. \tag{10}$$

**Constant Packet Size, Variable Transmission Rate.** In this case, the transmission time is dependent on the physical transmission rate. Let $g(t)$ be the PDF of the transmission rate and $f(t)$ the PDF of the longest transmission time in a sample. In [10] it is shown that the PDF of the product $g(t)f(t)$ of dependent random variables $g(x)$ defined in interval $(c, d)$ and $f(x)$ defined in the interval $(a, b)$ is

$$\mathrm{PDF} = \begin{cases} \int_a^{\frac{\nu}{c}} g\left(\frac{\nu}{x}\right) f(x) \frac{1}{x} \mathrm{d}x, \ a \cdot c < \nu < a \cdot d \\ \int_{\frac{\nu}{d}}^{b} g\left(\frac{\nu}{x}\right) f(x) \frac{1}{x} \mathrm{d}x, \ \ a \cdot d < \nu < b \cdot d. \end{cases} \tag{11}$$

The average throughput is

$$S_{\max} = \frac{\mathrm{E}[\nu]}{T_{\mathrm{meanmax}}} = \frac{\int_0^\infty \nu p(\nu) \mathrm{d}\nu}{T_{\mathrm{meanmax}}}. \tag{12}$$

*Uniform distribution.* The PDF of the transmission rate is

$$g(t) = \frac{1}{t^2 (T_{\max} - T_{\min})}. \tag{13}$$

The PDF of the longest transmission time in a sample is

$$f(t) = n \frac{(t - T_{\min})^{n-1}}{(T_{\max} - T_{\min})^n}. \tag{14}$$

The PDF of the product $g(t)f(t)$ of the above random variables is

$$p(\nu) = \begin{cases} \dfrac{(\upsilon T_{\max} - T_{\min})^n (T_{\min} + n T_{\max} \nu)}{\nu^2 (n+1)(T_{\max} - T_{\min})^{n+1}}, & \dfrac{T_{\min}}{T_{\max}} < \nu < 1, \\ \dfrac{T_{\min} + n T_{\max}}{(n+1)(T_{\max} - T_{\min}) \nu^2} & 1 < \nu < \dfrac{T_{\max}}{T_{\min}} \\ -\dfrac{T_{\min}^{n+1} (\nu - 1)^n (n\nu + 1)}{(n+1)(T_{\max} - T_{\min})^{n+1} \nu^2}, & \end{cases} \tag{15}$$

The throughput $S_{\max}$ with packet aggregation is then

$$S_{\max} = \frac{\mathrm{E}(\nu)}{T_{\text{meanmax}}} = \frac{\int_1^{\frac{T_{\max}}{T_{\min}}} \nu p(\nu) \mathrm{d}\nu + \int_{\frac{T_{\min}}{T_{\max}}}^1 \nu p(\nu) \mathrm{d}\nu}{T_{\text{meanmax}}} \tag{16}$$

which can be solved analytically, however the expression is large and not shown here due to space reasons.

Without packet aggregation, the throughput is calculated as

$$S_0 = \text{PacketSize}/T_{\text{meanmax}}, \tag{17}$$

where $T_{\text{meanmax}}$ can be calculated according to the left fraction of expression (5).

*Exponential distribution.* In this case, the PDF of the transmission time can be written as

$$p(t) = \lambda e^{-\lambda(t-s)}, s < t < m, \tag{18}$$

where $s$ is the min. and $m$ the max. transmission time.

The PDF of the transmission rate is then

$$g(t) = \frac{\lambda e^{\lambda(s-\frac{1}{t})}}{t^2}, \frac{1}{m} < t < \frac{1}{s}. \tag{19}$$

The PDF of the largest element in a sample is

$$f_{\max}(t) = n(1 - e^{-\lambda(t-s)})^{n-1} \lambda e^{-\lambda(t-s)}. \tag{20}$$

The PDF of the product between the largest element in a sample and the transmission rate for $n = 8$ is

$$h(\nu) = \begin{cases} \frac{8\lambda^2}{\nu^2} \int_s^{\nu \cdot m} x \left(1 - e^{-\lambda(x-s)}\right)^7 e^{\lambda\left(2s - x - \frac{x}{v}\right)} \mathrm{d}x \\ \frac{8\lambda^2}{\nu^2} \int_{\nu \cdot s}^{m} x \left(1 - e^{-\lambda(x-s)}\right)^7 e^{\lambda\left(2s - x - \frac{x}{v}\right)} \mathrm{d}x \end{cases} \tag{21}$$

for $\frac{s}{m} < \nu < 1$ and $1 < \nu < \frac{m}{s}$, respectively.

The average throughput with packet aggregation is

$$S_{\max} = \frac{\mathrm{E}[\nu]}{T_{\text{meanmax}}} = \frac{\int_{\frac{s}{m}}^{l} \nu h(\nu) \mathrm{d}\nu + \int_{\frac{m}{s}}^{l} \nu h(\nu) \mathrm{d}\nu}{T_{\text{meanmax}}}. \tag{22}$$

The average throughput without packet aggregation $S_0$ can be calculated according to (7) where

$$T_{\text{meanmax}} = \frac{761}{280 \cdot \lambda} + s. \tag{23}$$

**Transmission Rate and Transmission Time Variable and Independent.** Finally, it is assumed that the transmission rate and time behave according to the same distribution function, but both parameters are independent. This means that

$$\text{MeanPacketSize} = \text{MeanTxRate} \cdot \text{MeanTxTime}. \tag{24}$$

The throughput with packet aggregation $S_{\text{max}}$ is equal to the mean transmission rate independent of the transmission rate distribution.

*Uniform distribution.* With $T_{\text{min}}$ and $T_{\text{max}}$ as described earlier and $R_{\text{min}}$, $R_{\text{max}}$ being the minimum and maximum rate, the mean throughput is

$$S_0 = \frac{(n+1)(R_{\text{min}} + R_{\text{max}})(T_{\text{min}} + T_{\text{max}})}{4(nT_{\text{max}} + T_{\text{min}})} \tag{25}$$

and

$$\Delta S = \frac{R_{\text{min}} + R_{\text{max}}}{2} \cdot \left( \frac{n \cdot T_{\text{max}} + T_{\text{min}}}{n+1} - \frac{T_{\text{max}} + T_{\text{min}}}{2} \right). \tag{26}$$

*Exponential distribution.* The throughput without packet aggregation is

$$S_0 = \frac{280}{761} \text{MeanTxRate}. \tag{27}$$

With $S_{\text{max}} = \text{MeanTxRate}$, the increment of the throughput $\Delta S$ can then easily be calculated.

# 4 Simulation Results

## 4.1 Comparison with the Analytical Calculations

The theoretical deductions are compared with simulations which model the scenario of an access point serving $n = 8$ stations as it already has been described at the beginning of sect. 3. The simulator implements the generation and transmission of packets with given distributions of packet size and transmission time. The measurement values such as gap size and throughput are obtained by statistical evaluation.

First, the mean gap is determined by measuring the length of the longest packet and comparing it with the length of packets which are transmitted faster. Fig. 1 below shows the analytic results (given by the lines) and the simulation results (given by the circles). The relationship between the maximum resp. mean transmission time and the mean gap size is linear for both distribution functions. The analytic and the simulation results match with negligible difference.

Figure 2 shows the graphs for constant packet size and uniform resp. exponential distribution of the transmission time. The transmission rate is in this case dependent on the previously mentioned parameters. This scenario approximates the case that all users run applications such as file downloads or CBR video

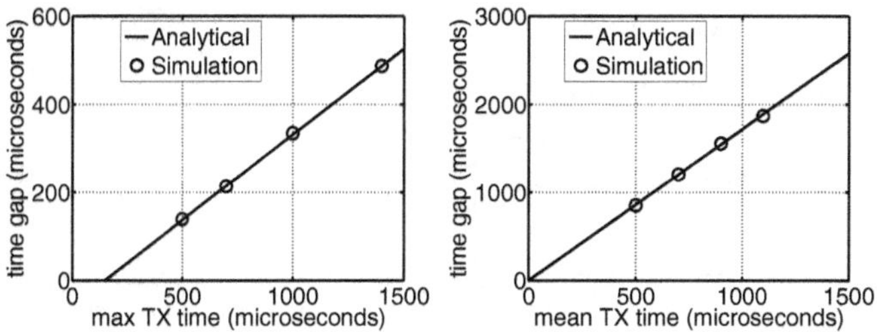


**Fig. 1.** Gap size for uniform and exponential distribution

transmissions where the packet size does not change. The achievable throughput is reduced when the transmission time is increased. The shape of the graph is similar for uniform and exponential distributions; they differ in the absolute values yielded for a particular packet size. The enhancement ratio of the throughput is increased the longer the transmission time becomes, which means it is in particular effective in case of bad channel conditions.

In Fig. 3, the graphs for constant transmission rate and uniform/exponential distributions of the transmission time are shown. In this case, the packet size is a dependent variable. This scenario would appear in reality if the users run different types of applications with variable packet size, but all have similar channel conditions. The graphs for the distributions without packet aggregation have different shapes: for the uniform distribution, it is reciprocal; for the negative exponential distribution, it is constant. Since the TX rate is assumed to be constant in this case, this means that the amount of enhancement increases with growing packet sizes. For the exponential distribution, the analytical results are slightly lower than the simulated ones. Due to the numeric processing, large values for the transmission time are not considered so that the numerically calculated packet length is slightly shorter then the analytical one which results in a higher throughput. Figure 4 shows the graphs for independent distribution functions for the transmission rate and the transmission time. In practice, this is the general case where users run different types of applications and face different channel conditions. The shapes of the result graphs are similar to the previous

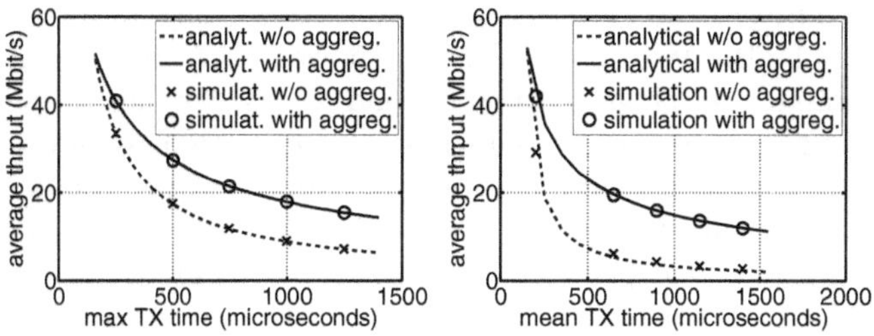


**Fig. 2.** Throughput for constant packet size and uniform/exp. distribution for TX time

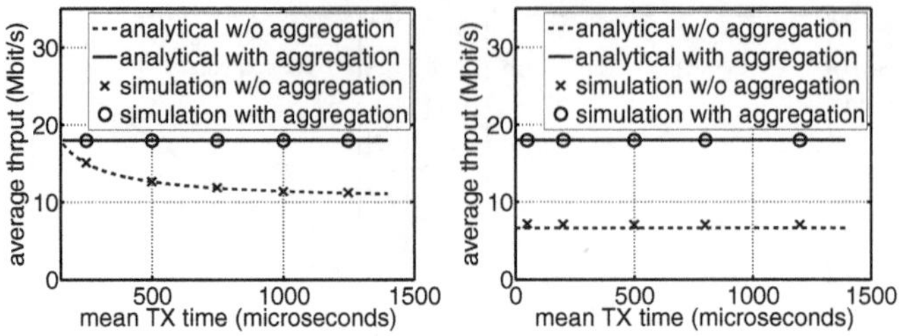


**Fig. 3.** Throughput for constant TX rate and uniform/exp. distribution of TX time

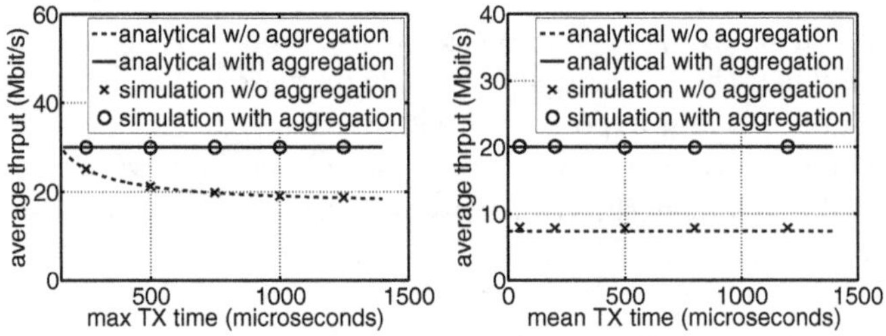


**Fig. 4.** Throughput for independent uniform/exp. distribution of TX rate and TX time

case where the transmission rate was constant; the same applies for the amount of enhancement due to packet aggregation. For the exponential distribution, the numerically calculated values are slightly higher than the analytical values due to same reason as in Fig. 3.

### 4.2 Realistic Channel Model

For comparison, a scenario with a realistic channel model is also considered. Like in the previous analytical and simulated scenarios, an access point serves $n = 8$ mobile stations using packet aggregation where each station receives one data flow. The load generator provides CBR distributed traffic for the flows 1, 2 and 4 and Poisson traffic for the other flows; the traffic load is configured individually for each flow. The packets for each flow are stored in the respective queue until they are served by the MAC scheduler. The flows 1 to 3 are time-critical due to a delay constraint; the scheduler tries to send a packet within max. 15 ms after arrival in the queue. The other flows are not delay-constrained, they are controlled according to the target throughput which corresponds to the offered load.

The channel capacities are determined based on the IEEE 802.11 TGn radio channel model proposed in [3] which is deployed here to implement a MIMO transmission with $M = 2$ transmit antennas at the base station and $N = 2$ receive antennas at each of the mobile stations, $K = 8$ users and $L = 52$ subcarriers. The model considers an indoor environment with moving obstacles which scatter the radio signal while it propagates from the sender to the receiver and result in varying

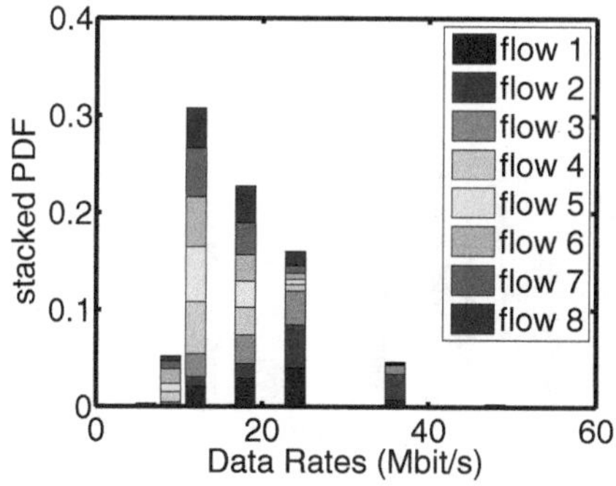


**Fig. 5.** Stacked PDF of the channel capacities for the different flows in case of realistic channel

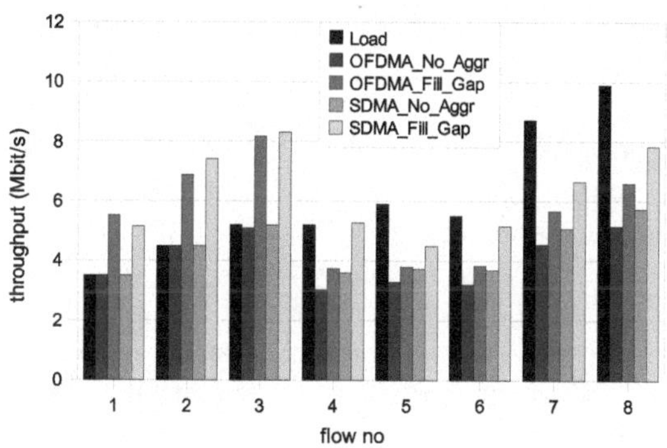


**Fig. 6.** Thrp. for the individual flows in case of realistic channel, constant packet size

channel conditions. An OFDMA or SDMA transmission is considered where each subcarrier is allocated to a particular user or subcarriers are shared between users based on spatial diversity. An important difference between the idealised scenarios discussed above and the realisitic channel model is that users in avarage may see different channel conditions regarding the mean capacity.

In Fig. 5, 6 and 7, measurements for a real channel are shown as it was described in the previous paragraph. The available data rates can be selected as in IEEE 802.11a or g systems between 6 and 54 Mbit/s. Fig. 5 shows the stacked distribution function for the relative amount of packets which was transmitted at a certain data rate. The length of each bar shows the total probability that a certain data rate was used, the lengths of the sections inside each bar show the probability for a certain flow. The graph shows the behavior of the QoS-aware scheduler: the time-critical flows 1 to 3 are assigned more often to higher bitrates than the non-time-critical flows.

Figures 6 and 7 show the per-flow throughput for data flows with constant and variable packet size. Each figure shows results which were achieved by using

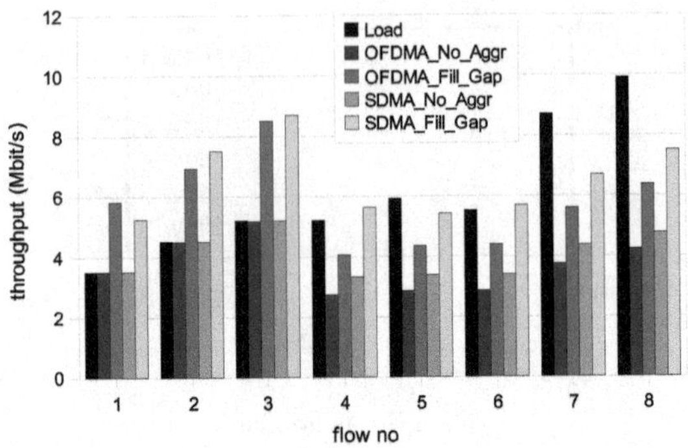


**Fig. 7.** Thrp. for the individual flows in case of realistic channel, variable packet size

OFDMA and SDMA as the method for the parallel transmission of packets. For each flow, the left bar denotes the offered load, the bars OFDMA_No_Aggr and SDMA_No_Aggr denote the throughput in case that no packet aggregation is used, respectively for OFDMA and SDMA. The bars OFDMA_Fill_Gap and SDMA_Fill_Gap show a hypothetical case, namely the throughput which could be achieved if the remaining gaps for each flow would be filled entirely with data, regardless of the offered traffic load. For the time-critical flows, this throughput is higher than the load, because for these flows the queues are usually empty due to the quick serivce of the packets. Therefore the gaps between the end of a packet transmission and the start of the next transmission are not filled in case of normal operation. Filling the gaps hence means that additional data beyond the offered load is transferred.

The flows 1 to 3 are real-time flows where the throughput must be kept. The remaining capacity is distributed among the other flows. The figures show that the time-critical flows are always served according to their needs. The throughput of the non-time-critical flows is increased in all scenarios when packet aggregation is used.

## 5 Conclusion and Outlook

A theoretical deduction was given how packet aggregation can enhance the throughput of a cross-layer scheduler with parallel transmissions. The achievable throughput increase was determined for different scenarios where the distribution of the packet size and the transmission time are either constant or according to a given distribution function. It was shown by analytical calculations and validated by simulations that packet aggregation significantly enhances the throughput. These results were compared with simulations based on a realistic channel model and with mixed time-critical and non-time-critical traffic which also showed a throughput improvement.

It was assumed that each flow fills the gap of air time completely to determine the theoretical upper boundary of the enhancement by packet aggregation. In practice, the packets to be aggregated have given sizes so that the gap might not be fully closed. Furthermore, it was assumed that each user always has data to be transmitted. In practice, a user might have a relatively low data rate so that there is no transmission requirement even if resources are available. Hence there is the need for more realistic simulations where the filling state of the queues for the different users are considered.

## References


1. Könsgen, A., Herdt, W., Timm-Giel, A., Görg, C.: A Crosslayer Two-Stage Scheduler for Wireless LANs. In: Mobile and Wireless Communications Summit, Budapest, Hungary (2007)
2. Könsgen, A., Herdt, W., Timm-Giel, A., Wang, H., Görg, C.: An Enhanced Crosslayer Two-Stage Scheduler for Wireless LANs. In: Int. Symposium on Personal and Indoor Wireless Comm. (PIMRC), Athens, Greece (2007)
3. Kermoal, J.P., Schumacher, L., Pedersen, K.I., Mogensen, P.E., Frederiksen, F.: A Stochastic MIMO Radio Channel Model with Experimental Validation. IEEE Journal on Selected Areas in Communications. Work supported by IST project I-METRA IST-2000-30148 20(6) (2002)
4. Senst, A., Schulz-Ritticg, P., Croonen, D., Ascheid, G., Meyr, H.: A Wireless Revenue Based Scheduler with QoS Support. In: 41st Annual Allerton Conf. on Communications, Control and Computing, Monticello, IL, USA (2003)
5. Inan, I., Keceli, F., Ayanoglu, E.: An Adaptive Multimedia QoS Scheduler for 802.11e Wireless LANs. In: Proc. IEEE Int. Conf. on Communication, Istanbul, Turkey (2006)
6. Yuan, Y., Gu, D., Arbaugh, W., Zhang, J.: Achieving Packet-Level Quality of Service Through Scheduling in Multirate WLANs. In: Proc. Vehicular Technology Conference (VTC), Los Angeles, USA (Fall 2004)
7. Khan, S., Suhovniko, S., Steinbach, E.: Application-driven Cross-layer Optimization for Mobile Multimedia Communication using a Common Application Layer Quality Metric. In: Proc. 2nd Int. Symposium on Multimedia over Wireless (ISMW), Vancouver, CDN (2006)
8. Rom, R., Tam, H.P.: Analysis of Performance Trade-offs for an Adaptive Channel-aware Wireless Scheduler. Springer Wireless Network Journal (December 2007)
9. Liu, Q., Zhou, S., Giannakis, G.B.: Queuing with Adaptive Modulation and Coding Over Wireless Links: Cross-Layer Analysis and Design. IEEE Transactions on Wireless Comm. 4(3) (May 2005)
10. Glenn, A.G., Leemis, L.M., Drew, J.: Computing the distribution of the product of two continuous random variables. Computational Statistics and Data Analysis 44(3) (January 2004)

# Revisiting the Performance of Short TCP Transfers

Aymen Hafsaoui[1,⋆], Denis Collange[2], and Guillaume Urvoy-Keller[1]

[1] Eurecom, Sophia-Antipolis, France
{aymen.hafsaoui,guillaume.urvoy}@eurecom.fr
[2] Orange Labs, Sophia-Antipolis, France
denis.collange@orange-ftgroup.com

**Abstract.** Performance of short TCP transfers, e.g., Web browsing, has a direct impact on the way users perceive the health of their Internet access. It is a common belief that TCP performs better with large than with short transfers, as the latters are more likely to time-out and their duration is dominated by the RTT.

In this paper, we revisit the performance of short TCP transfers. We highlight the interplay between TCP and the application on top. We show that while losses can have a detrimental impact on short TCP transfers, the application significantly affects the transfer time of almost all short - and even long - flows in a variety of way. Indeed, the application can induce extremely large tear-down times and it can also slow the rate of actual TCP transfers or affect the ability of TCP to recover using Fast Retransmit/Fast Recovery. We illustrate our findings using several traces from realistic networks including DSL, wireless hotspot and a research lab traffic.

**Keywords:** TCP, passive measurements, short transfers, application impact.

## 1 Introduction

TCP is the dominant transport protocol currently implemented in the Internet and responsible of the majority of packets and flows sent. Recent measurement studies show that TCP accounts for 60% to 90% of today's Internet traffic [1]. It is used by a large range of applications, including web, email, peer-to-peer file sharing and the newly emerging trend of YouTube-like media streaming.

A large majority of TCP flows are short lived, also known as "mice". Mice can contribute up to 97% of total number of flows and 6% of global traffic [2].This phenomenon was attributed to the domination of Internet flows by web data transfers, which are characterized by short connections. More generally, the interactive traffic of the end users often corresponds to short TCP transfers and a change in their performance directly affects the way the user perceives her Internet access.

⋆ Corresponding author.

L. Fratta et al. (Eds.): NETWORKING 2009, LNCS 5550, pp. 260–273, 2009.

A closer look at TCP loss recovery mechanisms brings to light some phenomena which can badly impact short connections. TCP detects and recovers from losses using two basic types of mechanisms: retransmissions timeouts (RTO) and fast retransmit/recovery (FR/R). Normally, a sender must receive at least three duplicate acknowledgments (ACKs) before it triggers a fast retransmit [3]. Short flows in the slow start phase often do not have a congestion window large enough to generate three duplicate ACKs, making timeouts the only loss recovery mechanism available to a TCP sender.

Given the above statements, a commonly used definition for a short TCP transfer is a transfer that can not rely on FR/R to recover from a loss. We will use this definition as a starting point and show that the emergence of new mechanisms to speed up short transfers, like *Limited Transmit* [3] and larger initial congestion window [4] prevents the derivation of a universal threshold in number of packets.

The main contribution of this paper lies in the study of the interplay between TCP and the application on top of it. Indeed, the application can slow down a TCP transfer by: (i) being stalled waiting for data to be crafted by back-end servers or from the end user, (ii) shaping the traffic to a specific rate, or (ii) delaying the closing of the transfer. In addition, the application can worsen the impact of losses by preventing TCP from sending large enough bursts of packets. We adopt an application agnostic approach, i.e., we do not make any assumption on the way the application is working, to develop a set of techniques that delineate the impact of the application from other causes that explain a given transfer duration, including the data transfer itself and the recovery time if any.

We rely on a passive study of more than 35,000 TCP connections to assess the impact of the application and the recovery mechanisms of TCP. Those connections originate from a variety of environments: one trace from an ADSL platform of a European ISP collected in 2005, one wireless trace from a public hotspot captured in Portland in 2007 (publicly available on Crawdad [5] ) and one trace from a research lab (Eurecom) collected in 2008.

Overall, we find that while losses can significantly impact the performance of short TCP transfers, only a small fraction of the short flows actually experience losses. In contrast, the application tends to affect the vast majority of the transfers, resulting in a significant drop of performance as compared to a TCP transfers where all the bytes to be sent are present in the application buffer at the onset of the transfer.

The reminder of the paper is organized as follows: related work is reviewed in Section 2. We present the main characteristics of the traces we used in Section 3. Section 4 reports on how to identify short TCP connections. In Section 5, we focus on the many different ways an application can impact a TCP transfer. Finally, Section 6 concludes the paper.

## 2 Related Work

The study of short TCP connections, a.k.a mice, has been the focus of several studies over the past two decades. The exact definition of a short transfer varies from one publication to the other. Some works rely on a fixed threshold: 10 KB [6],[7], which corresponds to 7 segments with a typical maximum segment size (MSS) of 1460 bytes, 13.5 KB in [8], i.e., 9 segments, or 32 KB [9], which is chosen equal to the median size of HTTP responses with status code 200 (indicates that the client request was successfully received). In [2] authors define short connection as data transfer comprising a number of packets less than or equal to 20 packets, assuming that the maximum congestion size is 8 KB and delayed acknowledgment is turned off. Unlike previous studies, the authors in [10],[11] define short transfers as connections that never leave the slow start phase of TCP.

Modeling short TCP transfers latency has received considerable attention. Several approaches have been proposed that take into account RTT estimation and losses impact. In [7] Cardwell et al. compare analytic models to understand how well several TCP performance models fit TCP behavior under realistic loss rate in the Internet. They propose a first model when the loss rate is zero. When the loss rate is strictly positive, they adapt the model based on the well-known TCP throughput formula of [12] to the case of short flows. In [8], a recursive analytical model is proposed to predict the TCP performance of short lived flow in the presence of losses. Completion time is computed using the connection establishment time and the duration of previous data transfers.

More recently, the authors in [9] investigated the use of short transfers latency prediction techniques based on the TCP throughput formula proposed in [7] and historical observations, being done at the server end. They demonstrated using real traces that prediction based on previous transfers systematically outperforms the analytical approach. Hence, they propose an hybrid approach: an equation based estimation for the first-contact transfer and a smoothed mean of the client previous bandwidths for subsequent transfers.

Few works have focused on the interplay between the transport and the application layers. In [13], the authors analyze passively captured TCP connections of more than 128 packets. They propose a technique to break each connection into time intervals where the application explains the transfer rate or not, based on the silences and also the rate of PUSH flags observed. In contrast, we focus on small transfers and provide a deeper analysis of the impact of the application on the transfer time (Section 5.2) and also on the impact of the application on the recovery mechanisms of TCP (Section 5.1).

## 3 Data Sets

Table 1 summarizes the main characteristics of the packet level traces used in this paper. These traces were collected from several different environments: the network of a DSL ISP, a wireless hotspot in Portland and a research lab (Eurecom). Those traces are interesting because of their diversity in terms of access technology and also in terms of applications. For instance, p2p transfers are banned from the Eurecom network while it represents a large fraction of the bytes for the DSL trace. A wireless hotspot should differ from a DSL network

**Table 1.** Trace description

| | Capture day | Duration connection | No. of connection | Well-behaved connection | Size in MB | Size in packets |
|---|---|---|---|---|---|---|
| ADSL | 2005-05-31 | 1 min and 29 s | 37790 | 5873 | 357.51 | 743683 |
| Portland Hotspot | 2007-09-14 | 2 h and 20 min | 5051 | 3798 | 174.13 | 352569 |
| Research Lab | 2008-10-20 | 1 h and 1 min | 32153 | 26837 | 1567.42 | 2867321 |

in that users tend to focus more on interactive application in such environment and tend to refrain themselves from generating large transfers, e.g. application updates or p2p transfers.

### 3.1 Well-Behaved connections

While analyzing the performance of TCP transfers, we focused on the connections that correspond to valid and complete transfers. Specifically, well-behaved TCP connections must fulfill the following conditions: (i) A complete three-way handshake; (ii) At least one TCP data segment in each direction; (iii) The connection must finish either with a FIN or RESET flag.

When applying the above heuristics, we are left with a total of over 35,000 TCP connections when summing over the three traces (detailed values are given in Table 1). The DSL trace is the one offering the smallest fraction of well-behaved connections, 5873 over 37,790, because of a large number of unidirectional transfers (SYN without a reply). P2p applications tend to generate such abnormal connections (contacting a non available p2p server to download a content), as well as malicious activities.

Figure 1 depicts the cumulative distribution of well-behaved connection size using bytes and data packets of the 3 traces. We observe that the Eurecom

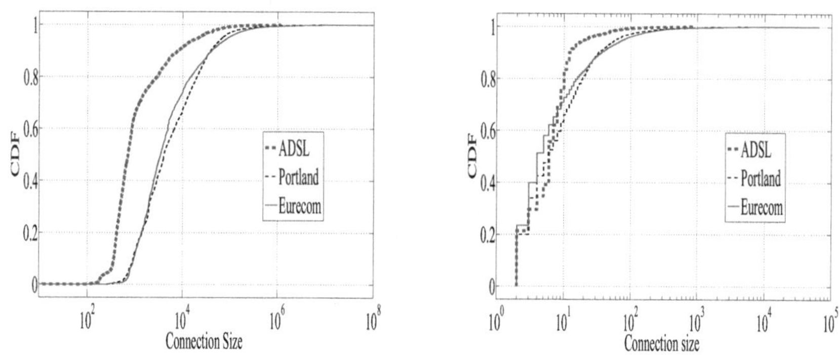


(a) Size in Bytes (b) Size in Packets

**Fig. 1.** Trace characteristics

and Portland traces offer a similar connection profile that significantly differs from the DSL trace. For instance, 65% of the DSL connections are less than 1 Kbytes and 25% are between 1 Kbytes and 1 Mbytes, unlike Portland and Eurecom traffic which offers larger values at similar connection percentiles. A reason behind this obervation is the small duration of the DSL trace. However, our focus is on short transfers, and from this perspective, the DSL trace offers valuable information.

When focusing on the performance of TCP transfers, the number of data packets to be transferred is a key element to consider. We can already observe from Figure 1 that irrespectively of the trace, a significant portion of connections (between 53% and 65%) have less than 7 data packets.

## 4 Short Transfers

### 4.1 Definition

In this section we introduce a first definition of a short TCP connection, which is commonly used in the literature.

*A short TCP connection is a connection unable to perform fast retransmit/ recovery (FR/R), after a packet loss detection.*

While simple, the above definition does not lead to a unique threshold value in terms of number of data packets for a short TCP transfer. Indeed, various TCP implementations and connection characteristics can affect this definition: the initial congestion window, the use of delayed ACK, the number of duplicate acks that triggers a FR/R. For instance, Windows Vista implements Limited Transmit, which means that only 2 duplicate ACKs are enough to trigger a fast retransmit. We estimated for the 3 traces, the number of segments observed in a duration equal to one RTT after the sending of the first data packet, and this for each direction - see Table 2. The obtained value provides a lower bound on the initial congestion window that the transport uses as the application may not provide TCP with enough data to send at the beginning of the transfer. This is especially true for the initiator side in the case of Web transfer where the GET might fit in a single data packet. Overall, we observe that values of 1 and 2 MSS (and possibly higher values) seem to be common initial congestion windows. Initial congestion windows larger than 2 MSS (we observed values up to 12 MSS) might be due to specific optimizations of operating systems that cache TCP level variables of previous transfers for a few minutes — see `http://www.csm.ornl.gov/~dunigan/netperf/auto.html`.

Given the estimated initial congestion window of Table 2, we report in Table 3 the main scenarios we focus on to find the threshold in terms of number of data packets that triggers a FR/R. A short connection is thus, for each scenario, one with a number of packets strictly smaller than the threshold. Those scenarios cover, to the best of our knowledge, all the most commonly encountered cases.

**Table 2.** Estimated initial congestion window

| Trace | Initiator | | | Remote party | | |
|---|---|---|---|---|---|---|
| | 1 pkt | 2 pkts | > 2 pkts | 1 pkt | 2 pkts | > 2 pkts |
| DSL | 99% | 1% | 0% | 80% | 18% | 2% |
| Portland | 82% | 17% | 1% | 64% | 24% | 2% |
| Eurecom | 90% | 10% | 0% | 65% | 24% | 1% |

**Table 3.** Minimum connection size to perform fast retransmit/recovery

| | Scenario 1 | Scenario 2 | Scenario 3 | Scenario 4 |
|---|---|---|---|---|
| initial cwnd | 1 | 1 | 2 | 2 |
| Delayed ACK | no | yes | yes | yes |
| Duplicate ACK | 3 | 3 | 3 | 2 |
| Minimum connection size (data packets) | 7 | 9 | 8 | 7 |

Based on the results presented in Table 3, we observe that:

- Different scenarios lead to different thresholds, from 7 to 9 data packets;
- A connection size with less than 7 data packets can not recover from packet loss using FR/R, whatever the exact scenario is;
- When considering a given scenario and a connection whose size is one packet over the threshold, we observe that this connection is able to perform a FR/R for only a single packet in its last round. The loss of any other packet will lead to timeout. A connection is thus not always able to perform FR/R if it is over the threshold.

Based on the result obtained from this section, we adopt a first definition of a short TCP transfer as a connection of size less than 7 data packets. This definition, while simple, relies on the implicit hypothesis that the application on top of TCP does not impact the way TCP sends packets. As we will see in Section 5, this assumption can be too strong in practice, as even long TCP transfers can be divided into short bursts (due to the application on top) that prevent TCP from relying on FR/R in case of losses.

### 4.2 Transfer Time Break-Down

To understand the factors that affect the performance of TCP transfers, we rely on the following decomposition of each transfer into 3 different phases:

**Set-up time.** This is the time between the first control packet and the first data packet. Since we consider only transfers which have performed a complete three-way handshake, the first packet is a SYN packet while the last one is a pure ACK in general. The connection set-up time is highly correlated to the RTT of the connection. For the three traces we consider, we have a correlation coefficient of 70% for the DSL trace, 60% for the Portland trace, and 39% for the Eurecom trace.

**Data transfer time.** This is the time between the first and the last data packet observed in the connection. Note that it includes loss recovery durations, if any.

**Tear-down time.** This is the time between the last data packet and the last control packet of the connection. We impose, as explained in Section 3.1, that at least one FIN or one RESET be observed, but there can be multiple combinations of those flags at the end of the transfer. Unlike set-up, tear down is not only a function of the RTT of the connection, but also a function of the application on top of TCP. For instance, the default setting of an Apache Web server is to allow persistent connection but with a keep alive timer of 15 seconds, which means that if the user does not post a new GET request after 15 seconds, the connection is closed. A consequence of the relation between the tear-down time and the application is a weak correlation between tear-down times and RTT in our traces: 40% for the DSL trace (which is still quite high), 0.7% for the Portland trace, and -2% for the Eurecom trace.

Using the above decomposition, we analyze, in the remaining of this article, the impact of losses (Section 4.3) and also of the application (Section 5) on the data transfer time.

### 4.3 Recovery Time

As explained above, the data transfer time possibly includes loss events. We estimate the time spent by TCP in recovering from losses using the *recovery time*. Specifically, for a given transfer, each time the sequence number in the stream of data packet decreases, we record the duration between this event and the observation of the first data packet whose sequence number is larger than the largest observed sequence number seen so far. For instance, assuming that we associate a unique sequence number to each packet, if we observe the sequence 1,2,3,4,7,6,5,6,8, we will record the duration between packet 7 and packet 8. This duration is added to the *recovery time* of the transfer. To filter out reorderings that occur at the network layer, we discard each recovery time smaller than one RTT. Rewaskar et al. [14] developed algorithms to assess whether an observed loss event can be attributed to a time-out or a FR/R. We were not able to use this technique as it requires to perform a passive OS finger printing of the sender of the data. However, in our traces, most losses occurred in the data stream issued by the remote party and not the local clients. While p0f (`http://lcamtuf.coredump.cx/p0f.shtml`), which is recommended in [14], is effective when used on SYN packets, it fails when working on SYN/ACK packets, which limits the applicability of the techniques proposed in [14].

Figure 2 presents the break-down of the small and large TCP transfers for the three traces. We first observe from Figure 2 that while set-up durations are consistently small for all traces and transfer sizes, tear-down take very high values, between 2.5 and 27.5 seconds on average. The tear-down phase in itself often represents the majority of the connection time. Note however, that the

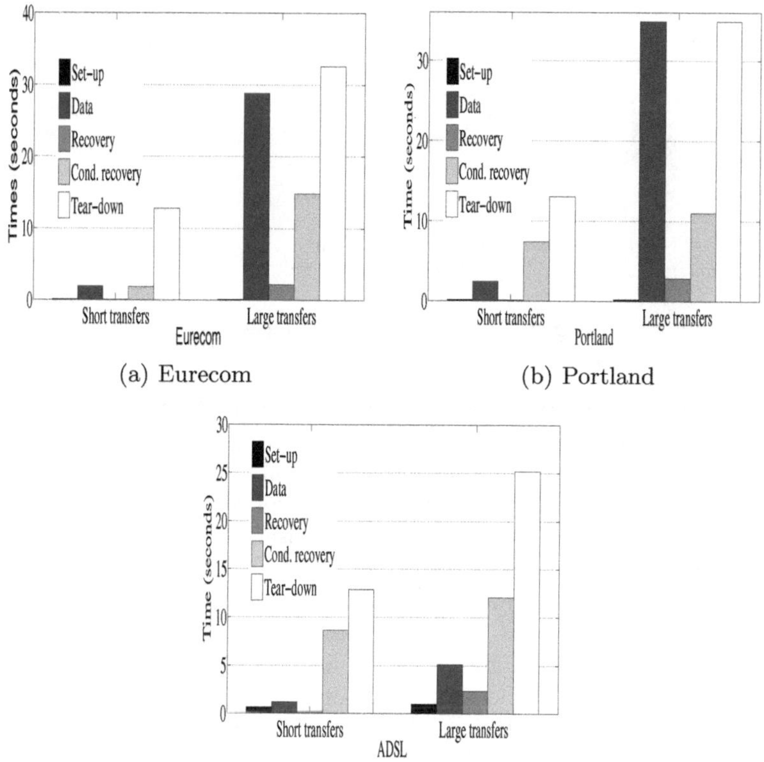


(a) Eurecom

(b) Portland

(c) ADSL

**Fig. 2.** Transfer time break-down

tear-down time should have no impact on the performance perceived from the application on top as the data transfer is completed.

As for losses, we present two distinct values for the recovery time: the average conditional recovery time and the average recovery time. The latter is computed over all transfers of the category while the former is computed only for the transfers that experience at least one recovery event. Since only a small fraction of the transfers experience losses (9.4% for DSL trace, 13.2% for Portland and 6.8% for Eurecom), the average conditional recovery time is often much larger than the average transfer time. This impact is clearly more pronounced for small than for large flows, over the three traces, most probably because of the predominance of time-outs for short transfers.

## 5 Application Impact

In this section, we are interested in assessing the impact of the application on the transfer time of a TCP connection. There are many ways by which the application can influence the pace at which data flows in a network. First, the user

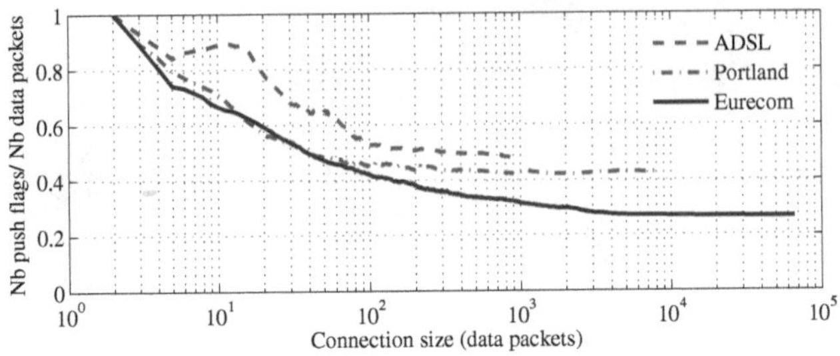


**Fig. 3.** Conditional ratio of push flags

might be involved in the transfer, as the case in a persistent HTTP connection, where the download of a new page is triggered by an HTTP Get message issued by the client browser. Second, the application might cap the rate at which information is sent to the TCP layer. This is typically what p2p applications do to limit the congestion on the uplink of the user. A third possibility is when the generation of data is done online. For instance, when querying Google for a specific keyword, several tens of machines are involved in this operation.

From the above discussion, we observe that the application may affect the transfer of data in many different ways. A first simple assessment that can be made to infer the impact of the application on a TCP transfer is to compute the fraction of packets with PUSH flags. The PUSH flag is a way for the application to specify that it has no more bytes to send at the moment and the current segment can be sent. We plot in Figure 3 the ratio of PUSH flags as a function of the transfer size for the three traces. We observe that the impact of application as captured by the PUSH flags decreases with increasing transfer size. For the short connections, the push flag ratio is extremely high, between 74% and 86%.

In the remaining of this section, we want to assess in more details the way the application influence the transfer time. We will first show that the application tends to fragment the transfer in small flights of packets that prevent TCP from relying on FR/R in cases of losses. In a second stage, we focus on the way the application forces TCP to pace the data.

### 5.1 Synchronism and Losses

For client/server applications, one often observes that even if the server is sending a large amount of bytes/packets, the actual exchange is fragmented: the server sends a few packets (hereafter called a train of packets), then waits for the client to post another request and then sends its next answer. If such a behavior is predominant in TCP transfers, it can have a detrimental impact if ever the train size is too small as it might prevent TCP from performing FR/R in cases of losses.

The question we raise is thus: are the two parties involved in a transfer synchronized or not? Proving synchronism requires an a priori knowledge of the application semantics. We can however prove that the synchronism hypothesis cannot be rejected as follows: for a given transfer, each time we observe a transition from one side sending packets, say A, to the other side sending packets, say B, we observe if the first packet from B acknowledges the reception of the last packet from A. If this is not the case, then there is no synchronism, otherwise, synchronism can not be rejected. Applying this methodology to the three traces, we obtained that for each trace, the fraction of connections for which synchronism could not be rejected was extremely high: 88.6% for the ADSL trace, 94.4% for the Portland trace and 95.3% for the Eurecom trace.

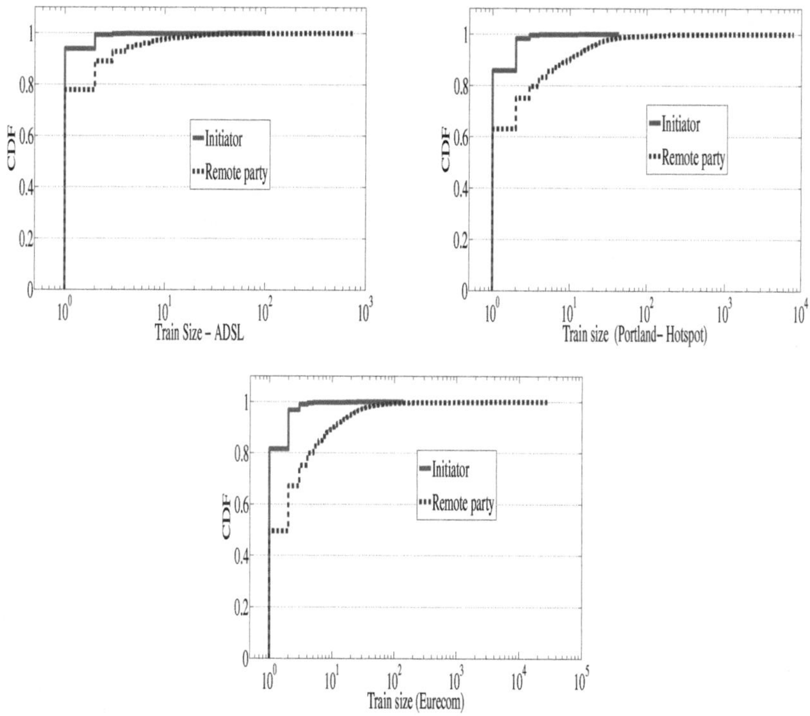


**Fig. 4.** Cumulative distribution of transmitted size blocs

For the connections for which synchronism could not be rejected, we looked at the distribution of the size of the trains of packets sent. We distinguished between the initiator of the connection and the remote party, as we expect the latter to be some kind of server that usually sends larger amount of packets than the former that simply posts requests. As illustrated by Figure 4:

- Trains size sent by the remote part are larger than those sent by the initiator, in line with our hypothesis that the remote party be a server;
- More than 97% of initiator trains are less than 3 data packets, which leaves TCP unable to trigger any Fast Retransmit, even if Limited Transmit is used;

- More than 75% of remote party trains are less than 3 data packets, which again leaves TCP unable to trigger the fast recovery/retransmit, even if Limited Transmit is used.

Taking a broader perspective, the fraction of connections that have a maximum train size of 3 packets is 85.2% for the DSL trace, 40.5% for the Portland trace and 54% for the Eurecom trace. Sizes of those connections remain quite in line with our definition of Section 4.1 as about 87% of those Eurecom and Portland connections have less than 7 packets. It falls to 62% for the DSL trace. For all 3 traces, we observe the vast majority (over 97%) of those connections have less than 20 packets.

### 5.2 Data Pacing

In this section, we focus on the transfers that obey to the definition of synchronism introduced in the previous section. For those transfers, we want to assess how the application[1] slows down the actual data transfer. To do so, we term A and B the two parties involved in the transfer (A is the initiator of the transfer) and we break down the data transfer times into a set of components (see Figure 5):

- $T^i_{\text{train time}}(A)$: time needed to transfer the $i$-th train of the initiator;
- $T^i_{\text{train time}}(B)$: is the time needed to transfer the remote party data train;
- $T^i_{\text{warm-up}}(A)$: time between receiving the last data packet from B and sending train $i$. The warm-up accounts either for the user thinking time or for some latency to generate the data at the server side for instance;
- $T^i_{\text{warm-up}}(B)$: time between receiving the last data packet from A and sending train $i$.

Note that to obtain accurate estimates of those durations that are related to the sender or receiver side, we have to shift in time the time-series of packets received at the probe. Specifically, we assume that a packet received from A at probe P was sent $\frac{RTT_{P-A}}{2}$ in the past and will be received $\frac{RTT_{P-B}}{2}$ in the future, where $RTT_{P-A}$ (resp. $RTT_{P-B}$) is the RTT between P and A (resp. B) . While doing this operation, we assume that the RTT of the transfer stays constant.

The above breakdown strategy results in a complete partition of the total transfer time. The application can impact both warm-up and train times. Concerning train times, we sum for each party, A or B, the total train times, from which we substract the recovery times if any. We term those values $T_{\text{train time}}(A)$ and $T_{\text{train time}}(B)$. We also record the total number of distinct data packets sent by A or B. We next compute the duration that an ideal TCP layer with an initial congestion of 1, delayed acknowledgment turned on, an infinite capacity, an RTT equal to $RTT_{A-B}$ and the same number of packets to send as A or B would take to complete the transmission of all those packets. We term those duration $T_{\text{theory}}(A)$ and $T_{\text{theory}}(B)$. The difference between theoretical quantities

[1] We consider the application in a broad sense, including the user interactions.

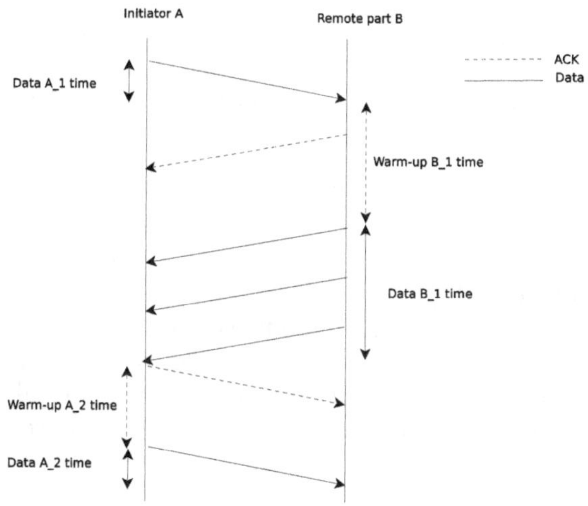


**Fig. 5.** Connections data time

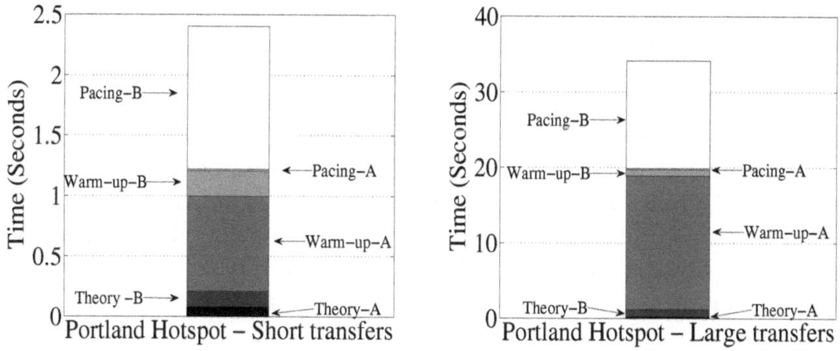


**Fig. 6.** Application impact for the Portland trace

and the total train time, $T_{\text{train time}}(A) - T_{\text{theory}}(A)$ and $T_{\text{train time}}(B) - T_{\text{theory}}(B)$, represent estimates of the delay introduced by the application on top of TCP. We term them as pacing times in the remaining of this section.

Figure 6 presents the result of applying the above methodology to the Portland trace. The two other traces offer qualitatively similar results. We observe when looking at Figure 6 that the warm up time of A (initiator) and the pacing time of B (remote party) represent the largest shares of the train time of A and B respectively. A possible explanation behind this observation is that the "average" connection features characteristics close to a client/server application with a large thinking-time of the user, that leads to large warm-up values for A, and a server whose rate is limited either by some back-end server or the use of a rate policy. A precise assessment of the causes behind those phenomena clearly calls for more advanced studies, that we leave for future work. For the time being, the major lesson learned from this study is that the application slow down most transfers in many different ways and this impact is observable for both small and large transfers. This is somehow in contrast to losses, which can have a more detrimental impact, but only for a minority of transfers.

## 6 Conclusion

In this paper, we have analyzed on three different traffic traces the performance limitations of short and of interactive TCP transfers.

Short transfers sending less than seven packets are not able to apply Fast Retransmit. Thus, they are really sensitive to loss events in the network. These short transfers represent the majority of transfers. We have also observed very long tear-down delays, between the last data packet of the connection and the last control packet. This tear-down delay does not influence the user perception, but it may affect the measurement of response times of short transfers in network management functions.

The sensitivity to loss concerns also many long transfers as many of them are a sequence of alternate exchanges and the vast majority of these bursts are less than 3 packets. Such a feature has a direct influence on the ability of TCP to recover from a loss using Fast Retransmit.

We have also highlighted that the delay to transfer a burst is usually much larger than the pure transmission time. Causes behind these slow downs can be found at the sender, e.g., rate shaping, and also at the receiver side, e.g., thinking time. To the best of our knowledge, this work is the first of its kind to pinpoint and quantify the impact of the application on top of TCP. An important lesson learned from this study is that while losses can have a highly detrimental impact on the transfer times, losses occur in fact (and hopefully) very rarely. In contrast, the application affects almost all flows and leads to a substantial slow down of the transfers.

## References

1. Fomenkov, M., Keys, K., Moore, D., Claffy, K.: Longitudinal study of Internet traffic in 1998-2003. Technical Report, Cooperative Association for Internet Data Analysis CAIDA (2003)
2. Ben Azzouna, N., Guillemin, F.: Analysis of ADSL traffic on an IP backbone link. In: IEEE GLOBECOM, San Francisco (2003)
3. Allman, M., Balakrishnan, H., Floyd, S.: Enhancing TCP's Loss Recovery Using Limited Transmit. RFC:3042 (2001)
4. Allman, M., Floyd, S., Partridge, C.: Increasing TCP's Initial Window. RFC:3390 (2002)
5. Kotz, D., Henderson, T., Abyzov, I.: CRAWDAD data set dartmouth/campus, `http://crawdad.cs.dartmouth.edu`
6. Ayesta, U., Avrachenkov, K.: The Effect of the Initial Window Size and Limited Transmit Algorithm on the Transient Behavior of TCP Transfers. ITC-SS15, Germany, Wuerzburg (2002)

7. Cardwell, N., Savage, S., Anderson, T.: Modeling TCP Latency. In: IEEE INFOCOM, Tel-Aviv (2000)
8. Melia, M., Stoica, I., Zhang, H.: TCP Model for short lived flows. IEEE Communications Letters (2003)
9. Arlitt, M., Krishnamurthy, B., Mogul, J.C.: Predicting short-transfer latency from TCP arcana: A trace-based validation. In: ACM SIGCOMM, Philadelphia (2005)
10. Ebrahim-Taghizadeh, S., Helmy, A., Gupta, S.: CP vs. TCP: a Systematic Study of adverse Impact of Short-lived TCP Flows on Long-lived TCP Flows. In: IEEE INFOCOM, Miami (2005)
11. Barakat, C., Altman, E.: Performance of short TCP transfers. In: Pujolle, G., Perros, H.G., Fdida, S., Körner, U., Stavrakakis, I. (eds.) NETWORKING 2000. LNCS, vol. 1815, pp. 567–579. Springer, Heidelberg (2000)
12. Padhye, J., Firoiu, V., Towsley, D., Kurose, J.: Modeling TCP Throughput: A Simple Model and its Empirical Validation. In: ACM SIGCOMM, Vancouver (1998)
13. Siekkinen, M., Urvoy-Keller, G., Biersack, E.W.: On the interaction between internet applications and TCP. ITC-20, Ottawa (2007)
14. Rewaskar, S., Kaur, J., Smith, F.D.: A Passive State-Machine Approach for Accurate Analysis of TCP Out-of-Sequence Segments. ACM Computer Communication Review (2006)

# Why Are Peers Less Stable in Unpopular P2P Streaming Channels?

Zimu Liu[1], Chuan Wu[2], Baochun Li[1], and Shuqiao Zhao[3]

[1] Department of Electrical and Computer Engineering, University of Toronto
{zimu,bli}@eecg.toronto.edu
[2] Department of Computer Science, The University of Hong Kong
cwu@cs.hku.edu
[3] Multimedia Development Group, UUSee Inc.
zhaosq@uusee.com

**Abstract.** In large-scale P2P live streaming systems, it is shown that peers in an unpopular channel often experience worse streaming quality than those in popular channels. In this paper, by analyzing 130 GB worth of traces from a large-scale P2P streaming system, UUSee, we observe that a large number of "unpopular" channels, those with dozens or hundreds of concurrent peers, tend to experience inferior streaming quality. We also notice a short lifespan in these channels, which further exacerbates streaming quality. To derive useful insights towards improving streaming performance, we seek to thoroughly characterize important factors that may cause peer volatility in unpopular channels. Specifically, we conduct a comprehensive statistical analysis on the impact of various factors on peer lifespan, using survival analysis techniques. We found that the initial buffering level, the variance of peer indegree, and the peer joining time all have important effects on the lifespan of peers.

**Keywords:** Measurement, Peer-to-Peer, Live Streaming.

## 1 Introduction

Real-world live P2P multimedia streaming systems have been successfully deployed in the Internet at a large scale, with hundreds of channels and hundreds of thousands of users at any given time. With the large number of concurrent channels, practical experiences have revealed the widely uneven distribution of peers across different channels: there may be thousands of concurrent users watching a popular channel, and no more than a few hundred of peers in an unpopular channel. These unpopular channels, usually representing the majority of the available channels in the streaming system, generally experience lower streaming quality, as compared to large popular channels. While many research efforts have been made to guarantee the performance of large popular channels, *e.g.*, to accommodate a flash crowd scenario where a large number of peers join in a short period of time, little attention has been devoted to the improvement of streaming quality in unpopular channels.

L. Fratta et al. (Eds.): NETWORKING 2009, LNCS 5550, pp. 274–286, 2009.

In this paper, we focus on unpopular channels in large-scale P2P streaming systems. Using more than 130 GB worth of run-time traces from hundreds of streaming channels in a large-scale real-world P2P live streaming system, UUSee [1] (among the top three commercial systems in mainland China, along with PPLive and PPStream), we have investigated the distribution of peer population and streaming quality across different channels, and observed inferior streaming qualities that are empirically experienced by unpopular channels. We have further discovered a short peer lifespan (severe peer volatility) in the unpopular channels, which reveals a less than desirable situation that may lead to a downward spiral of peer population: On one hand, the low streaming quality in an unpopular channel may lead to short peer stay in the channel; on the other hand, the more severe peer churn further exacerbates the streaming quality of existing peers. To promote peer stability for a better streaming quality, it is critical to thoroughly understand and characterize the important factors that may have caused the peer volatility in unpopular channels.

Towards this objective, we conduct a comprehensive and in-depth statistical analysis using the UUSee traces. Our objective is very clear: we wish to identify critical performance metrics as risk factors that may influence the lifespan of peers, in order to derive useful insights towards the improvement of stability of peers in unpopular channels. To achieve this, we have parsed and imported all run-time traces into a database, where we apply survival analysis techniques such as the Cox proportional hazards model and the Mantel-Haenszel test to discover such influential factors. We have found that the initial buffering level, peer indegree and peer joining time all have important effects on the lifespan of individual peers. We are able to discover these influential performance factors from the real-world traces, which has not been possible in the existing literature. Based on the results of our regression analysis, we have derived a number of useful insights to guide the design of better P2P streaming protocols that promote the stability of peers in unpopular channels.

The remainder of this paper is organized as follows. In Sec. 2, we present our research methodologies with respect to collecting and parsing UUSee run-time traces. In Sec. 3, we show our survival analysis of the traces to identify influential factors on the peer lifespan. In Sec. 4, we model the impact of influential factors using the Cox regression model. In Sec. 5, we discuss the implication of our model and its usage in promoting peer stability. We discuss related work and conclude the paper in Sec. 6 and Sec. 7, respectively.

## 2 Trace Overview: A First Glance at Unpopular Channels

### 2.1 Collecting Real-World Traces

Starting from 2006, we have been monitoring the performance statistics of a real-world commercial P2P streaming system, offered by UUSee Inc. [1]. Similar to all current-generation mesh pull-based P2P streaming protocols, UUSee's streaming protocol design is based on the principle of allowing peers to serve each other by exchanging blocks of data, that are received and cached in their local playback buffers. To dynamically monitor the entire system, we have implemented detailed measurement and reporting capabilities within the UUSee client application. Each peer collects a set of its vital statistics, and encapsulates them into "heartbeat" reports to be sent to the tracking servers every 5 minutes via UDP. The statistics include its IP address, the channel it is watching, its buffer availability map, the number of consecutive blocks in its current playback buffer (henceforth referred to as the *buffering level*), instantaneous aggregate download and upload throughput from and to all partners, as well as its download and upload bandwidth capacities.

Though we have been continuously monitoring the performance of UUSee, the study in this paper features a most recent set of run-time traces, collected between Thursday, May 29, 2008 (GMT+8) and Monday, June 2, 2008 (GMT+8), which contains continuous-time snapshots of the streaming system throughout the period, featuring over 16 million peer sessions. We believe these recent traces best captured the up-to-date characteristics of peers in the millions-of-users scale, to which the application has expanded over the years. Here, a *peer session* refers to the lifespan between the joining and the departure of a peer.

### 2.2 Observations on Unpopular Channels

Why do we need to investigate the streaming performance in unpopular channels and popular channels distinctively in such a large-scale system? First of all, we observe that the popularity differs significantly across channels: a small number of most popular channels ($\approx 2\%$) with an average peer population over 5000, a small percentage of less popular channels ($\approx 31\%$) with a population in the range of 500 to 5000, and the majority of UUSee channels accommodating a peer population less than 500 ($\approx 67\%$). Fig. 1 plots the correlation between the streaming quality and peer population in all UUSee channels in two representative snapshots, 9 a.m. on May 30 and 9 p.m. on May 30. Here, we evaluate the streaming quality in a channel at each time as the *percentage of high-quality peers in the channel*, where a high-quality peer has a buffering level of more than 80% of the total size of its playback buffer (buffer size in UUSee is 500 media blocks). The criterion of the buffering level (*i.e.*, the number of consecutive blocks in the playback buffer of a peer, starting from the current playback position) has been extensively used in the actual UUSee streaming protocol to evaluate the current streaming quality of a peer. Accordingly, we also use the peer buffering level as our basic streaming quality metric, based on the rationale that the more blocks a peer has cached in its buffer, the higher chance it has to enjoy a smooth playback. We can observe from Fig. 1 that unpopular channels generally represent worse streaming quality, as compared to large popular channels. The less than satisfactory streaming performance in unpopular channels—which represent the majority of streaming channels in UUSee—exposes a critical challenge in improving the performance of real-world streaming systems: How shall we boost the streaming quality of unpopular channels?

In P2P streaming systems, peer instability represents a "killer" factor that negatively affects the achievable streaming quality. It is even more so in unpopular

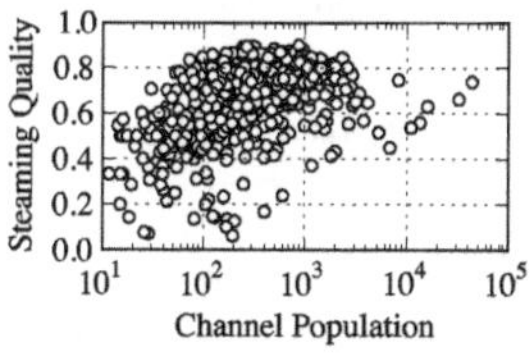


(a) 9 a.m., May 30

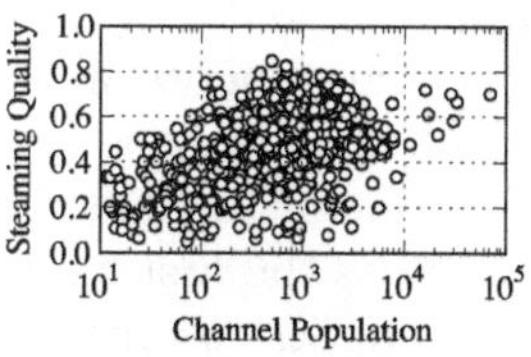


(b) 9 p.m., May 30

**Fig. 1.** Correlation between channel population and streaming quality

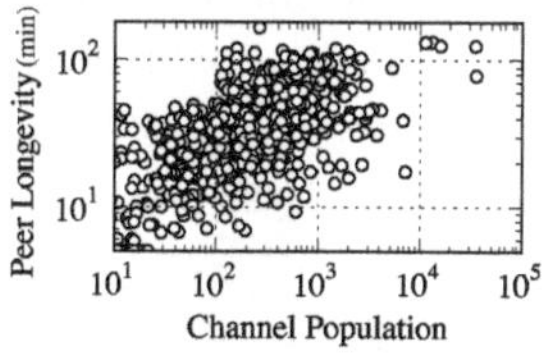


(a) 9 a.m., May 30

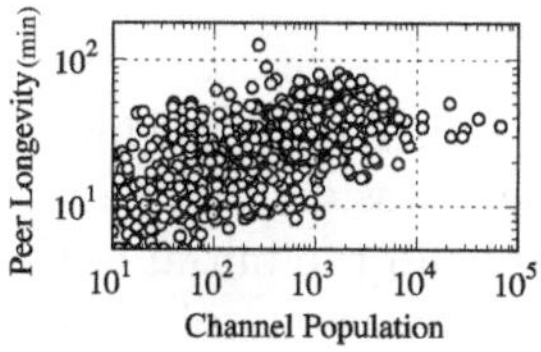


(b) 9 p.m., May 30

**Fig. 2.** Correlation between channel population and peer longevity

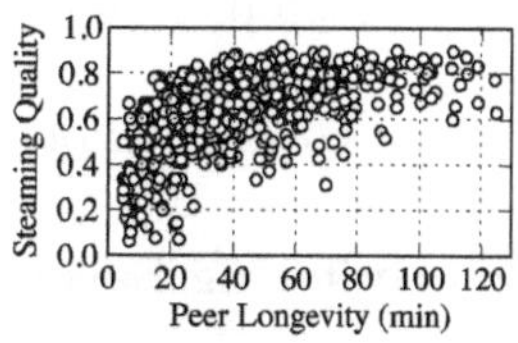


(a) 9 a.m., May 30

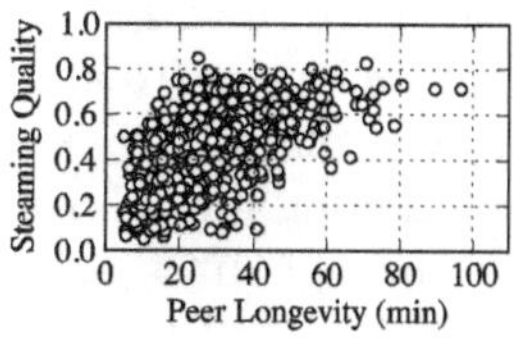


(b) 9 p.m., May 30

**Fig. 3.** Correlation between peer longevity and streaming quality

channels, as observed by our trace studies in Fig. 2 and Fig. 3. We found that the peer lifespan (also referred to as peer *longevity*) tends to be shorter in the unpopular channels from Fig. 2, while in most cases the more severe peer churns further exacerbate the streaming quality in those channels, as shown in Fig. 3. All these observations have pointed to the following fact: To promote the streaming quality in unpopular channels, a key step is to improve the peer stability in these channels, by promoting peer online times. In order to promote peer stability, we find it important to obtain a thorough and in-depth understanding of the critical factors that influence peer online times in unpopular streaming channels, which constitutes the major objective of our study in this paper.

# 3 Deciphering Peer Instability in Unpopular Channels

## 3.1 Survival Analysis and Censoring

Survival analysis [2] represents a set of statistical methods for the analysis of death or failure events and involves the modeling of time to event data, *i.e.*, the survival time. In our analysis of peer longevity in each streaming channel, a peer's departure represents a *failure* or *death* event, and the time between its joining and departure, the peer longevity, is the survival time to be considered. In survival analysis, a *survival function* is frequently used to describe the probability that an individual survives to a specific time $t$. Let a random variable $T$ represent the longevity of a peer session, the survival function is defined as: $S(t) = \mathbf{Pr}(T > t) = 1 - \mathbf{Pr}(T \leq t) = 1 - F(t)$, where $F(t)$ is the cumulative distribution function (CDF) of the longevity. A standard estimator of the survival function, based on a number of measured survival times, is proposed by Kaplan and Meier, referred to as the product-limit estimator or the K-M estimator [2]. We now seek to investigate critical factors that influence peer longevity in unpopular channels based on correlation plotting and survival function K-M estimator.

### 3.2 Buffering Level

Intuitively, the higher buffering level a peer experiences, the smoother its streaming is, and the more likely it will stay longer in the channel. Therefore, we start by investigating: *Do peer longevity patterns differ significantly under different buffering levels?* To answer this question, we explore the relevance between peer longevity and various statistical metrics of the buffering level, including the average buffering level during a peer session, the standard deviation of the buffering level throughout a peer session and the initial buffering level, as the first buffering level measured when a peer starts its playback. As a statistics to represent the distribution of peer longevity among a group of sessions, we define an EDR($t$) function, *i.e.*, Early Departure Rate function, as the percentage of peers whose lifespan is less than or equal to $t$ minutes within a group. In each sub-figure in Fig. 4, we plot the EDR(15 min) of each session group categorized according to different levels of the respective buffering level metric, as well as the smoothed lowess curves. Note that in all our studies hereinafter, we use peer session data from all the unpopular channels, *i.e.*, channels with less than 500 peers most of the time, in order to derive insights useful for their performance enhancement.

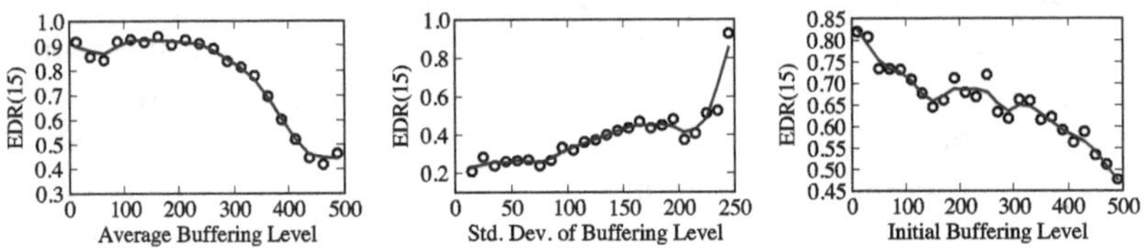


(a) (b) (c)

**Fig. 4.** Correlation of EDR(15 min) with metrics of buffering level

Fig. 4(a) reveals a negative correlation between the early departure rate and the average buffering level within the buffering level range of 200 – 475, showing that the departure rate is higher (peer lifespan is shorter) when the average buffering level is lower in this majority range. The positive correlation between the standard deviation of buffering level and early departure rate, as shown in Fig. 4(b), meets our expectation that the less stable the buffering level is, the shorter the peers are staying. We further investigate the tolerance of peers towards the initial buffering level, by plotting the correlations in Fig. 4(c). A strong negative correlation is observed in Fig. 4(c) between the early departure rate and initial buffering level within the buffering level range of 0 – 120 and 320 – 500, respectively. This reveals that at the two ends of the spectrum, an excellent initial buffering level brings a longer peer online time, and a very poor initial buffering level will almost definitely result in an early departure. In our study, we have varied $t$ in the EDR($t$) function from 5 minutes to 60 minutes, and made similar observations.

### 3.3 Peer Incoming Degree

In P2P streaming, the number of supplying peers a peer can obtain in a streaming channel, *i.e.* its incoming degree or indegree, and how stable these incoming connections are, affect the streaming quality it obtains, and thus affect the longevity of the peer in the channel. To investigate such impact of peer indegree, we plot in Fig. 5 the EDR(15 min) values of session groups at different levels of the average indegree during a peer session and the standard deviation of the indegree throughout a peer session, respectively.

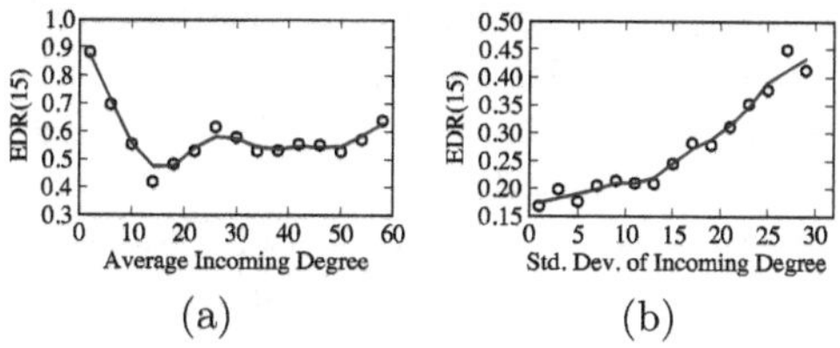


**Fig. 5.** Correlation of EDR(15 min) with metrics of peer indegree

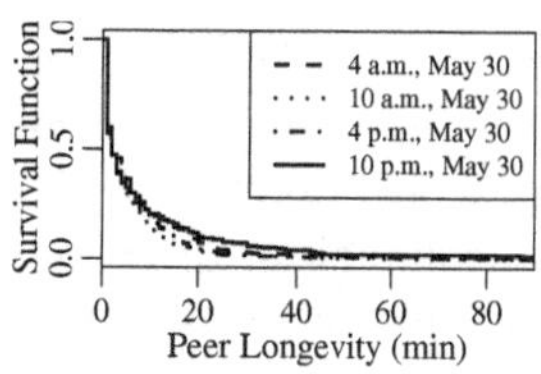


**Fig. 6.** Detecting the time of the day effect

Fig. 5(a) shows an interesting phenomenon: When the peer indegree is at a smaller value ($< 20$), a negative correlation exists between the average indegree and the early departure rate, meaning that the more suppliers a peer has, the longer it may stay; however, when the indegree goes up, a positive correlation result, showing that the departure rate is high even when peers know many others in the same channel. To explain the latter part of the observation, we have further observed that the majority of peers in unpopular channels in UUSee have an indegree lower than 30, and only a few may have large indegree up to one hundred. Interesting enough, the peers with large indegrees are generally those with poor buffering levels, which thus strive to find more possible suppliers, but are nevertheless unable to get a satisfactory streaming quality. Fig. 5(b) plots a positive correlation between the standard deviation of indegree and the early departure rate. This reveals the following: when the number of incoming connection fluctuates significantly at a peer, the peer may not be experiencing smooth streaming, and thus is more prone to early departure.

### 3.4 Time Effects

Intuitively, the time when a peer surfs the Internet also influences its viewing behavior. We then explore any possible effect of the time of the day, using sessions starting at different times on a same day and classify sessions according to the hours they start. Fig. 6 exhibits visible differences among survival curves for session groups of four different starting times on May 30. We further statistically validate our observations using the Mantel-Haenszel test [2], also referred to as the log-rank test. The log-rank test is commonly applied to test the null hypothesis that a set of survival functions are statistically equivalent, in which the null hypothesis is rejected if the result $p$-value is lower than the significance level of 0.05. The log-rank test result of $p \approx 0$, rejects the null hypothesis that survival functions are equivalent and validates our observations.

## 4 Modeling Peer Longevity: Cox Regression

### 4.1 The Cox Regression Model

The Cox proportional hazards model [2] is a classical regression model for the analysis of survival times with respect to their relationship with covariates (which are the terminologies in Cox modeling for influential factors). It models the relationship between the covariates and survival times based on the *hazard function*. A hazard function $\lambda(t)$, also referred to as the *hazard rate*, represents the instantaneous failure rate for a session that has survived to time $t$. Let $T$ denote the duration of a survival session. The hazard function is defined as:

$$\lambda(t) = \lim_{\Delta t \to 0} \frac{\mathbf{Pr}(t \leq T \leq t + \Delta t | T \geq t)}{\Delta t}.$$

In Cox regression modeling, it models the hazard rate at time $t$ for a session with covariate vector $\boldsymbol{z} = (z_1, \ldots, z_p)$ as a function of a baseline hazard function and the influential factors. The basic Cox model is:

$$\lambda(t; \boldsymbol{z}) = \lambda_0(t) \exp(\boldsymbol{\beta}^T \boldsymbol{z}) = \lambda_0(t) \exp(\sum_{k=1}^{p} \beta_k z_k), \tag{1}$$

where $\lambda(t; \boldsymbol{z})$ is the hazard rate at time $t$ for a session with covariate vector $\boldsymbol{z}$; $\lambda_0(t)$ is an arbitrary non-negative baseline hazard function, which is computed during the regression process; and $\boldsymbol{\beta} = (\beta_1, \ldots, \beta_p)$ is a column $p$-vector of coefficients corresponding to the covariates in $\boldsymbol{z}$. A major property of the Cox model is that given two sessions with covariate vector $\boldsymbol{z}_1$ and $\boldsymbol{z}_2$, the ratio of their hazard rate is independent of the time:

$$\frac{\lambda(t; \boldsymbol{z}_1)}{\lambda(t; \boldsymbol{z}_2)} = \frac{\lambda_0(t) \exp(\boldsymbol{\beta}^T \boldsymbol{z}_1)}{\lambda_0(t) \exp(\boldsymbol{\beta}^T \boldsymbol{z}_2)} = \exp(\boldsymbol{\beta}^T (\boldsymbol{z}_1 - \boldsymbol{z}_2)).$$

Such a property also imposes the *proportional hazards assumption* for applying the Cox model, that the hazard rate ratio for any two sessions should be always proportional, *i.e.*, dependent *only* on their covariate values.

In our regression modeling, the potential covariates are selected corresponding to the influential factors we have observed. The potential covariates, along with their description and type, are listed in Table 1. In order to use the Cox regression model in (1) to formulate the relationship among these covariates and the hazard

**Table 1.** Potential covariates in Cox Regression Model

| Covariate | Description | Type |
|---|---|---|
| BUFAVG | Average buffering level of the session | Continuous |
| BUFSTD | Standard deviation of buffering level during the session | Continuous |
| BUFINIT | Initial buffering level of the session | Continuous |
| INDAVG | Average incoming degree during the session | Continuous |
| INDSTD | Standard deviation of incoming degree during the session | Continuous |
| TOD | Joining time of the day ($\text{TOD} \in \{0, 1, \ldots, 23\}$, corresponding to the hours of the day) | Categorical |

rate, we first need to check if the proportional hazards assumption is satisfied, and may adjust the form of the covariates in Table 1, in order to meet the proportional requirement. Once the assumption check is passed, we proceed to derive the values of regression coefficients $\beta_k, k = 1, \ldots, p$ in the model. Using the Cox model, we can then estimate the probability that a session lasts to any specific time $t$ (*i.e.*, the survival curve of the session), given the values of the covariates for the session and using the derived coefficients.

### 4.2 Proportional Hazards Assumption Check

**Categorical Factor.** One approach to check the proportional hazards assumption for a categorical covariate, *i.e.*, whether or not the hazard ratio of sessions with different values of a categorical factor is a constant, is to group the sessions based on the values of the corresponding categorical factor, and plot the values of $-\log(\hat{S}(t)))$ against $t$ for each session group [2], where $\hat{S}(t)$ is the estimated survival function of the group. The plot for the categorical covariate TOD in our model is shown in Fig. 7(a). If the hazard ratios do not change with time, the curves in the figure should be approximately *parallel*, *i.e.* there is an approximate constant vertical distance between each pair of them at all times. However, in Fig. 7(a), we observe that the curves intersect with each other, indicating the violation of the proportional assumption for TOD.

Given the non-proportionality of the categorical factor, we modify our model in (1) to the stratified Cox model [2], in order to accommodate the categorical factor. The stratified Cox model extends the basic Cox model by incorporating strata, where each stratum corresponds to one hazard rate function, that models the hazard rate of sessions corresponding to one specific value of each categorical factor. The stratified Cox model with $n$ strata ($i = 1, \ldots, n$) is given by:

$$\lambda_i(t; \boldsymbol{z}) = \lambda_{0,i}(t) \exp(\boldsymbol{\beta}^T \boldsymbol{z}), i = 1, \ldots, n. \tag{2}$$

In our modeling, we have one categorical variable with 24 possible values, and thus the total number of strata $n$ is 24. We note that in such a stratified Cox model, each stratum may have different baseline hazard functions, but all strata share the same coefficient vector $\boldsymbol{\beta}$ as all other non-stratified factors are required to have a constant influence to the hazard functions.

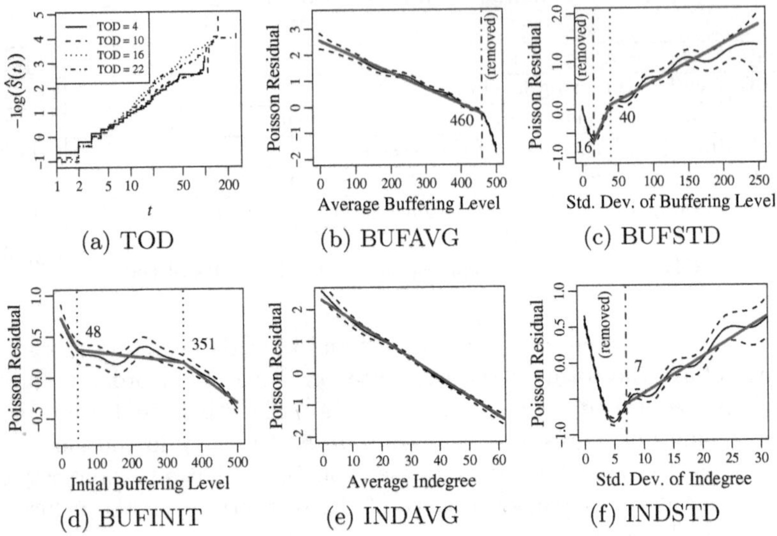


**Fig. 7.** Proportional hazards assumption check for covariates

**Continuous Factors.** For a continuous covariate, the proportional hazards assumption implies that it should have a linear influence on the hazard ratio, *i.e.*, the hazard ratio between a session with BUFINIT = 300 and one with BUFINIT = 340 should be the same as that between a session with BUFINIT = 400 and one with BUFINIT = 440. The approach to conduct such an assumption check is to plot the Poisson residual curve [2] for each continuous factor. The Poisson residual of a covariate reflects the *impact* of this specific factor in the hazard rate function: a positive Poisson residual implies a positive impact, *i.e.*, the hazard rate is greater with a larger value of the covariate, while a negative Poisson residual indicates a negative impact, vice versa. The Poisson residual curve should be approximately linear if the hazard ratio between any two sessions with two specific values of the factor is a time-independent constant. We plot the Poisson residual curves (black solid lines) for all our continuous factors in Fig. 7(b)–(f), along with their standard error confidence bands (black dashed lines), and the linear approximation lines (in red). The plots show that many of the Poisson residual curves are not satisfactorily linear, reflecting violation of the proportionality in certain value ranges of the factors. We thus seek to make necessary adjustments for the form of the covariates, such that all the new covariates have a linear influence on the hazard ratio.

Fig. 7(b) shows an approximated linear curve in the majority range of the average buffering level, except in the range of 460 to 500. To include BUFAVG into the Cox model, we only keep its value range of $[0, 460)$, *i.e.*, we exclude sessions with BUFAVG in the range of $[460, 500]$ when we derive the model coefficients, which nevertheless only represent a small portion ($\approx 7\%$) of all the sessions based on our measurement study. The major part of the Poisson curve in Fig. 7(c) can be approximated by two linear segments.To include BUFSTD into our Cox model, we exclude sessions with BUFSTD in the range of [0, 16] (which only represent a few extremely short sessions), and include a new variable BUFSTD_L to describe the section of BUFSTD with larger values, corresponding to the range of the second linear segment we approximate in Fig. 7(c):

$$\text{BUFSTD\_L} = \begin{cases} \text{BUFSTD} - 40 & \text{if BUFSTD} \geq 40 \\ 0 & \text{otherwise} \end{cases}$$

The Poisson curve in Fig. 7(d) can be approximated by three line segments connected at two knots at 48 and 351, respectively. Since each section of BUFINIT includes a substantial number of sessions, we include two new covariates, BUFINIT_M and BUFINIT_L, to describe the sections of BUFINIT corresponding to linear segments in the middle and to the right, respectively:

$$\text{BUFINIT\_M} = \begin{cases} \text{BUFINIT} - 48 & \text{if BUFINIT} \geq 48 \\ 0 & \text{otherwise;} \end{cases}$$

$$\text{BUFINIT\_L} = \begin{cases} \text{BUFINIT} - 351 & \text{if BUFINIT} \geq 351 \\ 0 & \text{otherwise.} \end{cases}$$

In Fig. 7(e), the Poisson curve of INDAVG can be nicely fitted by one line, revealing the proportionality of the factor on the hazard ratio. In Fig. 7(f), we remove the leftmost part (corresponding to INDSTD in the range of $0 - 7$ with a few sessions), and fit the rest of the curve with a line.

After the adjustment, a covariate vector $\boldsymbol{z} = (z_1, \ldots, z_p)$ with $p = 8$ components is used in our stratified Cox model. The covariates are summarized in Table 2. We note that in Fig. 7(b)–(f), all the linear approximation lines fall within the confidential bands of the original Poisson residual curves, indicating that the Poisson curves in those sections can be effectively approximated by the linear segments. Therefore, after the adjustment, all the covariates we now use in the Cox modeling satisfy the proportional hazards assumptions.

### 4.3 Estimation of the Coefficients and Model Validation

We next use a specific Cox regression technique proposed by Andersen and Gill [3], to estimate the stratified baseline functions $\lambda_{0,i}, i = 1, 2, \ldots, 24$ and the coefficient vector $\boldsymbol{\beta}$ in our stratified Cox model in (2).

Table 2 gives the coefficients of the covariates along with their standard errors, estimated using information of 12866 session from 20 unpopular channels in our traces. The 20 channels, whose average concurrent population varies from 48 to 457, are randomly chosen from all 530 unpopular channels contained in our traces. The purpose for such sampling is not only to expedite the speed of the regression process, but also to exhibit the usefulness of our model, trained using only a limited set of samples, as is to be illustrated in the following subsection. We have also computed the $p$ values to test the significance of all the coefficients, which are all far below 0.05, suggesting the significance of the covariates.

**Table 2.** Covariates and Coefficients for the Cox Model in (2)

| Covariate | $\beta$ | Std. Err. | Covariate | $\beta$ | Std. Err. |
|---|---|---|---|---|---|
| BUFAVG | −0.0074 | 1.7e-3 | BUFINIT | 0.011 | 1.8e-3 |
| BUFSTD | 0.059 | 2.9e-3 | BUFINIT_M | −0.012 | 1.9e-3 |
| BUFSTD_L | −0.044 | 2.9e-3 | BUFINIT_L | −0.0029 | 4.9e-4 |
| INDAVG | −0.051 | 1.9e-3 | INDSTD | 0.046 | 2.1e-3 |

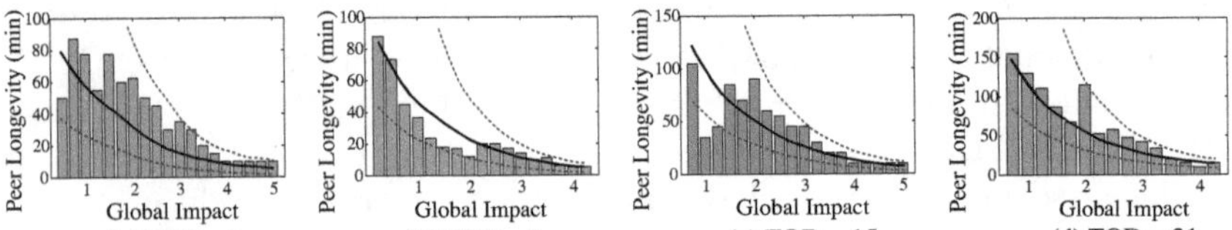


**Fig. 8.** Regression model validation by prediction of the peer longevity

With the Cox regression model established, we can now derive the survival curve of a session with covariate vector $\boldsymbol{z}$ in a certain category of TOD. The estimator of the survival function with covariate vector $\boldsymbol{z}$ at time $t$ is given by $S_i(t;\boldsymbol{z}) = \exp(-\int_0^t \lambda_i(u;\boldsymbol{z})du)$, where $\lambda_i(u;\boldsymbol{z})$ is the stratum corresponding to a specific category of TOD of sessions. We may use the expected session time of the survival curve corresponding to a session with $\boldsymbol{z}$, as the most probable duration of the session. In this way, given a covariate vector $\boldsymbol{z}$ and the corresponding TOD, we are able to predict the most probable duration of a session using our stratified Cox model. Recall that our regression model is trained using only a limited set of session data from 20 randomly selected unpopular channels. We now evaluate its accuracy in estimating the duration of sessions in other channels.

We calculate the overall influence of the continuous covariates by $\beta^T\boldsymbol{z}$, referred to as the *global impact* to the hazard rate. We group all the sessions in the unpopular channels in UUSee (other than the 20 channels for the regression), by their values of global impact and TOD, and plot the median session duration for session groups at different levels of their global impact in different cases of TOD, as shown by the bar plotting in Fig. 8. In solid curves, we also plot the session duration predicted using our Cox model in (2) at each global impact level, and the 75% confidence intervals (in dashed lines). The actual median session durations in all four figures fall into the confidence intervals, which validates the usefulness of our regression model—derived using a small portion of session data—in accurately capturing the session duration patterns for unpopular channels.

# 5 Model Implications to Unpopular Streaming Channels

## 5.1 Impact of the Streaming Quality Factors

Three streaming quality factors are involved in our Cox regression model. We seek to investigate the relative significance of their impact on peer longevity, by calculating an impact value, $\beta_k z_k - \min_{\boldsymbol{y}}(\beta_k y_k)$, for each individual session with a specific $z_k$, with respect to the three streaming quality factors of average buffering level, standard deviation of buffering level and initial buffering level, respectively. Here, $\min_{\boldsymbol{y}}(\beta_k y_k)$ is the minimal impact value for the corresponding streaming quality factor over all the sessions. We then compute the ratio of the impact values of the three streaming quality factors for each session, and derive the average ratio across all sessions. The normalized average ratio in the percentage format is 16% : 33% : 51%, which exhibits the relative level of user's *intolerance* to the three streaming quality factors, respectively.

An intriguing discovery is that the initial buffering level is the most important streaming quality factor affecting peer longevity. Using the coefficients in Table 2, we derive the coefficient $\beta$ corresponding to the initial buffering level factor in the range of 48 to 351 is $\beta_{\text{BUFINIT}} + \beta_{\text{BUFINIT\_M}} = -0.001$, and the coefficient corresponding to the range of 351 and 500 is $\beta_{\text{BUFINIT}} + \beta_{\text{BUFINIT\_M}} + \beta_{\text{BUFINIT\_L}} = -0.0039$. The more negative coefficient in the latter case illustrates that when the initial buffer is relatively full, a small increase of buffering level induces more significant decrease of failure probability, *i.e.*, more evident effect in keeping peers longer in the system. Therefore, to promote the stability of high-contribution peers, efforts should be made to guarantee them a high initial buffering level.

### 5.2 Impact of the Incoming Degree Factors

Following a similar methodology, we further compare the impact of the average peer indegree and the standard deviation of peer indegree, and derive a user's intolerance ratio of 38% : 62% to the two factors. It confirms that in unpopular channels, peers are much less tolerant to neighbor churns than the average level of neighbor numbers. Therefore, the P2P protocol should always try to find stable good neighbors for each peer, the number of which may be small, but is much desirable than a large number of transient neighbors.

## 6 Related Work

With respect to P2P measurements related to peer longevity, Hei *et al.* [4] characterized the distribution of peer life time in PPLive, and exhibited different life-time patterns among popular and unpopular channels. Li *et al.* [5] have also observed a heavy-tailed peer lifetime distribution in their measurement study of Coolstreaming. Based on simulations, Tang *et al.* [6] have shown that the longer peers stay, the better the overall streaming quality is in a streaming channel. Focusing on P2P applications other than live streaming, Stutzbach *et al.* [7] have characterized peer arrivals and departures in three popular P2P file-sharing systems (BitTorrent, Kad and Gnutella). Chen *et al.* [8] have investigated the influence of network QoS metrics on peer session lengths in a P2P VoIP application, Skype. Our work distinguishes itself from all the existing measurement work, by focusing on a thorough understanding of the causes to peer lifetime patterns in unpopular channels, in order to improve their streaming quality.

## 7 Concluding Remarks

This paper focuses on improving the streaming quality in the large number of unpopular channels in real-world P2P live streaming systems. Utilizing over 130 GB worth of traces from a large-scale commercial system, UUSee, we thoroughly characterize the important factors that influence peer longevity. Our key contributions include: *first,* we successfully identify the key factors that decide the duration of peer sessions, including the initial buffering level, incoming degree and peer joining time; *second,* we model their relationship into a Cox regression model, using a survival analysis approach; *third,* we discuss implications of our model and derive a number of useful insights to promote peer stability in unpopular channels. As important applications of our Cox model, we can compute and compare the relative stability of peers during the peer selection process, and can promote the online time of high-contribution peers by guaranteeing them a better initial buffering level and stable download bandwidth. All these assist in improving the stability and streaming quality in unpopular channels.

## References

1. UUSee, `http://www.uusee.com/`
2. Machin, D., Cheung, Y.B., Parmar, M.: Survival Analysis: A Practical Approach, 2nd edn. John Wiley and Sons Ltd., Chichester (2006)
3. Andersen, P., Gill, R.: Cox's regression model for counting processes: A large sample study. Annals of Statistics 10(4), 1100–1120 (1982)
4. Hei, X., Liang, C., Liang, J., Liu, Y., Ross, K.: A measurement study of a large-scale P2P IPTV system. IEEE Trans. Multimedia 9(8), 1672–1687 (2007)
5. Li, B., Xie, S., Keung, G., Liu, J., Stoica, I., Zhang, H., Zhang, X.: An empirical study of the Coolstreaming+ system. IEEE JSAC 25(9), 1627–1639 (2007)
6. Tang, Y., Sun, L.F., Zhang, K.Y., Yang, S.Q., Zhong, Y.Z.: Longer, better: On extending user online duration to improve quality of streaming service in P2P networks. In: Proc. IEEE ICME 2007, July 2007, pp. 2158–2161 (2007)
7. Stutzbach, D., Rejaie, R.: Understanding churn in peer-to-peer networks. In: Proc. ACM IMC 2006, pp. 189–202 (2006)
8. Chen, K.T., Huang, C.Y., Huang, P., Lei, C.L.: Quantifying Skype user satisfaction. In: Proc. ACM SIGCOMM 2006, pp. 399–410 (2006)

# Enhancing Application Identification by Means of Sequential Testing

Mohamad Jaber and Chadi Barakat

EPI Planète, INRIA, France
{mohamad.jaber,chadi.barakat}@sophia.inria.fr

**Abstract.** One of the most important challenges for network administrators is the identification of applications behind the Internet traffic. This identification serves for many purposes as in network security, traffic engineering and monitoring. The classical methods based on standard port numbers or deep packet inspection are unfortunately becoming less and less efficient because of encryption and the utilization of non standard ports. In this paper we come up with an online iterative probabilistic method that identifies applications quickly and accurately by only using the size of packets. Our method associates a configurable confidence level to the port number carried in the transport header and is able to consider a variable number of packets at the beginning of a flow. By verification on real traces we observe that even in the case of no confidence in the port number, a very high accuracy can be obtained for well known applications after few packets were examined.

**Keywords:** Internet traffic identification, statistical methods.

## 1 Introduction and Related Work

The identification of applications behind the Internet traffic is of major importance for network operators. On one side, this allows to treat flows in a different way based on their quality of service requirements. On the other side, it can serve for security reasons by blocking or looking closely at those users who run non legacy applications or utilize non-standard port numbers to counter security controls. Originally, this identification was considered to be straightforward by simply looking at the port number in the transport header. Legacy applications are supposed to use standard port numbers as, for example, port 80 for the WEB and port 25 for the SMTP. However, the Internet users tend more and more to use non-standard port numbers or to encrypt the payload of their packets including the transport header so that it cannot be read along the network path. This hiding of application information is done either to prohibit intermediate users and operators from reading packets and accessing private data, or more importantly, to make network operators believe that the application is legal while it is not [1] and [2]. A typical example is the case of users audio-conferencing over port 80 to be able to cross firewalls authorizing only the passage of WEB traffic. These manipulations of the transport information by end users make it hard to identify applications inside the network.

L. Fratta et al. (Eds.): NETWORKING 2009, LNCS 5550, pp. 287–300, 2009.

To counter this obstacle, there has been recently a trend to use traffic statistics for the identification of applications [3] and [4]. Indeed, it is known that different applications are governed by different end-to-end protocols and consequently they generate packets of different sizes and with different inter-packet times. Promising results have been recently found in this direction but we believe there is still room for more research to understand the capacity of the approach and of its limitations. For example, in [5] only the first four packets from a flow are jointly considered. This joint consideration of packet sizes prohibits the method from extending to more packets, otherwise the space of observations becomes complex to handle. Our idea in this paper is to separate the observations to allow more generality. Another example is the work in [6] which is relevant to our work, in the fact that it considers packets separately, however it uses a classification algorithm that is built mostly in an empirical way.

BLINC [7] is another solution for application identification that correlates flows as a function of the machines from which they originate and to which they are destined. Even though we think BLINC is a useful mechanism, in this paper we focus on the identification of applications from traffic properties independently of the relevance of machines for the different applications. The extension to the BLINC case is left for future research. We reconsider the application identification problem and we tackle it by a new statistical method that is able to extend to any number of packets per flow. In line with [6], we consider packets separately from each other, which has the main advantage of reducing the problem complexity at the expense of a small loss in performance caused by the correlation that might exist among packets. This separation is necessary to be able to consider more packets than the very few ones at the beginning of a flow. We introduce a new function that is able to quantify for each new flow the probability that a classification decision is wrong. We do this for different possible applications and for variable number of packets per flow. The function can be easily updated as long as new packets pop up from each flow. The classification is then simply to tag the flow with the most probable application. In this way we are able to evaluate with a high precision the probability of wrong or false decisions which allows a better understanding of the power of this approach and as a consequence, the proposition of more meaningful classification methods.

Another contribution of our work is that we associate a confidence levels to the port number carried in the transport header of packets. This level is to be set by the administrator. A low value of confidence level can be set in case of no confidence in the port number or a higher value can be set when the port number is a reliable information. Unfortunately this information on the port number has been largely overlooked in the literature. The port number has been either completely ignored or used partially as in [5] to identify flows after a first classification step. In our work, we choose to model this information by a configurable confidence level that can be set by administrators based on their experience about the traffic. The validation will show that even for a very low confidence level in the port number, we can get a very high accuracy. Clearly the accuracy increases when more confidence is associated to the port number.

The main contribution of this paper is then in the proposition of a new statistical method for application identification that is able to scale to more than the very few packets at the beginning of a flow. As we will show, this scaling is necessary since the more packets are monitored from a flow, the higher the precision of the classification. Our observations are made on real traces that we collected ourselves on the network of INRIA Sophia Antipolis in Spring 2008. We also consider the traces used in [6] for further validation. Over all these traces, one can notice the high performance of our method that is able, for example, to reach an accuracy of 97% for the first ten packets even without any confidence in the port number. Clearly, higher accuracy can be obtained if more weight is given to the value in the port number field. This accuracy is higher than what has been noticed so far, e.g. [5].

The remainder of the paper is structured as follows. In Section 2 we explain our methodology and we present our probability function. In Section 3 we describe the traces used to evaluate our method and in Section 4 we provide validation results and discussions. The paper is concluded in Section 5 with some perspectives on our future research.

## 2 Method Description and Assumptions

In this section we explain the details of our study and the assumptions we made. We are targeting a new statistical method for the identification of applications in the Internet traffic, which is able to classify flows early, on the fly, and with very high precision. A flow in our context is a TCP or a UDP connection defined by the 5-tuple information (IP addresses, port numbers and protocol). We want to be safe when our method affects a flow to an application while being able to identify the application before the end of the flow. We start by exploring the interesting method developed in [5] where the main idea was to identify traffic based on the size and the direction of the first four packets considered jointly. A precision of around 88% was announced in [5], but it has been also observed that if one uses more than four packets together, this will cause a loss in the identification accuracy. So we depart from the model in [5] but while considering more packets in order to understand the limitations of online statistical methods. Then, we make some further assumptions that will allow us to consider more packets while continuously increasing the accuracy. We couple this with the proposition of a new probability function to classify flows in a more accurate way. Our function is calculated on the fly, after a calibration phase by machine learning techniques to account for different application characteristics.

### 2.1 Joint Consideration of Packets

We begin to study the classification of traffic while using the size and the direction of the first N packets together (i.e. the first four together, the first five together, etc). By together we mean that the space in which we put ourselves is a multidimensional space where one dimension is associated to the size of each packet (+ and - to model the direction). Clusters are formed and associated to applications in this multidimensional space and new flows are classified accordingly. In this section, we anticipate the description of the clustering and classification procedure and of our traces used for validation to highlight an important limitation of a model considering packets jointly as in [5]. Then, we make and support the claim that studying packets separately from each other allows statistical traffic classification to scale to more packets and to provide more accuracy. We plot in figure 1 the global classification accuracy as a function of the number of packets considered per flow. This figure is an average over several standard applications existing in two of our traces (one line per trace). It shows that the precision of the classification increases with the number of packets until it reaches a maximum of 88% for four packets. At that point, the precision begins to decrease until it reaches 80% for 10 packets. After looking closely at the numerical results to understand the reasons behind this decrease in accuracy beyond four packets, we believe that this decrease is not because the packets five, six, etc are not distinctive of the different types of applications, but rather because we are using more dimensions during the classification and so the forming of clusters in the multidimensional space becomes more challenging. On one side, it is hard to find the optimal number of clusters to be used (the figure shows the results for 80 clusters which we found to give the best results for four packets). And on the other side, increasing the number of dimensions should be accompanied by an exponential increase in the number of clusters, which can become larger than what clustering algorithms (e.g. K-Means) can handle in practice. This is the main reason for which we propose to consider packets separately from each other as if they come from independent observations. Each packet (first, second, third, etc) is studied separately in its own low dimensional space, then the flow is classified using a probability function (kind of likelihood function) that combines the different observations resulting from its different packets. This assumption is supported next by the low level of correlation existing between packets of a flow. The main advantage of our approach is that it reduces the complexity of the multidimensional space needed for learning packet size characteristics when packets are considered jointly. We replace this multidimensional space by a separate low dimensional space per packet (one dimension per packet if the direction is represented by signs + and -). The benefit is clearly a less complex and a less erroneous learning method and a classification accuracy that keeps increasing as long as we add more packets from each flow (for now have a look at figure 10). In fact, the gain one gets from reducing the space complexity is much more important than what is lost by ignoring correlation among packets.

### 2.2 On the Auto-correlation of Sizes of Packets within a Flow

We don't claim that packet sizes are uncorrelated or they form independent observations. We only assume this independence to ease the classification of flows provided we have learned the individual characteristics of their packet sizes. Nevertheless, the low level of auto-correlation in the packet size process would

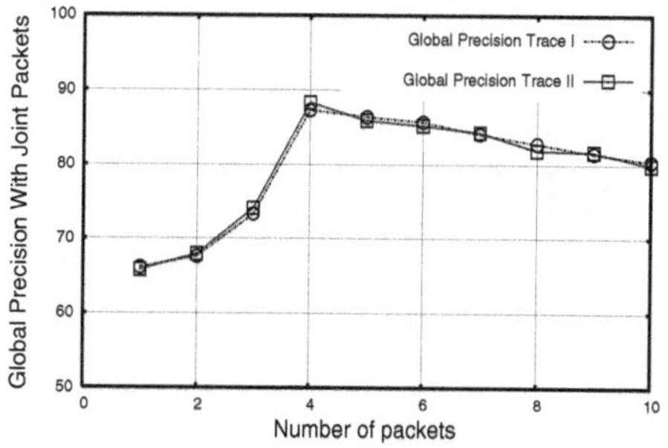


**Fig. 1.** Precision of traffic classification when using jointly the sizes of the first N packets of each flow

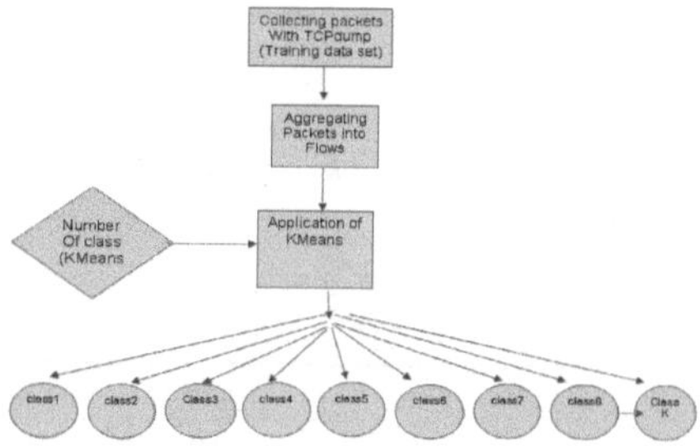


**Fig. 2.** Model building phase

help us making this assumption and would make our method even stronger. This is what we are going to check in this section.

We can evaluate the correlation between two random variables X and Y using the following correlation coefficient : $R(X,Y) = \frac{COV(X,Y)}{\sigma(X)*\sigma(Y)}$, where COV is the covariance function and $\sigma$ is the standard deviation. The common practice is to suppose a strong correlation between X and Y when $|R(X,Y)| \geq 0.7$ and a weak correlation when $|R(X,Y)| \leq 0.3$ . We measure the value of this coefficient for the first ten packets in each flow of Internet traffic. Several applications are considered : WEB, HTTPS, SSH, IMAP, SMTP, and POP3. Figures 3 and 4 show the correlation coefficient values between every two packets among the first ten. The X axis in the figures represents the lag. For example for lag 1, we plot the correlation value between the sizes of every two consecutive packets (packet 1 and 2, packet 2 and 3, etc). For lag 3 we consider all packets which are separated by two other packets from the same flow (packet 1 and 4, packet 2 and 5, etc), and so on. We can clearly see in these figures that the correlation value between any pair of packets, for all applications and for all lag values, is often smaller than 0.3 and in most cases even close to 0. This means that we can safely develop our method with the assumption that packet sizes are independent of each other, even if we know that this is not absolutely true. The correlation is low enough to make our method more scalable and more efficient than when considering packet sizes jointly in the learning and classification phases.

### 2.3 Our Method Description

In order to benefit from information carried by the first $N$ packets of a flow and to avoid problems during the clustering phase caused by the use of many parameters, we resort to an iterative packet-based approach where we use the size and the direction of every packet individually to calculate the likelihood of originating from this or that application, then we merge the results from all packets of a flow together using a probability function that we introduce to associate the flow to the most probable application. Note that we also considered

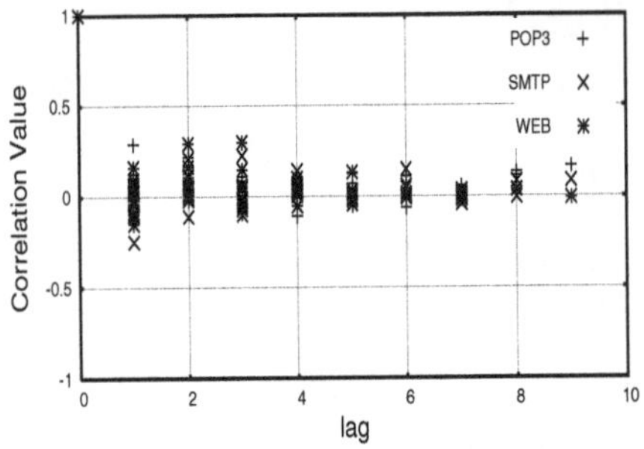


**Fig. 3.** Correlation values between the first ten packets (POP3, SMTP, WEB)

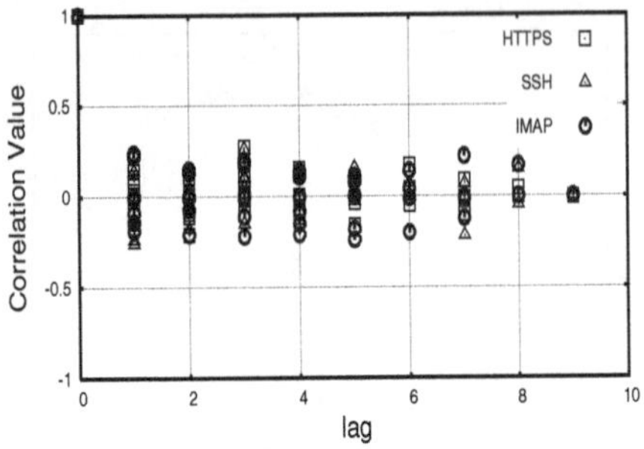


**Fig. 4.** Correlation values between the first ten packets (HTTPS, SSH, IMAP)

the time between packets, but since the results did not show any improvement in performance, we limit ourselves to showing a sample of these results and we focus more on the packet size. We believe this is an intuitive observation since the size of packets is defined by protocol and application behaviors, whereas the time between packets is mostly decided by network conditions.

Our method consists of three main phases: the model building phase, the classification phase and the application probability and labeling phase.

**Model building phase,** or learning phase, is very important in our work. In this phase, we construct a set of clusters or models (using a training data set) which we use later in the traffic classification phase. As we study each packet individually, we build a separate model for each packet using all flows in the training data set (model for the first packet, model for the second packet, etc). This model describes how the size of a packet, say the first one, varies over the different applications and how it occupies the different parts of the packet size space.

Let us describe the method that we use to generate models. We begin by explaining how we create good learning traces. Then, we describe how we represent spatially the flows in these traces. To cluster the flows in the space, we use the K-Means algorithm [8].

We choose the training traces such that they are representative of the applications to identify. To this end, we take a similar number of flows from all applications because if the number of flows is not the same, applications that predominate may bias the clustering and the classification afterwards. The number of flows is made equal by random selection from applications having more flows than necessary. To build the model for a given packet size (say the size of the i-th packet from a flow), we represent each flow as a point on an axis. This point has a positive coordinate if the packet is sent by the client and a negative coordinate if the packet is sent by the server. The coordinate of this point is equal to the size of the packet. When the time between packets is used, a flow is represented by two coordinates on two axis, one for the packet size and one for the time between this packet and the previous one (or the next one). Without loss of generality consider a maximum of ten packets per flow. At the end of the model building phase, we get ten models for the first ten packets from any flow. In each model, say for example there are 20 classes (or clusters). Note that 20 was shown to be a good tradeoff between complexity and precision [5]. Nevertheless, the performance of the method is of low dependence on this number of classes. Then, for each application and for each class we associate a specific probability proportional to the number of flows from this application present in the class. This probability models the likelihood that a packet from this application fills in this class in general. As we have the same number of flows from all applications in the training trace, we define the probability of class $i$ knowing the application $I$ noted by $Pr(i/I)$ as the number of flows that belong to the application $I$ in class $i$, $F_{I,i}$, divided by the total number of flows belonging to the application $I$ in the training trace, $F_I$. We have $Pr(i/I) = \frac{F_{I,i}}{F_I}$.

For example, in a class where we have 400 flows (300 WEB, 60 HTTPS and 40 SMTP), and we have a total of 8000 flows for each application in the training trace, we associate this class to these three applications following these probabilities:

$$Pr(i/\text{WEB}) = \frac{300}{8000}, Pr(i/\text{HTTPS}) = \frac{60}{8000}, Pr(i/\text{SMTP}) = \frac{40}{8000}.$$

**Classification Phase,** consists in using the models built in the learning phase to classify traffic online. Note that here we affect flow packets to classes and not to applications. Each time there is a new packet from a flow, it is classified independently of the other packets using the model corresponding to its position within the flow. For example, when we capture the first packet of a new flow, we affect the packet (and hence the flow) to one class of the model built for all packets which are first in their flows. The same for the second packet, and so on.

To carry this association flow-class in a given model, we calculate the Euclidean distance that separates the point corresponding to the new flow in the model space from the center of each class and we affect it to the closest one. We repeat this classification for all packets from a flow until we are satisfied or we reach some threshold. The way the satisfaction is measured is done via a new probability function to be described later. This function uses the per-packet and per-class probabilities calculated in the model building phase and identified during the classification phase. Clearly, the classification of a flow is different and independent from one packet to another, hence the result of the classification. For example, a new flow can be affected to the class number 5 using the first packet and to the class number 19 using the second one, and so on. It is this set of classification results that will decide on the most probable application to which the flow would belong.

**Application probability and labeling phase,** we affect here flows to applications while relying on the results of the classification phase. The classification phase indicates the probability that each packet of a flow belongs to this or that application. These are the functions $Pr(i/I)$ obtained from the classes in which the packets fall. We combine (on the fly) these probabilities together to obtain a new value evaluating how well we do if we associate the entire flow to this or that application. This leads to an online iterative application probability function that we use for assignment and identification. Let us define the following variables to be used next :

- $i$: used to denote classes.
- $I$: used to denote applications.
- $N$: the maximum number of packets tested from a flow.
- $k$: the test number $k$ (packet 1, packet 2, etc).
- $A$: the total number of applications.
- $\alpha_I$: the value (between 0 and 1) affected by the administrator to the confidence in the standarized port number relative to application $I$.
- $F_{I,i}$: the number of flows belonging to application $I$ and class $i$ in the training trace.
- $F_I$: the total number of flows belonging to application $I$ in the training trace.
- $Pr(I/Result)$: the probability that a flow belongs to application $I$ knowing the results of the classification phase (i.e. class $i(1)$ for the first packet, class $i(2)$ for the second packet and so on).
- $Pr(i(k)/I)$: the probability that the $k$-th packet of a flow from application $I$ falls in class $i$ this can be calculated from the training trace as $Pr(i(k)/I) = \frac{F_{I,i}}{F_I}$.
- $Pr(I)$: the probability that any flow randomly selected comes from application $I$, independently of any information on its packet sizes and times. The sum of these probabilities over all applications under consideration must be equal to 1. We integrate in this probability the confidence in the port number carried in the transport header of each packet. For example, for a given flow, and if the port number is equal to 80 (the standard WEB port number), we give $Pr(WEB)$ the value $\alpha_{80}$ set between 0 and 1 (specified by the network administrator as a function of how confident he is). This value models the probability that a flow carrying the port number 80 belongs to the WEB application. For all other applications we give $Pr(I)$ the same value that is equal to : $\frac{(1-\alpha_{80})}{(A-1)}$.
  Note that the administrator might not give any weight to the port number. This can be the case when he does not trust this information. Here, we give the $Pr(I)$ the same value for all applications independently of what is carried in the port number field. This specific value is equal to: $\frac{1}{A}$.

We aim at calculating the probability that a flow belongs to an application $I$ given the results of the classification phase applied to the first, let's say $N$, packets of the flow. Take for example the first two packets and their corresponding classes $i(1)$ and $i(2)$. The probability we are looking for can be written as follows:

$$Pr(\ I\ /(i(1) \cap i(2))) = \frac{Pr(I \cap (i(1) \cap i(2)))}{Pr(i(1) \cap i(2))} = \frac{Pr(I) * Pr((i(1) \cap i(2))/I)}{Pr(i(1) \cap i(2))}$$

$$= \frac{Pr(I) * Pr(i(1)/I) * Pr(i(2)/I)}{Pr(i(1) \cap i(2))} = \frac{Pr(I) * \prod_{k=1}^{2} Pr(i(k)/I)}{\sum_{I=1}^{A} Pr(I) * \prod_{k=1}^{2} Pr(i(k)/I)}$$

Now, we can generalize this expression to calculate the probability that a flow belongs to an application $I$, given the classification results for the first $N$ packets:

$$Pr(I/Result) = \frac{Pr(I) * \prod_{k=1}^{N} Pr(i(k)/I)}{\sum_{I=1}^{A} Pr(I) * \prod_{k=1}^{N} Pr(i(k)/I)}$$

We call this probability the assignment probability and we use it to decide on how well the profile of packet sizes of a new flow fits some application $I$. For each new flow and when we capture the first packet (except the SYN packet), we first classify the flow according to this packet and we calculate the probability that it belongs to each application. Then, we take the highest assignment probability and we compare it with a threshold $th$ specified by the network administrator. If this probability is greater than the threshold $th$, we label the flow by the application, otherwise we take and classify the next packet and we recalculate the assignment probability using the results of the classification phase obtained for the first and second packets separately. We check again the resulting probability and we keep adding more packets until the threshold is exceeded or a maximum allowed number of tests is reached.

## 3 Trace Description

We test the validity of our method by the help of two real traces (Table 1). The first trace, noted Trace I, is collected at the edge gateway of Brescia University's campus network in Italy (used and described in [6]). This trace is made up of three standard applications: HTTP (WEB), SMTP and POP3. The second trace, noted Trace II, is collected by us at the edge of INRIA Sophia Antipolis network in France during Spring 2008. Trace II is made up of five standard applications: HTTP (WEB), HTTPS, IMAP, SSH and SMTP. We divide every trace into a training trace and a validation trace. The training part is used to construct the models and the validation part is used to evaluate how well our iterative packet-based method behaves in identifying the application behind each flow. Note that we made sure that there are enough flows from each application in each trace so that our learning phase can provide representative models and our validation phase meaningful results. To well calibrate and evaluate our classification method, we need to know the real application associated with each flow. For the first trace, the authors use a method based on deep packet inspection to infer real applications. For the INRIA trace, we use tcpdump ([9]) to measure each application separately at the interface of the server reserved by INRIA to this application (for instance we collect the WEB flows from the interface of INRIA's WEB server, and so forth). Since servers at INRIA are dedicated to unique applications, we are sure this way about the real application behind each collected flow. Traffic coming from the different servers is then mixed together to form one large trace.

**Table 1.** Traces Description

| Trace Name | Place of capture | Applications | Time of capture |
|---|---|---|---|
| Trace I | Brescia University | HTTP, SMTP, POP3 | 2006 |
| Trace II | INRIA Laboratory | HTTP, SMTP, HTTPS, SSH, IMAP | Spring 2008 |

# 4 Experimental Results

In this section, we evaluate the overall effectiveness of our method. We define:

- **False Positive Ratio** as the number of flows classified by our method as belonging to an application $I$ without being in reality from this application, divided by the total number of flows not belonging to this application.
- **True Positive Ratio** as the number of flows classified by our method as belonging to an application $I$ and they belong really to this application, divided by the total number of flows belonging to this application.
- **Total Precision** as the average of the true positive ratio over all applications.

We present the results of our method as a function of the number of packets classified per flow and the weight affected to the port number (specified by the network administrator). For instance a port number weight equal to 0.5 means that the chance that the flow comes from the standard application associated to this port number, $Pr(I)$, is equal to 50% while the chance that the flow does not come from this standard application is also 50% (see definition of $Pr(I)$ in Section 2.3.3). Note that one can also reflect in $Pr(I)$ the proportion of flows from each application. Indeed, if the WEB for example forms the majority of the traffic, there is a high chance that a new flow belongs to WEB. In our validation, we don't account for this factor and we only calculate $Pr(I)$ using information on the port number.

To associate flows to applications we set the threshold $th$ to a high value equal to 0.99 and we fix a maximum number of tests that we vary. When the maximum number of tests is reached, we associate the flow to the application having the maximum assignment probability. This way the threshold $th$ can be seen as a way to leave early the identification procedure when we are almost sure about the flow, otherwise one has to wait the maximum number of tests to decide.

## 4.1 Assignment Probabilities

In Figure 5, we present an example of the average and the standard deviation of the assignment probabilities for the real POP3 flows as a function of the number of classified packets. We calculate the assignment probability for the two other applications as well when flows are really POP3 (this gives the three curves in the Figure).

We can clearly see how the assignment probability for the correct application keeps increasing while adding more packets and how its standard deviation reduces which means we become more and more sure for most flows. At the same

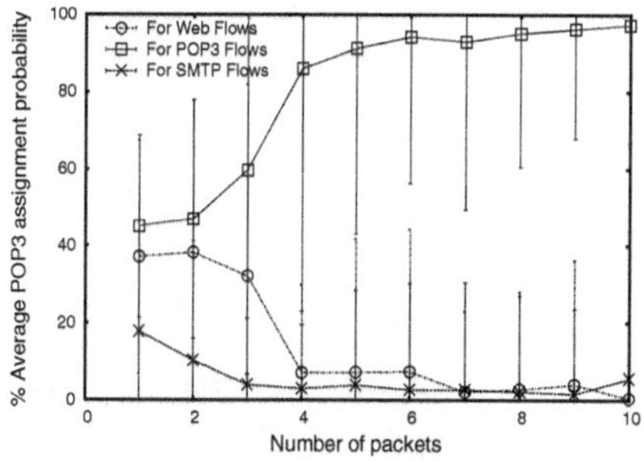


Fig. 5. Average and standard deviation of the POP3 assignment probabilities (Trace I)

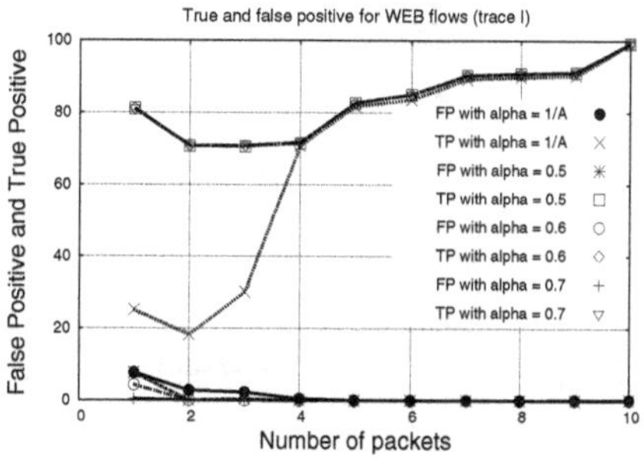


**Fig. 6.** Average True and false positive ratio for WEB flows (Trace I)

time, the assignment probability for the wrong applications drops to low values if not zero. We can also see in the Figure that until the fifth packet, there is an overlap between the standard deviations of the assignment probabilities for the different applications. This means that the first five packets do not bring enough information to make a clear separation between flows, most probably because some of them have common features. Starting from the sixth packet, we become able to separate between WEB, SMTP and POP3 flows with a high precision. The other flows present similar behaviors, so we conclude that one needs to take more than five to six packets to be able to separate flows.

## 4.2 Classification Accuracy

In Figure 6, we present the false positive ratio and the true positive ratio for the WEB application (Trace I) and this is for several values of the port number weight. We can observe that without using any weight for the port number ($\alpha_I$ set to $1/A$), we can obtain a true positive ratio that exceeds 99 % at the tenth packet. On the other hand the false positive ratio for the other applications drops to zero after testing four packets per flow. The same observation repeats in Figures 7 and 8, for the SMTP and POP3 applications (Trace I) but with a little difference that the false positive ratio is a little bit higher (1.6 % for SMTP and 2.3 % for POP3), and the true positive ratio is smaller than for HTTP at the tenth packet (95.5 % for SMTP, and 97.2 % for POP3). These values for the true positive ratio are still larger than previous ones in the literature to be considered as highly accurate. Note that one can easily improve this result by giving more weight to the port number.

In Figures 10 and 11 we present respectively the false positive ratio and the true positive ratio for the five applications: WEB, HTTPS, IMAP, SSH and SMTP in Trace II, this time without using any weight for the port number (worst case). It is clear that for all applications, the false positive ratio decreases with the number of classified packets and approaches zero at the tenth packet. For some applications as HTTP and HTTPS, it approaches zero even before (it seems hard for other applications to pass as HTTP or HTTPS).

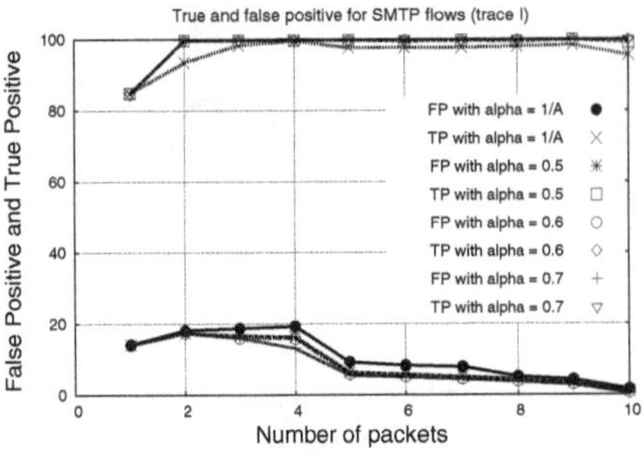


**Fig. 7.** Average True and false positive ratio for SMTP flows (Trace I)

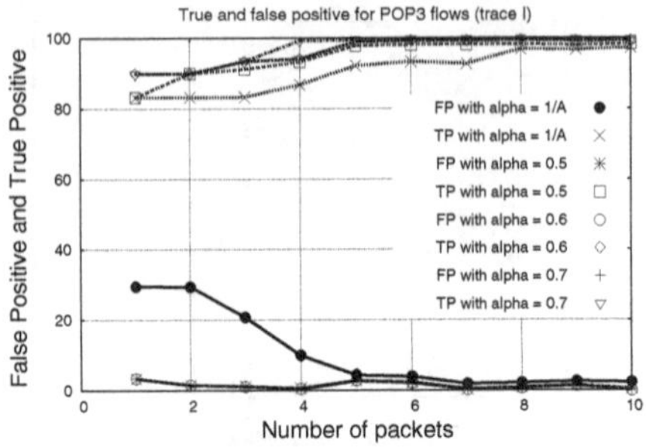


**Fig. 8.** Average True and false positive ratio for POP3 flows (Trace I)

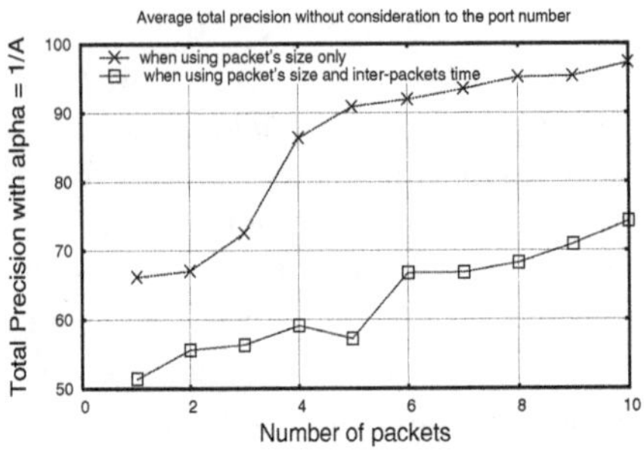


**Fig. 9.** Average total precision when using packet's size only and when using packet's size and inter-packet time (over all traces)

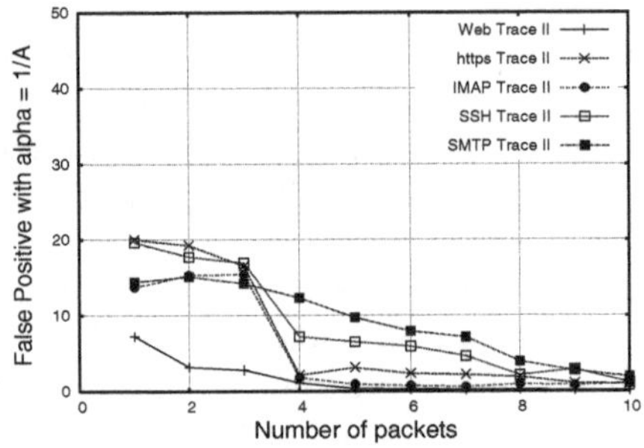


**Fig. 10.** Average false positive ratio for the different applications (Trace II)

The true positive ratio in its turn keeps increasing with the number of classified packets and approaches 100% (around 97% on average at the tenth packet). Again some applications approach 100% faster than others (case of SMTP). These results on Trace II (similar results obtained for Trace I but not shown for lack of space) confirm the strength of our method in reaching high accuracy levels and in being able to scale with the number of classified packets. Our results also show how well the packet size reflects the behavior of different applications either in terms of false positive ratio or true positive ratio.

In Figure 12 we present the average total precision of our method over all traces and for several values of the port number weight. This is an average over all true positive ratios. We can see that without using any weight for the port number, the precision of our method increases over 90% after the sixth packet and reaches 97% for ten packets. Clearly, one can get higher precision by assigning more weight to the port number carried in the packet headers.

In Figure 9 we present a comparison of the average precision of our method between only using the size of the first ten packets and using the size of packets and the inter-packet times (without considering the port number weight).

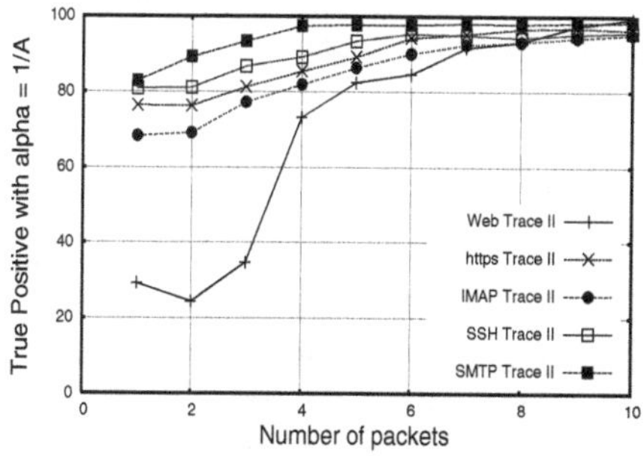


**Fig. 11.** Average true positive ratio for the different applications (Trace II)

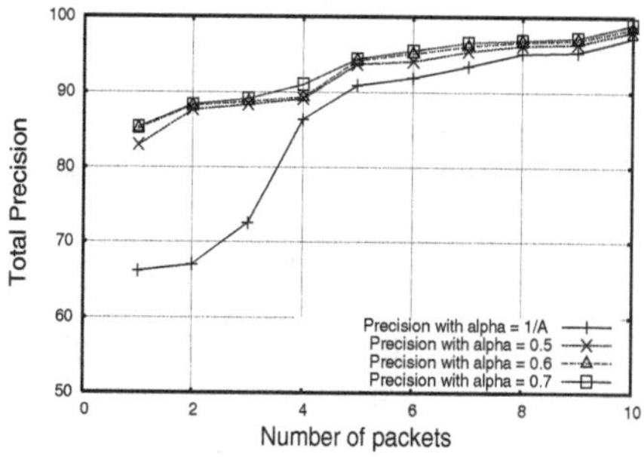


**Fig. 12.** Average total precision over all traces

Unfortunately, the Figure shows that using the inter-packet times makes the precision degrades to around 74% after then packets. It seems that the times between packets bring to the system wrong information not specific to each application because of their strong dependency on network conditions. These results indicate that at least in our setting, one cannot use the inter-packet times to differentiate between applications.

## 5 Conclusions

In this paper we have proposed and applied a statistical traffic classification technique based on the learning and analysis of the size and the direction of the first packets of flows. By considering packets separately from each other and with the help of a new probability function that combines the observations made on the different packets from a flow, we were able to obtain a high classification accuracy that kept improving by adding more packets from each flow, and that was able to reach high levels of around 97% and even more. Six applications were considered and validation was done on real traces. The validation over other applications and the consideration of other factors as the popularity of servers and hosts (BLINC [7]) are important steps for our future research.

## Acknowledgments

We would like to thank the authors of [6] for making their traces available to us, and the anonymous reviewers for their useful comments that helped to improve the paper.

## References

1. Moore, A., Papagiannaki, K.: Toward the accurate identification of network applications. In: Dovrolis, C. (ed.) PAM 2005. LNCS, vol. 3431, pp. 41–54. Springer, Heidelberg (2005)

2. Sen, S., Spatscheck, O., Wang, D.: Accurate, scalable in-network identification of p2p traffic using application signatures. In: WWW 2004 Conference, Philadelphia, USA (May 2004)
3. Erman, J., Mahanti, A., Arlitt, M.: Internet traffic identification using machine. In: Proc. of 49th IEEE Global Telecommunications Conference (GLOBECOM), San Francisco, USA (November 2006)
4. Moore, A., Zuev, D.: Internet traffic classification using bayesian analysis techniques. In: ACM Sigmetrics (2005)
5. Bernaille, L., Teixeira, R., Salamatian, K.: Early application identification. In: The 2nd ADETTI/ISCTE CoNEXT Conference, Lisboa, Portugal (December 2006)
6. Crotti, M., Dusi, M., Gringoli, F., Salgarelli, L.: Traffic classification through simple statistical fingerprinting. ACM-Sigcomm Computer Communication Review 37, 5–16 (2007)
7. Karagiannis, T., Papagiannaki, K., Faloutsos, M.: Blinc: Multilevel traffic classification in the dark. In: SIGCOMM 2005, New York, USA (August 2005)
8. Witten, I., Frank, E.: Data mining: Practical machine learning tools and techniques. Morgan Kaufmann, San Francisco (2005)
9. TCPdump: Collection of various patches that have been floating around for lbl's tcpdump and libcap programs (2007), http://www.tcpdump.org/

# The Illusion of Being Deterministic – Application-Level Considerations on Delay in 3G HSPA Networks

## (Work in Progress)

Joachim Fabini[1], Wolfgang Karner[2], Lukas Wallentin[1], and Thomas Baumgartner[2]

[1] Institute of Broadband Communications,
Vienna University of Technology
Favoritenstr. 9/E388, A-1040 Wien
`Joachim.Fabini@tuwien.ac.at`

[2] Mobilkom Austria AG, Obere Donaustr. 29, A-1020 Wien

**Abstract.** The delay experienced by mobile applications in HSPA networks depends to a large extent on highly dynamical global context like, e.g., cell load or algorithms and thresholds governing radio resource scheduling, and on local context like, e.g., user-generated load or load history. These complex uncertainty factors are outside of an applications sphere of influence and result in applications perceiving HSPA link behavior as non-deterministic and non-reproducible.

This paper analyzes accurate round-trip and one-way delay measurement results for three public HSPA networks to demonstrate the high degree of network non-determinism which mobile applications are likely to encounter in practice, particularly significant payload-dependence and halved delay on slightly increased user-generated link load. We argue that current HSPA radio link schedulers, relying on instantaneous user load as decision criterion for channel capacity allocation, neglect real-time application requirements. Cross-layer optimization is one solution which enables deterministic scheduler decisions based on application requirements.

**Keywords:** 3G, One-way Delay, HSPA, Measurements, User Experience.

## 1 Introduction

Next Generation Network (NGN) architectures aim at merging circuit switched (CS) voice networks and packet switched (PS) data networks into one common infrastructure. Central to these IP-based NGN architectures is their access-agnostic nature, meaning that applications need not be aware of the underlying access network technology. However, due to fundamentally different behavior of NGN access network technologies, e.g., with respect to delay and transfer rate, applications face difficulties in hiding access technologies from users. Moreover, high transfer capacities supported by technologies like HSPA in uplink and in

L. Fratta et al. (Eds.): NETWORKING 2009, LNCS 5550, pp. 301–312, 2009.

downlink direction force mobile operators to increasingly optimize radio resource usage of their networks, meaning that radio resources, e.g., dedicated channels, are allocated to users for a short time interval based on momentary user-generated load. Additional factors which may influence on this resource scheduling optimization include, among others, the user's mobile device type, radio provisioning, user load history, and current cell load generated by all users.

From a single mobile application's perspective the combination of various coding schemes and network scheduling parameters along with the uncertainty factors mentioned earlier yields an unpredictable mobile network behavior with respect to delay and transfer rate. This holds true for static measurements and the more for scenarios involving terminal mobility. Mobile applications, specifically real-time applications relying on deterministic network conditions, must handle delay which varies significantly, depending on the application's own traffic pattern, as well as on other background traffic generated by the mobile user.

Main aim of this paper is to raise NGN application developers' awareness concerning the peculiarity of mobile access networks. Specifically, the measurement which we present in this paper demonstrate that it is grossly negligent to infer from application tests in core- or fixed access networks onto application behavior in mobile access networks. The measurement results presented in this paper question the relevance of many scientific publications which rely on simplifying assumptions like, e.g., "the round-trip delay of UMTS is 150 ms". Our measurement results demonstrate that delay in today's mobile access networks is payload dependent and can effectively halve on slight increase in user-generated network load. Mobile real-time applications must, therefore, be prepared and tested to appropriately handle this special and apparent paradox behavior.

### 1.1 Related Work

There are only few publications analyzing 2.5G and 3G packet-switched performance aspects based on real network measurements. The authors of [1] use ICMP echo messages to infer on conversational and delay aspects of GPRS and EDGE networks. [2] analyzes impact of network load on TCP RTT in live GPRS and UMTS networks, whereas [3] proposes the use of passive monitoring for optimizing mobile network performance. In recent work, [4] relies on synchronized packet-level captures in the core network to infer on one-way delay. In [5] we have presented early GPRS and UMTS network measurement results as well as the impact of application-layer protocols like HTTP, POP3, SMTP or FTP on user-perceived performance, while [6], [7] and [8] present previous results of payload-dependent access network measurement and access network emulation results.

This paper's main contribution is to raise mobile application developer's awareness concerning the huge dependency of delay and, generally, of the behavioral determinism in mobile networks on user-generated traffic load and traffic patterns. The measurement results presented in this paper quantify for the first time application-perceived uncertainty of HSPA networks based on empirical measurement results and comparison of several public HSPA networks.

### 1.2 Structure of This Paper

In this paper we present payload-dependent delay measurement results for round-trip-, uplink- and downlink delay in public HSPA networks to point out the impact of resource allocation in mobile networks on real-time applications expecting deterministic traffic conditions. The remainder of this paper is structured as follows: Section 2 proposes a measurement methodology for round-trip- and one-way delay assessment based on round-trip delay decomposition into request and reply, followed by Section 3 which presents measurement results for this methodology in three distinct public HSPA networks depending on measurement configuration. In Section 4 we discuss the impact of our measurement results on NGN applications and NGN technologies, present methodological implications and conclude with an outlook on future work.

## 2 Measurement Methodology

The proposed measurement methodology for assessing HSPA network delay relies on two randomness factors, specifically on Internet Control Message Protocol (ICMP) packets having *random payload size* which are sent at *random start times*. Whereas random payload sizes distribute the impact of temporal network overload over the whole measurement payload space, start time randomness as proposed by the IP Performance Metrics (IPPM) framework in RFC 2330 eliminates correlations between send time and periodical network behavior.

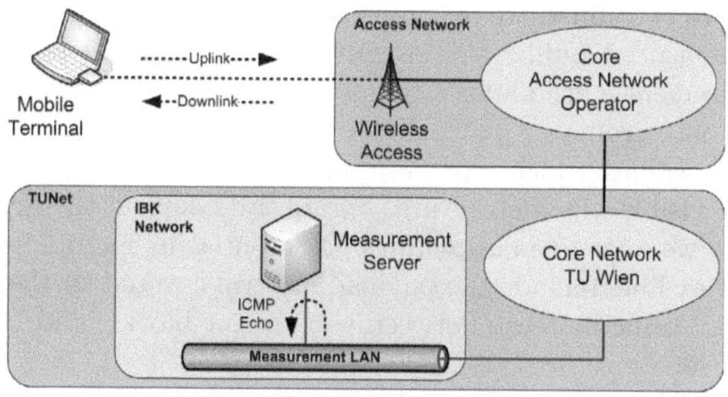


**Fig. 1.** Generic setup for automated round-trip delay measurement

The measurement setup in Figure 1 depicts the mobile client, a laptop computer connected to public HSPA networks using a Huawei E870 HSPA modem, and the measurement server which is connected to the Institute of Broadband Communication's switched Ethernet network. Depending on the measurement methodology, ICMP messages are sent either by the mobile client and reflected by the measurement server or vice-versa. For one-way measurements the clocks of mobile client and measurement server are continuously and accurately synchronized against a common global time base, whereas tcpdump [9] processes log incoming and outgoing packets along with their headers and timestamps on both hosts. Correlating the information stored in the two tcpdump trace files yields accurate uplink and downlink delay values, the precision being better than 0.5 ms. However, only ICMP *request* messages have been used for one-way delay computation.

The mobile terminal's position was fixed during all measurements, this static measurement regime minimizing the impact of variable measurement parameters like, e.g., of varying radio conditions and handovers because of terminal motion onto measurement results.

## 3 Measurement Results

Any single dot in the diagrams presented in this section represents the value of one ICMP round-trip- or one-way delay measurement, one diagram displaying approximatively 10,000 single measurements. Measurement ICMP payload size is selected uniformly distributed between 12 bytes and 1450 bytes, the start time of any ICMP-packet being chosen randomly between 50 ms and 500 ms following the last transmission.

### 3.1 Round-Trip Delay Measurement Results

Figure 2 depicts round-trip delay measurement results using the random measurement methodology presented in Section 2 for three public Austrian HSPA networks labeled as Operator A to Operator C.

Figures 2(a) to 2(c) illustrate that payload-dependent ICMP response patterns of the three measured public HSPA networks differ significantly. Due to the random measurement methodology, temporary changes in coding and scheduling influence on a wide range of payload sizes. All three diagrams show clustered groups of delay values which can be approximated by straight lines, some of them having a visible block structure. Figure 2(d) depicts the groups to which in the following we refer to as *delay lines*. Differentiating factors, which uniquely identify any delay line, include the payload interval covered by the delay line, its block size, delay value increase between subsequent blocks, and the delay line's initial delay offset.

For instance Operator A's diagram shows three main delay lines, as depicted by straight lines and corresponding labels in Figure 2(d). The first one, *Delay line 1*, positioned at the top left is composed of 40-byte wide payload blocks. It starts at an initial delay offset of approximatively 75 ms for 12 bytes ICMP payload size and ends at a delay value of 200 ms for 439 bytes payload size. The second delay line, *Delay line 2* starts at a delay value of 160 ms for 440 bytes payload but uses a block size of 160 bytes. The third and lowest one of these three main delay groups, labeled *Delay line 3* spans the entire payload range. It starts at a delay value of 75 ms for 12 bytes ICMP payload and terminates at 250 ms delay for 1450 bytes payload. Sparse measurement results below *Delay line 3* indicate the existence of a fourth, dotted, high-performance and low-delay *Delay line 4* which, however, is only rarely used for the tested load profile.

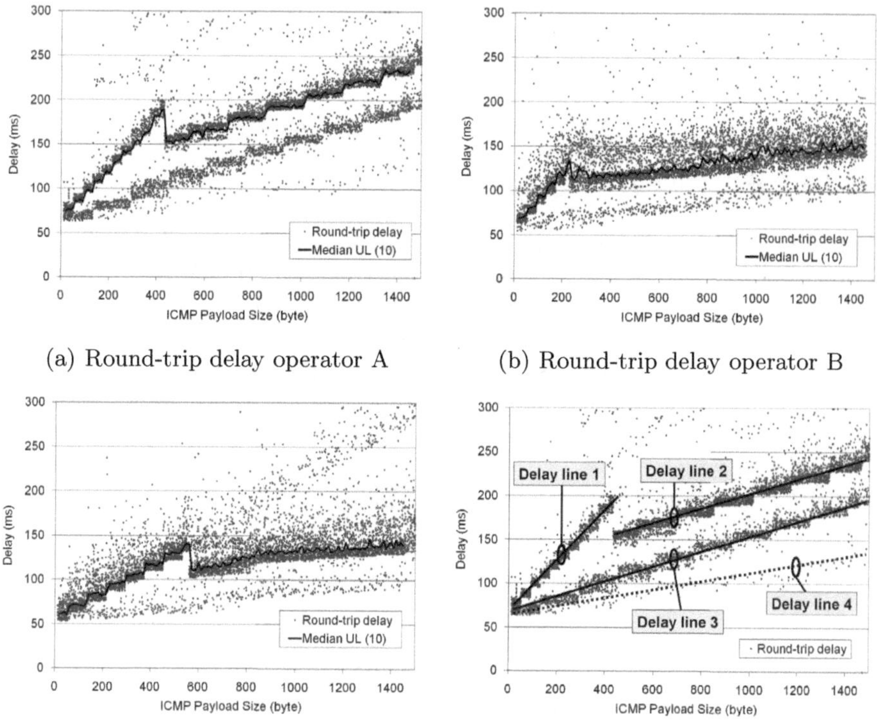


(a) Round-trip delay operator A

(b) Round-trip delay operator B

(c) Round-trip delay operator C

(d) Round-trip delay lines operator A

**Fig. 2.** HSPA Round-trip delay measurement results (measurement load only)

Due to the average function's sensitivity to outliers we have selected the median as representative function for the "common" network behavior. To compute the diagram's median curve, labeled in the diagram's legend as *Median(10)*, we have segmented the entire measurement payload range into sets which cluster 10 adjacent payload sizes. The diagram displays the median delay values for these sets, positioned in the center of the respective 10 payload values, interconnecting these points by straight lines.

The round-trip delay diagrams for Operator B in Figure 2(b) and for Operator C in Figure 2(b) exhibit different patterns, although the basic diagram structure is similar to Operator A. All diagrams start with steeply increasing low-blocksize delay lines and continue with larger block sizes. However, the "typical" delay at large payload sizes is significantly lower for Operator B and Operator C (150 ms) than for Operator A (200 ms to 250 ms).

Fundamental to the diagram discussion is the statement that the slope value of these delay lines equals the inverse of the channel's transfer capacity. Therefore, delay lines having a high slope value represent low-bandwidth channels whereas low slope values correspond to high-bandwidth channels. This finding raises an additional question, namely how can it happen that the diagrams in Figure 2 show two delay lines having equal transfer capacity but a delay offset of more than 50 ms. This analysis requires decomposition of any HSPA round-trip delay measurement value in its one-way delay components, which we present exemplary for Operator A.

### 3.2 One-Way Delay Measurement Results

Figure 3 depicts uplink delay and downlink delay measurement results and histograms for Operator A. The diagrams evidence that the HSPA uplink originates the multi-layering which we have noticed in the round-trip delay diagram in Figure 2(a). Therefore we can infer that the delay lines can be mapped to specific HSUPA uplink grants. The first (top leftmost) delay line uses E-TFCI 4 (Enhanced Dedicated Transport Channel Transport Format Combination Indicator) carrying 372 bits of RLC payload, whereas the lower delay line can be mapped to E-TFCI 15 which supports 1362 bits of payload for 10 ms Transmit Time Interval (TTI) according to Annex B.4 - Table 1 of TS 25.321 [10]. E-TFCI

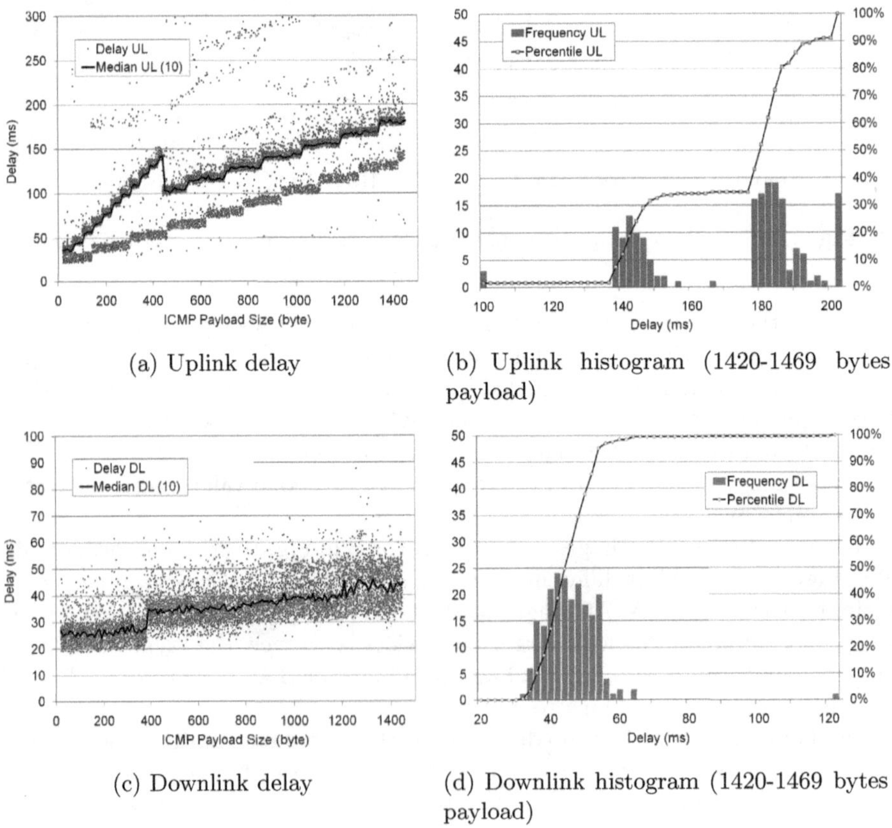


(a) Uplink delay

(b) Uplink histogram (1420-1469 bytes payload)

(c) Downlink delay

(d) Downlink histogram (1420-1469 bytes payload)

**Fig. 3.** HSPA one-way delay measurement result and histogram Operator A payload includes higher layer headers and therefore is larger than the IP layer block sizes which have been presented in section 3.1.

Our interpretation of the diagram in Figure 3(a) is that the mobile network grants E-TFCI 15 whenever a packet with payload larger than 440 bytes is to be sent while the current grant is E-TFCI 4. In this case the terminal starts transmitting using E-TFCI 4 while switching to E-TFCI 15 and continues using E-TFCI 15 afterwards. This "slow-start" causes the delay- and payload offset of the upper right E-TFCI 15-delay-line relative to the lower E-TFCI 15 layer. We also note that the terminal never obtains a higher grant if it uses E-TFCI 4 and a packet with payload size of up to 440 bytes must be sent.

The histogram analysis of uplink delay measurement results reveals that the upper left and right delay lines account for the majority of all measurement values. Figure 3(b) depicts exemplary the uplink delay histogram for the payload interval between 1420 and 1469 bytes. The percentile curve points out that approximatively 30% of all measurement values belong to the lower delay line and 55% to the upper line, this ratio being representative for the entire measured payload interval. Knowing that the delay between two subsequent measurement packets is chosen randomly between 50 ms and 500 ms one likely explanation for this distribution is that Operator A optimizes the network aggressively, downgrading terminal grants from E-TFCI 15 to E-TFCI 4 shortly after the packet has been transmitted. In a less aggressive optimization case, i.e., when the E-TFCI 15 grant persists until the next packet is ready to send, the majority of delay values would have been located in the lower (E-TFCI 15) delay layer.

Explanations analogous to the ones for Operator A, though with other grants and payload limits, apply to the diagrams for Operator B and Operator C which we have presented in Figures 2(b) and 2(c). The low delay at large payload sizes indicates that both operators use higher grants than E-TFCI 15 for large packets. Operator B uses the same E-TFCI 4 grant at low packet sizes but a significantly smaller threshold than Operator A (approximatively 230 byte) which triggers the switch to a higher grant. Finally, Operator C uses a higher grant than Operator A and Operator B (most likely E-TFCI 7 or E-TFCI 8) for packets up to 560 bytes in size.

### 3.3 Round-Trip Delay Measurement Results (Background Load)

Assuming that user-generated load – i.e., the load due to measurement packets in the case of measurements – biases on the transfer capacity which mobile networks allocate to users we have repeated the HSPA measurements with deterministic background load. The modified measurement methodology uses identical setup and methodology as presented in section 2 but adds two constant bit rate (CBR) flows between the mobile terminal and the measurement server, one flow loading the uplink and one the downlink. Main aim of this user load is to prevent the network from withdrawing higher grants suddenly after a measurement packet has been sent while minimizing collisions with measurement packets.

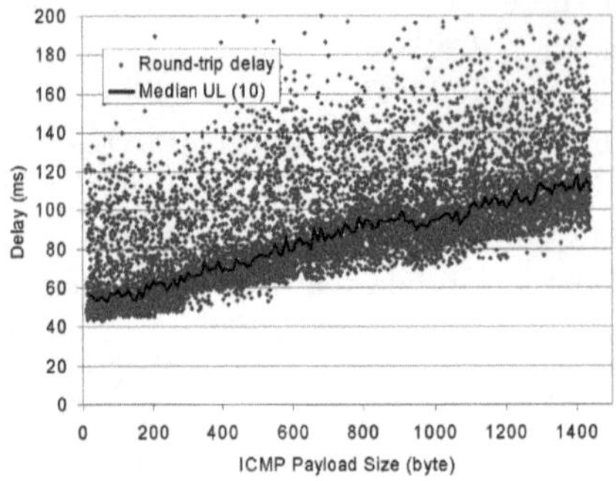


(a) Round-trip delay operator A (load)

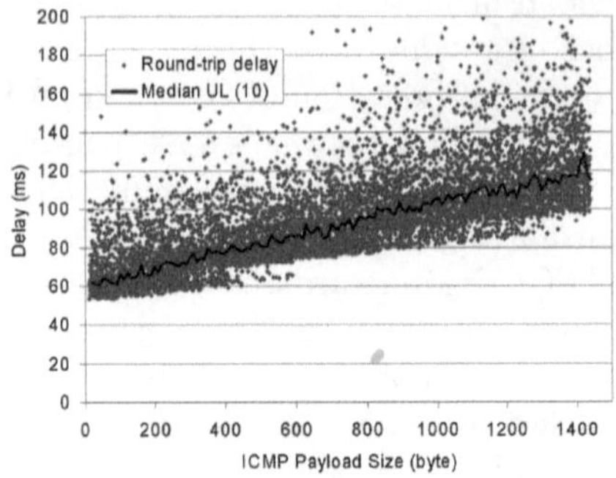


(b) Round-trip delay operator B (load)

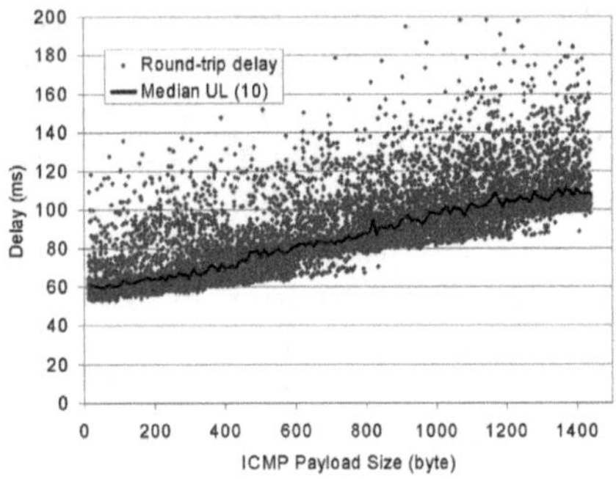


(c) Round-trip delay operator C (load)

**Fig. 4.** HSPA Round-trip delay measurement results (deterministic background load)

Figure 4 presents round-trip delay measurement results for the deterministic background load scenario. Uplink and downlink have been loaded using two periodic 50 kbit/s data rate 100 byte payload UDP flows generated by IPerf [11], one in each direction. Although the number of outliers in the background-loaded measurement diagrams increases visibly when compared to the original diagrams in Figure 2, the delay pattern in the background load case in Figure 4 is more deterministic. Specifically, the multi-layering clearly visible in Figure 2(a) disappears when measuring using deterministic background load.

Main result of this new measurement methodology is that the median round-trip delay value *decreases* significantly for all three operators throughout the entire measured payload interval. Most prominent, for Operator A the delay difference is substantial when comparing Figure 2(a) against 4(a). The median round-trip delay value at large payload sizes effectively halves from 225 ms to 110 ms when adding background load to the network. For Operator B and Operator C the decrease in round-trip delay is smaller, though still significant. Round-trip delay drops from 150 ms to 120 ms for Operator B and from 140 ms to 110 ms for Operator C, taking into account the median values at large payload sizes.

### 3.4 One-Way Delay Measurement Results (Background Load)

Analogous to the non-loaded measurements we have decomposed round-trip delay measurement values into their uplink and downlink delay components, enabling accurate analysis of performance improvements. All one-way diagrams which we have presented for the non-loaded case in Figure 3 are shown in Figure 5 for the background-load measurement scenario.

A comparison of uplink delay diagrams and their median delay curves in Figures 3(a) and 5(a) illustrates that adding background load to measurements decreases the uplink delay significantly. The histograms showing delay frequency for large payload sizes (1420-1459 bytes) in Figures 3(b) and 5(b) confirm this finding. Whereas in Figure 3(b) the percentile curve starts to increase at 140 ms and reaches 90% close to a delay value of 200 ms, the percentile curve in Figure 5(b) starts at 50 ms and crosses 90% at an uplink delay value of 110 ms. However, the uplink jitter in the background load case is significant and affects more than 12% of all measurement values for large payload sizes.

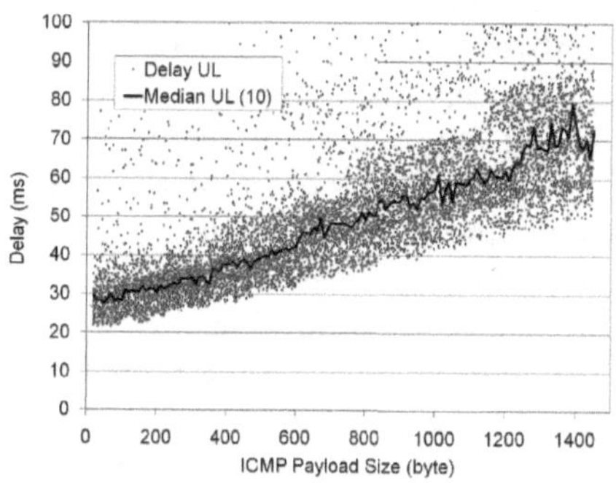


(a) Uplink delay (load)

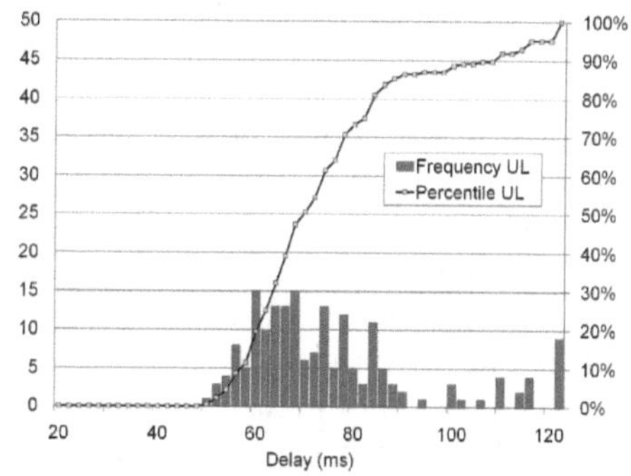


(b) Uplink histogram 1420-1469 bytes (load)

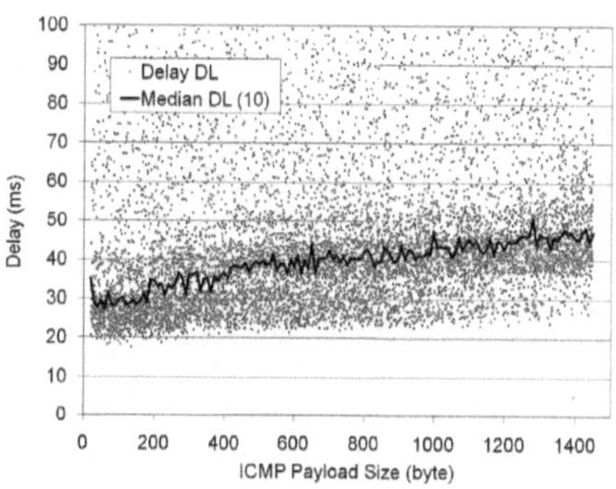


(c) Downlink delay (load)

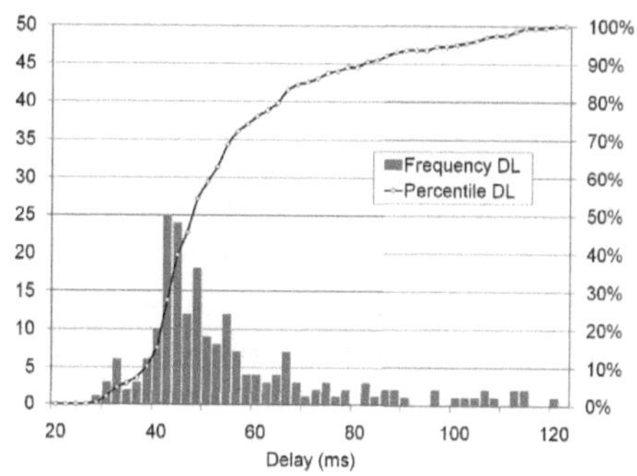


(d) Downlink histogram 1420-1469 bytes (load)

**Fig. 5.** HSPA one-way delay measurement results Operator A (background load)

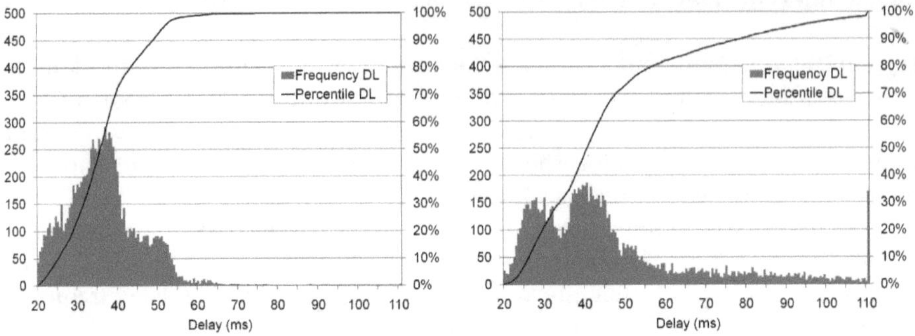


(a) HSPA downlink histogram (b) HSPA downlink histogram (load)

**Fig. 6.** HSPA downlink delay histograms for Operator A

Non-regarding this, the difference in uplink delay median values for the payload interval mentioned above is substantial: adding background load decreases the uplink delay median value from 182 ms in the non-loaded case to 72 ms in the loaded case. However, the measurement results indicate that background load does not improve HSPA downlink performance, on the contrary. A comparison of HSPA downlink delay diagrams shown in Figure 3(c) for the non-loaded case and in Figure 5(c) for the background load case points out that the median value curve is affected mainly for low payload values up to 400 bytes, for which median delay increases by 5 ms to 10 ms. Detrimental is the huge jitter which in the background load scenario affects more than 15% of all downlink packets and spreads over the entire measured payload interval. This finding is confirmed by the evolution of percentile curves in Figure 6(a) for the non-loaded case and Figure 6(b) for the background load scenario, the diagrams depicting downlink delay histograms for the entire measured payload interval.

## 4 Conclusions and Future Work

From an application perspective HSPA networks behave non deterministic. The measurement results presented in this paper demonstrate that the HSUPA uplink delay *decreases* significantly with *increasing* uplink load and associated higher grants in terms of radio resources. This radio link scheduling strategy is at least questionable from a fairness perspective, resulting in delay penalties for applications which manage their resources economically. However, additional load does not have a positive effect on downlink delay. Inherent reason for this asymmetric load influence on performance is the radio link scheduler's location in the NodeB. Due to the NodeB's low-latency fixed network connectivity towards RNC and core network the scheduler can react almost instantaneously by allocating appropriate radio resources for packets in the *downlink*, whereas for *uplink* transmission it is the mobile terminal's task to request uplink radio resource allocation according to the momentary load. Uplink resource requests therefore require an additional radio access round-trip delay until the mobile terminal receives the notification whether a specific request can be fulfilled or not fulfilled, depending on the radio scheduler policy and momentary context.

Underlying reason to applications perceiving radio link scheduler decisions as non-deterministic is eventually missing context information in both, the mobile terminal and in the radio link scheduler. Mobile applications which do posess a priori or momentary context information – e.g., having configured a specific real-time traffic profile, stating what amount of data is to be sent at which point in time – currently have no means to communicate this context information to lower layers or the radio link scheduler. In other words, data at IP-layer is considered by lower layers and by the radio link scheduler as being opaque, all packets having same priority and importance. It is therefore legitimate to question the traditional OSI layering model which does not foresee cross-layer information exchange. We expect context propagation in mobile access networks to be implemented using cross-layer optimization in the near future, marking IP packets by appropriate priorities in terms of delay, loss, and priority requirements, to enable and support appropriate radio link scheduler decisions.

Our measurements demonstrate that randomness in terms of payload and packet send time can unleash mobile operator specific configuration parameters and thresholds. Active measurements can therefore serve as a simple, straightforward tool to optimize own mobile networks or to reverse-engineer highly sensitive configuration parameters of competing network operators.

Seen from standardization perspective, the measurement results presented in this paper question the meaningfulness of compression techniques proposed for NGNs like, e.g., Signaling Compression [12]. The results demonstrate that these techniques, which are supposed to decrease delay in narrow-band mobile networks and are mandatory part of NGNs like, e.g., the IP Multimedia Subsytem (IMS), might show the contrary effect of increasing delay because of their reduction of user-generated network load.

Even more important, this paper's results point out a major drawback of existing standards like the ETSI Technical Specification TS 102 250 Part1 to Part 6, which target the definition of fair metrics for mobile network measurements but completely disregard measurement result dependence on measurement methodology. Facing the huge influence of measurement methodology, specifically of traffic profiles, on delay results, we conclude that these standards must be substantially revised to fulfill basic fairness requirements. As future work we therefore plan to contact ETSI and contribute our findings in order to enhance TS 102 250 with respect to measurement methodology specification.

## Acknowledgment

The authors would like to thank mobilkom austria AG for technical and financial support of this work. The views expressed in this paper are those of the authors and do not necessarily reflect the views of mobilkom austria AG.

## References

1. Ball, C., Masseroni, C., Trivisonno, R.: Introducing 3G like conversational services in GERAN packet data networks. In: IEEE 61st Vehicular Technology Conference (VTC 2005), Stockholm, Sweden, May 2005, pp. 2186–2191 (Spring 2005)
2. Vacirca, F., Ricciato, F., Pilz, R.: Large-scale RTT measurements from an operational UMTS/GPRS network. In: Proceedings of the First International Conference on Wireless Internet (WICON 2005), Budpest, Hungary, July 2005, pp. 190–197 (2005)
3. Ricciato, F.: Traffic monitoring and analysis for the optimization of a 3G network. IEEE Wireless Communications 13(6), 42–49 (2006)
4. Ricciato, F., Hasenleithner, E., Romirer-Maierhofer, P.: Traffic analysis at short time-scales: an empirical case study from a 3G cellular network. IEEE Transactions on Network and Service Management 5(1), 11–21 (2008)
5. Reichl, P., Jordan, N., Fabini, J., et al.: Wireless Inter-System Quality-of-Service: A Practical Performance Analysis of 3G and Beyond. In: Kommunikation in Verteilten Systemen (KiVS 2005), Kaiserslautern, Germany, March 2005, pp. 230–241 (2005)
6. Fabini, J., Reichl, P., Egger, C., et al.: Generic Access Network Emulation for NGN Testbeds. In: 4th International Conference on Testbeds and Research Infrastructures for the Development of Networks and Communities (TridentCom 2008), Innsbruck, Austria, March 2008, p. 10 (2008)
7. Fabini, J., Reichl, P., Poropatich, A.: Measurement-Based Modeling of NGN Access Networks from an Application Perspective. In: Proceedings of the 14th GI/ITG Conference on Measurement, Modeling, and Evaluation of Computer and Communication Systems (MMB 2008), Dortmund, Germany, April 2008, pp. 45–59 (2008)
8. Fabini, J., Reichl, P., Poropatich, A.: A Generic Approach to Access Network Modeling for Next Generation Network Applications. In: Proceedings of the 4th International Conference on Networking and Services (ICNS 2008), Gosier, Guadeloupe, March 2008, pp. 254–260 (2008)
9. Lawrence Berkeley National Laboratory, University of California, Berkeley, CA: TCPDump Monitoring Tool (December 2007), `http://www.tcpdump.org/`
10. 3GPP: 3rd Generation Partnership Project: 3GPP TS 25.321, Medium Access Control (MAC) protocol spec., version 6.15.0 (March 2008), `http://www.3gpp.org/`
11. National Laboratory for Applied Network Research (NLANR): IPerf Performance Measurement Tool (December 2007), `http://dast.nlanr.net/Projects/Iperf/`
12. Price, R., Bormann, C., Christofferson, J., et al.: IETF standards track RFC 3320: Signaling Compression (SigComp) (January 2003)

# Phoenix: Towards an Accurate, Practical and Decentralized Network Coordinate System

Yang Chen[1], Xiao Wang[1], Xiaoxiao Song[1], Eng Keong Lua[2], Cong Shi[3], Xiaohan Zhao[1], Beixing Deng[1], and Xing Li[1]

[1] Tsinghua National Laboratory for Information Science and Technology
Department of Electronic Engineering, Tsinghua University, Beijing 100084, China
[2] College of Engineering, Carnegie Mellon University, Pittsburgh, PA 15213
[3] College of Computing, Georgia Institute of Technology, Atlanta, GA 30332
chenyang04@mails.tsinghua.edu.cn

**Abstract.** Network coordinate (NC) system allows efficient Internet distance prediction with scalable measurements. Most of the NC systems are based on embedding hosts into a low dimensional Euclidean space. Unfortunately, the accuracy of predicted distances is largely hurt by the persistent occurrence of Triangle Inequality Violation (TIV) in measured Internet distances. IDES is a dot product based NC system which can tolerate the constraints of TIVs. However, it cannot guarantee the predicted distance non-negative and its prediction accuracy is close to the Euclidean distance based NC systems. In this paper, we propose Phoenix, an accurate, practical and decentralized NC system. It adopts a weighted model adjustment to achieve better prediction accuracy while it ensures the predicted distances to be positive and usable. Our extensive Internet trace based simulation shows that Phoenix can achieve higher prediction accuracy than other representative NC systems. Furthermore, Phoenix has fast convergence and robustness over measurement anomalies.

**Keywords:** P2P, Network Coordinate System, Triangle Inequality Violation, Dot Product, Weighted Model.

## 1 Introduction

Network Coordinate (NC) system is an efficient and scalable mechanism to predict distance (Round Trip Time) between any two Internet hosts without explicit measurements. In most of the NC systems, each host is assigned a set of numbers called coordinates to represent its position in the Euclidean space, and the distance between any two hosts can be predicted by their coordinates using Euclidean distance. NC system reduces the active probing overhead significantly, which is especially beneficial to large-scale distributed applications. To date, NC systems are widely used in different Internet applications such as application layer multicast [1], locality-aware server selection [1], distributed query optimization [2], file-sharing via BitTorrent [3], network modeling [4], compact routing [5], and application layer anycast [6].

L. Fratta et al. (Eds.): NETWORKING 2009, LNCS 5550, pp. 313–325, 2009.

Unfortunately, the Euclidean distance based NC systems have a common unremediable drawback, i.e., the predicted distances among each three hosts must satisfy the triangle inequality. Lots of existing studies report the existence of Triangle Inequality Violations (TIV) in the Internet delay structure [8,7,10,22,25,32]. As a result, these distances cannot be predicted accurately by using Euclidean distance based NC even if we increase the dimension of the space.

A dot product based NC system named IDES is proposed in [9] . The key idea of this system is that a large distance matrix can be approximately factorized into two smaller matrices by methods like Singular Value Decomposition (SVD) or Non-negative Matrix Factorization (NMF) [11]. This results in a compressed version of the Internet distance matrix. In contrast to the Euclidean distance based NC systems, the distances predicted by IDES do not have to satisfy the triangle inequality. However, IDES is still not an ultimate solution. First, unlike any other existing NC system, IDES will give negative predicted distance. This will cause the malfunction of the system because the distance (Round Trip Time) can not be negative. In addition, the prediction accuracy of IDES is close to typical Euclidean distance based NC system such as GNP [18] according to the experiments in [1].

In this paper, we propose an accurate, practical and decentralized NC system named Phoenix. Phoenix is also based on dot product, but remedies IDES's flaws. Phoenix can achieve much higher prediction accuracy than other typical representative NC systems such as IDES and Vivaldi [19]. The key contributions of this paper are twofold. (1) We propose a weight calculation algorithm to distinguish referred NCs with high errors and low errors. With the error propagation eliminated, Phoenix demonstrates the advantage of dot product based NC systems. Our extensive Internet trace based simulation results show that Phoenix can achieve much higher prediction accuracy than state-of-the-art methods. Our simulation results also demonstrate Phoenix's fast convergence and robustness over measurement anomalies. (2) Compared with IDES, Phoenix not only performs better in prediction accuracy but also guarantees the predicted distance non-negative. The results show that the implementation of Phoenix is an accurate solution to build a practical NC system.

The rest of this paper is organized as follows. In Section 2, we review the related work. The design of our accurate, practical and decentralized NC system - Phoenix is proposed in section 3. We evaluate the performance of Phoenix and compare it with two representative NC systems with extensive simulation results in Section 4. We conclude the whole paper in Section 5.

## 2 Related Work

### 2.1 Euclidean Distance Based Network Coordinates and Triangle Inequality Violation (TIV)

Suppose there are $N$ Internet hosts. Let $S$ be the set of these $N$ hosts. Let $D$ be the $N \times N$ distance matrix among the hosts in $S$. Thus $D(i, j)$ represents the measured Round Trip Time (RTT) between host $i$ and host $j$.

Basically, NC is an embedding of these $N$ hosts into m-dimensional Euclidean space $R^m$. Defining $x_i$ as the NC of host $i$, we have $x_i = (r_1^i, r_2^i, ..., r_m^i), r_k^i \in R, 1 \leqslant k \leqslant m$. Then $x_i$ and $x_j$ can be used to predict the RTT between host $i$ and host $j$. We use $D^E(i,j)$ to represent this predicted RTT. The definition of $D^E(i,j)$ is as follows.

$$D^E(i,j) = \|x_i - x_j\| = \sqrt{\sum_{1 \leq k \leq m} (r_k^i - r_k^j)^2} \tag{1}$$

Several NC systems have been proposed in the literature, and they can be categorized into two classes [8], namely centralized NC systems and decentralized NC systems. Centralized NC systems such as GNP [18], Virtual Landmarks [14] require a fixed set of dedicated landmarks to orient the NC calculation of the whole system, which will be a bottleneck of the system. Therefore, decentralized NC systems such as Vivaldi [19], NPS [26], PIC [31] were proposed to make NC system work well on large-scale applications. In this paper, we compare our system with Vivaldi since it's the representive Euclidean distance based NC system due to its clean and decentralized implementation. It is deployed in many well-known Internet systems, such as Bamboo DHT [28], Stream-Based Overlay Network (SBON) [2] and Azureus BitTorrent [3].

The prediction accuracy of an NC system is often denoted by the relative error (RE) of predicted distance over the real RTT measured on Internet. Relative Error (RE) of the distance between host $i$ and host $j$ is defined as [10,12,13,14, 15,16,17]

$$RE = \frac{| D^E(i,j) - D(i,j) |}{D(i,j)} \tag{2}$$

Smaller RE indicates higher prediction accuracy. When measured distance equals to predicted distance, the RE value will be zero.

Suppose there are three hosts, A, B and C. Let's consider the triangle ABC. Suppose AB is the longest edge of the triangle. If $D(A,B) > D(A,C) + D(C,B)$, then ABC is called a TIV, due to the violation of the triangle inequality. As mentioned in [25], any three hosts with TIV cannot be embedded into Euclidean space within some level of accuracy, for the distances among them in Euclidean space must obey triangle inequality. However, the TIV is natural and persistent in Internet [25]. Therefore the existence of TIV causes a serious problem for every Euclidean distance based NC systems [7,8]. In other words, it hurts the accuracy of Euclidean distance based NC a lot.

## 2.2 Dot Product Based NC and IDES

Suppose there are $N$ Internet hosts. Let $D$ be the $N \times N$ distance matrix among these $N$ hosts. $D(i,j)$ represents the measured RTT between host $i$ and host $j$. This $N \times N$ matrix can be factorized into two smaller matrices. $D \approx XY^T$ where $X$ and $Y$ are $N \times d$ matrices ($d << N$).

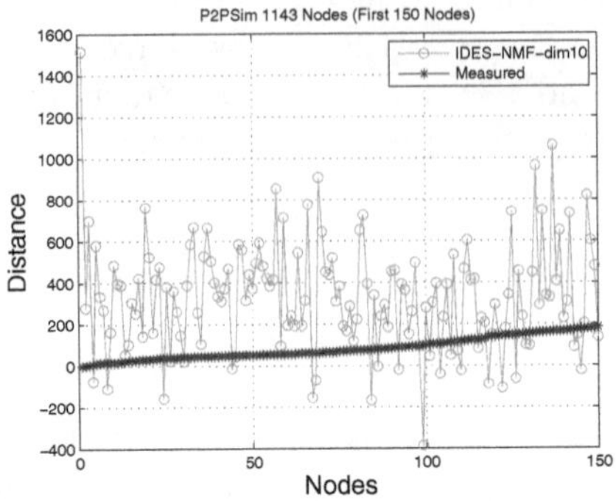


**Fig. 1.** Maximum RRL using IDES-NMF

In dot product based NC system, for host $i$, $\boldsymbol{X_i}$ is the *outgoing vector* and $\boldsymbol{Y_i}$ is the *incoming vector*. The predicted distance from host $i$ to host $j$ is simply the dot product between the outgoing vector of host $i$ and the incoming vector of host $j$ as follows

$$D^E(i,j) = \boldsymbol{X_i} \cdot \boldsymbol{Y_j} = \sum_{k=1}^{d} X(i,k) \cdot Y(j,k) \tag{3}$$

IDES [9] is the representative NC system based on dot product. There are several serious problems with IDES. First, IDES's predicted distance can be negative, which is very harmful to many practical applications. In a simple example, we apply maximum RRL test [10] using the P2PSim latency data set [9] for the IDES(NMF) methods in 10-dimensional space (with 15 landmarks). We use IDES simulator which is available at the author's homepage [29]. In this test, we select the site with maximum RRL value as the target site. Fig.1 demonstrates the results. The x-axis enumerates sites while the y-axis corresponds to the RTT distance of each site from the selected site. The signature plot marked with +'s indicates RTTs in the original distance matrix, and the sites on the x-axis have been sorted to ensure that this plot is in ascending order of RTTs. The signature plot marked with o's is the predicted distances given by IDES. It is obvious that IDES results in negativity in distances. This is not realistic or relevant since the distance stands for RTT on Internet. Despite the small percentage of negative distances, their impact on the whole application will be very severe. For example, a typical NC application in overlay multicast is the construction of Minimum Spanning Tree (MST) using NC predicted distance [1]. If we use Dijkstra algorithm to construct the MST, even one negative distance can fail the algorithm.

Furthermore, estimation error will be propagated in IDES's distance prediction. IDES uses the least square error for NC calculation. For a certain host, it uses the NCs of reference hosts and its distances to reference hosts to calculate its own NC. In the NC calculation, this host gives equal confidence to each referred

NC in both basic IDES and decentralized IDES. However, some NCs are very inaccurate due to different factors, such as network congestion or error propagation. Once these inaccurate NCs are referred, their errors will be propagated to the new hosts. Thus, it is not surprising that IDES doesn't show significant improvement on the prediction accuracy over Euclidean distance based NC [1].

In IDES, we can add non-negative constraint in the NC calculation of ordinary hosts to avoid negative predicted distance. However, the prediction accuracy still cannot be improved [9]. In next section, we will propose our solution, an accurate, practical and decentralized NC system called Phoenix. Phoenix will not only guarantee the predicted distance non-negative but also improve the prediction accuracy a lot.

# 3 Design of Phoenix

## 3.1 System Architecture

In this section, we propose the design of Phoenix, a practical dot product based NC system. Our design focuses on the most important aspects for NC systems, i.e., accuracy, decentralization and practicability. Basically, NC is used for Internet distance prediction; so the prediction accuracy is principally important. Moreover, NC is widely used in distributed applications, there may be thousands of hosts in the swarm; therefore, NC system should be decentralized. Last but not the least, NC system should be practical. In other words, it should never give negative predicted distances.

Phoenix maps each host to two $d$-dimensional row vectors - an incoming vector and an outgoing vector. The predicted distance from host $i$ to host $j$ is simply the dot product between the outgoing vector of host $i$ and the incoming vector of host $j$. In contrast to IDES, all the elements in these two vectors are non-negative, which guarantees Phoenix never gives negative predicted distance.

Unlike GNP [18] or other centralized NC systems, Phoenix system has no fixed network infrastructure or distinguished hosts to serve the whole system. Any host with calculated NC can serve as a *reference host* to orient the new host to participate in. Thus Phoenix is efficient for large scale applications since the communication and computation overhead will be distributed evenly to all hosts in the system. After a new host joins the system, it can pick any host with calculated NC as one of its reference hosts. Let $N$ be the set of hosts whose NCs have been calculated. When a new host $H_{new}$ joins the system, it can select any $m$ reference hosts randomly from the set $N$ with size $m < |N|$ and starts its NC update procedure. In every round, $H_{new}$ measures its RTTs to these $m$ hosts as well as retrieves the incoming and outgoing vectors of these $m$ hosts. Then its NC can be calculated and updated periodically. We will propose the detailed NC calculation algorithm in section 3.2.

For the first $m$ hosts of the Phoenix system, the NC calculation is a bit different. If $|N| \leq m$, the new host $H_{new}$ will be considered as one of the *early hosts*. These early hosts will probe each other to obtain the $|N| \times |N|$ distance matrix. The system will use NMF algorithm [11] to get the incoming vectors and the outgoing vectors of these early hosts.

### 3.2 Weighted Non-negative Least Squares NC Calculation

In Phoenix, we use a weighted non-negative least squares module to calculate the NC. The intuition behind this weighted NC calculation is as follows. The more accurate the referred NC (vector) is, the higher confidence (weight) should be given to this NC. In contrast, some referred NCs with abnormal high error will not be considered for NC calculation. Similar intuition is also used in decentralized Euclidean distance based NC systems such as [21].

For a host $H_{new}$, it has $m$ reference hosts namely $R_1, R_2, ..., R_m$. The outgoing vectors of these $m$ reference hosts are $X_1, X_2, ..., X_m$ and the incoming vectors of these $m$ reference hosts are $Y_1, Y_2, ..., Y_m$. We define $w_{X_i}$ as the weight of vector $X_i$ and $w_{Y_i}$ as the weight of vector $Y_i$. In our design, for all $X_i$ and $Y_i$, we have $0 \leqslant w_{X_i} \leqslant 1$ and $0 \leqslant w_{Y_i} \leqslant 1$.

Suppose $D_i^{out}$ is the distance from the host $H_{new}$ to reference host $R_i$, $D_i^{in}$ is the distance from reference host $R_i$ to the host $H_{new}$. The solution of the $X_{new}$ and $Y_{new}$ with the weighted non-negative least squares error is as follows.

$$\boldsymbol{X_{new}} = arg \min_{U \in R^{+d}} \sum_{i=1}^{m} w_{Y_i} (D_i^{out} - \boldsymbol{U} \cdot \boldsymbol{Y_i})^2 \tag{4}$$

$$\boldsymbol{Y_{new}} = arg \min_{U \in R^{+d}} \sum_{i=1}^{m} w_{X_i} (D_i^{in} - \boldsymbol{X_i} \cdot \boldsymbol{U})^2 \tag{5}$$

We define an $m \times 1$ matrix $D_W^{out} = [\sqrt{w_{Y_1}} D_1^{out} \ \sqrt{w_{Y_2}} D_2^{out} \ ... \ \sqrt{w_{Y_m}} D_m^{out}]^T$ and an $m \times d$ matrix $Y_W$ with $\sqrt{w_{Y_i}} Y_i$ as its row vectors. Therefore the Eq.(4) can be solved as follows.

$$\boldsymbol{X_{new}} = arg \min \|Y_W \boldsymbol{X_{new}^T} - D_W^{out}\|, \ s.t. \boldsymbol{X_{new}} \geqslant 0. \tag{6}$$

Likewise,we define an $m \times 1$ matrix $D_W^{in} = [\sqrt{w_{X_1}} D_1^{in} \ \sqrt{w_{X_2}} D_2^{in} \ ... \ \sqrt{w_{X_m}} D_m^{in}]^T$ and an $m \times d$ matrix $X_W$ with $\sqrt{w_{X_i}} X_i$ as its row vectors. Therefore the Eq.(5) can be solved as follows.

$$\boldsymbol{Y_{new}} = arg \min \|X_W \boldsymbol{Y_{new}^T} - D_W^{in}\|, \ s.t. \boldsymbol{Y_{new}} \geqslant 0. \tag{7}$$

Eq. (6)-(7) are non-negative least squares constraints problems, they can be solved by the algorithm proposed in [20]. In our simulator [33] written in Matlab 6.0, the function called *lsqnonneg* is used to solve the nonnegative least-squares constraints problem.

In Phoenix, we calculate the weight as follows. If $H_{new}$ is a newly joined host, we set all $w_{X_j} = w_{Y_j} = 1$ and calculate its initial NC. From this round, it applies Eq.(8)-(11) to determine the weights and updates its NC periodically.

For each reference host $j$ of $H_{new}$, we define the error of each vector as follows.

$$\begin{aligned} E_{X_j} &= \|X_j \cdot Y_{H_{new}} - D(j, H_{new})\| \\ E_{Y_j} &= \|X_{H_{new}} \cdot Y_j - D(H_{new}, j)\| \end{aligned} \tag{8}$$

Thus the median values of both the error of $E_{X_i}(1 \leq i \leq m)$ and $E_{Y_i}(1 \leq i \leq m)$ are

$$\begin{aligned} M_X &= median_i(E_{X_i}) \\ M_Y &= median_i(E_{Y_i}) \end{aligned} \tag{9}$$

Then we can get the $w_{X_j}$ and $w_{Y_j}$ as follows.

$$w_{X_j} = \begin{cases} 1, & \text{if } E_{X_j} < M_X; \\ M_X/E_{X_j}, & \text{if } M_X < E_{X_j} < C * M_X; \\ 0, & \text{otherwise.} \end{cases} \tag{10}$$

$$w_{Y_j} = \begin{cases} 1, & \text{if } E_{Y_j} < M_Y; \\ M_Y/E_{Y_j}, & \text{if } M_Y < E_{Y_j} < C * M_Y; \\ 0, & \text{otherwise.} \end{cases} \tag{11}$$

$C$ is set as 5 in our Phoenix implementation.

**Algorithm 1.** Phoenix Algorithm

```
Connect_to_Rendezvous_Point(RP)
Get_Reference_Host_Candidates(RP)
Connect_to_Reference_Hosts()
round = 1
while forever do
  Get(d(·), X(·), Y(·))
  if round = 1 then
    w_{X_j} = w_{Y_j} = 1
    [X_new, Y_new] = NC_Calculation(d(·), X(·), Y(·), w_X(·), w_Y(·))
  end if
  [w_X(·), w_Y(·)] = Weight_Calculation(D(·), X(·), Y(·), X_new, Y_new)
  [X_new, Y_new] = NC_Calculation(D(·), X(·), Y(·), w_X(·), w_Y(·))
  Wait(Update_Interval);
  round = round + 1
end while
```

Algorithm 1 shows the procedure of a new host joining the Phoenix NC system (Suppose this host is not one of the early hosts). This new host first contacts the Rendezvous Point (RP) of the Phoenix system like all other P2P schemes. After obtaining a list of reference host candidates, it contacts them and select $m$ hosts out of them as its reference hosts. Then it starts its NC calculation and updates its NC periodically using the weighted NC calculation model. In every round, the new host measures its RTTs to its reference hosts as well as retrieves their NCs before the weighted NC calculation. Specifically, in its first round, it will calculate its initial NC using non-negative least squares (without introducing weight), then the weighted model can be applied from then on.

**Table 1.** TIV of the Data Sets

| DataSet | *Hosts* | *triples in TIVs* | *pairs in TIVs* |
|---|---|---|---|
| AMP | 110 | 4.29% | 48.07% |
| PlanetLab | 169 | 25.71% | 86.53% |
| King | 1740 | 12.32% | 85.52% |
| P2PSim | 1143 | 17.10% | 97.33% |
| Meridian | 2500 | 23.50% | 96.55% |

# 4 Performance Evaluation

## 4.1 Setup of the Experiment

In our experiment, we compare Phoenix with IDES and Vivaldi. All of these three systems use 10-dimensional coordinates. In Phoenix, each host has $m$ reference hosts. Likewise, there are $m$ randomly selected landmarks in IDES. In Vivaldi, $c_c$ and $c_e$ are set to 0.25 as an empirical value and each host has $m$ neighbors. In our experiments, we set $m$ as 32. 10 runs are performed on each data set and the average results are reported.

We use five typical data sets from real Internet measurement to study the prediction accuracy of different NC systems. The first data set is the AMP data set [9], which includes the RTTs among 110 Internet hosts. The hosts are mainly at NSF supported HPC sites, with about 10% outside the US. The second data set is the PlanetLab data set [9], which includes the RTTs among 169 PlanetLab [23] hosts all over the world. The third data set is King data set which includes the RTTs among 1740 Internet DNS servers [19]. The fourth data set is P2PSim data set, which includes the RTTs among 1143 Internet DNS servers. These DNS servers were obtained from an Internet scale Gnutella network trace [9]. The fifth data set is Meridian data set which is from the Cornell Meridian project [27]. It measures the pairwise RTTs between 2500 hosts.

The effect of TIV is studied on these five data sets with two metrics proposed in [30]. The triple of hosts violating the triangle inequality is called a bad triangle. The first metric is *triples in TIVs*, which is defined as the percentage of triples that form bad triangles. The other metric is *pairs in TIVs*, which is defined as the percentage of pairs of hosts that are long sides in bad triangles (i.e., pairs that have an alternate shorter path). Table.1 shows the results. It is obvious that TIVs are quite different among these data sets. Thus we can evaluate our Phoenix NC system in different Internet delay structures.

## 4.2 Evaluation Results on Prediction Accuracy

In this subsection, we compare the REs of IDES, Vivaldi and Phoenix. More precisely, we evaluate both IDES(SVD) and IDES(NMF). Besides the complete

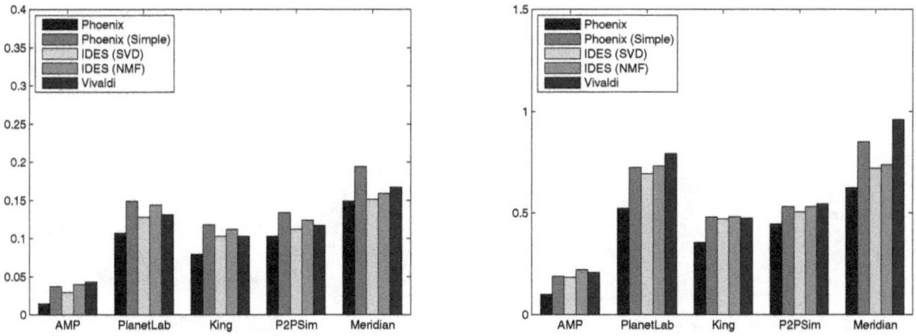


(a) Median Relative Error (b) 90th Percentile Relative Error

**Fig. 2.** Prediction Accuracy Comparison

Phoenix, we also show the results of Phoenix(Simple). In the NC calculation procedure, Phoenix(Simple) doesn't use weighted model. It just simply sets all the weight to be 1. The comparison between Phoenix and Phoenix(Simple) demonstrates the improvement of the weighted model. Thus we have five different implementations of NC systems in this comparison.

Fig.2 shows the comparison between these five NC implementations under five different data sets. As in [18, 9], more attention is paid to the 90th Percentile Relative Error (NPRE) since it can guarantee 90% of the hosts have lower RE values than it. In all these five data sets, Phoenix performs the best. The performance of Phoenix(Simple) is close to IDES(SVD) and IDES(NMF) because weighted model hasn't been applied on it. But it's still an improvement over IDES since it will never give negative predicted distance. Compared with Vivaldi, the representative Euclidean distance based NC, Phoenix can reduce the NPRE by between 18.34% (P2PSim data set) and 52.17% (AMP data set). Our simulation results demonstrate that Phoenix can achieve high prediction accuracy in a decentralized and practical way.

### 4.3 Convergence Behavior of Phoenix

In this subsection, we study Phoenix in terms of the number of rounds (samples) required for convergence under a flash-crowd scenario, i.e., all hosts join simultaneously. We define *median prediction error* as $median_{i,j}(\|D^E(i,j) - D(i,j)\|)$. As in [19], we plots the median prediction error as a function of NC update rounds used per host in Fig.4(a). We can see in all the five data sets, the convergence of Phoenix is fast according to our simulation results. Basically, Phoenix will converge in less than 10 rounds.

We compare Phoenix with Vivaldi in terms of the number of samples required for convergence. For a Vivaldi host, in each update round it probes one of its neighbors and retrieves the NC of this neighbor. We regard this process as one sample. For a Phoenix host, it has 32 neighbors and each neighbor has incoming

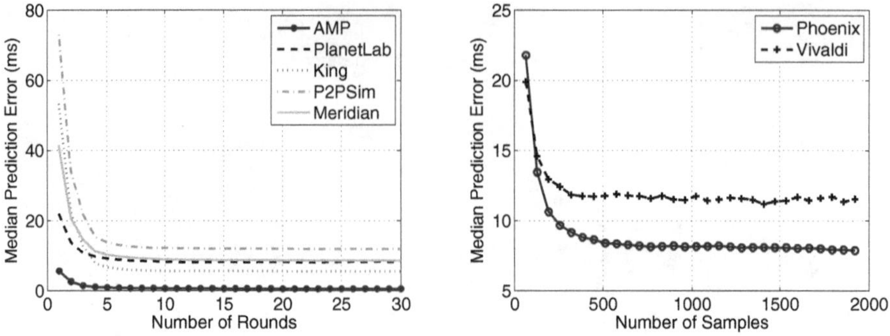


(a) Median Prediction Error

(b) Phoenix versus Vivaldi

**Fig. 3.** Convergence Behavior of Phoenix

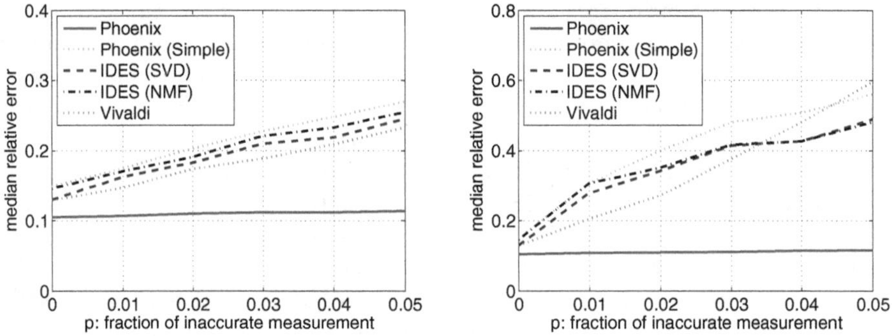


(a) $\lambda = 3$

(b) $\lambda = 10$

**Fig. 4.** The Impact on Measurement Anomalies

vector and outgoing vector, thus each update round needs 64 samples. We have done the comparison on all the five data sets and have drawn similar conclusions. Due to space limitation, we only show the results on PlanetLab data set. In Fig.4(b), we can see only in first round (first 64 samples), the median prediction error in Phoenix is 9% larger than that of Vivaldi. Thereafter, Phoenix converges very fast. The median prediction error in Phoenix needs less than 200 samples to converge to 12ms whereas in Vivaldi it needs about 300 samples. Also, we can find that the final median prediction error of Phoenix is about 31% smaller than Vivaldi. Thus the convergence of Phoenix is very fast and effective.

### 4.4 Robustness over Measurement Anomalies

In previous study, we assume that all the measurement results from the data sets are accurate. However, in the real world, measurement may contain various kinds of errors [9]. A robust NC system will still give accurate prediction under a small amount of measurement anomalies. The robustness of these systems are evaluated by the experiment proposed in [9]. For a certain data set, we randomly pick $p$ percent links of the data set and increase the RTTs of these links to $\lambda$ times of the original values. Then we run NC system on the modified data set. When calculating the RE after the simulation, we compare the predicted distances to the actual network distances (i.e. the original value without injected errors). We vary the $p$ value and see the evolution of median RE. As in [9], $\lambda$ value is set as 3 and 10.

We have done the experiment on all the five data sets and have drawn similar conclusions. Due to space limitation, we only show the results on PlanetLab data set in Fig. 4. In IDES, Vivaldi and Phoenix(Simple) system, the amount of median RE increases when the amount of inaccurately measured links increases. The higher the degree of measurement error $\lambda$ is, the faster the median RE increases along with $p$. In contrast, the results in Phoenix are quite different, while median RE only increases slightly along with $p$. Thus Phoenix is very robust to small amount of measurements anomalies. The difference between Phoenix and Phoenix(Simple) demonstrates that the weighted model can eliminate the impact of measurement anomalies greatly.

## 5 Conclusion and Future Work

In this paper we proposed the design and implementation of a decentralized dot product based NC system, Phoenix. Phoenix employs a weighted NC calculation model to reduce the effect of error propagation and it's a practical system which never gives negative predicted distance. Extensive simulation results with real Internet traces show that Phoenix achieves much higher prediction accuracy than state-of-the-art NC systems in different typical Internet data sets. We have also demonstrated that the convergence of Phoenix is fast and it performs robustly over small amount of measurement anomalies. Compared with IDES, Phoenix not only has better prediction accuracy but also can guarantee all the predicted distances non-negative. In short, Phoenix is an accurate, practical and decentralized solution to scalable Internet distance prediction.

Our future work of Phoenix is wide-area deployment. We will improve Phoenix in large scale Internet experiments and we believe Phoenix will be a robust, accurate and widely-used NC system.

## Acknowledgment

This work is supported by the National Science Foundation of China (No.60473087, No.60703052, No.60850003) and National Basic Research Program of China (No.2007CB310806). Thanks to Mr. Zengbin Zhang from Tsinghua University for his comments and suggestions.

## References

1. Zhang, R.M., Tang, C.Q., Hu, Y.C., et al.: Impact of the Inaccuracy of Distance Prediction Algorithms on Internet Applications: an Analytical and Comparative Study. In: Proc. of IEEE INFOCOM (2006)
2. Pietzuch, P., Ledlie, J.: Network-aware operator placement for stream-processing systems. In: Proc. of ICDE (2006)
3. Azureus bittorrent, `http://azureus.sourceforge.net/`
4. Zhang, B., Ng, T.S.E., Nandi, A., et al.: Measurement-Based Analysis, Modeling, and Synthesis of the Internet Delay Space. In: Proc. of ACM IMC (2006)
5. Abraham, I., Malkhi, D.: Compact routing on euclidian metrics. In: Proc. of ACM PODC (2004)
6. Wang, G., Chen, Y., Shi, L., Lua, E.K., Deng, B.X., Li, X.: Proxima: Towards Lightweight and Flexible Anycast Service. In: Proc. of IEEE INFOCOM Student Workshop (2009)
7. Lee, S., Zhang, Z., Sahu, S., et al.: On Suitability of Euclidean Embedding of Internet Hosts. In: Proc. of ACM SIGMetrics/Performance (2006)
8. Wang, G., Zhang, B., Ng, T.S.E.: Towards Network Triangle Inequality Violation Aware Distributed Systems. In: Proc. of ACM IMC (2007)
9. Mao, Y., Saul, L., Smith, J.M.: IDES: An Internet Distance Estimation Service for Large Network. IEEE Journal on Selected Areas in Communications, JSAC (2006)
10. Lua, E.K., Griffin, T., Pias, M., et al.: On the Accuracy of Embeddings for Internet Coordinate Systems. In: Proc. of ACM IMC (2005)
11. Lee, D.D., Seung, H.S.: Learning the parts of objects by non-negative matrix factorization. Nature 401(6755), 788–791 (1999)
12. Pias, M., Crowcroft, J., Wilbur, S., et al.: Lighthouses for Scalable Distributed Location. In: Proc. of IPTPS (2003)
13. Zhang, R., Hu, Y.C., Lin, X., et al.: A Hierarchical Approach to Internet Distance Prediction. In: Proc. of IEEE ICDCS (2006)
14. Tang, L., Crovella, M.: Virtual Landmarks for the Internet. In: Proc. of ACM IMC (2003)
15. Tang, L., Crovella, M.: Geometric Exploration of the Landmark Selection Problem. In: Barakat, C., Pratt, I. (eds.) PAM 2004. LNCS, vol. 3015, pp. 63–72. Springer, Heidelberg (2004)
16. Chen, Y., Xiong, Y., Shi, X., et al.: Pharos: Accurate and Decentralised Network Coordinate System. IET Communications (to appear)
17. Chen, Y., Zhao, G., Li, A., et al.: Handling Node Churn in Decentralised Network Coordinate System. IET Communications (to appear)
18. Ng, T.S.E., Zhang, H.: Predicting Internet Network Distance with Coordinates-Based Approaches. In: Proc. of IEEE INFOCOM (2002)
19. Dabek, F., Cox, R., Kaashoek, F., Morris, R.: Vivaldi: A Decentralized Network Coordinate System. In: Proc. of ACM SIGCOMM (2004)
20. Lawson, C.L., Hanson, R.J. (eds.): Solving Least Squares Problems, ch. 23, p. 161. Prentice-Hall, Englewood Cliffs (1974)
21. Lehman, L., Lerman, S.: A Decentralized Network Coordinate System for Robust Internet Distance Prediction. In: Proc. of ITNG (2006)
22. Ledlie, J., Gardner, P., Seltzer, M.: Network Coordinates in the Wild. In: Proc. of NSDI (2007)
23. PlanetLab, `http://www.planet-lab.org/` (accessed, November 2008)

24. Shavitt, Y., Tankel, T.: Big-Bang Simulation for Embedding Network Distances in Euclidean Space. In: Proc. of INFOCOM (2003)
25. Zheng, H., Lua, E.K., Pias, M., et al.: Internet Routing Policies and Round-Trip-Times. In: Proc. of the Passive Active Measurement Workshop (2005)
26. Ng, T.S.E., Zhang, H.: A Network Positioning System for the Internet. In: Proc. of USENIX Annual Technical Conference (2004)
27. Wong, B., Slivkins, A., Sirer, E.G.: Meridian: A Lightweight Network Location Service without Virtual Coordinates. In: Proc. of ACM SIGCOMM (2005)
28. Rhea, S., Geels, D., Roscoe, T., Kubiatowicz, J.: Handling Churn in a DHT. In: Proc. of the USENIX Annual Technical Conference (2004)
29. IDES Simulator, http://www.research.att.com/~maoy/ides-0.1.tar.gz
30. Lumezanu, C., Levin, D., Spring, N.: PeerWise Discovery and Negotiation of Faster Paths. In: Proc. of HotNets (2007)
31. Costa, M., Castro, M., Rowstron, A., et al.: PIC: Practical Internet Coordinates for Distance Estimation. In: Proc. of IEEE ICDCS (2004)
32. Lua, E.K., Griffin, T.: Embeddable Overlay Networks. In: Proc. of IEEE ISCC (2007)
33. Phoenix Simulator, http://www.net-glyph.org/~chenyang/Phoenix-sim.zip

# Topology Dynamics in a P2PTV Network

Siyu Tang, Yue Lu, Javier Martín Hernández, Fernando Kuipers, and Piet Van Mieghem

Delft University of Technology, P.O. Box 5031, 2600 GA Delft, The Netherlands
{S.Tang,Y.Lu,J.MartinHernandez,F.A.Kuipers,P.F.A.VanMieghem}@tudelft.nl

**Abstract.** In recent years, a number of commercial peer-to-peer TV (P2PTV) applications have been launched. Yet, their mechanisms and characteristics are unknown. In this paper, we study SopCast, a typical proprietary P2PTV system. Treating SopCast as a black box, we perform a set of experiments that are suitable to analyze SopCast in depth. We attempt to disclose the SopCast protocol. The dynamic nature of the SopCast overlay, in terms of node degree, is also addressed in this paper. Our approaches in analyzing the SopCast mechanism and characterizing its topological properties reveal important design insights in SopCast, and may help to better understand similar P2PTV systems.

**Keywords:** Topology dynamics, SopCast, P2PTV.

## 1 Introduction

The success of peer-to-peer (P2P) file-sharing systems has spurred the deployment of P2P technologies in many other bandwidth-intensive large-scale applications. Peer-to-Peer Television (P2PTV) has become a popular means of streaming audio and video content over the Internet. Example applications are CoolStreaming [8], TVAnts[1], TVU[2], SopCast[3], etc. It is important to evaluate the traffic impact of such applications, while modeling their behavior. However, P2PTV streaming systems, such as SopCast, are developed for commercial purposes: thus, very little is known about their architectures. Some papers claim that SopCast is based on similar principles as those underlying CoolStreaming, e.g. [2], some refer to it as a BitTorrent-based P2PTV system, e.g. [1], but all without substantiating their claims. Furthermore, SopCast traffic is encoded, which makes understanding the protocol even more challenging. In this paper, we will investigate the SopCast P2PTV system by answering the following two questions. *What is the operational mechanism in SopCast? How are topology dynamics reflected in the SopCast overlay?*

The rest of the paper is organized as follows. In Section 2, related work is discussed. Section 3 describes the measurement settings in PlanetLab[4] network.

[1] http://tvants.en.softonic.com/
[2] http://www.tvunetworks.com/
[3] http://www.sopcast.org/
[4] http://www.planet-lab.org/

L. Fratta et al. (Eds.): NETWORKING 2009, LNCS 5550, pp. 326–337, 2009.

Based on the experiments conducted in Section 3, we present our understanding of the SopCast protocol in Section 4. Our methodology of modeling the time-variant SopCast overlay and the derived results are provided in Section 5. In Section 6, we conclude the paper.

## 2 Related Measurement Studies Performed with SopCast

Ali *et al.* [6] evaluated the performance of both PPLive[5] and SopCast. They conducted experiments on a single host joining a system. The systems were run under different environments and the collected data was further analyzed to give insight into SopCast's operation. Silverston *et al.* [7] analyzed the different traffic patterns and underlying mechanisms of several P2PTV applications. Their study is based on a single measurement day. The dataset collected in this paper is from two personal computers that are directly connected to the Internet from their campus network. Sentinelli *et al.* [1] performed two sets of experiments on SopCast. There were in total 27 peers connected to a popular SopCast channel in the first experiment, while the second experiment was run on the PlanetLab network with 1 node acting as the source and 10 nodes performing as peers.

Most of the previous work is executed from a single point of observation [7], or from a small number of vantage points [1], [6]. Thus, results from these papers cannot accurately reflect the performance of the entire SopCast network. Furthermore, characterizing dynamic peer-to-peer networks has been considered a research field with a lot of contentions. The major dispute is on the accuracy of the captured topology snapshots in reflecting the actual peer-to-peer overlay. Stutzbach *et al.* [3] quantified the effects of the crawling duration and the ratio of unreachable peers on deriving network characterizations.

In this paper we study the SopCast protocol and provide our solutions to evaluating the topological properties (with respect to the degree distribution) and dynamics in a P2PTV network.

## 3 Experimental Settings

We have used PlanetLab for our experiments, because it allows us to evaluate the entire constructed overlay. The experiment consists of two types of nodes.

A standard personal computer located in our campus network, which acts as the *source provider*[6] (SP). With the SP, we registered a dedicated channel to the SopCast network. In this channel, a small cartoon movie with a duration of 2 minutes and size of 3.5 MBytes is continuously broadcast in a loop. Thus, our experiment resembles a streaming system. The SP runs Windows XP. It is equipped with an Intel Pentium 2.4 GHz processor, 512 MB RAM and a 10/100 FastEthernet network interface, which is further connected through a router

[5] http://www.pplive.com/

[6] The source provider is the node who broadcasts the entire video by using SopCast software..

to the Internet. The second type of nodes are PlanetLab nodes that act as SopCast peers viewing the TV channel released by us. Each of the 51 PlanetLab nodes under consideration runs the following software: (1) SopCast Client (Linux version), with command line control; (2) Tcpdump[7] to enable passive monitoring on the traffic transmitted at the SopCast peers; and (3) Perl[8] Scripts to remotely control the PlanetLab nodes.

We conducted a single experiment on 11:00 am, August $22^{nd}$, 2008. The experiment lasted for roughly 40 minutes. The 51 PlanetLab nodes were controlled in such a way that they joined and left the network simultaneously. We collected the traffic log files captured by tcpdump from all the 51 peers. Traffic collection at the SP is accomplished with Ethereal [4]. The collected trace files from the 51 PlanetLab nodes are further processed and analyzed with AWK[9] scripts.

## 4 Dissecting the SopCast Protocol

SopCast is a proprietary P2PTV application and consequently the SopCast website only provides limited information about SopCast's video delivery mechanism. Although the work presented in this paper has been conducted in a thorough and careful way, without having the source code of SopCast, the claims that we have made are based on our investigation of SopCast. They are not the exact description of the protocol. In this paper, we only present our conclusions on the peer communication scheme and video delivery rule in Sopcast, because they play important roles in determining the topological properties in SopCast.

### 4.1 Identification of SopCast Packets

Tcpdump reveals that SopCast relies on UDP. Since we are not able to decode the SopCast traffic, it is not possible to tell exactly what kind of messages are being exchanged in the captured trace files. To figure out the packet functionalities, we studied the packet lengths and the corresponding delivery patterns. Table 1 presents our findings.

**Table 1.** Summary of SopCast traffic

| Type | Size (bytes) | Functionality |
|---|---|---|
| Video packet | 1320 | Maximum size of the video packets |
| | 377, 497, 617, 1081, 1201 | Video fragments |
| Control packet | 52 | HELLO packet to initiate link connections |
| | 80 | Confirmation on receiving the HELLO packet |
| | 28 | Acknowledgement |
| | 42 | Keep-alive message with neighbors |
| | 46 | Video requesting packet |

[7] http://www.tcpdump.org/
[8] http://www.perl.org/
[9] AWK is a general programming language that is designed for processing text-based data, either in files or data streams.

### 4.2 Neighbor Communication in SopCast

Assuming that a SopCast peer has retrieved a random list of peers (a *peerlist*) in the network, it will start to choose some peers with whom connections are established. If two peers exchange a 42-byte control packets with each other, we refer to these peers as being *neighbors.*

Communication between two peers in SopCast is always initiated by a $52-80$ byte packets pair. Once the connection is established, the pair of peers keeps exchanging a sequence of 42-byte packets with each other. The 42-byte packets are transmitted with a high frequency, roughly every second. We denote this packet as a *keep-alive* packet. With the keep-alive packets, a peer announces its existence in the network. The purpose is to maintain neighbor relation with others via the decentralized communication between peers. Peers can lose neighbors. In case a neighbor does not respond to the keep-alive packets, the peer stops contacting this neighbor until it chooses the neighbor again. A graphic illustration of the neighbor communication scheme is shown in Fig. 1(a).

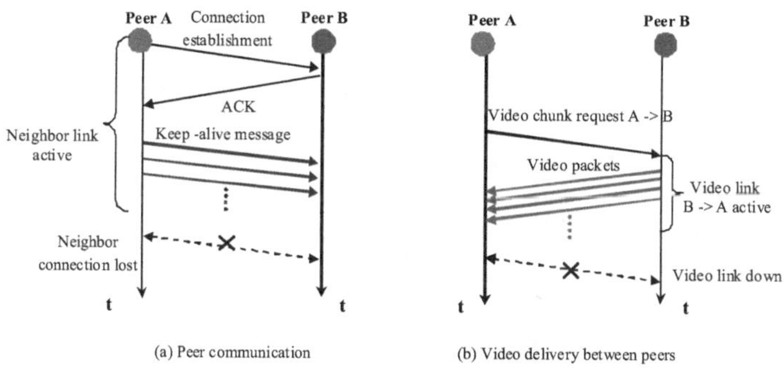


**Fig. 1.** Diagram of (a) neighbor communication and (b) video delivery in SopCast

### 4.3 Video Delivery in SopCast

Video delivery in SopCast is chunk-based. The TV content in SopCast is divided into *video chunks* or *blocks* with equal sizes of 10 kbyte. This finding is in line with the video chopping algorithm implemented in many existing P2P systems, although the chunk size may be different[10]. If a peer is providing video packets to its neighbors, we refer to the peer as a *parent.* A *child* is defined if a peer is receiving video packets. A peer is allowed to have multiple parents and multiple children. A parent-child relation can be established only when two peers are neighbors. A peer is free to request multiple video blocks from its parents.

A peer in SopCast never voluntarily delivers video streams to its neighbors. To download video packets, a child always needs to request them from its parent(s) via a *video request packet* with the size of 46 bytes, see Fig. 1(b). After receiving the request, its parent(s) will deliver a series of video packets to the child. In case

[10] In Bittorrent, the default size of the chopped block is 256 kbytes.

the child needs more blocks, it sends another request. The requested blocks are treated as a large datagram during transmission. Due to the IP fragmentation principle, a large datagram such as 10 kbytes should be segmented into smaller pieces in order to pass over the Ethernet. SopCast sets the maximum size of the video packet to be transmitted to 1320 bytes. A generalization of the video block fragmentation rule is

$$n \times 10 \ kbyte = x \times 1320 \ bytes + y \tag{1}$$

where $n$ is the number of requested video blocks, $x$ is the number of 1320-byte video packets, and $y$ is the size of the smaller fragment in bytes (377, 497, etc.), as presented in Table 1.

# 5 Topological Properties of SopCast

## 5.1 A Two-Layer SopCast Overlay

The topology of the SopCast overlay network can be generated with peers as nodes and connections as links. The SopCast overlay is constructed with decentralized peers. We study the SopCast overlay as a two-layer architecture consisting of the *neighbor graph* $G_N$ and the *video graph* $G_V$. Both graphs are formed with directed links. We define a peer as a node that is active in both downloading and uploading processes. The SP is not included in the two layers, because it only supports a number of children with video downloading. Notice that not all neighbors will share video streams with each other. A straightforward relation between the two layers is $G_V \subseteq G_N$.

The neighbor graph is formed by a set of nodes with neighbor relations. A link in the graph is denoted as a *neighbor link*. A neighbor link is established if a node pair is regularly sending keep-alive messages. If the keep-alive messages are not observed for some time, the link is considered to be inactive. The outgoing degree $D_{out}$ of $G_N$ defines the number of neighbors that a random peer has in the network. The video graph consists of peers that are transmitting video blocks. The incoming degree $D_{in}$ of the video graph indicates the number of parents that a random peer has. The outgoing degree determines the number of children that an arbitrary peer supports. A *video link* is activated if there are video packets being transmitted from the parent to the child.

## 5.2 Reflecting the Dynamic Overlay

In a high churning network, such as SopCast, we are not interested in taking *instantaneous* snapshots. Because an instantaneous snapshot is taken at an exact time point (in the order of microseconds), thus capturing only a few nodes and links. In this paper, we study the network dynamics by continuously taking network snapshots with the *duration* $\tau$ as time evolves and show them as a time series. A *snapshot* captures all participating peers and their connections within a particular time interval, from which a graph can be generated. The snapshot duration may have minor effects on analyzing slowly changing networks, such as Internet on Autonomous System (AS) level topologies. However, in a P2P streaming system, the characteristics of the network topology vary greatly with respect to the time scale of the snapshot duration, as addressed in [3].

We define an active (neighbor or video) link if two peers in SopCast are continuously transmitting keep-alive messages, or video packets. Often, we could notice temporary ceasing of the connection between two peers. After a period which ranges from several seconds to hundred seconds, they contact each other again. If the time period that two peers stop communication is in the order of hundred seconds, we can confirm a connection closure on the (neighbor or video) link. Whereas, due to transmission, or processing delay, or network congestion, a peer may send the keep-alive packets, or video packets with longer delay. Hence, such a link should be considered as active even though packet delivery is not observed for some small time.

To define the activation period of a neighbor link in SopCast, we have processed the traces and obtained the time interval between two consecutive keep-alive messages between all node pairs. From the parsed dataset, the corresponding probability density function (pdf) is plotted in Fig. 2 (left figure). This statistical analysis suggests that, with high probability (more than 95% of the cases), a peer sends two consecutive keep-alive packets to its neighbors within 2.5 seconds. Thus, we consider the threshold of $\tau_N \sim 2.5$ s as the activation period of a neighbor link. If two peers have not contacted each other within this interval, termination of a link connection is assumed. With the same approach, the pdf of the time interval between two consecutive video requests for all node pairs is plotted. Indicated in Fig. 2 (right figure), the activation period of a video link is around $\tau_V \sim 2.0$ s. If a child $A$ requests video blocks from its parent $B$ within 2.0 s from the previous request, the video link is considered to be active.

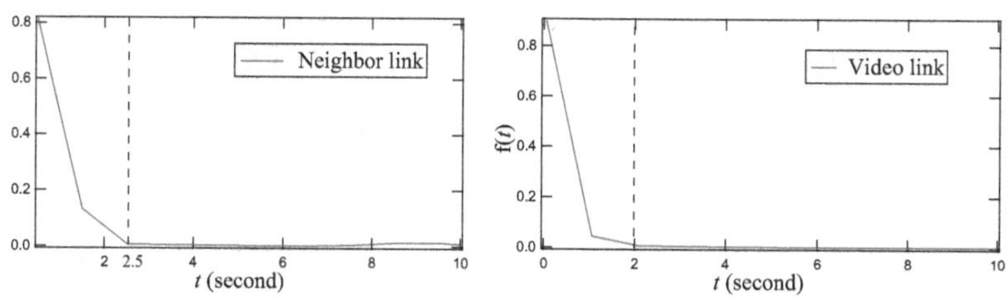


**Fig. 2.** Pdf of the activation period of the neighbor link (left) and the video link (right). Threshold of activation period for the neighbor link is $\tau_N \sim 2.5$ s, and$\tau_V \sim 2.0$ s for the video link.

## 5.3 Topology Evolution of the Neighbor Graph

SopCast deploys certain principles to discover neighbors by exchanging peerlists. A peer is consequently expected to be able to contact, at least a portion of the neighboring peers in the network. We take snapshots of the neighbor graph

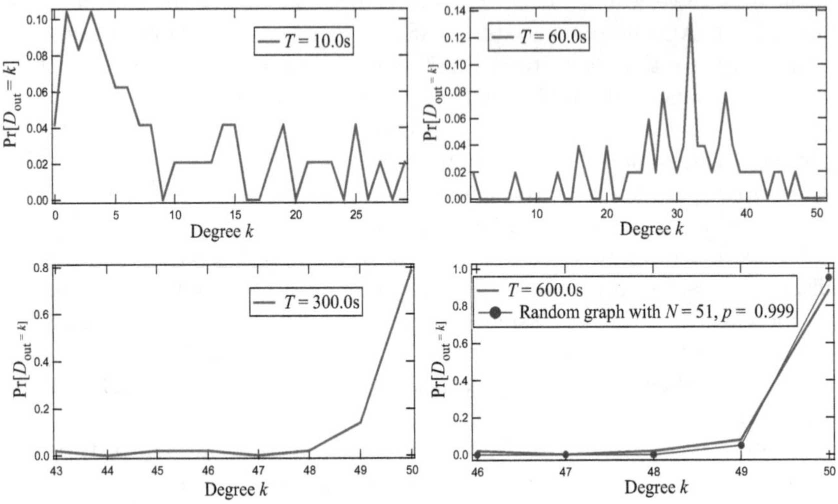


**Fig. 3.** Topology evolution of the neighbor graph as a function of time series. The outgoing degree defines the number of neighboring connections a node once maintained. In the bottom-right sub-figure, $Pr[D_{out} = k]$ is fitted with the degree distribution of a random graph with link density $p = 0.999$.

during the time interval of $[0, T]$, with $T = 10$, 60, 300, 600 seconds respectively, so that all the participating peers and links that have once been active from the beginning of the experiment till the time $T$ have been captured. All the historical neighboring connections are accumulated in these snapshots.

With the captured snapshots, four directed graphs are obtained. Multiple lines between two peers are removed. We plot the pdf of the outgoing degree ($D_{out}$) of the four graphs in Fig. 3. We notice that peers discover their neighbors quite fast in SopCast. After 300 seconds, the neighbor topology already converges to a full mesh, which means that SopCast peers have the ability to find almost all the other neighbors within a relatively short period of time. The distinct property of neighbor discovery is better illustrated in Fig. 4, in which the evolution of the average outgoing degree of all peers is plotted as a function of time. The average outgoing degree grows rapidly during the first 5 minutes of the experiment[11]. The standard deviation of the average outgoing degree is also shown. In the first several minutes, peers tend to behave differently. Some of them contacted with more neighbors, while some only meet a few. This is the reason that a large standard deviation appears at the beginning of the experiment. As time evolves, the activity of neighbor discovery gradually converges at different peers. After 5 minutes, the standard deviation has reduced substantially.

The activation period of a neighbor link was found to be 2.5 s in the previous subsection. By taking network snapshots with this interval, we could obtain some

[11] Compared to the entire duration of the experiment (40 minutes), this period can be considered to be short.

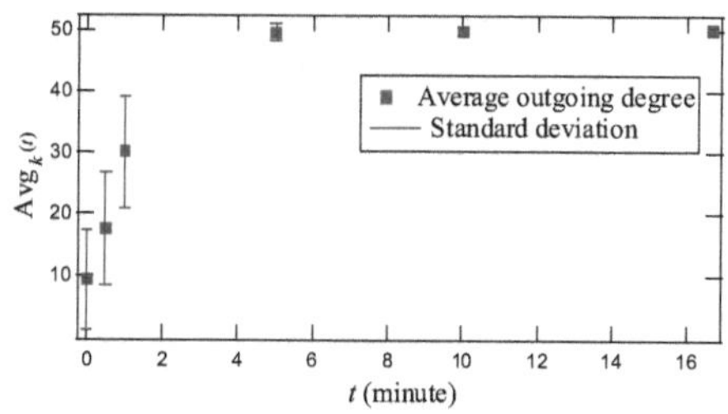


**Fig. 4.** The average outgoing degree of the neighbor graphs as a function of time

insights on the number of neighbors that a peer communicates during this period. The neighbor graph is examined during the time interval of [5 min, 35 min]. This is because, in the first 5 minutes, peers are trying to discover more neighbors, thus the peerlist may not be completely filled in. The last 5 minutes are also excluded, because two PlanetLab nodes came off-line unexpectedly. By taking snapshots every consecutive 2.5 s, 720 network topologies are obtained, which generate 28050 samples of the outgoing degree in total. The derived pdf of the outgoing degree is well fitted with a Gaussian distribution. The average number of contacted neighbors within 2.5 seconds is around 37.73. We also extended the snapshot duration to 10 seconds as a comparison. The average outgoing degree increases slightly to 41.60. This observation reveals relatively stable connections between peers. If a peer frequently removes its neighbors in the peerlist and substitutes them with new ones, the average outgoing degree of the neighbor graph would be larger than 41.60 in the longer snapshot interval of 10 s.

### 5.4 Topology Dynamics of the Video Graph

The SopCast protocol specifies a set of streaming delivery rules that are used to guide peers to actively connect to others, or allow other peers to establish connections to themselves. We aim to understand how SopCast coordinates peers to share video streams from the perspective of a network topology.

We study the video graph when peers have chances to contact with almost every neighbor in the network, e.g. after 5 minutes of the experiment in Fig. 3, because a peer is expected to keep the best performing neighbors in its peerlist and the network is considered to be more stable from this time. Hence, the topology of the video graph is also examined during the time interval of [5 min, 35 min].

Recall that the activation period of a video link was found to be 2.0 seconds. We divide the time period of [5 min, 35 min] to sequential time slots of 2.0 seconds, resulting in 900 network snapshots of $G_V$. The 900 snapshots provide us with 32320 samples of the incoming degree ($D_{in}$), and 27697 samples of the outgoing degree ($D_{out}$). The incoming and outgoing degree distributions are obtained by making the histogram of the 32320 in-degree samples and 27697 out-degree samples respectively. In Fig. 5, SopCast peers only retrieve video packets from, on average, 1.86 neighbors during 2.0 seconds. In rare cases, a node may

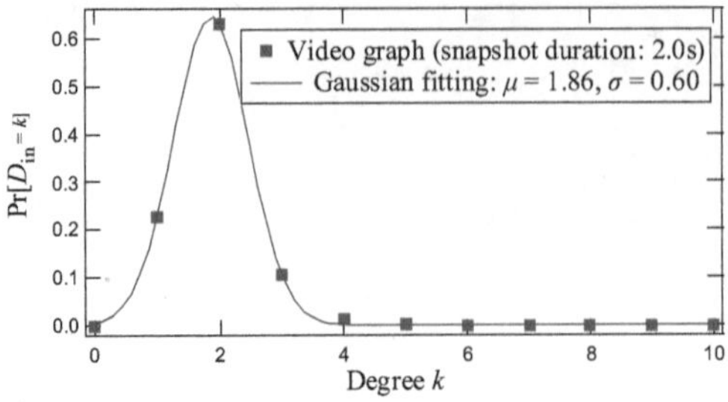


**Fig. 5.** Pdf of the incoming degree (number of parents) of the video graph on a linear scale. Snapshot duration of the video graph is set to 2.0 s. A Gaussian fitting describes the pdf of $D_{in}$ well, which suggests us to model the incoming degree of $G_V$ as a random graph with very low link density $p \ll 1.0$.

contact more than 10 peers to download streams. The observation suggests that there might be a threshold $\gamma$ of the maximum number of parents that a node is allowed to connect with, most likely with $\gamma = 4$.

The outgoing degree distribution of the video graph shows different behaviors. A supernode structure emerges when looking at the curve plotted on a log-log scale in Fig. 6 (left). A *supernode* in a P2P network refers to a node that can offer good uploading bitrates and establishes many connections with different children. It seems that SopCast peers are allowed to select several parents to establish downloading connections, while they will only choose the ones that provide better uploading performance, which leads to the existence of supernodes. In this section, we only inspect the supernode structure in terms of node degree. In Section 5.5, we will show the supernode structure from the perspective of uploading throughput. Interestingly, $\Pr[D_{out} = k]$ on a log-lin scale also fits well with a line curve. A similar observation was found by Janic *et al.* [5]. We think the network size is not large enough to warrant a definite conclusion. However, with the PlanetLab network, a large networking scale, e.g. $N \gg 100$ was not possible during the period in which we performed the experiments. We have carried out another experiment with 132 PlanetLab nodes, which gave us similar results as in Fig. 6.

In the following, we will substantiate our statement on the importance of the snapshot duration for reflecting the network properties. By increasing the snapshot duration to 10 s, a dominating power-law distribution of the outgoing degree is still observed. When enlarging the time scale of the snapshot duration to 100 s, the power-law behavior of $D_{out}$ is replaced by a Gaussian distribution. The discrepancy of the distribution in $D_{out}$ with snapshot duration of 100 s is mainly caused by the fact that a peer may support different children over time and these changes are accumulated in the captured snapshots. The divergency observed with different snapshot durations suggests us to model the video graph with respect to different time scales that are used to capture network snapshots. Besides, the accuracy of the obtained snapshots should be discussed carefully.

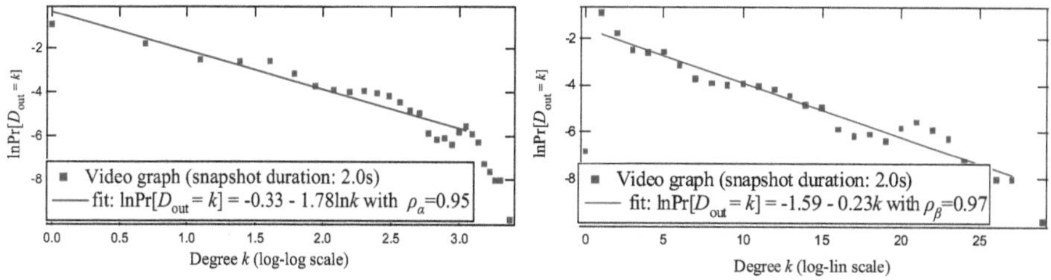


**Fig. 6.** The outgoing degree (number of children) distribution of the $G_V$ on both log-log scale (left) and log-lin scale (right). The linear correlation coefficients $\rho_\alpha$ on the log-log scale and $\rho_\beta$ on the log-lin scale are used to reflect the fitting quality.

In our case, we consider a snapshot duration of 2.0 seconds as the accurate parameter to characterize the essential properties of the video graph.

### 5.5 SopCast Network Load Distribution

A critical issue of a peer-to-peer streaming system is to fairly distribute the network load to peers participating in the video delivery process. It is very likely that the so called tit-for-tat principle is not implemented in SopCast. A peer can request video packets from one of its neighbors, without offering any video blocks to this neighbor. For instance, we have noticed constant uploading from PlanetLab node *freedom.informatik.rwth-aachen.de* to *planetlab1.pop-mg.rnp.br*, but no uploading activity in the reverse direction.

In Fig. 7, we present the total downloading and uploading throughput at every single peer. Peers tend to have uniform downloading throughput, as the black dotted line shows. However, they behave differently regarding their uploading performance. There are several nodes downloading much less than others, which may be caused by bandwidth limitations[12] at these nodes.

In SopCast, a node can download video streams either from the SP, or from other peers. We noticed that SopCast has limited the maximum number of children connected to the SP simultaneously to 11. The downloading throughput from decentralized peers and the SP at every single node are also presented in Fig. 7. We notice that the SP in SopCast does not change its children very often. During the entire experiment, only 15 nodes downloaded video blocks from the SP. These 15 peers act as the major force (supernodes) to provide video blocks to other peers and contribute the most in terms of uploading throughput. The burden of the SP is consequently alleviated.

We have indicated the existence of supernodes in terms of node degree and uploading throughput. We also found a strong correlation between the outgoing degree and the uploading throughput of a supernode, namely, a peer that establishes many connections with its children uploads more video blocks. Our observations suggest a hierarchical structure of the video graph. Although the

[12] Bandwith of a PlanetLab node can be limited by PlanetLab administrator.

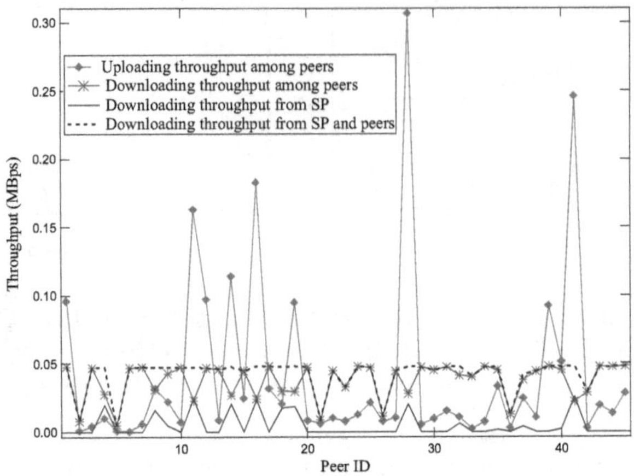


**Fig. 7.** Uploading and downloading throughput at every SopCast node during the entire experiment (40 minutes). The $x$-axis presents the identification number (ID) of each individual peer. The uploading throughput (downloading throughput) at a single peer is calculated based on the amount of data uploads (downloads) to all the children (parents) the peer has.

SP performs a critical role in broadcasting the video streams, it is not the major resource from which peers can download video blocks. The peers that are directly connected to the SP act as the supernodes in SopCast. They support many children, and contribute to the largest uploading bandwidth in the network. The other peers download video blocks from the supernodes.

## 6 Conclusions

In this paper, we have performed a series of experiments in a controlled environment with a set of PlanetLab nodes running the SopCast application. Our aim was to understand the mechanisms in SopCast and further to characterize the topological properties of this dynamic P2PTV overlay. Via passive measurements, we discovered the basic mechanism deployed in the SopCast protocol and modeled the SopCast network with a two-layer architecture.

Based on our measurements and analysis, our main conclusions are: 1) SopCast traffic can be categorized in control packets and video packets. The control packets fulfil different functionalities, which coordinate the operation of the SopCast application. The video packets deliver the TV content. 2) Communication between SopCast peers is decentralized. Peers discover their neighbors very fast. The constructed neighbor graph is considered to be resilient, since it is close to a full mesh. 3) Video delivery in SopCast is chunk-based. Each video chunk has equal length of 10 kbytes. A peer is free to request multiple blocks from its parent(s). 4) The incoming degree distribution of the video graph can be modeled as a random graph with very low link density $p \ll 1.0$. While a potential power law behavior is observed with the outgoing degree distribution of the video graph. A node with good uploading performance can act as a supernode in SopCast. A supernode sacrifices a large amount of upload bandwidth to its many children. 5) An adequate snapshot duration is important in understanding the actual topology changes of P2PTV networks. In SopCast, we have found $\tau_N \sim 2.5$ s, and $\tau_V \sim 2.0$ s as the reference snapshot duration in the neighbor graph and video graph respectively.

## References


1. Sentinelli, A., Marfia, G., Gerla, M., Kleinrock, L., Tewari, S.: Will IPTV Ride the Peer-to-Peer Stream. IEEE Communications Magazine 45(6), 86–92 (2007)
2. Purandare, D., Guha, R.: An Alliance based Peering Scheme for Peer-to-Peer Media Streaming. In: Proc. of the Workshop on Peer-to-Peer Streaming and IP-TV, in conjunction with ACM SIGCOMM, Kyoto, Japan, August 27-31 (2007)
3. Stutzbach, D., Rejaie, R., Sen, S.: Characterizing Unstructured Overlay Topologies in Modern P2P File-Sharing Systems. IEEE/ACM Transactions on Networking 16(2) (April 2008)
4. Orebaugh, A., Ramirez, G., Burke, J., Pesce, L.: Wireshark & ethereal network protocol analyzer toolkit (jay beale's open source security). Syngress Publishing (2006)
5. Janic, M., Van Mieghem, P.: On properties of multicast routing trees. International Journal of Communication Systems 19(1), 95–114 (2006)
6. Ali, S., Mathur, A., Zhang, H.: Measurement of Commercial Peer-to-Peer Live Video Streaming. In: Proc. of the Workshop In Recent Advances in Peer-to-Peer Streaming (August 2006)
7. Silverston, T., Fourmaux, O.: Measuring P2P IPTV Systems. In: Proc. of Network & Operating Systems Support for Digital Audio & Video (NOSSDAV 2007) (June 2007)
8. Zhang, X., Liu, J., Li, B., Yum, P.: DONet/Coolstreaming: A datadriven overlay network for live media streaming. In: Proc. of IEEE INFOCOM (March 2005)

# Collaboration in BitTorrent Systems

Rafit Izhak-Ratzin

Computer Science Department, University of California - Los Angeles
(310) 825-3886, 4732 Boelter Hall, Los Angeles, CA 90095
rafiti@cs.ucla.edu

**Abstract.** Recent research efforts have shown that the popular BitTorrent protocol does not strictly enforce fairness and allows free-riding, mainly via optimistic unchokes.

This paper proposes a BitTorrent-like protocol, that encourages peers of similar upload bandwidth to be *buddies*— peers collaborating for mutual benefit. Buddy peers mostly satisfy their download needs through their buddies and perform optimistic unchokes only when absolutely necessary. As a result, the buddy protocol improves fairness via explicit cooperation between buddies, and limits bandwidth spent on random optimistic unchokes, leading to a system more robust against free-riders.

We implemented the buddy protocol on top of an existing BitTorrent implementation and ran experiments on a controlled PlanetLab testbed to evaluate its impact. Our results show that the buddy protocol promotes fairness, discourages free-riding, and improves the robustness of the system as compared to regular BitTorrent. It also provides incentives to be adopted by all the peers in the system.

**Keywords:** BitTorrent, Buddies, fairness, incentives.

## 1 Introduction

The BitTorrent protocol [1], one of the most popular protocol for peer-to-peer content distribution, aims to provide fairness among peers by applying the tit-for-tat (TFT) unchoking mechanism to incentivize contribution and discourage free-riding. Despite this, previous studies [2,3,4] showed that BitTorrent suffers from unfairness, particularly for high capacity leechers. Other studies [5,6,7] showed that free-riding opportunity exists in BitTorrent. One of the main contributor to this lack of fairness is the *optimistic unchokes* mechanism, another unchoking mechanism that is used in BitTorrent. As this mechanism facilitates continuous discovery of better peers to interact with, it also facilitates upload imbalance, dynamically creates a major opportunity for peers to obtain data without uploading in exchange. However, it is also been shown to greatly contribute to the robustness of the protocol [5], making it hard to be replaced.

In this paper we propose an alternative mechanism that dynamically creates *buddies*—pairs of peers having similar upload bandwidth that collaborate for mutual benefit. The buddy relation is based on the upload history during the

L. Fratta et al. (Eds.): NETWORKING 2009, LNCS 5550, pp. 338–351, 2009.

relation existence, and not only on the last round of contribution as in regular BitTorrent's TFT mechanism. Peers in buddy mode aim to satisfy their download needs through their buddies and perform optimistic unchokes only if absolutely necessary. Hence, the buddy mechanism mostly obviates the need for optimistic unchokes, while at the same time improves fairness and system robustness.

We implemented our buddy protocol on top of an existing BitTorrent implementation. We ran experiments with different configurations on controlled PlanetLab testbed in order to evaluate the impact of the buddy protocol on different level of contributing leechers and free-riders in comparison to the regular BitTorrent. In the experiments, our buddy protocol exhibits the following properties:

1. It promotes fairness as high capacity leechers who suffer from the unfairness in regular BitTorrent improve their download rate, and low capacity leechers who used to benefit from this unfairness are slowed down.
2. It provides a clear incentive for all level of contributing leechers to adopt it, even to those low capacity leechers who are slowed down due to it.
3. It discourages free-riding by limiting the optimistic unchokes, in comparison to the regular BitTorrent, thus directly hurting free-riders performance.
4. It improves the system robustness as increasing free-riding has less impact on the contributors performance, as compared to the regular BitTorrent.

We devolop an analytical model that explains leechers interactions in the presence of the buddy protocol and justifies our design choices. Based on this model and game theory, we prove the existence of Nash Equilibrium when all the contributing leechers adopt the buddy protocol.

The rest of this paper is structured as follows. In section 2 we provide a brief description of BitTorrent. In section 3 we present the design of our buddy protocol, while in section 4 we describe the analytical model that has guided this design. Implementation details are discussed in Section 5, and then the results of our PlanetLab experiments are presented in Section 6. Finally, section 7 sets our approach in the context of related work, and section 8 concludes.

## 2 BitTorrent Overview

The BitTorrent protocol is a popular peer-to-peer content distribution protocol that has been demonstrated to scale well with many participating clients.

Before the distribution process, the content provider divides the data content into multiple *pieces* which further divided into multiple *subpieces*. It then creates a *metainfo file*, containing information necessary for initiating the download process. This information includes the address of the *tracker*, a centralized coordinator that facilitates peer discovery.

In order to join a *torrent*(or swarm)—a group of peers participating in download of a content—a client downloads the metainfo file. It then connects the tracker from which it receives a *peer set* of randomly selected peers currently transferring the content pieces. These might include both *leechers*, who are still in the downloading process, and *seeds*, who have the entire content and are providing it to others. The new client can then connect peers in this set and request pieces.

The decision to whom to upload data is made independently by the leecher via the *choking algorithm*, applying the *tit-for-tat* mechanism that gives preference to those leechers who upload data to the given leecher at the highest rate. More specifically, once every *rechoke period*, typically 10 seconds, a leecher checks the current download rates of all leechers in its peer set. Then it selects the fastest uploaders and only uploads to those, while choking the rest for the duration of the rechoke period. This unchoking mechanism is called *regular unchoke.* The number of unchoked peers (slots) depends on the implementation and might be fixed or changed dynamically as a function of the available upload bandwidth. A different unchoke strategy is followed by seeds since they do not need to download any pieces. The most common one is unchoke leechers in a round-robin aiming to distribute data uniformly. We used this strategy in our experiments.

Beside the regular unchokes, leechers reserve a portion of their available bandwidth for sending pieces to random peers, a so-called *optimistic unchoke* mechanism. This mechanism enables the continuous discovery of better partners, and it also enables newcomer leechers, to download some data and start exchanging pieces. Optimistic unchokes are rotated randomly among the peer set typically once every three rechoke periods allowing enough time for leechers to demonstrate cooperative behavior.

### 2.1 Motivation

In this section we look at some problems in BitTorrent that motivate our design. Figure 1 shows the impact of optimistic unchokes on the expected download rate as a function of the P2P observed bandwidth distribution given in [2]. We assume that one unchoke slot is used by the optimistic unchokes mechanism, and that the regular unchoke mechanism works perfectly, i.e. the download rate and the upload rate through regular unchoke are equal. Clearly, we can see that the sub linear behavior of the expected download rate leads to unfairness for high capacity leechers, where low capacity leechers benefit from this unfairness. Note that in the observed bandwidth distribution the majority of the leechers are low capacity leechers with 88% of the leechers having less than 300KB/s upload capacity. This result motivates us to look for a replacement for the optimistic unchoke mechanism.

As for the regular unchoke mechanism, the number of unchoke slots that a leecher uses for regular unchoke can be fixed or dynamic as a function of the leecher's ability to saturate its upload capacity. This again might lead to unfairness, since typically, in real torrents, the download capacity of a leecher might be greater than the upload capacity. Thus, high capacity leechers might upload in full capacity, but not be able to download as much due to upload constraints of the downloading leechers and limited number of unchoke slots.

In addition, the choking decision in based on leechers' upload rate from the last rechoke period only, ignoring longer history that is available.

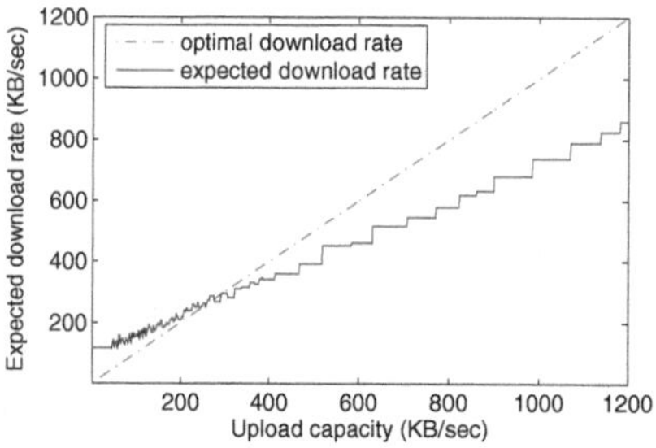


**Fig. 1.** Expected download rate vs. upload capacity for partial bandwidth spectrum

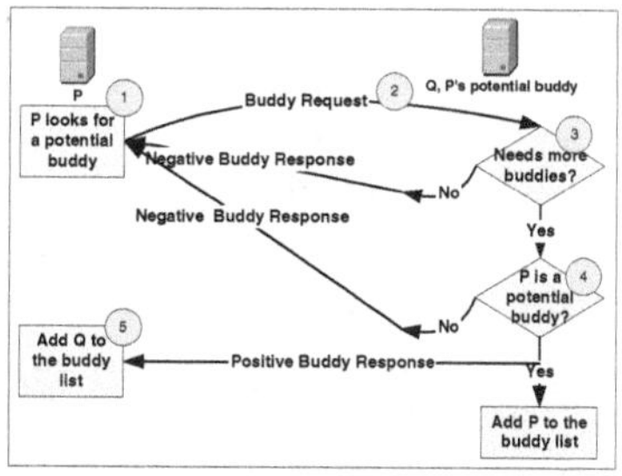


**Fig. 2.** Buddy formation mechanism

# 3 Design

We now turn to describe the design of our buddy protocol, which augments BitTorrent with the notion of *buddies*—pairs of peers having similar upload capacity, collaborating for mutual benefit. The main goals of the protocol are reducing the unfairness that BitTorrent suffers from, and discouraging free-riding. The protocol achieves these goals via explicit cooperation, as well as the tit-for-tat choking mechanism already employed by BitTorrent. In addition, it significantly limits the optimistic unchokes, where high capacity clients are forced to peer with those with low capacity while searching for better peers, an acknowledged opportunity for free-riding in BitTorrent [7].

The protocol comprises three main parts: the process by which buddy relations are formed, the process by which buddy relations are monitored, and the algorithm peers use to decide who to unchoke. We describe these in turn.

## 3.1 Buddy Formation

A leecher $P$ that is running in buddy mode, is willing to maximize the number of buddies, who make it their priority to serve $P$, as $P$ serves them in return. $P$ reserves an unchoked slot to each buddy it can upload data to in order to minimize buddy chokes, which can lead to termination of buddy relation. At the same time, $P$ reserves one unchoke slot for regular unchoke in order to avoid isolating group of buddies from the rest of the swarm. Therefore, $P$ can have maximum $n_s - 1$ buddies, given that $n_s$ is the number of unchoke slots in $P$.

Figure 2 shows the buddy formation process. $P$ will run this process once every rechoke period, only if it has not reached the maximum number of buddies.

**Step 1.** Aiming to find a potential buddy with similar upload as its own, $P$ will estimates the upload bandwidth of peers it interacted with, that are not its buddies, using its own history about past download interactions with these peers. If $P$ is not able to find a potential buddy $Q$ due to its upload capacity that is too high(low) compared to the leechers it has interacted with, it may

gradually increase(decrease) the number of unchoke slots $n_s$, while keeping maximum(minimum) $n_s$ size constraints satisfied.
**Step 2.** Once $P$ has found a potential buddy $Q$, $P$ sends a buddy request to $Q$, asking $Q$ to be its buddy.
**Step 3.** Upon receiving a buddy request from $P$, $Q$ checks that it has not reached the maximum number of buddies, otherwise it sends a negative buddy response.
**Step 4.** If $Q$ can have more buddies, it estimates $P$ upload capacity. If it finds that $P$ has similar upload capacity as its own, $Q$ sends a positive buddy response to $P$ and tags $P$ as buddy. Otherwise it sends a negative buddy response to $P$.
**Step 5.** Upon receiving a positive buddy response from $Q$, $P$ tags $Q$ as buddy.

In order to keep the design simple the look-up for buddies does not require a view of all the network peers, thus, we expect the improvement to be suboptimal.

### 3.2 Monitoring Buddies

Every rechoke period, a peer in buddy mode checks the upload rate of its buddies. It looks separately at the history of each buddy, since the buddy connection was established, and checks that the estimated upload rate of the buddy remains similar to its own. If this is not the case, the peer removes the buddy that uploads slower than expected from its buddy list. Thus the longer the connection the more reliable it is in comparison to the regular unchoke mechanism. A buddy can not gain much from cheating by uploading less than agreed, since it will be disconnected quickly. Thus, it will not be able to download much before being caught, and will lose an established buddy connection. Notice that a stronger punishment can be considered by keeping the cheating buddies' history and banning them from download or have a buddy relation for a time.

### 3.3 Unchoking Decisions

During the uploading process, a leecher decides periodically which peers to unchoke. In regular BitTorrent protocol, this unchoking decision is made using two separate mechanisms, the regular unchoke mechanisms and the optimistic unchoke mechanism. The former guides a leecher to unchoke those peers that upload data to the given peer at the highest rates, while the latter selects the unchoked peers at random, independently of their contributions.

Our buddy protocol extends the two given mechanisms with a third, *buddy unchoke* mechanism, which dictates that a leecher always unchokes all its buddies. The rationale is that a leecher that has buddies should opt for uploading to its buddies whenever possible, expecting its buddies to do the same. Hence, since buddies have similar upload rate, this guarantees similar upload as download rate. At the same time, it performs at least one regular unchoke to a peer that is not its buddy to avoid isolating group of buddies from the rest of the swarm. Uploading via optimistic unchokes is performed only as a last resort, and it may happen only when a leecher has no data other buddies are missing or does not reach the maximum number of buddies. The combined unchoke decision algorithm is as follows.

**Step 1 (buddy unchokes).** A leecher in buddy mode strives to satisfy as many of its buddies as possible. We denote the number of unchoke slots reserved for buddies as a function of time with $n_B(t)$, where $0 \leq n_B(t) \leq n_s - 1$, $n_B(t)$ is bounded by $n_s - 1$ devoting at least one slot for regular unchokes.
**Step 2 (optimistic unchokes).** Given that $P_0$ is the probability to perform optimistic unchokes in regular BitTorrent, a leecher in buddy mode performs optimistic unchokes with probability:

$$P_B(t) = P_0 \frac{n_s - n_B(t) - 1}{n_s - 1} \tag{1}$$

**Step 3 (Regular unchokes).** $n_O(t)$ denotes the number of optimistic unchokes slots. Thus, for the remaining $(n_s - n_B(t) - n_O(t))$ unchoke slots, a leecher selects those peers that were not selected already, having the highest uploading rates.

Note that for independent leechers that run the regular BitTorrent $P_B = P_0$, since $n_B = 0$. However, $P_B$ decreases for buddies as more buddies are served. Those peers who serve $n_s - 1$ buddies perform no optimistic unchokes ($n_B = n_s - 1 \rightarrow P_T = 0$). They do not need to continue peers discovery.

Based on this unchoking algorithm design, we expect the following effects:

1. High capacity buddy peers' performance should be improved, as there are now other high capacity peers who make it their priority to serve them.
2. Reduction in the total number of optimistic unchokes, thus directly hurting the free-riders, who critically depend on these unchokes to obtain data.

These hypotheses validated by our experimental evaluation results. In a sense, *BitTorrent's optimistic unchoke mechanism is mostly replaced by buddies* as the means of discovering better partners. Hence, buddies no longer need to explore the network as much, having the guarantee of other leechers with similar upload capacity willing to upload to them.

## 4 Analytical Model

This section presents an analytical model that guides the protocol design presented in the previous section. It describes peer interactions in the presence of buddies as a game with rational players, and shows that there is a Nash-Equilibrium when all the peers in the system are contributing buddies.

### 4.1 The Game

We model the process of distributing a given content between incentive leechers in the P2P system as an infinitely repeated *game* [8], with rational participants. To simplify this process, we do not consider long-term punishment, although this is a design issue that is part of future work. The players repeatedly engage in rounds. In each round they choose their actions simultaneously, where the actions are determined by the players' strategy choices. A player can choose one of the three following strategies.

1. **an independent peer:** a contributor who runs the regular BitTorrent
2. **a free-rider:** a player who does not upload data to others
3. **a buddy-aware peer:** a player who follows the buddy-enhanced protocol

### 4.2 Leecher Utility Function

The leecher utility function is defined as the average download rate from all other leechers, thus leecher i's utility function is $U_i = \overline{d}_i$. We will ignore the download from the seeds in the analysis, since a seed upload process does not depend on the existence of buddies.

In [9], Fan *et al.* have already expressed the leecher utility function for the regular BitTorrent protocol as a combination of the average download rate that is obtained through the regular unchoke and the optimistic unchoke mechanisms. The buddy-enhanced BitTorrent protocol adds yet another term for the average download rate that is obtained through the buddy unchoke policy. Thus, the utility function of a leecher in the buddy-enhanced BitTorrent will be

$$\overline{U}_i = \overline{d}_i(regular) + \overline{d}_i(buddies) + \overline{d}_i(optimistic) \tag{2}$$

We assume that all peer $i$'s buddies have a similar upload rate of $\frac{u_i}{n_s}$ to $i$, given that $u_i$ is $i$'s upload rate. That is due to the way the buddy connections are created and maintained. In addition, we assume that the download bandwidth of a leecher is at least as much as its own upload bandwidth; this is typically the case in real torrents.

Each peer that has $n_B$ buddies enjoys a download rate of $\frac{u_i}{n_s}$ from each of its buddies. As a result, the buddy policy yield will be

$$\overline{d}_i(buddy) \approx n_B \cdot \frac{u_i}{n_s} \tag{3}$$

$P_{db} = \frac{n_B}{n_s}$ describes the download rate a peer enjoys from its buddies, as it also represents the percentage of upload bandwidth it spends on those buddies.

In [9], Fan *et al.* have shown mathematically that the download bandwidth of a leecher is approximately equal to its upload bandwidth, for regular unchokes, i.e., $d_i \approx u_i$. The percentage of the upload bandwidth dedicated to regular unchokes is $1 - P_{db} - P_B$. Thus, the regular unchoke yield will be

$$\overline{d}_i(regular) \approx (1 - P_{db} - P_B) \cdot u_i \tag{4}$$

Therefore, from Equations (2), (3), and (4), the utility function of buddy $i$ is:

$$U_i \approx (1 - P_B) \cdot u_i + \overline{d}_i(optimistic). \tag{5}$$

The utility function of a free-rider who uploads nothing is:

$$U_{FR} = \overline{d}_{FR}(optimistic) \tag{6}$$

The utility function of an independent peer who runs the regular BitTorrent is:

$$U_{IL} \approx (1 - P_0) \cdot u_i + \overline{d}_{IL}(optimistic). \tag{7}$$

Note that since the download through optimistic unchokes can be received from any peer in $i$'s peer set as it is based on random selection among all peers, the optimistic unchokes yield is not sensitive to the strategy choice of peer $i$. Hence, the unfairness in the regular BitTorrent is a function of the upload capacity and $P_0$, and it appears when the amount of upload that is given to optimistic unchokes is not the same as the amount of download that is received through optimistic unchokes, i.e., $P_0 \cdot u_i \neq d_i(optimistic)$.

In the buddy-enhanced BitTorrent, the unfairness is a function of the upload capacity, and $P_B$. $P_B$ gets closer to 0, as the number of buddy connections increases, which leads to reduction in optimistic unchokes upload. Thus, peers will benefit more by maximizing the number of buddy connections. Morover, increasing in number of buddy connections in the system will reduce the average download that is received through optimistic unchokes. As for the system, if every peer will choose the buddy-aware strategy, thus minimizing $P_B$ and devoting less upload to the unfairness inequality, the unfairness in the system will decrease.

### 4.3 Nash Equilibrium

**Theorem 1.** *The game in a buddy-enhanced BitTorrent system has a Nash Equilibrium when all the rational leechers in the game are buddy-aware peers.*

*Proof.* In order to prove this, we need to show that the utility of a buddy-aware peer is not lower than any other strategy's utility. Thus, if all the peers choose the buddy-aware strategy, none of the peers have the incentive to change their strategy. Equation 1 shows that $P_B \leq P_O$. Therefore, $U_i(IL) \leq U_i$(buddy-aware). Both utilities are equal when the buddy-aware peer is not able to find any matching buddy through the whole downloading process. Otherwise, the utility of the buddy-aware peer will be always greater than the utility of an independent peer. A free-rider's utility will be $U_i(FR) \leq U_i$(buddy-aware). The utility of a free-rider will be equal to the utility of a buddy-aware $i$ only in the case when all the peers in the network do not upload to $i$ through the regular unchokes or the buddy unchokes policies, during the whole downloading process.

We can thus conclude that the buddy-enhanced system has a Nash Equilibrium when all the rational peers are buddy-aware peers.

## 5 Implementation

The analytical model shows that the buddy-enhanced protocol has the potential to reduce unfairness and free-riding in a BitTorrent system. In order to examine this claim, we modified an existing open-source BitTorrent client to implement our buddy protocol. This section discusses interesting aspects of our prototype.

Our buddy-enhanced BitTorrent client was implemented on top of Enhanced CTorrent, version 3.2 [10]. The modified client can run in *buddy mode*, or in *regular mode*, in which it simply runs the regular BitTorrent. We also added functionality for a buddy peer to maintain state about its buddies.

Rechoke period in Enhanced CTorrent takes 10 seconds, and optimistic unchokes are rotated every 3 rechoke periods. The number of unchoke slots is dynamic with a minimum of 4 slots. This number may increase if a client does not saturate its capacity. In buddy mode, this number might change also if the upload capacity of a peer is too low/high compared to the peers it interacts with. We did not limit the maximum number of slots.

In regular mode, one unchoke slot is always devoted to optimistic unchokes, while in buddy mode, our unchoking mechanism determines the number of optimistic unchokes. In particular, every 3 rechoke periods a single optimistic unchoke is performed with probability given by Equation 1.

Before running in buddy-mode, a peer needs to collect information about the upload rate of itself and other peers in its peer set. The first 3 minutes (6 optimistic unchoke periods) after a client starts uploading are used for that, only after this period the peer is able to start running in buddy mode.

No changes were required for the tracker.

## 6 Experimental Evaluation

We ran extensive experiments on a controlled testbed, validating our protocol properties. Here we discuss our experimental methodology and results.

### 6.1 Methodology

All our experiments were ran on the PlanetLab experimental platform [11], utilizing nodes that are located across the globe. We executed all experiment runs consecutively in time on the same set of machines. Unless otherwise specified, we use the default BitTorrent client and seeding behavior.

All peers started the downloading process simultaneously, emulating a flash crowd scenario. The initial seeds stayed connected through the whole experiment. Leechers left the network once they have downloaded the entire content.

We artificially constrained the upload capacity of the nodes according to the bandwidth distribution for typical BitTorrent leechers as presented in [2]. This distribution is based on empirical measurements of BitTorrent swarms including more than 300,000 unique BitTorrent IPs. Not all nodes were capable of matching the required upload capacity that drawn from the empirical distribution. Thus, we scaled by 1/20th the upload capacity and other relevant experimental parameters such as file size. We did not define any download limits.

Unless otherwise specified, the experiments host 104 PlanetLab nodes, 100 leechers and 4 seeds with combined capacity of 128 KB/s serving a 30 MB file.

### 6.2 Results

We look at comparative results for leechers with different contribution levels, in order to validate our design.

**Leecher Performance - System without free-riders.** In this subsection we discuss two aspects that have been raised in this study: (1) Does the buddy-enhanced protocol indeed promote fairness? (2) Do all strategic leechers (having different upload capacity) have incentives to adopt the buddy-enhanced protocol.

Figure 3 shows leechers' download completion time comparing a regular BitTorrent system to a buddy-enhanced system where all leechers run in buddy mode. For each group of leechers having similar upload capacity, we draw separate boxplots [12] for the different scenarios. The top and the bottom of the box represent respectively the 75th and the 25th percentile download rates over all 7 runs of the experiment. The marker inside the box represents the median, while the vertical lines extending above and below the box represent respectively, the maximum and minimum values within the ranges of 1.5 time the box height from the box boarder. Potential outliers are marked individually with "+" sign.

We observe a clear difference between high capacity leechers which are the fastest 40% leechers and low capacity leechers which are the slowest 60% leechers. High capacity leechers improved their download time tremendously. Leechers with upload capacity of at least 5kB/sec improved their download time by 9%-35% as measured by the median. This happened due to downloading consistently from leechers with similar upload rate, as well as limiting in optimistic unchokes, and increasing number of unchoke slots for high capacity leechers.

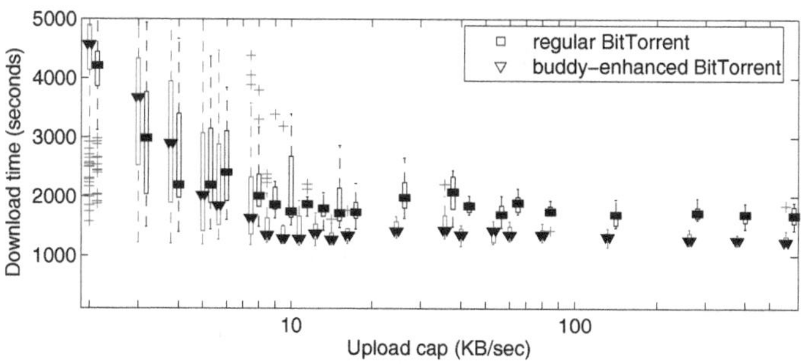


**Fig. 3.** Download completion time for leechers

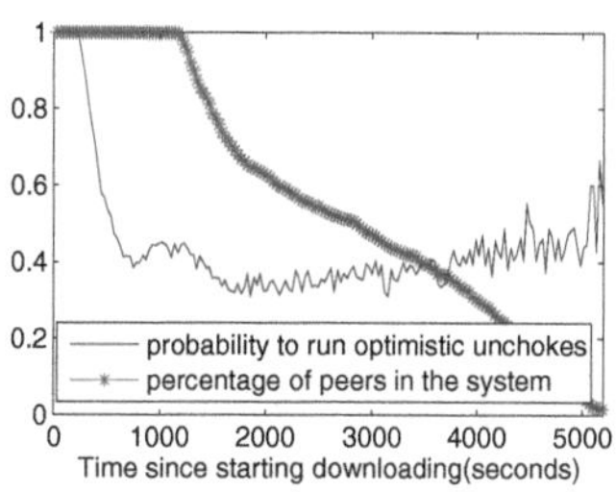


**Fig. 4.** The probability for optimistic unchokes in buddy-enhanced system

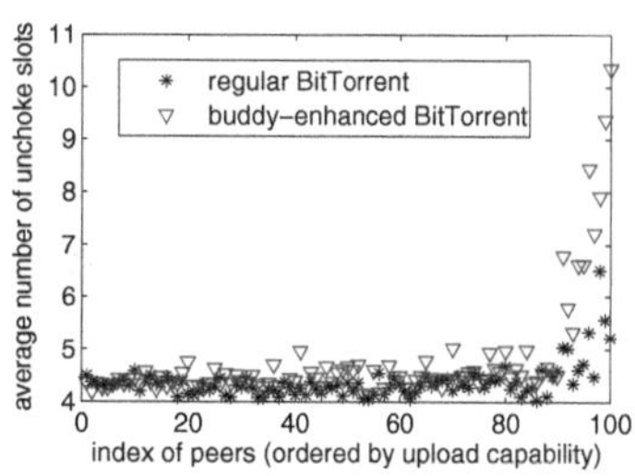


**Fig. 5.** The average unchoke slots for leechers ordered by upload capacity

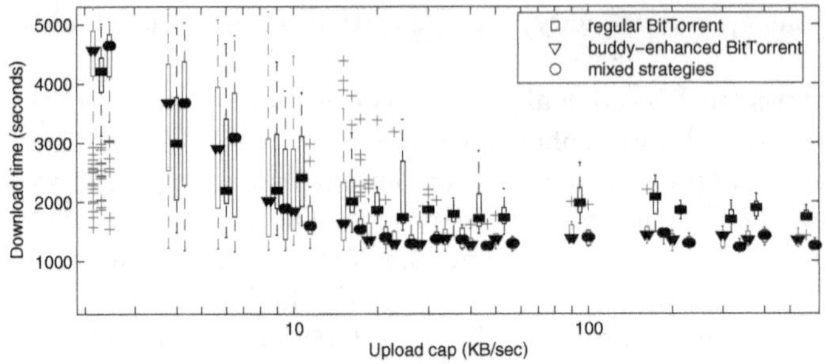


**Fig. 6.** Download completion time for leechers

Figure 4 confirms the reduction in optimistic unchokes by showing the average percentage of optimistic unchokes performed in the buddy enhanced system as a function of optimistic unchokes time periods. In regular BitTorrent, leechers run the optimistic unchokes mechanism every optimistic unchokes period, thus, this percentage value always equal to 1. In the buddy enhanced protocol, the optimistic unchokes reduced by $\sim$ 60% for most of the downloading process. Figure 5 confirms the increasing number of unchoke slots for the fastest 10% leechers in the buddy-enhanced protocol.

As a result of the aforementioned, low capacity leechers downloaded less from high capacity leechers, slowing down by 9%-32% as shown in Figure 3.

In summary, we see that *the buddy-enhanced protocol indeed promotes fairness and provides clear incentives for high capacity leechers to run in buddy mode*, as a high capacity buddy leecher achieves improved performance compared to an independent one. However, it is not clear that the low capacity leechers have the same incentives, since their download time performance suffers.

Hence, we ran another experiment, where low capacity peers ran in independent mode, and high capacity peers ran in buddy mode. Figure 6 that compares our previous results with the "mixed strategies" results shows that by changing strategy to independent, low capacity peers suffered even more, since as independents they were also losing part of their bandwidth for optimistic unchokes without getting anything in return. Moreover, the high capacity peers performance were not

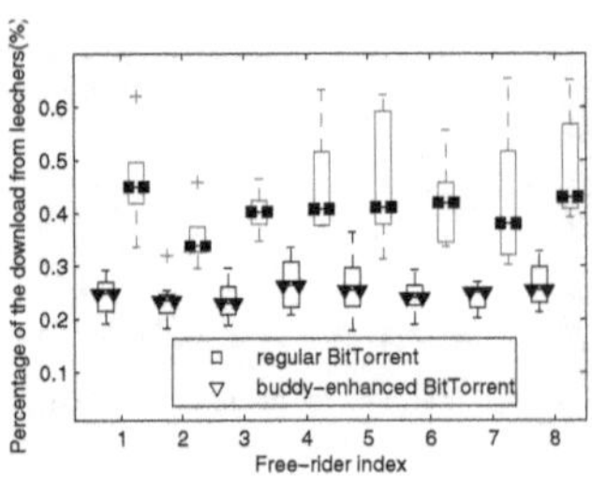


**Fig. 7.** FRs' download from leechers

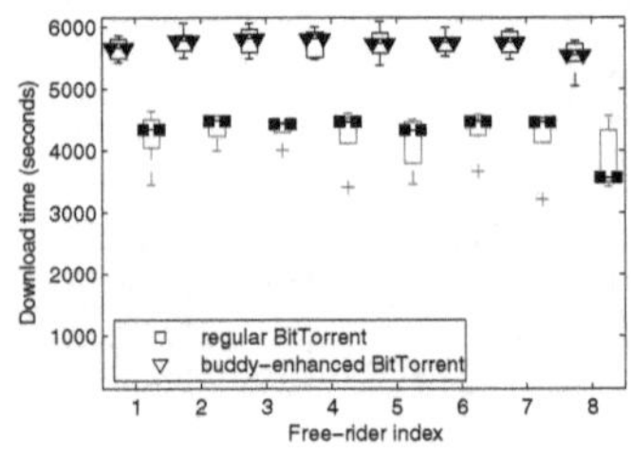


**Fig. 8.** FRs download time

very sensitive to the change, most of them slightly improved their performance in comparison to all are buddies results. This is because in the mixed strategies experiment high capacity peers enjoy increasing amount of optimistic unchokes from low capacity peers needing to give nothing in return.

The experiments validate our hypothesis in section 4 showing that it is to the benefit of all the leechers to run in buddy mode. Furthermore, running in buddy mode motivates leechers to contribute and improves fairness in the network.

**Leecher Performance - System with Free-rider.** As this section focuses on the impact of free-riders, we added 8 free-riders , having a total of 112 peers in the experiment: 100 contributors and 4 seeds as described in previous experiments and 8 free-riders that download as much as possible, but do not upload any data.

With free-riders in the network high capacity leechers in buddy mode improved their download time performance by 17%-54% as measured by the median in comparison to regular BitTorrent, while low capacity leechers were slowed down by 16%-59%. We got similar results to those shown in Figure 6, however, the different between the two protocols performance is more pronounce. That is since contributors in buddy mode limit the optimistic unchokes, hence limit the possibility to be affected by free-riders. Figure 7 validates this hypothesis, showing that the percentage of content free-riders download from leechers decreases from $\sim$ 45% in regular BitTorrent to $\sim$ 25% in buddy-enhanced system.

Figure 8 shows the download completion times of free-riders, showing that *the existence of buddies slowed down free-riders by 29-56%.* As in Figure 7, this is because free-riders can not take advantage of optimistic unchokes as in regular BitTorrent and thus have to depend mainly on the seeds for downloading data.

These results also indicate the increased robustness of the system to free-riding. We also investigate the system robustness by increasing the number of free-riders in the system to 16, 20, 24, however, due to lack of space we can not present in detail these results. In summary, these results show that the buddy enhanced system is less sensitive to the increasing number of free-riders, an indication to a more robust system.

## 7 Related Work

There is large amount of works that model the behavior and analyze the performance of BitTorrent, since Bram Cohen, the protocol's creator, first described the protocol's main mechanisms and their design rationale [1].

In [13], Qiu *et al.* presented a fluid analytical model of BitTorrent systems. They studied the choking algorithm and its effect on peer performance. They indicated that the optimistic unchokes mechanism may provide a way for leechers to free-ride. More recently, in [9], Fan *et al.* model BitTorrent capturing the fundamental trade-off between performance and fairness. Our analytical model is inspired by their work. Furthermore, in [14], Legout *et al.* observed that the incentives of tit-for-tat tend to cause leechers to cluster together with other similar upload bandwidth leechers. Our protocol facilitates this explicitly by having buddies, peers with similar bandwidth to collaborate.

Other researchers have studied the feasibility of free-riding behavior, when peers download without uploading, attempting to deceive the protocol mechanisms. The first to pinpoint that effective free-riding in BitTorrent is feasible were Shneidman *et al.* [15]. Liogkas *et al.* [5] designed and implemented 3 exploits that allow free-riders to obtain high download rates under specific circumstances. Sirivianos *et al.* [7] showed that maintaining a larger-than-normal view of the system, provides a free-rider a much higher probability to receive data from seeds and via optimistic unchokes. Finally, in [2], Piatek *et al.* also observed the unfairness. They proposed a new choking mechanism that reallocates upload bandwidth and aims to maximize peer download rates. The aforementioned studies identify one of the most important vulnerabilities that our work directly addresses, namely exploiting optimistic unchokes to download data for free.

Researchers [16, 17, 18, 19] have recognized the value of peers cooperating to download content. However all these works do not address the incentives of the volunteer peers to voluntarily download popular content pieces and upload them to others. Our proposed protocol, on the other hand, addresses this by providing a clear incentive for leechers to increase their buddy connections and upload to their buddies. This leads to reduction in enforced altruism and therefore improves fairness, and limits free-riding.

The most related protocol to our proposed protocol is the protocol that embeds teams [20], groups of peers that collaborate to improve their upload rate. There is a trade-off between the team protocol and the presented buddy protocol, which is a simplicity for the price of less improvement. The main different is that teams are formed and managed by the tracker, a centralized authority, therefore it achieves a faster convergence and a larger improvement in fairness and free-riding reduction. However, in the presented work the tracker does not involve in the buddy related part of the protocol. Hence, here we suggest a more distributed protocol, involves less management overhead, and is simpler to implement, which make it more realistic to be adopted.

## 8 Conclusion

In this paper we present an extension to the BitTorrent protocol, an alternative buddy mechanism that relies on continuous history to match similar upload capacity peers. It mostly replaces the optimistic unchokes, partially replaces the original TFT mechanism, and add more dynamic to the number of unchoke slots. Experimental results on a controlled testbed demonstrate that our buddy protocol promotes fairness compared to the regular BitTorrent, and provides clear incentives to be adopted, even by those low capacity peers that were slowed down due to the presence of buddies. In the presence of buddies, free-riders achieve lower download rates, compared to the regular BitTorrent. Furthermore, the system with buddies is more robust being less sensitive to increasing number of free-riders in the network.

## Acknowledgments

We thank Nikitas Liogkas, Rupak Majumdar, Mihaela van der Schaar, Giovanni Pau, and Cesar Marcondes for helpful discussions and comments.

## References

1. Cohen, B.: Incentives Build Robustness in BitTorrent. In: P2PEcon 2003 (2003)
2. Piatek, M., Isdal, T., Anderson, T., Krishnamurthy, A., Venkataramani, A.: Do incentives build robustness in BitTorrent?. In: NSDI 2007 (2007)
3. Bharambe, A., Herley, C., Padmanabhan, V.: Analyzing and improving a bittorrent network's performance mechanisms. In: INFOCOM (2006)
4. Guo, L., Chen, S., Xiao, Z., Tan, E., Ding, X., Zhang, X.: Measurements, analysis, and modeling of BitTorrent-like systems. In: IMC 2005 (2005)
5. Liogkas, N., Nelson, R., Kohler, E., Zhang, L.: Exploiting BitTorrent For Fun (But Not Profit). In: IPTPS 2006 (2006)
6. Locher, T., Moor, P., Schmid, S., Wattenhofer, R.: Free Riding in BitTorrent is Cheap. In: HotNets-V (2006)
7. Sirivianos, M., Park, J.H., Chen, R., Yang, X.: Free-riding in BitTorrent Networks with the Large View Exploit. In: IPTPS 2007 (2007)
8. Osborne, M.J., Rubinstein, A.: A Course in Game Theory. MIT Press, Cambridge (1994)
9. Fan, B., Chiu, D.M., Lui, J.C.: The Delicate Tradeoffs in BitTorrent-like File Sharing Protocol Design. In: ICNP 2006 (2006)
10. Enhanced CTorrent: `http://www.rahul.net/dholmes/ctorrent:`
11. Bavier, A., Bowman, M., Chun, B., Culler, D., Karlin, S., Muir, S., Peterson, L., Roscoe, T., Spalink, T., Wawrzoniak, M.: Operating System Support for Planetary-Scale Network Services. In: NSDI 2004 (2004)
12. Robert McGill, J.W.T., Larsen, W.A.: Variations of box plots. The American Statistician 32, 12–16 (1978)
13. Qiu, D., Srikant, R.: Modeling and Performance Analysis of BitTorrent-Like Peer-to-Peer Networks. In: SIGCOMM 2004 (2004)
14. Legout, A., Liogkas, N., Kohler, E., Zhang, L.: Clustering and Sharing Incentives in BitTorrent Systems. In: SIGMETRICS (2007)
15. Shneidman, J., Parkes, D.C., Massouliè, L.: Faithfulness in Internet Algorithms. In: PINS 2004 (2004)
16. Wang, J., Yeo, C., Prabhakaran, V., Ramchandran, K.: On the role of helpers in peer-to-peer file download systems: design, analysis, and simulation. In: IPTPS 2007 (2007)
17. Wong, J.H.T.: Enhancing Collaborative Content Delivery with Helpers. M.sc thesis, University of British Columbia (September 2004)
18. Garbacki, P., Iosup, A., Epema, D., van Steen, M.: 2Fast: Collaborative downloads in P2P networks. In: P2P 2006 (2006)
19. BTSlave protocol page, `http://btslave.sourceforge.net`
20. Izhak-Razin, R., Liogkas, N., Majumdar, R.: Team incentives in bittorrent systems. In: TR 090002, UCLA CSD

# Detecting Triangle Inequality Violations in Internet Coordinate Systems by Supervised Learning
## (Work in Progress)

Yongjun Liao[1,⋆], Mohamed Ali Kaafar[2], Bamba Gueye[1], François Cantin[1], Pierre Geurts[3], and Guy Leduc[1]

[1] Research Unit in Networking (RUN), University of Liège, Belgium
{liao,gueye,cantin,leduc}@run.montefiore.ulg.ac.be
[2] INRIA, France
mohamed-ali.kaafar@inria.fr
[3] Systems and Modeling, University of Liège, Belgium
p.geurts@ulg.ac.be

**Abstract.** Internet Coordinates Systems (ICS) are used to predict Internet distances with limited measurements. However the precision of an ICS is degraded by the presence of Triangle Inequality Violations (TIVs). Simple methods have been proposed to detect TIVs, based e.g. on the empirical observation that a TIV is more likely when the distance is underestimated by the coordinates. In this paper, we apply supervised machine learning techniques to try and derive more powerful criteria to detect TIVs. We first show that (ensembles of) Decision Trees (DTs) learnt on our datasets are very good models for this problem. Moreover, our approach brings out a discriminative variable (called $OREE$), which combines the classical estimation error with the variance of the estimated distance. This variable alone is as good as an ensemble of DTs, and provides a much simpler criterion. If every node of the ICS sorts its neighbours according to $OREE$, we show that cutting these lists after a given number of neighbours, or when $OREE$ crosses a given threshold value, achieves very good performance to detect TIVs.

**Keywords:** Internet Coordinate System, Triangle Inequality Violation, Supervised Learning, Decision Trees.

## 1 Introduction

Internet Coordinate Systems (ICS) have been widely used in large-scale network applications such as peer-to-peer file sharing [1], dynamic server selection [2] and overlay multicast [3]. The success roots in the embedding of Internet nodes

⋆ This work has been partially supported by the EU under projects FP7-Fire ECODE and FP6-FET ANA. M.A. Kaafar was also partly supported by the University of Liège. F. Cantin is a Research Fellow of the Belgian Fund for Research in Industry and Agriculture (FRIA). P. Geurts is a Research associate of the FRS-F.N.R.S. (Belgium).

L. Fratta et al. (Eds.): NETWORKING 2009, LNCS 5550, pp. 352–363, 2009.

into a metric space or a coordinate system. When the coordinates of the nodes are computed, the prediction of the distances (delays) between two nodes is a straightforward application of a distance function where no explicit communication between them is required. This significantly reduces the overhead of active probing and largely improves the efficiency of the network.

Most ICS algorithms collect, at each node, a small number of distance measurements and optimize the differences between the distances computed from the estimated coordinates of the nodes and the measured distances [4, 5, 6]. As a result, all nodes fall into a metric space where the triangle inequality holds. However, in reality, the triangle inequality is often violated due to the Internet's structure and routing policies. In [7,8,9,10], it has been shown that the violations of triangle inequality (TIVs) in the Internet delay space are not rare and that their impacts on ICS are not negligible. Indeed, a TIV indicates the existence of a shorter path from the source node to the destination node via an intermediate node than a direct path.

The importance of building TIV-aware systems has been addressed in e.g. [9, 10, 11]. In [11], by observing that TIV edges are often underestimated to some degree, a TIV alert mechanism was incorporated into ICS's neighbor selection procedure, which improved the precision of distance prediction. In our previous works [9, 10], we confirmed the significance of the impact of TIVs on ICS and proposed a different mechanism based on the observation that the variance of TIV edges is often smaller than that of non-TIV edges. A comparative study showed that the two mechanisms performed inconsistently on different networks: sometimes ours is better, sometimes the mechanism in [11] is better. This makes us wonder if there exists more discriminative and more consistent variables for TIV detection.

This paper introduces a systematic approach to detecting TIVs in ICS. Unlike previous heuristics-based work, we look for discriminative variables by using supervised learning methods, the use of which has become a trend in the field of networking [12, 13]. In particular, we collect training data from popular ICS algorithms such as Vivaldi [5] and extract as many variables of different kinds as possible. A classical supervised learning method, namely decision tree, is then applied. A discriminative variable (called $OREE$) is found that outperforms other competing variables consistently. The learned $OREE$ has a clear physical meaning that explains its discriminativeness. Due to the diverse characteristics of the TIV distributions, the decision trees learned on different networks have large variations. We then design a simple distributed algorithm for TIV detection based on $OREE$. Our detection algorithm is tested on different networks and convincing results are achieved.

The rest of the paper is structured as follows. Section 2 analyzes the characteristics of TIVs in Internet delay space. Section 3 introduces the supervised learning of $OREE$, the discriminative variable, for TIV detection, including the collection of training data, the building of decision trees and some experiments and observations. Section 4 presents a distributed detection algorithm based on $OREE$. Section 5 concludes our approach and discusses future work.

## 2 TIVs in Delay Space

Let $A$, $B$ and $C$ be three nodes that form a triangle. Denote $d(X,Y)$ the distance or delay between $X$ and $Y$. Without loss of generality, $d(A,B)$ is the largest edge in the triangle. Then, we claim that the triangle inequality is violated if $d(A,B) > d(B,C) + d(A,C)$, in short, TIV. In case of TIV, the largest edge $d(A,B)$ is referred to as the TIV base. Two criteria are defined to model the severity of TIVs:

$$\textbf{absolute severity:} \qquad G_a = d(A,B) - (d(A,C) + d(C,B)),$$

$$\textbf{relative severity:} \qquad G_r = \frac{d(A,B) - (d(A,C) + d(C,B))}{d(A,B)}.$$

These criteria also reflect the possible gains one could achieve by detecting TIVs. For instance, $G_a = 10ms$ suggests that, instead of the direct path from $A$ to $B$, taking a path via $C$ could possibly be $10ms$ faster. Therefore, larger (positive) $G_a$ and $G_r$ mean not only more severe violation but also more possible gains and hence, are more interesting. In this paper, we focus on TIVs that satisfy both $G_a > 10ms$ and $G_r > 0.1$ and consider the others as non-TIV.

It is expected that the distribution of TIV characteristics vary in different networks. We verify it by analyzing two network data sets, P2psim of 1740 nodes [14] and Meridian of 2500 nodes [15]. To this end, the whole range of the delay space is divided into equal bins of $5ms$ and the TIV bases that fall into each bin are counted and normalized. Figure 1 shows the resulting histograms of delays both for TIVs and non-TIVs in P2psim and Meridian respectively. The distributions are quite different. In Meridian, it is sharply peaked while it is much more spread in P2psim. This large difference could potentially mean that it would be difficult to learn a general model for TIV detection that works well over different networks. Our experiments in the next section will try to answer that question.

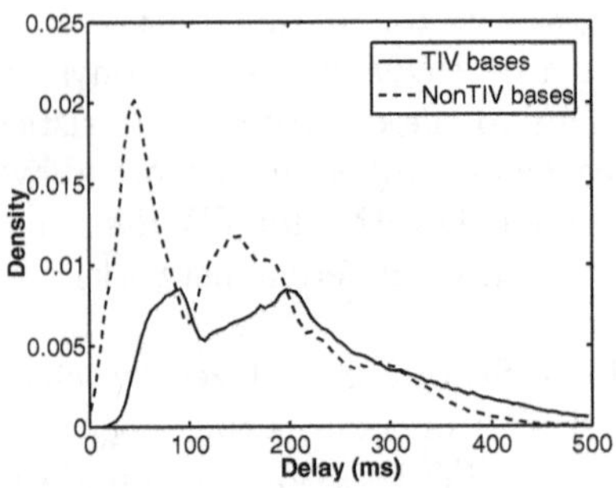

(a) P2psim

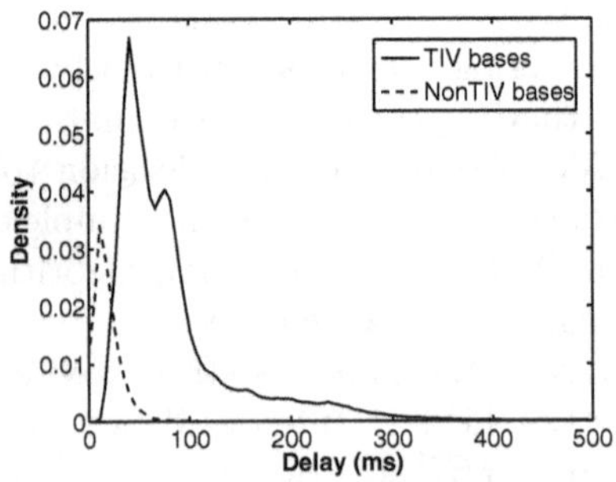

(b) Meridian

**Fig. 1.** TIV and non-TIV distributions in P2psim and Meridian. 42% of the edges in P2psim and 83% in Meridian are TIV bases.

# 3 Finding Discriminative Variables for TIV Detection

Our objective is to find variables that are consistently discriminative for TIV detections regardless of the networks. Supervised learning methods are promising techniques that allow us to achieve this goal.

Generally speaking, supervised learning methods take $N$ input/output pairs, $< x_1, y_1 >, < x_2, y_2 >, ..., < x_N, y_N >$, where $x_i$ is the vector of the input variables and $y_i$ is its output value or label, and try to reveal the relationships between the inputs and the outputs. In other words, the goal of supervised learning is to learn a function $f(x)$ from a set of training data that predicts, at best, the output $y$ for any new unseen input $x$.

This section details the use of a particular supervised learning method, namely decision tree, in finding discriminative variables for TIV detection.

## 3.1 Collection of Training Data

In this paper, we focus on a classical ICS algorithm, Vivaldi [5], which approximates a network by a system of springs and seeks to minimize its energy. The minimization is fully distributed and iteratively done at each node. Following [5], each node communicates with 32 neighbors, among which 16 are the nearest neighbors and the other 16 are randomly selected. In each iteration, each node updates its own coordinates by minimizing the embedding error with respect to its neighbors. After some iterations depending on the complexity of the network, the coordinates of the nodes become stable, i.e., the embedding error of the whole network is smaller than a threshold, with a small amount of jitter.

To collect training data, we ran Vivaldi on P2psim and Meridian respectively and recorded the coordinates of the nodes, after they become stable, at $K$ different times or ticks. Unless otherwise stated, $K$ will be fixed to 100 in all our experiments. Its effect will be discussed in Section 3.3. From the coordinates obtained, we computed the time-varying distances between each pair of neighboring nodes. We denote the measured distance by $d$ and the estimated distance by $\hat{d}$. Thus, for each edge, we have $\{d, \hat{d}_1, \hat{d}_2, \ldots, \hat{d}_K\}$. For the $K$ estimated distances, we further calculated some statistics including $\hat{d}_{max} = max\{\hat{d}_1, \ldots, \hat{d}_K\}$, $\hat{d}_{min} = min\{\hat{d}_1, \ldots, \hat{d}_K\}$, $\hat{d}_{mean} = mean\{\hat{d}_1, \ldots, \hat{d}_K\}$, $\hat{d}_{median} = median\{\hat{d}_1, \ldots, \hat{d}_K\}$ and $\hat{d}_{std} = standard\_deviation\{\hat{d}_1, \ldots, \hat{d}_K\}$.

The input variables to the machine learning algorithm consist of various combinations of these statistical variables and the measured distances. Note that the input variables are normalized in different ways in order to get rid of the influence of particular network conditions. Currently, 64 input variables for each edge are used. A few examples are

$$\frac{\hat{d}_{max} - \hat{d}_{min}}{d}, \frac{\hat{d}_{mean} - \hat{d}_{median}}{d}, \frac{\hat{d}_{mean} - d}{\hat{d}_{max}}, \frac{\hat{d}_{mean}}{\hat{d}_{std}}, \frac{\hat{d}_K}{d}, \frac{\hat{d}_{std}}{d}.$$

Note that the last two variables, $\frac{\hat{d}_K}{d}$ and $\frac{\hat{d}_{std}}{d}$, are equivalent to those used in [11] and [10] respectively. The former, $\frac{\hat{d}_K}{d}$, is a measure of relative estimation error, denoted by $REE$, while the latter, $\frac{\hat{d}_{std}}{d}$, is the standard deviation of the relative estimation error, denoted by $std\_REE$.

For supervised learning, the outputs or labels of the edges are needed. Here, the outputs are simply TIV or non-TIV, which can be easily obtained from the measured distances. To assess the effect of the random neighbor selection procedure in Vivaldi, we ran the simulation on each network for 10 times. In each simulation, we collected $1740 \times 32 = 55680$ edges from P2psim and $2500 \times 32 = 80000$ from Meridian, among which average 23% and 42% are TIV bases respectively. Note that the numbers of TIV bases vary only slightly in different simulations.

### 3.2 Decision Tree

Decision Tree (DT) is one of the most popular supervised learning algorithms. Each decision tree is a classifier in the form of a tree structure, where each interior node specifies a binary test carried on a single input variable and each terminal node is labeled with the value of the output. In the learning phase, a decision-tree builder recursively splits the training samples with binary tests, trying to reduce as much as possible the uncertainty about the output classification in the resulting subsets of the samples. The splitting of a node is stopped when the output in it is homogeneous or some other stopping criterion is met. During learning, a byproduct is the ranking of the input variables according to their importance, which is often used to find discriminative variables. In the classification phase, we start from the root of the tree and move through it until a terminal node, where the classification result is provided. See [16] for more details about decision trees.

In many applications, single decision trees are greatly improved by ensemble methods like boosting or random forests that combine the predictions of several trees (see e.g. [13]). However, in our experiments, we did not notice any significant improvement with these methods. We thus restrict our discussion below to standard (single) decision trees.

### 3.3 Experiments and Observations

In this paper, we learned our decision trees using a data mining software called PEPITO [16] which integrates most popular machine learning algorithms. The training data collected in Section 3.1 was randomly divided into disjoint learning and test sets of roughly equal sizes. In other words, we used half of the data to build the trees and the other half to evaluate the trees. Two different trees were separately built for P2psim and Meridian, shown in Figure 2[1]. There are a few interesting observations to be noticed.

[1] Note that the decision trees learned are essentially invariant from one simulation to another. The invariance can be extended to other results in this paper. Thus, unless otherwise stated, the results of all the experiments in the paper are derived from one simulation.

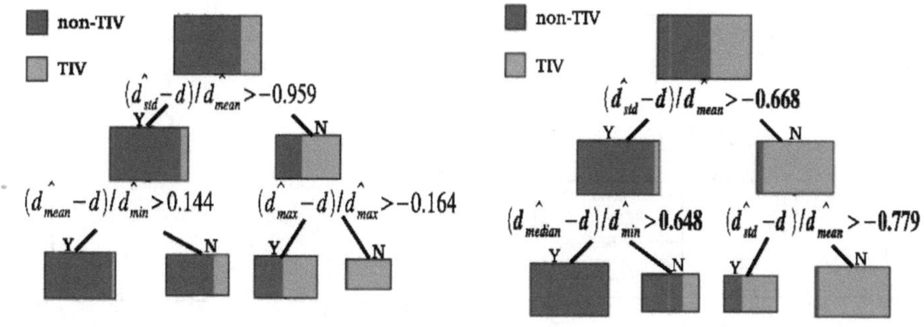


**Fig. 2.** Top three levels of the decision trees built on P2psim and Meridian data set. The colour in the rectangular boxes reflects the proportions of TIV and Non-TIV classes.

**The most discriminative variable.** By examining the root nodes of both trees in Figure 2, it can be seen that the first test is on the same variable, $\frac{\hat{d}_{std}-d}{\hat{d}_{mean}}$, which appears to be the most discriminative. For the reason that will be clear in the next section, we name this variable $OREE$. On both trees, when the value of $OREE$ is smaller than the cut point, the edge is more likely to be a TIV base, and vice versa. Figure 3(a) and 3(c) show the ROC [2] curves of $OREE$, $REE$ and $std\_REE$. Note that the ROC curve of $OREE$ is consistently the highest while the other two variables perform inconsistently: $REE$ is more discriminative on P2psim but less on Meridian than $std\_REE$. Figure 3(b) and 3(d) show the mean ROC curve of $OREE$ and the covariances at some selected thresholds obtained over the 10 simulations. These figures clearly highlight that the results are very stable from one simulation to another.

**The analysis of the most discriminative variable.** The most discriminative variable $\frac{\hat{d}_{std}-d}{\hat{d}_{mean}}$ consists of 2 parts, $\frac{\hat{d}_{std}}{\hat{d}_{mean}}$ and $\frac{d}{\hat{d}_{mean}}$. The former can be considered as a relative oscillation measure and the latter is a relative error measure. Thus, this variable is named by $OREE$ representing Oscillation and Relative Estimation Error. Intuitively, the oscillation of the estimated distances and the embedding error are all the information one can extract and exploit for TIV detection. In fact, the ROC curves of $OREE$ are almost identical to those of the decision trees in both networks, shown in Figure 3(a) and 3(c). This suggests that the other variables provide little extra information so that their corresponding nodes in the trees can be safely pruned with no performance degradation. Figure 4 shows the TIV and non-TIV distributions of P2psim and Meridian with respect to $OREE$. It can be seen that the overlap of the two distributions in P2psim is much larger than in Meridian, which is the reason why the ROC curves in Meridian are much higher than those in P2psim.

[2] A ROC (Receiver Operating Characteristic) curve is a graphical plot of the true positive rates (TPR) versus the false positive rates (FPR) for a binary classifier as its discrimination threshold is varied [17].

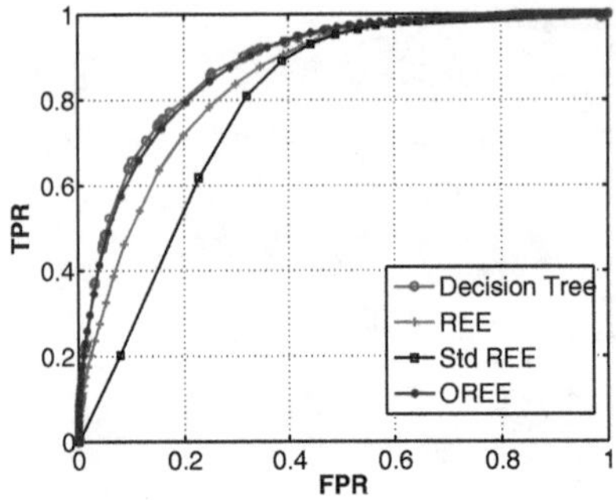


(a) P2psim

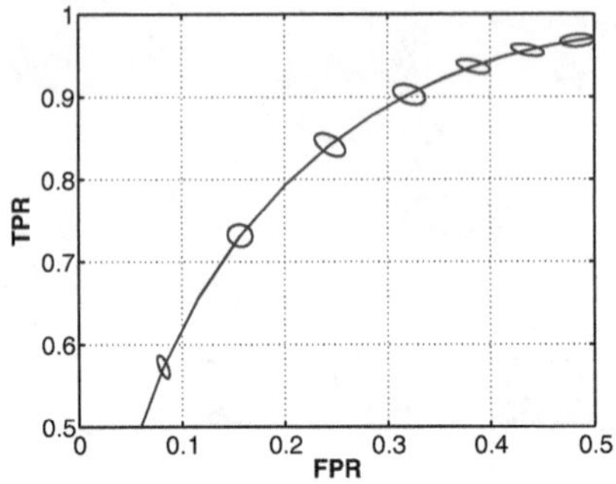


(b) 10 simulations on P2psim

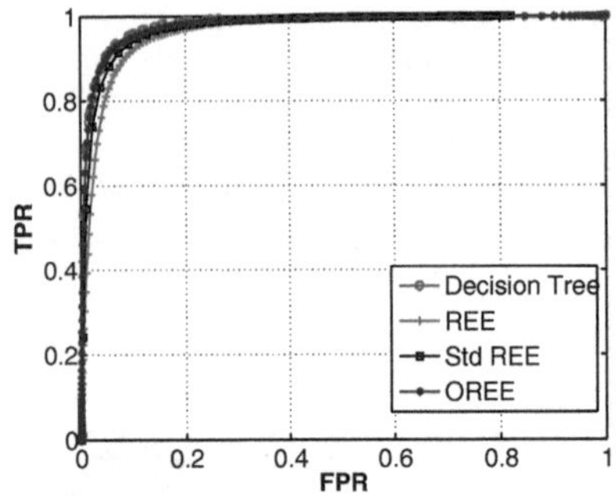


(c) Meridian

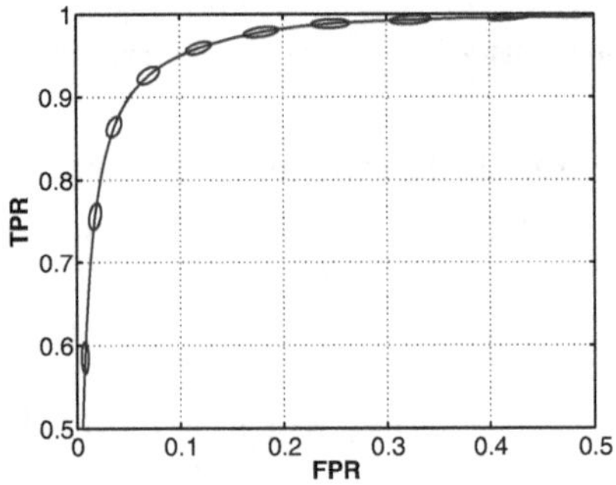


(d) 10 simulations on Meridian

**Fig. 3.** ROC curves of the decision trees, obtained by varying the threshold of the probability of TIV, and of three variables including *OREE*, *REE* and *std_REE*, obtained by thresholding the values of the corresponding variables. The right figures show the mean ROC curves of OREE and the covariances at some selected thresholds (the covariance ellipses are scaled by 3 times for the sake of visibility).

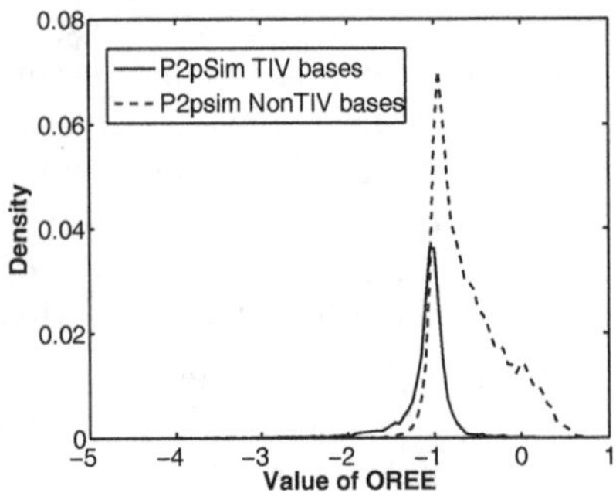


(a) P2psim

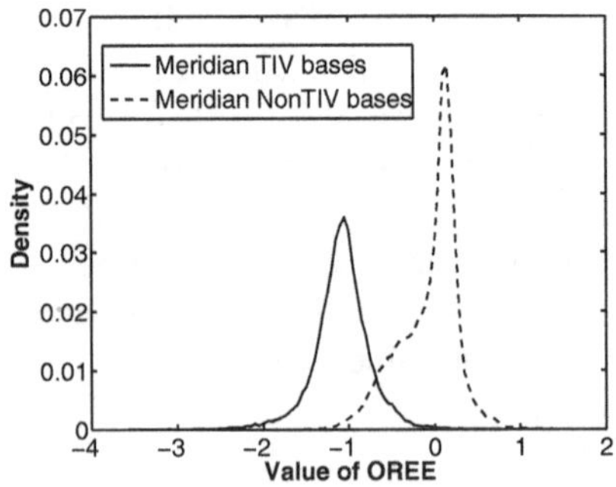


(b) Meridian

**Fig. 4.** TIV and non-TIV distributions in P2psim and Meridian with respect to *OREE*. It can be seen that detecting TIVs in Meridian is much easier than in P2psim and that there is a shift between the optimal cut points for the two networks.

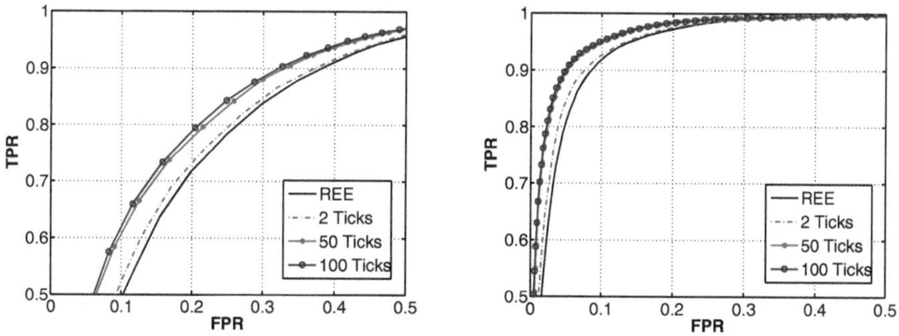


(a) P2psim (b) Meridian

**Fig. 5.** ROC curves of using different numbers of ticks to compute *OREE*

**Table 1.** Error rates of applying decision trees on P2psim and Meridian data sets

| Test Set \ Learning Set | P2psim | Meridian | P2psim+Meridian |
|---|---|---|---|
| P2psim | 15.6% | 43.7% | 20.5% |
| Meridian | 28.5% | 8.4% | 9.4% |

**How many ticks to be used.** The parameter $K$, i.e. the number of ticks that are used to compute $OREE$, might affect the accuracy of our model. Figure 5 shows ROC curves obtained with different values of $K$. It can be seen that a noticeable improvement can be achieved even when $K = 2$. With a larger $K$, the statistical variables, mean and standard deviation, become more stable and the performance of the detection is improved. However, no or minor differences are obtained when $K$ goes from 50 to 100, showing that the optimal value of $K$ should probably be smaller than 100. We nevertheless set $K = 100$ in all our experiments since it doesn't cause any side effect.

**The variations of the decision trees.** The two decision trees in Figure 2 have non-negligible variations. Even for the two root nodes that correspond to the same variable, a shift is found between the optimal cut points in them, which introduces errors when the tree learned on one network is applied to another. Table 1 shows the error rates when learning and evaluating on the same data set and on different data sets. Note that the first row specifies the data sets on which learning is carried while the first column gives the data sets on which evaluation is applied. We see that the variability of the trees indeed leads to high error rates in cross-testing. The tree learned from combining data of both networks gives better average results but is still inferior to the trees learned on each particular network. In practice, even if we can collect data and learn a specific classifier for each network, the TIV distribution in the network often evolves in time due to e.g. changes in routing or network upgrade. The next section introduces a simple but effective method for distributed TIV detection based only on $OREE$.

## 4 Distributed TIV Detection Based on *OREE*

We aim to embed the TIV detection algorithm into the ICS algorithm, Vivaldi in our case, so that the detection can be done at each node, i.e., each node classifies the edges with its 32 neighbors into TIV bases and non-TIV based on OREE variable.

### 4.1 Detection Algorithm

Recall that when the value of $OREE$ is smaller than a cut point, the edge will be detected as TIV, and vice versa. This makes us guess that the probability of an edge being a TIV increases when the value of $OREE$ decreases. To verify it, we sort the 32 edges connected to each node by the value of $OREE$ in ascending order and count the number of TIV bases appearing at each position from 1 to 32. These numbers are normalized by the total number of nodes so that we have the proportion of TIVs at each position. Ideally, we would like the proportions of TIVs at one side to be close to 1 and at the other side to 0. Figure 6 shows the proportion of TIVs at each position sorted by $OREE$, $REE$ and $std\_REE$ respectively.

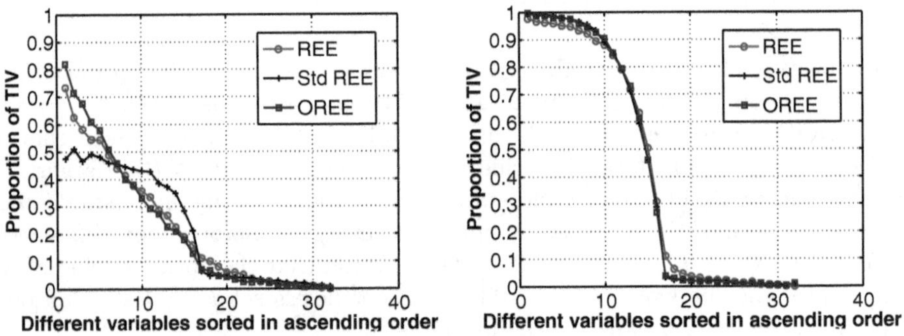


(a) P2psim (b) Meridian

**Fig. 6.** Proportions of TIVs at each ordering position sorted respectively by $OREE$, $REE$ and $std\_REE$ on P2psim and Meridian data sets. It can be seen that the TIV proportions of $OREE$ are larger on the left side and smaller on the right side, which further justifies the efficiency of $OREE$ for TIV detection.

It is clear that the proportion of TIV decreases monotonously with the ordering position. For $OREE$, the probability drops below 0.5 at the 6th position for P2psim and at the 14th position for Meridian. This suggests a simple detection method that, at each node, after the 32 edges are sorted in ascending order, fixes a cut point at the $n$th position, where the probability of TIV at the $n$th position is larger than a pre-defined threshold that should be at least above 0.5.

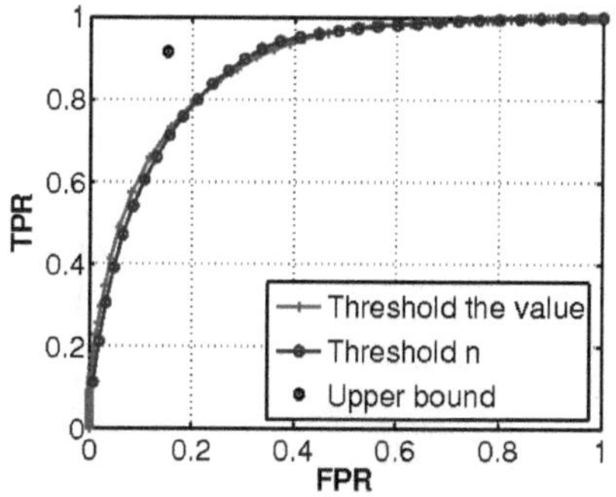


(a) P2psim

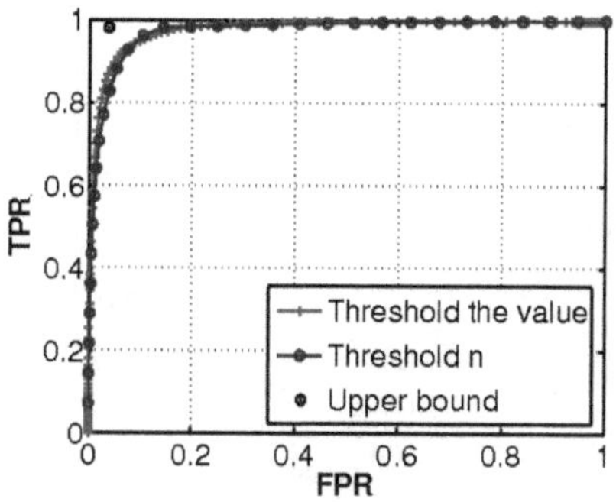


(b) Meridian

**Fig. 7.** Roc curves of thresholding $OREE$ and threshold $n$ on P2psim and Meridian. The circles, in the upper left corner, represent the upper bounds of the detection rates by choosing an optimal, different threshold for each node.

## 4.2 Experiments

We first compared our detection algorithm that thresholds $n$ with an alternative method that directly thresholds the value of $OREE$. A ROC curve can be obtained for either method by varying its threshold, shown in Figure 7. It can be seen that both methods work identically well as their ROC curves coincide, which is not surprising as the difference between them is subtle. In a distributive manner, a common fixed threshold on the value of $OREE$ is equivalent to a different threshold on $n$ chosen at each node. On the other hand, a common fixed threshold on $n$ for all nodes is equivalent to a different threshold on the value of $OREE$ at each node. In a sense, these are dual methods.

The two methods share a common problem. For decentralized detection, the optimal fixed cut point either for the value of $OREE$ or for $n$ can't be found out easily. Nevertheless, the advantages of tuning $n$ are twofold. First, $n$ has a fixed range of $[1, 32]$ while the value of $OREE$ is theoretically unbounded. Then, from the ROC curves in Figure 7, it can be observed that the detection rates are influenced by $n$ more evenly than by the value of $OREE$. Figure 8 shows the plot of recall versus precision by varying $n$. It can be seen that tuning $n$ affects recall and precision in the opposite directions. A large $n$ improves the recall but decreases the precision, and vice versa. In practice, the optimal $n$ should be set to trade the precision for the recall in order to meet the specific requirements of different applications. For instance, when the cost that a non-TIV edge is erroneously detected as a TIV edge is large, a conservative, small $n$ should be used, and vice versa.

Since we know whether an edge is TIV or non-TIV in the experiments, it is possible to calculate the optimal cut point at each node. To test this idea, we tried out, for each node, all possible $n$ from 1 to 32 and recorded the particular $n$ that maximizes the information gain [18]. Although these optimal cut points can't be computed in practice, they give the upper bound of the detection rates

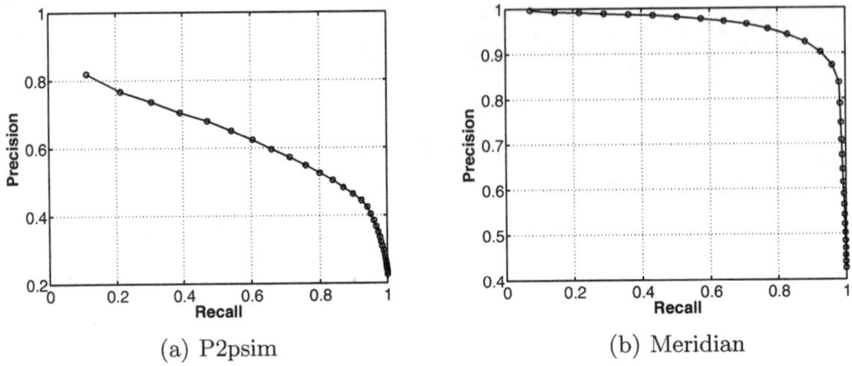


**Fig. 8.** Precision versus recall of our detection algorithm by thresholding $n$. Precision is the ratio between the true positive and the total positive detected by the classifier, and recall is another name of true positive rate.

by using $OREE$, plotted as a circle in Figure 7, and show how much room we have to further improve our detection algorithm.

## 5 Conclusions and Future Work

This paper presents a novel approach to detecting triangle inequality violations in Internet Coordinate Systems. The strength of our approach lies in the use of supervised learning methods, particularly decision trees, that successfully find a discriminative variable for TIV detection. A simple but effective method is developed that detects TIVs at each node based on the learned variable. This allows the detection algorithm to be embedded in the ICS algorithms.

One shortcoming of the detection algorithm is that a fixed cut point needs to be set for all nodes. Clearly, the performance of detection can be further improved by choosing an optimal cut point for each node autonomously. Another limitation is that the TIV detection is treated as a classification problem that classifies edges between nodes as TIV or non-TIV. In practice, it would be more useful to reason on the severity of the violations, which is a continuous value instead of a discrete level. This requires to solve a regression problem, which is expected to be more difficult than the classification problem addressed in this paper.

## References


1. Azureus Bittorrent, `http://azureus.sourceforge.net`
2. Ratnasamy, S., Handley, M., Karp, R., Shenker, S.: Topologically-aware overlay construction and server selection. In: Proc. IEEE INFOCOM, New York, NY, USA (June 2002)
3. Zhang, R., Tang, C., Hu, Y.C., Fahmy, S., Lin, X.: Impact of the inaccuracy of distance prediction algorithms on internet applications: an analytical and comparative study. In: Proc. IEEE INFOCOM, Barcelona, Spain (April 2006)

4. Ng, T.S.E., Zhang, H.: A network positioning system for the internet. In: Proc. of USENIX Annual Technical Conference (June 2004)
5. Dabek, F., Cox, R., Kaashoek, F., Morris, R.: Vivaldi: A decentralized network coordinate system. In: Proc. ACM SIGCOMM, Portland, OR, USA (August 2004)
6. Costa, M., Castro, M., Rowstron, R., Key, P.: Pic: practical internet coordinates for distance estimation. In: Proc. the ICDCS, pp. 178–187 (2004)
7. Zheng, H., Lua, E.K., Pias, M., Griffin, T.: Internet Routing Policies and Round-Trip-Times. In: Proc. the PAM Conference, Boston, MA, USA (April 2005)
8. Lee, S., Zhang, Z., Sahu, S., Saha, D.: On suitability of euclidean embedding of internet hosts. SIGMETRICS 34(1), 157–168 (2006)
9. Kaafar, M.A., Gueye, B., Cantin, F., Leduc, G., Mathy, L.: Towards a two-tier internet coordinate system to mitigate the impact of triangle inequality violations. In: Das, A., Pung, H.K., Lee, F.B.S., Wong, L.W.C. (eds.) NETWORKING 2008. LNCS, vol. 4982, pp. 397–408. Springer, Heidelberg (2008)
10. Kaafar, M.A., Cantin, F., Gueye, B., Leduc, G.: Detecting triangle inequality violations for internet coordinate systems. In: Proc. International Workshop on the Network of the Future, Dresden, Germany (June 2009)
11. Wang, G., Zhang, B., Ng, T.S.E.: Towards network triangle inequality violation aware distributed systems. In: Proc. the ACM/IMC Conference, San Diego, CA, USA, pp. 175–188 (October 2007)
12. Littman, M.L., Ravi, N., Fenson, E., Howard, R.: Reinforcement learning for autonomic network repair. In: 1st International Conference on Autonomic Computing, ICAC (2004)
13. Khayat, I.E., Geurts, P., Leduc, G.: Machine-learnt versus analytical models of TCP throughput. Computer Networks 51(10), 2631–2644 (2007)
14. A simulator for peer-to-peer protocols, `http://www.pdos.lcs.mit.edu/p2psim/index.html`
15. Wong, B., Slivkins, A., Sirer, E.: Meridian: A lightweight network location service without virtual coordinates. In: Proc. the ACM SIGCOMM (August 2005)
16. PEPITO: A data mining software, `http://www.pepite.be`
17. Fawcett, T.: Roc graphs: Notes and practical considerations for researchers. Technical report, HP Laboratories (March 2004)
18. Breiman, L., Friedman, J.H., Olshen, R.A., Stone, C.J.: Classification and Regression Trees. Wadsworth International, California (1984)

# Analysis of the Effects of XLFrames in a Network*

Dinil Mon Divakaran[1], Eitan Altman[2], Georg Post[3], Ludovic Noirie[3], and Pascale Vicat-Blanc Primet[1]

[1] INRIA / Université de Lyon / ENS Lyon,
LIP, ENS Lyon, Lyon - 69007, France
{Dinil.Mon.Divakaran,Pascale.Primet}@ens-lyon.fr
[2] INRIA
Eitan.Altman@sophia.inria.fr
[3] Alcatel-Lucent Bell Labs
{Georg.Post,Ludovic.Noirie}@alcatel-lucent.fr

**Abstract.** The phenomenal increase in network capacity to hundreds and thousands of Gbits/s in the core as well as Gbits/s at the access, is soon to witness stupendous amounts of packets that have to be processed and switched at amplifying line rates. Looking into the future, we address the need for the integration of packets of larger size, called XL-Frames (XLFs), into the future Internet framework. This paper analyses the effects of introducing XLFs in a network that has both packets and XLFs. We evaluate the gains in terms of processing power and throughput. As we observe that XLFs have an impact on loss rate and fairness, we study how, with minimal efforts at routers while keeping the existing protocols (TCP/UDP, IP), XLFs may integrate in the current scenario.

**Keywords:** Future Internet, Packet size, IP network.

## 1 Introduction

The phenomenal growth of the Internet has been accompanied by the increase in network capacity at a remarkable rate. With the core of the Internet moving from capacities of tens of Gbits/s to hundreds and thousands of Gbits/s, it is just a matter of time before the access links will have matching capacities to flood the network with mammoth data. Such an unprecedented growth brings along multiple challenges at end-hosts, switches and routers.

In this paper, we explore the concept of large packets, called *XLFrames* (XLFs), challenged by the problems facing future Internet. We perform studies to analyse the effect of introducing XLFs in a network with standard packets. Indeed, we foresee a future Internet that has classical packets as well as XLFs. Some of the reasons for keeping standard packets along with XLFs are: (i) Not

* This work was done in the framework of the INRIA and Alcatel-Lucent Bell Labs Joint Research Lab on Self Organized Networks.

L. Fratta et al. (Eds.): NETWORKING 2009, LNCS 5550, pp. 364–377, 2009.

all end-hosts will have the ability to send XLFs, in particular those fixed and mobile devices whose access rates stay below 100 Mbps. (ii) A TCP connection has short packets (iii) Some delay-constraint applications (say voice) might stick to packets (iv) Small flows, e.g HTTP requests, are expected to use packets as there might not be sufficient data to transfer.

We discuss our motivation and related works in Section 2. In Section 3, we analyse the gains in processing and throughput. The analyses on loss rates and unfairness are studied in Section 4. Therein, we present how existing standard mechanisms can be used to tackle these issues. The effect of having XLFs in wireless networks is studied in Section 5. We conclude in Section 6.

## 2 Motivations and Related Work

Huge amounts of traffic implies processing of large number of packets per unit time at terminals and network nodes. The increase in processing per unit of time also increases the power consumption at the equipments. The computing power required would increase too much unless some drastic measures are taken [1]. *Therefore minimizing power consumption at equipments is one important criteria that any research on Internet architecture should focus on* [2].

The growing line rate raises another major concern. The increase in processing power and memory speed is slower compared to the increase in transmission rates. For example, arrival of consecutive minimum-sized packets (say, 64B) at a router with 40 Gbps line cards, puts an upper bound of 12.8 $ns$ on processing time and memory access time. Parallel, pipelined processing and the use of fast SRAM can keep up with reduced inter-packet time only at increased complexity and cost. As the bottleneck moves from transmission capacity to processing power and/or memory access time, the maximum throughput achievable by a flow becomes less than the line capacity. *Achieving maximum throughput is another goal*

The cost of performing packet-level functions at line rate can be partitioned into two: per-byte processing cost and per-packet processing cost. Storing and retrieving packets to and from memory obviously involves per-byte cost (in fact, it depends on the word size). Functions such as route lookup, classification, arbitration, scheduling etc. have per-packet costs. Besides, flow-level functions also add to per-packet cost. If the trend towards flow-aware networking is anything to go by [3], additional flow-level functions (such as flow table lookup, estimation of flow parameters, flow policing etc.) will add to per-packet cost. *Per-packet processing cost reduces if the number of packets that need to be processed per unit time is reduced.*

At an end-host, the costs in terms of protocol processing and interrupt handling for packets make it challenging to achieve 100% throughput at high line rates [4]. According to a study [5], it is hardly possible to achieve 1 $bps$ for every 1 $Hz$. The authors also note that packet transmission/reception form a substantial part (28 - 40%) of commercial server workloads. Though methods like TOE (TCP Offload Engine) and packet coalescence have been proposed, it is not clear if these are long term solutions. To achieve high throughput, and at the same time, to save CPU cycles for other applications, *it is necessary to minimize the processing cost involved in protocol processing and interrupt handling.*

We argue that all the above points are sound motivating factors to break the barrier of traditional packet size; and look into a future where data is transported in *XLFs*.

### 2.1 Related Work

The concept of large packet sizes has been floating around for a decade [6,7]. *Jumbo frames* was introduced to solve some of the end-host related problems. Now, we have more reasons, mainly from the network perspective (as described above), to pursue the idea of large packet size. The limiting factor of the MTU comes from the early Ethernet designs. But today, we have Gigabit Ethernet NICs that support packet sizes of 9000B and even larger [8,9,10]. Research networks such as Internet2 and GEANT also support XLFs. Wang *et al.* showed that in Ethernet-based storage area networks, the use of XLFs reduce CPU utilization and interrupts notably, while improving the throughput during data transfers [11]. The use of larger MTUs also contributes to high throughput achieved on iSCSI storage networks [12].

Researchers have also put forward the idea of aggregating packets in the network, at the edges. In [13], the authors propose to dynamically encapsulate packets into large packet at the ingress of a domain, and sent out to the network with an additional new header. At the egress of the domain the original packets are decapsulated from the large packet and transmitted to the destination. A timer is used to decide on the number of packets that will be encapsulated. This is similar to the burst assembly process in OBS (Optical Burst Switching), one of the paradigms in optical networks [14]. The burst assembly process is known to have a significant impact on TCP performance [15]. Short assembly times hinder the congestion window growing pace, and long assembly time causes unacceptable delays. In addition, the burstification might also lead to synchronization of TCP flows, resulting in inefficient bandwidth utilization.

## 3 Benefits of Introducing XLFs in a Packet Network

In this section, we analyse the benefits of XLF introduction and also provide an insight into the main drawback. We focus on the following questions: (i) What is the gain in terms of processing power? (ii) What is the gain in terms of throughput? We study these two questions first through simulations.

### 3.1 Simulation Setup

We simulate the multiplexing of TCP flows with small and large packets using NS-2; for all the simulations, we consider a dumbbell topology, connecting $n$ *src-dst* pairs (see Fig. 1). Packet size is 1500B. XLF size is in number of packets; i.e.,

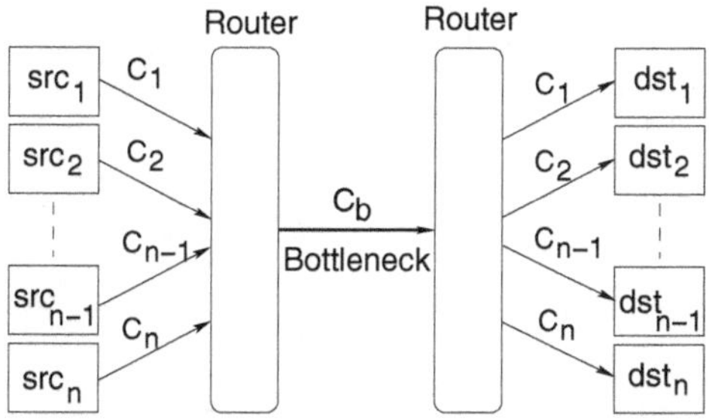


**Fig. 1.** Topology considered for simulations

an XLF of size $F$ has $F \times 1500$B in it. We use *data units* to refer to a transporting unit independent of its size (a data unit can be a packet or an XLF). Fixing the bottleneck capacity as $C_b$, the capacities of links from *src* nodes to the bottleneck, and from the bottleneck to *dst* nodes, are set depending on the level of congestion desired for the scenario. The congestion factor $\rho = \sum_{i=1}^{n} C_i/C_b$, $C_i$ being the capacity of link $i$.

The size of the bottleneck queue is set in bytes, as the bandwidth delay product ($BDP$) for 100 ms. The buffer size at each source is $1.2 \times BDP$. TCP window size is set high enough so as to be limited only by the network (and not by the end-hosts). TCP Sack is used in all scenarios, with no delayed ACKs. Since simulating events using links with 10 Gbps or higher has practical difficulties, we set $C_b = 1$ *Gbps* for simulations. Simulations are run for 540s . For all measurements, we ignore the first 60s of the simulation to avoid transient states. The metrics measured are throughput and drop rate of individual and aggregate flows, all measurements being made at the bottleneck link. For a single flow, the throughput is measured as the number of bytes transmitted during an interval; whereas the drop rate equals the number of data units dropped (due to the queue being full) over the total number of data units sent.

### 3.2 Processing Gain

The reduction in the number of data units at an equipment results in processing gain. A recent study shows, the packet size distribution is no more trimodal, but rather bimodal, with nearly 50% of packet lengths between 40 and 100B, and around 40% between 1400 and 1500 bytes [16]. Since it is also well-known that TCP contributes more than 90% in bytes as well as packets seen in the Internet traffic, it can be concluded that a majority of the small-sized packets are TCP ACKs (around 40% of IP packets). The use of XLFs for large flows reduces the number of data units as well as ACKs.

We illustrate this using numerical analysis. TCP flow sizes were generated using Pareto distribution (with $\alpha = 1.2$) for varying mean flow sizes. Number of data units required to transfer each flow was estimated for a given flow size (this included TCP control packets too). In packet-switched architecture, all flows are switched in packets of size 1500B. In order to estimate the number of data units

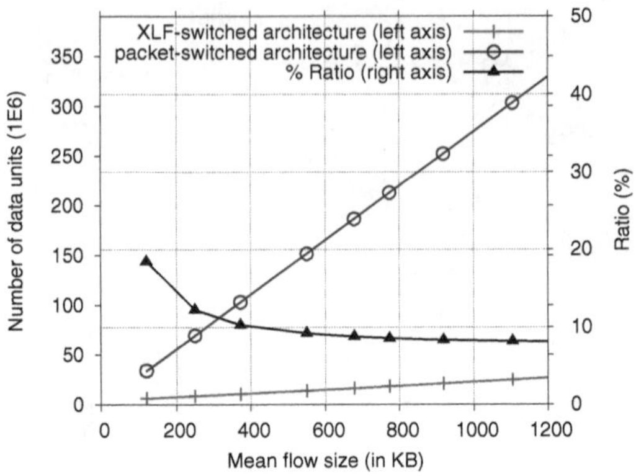


**Fig. 2.** Comparison of packet counts

required to transfer flows in an *XLF-switched architecture*, we assume flows of size greater than 50kB to be switched in packets of size 19500B ($F = 13$) and the rest in 1500B packets. The volume of bytes contained in flows of size greater than 50kB is 90% (of a total of 20GB) when the mean flow size is 100kB, and more than 99% (of a total of around 200GB) when the mean flow size is 1MB. Fig. 2 compares the number of data units in XLF-switched and packet-switched architectures. The ratio of data units in XLF-switched architecture to that of packet-switched architecture is also shown in the same figure, but on the right axis. For a distribution of flow sizes with a mean of 1MB, the total number of data units in XLF-switched architecture is just around 8% of that required in packet-switching architecture.

The reduction in the number of packets also reduces the number of arbitration decisions required to switch packets from the inputs to the outputs. Though this is not possible in traditional switches using cell-based switching (where variable length packets are segmented into fixed size cells), it should be noted that with the feasibility of adding buffers in a crossbar switch, recent research works have explored asynchronous buffered crossbar architecture, proposing it as the next level of crossbar switches that can scale to hundreds of ports [17,18]. As the units being switched in such architectures are variable-size packets, the gain due to XLFs is directly proportional to the reduction in the number of packets.

### 3.3 Throughput Gains

To estimate the gain in throughput when a flow is switched in XLFs, consider packet processing time in high-capacity switches. For example, processing 64B packets at 100 Gbps represents about 200M packets/s, whereas with 1500B packet size, the required speed decreases to 8.3M packets/s, and with even larger packets of, say 19500B, the required speed is only 640k packets/s. Of course, real traffic is a mix of different packet sizes, but the mean time between packet operations will increase with average size. With appropriate implementation of electronics, significant saving of electrical power can be achieved in the packet-header processing parts of an equipment.

We multiplex two TCP flows, one in packets and the other in XLFs, using NS-2 for a processing delay equal to the transmission delay. $C_b = 1$ Gbps, and $\rho = 1.2$; that is $C_1 = C_2 = 600$ Mbps. The results are displayed in Fig. 3. Observe that the aggregate throughput achieved is minimum when $F = 1$ (packet size = XLF size = 1500B). Higher link utilization is achieved with larger XLFs. In fact, even the packet-switched flow achieves higher throughput when the other flow is switched in XLFs. The drop rates are as seen in Fig 3(b). As expected, drop rates of XLFs is higher than that of packets. The evolutions of TCP congestion window (*cwnd*) for both flows are plotted in Fig. 3(c). This gives insights into the higher drop rates experienced by XLFs. XLFs of size $F$ arriving at a queue with space for just $F-1$ packets get rejected, whereas, packets are still accepted. Additionally, the slowing down of the XLF-switched flow makes more space in the queue, resulting in the increase in *cwnd* of packet-switched flow.

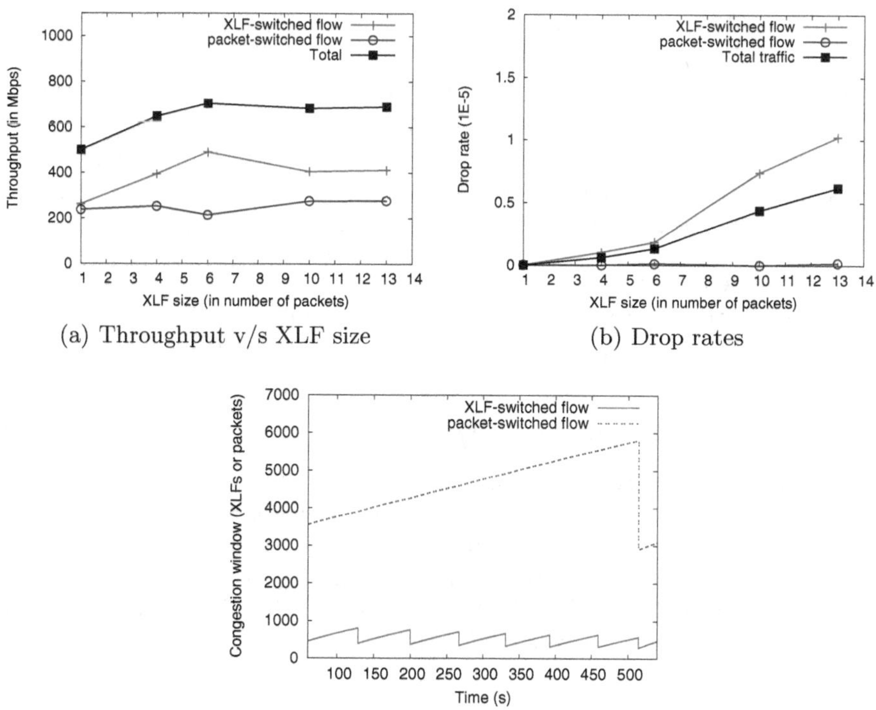


(a) Throughput v/s XLF size

(b) Drop rates

(c) Evolution of *cwnd*, $F = 13$

**Fig. 3.** Flows through Droptail queue, $\rho = 1.2$

# 4 Adapting the Protocols and Mechanisms to XLFs

In the previous section, simulation results highlighted the impact of packet size on the loss rate and the resulting unfairness. In this section, we explore how, without changing the existing protocols (TCP/UDP, IP), XLFs integrate in the current scenario.

## 4.1 XLF and Loss Probability

To understand the loss rate issue, we first analyse the losses at a queue that has arrivals in packets as well as XLFs using a Markov model. $\lambda_p$ is the packet arrival rate, and $\lambda_f$ is the XLF arrival rate; both arrival processes are assumed to be Poisson in nature. $\lambda = \lambda_p + \lambda_f$. The service time of packets is exponentially distributed with mean rate $\mu$. The Markov model is as given in Fig. 4.

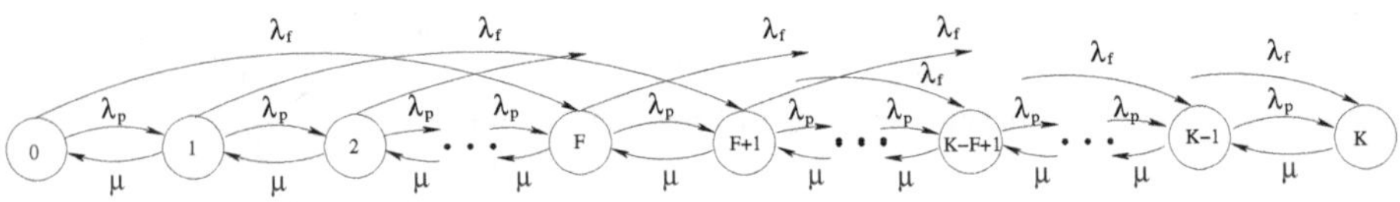

**Fig. 4.** Markovian queue with packet and XLF arrivals

The queue is measured in packets. An XLF arrival brings $F$ packets. Hence an arrival of an XLF when the state is $i$, results in a transition to state $i + F$, provided $i + F < K$, where $K$ is the queue size in packets. Similarly, it takes $F$ stages of services to completely serve an XLF. If $\pi_i$ is the probability for having $i$ packets in the queue, the steady-state equations are given by,

$$\begin{aligned}
\pi_0 \lambda &= \pi_1 \mu \\
\pi_i(\lambda + \mu) &= \pi_{i-1}\lambda_p + \pi_{i+1}\mu \qquad 1 \le i \le F-1 \\
\pi_i(\lambda + \mu) &= \pi_{i-F}\lambda_f + \pi_{i-1}\lambda_p + \pi_{i+1}\mu \qquad F \le i \le K-F \\
\pi_i(\lambda_p + \mu) &= \pi_{i-F}\lambda_f + \pi_{i-1}\lambda_p + \pi_{i+1}\mu \qquad K-F+1 \le i \le K-1 \\
\pi_K \mu &= \pi_{K-1}\lambda_p + \pi_{K-F}\lambda_f
\end{aligned} \tag{1}$$

No more XLFs can be accepted if the queue is any state $k$ such that $k \ge K-F+1$. Packet loss probability is $\pi_K$. Fig. 5 shows the loss probabilities for packets and XLFs for increasing XLF size. The load brought by the packets was set equal to that brought by XLFs, and they were kept constant so as to have a constant load for varying XLF size. The buffer size was set to 200 packets. The losses are seen to increase; and more importantly, the losses experienced by XLFs are increasing by orders of magnitude.

## 4.2 XLFs and Unfairness

As the introduction of XLFs in packet networks increases the losses of both packets and XLFs, we analyse the effects on individual and aggregated throughput when classical queuing and scheduling mechanisms are used. We simulate

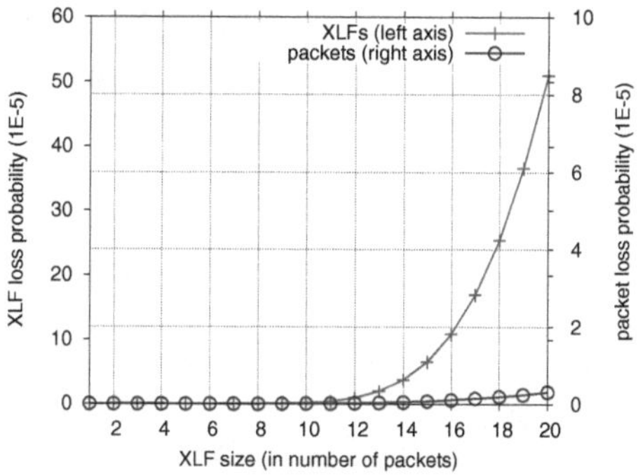


**Fig. 5.** Loss probabilities as a function of XLF size

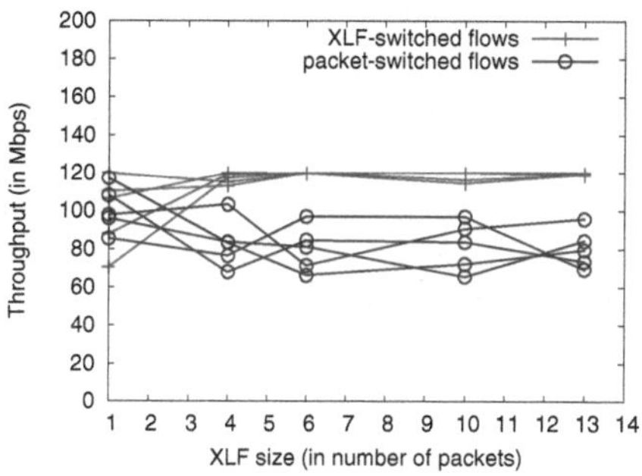


(a) Five XLF-switched flows

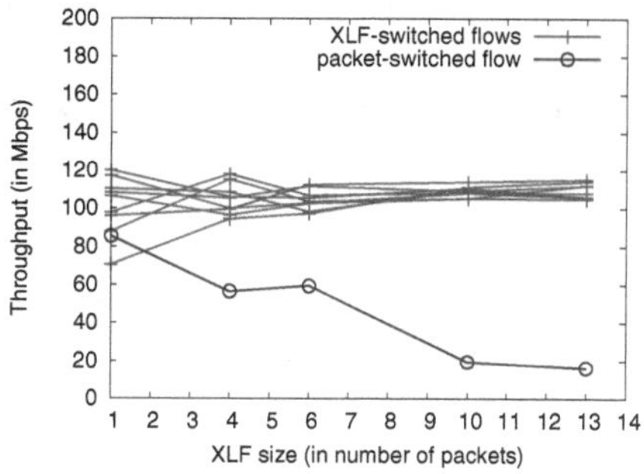


(b) Nine XLF-switched flows

**Fig. 6.** Throughput of flows through Droptail queue, $\rho = 1.2$

the basic scenario using Droptail queues and FCFS discipline. We assume that the effect of processing delay (on throughput) is negligible due to pipelining, as is the case today. We consider 10 parallel flows, each being switched either in packets or in XLFs. $\rho = 1.2$, $C_b = 1$ *Gbps*. The aggregate throughput achieved is always more than 999 Mbps, and hence not shown.

Fig. 6 displays the throughputs of 10 TCP flows. In Fig. 6(a), five of them are switched in XLFs and the rest in packets, whereas nine of the 10 are XLF-switched flows in Fig. 6(b). The drop rates for individual flows and aggregate traffic for the scenario corresponding to Fig. 6(b) are as seen in Fig. 7. It is to be noted that packet-switched flow gets lesser bandwidth as the number of XLF-switched flow increases, causing unfairness. Besides, XLFs experience higher drop rates.

## 4.3 Using Deficit Round Robin (DRR) for Achieving Fairness

To achieve fairness (in terms of throughput) among competing flows of very different packet sizes, we propose to use DRR (instead of FCFS) for scheduling

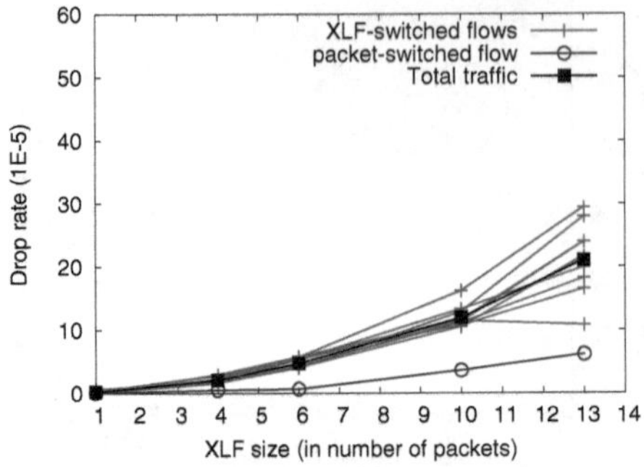


**Fig. 7.** Drop rates. Nine XLF-switched flows.

the data units. We simulate the same scenarios (as in the previous section) using DRR for scheduling packets and XLFs. The results for nine XLF-switched flows and one packet-switched flow are plotted in Fig. 8. We observe from Fig. 8(a) that each flow gets equal share of the bottleneck bandwidth. The results are similar for varying number of XLF-switched flows, though not plotted here. The drop rates of XLFs are seen to be high. Packet drop rate remains more or less a constant as expected from the standard approximation (see discussion below).

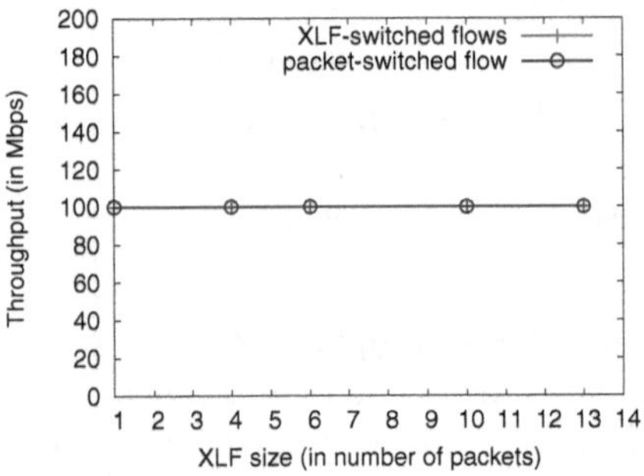


(a) Individual throughput

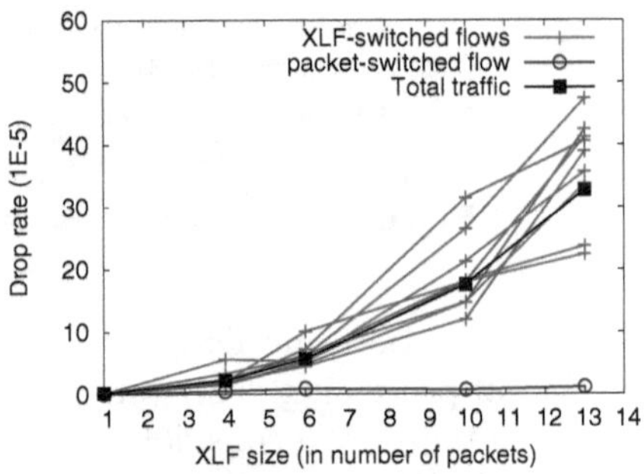


(b) Drop rates of each flow

**Fig. 8.** Throughput and drop rates using DRR, $\rho = 1.2$

Observing the ratio of XLF drop rate to packet drop rate, say $\beta$, we notice that $\beta$ has higher values in simulations using DRR compared to those without DRR. In the standard approximation of throughput [19],

$$\phi \approx \frac{C \times S}{RTT \times l^k} \tag{2}$$

where $l$ is the packet loss event rate, and $C$ a constant. $k$ is usually around 0.5. Using subscripts $f$ and $p$ for XLF and packet respectively, and $\beta'$ to represent the ratio of XLF to packet loss event rates ($l_f : l_p$), $\beta' = (\frac{F}{\phi_f/\phi_p})^2$. Note that the throughput formula uses loss event rate, where a loss event corresponds to one or more packet losses within a single RTT. As the loss (drop) probability increases, the probability that the losses are in a single RTT is higher for small packets. Hence $\beta' \geq \beta$. Nevertheless, this explains the high values for $\beta$ with DRR scheduling: as $\phi_p = \phi_f$, the ratio grows with the square of XLF size, a trend well observed in Fig. 8(b).

For the Droptail, assume that the drop probability per arriving bit is constant and so the packet loss rate is proportional to the size $S$ of a packet. From the above throughput approximation, it follows immediately that, $\phi \propto S/sqrt(S) = sqrt(S)$, explaining the observed unfairness of the Droptail model: flows receive throughput in proportion to the square root of their packet sizes. A simulation with 50 flows was performed. The dumbbell topology had 25 branches. There were flows in both directions, and packet sizes were linearly spaced from 1000B to 20000B. Each source had 1 Gbps capacity, and so did the bottleneck. Fig. 9 shows the throughput and loss behaviour for flows with different packet sizes.

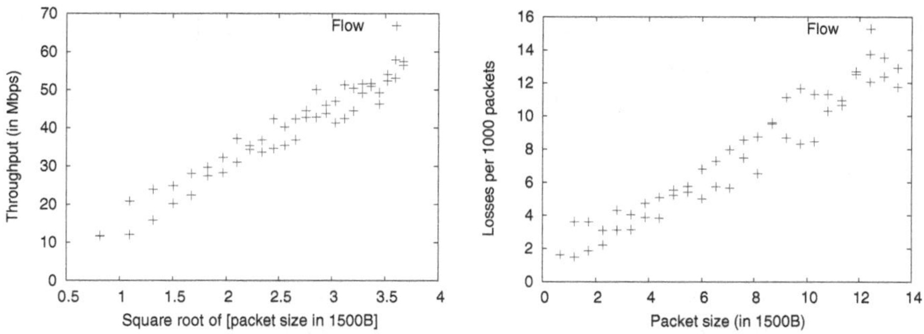


(a) Throughput behaviour (b) Loss behaviour

**Fig. 9.** Analysis of 50 flows through a Droptail queue

### 4.4 Loss Reduction Using ECN (Explicit Congestion Notification)

This section motivates the use of ECN, an IETF standard option for packet marking in routers, to reduce the high drop rates experienced by XLFs. At ECN-enabled routers, instead of being dropped, packets are marked before the queue starts to overflow. This makes it necessary to estimate the queue length, and accordingly mark the packets with some probability. At the same time, we also use DRR scheduling to ensure fairness among flows with varying packet lengths. To simulate this setup, we incorporate code in NS-2 to probabilistically mark the packets, depending on the length of the common queue[1]. Fig. 10 compares the drop rates experienced by flows switched in XLFs and packets, when nine out of ten flows were XLF-switched flows. Evidently, the drop rates have gone down very low. Obviously, due to the DRR scheduling there was also no unfairness among flows.

[1] Even though NS-2 has RED implementation, its design rules out the use of RED and DRR scheduling at the same time.

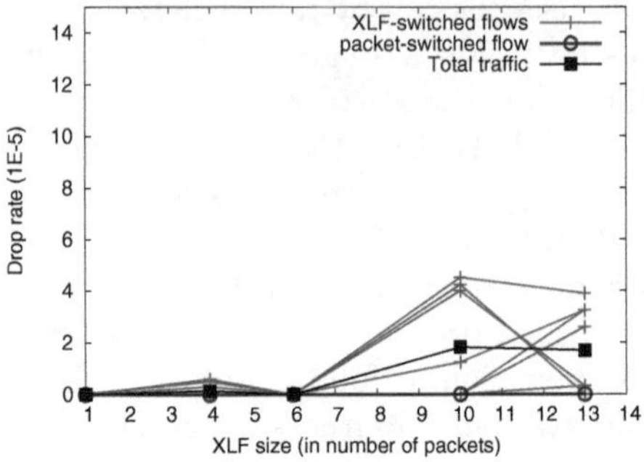


**Fig. 10.** Drop rates using ECN with DRR. Nine XLF-switched flows.

## 5 XLFs in Wireless Networks

Nowadays, access to the network is often through some wireless channels that introduce random drop of packets. We show below that if TCP is used, then its throughput may be very sensitive to the choice of packet size, making the access point inefficient. Assume that the probability for an erroneous bit is $p$. The probability to lose a packet of size $N$ is

$$P_l(N) = 1 - (1-p)^N \tag{3}$$

Throughput in packets/s is proportional to $1/\sqrt{(P_l(N))}$, and the goodput is thus proportional to $(1 - P_l(N))/\sqrt{(P_l(N))}$; For small $p$,

$$P_l(N) = 1 - [(1-p)^{(N/p)}]^p \sim 1 - \exp(-Np) \tag{4}$$

To obtain the best packet size $N$ we need to maximize

$$Goodput(N) = \frac{N \exp(-Np)}{\sqrt{1 - \exp(-Np)}} c \tag{5}$$

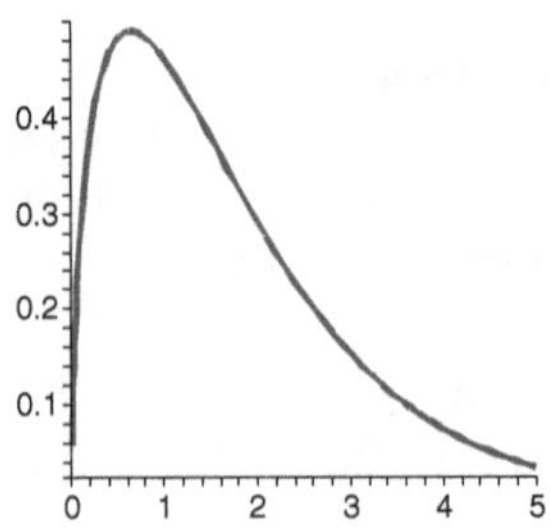


**Fig. 11.** Normalized goodput as a function of $\theta$

where $c$ is a constant. We write

$$Goodput = \frac{\theta \exp(-\theta)}{p\sqrt{1-\exp(-\theta)}}c \tag{6}$$

where $\theta = Np$. Fig. 11 shows the goodput in units of length per second, where we take a unit to be $c/p$, as a function of the normalized packet length $\theta$. The function has a unique maximum at $\theta = 0.6438$ obtained by differentiating the Goodput with respect to $\theta$ and equating to zero. We obtain,

**Lemma 1.** *The optimal throughput is obtained for the packet size $N = 0.6438/p$, where $p$ is the bit-error rate.*

Hence we propose, packets be put into XLFs after the access which suggests using Split TCP, or that the lower layers should include coding or ARQ so as to decrease the bit error rate.

### 5.1 Using Block FEC

Block FEC can be used to reduce the loss rates in wireless networks. We use the Gilbert loss model and the FEC model described in [20]. A link-level packet is sent as multiple smaller units called transmission units (TUs in short). Let $p$ be the probability of losing a TU, and let $N$ be the packet size in TUs. $N = K + R$, where $R$ is the number of redundant TUs and $K$ is the number of *useful* TUs. In block FEC, such a packet of $N$ TUs is lost, only if more than $R$ TUs are lost. The probability of losing more than $R$ TUs is,

$$P_{(R,N)} = 1 - \sum_{i=0}^{R} \binom{N}{i} p^i (1-p)^{N-i}$$

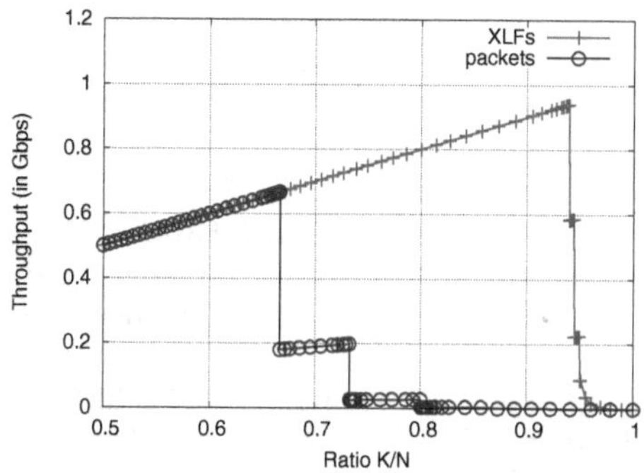


**Fig. 12.** Throughput as a function of K/N

For the throughput of a TCP flow, we use the formula, $\phi = \frac{S}{RTT}\sqrt{\frac{3}{2P}}$, where $P$ is the probability of losing a packet. The throughput under the FEC scheme with parameter $R$ is,

$$\phi_{(R,N)} = min\left(\frac{S}{RTT}\sqrt{\frac{3}{2P_{(R,N)}}}, \frac{K}{N}C\right)$$

where $C$ is the capacity of the (bottlenecked) wireless link. Fig. 12 shows the throughput as a function of the ratio of $K/N$. The transmission unit is assumed to be 100 bytes. $N$ is then set to 15 (TUs) for packets, and to 200 for XLFs. The capacity $C$ was set to 1 Gbps. We see that, it is possible to achieve higher throughput with considerably larger K/N ratio for XLFs.

## 6 Conclusions

Through arguments, analyses and simulations, this article reasoned the need for having XLFs in future Internet. Numerical analysis showed that the number of data units at an equipment can be brought down to as low as 8% by the use of XLFs along with packets, thus bringing savings in processing resources and electrical power. In the case where processing is the bottleneck, achievable throughput is close to the line rate when XLFs are used. Increase in packet size by an order also decreases the speed required to process packets by an order. Though XLFs introduce unfairness without any fair queueing algorithm, simple round robin mechanisms like DRR has been illustrated as solving this issue. The high drop rates experienced by XLFs is brought down by using ECN feature. Finally, we analyse the redundancy that has to be added in the wireless access network integrating XLFs. These analyses have shown that XLFs will bring a lot of benefits and can be integrated with light adaptation of current mechanisms and no changes to the standard protocols.

## References


1. Ye, T.T., Micheli, G.D., Benini, L.: Analysis of power consumption on switch fabrics in network routers. In: DAC 2002: Proc. of the 39th Conf. on Design automation, pp. 524–529 (2002)
2. Telcos demand greener network equipment (2008), `http://www.reuters.com/article/idUSN1847837420080619`
3. Bonald, T., Oueslati-Boulahia, S., Roberts, J.: IP traffic and QoS control: the need for a flow-aware architecture. In: World Telecommunications Congress (September 2002)
4. Feng, W., Balaji, P., Baron, C., Bhuyan, L.N., Panda, D.K.: Performance Characterization of a 10-Gigabit Ethernet TOE. In: HOTI 2005: Proc. of the 13th Symposium on High Perf. Interconnects, pp. 58–63 (2005)
5. Makineni, S., Iyer, R.: Architectural Characterization of TCP/IP Packet Processing on the Pentium®M Microprocessor. In: HPCA 2004, pp. 152–161 (2004)
6. Dykstra, P.: Gigabit ethernet jumbo frames and why you should care (1999), `http://sd.wareonearth.com/~phil/jumbo.html`
7. Mathis, M.: Pushing up the Internet MTU. Internet2/NLANR Joint Techs Meeting, Miami, Florida (March 2003)
8. `http://www.intel.com/support/network/sb/cs-001911.htm`

9. Myri-10G: Myrinet Converges with Ethernet, http://www.myri.com/news/051121/Myricom_SC05_Myri-10G.pdf
10. http://kbserver.netgear.com/kb_web_files/n101539.asp
11. Wang, W.Y.H., Yeo, H.N., Zhu, Y.L., Chong, T.C., Chai, T.Y., Zhou, L., Bitwas, J.: Design and development of Ethernet-based storage area network protocol. Computer Communications 29(9), 1271–1283 (2006)
12. Simitci, H., Malakapalli, C., Gunturu, V.: Evaluation of SCSI over TCP/IP and SCSI over fibre channel connections. In: Hot Interconnects 2001, vol. 9, pp. 87–91 (2001)
13. Salyers, D., Jiang, Y., Striegel, A., Poellabauer, C.: JumboGen: dynamic jumbo frame generation for network performance scalability. SIGCOMM CCR 37(5), 53–64 (2007)
14. Qiao, C., Yoo, M.: Optical burst switching (OBS) - a new paradigm for an optical Internet. J. High Speed Netw. 8(1), 69–84 (1999)
15. Vlachos, K.: Burstification effect on the TCP synchronization and congestion window mechanism. In: BROADNETS, pp. 24–28 (2007)
16. John, W., Tafvelin, S.: Analysis of Internet backbone traffic and header anomalies observed. In: IMC 2007, pp. 111–116 (2007)
17. Katevenis, M., Passas, G., Simos, D., Papaefstathiou, I., Chrysos, N.: Variable packet size buffered crossbar (CICQ) switches. In: ICC 2004, June 2004, vol. 2, pp. 1090–1096 (2004)
18. Chrysos, N., Katevenis, M.: Scheduling in non-blocking buffered three-stage switching fabrics. In: IEEE INFOCOM (April 2006)
19. Mathis, M., Semke, J., Mahdavi, J.: The macroscopic behavior of the TCP congestion avoidance algorithm. SIGCOMM CCR 27(3), 67–82 (1997)
20. Barakat, C., Altman, E.: Bandwidth tradeoff between tcp and link-level fec. Comput. Netw. 39(2), 133–150 (2002)

# Complementing TCP Congestion Control with Forward Error Correction

Vicky Sharma[1], Kadangode Ramakrishnan[2], Koushik Kar[3], and Shivkumar Kalyanaraman[4]

[1] Rensselaer Polytechnic Institute, Troy, NY 12180
sharmv@rpi.edu
[2] AT&T Labs Research, Florham Park, NJ 07932
kkrama@research.att.com
[3] Rensselaer Polytechnic Institute, Troy, NY 12180
koushik@ecse.rpi.edu
[4] IBM Research, Bangalore, India
shivkumar-k@in.ibm.com

**Abstract.** In this paper, we examine an emerging combination of challenges for TCP: increasingly bursty background traffic that is not subject to flow and congestion control, higher bandwidth networks and small buffers at network routers. As a result, TCP experiences short-term bursty packet losses that may not reflect long-term congestion. In this work, we propose a balanced approach that uses TCP congestion control mechanisms including Explicit Congestion Notification (ECN) to identify and overcome congestion, supported by adaptive FEC for short-term packet loss recovery due to bursty flows. We demonstrate the effectiveness of using FEC with TCP SACK congestion control by showing that such an approach improves performance for TCP flows for the emerging bursty traffic, small buffer, high bandwidth-delay product environments.

**Keywords:** Next-generation transport, Congestion control, Loss recovery.

## 1 Introduction

TCP has been remarkably successful as a transport protocol, and has evolved over the years to deliver the two core functions of reliability and congestion control. In this paper, we examine potential ways in which TCP may evolve to address a new combination of emerging challenges to TCP's performance.

The first challenge for TCP is to co exist with the exploding growth of multi media communications, including streaming and interactive video applications. Multimedia applications predominantly use RTP/UDP, and manifest themselves as bursty background traffic to regular TCP flows.

The second challenge is the increasing capacity of the end to end communication path coupled with smaller buffers at core routers (relative to the bandwidth delay product). The purpose of buffering is to absorb short term burstiness, so

L. Fratta et al. (Eds.): NETWORKING 2009, LNCS 5550, pp. 378–391, 2009.

that any loss signal received by the end system is a reasonably unambiguous indication of medium to long term congestion, in terms of the round trip time and the time scales of end to end congestion control, and not burstiness. However, small buffers when subjected to background traffic that is highly bursty result in buffer overflows that are correspondingly bursty. Packet loss is transient and short lived, operating in time scales well beyond the dynamic range of end to end feedback based congestion control mechanisms. Importantly, such packet loss patterns significantly blur the distinction between medium term congestion and short term burstiness. TCP will increasingly confuse short term burstiness for medium term congestion, and cut its transmission window multiplicatively (following the traditional AIMD principle, and even with the newer versions that are more aggressive in their increase policies) each time, thereby paying a significant performance penalty. Worse yet, TCP may also timeout during larger bursts when a complete window of packets or a retransmission is lost. Such timeouts are a more severe response to the bursty cross traffic than is "truly necessary", and recovery from such "mistakes" takes several round trip times.

The right response to these short term bursty losses is to combat it with short term recovery mechanisms that don't necessarily operate in the larger multiple RTT time scales of feedback based congestion control schemes. Approaches that would help TCP recover from short term transient losses can greatly improve TCP's goodput and delay performance in these situations. Forward Error Correction (FEC) coding presents an alternative. If packets at the transport layer are FEC coded, the receiver would be able to recover the data even when some packets have been lost in transmission. This "insurance" like property of FEC coding results in a reduction of packet re transmissions and consequently, saves time and bandwidth.

The role of FEC coding in recovering from losses due to noisy/lossy links has been explored earlier [1],[2],[3]. In prior work, we developed enhancements to TCP like transport protocols MPLOT and LT TCP in [1] and [2] (for multiple paths and single path respectively) that employed FEC coding at the transport layer as a means to counter packet losses from noisy/lossy wireless links. Our protocols distinguished congestion losses from link losses by using the Explicit Congestion Notification (ECN) [4] as definitive indicator of congestion. Another useful feature of these protocols was the dynamic adaptation of FEC coding to match the losses incurred in the network. This helped in minimizing bandwidth wastage due to coding overhead. The end to end packet delay was also reduced due to the reduction in re transmissions needed to recover from losses. We showed in [1],[2] that such features are very effective in recovering from losses due to noisy links. As a result, the enhanced transport achieves a significantly higher goodput and lower delay than existing TCP protocols.

In this paper, we are interested in the question whether adaptive FEC mechanisms have a role beyond lossy wireless environments, and in the overcoming the challenge of short term transient losses in high bandwidth delay product networks with small buffers, combined with the presence of non congestion controlled traffic such as UDP. We show that adaptive FEC mechanisms can be used to recover more efficiently from bursty losses in such situations. Beyond loss recovery, we are also interested in complementing congestion control by distinguishing true congestion from short term bursty losses. In this paper, we propose a balanced approach for TCP congestion control evolution: to use explicit congestion notification (ECN) to respond to congestion, supported by adaptive FEC for short term packet recovery. While fixed coding overhead may be undesirable as it reduces goodput, we show that adaptive coding feature of MPLOT and LT TCP, on the other hand, actually boosts goodput. We demonstrate that this approach results in a much smoother degradation in goodput with increasing packet losses as compared to conventional TCP SACK. This is one of our fundamental goals to achieve a much more gradual degradation in goodput as compared to the highly non linear (exponential) decay in goodput seen by TCP applications as the level of transient overload increases. It is worth noting that our scheme does not make TCP more aggressive, or send beyond the constraints placed by congestion control mechanisms such as self clocking, congestion window, AIMD, timeout window reduction etc. We use FEC primarily as a reliability technique to enhance TCP SACK, and it has collateral benefits of helping in "brittle" phases of congestion control caused by burstiness and small buffers. In essence, our paper makes the case that FEC as a reliability technique has broader applicability beyond lossy wireless environments, and can complement TCP SACK's congestion control mechanisms for emerging bursty, small buffer, high bandwidth delay product environments.

The paper is structured as follows: in the next section, we discuss the motivation for our work along with the related work, followed by a brief discussion on transport layer FEC coding and how TCP SACK can be modified to make effective use of FEC. We then discuss the results of our simulation based evaluation and conclude by summarizing our results.

## 2 Motivation and Related Work

Liu *et al.* [5] have observed that medium/high speed streams exhibited high burstiness due to packet accumulation. One stream was observed to peak at 600 Mb/s for a few micro-seconds when it transferred only 3.1 Mb/s in 3 seconds. Fitzek *et al.* [6] observed peak-to-mean values between 15 and 45 for bit-rates of audio/video streams. These studies indicated a high burstiness in traffic patterns for new streaming applications.

Appenseller *et al.* [7] showed that router buffer sizes can be much smaller than the end-to-end bandwidth-delay product. Their theoretical results showed that the router buffer size is bounded by $1/\sqrt{n}$ of the bandwidth-delay product when a link is being used by $n$ identical TCP-like flows. Mascolo *et al.* [8] take this study a step further by proposing a TCP congestion control framework aimed at minimizing end-to-end bandwidth loss due to congestion. However, the results in [7],[8] were based on the assumption that only identical TCP-like flows use the network. Furthermore, in deriving bounds on the buffer sizes, these works only focus on maintaining a desired throughput value, and ignore the goodput metric which is more relevant in practice.

A bursty unresponsive flow (e.g. UDP) can quickly overflow small queue buffers leading to congestion losses for TCP flows. Consequently, TCP flows would reduce their packet rates, thereby reducing throughput. Due to packet losses, a TCP flow can take multiple Round Trip Times (RTTs) to recover lost packets by re-transmissions. Due to the nature (AIMD policy) of TCP congestion control, after a packet loss has occurred, it can take a long time for a TCP flow to build up it's throughput and goodput to levels before the congestion losses. Consequently, a TCP flow will potentially lose significant amounts of bandwidth and incur large delays due to the bursty losses caused by an unresponsive flow.

In presence of such bursty losses, therefore, it serves TCP well to recover from losses quickly to reclaim lost bandwidth. FEC has been used to recover from packet losses due to noisy links in [1],[2] and [3]. Hayasaka *et al.* [9] use FEC to recover lost video packets due to congestion. Their solution needs a long buffering time to transmit FEC packets before video transmission, however. Yu *et al.* [10] use simplistic M/M/1 queueing models to derive optimal block size and code-rate for FEC to recover from congestion losses, but only under very idealistic conditions and assumptions (like all flows use TCP and are identical) which may not hold in reality. Li *et al.* [11] use fixed coding overhead and flexible packet scheduling to counter packet losses from noisy links. The packet scheduling is varied to control the average loss-rate. A fixed coding rate approach may however reduce goodput due to unnecessary coding overhead, as we demonstrate later in this paper. Nguyen *et al.* use an exhaustive line-search to schedule a block of coded packets for minimizing the likelihood that the receiver would not be able to recover lost data. Here the authors only consider bandwidth constraints in their work, ignoring any increments in delays incurred in the process.

Ahlswede *et al.* [12] present network-coding as a possible solution to recover from losses in multi-cast applications. Application of network-coding has not been applied/studied for unicast TCP flows extensively, however. There have been a few attempts to study the effects of network-coding on TCP flows, but in the context of wireless networks. For example, Huang *et al.* [13] and Ghaderi *et al.* [14] study the impact of network-coding on unicast TCP flows running over wireless mesh networks, and conclude that significant gains cannot be attained unless the MAC layer is significantly changed.

We note that FEC has been primarily used to recover from non-congestion losses. The overhead required by FEC is an undesirable feature which can potentially reduce goodput (compared to the case with no coding) if FEC provisioning is not done appropriately. In the next section we show how FEC can be used intelligently to employ it's loss-tolerance properties while limiting the overhead incurred to negligible levels in presence of bursty non-responsive flows. The basic idea is to employ FEC only when losses are incurred, and in proportion to the amount of losses incurred. We then show through simulations that our proposal benefits TCP by allowing it to recover from losses and obtain more goodput than the conventional TCP-SACK.

# 3 Packet Recovery Using FEC Coding

## 3.1 FEC Coding at the Transport Layer

In this paper, we use the term FEC to refer to erasure codes which have excellent loss tolerance properties. In brief, a $(n, k)$ block FEC code adds $(n-k)$ redundant units to $k$ data units in such a fashion that the original $k$ data units can be recovered from any of the $k$ units. This implies that a few FEC coded packets can be lost during transmission, but the receiver would still be able to decode and recover the original data packets from the remaining packets received.

Since our goal is to complement TCP congestion control with FEC, we propose to code packets (or TCP segments) at the transport layer itself. As a result, data can be recovered at the receiver even when a few packets are lost due to short-term congestion. FEC would reduce the number of re-transmissions, thus increasing TCP goodput during lossy phases. Reduction in re-transmissions also reduces the delay incurred in recovering from packet losses, allowing the TCP flow to quickly build up to throughput and goodput levels prevailing before the losses.

The performance of an $(n, k)$ FEC block code depends on two key parameters – (i) code-rate $\frac{n}{k}$, and (ii) block size $n$. The choice of these two parameters significantly influences the goodput and delay properties of the transport protocol.

The degree of loss-tolerance attained by FEC coding is directly proportional to the code-rate. A higher code-rate also indicates a higher coding overhead. Hence, there exists a trade-off between degree of loss-tolerance and coding overhead. A fixed code-rate is clearly not optimal because it would consume bandwidth even in zero-loss conditions, reducing goodput. A fixed code-rate is also useless if packet losses exceed the degree of loss-tolerance provided by the code.

Clearly, we require a code-rate that would adapt to variations in network conditions. An adaptable code-rate would only provide loss-tolerance during lossy conditions (more specifically, provision enough FEC based upon an estimate of loss statistics), thereby reducing bandwidth wastage. Therefore, the transport protocol must be able to estimate network losses as well as establish requirements for the degree of loss-tolerance required from the FEC code. We show in [1] that in order to decode any block without additional re-transmissions with high probability, a code-rate that is inversely proportional to the sum of average packet loss-rate and packet loss-rate deviation is appropriate. The likelihood of decoding bounds the expected goodput from below, providing a minimum performance guarantee.

The choice of block size determines the average delay experienced by the application. Since the receiver must at least wait for the amount of time it takes to transmit the complete block of $n$ packets, a larger block size would lead to an increase in delay. As a block must consist of an integral number of packets, the desired code rate often imposes restrictions on the block size. For instance, a code-rate of 1.01 can be realized exactly only by a block of size 101 (as a (101, 100) code, for example), or multiples of it.

Lundqvist *et al.* [3] and our prior work [1] argue for a maximum block size equal to the window size (which is a measure of the connection's bandwidth-delay product) to attain the optimal trade-off between goodput and delay.

### 3.2 TCP-SACK Modifications for Effective Use of FEC

We modify the standard TCP-SACK to employ FEC coding as mentioned in the previous section. We denote this version of TCP as *TCP-cd (coded TCP).*

The focus of TCP-cd design is to emphasize that FEC, a mechanism for reliability and loss recovery, when integrated into TCP, also has collateral benefits by reducing the "brittleness" of TCP congestion control. The brittleness is observed particularly in small-window phases and during transient burst losses, in addition to other situations. The "insurance" provided by FEC helps provide additional resilience under these conditions, and specifically leads to reduced latency during the loss recovery phase and reduced fall back to timeouts that hurt performance.

TCP-cd calculates the packet loss rate from loss measurements using Exponentially Moving Weighted Average (EWMA), and then calculates the variance of packet losses accordingly. We use the method described in [1] for the MPLOT protocol to measure and update packet loss statistics. TCP-cd then scales the TCP congestion window by the inverse of the packet delivery rate, i.e., (1 - packet loss rate). This extra redundancy ensures that the "loss adjusted" congestion window remains at the level as originally targeted, on an average. TCP-cd encodes the data packets with a $(n, (1 - \mu - \sigma)n)$ block code where $n$ is the scaled window size, $\mu$ is the average packet loss rate and $\sigma^2$ is the packet loss variance. TCP-cd dynamically adapts the code-rate and block size as the measured/estimated loss rate varies. Unlike some implementations of TCP, the window is not reduced simply based on the receipt of three duplicate acknowledgements. Instead, we use ECN to detect congestion and reduce the window. TCP-cd is a simpler version of MPLOT [1] in the sense that it only includes FEC coding from MPLOT while excluding any capabilities to take advantage of out-of-order transmission, packet size adjustment and other policies employed by MPLOT to take advantage of multiple paths.

In order to estimate the effect of FEC, we also simulate another variant of TCP-SACK – *TCP-la (loss aware TCP)*, that measures the packet loss statistics like TCP-cd but does not employ FEC coding. Thus in TCP-la, packet loss measurements are only used to scale up the congestion window as described above. We simulate TCP-SACK, TCP-la and TCP-cd under different conditions to measure the throughput, goodput, packet losses and timeouts. The simulations in the next section show that TCP-cd achieves better goodput than TCP-SACK and TCP-la even though all the three protocols have the same throughput.

## 4 Simulation Results

In this section, we present results from simulation experiments which show that intelligent use of FEC coding, as discussed in section 3.2, helps in substantially improving the performance of TCP by increasing end-to-end goodput. The effect of bursty packet losses due to short-term transient congestion are overcome without triggering the more severe congestion response of TCP. Through simulations, we aim to investigate the following issues: (i) whether and by how much does coding help in improving TCP throughput and goodput in the presence of bursty unresponsive flows, (ii) how buffer sizes at network core routers impact the effectiveness of the use of FEC, and (iii) the extent of burstiness (in terms of the the peak-to-mean ratio of the bursts) for which of FEC is effective in improving TCP goodput.

Our ns-2.30 simulations use the typical single bottleneck dumb-bell topology. We use varying numbers of different types of sources (responsive (TCP), non-responsive (UDP), with/without congestion control) for assessing the contribution of FEC coding in recovering from congestion losses. We use a bottleneck link bandwidth of 10 Mb/s and a Round Trip Time (RTT) of 40 ms in the simulations, unless specified otherwise. To understand the impact of our work for networks with high bandwidth-delay product environments, we examine our results with varying buffer sizes, as a function of the bandwidth-delay product, at the bottleneck.

### 4.1 Complementarity of FEC and Congestion Control

We first investigate the effectiveness of FEC coding in attaining higher goodput in the presence of losses due to buffer overflows, both with and without congestion control. To first understand the role of coding in improving goodput when a transport protocol does not use flow and congestion control, we simulate an end-end reliable transport protocol similar to TCP except that the sources have no window flow control and therefore transmit at a constant packet rate (e.g., UDP with reliability built-in). We simulate 10 such sources over a single-bottleneck topology and vary the bottleneck bandwidth from 6 Mb/s to 15 Mb/s. Each source transmits application (data) packets at 1 Mb/s. If the packet flow is $(n, k)$ coded, then the sending rate is increased to $\frac{n}{k}$ Mb/s to keep the rate of data packets the same. For the coded flows, a block size of 10 packets is used. The queue is a simple drop-tail queue with a buffer size equal to the bandwidth-delay product.

Figure 1(a) shows the cumulative goodput achieved by all the sources as the *traffic overload*, defined as the difference between the aggregate application traffic rate and the bottleneck bandwidth, is increased. The uncoded flow experiences losses only when the bottleneck is overloaded (i.e., the bottleneck bandwidth is 10 Mb/s or lower). However, for the coded flows, packet loss occurs earlier, once the sending rate exceeds $10\frac{k}{n}$ Mb/s for an $(n, k)$ coded flow (i.e., the aggregate sending rate exceeds the bottleneck bandwidth). Furthermore, the uncoded flows exhibit higher goodput as compared to coded flows when the traffic overload is less than 1 Mb/s. This is intuitively expected, since the code-rates used for the coded flows in the three cases shown in the simulations are at least 10%. However, as the traffic overload increases, the coded flows experience a more gradual decay in goodput than uncoded flows, thereby achieving higher goodputs

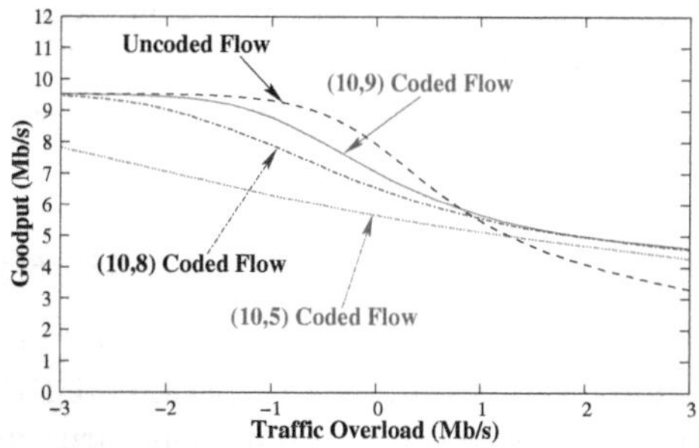


(a) Effect of coding on goodput of flows with no congestion control.

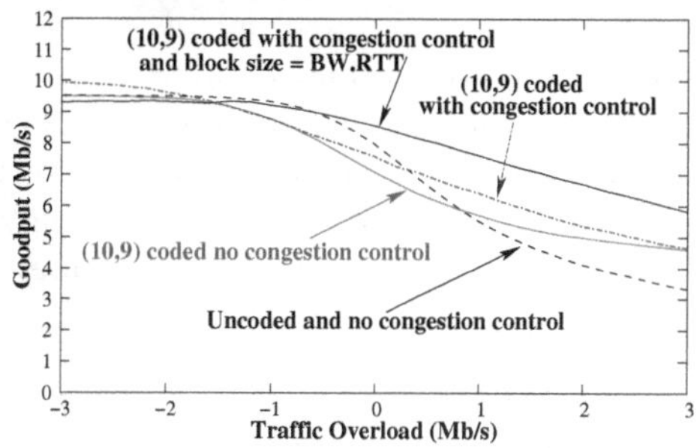


(b) Effect of coding and congestion control on goodput.

**Fig. 1.** The use of FEC coding in congested networks can have a significant effect on goodput. The combination of coding and congestion control leads to more gains.

at higher levels of congestion. These results show that FEC coding can certainly help reduce packet loss rates at high congestion levels. In addition, the extent of coding overhead (low with (10,9); high with (10,5)) has a significant impact on the goodput.

In the next set of simulations (results shown in figure 1(b)), we introduce a rudimentary congestion control function (similar to the AIMD algorithm in TCP) along with FEC coding to gauge their combined effect, and compare the performance of coded flows with the uncoded case subject to similar congestion control. For coded flows, we only consider the $(10, 9)$ block code case. We use two block sizes: one a fixed block size of 10 as before, and another where the block size is set to the bandwidth-delay product. Note that the two curves corresponding to no congestion control are the same as in figure 1(a), and are included here for comparison. We observe that the goodput is substantially improved with the introduction of the simple congestion control mechanism. However, performance of the coded case improves more than the uncoded case. This is particularly true if the block size is set to the bandwidth delay product, which attains significantly higher goodput compared to the uncoded case. More importantly, the reduction in goodput for coded flows is linear (gradual) in nature as compared to uncoded flows which experience a more severe non-linear reduction in goodput. These results imply that FEC coding can work well with congestion control mechanisms in improving goodput by reducing losses, particularly if the block size is set to the bandwidth-delay product. We explore this issue in more detail in the rest of our simulation experiments.

## 4.2 Benefits of FEC with TCP Flows under Bursty Losses

In order to represent the conditions of unresponsive and bursty flows that expose TCP flows to congestion losses, we simulate a periodic UDP flow that transmits at a peak rate for a certain time-period and then turns off for an equal time period (50% duty cycle). We use an ON-period of 50 RTT (2000 ms) and vary the peak rates from 3 Mb/s (low congestion) to 15 Mb/s (high congestion) in our simulations.

We first compare the throughput values (aggregate rate of packets sent, including retransmissions and in the coded case, redundancy packets) obtained by the three TCP variants for different UDP rates and router buffer sizes, and establish the fact that our use of coding stays within the window flow control limits of TCP. Figure 2 shows the average throughput values (as a percentage of

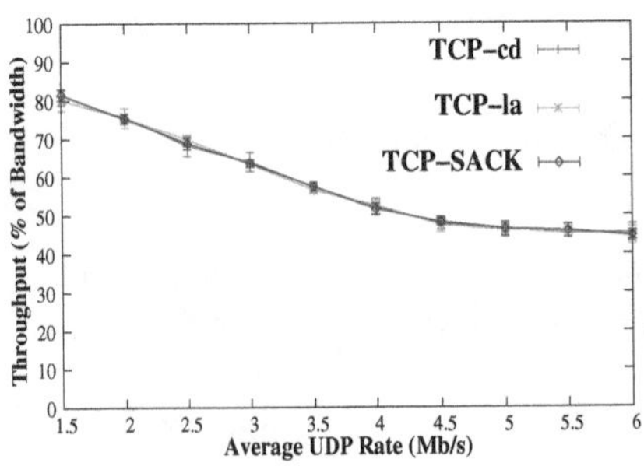


**Fig. 2.** Throughput when router buffer size equals bandwidth-delay product

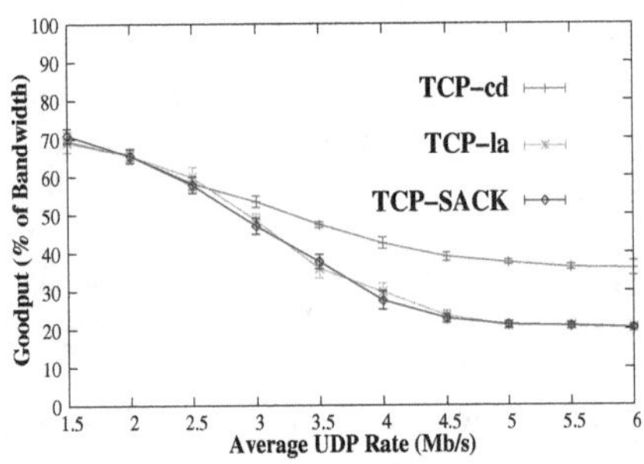


**Fig. 3.** Goodput when router buffer size equals bandwidth-delay product

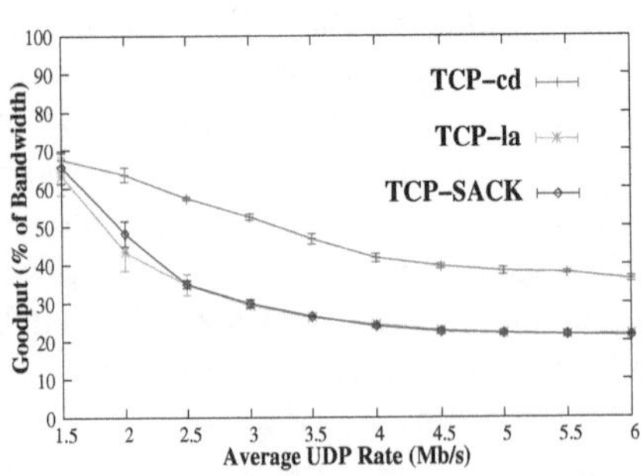


**Fig. 4.** Goodput when router buffer size equals *half* of the bandwidth-delay product

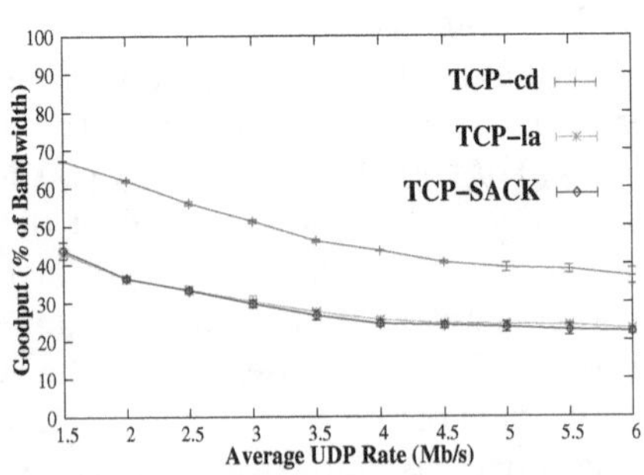


**Fig. 5.** Goodput when router buffer size equals *quarter* of the bandwidth-delay product

the bottleneck bandwidth) along with their 95% confidence intervals for TCP-cd, TCP-la and TCP-SACK, when the buffer size at the routers is equal to the bandwidth-delay product. We note that the throughput values of the three algorithms are roughly the same. (This trend is also observed for buffer sizes of $BW.RTT/2$ and $BW.RTT/4$ – not shown due to space limitations.) The similar throughput for the three algorithms is because they use the same TCP congestion control mechanism (AIMD algorithm with similar parameters), and therefore increase and decrease window sizes similarly in response to congestion losses. However, for the same throughput, we show below that TCP-cd is able to attain much higher goodput by being able to convert a larger fraction of the throughput to goodput, as compared to TCP-SACK and TCP-la.

By using dynamic coding, we limit the coding overhead appropriately (as compared to using a fixed coding rate), and adapt the coding rate nicely to match the time-varying loss rate on the path of a flow. We now measure how dynamic adaptive coding affects goodput in presence of TCP congestion control. The average goodputs (as a percentage of the capacity of the bottleneck) with their 95% confidence intervals are shown in figures 3-5, for the three variants and buffer sizes of $BW.RTT$, $BW.RTT/2$ and $BW.RTT/4$, respectively. Note that the bottleneck capacity is 10 Mb/s; thus even at an average UDP rate of 6 Mb/s, only 60% of the bottleneck bandwidth is used by the UDP flow.

For TCP-SACK and TCP-la, the goodput reduces by almost 75% as the average UDP rate increases from 1.5 Mb/s to 6 Mb/s. This shows the severe impact that bursty flows can cause to goodput, even if they do not lead to long-term congestion. In contrast, the goodput for TCP-cd decays by about 60% across the same range of average UDP rates. This shows that use of FEC helps in rapid recovery from losses, thereby reducing re-transmissions and increasing goodput. This behavior is consistent across all values of buffer sizes simulated. Comparing the throughput and goodput values, we note that TCP-SACK is only able to convert 30% of its throughput to goodput (at an average UDP rate of 6 Mb/s) while TCP-cd can convert 66.7% of throughput to goodput at the same average UDP rate. This is more than a 100% improvement in converting throughput to goodput. Since the goodput performance of TCP-la and TCP-SACK is very similar, this implies that the performance improvement observed for TCP-cd is exclusively due to the effective use of coding, and not due to the loss-rate based scaling of the congestion window. Comparing the three figures (the three different bandwidth-delay product cases) we observe that the performance difference between TCP-cd and the other two TCP versions is more pronounced at lower values of the average UDP rates as the buffer size becomes smaller. This is particularly important as we go to higher bandwidth-delay product environments, where correspondingly the amount of buffering available at routers may be smaller.

To understand the cause for the improved goodput with TCP-cd, we examine the packet loss and timeout behavior of the different alternatives. The packet loss rate increases with an increase in average UDP rate as expected. In addition to the reduction in window size as a result of TCP's loss-based congestion response, there is also a likelihood of TCP flows suffering timeouts. We look at the average

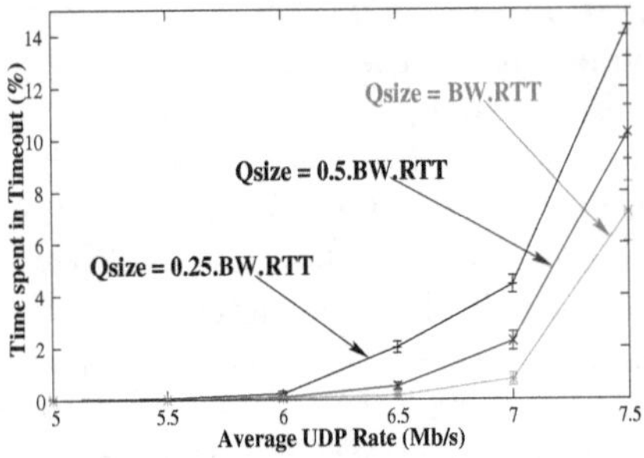


(a) TCP-SACK

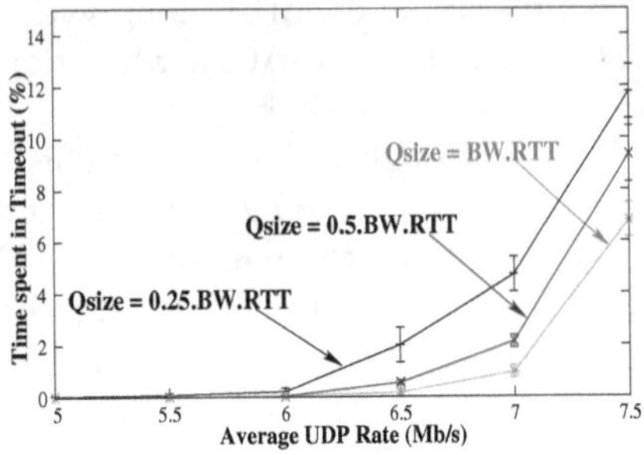


(b) TCP-cd

**Fig. 6.** The time spent in timeout by TCP-SACK and TCP-cd

time spent in timeouts by TCP-SACK and TCP-cd respectively in figures 6(a) and 6(b), for the three different buffer sizes. The timeouts become significant when the average UDP rate exceeds about 60% of the bottleneck capacity (which corresponds to the peak UDP rate exceeding 100% of the bottleneck capacity). However, while there is no dramatic difference between TCP-SACK and TCP-cd in terms of time spent in timeouts, TCP-cd does in fact see improvements in the amount of timeouts experienced with smaller router buffers.

Figure 7(a) overlays the variation in UDP traffic and the corresponding average packet loss rate measured at a TCP receiver when the UDP flow transmits at a peak rate of 10 Mb/s on a path with a bandwidth 10 Mb/s. TCP experiences losses when the UDP flow turns on, resulting in TCPs reducing their window, thereby reducing throughput and goodput as well. Packet retransmission and the potential of retransmitted packets being lost result in further reduction in goodput. In figure 7(a), which shows the loss-rates for TCP-SACK, the average packet loss is high, frequently experiencing 100% packet loss (resulting in timeouts). We now look at the average loss rate measured by a TCP-cd receiver in figure 7(b). While the packet loss rate is still high, it is considerably lower than the loss-rate observed for TCP-SACK. FEC coding thus enables a TCP flow to tolerate the co-existence of non-cooperating flows and overcome the effect of transient packet losses. Even under heavy congestion useful information is delivered through the use of coding.

We now study the impact of increasing the burstiness of UDP flows on TCP-SACK and TCP-cd. For this purpose, we maintain the peak rate of the UDP flow equal to the bottleneck bandwidth of 10 Mb/s and vary the ON-period for the UDP flow. The OFF-period of the UDP flow is fixed (at 2 secs), and we observe the throughput and goodput achieved by TCP-SACK and TCP-cd, while varying the ON-period from 0.4 secs to 20 secs, resulting in a peak-to-mean rate ratio of the UDP flow varying from 1.1 to 6. We present the throughput and goodput values for the two protocols in figures 8 and 9 respectively, where the router buffer size equals half the bandwidth-delay product.

We note from figure 8 that the throughput achieved by TCP-cd is the same as TCP-SACK, across the range of UDP burstiness values considered. In contrast,

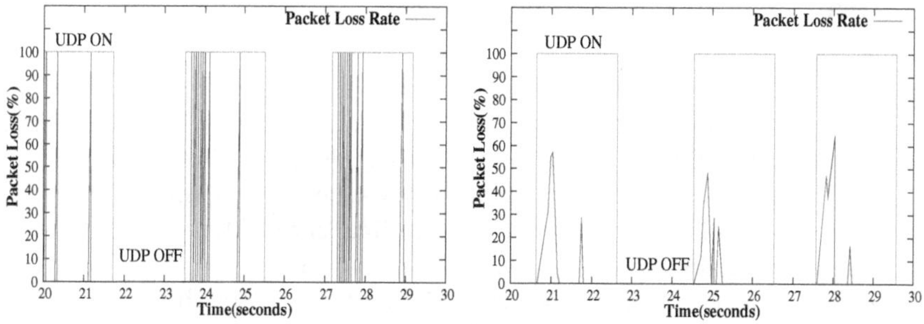


(a) Packet loss seen by a TCP-SACK flow. (b) Packet loss seen by a TCP-cd flow.

**Fig. 7.** Comparison of packet losses for TCP-cd and TCP-SACK when UDP is transmitting at peak rate of 10 Mb/s and the buffer size is quarter of the bandwidth-delay product. The losses seen by TCP-cd are significantly less than those of TCP-SACK.

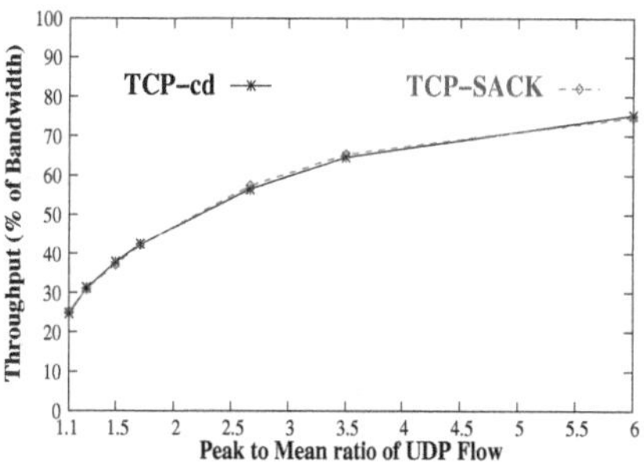


**Fig. 8.** Throughput comparion for different levels of burstiness of UDP flow (buffer=half BW-Delay product)

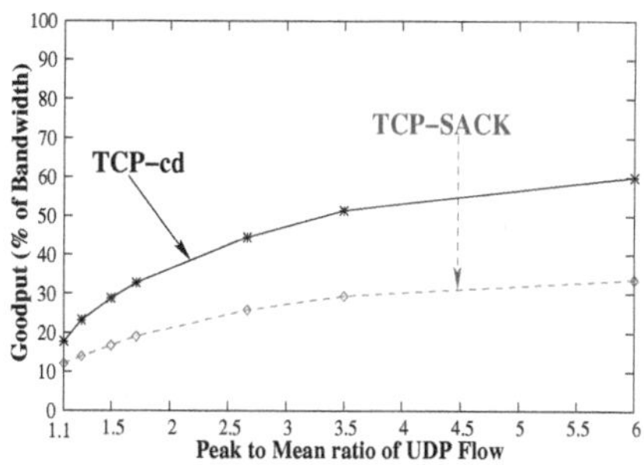


**Fig. 9.** Goodput comparison for different levels of burstiness of UDP flow (buffer=half BW-Delay product)

the goodput achieved by TCP-cd gets significantly better as the burstiness of UDP flow increases, as we observe in figure 9. For a peak-to-mean ratio of 6, the goodput achieved by TCP-cd is twice that of TCP-SACK. As the burstiness reduces, the UDP flow acts more like a constant rate flow, reducing the gains made by FEC.

In summary, we conclude that the use of adaptive FEC along with TCP congestion control can lead to significant goodput gains in scenarios that are typically observed in current networks, where there is likely to exist significant amounts of cross-traffic from non-cooperative UDP flows. The goodput improvement is more pronounced as the burstiness of the unresponsive flows, and hence the congestion losses, increases. This becomes particularly important in higher bandwidth-delay product environments, especially with routers having small buffers. Use of adaptive FEC coding helps in significantly extending the dynamic range of operation of feedback based congestion control mechanisms.

## 5 Summary and Conclusions

As networks evolve to higher bandwidth-delay product environments and have to carry traffic that is unresponsive to congestion control, the capabilities of congestion control will be pushed beyond their designed limits. In this paper, we demonstrated that existing TCP congestion control mechanisms can be effectively complemented by packet level FEC coding to quickly recover from short-term congestion losses caused by bursty unresponsive flows. The approach we propose is particularly useful in to help TCP perform reasonably well even when traffic is bursty causing transient overloads, buffer sizes are small and feedback delays are large, thus extending the dynamic range of traditional feedback-based congestion control protocols. The extent of goodput benefit provided by FEC coding increases as the router buffer size reduces. This is a particularly important as it allows routers to retain relatively small buffers for the emerging high bandwidth-delay environments, thereby reducing queueing delay while FEC coding protects goodput against short-term queue overflows. We also showed that the performance gains delivered by packet level FEC increases as the burstiness in the traffic and the associated packet losses due to congestion increase. The significant gains in goodput are achieved without a corresponding increase in throughput, as the overhead of FEC coding is minimized by adapting the code-rate to network losses. Our scheme is able to use the loss-tolerance property of FEC to attain higher goodput, while right-sizing the overhead associated with FEC.

In summary, intelligent/adaptive FEC coding can complement TCP's congestion control to effectively counter losses caused by the combination of bursty unresponsive flows and router buffer sizes that are small compared to the bandwidth delay product in high-speed networks.

## References

1. Sharma, V., Kalyanaraman, S., Kar, K., Ramakrishnan, K., Subramanian, V.: MPLOT: A Transport Protocol Exploiting MultiPath Diversity using Erasure Codes. In: Proc. IEEE INFOCOM (2008)
2. Subramanian, V., Kalyanaraman, S., Ramakrishnan, K.: An end-to-end transport protocol for extreme wireless environments. In: Proc. IEEE Military Communications Conference (MILCOM 2006), Washington D.C, USA (October 2006)
3. Baldantoni, L., Lundqvist, H., Karlsson, G.: Adaptive end-to-end FEC for improving TCP performance over wireless links. In: IEEE International Conference on Communications, vol. 7, pp. 4023–4027 (2004)
4. Ramakrishnan, K., Floyd, S., Black, D.: RFC 3168 - The Addition of Explicit Congestion Notification (ECN) to IP (2001)
5. Liu, Z., Hu, C., Zheng, K., Chen, S., Liu, B.: A trace study for characteristics of packet arrivals. In: IEEE 7th Malaysia International Conference on Communication & 13th IEEE International Conference on Networks, MICC-ICON, vol. 1 (November 2005)

6. Fitzek, F., Zorzi, M., Seeling, P., Reisslein, M.: Video and audio trace files of pre-encoded video content for network performance measurements. In: First IEEE Consumer Communications and Networking Conference (CCNC), pp. 245–250 (January 2004)
7. Appenzeller, G., Keslassy, I., McKeown, N.: Sizing router buffers. SIGCOMM Computer Communication Review 34(4), 281–292 (2004)
8. Mascolo, S., Vacirca, F.: Congestion control and sizing router buffers in the internet. In: IEEE Conference on Decision and Control & European Control Conference (CDC-ECC), pp. 6750–6755 (December 2005)
9. Hayasaka, M., Gamage, M., Miki, T.: An efficient loss recovery scheme for on-demand video streaming over the internet. In: The 7th International Conference on Advanced Communication Technology (ICACT), vol. 2, pp. 1301–1306 (2005)
10. Yu, X., Modestino, J., Bajic, I.: Performance analysis of the efficacy of packet-level fec in improving video transport over networks. In: IEEE International Conference on Image Processing (ICIP), vol. 2, pp. II–177–180 (September 2005)
11. Li, Y., Zhang, Y., Qiu, L., Lam, S.: Smarttunnel: Achieving reliability in the internet. In: Proc. of IEEE INFOCOM (2007)
12. Ahlswede, R., Cai, N., Li, S.Y.R., Yeung, R.W.: Network information flow. IEEE Transactions on Information Theory 46(4), 1204–1216 (2000)
13. Huang, Y., Ghaderi, M., Towsley, D., Gong, W.: TCP performance in coded wireless mesh networks. In: IEEE SECON (2008)
14. Ghaderi, M., Towsley, D., Kurose, J.: Reliability gain of network coding in lossy wireless networks. In: IEEE INFOCOM (2008)

# Collateral Damage: The Impact of Optimised TCP Variants on Real-Time Traffic Latency in Consumer Broadband Environments

Lawrence Stewart, Grenville Armitage, and Alana Huebner

Swinburne University of Technology
Melbourne, Australia
{lastewart,garmitage}@swin.edu.au, alanahuebner@gmail.com

**Abstract.** In recent years a number of TCP variants have emerged to optimise some aspect of data transport where high delay-bandwidth product paths are common. We evaluate a different scenario - latency-sensitive UDP-based traffic sharing a consumer-grade 'broadband' link with one or more TCP flows. In particular we compare Linux implementations of NewReno, H-TCP and CUBIC. We find that dynamic latency fluctuations induced by each TCP variant is a more significant differentiator than 'goodput' (useful throughput), and that CUBIC induces far more latency than either H-TCP or NewReno when multiple TCP flows are active concurrently. This potential for 'collateral damage' should influence future efforts to re-design TCP for widespread deployment.

**Keywords:** TCP, congestion control, latency, interactive, broadband.

## 1 Introduction

Transmission control protocol (TCP) [1] deserves significant credit for the internet's wide-spread utility over the past 25+ years. The relatively modern *NewReno* variant of TCP [2] balances two key goals: Provide reliable transfer of byte-streams across the IP layer's unpredictable packet-based service, and minimise congestion inside end hosts and the underlying IP network(s) while maximising performance [3]. As the dominant transport protocol for internet-based applications [4], maximisation of TCP performance has been an active and challenging area for academic and industry research into congestion control (CC) techniques [5]. A common focus has been on the interactions between multiple TCP sessions when sharing a common congestion point or path segment, particularly on long paths with high link speeds.

This paper looks at an increasingly important new scenario - TCP flows sharing consumer-grade 'broadband' links with non-reactive, latency-sensitive UDP-based applications such as Voice over IP (VoIP) and online games. In particular we focus on the impact of the Linux implementations of NewReno, *CUBIC* [6,7] (currently the default CC algorithm for Linux TCP connections), and *H-TCP* [8,9] variants of TCP. Each variant provides different dynamic response to congestion within an IP network, which has a direct impact on the

L. Fratta et al. (Eds.): NETWORKING 2009, LNCS 5550, pp. 392–403, 2009.

latency experienced by other flows sharing a congestion point. We characterise the extent to which both CUBIC and H-TCP induce greater median latency than NewReno yet provide little nett gain in *goodput* (useful throughput as measured by the application on top of TCP).

Section 2 begins with an over-view of issues surrounding TCP CC research. Section 3 describes our testbed, instrumented to allow precise tracking of latency versus TCP's congestion window size and congestion events. In section 4 we illustrate the range of latencies, and frequency of latency fluctuations, experienced by unrelated UDP flows sharing a congestion point with long-lived NewReno, CUBIC and H-TCP flows.

## 2 Background

The evolution of the internet's IP-based network and underlying infrastructure has exceeded initial architectural design assumptions and expectations in a number of areas. Since congestion control (CC) was first proposed [10] and subsequently mandated [11], there has been significant ongoing research to ensure CC kept pace with the underlying network it was designed to efficiently utilise and protect from congestion collapse.

The Internet Research Task Force's (IRTF) Internet Congestion Control Research Group (ICCRG) [12] and Transport Modeling Research Group (TMRG) [13] have been established to shepherd the development, evaluation and (where applicable) standardisation of improved and altogether new CC mechanisms and schemes, with a focus on transport protocols.

A particularly large body of work has developed around improvements to the Additive Increase Multiplicative Decrease (AIMD) congestion window (cwnd) scaling factors and congestion detection mechanisms used by the defacto-standard NewReno CC algorithm. Wireless environments (where packet loss is not indicative of congestion) and large bandwidth-delay product (BDP) paths (which take multiple minutes to re-probe network capacity after congestion back-off) are some of NewReno's current problem areas where it is failing to meet the goal of efficient network utilisation.

A raft of new protocols have been proposed in recent years, which typically aim to solve a weakness in NewReno under a specific subset of network scenarios. The resulting proposals are being discussed, evaluated and refined within the context of the ICCRG and TMRG, with an eye to eventual publication as an experimental or fully fledged Internet Engineering Task Force (IETF) standard.

Evaluation of TCP related CC mechanisms is itself an active area of research. Steps are being taken to develop a set of public baseline test scenarios and metrics with which to compare new CC algorithms [14]. Evaluation by individual algorithm implementers thus far has largely focused on aspects such as intra-protocol fairness, inter-protocol fairness with NewReno, throughput and convergence.

While there are good reasons to maintain multiple CC implementations for research purposes and specific application use cases, the need for a suitable default for the average network stack is important. It is therefore crucial to test real-world implementations of proposed protocols in common network scenarios that a standardised protocol would likely be deployed in.

To date, we have found very little prior work [15,16] investigating home broadband scenarios and the behaviour of emerging TCPs, or TCPs interacting with non-congestion reactive, latency-sensitive traffic in this environment. The continuing convergence towards IP based entertainment, information access and communication service delivery ensures this is an area of increasing importance.

Gaming and voice over IP are two such services of interest, both typically delivered using constant rate, non-congestion reactive flows of small UDP packets. Online multiplayer computer games, and the popular first person shooter genre in particular, have well known latency sensitivity requirements for effective game play [17]. Voice over IP is another real-time service with well characterised latency tolerances, previously examined in the context of traditional telephony [18,19] and now IP networks more recently [20]. The requirements of these increasingly important applications must be taken into consideration within the context of transport protocol research and development.

## 3 Experimental Methodology

Figure 1 shows our experimental topology - one congestion point (a FreeBSD router[1] with configurable forwarding latency and instrumented to log actual queue utilisation over time), four Debian Linux hosts[2] acting as TCP or UDP sources and sinks, and a precision traffic capture tool[3] to calculate one way delay (OWD) through the router. Our simple *dumbbell* testbed is not topologically equivalent to the actual network paths traversed by typical consumer traffic. Nevertheless it is suitable for this paper's focus on the interactions between TCP and UDP flows.

As consumer broadband links vary widely around the world we settled on emulating 1Mbps links with drop-tail queues of 60000 bytes in each direction (based on previously published estimations of buffering in consumer routers [21,22]). The drop-tail queues create the bottleneck shared by all traffic traversing the router, and $RTT/2$ of delay is added in each direction using *dummynet* [23]. We experimented with total RTTs of 50ms and 100ms to emulate delays conceivably experienced by typical consumer activities. Configured latency is accurate to within 0.5ms as the router's kernel was set to tick at 2000Hz ($kern.hz = 2000$).

Hosts A, B, C and D are instrumented with Web100 [24,25] - tracking TCP connection parameters (such as *cwnd*) by polling every 1ms over the lifetimes of active TCP sessions.

[1] FreeBSD 7.0-RC1 on a 2.80GHz Intel Celeron D (256K L2 Cache), 512MB PC3200 DDR-400 RAM, with two Intel PRO/1000 GT 82541PI PCI gigabit Ethernet cards as forwarding interfaces.

[2] A 2.6.25 kernel ticking at 1000Hz, each one a 1.86GHz Intel Core2 Duo E6320 (4MB L2 Cache) CPU, 1GB PC5300 DDR2 RAM and Intel PRO/1000 GT 82541PI PCI gigabit Ethernet NIC.

[3] Two Endace DAG 3.7GF gigabit ethernet capture card ports.

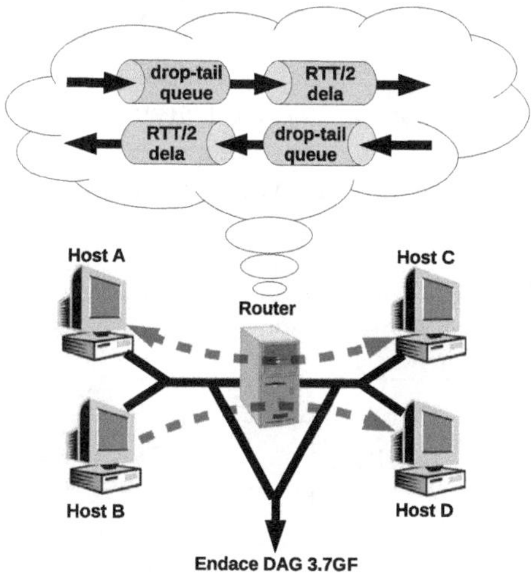


**Fig. 1.** Testbed for measurement of shared, congestion-induced queuing delays

Bulk TCP traffic was generated using Iperf [26], with data flowing from Host A to C and ACKs from Host C to A. Uni-directional UDP traffic from Host B to D (186byte IP packets every 20ms, emulating non-reactive VoIP traffic) was generated using Tcpreplay [27]. No traffic flowed in the reverse direction from Host D to B.

Load-time tunable variables of the Linux CUBIC and H-TCP implementations were left at their default values, except as noted here. For CUBIC, we set "max increment" to 100[4]. To ensure H-TCP's consistency with the published internet draft [8] we disabled "adaptive backoff" and "adaptive reset", but left "RTT scaling" on[5].

Trials were run five times for each combination of TCP algorithm, testbed RTT and number of TCP flows. Trials lasted for at least three minutes or twenty congestion epochs, which ever was longer. For each group of five runs we discarded the highest and lowest results and took the average of the remaining three.

## 4 Results

Our initial experiments focus on measuring the following characteristics: TCP goodput, latency as experienced by the UDP flow over time, the frequency of congestion events as observed in the router's bottleneck queue and the number of packets retransmitted by TCP.

[4] This variable has been removed in more recent CUBIC specifications and setting it to 100 effectively disabled any effect it might have in our test scenarios.

[5] The H-TCP code also ran with a patch functionally equivalent to [28] that fixed a visible bug in H-TCP's behaviour.

### 4.1 Goodput Achieved by NewReno, CUBIC and H-TCP

Goodput for each trial run was calculated using tcptrace [29], providing a measure of the application level (or "usable") bitrate achieved during each trial. All three TCP variants exhibited much the same average goodput when RTT was both 50ms and 100ms. Table 1 lists the average aggregate goodput across all flows.

**Table 1.** Aggregate 'goodput' of one, two and five concurrent flows - NewReno, H-TCP or CUBIC congestion control, 50ms or 100ms RTT, 1Mbit/sec bottleneck

| RTT (ms) | Algorithm | No. Flows | | | | | |
|---|---|---|---|---|---|---|---|
| | | 1 | | 2 | | 5 | |
| | | Goodput Statistics (KB/s)[a] | | | | | |
| | | $\mu$ | $\sigma$ | $\mu$ | $\sigma$ | $\mu$ | $\sigma$ |
| 50 | NewReno | 111.8 | 0.01 | 112.2 | 0.19 | 112.2 | 0.10 |
| | H-TCP | 111.9 | 0.06 | 112.5 | 0.11 | 114.2 | 0.50 |
| | CUBIC | 111.9 | 0.04 | 112.6 | 0.24 | 114.5 | 0.39 |
| 100 | NewReno | 111.9 | 0.01 | 112.2 | 0.01 | 112.5 | 0.23 |
| | H-TCP | 111.8 | 0.04 | 112.7 | 0.19 | 114.0 | 0.43 |
| | CUBIC | 111.9 | 0.04 | 112.7 | 0.25 | 114.3 | 0.39 |

[a]Where 1KB = 1000 bytes.

Consumers constrained by 1Mbit/sec links and 50ms or 100ms paths were not the original design target for TCP variants such as CUBIC and H-TCP. Indeed, H-TCP and CUBIC were intended to be NewReno-like over links with low bandwidth delay product (BDP), so our goodput results are not unexpected.

However, our experiments reveal that H-TCP and CUBIC diverge from NewReno in significant ways when we consider the latency caused by each TCP variant's congestion control behaviour in low BDP environments.

### 4.2 Latency Induced by NewReno, CUBIC and H-TCP

Latency-sensitive applications such as VoIP and online games typically generate UDP flows that have simplistic (or non-existent) reaction to congestion within the underlying IP network path. Such applications are increasingly important in the consumer market, and TCP's cyclical variation of *cwnd* is known to cause bottleneck queuing delays to fluctuate regularly. Our experiments suggest that this 'induced latency' should be considered an important differentiator between TCP variants that expect deployment in consumer contexts.

Illustrative examples of the cyclical variation in cwnd, one way delay (OWD) and queue length induced by TCP are provided by Figures 2a and 2b. These figures clearly highlight the relationship between cwnd, queue occupancy and induced OWD during three consecutive congestion epochs of a single H-TCP flow sharing the bottleneck link with a single UDP flow over a 100ms RTT path.

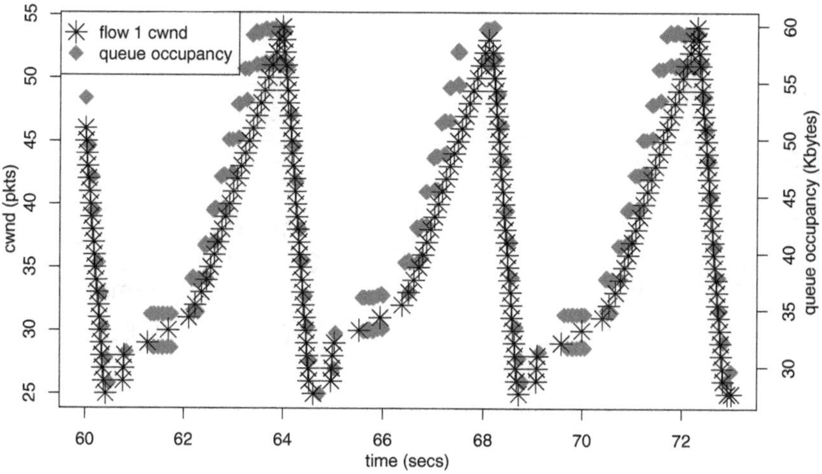


(a) Cwnd and Queue Occupancy vs Time

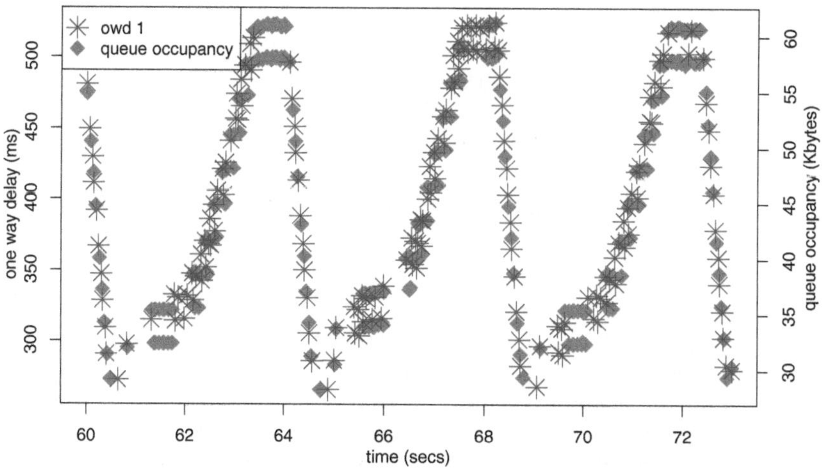


(b) Induced OWD and Queue Occupancy vs Time

**Fig. 2.** Cwnd, Induced OWD and Queue Occupancy vs Time across three consecutive congestion epochs of a single H-TCP flow. Path is 1Mbit/sec @ 100ms RTT.

More usefully, Figures 3a, 3b and 3c show the CDF of OWD experienced by our single UDP flow over at least twenty congestion epochs when the bottleneck link is shared with one, two and five TCP flows respectively. Plots for 100ms looked identical. In this case the path has a 50ms baseline RTT (25ms OWD), thereby fixing the best case OWD at 25ms, which is visible in the figures. The upper bound is the maximum queuing delay (480ms, being 60000 bytes at 1Mbps) plus the path's intrinsic 25ms OWD. A statistically insignificant number of UDP packets were dropped during queue full events, and these are excluded from the latency statistics presented here. NewReno, H-TCP and CUBIC clearly interact very differently

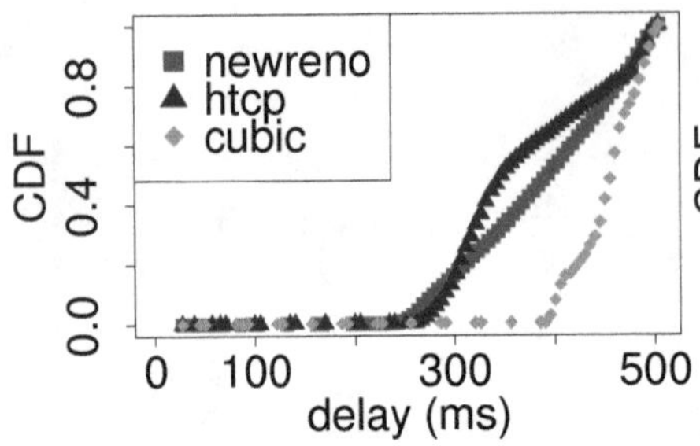


(a) One TCP flow

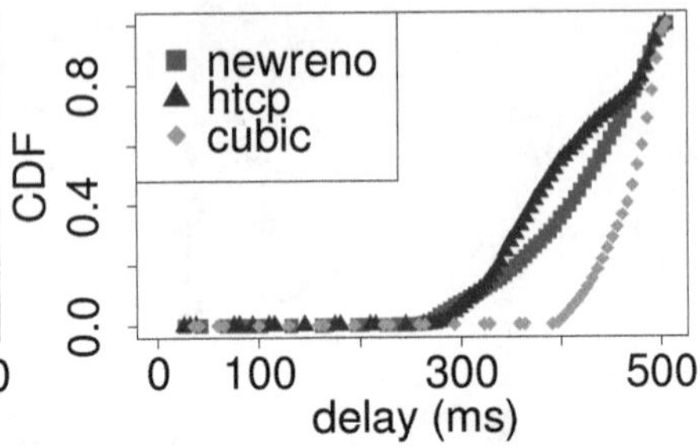


(b) Two TCP flows

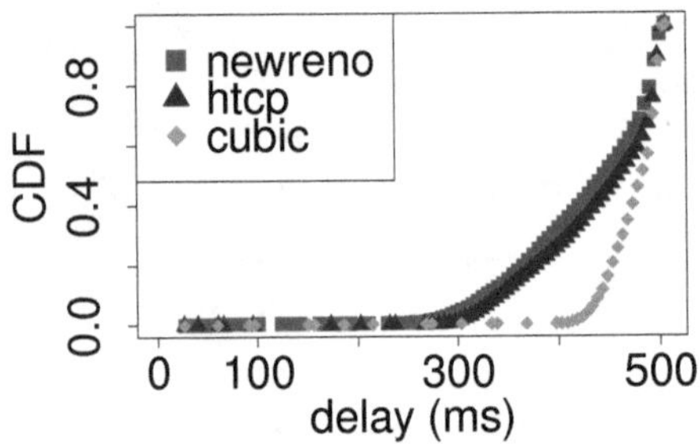


(c) Five TCP flows

**Fig. 3.** Cumulative distribution of OWD for NewReno, H-TCP and CUBIC. Path is 1Mbit/sec @ 50ms RTT.

with the bottleneck queue over time - behaviours directly attributable to their particular algorithms for managing the TCP congestion window.

H-TCP mimics NewReno cwnd growth for the first second after congestion, and then follows a parabolic growth function proportional to $x^2/2$ until the next congestion event (where $x$ is related to the time since last congestion).

CUBIC follows a cubic growth function proportional to $0.4x^3$ (where $x$ is also related to the time since last congestion). Unlike H-TCP, CUBIC takes NewReno cwnd growth as a baseline. The algorithm switches to the NewReno growth function if the cubic growth function calculates a value less than NewReno would have achieved in the same amount of time since congestion.

The difference between the growth functions themselves is also pertinent. Unlike parabolas, cubic functions grow quickly in their concave region, gradually slowing as they reach an inflection point before switching to convex mode operation and grow quickly again. This behaviour is particularly noticeable in Figure 3a. The CUBIC data shows a steep increase in latency at 400ms (concave growth), which slows between 440ms-480ms (near inflection point) before starting to grow again (convex growth).

As more TCP flows are added (Figures 3b and 3c) the CUBIC flows continue to induce significant additional delay whereas the H-TCP flows tend to induce NewReno-like latency. (More concurrent flows create more frequent congestion

**Table 2.** Latency induced by NewReno, H-TCP or CUBIC congestion control on one, two or five concurrent flows, 50ms or 100ms RTT, 1Mbit/sec bottleneck

| RTT (ms) | Algorithm | No. Flows | | | | | | | | |
|---|---|---|---|---|---|---|---|---|---|---|
| | | 1 | | | 2 | | | 5 | | |
| | | Induced Latency Statistics (ms) | | | | | | | | |
| | | Median | $\mu$ | $\sigma$ | Median | $\mu$ | $\sigma$ | Median | $\mu$ | $\sigma$ |
| 50 | NewReno | 393.2 | 387.7 | 74.2 | 422.8 | 410.1 | 67.1 | 445.5 | 427.3 | 68.2 |
| | H-TCP | 347.0 | 371.1 | 73.5 | 390.8 | 397.8 | 66.3 | 459.6 | 439.2 | 59.7 |
| | CUBIC | 456.2 | 452.2 | 35.5 | 474.1 | 465.7 | 31.9 | 485.0 | 476.4 | 26.1 |
| 100 | NewReno | 409.8 | 402.4 | 81.0 | 442.4 | 427.6 | 73.7 | 466.0 | 447.1 | 74.1 |
| | H-TCP | 353.3 | 383.6 | 85.1 | 405.5 | 414.8 | 71.5 | 479.0 | 459.1 | 62.9 |
| | CUBIC | 487.8 | 476.7 | 42.0 | 494.5 | 485.2 | 37.8 | 504.3 | 494.7 | 32.1 |

events, and the individual H-TCP flows spend proportionally more time exhibiting NewReno-like cwnd growth.) Table 2 quantifies the median, mean and standard deviation of latencies induced during the separate trials of NewReno, H-TCP and CUBIC congestion control with one, two or five concurrent flows.

### 4.3 Impact of Congestion Events

Figure 4 plots the average number of queue full events per minute, observed during a sixty second period in the middle of each trial. A queue full event occurs when the dummynet bottleneck queue reaches capacity and drops packets until it is able to accept a new packet.

In the one and two TCP flow cases, H-TCP's parabolic cwnd growth function causes it to consistently overshoot the queue capacity by some margin, resulting in the larger number of queue full events. However for the five flow case, H-TCP's tendency to remain in NewReno mode due to the increased frequency of congestion events results in fewer queue full events. CUBIC on the other hand

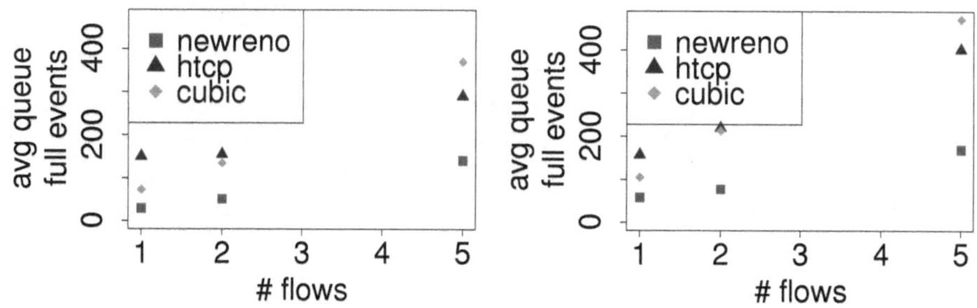


(a) 50ms (b) 100ms

**Fig. 4.** Average number of "queue full" events per minute

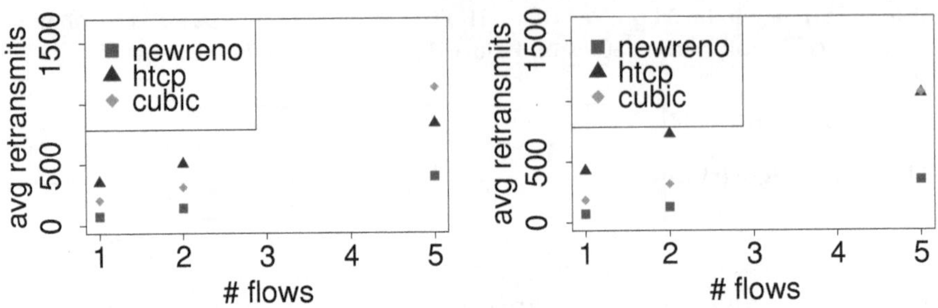


(a) 50ms (b) 100ms

**Fig. 5.** Average number of retransmitted packets

demonstrates no such restraint, which results in faster increases in queue full events per flow than H-TCP or NewReno.

Retransmitting packets due to network and receiver losses should ideally be kept to a minimum to ensure efficient network resource utilisation. In the low BDP bottleneck scenarios under investigation, retransmissions are almost entirely caused by a flow's cwnd overshooting the path capacity and subsequently overflowing the bottleneck queue. A TCP's cwnd growth function therefore plays an integral part in balancing protocol scalability (quickly probing the path to utilise available bandwidth) and stability (minimising the time a connection is not sending user data).

Figures 5a and 5b plot the average aggregate number of retransmissions for the 50ms and 100ms trial sets respectively[6]. Because they are optimised to aggressively probe for network capacity for benefit in high BDP environments, H-TCP and CUBIC both trigger significantly more retransmissions than NewReno.

These results also align with those of Figure 4, as we would expect queue full events and retransmissions due to lost packets to be closely correlated.

## 5 Conclusion and Further Work

We have performed a comparison of the Linux implementations of NewReno, H-TCP and CUBIC TCP over an emulated path that (roughly) approximates a congested consumer broadband link. First we observed that 'goodput' (useful throughput) was essentially equivalent for each TCP variant when pushing data over a 50ms or 100ms RTT path with a 1Mbps bottleneck link speed and 60000 byte queue. Then we explored the latency that would likely be experienced by VoIP-like UDP traffic sharing such a congested 'consumer' link with each TCP variant.

CUBIC was observed to induce noticeably more latency than either H-TCP or NewReno when one or more active TCP flows congest a link. With one or two concurrent flows, it is even possible for H-TCP to induce less latency than

[6] Calculated using tcptrace and summed for each test run.

both CUBIC and NewReno. For example, two TCP flows sharing a 50ms RTT path with a 60000 byte queue at a 1Mbps bottleneck induced median latencies of 422.8ms, 390.8ms and 474.1ms for NewReno, H-TCP and CUBIC respectively. With five concurrent flows the median induced latencies rose to 445.5ms, 459.6ms and 485.0ms respectively.

The impact of induced latency on VoIP and online game applications is a form of 'collateral damage'. We believe future efforts to re-design TCP for widespread deployment should aim to minimise this impact, rather than aiming for fairness between TCP flows or maximisation of goodput in long, fast networks. Additionally, the largest difference between the latency induced by the optimised variants stems from differences in their "low speed" compatibility modes. Algorithm developers should consider these findings in the future refinement of compatibility modes, as there are likely gains to be had by defining better behaving options for broadband-like environments.

The current behaviour of the tested common TCP variants in broadband environments strongly indicates that CPE manufacturers need to provide better mechanisms to appropriately manage their device's queues. The current trend towards large unmanaged (by default) queues in CPE is demonstrably bad news for consumers.

Our analysis suggests a number of areas for future work. In order to rule out implementation effects, a similar suite of tests should be performed using independent implementations of NewReno, H-TCP and CUBIC (ideally under a different operating system, such as FreeBSD). Extending the evaluation to a wider set of TCP variants and further exploring useful metrics to differentiate the impact of TCPs in the home broadband environment is required. Finally, exploring simple ways in which CPE manufacturers can easily address the issues raised by this research would result in some material benefit to consumers. For example, evaluating the appropriate tuning and impact of various queue management schemes with a goal of making recommendations to CPE manufacturers would be a worthwhile outcome.

## Acknowledgments

This work has been made possible in part by a grant from the Cisco University Research Program Fund at Community Foundation Silicon Valley.

## References

1. Postel, J.: Transmission Control Protocol. RFC 793, Standard, Updated by RFC 3168. (September 1981), http://www.ietf.org/rfc/rfc793.txt
2. Floyd, S., Henderson, T., Gurtov, A.: The NewReno Modification to TCP's Fast Recovery Algorithm. RFC 3782 (Proposed Standard) (April 2004), http://www.ietf.org/rfc/rfc3782.txt
3. Allman, M., Paxson, V., Stevens, W.: TCP Congestion Control. RFC 2581 (Proposed Standard), Updated by RFC 3390 (April 1999), http://www.ietf.org/rfc/rfc2581.txt

4. Fomenkov, M., Keys, K., Moore, D., Claffy, K.: Longitudinal study of Internet traffic in 1998-2003. In: Winter International Symposium on Information and Communication Technologies (WISICT), Cancun, Mexico (January 2004), http://www.caida.org/publications/papers/2003/nlanr/nlanr_overview.pdf
5. Floyd, S.: Congestion Control Principles. RFC 2914 (Best Current Practice) (September 2000), http://www.ietf.org/rfc/rfc2914.txt
6. Rhee, I., Xu, L., Ha, S.: CUBIC for Fast Long-Distance Networks. Technical report, North Carolina State University (August 2008), http://tools.ietf.org/id/draft-rhee-tcpm-cubic-02.txt
7. Ha, S., Rhee, I., Xu, L.: CUBIC: A New TCP-Friendly High-Speed TCP Variant. ACM SIGOPS Operating System Review 42(5), 64–74 (2008), http://netsrv.csc.ncsu.edu/export/cubic_a_new_tcp_2008.pdf
8. Leith, D.: H-TCP: TCP Congestion Control for High Bandwidth-Delay Product Paths. Technical report, Hamilton Institute (April 2008), http://tools.ietf.org/id/draft-leith-tcp-htcp-06.txt
9. Li, Y.T., Leith, D., Shorten, R.N.: Experimental evaluation of tcp protocols for high-speed networks. IEEE/ACM Trans. Netw. 15(5), 1109–1122 (2007), http://dx.doi.org/10.1109/TNET.2007.896240
10. Jacobson, V.: Congestion avoidance and control. In: SIGCOMM 1988: Symposium proceedings on Communications architectures and protocols, pp. 314–329. ACM, New York (1988), http://doi.acm.org/10.1145/52324.52356
11. Braden, R.: Requirements for Internet Hosts - Communication Layers. RFC 1122, Standard, Updated by RFC 1349 (October 1989), http://www.ietf.org/rfc/rfc1122.txt
12. Internet Research Task Force: (Internet Congestion Control Research Group), http://www.irtf.org/charter?gtype=rg&group=iccrg (accessed November 8, 2008),
13. Internet Research Task Force: (Transport Modeling Research Group), http://www.irtf.org/charter?gtype=rg&group=tmrg (accessed November 8, 2008)
14. Andrew, L., Marcondes, C., Floyd, S., Dunn, L., Guillier, R., Gang, W., Eggert, L., Ha, S., Rhee, I.: Towards a Common TCP Evauation Suite. In: Sixth International Workshop on Protocols for Fast Long-Distance Networks, Manchester, GB (March 2008), http://www.hep.man.ac.uk/PFLDnet2008/paper/08_Lachlan_pfldnet2008.pdf
15. Andrew, L., Atov, I., Kennedy, D., Wydrowski, B.: Evaluation of FAST TCP on Low-Speed DOCSIS-based Access Networks. In: IEEE TENCON 2005, Melbourne, Australia (November 2005), http://ieeexplore.ieee.org/iel5/4084859/4084860/04085219.pdf?tp=&arnumber=4085219&isnumber=4084860
16. Armitage, G., Stewart, L., Welzl, M., Healy, J.: An Independent H-TCP Implementation Under FreeBSD 7.0: Description and Observed Behaviour. SIGCOMM Comput. Commun. Rev. 38(3), 27–38 (2008), http://doi.acm.org/10.1145/1384609.1384613
17. Armitage, G.: An experimental estimation of latency sensitivity in multiplayer quake 3. In: 11th IEEE International Conference on Networks (ICON 2003), Sydney, Australia, pp. 137–141 (2003), http://dx.doi.org/10.1109/ICON.2003.1266180

18. Helder, G.K.: Customer evaluation of telephone circuits with transmission delay. Bell System Technical Journal 45, 1157–1191 (1966)
19. Kitawaki, N., Itoh, K.: Pure delay effects on speech quality in telecommunications. IEEE Journal on Selected Areas in Communications 9(4), 586–593 (1991)
20. Markopoulou, A., Tobagi, F., Karam, M.: Assessing the quality of voice communications over internet backbones. IEEE/ACM Transactions on Networking 11(5), 747–760 (2003)
21. Claypool, M., Kinicki, R., Li, M., Nichols, J., Wu, H.: Inferring Queue Sizes in Access Networks by Active Measurement. In: Passive and Active Measurement Workshop, Antibes Juan-les-Pins, France (April 2004), `http://www.pamconf.org/2004/papers/209.pdf`
22. Dischinger, M., Haeberlen, A., Gummadi, K.P., Saroiu, S.: Characterizing residential broadband networks. In: IMC 2007: Proceedings of the 7th ACM SIGCOMM conference on Internet measurement, pp. 43–56. ACM, New York (2007), `http://doi.acm.org/10.1145/1298306.1298313`
23. Rizzo, L.: Dummynet: a simple approach to the evaluation of network protocols. ACM SIGCOMM Computer Communication Review 27(1), 31–41 (1997), `http://doi.acm.org/10.1145/251007.251012`
24. Mathis, M., Heffner, J., Reddy, R.: Web100: extended tcp instrumentation for research, education and diagnosis. SIGCOMM Comput. Commun. Rev. 33(3), 69–79 (2003)
25. Unknown (The Web100 Project), `http://web100.org/` (accessed November 19, 2008)
26. Unknown: (Iperf), `http://sourceforge.net/projects/iperf` (accessed November 19, 2008)
27. Turner, A.: Tcpreplay, `http://tcpreplay.synfin.net/` (accessed December 4, 2008)
28. Leith, D.: [2.6.27] tcp_htcp: last_cong bug fix, `http://patchwork.ozlabs.org/patch/8341/` (accessed December 4, 2008)
29. Ostermann, S.: tcptrace, `http://www.tcptrace.org/` (accessed December 4, 2008)

# Why Rely on Blind AIMDs?
## (Work in Progress)

Ioannis Psaras, Mehrdad Dianati, and Rahim Tafazolli

Center for Communication Systems Research (CCSR)
Dept. of El. Eng., University of Surrey
Guildford, GU2 7XH, Surrey, UK
(i.psaras,m.dianati,r.tafazolli)@surrey.ac.uk

**Abstract.** We are motivated by the fact that fixed Increase rates and Decrease ratios for AIMD cannot adjust TCP's performance to the Internet's diverse networking conditions. Indeed, we find that fixed values for the increase/decrease factors of AIMD restrain flexibility, which is a fundamental property of transport protocols in order to guarantee utilization and fairness in Modern and Future internetworks. We propose a new paradigm for hybrid AIMD designs that has the potential to adjust TCP's behavior according to network conditions.

The proposed *Multi-Rate AIMD* (MR-AIMD) increases additively the Additive Increase factor of AIMD in case of positive feedback and decreases multiplicatively (the AI factor) in case of negative feedback *and* Explicit Congestion Notifications. In other words, MR-AIMD takes into account ECN signals in order to quantify the level of network contention and adjusts its response accordingly.

We show that MR-AIMD reduces retransmission effort significantly, when contention is high, becomes aggressive when contention decreases and tolerates against random, transient errors due to fading channels.

**Keywords:** TCP, AIMD, Multi-Rate AIMD, Congestion Control, ECN.

## 1 Introduction

The huge expansion of the Internet and the explosion of its applicability in our every day life has triggered extensive research in various fields of the networking technology. Clearly, TCP and its supporting AIMD algorithm is one of the most overworked topics during the last 15 years. Research efforts have focused on faster convergence to fairness [1] and efficiency [2], transmission over wireless lossy links [3], fast exploitation of high-speed links [4], [5], [6], [7], [8], [9] and optimization for web traffic [10], just to name a few.

Many researchers approached the above issues from a non-end-to-end point of view. Efforts on that direction focused on cooperation techniques between mobile hosts and base stations to improve TCP's performance over wireless media (e.g., [11]), or sophisticated AQM techniques to provide preferential treatment for short (web) flows (e.g., [12], [13], [14]).

L. Fratta et al. (Eds.): NETWORKING 2009, LNCS 5550, pp. 404–415, 2009.

Apart from their design-specific goals (i.e., tolerate wireless loss, converge to fairness, treat preferentially short flows), the above-mentioned approaches always target *full resource utilization*. Here, we argue that *full utilization does not necessarily translate into efficient utilization*. That is, when contention is low, the transport protocol should exploit all available resources. However, when contention increases, demand exceeds resource supply and therefore, full utilization should not be a matter of concern anymore. Instead, *efficient* utilization should become the challenge to deal with.

We consider that an efficient transport and congestion control mechanism should be aggressive when contention is low (in order to exploit available resources and tolerate against wireless errors) and conservative when contention increases (in order to reduce retransmission effort and decongest the buffers' queues). We argue that the fixed increase rates and decrease ratios for AIMD restrain flexibility and therefore, fail to provide *efficient* resource utilization. Motivated by similar studies such as TCP-SIMD [2], which however lacks the potential to tolerate against wireless errors and AIRA [15], [16], we attempt to design a hybrid congestion controller for future internetworks, which takes advantage of ECN signals. The Explicit Congestion Notification mechanism has been shown to provide some benefits for web traffic (e.g., [14], [17]). However, not many studies elaborated on the potential benefits that ECN can provide to long flows, or on its properties as an error discriminator.

We investigate the properties of a *Multi-Rate*, AIMD-based, Additive Increase (AI) factor. Briefly, the algorithm operates as follows: upon successful delivery of *cwnd* number of packets to the receiver side, not only the *cwnd*, but also the *the Additive Increase factor* increases Additively (i.e., $a \leftarrow a + \frac{a'}{cwnd}$), while on the face of loss, *the Additive Increase factor* is Multiplicatively Decreased (i.e., $a \leftarrow a - b \cdot a$). Decisions as to whether the AI factor should be increased or decreased and by how are based on AQM techniques, namely ECN.

The novelty of the proposed algorithm lies on its ability to adjust according to network conditions. MR-AIMD becomes aggressive when contention is low, although we explicitly note that it does not target high-speed environments; conservative when contention increases and tolerant against wireless errors, since it exploits ECN signals. We also note that although ECN is not famous for its capability as an error discriminator, our initial results show that there exists a lot of space for exploitation of such a system property.

## 2 Motivation: Blind AIMD

Deployment of AIMD is associated with two operational standards: (i) the fixed increase rate and decrease ratio and (ii) the corresponding selection of appropriate values.

Recent research has focused on altering the values for the *Additive Increase*, $a$, and *Multiplicative Decrease*, $b$, factors, in order to achieve fast bandwidth exploitation (e.g., [4], [5], [6], [7], [8], [9]) or faster convergence to fairness (e.g., [2], [1] and references therein), but has not questioned really the validity and efficiency of fixed rates throughout the lifetime of participating flows.

In this context, research efforts cannot address questions such as: *Why do flows increase their rate by "a" packets instead of "2a" packets, even when half users of a system leave and bandwidth becomes available?*

## 2.1 Congested Wired Network

One possible justification for not highlighting the above research direction is that:

*The Additive Increase factor of AIMD does not contribute to the long-term Goodput performance of TCP, when losses are due to buffer overflow.*

In Figure 1, we present the *cwnd* evolution for two TCP flavors: Figure 1(a), where $a = 1$ (regular TCP) and Figure 1(b), where $a = 0.5$. The area underneath the solid *cwnd* lineplot (Area 1 and 2) represents the Goodput[1] performance of the protocols. In Figure 1(c), we show that both protocols achieve the same Goodput performance, since $A1 = A2$ and $A3 = A4$[2]. However,

*Additive Increase affects significantly the Retransmission Effort of flows, which impacts overall system behavior as well.*

For example, TCP $a = 1$, in Figure 1, experiences 4 congestion events, while TCP $a = 0.5$ experiences only 2. Assuming that each congestion event is associated with a fixed number of lost packets, regular TCP (i.e., $a = 1$) will retransmit twice as many packets as TCP with $a = 0.5$, without any gain in Goodput.

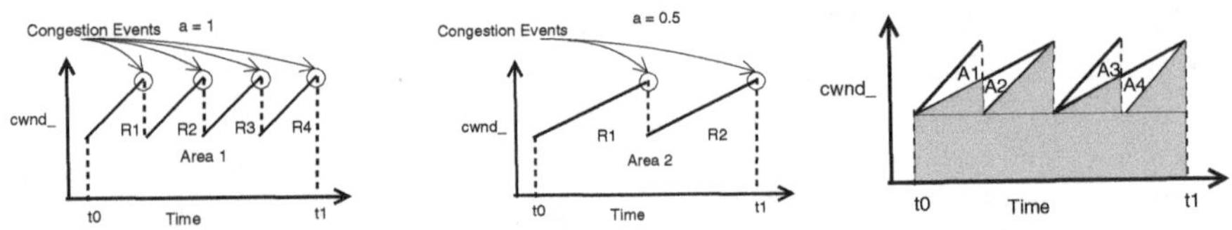


(a) Increase Factor = 1 (b) Increase Factor = 0.5 (c) Difference

**Fig. 1.** Different Increase Factors

We verify the above observations through simulations (using ns-2). We simulate TCP-SACK flows, for 200 seconds, over a single bottleneck dumbbell network topology (Figure 2); the backbone link transmits 1Mbps, its propagation delay is 20ms and the RED Router has buffer capacity equal to 25 packets.

---

[1] We define the system Goodput as $\frac{Original\ Data}{Connection\ Time}$, where *Original Data* is the number of bytes delivered to the high level protocol at the receiver (i.e., excluding retransmissions and the TCP header overhead) and *Connection Time* is the amount of time required for the data delivery. Instead, system Throughput includes retransmitted packets and header overhead (i.e., $\frac{Total\ Data}{Connection\ Time}$).

[2] In Figure 1(c) grey areas are common for both protocols; white areas are equal ($A1$ is similar to $A2$ and $A3$ is similar to $A4$).

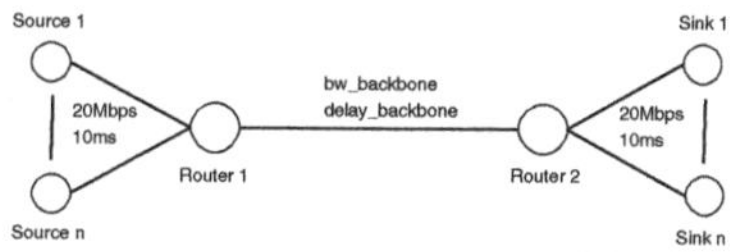


**Fig. 2.** Dumbbell Network Topology

**Table 1.** TCP Performance - Different Increase Factors

| **Wired** | Goodput | Retransmissions | | **Wireless** | Goodput | |
|---|---|---|---|---|---|---|
| | 2/4 flows | 2 flows | 4 flows | | 2 flows | 4 flows |
| | | | | $a = 5$ | 114.8 KB/s | 238.4 KB/s |
| $a = 2$ | 118.9 KB/s | 742 pkts | 1653 pkts | $a = 2$ | 81.3 KB/s | 179.9 KB/s |
| $a = 1.5$ | 118.9 KB/s | 548 pkts | 1777 pkts | $a = 1.5$ | 73.5 KB/s | 150.5 KB/s |
| $a = 1$ | 118.8 KB/s | 278 pkts | 704 pkts | $a = 1$ | 63.1 KB/s | 127.9 KB/s |
| $a = 0.5$ | 118.9 KB/s | 172 pkts | 452 pkts | $a = 0.5$ | 44.2 KB/s | 89 KB/s |

Clearly, there is a tradeoff between *Aggressiveness* and *Retransmission Effort* (see Table 1). The degree of *Aggressiveness* that a transport protocol can achieve is tightly associated with its Retransmission Effort. The higher the Additive Increase factor, the more the retransmission effort of the transport protocol (see for example, the 4 flow, wired scenario in Table 1). Note that further increasing the level of contention may even degrade the system Goodput performance, due to timeout expirations [18], which are not considered in Figure 1.

*In high contention scenarios, the higher the Additive Increase factor, the more retransmissions it causes, with zero gains in system Goodput.*

## 2.2 Wireless, Mobile Computing

On the contrary, losses due to congestion may not always be the case. The evolution of mobile, wireless networking calls for further investigation and adjustment of transport layer algorithms to deal with losses due to wireless, fading channels as well. In this context,

*The Additive Increase factor of AIMD may very well impact TCP's Goodput performance, when contention is low and losses are due to wireless errors.*

We repeat the previous simulation to verify the above statement. The backbone link can now transfer 10Mbps (instead of 1Mbps) and we additionally insert 0.3 Packet Error Rate (PER) to emulate losses due to fading, wireless channels. The results are presented on the right side of Table 1. We observe that in case of low contention and transient errors due to fading channels, higher values for the Additive Increase factor of AIMD can boost TCP's performance significantly.

*In low contention scenarios, where transient losses happen due to fading channels, the higher the Additive Increase factor, the more the Goodput gains for TCP.*

### 2.3 Bandwidth Exploitation Properties

Today's Internet application and infrastructure heterogeneity demands for responsive transport protocols, which are able to exploit extra available bandwidth rapidly, in case of contention decrease; at the same time the transport layer protocol should be able to adjust its transmission rate downwards in case of incoming flows, in order to (i) leave space for the new flows and (ii) not overflow the network. That said, fixed Additive Increase provides fixed transmission rate acceleration both in case of contention decrease and in case of extra bandwidth constraints. We argue that such behavior is undesirable indeed, since it leads to slow bandwidth exploitation, when bandwidth becomes available, while it requires significant retransmission effort when bandwidth constraints prevail.

*In case of contention decrease / increase scenarios, fixed acceleration leads to either slow resource exploitation or high retransmission effort, respectively.*

## 3 MR-AIMD: Multi-Rate AIMD

### 3.1 The Algorithm

Regular TCP increases its congestion window by 1 packet, upon successful transmission of *cwnd* number of packets (i.e., $cwnd \leftarrow cwnd + \frac{a}{cwnd}$ on every ACK), while negative feedback (i.e., three duplicate ACKs), which is interpreted as network congestion, triggers *multiplicative cwnd* decrease (i.e., $cwnd \leftarrow cwnd - b \cdot cwnd$), where $a = 1$ and $b = 0.5$, according to [19].

As an initial approach to a "non-blind", dynamically adjustable increase factor, we attempt to graft the basic AIMD functionality *into the Additive Increase factor of TCP*. More precisely, the *Multi-Rate AIMD* algorithm increases the *cwnd* value according to:

$$cwnd \leftarrow cwnd + \frac{a \leftarrow a + \frac{a'}{cwnd}}{cwnd} \qquad (1)$$

The initial value for $a$ is 1, while for $a'$ is 0.5. The proposed algorithm makes use of Active Queue Management (AQM) techniques in order to regulate the Additive Increase rate, $a'$. In particular, the algorithm uses the Explicit Congestion Notification (ECN) bit. There are two salient points that need to be clarified at this point regarding the cooperation scheme between the MR-AIMD sender and ECN: (i) an ECN marked packet triggers adjustment of the Additive Increase rate ($a'$) *only* (i.e., the flow's *cwnd* is *not* reduced), and (ii) an ECN marked packet decelerates MR-AIMD's rate $a'$ to 0.005 (instead of its initial value 0.5).

Modification of the ECN algorithm exhibits a number of desirable properties: (i) it smooths TCP's transmission rate and (ii) it avoids (to an extend) TCP's drastic rate fluctuations, whenever deemed appropriate according to the proposed algorithm.

For the AI rate adaptation upon arrival of an ECN marked packet, we reason as follows: Once set, by the intermediate router, the ECN bit may trigger one of three possible responses: (i) Additive Increase, (ii) Multiplicative Decrease or (iii) stabilization *of the AI rate, $a'$*. The current implementation of MR-AIMD acts according to choice (i) (i.e., AI of rate $a'$). The rationale behind our choice is as follows: multiplicative decrease of the AI rate results in very low values for $a'$ and therefore, conservative behavior. On the other hand, rate stabilization, through choice (iii), may result in system instability and flow unfairness, in the long term. Due to space limitations, we do not elaborate further on this issue, but we report that initial results verify our decisions for increased system performance (see Section 4).

The Additive Increase factor, $a$, decreases multiplicatively, according to Equation 2, upon a triple duplicate ACK event:

$$a \leftarrow a - b \cdot a. \tag{2}$$

The Multiplicative Decrease factor of MR-AIMD, $b$, is set to 0.5 similarly to TCP-AIMD (Equation 2), in order to guarantee fairness and stability [19]. Furthermore, upon a timeout event, MR-AIMD reduces the Additive Increase rate, $a'$, to its initial value, 0.5, in order to account for increased levels of network contention. The rest of TCP's functions remain unchanged (e.g., RTO back-off, *cwnd* adjustments etc.). Although we do not elaborate on the convergence properties of MR-AIMD here, we report that according to our initial simulation results the algorithm indeed converges to fairness and stability, due (i) to its Multiplicative Decrease properties and (ii) to its ability to reduce retransmission effort, which implicitly increases system stability. We refer the reader to [16] for a more complete discussion on that topic.

### 3.2 Discussion

We assume that TCP's operational space, with regard to the Additive Increase and Multiplicative Decrease factors ALPHA and BETA, is represented by four basic domains: (i) conservative, (ii) aggressive, (iii) smooth and (iv) responsive (see Figure 3). The current, blind TCP-AIMD implementation covers a single point, only, within TCP's operational space (see Figure 3(a)). Clearly, the fixed increase/decrease parameters deal with none of the four operational domains, efficiency-wise and moreover, *any* pair of *fixed* increase/decrease parameters can deal with *one* operational domain *only*. We argue that such settings form an inflexible, conservative, worst-case approach to TCP's operational properties. For instance, a sophisticated transport layer algorithm should adjust according to network conditions: it should become conservative when contention is high, aggressive in case of transient wireless errors, responsive in case of contention increase/decrease and smooth in case of (relatively) static network load.

The proposed scheme extends TCP's functionality to operate along the x-axis of TCP's operational space. This way, several desirable properties are added to TCP's inherent functionality. For example, we show that careful design can lead to more aggressive transmission when contention is low and losses happen on the wireless portion of the network, while conservative transmission, when contention

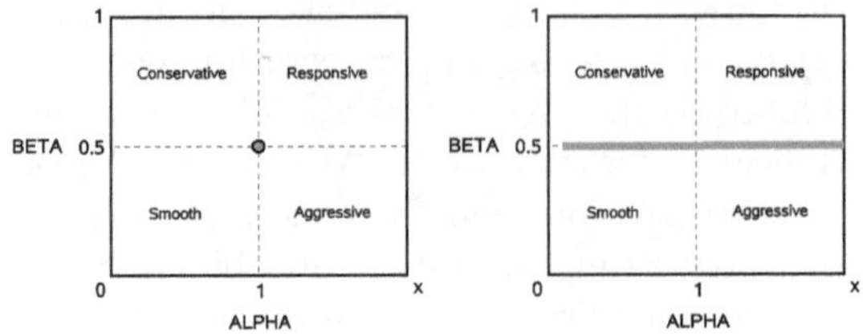


(a) TCP-AIMD (b) MR-AIMD

**Fig. 3.** x-AIMD Operational Space

increases, can account for reduced retransmission overhead. The current proposal constitutes a first step on the further extension of TCP's functionality, in order to exploit the whole spectrum of possible behaviors (i.e., utilization of the y-axis as well). We note that although there have been some proposals on the same direction (e.g., [8], [20]), these proposals target high-speed environments and therefore have different design goals. Hence, we do not attempt to compare MR-AIMD with those approaches.

We note that MR-AIMD does not target high-speed environments. Although the MR-AIMD's transmission rate may increase compared to regular TCP, its operational properties are not intended to exploit high-speed links. Instead, the proposed algorithm attempts to deal with the application diversity and infrastructure heterogeneity of present and future internetworks.

## 4 Preliminary Results

We use the SACK version of TCP with the timestamps option enabled. The simulation scenarios are similar to the ones presented in Section 2. That is, we use the dumbbell network topology[3], the queuing policy is RED and the buffer size is set according to the bandwidth-delay product of the outgoing link.

### 4.1 Congested Wired Networks

Initially, we simulate a wired network where the backbone link transfers 48 Mbps and induces propagation delay of 40ms. We repeat the simulation for increasing number of participating flows (from 10 to 100), to capture the performance of the proposed algorithm relatively with the level of contention.

We observe that when contention is low (e.g., 10 flows over 48Mbps) the proposed algorithm achieves the same Goodput performance as regular AIMD (see Figure 4(a)); the retransmission effort graph (Figure 4(b)) reveals that for

[3] We have experimented with diverse-RTT topologies as well and we report that RED's inherent property to penalize higher-bandwidth flows, alleviates RTT-unfairness at least for RTT-differences in the order of 100ms or less. For the sake of simplicity and given the space limitations, we present results for equal-RTT flows only.

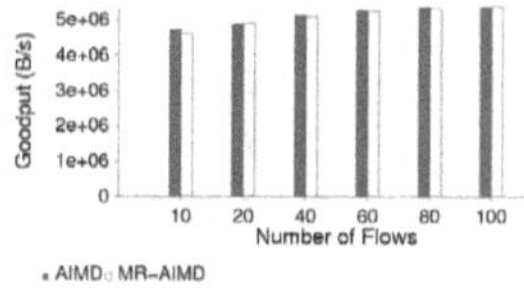


(a) System Goodput

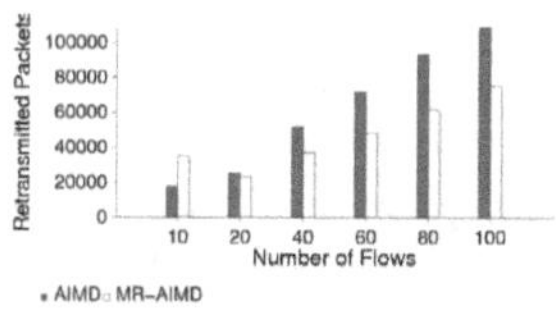


(b) Retransmitted Packets

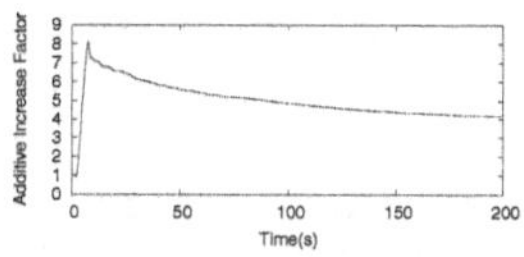


(c) 10 Flows Experiment

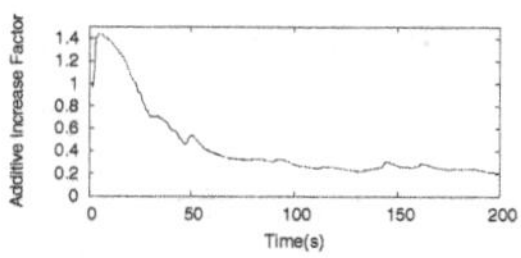


(d) 100 Flows Experiment

**Fig. 4.** Performance over Congested Wired Networks

low contention environments, the proposed AIMD operates aggressively. In Figure 4(c), we graph the average Additive Increase factor for a random flow, when 10 flows compete. This Figure verifies the aggressive behavior of MR-AIMD, when contention is low. We note that according to our solution framework this behavior is desirable indeed. That is, when contention is low we want the algorithm to be aggressive, ready to exploit extra bandwidth that may potentially become available due to flows that end their tasks and leave the system.

As contention increases, however, losses due to buffer overflow become more frequent, leading to multiplicative decreases of both the *cwnd* and the Additive Increase factor. In turn, smaller increase rates lead to reduced retransmission effort (see Figure 4(b)). In Figure 4(d), we present the average Additive Increase factor of MR-AIMD, for a random flow, when the total number of participating flows is 100. Indeed, we see that the average value of MR-AIMD's Additive Increase factor is below 1, allowing for less aggressive transmission, since the level of network contention so permits.

Overall, we see that the proposed algorithm adjusts efficiently to the level of network contention, taking advantage of its dynamic increase/decrease acceleration properties to utilize resources accordingly. In particular, when contention is low the algorithm is aggressive, ready to utilize rapidly extra available bandwidth, while when contention increases the algorithm lowers the transmission rate to reduce retransmission effort.

## 4.2 Wireless Networks

We repeat the simulation presented in Section 2.2 to observe the performance of the proposed algorithm over lossy links. In the current setup the backbone link transfers 48Mbps with 40ms propagation delay, while the PER is 0.3.

Figure 5(a) depicts the outcome of the simulation. We see that the proposed algorithm is tolerant against random, transient errors caused by wireless, fading channels. MR-AIMD accelerates transmission faster than conventional AIMD, becoming more aggressive, when conditions permit, which is another desirable

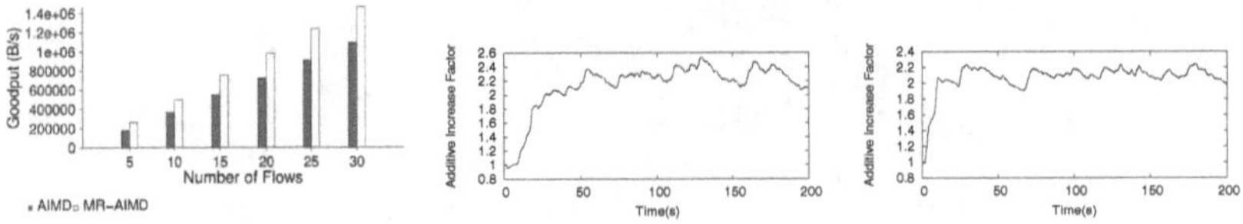


(a) System Goodput (b) 5 Flows Experiment (c) 30 Flows Experiment

**Fig. 5.** Performance over Wireless Networks

property in case of wireless errors [3]. Indeed, we see in Figures 5(b) and 5(c) that the Additive Increase factor is far above 1, allowing for speedy transmission and up to, approximately, 30% higher Goodput performance (Figure 5(a)) in case of errors due to wireless, lossy links.

### 4.3 Mixed Wired-Wireless Environments

We perform one more experiment in order to verify that MR-AIMD adjusts correctly in mixed wired-wireless environments, where the level of contention may vary. The simulation environment is the same as previously, but the backbone link can now transfer 24Mbps. We repeat the simulation for increasing number of participating flows, from 5 to 100. Indeed, we see in Figure 6(a) that when contention is low MR-AIMD exploits the available resources, tolerates against random link errors and accelerates transmission (see Figure 6(c)), increasing the overall system Goodput (see Figure 6(a), flows 5-80). In contrast, when contention increases (i.e., 80 and 100 participating flows), MR-AIMD reduces its transmission rate, through lower Additive Increase factors (Figure 6(d)), although some errors may still be due to fading channels (i.e., we consider buffer overflow to be a more important factor for rate reduction than wireless errors). By doing so, MR-AIMD achieves the same Goodput performance as conventional AIMD, but reduces the retransmission effort of the transport protocol (Figure 6(b)). The present experiment verifies the hybrid behavior of MR-AIMD, which based on ECN signals adapts appropriately to the network conditions. Although ECN is not famous as an error discriminator, our initial results show that ECN-capable transports may be benefited, at least to an extend, from its operation as such. Further experimentation is needed in order to uncover ECN's capabilities regarding its accuracy on that direction.

### 4.4 Bandwidth Exploitation Properties

We attempt to briefly assess the bandwidth exploitation properties of the proposed algorithm. The Additive Increase factor of MR-AIMD progresses in time according to:

$$a_n = a_{n-1} + a' \cdot n, \ where \ n \geq 1. \tag{3}$$

In Equation 3, $n$ stands for the number of RTTs, and $a'$ is the acceleration factor of MR-AIMD (i.e., either 0.5 or 0.005). The initial value for $a_n$, for a new

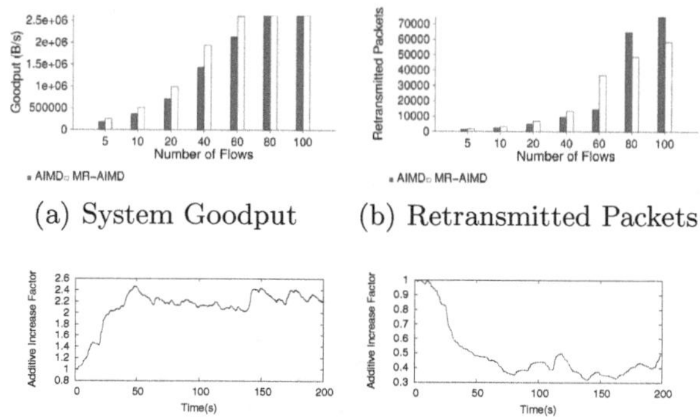


(a) System Goodput (b) Retransmitted Packets

(c) 5 Flows Experiment (d) 80 Flows Experiment

**Fig. 6.** Performance over Mixed Wired-Wireless Networks

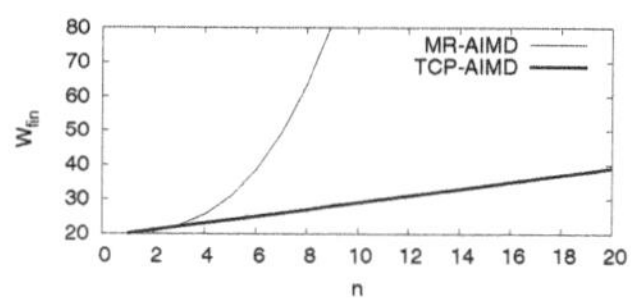


**Fig. 7.** Extra Bandwidth Exploitation Properties

connection for example, is 1. Otherwise, for an existing connection, the initial value of $a_n$ depends on the algorithm's state (i.e., AI through Equation 1, or MD through Equation 2).

In turn, MR-AIMD's *cwnd* after $n$ RTTs is given by:

$$W_{fin} = W_{init} + \sum a_n, \tag{4}$$

TCP's *cwnd* after $n$ RTTs is given by:

$$W_{fin} = W_{init} + a \cdot n, \tag{5}$$

where $W_{init}$ is the initial *cwnd* and $W_{fin}$ is the *cwnd* after $n$ RTTs. Obviously, for TCP-AIMD $a = 1$, while for MR-AIMD $a'$ is either equal to 0.5 or 0.005.

We assume a contention decrease event, where a number of participating flows leave the system when $W_{init} = 20$. From that point onwards, the rest of the flows have to exploit the extra bandwidth as fast as possible. We assume that since contention has decreased there are no ECN signals to the TCP sender (at least up to a certain point where contention becomes high due to the increased congestion windows of the rest of the participating flows).

As expected, we see in Figure 7 that MR-AIMD has the potential to exploit extra available network resources rapidly, without threatening the system's stability. That is, although initially the algorithm appears aggressive (MR-AIMD doubles its window within 6 RTTs, while TCP-AIMD needs 20 RTTs), it will immediately slow down, when ECN marked packets indicate incipient congestion. Due to limited space, we do not elaborate further on that issue here.

## 5 Conclusions

We argue that a blind Additive Increase factor for AIMD limits TCP's performance in terms of efficient resource utilization. We proposed a rather simple but novel approach towards a new design space for transport layer internetworking. Although the proposed settings are chosen based on experimental evaluations only, they seem to boost TCP's performance significantly. Moreover, additional modifications can easily be incorporated. For example, we did not evaluate here the properties of MR-AIMD with regard to the RTT-unfairness problem of TCP. Although one may argue that the proposed algorithm, in its current form, may extend TCP's inability to treat diverse RTT flows fairly, simple modifications can improve TCP's performance on that direction as well. For instance, MR-AIMD's Additive Increase function may be complemented with a fraction of the flow's measured RTT sample (i.e., $a \leftarrow a + \frac{c \cdot a'}{cwnd}$, where $c$ is the flow's latest measured RTT sample). We note, however, that since MR-AIMD requires AQM techniques, namely RED with ECN to be implemented in the intermediate router [21], RTT-unfairness issues are partially eliminated due to RED's inherent properties, as our initial results (not included here) indicate.

## Acknowledgments

This work is supported in part by the European Commission through the *4WARD* project in the $7^{th}$ *Framework Programme*. The views and conclusions contained herein are those of the authors and should not be interpreted as necessarily representing the *4WARD* project or the European Commission.

## References

1. Lahanas, A., Tsaoussidis, V.: Exploiting the Efficiency and Fairness Potential of AIMD-based Congestion Avoidance and Control. Computer Networks 43(2), 227–245 (2003)
2. Jin, S., Guo, L., Matta, I., Bestavros, A.: TCP-friendly SIMD Congestion Control and Its Convergence Behavior. In: Proc. of 9th IEEE ICNP, Riverside, CA (2001)
3. Tsaoussidis, V., Matta, I.: Open issues on TCP for Mobile Computing. Wiley, WCMC 2(1) (Feburary 2002)
4. Floyd, S.: HighSpeed TCP for Large Congestion Windows, RFC 3649 (2003)
5. Kelly, T.: Scalable TCP: Improving Performance in Highspeed Wide Area Networks. ACM SIGCOMM CCR 33(2), 83–91 (2003)
6. Wei, D.X., et al.: FAST TCP: motivation, architecture, algorithms, performance. IEEE/ACM Transactions on Networking (2007)
7. Rhee, I., Xu, L.: CUBIC: A New TCP-Friendly High-Speed TCP Variant. In: Proceedings of PFLDnet (2005)

8. Tan, K., et al.: A Compound TCP Approach for High-Speed and Long Distance Networks. In: Proc. of INFOCOM 2006, Barcelona, Spain (2006)
9. King, R., Baraniuk, R., Riedi, R.: TCP-Africa: An Adaptive and Fair Rapid Increase Rule for Scalable TCP. In: Proceedings of INFOCOM 2005, vol. 3, pp. 1838–1848 (2005)
10. Le, L., Aikat, J., Jeffay, K., Smith, F.D.: The effects of active queue management on web performance. In: SIGCOMM 2003, pp. 265–276 (2003)
11. Bakre, A., Badrinath, B.R.: I-TCP: Indirect TCP for mobile hosts. In: Proc. 15th Int'l Conf. on Distributed Computing Systems, pp. 136–143 (1995)
12. Mahajan, R., Floyd, S., Wetherall, D.: Controlling high-bandwidth flows at the congested router. In: ICNP 2001, pp. 192–201 (2001)
13. Guo, L., Matta, I.: The war between mice and elephants. In: Proceedings of ICNP (October 2001)
14. Le, L., Aikat, J., Jeffay, K., Smith, F.D.: Differential congestion notification: Taming the elephants. In: ICNP 2004, pp. 118–128 (2004)
15. Psaras, I., Tsaoussidis, V.: AIRA: Additive Increase Rate Accelerator. In: Proc. of IFIP Networking 2008, Singapore, pp. 691–702 (May 2008)
16. Psaras, I., Tsaoussidis, V.: On the Properties of an Additive Increase Rate Accelerator. Computer Networks (to appear)
17. Kuzmanovic, A.: The power of explicit congestion notification. In: SIGCOMM 2005, pp. 61–72 (2005)
18. Psaras, I., Tsaoussidis, V.: Why TCP timers (still) don't work well. Computer Networks 51(8), 2033–2048 (2007)
19. Jacobson, V.: Congestion avoidance and control. In: Proc. of SIGCOMM, pp. 314–329 (1988)
20. Shorten, R.N., Leith, D.J.: H-TCP: TCP for high-speed and long-distance networks. In: Proc. PFLDnet, Argonne (2004)
21. Medina, A., Allman, M., Floyd, S.: Measuring the evolution of transport protocols in the internet. SIGCOMM Comput. Commun. Rev. 35(2), 37–52 (2005)

# SBA: A Simple Backoff Algorithm for Wireless Ad Hoc Networks

Tahiry Razafindralambo[1] and Isabelle Guérin Lassous[2]

[1] Inria Lille - Nord Europe / CNRS / Université Lille 1
tahiry.razafindralambo@inria.fr
[2] Université de Lyon / LIP / Inria
isabelle.guerin-lassous@ens-lyon.fr

**Abstract.** The performance of ad hoc networks based on IEEE 802.11 DCF degrade when congestion increases. The issues concern efficiency and fairness. Many solutions can be found at the MAC layer in the literature, but very few solutions improve fairness and efficiency at the same time. In this paper, we design a new backoff solution, called SBA. SBA uses only local information and two contention window sizes. By simulations, we compare SBA with IEEE 802.11 and several alternatives to 802.11 in ad hoc networks. We show that SBA achieves a good trade-off between fairness, simplicity and efficiency.

**Keywords:** Ad hoc Networks, MAC protocol, Backoff Algorithm.

## 1 Introduction

An ad hoc network is a collection of nodes that communicate via wireless links in a multihop fashion without any fixed infrastructure or centralized servers. An ad hoc network has many practical applications including emergency/rescue operations, military operations and personal area networking.

One challenge in designing protocols for such a network is Medium Access Control. The IEEE 802.11 standard provides a communication mode called DCF (Distributed Coordination Function) that is fully distributed and usable in ad hoc networks [1]. However, the MAC protocol described in DCF exhibits some performance issues in terms of efficiency (often given as the overall throughput in the network) and fairness [2]. If some stations cannot access to the medium or cannot send its packets successfully, these nodes can disconnect the network. Such a disconnection may affect the behaviour of the whole network or prevent some nodes from providing a given service. On the other hand, if fairness and efficiency are achieved, QoS guarantees may be expected and/or provided by/for upper layer and thus for the user application.

Many works have tackled the problem of fairness and efficiency in ad hoc networks. Some approaches modify the BEB (Binary Exponential Backoff) algorithm of IEEE 802.11 to provide better performance, whereas ohers introduce new mechanisms. Most ot the existing solutions provide either efficiency or fairness, but very few solutions consider both of them.

L. Fratta et al. (Eds.): NETWORKING 2009, LNCS 5550, pp. 416–428, 2009.

In this article, we propose a new approach for a backoff-based algorithm for ad hoc networks, called hereafter SBA for *Simple Backoff Algorithm*. Our solution relies only on *local information*, as successful transmissions and collisions undergone by each station without taking advantage of the carrier sensing mechanism nor the packets that can be decoded. Contrary to the BEB algorithm, SBA has only two distinct contention window sizes. SBA also uses the same contention window for all the packets it has to send in a given time interval. At the end of this interval, some computations are done to choose the contention window for the next interval. The first results show that SBA is a good candidate in terms of trade-off between fairness, simplicity and efficiency.

After a state-of-the-art given in Section 2, SBA is described in Section 3. Then, the protocol is evaluated by simulation. The results are given in Section 4. We compare SBA with 802.11 and some other fair MAC protocols. Finally, we give a preliminary study on the complexity of these protocols in Section 5.

## 2 Related Work

Due to space limitation, we only cite works that focus on the fairness issue in ad hoc networks and that design single channel and single interface MAC protocols.

There are mainly two categories of MAC protocols in the literature. The first one is based on the backoff mechanism: the protocol uses the contention window to modify the behaviour of the protocol. The use of the contention window can be a modification of the window size or of the way of increasing/decreasing it. The protocols that belong to this category are mainly based on the DCF of IEEE 802.11 for the medium access method. The second category consists of protocols that do not use the backoff algorithm to modify their behaviour. This does not mean that there is no backoff algorithm but the main changes and evolutions inside the protocol are not made via the backoff.

**Backoff-based approach.** The protocols in this category can be also divided into two classes. The first class, called BEB-like algorithms, contains the backoff algorithms that behave like the BEB algorithm: this means that, directly after a successful transmission, the contention window is decreased, and after a collision, the contention window is increased. The MILD (Multiplicative Increase Linear Decrease) algorithm [3] and the DIDD (Double Increase Double Decrease) algorithm [4] belong to this category. Surprisingly, in the literature, most of these algorithms are mainly designed for single-hop networks, and very few of them are tested in ad hoc networks conditions.

The protocols of the second class modify the contention window based on some computations made by the station. These computations use some information gathered by each station from its neighbourhood. For example, in MBFAIR [5], the algorithm computes what the authors call a fair share and modifies the contention window size accordingly. The fair share is computed based on some information received from the two-hop neighborhood. The protocols described in [6,7] also belong to this category.

**Other approach or non backoff-based approach.** In this category, there are also two main classes of protocols. The first class uses some information other than the one initially provided by 802.11. In the protocol EHATDMA [8], a handshake in EHATDMA can be initiated by the sender or the receiver. The protocols of this category use broad information to provide fairness for all stations.

The protocols of the second class do not use any extra information except the information gathered from the active carrier sensing mechanism. For instance, the protocols PNAV [9] and MadMac [10] belong to this category. MadMac and PNAV insert a waiting time before sending a packet with 802.11. The waiting time is inserted depending on the activity perceived on the medium.

## 3 SBA: A Simple Backoff Algorithm

### 3.1 Basic Issues with 802.11

We describe here some basic ad hoc configurations that present performance issues when 802.11 is used. These configurations are often sub-configurations of larger ad hoc topologies. We think that a good understanding of these basic scenarios is important because the rules of our algorithm SBA are based on these issues. We describe three configurations: hidden terminals, asymmetric hidden terminals and the three pairs.

*a)* The hidden terminals configuration is given on Figure 1(a): the two transmitters are independent and try to send packets to the same receiver. The collisions due to concurrent transmissions lead to a throughput degradation and a short term unfairness as described in [11]. The collisions may be reduced with the RTS/CTS mechanism of 802.11. *b)* The asymmetric hidden terminals configuration is given on Figure 1(b): the transmission of the upper emitter always collides and the transmission of the lower emitter always succeeds. This asymmetry leads to a long term unfairness between the two transmitters when 802.11 is used and the RTS/CTS mechanism does not solve the problem [3]. *c)* The three pairs scenario is presented on Figure 1(c): there is no collision issue, but unfairness arises with 802.11 because the central pair cannot access the medium due to asymmetric contention [12].

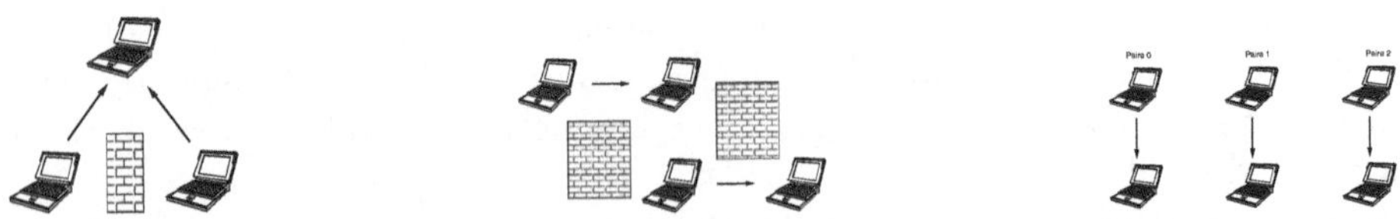

(a) The hidden terminals. (b) The asymmetric hidden terminals. (c) The three pairs.

**Fig. 1.** Basic configuration issues with IEEE 802.11 DCF

### 3.2 Principles

In SBA, we keep the main principles of the MAC protocol of IEEE 802.11 DCF: a packet is emitted on the radio medium after a DIFS and a backoff time (during which the medium must be free) and in unicast mode, the data packet is acknowledged after a SIFS. The main differences between SBA and 802.11 are the number of backoff stages and the conditions that determine the contention window size to use. Due to space limitations, we do not describe IEEE 802.11 DCF (see [1] for more details).

**Only two states.** Compared to 802.11, SBA has only two different contention window sizes, one large called hereafter $CW_{max}$ and one small called $CW_{min}$.

**Local information.** Our protocol relies only on local information. What we mean by local information is information that a station can derive from its own experience without using any measure from the carrier sensing mechanism[1] and without exploiting the content of the packets it receives nor the packets it can listen on the radio medium. Therefore the used information is collisions and successful transmissions undergone by the station, as done in the DCF mode. There are many advantages in using such local information. For example, such information is always available and reliable and if no message is exchanged the protocol overhead is reduced. Not using the carrier sensing mechanism, for computing the medium occupancy ratio for instance, can be also of some interest concerning the energy consumption. But using only local information has also some drawbacks. For example, when the number of stations is not known, it is difficult for the protocol to have an optimal behaviour. However locality is an important property for ad hoc networks protocols, especially for MAC protocols.

**How to use local information?** The BEB algorithm of IEEE 802.11 is said to be aggressive due to the adaption of the contention window size after a successful transmission. Many BEB-like algorithms try to cope with this aggressive behaviour by a slow reduction of the contention window size. As SBA has only two distinct contention window sizes, we do not adapt the contention window size after each transmission, in order to reduce the oscillations between these two states. SBA uses the local information gathered during a given time interval to adapt the contention window for the next time interval. The length of this interval must not be too short to avoid possible oscillation and not be too long to increase the reactivity of the algorithm. The time interval used in SBA is fixed and the same for all stations. This interval is called $\Delta$ hereafter.

**Computing medium state probabilities.** The medium can have different states. From the station point of view, the medium can be occupied by other stations, or occupied by its transmission resulting in a successful transmission or in a collision, or the medium can be idle. We denote the probability for the

[1] Note that SBA uses the carrier sensing mechanism for accessing the medium to check whether the medium is free or not, but does not make any measure, as the medium occupancy ratio for instance, with this mechanism.

medium to be in each state by $P[occ]$, $P[suc]$, $P[col]$ and $P[free]$ respectively. These four probabilities cannot be accurately computed by using only local information as defined previously. However they can be approximated with this local information. The local information, known and used by a station, is the length and the time needed to transmit each packet. The value of $P[suc]$ for a given interval is thus given by:

$$P[suc] = \frac{T_{suc}}{\Delta} \quad (1)$$

where $T_{suc}$ is the time spent by a station in successful transmissions. In the same way, we can compute $P[col]$ based on $T_{col}$, where $T_{col}$ is the time spent by a station in collisions. Another local information used by a station is the number of successful transmissions and the number of collisions it undergoes during $\Delta$, and denoted by $N_{suc}$ and $N_{col}$ (resp.). As SBA uses only one contention window size during a given time interval, it is possible to approximate, during a given $\Delta$ and without sensing the medium, the period during which the medium is idle, denoted by $P[free]$, by:

$$P[free] = \frac{(N_{col} + N_{suc}) \times (cw + DIFS)}{\Delta} \quad (2)$$

where $cw$ is the mean backoff time. This value is an approximation, because, we use the mean backoff time and because it only takes into account the free periods just before the transmissions. Based on previous equations, we can compute $P[occ]$ by using the following relation:

$$P[occ] = 1 - (P[suc] + P[free] + P[col]) \quad (3)$$

Note also that when the network is unsaturated, the value of $P[occ]$ ($P[free]$ resp.) is not the real occupation (idle period resp.) of the medium. This behaviour is due to the fact that the station does not use the carrier sensing mechanism to define free periods and thus occupation periods, but only local information. However, we will see that this approximation ensures good performances to SBA in unloaded networks.

### 3.3 SBA: The Protocol

The code of SBA is presented in Algorithm 1. As we have only two states, the algorithm can be seen as the transition between these two states. Algorithm 1 shows how the statistics gathered during the $\Delta$ period are used to switch from one state to another.

Algorithm 1 consists of two main cases given in Line 1 and Line 9. If the condition of Line 1 is true, it means that the station's occupancy with successful transmissions is lower than the other stations' occupancy and then its contention window is set to $CW_{min}$ (Line 2), otherwise it is set to $CW_{max}$ (Line 10). This condition ensures that a node does not monopolize the medium as in, for instance, the three pairs or asymmetric hidden nodes scenarios. If a node thinks that it uses too much the radio medium, then it slows down its emissions by using $CW_{max}$.

**Algorithm 1.** SBA

```
1: if (P[suc] ≤ P[occ] + P[free]) then
2:    CW ← CW_min
3:    if (P[col] > r && rand{0,1}=1) then
4:       CW ← CW_max
5:    end if
6:    if ((P[free] ≤ s && P[col] > 0) || (N_suc + N_col = 0)) then
7:       CW ← CW_max
8:    end if
9: else
10:    CW ← CW_max
11: end if
```

The condition of Line 1 does not take into account the medium load. The medium load is considered in the condition of Line 6 with two cases. If the medium is heavily loaded, then free periods are reduced and the collision probability is greater than 0 and/or the station can not access to the medium. When these two conditions are satisfied, the station uses $CW_{max}$.

The condition of Line 3 uses a random variable to create asymmetry when the probability of collisions is greater than $r$. In this case, a random variable is drawn, and depending on its value, the contention window is set to $CW_{max}$ or left to $CW_{min}$. When the collision probability is very high, it means that there is problem of symmetry in the network. This symmetry appears when the stations use the same contention window size, especially $CW_{min}$. If these configurations lead to a high collision probability, this is probably due to the value of $CW_{min}$ which cannot solve the collisions. In this case, this random drawing tries to break the symmetry and divides by two the number of stations that can potentially use $CW_{min}$, which may reduce the number of collisions in the next period. This condition should solve hidden nodes configurations.

## 4 Simulation Results

In this section we present the simulation results of SBA. The simulations are realized on different ad hoc scenarios. We selected scenarios presented in the literature, most of them can be found in [2].

### 4.1 Simulation Parameters

The simulations presented in this section were run with NS-2 [13]. We modified the simulator to reflect the DSSS modulation described in 802.11b. We also removed the overhead due to the routing protocol and the ARP protocol in order to better understand the behavior of SBA. Table 1 gives the summary of the simulation parameters we use by defaults[2]. These parameters are used in all simulations if changes are not specified.

[2] Due to space limitation, we do not discuss the possible values for the SBA parameters. The results are given with a single set of values.

**Table 1.** Summary of the simulation parameters

| | | | |
|---|---|---|---|
| SIFS | $10\mu s$ | Routing protocol | static |
| DIFS | $50\mu s$ | $\Delta$ | $0.2s$ |
| backoff slot | $20\mu s$ | $s$ | 0.15 |
| $CW_{min}$ | 31 | $r$ | 0.5 |
| $CW_{max}$ | 1023 | Traffic | backlogged / CBR / UDP |
| Physical header | $192\mu s$ | Packet size | random [600;1400] bytes |
| Data rate | 11 Mbps | Synchronization | random |
| ARP protocol | disabled | Mobility | none |
| Transmission range | 100 m | Carrier Sensing range | 200 m |

We compare SBA to other MAC layer solutions, including the IEEE 802.11 DCF, MBFAIR [5], PNAV [9], MadMac [10] and DIDD [4]. These protocols belong to different categories of MAC protocols presented in Section 2. We also evaluate SBA when the $\Delta$ periods of all the nodes are synchronized, called *SBA sync* hereafter. Of course, it is difficult to achieve such a synchronization in ad hoc networks, but this evaluation shows the upper bound on SBA performance. The studied results are the global throughput and the fairness of these protocols. The global throughput, *i.e.* the sum of the throughputs of all the stations, corresponds to the efficiency of the protocol. We measure fairness, according to the Jain fairness index defined in [14]. When this index is close to 1, it means that the protocol is fair according to a fairness scheme. We use the Max-Min fairness scheme as a baseline comparison, as many studies on fairness in ad hoc networks do[3].

### 4.2 Performance Results

#### Specific ad hoc scenarios

*The 3 pairs.* The results of efficiency and fairness for the three pairs scenario, depicted in Figure 1(c), are given in Figure 2. In this scenario, the protocol is fair (under a Max-Min fairness scheme) when the throughputs obtained by the three pairs are roughly the same.

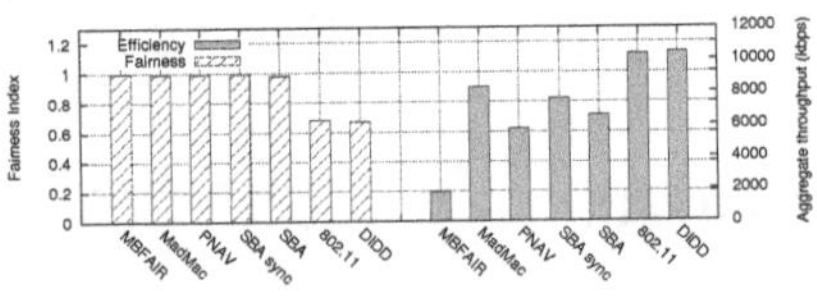


**Fig. 2.** Simulation results for the three pair scenario

This figure shows that the fairness index of SBA is close to 1, therefore SBA is fair, like MBFAIR, MadMac and PNAV. In this scenario, the three pairs alternate the use of $CW_{min}$ and $CW_{max}$. This alternation allows the central pair to access the medium more often than with 802.11, which increases the throughput of

[3] The discussion on this choice is out of the scope of this article.

the central pair and decreases the throughputs of the two external pairs. The flows throughput are roughly equal, which is not the case for 802.11 (or DIDD). On the other hand, the throughput reduction of the external pairs reduces the global throughput of the network. Therefore, SBA is less efficient than 802.11 (or DIDD).

We also see that MadMac has the better global throughput among the fair MAC protocols. This throughput is the maximum throughput that can be achieved under a Max-Min fairness restriction. This maximum throughput is achieved with MadMac by synchronizing the 3 pairs by using carrier sensing information. In SBA, we do not use such a synchronization scheme and we can see that the performance of SBA is smaller but not so far from the fair capacity. We can also notice that SBA is more efficient than MBFAIR and PNAV.

*The hidden terminals.* Figure 3 shows the fairness index and the global throughput obtained in the hidden terminals scenario presented in Figure 1(a). As the stations have a symmetrical behaviour, the two stations have statistically the same behaviour (see [11] for more details). Thus, there is no long term unfairness issue with this scenario, as the different fairness indexes show. We can see from this figure that, when SBA is randomly asynchronous, the global throughput is roughly the same as with 802.11. But when the stations are fully synchronized, the global throughput of SBA is better than the throughput of 802.11 with or without the use of RTS/CTS mechanisms. The throughput of MadMac is better than the throughput of SBA, because MadMac synchronizes the hidden terminal by using the acknowledgement to identify the hidden terminals problem. Note that the performance of MadMac degrades when acknowledgements are not used. We show here that SBA can provide a good fairness-efficiency trade-off without any information.

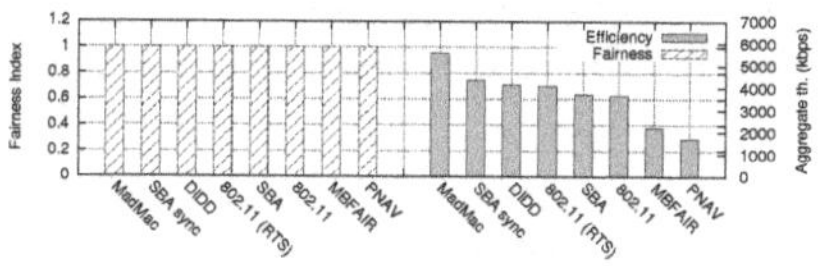


**Fig. 3.** Simulation results for the hidden terminals scenario

*The asymmetric hidden terminals.* Figure 4 shows the simulation results on the asymmetric hidden terminals, given in Figure 1(b). We can see that when the stations are synchronized, the throughput of SBA is close to the throughput of 802.11 and the fairness index of SBA is close to 1, which means that the two emitters have roughly the same throughput. The figure also shows that even if the stations are not synchronized, the performances of SBA are either better or close to the performances of the other fair protocols. These good performances are due to the alternation between the use of the contention window sizes. Concerning 802.11 (or DIDD), it is more efficient than all the other tested protocols, but much less fair.

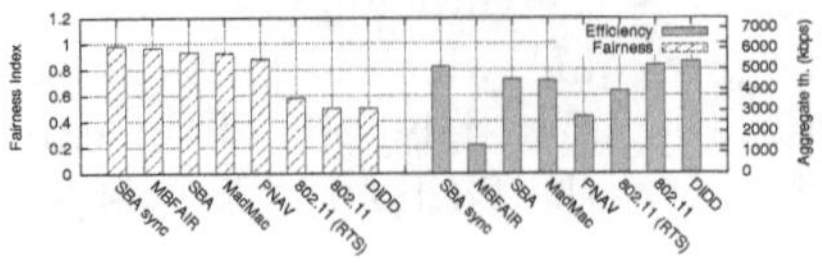


**Fig. 4.** Simulation results for the asymmetric hidden terminals scenario

**Random topologies.** In this subsection, we present the results on random topologies. In these topologies, stations' positions are uniformly and randomly chosen in a square of $500m \times 500m$ size. The communication range of the station is set to $150m$. Sources and destinations of flows are also randomly chosen. It is worth noting that, in these simulations, the same topology and flows configuration are tested for each protocol. The results given in this section are thus topology dependent.

Due to the random aspect of these scenarios, the performance metrics we use are different. For efficiency, we compute the global throughput of the network, the mean throughput of each flow and the corresponding 95% confidence interval. For fairness, we measure the $min - max$ ratio which is the ratio between the minimum throughput achieved in the network over the maximum throughput. This metric has the advantage, compared to the usual variance, of being independent from the number of considered emitters and does not mask the situation in which a few stations are starved. If the $min - max$ ratio is close to 1 this means that all emitters have the same throughput. If the $min - max$ ratio is equal to 0, it means that at least one flow is starved in the networks. We also use the variation coefficient which gives the dispersion around the mean value. The variation coefficient is the ratio of the standard deviation over the mean. This metric is independent from the mean value unlike the usual variance. The higher the value of the coefficient, the higher the dispersion around the mean value. This metric gives an intuition on how the flows are distributed around the mean value.

Figure 5 shows the simulations results in a random scenario with 200 stations and 150 flows. This figure shows that SBA is slightly more efficient than MadMac, PNAV and MBFAIR. We can also see that no flow is starved with SBA, unlike the other protocols for which the $min - max$ ratio is equal to 0. The coefficient variation also shows that the dispersion around the mean throughput is smaller for SBA, which means that the flows' throughputs are close. This figure shows that in this case, SBA is the best trade-off between fairness and efficiency.

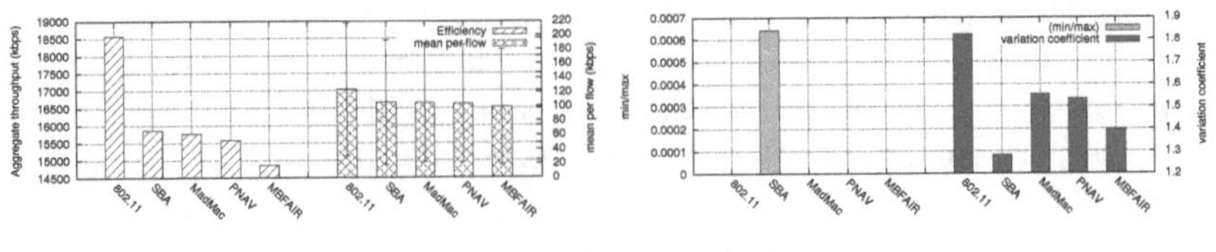


**Fig. 5.** Simulation results for a random scenario with 200 stations and 150 flows

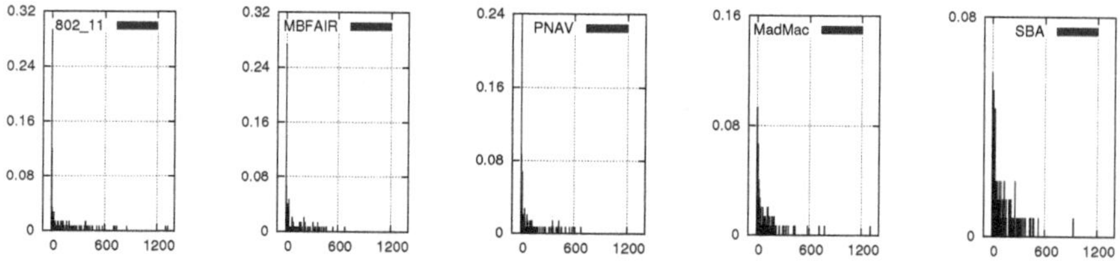


**Fig. 6.** Throughputs distribution for a random scenario with 200 stations and 150 flows (throughput vs percentage of flows achieving this throughput)

To better see the difference between SBA and the other protocols, we plotted the distribution of the flows throughputs.

Figure 6 shows the throughputs distribution of each protocol. The $x$-axis gives the throughput in *kbps* and the $y$-axis the proportion of flows that get this throughput. The size of each chart is equal to 3 *kbps*. We can see that with SBA, the number of flows that have a throughput close between 0 and 3 *kpbs* is less than 6%. Moreover, the $min - max$ ratio tells us that no flow is starved with SBA. On the other hand, with 802.11, more than 28% of flows have a throughput between 0 and 3 *kpbs*. This value is more than 24% for PNAV and MBFAIR and more than 8% for MadMac. These charts also show that the difference between the aggregated throughputs of 802.11 and SBA comes from the fact that with 802.11 a little proportion of flows can get more that 1200 *kbps*. With SBA, there is no flow with such a throughput, the higher throughput is around 920 Kbps. One interesting point is that even if a small percentage of flows have a throughput around 1300 Kbps with MadMac, the global throughput of MadMac is lower than the SBA's one.

It is worth noting that we ran many other random simulations and the results are always the same. We mean here that SBA exhibits a good trade-off between fairness and efficiency among the tested protocols.

## 5 Complexity

The complexity of MAC protocols for ad hoc networks is hard to quantify. In this section, we define some parameters to quantify complexity and then we provide the measure of each indicator for the protocols we compare in the previous section.

### 5.1 Complexity Parameters

**(HD) History dependence.** With this parameter, we try to provide the age of the information used by each protocol.

**(Dist) Distance of information.** Given in number of hops: it indicates the locality degree of the protocol.

**(E) Event.** It indicates the event(s) from which the protocol reacts. We have identified different events: (a) *successful transmission and/or collision*: such an event gives, to a node, the status of the packet it has just sent; (b) *carrier sensing detection*; (c) *reception of control packets*: the protocol can use control packets to adapt its behavior; (d) *reception of data packets* from which the protocol can extract and use useful information like packet source, packet size, etc.

**(NV) Number of variables.** We give here the maximum number of variables that are modified by a node after each different event given previously and used by the protocol.

### 5.2 Complexity Results

Table 2 gives the complexity parameters for the MAC protocols we evaluate in Section 4. For the **HD** parameter, $\epsilon$ means that the history length is very short, *i.e.* the protocol adapts its behavior just after an event. For example, IEEE 802.11 DCF only needs the transmission status of the last sent packet. Stochastic process means that the history length depends on a stochastic process which can be probabilistically long even if it can only depend on the previous status of the protocol.

We give the **NV** parameter in respect to the format of the parameter (E): for instance $(x_a, x_b)$ for **NV** corresponds to at most $x_a$ modified variables after event (a) and $x_b$ modified variables after event (b) if **E** is denoted by (a,b). With 802.11, each node stops its backoff decrease after event (a) or changes its contention window size or not after event (b). When RTS/CTS are activated, these packets are also used to update the NAV, which corresponds to one variable modification after event (c). With SBA, $T_{succ}$ or $T_{col}$ and $N_{succ}$ or $N_{col}$ are updated after event (b). At the end of each interval, the four different probabilities, as the new contention window size, are computed, and $\Delta$ is reset. This is denoted by 6 before each $\Delta$ in the table. With MadMac, more variables are maintained and modified: after event (a), the backoff decrease is stopped, but some variables are updated to indicate the share of the medium as the status of a possible multiple hidden nodes configuration; after event (b), the contention window size may be updated, but MadMac also counts, via some variables, the number of collisions as the number of successive sent packets and infers, via other variables, whether the node is within a multiple hidden nodes configuration. At the end of the $\Delta$ period, some of these variables are reset. With PNAV, a stochastic process is used to determine if the NAV is used or not. For that, different variables are updated after events (a,b). With MBFAIR, four variables are updated after event (b), (c) or (d): thanks to the sent packets or the received packtets, each node can compute its medium's occupancy of the medium, the medium's occupancy of the other stations, its fair share and the contention window size to use. With the receipt of RTS/CTS, each node can also update its NAV.

It is difficult to order these algorithms in terms of complexity. It is clear the IEEE 802.11 without RTS/CTS uses very few information and very few variables. If we consider that the use of control packets introduces more complexity in the protocol than local operations, then SBA is rather simple and requires less variables than MadMac.

**Table 2.** Complexity results

| | HD | Dist | E | NV |
|---|---|---|---|---|
| 802.11 | $\epsilon$ | $\emptyset$ | (a,b) | (1,1) |
| 802.11 (RTS/CTS) | $\epsilon$ | 2 | (a,b,c) | (1,1,1) |
| SBA | $\Delta$ | $\emptyset$ | (a,b) | (1,2) + 6 before each $\Delta$ |
| MadMac | $2 \times \Delta$ | $\emptyset$ | (a,b) | (4,6) + 7 before each $2\Delta$ |
| PNAV | stochastic process | $\emptyset$ | (a,b) | (2,6) |
| MBFAIR | $\infty$ | 2 | (a,b,c,d) | (1,4,5,4) |

## 6 Conclusion and Future Works

In this paper, we have presented a new backoff algorithm for ad hoc networks: only two contention window sizes are used and the choice of the contention window depends on computations based on local information only. Compared to existing backoff algorithms, SBA is designed for ad hoc networks. The results presented in this paper show that SBA is fairer than 802.11. Of course, the fairness induced by SBA reduces the global throughput (compared to 802.11), but this decrease is often lower than the one achieved by some other fair MAC protocols. Given the first obtained results, we think that SBA is a good candidate in terms of trade-off between fairness, simplicity and efficiency.

The next step of this work is to modify the value of each parameter of SBA such as $CW_{min}$, $CW_{max}$ and the conditions presented in the algorithm to optimize the performances of SBA depending on the requirement of the networks.

## References

1. IEEE, between Systems, I.E.: Local and Metropolitan Area Network – Specific Requirements –Part 11: Wireless LAN Medium Access Control (MAC) and Physical Layer (PHY) Specifications. The Institute of Electrical and Electronics Engineers (1997)
2. Chaudet, C., Dhoutaut, D., Guérin-Lassous, I.: Performance Issues with IEEE 802.11 in Ad Hoc Networking. IEEE Communications Magazine 43(7), 110–116 (2005)
3. Bharghavan, V., Demers, A., Shenker, S., Zhang, L.: MACAW: a media access protocol for wireless LAN's. In: ACM SIGCOMM Computer Communication Review, London, United Kingdom, pp. 212–225 (1994)
4. Chatzimisios, P., Boucouvalas, A.C., Vitsas, V., Vafiadis, A., Oikonomidis, A., Huang, P.: A simple and effective backoff scheme for the IEEE 802.11 MAC protocol. In: Cybernetics and Information Technologies, Systems and Applications (CITSA), Orlando, Florida, USA (July 2005)
5. Fang, Z., Bensaou, B., Wang, Y.: Performance evaluation of a fair backoff algorithm for IEEE 802.11 DFWMAC. In: ACM Symposium on Mobile Ad Hoc Networking and Computing (MOBIHOC), Lausanne, Switzerland, pp. 48–57 (2002)
6. Nandagopal, T., Kim, T., Gao, X., Bharghavan, V.: Achieving MAC Layer Fairness in Wireless Packet Networks. In: ACM Conference on Mobile Computing and Networking (MOBICOM), Boston, Massachusetts, United States, pp. 87–98 (2000)

7. Luo, H., Lu, S., Bharghavan, V.: A new model for packet scheduling in multihop wireless networks. In: ACM Conference on Mobile Computing and Networking (MOBICOM), Boston, Massachusetts, United States, pp. 76–86 (2000)
8. He, J., Pung, H.K.: Fairness of Medium Access Control Protocols for Multi-hop Ad Hoc Wireless Networks. Computer Networks 48(6), 867–890 (2005)
9. Chaudet, C., Chelius, G., Meunier, H., Simplot-Ryl, D.: Adaptive Probabilistic NAV to Increase Fairness in Ad Hoc 802.11 MAC. Ad Hoc and Sensor Wireless Networks: an International Journal (AHSWN) 2(2) (June 2005)
10. Razafindralambo, T., Guérin-Lassous, I.: Increasing Fairness and Efficiency using the MadMac Protocol in Ad Hoc Networks. Ad Hoc Networks Journal, Elsevier Ed. 6(3), 408–423 (2008)
11. Li, Z., Nandi, S., Gupta, A.K.: Modeling the Short-Term Unfairness of IEEE 802.11 in Presence of Hidden Terminals. In: Mitrou, N.M., Kontovasilis, K., Rouskas, G.N., Iliadis, I., Merakos, L. (eds.) NETWORKING 2004. LNCS, vol. 3042, pp. 613–625. Springer, Heidelberg (2004)
12. Dhoutaut, D., Lassous, I.G.: Impact of Heavy Traffic Beyond Communication Range in Multi-Hops Ad Hoc Networks. In: Third International Network Conference, Plymouth, UK (2002)
13. NS-2: The Network Simulator, `http://www.isi.edu/nsnam/ns/`
14. Jain, R., Durresi, A., Babic, G.: Throughput Fairness Index: An Explanation. In: ATM Forum Document Number: ATM Forum/990045 (February 1999)

# Hashing Backoff: A Collision-Free Wireless Access Method

Paul Starzetz, Martin Heusse, Franck Rousseau, and Andrzej Duda

Grenoble Informatics Laboratory (LIG)*
681 rue de la Passerelle, BP 72
38402 Saint Martin d'Hères Cedex, France
{Paul.Starzetz,Martin.Heusse,Franck.Rousseau,Andrzej.Duda}@imag.fr

**Abstract.** In this paper, we propose *Hashing Backoff*, an access method in which stations select backoff values by means of asymptotically orthogonal hashing functions, so that contending stations converge to a collision-free state. This solution is a half-way between TDMA, CDMA, and random access. Our simulations show that it presents significant improvement over *Idle Sense*, the access method with much better performance that the standard 802.11 DCF. The fact that the proposed method focuses on reducing collisions makes it particularly interesting for some specific applications such as sensor networks in which eliminating collisions leads to energy savings.

**Keywords:** 802.11, random access methods, collision-free access method.

## 1 Introduction

Sharing a common radio channel requires an access method to define how contending stations can access the channel. Some existing access methods use fixed channel allocation based on various multiplexing schemes: time (TDMA), frequency (FDMA), or code (CDMA), however they require some kind of synchronization between stations usually done by a centralized coordinator that adapts allocation to varying traffic conditions. Other channel allocation schemes rely on dynamic random access such as the familiar CSMA/CA (Carrier Sense Multiple Access/Collision Avoidance) that allows stations to operate in a fully distributed way and to dynamically adapt to changing load. However, this kind of access methods results in possible collisions that significantly limit performance. The current IEEE 802.11 DCF (Distributed Coordination Function) access method [1] is based on the principles of CSMA/CA in which stations independently choose a backoff interval before transmission—a uniformly distributed number of slots in a contention window to avoid starting transmission at the same instant and colliding.

* LIG is a joint research laboratory of CNRS (*Centre National de la Recherche Scientifique*), Grenoble INP (*Institut Polytechnique de Grenoble*), INRIA (*Institut National de Recherche en Informatique et Automatique*), UJF (*Université Joseph Fourier*), and UPMF (*Université Pierre-Mendès-France*).

L. Fratta et al. (Eds.): NETWORKING 2009, LNCS 5550, pp. 429–441, 2009.

Most of the recent work on wireless access methods has focused on proposing new access methods such as TCF (*TDM-based Coordination Function*) [2] inspired by TDMA or enhancing the operation of 802.11 DCF, e.g. choosing the right parameters of the contention window $CW_{min}$ and $CW_{max}$, improving collision resolution (*Fast Collision Resolution* [3]), or tuning control algorithms of the random backoff (*Slow Decrease* [4], *Asymptotically Optimal Backoff (AOB)* [5], *Idle Sense* [6]). To improve throughput and fairness in 802.11-like wireless networks *AOB* and *Idle Sense* propose to adjust the contention window of stations based on the observed average number of idle slots. The value of contention windows increases with the number of active stations, which results in less collisions.

In the context of recent research on sensor networks, wireless location systems, and power-saving in ad hoc networks, Tay *et al.* have proposed CSMA/$p^*$, a protocol to minimize collisions between contending stations [7]. It is non-persistent carrier sense multiple access (CSMA) with a non-uniform probability distribution that nodes use to randomly select contention slots.

In this paper, we propose a CSMA/CA access method that goes in a similar direction—*Hashing Backoff*, a method that is a half-way between TDMA, CDMA, and random access. Similarly to IEEE 802.11 DCF, it is fully distributed and independent of $N$, the number of contending active wireless stations, i.e. stations do not need to know $N$ to achieve the collision-free state. It is also somehow similar to CDMA, because it uses orthogonal backoff sequences to attain the collision-free state. Under favorable conditions (low error rates), it converges to the collision-free state by leveraging the information conveyed by collisions on the state of colliding stations. For higher error rates, the performance of the method is limited by the scheme used to control contention windows, which is based on *Idle Sense.* This means that we obtain better performance than *Idle Sense* in favorable conditions and similar performance otherwise.

Although we compare the proposed access method with the IEEE 802.11 DCF, our scheme can successfully apply to sensor networks for saving energy lost in collisions. Consider for instance sensor MAC access schemes based on *Preamble Sampling*: sensor nodes sleep for long periods of time instead of being permanently active and periodically wake up to check if there is an ongoing transmission. To send a frame, a node transmits a preamble before each data frame. The preamble is long enough to make sure that all potential receivers will wake up and get their data. Collisions may arise when several nodes wake up during a preamble, get the data frame, and reply. Usually, they use a contention window to avoid collisions, but our method can further improve energy saving by eliminating collisions.

The rest of the paper is organized as follows. We first define *Hashing Backoff*, the access method that converges to a collision-free state (Section 2). We present simulation results that illustrate its performance (Section 3). Finally, we briefly review the related work (Section 4) and present some conclusions (Section 5).

## 2 Hashing Backoff Access Method

We adopt common assumptions for 802.11-like wireless networks: we assume a certain number of wireless stations sharing a common radio channel (this assumption corresponds to the infrastructure mode with an access point acting as a bridge and to the ad hoc way of operation in which several stations directly communicate with close neighbors). As in the basic DCF method of 802.11, we do not deal with spatial problems related to the relative positions of stations—we assume that other methods such as RTS/CTS can help to cope with spatial problems such as hidden, blocked, exposed, and masked stations. We also assume that radio devices are half-duplex, they use carrier sense (there is no busy tone), and collisions can be resolved only after a frame transmission (they are detected by the absence of the corresponding ACK frame). Under such assumptions, a collision occurs when two or more contending stations choose the same slot for a transmission.

### 2.1 Distributed Hash Table View

Let us consider a wireless network with $N$ active contending wireless stations. We observe that the pattern of channel access is a repeating sequence of some idle time slots terminated by a collision or a successful transmission. A similar problem arises when we want to insert key values into a *hash table* of a given size—it results in a successful insertion or in a collision. We can thus analyze an access method by analogy with key insertion: each time a wireless station wants to send a frame, it selects a backoff value $b_j^i$ that corresponds to hashing of some station key value into a slot of a virtual hash table (in this view, the contention window can be viewed as the hash table). Thus, we can formalize the backoff as:

$$b_j^i = H^i(i, j, k_j^i, CW), i = 1 \dots N, j = 1 \dots \infty, \tag{1}$$

where $i$ enumerates stations, $j$ is the contention cycle number[1], $k_j^i$ is the key used as the input to the hash function $H^i$ of station $i$, and finally $CW$ is the table size measured in slots. Once each station has selected a slot in the hash table, contention proceeds as in IEEE 802.11: stations wait counting down time slots until the station with the shortest backoff wins and sends its frame. Other stations lose contention and the cycle repeats—stations choose another backoff according to Eq. 1. The state of the hash table just after the selection of backoffs determines which station will send a frame (the one with the smallest backoff) and after what delay (its $b_j^i$ measured in time slots).

Note that collisions only happen if two stations select the same smallest contention-winning table slot, whereas in the classical table hashing scheme, collisions may happen at any table slot. So, in the access method above, only

[1] We use the term *contention cycle* to denote repeating transmission attempts. Even if a contention cycle has index $j$, it does not mean that stations are synchronized in any way.

the probability density function of collisions at the contention-winning slot is important. Therefore, we expect that hash function $H^i$ minimizing the number of collisions is different from the uniform distribution, the latter being the optimal choice for the classical hashing problem.

Our idea for a collision-free access method is to construct on the fly orthogonal hash functions for each station by taking into account the inferred state of other stations after each collision. Such a scheme should ideally achieve the following objectives:

- it should adapt to the varying number of active contending wireless stations,
- it must cope with varying load of wireless stations, that is, maintain sufficient orthogonality if a station is not always backlogged, at least for short time periods (e.g. few contention cycles),
- and finally, it should obtain better performance than the standard 802.11 DCF access method and its improvements, or in the worst case, just degrade to their performance.

We start with the case of a network composed of a known number of active contending stations $N$ and then, we relax this assumption so that the proposed access method does not require the knowledge of $N$.

### 2.2 Simplified Case: Fixed Contention Window

We begin with a simplified case of $N$ stations that use a fixed contention window of $CW$ slots. We can formalize backoff generation in the IEEE 802.11 DCF by station $i$ in contention cycle $j$ as:

$$b_j^i \;=\; H^i(i,j,k_j^i,CW) \;=\; rand(CW), \tag{2}$$

where $rand(CW)$ denotes a discrete uniformly distributed function with the image $[0 \ldots CW - 1]$. In *Hashing Backoff*, we define $b_j^i$ so that it forms an *orthogonal hash function* set:

$$b_j^i \;=\; H^i(i,j,k_j^i,CW) \;=\; a_j^i + m \cdot rand(n) \tag{3}$$

with

$$m \cdot n = CW,\; a_j^i = 0 \ldots m-1,\; m \geq N. \tag{4}$$

The idea is to divide the total contention window $CW$ into $m$ disjoint "comb-like" subsets of $n$ slots each as illustrated in Figure 1. Each subset of slots can be selected by an appropriate offset $a_j^i$. Note also that for $m=1$, Eq. 3 reduces to Eq. 2. In contention cycle $j$ each station selects the backoff value $b_j$ according to Eq. 3 and counts down to zero. If the channel is still free, it transmits the frame. When the channel is sensed busy while the station counts down, the backoff value is discarded and the cycle repeats by generating a new backoff value. It is clear that for each station having a different offset value $a_j^i$, there will be no collision at all at the winning (minimal) slot, since Eq. 3 yields orthogonal sequences of random numbers for different offsets $a_j^i$.

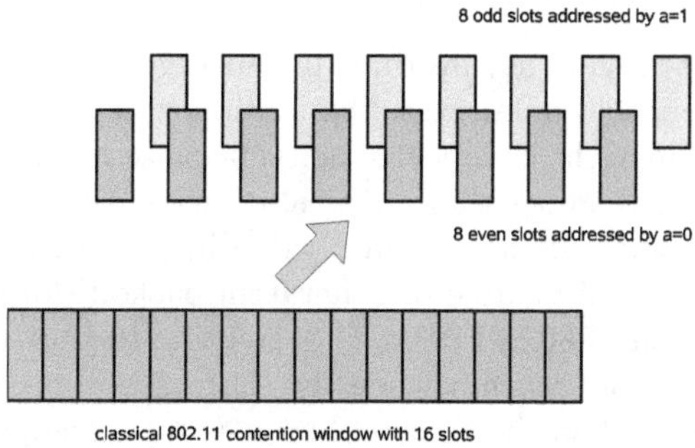


**Fig. 1.** Principle of the split of the contention window into orthogonal sequences, $m = 2$, $n = 8$

Initially, each station selects its offset $a_j^i$ as a random value from interval $0 \ldots m-1$ and keeps it constant until a collision occurs. Then, each colliding station simply reselects a random value for $a_j^i$ and repeats this procedure after each collision until no more collisions occur. This procedure guarantees asymptotic convergence to a collision-free state given that the number of active stations $N$ is less or equal to modulus $m$. We can prove asymptotic convergence through the following reasoning. We number collision events with index $n_c = 0 \ldots \infty$. Reselecting offset values $a^i$ by colliding stations can be considered as redistributing all $a^i$ values into available $m$ subset values (stations that do not collide keep their previous values). There are $x = m^N$ different combinations of how $N$ $a^i$ values of the stations can be distributed and there are $y > 0$ combinations for which each station has a different value of $a^i$. Thus, after each reselection of $a^i$ values, each station has a different $a^i$ value (orthogonal hash functions) with probability $p = y/x > 0$ and with probability $1 - p < 1$ at least two stations have identical $a^i$ values. Then, we can conclude that starting with non-orthogonal hash functions, probability $P(n_c)$ of still having non-orthogonal hash functions after $n_c$ collisions is bounded by:

$$P(n_c) = (1-p)^{n_c} \tag{5}$$

and since $1 \geq p > 0$ we have:

$$\lim_{t \to \infty} (1-p)^{n_c} = 0, \tag{6}$$

which guarantees asymptotic convergence of the algorithm. Probability $P(n_c)$ is an upper bound, because only stations that collided will reselect a new value of $a^i$ while other stations keep their already orthogonal hash functions, thus the actual convergence is faster.

The convergence to the collision-free state will be however slowed down by frame errors: a transmitting station considers a failed transmission as a collision, which forces the reselection of $a^i$.

### 2.3 Improving Fairness with Orthogonal Residual Backoff

While the proposed method asymptotically converges to a collision-free state, it suffers from unacceptable unfairness: the station with the smallest value of $a^i$ obtains a higher share of the channel capacity, because it has a higher probability of using smaller backoff values than other stations and thus winning contention more often. For instance, if the minimal value in the current set of all $a^i$ is 0, then only this station will be able to generate a backoff slot of value 0.

The residual backoff of 802.11 consists of freezing the current value of backoff $b^i_j$ of stations that do not transmit when the channel becomes busy; they resume the count down again when the channel becomes free. The procedure has a nice property of improving short time fairness, because even stations that choose large values of backoff will eventually gain the access to the channel after few contention cycles. However, we cannot directly reuse the residual backoff scheme of 802.11, because the set of residual backoff timer values $S_{j+1}$ in the next contention cycle $j+1$ will not be anymore distinct. To illustrate this, let us denote by $r^i_j = rand(n)$ the random value drawn by station $i$ in cycle $j$. Station $l$ that transmits a frame, draws a new backoff value according to Eq. 3 and all other stations apply the principle of the residual backoff:

$$b^i_{j+1} = \begin{cases} a^i_j + m \cdot r^i_j - (a^l_j + m \cdot r^l_j) - 1, & \text{for } i \neq l, \\ a^l_{j+1} + m \cdot r^l_{j+1}, & \text{for } i = l, \end{cases} \tag{7}$$

where $-1$ comes from the fact that as we count the first slot from 0, the residual backoff subtracts 1 from the timers of other stations, if the sending station has selected slot 0 and transmitted after sensing the channel idle for one slot. This relation can also be rewritten as:

$$b^i_{j+1} = \begin{cases} (a^i_j - a^l_j - 1) + m \cdot (r^i_j - r^l_j), & i \neq l, \\ a^l_{j+1} + m \cdot r^l_{j+1}, & i = l, \end{cases} \tag{8}$$

which is obviously not orthogonal anymore in cycle $j+1$, for example $a^i_j - a^l_j - 1 = a^l_j$ (e.g. for two stations, $a^1 = 5$, $a^2 = 2$, $a^1 - a^2 - 1 = 2$).

To solve this problem, we propose to apply the residual backoff procedure not only to the backoff timer value selected from Eq. 3, but also independently (and modulo $m$) to all offsets $a^i_j$ of contending stations, which can be written for the next cycle as:

$$b^i_{j+1} = \begin{cases} a^i_j + m \cdot r^i_j - (a^l_j + m \cdot r^l_j) - 1, & \text{for } i \neq l, \\ (a^l_j - (a^l_j + m \cdot r^l_j) - 1) + m \cdot r^l_{j+1} & \text{for } i = l. \end{cases} \tag{9}$$

In modulo-$m$ arithmetic (which applies to the new value of $a^i_j$) we have $m \cdot r^l_j \ mod \ m = 0$ and $-1 \ mod \ m = (m-1)$, which means that $a^l_j$ wraps around to $m-1$. Thus, the previous relation can be rewritten as:

$$b^i_{j+1} = \begin{cases} (a^i_j - a^l_j - 1) + m \cdot (r^i_j - r^l_j), & \text{for } i \neq l, \\ (m-1) + m \cdot r^l_{j+1} & \text{for } i = l, \end{cases} \tag{10}$$

which now is obviously orthogonal in cycle $j+1$ for an initially orthogonal set of offsets $a_j^i$: since station $l$ was the last contention winner $a_j^i > a_j^l \ \forall i \neq l$, it follows from $0 \leq a_j^i < m$ that $a_j^i - a_j^l - 1 \neq m-1$ so the orthogonality is preserved.

### 2.4 Dynamic Adaptation of the Contention Window and the Modulus

The last thing that we need to consider is to propose an algorithm for a dynamic choice of contention window $CW$ and modulus $m$ in function of the number of active contending stations $N$. The algorithm should not rely on the knowledge of $N$ by contending stations.

We propose to use contention control of *Idle Sense* [6] in which each station estimates $n_i$, the number of consecutive idle slots between two transmission attempts and uses it to adjust its contention window $CW$ to reach a target value $n_{\mathrm{i}}^{\mathrm{target}}$ (we keep the notation of *Idle Sense* with $n_i$ denoting the number of idle slots). The value of $n_{\mathrm{i}}^{\mathrm{target}}$ is computed numerically for a given variant of IEEE 802.11 from the parameters of the PHY and MAC layers (its value is 5.68 for IEEE 802.11b and 3.91 for IEEE 802.11g [6]). Dynamic adaptation follows the AIMD (*Additive Increase Multiplicative Decrease*) principle [8] applied to contention window $CW$:

- if $n_i \geq n_i{}^{target}$, $CW \leftarrow \alpha \cdot CW$,
- if $n_i < n_i{}^{target}$, $CW \leftarrow CW + \beta$,

where $\alpha$ and $\beta$ are the parameters of the AIMD scheme chosen for the implementation of *Idle Sense* after tuning and measurements [9] (the parameters result in the best tradeoff between accuracy and convergence speed):

- $\frac{1}{\alpha} = 1.0666$,
- $\beta = 6.0$.

As we need to choose modulus $m$ in function of $N$, the number of contending stations, we propose to deduce an estimate of $N$ from $CW_{IS}$, the contention window controlled through dynamic adaptation of *Idle Sense*. Recall that the contention window used by *Idle Sense* is maintained proportional to the number of contending stations: $\zeta = NP_{\mathrm{e}}$, where $\zeta$ is a constant and $P_{\mathrm{e}} = 2/(CW_{IS}+1)$ is the transmission attempt probability in a given slot [6]. However, $CW_{IS}$ varies fast to keep track of the number of contending stations (a station refreshes $CW_{IS}$ every *maxtrans* transmissions, by default 5) so we cannot directly use it to adjust modulus $m$. Instead, we propose to smooth $CW_{IS}$ out by means of EWMA (Exponentially Weighted Moving Average) and relate to modulus $m$ through the following formula:

$$CW_{SM} = q \cdot CW_{SM} + (1-q) \cdot CW_{IS}, \tag{11}$$

$$e = \max\{3,\ round[\log_2(CW_F)] - 1\}, \tag{12}$$

$$m = 2^e, \tag{13}$$

where $CW_{SM}$ is the smoothed value of the current contention window. We have tuned this relation by running simulations described in the next section and we obtain good results.

In addition to that, *Hashing Backoff* uses the current estimate of contention window $CW_{IS}$ to support $m$ orthogonal sequences, so the adaptation scheme needs to round the value of $CW_{IS}$ to the nearest multiple of modulus $m$:

$$CW_{HB} = m \cdot round(CW_{IS}/m). \tag{14}$$

# 3 Performance Comparisons

To evaluate the performance of the proposed access method, we have developed a discrete-event simulator in Java. Our simulator implements the standard 802.11 DCF method (without RTS/CTS), *Idle Sense*, and *Hashing Backoff*. The simulator focuses on the MAC layer performance and takes into account the influence of an imperfect physical layer by simulating frame losses at a given frame error rate. Before the present study, we validated the simulator by comparing its results with other simulators [10] and measurements on a 802.11b platform [9]. A simulation runs for $10^6$ transmissions to obtain a relative precision of the order of $10^{-3}$ (confidence intervals are too small to show in figures).

We present performance results for the PHY and MAC parameters of 802.11b and frame data size of 1500 Bytes. Stations behave like greedy (backlogged) sources transmitting at the maximal bit rate—they always have a frame to send. Considering such saturated conditions enables us to examine performance limits of the proposed method.[2]

We start with the detailed results for fixed values of modulus $m$ and then present the results for dynamic adaptation to the number of contending stations.

## 3.1 Fixed Modulus, $m = 8$

Recall that when modulus $m$ is fixed, *Hashing Backoff* is intended to operate for the number of contending stations $N \leq m$, because it allows for $m$ orthogonal sequences resulting in no collisions after convergence. Figure 2 presents a comparison of throughput for $m = 8$. We can see a significant improvement over 802.11 DCF and a slight improvement over *Idle Sense*. In this last case, the difference is small, because *Idle Sense* already provides an improvement over DCF and for $N \leq 8$, the collision rate remains small (under 10%).

To characterize the delay, we report on some parts of the transmission delay: it is composed of some number of backoff slots a station needs to wait before a transmission attempt, some unsuccessful transmissions due to collisions or frame

[2] Note that the network with a given number of greedy stations corresponds to a network with a much higher number of stations generating intermittent traffic, the increase factor depending on the proportion of active to idle periods. For instance, 10 greedy stations may correspond to 100 non saturated stations amongst which 10% are active and contending for channel access during a given time period.

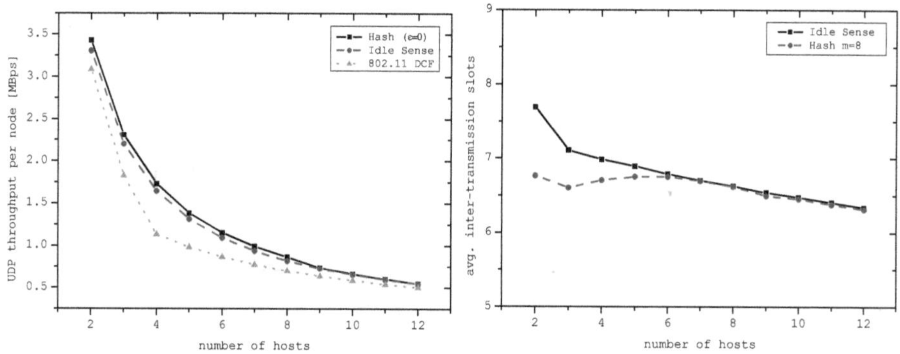


(a) throughput (b) average number of slots

**Fig. 2.** Throughput for DCF, *Idle Sense* and *Hashing Backoff* and the average number of slots before a transmission, $m = 8$, no frame errors

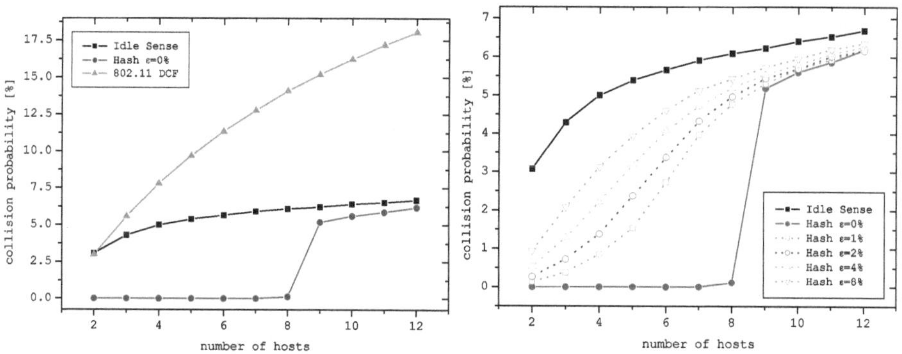


(a) no frame errors (b) varying frame error rate

**Fig. 3.** Collision probability, $m = 8$

errors, and a final successful transmission. As for *Hashing Backoff* the collision rate tends to zero, the only element that influences the delay is the number of slots to wait before a transmission, a successful transmission being the same for the compared access methods.

Figure 2 shows the average number of slots a station waits before a transmission. We can see that for $N \leq m$, *Hashing Backoff* results in a lower value than that of *Idle Sense*, which implies that the delay of *Hashing Backoff* is better, because unlike *Idle Sense*, it does not experience collisions. Consequently, *Hashing Backoff* also performs better than DCF, because *Idle Sense* provides shorter delays than DCF (cf. measurement based comparisons [9]).

Figure 3 presents collision probability. We can see from the figure that *Hashing Backoff* achieves its objective—the collision rate is closed to 0 for the number of stations up to $m$. Then it increases, but even for a larger number of stations, it

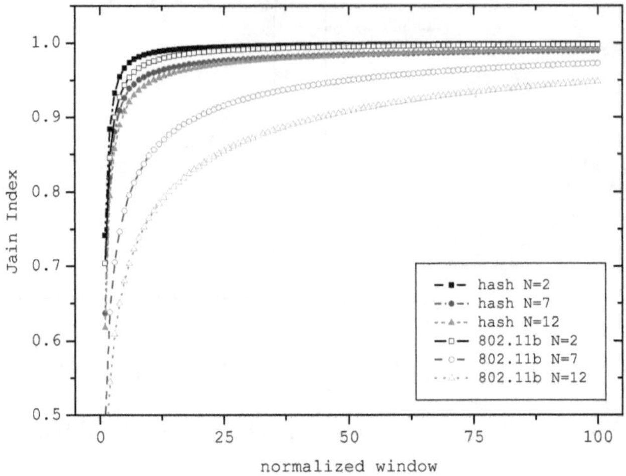


**Fig. 4.** Jain fairness index for *Hashing Backoff* and DCF, $N$ contending stations, $m = 8$

remains lower than the collision probability for *Idle Sense*. When transmission conditions degrade and the frame error rate increases, the collision probability of *Hashing Backoff* becomes greater than zero, however it remains much better than that observed in *Idle Sense*. Note that the comparison is only with *Idle Sense* without frame errors to keep the figure legible, a fair comparison would require comparing the methods for the same frame error rate.

Figure 4 presents a comparison of short term fairness for DCF and *Hashing Backoff*. We use the sliding window method that considers patterns of transmissions and computes the average *Jain fairness index* in a window of an increasing size [11]. Perfect fairness is achieved for $F_J(w) = 1$ and perfect unfairness for $F_J(w) = 1/N$. We can see that fairness of *Hashing Backoff* is much better than DCF.

### 3.2 Dynamic Adaptation of Modulus $m$

Finally, we evaluate the dynamic scheme that adapts modulus $m$ to the current number of contending stations (up to 25 greedy stations, which is a fairly large scale for 802.11 networks). Figure 5 shows the collision probability in this case and compares its behavior with the performance of fixed size modulus ($m = 8, 16$). We can observe that the collision probability is almost zero for a large range of the number of contending stations. The figure also presents the collision probability for an increased frame error rate and shows that even in the case of a 4% error rate, the performance of *Hashing Backoff* is still much better than that of *Idle Sense* without frame errors.

Figure 5 also shows the average number of slots a station waits before a transmission for *Hashing Backoff* with dynamic adaptation. We can see that it

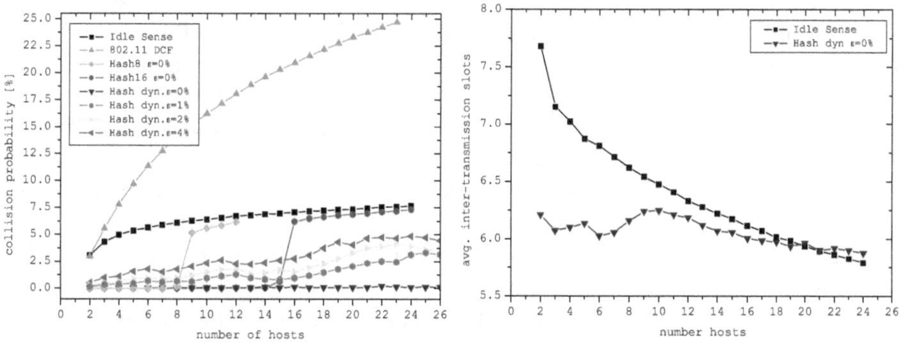


(a) collision probability (b) number of slots

**Fig. 5.** Collision probability and average number of slots before a transmission for *Hashing Backoff* with dynamic adaptation of modulus $m$, varying frame error rate $\epsilon$

is lower for *Hashing Backoff* than for *Idle Sense*, which implies that *Hashing Backoff* presents shorter delays.

### 3.3 Discussion

The results show that *Hashing Backoff* presents better performance than *Idle Sense* with respect to all performance indices and much better improvement over the 802.11 DCF. This comes from the reduction of collisions through the use of orthogonal hash functions that separate sets of slots chosen by different stations. This feature is even more important for high rate variants such as 802.11g in which the duration of the slot time compared to the average communication time is increased (in 802.11g, the slot time is divided by 2, but the bit rate is increased by the factor of 5).

The physical layer capture effect [12] has a positive impact on our method: only the station that perceives a collision adjusts its backoff while tending to the collision free state.

Note that *Hashing Backoff* with dynamic adaptation of the contention window and the modulus supports varying numbers of contending stations that can join and leave the network. Contention control borrowed from *Idle Sense* leads to the operation that does not require the knowledge of the number of contending stations.

In the case of an increased frame error rate, the performance of the method degrades—the collision probability of *Hashing Backoff* becomes greater than zero, however it remains much better than that observed in *Idle Sense*.

## 4 Related Work

This section only addresses the work closely related to *Hashing Backoff*. In particular, we compare our method with other proposals that try to reduce collisions of random access methods.

The *Binary Countdown Method* [13] can reduce collision overhead. As collisions significantly limit throughput, the method is more efficient than the standard 802.11 DCF. However, it requires a control channel for transmitting management messages to schedule each transmission. As the channel consumes 20% of the available bandwidth, it is not a method that compares favorably with our approach.

CSMA/$p^*$ lowers the collision rate, because stations use a nonuniform probability distribution to randomly select backoff contention slots [7]. The optimal distribution derived for a known number of contending stations N is highly nonuniform—the probability mass is concentrated on the largest values of the contention window. The reduction of the collision rate is thus obtained at the expense of a longer average wait before a transmission.

TCF (*TDM-based Coordination Function*) is an original approach, much different from the contention based methods [2]. It eliminates contention periods by allocating the channel dynamically using a TDMA scheme. The method offers high throughput and good fairness if the number of contending stations remains stable; otherwise, it presents similar problems as other proposals, because the phase allowing stations to join is based on contention.

## 5 Conclusion

In this paper, we have proposed a view that considers collision avoidance as a problem similar to distributed hashing. This formalization leads to the definition of *Hashing Backoff*, an access method in which stations choose backoff values based on hashing functions that are asymptotically orthogonal, so that stations converge to a collision-free state. We achieve orthogonality by taking advantage of the information conveyed by collisions.

We have validated the performance of *Hashing Backoff* by simulation and our experiments show that it presents significant improvement over *Idle Sense*, the access method with much better performance that the standard 802.11 DCF. The fact that the proposed method focuses on reducing collisions makes it particularly interesting for some specific applications such as sensor networks in which eliminating collisions leads to energy savings.

The dynamic adaptation scheme allows *Hashing Backoff* to efficiently operate in the presence of joining and leaving stations. As we dynamically estimate the current number of contending stations and adjust modulus $m$ accordingly, a joining station will possibly begin with a backoff hashing function that generates collisions, but after some number of collisions, it will change its hashing function and finally attain the collision free state.

## Acknowledgments

This work was partially supported by the European Commission project WIP under contract 27402, the French Ministry of Research projects AIRNET under contract ANR-05-RNRT-012-01, and ARESA under contract ANR-05-RNRT-01703.

## References


1. IEEE: IEEE 802.11: Wireless LAN Medium Access Control (MAC) and Physical Layer (PHY) Specifications (1999)
2. Lim, C., Choi, C.: TDM-based Coordination Function in WLAN for High Throughput. In: Proc. IEEE GLOBECOM (2004)
3. Kwon, Y., Fang, Y., Latchman, H.: A Novel MAC Protocol with Fast Collision Resolution for Wireless LANs. In: Proc. of INFOCOM 2003, San Francisco, USA (March- April 2003)
4. Ni, Q., Aad, I., Barakat, C., Turletti, T.: Modeling and Analysis of Slow CW Decrease for IEEE 802.11 WLAN. In: Proc. of PIMRC 2003 (2003)
5. Bononi, L., Conti, M., Gregori, E.: Runtime Optimization of IEEE 802.11 Wireless LANs Performance. IEEE Trans. Parallel Distrib. Syst. 15(1), 66–80 (2004)
6. Heusse, M., Rousseau, F., Guillier, R., Duda, A.: Idle Sense: An Optimal Access Method for High Throughput and Fairness in Rate Diverse Wireless LANs. In: Proc. of ACM SIGCOMM 2005 (August 2005)
7. Tay, Y., Jamieson, K., Balakrishnan, H.: Collision-Minimizing CSMA and its Applications to Wireless Sensor Networks. IEEE Journal on Selected Areas in Communications (August 2004)
8. Chiu, D., Jain, R.: Analysis of the Increase and Decrease Algorithms for Congestion Avoidance in Computer Networks. Journal of Computer Networks and ISDN 17(1) (June 1989)
9. Grunenberger, Y., Heusse, M., Rousseau, F., Duda, A.: Experience with an Implementation of the Idle Sense Wireless Access Method. In: Proc. of CoNEXT 2007, New York (2007)
10. Lopez-Aguilera, E., Heusse, M., Rousseau, F., Duda, A., Casademont, J.: Performance of Wireless LAN Access Methods in Multicell Environments. In: Proc. of IEEE GLOBECOM 2006 (November-December 2006)
11. Jain, R., Chiu, D., Hawe, W.: A Quantitative Measure of Fairness and Discrimination for Resource Allocation in Shared Computer Systems. DEC Research Report TR-301 (September 1984)
12. Kochut, A., Vasan, A., Shankar, U., Agrawala, A.: Sniffing Out the Correct Physical Layer Capture Model in 802.11b. In: Proc. of ICNP 2004, Berlin, Germany (October 2004)
13. You, T., Yeh, C.H., Hassanein, H.: A New Class of Collision-Prevention MAC Protocols for Ad Hoc Wireless Networks. In: Proc. of ICC 2003, San Francisco, USA (May 2003)

# Optimal Placement of Multiple Interconnected Gateways in Heterogeneous Wireless Sensor Networks*

Antonio Capone, Matteo Cesana, Danilo De Donno, and Ilario Filippini

Dipartimento di Elettronica e Informazione, Politecnico di Milano
Piazza L. da Vinci 32, 20133 Milan, Italy
surname@elet.polimi.it

**Abstract.** Data collected by sensors often have to be remotely delivered through multi-hop wireless paths to data sinks connected to application servers for information processing. The position of these sinks has a huge impact on the quality of the specific Wireless Sensor Network (WSN). Indeed, it may create artificial traffic bottlenecks which affect the energy efficiency and the WSN lifetime. This paper considers a heterogeneous network scenario where wireless sensors deliver data to intermediate gateways geared with a diverse wireless technology and interconnected together and to the sink. An optimization framework based on Integer Linear Programming (ILP) is developed to locate wireless gateways minimizing the overall installation cost and the energy consumption in the WSN, while accounting for multi-hop coverage between sensors and gateways, and connectivity among wireless gateways. The proposed ILP formulations are solved to optimality for medium-size instances to analyze the quality of the designed networks, and heuristic algorithms are also proposed to tackle large-scale heterogeneous scenarios.

**Keywords:** Wireless Sensor Networks, Gateway Placement, Optimization.

## 1 Introduction

Wireless Sensor Networks (WSN) have recently emerged as an ideal solution to a large number of applications where the goal is collecting measurements of a physical parameters (temperature, humidity, light intensity, etc.) or detecting events in the covered area (intrusion, wild fire, etc.) [1]. Sensors are usually low-cost battery-operated devices geared with sensing, processing and communication functionalities. Obviously, the limited energy availability deeply impacts on the design of WSN and in particular on communication protocols running at different levels [2,3,4] as well as on network deployment [5,6] and topology.

* This paper has been partially supported by research project PRIN SESAME.

L. Fratta et al. (Eds.): NETWORKING 2009, LNCS 5550, pp. 442–455, 2009.

In many scenarios, data collected by sensors must be delivered to application servers for processing. These servers can be reached via a sink node that is connected to a local or geographical network through a communication interface with a different technology with respect to the WSN. Since the transmission range of sensor nodes is often much smaller than the area covered by the WSN, sensors cooperate to deliver information to the sink node through multi-hop paths.

Multi-hop information delivery to sink nodes requires multiple transmissions that consume energy in intermediate nodes and heavily limit lifetime of those nodes that are used more frequently. Even if many routing approaches have been proposed to balance the load among nodes and prolong the overall network lifetime, it is quite evident that nodes closer to the sink are anyway the most critical ones since they are required to relay all traffic generated by the other nodes and headed to the sink. For this reason, several methods have been proposed for optimizing the sink position in WSN design [13,15].

However, a single sink node collecting all data from the network is not for sure an efficient solution when dealing with large size WSNs, like e.g. in applications for monitoring natural areas. Indeed, as the network grows, the amount of information that must be delivered to the sink by surrounding nodes increases creating traffic and energy consumption bottlenecks. A promising alternative approach is based on the use of additional concentration nodes, *gateways*, that allow to spread the load and also to keep multi-hop paths within a reasonable length [8,9,10]. Also in this case gateway positioning is a key element for designing efficient networks. The entire set of sensor nodes need to be divided into independent subsets that send their traffic to a gateway, giving rise to a clustering problem.

Generally speaking, gateways are more powerful and expensive devices with respect to sensors and a reasonable objective is minimizing their number while matching constraints on energy efficiency and network quality [8,10,12]. However, this approach focuses on the WSN only neglecting the problem of interconnecting the gateways to the sink or more in general to the network where application servers are located. According to the specific application scenario, several solutions can be adopted for this problem, including the use of direct geographical links among the gateways (wired links, satellite links, etc.). However, this may be quite expensive and inefficient, since the traffic collected at each gateway is usually much smaller than the link capacity while its cost (interface, installation, and operation) often high.

For this reason, in this paper we consider an alternative network scenario where gateways are inter-connected through a Wireless Mesh Network (WMN) and act as Wireless Routers (WRs) forwarding traffic along multi-hop paths toward a sink node that is the only one equipped with a geographical link (see Figure 1). The resulting architecture is an heterogenous multi-hop network with different devices (sensors, MRs, and access point) and wireless links (low-power low-rate links for sensors, and high rate links for mesh routers) that can be

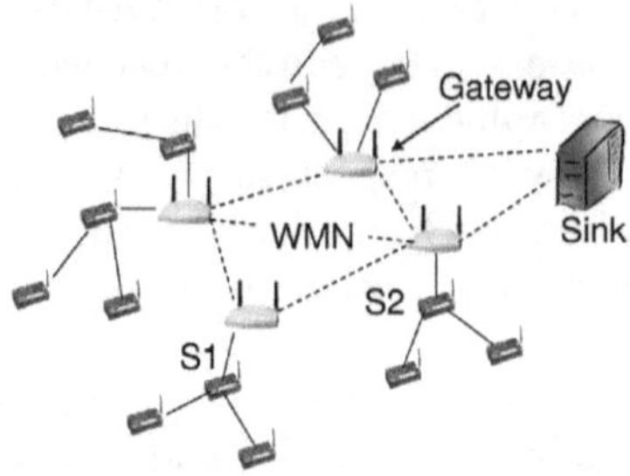


**Fig. 1.** Reference Network Scenario

adapted to the specific application scenario using available technologies (like e.g. ZigBee, WiFi, WiMax, etc.).

In this work we investigate the joint problem of selecting/positioning the gateways and designing the WMN that interconnects them. We provide an optimization framework that minimizes installation costs and maximizes the energy efficiency, while considering both multi-hop coverage and connectivity constraints. Coverage considers the availability of a multi-hop path between each sensor node and at least a gateway, while connectivity among gateways must be ensured by properly designing the WMN. The proposed optimization framework is independent on the specific routing strategy in the WSN; indeed, routes in the WSN (from sensors to gateways) are assumed to be given. This allows to use any of the routing mechanisms that have been proposed for WSNs [3] and to plug it into our optimization framework just modifying a separate software module that compute traffic loads in the WSN. We note here that routing could be easily optimized together with gateway positions by including further degrees of freedom (variables) in the optimization framework [8,12]. However, we believe that our approach is more practical since in real-life wireless sensor networks the network installer can hardly modify/optimize the routing schemes running in the specific WSNs technology.

We propose an Integer Linear Programming (ILP) framework that allows to solve to optimality reasonable size networks and to evaluate the impact of system parameters, including the routing strategy, on the obtained solutions. We also propose an heuristic approach that allows to achieve good quality solutions (within 10% from the optimum in the considered scenarios) in short computing time, tackling also the optimization of large-scale networks.

The paper is organized as follows. In Section 2 we provide a detailed description of the the Multiple Interconnected Gateway Placement (MIG-P) problem and we present our ILP framework to formulate the problem. Section 3 provides numerical results to get insight on the characteristics of the optimal solutions obtained through the ILP framework. The heuristic algorithms proposed for tackling large size instances and results to assess their performance are presented in Section 4. We review and discuss previous works on this topic in Section 5 and conclude the paper in Section 6.

## 2 The Multiple Interconnected Gateway Placement Problem

We consider a heterogeneous network scenario where clouds of Wireless Sensor Networks (WSNs) for data collection are connected to a sink node through a wireless mesh backbone of gateway devices. We will use gateways and WRs as synonyms throughout the paper. Each WR acts as a traffic concentration point for all the sensor nodes belonging to the corresponding WSN. In such network scenario, the number and the positions of the WRs obviously impact on the wireless sensor network lifetime. In fact, roughly speaking, placing a WR in a given position increases the load of all the sensors in the one-hop neighborhood, which are forced to relay to the gateway also the traffic coming from the leaves sensors. We will call these sensor nodes *critical nodes* with respect to a given WR (e.g., nodes S1 and S2 in Fig. 1 are critical nodes).

In this work, we aim at optimizing the overall network topology by properly selecting number and locations of WRs in such a way that installation cost are minimized and energy efficiency (network lifetime) maximized, under the constraints that each sensor must be connected through a multi-hop path to the closest WR (to limit energy consumption due to packet forwarding), and that a multi-hop path does exist also between each WR and the traffic sink. Therefore, the optimization framework jointly considers the selection/positioning of WRs, as well as the design of the WMN among WRs.

We take a constructive approach in presenting the mathematical programming framework. Indeed, we introduce first a *basic version* of the MIG-P problem which considers only installation cost minimization while ensuring coverage within a maximum number of hops of all the sensors and connectivity between gateways and sink node; finally, we extend the basic version to an *advanced version* which also minimizes the load of critical nodes.

The WSN can be represented as a set of test points $\mathcal{S} = \{1..N_S\}$ in the network, whereas a set of candidate sites, denoted by $\mathcal{W} = \{1..N_W\}$, defines the potential positions where WRs can be installed. Each WR can establish a wireless link with any other WR at a distance smaller than its communication range $R_W$ on the backbone, as well as with any sensor at distance lower than $R_C$ in the WSN[1]. According to that, we define an adjacency matrix that summarizes connectivity among candidate sites:

$$I_{ij} = \begin{cases} 1 \text{ if } i \in \mathcal{W}, j \in \mathcal{W}, dist(i,j) \leq R_W \\ 0 \text{ otherwise} \end{cases}$$

where $dist(\cdot)$ is the Euclidean distance between the positions of the two involved devices.

The routing pattern is an input parameter of the optimization problem. Given a candidate site, the specific routing strategy adopted in the WSN determines

[1] The link definition on the basis of a communication range is only for the sake of presentation. We note here that the proposed framework holds for any type of connectivity matrix.

the assignment of sensor nodes to gateways, as well as the set of multi-hop paths used for collecting data. To account for the pre-computed routes, we introduce matrix $R$ which represents the pre-computed routing trees rooted at each WR:

$$R_{ij}^c = \begin{cases} 1 \text{ if link } (i,j),\ i,j \in \mathcal{S}, \text{ belongs to the tree rooted at WR } c \\ 0 \text{ otherwise} \end{cases}$$

To further represent the concept of multi-hop coverage, we introduce the following binary matrix $C$, which indicates whether each sensor is connected through a multi-hop path to a given WR:

$$C_{ij} = \begin{cases} 1 \text{ if there exists a route from sensor node } i \text{ and WR } j \\ \quad \text{with no more than } MAX_{HOPS} \text{ hops} \\ 0 \text{ otherwise} \end{cases}$$

being $MAX_{HOPS}$ is the maximum allowed number of hops of a route.

Finally, for each sensor node $i$ we define the ordered vector $O_i$ of the reachable WRs. The ordering relation is such that given two WRs, one at the $j$-th place, $O_i(j)$, and one at the $k$-th, $O_i(k)$, if $j < k$ then $dist(i,j) \leq dist(i,k)$. Sets $\mathcal{J}_i$ are index sets of vectors $O_i$.

We introduce the following sets of decision variables defining WR installation, node-gateway assignment and routing in the WMN. In detail:

$$y_i = \begin{cases} 1 \text{ if a WR is installed in candidate site } i \\ 0 \text{ otherwise} \end{cases}$$

$$x_{ij} = \begin{cases} 1 \text{ if sensor node } i \text{ is assigned to WR } j, \\ \quad \text{that is, node } i \text{ routes its traffic toward WR } j \\ 0 \text{ otherwise} \end{cases}$$

$f_{ij}$ are integer flow variables for links $(i,j)$, $i,j \in \mathcal{S}$ in the WMN formed by installed WRs, while the $f_w$ represents the overall traffic entering the sink node.

The proposed ILP model formulation for the basic version of the MIG-P problem is:

$$\min \sum_{i \in \mathcal{W}} y_i \tag{1}$$

$s.t.$

$$\sum_{j \in \mathcal{W}} x_{ij} = 1 \qquad \forall\, i \in \mathcal{S} \tag{2}$$

$$x_{ij} \leq C_{ij} y_j \qquad \forall\, i \in \mathcal{S},\ j \in \mathcal{W} \tag{3}$$

$$\sum_{i \in \mathcal{S}} x_{ij} + \sum_{l \in \mathcal{W}} (f_{lj} - f_{jl}) = \begin{cases} f_w, \text{ if } j = SINK \\ 0, \quad \text{otherwise} \end{cases} \qquad \forall\, j \in \mathcal{W} \tag{4}$$

$$f_{ij} \leq N_S I_{ij} y_i,\ f_{ij} \leq N_S I_{ij} y_j \qquad \forall\, i,\ j \in \mathcal{W} \tag{5}$$

$$y_{O_i(k)} + \sum_{h \in \mathcal{J}_i, h > k} x_{iO_i(h)} \leq 1 \qquad \forall\, i \in \mathcal{S},\ k \in \mathcal{J}_i \tag{6}$$

$$y_i \in \{0,1\},\ x_{hj} \in \{0,1\},\ f_{ij} \in \mathbb{Z}^+,\ f_w \in \mathbb{Z}^+ \qquad \forall i,j \in \mathcal{W},\ \forall h \in \mathcal{S} \tag{7}$$

where $SINK$ is the WR chosen as sink node (MAP). Objective function (1) states the minimization of the WR installation cost. In this formulation all candidate sites have the same installation cost, however, it can be easily extended to consider a different cost for each site. Constraints (2) and (3) impose that each sensor node is covered by an installed WR. Constraints (4) are flow balance constraints defining the routing among WRs, while constraints (5) allow the traffic flowing through a link only if the two extreme WRs are installed. Constraints (6) force each sensor node to be assigned to the closest installed WR. Finally, constraints (7) define the domain of each variable.

The second version of the MIG-P problem directly accounts for load balancing and energy efficiency, by minimizing the number of concurrent flows travelling through of the most loaded nodes around the WRs (*critical nodes*). The set of critical nodes can be denoted by $\mathcal{S}^C \subset \mathcal{S}$, whereas $\mathcal{W}_i^C \subset \mathcal{W}$ is the set of candidate sites for which a sensor node $i$ is a critical node. The number of paths through a critical node $i$ is given by variable $p_i, \forall i \in \mathcal{S}^C$. The new ILP model consists in replacing objective function (1) with the following:

$$\min \sum_{i \in \mathcal{W}} y_i + \alpha \sum_{j \in \mathcal{S}^C} p_j \tag{8}$$

and including the additional constraints:

$$\sum_{j \in \mathcal{S}: j \neq i} R_{ji}^c x_{jc} \leq M_P x_{ic} + p_i \qquad \forall i \in \mathcal{S}^C, c \in \mathcal{W}_i^C \tag{9}$$

$$p_i \in \mathbb{Z}^+ \qquad \forall i \in \mathcal{S}^C \tag{10}$$

where $M_P$ is a threshold on the number of paths going through a critical node: when this number is greater than $M_P$ the critical nodes is considered overloaded. These constraints enforce the number of actually used routing paths through a critical node $i$ (left term) to be less or equal to the overload threshold $M_P$. Note that (9) does not define a hard constraint, but, on the other hand, if the overload threshold is violated, the variable $p_i$ records the violation.

The new objective function (8) aims at minimizing both the number of installed WRs and the overload of critical nodes. The scalar $\alpha$ is a conversion parameter between the cost of installing a WR and the overload of critical nodes. The amount of overload recorded in $p_i$ penalizes the objective function and can be actually mapped into the cost of additional power supply for node $i$ (f.i., recharging costs, high capacity batteries, etc.). We have experimentally set $\alpha$ equal to 0.1.

## 3 Optimal Gateway Placement

We have tested the quality of the planned heterogeneous WSNs under two different routing algorithms: a classical min-hop shortest path, and an ideal energy-aware routing. In the latter case, routing paths are obtained solving a Minimum Cost Flow problem towards each candidate site, where the the best routing paths from every sensor to a given candidate site are determined according to the following objective function: $\min \sum_{(i,j) \in \mathcal{L}} dist(i,j) flow_{ij} + \beta m^C$, where $\mathcal{L}$ is the set of wireless links among sensor nodes and the gateway located at the site, $flow_{ij}$ are variables defining the number of node-gateway paths through link $(i, j)$ and $m_C$ expresses the number of routing paths going through the most loaded critical node. Note that the obtained routes preserve a shortest path nature but, at the same time, guarantee an even load distribution among critical nodes. Once the routing is defined, matrices $R$ and $C$ can be fed into the formulations to be solved in reference scenarios.

We consider here randomly deployed sensors and candidate sites over a square area with edge in the range 250m-350m. The communication ranges for sensors and WRs are 20m, and 100m, respectively. The maximum route length parameter $MAX_{HOPS}$ varies from 3 to 5. The problem is modelled with AMPL [21] and solved with the CPLEX 10.0 solver [22] on INTEL 3.2GHz machines with 2GB of RAM operated by Linux. Each result has been averaged on 10 instances.

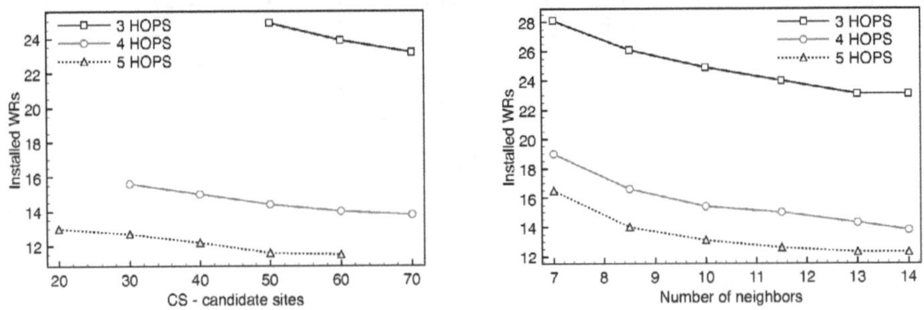


(a) Candidate sites (b) Deployment density

**Fig. 2.** Installed MRs vs. the number of candidate sites (a) and deployment (b). Area 300m x 300m and $MAX_{HOPS} = 3, 4, 5$. a) 1000 sensors; b) from 500 to 1000 sensors, 60 candidate sites.

Figure 2(a) shows the average number of installed WRs when varying the number of candidate sites and the maximum path length. As the number of candidate sites increases, the solution space enlarges and, therefore, the solver can select better options reducing the overall installation cost. As for the maximum path length, achieving coverage is more difficult with shorter paths and cost increases since additional WRs must be installed to provide multi-hop coverage to isolated sensors. We observe that the network installation cost halves when passing from $MAX_{HOPS} = 3$ to $MAX_{HOPS} = 5$. Figure 2(b) shows the average number of installed WRs vs. the density of sensor nodes (expressed as number of neighbors). As expected, the installation cost decreases as the sensor density increases since more multi-hop paths between sensors and WRs area available and coverage is easier.

In order to assess the importance of modelling the energy efficiency directly into the optimization problem, we compare the solutions obtained with the basic version of the gateway placement problem against the advanced one. Some results on the load distribution of critical nodes are reported in Figures 3(a) and 3(b) which refer to the minimum hop routing, and the ideal energy-aware routing algorithm, respectively. Plots show the percentage of critical nodes crossed by the number of routing paths reported on x-axis. For a given critical node, the number of crossing paths corresponds to the number of incoming traffic flows to be forwarded to the closest installed WR. We observe that the advanced formulation provides a larger fraction of critical nodes with less than $M_P$ paths ($M_P = 3$ in the figure), that is, a larger fraction of not overloaded critical nodes. The gap between the two curves is larger when the energy-aware routing is used as it allows to spread the load among critical nodes.

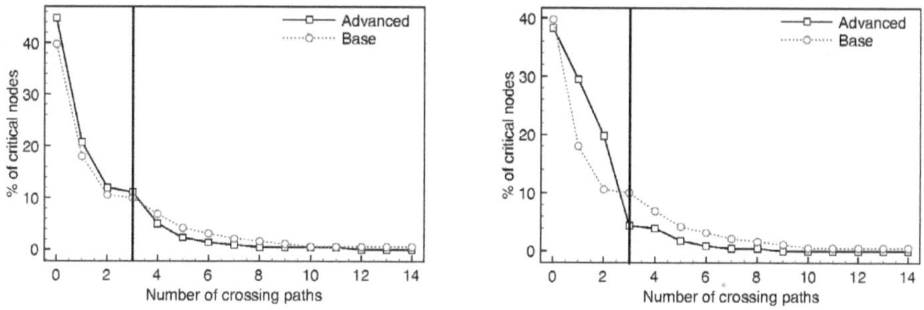


(a) Min-hop routing (b) Energy-aware routing

**Fig. 3.** Comparison of the basic and advanced formulations according to the load distribution among critical nodes for both min-hop and energy-aware routing. Area 300m x 300m, 1000 sensor nodes, 70 candidate sites, and $MAX_{HOPS} = 4$.

**Table 1.** Comparison among different problem formulations. Area 300m x 300m, 1000 sensor nodes, 70 candidate sites, and $MAX_{HOPS} = 3$.

| Formulation | Routing | Average Load of critical nodes | Installed WRs |
|---|---|---|---|
| Basic | Min-hop | 2.75 | 23.2 |
| Advanced $M_P = 3$ | Min-hop | 2.37 | 25.2 |
| Advanced $M_P = 0$ | Min-hop | 2.23 | 29.2 |
| Advanced $M_P = 3$ | Energy-aware | 1.97 | 26.9 |

In Table 1 we summarize the average results for the number of installed WRs and the load of critical nodes with possible combinations of problem versions and routing algorithms. Note that the results with $M_P = 0$ refer to the minimization of the total load of critical nodes, and not only the load of the overloaded ones. Results confirm that the best solutions are obtained with the advanced problem formulation and the energy-aware routing. Moreover, with min-hop routing the lower routing efficiency needs to be compensated by a high number of gateways.

Figures 4(a) and 4(b) report two snapshots of planned networks with different $MAX_{HOPS}$. We reported the Voronoi diagrams representing the WSNs clouds

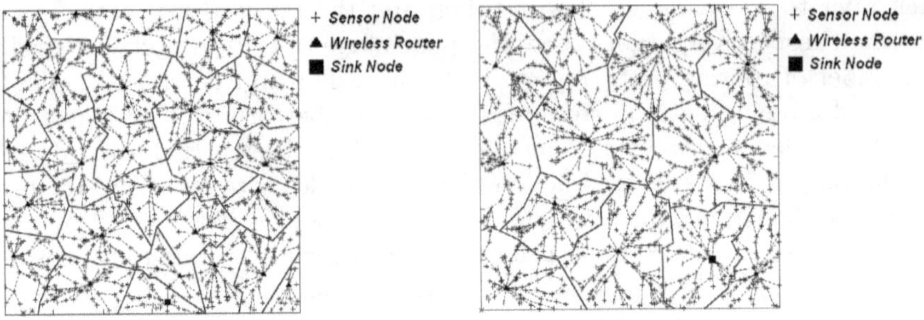


(a) $MAX_{HOPS} = 3$ (b) $MAX_{HOPS} = 5$

**Fig. 4.** Network Planned with the advanced version model. Area $300m$ x $300m$, 1000 sensor nodes, 70 candidate sites.

around each installed gateway that allow to easily observe that increasing the $MAX_{HOPS}$ leads to larger clusters.

## 4 Heuristic Algorithms

It is easy to show that the MIG-P problem is NP-hard since it contains the Minimum Cardinality Set Covering problem as sub-problem. Even if we have shown that the problem can be solved for reasonable size instances, we have also designed heuristics to cope with large networks. We present first an algorithm for the basic version of the MIG-P problem, and then we extend it to solve the advanced version.

Our heuristic approach is based on a continuous relaxation of the corresponding ILP model, and it can be summarized in the following three steps.

*STEP 1 - Continuous relaxation.* The ILP model (1)-(7) is solved relaxing the integrality of variables $y_j$ (WR installation) and $x_{ij}$ (node-WR assignment). Experimentally, we noted that about $60 - 70\%$ of relaxed variables are integer (with $10^{-9}$ precision), thus, we set the value of these variables to 1 or 0. Other variables are rounded in the following step.

*STEP 2 - Variable rounding.* It consists of a greedy rounding over variables $y_j$:

- We generate an ordered list of candidate sites associated to fractional variables $y_j$. The top-list candidate site is the one covering the largest number of sensor nodes.
- We scroll down the list starting from the top. For each site, we round to 1 the associated variable until full coverage is achieved. At this point the coverage requirement is satisfied, but the connectivity one may be not.
- We generate a new ordered list of the candidate sites not yet considered: the first item is the candidate site with the smallest number of installed WRs in its neighborhood. In other words, the head of the list is the most isolated candidate sites.

- Again, we scroll the list rounding to 1 the associated variables, until the WMN is connected. At the end of this step coverage and connectivity requirements are satisfied, thereby remaining fractional variables are set to 0. The result is an initial feasible solution to be refined in STEP 3 removing WRS in excess.

*STEP 3 - Refinement* This step removes as many installed WRs as possible without violating coverage and connectivity constraints. The installed WRs are ordered according to the non-increasing number of neighboring WRs. We scroll the list iteratively removing WRs. At each WR, if the solution is still feasible after its removal, we keep the new solution, otherwise, we re-insert the WR. When the end of the list is reached, the last feasible solution is the final one. Finally, the node-WR assignment is obtained running the ILP model where variable $y_i$ are now parameters set according to the above-described algorithm.

Figures 5(a) and 5(b) show the average optimality gap of the heuristic algorithm under different numbers of sensor nodes and candidate sites. As expected, the quality of the heuristic solution slightly decreases when the solution space increases (candidate sites number increases). However, the largest gap is way below 10%.

As for the advanced version of the problem, the workflow of the heuristic does not change: STEP 1 solves the relaxed ILP model for the advanced formulation, while STEP 2 and STEP 3 provide a feasible initial solution. Only one further step, STEP 4, has been added to the previous framework.

*STEP 4 - WR re-positioning* The step that most significantly impacts on the result of the algorithm is STEP 1, which fixes values of 60 – 70% of the overall variables. However, since assignment variables $x_{ij}$ are relaxed during STEP 1, constraints (9) on the load of critical nodes cannot be properly evaluated. Consequently, after STEP 3, we have solutions with a small number of installed WRs, but with a poor quality in terms of energy consumption. The aim of this STEP 4 is to re-position installed WRs in order to decrease the load of critical nodes. We iteratively perform the following move:

- We place each installed WR in one of the free candidate sites ($y_j = 0$) in its neighborhood. At each iteration, coverage and connectivity requirements are verified. If the solution is feasible and it decreases the load of critical nodes (the second term of the objective function (8)) with respect to the last feasible solution, then the new solution is kept.

This move is repeated until an improving solution cannot be found or a maximum number of iterations is reached.

We remark here that STEP 4 does not change the number of installed WRs which are just re-positioned in order to achieve a better load balancing among critical nodes. The optimality gap of this approach, not reported here for the sake of brevity, has a very similar behavior to the one reported in Figures 5(a) and 5(b). Again, the gap is upper bounded by 10% in all the cases under analysis.

We have tested the heuristic approaches also in large scale heterogeneous WSNs with 1000-3000 deployed sensor nodes and 50-200 available candidate

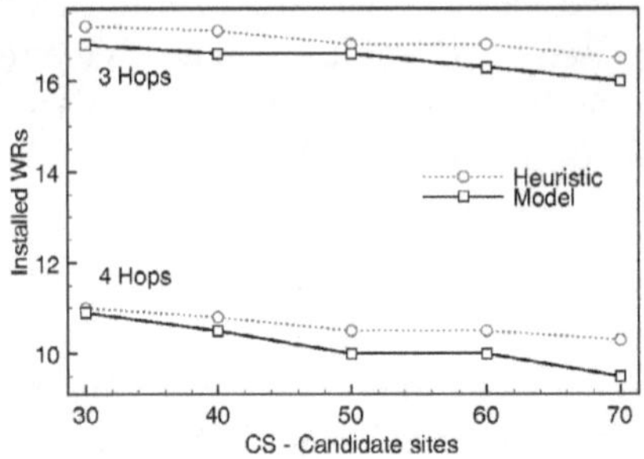


(a) Area 250m x 250m, 700 sensor nodes and $MAX_{HOPS} = 3, 4$

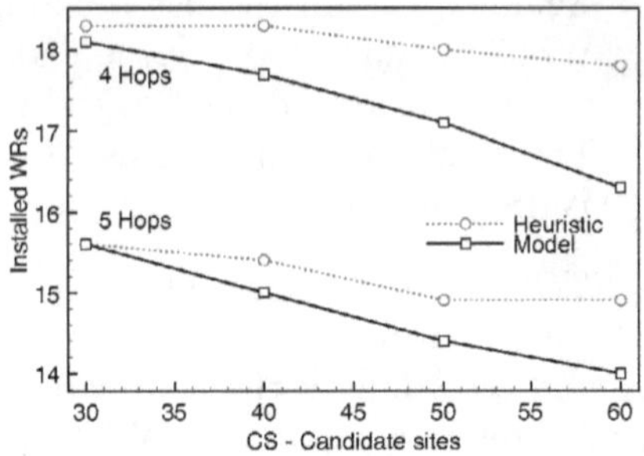


(b) Area 300m x 300m, 1000 sensor nodes and $MAX_{HOPS} = 3, 4$

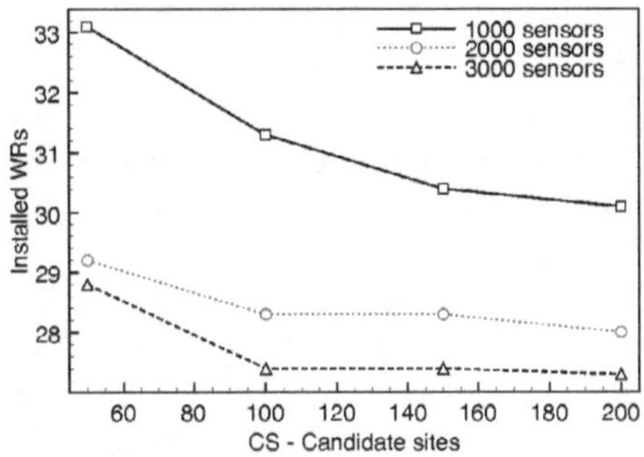


(c) Area 350m x 350m and $MAX_{HOPS} = 3$

**Fig. 5.** Installed WRs as the number of candidate site changes using heuristic algorithms

sites in a square area 350m x 350m, which corresponds to a density of 10-30 neighbors within the communication range. The corresponding numerical results are shown in Figure 5(c) and confirm the same performance trend highlighted by the optimal solutions.

## 5 Related Work

The Gateway Placement problem has been extensively studied in literature. The whole set of works can be divided into two classes. One is the class of *Static Base Station Positioning* problems where the WSN is deployed, gateways are installed and their positions never change for the entire network lifetime. The other class, the class of *Dynamic Base Station Positioning* problems is characterized by the gateway re-positioning during the network operation. The latter type of problems investigate the way of moving gateways to react in front of topology changes, failures, new QoS parameters or simply to balance the energy consumption among sensor nodes. Even though many works of this class exist [11,18,20], this type of problems is out of the scope of our work.

We focus on the class of Static Base Station Positioning problems. Some approaches consider the best placement of a single sink node. In [13] authors propose an iterative repositioning geometric algorithm to find the best sink location in terms of minimum overall transmission energy. However, they assume the gateway is always reachable within a single hop. General sensor-sink paths are considered in [15] where an approximated algorithm aims to maximize the lifetime of a single-sink WSN.

When more gateways are available the problem gets harder because both positions and number of installed gateways must be chosen. The installation of multiple gateways induces the partition of the WSN into clusters, each one assigned to a gateway. Authors in [16] study the problem of positioning the smallest number of gateways in such a way sensors are covered in a single hop by gateway and the set of gateways forms a connected network. They propose a solution approach that divides the deployment area in square cells, finds the best single gateway placement for each cell, and then connects cells installing, if required, new gateways. It does not include any energy issue and lack in multi-hop paths between sensors and gateways.

In [10] heuristic algorithms based on local search and greedy techniques are proposed. Their objective is finding the best sensor nodes to be selected as gateways. The selection of the fixed number of gateways aims to minimize the overall energy consumption. Authors in [9] propose ILP models. They define several versions of the Gateway Placement problem that minimize the number of gateways with a coverage requirement or minimize path lengths from nodes to gateways with an installation budget and coverage constraints. However, these approaches do not guarantee connectivity among gateways and/or pay not attention to the load of critical nodes.

The work in [17] present ILP models where energy issues are expressed as limits on the cluster size and the number of hops between a node and a gateway. Authors in [8] propose an ILP model. The objective is minimizing the maximum number of transmitted/received packets at each sensor node. Given a fixed number of gateways, the solution provides optimum gateway placement and flows along links. The work in [12] presents Linear Programming (LP) models that maximize the total throughput during the network lifetime with fairness guarantees. Solutions provide gateway positions and data rate values. The last three modeling aproaches, besides not address gateway connectivity, force the selection of optimized routing paths.

Some further works start from a planned network and try to improve the topology according to some energy-aware principles. Although they do not consider an actual placement problem, they are interesting because of their approaches to energy-aware sensor clustering. Authors in [14] consider the problem of adding a given amount of energy to the WSN, expressed as gateways to be properly located in the area, in order to extend the network lifetime. In [7] a node-gateway assignment algorithm is proposed. It aims at distributing evenly the number of nodes among gateways. In [19], instead, the assignment algorithm is a genetic algorithm aiming at minimizing the total number of hops within the network.

## 6 Conclusion

In this paper we have proposed an optimization framework for the design of WSNs where gateways must be placed in the area and interconnected with the sink through the wireless links of a WMN. The resulting network architecture is an heterogeneous multi-hop wireless networks where short-range low-rate links interconnect sensor nodes, while long-rage high-rate links create a wireless backbone among gateways.

The objective is that of optimizing a tradeoff between the network installation cost (the number of gateways if their installation cost is the same) and the energy consumption which is modelled considering the load of nodes directly connected to the gateways (critical nodes). The gateway selection is subject to multi-hop sensor coverage constraints that guarantee that each sensor can reach a gateway within a maximum number of hops. Moreover, the wireless network among gateways must be connected so as to allow them to deliver data to the sink.

The proposed solution approach is based on ILP models that allow to obtain optimum solutions for instances of reasonable size. Our problem formulation can consider any type of routing mechanism in the WSN and just account for its effect on the network energy efficiency. This is a key issue since in most application scenarios it is absolutely not possible to enforce a routing scheme optimized during the network design phase as it is constrained by the wireless technology and protocols adopted.

We have analyzed the effect of several parameters on the optimal solutions obtained via the ILP formulations and showed that the coverage constraints and the routing scheme can greatly impact the overall network cost and energy efficiency.

Furthermore, to tackle large problem instances we developed heuristic algorithms based on the continuous relaxation of ILP models. We showed that our heuristics provide good sub-optimal solutions (within 10% from the optimum) in short time and allow planning very big networks with thousands of nodes.

## References

1. Akyildiz, I.F., Su, W., Sankarasubramaniam, Y., Cayirci, E.: Wireless Sensor Networks: A Survey. Comp. Netw. 38, 393–422 (2002)
2. Demirkol, I., Ersoy, C., Alagoz, F.: MAC protocols for wireless sensor networks: a survey. IEEE Comm. Mag. 44(4), 115–121 (2006)
3. Akkaya, K., Younis, M.: A survey on routing protocols for wireless sensor networks. Ad Hoc Networks 3(3), 325–349 (2005)
4. Campelli, L., Capone, A., Cesana, M., Ekici, E.: A Receiver Oriented MAC Protocol for Wireless Sensor Networks. In: IEEE MASS 2007, Pisa, Italy (October 2007)
5. Cardei, M., Wu, J.: Energy-efficient coverage problems in wireless ad-hoc sensor networks. Comp. Comm. 29(4), 413–420 (2006)

6. Amaldi, E., Capone, A., Cesana, M., Filippini, I.: Coverage planning of Wireless Sensors for Mobile Target Detection. In: IEEE MASS 2008, Atlanta, Georgia (October 2008)
7. Gupta, G., Younis, M.: Load-Balanced Clustering of Wireless Sensor Networks. In: IEEE ICC 2003, Anchorage, Alaska (May 2003)
8. Gandham, S.R., Dawande, M., Prakash, R., Venkatesan, S.: Energy Efficent Schemes for Wireless Sensor Networks with Multiple Mobile Base Stations. In: IEEE GLOBECOM 2003, San Francisco (December 2003)
9. Wong, J.L., Jafari, R., Potkonjak, M.: Gateway placement for Latency and Energy Efficient Data Aggregation. In: IEEE LCN 2004, Tampa, Florida (November 2004)
10. Bogdanov, A., Maneva, E., Riesenfeld, S.: Power-Aware Base Station Positioning for Sensor Networks. In: IEEE INFOCOM 2004, Hong Kong (March 2004)
11. Akkaya, K., Younis, M., Bangad, M.: Sink Repositioning for Enhanced Performance in Wireless Sensor Networks. Comp. Netw. 49(4), 512–534 (2005)
12. Kim, H., Seok, Y., Choi, N., Choi, Y., Kwon, T.: Optimal Multi-sink Positioning and Energy-efficient Routing in Wireless Sensor Networks. In: IEEE ICOIN 2005, Jeju Island (Febuary 2005)
13. Pan, J., Cai, L., Hou, Y.T., Shi, Y., Shen, X.: Optimal Base-Station Locations in Two-Tiered Wireless Sensor Networks. IEEE Trans. on Mob. Comp. 4, 458–473 (2005)
14. Hou, Y.T., Shi, Y., Sherali, H.D., Midkiff, S.F.: On Energy Provisioning and Relay Node Placement for Wireless Sensor Networks. IEEE Trans. on Wireless Comm. 4, 2579–2590 (2005)
15. Efrat, A., Har-Peled, S., Mitchell, J.S.B.: Approximation Algorithms for Two Optimal Location Problems in Sensor Networks. In: IEEE BROADNETS 2003, Boston (October 2005)
16. Tang, J., Hao, B., Sen, A.: Relay node placement in large scale wireless sensor networks. Comp. Comm. 29(4), 490–501 (2006)
17. Aoun, B., Boutaba, R.: Clustering in WSN with Latency and Energy Consumption Constaints. Journal of Network and Systems Management 14(3), 415–439 (2006)
18. Youssef, W., Younis, M., Akkaya, K.: An Intelligent Safety-Aware Gateway Relocation Scheme for Wireless Sensor Networks. In: IEEE ICC 2006, Istanbul, Turkey (June 2006)
19. Youssef, W., Younis, M.: Intelligent Gateways Placement for Reduced Data Latency in Wireless Sensor Networks. In: IEEE ICC 2007, Glasgow, Scotland (June 2007)
20. Akkaya, K., Younis, M., Youssef, W.: Positioning of Base Stations in Wireless Sensor Networks. IEEE Comm. Mag. 45(4), 96–102 (2007)
21. Fourer, R., Gay, D.M., Kernighan, B.W.: AMPL, A modeling language for mathematical programming (1993)
22. ILOG CPLEX 10.0 user's manual, `http://www.ilog.com/products/cplex/`

# Extended Cooperative Balanced Space-Time Block Coding for Increased Efficiency in Wireless Sensor Networks*

## (Work in Progress)

Ali Ekşim[1] and Mehmet Ertuğrul Çelebi[2]

[1] Tubitak-UEKAE, P.K. 74, 41470, Gebze, Kocaeli, Turkey
alieksim@uekae.tubitak.gov.tr
[2] Department of Electronics and Communication Engineering,
Istanbul Technical University, Maslak, Istanbul, Turkey, 34469
celebi@ehb.itu.edu.tr

**Abstract.** Diversity techniques for communications among sensors are very effective tool to increase reception quality and battery lifetimes. A well-known method to increase diversity in cooperative communications is sensor (relay) selection. However, sensor selection method may lead to the selection of the same (near) sensor for transmission over a long period. One of the alternative techniques to sensor selection is cooperative balanced space-time block coding which utilizes every sensor in sight, thus, distributes the energy consumption among many sensors. Furthermore, it guarantees full diversity for any number of relay sensors. In this work, we consider dual-hop amplify-and-forward wireless sensor network and extend the cooperative balanced space-time block code family to improve its performance. In the proposed scheme, a larger number of codes can be generated for improved coding gain, and better signal-to-noise ratio improvement can be obtained compared to sensor selection schemes.

**Keywords:** Balanced space-time block codes, Cooperative communications, Relay selection, Wireless sensor networks.

## 1 Introduction

Recently, the desire for connectivity has caused an exponential growth in wireless sensor networks. These networks support a wide variety services while maintaining low power consumption and increased spectrum efficiency. On the other hand, they suffer from fading which causes severe variation in signal attenuation. These constraints can be met via diversity techniques which combine independent fading copies of transmitted signal at the receiver [1]. One of the well-known diversity methods in MIMO systems is transmit antenna diversity which does not especially increase transmission bandwidth and power. Space-time coding methods have been developed to utilize the transmit diversity potential [2-5].

* The work of Ali Ekşim is supported partially by the European Comission (EC), FP-7 Project ICE, Project No. 206546. The work of Mehmet Ertuğrul Çelebi is supported partially by the Scientific and Technological Research Council of Turkey (TUBITAK), Project No.107E022.

L. Fratta et al. (Eds.): NETWORKING 2009, LNCS 5550, pp. 456–467, 2009.

One of the space-time coding scheme is Orthogonal Space-Time Block Codes (OSTBCs) which provide full diversity advantages with a low decoding complexity. The transmitted symbols are decoded separately using linear processing [5]. However, full diversity and full rate for more than two antennas cannot be achieved with OSTBCs. Several quasi-orthogonal STBCs that provide full rate at the expense of some loss in diversity [6],[7] and OSTBCs that provide full diversity with some loss in code rate [5], [8] have been proposed in the literature. In [4], full rate Balanced Space-Time Block Codes (BSTBCs) have been proposed which achieve full diversity for arbitrary number of transmit antennas when one or more feedback bits are transmitted via feedback channel.

Due to insufficient antenna space, cost and hardware limitations, wireless sensors may not be able to support multiple transmit antennas. To overcome this difficulty, recently, with the increasing interests in sensor networks, researchers have been looking for methods to exploit spatial diversity using the antennas of different units in the network. This type of diversity is called *cooperative diversity* [9-13] where virtual antenna arrays can be formed to overcome the drawback of channel correlation and space limitations of mobile units. Cooperative diversity reduces the required transmit power which leads to longer battery life, and increases capacity in interference limited systems.

In [14], Cooperative Balanced Space-Time Block Codes (CBSTBCs) have been proposed where the number of relay sensors can be arbitrary. CBSTBCs cause uniform energy consumption over relay sensors and increase the wireless sensor network lifetime. However, in the proposed scheme, the number of CBSTBCs is limited. For example, if three or four sensors are active in the environment, the number of available codes is only six and twenty-eight, respectively. It can be seen that extension of available coding family increase the efficiency.

In this paper, we propose the Extended Cooperative Balanced Space-Time Block Codes (ECBSTBCs) scheme. With the help of the feedback information, arbitrary number of ECBSTBCs can be utilized. The reason for this extension is to improve upon relay (sensor) selection methods [17] in terms of bit-error performance and energy consumption. In this regard, in the second section, the system model is described, in the third section, the ECBSTBCs are explained, in the fourth section, performance analysis is presented, and in the last section, the results of the paper and the conclusion are given.

The following notation used in the paper: * denotes the conjugate operation; $Re\{.\}$ and $Im\{.\}$ are the real and imaginary part of the argument, respectively. diag$\{d_1,..,d_N\}$ is the $N$x$N$ diagonal matrix whose $i$th diagonal entry is $d_i$. $P(s_k \rightarrow \hat{s}_k)$ is the Pairwise Error Probability (PEP) of deciding in favor of symbol $\hat{s}_k$ when $s_k$ is transmitted. The operator $E(.)$ and exp(.) indicate the statistical expectation and the exponential of the elements of the argument, respectively. The operator $ceil\{.\}$ rounds to the largest integer less or equal than its argument.

## 2 System Model

The wireless sensor network consists of one source sensor, one fusion center and $N$ relay sensors which are located randomly and independently. All sensors are equipped with a single antenna. They cannot transmit and receive at the same time. The relay sensors utilize amplify and forward cooperative protocol [9]. All channels are assumed frequency flat Rayleigh fading channel where channel gains are circularly complex Gaussian random variables and statistically independent from each other. The channels are quasi-static, namely, the fading coefficients remain constant over the duration of one frame. $h_{sri}$ is the channel coefficient from the source sensor to the $i$th relay sensor and $h_{rid}$ is the channel coefficient from the $i$th relay sensor to the fusion center where $i$=1, 2,.., $N$.

The fusion center is assumed to have perfect knowledge of the relay-fusion center channels. It is also assumed that the fusion center has no knowledge of the source sensor-relay sensor channels, since perfect knowledge of the source sensor-relay sensor channels at the fusion center requires considerable protocol overhead. Each sensor is assumed to have perfect knowledge of its own source sensor-relay sensor channel. $T$ is the coherence interval, $P_1$ is the average transmit power of the source sensor, and $P_2$ is the average total transmit power of the relay sensors. The source sensor data bits are mapped by streams of $y$ bits into $M$-PSK symbols where $M=2^y$.

## 3 Extended Cooperative Balanced Space-Time Block Codes

The ECBSTBCs can be obtained with an OSTBC multiplied by an extension matrix. Since Alamouti`s code is the only orthogonal code with rate one and minimum delay, the ECBSTBCs can be obtained as an extension of the Alamouti`s code [15]

$$C=XW. \tag{1}$$

Here $\boldsymbol{X}$ is the Alamouti`s code matrix, $\mathbf{W}$ is a 2x$N$ ($N$>2) matrix whose columns are 2x1 standard basis vectors, and the rank of $\mathbf{W}$ must be 2. The following example shows how to generate the ECBSTBCs for three relay sensors. Consider the ECBSTBC pair with transmission matrix

$$C_1 = \begin{bmatrix} s_1 & s_2 & as_2 \\ -s_2^* & s_1^* & as_1^* \end{bmatrix}. \tag{2}$$

where $a=e^{j2\pi m/q}$, $q$ is the extension level and $m$=0, 1,…$q$-1. The columns and rows of $C_1$ denote symbols transmitted from three relay sensors in two signaling intervals, respectively. $C_1$ is obtained from the Alamouti code using (1) where

$$X = \begin{bmatrix} s_1 & s_2 \\ -s_2^* & s_1^* \end{bmatrix} \quad W = \begin{bmatrix} 1 & 0 & 0 \\ 0 & 1 & a \end{bmatrix}. \tag{3}$$

In this fashion, arbitrary number of the ECBSTBCs can be generated by increasing the extension level. In CBSTBC, the number of codes is limited since $q$ is equal to 2 and $a$=±1. It can be shown that the number of possible ECBSTBCs is $q^{N-2}\left(2^{N-1}-1\right)$.

For that reason, the fusion center needs $N$+$d$ feedback bits ($N$≥3) to select any possible ECBSTBCs where $d$=ceil{($N$-2)log$_2q$}-1. $N$-2 feedback bits are needed to achieve full diversity as in CBSTBCs [14]. The rest of the $d$-1 feedback bits provide coding gain. In CBSTBCs, the extension level is only two, then, the number of available codes is $2^{N-2}\left(2^{N-1}-1\right)$ [14].

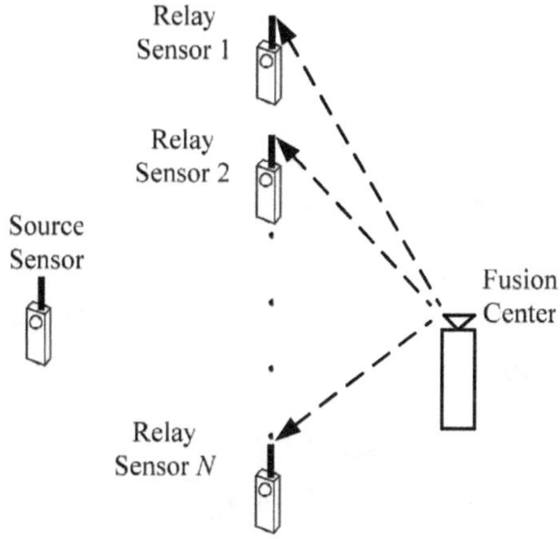


**Fig. 1.** Frame Initialization phase of the ECBSTBCs

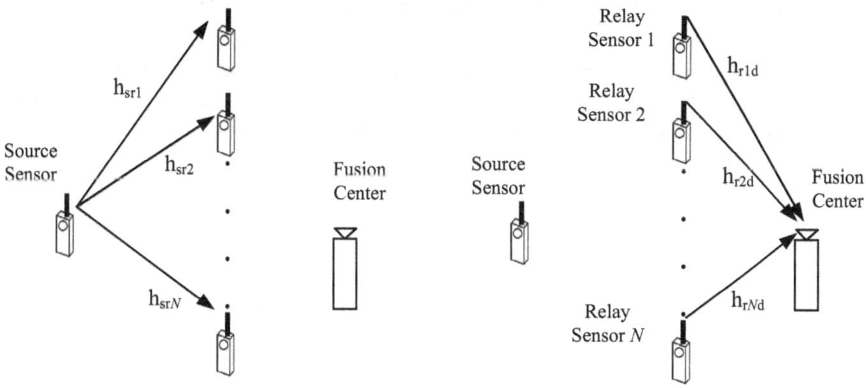


**Fig. 2.** Broadcast and Cooperation phases of the ECBSTBCs

The ECBSTBCs can be used in wireless sensor networks. The ECBSTBCs contain two phases: Broadcast and cooperation. There are many broadcasts and cooperation phases respectively within a frame. Additionally, each frame includes an initialization phase. In the initialization phase, which occurs at the beginning of the each frame, the fusion center informs the relay sensors about which ECBSTBC would be utilized within the frame as shown in Figure 1. The selected code is fixed over one frame. Figure 2 depicts the broadcast and cooperation phases of the ECBSTBCs. In the broadcast phase, the source sensor transmits symbols $s_1$ and $s_2$ to the relay sensors and the received signals at the sensor relays can be given as follows

$$\begin{aligned} r_{ri,1} &= \sqrt{P_1} h_{sri} s_1 + n_{ri,1} \\ r_{ri,2} &= \sqrt{P_1} h_{sri} s_2 + n_{ri,2} \ . \end{aligned} \tag{4}$$

Here $r_{ri,j}$ is the received signal by the $i$th relay sensor at the $j$th symbol interval, $n_{ri,j}$ is the additive white Gaussian noise at the $i$th relay sensor at the $j$th symbol interval. Each cooperating relay sensor makes a basic linear transformation to normalize the received signal.

$$\overline{r}_{ri,1} = \frac{r_{ri,1}}{|r_{ri,1}|} \qquad \overline{r}_{ri,2} = \frac{r_{ri,2}}{|r_{ri,2}|}. \tag{5}$$

Here $\overline{r}_{ri,k}$ is the normalized received signal by the $i$th relay sensor at the $j$th symbol interval. In the cooperation phase of the ECBSTBCs, the fusion center receives the signal, $\boldsymbol{r_D}$,

$$\boldsymbol{r_D} = \sqrt{\frac{P_2}{N}}\boldsymbol{C}\boldsymbol{h}_{rd} + \boldsymbol{n_D}. \tag{6}$$

Here $\boldsymbol{h}_{rd}$ is the channel coefficient vector that contains path gains from the relay sensors to the fusion center, $\boldsymbol{n_D}$ is additive white Gaussian noise vector whose components are complex zero-mean with variance $\sigma^2$, and $\boldsymbol{C}$ is the ECBSTBC matrix.

In [16], distributed space-time coding scenario is analyzed. The optimum power allocation between the source and the relays is chosen to minimize the PEP [16]. For large number of relays which utilize the amplify-and-forward protocol, the optimum power allocation is when the source consumes half of the power and the relays consume the other half. Following a similar analysis carried out in [16] an upper bound for the ECBSTBCs PEP expression is given as,

$$P(s_k \rightarrow \hat{s}_k) \leq E\left(\det{}^{-1}\left[I_N + \frac{PT}{8\left(N + \sum_{i=1}^{N}|h_{rid}|^2\right)}|s_k - \tilde{s}_k|^2 H_{rd}\right]\right). \tag{7}$$

where $P_T$ is the total power used in the network and $H_{rd}$=diag$\{|h_{r1d}|^2, |h_{r2d}|^2,\ldots, |h_{rNd}|^2\}$. Then, the achievable diversity of the ECBSTBCs is min$\{T,N\}$(1-loglog$P$/log$P$) [16]. With the assumption that $T$ is greater or equal to $N$ and $P$ is very large, the diversity gain about $N$ is obtained [16].

## 3.1 Three Relay Sensors

Due to energy efficiency, when three relay sensors are active in the wireless environment, then, $C_1$, $C_2$ and $C_3$ are available ECBSTBC matrices. These matrices are

$$C_1 = \begin{bmatrix} s_1 & s_2 & as_2 \\ -s_2^* & s_1^* & as_1^* \end{bmatrix} \; C_2 = \begin{bmatrix} s_1 & s_2 & as_1 \\ -s_2^* & s_1^* & -as_2^* \end{bmatrix} \; C_3 = \begin{bmatrix} s_1 & as_1 & s_2 \\ -s_2^* & -as_2^* & s_1^* \end{bmatrix}. \tag{8}$$

Here $a$ is the coefficient as defined previously. The fusion center selects the ECBSTBC $C_j$, j=1,2,3 and the feedback bit $a$ that gives the maximum coding gain. Two bits of feedback is needed to select the ECBSTBC matrices and $k$ bit of is needed to select the feedback bit $a$ where $k$=ceil$\{\log_2 q\}$.

The decoding of the ECBSTBCs is similar to CBSTBCs [14]. Assume that the $C_1$ matrix gives maximum coding gain. The received signals at fusion center are given as

$$r_{D,1} = \sqrt{\frac{P_1}{3}}\left[h_{r1d}\overline{r}_{r1,1} + h_{r2d}\overline{r}_{r2,2} + ah_{r3d}\overline{r}_{r3,2}\right] + \eta_1$$
$$r_{D,2} = \sqrt{\frac{P_1}{3}}\left[-h_{r1d}\overline{r}^*_{r1,2} + h_{r2d}\overline{r}^*_{r2,1} + ah_{r3d}\overline{r}^*_{r3,1}\right] + \eta_2. \quad (9)$$

Here $\eta_1$ and $\eta_2$ are noise at the fusion center. The fusion center estimates $s_1$ and $s_2$ by linear processing

$$\hat{s}_1 = h^*_{r1d} r_{D,1} + (h_{r2d} + ah_{r3d}) r^*_{D,2}$$
$$\hat{s}_2 = (h^*_{r2d} + ah^*_{r3d}) r^*_{D,1} - h_{r1d} r^*_{D,2}\ . \quad (10)$$

Substituting $r_{D,1}$ and $r_{D,2}$ in Eq.10,

$$\hat{s}_1 = \sqrt{\frac{P_2}{3}}\left[\left(|h_{r1d}|^2 + |h_{r2d}|^2 + |h_{r3d}|^2 + 2\max\left(\mathrm{Re}\{ah^*_{r2d}h_{r3d}\}, \mathrm{Re}\{ah^*_{r1d}h_{r3d}\}, \mathrm{Re}\{ah^*_{r1d}h_{r2d}\}\right)\right)\right] s_1 + n_1\ .$$
$$\hat{s}_2 = \sqrt{\frac{P_2}{3}}\left[\left(|h_{r1d}|^2 + |h_{r2d}|^2 + |h_{r3d}|^2 + 2\max\left(\mathrm{Re}\{ah^*_{r2d}h_{r3d}\}, \mathrm{Re}\{ah^*_{r1d}h_{r3d}\}, \mathrm{Re}\{ah^*_{r1d}h_{r2d}\}\right)\right)\right] s_2 + n_2 \quad (11)$$

where $n_1$ and $n_2$ are the noise terms which include the observation noise of the both at the relay sensors and at the fusion center. The contribution of the $2\max\left(\mathrm{Re}\{ah^*_{r2d}h_{r3d}\}, \mathrm{Re}\{ah^*_{r1d}h_{r3d}\}, \mathrm{Re}\{ah^*_{r1d}h_{r2d}\}\right)$ in Eq.11 will always be positive and the gain will be greater than the sum of the magnitude squares of all path gains $(|h_{r1d}|^2 + |h_{r2d}|^2 + |h_{r3d}|^2)$.

### 3.2 Four Relay Sensors

It is assumed that four relay sensors are active in wireless environment, then, 7 different ECBSTBC matrices can be obtained. These matrices are

$$C_1 = \begin{bmatrix} s_1 & as_1 & bs_1 & s_2 \\ -s_2^* & -as_2^* & -bs_2^* & s_1^* \end{bmatrix} C_2 = \begin{bmatrix} s_1 & as_1 & s_2 & bs_1 \\ -s_2^* & -as_2^* & s_1^* & -bs_2^* \end{bmatrix}$$
$$C_3 = \begin{bmatrix} s_1 & s_2 & as_1 & bs_1 \\ -s_2^* & s_1^* & -as_2^* & -bs_2^* \end{bmatrix} C_4 = \begin{bmatrix} s_1 & s_2 & as_2 & bs_2 \\ -s_2^* & s_1^* & as_1^* & bs_1^* \end{bmatrix}$$
$$C_5 = \begin{bmatrix} s_1 & as_1 & s_2 & bs_2 \\ -s_2^* & -as_2^* & s_1^* & bs_1^* \end{bmatrix} C_6 = \begin{bmatrix} s_1 & s_2 & as_1 & bs_2 \\ -s_2^* & s_1^* & -as_2^* & bs_1^* \end{bmatrix} \quad (12)$$
$$C_7 = \begin{bmatrix} s_1 & s_2 & as_2 & bs_1 \\ -s_2^* & s_1^* & as_1^* & -bs_2^* \end{bmatrix}.$$

where $a = e^{j2\pi m/q}$ and $b = e^{j2\pi m/q}$. The fusion center selects the ECBSTBC matrix $C_j$, $j$=1,.,7 and the coefficients $a$ and $b$ that yields the maximum coding gain. Three bits of feedback is needed to select the ECBSTBC matrix and $(N\text{-}2)k$ bits of feedback is needed to select feedback $a$ and $b$ where $k = \mathrm{ceil}\{\log_2 q\}$.

The decoding of the ECBSTBCs is similar to CBSTBCs [14]. After combining and linear processing of received signals, the estimates of the PSK symbols $s_1$ and $s_2$ can be expressed as follows

$$\hat{s}_1 = \frac{\sqrt{P_2}}{2}\left[\left(\left|h_{r1d}\right|^2+\left|h_{r2d}\right|^2+\left|h_{r3d}\right|^2+\left|h_{r4d}\right|^2 + 2\max\begin{pmatrix} \left[\mathrm{Re}\{ah_{r1d}^{*}h_{r2d}\}+\mathrm{Re}\{bh_{r1d}^{*}h_{r3d}\}+\mathrm{Re}\{ab^{*}h_{r2d}h_{r3d}^{*}\}\right], \\ \left[\mathrm{Re}\{ah_{r1d}^{*}h_{r2d}\}+\mathrm{Re}\{bh_{r1d}^{*}h_{r4d}\}+\mathrm{Re}\{ab^{*}h_{r2d}h_{r4d}^{*}\}\right], \\ \left[\mathrm{Re}\{ah_{r1d}^{*}h_{r3d}\}+\mathrm{Re}\{bh_{r1d}^{*}h_{r4d}\}+\mathrm{Re}\{ab^{*}h_{r3d}h_{r4d}^{*}\}\right], \\ \left[\mathrm{Re}\{ah_{r2d}^{*}h_{r3d}\}+\mathrm{Re}\{bh_{r2d}^{*}h_{r4d}\}+\mathrm{Re}\{ab^{*}h_{r3d}h_{r4d}^{*}\}\right], \\ \left[\mathrm{Re}\{ah_{r1d}^{*}h_{r2d}\}+\mathrm{Re}\{bh_{r3d}^{*}h_{r4d}\}\right], \\ \left[\mathrm{Re}\{ah_{r1d}^{*}h_{r3d}\}+\mathrm{Re}\{bh_{r2d}^{*}h_{r4d}\}\right], \\ \left[\mathrm{Re}\{ah_{r2d}^{*}h_{r3d}\}+\mathrm{Re}\{bh_{r1d}^{*}h_{r4d}\}\right] \end{pmatrix}\right)\right] s_1 + \omega_1 \tag{13}$$

$$\hat{s}_2 = \frac{\sqrt{P_2}}{2}\left[\left(\left|h_{r1d}\right|^2+\left|h_{r2d}\right|^2+\left|h_{r3d}\right|^2+\left|h_{r4d}\right|^2 + 2\max\begin{pmatrix} \left[\mathrm{Re}\{ah_{r1d}^{*}h_{r2d}\}+\mathrm{Re}\{bh_{r1d}^{*}h_{r3d}\}+\mathrm{Re}\{ab^{*}h_{r2d}h_{r3d}^{*}\}\right], \\ \left[\mathrm{Re}\{ah_{r1d}^{*}h_{r2d}\}+\mathrm{Re}\{bh_{r1d}^{*}h_{r4d}\}+\mathrm{Re}\{ab^{*}h_{r2d}h_{r4d}^{*}\}\right], \\ \left[\mathrm{Re}\{ah_{r1d}^{*}h_{r3d}\}+\mathrm{Re}\{bh_{r1d}^{*}h_{r4d}\}+\mathrm{Re}\{ab^{*}h_{r3d}h_{r4d}^{*}\}\right], \\ \left[\mathrm{Re}\{ah_{r2d}^{*}h_{r3d}\}+\mathrm{Re}\{bh_{r2d}^{*}h_{r4d}\}+\mathrm{Re}\{ab^{*}h_{r3d}h_{r4d}^{*}\}\right], \\ \left[\mathrm{Re}\{ah_{r1d}^{*}h_{r2d}\}+\mathrm{Re}\{bh_{r3d}^{*}h_{r4d}\}\right], \\ \left[\mathrm{Re}\{ah_{r1d}^{*}h_{r3d}\}+\mathrm{Re}\{bh_{r2d}^{*}h_{r4d}\}\right], \\ \left[\mathrm{Re}\{ah_{r2d}^{*}h_{r3d}\}+\mathrm{Re}\{bh_{r1d}^{*}h_{r4d}\}\right] \end{pmatrix}\right)\right] s_2 + \omega_2 \tag{14}$$

where $\omega_1$ and $\omega_2$ are the noise terms which include the observation noise of both at the relay sensors and at the fusion center. The contribution of the coding gain in Eq.13-14 will always be positive and the gain will be greater than the sum of the magnitude squares of all path gains ($\left|h_{r1d}\right|^2+\left|h_{r2d}\right|^2+\left|h_{r3d}\right|^2+\left|h_{r4d}\right|^2$).

## 4 Performance Evaluations

In the cooperative communication transmitting only from selected relays is called distributed transmit antenna selection (DTAS) [17] which is an alternative approach to the ECBSTBCs. The criterion in selecting a single relay sensor is the best instantaneous relay-fusion center channel gain [18], and this is called as transmit sensor selection (TSS *N*:1). To maximize SNR at the fusion center, two relay sensors are chosen out of all the available transmit relay sensors and then the relay sensors transmit the received signals using the Alamouti scheme [19]. In the simulations, the best relay pair which has best instantaneous relay-fusion center channel pair is selected. This is called as transmit sensor selection with Alamouti (TSS $N$:2).

The bit error probabilities of the proposed ECBSTBC sets are evaluated for quaternary phase-shift keying (QPSK) modulation by computer simulations. A frame of 130 symbols is used. For meaningful comparison, the total transmission power and bandwidth are fixed. Namely, in the broadcast phase; the source sensor transmits with full power but in the cooperation phase, the power is divided equally among relay sensors.

For comparison purposes, the bit error rate (BER) curves of the transmit sensor selection with Alamouti's scheme and transmit sensor selection are also included in Figs. 3–6.

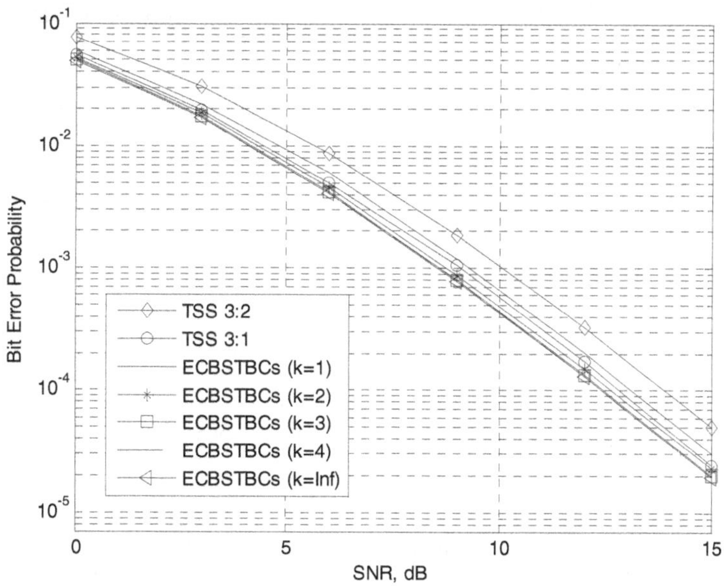


**Fig. 3.** Bit-error probability performance of sensor networks when three relay sensors are active and the source sensor-relay sensors' channels are ideal

Figure 3 presents the bit error probabilities of the ECBSTBC with one to four bits extension of feedback and ideal (infinite) feedback for three relay sensors. We assume that the source-relay sensor channels are ideal to be able to investigate pure coding performance. Compared to the transmit sensor selection with Alamouti's scheme (TSS 3:2), the ECBSTBC with one bit extension (ECBSTBCs (k=1)) provides an SNR advantage of approximately 0.74dB for a BER value of $P_b=10^{-4}$. However, transmit sensor selection (TSS 3:1) provides approximately 0.31dB better performance than ECBSTBC with one bit extension due to limited coding gain. If we extend feedback by two bits (ECBSTBCs (k=2)), the ECBSTBCs provide approximately 0.21dB better performance than the TSS. Thus, on the basis of the same SNR, the ECBSTBCs outperform the transmit sensor selection. If three bits of extension

(ECBSTBCs (k=3)) is available for feedback, an additional 0.13dB coding gain is available. Performance of four bits extension of feedback (ECBSTBCs (k=4)) is similar to ideal feedback (ECBSTBCs (k=Inf.)). As expected, when the number of extension level is increased, the bit error performance of ECBSTBC is improved.

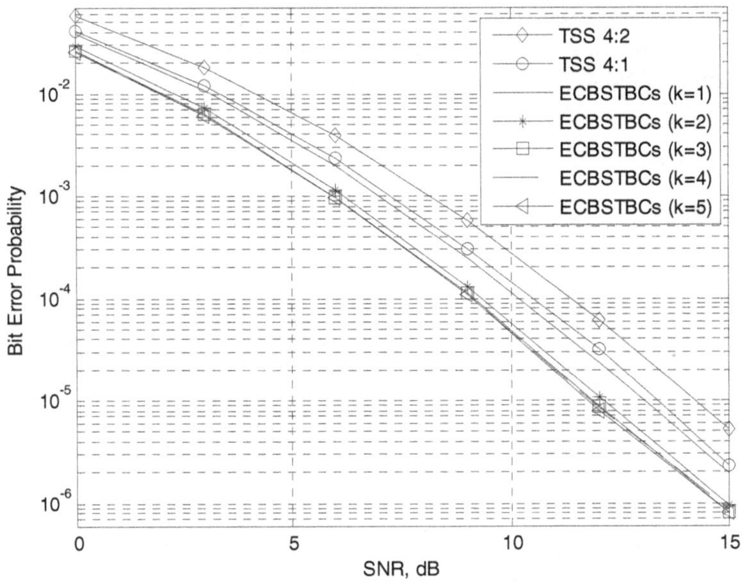


**Fig. 4.** Bit-error probability performance of sensor networks when four relay sensors are active and the source sensor-relay sensors' channels are ideal

Figure 4 presents the bit error probabilities of the ECBSTBC with one to five bits extension of feedback bits for four relay sensors. Compared to the TSS with Alamouti's scheme (TSS 4:2), the TSS (TSS 4:1) provides an SNR advantage of approximately 0.89 dB for a BER value of $Pb=10^{-5}$. However, the ECBSTBCs with one bit extension (ECBSTBCs (k=1)) provides approximately 0.29dB better performance than the TSS. If we extend the feedback information by two bits (ECBSTBCs (k=2)), the ECBSTBC provides approximately 0.94dB better performance than the ECBSTBCs with one bit extension. If three bits extension of feedback is available, an additional 0.23dB coding gain is obtained. Four bits and five bits extension of feedback yields approximately similar performance. As expected, as the number of available codes is increased the coding gain is increased.

In the sequel, we assumed that the source-relay channels are not ideal, nevertheless, are better quality than relay-fusion center channels whose difference is quantified by differential signal-to-noise ratio (DSNR). In the Figure 5-6, DSNR is assumed to be 15dB for three and four relay sensors. Once again, the ECBSTBCs provide better performance than the sensor selection schemes and the CBSTBCs.

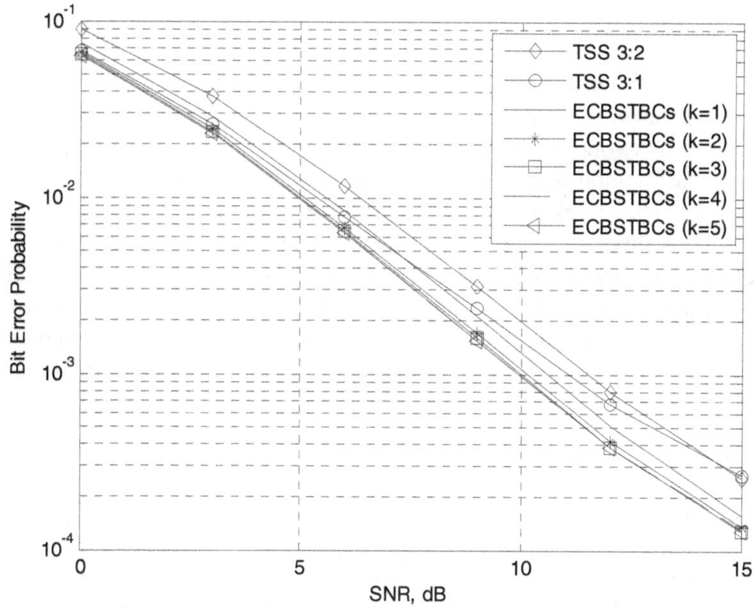


**Fig. 5.** Bit-error probability performance of sensor networks when three relay sensors are active and the source sensor-relay sensors' channels are 15dB DSNR

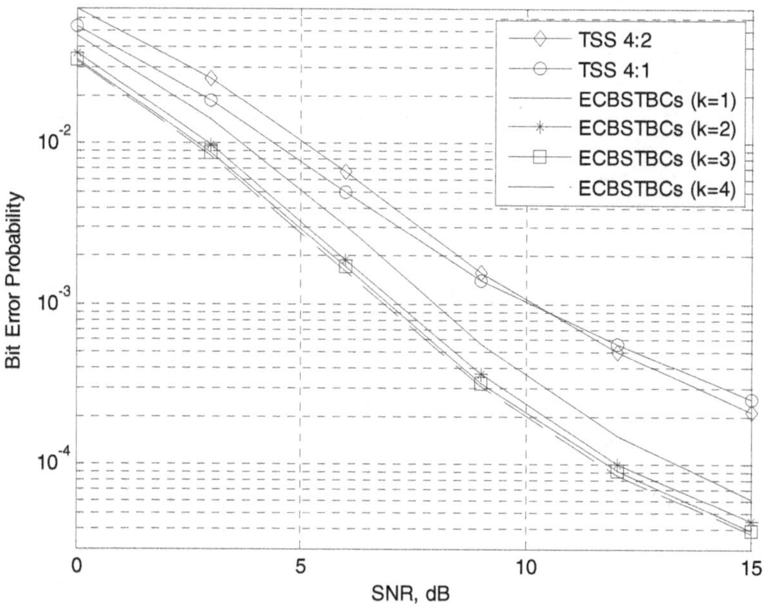


**Fig. 6.** Bit-error probability performance of sensor networks when four relay sensors are active and the source sensor-relay sensors' channels are 15dB DSNR

## 5 Conclusions

Network lifetime extension is better achieved via uniform energy distribution over a wireless sensor network. One of the promising methods for effective communication in wireless sensor networks is the CBSTBC which provides full diversity and distributes the energy consumption over a larger part of the network as opposed to sensor selection methods. In this work, we improve the performance of the CBSTBCs by means of extending the number of available codes to provide improved coding gain. Simulations, demonstrated that four bit extension of feedback gives approximately ideal feedback performance. In addition, the ECBSTBCs have better signal-to-noise ratio improvement compared to sensor selection schemes. Consequently, we conclude that the ECBSTBC is a useful tool to increase the QoS of wireless sensor networks.

## References


1. Vucetic, B., Yuan, J.: Space-Time Coding. John Wiley & Sons, Chichester (2003)
2. Liu, L., Lim, M.-S.: Selective receiver switching scheme for space time block coding with full code rate and non-orthogonal design. IET Commun. 2(5), 664–672 (2008)
3. Yousafzai, A.K., Nakhai, M.R.: Reduced complexity detection technique for layered space time block coded multiple-input multiple-output orthogonal frequency division multiplexing. IET Commun. 3(1), 115–122 (2009)
4. Çelebi, M.E., Şahin, S., Aygölü, Ü.: Balanced space-time block coding: An efficient way to increase diversity with feedback. IEE Proc-Commun. MIMO Wireless and Mobile Commun. 153(4) (2006)
5. Tarokh, V., Jafarkhani, H., Calderbank, A.R.: Space-time block codes from orthogonal designs. IEEE Transactions on Information Theory 45, 1456–1467 (1999)
6. Jafarkhani, H.: A quasi-orthogonal space-time block code. IEEE Trans. Commun. 49, 1–4 (2001)
7. Tirkkonen, O., Hottinen, A.: Complex space-time block codes for four Tx antennas. In: Proc. GLOBECOM, pp. 1005–1009. IEEE Press, San Fransisco (2000)
8. Su, W., Xia, X.G.: On space-time block codes from complex orthogonal designs. Wirel. Pers. Commun. 25, 1–26 (2003)
9. Laneman, J.N., Wornell, G.W., Tse, D.N.C.: An efficient protocol for realizing cooperative diversity in wireless networks. In: Proc. IEEE ISIT, Washington D.C. IEEE Press, Los Alamitos (2001)
10. Sendonaris, A., Erkip, E., Aazhang, B.: User cooperation diversity part I: System description. IEEE Trans. Commun. 51(11), 1927–1938 (2003)
11. Sendonaris, A., Erkip, E., Aazhang, B.: User cooperation diversity part II: Implementation aspects and performance analysis. IEEE Trans. Commun. 51(11), 1939–1948 (2003)
12. Nabar, R.U., Bolcskei, H., Kneubuhler, F.W.: Fading relay channels: Performance limits and space-time signal design. IEEE J. Select. Areas Commun., 1099–1109 (2004)
13. Tang, Y., Valenti, M.C.: Coded transmit macrodiversity: Block space-time codes over distributed antennas. In: IEEE Veh. Technol. Conference, pp. 1435–1438. IEEE Press, Los Alamitos (2001)
14. Ekşim, A., Çelebi, M.E.: Diversity enhancement with cooperative balanced space-time block coding. In: Proc. of the 18th Annual IEEE International Symposium on Personal, Indoor and Mobile Radio Communications (PIMRC 2007). IEEE Press, Athens (2007)

15. Alamouti, S.M.: A simple transmit diversity technique for wireless communications. IEEE J. Select. Areas Commun. 16, 1451–1458 (1998)
16. Jing, Y., Hassibi, B.: Distributed space-time coding in wireless relay networks. IEEE Trans. on Wirel. Commun. 5(12) (2006)
17. Michalopoulos, D.S., Karagiannidis, G.K., Tsiftsis, T.A., Mallik, R.K.: Distributed transmit antenna selection (DTAS) under performance or energy consumption constraints. IEEE Trans. Wireless Commun. 7, 1168–1173 (2008)
18. Luo, J., Blum, R.S., Greenstein, L.J., Haimovich, A.M.: Link-failure probabilities for practical cooperative relay networks. In: Proc. of the IEEE vehicular Technology Conference (VTC 2005 Spring). IEEE Press, Stockholm (2005)
19. Gore, D., Paulraj, A.: MIMO antenna subset selection with space-time coding. IEEE Trans. Signal Process 50, 2580–2588 (2002)

# Congestion and Flow Control in the Context of the Message-Oriented Protocol SCTP

Irene Rüngeler[1], Michael Tüxen[1], and Erwin P. Rathgeb[2]

[1] Münster University of Applied Sciences
Department of Electrical Engineering and Computer Science
Bismarckstrasse 11
D-48565 Steinfurt, Germany
{i.ruengeler,tuexen}@fh-muenster.de
[2] University of Duisburg-Essen
Ellernstrasse 29
D-45326 Essen, Germany
erwin.rathgeb@iem.uni-due.de

**Abstract.** Congestion and flow control are key mechanisms used to regulate the load in modern packet networks. The new IETF Stream Control Transmission Protocol (SCTP) inherited these algorithms from the Transmission Control Protocol (TCP). Although the principles used are the same, some issues arise from the fact that SCTP operates message-oriented whereas TCP operates byte-stream oriented. SCTP also supports bundling of multiple small user messages into one SCTP packet. As a consequence, the overall overhead of an SCTP packet depends on the user message size and the number of user messages that are bundled into the packet. RFC 4960 defining SCTP does not specify whether the message specific headers have to be considered when updating the parameters for congestion control. We will show that neglecting the additional headers when calculating outstanding bytes can lead to unfairness towards TCP connections. We will also show that incorrect handling of the additional memory needed to process each message in the flow control calculations will lead to an exhaustion of the receiver window resulting in a huge amount of unnecessary retransmissions. Based on experiments with the flow control of the SCTP implementations available in several operating systems, we will identify the issues and analyze them by using simulations. As a result, we will present solutions that will lead to fairness towards TCP and reduce the number of retransmissions substantially. Although we will focus on SCTP, the results are also true for other message-oriented protocols using bundling.

**Keywords:** Congestion Control, Flow Control, Message Orientation, SCTP.

## 1 Introduction

Although the Transmission Control Protocol (TCP) is still the main transport protocol for IP-based networks, the Stream Control Transmission Protocol (SCTP) [1]

L. Fratta et al. (Eds.): NETWORKING 2009, LNCS 5550, pp. 468–481, 2009.

gains more and more importance being integrated in the major operating systems. In contrast to TCP, SCTP is not byte-stream oriented but message-oriented. It also supports bundling of multiple small user messages into one SCTP packet to increase transport efficiency.

In order to prevent congestion, SCTP adopted the window based algorithms for flow control and congestion control from TCP. Although the mechanisms seem to be the same for both protocols, the message orientation of SCTP and thus the data processing on a per message basis requires specific consideration.

In TCP, the header size is relatively small and almost constant (between 20 and 60 bytes depending on the TCP options used). Therefore, it is not considered for the flow and congestion control computations. In SCTP, however, every packet requires a common header and an additional header of 16 bytes for each user message. When bundling several small messages into one SCTP packet, this additional overhead is variable and can have a significant impact on the flow and congestion control calculations.

RFC 4960 defining SCTP gives no clear recommendation how to handle the message overhead, and it is up to the implementer of the protocol, if it is included in the calculations or not. As a consequence, implementations like Solaris include the additional headers whereas in FreeBSD they are excluded.

The choice on handling the message overhead in the calculation of the amount of outstanding (not yet acknowledged) data for congestion control can lead to unfairness towards competing TCP connections, if small messages are sent. Therefore, after a short introduction of the relevant SCTP features in Section 2, we will investigate this issue in detail in Section 3 by using simulations. We will show that only including the message headers allows to achieve the mandatory TCP-friendliness.

Inappropriate handling of the overhead, when calculating the amount of data a (slow) receiver is still able to accept, can lead to memory exhaustion at the receiver before the sender is prevented from sending. This leads to packet loss and consequently to unnecessary retransmissions. Based on experiments with the implementations of the Linux 2.6.25 kernel, FreeBSD release 7.0 and Solaris 10, we will highlight the flow control issues in Section 4 and propose a way to implement SCTP flow control such that no unnecessary retransmissions are triggered. This includes a correct announcement of the receiver window as well as applying the silly window syndrome (SWS) avoidance and enabling the Nagle algorithm. The proposal will be validated by simulations.

The simulations in this paper have been performed with the OMNeT++ simulation environment [2] and an extended version of the SCTP simulation model [3] we contributed to the INET framework [4].

## 2 SCTP, a Message-Oriented Protocol

With the emergence of IP-based networks as universal platform for communication services the need arose to send telephony signaling data across the internet. Signaling data feature relatively small single messages that have to be sent reliably and some of them also in the correct sequence. To fulfill this task, the new protocol SCTP was designed as a reliable message-oriented protocol and finally adopted by the IETF as official standard in RFC 2960 in 2000. After some modifications, RFC 4960 [1] adopted in September 2007, is the current SCTP specification.

An SCTP packet consists of a 12 byte common header and a number of so called *chunks*. Each chunk has a chunk header that varies with the chunk type. The `DATA`-chunk header is 16 bytes long. Its payload has to be 32 bit aligned. Each chunk is identified by a transmission sequence number (TSN).

As SCTP is a reliable transport protocol, the arrival of user data has to be acknowledged. This is handled by `SACK`-chunks. Here the cumulative TSN ack parameter indicates the highest TSN received in sequence. The acceptance of additional chunks is reflected in gap ack blocks. Another important parameter of the `SACK`-chunk is the advertised receiver window (arwnd), that announces the amount of bytes the receiving endpoint can still accept.

More information about other distinctive features of SCTP like multi-homing or multi-streaming is provided in RFC 3286 [5].

# 3 SCTP Congestion Control

Congestion control is a mechanism to control the traffic of a network. The goal is to prevent senders from blocking links by reducing the rate of sending packets. Although in the next subsections we will focus on the congestion control mechanism integrated in the examined implementations, the considerations are also true for other congestion control algorithms and other message-orientated protocols using bundling.

## 3.1 The Congestion Window

The congestion window limits the number of bytes the sender is allowed to transmit before waiting for a new acknowledgement. That means, that not more than cwnd bytes may be outstanding.

The congestion control mechanism is divided into two phases. The first one is called *slow start*. It operates for cwnd values less than or equal to the slow start threshold, which is set to an arbitrary value (mostly the advertised receiver window of the peer during association setup) at the beginning of an association. *Slow start* is characterized by an exponential increase of the congestion window. Every time an incoming `SACK`-chunk announces that the cumulative TSN ack parameter has advanced and the cwnd is fully utilized, i.e. the number of outstanding bytes is greater than cwnd, the minimum of the path MTU and the acknowledged bytes is added to cwnd.

When cwnd exceeds the slow start threshold, *congestion avoidance* makes for a linear increase of cwnd. As the growth of cwnd can lead to an excessive injection of data into the network, packet loss is the consequence. While fast retransmissions result in halving the congestion window, a timer based retransmission leaves cwnd at the size of the path MTU and in *slow start* again. Thus cwnd follows usually a zigzag curve in the lifetime of a connection.

## 3.2 Counting Outstanding Bytes

As pointed out, cwnd has an influence on the network load and thus on the throughput. Therefore, the way the outstanding bytes, that limit cwnd, are counted, is important and should be examined.

Looking at an SCTP packet containing several data chunks, the amount of user data can vary significantly with the size of the individual chunks (i.e. messages) assuming the same packet length.

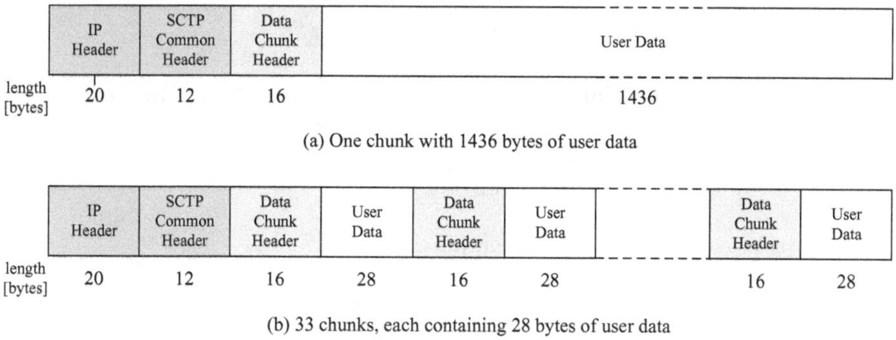


**Fig. 1.** SCTP packet format

In Figure 1(b) the packet contains 33 DATA-chunks with 28 bytes of user data each, adding up to 924 bytes of user data compared to 1436 bytes in the packet in Figure 1(a). Both packets have a size of 1484 bytes. Whereas the overhead is just 1 % in (a) the headers add up to 36 % in (b) and can be more than 60 % for even smaller user message sizes.

Therefore, we have to distinguish between the amount of data that is injected into the network and the user data that arrive at the application layer. Whereas the first has a direct impact on the network load, the second results in the goodput. Both depend on the number of packets (1), that are allowed by the cwnd.

$$NoOfPackets = \left\lceil \frac{cwnd}{Size_P} \right\rceil \tag{1}$$

Calculating the size of a packet ($Size_P$), the headers for IP ($H_{IP}$) and SCTP ($H_{SCTP}$) have to be considered as well as the size of the DATA-chunks ($Size_{Chunk}$).

$$Size_P = H_{IP} + H_{SCTP} + CPP \cdot Size_{Chunk} \tag{2}$$

The number of the chunks per packet ($CPP$) is calculated as

$$CPP = \left\lfloor \frac{MTU - H_{IP} - H_{SCTP}}{UMS + P_{UMS} + H_{Chunk}} \right\rfloor \tag{3}$$

The average user message size ($UMS$) per packet and the corresponding padding bytes ($P_{UMS}$) feature the variable parts of the packets.

$Size_P$ is a variable used for calculating the bytes allowed by the cwnd, and is not necessarily equivalent to the real size of a packet. If talking about $Size_P^+$, the bundled chunks are calculated with header $Size_{Chunk}^+ = UMS + P_{UMS} + H_{Chunk}$ and for $Size_P^-$ is $Size_{Chunk}^- = UMS$ without header. To compute the number of bytes that are induced into the network and which arrive at the receiver, four different cases are possible:

– Network load taking the header into account

$$Bytes_{SCTP}^+ = \left\lceil \frac{cwnd}{Size_P^+} \right\rceil \cdot Size_P^+ \tag{4}$$

– Bytes at the application layer if the header had been taken into account

$$Bytes_{App}^+ = \left\lceil \frac{cwnd}{Size_P^+} \right\rceil \cdot CPP \cdot Size_{Chunk}^- \tag{5}$$

– Network load without taking the header into account

$$Bytes_{SCTP}^- = \left\lceil \frac{cwnd}{Size_P^-} \right\rceil \cdot Size_P^+ \tag{6}$$

– Bytes at the application layer if the header had not been taken into account

$$Bytes_{App}^- = \left\lceil \frac{cwnd}{Size_P^-} \right\rceil \cdot CPP \cdot Size_{Chunk}^- \tag{7}$$

After the initialization of a connection, the initial window is specified in [6] to be

$$\min(4 * \text{MSS}, \max(2 * \text{MSS}, 4380 \text{ bytes})) \tag{8}$$

Assuming a path MTU of 1500 bytes, the maximum segment size (MSS) equals 1460 bytes, which is 1500 bytes minus 20 bytes for the IP header and 20 bytes for the TCP header. Thus the initial cwnd is 4380 bytes.

As one property of fairness is the evenly distribution of the link bandwidth, we will look at the behavior of associations (terminology for SCTP connections) with and without header inclusion in more detail.

### 3.3 TCP-Friendliness

When SCTP was designed, one of the major goals was to guarantee TCP-friendliness. In RFC 2309 [7] a *TCP-friendly* or *TCP-compatible* flow is defined as follows: "A TCP-compatible flow is responsive to congestion notification, and in steady state it uses no more bandwidth than a conforming TCP running under comparable conditions."

Since TCP is a byte-stream oriented protocol and typically the Nagle algorithm is not disabled, all packets are filled with enough user data to result in full sized link layer frames, if sufficient data are provided in the send queue. The overhead consists of the IP-header and the TCP-header, which is independent from the user message sizes.

Although SCTP and TCP implementations, which we inspected for the differences in the handling of header bytes, are readily available, we will base our solutions on simulation results. Since we found that some implementations have bugs substantially influencing the measurement results we decided to use a simulation for the measurements instead of waiting for the bugfixes to be included in the implementations.

### 3.4 The Simulation Scenario

We have used the INET framework [4] as a simulation tool. Although TCP is integrated in the framework, not all optional TCP features that are common nowadays, like Appropriate Byte Counting (ABC) [8] or delayed acknowledgements [9], are implemented. However, some of these features are mandatory for SCTP and are, therefore, implemented in the INET SCTP model, contributed by us [3]. Hence, a meaningful comparison between SCTP and TCP is not possible with INET. Nevertheless, we wanted to examine TCP-friendliness for flows with and without counting the header. Therefore we used an SCTP association transporting user data messages of 1452 bytes length to mimic the behavior of a state-of-the-art TCP connection. From a congestion control perspective, such an SCTP association behaves identical to a TCP connection. When talking about including or excluding the header, we always refer to the DATA-chunk header of 16 bytes.

The scenario for the simulation consists of an SCTPClient, connected to an SCTPServer, and an TCP-like client, connected to a TCP-like server. The connections have to share a bottleneck link between two routers with a data rate of 1 Mbps. The SCTP client sends data with configurable user message sizes from 12 to 204 bytes to the SCTP server. The TCP-like client only sends full packets with a payload of 1452 bytes, the headers are not included. Including them does not change the result, since the difference is neglectable for large user messages.

To test the behavior with and without counting the header bytes, the SCTP simulation has been extended by two parameters, *osbWithHeader* and *padding*. They are boolean variables that can be set to true, if the header and the padding bytes should be taken into account for the congestion control calculations.Tests showed that the influence of the padding bytes is not significant. Therefore, all described simulations were run with either both variables true or false.

### 3.5 Fairness on the Transport Layer

The SCTP association and the "TCP-like" association have to share the bandwidth equally. This means that all bytes that have been sent over the network have to be counted, including the retransmitted bytes. To assure that the same time interval is chosen and the associations have reached a steady state, a start and stop time can be configured for counting the bytes that have arrived at the server. The timers were set for the measurement to start after 50 secs and continue for 400 secs. As the ratio of additional header bytes to the user message

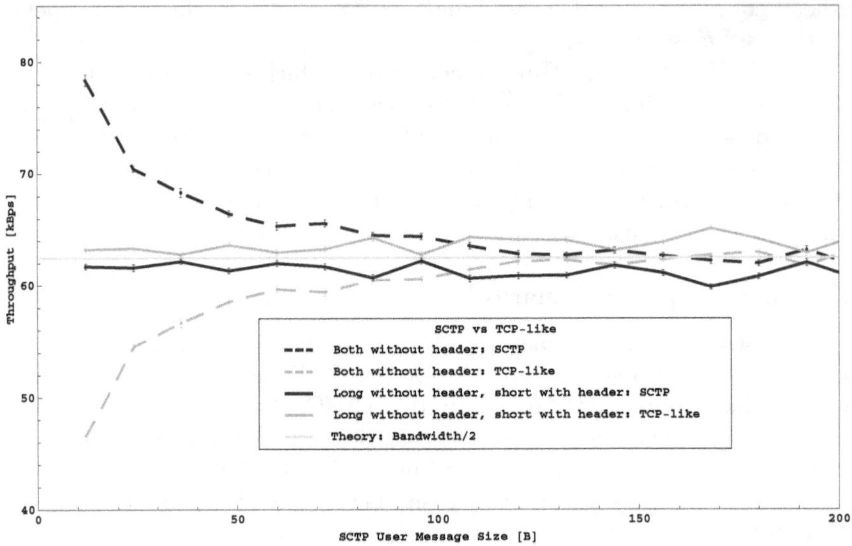


**Fig. 2.** Throughput on the transport layer

size is only significant for small payload sizes, we chose user messages from 10 to 200 bytes length in 10 byte intervals. Each simulation run was repeated 100 times with different seeds for the random numbers to ensure validity. Figure 2 shows the throughput on the transport layer. The vertical bars represent the 95 % confidence intervals. It is obvious that the associations that calculate the outstanding bytes with header share the link, symbolized by the straight line, equally, whereas the behavior is not fair for small message sizes, if the header is not taken into account. The SCTP client utilizes the link much more intensively than the TCP-like client, thus taking bandwidth from the other connection.

### 3.6 Fairness on the Application Layer

The behavior on the transport layer has an influence on the throughput on the application layer (goodput). Therefore, we chose the same setup as in the last section and counted the bytes that arrived at the user level of the servers during a predefined time period and calculated the throughput. Figure 3 shows the graphs when the header is not taken into account. Although the TCP-like client achieves a higher throughput than the SCTP client using the different message sizes, the throughput is much lower than it should be. As Figure 2 indicated, the SCTP client takes over so much bandwidth that the TCP goodput is considerably reduced. The two theoretical graphs show the ideal case, where the SCTP client (lowest graph) and the TCP-like client (top graph) share the link equally. The zigzagging of the lowest graph results from padding, i.e. the necessity to add 1 to 3 bytes to the payload to get it 32 bit aligned. The graphs

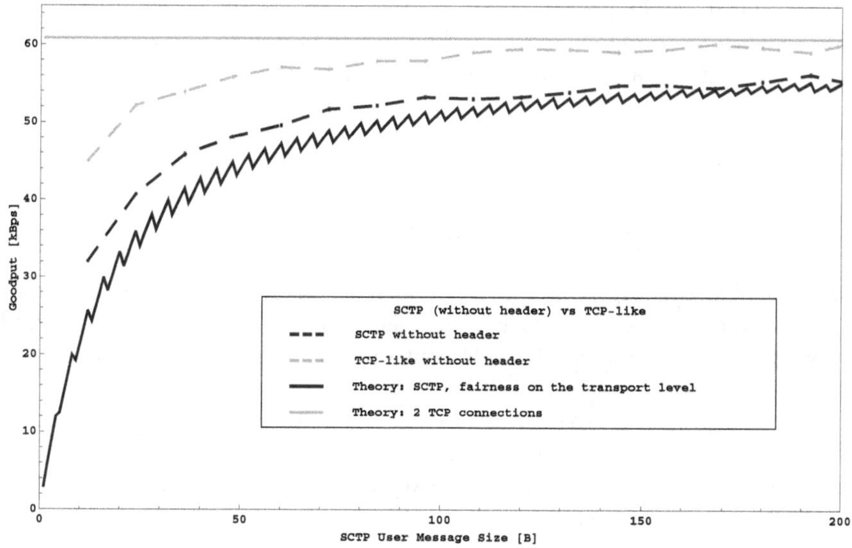


**Fig. 3.** Goodput, without considering the header

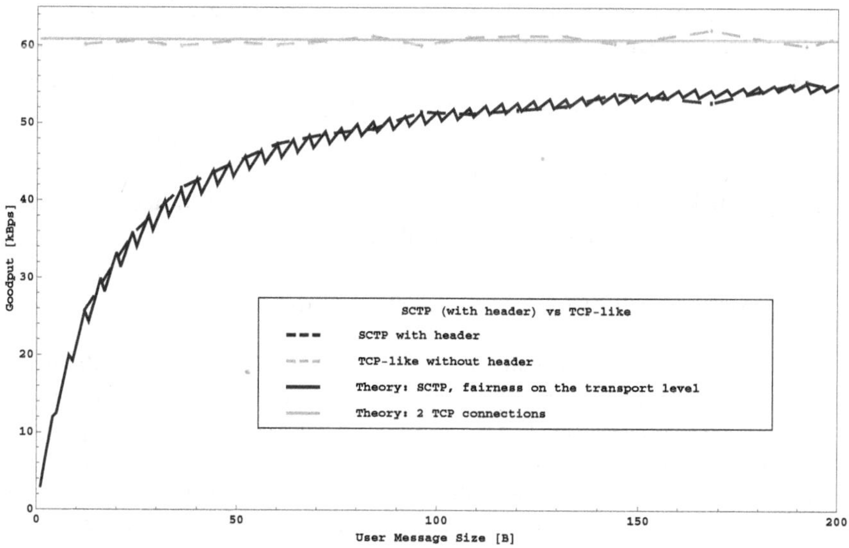


**Fig. 4.** Goodput, if the header is taken into account

in Figure 4 illustrate the results, if the header is taken into account. Now the curves show the desired behavior and fit the theoretical graphs.

As a result it can be postulated, that all implementations of message-oriented protocols with bundling should take the headers into account, when calculating the outstanding bytes, in order to be TCP-compliant.

## 4 Flow Control

### 4.1 The Concept of SCTP Flow Control

Flow control like congestion control is a mechanism to influence the amount of data injected into the network. Whereas the congestion control protects the network from a fast sender, the flow control should prevent the overload of the receiver.

To achieve this, the advertised receiver window parameter (arwnd) is used to announce the amount of data that the receiver is willing to accept. During the setup of the association the hosts exchange their initial arwnd in the `INIT` and `INIT_ACK`-chunk. Upon arrival of a `DATA`-chunk, arwnd is decremented by the message size. When the data is delivered to the upper layer, arwnd can be incremented again. When the receiver sends a `SACK`-chunk to acknowledge data, it includes the actual value of the arwnd. The sender tries to keep track of the size of its peer's arwnd by trying to predict the window size. It takes the value of the announced arwnd as basis, reduces it by the number of outstanding bytes, i.e. the data that are assumed to be in flight. If the peer's arwnd reaches zero, only one `DATA`-chunk may be sent to probe the window, which is similar to the Zero Window Probing mechanism in TCP (see [10]). But before zero is reached the silly window syndrome (SWS) avoidance algorithm (see [11]) has to be applied. FreeBSD achieves this by announcing a window of 1, if its size has dropped below two MTUs. During that time data is still accepted, but the sender is warned to reduce the amount of data.

### 4.2 SCTP Flow Control in Implementations

As flow control is a major feature of SCTP, it is supported in all available implementations. The advertised receiver window corresponds to the important resource receiver window, but the sizes are not necessarily the same. Upon arrival of a packet, the kernel has to provide memory for the storage of each chunk. Besides the actual user message, information has to be stored, like the stream sequence number, the TSN and so on. The amount of memory needed depends on the operating system.

We examined the change of the advertised receiver window for the Linux Kernel 2.6.25, OpenSolaris 10 and FreeBSD version 7.0. We distinguished two scenarios to see whether the implementations behaved in a different way, when gaps were reported or not. First, the application at the receiver was prevented from reading. Second, the first Transmission Sequence Number (TSN) was left out. Thus a gap was created preventing SCTP from pushing data to the upper layer.

The experiment was performed by using an SCTP testtool [12], to generate SCTP packets. Test scripts were programmed with the Guile scheme implementation [13] to create the desired message flow on the sending side.

FreeBSD behaved differently in the two scenarios. In the first one, the arwnd was reduced by the payload size plus an overhead of 256 bytes, which is equal to the memory that the kernel allocates for a chunk. In the second case the arwnd was only decremented by the payload size. For small message sizes a limit of the maximum number of chunks that were accepted was observed. When this limit of 3200 chunks was reached, the arwnd was not reduced any more, and newly arrived packets were dropped. This limit is a means for the kernel to protect resources. It can be configured by the network administrator, if necessary. Hence, the number of chunks accepted can be computed by

$$n = \max(3200, \left\lceil \frac{arwnd}{256 + UMS} \right\rceil) \tag{9}$$

Linux showed the same behavior in both scenarios. The arwnd is always reduced by the user message size (UMS). For messages smaller than 176 bytes, only

$$n = \left\lceil \frac{arwnd * 2}{176 + UMS} \right\rceil \tag{10}$$

chunks are accepted. Then the arwnd is not reduced any more.

OpenSolaris decrements the arwnd by the UMS until the next message does not fit any more. Thus the window is reduced to a value smaller than UMS and the number of accepted messages equals

$$n = \left\lfloor \frac{arwnd}{UMS} \right\rfloor \tag{11}$$

Neither in the Linux nor in the OpenSolaris SCTP kernel the silly window syndrome avoidance principle is implemented.

## 4.3 Simulation Results

Keeping in mind that all implementations need extra memory to store the received user data, and that the arwnd is coupled with the receiver window, we wanted to examine the effects of the different implementation dependent algorithms and their impact on interoperability. Therefore, we extended the simulation by a parameter for the additional memory needed per chunk. As seen in Linux and partly in FreeBSD, the receiver reduces its receiver window by the UMS plus the additional memory, but announces an arwnd, that is only decremented by the UMS. The sender, not knowing that the arwnd does not report the true value, tries to keep track of its peer's window and adjusts the value every time a SACK-chunk arrives.

To simulate this behavior and examine its impact on the network load, we configured a slow receiver by distributing the reading intervals exponentially with a mean of $\frac{5000\text{bytes}}{UMS}$ secs. After each interval, one message was read, so that

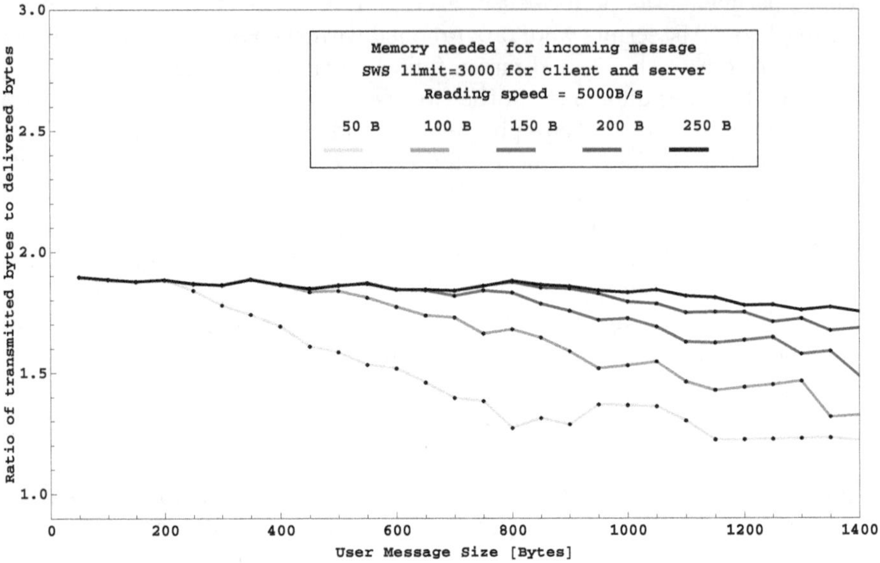


**Fig. 5.** Ratio of transmitted to delivered bytes for a varying amount of additional memory

approximately 5000 bytes were read per second independent from the UMS. Figure 5 shows the results for 50 to 250 bytes overhead. Here and in the next simulations, each run was repeated 10 times. The black dots represent the 95% confidence intervals. As a rate of the network load we calculated the ratio of the transmitted bytes, meaning the data sent over the network including all retransmissions, to the data that reached the upper layer. For an overhead of 250 bytes, which is even less than the memory needed by FreeBSD, almost twice the necessary data is transmitted. The other graphs show that the number of retransmissions is less for larger message sizes. Nevertheless, the ideal ratio of 1 is never reached. Noteworthy is also the slight inclination of the lowest graph for larger message sizes. This can be explained as follows. When an arwnd of 0 is announced, the sender is allowed to send zero window probes in the absence of outstanding data. Zero window probes consist of one `DATA`-chunk. The method that was chosen to simulate a slow receiver implies that the reading intervals are much smaller for small message sizes than for bigger ones. Thus the probability that data has been pushed and a new chunk can be accepted is higher for smaller messages. As a consequence the larger messages are more likely to be dropped.

Another difference between the operating systems is the implementation of the silly window syndrome avoidance algorithm. Of the three operating systems only FreeBSD has integrated this feature. In the following, we will examine the impact of its availability.

We varied the presence of SWS avoidance for sender and receiver and plotted again the ratio of transmitted to delivered bytes. We chose an additional memory of 50 bytes, because the measurements with the more realistic memory size of 250 bytes revealed a worse ratio, that made a graphical judgment almost impossible. Figure 6 shows the results for this scenario. As the measurements of Figure 5 were taken with SWS enabled for sender and receiver, the lowest graph of Figure 6 is equal to the 50 bytes graph of Figure 5. It is well to be seen, that the absence of the SWS avoidance algorithm on the sending side has a negative impact on the network load, whereas the implementation on the receiving side is not as important. The network load can be reduced by up to 20% , if the sender follows the principals of SWS avoidance.

For smaller chunks bundling results in a better payload to header ratio. The Nagle algorithm (see [14]) is a feature, first introduced in TCP, to prevent the sending of small packets, if there are still data in flight. For SCTP this means, that chunks have to be bundled, until the next chunk does not fit in the packet any more, unless there are no bytes outstanding. Applying this algorithm leads to delaying the sending of data. To examine the impact of the Nagle algorithm on the network load, we carried out the same runs as in Figure 6 and disabled the execution of the Nagle algorithm. The results showed that the number of retransmissions stayed at 80% to 90%, if the SWS avoidance algorithm was not applied on the client.

### 4.4 Solutions

Our first idea was to notify the sender about the amount of additional memory needed. Thus the sender was to be able to predict the reduction of the arwnd

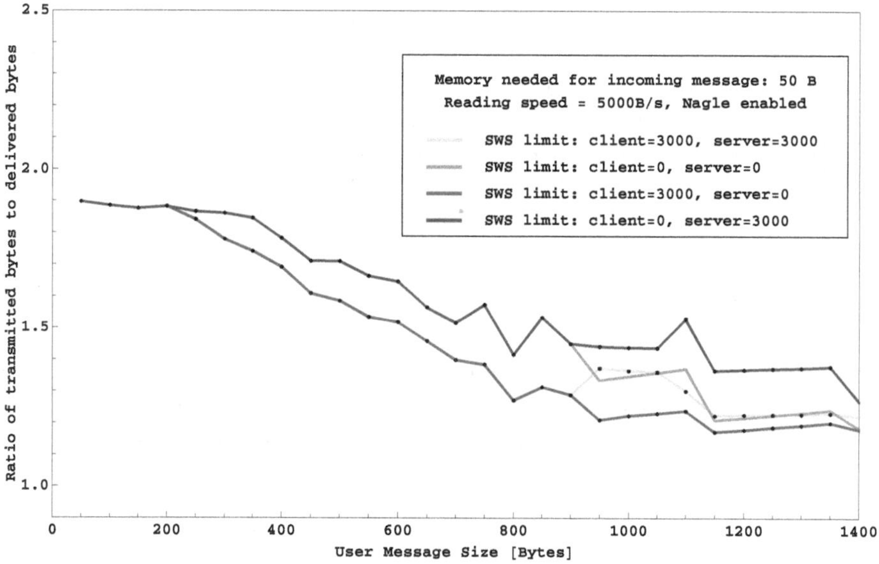


**Fig. 6.** Ratio of transmitted to delivered bytes in the absence or presence of the SWS avoidance algorithm

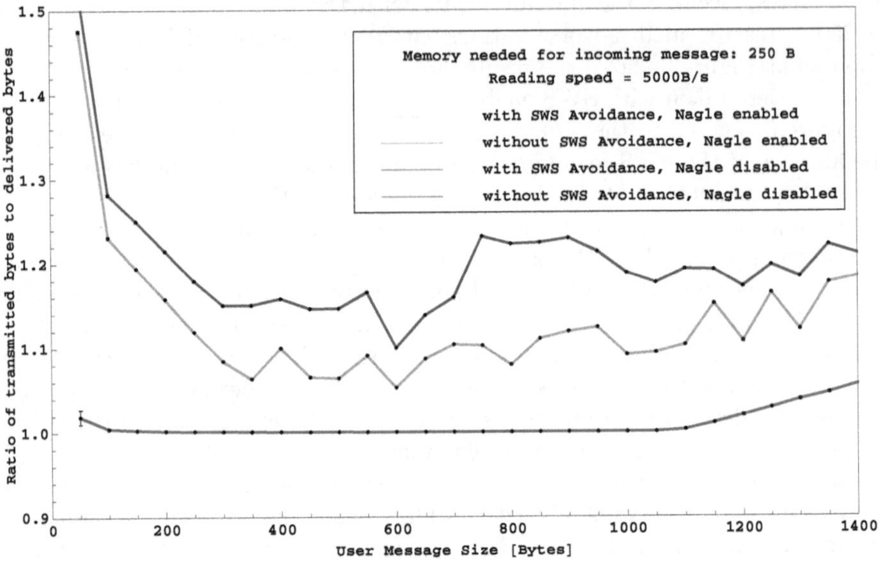


**Fig. 7.** Ratio of transmitted to delivered bytes, if the size of the real receiver window is announced

more exactly. However, simulation runs with this feature did not lead to significantly better results.

The best outcomes were achieved by "telling the truth". Like in FreeBSD, when the receiver did not read, the arwnd was reduced by the payload and the additional memory. Even if the sender cannot follow the peer window closely, the regular updates are enough to guide the sender.

Figure 7 shows the transmitted to delivered bytes ratio for the cases that SWS avoidance is present and Nagle enabled. It is well to be seen, that the application of the Nagle algorithm only has an impact on the number of retransmissions, if SWS avoidance is not present. In the other case, both graphs are the same. Thus, if SWS avoidance is applied, there are almost no retransmissions needed. The reason for the increase of the graph for larger user message sizes has been explained in the last subsection.

As a consequence, the strategy for implementors to avoid retransmissions in case of flow control is to set the advertised receiver window to the real value, so that it reflects the receiver window. Thus it is not possible that the receiver window runs out of memory before the arwnd reaches zero.

## 5 Conclusion

In this paper we discussed the influence of the message orientation on the implementation of congestion and flow control, using the example of SCTP.

We pointed out that the way the outstanding bytes are counted, has an impact on the fairness towards TCP connections. Therefore we recommended that the outstanding bytes should be calculated by taking the data message headers into account.

Looking at the different implementations and their way to apply flow control, we encountered problems, if the receiver window was exhausted faster than the advertised receiver window. The absence of the SWS avoidance algorithm still worsened the transmitted to delivered bytes ratio. Simulation results revealed that a solution to this problem is the announcement of the value of the actual receiver window in the advertised receiver window parameter. The application of the SWS avoidance algorithm even led to runs without retransmissions.

## References


1. Stewart, R.: Stream control transmission protocol. RFC 4960 (September 2007)
2. Varga, A., Hornig, R.: An Overview of the OMNeT++ Simulation Environment. In: International Conference on Simulation Tools and Techniques for Communications, Networks and Systems (SimuTools) (March 2008)
3. Rüngeler, I., Tüxen, M., Rathgeb, E.: Integration of SCTP in the OMNeT++ simulation environment. In: Proc. of the 1st international conference on Simulation tools and techniques for communications, networks and systems workshops (2008)
4. Varga, A., Hornig, R.: INET Framework Documentation (2009), `http://github.com/inet-framework/inet`
5. Ong, L., Yoakum, J.: An Introduction to the Stream Control Transmission Protocol (SCTP). RFC 3286 (May 2002)
6. Allman, M., Floyd, S., Partridge, C.: Increasing tcp's initial window. RFC 3390 (October 2002)
7. Braden, B., et al.: Recommendations on Queue Management and Congestion Avoidance in the Internet. RFC 2309 (April 1998)
8. Allman, M.: TCP Congestion Control with Appropriate Byte Counting (ABC). RFC 3465 (February 2003)
9. Allman, M., Paxson, V., Stevens, W.: TCP Congestion Control. RFC 2581 (April 1999)
10. Postel, J.: Transmission Control Protocol. RFC 793 (September 1981)
11. Clark, D.: Window and Acknowledgement Strategy. RFC 813 (July 1982)
12. Tüxen, M.: SCTP Testtool, `http://sctp.fh-muenster.de/sctp-testtool.html`
13. Free Software Foundation: Guile, `http://www.gnu.org/software/guile/guile.html`
14. Nagle, J.: Congestion Control in IP/TCP Internetworks. RFC 896 (January 1984)

# Compound TCP with Random Losses

Alberto Blanc[1], Konstantin Avrachenkov[2], Denis Collange[1], and Giovanni Neglia[2]

[1] Orange Labs, 905 rue Albert Einstein, 06921 Sophia Antipolis, France
{alberto.blanc,denis.collange}@orange-ftgroup.com
[2] I.N.R.I.A. 2004 route des lucioles, 06902 Sophia Antipolis, France
{k.avrachenkov,giovanni.neglia}@sophia.inria.fr

**Abstract.** We analyze the performance of a single, long-lived, Compound TCP (CTCP) connection in the presence of random packet losses. CTCP is a new version of TCP implemented in Microsoft Windows to improve the performance on networks with large bandwidth delay-products. We derive a Markovian model for the CTCP sending window and compute the steady state distribution of the window and the average throughput of a CTCP connection. We observe that the previous approximation, using a "typical cycle," underestimates the average window and its variance while the Markovian model gives more accurate results. We use our model to compare CTCP and TCP Reno. We notice that CTCP gives always a throughput equal or greater than Reno, while relative performance in terms of jitter depends on the specific network scenario: CTCP generates more jitter for moderate-high drop rate values, while the opposite is true for low drop rate values.

**Keywords:** TCP, Compound TCP, Bernoulli Losses, Markov Model.

## 1 Introduction

With the increasing popularity of faster access links like Fiber To The Home [1], the current standard TCP is not always ideal. As indicated by Floyd [2] the current standard is not able to reach high rates in realistic environments, i.e. with typical packet loss rates. Many new transport protocols have been proposed and are currently being studied to replace it. Some of them are already implemented in the latest versions of some operating systems, like Compound TCP (CTCP) on Windows, and Cubic (and others) on Linux. For a survey and comparative analysis of several high speed TCP versions see, for example, [3,4,5]. For the next few years, the new high speed TCP versions will play an increasing role in resource sharing among flows in the Internet. Yet the behavior, the performance, and the impact on the network of these protocols are not well-known. In particular, there is no comprehensive analytical study of CTCP.

CTCP has been presented by Microsoft Research in [6] and [7] in 2006. It is currently submitted as a draft to the IETF Network Working group with minor differences [8]. CTCP is enabled by default in computers running Windows Server 2008 and disabled by default in computers running Windows Vista [9]. It

L. Fratta et al. (Eds.): NETWORKING 2009, LNCS 5550, pp. 482–494, 2009.

is also possible to add support for CTCP to Windows XP. An implementation of CTCP, based on [7,8], is also available for Linux [10]. The main objective of the authors of CTCP [7] is to specify a transport protocol which is efficient, using all the available bandwidth, fair and conservative, limiting its impact on the network. They propose to combine the fairness of a delay-based approach with the aggressiveness of a loss-based approach. As the proposal of CTCP is still recent, there are only a few published evaluations of it. The only analytical model of CTCP in [7] is based on a de facto deterministic model.

In the present work we study the performance of CTCP under random losses. This model has been widely used in the literature (e.g. [11,12,13]) to analyze the influence of random traffic fluctuations on the performance of TCP Reno. While it is not necessarily the most sophisticated model, it does capture the behavior of TCP Reno in several real cases (see, for example, [11,13]). Whether the same holds true for more recent versions of TCP is an open question, to the best of our knowledge. The present work is just a first attempt at studying the influence of random losses on long-lived CTCP connections. These losses can be caused by a large number of connections sharing the same bottleneck link or by transmission errors in wireless networks. As new physical layer rates keep increasing thanks to new technologies like WiMax, wireless networks can have larger bandwidth delay products, reaching values for which the behavior of new versions of TCP differs from Reno. (In the case of CTCP if the window is less than 41 packets CTCP behaves like Reno.)

The outline of the paper and of our results is as follows. In Section 2 we give a brief overview CTCP. In Section 3 we present a two-dimensional Markov chain model for CTCP. With the help of this model we obtain the long-run average throughput of CTCP and the distribution of its congestion window at congestion events. Then, in Section 3.2 we use Palm calculus to obtain the distribution of the congestion window at arbitrary time moments. In Section 3.3 we propose some heuristics and compare them with the accurate two-dimension Markov chain model. Finally, in Section 4 we provide numerical and simulation results which confirm our theoretical findings. In particular, we conclude that CTCP provides a higher throughput than TCP Reno and , even if more aggressive, it causes less traffic jitter than TCP Reno on high speed links with small random losses.

Due to space constraints, some technical details are provided in a companion technical report [14].

## 2 Compound TCP Overview

In this section we give a very brief overview of CTCP (see [7] for a complete description). The main idea of CTCP is to quickly increase the window as long as the network path is not fully utilized, then to keep it constant for a certain period of time and finally to increase it by one MSS (Maximum Segment Size) per round trip time just like TCP Reno. We will often use the term "phases" for these three different behaviors.

During phase 1 the sender computes the sending window ($w$) at the $(i+1)$-th round trip as $w_{i+1} = w_i + \alpha w_i^k$ (as suggested in [7] we use $\alpha = 1/8$ and $k = 3/4$). At each round trip the sender estimates the bandwidth-delay product and the amount of data backlogged in the network using the same method adopted by TCP Vegas [15]. If the amount of backlogged data is greater than a certain threshold ($\gamma$, usually set to 30 [7,16]) the sender switches to phase 2 and keeps the window constant. This constant value corresponds to the sum of the estimated bandwidth-delay product and the estimated amount of backlogged data. We consider an ideal behavior of CTCP, assuming that such estimates are correct. In such case the window in phase 2 is equal to $\theta \triangleq \mu\tilde{\tau} + \gamma$, where $\mu$ is the capacity of the bottleneck link and $\tilde{\tau}$ is the round trip propagation delay. In reality any queue size estimate available at the sender is outdated due to feedback delays, this fact combined with the CTCP algorithm presented in [7] causes the window to oscillate during this phase as we analyzed in [17]. The length of phase 2 is dictated by the "congestion component" of the window. In fact in CTCP the congestion window $w$ is the sum of two components: the delay window $w_d$ and the congestion window $w_c$. The congestion component is incremented by one every round trip (just like the TCP Reno congestion window) and when this component reaches $\theta$ phase 2 ends. The delay component is set such that the value of the total window ($w_i = w_{c_i} + w_{d_i}$) at the $i$-th round trip time follows the following evolution in absence of packet losses:

$$w_i = \begin{cases} w_{i-1} + \delta_i & \text{, if } w_{i-1} + \delta_i < \theta \\ \theta & \text{, if } w_{i-1} + \delta_i \geq \theta \text{ and } w_{c_0} + i < \theta \\ w_{i-1} + 1 & \text{, otherwise} \end{cases} \tag{1}$$

where $\delta_i = \max\left\{\left\lfloor \alpha w_{i-1}^k \right\rfloor, 1\right\}$. Figure 1 shows the three phases of the window evolution. When a loss occurs both the total window and the congestion window are halved.

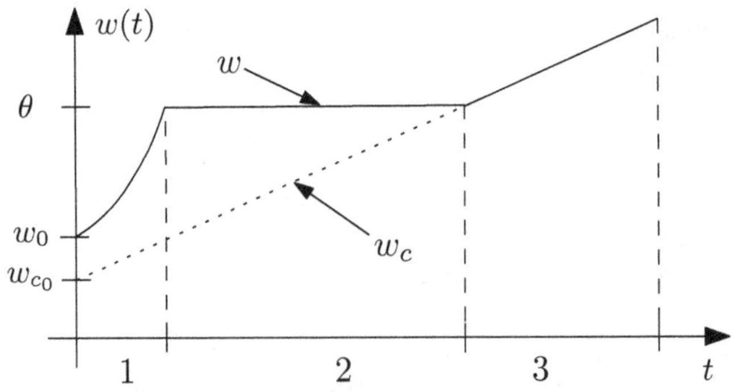


**Fig. 1.** The evolution of $w$ and $w_c$ in CTCP

## 3 Performances with Random Losses

We consider a single long lived TCP compound flow using a path with $\mu\tilde{\tau}$ bandwidth delay product and buffer size equal to $b$. The flow will experience a loss every time that its window size reaches the value $\mu\tilde{\tau} + b$. For this reason, we can consider $w_{max} = \mu\tilde{\tau} + b$ as an upper bound for the the window size. Beside the deterministic losses due to buffer overflow, we consider also that each packet can be dropped with some probability $p$, independently from all other packets, i.e. according to a Bernoulli process. In what follows we derive the throughput and the window distribution in steady state. We are going to assume that $w$ can only take integer values.

### 3.1 Throughput Calculation

We define a cycle as the time interval between two consecutive losses. We denote as $w_{c_n}^t$ (respectively $w_{d_n}^t$) the congestion (respectively delay) window at the begin of the $(n+1)$-th round trip time of the $t$-th cycle. We will omit the superscript $t$ whenever it is clear which cycle is being considered.

We observe that in our framework the evolution of the window in each cycle $t$ depends from previous cycles only through the window value at the begin of the cycle, or, more precisely, through the two initial values $w_{c_0}^t$ and $w_{d_0}^t$, which can be determined by the final value of the windows at the $t-1$-th cycle. This also implies that it is possible to use the renewal reward theorem to compute the average throughput as (see [13]):

$$E[\lambda] = \frac{E[S]}{E[T]} \tag{2}$$

where $\lambda$ is the throughput (in MSS/s), $S$ is the total number of packets sent during a cycle and $T$ is the duration of the cycle.

Both $E[T]$ and $E[S]$ can be evaluated starting from the knowledge of the distribution size of the two (correlated) random variables $W_{c_0}$ and $W_{d_0}$. We denote $g(w_{c_0}, w_{d_0})$, the probability mass function (pmf) of these random variables. We first show how to derive the distribution of cycle duration from $g(w_{c_0}, w_{d_0})$ and then derive the pmf $g()$ itself. $E[S]$ can be evaluated similarly to $E[T]$.

A cycle has length equal to $n$ if there is a loss at the $n$-th round trip time. As we are assuming that there is a loss whenever $w = w_{\max}$, all cycles have a finite length. Let $m(w_0) = \min\{n | w_n \geq w_{\max}\}$ be the maximum possible length (in round trips) of a cycle starting at $w_0$. For $n < m(w_0)$ the probability of a cycle having length equal to $n$ can be derived from the Bernoulli loss process as:

$$\begin{aligned} P[T = n | W_{c_0} = w_{c_0}, W_{d_0} = w_{d_0}] &= (1-p)^{V_{n-1}(w_0)} - (1-p)^{V_n(w_0)} \\ &\triangleq a_n(w_{c_0}, w_{d_0}), \end{aligned} \tag{3}$$

where

$$\mathrm{V}_n(w_0) \triangleq \sum_{i=0}^{n-1} w_i,\ V_0 \triangleq 0,$$

and $w_i$ is computed as in (1) so that $V_n$ is the number of packets sent during the $n$-th round trip of a cycle starting with $w = w_0$. Both $V_n$ and $a_n$, as most quantities used in this section, depend on the initial window $w_0 = w_{c_0} + w_{d_0}$ as highlighted by the notation $a_n(w_0)$ and $V_n(w_0)$, even though we will also use the simplified notation $a_n$ and $V_n$.

The probability that a loss occurs at the $m(w_0)$-th round trip can be evaluated simply considering that $\sum_{i=1}^{m} P[T = i] = 1$:

$$P[T = m_(w_0)|W_{c_0} = w_{c_0}, W_{d_0} = w_{d_0}] = 1 - \sum_{i=1}^{m_(w_0)-1} a_i(w_{c_0}, w_{d_0}).$$

Finally, being that the support of the discrete random variables $W_{c_0}$ and $W_{d_0}$ is finite, we can use a finite sum to compute $P[T]$:

$$P[T = n] = \sum_{w_{c_0}, w_{d_0}} a_n(w_{c_0}, w_{d_0}) g(w_{c_0}, w_{d_0}). \tag{4}$$

In order to compute $g(w_{c_0}, w_{d_0})$ we model the evolution of the window (at the beginning of each cycle) with a Markov chain. The evolution of the window of TCP Reno at the begin of a cycle ($w_0^t$) has been modeled in other works as a Markov chain (see, for example, [12,13]). In fact $w_0^t$ is equal to half of the window value at the end of the $(t-1)$-th cycle, which depends only on $w_0^{t-1}$ and on packet loss probability $p$.

In order to model CTCP we use a two-dimensional discrete Markov chain $X_t$ to account for $w_{c_0}^t$ and $w_{d_0}^t$. For each state $(i, j)$ the first index represents $w_{c_0}^t$ and the second $w_{d_0}^t$. For any pair of states it is possible to compute the transition probability as:

$$P[X_{t+1} = (k, l)|X_t = (i, j)] = \sum_{n \in B} a_n(i, j)$$

with $w_{c_0}^t = i$, $w_{d_0}^t = j$, $B = \{n| \lfloor w_{c_n}^t/2 \rfloor = k, \lfloor w_{d_n}^t/2 \rfloor = l\}$, where $w_{c_n}^t$ and $w_{d_n}^t$ are evaluated according to (1). The sum on the right hand side is needed because different pairs $(w_{c_n}^t, w_{d_n}^t)$ can originate, after a loss, the same pair $(w_{d_0}^{t+1}, w_{d_0}^{t+1})$ as we use integer values for the window. As $w \leq w_{\max}$ we have that $w_{c_0} \leq \lfloor w_{\max}/2 \rfloor \triangleq N$ and, if $\theta$ is the value of the window during the constant window phase, $w_{d_0} \leq \lfloor \theta/2 \rfloor \triangleq M$ (as $w_d \leq \theta$). Combining these two bounds we obtain that the number of states in the Markov chain is $NM$. Using the ARPACK implementation of the Arnoldi method [18] it is possible to efficiently calculate the steady state distribution of the Markov chain $X_t$ even for large values of $NM$. The more time consuming step is actually to compute the transition matrix for $X$. The complexity of the algorithm we used is $O(MN^2)$; we believe that it is not possible to decrease the complexity of the algorithm given that it has to compute all the possible transitions and these grow like $MN^2$. Note that the number of possible transitions for a Markov chain with $NM$ states is $N^2M^2$ therefore we already take into account that, in this case, some transitions are not possible.

Once the steady state distribution of the Markov chain $X_t$ $(g(w_{c_0}, w_{d_0}))$ is derived, we can use it to compute $E[T]$, $E[S]$ and then the average throughput using (2).

We observe that so far we have implicitly measured $T$ in terms of number of round trip times, but we can slightly change our model in order to consider real time duration (seconds) taking into account also queuing delay variation due to the TCP flow itself [14].

### 3.2 Steady State Distribution of the Window

In the previous section we have described how to compute the steady state distribution of $w_{c_0}$ and $w_{d_0}$, and consequently also the value of the window $w_0 = w_{c_0} + w_d$ *at the beginning of each cycle.* In this section we are interested in the steady state distribution of the window as a function of time. We denote as $Y_n$ the value of the window at the begin of the $(n-1)$-th round trip time. Note that $Y_n$ is different both from $X_t$, which is the value of the window after a packet loss, and from $w_n^t$, which is the value of the window at the begin of the $(n-1)$-th round trip time in the $t$-th cycle. Clearly $X_t$ represents a subsequence of the sequence $Y_n$. We observe that $Y_n$ can also be modeled as a discrete time Markov chain where a transition occurs every round trip. Also this Markov chain is ergodic, hence it admits a steady state and we assume that it is in steady state at time 0.

Using Palm calculus, we first compute $P[Y_n = k]$ starting from $P[W_0 = w_0]$, where $Y_n$ represents the window after $n$ round trip times starting from some arbitrary value (given that all the Markov chains involved are ergodic, the initial value is irrelevant).

Let $Z_n$ be the (discrete) time of the $n$-th packet drop after time 0. Using the intensity and inversion formulas of Palm calculus [19] we can compute $P[Y_n = k]$ as a function of $P[W_0 = w_0]$:

$$P[Y_n = k] = E\left[1_{\{Y_n=k\}}\right] = P[Z_0 = 0]E^0\left[\sum_{s=1}^{Z_1} 1_{\{Y_s=k\}}\right] \tag{5}$$

$$= \eta E^0\left[\sum_{s=1}^{Z_1} 1_{\{Y_s=k\}}\right] \tag{6}$$

$$= \eta \sum_{w_0,l}\left[P[W_0 = w_0]P[Z_1 = l|W_0 = w_0]\sum_{s=1}^{l} 1_{\{Y_s=k|W_0=w_0\}}\right] \tag{7}$$

where $E^0$ is the Palm expectation, $1_{\{Y_n=k\}}$ is the indicator function for the event $Y_n = k$ and $\eta$ is the intensity of the process $Z_n$. The second (5) and third (6) equalities follow from the inversion and intensity formulas, respectively, while (7) follows from the total probability theorem, conditioning on all the possible values of $W_0$ and $l$. Given that $Z_n$ is an ergodic process $P[Z_0 = 0]$ can be computed, using the intensity formula, as the inverse of the expected value of $T \overset{\mathrm{d}}{=} Z_n - Z_{n-1}$ that is as the average length of a cycle (in round trips) so that $\eta = 1/E[T]$ where $E[T]$ can be computed using (4).

As for the calculations in section 3.1 we can evaluate the distribution of the window taking into account the effect of queueing delays as well [14].

### 3.3 A Simple Approximation and the Deterministic Response Function

The method described in the previous section provides an exact solution, but it can be computationally expensive for medium and large values of $w_{\max}$ and $\theta$. Using the same method as in [12] it is possible to quickly find an approximate solution for the average window size. The idea is to consider a sequence of "typical" or "average" cycle. If $p$ is the probability that a packet is dropped, the average cycle has exactly $1/p$ packets (provided the probability of reaching $w_{\max}$ is negligible). Let us denote $w_0$ the initial window of an average cycle and $w_n(w_0)$ the final window, after $n$ round trip times during which $1/p$ packets are transmitted. Being that the following average cycle has to be identical and that CTCP, as Reno, halves the window at the end of each cycle, it has to be:

$$w_n(w_0) = 2w_0. \tag{8}$$

Imposing the constraint $V_n(w_0) = 1/p$ ($V_n(w_0)$ is defined in section 3.1), we can identify the unique possible value of $w_0$ and then the unique possible window evolution corresponding to $p$. Once the window evolution is known, the average window can be obtained and can be plotted as a function of the drop rate.

As previously noted in [11] this approach corresponds to the calculation of what is usually called the "deterministic response function." This function is often used in the literature on TCP. For example, [2] defines the response function as "the function mapping the steady-state packet drop rate to TCP's average sending rate in packets per round-trip time." Where the drop rate is simply the inverse of the number of packets sent during each cycle. In this case the term "drop rate" always refers to this quantity and not to any probabilistic model, while this usage is not the most appropriate one it is, nonetheless, common in the literature. From another point of view, we observe that the average cycle corresponds to the actual evolution of the window under our loss model, when there is no random loss, but the bandwidth delay product and the buffer size are such to cause a deterministic loss every $1/p$ packets.

When the window update rule operates on a round trip time basis - as in CTCP but also in Reno and HighSpeed for example - the deterministic response function does not depend on physical parameters like capacity or propagation delay. In other words it can be considered as an intrinsic property of the specific growth function used to increment the window. On the contrary, for TCP Cubic the growth of the window depends on real time and hence also on link capacities and propagation delays, so that a comparison with the TCP versions indicated above would need special care.

Figure 2 shows the response function for different values of $\theta$, between 100 MSS and 1000 MSS (Maximum Segment Size). The two lines correspond to two different ways to express the window evolution in (8). The dotted line corresponds to a fluid model where the growth of the window is approximated with a continuous function (see [20]). More precisely

$$w_n(w_0) = \left((1-k)\alpha n + w_0^{*1-k}\right)^{\frac{1}{1-k}}.$$

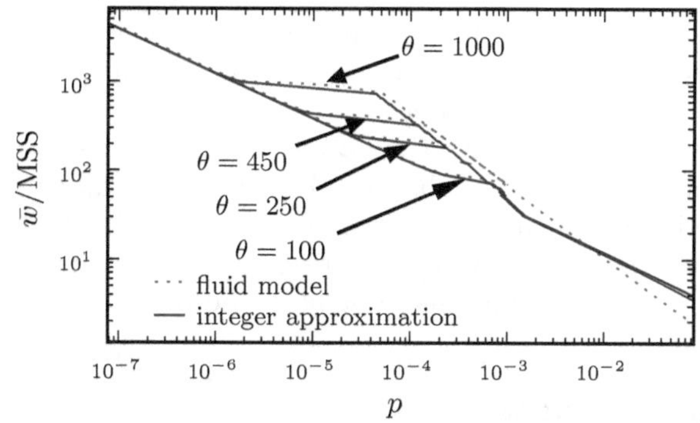


**Fig. 2.** The deterministic response function for $\theta = 100, 250, 450, 1000$

The solid line, instead, considers only integer values of the window and computes the increment of the window as according to (1) with $w_{c_0} = w_0$. In this case the last round trip time of the cycle -the $n$-th one- is evaluated as $n = \min\{i : w_i \geq 2w_0\}$. In both cases we have considered $w_0$ as the independent value. For each value of $w_0$ we have then computed $S$, the total amount of packets sent during a cycle starting with $w = w_0$ and then we have plotted the average window as a function of $1/S = p$. Clearly the total number of packets sent during such a cycle is a monotonically increasing function of $w_0$ (as the TCP window is monotonically increasing during each cycle) so that increasing values of $w_0$ correspond to increasing values of $S$, and decreasing values of $p = 1/S$.

Regarding the integer approximation we observe that, for $\alpha = 1/8$ and $k = 3/4$ (values suggested in [7]), $\alpha w^k \geq 2$ only if $w > 30$. That is the CTCP window grows by one each round trip, the same as Reno, as long as $w < 30$. This is somewhat consistent with what suggested in [7] where the authors call for the delay component to be used (that is to increment the window by $\alpha w^k$) only if the window is larger than *lowwnd* which they set equal to 41. This observation explains why for small values of $w$ the fluid and integer approximation are different, with the integer approximation giving a larger average value of the window.

In the case of the integer approximation and for large drop rates (more than $10^{-3}$) the response function is the same as Reno. The same is true, regardless of the approximation, for small drop rates (less than $10^{-6}$) given that, in this latter case, the evolution of the window is the same in CTCP and Reno. For drop rates roughly between $4 \cdot 10^{-4}$ and $10^{-3}$ only the rapid increase phase of CTCP is used so that the response function is steeper. For drop rates between $10^{-6}$ and $4 \cdot 10^{-4}$ the "constant" window phase is used along with all the other phases and this explains why the response function increases more slowly until it reaches the Reno response function.

While for Reno and HighSpeed TCP the response function is only a function of the drop rate, for CTCP the response function is also a function of $\theta$ (the value of the window during the constant window phase). The larger the value of $\theta$ the longer the rapid increase phase can be. In the extreme case of $\theta = \infty$ the other phases would not take place at all and the response function would be much steeper. Note that in [7] the authors compute the response function for exactly this case ($\theta = \infty$). Given that they are interested in limiting the aggressiveness of CTCP, they are in effect considering the worst case, by only considering the rapid increase phase. At the same time it can be argued that, as a consequence, CTCP is *less* aggressive than HighSpeed TCP as, for small values of $p$, CTCP is as aggressive Reno, which is less aggressive than HighSpeed TCP.

Finally it is worth noting that, as $p$ decreases, the difference between the fluid and integer models for the rapid increase phase becomes negligible.

## 4 Numerical Results

Using the models presented in the previous sections we can compute the average throughput and the average window size for different values of the drop probability $p$. Figure 3 shows the CTCP response function computed using the two deterministic models presented in section 3.3 and the probabilistic model discussed in section 3.1, for $\theta = 250\,\text{MSS}$. For the same "drop probability" the deterministic model with integer approximations gives a smaller average window than the probabilistic model, which uses the very same integer approximations, as already observed in [11]. The fluid model, instead, does agree with the probabilistic model, for larger values of $p$. As discussed in the previous section, the fluid model overestimates the window growth for certain values of $p$ (roughly between $10^-4$ and $10^-2$). We cannot explain why this overestimation is almost in agreement with the probabilistic model and we believe that it is just a coincidence.

The probabilistic and deterministic models do have almost the same values for values of $p$ around $2 \cdot 10^-4$. This can be explained by the fact that, for these values of $p$, most of the drops take place during the "constant window" phase. In this case even if the cycles have a different length the average window size is the same: random drops do change the cycle length but not the average window size.

One should take some care in comparing these two models: in the case of the deterministic model the buffer size at the bottleneck is fixed (so that all the

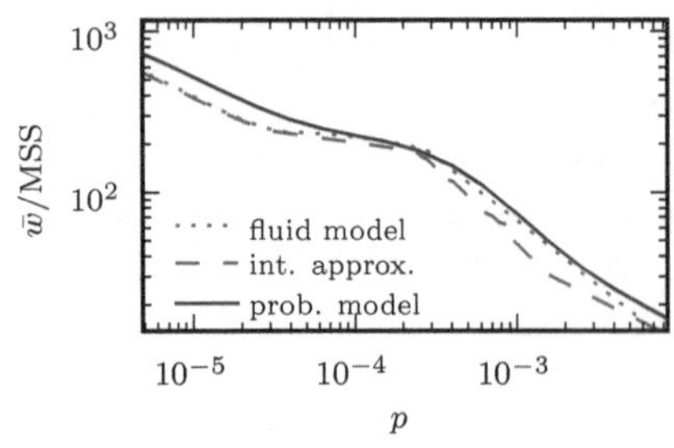


**Fig. 3.** CTCP response function

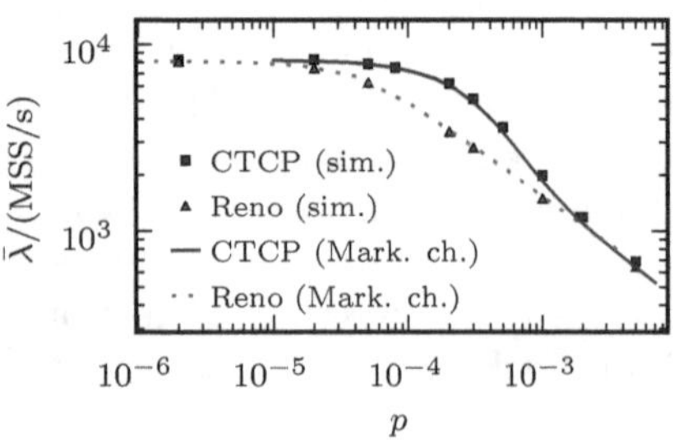


**Fig. 4.** Average Throughput

cycles have the same size) while in the case of the probabilistic model the buffer size at the bottleneck link is much larger (in theory infinite, set to 1600 MSS for the numerical results) and allows the window to reach larger values. If we used the same buffer size in both models the average window would be smaller in the probabilistic case.

Figure 4 shows the average throughput for CTCP and TCP Reno. For CTCP we have used the probabilistic model introduced in section 3.1 while for Reno we have used an equivalent model but with a "one dimensional" Markov chain as Reno does not have two components in the congestion window. The squares and triangles in Figure 4 correspond to simulation results. For both versions of TCP there is a good match between the probabilistic model and the simulations, obtained using ns-2 (version 2.33) with a Linux implementation of CTCP [10]. The simulations use the classical "dumbbell" topology with a single source and a single destination. There is a single TCP connection with no cross traffic going through a bottleneck of 100 Mb/s with propagation delay of 26.4 ms and where each packet of 1500 B is dropped with probability $p$. The duration of each simulation is 200 000 s. As the difference between different simulation runs is very small (less than 1%) we did not plot errorbars. As the simulations and the model share the same assumptions they can only be used as a sanity check. While assessing the influence of the assumptions used in the model and how realistic they are is a very interesting problem it is outside the scope of this work.

In Figure 4 $w_{\max} = 370$ MSS, the bandwidth-delay-product is 220 MSS and the buffer size is 150 MSS, while in Figure 3 $w_{\max} = 1600$ MSS. This explains why in Figure 4 the throughput is constant for small drop probabilities (most of the packets are dropped when $w = w_{\max}$) a similar behavior takes place for larger values of $w_{\max}$ but for drop probabilities smaller than those included in Figure 3.

Figures 5 and 6 show the distributions for $w_0$, $w_{c_0}$ and $w_{d_0}$ when $\theta = 430$ MSS and $w_{\max} = 600$ MSS for $p = 3 \cdot 10^{-4}$ and $p = 5 \cdot 10^{-6}$ respectively. The dotted lines represent simulation results, confirming that there is a good match with the Markov model. Figure 5 corresponds to the case where most of the packet are dropped during phase 1 when the window is growing quickly. In this case the

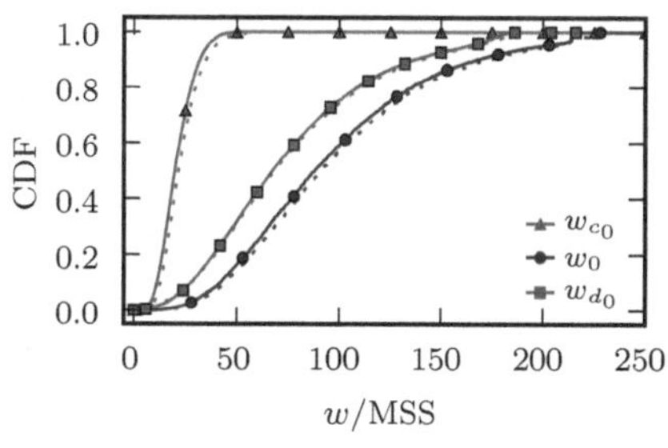


**Fig. 5.** Distribution of $w_0$,$w_{c_0}$ and $w_{d_0}$ ($\theta = 430$ MSS, $w_{\max} = 600$ MSS, $p = 3 \cdot 10^{-4}$)

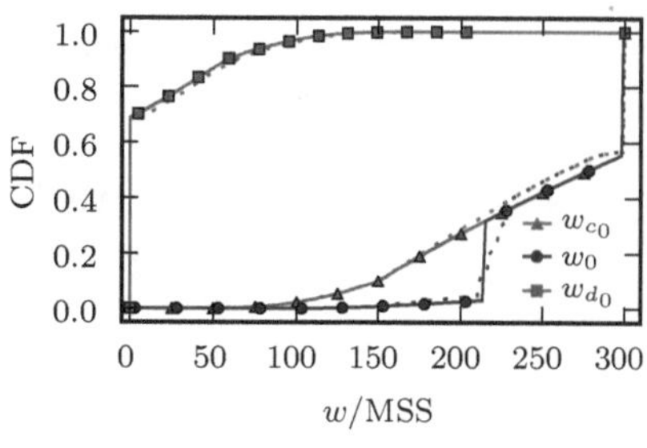


**Fig. 6.** Distribution of $w_0$,$w_{c_0}$ and $w_{d_0}$ ($\theta = 430$ MSS, $w_{\max} = 600$ MSS, $p = 5 \cdot 10^{-6}$)

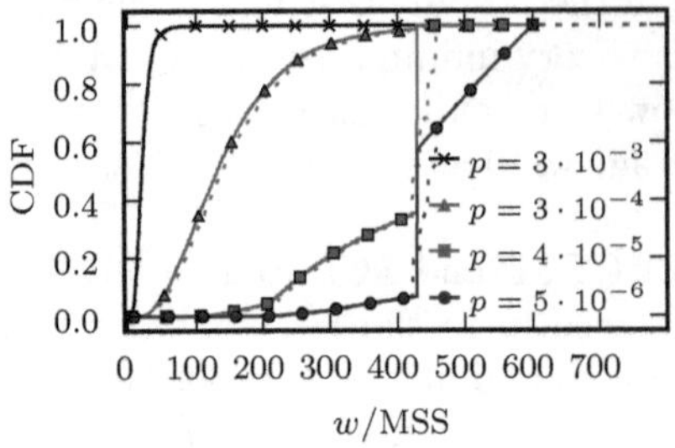


**Fig. 7.** Distribution of $Y_n$ ($\theta = 430\,\text{MSS}$, $w_{\max} = 600\,\text{MSS}$)

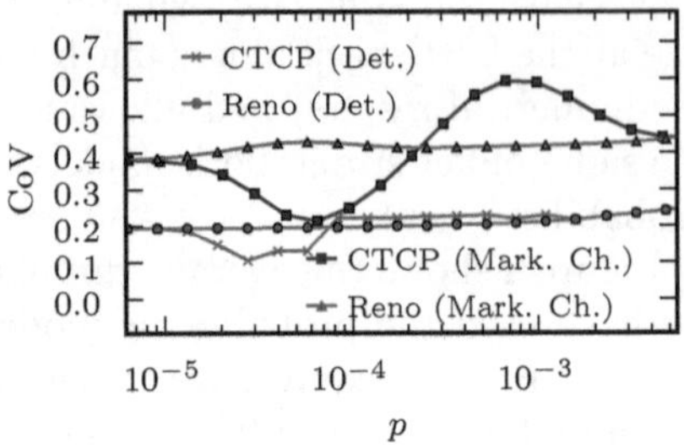


**Fig. 8.** Coefficient of variation of $Y_n$ for CTCP and Reno

distribution of $w_{c_0}$ is greater than the distribution of $w_{d_0}$, so that, on average, $w_{c_0} < w_{d_0}$. Figure 6 corresponds to the case where a significant fraction of packets is dropped during the Reno phase (almost 50%). The jump at $w_0 = 215\,\text{MSS}$ represents the packets dropped during the constant window phase (with the simulations having smaller jumps caused by the oscillations of the window). The jump at $w = 300\,\text{MSS}$ correspond to the case when packets are dropped due to a buffer overflow (recall that in this case $w_{\max} = 600\,\text{MSS}$ and for the simulations the buffer size $b$ is 200 MSS and $\mu\tilde{\tau} = 400\,\text{MSS}$). In this case the distribution of $w_{c_0}$ is smaller than the distribution of $w_{d_0}$, the opposite of what happens in the previous case.

Figure 7 shows the steady state distribution of $Y_n$ for different values of $p$ with $\theta = 430\,\text{MSS}$ and $w_{\max} = 2000\,\text{MSS}$. Again the dotted lines represent the same distribution for the corresponding ns-2 simulations. As expected, with increasing drop probabilities, each distribution is strictly greater than all the previous ones. In Figures 6 and 7 while the probabilistic models have a sharp jump for $w = 215\,\text{MSS}$ and $w = 430\,\text{MSS}$ the simulations (dotted lines) have smaller jumps. This is caused the oscillations of the sending window during phase 2 in the simulations. As mentioned in section 2 this is consistent with the CTCP algorithm but it is not taken into account by the probabilistic model. While it is, at least in principle, possible to incorporate this aspect into the model, we prefer using a simpler model with a constant value during phase 2 given that the differences between this simplified model and the simulations are not significant (especially as far as the throughput is concerned).

Figure 8 show the coefficient of variations (CoV) for CTCP ($\theta = 250\,\text{MSS}$, $w_{\max} = 2000\,\text{MSS}$) and Reno for $\mu\tilde{\tau} = 220\,\text{MSS}$. In this cases the difference between the deterministic and probabilistic model is more pronounced than in the case of the average window (response function). This can be explained by the fact that the average window depends only the first moment of $Y_n$ while the CoV depends on the second moment as well. For the CoV, in particular, the difference between the two models is significant. For small values of $p$ the CoV of CTCP is smaller than Reno but for larger values of $p$ the opposite is true indicating that for $p > 10^{-4}$ CTCP might not be the best solution.

## 5 Conclusions and Future Work

In this paper we have presented a Markovian model of CTCP under random losses. This kind of model is a first attempt to roughly assess the impact of varying network conditions on a CTCP connection. The network is seen as a black box randomly dropping packets, due to buffer overflows. This model could also be used to describe the impact of transmission errors in some "challenging environments" (e.g. wireless networks) as expected from new TCP versions [21].

In this first analysis, we have assumed that the loss arrivals follow a simple Bernoulli process. We have computed the distribution of the sending window on loss events with a Markovian model, and then the average throughput. Using Palm Calculus we have computed the steady state distribution of the window. Its value has a direct influence on the buffer occupancy and on the jitter experienced by all the flows sharing the same bottleneck link.

This analysis can be extended in many ways. The Bernoulli loss process could be replaced with a more bursty and realistic process. In this case, multiple losses could take place during the same round-trip time, and the recovery time could be longer. We could also consider time-outs in the case of high loss rates, as in [13]. A similar analytical study could also be applied to the other TCP versions, currently under standardization, comparing their efficiency and their robustness.

## References


1. FFTH: Fiber to the home council (2008), `http://www.ftthcouncil.org/`
2. Floyd, S.: HighSpeed TCP for Large Congestion Windows. RFC 3649 (2003) (Experimental)
3. Kherani, A., Prabhu, B., Avrachenkov, K., Altman, E.: Comparative study of different adaptive window protocols. Telecomm. Systems 30, 321–350 (2005)
4. Li, Y., Leith, D., Even, B.: Evaluating the performance of TCP stacks for high-speed networks. In: Proc. 4th Int. Workshop on Protocols for FAST Long-Distance Networks (2006)
5. Li, Y., Leith, D., Shorten, R.: Experimental evaluation of tcp protocols for high-speed networks. IEEE/ACM Trans. Netw. 15, 1109–1122 (2007)
6. Tan, K., Song, J., Zhang, Q., Sridharan, M.: Compound TCP: A scalable and TCP-friendly congestion control for high-speed networks. In: Proc. 4th Int. Workshop on Protocols for FAST Long-Distance Networks (2006)
7. Tan, K., Song, J., Zhang, Q., Sridharan, M.: A compound tcp approach for high-speed and long distance networks. In: INFOCOM (2006)
8. Sridharan, M., Tan, K., Bansal, D., Thaler, D.: Compound TCP: A new TCP congestion control for high-speed and long distance networks. Internet draft, Internet Engineering Task Force (2007) (work in progress)
9. Davies, J.: Performance enhancements in the next generation TCP/IP stack. The Cable Guy (2007), `http://www.microsoft.com/technet/community/columns/cableguy/cg1105.mspx`
10. Andrew, L.: Compound TCP Linux module (2008), `http://netlab.caltech.edu/lachlan/ctcp/`
11. Altman, E., Avrachenkov, K., Barakat, C.: A stochastic model of tcp/ip with stationary random losses. IEEE/ACM Trans. Netw. 13(2), 356–369 (2005)

12. Lakshman, T., Madhow, U.: The performance of tcp/ip for networks with high bandwidth-delay products and random loss. IEEE/ACM Transactions on Networking 5, 336–350 (1997)
13. Padhye, J., Firoiu, V., Towsley, D., Kurose, J.: Modeling tcp reno performance: a simple model and its empirical validation. IEEE/ACM Transactions on Networking 8, 133–145 (2000)
14. Blanc, A., Avrachenkov, K., Collange, D., Neglia, G.: Compound tcp with random losses. INRIA Research Report 6736, INRIA (2008), `http://hal.inria.fr/docs/00/34/98/45/PDF/RR-6778.pdf`
15. Brakmo, L.S., Peterson, L.L.: Tcp vegas: end to end congestion avoidance on a global internet. IEEE JSAC 13, 1465–1480 (1995)
16. Tan, K., Song, J., Sridharan, M., Ho, C.: CTCP-TUBE: Improving TCP-friendliness over low-buffered network links. In: Proc. 6th Int. Workshop on Protocols for FAST Long-Distance Networks (2008)
17. Blanc, A., Collange, D., Avrachenkov, K.: Oscillations of the sending window in Compound TCP. In: Proc. 2nd NetCoop Workshop (2008)
18. Sorensen, D., Lehoucq, R., Yang, C., Maschhoff, K.: ARPACK (2008), `http://www.caam.rice.edu/software/ARPACK/`
19. Boudec, J.Y.L.: Understanding the simulation of mobility models with palm calculus. Perform. Eval. 64, 126–147 (2007)
20. Blanc, A., Collange, D., Avrachenkov, K.: Modelling an isolated Compound TCP connection. Tech. Report 6736, INRIA (2008), `http://hal.inria.fr/docs/00/35/00/41/PDF/RR6736.pdf`
21. Floyd, S., Allman, M.: Specifying New Congestion Control Algorithms. RFC 5033 (Best Current Practice) (2007)

# Preventing the Unnecessary Propagation of BGP Withdraws*

Virginie Van den Schrieck[1], Pierre Francois[1], Cristel Pelsser[2], and Olivier Bonaventure[1]

[1] Universite catholique de Louvain (UCL), CSE Dept
Place Sainte-Barbe 2
1348 Louvain-la-Neuve, Belgium
firstname.lastname@uclouvain.be
[2] NTT Corporation, NTT Network Service Systems Laboratories
9-11, Midori-Cho 3 Chrome
Musashino-shi, Tokyo 180-8585, Japan
pelsser.cristel@lab.ntt.co.jp

**Abstract.** Due to the way BGP paths are distributed over iBGP sessions inside an Autonomous System (AS), a BGP withdraw that follows a failure may be propagated outside the AS although other routers of the AS know a valid alternate path. This causes transient losses of connectivity and contributes to the propagation of a large number of unnecessary BGP messages. In this paper, we show, based on RouteViews data, that a significant number of BGP withdraws are propagated even though alternate paths exists in another border router of the same AS. We propose an incrementally deployable solution based on BGP communities that allows the BGP routers of an AS to suspend the propagation of BGP withdraws when an alternate path is available at the borders of their AS.

**Keywords:** BGP, Internet, Churn.

## 1 Introduction

The Border Gateway Protocol (BGP) [1] plays a key role in today's Internet as it allows Internet Service Providers (ISPs) and enterprise networks to announce routes towards their IP prefixes. During the last years, network operators and researchers have been concerned by the limits to the scalability of the Internet architecture and BGP in particular [2]. An important problem that affects BGP is the BGP churn, i.e. the number of BGP messages exchanged among BGP routers.

* The research results presented herein have received support from Trilogy (http://www.trilogy-project.eu), a research project (ICT-216372) partially funded by the European Community under its Seventh Framework Programme. The views expressed here are those of the author(s) only. The European Commission is not liable for any use that may be made of the information in this document.

L. Fratta et al. (Eds.): NETWORKING 2009, LNCS 5550, pp. 495–508, 2009.

BGP is a path vector protocol with two different types of BGP messages : updates and withdraws. A BGP update is used to advertise a path towards a prefix or a change in a previously announced path towards a prefix. A BGP withdraw indicates that a previously announced prefix becomes unreachable. BGP routers exchange BGP messages over a BGP session.

An analysis of the BGP messages exchanged in the global Internet shows that their number is very high [2]. This BGP churn causes high-CPU load on smaller BGP routers. Several causes have been identified. First, some interdomain links are unstable and fail frequently [3,4,5]. Each of these failures causes the transmission of a number of BGP withdraws. Second, as BGP relies on path vectors, it suffers from the path exploration problem when a route becomes unavailable[6]. When a route fails, a new BGP convergence starts. During this convergence, routers may advertise paths that they consider valid although they are also affected by the failure. These paths will be withdrawn later causing another exchange of BGP messages. The MRAI timer [1] and the route flap damping mechanism [7] may further delay this convergence. Third, due to their routing policies and internal BGP organization [8], some BGP routers from large ASes may transiently send BGP withdraws although alternate paths are available inside the AS, because those alternate paths are not known by all the routers of the AS [9].

In this paper, we analyse Route Views data to show that an important number of BGP withdraws are probably due to an insufficient alternate paths propagation in iBGP. We propose an incrementally deployable solution that can be used in an AS to ensure that its BGP routers will not propagate a withdraw to neighboring ASes when an alternate path is already known by another BGP router inside this AS. This problem has been identified in [8] as one of the main factors that causes interdomain transient losses of connectivity.

The paper is organised as follows. First, we explain in Sect. 2 how iBGP is used in large ASes. In the next section, we evaluate the number of BGP withdraws that are transienty sent by routers of large ASes although an alternate path is known by another router of this AS. The fourth section details the impact of the internal BGP organization on the propagation of those withdraws. In the fifth section, we present our solution for preventing those unnecessary withdraws. The penultimate section is a review of the related work, and we finish by a conclusion.

## 2 iBGP Organizations

There are two types of BGP sessions : external BGP (eBGP) sessions between routers belonging to two different Autonomous Systems (AS) and internal BGP (iBGP) sessions established between routers belonging to the same AS. Over an eBGP session, a router announces one path towards each prefix according to its routing policies. Common policies are Customer-Provider and Shared-Cost [10] : Routes learned from customers are advertised to all peers while routes learned from providers and shared-cost peers are only advertised to customers. When an AS contains more than one BGP router, its BGP routers must exchange BGP routes among themselves over iBGP sessions. If the AS is small, these iBGP sessions are usually organised as a full-mesh, i.e. each BGP router has one iBGP session with each other router of the AS. Figure 1(a) shows an AS with a full-mesh of iBGP sessions. Over an iBGP session, each router only advertises the best routes that it learned over eBGP sessions. A BGP router does not advertise over an iBGP session a route that it learned over another iBGP session. The main drawbacks of using a full-mesh of iBGP sessions are that $\frac{n\times(n-1)}{2}$ iBGP sessions need to be established in an AS with $n$ BGP routers. Moreover, each BGP router receives all the best eBGP paths learned by the AS. This increases the load on all routers inside the AS as they need to process and store a large number of paths.

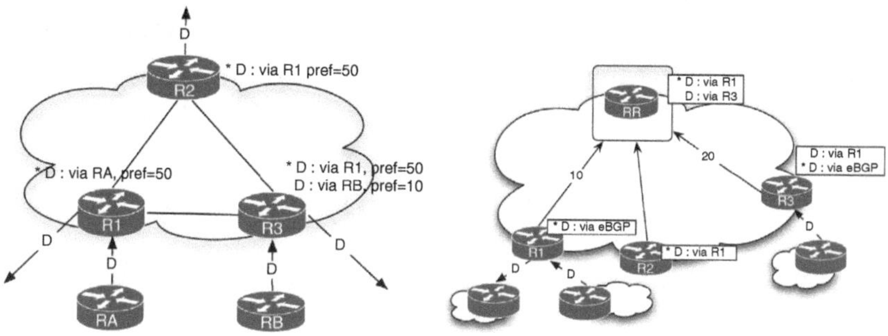


(a) Full-Mesh (b) Route Reflection

**Fig. 1.** iBGP organizations

In large ASes containing hundreds or even thousands of BGP routers, it is impossible to use a full-mesh of iBGP sessions. Such large ASes can rely on two possible iBGP organisations : confederations and route reflectors (RR). We focus on route reflection, which is the most widespread iBGP organization, but our solution is also applicable to confederations. A RR is a BGP router that is allowed to advertise over some iBGP sessions the routes that it received over other iBGP sessions. Figure 1(b) show an AS using one route reflector. In most ASes, each edge router has one iBGP session with two different RRs for redundancy. In medium-sized networks, a full-mesh of iBGP sessions is used among the RRs. In larger networks, a hierarchy of RRs is used and only the top-level RRs are interconnected by using a full-mesh of iBGP sessions [9].

We evaluated the cost of using an iBGP full-mesh in GEANT, the pan-european research network, and in a Tier-1 ISP. Geant has 23 routers, each of them having 22 iBGP sessions and about four routes for each prefix in their BGP tables. The Tier-1 ISP uses a two-level hierarchy of Route Reflectors. If it used an iBGP full-mesh instead, the number of iBGP sessions to be maintained by border routers would increase by a factor of 50. With Route Reflection, those routers are RR-clients of two RRs and typically receive two paths for each prefix, one from each of their Route Reflectors. Those two paths are often identical [9], i.e they have the same nexthop address. With a full-mesh, the total number of paths that they have to maintain in their routing tables would be three times higher than with Route Reflection, because in that AS, on average, six different paths are known for each prefix. The number of BGP messages would also be accordingly higher.

When an interdomain link fails, the failure is notified either by the IGP protocol or by the sending of BGP withdraws. The first case occurs when the nexthop address of those BGP routes is the interface to the router from the neighboring AS. It is thus advertised by the IGP protocol. When the link fails, that nexthop address becomes unreachable in the IGP and the corresponding BGP routes are removed from the BGP tables. If the nexthop address of the eBGP learned routes is the loopback of a local BGP router, the failure is not learned via the intradomain protocol but from BGP withdraws received over iBGP sessions. This is for example the case in Fig. 1(a) : For $R2$, the nexthop of destination D is local router $R1$ and not router $RA$. Failure detection is faster in the first case, as it relies upon the convergence of the intradomain protocol.

## 3 Evaluation of the Number of iBGP-Caused Withdraws

The propagation of unnecessary BGP withdraws by an AS is responsible for transient losses of connectivity [8,11]. In this section, we present an analysis of RouteViews BGP feeds that evaluates the number of occurences of such BGP withdraws.

We define a withdraw as iBGP-caused if it was sent by a router of some AS but at least one other router of the same AS did not send a withdraw for the same destination during the same period of time. Indeed, if the second router does not send a withdraw, this means that either it uses another path, or that it knew an alternate path to replace the withdrawn one. In such a situation, at least two paths are available inside the AS, but the alternate path is not known by all routers.

For our evaluation, we took the BGP data from the first two weeks of October 2008 on the RouteViews Oregon collector, and considered the BGP messages received from pairs of routers belonging to the same AS. First, we filter all BGP messages received during reboot periods using the BGPMCT algorithm [12]. Second, we classify a BGP withdraw for a destination as iBGP-caused if it is seen on one session with an AS while the other router of this AS has a stable route, i.e. no withdraw for that destination is seen on the session with the other router during 30 seconds before and after the withdraw. This is an upper bound to the propagation time of the withdraw for the path inside an AS, if we assume that, at worst, the withdraw has to cross a whole two-levels iBGP hierarchy between two edge routers, which gives 5 BGP hops.

Figure 2 shows that most of the routers send several thousands of iBGP-caused withdraws per day. On a per-hour basis, results show peaks of more than 2000 iBGP-caused withdraws per hour. Variations between the results for different routers are probably due to different iBGP configurations, but we don't have

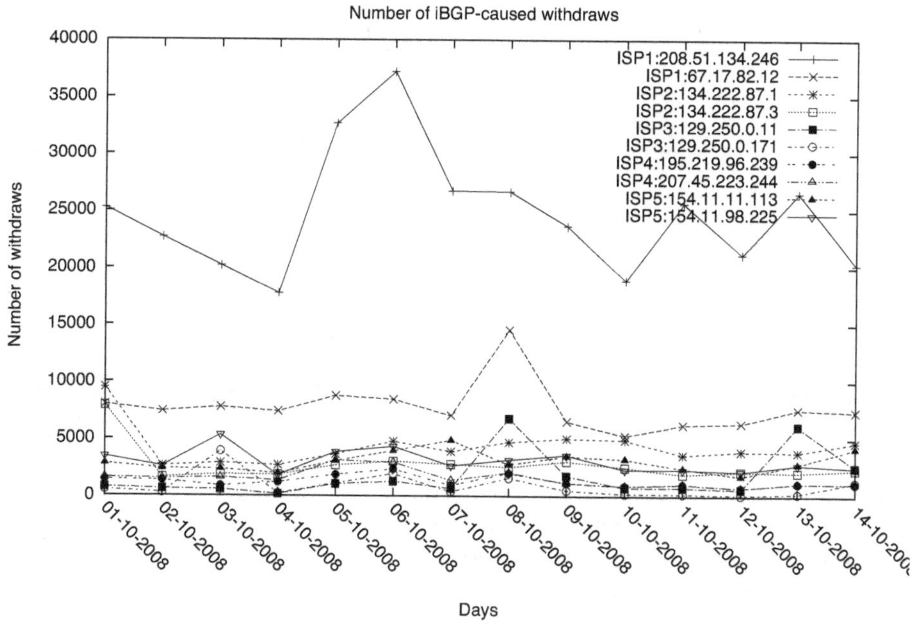


**Fig. 2.** Number of iBGP-caused BGP withdraws

information about the organizations of the observed ASes. Still, this analysis shows that for all the routers that we analysed, the number of iBGP-caused withdraws is important, and reducing this particular churn would help reducing transient losses of connectivity in the Internet.

## 4 iBGP Organization and Withdraws Propagation

In this section, we explore the BGP withdraw propagation, and identify that path diversity is the key for blocking this propagation. First, we analyse the problem at the AS level, then we focus on the influence of the iBGP organization on withdraw propagation.

To prevent the unnecessary announcement of a local failure to the entire Internet, the corresponding BGP withdraw must be stopped as close as possible to the failure. We call Withdraw-Blocking a router or an AS that is able to stop the propagation of a withdraw message.

**Definition 1.** *An AS is said to be* ***Withdraw-Blocking*** *for a destination D if that AS advertises D on at least one eBGP session and does not propagate a BGP withdraw to a neighbor not advertising D itself, upon reception of a BGP withdraw for its primary path towards that destination.*

On the topology of Fig. 3, $AS3$ is Withdraw-Blocking for destination D. For example, if the link between $AS1$ and $AS2$ fails, the withdraw is propagated by

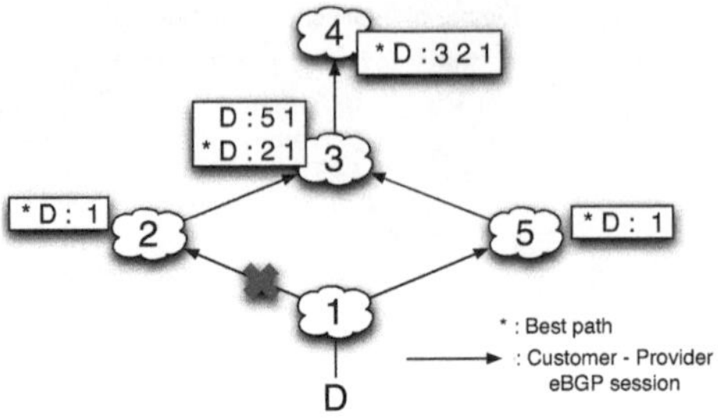


**Fig. 3.** Withdraw blocking AS

*AS*2 to *AS*3. *AS*3 knows the alternate path via *AS*5, such that it can advertise this alternate path to *AS*4 instead of a propagating the withdraw. A withdraw is still sent to *AS*5, but as this neighbor uses and advertises the alternate path, this withdraw won't result in any connectivity loss.

An AS must know an alternate path to reach the destination in order to be withdraw-blocking. However, this is not sufficient, as the AS must forward this alternate path to its neighbor to replace the withdraw message. Policies can prohibit the announcement of the alternate path on some eBGP sessions [10]. Therefore, we define a new property for an alternate path :

**Definition 2.** *Let $S_{P_X}$ be the set of eBGP sessions on which a path $P_X$ would be advertised if it were the only path available in the AS. A path $P_A$ to destination D is **export-policy compliant** (EPC) with another path $P_B$ to the same destination if $S_{P_B}$ is included in $S_{P_A}$.*

In the following subsections, we give conditions for an AS to be withdraw-blocking, first at the AS level (i.e. with ASes composed of a single router). Then, we describe the conditions in the context of ASes composed of multiple routers.

### 4.1 Withdraw Propagation Prevention with Blackboxed ASes

When modeling an AS as a single router, that "router" knows about all the paths to a given destination learned by the AS. In this case, the conditions to be Withdraw-Blocking are the following :

**Theorem 1.** *An AS is withdraw-blocking for a destination if and only if it knows an export policy compliant alternate path for its primary path to that destination.*

When the policies used in the AS are the classical routing policies [10], this theorem can easily be proven :

*Proof.* Two cases must be considered. First, if a withdraw is received from a provider or a shared-cost peer, the only sessions on which the AS was advertising the destination D are customer sessions. In this case, any other path is export-policy compliant, and the AS advertises the alternate path to its customers The second case is when the withdraw is received from a customer. If there exists an alternate path via a peer or a provider, the alternate path is advertised to the customers, but as it is not export-policy compliant, it cannot be advertised over session with peers or providers. The destination is thus withdrawn on those sessions. If the alternate path is learned from the same or another customer, this path is export-policy compliant. As customer path should be preferred over peers and providers paths [10], it will be selected as best once the primary path is withdrawn. The propagation of this customer alternate path is not constrained by policies and an update is sent instead of a withdraw.

### 4.2 Withdraw Propagation Prevention at the Router Level

In the previous subsection, we analysed the withdraw-blocking property of ASes containing only one router. Real ASes usually contain multiple routers, connected together by iBGP sessions. When two paths for a given destination are available inside an AS, they are learned over two different eBGP sessions. However, due to the iBGP organization, there can be several routers in the AS that are not necessarily aware of the alternate path [9]. Thus, even if the AS is Withdraw-Blocking, some of its routers may not be Withdraw-Blocking themselves.

For example, consider an AS with a full-mesh of iBGP sessions. If a destination D is learned over two eBGP sessions on two different routers with the same local preference, MED and AS path length, all routers receive both paths. No withdraw is propagated in case of failure of one of those paths. However, if one path has a higher local preference than the other, the alternate path is hidden at the router that received it, because that router prefers the other path. On Fig. 1(a), the path via $RB$ is not advertised by $R3$. If the best path fails, i.e. the path via $RA$, $R1$ and $R2$ will send withdraws for D on their eBGP sessions.

If Route Reflection is used, the situation is even more problematic [9]. In this case, even if two paths have the same local preference, one of them can be stuck in the Route Reflector, as shown on Fig. 1(b). All RR clients except $R3$ only know the path via $R1$. At least $R1$ will send a withdraw outside the AS if its primary path fails. If the failure is learned via BGP, the BGP withdraw can be blocked by the Route Reflector, because it knows the alternate path via $R3$ and sends to $R2$ and $R3$ an update containing that path instead of a withdraw. However, if the information that the nexthop is not reachable anymore is propagated inside the AS by the intra-domain protocol faster than the BGP messages, all routers that do not have any alternate path will send a withdraw outside the AS even if the Route Reflector does not send withdraws.

Based on this example, we can extend Theorem 1 to ASes containing several routers to give a sufficient condition for the Withdraw-Blocking property :

**Theorem 2.** *An AS is withdraw-blocking for destination D if all routers of the AS know at least one alternate path to D that is export-policy compliant with their primary path.*

*Proof.* If all routers of the AS have at least one export-policy compliant alternate path, any router that receives a withdraw is able to send an update with the alternate path instead of propagating the withdraw for the primary path on its eBGP sessions. Withdraw propagation is then blocked directly at the border of the AS. Also, when an external nexthop fails, all routers that learn the failure via the IGP have an alternate path, and none will send a withdraw.

Autonomous systems often connect with each other using multiple links [13]. Inside an AS, export-policy compliant paths are then usually available for destinations advertised by multi-connected neighbors. That makes the AS Withdraw-Blocking for these neighbors. However, as explained earlier, local preference settings or iBGP organization prevent this diversity to be propagated to all routers of the AS. It has been evaluated that, in a Tier-1 AS using a two levels' hierarchy of Route Reflectors, most routers typically know only a single path towards a destination [9]. In such an AS, a withdraw can easily be propagated outside the AS even if diversity is available.

## 5 Blocking Withdraw Propagation Outside an AS

Avoiding unnecessary BGP withdraw propagation should allow ISPs to improve the stability of the prefixes advertised by their customer in case of link failure. Furthermore, providing a solution to this problem is affordable, as it can be tackled within the iBGP organization.

When a full-mesh of iBGP sessions is used, it is easy to provide diversity to all routers. Diversity is stuck in a router when there is a better iBGP path (i.e. higher local preference, lower MED or shorter AS path) in the AS. If the advertisement rule is modified such that a router announces its best eBGP-learned path for each destination to its iBGP peers, up to one path per router is propagated in the AS. This mechanism is called Best-External [14].

When an AS uses Route Reflection, it is also possible to prevent withdraw propagation. Using Best-External with Route Reflection is not sufficient, although the best external paths are advertised to Route Reflectors. They do not propagate this diversity further in the network because they only advertise one path per prefix. Modifying the BGP protocol for advertising several paths for each prefix is a solution that has been proposed at the IETF [15]. This allows for a perfect diversity propagation thus achieving the Withdraw-Blocking property, but as it implies increased memory usage and number of BGP messages for exchanging those paths, it is not suitable for ASBR with limited resources.

We propose a lighter solution that would allow routers to propagate the information about the existence of alternate paths without modifying BGP itself. Upon reception of a BGP withdraw, a router that knows that an alternate path exists in the AS can wait until iBGP has converged before sending a withdraw over its eBGP sessions. The AS becomes thus Withdraw-Blocking without requiring its routers to store all BGP routes learned by the AS in their memory.

The principle of our solution is that, whenever a router knows an alternate path, it tags a special BGP community `PATH_DIVERSITY` to the primary path when it advertises it to its iBGP peers, including the one from which the path has been learned if that path comes from iBGP. This is needed because the router that sent the primary path also needs to learn the existence of the backup path.

The primary path is then propagated in the AS with the `PATH_DIVERSITY` community. Legacy BGP routers that do not support the community simply propagate the path with the community following classical iBGP rules, without taking its signification into account. The `PATH_DIVERSITY` community is removed when the path is advertised over eBGP sessions.

When the primary path is withdrawn, all routers that support the community and do not have an alternate path themselves will not propagate the withdraw on their eBGP session. Instead, they start a timer and re-advertise the route for the withdrawn path with a local-preference of 0 in iBGP. We thus allow the path to stay temporarily in the routing table of the router. This does not prevent traffic losses, as explained later, but blocks withdraw sending. Also, the routers that do not support the `PATH_DIVERSITY` community will not remove the primary path when receiving the advertisement with the low local-preference and won't send BGP withdraws before receiving the alternate path. With this local-preference value, alternate paths will be preferred over the primary by routers that know them and they will be propagated in the AS [16].

If the timer expires and no alternate path has been received, the router sends the BGP withdraw on its eBGP sessions. The timer is needed if the alternate path is withdrawn shortly after the primary, which can happens when both paths are impacted by the same failure. In that case, the alternate path cannot be propagated in the AS even if the `PATH_DIVERSITY` community has been tagged to the primary path. The timer prevents the BGP convergence to be blocked, waiting for an alternate path that doesn't exist anymore. A suitable value for the timer should be established by evaluating the iBGP convergence time. This value will typically depend on the type of iBGP organization of the AS, and the number of primary paths that can be impacted by a given eBGP link failure.

In the example of Fig. 1(b), R3 tags the path learned via $R1$ with the community, and sends it back to the Route Reflector, which in turn advertises it to all its clients including $R1$. Thanks to this community, $R1$ knows that there exists an alternate path, and if the primary path is withdrawn, it will not send a BGP withdraw on its eBGP sessions. Instead, it will wait until the Route Reflector advertises the alternate path via $R3$.

However, as such, the mechanism is not sufficient to ensure the Withdraw-Blocking property of the AS. Indeed, the first alternate path received during the convergence is not necessarily export-policy compliant with the primary one. In this case, the router will have to send a BGP withdraw on the eBGP sessions over which that alternate path cannot be advertised. We refine our solution to face this issue by relying on two community values, `EPC_DIVERSITY` and `NON_EPC_DIVERSITY`. The procedure for tagging those diversity communities is explained in Algorithm 1 : A router tags a path with either community depending on whether its alternate path is export-policy compliant with the primary or not. The export-policy compliance can be easily computed by a router if the

**Algorithm 1.** Update reception

```
if BGP path received then
    Run decision process
    {C}heck diversity
    if alternate path exists in Adjribins then
        Tag diversity community to the best path, depending on the policies applied to the alternate path.
    end if
    if Best path changed then
        Propagate new best path on eBGP sessions and iBGP sessions (including the originator of the path)
    end if
    if Best path unchanged, but a community has been tagged then
        Advertise on iBGP sessions, including originator of the best path.
    end if
end if
```

**Algorithm 2.** Withdraw reception

```
if BGP withdraw received from eBGP session or BGP nexthop becomes unreachable then
    Run decision process
    if best path unchanged then
        check diversity, update communities and readvertise over iBGP if needed
    else
        if best path is tagged with EPC_DIVERSITY then
            if Export-policy compliant path available in Adjribins then
                Propagate alternate path as new best path over iBGP sessions and eBGP sessions
            else
                Set local-preference of the path to 0
                Wait until export-policy compliant path is received, or timer expires
                if Timer expires then
                    Propagate withdraw
                else
                    Propagate alternate path as new best path over iBGP sessions and eBGP sessions
                end if
            end if
        else if best path is tagged with NON_EPC_DIVERSITY then
            if alternate path is available in Adjribins then
                Propagate alternate path over iBGP sessions and over policy-compliant eBGP sessions
                Propagage withdraw over non policy-compliant eBGP sessions
            else
                Set local-preference of the path to 0
                Wait until any alternate path is received, or timer expires
                if Timer expires then
                    Propagate withdraw
                else
                    Propagate alternate path as new best path over iBGP sessions and over policy-compliant eBGP sessions
                    Propagage withdraw over non policy-compliant eBGP sessions
                end if
            end if
        else
            Act as usual
        end if
    end if
end if
```

paths are tagged with a community that identifies their origin [17], i.e. if they come from a customer or from a peer or provider. This is a good practice rule that is often used. The router then readvertises the path to its iBGP neighbors, including to the one from which it was learned.

Algorithm 2 is applied when a router receives a BGP withdraw. The principle is that when a router receives a BGP withdraw for a path tagged with one of the communities, and for which it does not have an alternate path, it does not send any BGP withdraw over eBGP sessions. Instead, it waits until it receives that alternate path. If, during the convergence, a first alternate path that is not export-policy compliant is learned while the path is tagged with EPC_DIVERSITY, the router still waits for the export-policy compliant path instead of sending withdraws. When common policies are used [10], the export-policy compliant path will finally be selected as best, and no BGP withdraw is sent over eBGP sessions.

On Fig. 4, RR1 knows an export-policy compliant path via *RB*, so it adds the community EPC_DIVERSITY to the BGP route. RR2 also has diversity for that path. As its alternate path is not export-policy compliant (this is a path received from a provider while the primary comes from a customer), *RR*2 also tags the community NON_EPC_DIVERSITY. All routers know that diversity is available, and the AS is Withdraw-Blocking. For example, if the link between *RA* and *R*1 fails, *R*3 has no diversity but knows that an export-policy compliant path is available, so it does not advertise a withdraw to its eBGP peer. When *RR*2 learns the failure via the IGP, it will not yet send an update for D with the path via *RC*, because it is not export-policy compliant. Instead, it waits until it receives the export-policy compliant path. Eventually, *RR*2 then *R*3 will learn the alternate path via *RB*, and *R*3 can send an update with the export-policy compliant path on its eBGP session.

**BGP convergence** Using those communities slightly increases the number of BGP messages exchanged during the initial convergence, as an additionnal update is emitted when the route is tagged with a diversity community. In the worst case, two additionnal updates will be emitted by a router, one when a non export-policy compliant alternate path is known to exist, and a second when

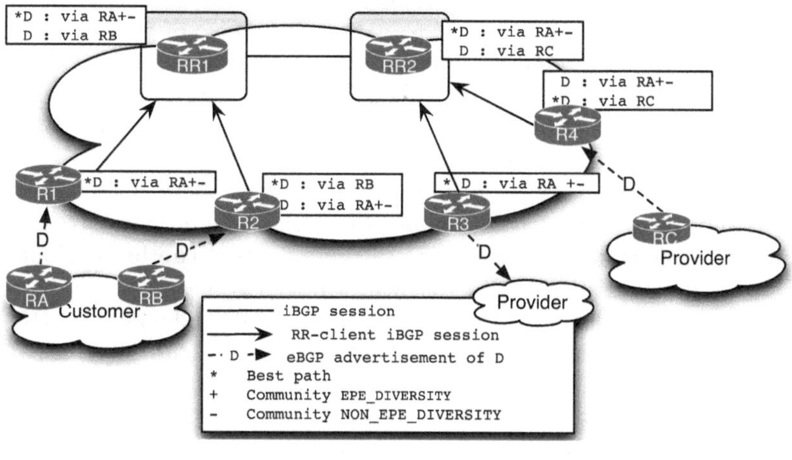


**Fig. 4.** Announcing diversity in a community

the existence of an export-policy compliant alternate path is learned. Also, upon failure of an alternate path, a few BGP messages are also exchanged to update the communities of the primary path. However, those BGP messages will not be announced outside the AS, hence the small message overhead is limited to the AS.

The diversity communities also do not impact routing stability : Tagging the diversity communities is a deterministic process that does not lead to routing loops. Indeed, when the paths to a destination are stable, once a diversity community has been tagged to a route, it is not removed as long as there is a corresponding alternate path in the AS. Sending the tagged path back to the sender or the original path also does not result in routing loops. As a path is at most tagged twice, it is sent back to the sender at most twice and then the routing state becomes stable. Also, if the alternate path fails, the router that tagged the primary path stops re-advertising it with the backup community, and the tagged path is replaced by the original path in all routers after iBGP convergence.

**Impact on the data plane.** When a router receives a withdraw for a destination and waits for the alternate path, it doesn't know any other nexthop to which send the traffic until it receives the new path. The traffic might then be dropped during the iBGP convergence. However, even if it cannot prevent the local loss of connectivity, waiting for the new path before sending the withdraw on eBGP reduces the losses of connectivity that would occur further in the Internet due to the unnecessary withdraw propagation.

## 6 Related Work

As explained in the introduction, the churn that affects BGP has several causes. The first one is the path vector nature of BGP. This, combined with routing policies, leads to path exploration as explained by Gao et al. among others in [18]. Several solutions have been proposed to reduce the impact of path exploration in the global Internet. These solutions rely on the utilization of new BGP attributes. The most complete ones are BGP RCN proposed by Pei et al. [19] and EPIC proposed by Chandrashekar et al. [20], but are not currently deployed. Our solution complements these approaches.

The current iBGP organizations are known to be imperfect [21]. Several researchers have proposed solutions to improve them. One approach is to centralise all routing decisions in a Routing Control Platform that can be considered as a super Route Reflector RCP [22]. Another approach is to extend iBGP to allow each router/RR to advertise several paths towards each destination [15]. This iBGP extension allows all BGP routers inside an AS to learn several paths towards each destination and thus block the propagation of withdraws, achieving the same objective as our communities. Furthermore, as the backup paths are propagated and not only the information about their availability, connectivity losses are also prevented. This extension also allows for other utilizations of additionnal paths, such as multipath routing. This is definitely a very promising solution for the future, but currently, as this mechanism increases the number of BGP messages and the memory required to store all the additional paths, all routers cannot support it. In the meantime, our communities can be used to prevent withraw propagation, as they do not require any change to the BGP syntax and could be deployed incrementally .

## 7 Conclusion

In this paper, we have first explained that the iBGP organisation used in large ASes reduces the number of paths learned by each router. When a link fails or a path is withdrawn, BGP routers inside an AS may send an unnecessary BGP withdraw. This causes transient losses of connectivity. We proposed a solution that allows routers to know if there is an alternate path for each prefix inside the AS. When a link fails or a BGP withdraw is received, BGP routers will block the propagation of withdraws for prefixes for which an alternate path is known in the AS.

Our further work is to evaluate the convergence time of iBGP with and without using our solution.

## References


1. Rekhter, Y., Li, T., Hares, S.: A border gateway protocol 4 (BGP-4). Internet RFC4271 (2006)
2. Meyer, D., Zhang, L., Fall, K.: Report from the IAB workshop on routing and addressing. RFC4984 (2007)
3. Bonaventure, O., Filsfils, C., Francois, P.: Achieving sub-50 milliseconds recovery upon BGP peering link failures. IEEE/ACM Trans. Netw. 15, 1123–1135 (2007)
4. Wu, J., Mao, Z.M., Rexford, J., Wang, J.: Finding a needle in a haystack: pinpointing significant BGP routing changes in an IP network. In: NSDI 2005, Berkeley, CA, USA, pp. 1–14. USENIX Association (2005)
5. Wang, F., Gao, L., Wang, J., Qiu, J.: On understanding of transient interdomain routing failures. In: ICNP 2005, Washington, DC, USA, pp. 30–39. IEEE Computer Society, Los Alamitos (2005)
6. Oliveira, R., Zhanf, B., Pei, D., Izhak-Ratzin, R., Zhang, L.: Quantifying path exploration in the internet. In: Internet Measurement Conference, Rio de Janeiro, Brazil (2006)
7. Mao, Z.M., Govindan, R., Varghese, G., Katz, R.: Route flap damping exacerbates internet routing convergence. In: ACM SIGCOMM 2002 (2002)
8. Wang, F., Mao, Z.M., Wang, J., Gao, L., Bush, R.: A measurement study on the impact of routing events on end-to-end internet path performance. In: SIGCOMM 2006, pp. 375–386. ACM, New York (2006)
9. Uhlig, S., Tandel, S.: Quantifying the impact of route-reflection on BGP routes diversity inside a tier-1 network. In: Boavida, F., Plagemann, T., Stiller, B., Westphal, C., Monteiro, E. (eds.) NETWORKING 2006. LNCS, vol. 3976, pp. 1002–1013. Springer, Heidelberg (2006)
10. Gao, L., Rexford, J.: Stable internet routing without global coordination. IEEE/ACM Trans. Netw. 9, 681–692 (2001)

11. Kushman, N., Kandula, S., Katabi, D., Maggs, B.: R-BGP: Staying Connected in a Connected World. In: NSDI 2007, Cambridge, MA (2007)
12. Zhang, B., Kambhampati, V., Lad, M., Massey, D., Zhang, L.: Identifying bgp routing table transfers. In: MineNet 2005: Proceedings of the 2005 ACM SIGCOMM workshop on Mining network data, pp. 213–218. ACM, New York (2005)
13. Feamster, N., Mao, Z.M., Rexford, J.: BorderGuard: Detecting Cold Potatoes from Peers. In: Internet Measurement Conference, Taormina, Italy (2004)
14. Marques, P., Fernando, R., Chen, E., Mohapatra, P.: Advertisement of the best-external route to iBGP. Internet Draft, draft-marques-idr-best-external-00, work in progess (2008)
15. Walton, D., Retana, A., Chen, E.: Advertisement of Multiple Paths in BGP. Internet draft, draft-walton-bgp-add-paths-05.txt, work in progress (2006)
16. Francois, P., Decraene, B., Pelsser, C.: Graceful eBGP Session Shutdown. Internet draft, draft-francois-bgp-gshut-00.txt, work in progress (2008)
17. Meyer, D.: BGP Communities for Data Collection. RFC 4384 (Best Current Practice) (2006)
18. Gao, L., Wang, F.: The extent of as path inflation by routing policies. In: Global Internet 2002 (2002)
19. Pei, D., Azuma, M., Nguyen, N., Chen, J., Massey, D., Zhang, L.: BGP-RCN: Improving BGP convergence through Root Cause Notification. Computer Networks 48, 175–194 (2005)
20. Chandrashekar, J., Duan, Z., Zhang, Z., Krasky, J.: Limiting path exploration in BGP. In: IEEE INFOCOM, Miami, Florida (2005)
21. Griffin, T., Wilfong, G.: On the correctnes of iBGP configuration. In: SIGCOMM 2002, Pittsburgh, PA, USA, pp. 17–29 (2002)
22. Caesar, M., Caldwell, D., Feamster, N., Rexford, J., Shaikh, A., van der Merwe, J.: Design and implementation of a routing control platform. In: NSDI 2005, Berkeley, CA, USA, pp. 15–28. USENIX Association (2005)

# Backup Path Classification Based on Failure Risks for Efficient Backup Path Computation
## (Work in Progress)

Mohand Yazid Saidi[1], Bernard Cousin[1], and Jean-Louis Le Roux[2]

[1] Université de Rennes I, IRISA, 35042 Rennes Cedex, France
{msaidi,bcousin}@irisa.fr
[2] France Télécom, 2 Avenue Pierre Marzin, 22300 Lannion, France
jeanlouis.leroux@orange-ftgroup.com

**Abstract.** We propose a new approach exploiting the failure risk (node, link or Shared Risk Link Group) structures to enhance the backup path computation. Upon failure, our approach classifies the backup paths into two categories: operative backup paths and inoperative backup paths. An operative backup path is an active backup path which really receives traffic of some affected communications while an inoperative backup path does not receive any traffic.

With the observation that only the operative backup paths really participate to the recovery procedure, we enhance the backup path computation (1) by limiting the concurrence for the protection bandwidth allocations to the operative backup paths (instead of all the active backup paths like in the classical approaches) and (2) by reducing the set of failure risks that a backup path must bypass.

Simulations show that our approach improves the protection bandwidth allocations and decreases the ratio of rejected backup paths.

**Keywords:** network, survivability, local protection, Shared Risk Link Group (SRLG), resource optimization, MPLS, backup path computation.

## 1 Introduction

With the explosion of the number of real-time network applications which are sensitive to the disruption time of communications, local proactive protection techniques [1,2] are more and more deployed and used to ensure service continuity. Indeed, the proactive protection techniques permit to achieve fast recovery from failures by pre-computing and generally pre-configuring local backup paths capable to receive and reroute the traffic of affected communications upon failure.

To provide local protection for communications, two types of backup paths are defined [3]: Next HOP path (or NHOP path) and Next Next HOP path (or NNHOP path). A NHOP path (resp. NNHOP path) is a backup path protecting against a link failure (resp. a node failure); it is setup between a primary router called Point of Local Repair (PLR) and one primary router downstream to the PLR (resp. to the PLR next-hop) called Merge Point (MP). Such backup

L. Fratta et al. (Eds.): NETWORKING 2009, LNCS 5550, pp. 509–520, 2009.

path bypasses the link (resp. the node) downstream to the PLR on the primary path. When a link failure (resp. node failure) is detected by a router, this later activates[1]locally all its NHOP and NNHOP (resp. its NNHOP) backup paths by switching traffic from the affected primary paths to their backup paths.

In order to guarantee the recovery success from any failure, enough resources (bandwidth) must be pre-allocated to the backup paths to reroute the traffic of affected communications. Due to the high number of backup paths (number which can be very higher than that of primary protected paths), the backup path bandwidth pre-allocation can lead to a rapid decrease of the network available bandwidth which, in its turn, can block (or prevent) the establishment of new communications. To decrease the blocking risks, numerous works consider only single physical failures [4,5,6,7,8,9]. With such practical hypothesis, the bandwidth allocation could be performed efficiently since the bandwidth can be shared between some backup paths. Indeed, in such a case, the backup paths which protect against different failure risks are never active at the same time and as a result, they can share the bandwidth on their common links. For instance, to decrease the amount of bandwidth allocated to the backup paths, several classical approaches [4,5,6,7,8,9] suggest to determine the cumulative bandwidth of the backup paths which would be activated on each link, to recover quickly from any possible failure. As only the activated backup paths can really use their resources, the classical approaches propose to allocate the maximum between the cumulative bandwidths of the backup paths which could be active at the same time on each link.

To deal with a *physical failure* in a *logical layer* (Network Layer), three types of failure risks are defined: link, node and Shared Link Risk Group (SRLG). The first type of failure risk corresponds to the risk of a logical link failure due to the breakdown of an exclusive physical component of the logical link. The second type of failure risk corresponds to the risk of a logical node failure due to the breakdown of an exclusive physical component of the logical node. Finally, a SRLG risk is a set of links that share a common physical component whose failure may impact all links in the set. For instance, two logical links using the same fiber (or sharing the same crossconnect) belong to the same SRLG. More details about the SRLG risk can be found in [10,7,11,12].

Contrarily to the protection against link and node failure risks which requires the setup of only one backup path, the protection against a SRLG risk requires the setup of several backup paths, one for each primary (logical) link belonging to the protected SRLG. Moreover, for fast recovery, all the backup paths which protect against the failure of links belonging to a failed SRLG will be activated simultaneously. With the observation that some activated backup paths don't really use their resources (bandwidth) upon a SRLG failure (because the traffic of the primary paths they protect was switched towards other backup paths bypassing their head-end routers), we propose in this article to enhance the protection quality and increase the bandwidth sharing by extending its ap-

[1] When a backup path $b$ protecting a primary path $p$ is activated, all the packets of $p$ which traverse the source router of $b$ are sent and redirected onto this backup path.

plication to some activated backup paths. In our approach, we exploit the SRLG structures to determine the active backup paths which do not really use their resources upon a failure. Such active backup paths are in reality *inoperative* (they do not receive/reroute any data flow) and thus, they can share the bandwidth with any other active or inactive backup path which is inoperative at that time. In addition to the bandwidth sharing improvement, we enhance the protection quality (protection rate) by decreasing the number of backup paths which protect against SRLG failure risks. In our proposition, more flexibility is provided for backup path selection since a backup path does not systematically bypass all the links sharing a SRLG with the protected link.

The rest of this article is organized as follows: In section 2, we review some works related to the bandwidth sharing. Then, we explain in section 3 the principles of the *failure risk-based backup path classification* (FRBPC) algorithm which enhances the backup path computation. In section 4, we present and analyze some simulation results and we finish, in section 5, by giving some conclusions.

## 2 Related Works

With the increasing interest for local proactive protection in the last decade, several works [4,5,1,2,7,8,9] are devoted to the determination of algorithms computing the backup paths. To minimize the quantity of bandwidth allocated on links while avoiding the bandwidth constraint violation (bandwidth insufficiency), the *backup path computation* (BPC) algorithms require the knowledge of some information like the primary and backup paths, the bandwidth allocations and the protected risks. This information enables the Backup Path Computation Element (BPCE) to deduce, for each new protection request, the additional bandwidth quantity which should be reserved and the bandwidth quantity which can be shared on each link to satisfy the new request.

Depending on the number of simultaneous failures that can be treated, the quantity of protection bandwidth reserved on each link can be high (when the number of simultaneous failures is large) or low (when the number of simultaneous failures is small). Indeed, the number of simultaneous failures that can be treated successfully determines all the failure scenarios, which in turn control the number and structures of the backup paths which provide the protection. Due to the rarety of multiple failures and to the difficulty to protect (in local and proactive manner) against this type of failure, and in order to increase the bandwidth availability (increase the bandwidth sharing), most of works in the literature consider only single failures [4,5,7,8,9]. With such type of failure (single failures), the quantity of protection bandwidth $Bk_\lambda$ that should be reserved on each unidirectional link $\lambda$, depends on the cumulative bandwidth of the paths which could be active at the same time after any single failure occurrence. It is computed as follows:

$$Bk_\lambda = \mathrm{Max}_r(\delta_r^\lambda) \tag{1}$$

where $\delta_r^\lambda$, called the *protection cost* of the risk $r$ on the unidirectional link $\lambda$, corresponds to the cumulative bandwidth of the backup paths ($BPaths$) which would be activated on the unidirectional link $\lambda$ upon a failure of the risk $r$, i.e.:

$$\delta_r^\lambda = \sum_{b \in BPaths:\, \lambda \in b} Act(b,r) \times BW(b) \tag{2}$$

where $BW(b)$ returns the bandwidth of $b$ and $Act(b,r)$ return 1 if the backup path $b$ is active upon a failure of the risk $r$. Otherwise, it returns 0.

When a new backup path $b$ is computing, only the links $\lambda$ verifying the following inequality can be used:

$$Pr_\lambda + \text{Max}_{r:\, Act(b,r)=1}(\delta_r^\lambda) + BW(b) \leq C_\lambda \tag{3}$$

where $Pr_\lambda$ is the cumulative bandwidth of the primary paths traversing the unidirectional link $\lambda$ and $C_\lambda$ is the capacity of the unidirectional link $\lambda$.

Once the links verifying the bandwidth constraints are selected according to (3), any BPC algorithm can be used to determine the backup path in computation. This approach increases the bandwidth availability by sharing the bandwidth between the backup paths. It is easy to be deployed in centralized environments where the unique BPCEs know the bandwidth information (protection costs, link capacities, cumulative primary bandwidths, etc.) required for the backup path computation. In distributed environments however, the advertisement of the bandwidth information required for the backup path computation is costly and could overload the network. Thus, using heuristics aggregating and/or reducing this bandwidth information before its advertisement in the network could give some interesting and practical solutions [4,5,7]. For instance, to decrease the size and frequency of the advertisement messages, the Kini's heuristic [4] suggests to approximate all the protection costs on a given unidirectional link by the highest protection cost on that link. In this way, a given unidirectional link $\lambda$ can be used to establish a new backup path $b$ if it verifies the following inequality: $Pr_\lambda + \text{Max}_r(\delta_r^\lambda) + BW(b) \leq C_\lambda$.

## 3 Failure Risk-Based Backup Path Classification Algorithm for Efficient Backup Path Computation

For fast recovery, each router detecting a failure on its interface activates locally all the backup paths which protect the primary paths traversing the failed interface (because it cannot distinguish quickly the different types of failures). Although active, some backup paths (*inoperative* backup paths) do not participate to the recovery of the affected communications because the traffic was already redirected by upstream routers onto other backup paths (*operative* backup paths) bypassing their head-end routers. Hence, to improve the bandwidth availability, we propose in this section to take into account the risk structures (specifically the SRLG structures) to determine the *operative* backup paths which *really participate* to the recovery. In our proposal, only the operative backup paths can be in concurrence for bandwidth allocation. Moreover, to provide more flexibility for the backup path selection, we restrict the set of risks protected by a backup path to the risks whose failure induces traffic to be switched onto this backup path.

### 3.1 Active Backup Paths vs. Operative Backup Paths

Due to the difficulty to distinguish quickly between the types of failure (node, link or SRLG) [13], each router detecting a failure on its outgoing interface activates all the backup paths which protect the primary paths traversing the affected interface. As a single physical failure can affect many logical links (cf. case of a SRLG failure), several backup paths protecting the same primary path can be activated upon a failure. In some cases, the head-end router of an activated backup path $b_1$ is bypassed by another activated backup path $b_2$ protecting the same primary path. In such a case, the backup path $b_1$ does not receive and reroute the traffic of the affected primary path; it is considered as *inoperative* since it does not really use its resources (particularly its bandwidth). Hence, the bandwidth allocated for such inoperative path can be freed and reallocated to other paths. Contrarily to the backup path $b_1$, the other backup path $b_2$ really participates to the recovery since it reroutes the traffic of the affected primary path. This path is considered as *operative*. Its resources (particularly the bandwidth) cannot be reallocated to other paths.

In figure 1 (a), two backup paths $b_E$ ($E \rightarrow F \rightarrow D \rightarrow C$) and $b_B$ ($B \rightarrow A \rightarrow E \rightarrow F \rightarrow D \rightarrow C$) are setup to protect the primary path $p$ ($E \rightarrow B \rightarrow C$) against the failure of the four following risks: node $B$, link $E$-$B$, link $B$-$C$ and SRLG $srlg = (B\text{-}C, B\text{-}E)$. When router $E$ (resp. router $B$) detects a failure on the interface leading to its adjacent router $B$ (resp. router $C$), it activates locally the backup path $b_E$ (resp. $b_B$) which protects the unique primary path traversing the failed interface. Hence, for the failure of node $B$ or the failure of link $E$-$B$ (resp. the failure of link $B$-$C$), traffic of the affected primary path $p$ will be switched onto the unique activated backup path $b_E$ (resp. $b_B$). As only one outgoing interface of the primary path routers can be affected upon a single link or node failure, we conclude that at most one backup path per primary path could be activated. As a result, all the backup paths activated to recover from a link or node failure *really receive* and *reroute* the traffic of the affected primary paths.

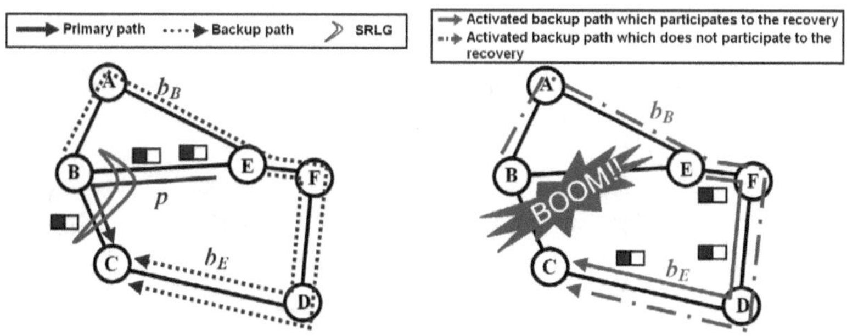


(a) Local protection of a primary path (b) Backup path activation and traffic rerouting

**Fig. 1.** Operative vs inoperative backup paths

With risks of type SRLG however, some activated backup paths do not receive or reroute the traffic of the affected primary paths. For instance, when the SRLG *srlg* in figure 1 (a) fails, all the end routers of the *srlg*'s links (i.e. routers $E$, $B$ and $C$) will detect a failure. As a result, all the backup paths protecting an affected primary path and whose head-end router is an end router of a link belonging to the failed SRLG will be activated. Typically, the backup path $b_E$ (resp. $b_B$) will be activated since it protects an affected primary path ($p$) and its head-end router $E$ (resp. $B$) is an end router of a link $E$-$B$ (resp. $B$-$C$) belonging to the affected SRLG *srlg*. As the traffic switching toward a backup path results in the bypassing of a primary path segment located between the head-end and the tail-end routers of the backup path, we deduce that only the backup path $b_E$ receives and reroutes the traffic of the affected primary path $p$ after the recovery from the failure of the SRLG *srlg* (cf. figure 1 (b)). Indeed, after the activation of the backup path $b_E$, the traffic of the primary path $p$ is forwarded on the path $E \rightarrow F \rightarrow D \rightarrow C$: the head-end router $B$ of the second activated backup path $b_B$ is bypassed and thus, no data flow traverses this backup path.

## 3.2 Decreasing the Bandwidth Allocation

To decrease the protection bandwidth reserved on a link, the bandwidth sharing should be extended to all the backup paths which cannot be operative at the same time. Concretely, if a backup path really receives traffic of an affected primary path upon a failure, the backup path is considered as operative and should be assigned a sufficient quantity of bandwidth to recover from the failure. However, if a backup path is inoperative upon a failure of a given risk, it will be assigned a null quantity of bandwidth since it does not receive any data flow.

In order to determine the exact set of operative backup paths $OPB_r$ upon a failure of a risk $r$, we consider the simple risks (node and link risks) and composite

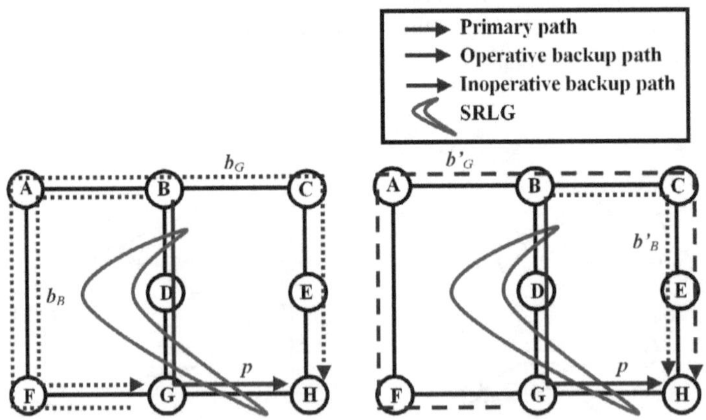


(a) Two operative backup paths upon the SRLG failure

(b) One operative backup path upon the SRLG failure

**Fig. 2.** Operative backup paths

risks (SRLGs). With a simple failure risk $r$, the operative backup path set $OPB_r$ is composed of all the activated backup paths upon a failure of $r$ (cf. section 3.1). With a composite risk *srlg*, a backup path $b$ protecting a primary path $p$ is in the operative backup path set $OPB_{srlg}$ if and only if:

1. The backup path $b$ protects against the failure of a link belonging to the SRLG *srlg*.
2. There is no backup path $b'$ ($b' \neq b$) such as:
   - $b'$ protects the primary path $p$ against the failure of a link belonging to the SRLG *srlg*,
   - the sub-path of $p$ located between the end routers of $b'$ contains, as transit router, the head-end router of $b$.

To better understand the process of the operative backup path determination upon a SRLG failure, let us consider an example. In figure 2, a primary path $p$ ($B \rightarrow D \rightarrow G \rightarrow H$) traversing the unique SRLG *srlg* = (*B-D*, *D-G*, *G-H*) of the network is established. To protect this primary path against the failure of link *G-H*, we setup the same NHOP backup path $G \rightarrow F \rightarrow A \rightarrow B \rightarrow C \rightarrow E \rightarrow H$ in both subfigures 2 (a) and 2 (b) ($b_G$ in subfigure 2 (a) and $b'_G$ in subfigure 2 (b)). To protect the primary path $p$ against the failure of node $D$ (and against the failure of link *B-D*), we used two different backup paths. In subfigure 2 (a), we setup the backup path $b_B$ ($B \rightarrow A \rightarrow F \rightarrow G$) and in subfigure 2 (b), we configured the backup path $b'_B$ ($B \rightarrow C \rightarrow E \rightarrow H$).

Upon a failure of the SRLG *srlg*, node $B$ and node $G$ activate the backup paths $b_B$ and $b_G$ in the subfigure 2 (a) (resp. the backup paths $b'_B$ and $b'_G$ in the subfigure 2 (b)) for recovery. In figure 2 (a), both the backup paths $b_B$ and $b_G$ become operative after the recovery from the SRLG failure. Indeed, the backup path $b_B$ (resp. $b_G$) protects the primary path $p$ against the failure of a *srlg*'s link *B-D* (resp. *G-H*) and its head-end router $B$ (resp. $G$) does not belong to the primary path segment located between the end routers $G$ and $H$ (resp. $B$ and $G$) of the unique other backup path $b_G$ (resp. $b_B$) protecting the primary path $p$ (against the failure of a link belonging to the SRLG *srlg*). In figure 2 (b) however, only the backup path $b'_B$ becomes operative (for the same reasons as $b_B$ in figure 2 (a)) upon the failure of the unique network SRLG *srlg*. The second backup path $b'_G$ is inoperative upon the failure of the SRLG *srlg* since there is another backup path $b'_B$ verifying these two conditions: (1) $b'_B$ protects the primary path $p$ (i.e. the same primary path as that protected by $b'_G$) against the failure of a link (*B-D*) belonging to *srlg*. (2) the sub-path ($B \rightarrow D \rightarrow G \rightarrow H$) of $p$ located between the end routers ($B$ and $H$) of $b'_B$ contains, as a transit router, the head-end router ($G$) of the backup path $b'_G$.

With the definition of the *protection price* $\gamma_r^\lambda$ as the cumulative bandwidth of the operative backup paths that traverse the unidirectional link $\lambda$ upon a failure of the risk $r$, we obtain:

$$\gamma_r^\lambda = \sum_{b \in BPaths:\, \lambda \in b} Op(b, r) \times BW(b) \qquad (4)$$

where $Op(b,r)$ return 1 if the backup path $b$ is operative upon a failure of the risk $r$. Otherwise, it returns 0.

As only the operative backup paths can be in concurrence for resources, we reduce and deduce the minimal protection bandwidth $Bk_\lambda$ required on a unidirectional link $\lambda$ as follows:

$$Bk_\lambda = \text{Max}_r(\gamma_r^\lambda) \tag{5}$$

To compute a new backup path $b$, only the unidirectional links ($\lambda$) verifying the following inequality can be used:

$$Pr_\lambda + \text{Max}_{r:\,Op(b,r)=1}(\gamma_r^\lambda) + BW(b) \leq C_\lambda \tag{6}$$

Since the set of the operative backup paths is included in the set of the activated backup paths, we deduce that all the protection prices are lower or equal to their corresponding protection costs ($\forall(r,\lambda) : \gamma_r^\lambda \leq \delta_r^\lambda$). As a result, we conclude that our approach permits to save much more bandwidth.

**Example:** Consider the link $B \rightarrow C$ in figure 2 (b).

Without the exploitation of the failure risk structures, we compute the minimal protection bandwidth $Bk1_{BC}$ allocated on the link $B \rightarrow C$ as follows:
$Bk1_{BC} = \text{Max}(\delta_{BD}^{BC}, \delta_D^{BC}, \delta_{GH}^{BC}, \delta_{srlg}^{BC}) = \delta_{srlg}^{BC} = 2 \times BW(p)$

With the FRBPC algorithm, we compute the minimal protection bandwidth $Bk2_{AB}$ allocated on the link $A \rightarrow B$ as follows:
$Bk2_{BC} = \text{Max}(\gamma_{BD}^{BC}, \gamma_D^{BC}, \gamma_{GH}^{BC}, \gamma_{srlg}^{BC}) = \gamma_{srlg}^{BC} = BW(p)$

Thus, we conclude that $Bk2_{BC} = Bk1_{BC}/2$

### 3.3 Providing Flexibility for the Backup Path Selection

In addition to the decrease of the protection bandwidth, our approach provides more flexibility for the backup path selection by reducing the set of risks that

---

**Algorithm 1.** Computation of a backup path $b$ with the FRBPC algorithm

---

**inputs**

A graph $G = (V, E)$ corresponding to the network topology. $V$ is the set of vertices (routers) and $E$ is the set of edges (links)

**begin_algorithm**

1. Determine the links verifying the bandwidth constraints.
   $E' \leftarrow \{\lambda \setminus \lambda \in E \wedge \forall\, r \in \textit{Risks}\text{: } Pr_\lambda + \text{Max}_{r \setminus Op(b,r)=1}(\gamma_r^\lambda) + BW(b) \leq C_\lambda\}$
2. Deduce the links and nodes which should be bypassed by $b$.
   $E" \leftarrow \{\lambda \setminus \exists\, (\lambda, r)\text{: } \lambda \in r \wedge Op(b,r) = 1\}$
   $V" \leftarrow \{n \setminus \exists\, (n, r)\text{: } n \in r \wedge Op(b,r) = 1\}$
3. Use any local protection technique (one-to-one backup or facility backup) and any path computation algorithm to determine the backup path $b$ on the graph $G' = (V \setminus V", E' \setminus E")$.

**end_algorithm**

---

must be protected by the backup paths. Concretely, with our FRBPC algorithm, the set of risks that must be bypassed by a backup path is reduced and composed only of risks whose failure operates that backup path. For instance, in subfigure 2 (b), any new NHOP backup path $b'_G$ protecting the primary path $p$ against the failure of the link $G-H$ is inoperative upon the failure of the SRLG *srlg*. As a result, any link of *srlg* (except the protected link $G-H$) can be utilized to build the new backup path $b'_G$. For instance, the backup path $G \to D \to B \to C \to E \to H$ can be selected (as $b'_G$) to protect the primary path $p$ against the failure of link $G-H$.

To summarize, the steps of algorithm 1 permit the computation of a backup path with our FRBPC algorithm. In the first step, the links verifying the bandwidth constraints are selected according to (6). In the second step, the set of risks whose failure operates the backup path which is being computed are determined. Finally, in the last step, we run any BPC algorithm on the network topology reduced to the links and nodes (1) verifying the bandwidth constraints (step 1) and (2) whose failure does not operate the backup path (step 2) which is being computed.

## 4 Analysis and Simulation Results

### 4.1 Simulation Model

In order to evaluate the performances of our FRBPC algorithm, we compared it to two classical approaches: Kini's heuristic [4] and the TDRA algorithm [8]. We have chosen the Kini's heuristic for its practicability whereas we opted for the TDRA algorithm for its efficiency to reduce the protection bandwidth allocation.

Two metrics are used for the comparison: ratio of rejected backup paths (*RRP*) and normalized SRLG protection bandwidth (*NSPB*).

The first metric *RRP* measures the ratio of backup paths that are rejected because of the lack of protection bandwidth on the links. It corresponds to the ratio between the number of backup path requests that are rejected and the total number of backup path requests. Formally, *RPR* is computed as follows: *RRP* = *#rejected protection requests / #protection requests*

The second metric *NSPB* measures the efficiency of the SRLG protection bandwidth allocations. For classical approaches (i.e. Kini's heuristic and TDRA algorithm), this metric is determined as the ratio between the sum of the SRLG protection costs and the cumulative bandwidth of the backup paths on all the links. For our FRBPC algorithm, this metric is determined as the ratio between the sum of the SRLG protection prices and the cumulative bandwidth of the backup paths on all the links. Note that, more high the *NSPB* is, less SRLG can be protected and more protection bandwidth is wasted.

To focus only on the impact of the compared methods on the ratio of rejected backup paths and on the normalized SRLG protection bandwidth, we splitted the capacity of each unidirectional link in two pools: primary pool and protection pool. The primary pool is used to allocate the bandwidth for the primary paths whereas the protection pool is used for backup path bandwidth allocations. In

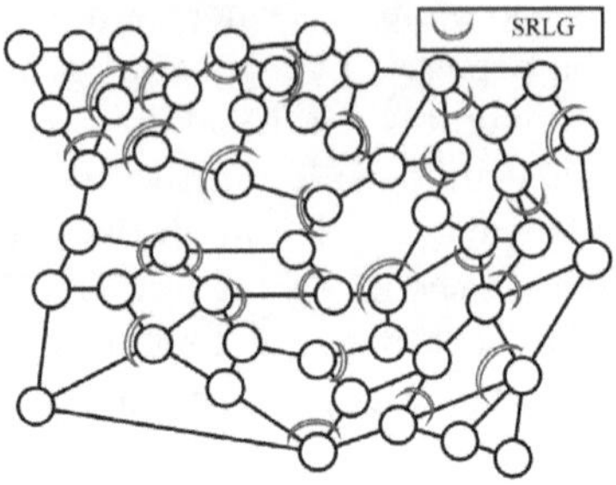


**Fig. 3.** Test topology (162 risks)

our simulations, we considered that the primary pool capacities are sufficient to satisfy all the requests of primary path establishment whereas we set the protection pool capacity ($PC_\lambda$) of each link $\lambda$ to 200 units. Hence, to ensure the respect of the bandwidth constraints, the protection bandwidth allocated on each unidirectional link should be always lower or equal to the corresponding protection pool capacity (i.e. $\forall\lambda : Bk_\lambda \leq PC_\lambda$).

The network topology used in our tests is depicted in figure 3. It is composed of 162 risks: 50 routers, 87 bidirectional links and 25 SRLGs (crescent-shaped in figure 3). The traffic matrix is generated randomly and consists of requests arriving one by one and asking for quantities of bandwidth uniformly distributed between 1 and 10. The head-end and tail-end routers of each primary path are chosen randomly among the network routers. Both the primary and backup path computations are based on the Dijkstra's algorithm.

At each establishment of 20 primary paths, the two metrics *RRP* and *NSPB* are computed for the compared methods.

### 4.2 Results and Analysis

Figure 4 depicts the evolution of *RRP* as a function of the number of primary paths setup in the network. This figure shows clearly that the *RRP* values of the FRBPC algorithm are lower and better (except for the 240 first primary paths where the *RRP* values of the three compared methods are null) than those of TDRA which are in turn lower than those of Kini's heuristic.

The wide difference in the *RRP* values between the Kini's heuristic and the FRBPC algorithm is essentially due to the partial knowledge of the protection bandwidth information with the Kini's heuristic (thus, the Kini's heuristic overestimates the bandwidth parameters required for the BPC) whereas the FRBPC algorithm (and the TDRA algorithm) utilizes and has a complete knowledge of the protection bandwidth parameter information. Concerning the comparison between the *RRP* values of TDRA and those of FRBPC, we note that the difference is large and considerable although it is not high in relation to the total number of protection requests. For instance, the difference varies between 4.5%

and 5.5% when the number of primary paths is between [600, 800]). When rejection of the protection requests is not allowed (as desired by the Internet service providers), the selection of FRBPC instead of TDRA permits to increase the number of protected primary paths from 240 to 400. Obviously, the positive difference between the *RRP* values of FRBPC and TDRA is totally due to the presence of SRLGs in the network.

In figure 5, the evolution of the normalized SRLG protection bandwidth (*NSPB*) as a function of the number of primary paths setup in the network is depicted. As we see, the application of the FRBPC algorithm instead of the TDRA algorithm and the Kini's heuristic permits to save up to 10% of the normalized SRLG bandwidth (i.e. for the 20 first primary paths, we have *NSPB (TDRA) / NSPB (FRBPC)* $\approx$ *NSPB (Kini) / NSPB (FRBPC)* $\approx$ 1.1). This difference in the *NSPB* values between FRBPC and TDRA (or Kini's heuristic) is due to the limitation of the concurrence for the protection bandwidth allocations (see section 3.2) and to the reduction of the risks to be bypassed by each backup path (see section 3.3) with FRBPC (contrarily to TDRA algorithm and Kini's heuristic which waste the protection bandwidth and bypass more risks).

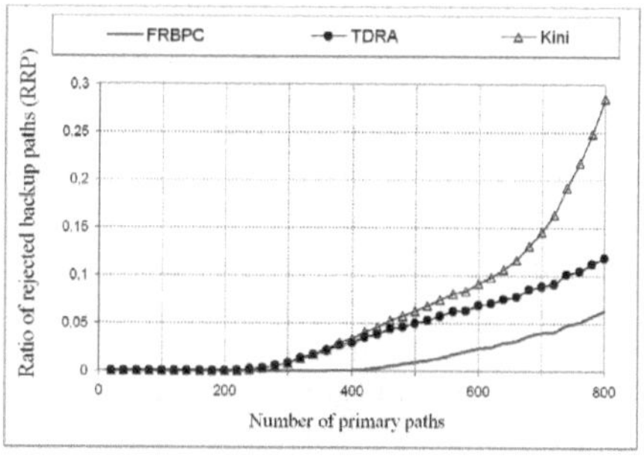


**Fig. 4.** Ratio of rejected backup paths (RRP)

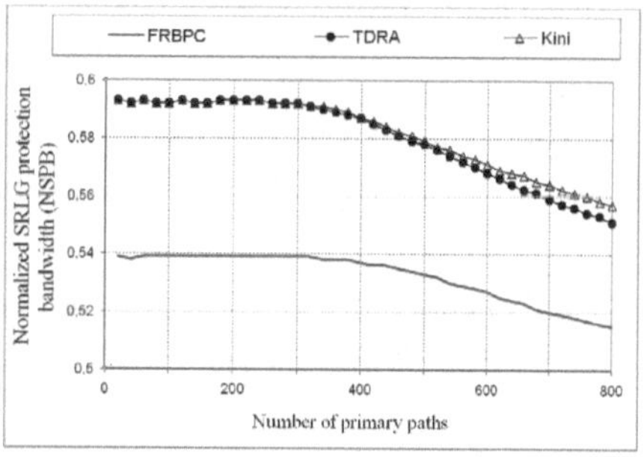


**Fig. 5.** Normalized SRLG protection bandwidth (NSPB)

## 5 Conclusion

In this paper, we shown that upon a SRLG failure some activated backup paths are *inoperative* (they don't receive traffic) and don't participate to the recovery process. As the *operative* state of a backup path can be determined beforehand by taking into account the risk structures (particularly the SRLG structures), we proposed a new algorithm, called Failure Risk-based Backup Path Classification (FRBPC) algorithm, decreasing the protection bandwidth allocations and providing more flexibility for the path selection.

Since it is useless to protect against the failure of a SRLG whose failure activates but does not operates a backup path, we proposed to restrict the set of SRLGs to be protected to those whose failure operates the backup path which is being computed. In this way, the amount of bandwidth required to protect against SRLG failures is decreased and much more flexibility is provided for the backup path selection. As a result, the reject probability of new protection requests is decreased.

Simulations results show that our failure risk-based backup path classification algorithm decreases the number of rejected backup paths and reduces the amount of protection bandwidth dedicated to the protection against the SRLG risks.

## References


1. Meyer, P., Van Den Bosch, S., Degrande, N.: High Availability in MPLS-based Networks. Alcatel telecommunication review (4th Quarter 2004)
2. Ramamurthy, S., Mukherjee, B.: Survivable WDM Mesh Networks (Part I - Protection). In: IEEE INFOCOM, vol. 2, pp. 744–751 (1999)
3. Pan, P., Swallow, G., Atlas, A.: Fast Reroute Extensions to RSVP-TE for LSP Tunnels. RFC 4090 (May 2005)
4. Kini, S., Kodialam, K., Lakshman, T.V., Sengupta, S., Villamizar, C.: Shared Backup Label Switched Path Restoration. Internet Draft draft-kini-restoration-shared-backup-01.txt, IETF (May 2001)
5. Kodialam, M.S., Lakshman, T.V.: Dynamic Routing of Locally Restorable Bandwidth Guaranteed Tunnels using Aggregated Link Usage Information. In: IEEE INFOCOM, pp. 376–385 (2001)
6. Mélon, L., Blanchy, F., Leduc, G.: Decentralized Local Backup LSP Calculation with Efficient Bandwidth Sharing. In: Proceeding of 10th International Conference on Telecommunications (February 2003)
7. Saidi, M.Y., Cousin, B., Le Roux, J.L.: A Distributed Bandwidth Sharing Heuristic for Backup LSP Computation. In: Global Telecommunications Conference (IEEE GLOBECOM 2007), Washington (USA), pp. 2477–2482 (November 2007)
8. Saidi, M.Y., Cousin, B., Le Roux, J.L.: Targeted Distribution of Resource Allocation for Backup LSP Computation. In: Seventh European Dependable Computing Conference (May 2008)
9. Vasseur, J.P., Charny, A., Le Faucheur, F., Achirica, J., Le Roux, J.L.: Framework for PCE-based MPLS-TE Fast Reroute Backup Path Computation. Internet Draft draft-leroux-pce-backup-comp-frwk-00.txt, IETF (July 2004)
10. Le Roux, J.L., Calvignac, G.: A Method for an Optimized Online Placement of MPLS Bypass Tunnels. Internet Draft draft-leroux-mpls-bypass-placement-00.txt, IETF (February 2002)
11. Kompella, K., Rekhter, Y.: Intermediate System to Intermediate System (IS-IS) Extensions in Support of Generalized Multi-Protocol Label Switching (GMPLS). RFC 4205 (October 2005)
12. Kompella, K., Rekhter, Y.: OSPF Extensions in Support of Generalized Multi-Protocol Label Switching (GMPLS). RFC 4203 (October 2005)
13. Aggarwal, R., Kompella, K., Nadeau, T., Swallow, G.: BFD For MPLS LSPs. Internet Draft draft-ietf-bfd-mpls-07.txt, IETF (June 2008)

# An Efficient Analytical Model for the Dimensioning of WiMAX Networks

Bruno Baynat[1], Georges Nogueira[1], Masood Maqbool[2], and Marceau Coupechoux[2]

[1] Universite Pierre et Marie Curie - Paris, France
{firstname.lastname}@lip6.fr
[2] Telecom ParisTech - Paris, France
{firstname.lastname}@telecom-paristech.fr

**Abstract.** This paper tackles the challenging task of developing a simple and accurate analytical model for performance evaluation of WiMAX networks. The need for accurate and fast-computing tools is of primary importance to face complex and exhaustive dimensioning issues for this promising access technology. In this paper, we present a generic Markovian model developed for three usual scheduling policies (slot sharing fairness, throughput fairness and opportunistic scheduling) that provides closed-form expressions for all the required performance parameters at a click speed. This model is compared in depth with realistic simulations that show its accuracy and robustness regarding the different modeling assumptions. Finally, the speed of our analytical tool allows us to carry on dimensioning studies that require several thousands of evaluations, which would not be tractable with any simulation tool.

**Keywords:** WiMAX, performance evaluation, dimensioning, analytical models.

## 1 Introduction

The evolution of last-mile infrastructure for wired broadband networks faces acute implications such as difficult terrain and high cost-to-serve ratio. Latest developments in wireless domain could not only address these issues but could also complement the existing framework. One of such highly anticipated technologies is WiMAX (Worldwide Interoperability for Microwave Access) based on IEEE standard 802.16. The first operative version of IEEE 802.16 is 802.16-2004 (fixed/nomadic WiMAX) [1]. It was followed by a ratification of mobile WiMAX amendment IEEE 802.16e in 2005 [2]. On the other hand, the consortium WiMAX Forum was found to specify profiles (technology options are chosen among those proposed by the IEEE standard), define an end-to-end architecture (IEEE does not go beyond physical and MAC layer), and certify products (through inter-operability tests).

Some WiMAX networks are already deployed but most operators are still under trial phases. As deployment is coming, the need arises for manufacturers

L. Fratta et al. (Eds.): NETWORKING 2009, LNCS 5550, pp. 521–534, 2009.

and operators to have fast and efficient tools for network design and performance evaluation. In [3] authors propose an analytical model for studying the random access scheme of IEEE 802.16d. Niyato and Hossain [4] formulate the bandwidth allocation of multiple services with different QoS requirements by using linear programming. They also propose performance analysis, first at connection level, and then, at packet level. In the former case, variations of the radio channel are however not taken into account. In the latter case, the computation of performance measures rely on multi-dimensional Markovian model that requires numerical resolutions. Not specific to WiMAX systems, generic analytical models for performance evaluation of cellular networks with varying channel conditions have been proposed in [5,6,7]. The models presented in these articles are mostly based on multi-class processor-sharing queues with each class corresponding to users having similar radio conditions and subsequently equal data rates. The variability of radio channel conditions at flow level is taken into account by integrating propagation models, mobility models or spatial distribution of users in a cell. In order to use classical PS-queues results, these papers consider implicitly that users can only switch class between two successive data transfers. However, as highlighted in the next section, in WiMAX systems, radio conditions and thus data rates of a particular user can change frequently during a data transfer. In addition, capacity of a WiMAX cell may vary as a result of varying radio conditions of users. As a consequence, any PS, DPS (discriminatory PS) or even GPS (generalized PS) queue is not appropriate for modeling these channel variations.

In this paper, we develop a novel and generic analytical model that takes into account frame structure, precise slot sharing-based scheduling and channel quality variation of WiMAX systems. Unlike existing models [5, 6, 7], our model is adapted to WiMAX systems' assumptions and is generic enough to integrate any appropriate scheduling policy. Here, we consider three classical policies: *slot sharing fairness*, *instantaneous throughput fairness*, and *opportunistic*. For each of them, we develop closed-form expressions for all performance metrics. Moreover, our approach makes it possible to take into account the so-called "outage" situation. A user experiences an outage, if at a given time radio conditions are so bad that it cannot transfer any data and is thus not scheduled. Once again, classical PS-like queues are not appropriate to model this feature.

The paper is organized as follows. Modeling assumptions are presented in Section 2. Section 3 presents the generic analytical model and its adaption to the three considered scheduling policies. Validation and robustness are discussed in Section 4. Section 5 finally gives an example of WiMAX dimensioning process.

## 2 Modeling Assumptions

The development of our analytical model is based on several assumptions related to the system, the channel, the traffic and the scheduling algorithm. We present here these assumptions. All of them will be discussed in Section 3.4, and, as will be developed in that section, most of them can be relaxed, if necessary, by slightly modifying the model. Wherever required, related details of WiMAX system are specified. Various notations are also introduced in this section.

A WiMAX time division duplex (TDD) frame comprises of slots that are the smallest unit of resource and which occupies space both in time and frequency domain. A part of the frame is used for overhead (e.g., DL_MAP and UL_MAP) and the rest for user data. The duration $T_F$ of this TDD frame is equal to 5 ms [2].

**System assumptions.** We consider a single WiMAX cell and focus on the downlink part which is a critical portion of asymmetric data traffic.

1. Overhead in the TDD frame is assumed to be constant and independent of the number of concurrent active mobile station (MS). As a consequence, the total number of slots available for data transmission in the downlink part is constant and will be denoted by $N_S$.
2. We assume that the number of MS that can simultaneously be in active transfer is not limited. As a consequence, any connection demand will be accepted and no blocking can occur.

One of the important features of IEEE 802.16e is link adaptation: different modulation and coding schemes (MCS) allows a dynamic adaptation of the transmission to the radio conditions. As the number of data subcarriers per slot is the same for all permutation schemes, the number of bits carried by a slot for a given MCS is constant. The selection of appropriate MCS is carried out according to the value of signal to interference plus noise ratio (SINR). In case of outage, i.e., if the SINR is too low, no data can be transmitted without error. We denote the radio channel states as: $MCS_k$, $1 \leq k \leq K$, where $K$ is the number of MCS. By extension, $MCS_0$ represents the outage state. The number of bits transmitted per slot by a MS using $MCS_k$ is denoted by $m_k$. For the particular case of outage, $m_0 = 0$.

**Channel assumption.** The MCS used by a given MS can change very often because of the high variability of the radio link quality.

3. We assume that each MS sends a feedback channel estimation on a frame by frame basis, and thus, the base station (BS) can change its MCS every frame. Since we do not make any distinction between users and consider all MS as statistically identical, we associate a probability $p_k$ with each coding scheme $MCS_k$, and assume that, at each time-step $T_F$, any MS has a probability $p_k$ to use $MCS_k$.

**Traffic assumptions.** The traffic model is based on the following assumptions.

4. All users have the same traffic characteristics. In addition, we don't consider any QoS differentiation here.
5. We assume that there is a fixed number $N$ of MS that are sharing the available bandwidth of the cell.

6. Each of the $N$ MS is assumed to generate an infinite length ON/OFF elastic traffic. An ON period corresponds to the download of an element (e.g., a web page including all embedded objects). The downloading duration depends on the system load and the radio link quality, so ON periods must be characterized by their size. An OFF period corresponds to the reading time of the last downloaded element, and is independent of the system load. As opposed to ON, OFF periods must then be characterized by their duration.
7. We assume that both ON sizes and OFF durations are exponentially distributed. We denote by $\bar{x}_{on}$ the average size of ON data volumes (in bits) and by $\bar{t}_{off}$ the average duration of OFF periods (in seconds).

**Scheduling assumption.** The scheduling algorithm is responsible for allocating radio resources to users. In wireless networks, scheduling may take into account their radio link quality. In this paper, we have considered three traditional schemes. The slot fairness scheduling allocates the same number of slots to all active users. The throughput fairness scheduling ensures that all active users have the same instantaneous throughput. The opportunistic scheduling gives all resources to active users with the best channel.

8. At any time and for all scheduling policies, if there is only one active user, we assume that the scheduler can allocate all the available slots for its transfer.

## 3 WiMAX Analytical Model

### 3.1 Markovian Model

A first attempt for modeling this system would be to develop a multi-dimensional Continuous Time Markov Chain (CTMC). A state $(n_0, ..., n_K)$ of this chain would be a precise description of the current number $n_k$ of MS using coding scheme $MCS_k$, $0 \leq k \leq K$ (including outage). The derivation of the transitions of such a model is an easy task. However the complexity of the resolution of this model makes it intractable for any realistic value of $K$. In order to work around the complexity problem, we aggregate the state description of the system into a single dimension $n$, representing the total number of concurrent active MS, regardless of the MCS they use. The resulting CTMC is thus made of $N + 1$ states as shown in Fig 1.

- A transition out of a generic state $n$ to a state $n + 1$ occurs when a MS in OFF period starts its transfer. This "arrival" transition corresponds to one MS among the $(N - n)$ in OFF period, ending its reading, and is performed with a rate $(N - n)\lambda$, where $\lambda$ is defined as the inverse of the average reading time: $\lambda = \frac{1}{\bar{t}_{off}}$.
- A transition out of a generic state $n$ to a state $n - 1$ occurs when a MS in ON period completes its transfer. This "departure" transition is performed with a generic rate $\mu(n)$ corresponding to the total departure rate of the frame when $n$ MS are active.

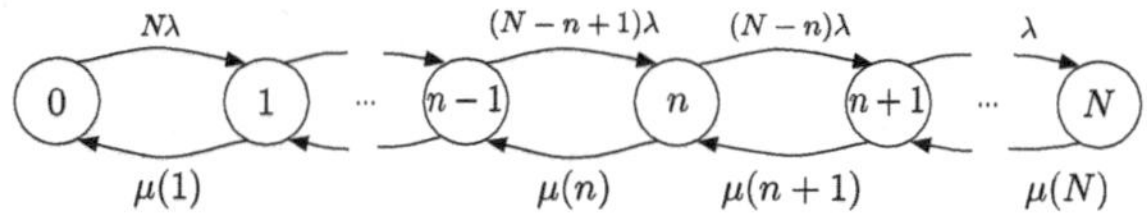


**Fig. 1.** General CTMC with state-dependent departure rates

Obviously, the main difficulty of the model resides in estimating the aggregate departure rates $\mu(n)$. In order to do so, we first express $\mu(n)$ as follows:

$$\mu(n) = \frac{\bar{m}(n)\, N_S}{\bar{x}_{on}\, T_F}, \tag{1}$$

where $\bar{m}(n)$ is the average number of bits per slot when there are $n$ concurrent active transfers. Obviously, $\bar{m}(n)$ depends on $K$, the number of MCS, and $p_k$, $0 \leq k \leq K$, the MCS vector probability. It also strongly depends on $n$, because the number of bits per slot must be estimated by considering all possible distributions of the $n$ MS between the $K+1$ possible MCS (including outage). It is worthwhile noting that the parameters $\bar{m}(n)$ finally depend on the scheduling policy, as it defines, at each time-step, the quantity of slots given to each of the $n$ MS with respect to the MCS they use.

In order to provide a generic expression of $\bar{m}(n)$, we define $x_k(j_0, ..., j_K)$ the proportion of the resource (i.e., of the $N_S$ slots) that is associated to a MS using $MCS_k$, when the current distribution of the $n$ MS among the $K+1$ coding schemes is $(j_0, ..., j_K)$. The average number of bits per slot, $\bar{m}(n)$, when there are $n$ active users, can then be expressed as follows:

$$\bar{m}(n) = \sum_{\substack{(j_0,...,j_K) = (0,...,0) | \\ j_0 + ... + j_K = n \\ j_0 \neq n}}^{(n,...,n)} \left( \sum_{k=1}^{K} m_k j_k x_k(j_0, ..., j_K) \right) \binom{n}{j_1, ..., j_K} \prod_{k=0}^{K} p_k^{j_k}, \tag{2}$$

where $\prod_{k=0}^{K} p_k^{j_k}$ is the probability of any distribution of the $n$ MS such that the number of MS using $MSC_k$ is $j_k$, and $\binom{n}{j_0,...,j_K}$ is the multinomial coefficient that takes into account all such possibles distributions.

## 3.2 Scheduling Policy Modeling

We now present the adaptation of the model, for the three specific scheduling policies we consider in this paper. For each of them we provide closed-form expressions for the average number of bits per slots, $\bar{m}(n)$.

**Slot sharing fairness.** Each time-step, the scheduler equally shares the $N_S$ slots among the active users that are not in outage. If, at a given time-step, there are $n$ active MS, each of the MS that are not in outage receives a portion $\frac{N_S}{n-j_0}$ of the whole resource. As a consequence, the proportion of the resource that is associated to a MS using $MCS_k$, is thus given by: $x_k(j_0, ..., j_K) = \frac{1}{n-j_0}$ for any $k \neq 0$. By replacing these proportions in generic expression (2) we obtain:

$$\bar{m}(n) = \sum_{\substack{(j_0, ..., j_K) = (0, ..., 0) | \\ j_0 + ... + j_K = n \\ j_0 \neq n}}^{(n, ..., n)} \frac{n!}{n - j_0} \left( \sum_{k=1}^{K} m_k j_k \right) \prod_{k=0}^{K} \frac{p_k^{j_k}}{j_k!}. \quad (3)$$

**Instantaneous throughput fairness.** The resource is shared in order to provide the same instantaneous throughput to all active users that are not in outage. This policy allows MS using MCS with a low bit rate per slot to obtain, at a given time-step, proportionally more slots compared to MS using a MCS with a high bit rate per slot. In order to respect instantaneous throughput fairness between all active users that are not in outage, the $x_k(j_0, ..., j_K)$ must be such that: $m_k x_k(j_0, ..., j_K) = C$ for any $k \neq 0$, where $C$ is a constant such that $\sum_{k=1}^{K} j_k x_k(j_0, ..., j_K) = 1$. By replacing the proportions $x_k(j_0, ..., j_K)$ in generic expression (2), the average number of bits per slot $\bar{m}(n)$ becomes:

$$\bar{m}(n) = \sum_{\substack{(j_0, ..., j_K) = (0, ..., 0) | \\ j_0 + ... + j_K = n \\ j_0 \neq n}}^{(n, ..., n)} \frac{(n - j_0)\, n! \prod_{k=0}^{K} \frac{p_k^{j_k}}{j_k!}}{\sum_{k=1}^{K} \frac{j_k}{m_k}}. \quad (4)$$

**Opportunistic scheduling.** All the resource is given to users having the highest transmission bit rate, i.e., the better radio conditions and then the better MCS. Without loss of generality, we assume here that the MCS are classified in increasing order: $m_0 < m_1 < ... < m_K$. And even if it is still possible to derive the average bit rates from generic expression (2), we prefer to give here a more intuitive and equivalent derivation.

We consider a system with $n$ current active MS. We denote by $\alpha_i(n)$ the probability of having at least one active user (among $n$) using $MCS_i$ and none using a MCS giving higher transmission rates (i.e., $MCS_j$ with $j > i$). As a matter of fact, $\alpha_i(n)$ corresponds to the probability that the scheduler gives at a given time-step all the resource to MS that use $MCS_i$. As a consequence, we can express the average number of bits per slot when there are $n$ active users as:

$$\bar{m}(n) = \sum_{i=1}^{K} \alpha_i(n) m_i. \quad (5)$$

In order to calculate the $\alpha_i(n)$, we first express the probability that there are no MS using a MCS higher than $MCS_i$ as: $p_{\leq i}(n) = \left(1 - \sum_{j=i+1}^{K} p_j\right)^n$. Then, we calculate the probability that there is at least one MS using $MCS_i$ conditioned by the fact that there are no MS using a better MCS: $p_{=i}(n) = 1 - \left(1 - \frac{p_i}{\sum_{j=0}^{i} p_j}\right)^n$. $\alpha_i(n)$ can thus be expressed as: $\alpha_i(n) = p_{=i}(n)\, p_{\leq i}(n)$.

### 3.3 Performance Parameters

The steady-state probabilities $\pi(n)$ can easily be derived from the birth-and-death structure of the Markov chain (depicted in Fig. 1):

$$\pi(n) = \frac{N!}{(N-n)!} \frac{T_F^n \rho^n}{N_S^n \prod_{i=1}^{n} \bar{m}(i)} \pi(0), \tag{6}$$

where $\rho$ is given by relation (7) and plays a role equivalent to the "traffic intensity" of Erlang laws [8], and $\pi(0)$ is obtained by normalization.

$$\rho = \frac{\bar{x}_{on}}{\bar{t}_{off}} \tag{7}$$

The performance parameters of this system can be derived from the steady-state probabilities as follows. The average utilization $\bar{U}$ of the TDD frame is:

$$\bar{U} = \sum_{n=1}^{N} (1 - p_0^n) \pi(n). \tag{8}$$

The average number of active users $\bar{Q}$ is expressed as:

$$\bar{Q} = \sum_{n=1}^{N} n\, \pi(n). \tag{9}$$

The mean number of departures $\bar{D}$ (MS completing their transfer) by unit of time, is obtained as: $\bar{D} = \sum_{n=1}^{N} \pi(n)\, \mu(n)$. From Little's law, we can derive the average duration $\bar{t}_{on}$ of an ON period (duration of an active transfer): $\bar{t}_{on} = \frac{\bar{Q}}{\bar{D}}$. We finally compute the average throughput $\bar{X}$ obtained by each MS in active transfer as:

$$\bar{X} = \frac{\bar{x}_{on}}{\bar{t}_{on}}. \tag{10}$$

### 3.4 Discussion of the Modeling Assumptions

Our Markovian model is based on several assumptions presented in Section 2. We now discuss these assumptions one by one (item numbers are related to the corresponding assumptions), evaluate their accuracy, and provide, if necessary and possible, extensions and generalization propositions.

1. DL_MAP and UL_MAP are located in the downlink part of the TDD frame. They contain the information elements that allow MS to identify the slots to be used. The size of these MAPs, and as a consequence the number $N_S$ of available slots for downlink data transmissions, depends on the number of MS scheduled in the TDD frame. In order to relax assumption 1, we can express the number of data slots, $N_S(n)$, as a function of $n$, the number of active users. This dependency can be easily integrated in the model by replacing $N_S$ by $\prod_{i=1}^{n} N_S(n)$ in relation (6), and $N_S$ by $N_S(n)$ in relation (1).

2. A limit $n_{max}$ on the total number of MS that can simultaneously be in active transfer, can be introduced easily if required. The corresponding Markov chain (Fig. 1) has just to be truncated to this limiting state (i.e., the last state becomes $\min(n_{max}, N)$). As a result, a blocking can occur when a new transfer demand arrives and the limit is reached. The blocking probability can be derived easily from the Markov chain [9].
3. Radio channel may be highly variable or may vary with some memory. Our analytical model only depends upon stationary probabilities of different MCS whatever be the radio channel dynamics. This approach is authenticated through simulations in Section 4.
4. More complex systems with multiple-traffic or differentiation between users would naturally result into more complex models. This is left for future work.
5. Poisson processes are currently used in the case of a large population of users, assuming independence between the arrivals and the current population of the system. As we focus in this paper on the performance of a single cell system, the potential population of users is relatively small. The higher the number of on-going data connections, the less likely the arrival of new ones. Poisson processes are thus a non-relevant choice for our models. Note however that if Poisson assumptions have to be made for connection demand arrivals, one can directly modify the arrival rates of the Markov chain (i.e., replace the state-dependent rates $(N - n)\lambda$ by some constant value, and limit the number of states of the Markov chain as explained above in point 2).
6. Each MS is supposed to generate infinite length ON/OFF session traffic. In [10], an extension to finite length sessions is proposed in the context of (E)GPRS networks, where each MS generates ON/OFF traffic during a session and does not generate any traffic during an inter-session. This work shows that a very simple transformation of traffic characteristics that increases OFF periods by a portion of the inter-session period, enables to derive the average performance from the infinite length session model. The accuracy of this transformation is related to the insensibility of the average performance parameters with regards to the traffic distributions (see next point). A similar transformation can be applied to our WiMAX traffic model.
7. Memoryless traffic distributions are strong assumptions that are validated by several theoretical results on PS-like queues. Several works on insensitivity have shown that the average performance parameters are insensitive to the distribution of ON and OFF periods [11,12,13]. In its generic form, our model is no longer equivalent to any PS-like queue, but we show in Section 4 by comparing our model to extensive simulations (using Pareto distributions), that insensibility still holds or is at least a very good approximation.
8. In some cellular networks (e.g. (E)GPRS), MS have limited transmission capabilities because of hardware considerations. This constraint defines a maximum throughput the network interface can reach or a maximum number of resource units that can be used by the MS. This characteristic has been introduced in the case of (E)GPRS networks [9] and consists in reducing the departure rates of the first states of the Markov chain. The same idea can be applied to our WiMAX model.

# 4 Validation

In this section we discuss the validation and robustness of our analytical model through extensive simulations. For this purpose, a simulator has been developed that implements an ON/OFF traffic generator and a wireless channel for each user, and a centralized scheduler that allocates radio resources, i.e., slots, to active users on a frame by frame basis.

## 4.1 Simulation Models

**System Parameters.** System bandwidth is assumed to be 10 MHz. The downlink/uplink ratio of the WiMAX TDD frame is considered to be 2/3. We assume for the sake of simplicity that the protocol overhead is of fixed length (2 symbols). Considering subcarrier permutation PUSC, the total number of data slots (excluding overhead) per TDD downlink sub-frame is $N_S = 450$.

**Traffic Parameters.** In our analytical model, we consider an elastic ON/OFF traffic. Mean values of ON data volume (main page and embedded objects) and OFF period (reading time), are 3 Mbits and 3 s respectively.

In the first phase (validation study), we assume that the ON data volume is exponentially distributed as it is the case in the analytical model assumptions. Although well adapted to Markov theory based analysis, exponential law does not always fit the reality for data traffic. This is the reason why we consider truncated Pareto distributions in the second phase (the robustness study). Recall that the mean value of the truncated Pareto distribution is given by equation $\bar{x}_{on} = \frac{\alpha b}{\alpha - 1}\left[1 - (b/q)^{\alpha-1}\right]$, where $\alpha$ is the shape parameter, $b$ is the minimum value of Pareto variable and $q$ is the cutoff value for truncated Pareto distribution. Two values of $q$ are considered: lower and higher. The mean value in both cases ($q = 300$ Mbits and $b = 611822$ bits for the higher cutoff and $q = 3000$ Mbits and $b = 712926$ bits lower cutoff) is 3 Mbits for the sake of comparison with the exponential model. The value of $\alpha = 1.2$ has been adopted from [14].

**Channel Models.** A generic method for describing the channel between the BS and a MS is to model the transitions between MCS by a finite state Markov chain (FSMC). The chain is discrete time and transitions occurs every $L$ frames, with $L\,T_F < \bar{t}_{coh}$, the coherence time of the channel. In our case, and for the sake of simplicity, $L = 1$. Such a FSMC is fully characterized by its transition matrix $P_T = (p_{ij})_{0 \le i,j \le K}$, where state 0 represents outage. Stationary probabilities $p_k$ provide the long term probabilities for a MS to receive data with MCS $k$.

In our analytical study, channel model is assumed to be memoryless, i.e., MCS are independently drawn from frame to frame for each user, and the discrete distribution is given by the $(p_i)_{0 \le i,j \le K}$. This corresponds to the case where $p_{ij} = p_j$ for all $i$. This simple approach, referred as the *memoryless channel model*, is the one considered in the validation study. Let $P_T(0)$ be the transition matrix associated to the memoryless model.

**Table 1.** Stationary probabilities

| Channel model | Memoryless | Average | Combined | |
|---|---|---|---|---|
| | | | good 50% MS | bad 50% MS |
| $a$ | 0 | 0.5 | 0.5 | 0.5 |
| $p_0$ | 0.225 | 0.225 | 0.020 | 0.430 |
| $p_1$ | 0.110 | 0.110 | 0.040 | 0.180 |
| $p_2$ | 0.070 | 0.070 | 0.050 | 0.090 |
| $p_3$ | 0.125 | 0.125 | 0.140 | 0.110 |
| $p_4$ | 0.470 | 0.470 | 0.750 | 0.190 |

**Table 2.** Channel parameters

| Channel state $\{0, ..., K\}$ | MCS and outage | Bits per slot $m_k$ |
|---|---|---|
| 0 | Outage | $m_0 = 0$ |
| 1 | QPSK-1/2 | $m_1 = 48$ |
| 2 | QPSK-3/4 | $m_2 = 72$ |
| 3 | 16QAM-1/2 | $m_3 = 96$ |
| 4 | 16QAM-3/4 | $m_4 = 144$ |

In the robustness study, we introduce two additional channel models with memory. In these models, the MCS observed for a given MS in a frame depends on the MCS observed in the previous frame according to the FSMC presented above. The transition matrix is derived from equation $P_T(a) = aI + (1-a)P_T(0)$ given that $0 \le a \le 1$. In this equation, $I$ is the identity matrix and parameter $a$ is a measure of the channel memory. A MS maintains its MCS for a certain duration with mean $\bar{t}_{coh} = 1/(1-a)$. With $a = 0$, the transition process becomes memoryless. On the other extreme, with $a = 1$, the transition process will have infinite memory and MS will never change its MCS. For simulations we have taken $a$ equal to 0.5, so that the channel is constant in average 2 frames. This value is consistent with the coherence time given in [15] for 45 Km/h at 2.5 GHz. We call the case where all MS have the same channel model with memory ($a = 0.5$), the *average channel model.* Note that the stationary probabilities of the average channel model are the same as those of the memoryless model.

As the channel depends on the BS-MS link, it is possible to refine the previous approach by considering part of the MS to be in a "bad" state, and the rest in a "good" state. Bad and good states are characterized by different stationary probabilities but have the same coherence time. In the so called *combined channel model*, half of the MS are in a good state, the rest in a bad state, and $a$ is kept to 0.5 for both populations.

Three models are thus considered: the memoryless, the average, and the combined channel models. Wireless channel parameters are summarized in Tab. 2. Considered MCS are given including outage, and for each of them, the number of bits transmitted per slot. Channel stationary probabilities are given in Tab. 1. The probabilities for the combined model are obtained by averaging corresponding values of good and bad model stationary probabilities.

### 4.2 Simulation Results

In this section, we first present a comparison between the results obtained through our analytical model and scheduling simulator. The output parameters in consideration are $\bar{U}$, $\bar{X}$, and $\pi(n)$ (see Section 3.3).

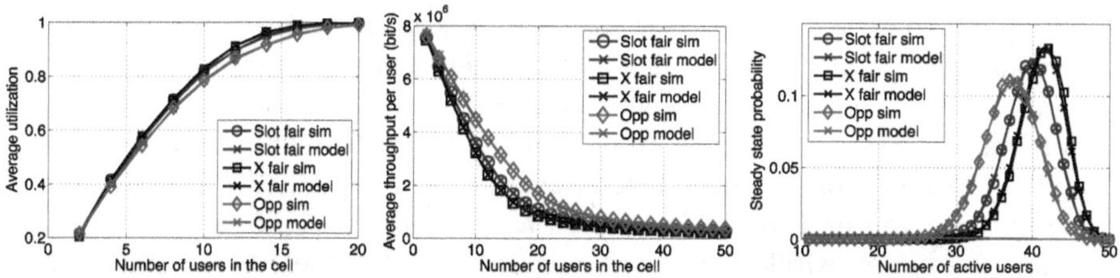


(a) Average utilization. (b) Average throughput per user. (c) Stationary probabilities for $N = 50$.

**Fig. 2.** Validation for the three scheduling policies with $\bar{x}_{on} = 3$ Mbits and $\bar{t}_{off} = 3$ s

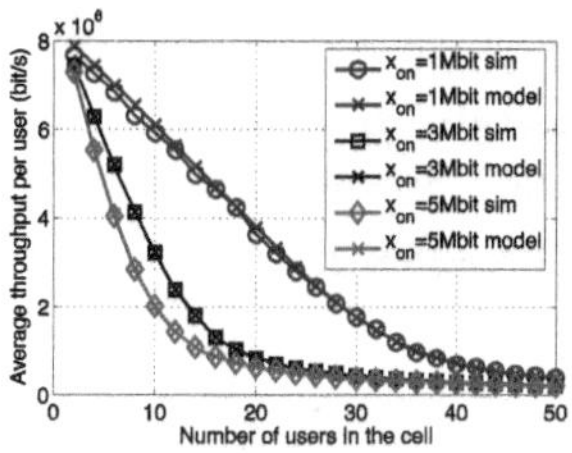


**Fig. 3.** Average throughput per user for different loads

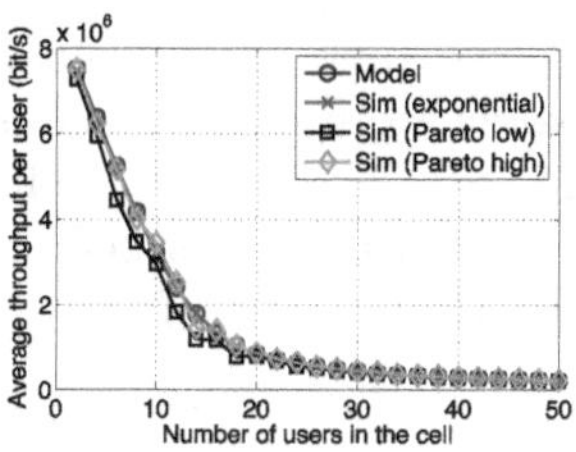


**Fig. 4.** Average throughput per user for different traffic distributions

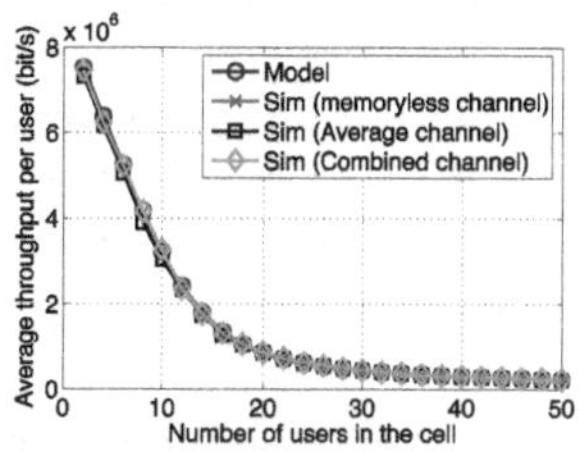


**Fig. 5.** Average throughput per user for different channel models

**Validation Study.** In this study, simulations take into account the same traffic and channel assumptions as those of the analytical model. However, in simulator MCS of users are determined on per frame basis and scheduling is carried out in real time, based on MCS at that instant. The analytical model on the other hand, considers stationary probabilities of MCS only.

Fig. 2(a, b) show respectively the average channel utilization ($\bar{U}$) and the average instantaneous throughput per user ($\bar{X}$) for the three scheduling schemes. It is clear that simulation and analytical results show a good agreement: for both utilization and throughput, the maximum relative error stays below 6% and the average relative error is less than 1%. Fig. 2(c) further proves that our analytical model is a very good description of the system: stationary probabilities $\pi(n)$ are compared with those of simulations for a given total number $N = 50$ of MS. Again results show a perfect match between two methods with an average relative error below 9%. At the end, Fig. 3 shows the validation for three different loads (1, 3 and 5 Mbps). Our model shows a comparable accuracy for all three load conditions with a maximum relative error of about 5%.

**Robustness Study.** In order to check the robustness of our analytical model towards distribution of ON data volumes, simulations are carried out for exponential and truncated pareto (with lower and higher cutoff). The results for this analysis are shown in Fig. 4. The average relative error between analytical results and simulations stays below 10% for all sets. It is clear that considering a truncated Pareto distribution has little influence on the design parameters.

Next we evaluate the robustness of our analytical model with respect to the channel model. We compare the analytical results with simulation for the three pre-cited channel models: memoryless, average and combined (with stationary probabilities given in Tab. 1). If we look at the plot of Fig. 5, we can say that even for a complex wireless channel, our analytical model shows considerable robustness with an average relative error below 7%. We can thus deduce that for designing a WiMAX network, channel information is almost completely included in the stationary probabilities of the MCS.

# 5 Network Design

In this section we provide some examples to demonstrate application of our model while considering throughput fairness scheduling. However, results can be obtained in the same manner for other scheduling schemes by using their respective average bits per slot $\bar{m}(n)$.

## 5.1 Performance Graphs

We first draw 3-dimensional surfaces where performance parameters are function of, e.g., $N$, the number of users in the cell and $\rho$, the combination of traffic parameters. For each performance parameter, the surface is cut out into level lines and the resulting 2-dimensional projections are drawn. The step between level lines can be arbitrarily chosen.

The average radio resource utilization of the WiMAX cell $\bar{U}$, and the average throughput per user $\bar{X}$ for any MS in the system are presented in Fig. 6 and 7 (corresponding to the radio link characteristics presented in Section 4). These graphs allow to directly derive any performance parameter knowing the traffic load profile, i.e., the couple $(N, \rho)$. Each graph is the result of several thousands of input parameter sets. Obviously, any simulation tool or even any multi-dimensional Markov chain requiring numerical resolution, would have precluded the drawing of such graphs.

## 5.2 Dimensioning Study

In this section, we show how our model can be advantageously used for dimensioning issues. Two examples, each respecting a certain QoS criterion, are given.

In Fig. 8 we find minimum number $N$ of MS in the cell to guarantee that the average radio utilization is over 50%. This kind of criterion allows operators to maximize the utilization of network resource in comparison with the traffic load of their customers. For a given traffic load profile and a given set of system parameters, the point of coordinates $(N_S, \rho)$ in the graph is located between two level lines, and the level line with the higher value gives the optimal value of $N$.

The QoS criterion chosen for second example is the user throughput. We have taken 50 Kbps, an arbitrary value of minimum user throughput. Next we

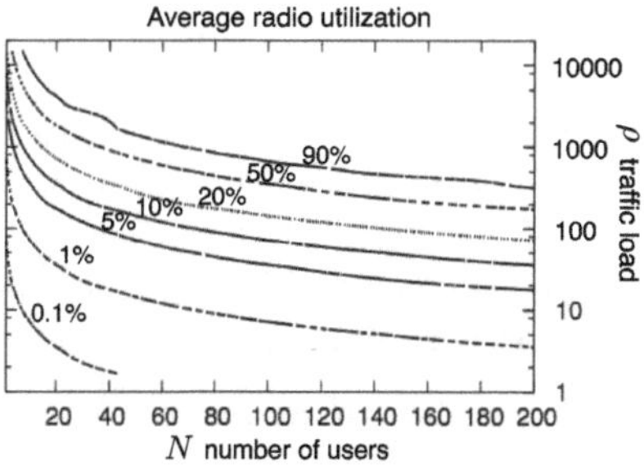


**Fig. 6.** Average utilization $\bar{U}$

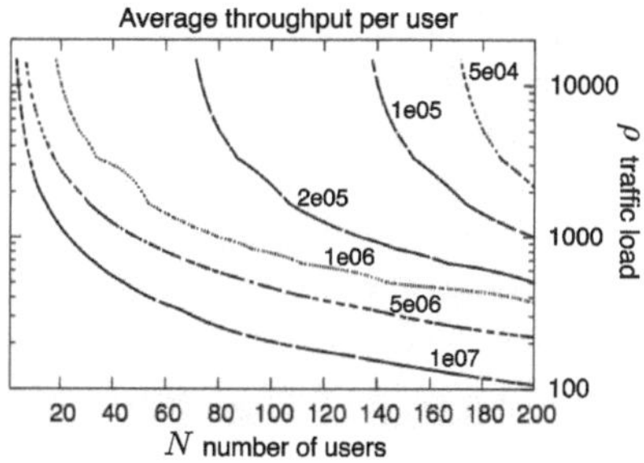


**Fig. 7.** Average throughput per user $\bar{X}$

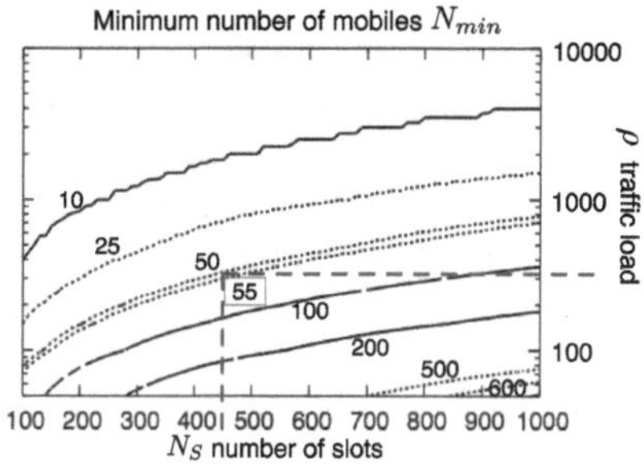


**Fig. 8.** Dimensioning the minimum value of $N$ for having $\bar{U} \geq 50\%$

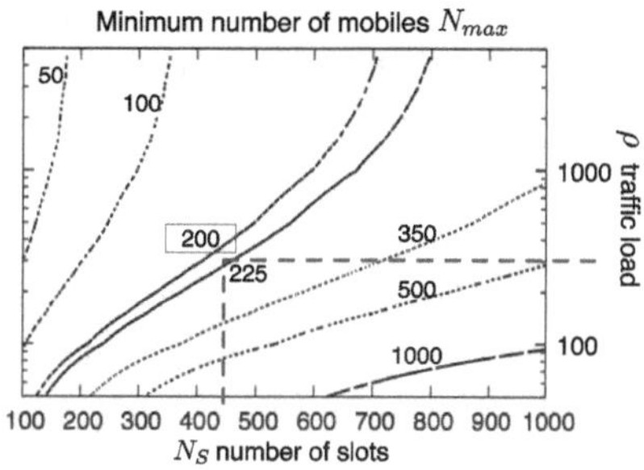


**Fig. 9.** Dimensioning the maximum value of $N$ for having $\bar{X} \geq 50$ Kbps per user

find the maximum number $N_{max}$ of users in the cell to guarantee the minimum throughput threshold. In Fig. 9, a given point $(N_S, \rho)$ is located between two level lines. The line with the lower value gives $N_{max}$. As explained before, the average throughput per user is inversely proportional to $N$.

The graphs of Fig. 9 and 8 can be jointly used to satisfy multiple QoS criteria. For example, if we have a WiMAX cell configured to have $N_S = 450$ slots and a traffic profile given by $\rho = 300$ (e.g., $x_{on} = 1.2$ Mbits and $t_{off} = 20$ s), Fig. 8 gives $N_{min} = 55$, and Fig. 9 gives $N_{max} = 200$. The combination of these two graphs recommend to have a number of users $N \in [55; 200]$ to guarantee a reasonable resource utilization and a minimum throughput to users.

## 6 Conclusion

As deployment of WiMAX networks is underway, need arises for operators and manufacturers to develop dimensioning tools. In this paper, we have presented novel analytical models for WiMAX networks and elastic ON/OFF traffic. The models are able to derive Erlang-like performance parameters such as throughput per user or channel utilization. Based on a one-dimensional Markov chain and the derivation of average bit rates, whose expressions are given for three main scheduling policies (slot fairness, throughput fairness and opportunistic scheduling), our model is remarkably simple. The resolution of model provides closed-form expressions for all the required performance parameters at a click-speed. Extensive simulations have validated the model's assumptions. The accuracy of the model is illustrated by the fact that, for all simulation results, maximum relative errors do not exceed 10%. Even if the traffic and channel assumptions are relaxed, analytical results still match very well with simulations that shows the robust nature of our model.

## References


1. Group, I.S.W.: IEEE Standard for local and metropolitan area networks - Part 16: Air Interface for Fixed Broadband Wireless Access Systems (2004)
2. Group, I.S.W.: Draft IEEE std 802.16e/D9. IEEE Standard for local and metropolitan area networks - Part 16: Air Interface for Fixed Broadband Wireless Access Systems (2005)
3. Vinel, A., Zhang, Y., Lott, M., Tiurlikov, A.: Performance analysis of the random access in IEEE 802.16. In: Proc. of IEEE PIMRC, pp. 1596–1600 (September 2005)
4. Niyato, D., Hossain, E.: A queuing-theoretic and optimization-based model for radio resource management in IEEE 802.16 broadband networks. IEEE ToC, vol. 55 (2006)
5. Borst, S.: User-level performance of channel-aware scheduling algorithms in wireless data networks. In: IEEE Infocom. (2003)
6. Bonald, T., Proutiere, A.: Wireless downlink channels: User performance and cell dimensioning. In: ACM Mobicom. (2003)
7. Liu, S., Virtamo, J.: Performance Analysis of Wireless Data Systems with a Finite Population of Mobile Users. In: 19th ITC. (2005)
8. Engset, T.O.: On the calculation of switches in an automatic telephone system. In: Tore Olaus Engset: The man behind the formula (1998)
9. Baynat, B., Eisenmann, P.: Towards an Erlang-Like formula for GPRS/EDGE network engineering. In: IEEE Int. Conf. on Communications (ICC) (June 2004)
10. Baynat, B., Boussetta, K., Eisenmann, P., Rached, N.B.: Towards an Erlang-Like formula for the performance evaluation of GPRS/EDGE networks with finite-length sessions. In: Proc. of 3rd IFIP-TC6 Networking Conference (May 2004)
11. Berger, A., Kogan, Y.: Dimensioning bandwidth for elastic traffic in high-speed data networks. IEEE/ACM Transactions on Networking 8, 643–654 (2000)
12. Fredj, S.B., Bonald, T., Proutiere, A., Regnie, G., Roberts, J.: Statistical bandwidth sharing: A Study of congestion at flow level. In: Proc. of ACM Sigcomm. (August 2001)
13. Heyman, D., Lakshman, T., Neidhardt, A.: New method for analyzing feedback protocols with applications to engineering web traffic over the internet. In: Proc. of the ACM Sigmetrics (June 1997)
14. Feldmann, A., Gilbert, A.C., Huang, P., Willinger, W.: Dynamics of IP traffic: A study of the role of variability and the impact of control. Computer Communication Review (October 1999)
15. Ramadas, K., Jain, R.: WiMAX System Evaluation Methodology. Technical report, Wimax Forum (January 2007)

# Performance Evaluation of Gradient Routing Strategies for Wireless Sensor Networks

Fadila Khadar and Tahiry Razafindralambo

INRIA Lille - Nord Europe, IRCICA/LIFL, CNRS UMR 8022, Univ. Lille 1
Parc Scientifique de la Haute Borne
50, avenue Halley - BP 70478
59658 Villeneuve d'Ascq, France
tahiry.razafindralambo@univ-lille1.fr, fadila.khadar@inria.fr

**Abstract.** We consider Wireless Sensor Networks (WSN) applications in which sensors have to send data to a unique sink in a multi-hop fashion. Gradient routing protocol is a scalable way to route data in these applications. Many gradient routing protocols exist, they mainly differ in their performances (delay, delivery ratio, etc.). In this paper, we propose an extensive performance evaluation study of some gradient routing protocols in order to give guidelines for WSN developers.

**Keywords:** Sensor Network, Routing protocol, Gradient.

## 1 Introduction

A Wireless Sensor Network (WSN) is a set of autonomous objects with limited processing and storage capabilities which cooperate in order to perform a common task. They are receiving more and more attentions due to their potential applications in various areas such as monitoring, security and data gathering. The communication paradigm in WSN is very characteristic since in most applications nodes send data only to a sink. Furthermore, they operate on limited capacity batteries. As a result, WSN protocols' design, especially routing protocol, is a challenging task.

The routing problem in WSN has been the subject of intense studies. One important difference between wired and wireless networks is the use of location for routing purposes. Position awareness improves the efficiency and scalability of routing protocols as it helps reducing the number of messages used for route discovery. This information could be obtained by GPS or by an internal service. GPS is not a pratical solution in WSNs, as sensors would have to be equipped with additional hardware, which would increase both cost and energy consumption. Yet, other solutions exist, such as localization protocols based on trilateration and triangulation ([1],[2]). Other solutions include the use of virtual coordinates, i.e., a coordinate system set up for routing purposes only. This is how work gradient routing protocols. They create a one-dimensional virtual coordinate system where the position of a node corresponds to its hop distance to the sink. This information is then used to efficiently route packets to the sink in

L. Fratta et al. (Eds.): NETWORKING 2009, LNCS 5550, pp. 535–547, 2009.

a multi-hop fashion. When receiving a packet, a node forwards it if it is closer to the sink than the previous sender. Since each sensor receiving the packet decides locally whether it should forward it, no state information about its neighborhood has to be kept, thus reducing routing complexity.

Many variants of gradient routing protocols have been proposed in the literature. They mostly differ in the way a node decides whether it should forward a received packet. As a result, their performances concerning the message delivery ratio, the end-to-end delay, and the number of nodes involved in the transmission of a message also differ.

We propose an extensive performance evaluation of different gradient routing algorithm strategies for wireless sensor networks. We identified three basic schemes: basic, probabilistic and unicast schemes. The evaluation of these strategies shows that there exists a trade-off between the four considered metrics (delay, delivery ratio, overhead and energy consumption).

This paper is organized as follows. In Section 2 we give a brief review of gradient-based routing protocols. In Section 3 we describe the assumptions we make. We then describe the algorithm we consider in Section 4. The simulation results are discussed in Section 5. Finally we conclude in Section 6.

## 2 Related Work

A number of routing protocols have been proposed for wireless sensor networks [3]. They aim at providing energy efficient, low delay and low overhead routing process. In this section we briefly describe data centric routing protocol in which the sink sends a request to certain regions (or all the network) and waits for data from the sensors which are located in these regions. For a more complete review of existing routing protocols for sensor networks please refer to [3] or other surveys.

The most classical protocols are flooding and gossiping [4] [5] [6] [7]. In flooding, each sensor broadcasts the packet to all its neighbors. The basic gradient protocol described in this paper is equivalent to flooding except that only a subset of sensors are allowed to broadcast packets. Flooding is very easy to implement and delivery ratio can be very high. However, it has several drawbacks, the most important one being the overhead generated by the transmission of a single message. The probabilistic flooding described in this paper is very similar to gossip algorithm especially the one described in [6]. Gossiping algorithms alleviate the problem of overhead compared to flooding algorithm. However delivery ratio can be very low.

Direct diffusion [8] is one of the first routing protocols that introduces gradient. In this protocol, an interest is flooded by the sink. The gradient is a reply link to a neighbor from which the interest was received. Rumor routing [9] is a variant of direct diffusion where flooding is not used.

In Gradient-based routing protocol [10], the authors propose a variant of Directed diffusion. The authors keep the number of hops when the interest is flooded into the network. Therefore each sensor can evaluate the number of hops to the sink. The difference between a node's height and the one of its neighbor is considered as the gradient on that link. A packet is forwarded on a link with the largest gradient. In this paper, we use this version of gradient routing protocol as a basis. We do not use the version proposed in [10] in this paper since we do not consider data aggregation. Moreover, we mainly focus on forwarding scheme and consider that sensors do not have any information on possible next hops for the forwarding scheme.

## 3 Background and Assumptions

We assume a large number of quasi-static wireless sensor nodes and time-driven applications: the main task of each sensor node is to periodically collect data from the sensing devices. These data are then sent to the sink in a multi-hop fashion due to a limited radio range. The data are sent toward the sink using a geographic routing based on gradient. The sensor nodes are assumed to use CSMA/CA MAC protocol such as 802.15.4 [11]. Acknowledgement scheme is used for unicast packet but broadcast packets are not acknowledge.

Each periodically collected data is stored into a packet P and broadcasted by the node. This packet also contains the depth of the node ($\texttt{P}.depth = node.depth$), the source of the packet ($\texttt{P}.dsrc = node.id$) and a sequence number ($\texttt{P}.seq = node.seq$) which is generated by the node.

Each node also keeps the list of sequence numbers of packets it has already forwarded. It is worth noting that our application does not need to store neighbors' list nor exchange discovery message (after the initialization phase) due to the stateless aspect of the presented protocols.

In this paper, the term "broadcast" stands for message propagation in a node's neighborhood and the term "flooding" refers to network-wide message propagation. In the sequel, given a node $n$, we use the denotation $n.x$ to refer to the variable $x$ at the node $n$ and given a message M and a name $x$ that identifies a field, we use the denotation $\texttt{M}.x$ to refer to the field $x$ in message M.

## 4 Algorithm Description

In this section, we describe the gradient establishment and the three algorithms that we evaluate. The gradient establishment or gradient construction is the scheme used to help routing process. The three algorithms are all based on the same gradient construction. The only difference between the algorithms is the way sensors forward each received packets to the sink.

### 4.1 Gradient Construction

A BUILD message is flooded in the network starting from the sink $s$. The BUILD message contains the source of the message $src$ which is the $id$ of the node and the $depth$ of the node. Initially, the sink sets $\texttt{BUILD}.depth = s.depth = 0$ and $\texttt{BUILD}.src = s$ . Each node $n$ that receives the BUILD message set its own $n.depth$

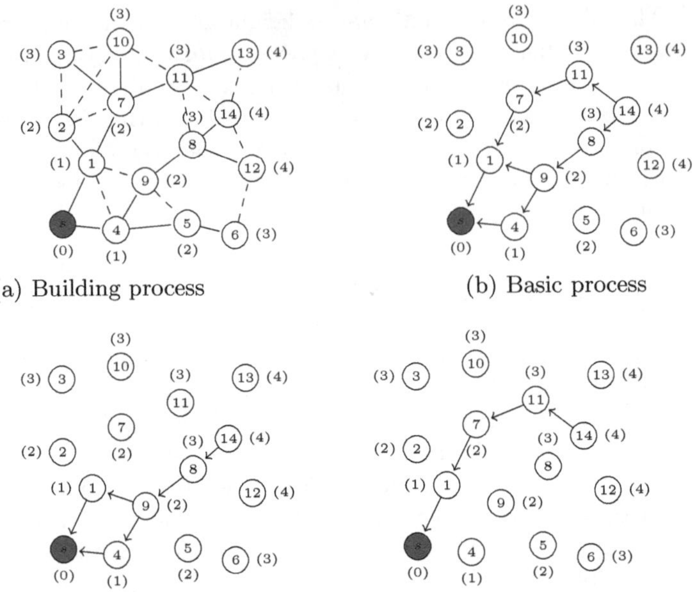

(a) Building process

(b) Basic process

(c) Probabilistic process

(d) Unicast process

**Fig. 1.** Illustration of the gradient routing. The sink is node $s$. In Figure 1(a) Lines represent `BUILD` message flooding used to set node depth. The number in brackets represents the node depth. Dashed lines represent communication links. Figure 1(b), 1(c) and 1(d) are examples of basic, probabilistic and unicast process based on the depth of each node.

to BUILD.*depth* + 1 and broadcasts the BUILD message with its own *depth* and also changes the BUILD.*src* field. A node can receive a BUILD message more than once. In this case, the node keeps the smallest *depth* received among all the BUILD messages but does not send the BUILD message if it has already sent one. See Figure 1(a) for the resulting process.

### 4.2 Basic Gradient Routing Protocol

Upon receiving a packet P which is not a BUILD message a node computes algorithm 1. It first checks if the depth in the packet field is greater than its own depth and if the sequence number of the packet is not in its packet sequence number list. If these conditions (Line 2 of Alg. 1) are met, the packet depth is changed to the actual node depth and the sequence number of the packet is added into the sequence list. If the node $n$ is not the sink the packet is broadcasted.

Figure 1(b) gives an example of the paths obtained by the basic gradient routing protocol. Here, node 14 broadcasts a message, only nodes 8 and 11 forward it since their depths are smaller than the depth of node 14.

**Algorithm 1.** Basic Gradient routing protocol – Forwarding process

**Message forwarding on node $n$:**

```
1: Reception of packet P
2: if (P.depth > n.depth) && (P.seq ∉ n.seqlist) then
3:    if (n ≠ sink ) then
4:       P.depth = n.depth ;
5:       n.seqlist ← P.seq
6:       broadcast packet P ;
7:    else
8:       Process packet P ;
9:    end if
10: else
11:    Drop P;
12: end if
```

### 4.3 Probabilistic Gradient Routing Protocol

The forwarding policy is the only difference between the basic gradient routing protocol and the probabilistic one. Indeed, in the probabilistic version, a packet P is forwarded based on a probability $p$. Line 3 of Alg. 1 is changed to:

[3-1] **if** $(n \neq sink \ \&\& \ rand(0,1) \leq p)$ **then**

The idea behind the probabilistic version of the gradient protocol is to reduce the overhead due to message forwarding while maintaining a good delivery ratio. Figure 1(c) gives an example of the paths obtained by the probabilistic gradient routing protocol. Here, node 14 broadcasts a message and only node 8 forwards it due to the proability and since its depth is smaller than the depth of node 14.

### 4.4 Unicast-Based Gradient Routing Protocol

The unicast version of the gradient routing is built to reduce the overhead due to forwarding while maintaining a good delivery ratio. However this version is resource costly for nodes that are in the forwarding path. It is also worth noting that this version strongly relies on gradient construction.

In this version of gradient routing, each node needs to store the packet source of the BUILD message it considers for its depth. This information is stored at each node in a variable called *hop* ($n.hop =$ BUILD$.src$). The forwarding process described in Alg. 1 is modified on Line 6 to implement the unicast version by the following line:

[6-1] unicast packet P to $n.hop$;

Figure 1(d) gives an example of the path obtained by the unicast-based gradient routing protocol. Here, the message follows a single path based on the depth of each node.

## 5 Performance Evaluation

We evaluate the performances of the gradient routing protocols through simulations using WSNet[1]. The performance metrics we use are delays, delivery ratio and overhead (number of duplicated messages received by the sink).

### 5.1 Simulation Settings

**Network Topology.** We consider scenario where static nodes are randomly deployed in a 1000 meters $\times$ 1000 meters flat square. The radio range is set to be 100 meters. The physical layer is modelled by the unit disk graph model. The total number of nodes varies from 100 to 400 and only connected networks are considered. We assume only one sink (at position (0,0)) initiates the gradient construction.

**Node Settings.** Sensors have to send data every 10 seconds with a random jitter to avoid synchronization. Each sensor has a buffer $B$ of size 10 to keep packets it has to forward. FIFO policy is used for packet transmission. It has also a table $S$ of size 20 in which it stores sequence numbers of packets it has already forwarded.

**Metrics of Interest.** We consider three metrics:

**Delay** Time between the sending of a packet and the first reception of it at the sink node.

**Delivery ratio** number of distinct received packets on number of distinct sent packets.

**Overhead** average number of times the same packet is received by the sink.

**Gradient Algorithms Settings.** For the probabilistic version of gradient routing, the probability of packet forwarding is set to 0.5. For the unicast version, data packets are acknowledged and retransmitted in case of failure following the scheme described in the 802.15.4 standard. Broadcast packets are also sent using the 802.15.4 standard.

As our focus is to evaluate the routing process we do not try to optimize the gradient construction. It is worth noting that some optimizations avoiding unstable or long links increase the performance of the routing protocol. These optimizations on gradient construction are left to future work.

The simulation results are divided into three parts presented in Figure 2 for part I, Figures 3, 4, and 5 for part II and Figure 6 and 7 for part III.

The first part (Section 5.2) shows a performance comparison between the three routing strategies depending on the number of nodes in the network. None of the studied schemes outperforms the other ones in all the metrics. The second part (Section 5.3) studies the behavior of each protocol according to nodes' depth. As expected, it shows that the delay increases when the depth increases. In the third part we evaluate the energy consumption of each algorithm and

[1] http://wsnet.gforge.inria.fr

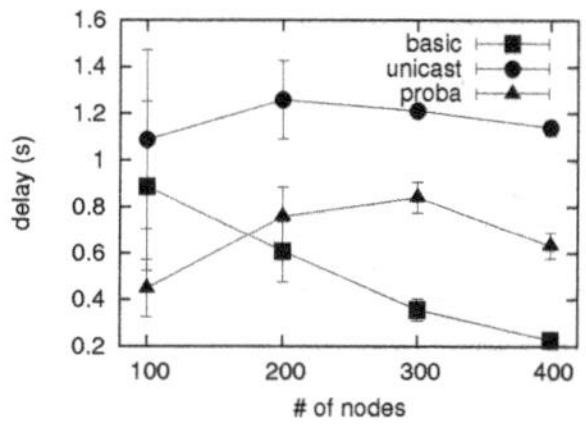


(a) Delay

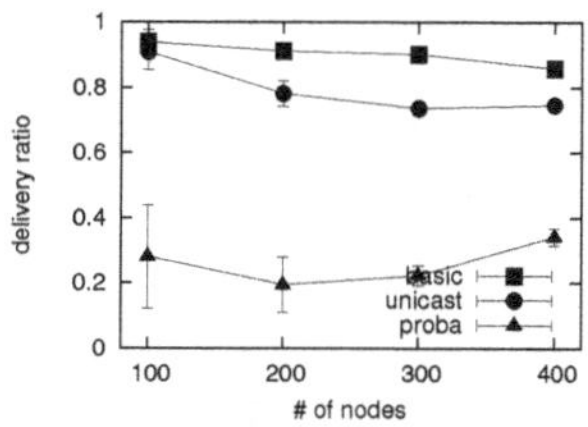


(b) Delivery ratio

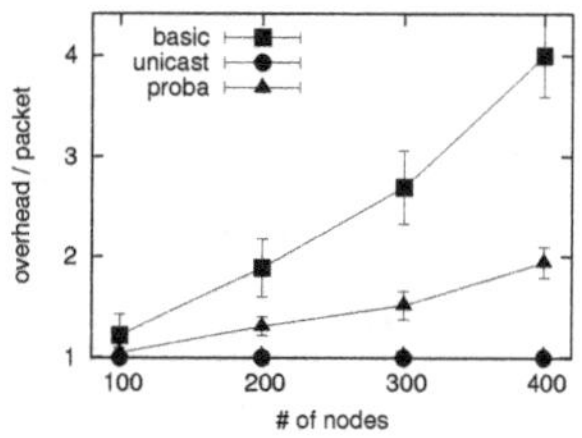


(c) Overhead

**Fig. 2.** Simulation results

plot the energy map at different simulation time. The results of this part show that the proabilistic version has the best performance among the three versions (Section 5.4).

### 5.2 Simulation Results: Density Effect (Part I)

**Delay.** Figure 2(a) plots the mean end-to-end delay for different number of sensors in the network. Surprisingly, when the number of nodes increases, the mean delays for the basic version of the gradient decreases. This is mainly due to the multiple path followed by each packet. This is confirmed by Figure 2(c) which shows the duplication ratio for each packet. Indeed, when the number of nodes increases, the number of possible forwarding nodes also increases. Therefore, packets can follow different paths using nodes that are not overloaded, decreasing the delay of each packet. The unicast version performs badly in term of delay. The mean delay remains constant when the number of nodes increases. This can be explained by the double effect increasing density has. On the one hand it increases the collision rate, leading to higher delays as packets must be retransmitted. On the other hand, delay decreases due to shorter paths in hop count. As a result, the end-to-end delay of the unicast version is stable.

We can notice that the probabilistic version is a shift of the basic version. Indeed, the mean delay for 400 nodes in *proba* is close to the one for 200 nodes of *basic* and the mean delay for *proba* 300 nodes is close to the one for 150 nodes of *basic*.

**Delivery ratio.** The Figure 2(b) depicts the delivery ratio for the three strategies. The basic scheme is more efficient than the others. The delivery ratio for *basic* is always above 85%. This is explained by the multiple paths a packet can follow. The delivery ratio is decreasing when the number of nodes increases. This is due to the increasing number of collisions when the node density increases. *proba* gives a low but increasing delivery ratio. When the number of nodes is low there may exist only one single path from a sensor to the sink. In this case, the probability that a packet is successfully received by the sink is $P_r = (p)^{hops}$ where $p$ is the forwarding probability, a parameter of *proba* and *hops* is the number of hops from a given sensor to the sink. When the number of nodes increases,

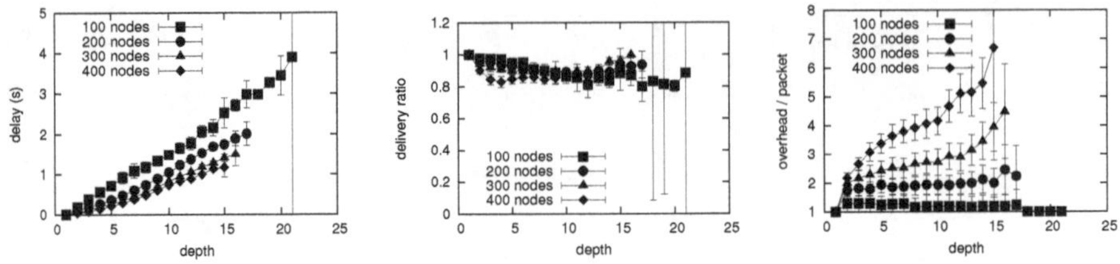


(a) Delay (b) Delivery ratio (c) Overhead

**Fig. 3.** Simulation results basic gradient

the probability that multiple paths exist increases and thus the delivery ratio increase for *proba*.

The delivery ratio of *unicast* is around 80%. This high value is due to the retransmission used at the MAC layer for unicast packets. However, when the number of nodes increases, the number of collisions also increases, which then reduces the delivery ratio.

**Overhead.** The overhead of each strategy is plotted in Figure 2(c). We can see that the overhead of *unicast* is 1, which means that each packet is received only once by the sink. This is the perfect behavior.

The overhead of *basic* is increasing with the number of nodes. This is mainly due to the multipath effect produced by the forwarding scheme. The overhead *proba* is half of the overhead of *basic* when the number of nodes is 300 and 400. It is also interesting to notice that when the number of nodes is low (100), the overhead is close to 1 for the three strategies, reflecting the fact that there is no multiple disjoint path from a sensor to a sink.

### 5.3 Simulation Results: Depth Effect (Part II)

**Basic.** The simulations for the basic gradient routing protocol are presented in Figure 3. The delay, delivery ratio and the overhead are plotted depending on the depth of the node initiating te message. Figure 3(a) shows that the delay

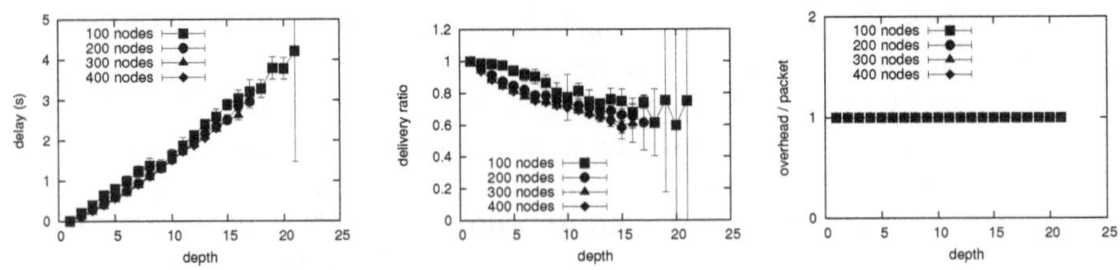


(a) Delay (b) Delivery ratio (c) Overhead

**Fig. 4.** Simulation results unicast gradient

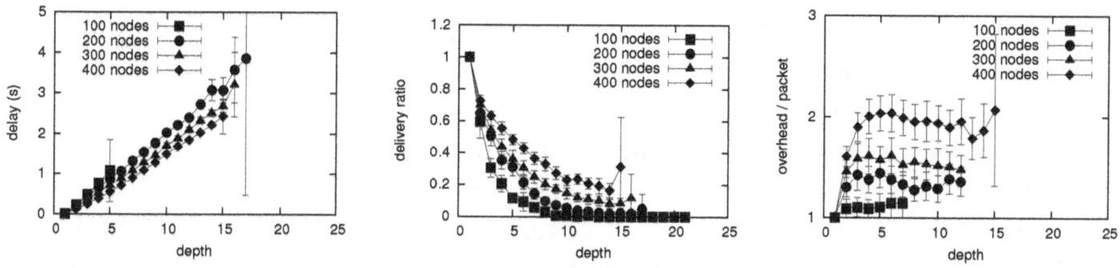


(a) Delay (b) Delivery ratio (c) Overhead

**Fig. 5.** Simulation results probabilistic gradient

increases when the depth increase, as expected. We can also notice that the maximum depth is lower when the number of nodes increases (21 for 200 nodes, and 15 for 400 nodes). This first result also indicates that for any depth the delay decreases when the number of nodes increases.

Figure 3(b) plots the delivery ratio depending on the depth of the originating nodes. The depth does not seem to strongly affect the delivery ratio. The latter remains roughly constant. This behavior is explained by the multiple paths followed by each packet. The overhead for each packet depending on the depth is drawn in Figure 3(c). We can see that for 100 nodes, the overhead is decreasing along with the depth. This is probably due to the small number of paths available when the source node has an higher depth. For the other densities, the overhead is increasing when the number of nodes increases. Again, this is due to the multiple paths a packet can follow when the depth increases. For 200 nodes, we can see that the overhead remains stable. For this density, only a constant number of multiple paths can be found.

**Unicast.** Figure 4 shows the simulation results for the unicast gradient routing protocol. The mean delay increases in the same way for each density when the depth increases ( Figure 4(a)). This is due to the fact that all packets follow only one path and that the delay for a one hop communication is constant. Figure 4(b) plots the delivery ratio depending on the depth. We can see that the delivery ratio decreases when the depth increases. This is mainly due to the buffer size of each node. In the unicast version, packets follow the same path which overloads the buffer. Packets may then be dropped. When the number of hops increases (depth), the probability for a packet to be dropped also increases, which explains this behavior. In Figure 4(c), we can see that the overhead is equal to 1 as unicast routing is used.

**Probabilistic.** The simulation results for probabilistic gradient routing are shown in Figure 5. The results are similar to the basic gradient results (Figure 5(a)). However, the maximum depth changes depending on the number of nodes. When the number of node is 100, the maximum depth is only 5. This means that for the probabilistic version, messages coming from higher depth do not reach the sink. This also confirms the fact that for 100 nodes the probability of having multiple paths is very low. A simple way to avoid this problem is to modify the forwarding probability depending on the node depth. With the probabilistic approach, the delivery ratio (see Figure 5(b)) decreases when the depth increases. This is foreseeable since for a given path the delivery probability decreases when the number of hops increases.

Figure 5(c) shows the overhead for each packet depending on the depth of the originating nodes. For all number of nodes, the overhead increases and then remains stable. This is due to the number of paths reduction caused by the probability of retransmission. It is worth noting that the shape of the *proba* curve of 400 nodes is similar to the shape of *basic* the curve for 200 nodes. This indicates that the effect of the probability is the same as the effect of reducing the number of nodes.

### 5.4 Energy Consumption (Part III)

In the previous simulation battery lifetime was set to infinity to evaluate the performance of each protocol. In this part we introduce results about energy consumption for the different algorithms. The energy model we use is very simple and linear: the energy cost of a transmission is 2, the energy cost of a reception is 1. Each sensor has initially 200000 units of battery.

The graph in Figure 6 shows the evolution of the percentage of dead nodes over time for the three algorithms. The unicast version seems to perform better than the other two versions after 160 secondes. Yet, one must be cautious with these results as they do not indicate the geographical repartition of dead nodes. Indeed, if all dead nodes are located near the sink, the network becomes useless. For better understanding the repartition of dead nodes across the network we use energy maps.

The energy maps of each algorithm at different step of the simulation (20s, 30s, 100s and 200s) for 200 sensors in a field of $1000m \times 1000m$ are drawn on Figure 7. The sink is at the bottom-left corner of each figure and has the

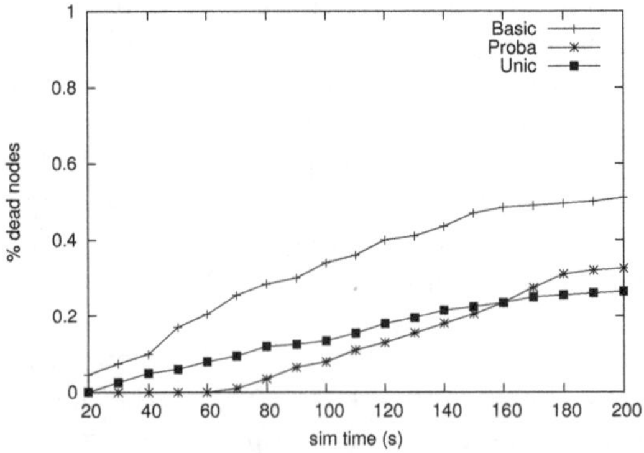


**Fig. 6.** Percentage of dead nodes over time for the 3 algorithms. After 160s the unicast version performs better, but the mean delay is higher (see Figure 2(a)).

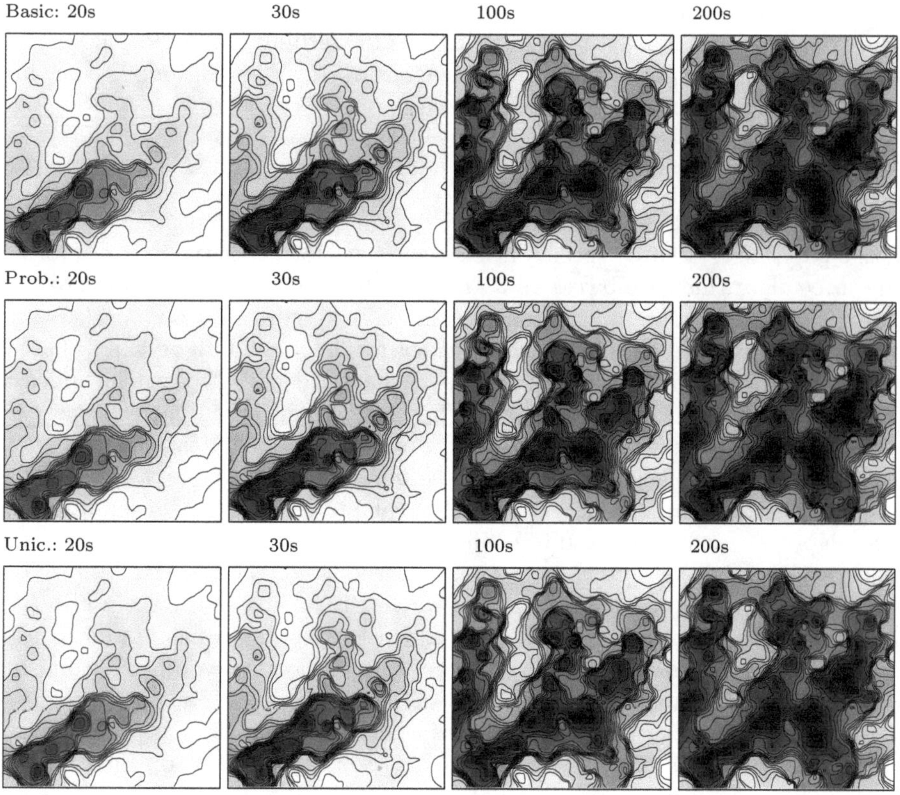


**Fig. 7.** Energy map after of each algorithm at 20, 30, 100, and 200 seconds of simulation. Black (blank) color means that sensors within this area have 0% (100%) battery left.

geographical coordinate (0,0). All other parameters are left unchanged. Black color means that sensors within the area have no battery left (0%). Blank color means that battery is full (100%).

We can conclude from Figure 7 that the worst algorithm is the basic version. Indeed, after 20s most of the nodes within the sink's neighborhood have 0% battery left. This is due to the high number of sensors, especially sensors close to the sink, involved in the reception and (re)transmission of packets. For the unicast version, after 30s, most of the nodes around the sink have 0% battery left. After 100s and for the unicast version, nodes within the paths followed by the packets have almost 0% battery left. The probabilistic version exhibits better performances compared to both unicast and basic versions. We can see that for the probabilistic version, most of the nodes within the sink neighborhood are dead after 200s.

These results demonstrate that there exists a trade-off between the network lifetime and the performances of the gradient routing algorithm. Note that the results may be different depending on the energy model used but the conclusion are still the same. It is also worth noting that the performance of the probabilistic algorithm can be enhanced by modifying the forwarding probability according to the remaining battery of each sensor such as in [12]. However, modifying this probability may affect other performance metrics.

## 6 Conclusion and Future Work

In this paper we evaluated three routing algorithms for wireless sensor networks. The three algorithms are based on a gradient that helps sensors to forward the data they receive to the sink. We first evaluated an algorithm where packets are forwarded by each node using local broadcast. The second algorithm is based probabilistic forwarding of the packet and the third one is based on unicast forwarding of the packet. We evaluated these algorithms based on three metrics namely the delay, the delivery ratio and the overhead. Our simulations show that the performances of each scheme are very different and that none of the studied scheme is better than the other. We also evaluate the energy consumption of each algorithm and these results confirm that there exists a tradeoff between delivery ratio and energy comsumption. The choice of the algorithm must be based on what is expected from the network. If reliability is the main criteria, the basic scheme should be used at the expense of energy consumption. If the lost of some messages is not critical, the unicast version is a good compromise. The next step of this work is to evaluate the impact of having a better gradient construction and maintenance on the performance of each routing protocol. Indeed, in the gradient construction we use gradient may not be optimal due to message loss. We also want to evaluate the effect of different parameters especially for the probabilistic version of the gradient or evaluate the combination of different strategies depending on the depth of the originating sensor.

## References

1. Niculescu, D., Nath, B.: Ad hoc positioning system (aps). In: GLOBECOM 2001, vol. 5, pp. 2926–2931. IEEE, Los Alamitos (2001)
2. Bulusu, N., Heidemann, J., Estrin, D.: Gps-less low-cost outdoor localization for very small devices. Personal Communications 7, 28–34 (2000)
3. Akkaya, K., Younis, M.: A survey on routing protocols for wireless sensor networks. Elsevier Ad Hoc Network Journal 3, 325–349 (2005)
4. Haas, Z., Halpern, J., Li, L.: Gossip-based ad hoc routing. IEEE/ACM Trans. Netw. 14(3), 479–491 (2006)
5. Maroti, M.: Directed flood-routing framework for wireless sensor networks. In: Jacobsen, H.-A. (ed.) Middleware 2004. LNCS, vol. 3231, pp. 99–114. Springer, Heidelberg (2004)
6. Barrett, C.L., Eidenbenz, S.J., Kroc, L., Marathe, M., Smith, J.P.: Parametric probabilistic sensor network routing. In: WSNA 2003, New York, NY, USA, pp. 122–131. ACM, New York (2003)
7. Hedetniemi, S.M., Hedetniemi, S.T., Liestman, A.L.: A survey of gossiping and broadcasting in communication networks. Networks 18(4), 319–349 (1988)

8. Intanagonwiwat, C., Govindan, R.D., Estrin, D.: Directed diffusion: a scalable and robust communication paradigm for sensor networks. In: MobiCom 2000, pp. 56–67 (2000)
9. Braginsky, D., Estrin, D.: Rumor routing algorthim for sensor networks. In: Proceedings of WSNA 2002, pp. 22–31. ACM, New York (2002)
10. Schurgers, C., Srivastava, M.: Energy efficient routing in wireless sensor networks. In: IEEE MILCOM 2001, vol. 1, pp. 357–361 (2001)
11. De Nardis, L., Di Benedetto, M.G.: Overview of the ieee 802.15.4/4a standards for low data rate wireless personal data networks. In: 4th Workshop on Positioning, Navigation and Communication, WPNC 2007, pp. 285–289 (March 2007)
12. Jaffres-Runser, K., Comaniciu, C.: A probabilistic interference and energy aware gradient broadcasting algorithm for wireless sensor networks. In: 3rd International Symposium on ISWPC 2008, pp. 1–5 (May 2008)

# Online Estimation of Available Bandwidth and Fair Share Using Kalman Filtering

Zdravko Bozakov and Michael Bredel

Institute for Communications Systems, Leibniz Universität Hannover, Germany
{zdravko.bozakov,michael.bredel}@ikt.uni-hannover.de

**Abstract.** In this paper we address the problem of online bandwidth estimation in wired and wireless LANs. To this end, we employ active probing, i.e. we continuously inject packet probes into the network. We present the key challenges and analyze the trade-offs between fast change detection and estimate smoothness. We show the benefit of using Kalman filtering to obtain optimal estimates under certain conditions and provide a procedure for parameterizing the filter with respect to specific use cases. Furthermore, we evaluate the influence of probing train length on the results. Based on our findings we developed a tool implementing the presented methodology. We support our theoretical results by a number of real-world measurements as well as simulations.

**Keywords:** measurements, online bandwith estimation, fair share, kalman filter.

## 1 Introduction

The idea of using end-to-end measurements to infer the capacity left over by cross traffic, also called available bandwidth, dates back to TCP congestion control [1] and packet pair probing [2]. In the field of wired networks a number of estimation methods, e.g. [3–7] exist that are well understood. The task of bandwidth estimation in wireless LANs, poses additional challenges [8, 9]. Nevertheless, some tools originating from the wired domain have been suggested for available bandwidth estimation in wireless networks [7, 9, 10]. Empirical evaluations for both domains can be found e.g. in [8, 11].

Active probing methods inject synthetic traffic into the network and attempt to infer the unused resources of the network path by analysing the packet dispersion, i.e, the changes in inter-packet gaps. Almost all tools calculate the average of several measurements and report a single bandwidth estimate. Under constant channel conditions, increasing probing intervals or the probing traffic intensity can improve the accuracy of the results. On the other hand, it is often necessary to probe bandwidth continuously in order to accurately detect changes over time. For minimal intrusiveness it is desirable to minimize probing traffic, which might negatively impact the sample accuracy. At the same time, the sampling rate must be high enough to guarantee that changes are detected sufficiently fast.

L. Fratta et al. (Eds.): NETWORKING 2009, LNCS 5550, pp. 548–561, 2009.

In this paper we analyze the constraints related to continuous probing and show how Kalman filtering can be used to improve estimates. We model the available bandwidth and the fair bandwidth share in wireless networks respectively as a time-varying process that is at least piecewise stationary. In wired networks this process is cross traffic dependent while in wireless domains it is also related to the number of nodes and the physical channel conditions. Using trains of packet probes we sample this process continuously to adapt to changes in the channel state, e.g. changing number of transmitting stations, rate adapting stations or cross traffic variations. Additionally, we assume that the maximum probing rate is limited. We show that the estimate variance, which we interpret as *measurement noise*, is related to the train length. Furthermore, we present a procedure for tuning the *process noise* which influences smoothness and agility of the Kalman estimate. We substantiate our analytical findings by an extensive set of measurements and simulations. Furthermore, we address the problem of finding a suitable number of packets per probing stream.

## 2 Related Work

In this section we discuss related work on bandwidth estimation in wireless and wired networks. We focus on active probing techniques rather than passive approaches which analyze cross traffic directly. In wireless networks for instance, the broadcast medium can be used to capture packets of other stations and derive bandwidth information. However, theses methods cannot obtain end-to-end information.

Available bandwidth estimation techniques attempt to infer the capacity left over by cross-traffic in a network path. Hence, the available bandwidth $AB_i$ of a single link $i$ in a time interval $[\tau, t)$ is defined as [12]

$$AB_i(\tau, t) = C_i(1 - u_i(\tau, t)) \tag{1}$$

where $C_i$ is the channel capacity and $u_i \in [0, 1]$ the utilization. The available bandwidth of the network path is given by the minimum available bandwidth over all single links $AB = min_i(AB_i)$.

In recent years a number of tools, which use active probing for available bandwidth estimation have been proposed, e.g. [3, 4, 10]. Probes consisting of packet pairs or packet trains are injected into the network at specified rate. The inter packet gaps at the path egress are then used to infer the available bandwidth. So-called packet chirps [4] represent a special class where packets are sent with a geometrically decreasing gap.

Most tools assume First In First Out (FIFO) multiplexing which is the typical scheduling policy in today's Internet. Thus, network flows obtain a share of the capacity which is proportional to their sending rate. Under the assumption of constant bit rate (CBR) probing rates this yields

$$\frac{r_{in}}{r_{out}} = \max\left(1, \frac{r_{in} + \lambda}{C}\right) = \begin{cases} 1 & , \text{if } r_{in} \leq C - \lambda \\ \frac{r_{in}+\lambda}{C} & , \text{if } r_{in} > C - \lambda \end{cases} \tag{2}$$

also referred to as the rate response curve [6, 13], where $r_{in}$ and $r_{out}$ are the input and output rates of the probing traffic, $\lambda$ is the cross traffic rate and $C$ is the channel capacity. Therefore, estimating the available bandwidth is equivalent to finding the break in the rate response curve. *Iterative* probing tools [14] aim to detect the break using a rate scan, i.e. by iteratively increasing their sending rate. While (2) is typically applied to packet trains, it can easily be mapped to a gap response curve, used for packet pairs [13] by using the expressions $g_{in} = L/r_{in}$ and $g_{out} = L/r_{out}$ where $L$ is a given packet size and $g_{in}$ and $g_{out}$ are the input and output gaps between successive packets.

The BART tool [7], a successor of DietTOPP [10], is an iterative probing method which uses a Kalman filter to obtain both, the end-to-end available bandwidth and the bottleneck capacity. However, the approach employs iterative probing while our work focuses on direct probing and an analytical derivation of the Kalman filter noise parameters.

If the link capacity $C$ is known in advance (2) can be solved for the cross traffic rate $\lambda$ if the probing rate is larger than the available bandwidth. In contrast to iterative probing this yields one estimate for every probe. Hence, rather than testing whether the probing rate is above or below the available bandwidth, the cross traffic itself is sampled. This approach is referred to as *direct* probing [14]. Let $r_{in} = C$ it follows from (2) that the available bandwidth can be calculated directly by [5, 9]

$$AB = C\left(2 - \frac{C}{r_{out}}\right) = C\left(1 - \frac{g_{out} - g_{in}}{g_{in}}\right). \tag{3}$$

In wireless networks the DCF aims to achieve per packet fairness for the medium access. Studies have related this fairness to the fair share achieved by Generalized Processor Sharing, e.g [15, 16]. Hence, the assumption of FCFS scheduling does not hold in wireless LANs. The fair share $fs$ can be computed recursively as a solution of

$$fs : \sum_{i=1}^{M} min\{r_i, fs\} = C \tag{4}$$

where $C$ is the capacity, $r_i$ is the sending rate of station $i$, and $M$ is the number of competing stations. Unlike in FCFS systems in fair queueing systems iterative probing reports the fair share, which represents an upper bound for the available bandwidth. However, using (3) in fair queueing systems yields neither the available bandwidth nor the fair share [8]. Thus, direct probing tools using this equation tend to fail in wireless networks. Nevertheless, direct probing using a rate above the fair share easily reports the fair share, since it is equal to the output rate.

A partially implicit assumption of many bandwidth estimation tools is a simplified network model: the network path is abstracted by a single tight link, cross traffic is viewed as constant rate fluid and channel conditions are invariant. At least the last point is not necessarily true in wireless networks due to interference of other stations, other external radio sources or rate adaption.

## 3 Filtering of Active Probing Measurements

In order to detect changes in the cross traffic process, we probe continuously over time. Therefore, we require an accurate and fast probing approach with low probing overhead. To this end, we use direct probing as it provides fast and fairly reliable samples in FIFO as well as DCF networks. We measure the dispersion of packets in a packet train of length $l+1$ with packet size $L$ and derive the averaged gap of the departures, i.e, packets at the receiver side by

$$g_d = \frac{d(l+1) - d(1)}{l} \tag{5}$$

where $d(n)$ represents the receiving time-stamp of the $n$th packet. In FIFO networks available bandwidth is derived by (3) whereas the fair share is computed by $fs = L/g_d$.

In a perfect constant fluid system, every single sample would lead to a correct bandwidth estimate. Unfortunately, estimates derived through injected packet probes are strongly influenced not only by packetized cross-traffic but also by issues such as interface buffers or timer inaccuracies. We regard the resulting probing train gap variations as *measurement noise*. We found that these effects are diminished when using larger packet trains. Most existing tools use averaging over several probe estimates to deal with measurement noise and obtain more accurate bandwidth estimates [13, 17]. In essence, all approaches utilize some form of filtering. For instance, the use of packet trains itself constitutes a low pass filter. A naive approach for on-line bandwidth estimation is to employ a moving average over the past $N$ samples. In all cases an open question is the suitable filter parametrization. Parameters include train lengths, probing intensity and filter lengths. In the following, we employ the Kalman filter, which known to produce optimal estimates for processes perturbed by Gaussian noise. In section 4.1 we argue, that active probing methods which use long trains of packet probes, produce estimates which are normally distributed around the true available bandwidth. Consequently, Kalman filtering is ideal for eliminating this type of measurement error. Additionally, we relate the Kalman filter to an exponentially weighted moving average filter (EWMA).

In order to correctly parametrize a probing technique, it is vital to consider its use-case. It is evident that in a scenario where the user is only interested in the average available bandwidth over several minutes different settings are required than if changes must be detected within seconds. To this end, we require the user to specify the minimum bandwidth change $B$ to be identified within a given time $T_s$. Furthermore, we assume that the maximum probing rate is constrained to a specific value $r_p$ (to e.g. 5% of the maximum link capacity), resulting in a maximum inter-train sending time $t_\Delta = L(l+1)/r_p$. Given these constraints, our goal is to derive optimal settings for the Kalman filter.

## 4 Bandwidth Estimation Using Kalman Filtering

As shown in [18] the channel access procedure used in the 802.11 DCF transforms the cross traffic process into an uncorrelated, approximately Gaussian distributed random process. Furthermore, in Sect. 4.1 we demonstrate that Poisson cross traffic observed over long timescales has the same properties. Therefore, we model perturbations within packet train samples as a Gaussian noise process. This makes the Kalman filter an ideal candidate for estimating the true channel bandwidth.

In the following, we will outline the working of the Kalman filter and present simplifications, which are applicable when the filter noise parameters are time invariant. Moreover, we will show how suitable noise values should be derived in the context of available bandwidth and fair share measurements. Furthermore, we will demonstrate the effect of the parameters on the filter performance and convergence speed.

It is well known that the Kalman filter is the *minimum mean squared error* (MMSE) estimator when the process and measurement noises are Gaussian. If the Gaussian assumption is dropped, the Kalman filter is still the *best linear unbiased estimator*. The filter can estimate the state $x_k$ at time $k$ of any system which can be represented in a state space framework

$$x_k = Ax_{k-1} + B_k u_k + \omega_k \tag{6}$$

$$z_k = Hx_k + \nu_k \tag{7}$$

where $A$ is a state transition matrix, $B$ is a control-input matrix associated with an external input $u_k$, and $\omega_k$ is a normally distributed process noise with $\omega_k \approx N(0, Q)$. The measurements $z_k$, linked to the system state through an observation matrix $H$ are perturbed by a normally distributed measurement noise $\nu_k \approx N(0, R)$.

The optimal state estimate is obtained by iteratively applying a set of equations [19] known as the time and measurement updates as samples $z_k$, e.g. packet train estimates in our specific use-case, become available. In our probing framework, the state estimate, i.e. the available bandwidth, is scalar, $A = 1$, $B = 0$ and $H = 1$. The reduced Kalman equations are then given by

$$G_k = \frac{P_{k-1} + Q}{P_{k-1} + Q + R} \tag{8}$$

$$\hat{x}_k = \hat{x}_{k-1}(1 - G_k) + G_k z_k \tag{9}$$

$$P_k = (1 - G_k)(P_{k-1} + Q) \tag{10}$$

Above $G_k$ is the Kalman gain and $P_k$ is the filter estimate error variance. Typically, the filter is initialized with a guess for the state estimate $x_0$ and large value $P_0$ representing the uncertainty associated with the guess.

It is evident from equations (8, 9, 10) that the error variance $P_k$ and the Kalman gain $G_k$ are independent of the current state estimate $\hat{x}_k$ and the measurement $z_k$. Moreover, for stationary noises the parameters $P_k$ and $G_k$ quickly converge to constant values $P_\infty$ and $G_\infty$ respectively [19]. The values of these steady-state parameters can be calculated offline analytically

$$P_\infty = \frac{Q}{2} \pm \frac{Q}{2}\sqrt{1 + 4R/Q} = P_\pm \tag{11}$$

$$G_\infty = \frac{P_+}{P_+ + R} = -\frac{P_-}{R} \tag{12}$$

where $P_\infty$ is positive. Evidently, the steady-state Kalman filter equations are equivalent to the recursive formulation of the exponentially weighted moving average (EWMA) filter with smoothing factor $G_\infty$. Using $z_k = x_k$ we get

$$\hat{x}_k = (1 - G_\infty)\hat{x}_{k-1} + G_\infty x_k \tag{13}$$

In contrast to classic EWMA filtering, where the smoothing factor is generally selected in an ad-hoc manner, the Kalman framework allows us to optimally parametrize the filter. If the Gaussian noise variances $Q$ and $R$ are known, the Kalman filter is optimal, i.e. no other filter can achieve a smaller MSE. For non-Gaussian noises, the filter is still the best linear unbiased estimator. In the following sections, we will outline a procedure for determining suitable process and measurement noises.

### 4.1 Measurement Noise Parametrization

We now derive the measurement noise associated with packet train sampling of a Poisson cross-traffic process. To calculate the dispersion and variance of $g_d$ caused by cross traffic in the wireless case we use [18]

$$d(l+1) - d(1) = (l+K)\left(\frac{L}{C} + \Delta\right) + \sum_{j=2}^{l+1} b(j) \tag{14}$$

where $K$ describes the random number of cross traffic packets transmitted between packet 1 and $l$ of the probing stream, $\Delta$ accounts for the protocol overhead and $b(j)$ for the DCF back-off procedure. For a wired FIFO system we set $\Delta$ and $b(j)$ to zero. The dominant source of randomness is given by the number of inter-transmitted packets $K$. Thus, the variation and the measurement noise mainly depend on $K$ and can be derived from its distribution.

To derive the conditional distribution of $P[K = k|l]$, i.e. that a cross traffic source transmits $k$ packets given a tagged station transmits $l$ packets in a DCF system, [18] uses probability theory and findings presented in [20] and [21]. The conditional distribution of $k$ under $l$ can be expressed as follows

$$\mathsf{P}[K = k|l] = \mathsf{P}\left[\sum_{j=1}^{k} b_1(j) \leq b_2\, l \text{ and } \sum_{j=1}^{k+1} b_1(j) > b_2\, l\right] \tag{15}$$

where $b_1$ are i.i.d. random variables representing the inter-arrival times of cross traffic whereas $b_2$ is the gap between two successive packets of a CBR probing stream with length $l$. Based on this model, the authors relate measurement noise to probing train length. Equipped with these findings, we derive a similar expression for cross traffic in FIFO networks.

**Lemma 1 (Poisson approximation).** *Let $\lambda$ be the average packet rate of Poisson cross traffic arrivals. Furthermore, let $l$ be the length and $r_p$ the rate of a probing stream. For $E[K] = l\lambda/r_p \gg 1$ (15) is approximately Gaussian where*

$$\mathsf{P}[K \leq k|l] \approx \mathsf{P}\left[N(0,1) \leq \frac{k - l\lambda/r_p}{\sqrt{l\lambda/r_p}}\right].$$

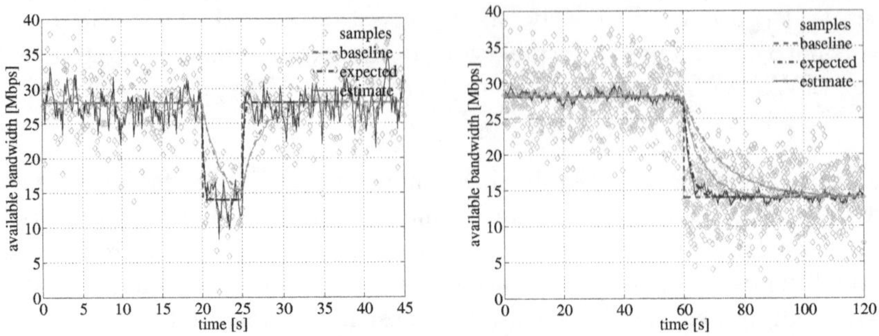


(a) Change tracking using different process noise values: Q=3.8, Q=0.38.

(b) Influence of process noise on convergence speed: Q=0.04 (T=10s), Q=0.0044 (T=30s), Q=0.0011 (T=60s).

**Fig. 1.** Simulation of estimates corrupted by Gaussian noise (R=16) with $t_\Delta = 0.1\ s$

*Proof.* Using the central limit theorem, the Poisson distribution can be approximated by a normal distribution for $E[K] \gg 1$ with mean $\mu = E[K]$ and variance $\sigma^2 = E[K]$, i.e. it becomes $N(E[K], E[K])$-distributed

$$\mathsf{P}[K = k] \approx \frac{1}{\sqrt{2\pi\, E[K]}} e^{-\frac{(k-E[K])^2}{2\,E[K]}}$$

To calculate $E[K]$, we assume the cross traffic arrivals as a Poisson process with an average packet rate $\lambda$. For a probing stream of length $l$ and probing rate $r_p$, we get

$$E[K] = \frac{l}{r_p}\lambda$$

that describes the expected average packet arrivals during a sample interval related to the probing stream. We can now use the normal distribution to calculate the conditional probability

$$\mathsf{P}[K \leq k|l] \approx \mathsf{P}\left[N\left(0, \frac{l}{r_p}\lambda\right) \leq k - \frac{l}{r_p}\lambda\right]$$

Finally, we use the fact, that if $X$ is $N(a\mu, a^2\sigma^2)$ then $Y = X/a$ is $N(\mu, \sigma^2)$ with $a^2 = l\lambda/r_p$ to standardize the result.

Combining (5) and (14) we find that in FIFO systems the variation of $g_d$ is given by the distribution $K/l$. Using Lemma 1 and $Var(aX+b) = a^2 Var(X)$ we calculate the standard deviation for Poisson cross traffic and derive the measurement noise

$$\sigma_{g_d} = \sqrt{\frac{\lambda}{r_p\, l}}\, \frac{L}{C} \tag{16}$$

### 4.2 Process Noise Parametrization

For optimal operation, the Kalman filter must also be supplied with the true variance $Q$ of the measured process. Essentially, the parameter $Q$ determines how quickly the filter considers new measurements to be reliable. However, generally the variance of the measured process is not known in advance. Nevertheless, we may choose a value based on knowledge of the type of changes we are interested in capturing. Even if the measured process is not normally distributed, the Kalman filter provides the lowest MSE achievable by a linear filter.

Consider the simulation scenario depicted in figure 1(a). After 20 s the fair share abruptly drops from 28 to 14 Mbps. Expressed as a gap length, we denote this change of $B = 14$ Mbps as $g_j = L/B$. The fair share remains in this state for $T_s = 5$ $s$. Samples corrupted by a Gaussian noise process with variance $R = 16$ are collected every $t_\Delta = 0.1$ s. To ensure that the filter follows cross-traffic variations, we must set $Q > 0$. It is evident that if $Q$ is too large, the estimate will be unnecessarily noisy. If, on the other hand, $Q$ is too small, the discontinuity will be over-smoothed. This might be acceptable if one is not interested in tracking such short variations. Let us however assume that the application relying on our bandwidth estimate can benefit from accurately identifying this short fair share jump.

We can optimally identify the discontinuity by calculating the variance $Q$ for the segment $T_s$ after the fair share change. As $T_s/t_\Delta = n_p$ samples are generated during $T_s$, $Q$ is calculated as follows

$$Q = \frac{g_j^2}{n_p} = \frac{g_j^2 L}{T_s r_p}(l+1) \tag{17}$$

Evidently $Q$ decreases linearly as the number of samples in the segment is increased, i.e. the number of packets per train is reduced. Using Eq. 17 the Kalman filter will yield the minimal MSE within the considered segment. Naturally the detection of shorter discontinuities will be sub-optimal. As a result, the parametrization can be viewed as a lower bound for the filter tracking ability.

### 4.3 Convergence Speed and Estimate Variance

We now consider the effects of the filter parametrization on the estimate. Firstly, we calculate the time needed for the filter to converge to a new value after an abrupt change. To this end, we examine the impulse response of the scalar steady-state Kalman filter where $y_k$ is the system output and $u_k$ is the step function

$$y_k = (1 - G_\infty)^k u_k = e^{-\alpha k} u_k \tag{18}$$

Evidently, the convergence speed to a new bandwidth is exponential. To calculate the convergence time, we make use of the fact that $e^{-\alpha k}$ decays to less than 1% of its initial value after $k = 5/\alpha$ time-steps.

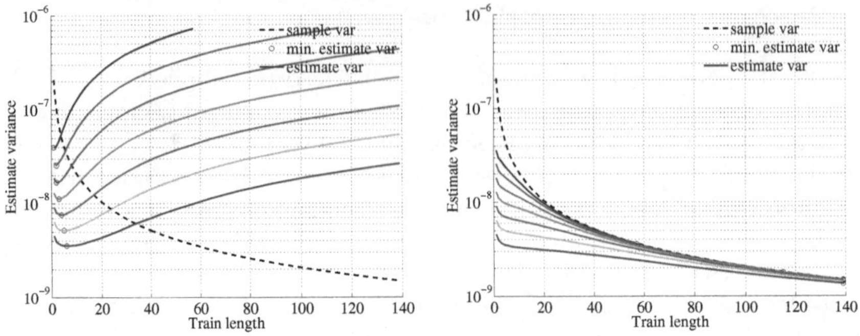


**Fig. 2.** Estimate variance for a 14 Mbit bandwidth change within $T_s = 0.5, 1, 2, 4 \ldots 32s$ (solid curves; top to bottom) and dependency on train length and measurement noise variance. Within the segment $T_s$ in (a) and for $t \gg T_s$ in (b).

The relationship[1] between the noise process variances $Q$ and $R$ to $\alpha$, and consequently $G_\infty$ is derived using

$$e^{-\alpha} = (1 - G_\infty) = 1 + \frac{P_-}{R} = 1 + \frac{Q}{2R} - \frac{Q}{2R}\sqrt{1 + 4R/Q} \tag{19}$$

$$e^{\alpha} = \frac{1}{(1 - G_\infty)} = \frac{R + P_+}{R} = 1 + \frac{Q}{2R} + \frac{Q}{2R}\sqrt{1 + 4R/Q} \tag{20}$$

$$\cosh(\alpha) = 1 + \frac{Q}{2R} \tag{21}$$

Moreover, taking into account that packet train probes are sent every $t_\Delta$ seconds, we can calculate the filter convergence time for a set of values $Q$,$R$

$$T = 5t_\Delta / \operatorname{arcosh}(1 + \frac{Q}{2R}) \tag{22}$$

Next, we derive the overall variance of the filtered estimate during the discontinuity period $T_s$. The EWMA filter is in essence an auto-regressive $AR(1)$ process with $a_1 = 1 - G$, driven by the process $G_\infty z$. The variance of the sample process $z$ is the sum of the process and measurement variances $\mathrm{Var}(z) = Q + R$, which are independent by definition. Therefore, $\sigma_z^2 = \mathrm{Var}(G_\infty z) = G_\infty^2(R + Q)$. Thus, the overall variance of the filtered estimate process is given by

$$\mathrm{Var}(\hat{x}_k) = \frac{\sigma_z^2}{1 - a_1^2} = \frac{G_\infty^2(Q + R)}{1 - (1 - G_\infty)^2} \tag{23}$$

Figure 2(a) illustrates the relationship between the estimate variance and train length for discontinuity periods of different lengths.

Assuming no further cross-traffic jumps, for times significantly longer than $T_s$ the effects of the jump on the variance become negligible, i.e. $Q$ tends towards

[1] For $R \gg Q$ the approximation $\alpha \approx \sqrt{Q/R}$ may be used.

zero resulting in $\mathrm{Var}(\hat{x}_k) = G_\infty^2 R/(1-(1-G_\infty)^2)$. The train length dependencies for this case are depicted in Figure 2(b).

For cases in which the variance of the cross-traffic is known in advance, we can readily employ equation 23 to determine the probing train length which produces a minimal MSE.

### 4.4 Packet Train Length Considerations

For the remainder of the paper, we focus on the wireless case, i.e. fair share estimation. The measurement noise variance derived in [18] is used.

As we showed above, increasing the packet train length $l$ reduces the variance of the bandwidth samples proportionally to $1/l$. Additionally, because the ratio between the bandwidth used per train and number of train gaps $(l+1)L/l$ is non-linear, probing bandwidth is wasted when using short packet trains. It is also clear that long train lengths, i.e. low sampling rates, will have a negative impact on the bandwidth change tracking accuracy.

Next, we analyze the optimal train length for a given $T_s$ and $B$. Figure 2(a) depicts the variance of the filtered samples for a process noise optimized to track changes of $B = 14$ Mbps within $T_s = 0.5, 1, 2, 4 \ldots 32s$. This variance is related to the MSE between the estimate and the actual bandwidth during the time $T_s$. It is comprised of the deviation during the convergence period $T$ and the measurement variance after convergence.

It is evident that for the depicted scenario, it is desirable to use short train lengths even if the resulting sample variance $R$ is increased: the overall segment MSE is minimized. For $t \gg T_s$, the filter has the largest variance improvement for short train lengths, as depicted in Figure 2(b). However, as the cross-traffic is assumed to be constant, we can use long packet trains sent at a low frequency to further improve the estimate.

Naturally, increasing the probing traffic intensity $r_p$ will yield even better results, as longer trains can be used, yielding a lower probe variance.

## 5 Experimental Evaluation of Bandwidth Estimation

Based on the findings in Sect. 4 we developed a modular, portable measurement framework called *WiProbe*. It implements direct probing for estimating the available bandwidth in the wired domain and the fair share in wireless networks. Kalman filtering is used to continuously remove measurement noise from the probes.

In order to evaluate our method and the implementation of our tool, we performed experiments in a controlled testbed environment containing both wired and wireless links. We investigate the performance of the Kalman filter and the effects of parametrization to provide an underpining of our theoretical findings in section 3 and section 4.

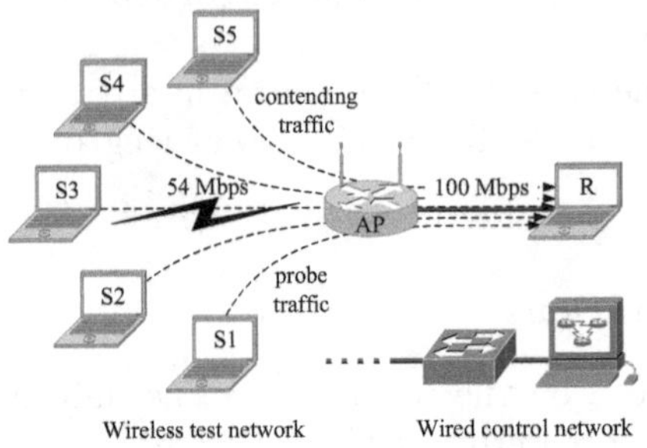


**Fig. 3.** Wireless testbed setup

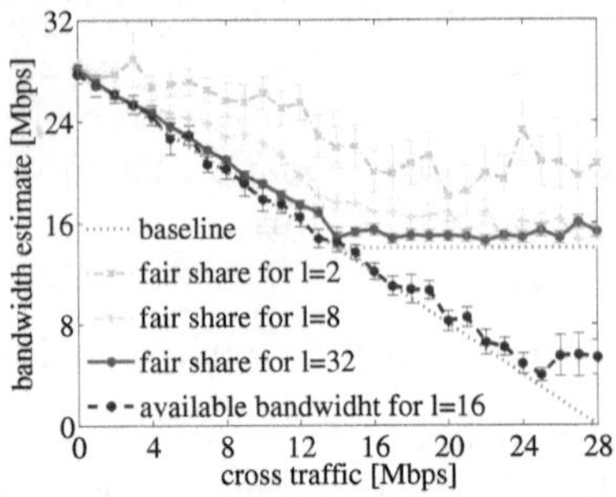


**Fig. 4.** Accuracy of WiProbe using the probe gap model for wired (FIFO) networks and the fair share model for wireless (DCF) networks at different train lengths

The wireless testbed[2] setup is depicted in Fig. 3. The distance between the wireless stations and the access point was between 0.5 m and 1.5 m. We switched off RTS/CTS, automatic rate adaption as well as packet fragmentation. The DCF is used for medium access. All nodes were connected to a separate switched Ethernet control network. We employed *SSHLauncher* [22] to atomate the experiment execution.

First, we focus on the accuracy of unfiltered packet train probes. We estimate the fair share by sending probing traffic from S1 to R and a single contending flow from S2 to R with an increasing rate $\lambda$ from 0 Mbps to 28 Mbps. The cross traffic is generated using the D-ITG [23] traffic generator. Fig. 4 shows the average of 25 fair share and bandwidth estimates with different probing train lengths gathered in approximately 1 second per rate for all rates of cross traffic. Furthermore, we show the corresponding confidence intervals at a confidence level of 0.95. As a reference, the true fair share of a new flow $fs = max\{C - \lambda, C/2\}$ and the available bandwidth are plotted. Due to large protocol overhead [8, 24], we use a nominal capacity $C = 28$ Mbps for a packet size of 1500 Bytes. It is obvious that at least for long probing trains, WiProbe accurately estimates the available bandwidth and the fair share in FIFO and DCF networks respectively. As stated in section 3, using shorter train lengths results in samples with a higher variance, as indicated by the larger confidence intervals in Fig. 4. Furthermore, the direct probing samples for FIFO networks and high cross traffic rates exhibit some inaccuracies due to packet loss caused by congestion.

[2] We used Lenovo ThinkPad R61i notebooks with 2.0 GHz, 2 GB RAM running Ubuntu Linux 8.04 with kernel version 2.6.24. We employed the internal Intel PRO/Wireless 4965 AGN IEEE 802.11g WLAN adapters. The access point is a Buffalo Wireless-G 125 series running DD-WRT version 24 RC-4. The switch used in the wired test network is a Netgear FS-108.

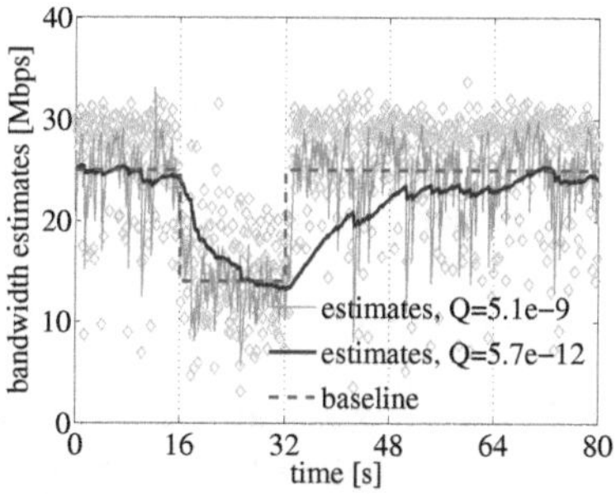


**Fig. 5.** $B = 14$ Mbit, $T_s = 16$ s: Optimal $Q = 5.1 \times 10^{-9}$ and non-optimal $Q = 5.7 \times 10^{-12}$ process noise

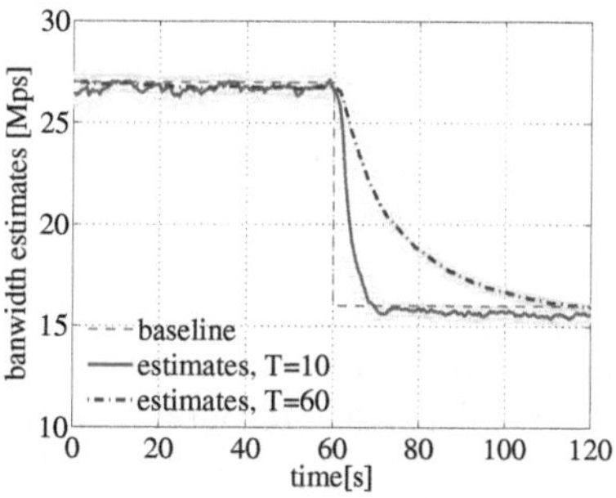


**Fig. 6.** Effect of process noise on estimate convergence speed using WiProbe

Fig. 5 depicts the effect of train length and process noise on the ability to track a fair share discontinuity. Based on the results plotted in Fig. 2(a) we probe using a train length of $l = 8$ to detect the change of 14 Mbits for $T_s = 16$s. Using the calculated, optimal process noise $Q = 7.1 \times 10^{-5}$, the jump is accurately tracked at the expense of a higher estimate noise. Nevertheless, the measurement noise is significantly lower when compared to the unfiltered case. Using a sub-optimal value for $Q$, the filtered estimate variance is reduced for longer timescales, however the estimate of the discontinuity is highly distorted.

The measurement results depicted in Fig. 6 confirm our theoretical findings from section 4.3. We were able to achieve a predetermined convergence time by selecting the process noise $Q$ according to Eq. 23. As expected, fast convergence times are associated with a larger estimate variance as indicated by the larger confidence intervals in Fig. 6.

## 6 Conclusion

In this paper we showed the benefit of employing Kalman filtering for improving estimates derived from continuous active probing of fair share or available bandwidth using a constant probing rate. Specifically, we showed how filter parameters should be chosen to fit a specific use case and calculated the effects of filtering on the estimate variance. Additionally, we showed the relationship between filter convergence time and the process and measurement noise parameters. Furthermore, we evaluated the influence of train lengths on the estimate's variance for wired and wireless setups. We conclude that depending on the time-scale and intensity of the cross traffic variations of interest, it can be beneficial to sample cross-traffic frequently using short packet trains. When tracking short-term changes is not a primary concern, long packet trains sent less frequently should be employed, as these provide estimates with the lowest variance.

## References


1. Jacobson, V.: Congestion avoidance and control. In: Proc. ACM SIGCOMM, pp. 314–329 (1988)
2. Keshav, S.: A control-theoretic approach to flow control. In: Proc. ACM SIGCOMM, pp. 3–15 (1991)
3. Jain, M., Dovrolis, C.: Pathload: A measurement tool for end-to-end available bandwidth. In: Proc. of PAM (2002)
4. Ribeiro, V., Riedi, R., Baraniuk, R., Navratil, J., Cottrell, L.: PathChirp: Efficient available bandwidth estimation for network paths. In: Proc. of PAM (2003)
5. Strauss, J., Katabi, D., Kaashoek, F.: A measurement study of available bandwidth estimation tools. In: Proc. ACM IMC, pp. 39–44 (2003)
6. Melander, B., Björkman, M., Gunningberg, P.: Regression-based available bandwidth measurements. In: Proc. of SPECTS (2002)
7. Ekelin, S., Nilsson, M., Hartikainen, E., Johnsson, A., Mangs, J.E., Melander, B., Björkman, M.: Real-time measurement of end-to-end available bandwidth using kalman filtering. In: Proc. IEEE/IFIP NOMS, pp. 5–17 (2006)
8. Bredel, M., Fidler, M.: A measurement study of bandwidth estimation in IEEE 802.11g wireless LANs using the DCF. In: Das, A., Pung, H.K., Lee, F.B.S., Wong, L.W.C. (eds.) NETWORKING 2008. LNCS, vol. 4982, pp. 314–325. Springer, Heidelberg (2008)
9. Li, M., Claypool, M., Kinicki, R.: WBest: A bandwidth estimation tool for multimedia streaming application over IEEE 802.11 wireless networks. Technical Report WPI-CS-TR-06-14, Department of Computer Science - Worcester Polytechnic Institute (2006)
10. Johnsson, A., Melander, B., Björkman, M.: Diettopp: A first implementation and evaluation of a simplified bandwidth measurement method. In: Proc. of SNCNW, p. 5 (2004)
11. Shriram, A., Kaur, J.: Empirical evaluation of techniques for measuring available bandwidth. In: Proc. IEEE INFOCOM (2007)
12. Jain, M., Dovrolis, C.: End-to-end available bandwidth: Measurement methodology, dynamics, and relation with TCP throughput. In: Proc. ACM SIGCOMM, pp. 295–308 (2002)
13. Liu, X., Ravindran, K., Loguinov, D.: A queuing-theoretic foundation of available bandwidth estimation: Single-hop analysis. IEEE/ACM Trans. Networking 15(4), 918–931 (2007)
14. Jain, M., Dovrolis, C.: Ten fallacies and pitfalls on end-to-end available bandwidth estimation. In: Proc. ACM IMC, pp. 272–277 (2004)
15. Vaidya, N.H., Bahl, P., Gupta, S.: Distributed fair scheduling in a wireless lan. In: Proc. of ACM MOBICOM, pp. 167–178 (2000)
16. Banchs, A., Perez, X.: Distributed weighted fair queuing in 802.11 wireless lan. In: IEEE ICC 2002, vol. 5, pp. 3121–3127 (2002)
17. Liu, X., Ravindran, K., Liu, B., Loguinov, D.: Single-hop probing asymptotics in available bandwidth estimation: sample-path analysis. In: Proc. of ACM SIGCOMM, ACM, pp. 300–313. ACM, New York (2004)
18. Bredel, M., Fidler, M.: Understanding fairness and its impact on quality of service in IEEE 802.11. In: IEEE Infocom (2009)
19. Anderson, B.D.O., Moore, J.B.: Optimal Filtering. Prentice-Hall, Englewood Cliffs (1979)

20. Berger-Sabbatel, G., Duda, A., Gaudoin, O., Heusse, M., Rousseau, F.: Fairness and its impact on delay in 802.11 networks. In: Proc. of IEEE GLOBECOM, pp. 2967–2973 (2004)
21. Berger-Sabbatel, G., Duda, A., Heusse, M., Rousseau, F.: Short-term fairness of 802.11 networks with several hosts. In: Proc. of IFIP MWCN, pp. 263–274 (2004)
22. Bozakov, Z., Bredel, M.: Sshlauncher.kom - a tool for experiment automation in distributed environments. Technical Report KOM-TR-2008-11, TU-Darmstadt, Merckstr. 25, 64283 Darmstadt (2008)
23. D-ITG: Internet Traffic Generator, `http://www.grid.unina.it/software/ITG`
24. IEEE Standard for Wireless LAN Medium Access Control (MAC) and Physical Layer (PHY) Spedifications, Std. 802.11-2007 (2007)

# A New Metric for Robustness with Respect to Virus Spread
## (Work in Progress)

Robert E. Kooij[1,2,3,*], Phillip Schumm[3], Caterina Scoglio[3], and Mina Youssef[3]

[1] Delft University of Technology, EEMCS, The Netherlands
[2] TNO ICT, The Netherlands
[3] Kansas State University, EECE, U.S.A.
robert.kooij@tno.nl, {pbschumm,caterina,mkamel}@ksu.edu

**Abstract.** The robustness of a network is depending on the type of attack we are considering. In this paper we focus on the spread of viruses on networks. It is common practice to use the epidemic threshold as a measure for robustness. Because the epidemic threshold is inversely proportional to the largest eigenvalue of the adjacency matrix, it seems easy to compare the robustness of two networks. We will show in this paper that the comparison of the robustness with respect to virus spread for two networks actually depends on the value of the effective spreading rate $\tau$. For this reason we propose a new metric, the viral conductance, which takes into account the complete range of values $\tau$ can obtain. In this paper we determine the viral conductance of regular graphs, complete bipartite graphs and a number of realistic networks.

**Keywords:** Robustness, virus spread, epidemic threshold, viral conductance.

## 1 Introduction

Our daily activities rely increasingly on complex networks. The power grid, the Internet, and transportation networks are examples of complex networks. In contrast to simple networks, such as regular or Erdös-Rényi random graphs [7], complex networks are characterized by a large number of vertices (from hundreds of thousands to billions of nodes), a low density of links, clustering effects, and power-law node-degree distribution [1], [20]. Being so large, complex networks are often controlled in a decentralized way and show properties of self-organization. However, even if decentralization and self-organization theoretically reduce the risk of failure, complex networks can experience disruptive and massive failure.

As an example of massive attacks, in 2001, Code Red, a computer virus that incapacitated numerous networks, resulted in a global loss of 2.6 billion US dollars. In 2004, the Sassar virus caused Delta airlines to cancel 40 transatlantic flights in addition to halting trains in Australia. Additionally, the US General Accounting Office

* Corresponding author: R.E. Kooij, TNO ICT, Brassersplein 2, 2612 CT Delft, the Netherlands.

L. Fratta et al. (Eds.): NETWORKING 2009, LNCS 5550, pp. 562–572, 2009.

estimated 250,000 annual attacks on Department Of Defense networks. Objectives of such attacks range from theft, modification, and destruction of data to dismantling of entire networks. In another example concerning the power grid, the Northeastern and Midwestern United States, and Ontario, Canada suffered a massive widespread power outage on August 14, 2003. Since our daily routines would cease if the technological information infrastructure disintegrates, thus, it becomes crucial to maintain the highest levels of robustness in complex networks.

Therefore, the first step is to assess the robustness of networks. Obviously the robustness of a network is depending on the type of attack we are considering. In this paper we will focus on the spread of viruses on networks.

The Susceptible-Infected-Susceptible (SIS) infection model, which arose in mathematical biology, is often used to model the spread of viruses [10], [9], [14], epidemic algorithms for information dissemination in unreliable distributed systems like P2P and ad-hoc networks [3], [8], and propagation of faults and failures in networks like BGP [4]. The SIS model assumes that a node in the network is in one of two states: infected and therefore infectious, or healthy and therefore susceptible to infection. The SIS model usually assumes instantaneous state transitions. Thus, as soon as a node becomes infected, it becomes infectious and likewise, as soon as a node is cured it is susceptible to re-infection. There are many models that consider more aspects like incubation periods, variable infection rate, a curing process that takes a certain amount of time and so on [6], [10], [19]. In epidemiological theory, many authors refer to an epidemic threshold $\tau_c$, see for instance [6], [2], [10] and [14]. If it is assumed that the infection rate along each link is $\beta$ while the curing rate for each node is $\delta$ then the effective spreading rate of the virus can be defined as $\tau = \beta/\delta$. The epidemic threshold can be defined as follows: for effective spreading rates below $\tau_c$ the virus contamination in the network dies out - the mean epidemic lifetime is of order log n, while for effective spreading rates above $\tau_c$ the virus is prevalent, i.e. a persisting fraction of nodes remains infected with the mean epidemic lifetime [9] of the order $\exp(n^\alpha)$. In the case of persistence we will refer to the prevailing state as a metastable state or steady state. It was shown in [18] and [9] that $\tau_c = 1/\rho(A)$ where $\rho(A)$ denotes the spectral radius of the adjacency matrix A of the graph. Recently, the epidemic threshold formula has also been verified by using the N-intertwined model [17], which consists of a pair of interacting continuous Markov chains.

It is common practice to use the epidemic threshold as a measure for robustness: the larger the epidemic, the more robust a network is against the spread of a virus, see [11]. Because the epidemic threshold is inversely proportional to the largest eigenvalue of the adjacency matrix, it seems easy to compare the robustness of two networks. We will show in this paper that the comparison of the robustness with respect to virus spread for two networks is not so straightforward. To be more precise, we will show that the comparison of networks depends on the actual value of the effective spreading rate $\tau$. For this reason we propose a new metric, the viral conductance, which takes into account the complete range of values $\tau$ can obtain.

The rest of this paper is organized as follows. In Section 2 we consider the spread of viruses on regular and complete bi-partite graphs and show the need for a new metric for robustness with respect to virus spread. We propose this new metric, the viral conductance, in Section 3. In Section 4 we suggest a heuristic for the computation of the viral conductance. We determine the viral conductance for some realistic networks in Section 5. The main conclusions are summarized in Section 6.

## 2 Virus Spread on Regular and Complete Bi-partite Graphs

In this section we will compare the fraction of infected nodes for two example networks and show that the value of the effective spreading rate $\tau$ determines for which network this fraction is higher. The example networks belong to the class of regular and complete bi-partite graphs, respectively.

### 2.1 Virus Spread on Regular Graphs

In this subsection we discuss the spread of viruses over a simpler network, i.e. the connected regular graph. This model is based on a classical result by Kephart and White [10] for SIS models.

We consider a connected graph on N nodes where every node has degree k. We denote the number of infected nodes in the population at time t by Y(t). If the population N is sufficiently large, we can convert Y(t) to y(t)=Y(t)/N, a continuous quantity representing the fraction of infected nodes. Now the rate at which the fraction of infected nodes changes is due to two processes: susceptible nodes becoming infected and infected nodes being cured. Obviously, the cure rate for a fraction i of infected nodes is $\delta y$. The rate at which the fraction y grows is proportional the fraction of susceptible nodes, i.e. 1-y. For every susceptible node the rate of infection is the product of the infection rate per node ($\beta$), the degree of the node (k) and the probability that on a given link the susceptible node connects to an infected node (y).

Therefore we obtain the following differential equation describing the time evolution of y(t):

$$\frac{dy}{dt} = \beta k y(1-y) - \delta y. \tag{1}$$

The steady state solution $y_\infty$ of Eq. (1) satisfies

$$y_\infty = \frac{\beta k - \delta}{\beta k} = 1 - \frac{1}{k\tau}. \tag{2}$$

Because an epidemic state only exists if $y_\infty > 0$, we conclude that the epidemic threshold satisfies

$$\tau_c = \frac{1}{k}. \tag{3}$$

Because for k-regular graphs the spectral radius of the adjacency matrix is equal to k, see [5], Eq. (3) is in line with the result by [18].

## 2.2 Virus Spread on Complete Bi-partite Graphs

In this subsection we will consider complete bi-partite graphs. A complete bi-partite graph $K_{M,N}$ consists of two disjoint sets $S_1$ and $S_2$ containing respectively M and N nodes, such that all nodes in $S_1$ are connected to all nodes in $S_2$, while within each set no connections occur. Fig. 1 gives an example of a complete bi-partite graph on 10 nodes.

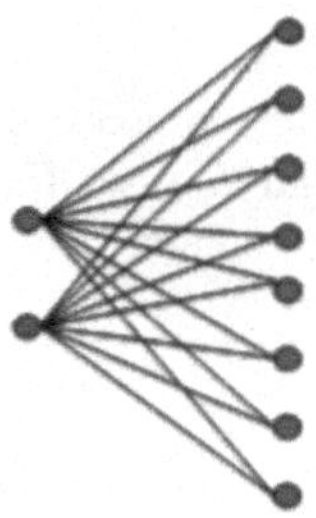

**Fig. 1.** Complete bi-partite graph $K_{2,8}$

Notice that (core) telecommunication networks often can be modeled as a complete bi-partite topology. For instance, the so-called double-star topology (i.e. $K_{M,N}$ withM = 2) is quite commonly used because it offers a high level of robustness against link failures. For example, the Amsterdam Internet Exchange (see www.ams-ix.net), one of the largest public Internet exchanges in the world, uses this topology to connect its four locations in Amsterdam to two high-density Ethernet switches. Sensor networks are also often designed as complete bi-partite graphs.

In [12] a model for virus spreading on the complete bi-partite graph $K_{M,N}$ was presented. Using differential equations and two-state Markov processes it was shown in [12] that, above the epidemic threshold $\tau_c = \frac{1}{\sqrt{MN}}$, the fraction of infected nodes for $K_{M,N}$ satisfies

$$y_\infty = \frac{(MN\tau^2 - 1)((M+N)\tau + 2)}{\tau(M+N)(M\tau+1)(N\tau+1)}. \tag{4}$$

It is easy to verify that for the case M = N, Eq. (4) reduces to Eq. (2), with k = N.

## 2.3 Comparing the Fraction of Infected Nodes for Two Graphs

In this subsection we consider two networks on 10 nodes, the Petersen graph (see Fig. 2) and $K_{2,8}$. Note that the Petersen graph is a regular graph where every node has 3 neighbours, i.e. k = 3 in the notation of section 2.1.

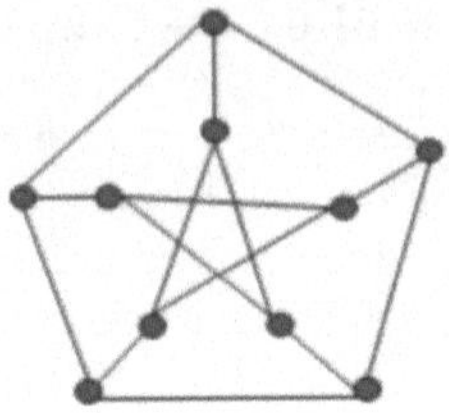

**Fig. 2.** Petersen graph

Using Eq. (2) and Eq. (4) we can compare the fraction of infected nodes $y_\infty$ at steady state for the Petersen graph and $K_{2,8}$, see Fig. 3.

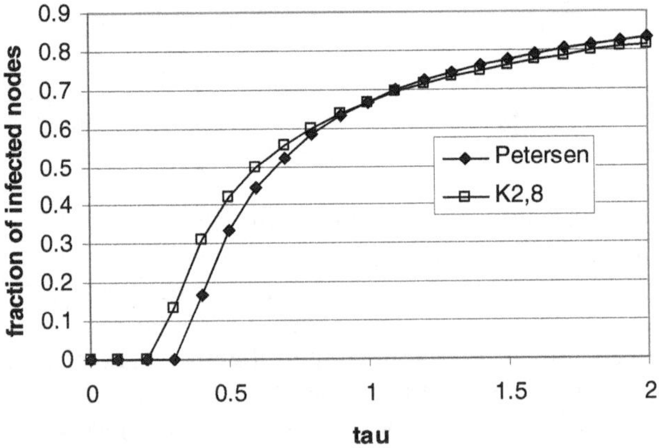


**Fig. 3.** Fraction of infected nodes for Petersen graph and $K_{2,8}$

If we only look at the epidemic threshold, we see that the Petersen graph outperforms $K_{2,8}$. However, for $\tau$ sufficiently large, $K_{2,8}$ performs better because then the fraction of infected nodes is lower than for the regular graph. Note that this effect, that regular graphs have a higher fraction of infected nodes than non-regular graphs for large values of $\tau$, was already observed in [16].

This section shows that in order to compare for two networks the robustness with respect to virus spread, it does not suffice only to look at the epidemic threshold.

## 3 Viral Conductance

In this section we propose a new metric for robustness with respect to virus spread that takes into account the complete range of $\tau$ values.

A natural way to take all values of $\tau$ into account is by considering the area under the curve that gives the fraction of infected nodes. However, because this will lead to divergent integrals, from now on, instead of considering the effective spreading rate $\tau$ we look at the reciprocal of $\tau$, that is the effective spreading rate $s = \delta/\beta$.

We are interested in $y_\infty(s)$, the fraction of infected nodes in steady state, as a function of the effective curing rate. Note that the behaviour of $y_\infty(s)$ around s = 0 reflects the behaviour of the original system for $\tau \rightarrow \infty$.

We are now in the position to suggest a new robustness measure with respect to virus spread that takes into account all values of $\tau$, and hence s.

**Definition.** The viral conductance V of a network G is given by

$$V(G) = \int_0^\infty y_\infty(s)ds, \tag{5}$$

where $y_\infty(s)$ denotes the fraction of infected nodes in steady state and $s = \delta/\beta$.

Note that it is possible to come up with other metrics that take the full range of $\tau$ values into account, but in our opinion the viral conductance is the simplest one. Determination of the operational meaning of the viral conductance is left for further study. We will now state some theorems for the viral conductance V(G).

**Theorem 1.** For regular graphs $H_k$, where every node has k neighbours, it holds that $V(H_k) = k/2$.

Proof. This follows directly from Eq. (2) and Eq. (5).

**Theorem 2.** For complete bi-partite graphs $K_{M,N}$, it holds that $V(K_{M,N}) =$

$$\frac{(M+N)\sqrt{MN} - MN + (M-N)(N\ln(M+\sqrt{MN}) - M\ln(N+\sqrt{MN}) + M\ln M - N\ln N)}{M+N}.$$

Proof. This follows from applying Eq. (5) to Eq. (4).

## 4 A Heuristic for the Viral Conductance

For general networks we cannot compute the fraction of infected nodes $y_\infty(s)$, and hence the viral conductance, explicitly. Therefore, in this section we propose a heuristic for the computation of the viral conductance for general networks.

We start with listing some properties of the fraction of infected nodes $y_\infty(s)$.

**Lemma 1.** For any connected graph G let A denote its adjacency matrix and $\rho(A)$ the largest eigenvalue of A. Then $y_\infty(\rho(A))=0$.

Proof. This is just the threshold theorem, see e.g. [17].

**Lemma 2.** Consider a connected graph on N nodes and denote the degree of node i by $d_i$. Then $y_\infty(s) = 1 - s\frac{1}{N}\sum_{i=1}^{N}\frac{1}{d_i} + O(s^2)$.

Proof. This follows from Section IV in [17].

**Conjecture.** For any connected graph G on N nodes, let A denote its adjacency matrix, $\rho(A)$ the largest eigenvalue of A, $d_i$ the degree of node i and $E[d_i]$ the mean nodal degree. Then $\lim_{s\uparrow\rho}\frac{dy_\infty}{ds} = -\frac{E[d_i]}{\rho^2}$.

Motivation. The conjecture is true for regular and complete bi-partite graphs and also is supported by numerical evidence.

We will use the above lemmas and conjecture to construct a heuristic for estimating the new robustness metric V.

We first approximate the curve $y_\infty(s)$ by two straight lines, $y_{\infty 1}(s)$ and $y_{\infty 2}(s)$ such that $y_{\infty 1}(s)$ is the linearization of $y_\infty(s)$ at $s = 0$ and $y_{\infty 2}(s)$ is the linearization of $y_\infty(s)$ at $s = \rho$.

From Lemma 2 and the conjecture it follows that

$$y_{\infty 1}(s) = 1 - \sigma s,\quad y_{\infty 2}(s) = \frac{E[d_i]}{\rho} - \frac{E[d_i]}{\rho^2} s$$

where $\sigma = \frac{1}{N}\sum_{i=1}^{N}\frac{1}{d_i}$. The lines $y_{\infty 1}(s)$ and $y_{\infty 2}(s)$ intersect at $s^* = \frac{\rho(\rho - E[d_i])}{\sigma\rho^2 - E[d_i]}$.

Hence we obtain for this heuristic

$$V_{PL} = \int_0^{s^*} y_{\infty 1}(s)ds + \int_{s^*}^{\rho} y_{\infty 2}(s)ds = \frac{\rho((\sigma\rho - 2)E[d_i] + \rho)}{2(\sigma\rho^2 - E[d_i])}.$$

Next we approximate the curve $y_\infty(s)$ by a non-linear curve $y_{\infty NL}(s)$ of the following form $y_{\infty NL}(s) = a + bs + cs^d$ such that $y_{\infty NL}(s)$ and $y_\infty(s)$ have the same linearization, both at $s = 0$ and at $s = \rho$. A straightforward calculation shows that

$$y_{\infty NL}(s) = 1 - \sigma s + (\sigma\rho - 1)\left(\frac{s}{\rho}\right)^d,$$

where $d = \frac{\sigma\rho^2 - E[d_i])}{\rho(\sigma\rho - 1)}$.

Note that for non-regular graphs $\rho > E[d_i]$, see e.g. [5], from which it can be deduced that $d > 1$, hence the derivative of $y_{\infty NL}(s)$ at $s = 0$ is finite.

Hence we obtain for this heuristic

$$V_{NL} = \int_0^{\rho} y_{\infty NL}(s)ds = \frac{\rho((\sigma\rho - 2)E[d_i] + \sigma\rho^2)}{2(2\sigma\rho^2 - E[d_i] - \rho)}.$$

Because $V_{NL}$ always seems to lead to an overestimation of V, while $V_{PL}$ underestimates V, as the final heuristic $V_H$ for the viral conductance we propose a weighted average of $V_{PL}$ and $V_{NL}$:

$$V_H = 0.95 * V_{PL} + 0.05 * V_{NL}.$$

The choice of the weight (0.95) is based upon the results of the next section. We will now validate the heuristic for the complete bi-partite graph $K_{M,N}$, for which we can determine the new robustness measure V explicitly, see Theorem 2.

The results are given in Table 1.

**Table 1.** Comparison viral conductance V with heuristic $V_H$ for $K_{M,N}$

| M | N | V | $V_H$ | Relative error |
|---|---|---|---|---|
| 10 | 90 | 11.38 | 10.23 | -10% |
| 30 | 70 | 21.85 | 21.49 | -2% |
| 50 | 50 | 25 | 25 | 0% |
| 10 | 990 | 20.14 | 14.08 | -30% |
| 100 | 900 | 113.77 | 102.34 | -10% |
| 250 | 750 | 200.24 | 194.71 | -3% |

Apart from the case $K_{10,990}$, the heuristic $V_H$ performs reasonably well.

# 5 Viral Conductance for Realistic Networks

In this section we will determine the viral conductance for a number of real-life networks and for some toy networks that are commonly used to model realistic networks.

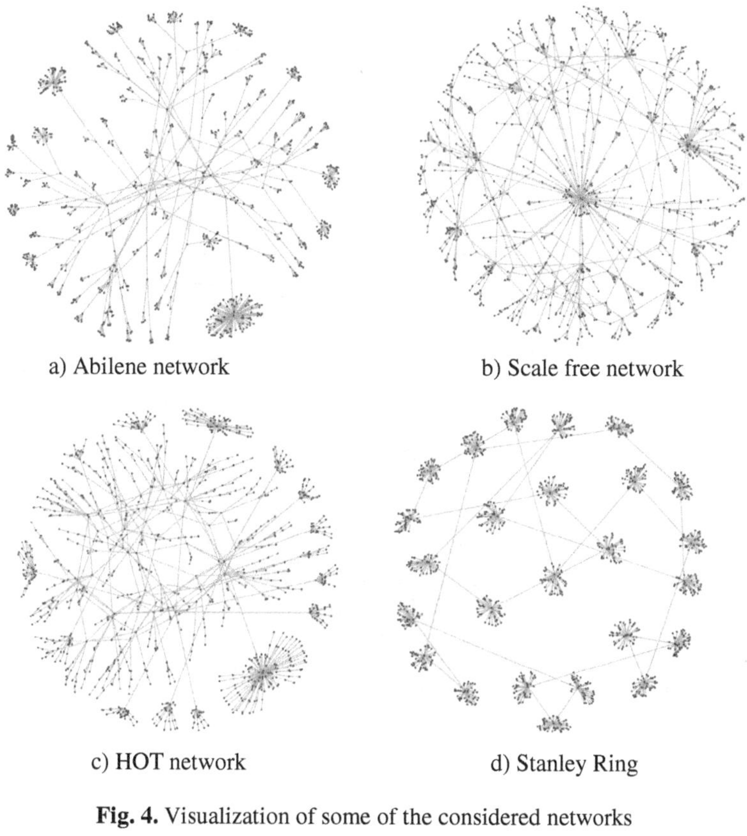


**Fig. 4.** Visualization of some of the considered networks

### 5.1 Considered Networks

The following networks are considered in this section: the Abilene backbone network, a scale free network, HOT (Heuristically Optimal Topology), the Erdös-Rényi graph, the Stanley Ring, the Stanley Mesh and a 2D-lattice. The Stanley Ring and Stanley Mesh were proposed in [15] to maximize network robustness against both random and targeted attacks while minimizing the network cost.

More information on the considered networks can be found in [12], [15] and [16].

All networks contain approximately 1000 nodes. This value reflects the number of nodes in current inter-domain network design and thus provides a comparatively real-life scenario. Some of the considered networks are visualized in Fig. 4.

### 5.2 Numerical Results

Because for general networks no explicit expression for the fraction of infected nodes $y_\infty(s)$ is available, we have used numerical analysis to determine the viral conductance for the considered networks. The first step is to obtain steady state values for the number of infected nodes, for a given value of the effective curing rate s. For this we use the discrete, deterministic expression for $p_{i,t}$, the probability that node i is infected at time t, as given in [18]. By summing over all nodes and after the appropriate re-scaling we obtain $y_\infty(s)$. This fraction of infected nodes is evaluated for 100 equidistant values of s, between 0 and $\rho(A)$, the spectral radius of the adjacency matrix A. Finally, the viral conductance V is determined by means of a simple triangular integration method. The results are given in Table 2.

**Table 2.** Viral conductance V for realistic networks

| Network | N | L | <d> | $\tau_c$ | V | $V_H$ | rel. error |
|---|---|---|---|---|---|---|---|
| Abilene | 886 | 896 | 2.02 | 0.11 | 1.43 | 1.46 | 3% |
| Scale free | 1000 | 1049 | 2.10 | 0.10 | 1.49 | 1.54 | 3% |
| HOT | 1000 | 1049 | 2.10 | 0.11 | 1.46 | 1.53 | 5% |
| Erdös-Rényi | 1000 | 2009 | 4.02 | 0.19 | 2.20 | 2.15 | -2% |
| Stanley Ring | 1000 | 1000 | 2.00 | 0.14 | 1.79 | 1.34 | -25% |
| Stanley Mesh | 1000 | 1275 | 2.55 | 0.04 | 2.81 | 1.96 | -30% |
| 2D-lattice | 900 | 1740 | 3.87 | 0.25 | 2.00 | 1.96 | -2% |

In Table 2 N denotes the number of nodes, L the number of links, <d> the average nodal degree and $\tau_c$ the epidemic threshold. $V_H$ denotes the viral conductance according to the heuristic proposed in Section 4. Several conclusions can be drawn from Table 2. We confine ourselves to mention just a few:

- Of the considered networks Abilene has the lowest viral conductance; hence it is the most robust with respect to virus spread.
- The threshold of the Erdös-Rényi graph is almost twice as high as that of Abilene, yet its viral conductance is about 50% higher.
- For the Stanley Ring and Mesh the heuristic $V_H$ leads to an underestimation of the viral conductance in the order of 30%. For the other considered networks the heuristic is very accurate.

- Rewiring the links of the Stanley Ring could lead to a reduction of the viral conductance of about 44%.

The last conclusion follows from the fact that every ring topology has a viral conductance of 1, see Theorem 1.

## 6 Conclusions

In this paper we have proposed the viral conductance as a new metric for robustness with respect to virus spread in networks. The viral conductance takes the complete range of values the effective spreading rate can attain into account. We have given an explicit expression for the viral conductance in case of regular and complete bi-partite graphs. For general networks we have proposed a heuristic. Next we have determined the viral conductance for a number of realistic networks, by means of numerical computation.

The main conclusions are the following:

- of the considered networks Abilene has the lowest viral conductance; hence it is the most robust with respect to virus spread;
- For the Stanley Ring and Mesh the heuristic $V_H$ leads to an underestimation of the viral conductance in the order of 30%. For the other considered networks the heuristic is very accurate..

The following issues will be subject of our future work:

- determination of operational meaning of the viral conductance;
- design of topologies, with given number of nodes and links, which minimize the viral conductance;
- construction of a more accurate heuristic for the computation of the viral conductance;
- computation of the viral conductance for networks of a larger scale.

## Acknowledgement

This research was partially supported by the Netherlands Organization for Scientific Research (NWO) under project number 643.000.503 and by the National Science Foundation (NSF) under award number 0841112.

## References

1. Albert, R., Jeong, H., Barabasi, A.: Error and attack tolerance of complex networks. Nature 406, 378–382 (2000)
2. Bailey, N.T.J.: The Mathematical Theory of Infectious Diseases and its Applications, 2nd edn. Charlin Griffin & Company, London (1975)

3. Chakrabarti, D., Leskovec, J., Faloutsos, C., Madden, S., Guestrin, C., Faloutsos, M.: Information Survival Threshold in Sensor and P2P Networks. In: Proc. IEEE INFOCOM 2007, Anchorage (2007)
4. Coffman Jr., F.G., Ge, Z., Misra, V., Towsley, D.: Network resilience: exploring cascading failures within BGP. In: Proc. 40th Annual Allerton Conference on Communications, Computing and Control (2002)
5. Cvetkovic, D.M., Doob, M., Sachs, H.: Spectra of graphs, Theory and Applications, 3rd edn. Johan Ambrosius Barth Verlag, Heidelberg (1995)
6. Daley, D.K., Gani, J.: Epidemic modelling: An Introduction. Cambridge University Press, Cambridge (1999)
7. Erdős, P., Rényi, A.: On Random Graphs. I. Publicationes Mathematicae 6, 290–297 (1959)
8. Eugster, P.T., Guerraoui, R., Kermarrec, A.M., Massoulié, L.: From Epidemics to Distributed computing. IEEE Computer 37(5), 60–67 (2004)
9. Ganesh, A., Massoulié, L., Towsley, D.: The Effect of Network Topology on the Spread of Epidemics. In: Proc. IEEE INFOCOM 2005, Miami (2005)
10. Kephart, J.O., White, S.R.: Direct-graph epidemiological models of computer viruses. In: Proc. IEEE Computer Society Symposium on Research in Security and Privacy, pp. 343–359 (1991)
11. Kooij, R.E., Jamakovic, A., Van Mieghem, P., van Dam, E.R.: Robustness of networks against the spread of viruses: the role of the spectral radius. In: Proceedings of the 13th Annual Symposium of the IEEE/CVT Benelux, Liége, Belgium (2006)
12. Newman, M.E.J., Barabási, A.-L., Watts, D.J.: The Structure and Dynamics of Networks. Princeton University Press, Princeton (2006)
13. Omic, J., Kooij, R.E., Van Mieghem, P.: Virus Spread in Complete Bi-partite Graphs. In: Proc. BIONETICS 2007, Budapest (2007)
14. Pastor-Satorras, R., Vespignani, A.: Epidemic Spreading in Scale-Free Networks. Physical Review Letters 86(14), 3200–3203 (2001)
15. Paul, G., Tanizawa, T., Havlin, S., Stanley, H.E.: Optimization of robustness of complex networks. The European Physical Journal B 38(2), 187–191 (2004)
16. Schumm, P., Scoglio, C., Easton, T., Gruenbacher, D.: Epidemic Spreading on Weighted Contact Networks. In: Proc. of BIONETICS 2007, Budapest, Hungary (2007)
17. Van Mieghem, P., Omic, J., Kooij, R.E.: Virus Spread in Networks. IEEE/ACM Transactions on Networking 17(1), 1–14 (2009)
18. Wang, Y., Chakrabarti, D., Wang, C., Faloutsos, C.: Epidemic spreading in real networks: An eigenvalue viewpoint. In: 22nd Symposium in Reliable Distributed Computing, Florence Italy, October 6-8 (2003)
19. Wang, Y., Wang, C.: Modeling the Effects of Timing Parameters on Virus Propagation. In: ACM Workshop on Rapid Malcode, Washington, DC, October 27 (2003)
20. Watts, D.J., Strogatz, S.H.: Collective dynamics of small-world networks. Nature 393, 440–442 (1998)

# Practical Random Linear Network Coding on GPUs*

Xiaowen Chu[1], Kaiyong Zhao[1], and Mea Wang[2]

[1] Department of Computer Science, Hong Kong Baptist University, Hong Kong
{chxw,kyzhao}@comp.hkbu.edu.hk
[2] Department of Computer Science, University of Calgary, Alberta, Canada
meawang@ucalgary.ca

**Abstract.** Recently, random linear network coding has been widely applied in peer-to-peer network applications. Instead of sharing the raw data with each other, peers in the network produce and send encoded data to each other. As a result, the communication protocols have been greatly simplified, and the applications experience higher end-to-end throughput and better robustness to network churns. Since it is difficult to verify the integrity of the encoded data, such systems can suffer from the famous pollution attack, in which a malicious node can send bad encoded blocks that consist of bogus data. Consequently, the bogus data will be propagated into the whole network at an exponential rate. Homomorphic hash functions (HHFs) have been designed to defend systems from such pollution attacks, but with a new challenge: HHFs require that network coding must be performed in GF($q$), where $q$ is a very large prime number. This greatly increases the computational cost of network coding, in addition to the already computational expensive HHFs. This paper exploits the potential of the huge computing power of Graphic Processing Units (GPUs) to reduce the computational cost of network coding and homomorphic hashing. With our network coding and HHF implementation on GPU, we observed significant computational speedup in comparison with the best CPU implementation. This implementation can lead to a practical solution for defending against the pollution attacks in distributed systems.

**Keywords:** applications and services, network coding, pollution attack, GPU computing.

## 1 Introduction

In recent years, peer-to-peer (P2P) content distribution applications (e.g., BitTorrent) and video streaming applications (e.g., ppLive) have become popular and constitute more than 30% of today's Internet traffic. The new coding technique, random linear network coding, is recently adopted by P2P applications [6-12], leading to simpler communication protocols, higher throughput, better resilience to network churns, and many more benefits to be discovered [6-12]. With network coding, the source

* This research was supported in part by Hong Kong RGC under grant HKBU 210406, and FRG grant HKBU FRG/07-08/II-36.

L. Fratta et al. (Eds.): NETWORKING 2009, LNCS 5550, pp. 573–585, 2009.

segments the to-be-distributed content into $n$ data blocks of equal size. Each peer (including the source) sends out encoded data blocks, each of which is a linear combination of the original data blocks. After receiving $n$ linearly independent encoded data blocks, a peer is able to decode the original data blocks by solving $n$ linear equations with $n$ variables. Since it is difficult to verify the integrity of the encoded data, such systems can suffer from the famous pollution attack, in which a malicious node can send bad encoded blocks that consist of bogus data. Consequently, the bogus data will be propagated into the whole network at an exponential rate. To defend against such attacks, homomorphic hash functions (HHFs) have been proposed to provide a mechanism for verifying the integrity of the encoded data blocks received from the network. In a nutshell, HHFs offer a nice property that the hash value of any encoded data block can be derived from the hash values of the original data blocks, based on which we can identify the bad encoded data blocks without decoding them [14] [15]. Hence, it can effectively prevent the propagation of the bogus data blocks.

The theoretical property of HHFs is very attractive to practical P2P applications. Unfortunately, the computational complexity posed by HHF is the stumbling stone to this realization. First, HHF itself is computationally expensive. On a contemporary CPU, say 3.0GHz Pentium 4 PC, we can only hash hundreds of kilobit per second [13]. Second, Homomorphic hashing requires extensive modular exponentiation operations over a very larger prime modulus $p$ (e.g., 1024 bit). This require that the data must be encoded in large finite field, GF($q$), where $q$ is a large prime number (e.g., 257 bit). This greatly increases the computational cost of network coding, nearly impractical on CPUs. Our imperative goal is to remove the barrier by reducing the computational cost. The key enabling technologies here are the modern GPUs and the CUDA programming model for non-graphical application development on GPUs.

On the hardware level, recent advances in GPUs open a new era of GPU computing [20]. For instance, NVIDIA's GTX 280 can achieve 933 GFLOPS of computing power, about 8 times faster than the Intel Harpertown 3.2GHz CPU. However, using GPU for non-graphic applications has been considered very difficult, mostly due to the limited API support. Nonetheless, the introduction of CUDA programming model makes it easier for software developers to develop non-graphic applications on GPUs [1]. In CUDA, GPU is treated as a dedicated coprocessor to the CPU, and multiple threads based on the same code can run simultaneously on the GPU, working on different data set. With supports from CUDA, it is now possible to implement network coding and HHFs on GPUs. In this paper, we propose to use GPU to accelerate random linear network coding as well as homomorphic hashing. We designed and developed massively parallel network encoding and decoding algorithms and homomorphic hashing. By carefully applying optimization techniques, we successfully achieved a significant performance boost: 95x speedup for network encoding, 33x speedup for network decoding, and 15x speedup for homomorphic hashing on a contemporary GPU. This makes network coding and HHF a practical solution for pollution attacks in P2P systems.

The rest of the paper is organized as follows. Sec. 2 provides background information on network coding, homomorphic hashing, the GPU architecture, and the CUDA programming model. Sec. 3 presents the parallel algorithms for random linear network coding in GF($q$). Sec. 4 presents the parallel homomorphic hash algorithm. Our experimental results are presented in Sec. 5, followed by the conclusions in Sec. 6.

## 2 Background and Related Work

This section provides the necessary background knowledge of network coding, homomorphic hash function, the GPU architecture, and the CUDA programming model.

### 2.1 Network Coding

Network coding has been originally proposed in information theory to achieve the optimal throughput in a multicast session [6]. Since then, it has been applied in various communication networks for better throughput and robustness to network dynamics. The essence of network coding is a paradigm shift to allow coding at intermediate nodes between the source and the receivers in one or multiple communication sessions. The seminal work of network coding has been studied in [6] [7] [9], which has shown that a multicast session can achieve the data rate of multicast upper bound if network nodes are allowed to perform coding. The framework of random network coding was proposed in [8], which makes network coding theory applicable to practical applications. Since then, there are quite a number of proposal to apply network coding in practical systems for performance enhancement. The Avalanche project by Microsoft Research applied random linear network coding in a P2P content distribution application [10]. Similarly, Lava incorporates random linear network coding into a live multimedia streaming system [12]. Network coding has also been applied in other fields, such as distributed storage systems [11] and wireless networks [21].

In network coding, each data block is treated as a vector of elements in the finite field, and an encoded block is simply a vector representing the linear combination of a set of data blocks (vectors) with randomly generated coefficients in the finite field. The network coding operations, encoding and decoding, are implemented in finite fields, i.e., prime fields GF($q$) or extension fields GF($q^r$), where $q$ is a prime number and $r$ is a positive integer. The computational performance of random linear network coding in GF($2^r$) has been previously studied in [17]. In [18] [22], GPUs have been used to accelerate the performance of network coding in GF($2^r$). To the best of our knowledge, this is the first paper studying the computational performance of random linear network coding in prime field GF($q$), which has a much higher demand of computing power than that of network coding in GF($2^r$).

### 2.2 Homomorphic Hashing

As discussed in Sec. 1, network coding enabled P2P applications are prone to the pollution attacks, in which a malicious peer can easily inject bogus data blocks into an encoded block without being noticed. When network coding is not deployed, peers will receive original data blocks from each other. Hence it is possible to use normal hash functions, such as SHA1, to verify the correctness of a data block by comparing the hash of each received data block to the corresponding hash provided by the source. With network coding, the effect of pollution attack becomes more serious and harder to detect [14] [15] [16] for two reasons: First, each bogus block can be encoded with regular data blocks before being propagated in the network. Second, the traditional hash functions, e.g., SHA1, are no longer practical since the encoded blocks received by each peer cannot be predetermined by the source. Some workarounds have been proposed to address these issues. In [14], a cooperative scheme is proposed, in which peers perform probabilistically block verification and inform others when a malicious node has been identified. However, this scheme cannot detect the bogus blocks at the earliest stage and could potentially have false alarms.

To this end, homomorphic hash functions (HHFs) are currently the best solution to address this security issue with network coding. HHFs have the property that the hash value of an encoded block can be constructed by the hash values of the original blocks. In other words, a peer only need to get the hash values of the original blocks from the source, it then can easily verify the integrity of an encoded block immediately after receiving the encoded block. Although the HHFs can theoretically resolve the pollution attack problem, it is technically not practical on today's desktop CPUs. A 3 GHz Pentium 4 CPU can only achieve around 300 Kbps of hashing throughput [13]. Furthermore, it requires network coding to be performed in GF($q$) with large value for $q$, which makes it computationally expensive.

### 2.3 GPU Computing and CUDA

GPUs are dedicated hardware for manipulating computer graphics. Due to the huge demand for computing for real-time applications and high-definition 3D graphics, GPUs have been evolved into highly paralleled multi-core processors. The NVIDIA GeForce GTX260 has 24 Streaming Multiprocessors (SMs), and each SM has 8 Scalar Processors (SPs). At any given clock cycle, all SPs of the same SM must execute the same instruction, but can operate on different data. Each SM has four different types of on-chip memory: constant cache, texture cache, registers, and shared memory. The properties of the different types of memories have been summarized in [1] [19]. A general optimization principle is that registers and shared memory should be carefully utilized to amortize the global memory latency cost.

The exceptional GPU computing power is very attractive to general-purpose system development. The first generation of GPU computing (namely GPGPU) requires that non-graphics application must be mapped through the graphics application programming interfaces, which is very challenging. In early 2007, one of the major GPU vendors, NVIDIA, announced a new general-purpose parallel programming model, Compute Unified Device Architecture (CUDA) [1], which extends the C programming language for general-purpose application development. Meanwhile, another GPU vendor AMD introduced Close To Metal (CTM) programming model that provides an assembly language for application development [2]. Intel is also planning to release Larrabee [3], a new multi-core GPU architecture specially designed for GPU computing. Currently, CUDA is the best available programming model, and is the most well accepted model by the research and development community. Since the release of CUDA, it has been used for speeding up a large number of applications [18-20] [22] [23]. For these reasons, we chose to use CUDA in our research. Nevertheless, out algorithms can be easily implemented on other GPU computing models.

In the CUDA model, the GPU is regarded as a coprocessor capable of executing a great number of threads in parallel. A single program consists of *host code* to be executed on CPU and *kernel code* to be executed on GPU. The kernel code is usually computational-intensive, data-parallel and multi-threaded. Threads are organized into *thread blocks*, where each block is associated with one SM. Threads belonging to the same thread block can share data through the shared memory and can perform barrier synchronization. CUDA does not provide any direct synchronization methods between threads that belong to different thread blocks, however. When a thread block terminates, a new thread block can be launched on the vacant SM.

# 3 Parallel Network Coding on GPUs

To facilitate the development of network coding, we have implemented a set of library functions of multiple-precision modular arithmetic on the CUDA platform. These library functions simplify the development of the network coding system and homomorphic hash functions. Our multiple-precision library includes the following functions: comparison, subtraction, modular addition, modular subtraction, multiplication, division, multiplicative inversion, Montgomery reduction, Montgomery multiplication.

Assume the original data to be distributed is divided into $n$ equally sized data blocks ($b_1$, $b_2$, ..., $b_n$), where each data block $b_i$ contains $m$ codewords $b_{i,k}$, $k \in \{1,\dots,m\}$. An encoded block $e_j$ is a linear combination of the $n$ original blocks and it also contains $m$ codewords $e_{j,k}$, $k \in \{1,\dots,m\}$. The linear relationship between $e_j$ and the original $n$ blocks is described by $e_j$'s *global coefficient vector* $(c_{j,1}, c_{j,2}, \dots, c_{j,n})$: $e_{j,k} = \sum_{i=1}^{n} c_{j,i} \cdot b_{i,k}$, $k \in \{1,\dots,m\}$. Obviously the encoding process is a vector-matrix multiplication. A peer can decode the original $n$ data blocks as soon as it has received $n$ linearly independent encoded data blocks ($e_1$, $e_2$, ..., $e_n$), by solving the set of linear equations $e_{j,k} = \sum_{i=1}^{n} c_{j,i} \cdot b_{i,k}$, $k \in \{1,\dots,m\}$, $j \in \{1,\dots,n\}$. In a P2P application with network coding, a peer receives encoded data blocks from upstream peers, and also creates new encoded data blocks by randomly and linearly combining its received encoded blocks, and then disseminates the new encoded blocks to its downstream peers.

## 3.1 Network Encoding in GF($q$)

When network coding is performed in GF($q$) where $q$ is a predefined large prime number, the encoding process will creates a sequence of encoded blocks $e_j$, each of which contains $m$ codewords $e_{j,k}$. $e_j$ is generated based on a random coefficient vector $c_j = (c_{j,1}, c_{j,2}, \dots, c_{j,n})$: $e_{j,k} = \sum_{i=1}^{n} c_{j,i} \cdot b_{i,k} \bmod q$, $k \in \{1,\dots,m\}$. Here $c_{j,i}$ are positive 32-bit integers. The encoding process includes two steps: (1) generating the coefficient vector $c_j$; (2) modular vector-matrix multiplication. The CUDA library provides a very high efficient random number generator using Mersenne Twister method, which can generate tens of millions of random numbers per second using GPU. As the time of generating $n$ random numbers are negligible as compared with the encoding time, we will focus on the vector-matrix multiplication operation, as shown in Figure 1(a). We implement the computing of each codeword $e_{j,k}$ by a CUDA thread. Hence encoding a single block requires $m$ threads. Each thread computes a dot product and then performs a modular operation, using our multiple-precision CUDA library. In order to fully exploit the computing power of GPUs, thousands of threads are a normal requirement. Therefore we propose a batched encoding approach for small values of $m$, which encodes $K$ blocks simultaneously, as shown in Figure 1(b). In this case, the number of threads equals $mK$.

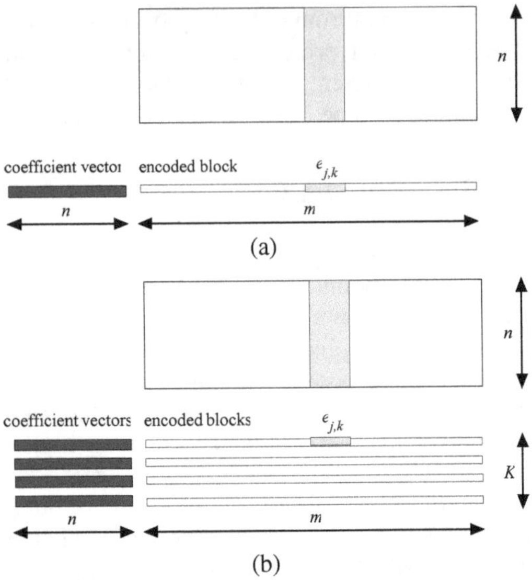


**Fig. 1.** Network encoding: (a) Encode a single block (b) Encode multiple blocks in a batch

### 3.2 Network Decoding in GF($q$)

The decoding process includes two steps: (1) matrix inversion; and (2) matrix multiplication. Matrix inversion for floating-point numbers on GPU has been recently studied in [23]. Our problem is very different because we are operating in GF($q$). We use Gauss-Jordan elimination for the matrix inversion, which brings a matrix to its reduced row echelon form. There is no stability issue because we are operating in finite field. To overcome the synchronization challenge, our parallel matrix inversion algorithm uses both CPU and GPU, as shown in Table 1. The non-parallel parts, e.g., finding the multiplicative inverse, are done by the CPU, whilst the parallel parts, e.g., reducing to row echelon form, are done by GPU.

The matrix multiplication can be implemented in a similar way as the vector-matrix multiplication. Each element in the output matrix is computed by one thread; hence the total number of threads is $n^2$, which is sufficient to fill the GPU cores since $n$ is normally no less than 64 in practice. Difference from the encoding process, the integer multiplications here are performed between two large integers, since the corresponding values of the coefficients grow as the blocks are being re-encoded at each peer. In this case, a straightforward parallel implementation cannot achieve satisfactory performance due to the global memory latency. GPU's on-chip shared memory can be exploited to amortize the global memory latency, and we propose to use a titled version of matrix multiplication, in which the matrix is divided into a number of sub-blocks [1] [19]. As illustrated in Figure 2, the computing of sub-block $B_{sub}$ is done by a thread block. The threads in this block cooperatively load the data from the two tiles in coefficient matrix and $E^T$ into shared memory. These threads compute the partial dot product in shared memory, and then continue with the next tile. The size of the tile should be controlled such that two tiles can be accommodated by the shared memory of a SM.

**Table 1.** Algorithm of Matrix Inversion in GF($q$)

**Algorithm 1.** Matrix Inversion in GF($q$)

```
INPUT: An n x n non-singular matrix M, an n x n unit matrix U
OUTPUT: the inverse of M
lead ← 0;
row ← n; col ← n;
for ( r = 0 to row - 1)
    i ← r;
    while M[i, lead] equals 0
        i++;
    Swap rows i and r of M and U;
    t ← multiplicative inverse of M[r, r];  /* on CPU */
    Multiply row r of M and U by t;          /* on GPU */
    for all rows j except row r of M and U
        For M, subtract M[j, lead] multiplied by row r from row j; /* on GPU */
        For U, subtract M[j, lead] multiplied by row r from row j; /* on GPU */
    end for
    lead++;
end for
return U
```

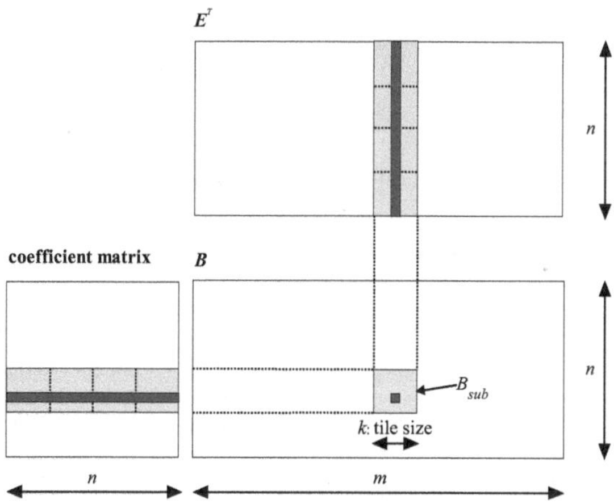


**Fig. 2.** Decoding: tiled matrix multiplication

## 4 Parallel Homomorphic Hashing on GPUs

The homomorphic hash function, $h(\cdot)$, proposed in [13] requires a set of hash parameters $G = (p, q, g)$. The parameters $p$ and $q$ are large prime numbers of order $\lambda_p$ and $\lambda_q$ chosen such that $q \mid p - 1$. The parameter $g$ is a vector of $m$ numbers, each of which can be written as $x^{(p-1)/q} \bmod p$ where $x \in Z_q$ and $x \neq 1$. The method of creating the parameter set can be found in [13]. Typical values of the parameters are summarized in Table 2. The homomorphic hash of a data block $b_i$ is then calculated as

$h(b_i) = \prod_{k=1}^{m} g_k^{b_{k,i}} \bmod p$. The hash values of the original data blocks $(b_1, b_2, \ldots, b_n)$ are $h(b_1)$, $h(b_2)$, …, $h(b_n)$, respectively. Given an encoded data block $e_j$ with global coefficient vector $(c_{j,1}, c_{j,2}, \ldots, c_{j,n})$, the homomorphic hash function $h(\cdot)$ can be shown to satisfy the following condition: $h(e_j) = \prod_{i=1}^{n} h^{c_{j,i}}(b_i) \bmod p$. This property can be used to verify the integrity of an encoded block, as illustrated in Figure 3. The content publisher first calculates the homomorphic hash values for each of the data blocks. The downloaders need to download a copy of these hash values for the purpose of verifying every single encoded data block.

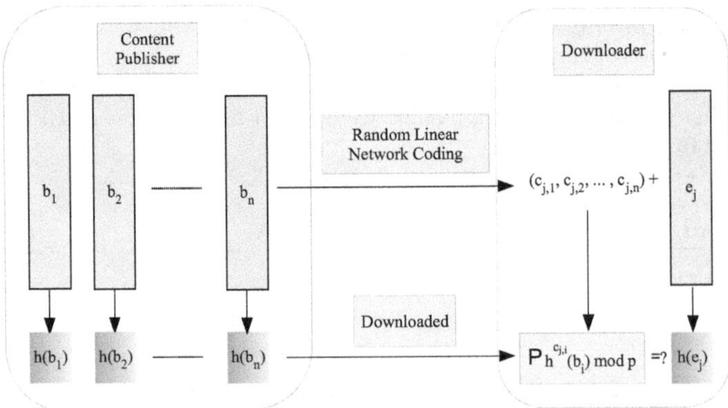


**Fig. 3.** Data verification using homomorphic hashing in network coded P2P applications

**Table 2.** Homomorphic hash function parameters

| **Name** | **Description** | **Typical Value** |
|---|---|---|
| $\lambda_p$ | Discrete log security parameter | 1024 bit |
| $\lambda_q$ | Discrete log security parameter | 257 bit |
| $p$, $q$ | Random primes, $\lvert p \rvert = \lambda_p, \lvert q \rvert = \lambda_q,\ q \mid p-1$ | |
| $m$ | Number of codewords per data block | 512 |
| $n$ | Number of data blocks | 128 |

As shown earlier, homomorphic hashing, i.e., $h(b_i) = \prod_{k=1}^{m} g_k^{b_{k,i}} \bmod p$, involves $m$ modular exponentiations and $m$-1 modular multiplications. The $m$-1 modular multiplications can be easily parallelized by a regular reduction process. The $m$ modular exponentiations can be very time consuming. We distribute the $m$ modular exponentiations to the GPU processing cores. The challenge is to implement modular exponentiation on GPU in the most efficient way. On the current CUDA platform, integer division and modulo operations are very costly. Therefore we choose to use the

Montgomery exponentiation algorithm (as shown in Table 3) which can decrease the number of division operations significantly.

**Table 3.** Algorithm of multiple-precision Montgomery exponentiation

**Algorithm 2.** Multiple-precision Montgomery Exponentiation

INPUT: integer $m$ with $n$ radix $b$ digits and $\gcd(m, b) = 1$, $R = b^n$, positive integer $x$ with $n$ radix $b$ digits and $x < m$, and positive integer e = $(e_t \cdots e_0)_2$ .

OUTPUT: $x^e$ mod $m$.

$\tilde{x} \leftarrow Mont(x, R^2 \ mod\ m)$ ;
$A \leftarrow R$ mod $m$;
**for** ( $i$ **from** $n$ **down to** 0)
  $A \leftarrow Mont(A,\ A)$ ;
  **if** $e_i == 1$
  **then** $A \leftarrow Mont(A,\ \tilde{x})$ ;
**end for**
$A \leftarrow Mont(A,\ 1)$ ;
**return** $A$;

# 5 Experimental Results

We have implemented the proposed network encoding/decoding and parallel homomorphic hashing algorithm using CUDA. For comparison purpose, we also implemented network coding and homomorphic hash function for CPU in C language, by utilizing the GNU MP arithmetic library, version 4.2.3 [4]. These implementations are running on an Intel Core2 CPU 1.6 GHz. We tested all our algorithms on XFX GTX280 graphic card with an NVIDIA GeForce GTX280, which has 240 processing cores. On GTX280, there are 30 Streaming Multiprocessors (SMs), and each SM has 8 Scalar Processors (SPs), 16384 32-bit registers and 16KB shared memory.

## 5.1 Performance of Encoding in GF($q$)

The throughput of encoding process is shown in Figure 4(a) in log-scale. In theory, the encoding time complexity is linear to the size of $n$. This is in accordance with our experimental results. The throughput of network encoding on CPU is very poor: only 10.3 Mbps for $n$ = 128. The GPU performance is very impressive: around 800 Mbps can be achieved for $n$ = 128 with a small batch size $K$ of 32. We observe that larger batch sizes can lead to better performance, until some threshold value has been met. In our testing environment, $K$ = 128 is the optimal setting. The speedup of GPU over CPU has been plotted in Figure 4(b), for different batch sizes and $n$. With a batch size larger than 32, the speedup is greater than 66x. The highest speedup of 95x is obtained when $n$ = 128 and $K$ = 128.

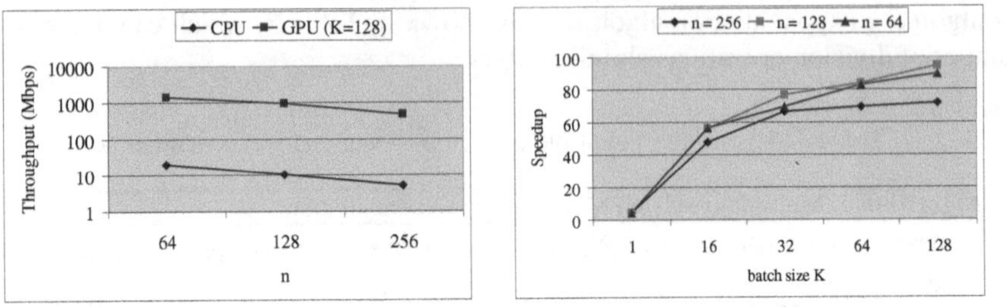


(a) (b)

**Fig. 4.** Performance of network encoding: (a) Encoding throughput (b) Speedup over CPU

## 5.2 Performance of Decoding

As mentioned before, the decoding process includes two steps: (1) matrix inversion; (2) matrix multiplication. The time used for matrix inversion is shown in Figure 5(a) using log-scale, and the speedups on GPU over CPU are plotted in Figure 5(b). It is a well known fact that matrix inverse using Gauss-Jordan elimination has a time complexity of $O(n^3)$. Our experimental results on CPU follow this pattern as well. The performance of parallel matrix inversion on GPU is a bit more complicated due to the kernel loading overhead and communication overhead between the CPU and GPU. Figure 5(b) shows that the speedup on GPU grows as the number of blocks, $n$, increase: 17 for $n$ = 128 and 24 for $n$ = 256. This is where the benefit of GPU manifests itself, i.e., the overheads are amortized by the increasing parallelism in the computation.

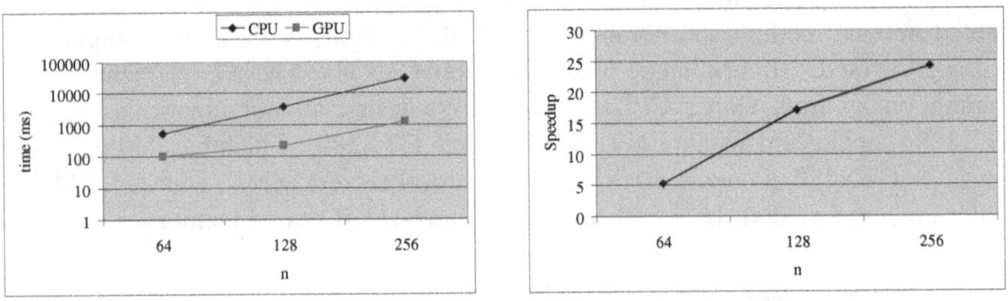


(a) (b)

**Fig. 5.** Performance of matrix inversion: (a) Matrix inversion time in *ms* (b) Speedup over CPU

The performance of the matrix multiplication process is shown in Figure 6. As expected, the throughput is much slower than the encoding process. Even so, the GPU can achieve 243 Mbps of throughput for $n$ = 128 when shared memory is utilized. The speedup on GPU over CPU ranges from 64x to 75x for $n$ = 64, 128, 256 respectively when shared memory is used.

The performance of the whole decoding process is shown in Figure 7, for $m$ = 512. Since the speedup of matrix multiplication is much larger than the speedup of matrix inversion, the speedup of the overall decoding process is limited by the performance of matrix inversion. The overall decoding throughput when $n$ = 128 is 58 Mbps which includes the matrix inversion and matrix multiplication. The decoding performance can be further enhanced by using a larger value of *m*, because the same matrix inverse operation is now used for a larger data volume. The speedup ranges from 15x to 33x for different values of *n*.

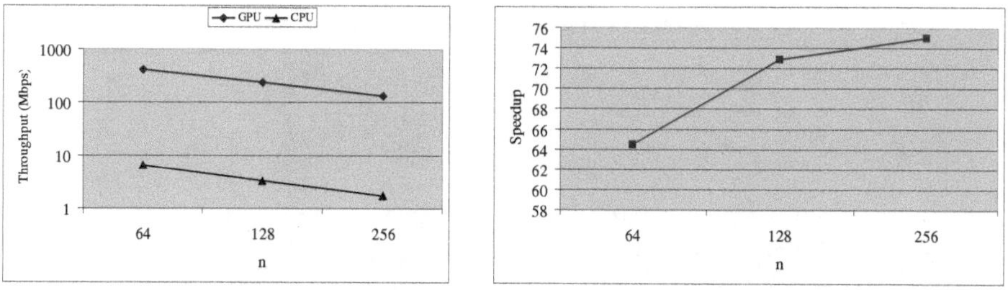


(a) (b)

**Fig. 6.** Performance of matrix multiplication: (a) Throughput (b) Speedup over CPU

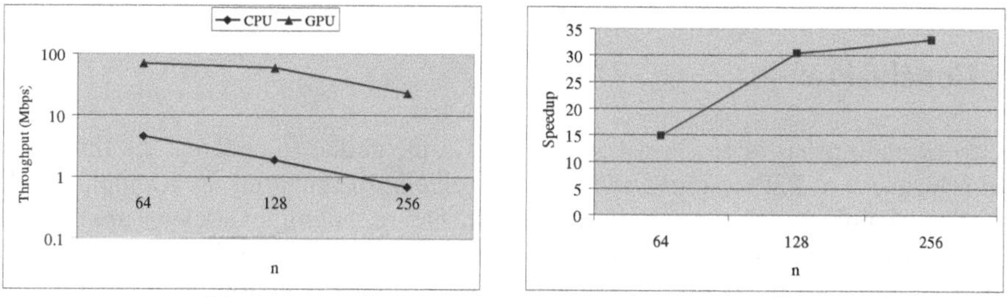


(a) (b)

**Fig. 7.** Performance of network decoding. (a) Throughput of decoding (b) Speedup over CPU.

## 5.3 Performance of Homomorphic Hashing

Our CPU version of homomorphic hashing achieves 130 Kbps of throughput, which is relatively lower than the results reported by [13] [14] due to our relatively lower CPU frequency.

The parallel homomorphic hashing uses Algorithm 2 to calculate exponentiations. The CUDA architecture requires a large number of threads to hide the memory latency and to fully utilize the computing power. The number of threads per thread block (denoted by ***TB***), and also the number of thread blocks (denoted by ***NB***), are the two main factors that affect the hashing throughput. We plot the throughput for different configurations in Figure 8. It is easy to observe that more threads per block can generally achieve better throughput. When the number of threads per block is fixed, the throughput can be improved by creating more thread blocks, until some threshold has been reached. Since our GPU has 30 SMs, the number of thread blocks should be a multiple of 30. Better throughput can be achieved if the following conditions are satisfied: (1) ***TB*** is a multiple of 32 (i.e., the *warp* size [1] [19]). In CUDA, a *warp* is formed by 32 parallel threads and is the scheduling unit of each SM. If the number of threads in a block is not a multiple of warp size, the remaining instruction cycles will be wasted. (2) ***NB*** is a multiple of 30 (i.e., the number of SMs); (3) ***TB*** x ***NB*** > 6144.

For example, if $m = 512$, we should perform the homomorphic hashing for 12 different data blocks simultaneously. The highest throughput of 1.9 Mbps is obtained when ***TB*** = 128 and ***NB*** = 90, which is 15 times faster than the CPU implementation.

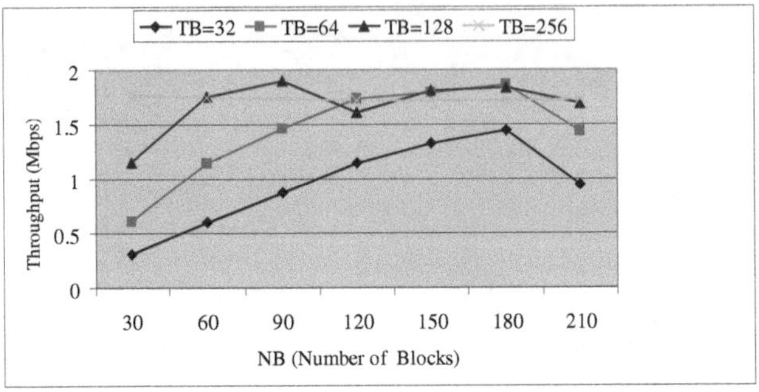


**Fig. 8.** Throughput of homomorphic hashing on GPU

## 6 Conclusions

Network coding has been shown as a powerful technique to enhance the throughput and robustness of P2P systems; and homomorphic hash functions are a supplementary tool for defending against the pollution attack. The remaining challenges are the computational requirement of network coding in prime field and the homomorphic hashing. This paper demonstrates a practical parallel implementation of network coding and homomorphic hashing using GPUs. Our experimental results show that the computational obstacle of network coding and homomorphic hashing can be overcome by designing efficient parallel algorithms and fully exploiting the computing power of contemporary GPUs that are widely available on today's desktop PCs.

## References

1. NVIDIA CUDA Compute Unified Device Architecture: Programming Guide, Version 2.0beta2 (June 2008)
2. AMD CTM Guide: Technical Reference Manual (2006), `http://ati.amd.com/companyinfo/researcher/documents/ATI_CTM_Guide.pdf`
3. Seiler, L., et al.: Larrabee: a many-core x86 architecture for visual computing. ACM Transactions on Graphics 27(3) (August 2008)
4. GNU MP Arithmetic Library, `http://gmplib.org/`
5. Brickell, E.F., Gordon, D.M., McCurley, K.S., Wilson, D.B.: Fast exponentiation with precomputation: algorithms and lower bound. In: Rueppel, R.A. (ed.) EUROCRYPT 1992. LNCS, vol. 658, pp. 200–207. Springer, Heidelberg (1993)
6. Ahlswede, R., Cai, N., Li, S.R., Yeung, R.W.: Network information flow. IEEE Transactions on Information Theory 46(4), 1204–1216 (2000)
7. Koetter, R., Medard, M.: An algebraic approach to network coding. IEEE/ACM Transactions on Networking 11(5), 782–795 (2003)

8. Ho, T., Koetter. R., Médard, M., Karger, D.R., Effros, M.: The benefits of coding over routing in a randomized setting. In: Proceedings of IEEE ISIT (2003)
9. Li, S.-Y.R., Yueng, R.W., Cai, N.: Linear network coding. IEEE Transactions on Information Theory 49, 371–381 (2003)
10. Gkantsidis, C., Rodriguez, P.: Network coding for large scale content distribution. In: Proceedings of IEEE INFOCOM 2005 (2005)
11. Dimakis, A.G., Godfrey, P.B., Wainwright, M.J., Ramchandran, K.: Network coding for distributed storage systems. In: Proceedings of IEEE INFOCOM 2007 (2007)
12. Wang, M., Li, B.: Lava: a reality check of network coding in peer-to-peer live streaming. In: Proceedings of IEEE INFOCOM 2007 (2007)
13. Krohn, M., FreedMan, M., Mazieres, D.: On-the-fly verification of rateless erasure codes for efficient content distribution. In: Proceedings of IEEE Symposium on Security and Privacy, Berkeley, CA (2004)
14. Gkantsidis, C., Rodriguez, P.: Cooperative security for network coding file distribution. In: Proceedings of IEEE INFOCOM 2006 (2006)
15. Li, Q., Chiu, D.-M., Lui, J.C.S.: On the practical and security issues of batch content distribution via network coding. In: Proceedings of IEEE ICNP 2006, pp. 158–167 (2006)
16. Yu, Z., Wei, Y., Ramkumar, B., Guan, Y.: An efficient signature-based scheme for securing network coding against pollution attacks. In: Proceedings of IEEE INFOCOM 2008 (April 2008)
17. Shojania, H., Li, B.: Parallelized progressive network coding with hardware acceleration. In: Proceedings of the 15th International Workshop on Quality of Service, IWQoS (2007)
18. Chu, X., Zhao, K., Wang, M.: Massively parallel network coding on GPUs. In: Proceedings of the 27th IEEE IPCCC (December 2008)
19. Ryoo, S., Rodrigues, C.I., Baghsorkhi, S.S., Stone, S.S., Kirk, D.B., Hwu, W.: Optimization principles and application performance evaluation of a multithreaded GPU using CUDA. In: Proceedings of ACM PPoPP 2008 (Feburary 2008)
20. Owens, J.D., Houston, M., Luebke, D., Green, S., Stone, J.E., Phillips, J.C.: GPU computing. In: IEEE Proceedings, May 2008, pp. 879–899 (2008)
21. Katti, S., Katabi, D., Balakrishna, H., Medard, M.: Symbol-level network coding for wireless mesh networks. In: Proceedings of ACM Sigcomm 2008 (August 2008)
22. Shojania, H., Li, B., Wang, X.: Nuclei: GPU-accelerated Many-core Network Coding. In: Proceedings of IEEE INFOCOM 2009 (April 2009)
23. Volkov, V., Demmel, J.W.: Benchmarking GPUs to tune dense linear algebra. In: Dolev, S., Haist, T., Oltean, M. (eds.) OSC 2008. LNCS, vol. 5172. Springer, Heidelberg (2008)

# Peer-assisted On-demand Video Streaming with Selfish Peers*

Niklas Carlsson[1], Derek L. Eager[2], and Anirban Mahanti[3]

[1] University of Calgary, Calgary, Canada
[2] University of Saskatchewan, Saskatoon, Canada
[3] NICTA, Sydney, Australia
niklas.carlsson@cpsc.ucalgary.ca, eager@cs.usask.ca,
anirban.mahanti@nicta.com.au

**Abstract.** Systems delivering stored video content using a peer-assisted approach are able to serve large numbers of concurrent requests by utilizing upload bandwidth from their clients to assist in delivery. In systems providing download service, BitTorrent-like protocols may be used in which "tit-for-tat" policies provide incentive for clients to contribute upload bandwidth. For on-demand streaming delivery, however, in which clients begin playback well before download is complete, all prior proposed protocols rely on peers at later video play points uploading data to peers at earlier play points that do not have data to share in return. This paper considers the problem of devising peer-assisted protocols for streaming systems that, similar to download systems, provide effective "tit-for-tat" incentives for clients to contribute upload bandwidth. We propose policies that provide such incentives, while also providing short start-up delays, and delivery of (almost) all video frames by their respective playback deadlines.

**Keywords:** BitTorrent-like systems, peer-assisted streaming, tit-for-tat.

## 1 Introduction

Peer-assisted content delivery techniques are increasingly being adopted by media companies. For example, in April 2008 it was reported that the BBC iPlayer service[1], which in part uses peer-assisted delivery, was being used for downloading more than one million BBC programmes each week.[2]

When a download-and-play approach is used (as in the BBC iPlayer service), BitTorrent-like protocols [1] may be used in which a tit-for-tat policy provides incentives for clients to contribute their upload bandwidth.[3] Tit-for-tat is effec-

* This work was supported by the Natural Sciences and Engineering Research Council (NSERC) of Canada and by the Informatics Circle of Research Excellence (iCORE) in the Province of Alberta.

[1] http://www.bbc.co.uk/iplayer/

[2] http://technology.timesonline.co.uk/tol/news/tech_and_web/article3716781.ece

[3] BBC iPlayer also offers a streaming service, but this does not use peer-assisted delivery.

L. Fratta et al. (Eds.): NETWORKING 2009, LNCS 5550, pp. 586–599, 2009.

tive in this context because the protocol's "rarest first" piece selection policy makes it highly likely that peers will have differing sets of file pieces and are thus able to carry out two-way mutually beneficial piece exchanges.

Clients may, however, prefer services in which they can begin viewing a video shortly after beginning download, with only a short start-up delay. In such on-demand streaming systems, once playback begins, reception of each subsequent frame must occur before the frame's play point if impairment in playback quality is to be avoided. Unfortunately, the in-order requirements of playback fundamentally conflict with the goal of high piece diversity, as needed for effective use of tit-for-tat. In fact, all prior peer-assisted protocols for on-demand video streaming have relied, for good performance including low start-up delay, on peers at later video play points uploading data to peers at earlier play points that do not have data to share in return.

This paper considers the problem of improving quality of service in peer-assisted on-demand streaming systems while retaining the basic tit-for-tat behaviour of BitTorrent-like protocols. Tit-for-tat is one of the cornerstone ideas supporting fairness and scalability in BitTorrent and has the advantage of being fully decentralized [1]. We attempt to achieve this goal through design of new, tit-for-tat compatible, peer and piece selection policies, focusing for the most part on policies to be used for server-to-peer uploads (for which tit-for-tat behavior is not an issue) rather than peer-to-peer uploads. We find that the best system performance is achieved with a peer selection (by the server) policy that preferentially allocates server bandwidth for uploads to peers at imminent risk of receiving data too late for playback, and secondly for uploads of rare pieces to newly arrived peers. Simulations of these new policies are used to evaluate their effectiveness. Our results show that substantial improvements in quality of service are feasible while ensuring that the piece diversity is sufficient for peers to effectively employ tit-for-tat.

The remainder of the paper is organized as follows. Related work is discussed in Section 2. Section 3 describes a baseline tit-for-tat based peer-assisted streaming protocol. New piece selection policies, and a server policy for prioritizing upload requests from peers, are discussed in Section 4. Section 5 describes the simulation model used for evaluating the new policies. Section 6 presents performance results. Conclusions are presented in Section 7.

## 2 Related Work

There exists a large literature on peer-assisted on-demand streaming systems. A significant portion of this literature has focussed on *explicit allocation* of peer upload bandwidth [2,3,4,5,6,7,8,9,10]. This literature, for example, includes tree-based cache-and-relay approaches [4,5], and work that considers the problem of determining the set of servers (or peers) that should serve each peer, and at what rate each server should operate [6,7]. In recent work, Parvez *et al.* show that systems in which pieces are retrieved in-order can significantly benefit from peer selection policies that bias uploads towards peers with fewer potential uploaders (i.e., peers requiring data closer to the end of the file) [9]. Such bias results in a natural "daisy-chain" effect wherein peers upload the most pieces to peers with slightly fewer in-order pieces than the peer itself has obtained thus far. Peer selection bias towards peers with similar playback points has also been used in an implementation of a video-on-demand system [10]. While effective in cooperative environments, we note that these techniques do not allow for effective use of tit-for-tat as younger peers typically are not able to upload to older peers.

Other work has proposed using feedback mechanisms to gain information about the upload contributions of peers at earlier playback points, and use this information to reward peers that forward pieces at a higher rate [11]. We note that such feedback-based schemes are significantly more sensitive to cheating peers than tit-for-tat techniques in which the peers themselves can effectively measure the rates they receive from other peers.

There has also been related work on piece selection policies that attempt to mediate the conflict between high piece diversity and the in-order requirements of playback (e.g., [12,13,14,15,16]). For example, Annapureddy *et al.* [12] propose splitting each file into fixed-sized segments, each consisting of some number of consecutive pieces. Segments are downloaded sequentially using a BitTorrent-like protocol. To increase the likelihood that peers downloading the same segment have pieces to exchange they propose using distributed network coding within segments, and pre-fetch some smaller number of pieces from future segments. Probabilistic piece selection policies have also been proposed [14,15,16]. Note that in order to achieve low start-up delays, these policies depend on older peers uploading to new peers that most likely do not have any needed pieces to offer in exchange.

Finally, we note that most prior work on peer-assisted video-on-demand has assumed that peers begin playback after buffering a *fixed* amount of data, or evaluate protocols with respect to the *lowest possible* start-up delay a peer could have chosen such that the (unknown *a priori*) download completion time of every piece is no later than its playback point. In contrast, we use a simple *online* rule (based on LTA [14]) to determine when playback can safely commence, and evaluate our policies with regards to both the *actual* start-up delay, as determined by this online rule, and the percentage of pieces that are not received by their playback point, given the chosen start-up delay.

## 3 Baseline Protocol Using Tit-for-Tat

We consider peer-assisted on-demand streaming systems with a single server and varying numbers of active peers, and focus on the delivery of a single video file. This file is divided into fixed-size pieces; each piece is further divided into subpieces as in other BitTorrent-like systems [17,18]. The unit of upload/download is the subpiece. When a peer acquires (all of) a piece, it can advertise this to other peers, and upload subpieces to other peers that do not yet have (all of) the piece. A peer may download multiple different subpieces of a piece in parallel from multiple other peers. We further assume that pieces are grouped into fixed-sized segments, as might be needed in systems utilizing coding, although the simulation results that we present here are for uncoded content delivery.

A peer is considered to be *interested* in another peer if the latter peer has at least one piece that the former peer has not yet received. A *piece selection* policy is used to select among the pieces that are available from a particular peer. In our baseline policy, the candidate pieces are ranked according to how soon they will be needed for playback, and a piece is selected by sampling from a Zipf probability distribution [14]. More specifically, with the Zipf($\theta$) policy, a peer $j$ about to request a piece from peer $i$ selects a piece $k$ from the set of pieces that $i$ has, but that $j$ does not have, with a probability proportional to $1/(k+1-k_0)^\theta$, where $k$ is the index of the piece, and $k_0$ is the index of the first piece that peer $j$ does not yet have. While the Zipf parameter $\theta$ can be tuned so that the policy is more or less aggressive with respect to its preference for earlier pieces,[4] for the results presented here $\theta$ is fixed at 1.25. The Zipf($\theta$) policy has been shown to achieve a good tradeoff between high piece diversity and sequential progress. In contrast to in-order policies (including segment-based in-order versions [12]), or proposals that make explicit allocation of peer upload bandwidth so that older peers upload to newer peers, Zipf-based policies allow for effective use of tit-for-tat.

As with BitTorrent, each peer establishes persistent connections with a large set of peers but only uploads to a limited number of peers at each time instance. A *peer selection* policy determines which peers to upload to, among the interested peers. A *tit-for-tat* peer selection policy is assumed, wherein upload priority at a peer is given to those other peers that are providing the highest download rates to that peer. Periodically, a new peer is chosen to upload to, in the chance that it may offer a better download rate than the current peers. With this *optimistic unchoke* component, a random selection is made among the interested peers to which the peer is not already uploading, if any. At the server, the baseline peer selection policy is random. The above policies are commonly used in simulations of BitTorrent-like systems and provide a baseline for comparison.

## 4 New Policies

### 4.1 Acquiring Rare Pieces

For good performance including low start-up delay, all previous policies for peer-assisted streaming delivery depend on older peers uploading to new peers that most likely do not have any needed pieces to offer in exchange. Such behavior is not a concern in tit-for-tat based download systems, since in that context rarest-first piece selection can be used, and new peers can quickly acquire pieces needed by many others. It *is* a potential concern, however, in tit-for-tat based streaming systems. Even with probabilistic piece selection policies such as that used in our baseline protocol, it may take a relatively long time for new peers to acquire pieces needed by many others. Selfish peers may therefore be motivated

[4] For example, note that $\theta = 0$ and $\theta \to \infty$ yield random and in-order piece selection, respectively.

to adopt optimistic unchoke policies discriminating against new peers. Also, the upload bandwidth of new peers may be underutilized, particularly in low to moderate request rate scenarios in which there is frequently only a single or a small number of concurrently active new peers. Finally, owing to use of tit-for-tat, new peers may receive a relatively small share of the aggregate upload bandwidth of other peers, resulting in a relatively low total download rate and lengthened start-up delays. We address this concern with *rare piece delivery to new peers* (RPNP), which entails two modifications to the baseline protocol.

First, when the server unchokes a "new" peer (specifically, a peer that has not yet begun playback), rather than using the Zipf($\theta$) piece selection policy used in the baseline protocol, the selected piece is the rarest piece not currently being uploaded by the server to some other peer. If there are multiple rarest pieces, ties are broken randomly except when only the server has these pieces, or when the server has sufficient upload bandwidth to serve every currently active peer at the play rate, in which case ties are broken using the Zipf($\theta$) policy.

Second, to more quickly disseminate rare pieces to new peers, the server gives upload priority to peers that have not yet begun playback.[5] Among such peers, higher priority is given to peers for which the server has uploaded less data. The server uploads to only the $n$ highest priority peers, where $n$ is the number of server upload connections; ties are broken randomly.

### 4.2 Prioritizing Urgent Piece Downloads

We further modify the baseline protocol so as to increase the likelihood that each piece is received by its scheduled playback point, by prioritizing delivery of the next required piece for any peer that has started playback and for which this next required piece is within either the current segment being played back, or the next segment (i.e., the peer is in a "low-buffer" state).

This prioritization is accomplished through two policy modifications. First, peers in the low-buffer state use in-order piece selection, rather than Zipf-based piece selection. Second, the server gives the highest upload priority to those peers that are in the low-buffer state. The remaining peers may be prioritized as in RPNP. Among those peers in the low buffer state, higher priority is given to peers for which the server has uploaded less data. As with RPNP, the server uploads to the $n$ highest priority peers, with ties broken randomly.

When used together with RPNP, we call this approach *urgent piece prioritization with rare piece delivery to new peers* (UP/RPNP). Note that neither RPNP nor UP/RPNP alter the tit-for-tat peer selection policy used by peers.

## 5 Simulation Model

We use an existing event-based simulator of BitTorrent-like systems [14]. For the results presented here it is assumed that: (i) all active peers have

[5] Both here, and in Section 4.2, we assume that the server is able to reliably identify peers to which it wishes to give preferential treatment. For example, the system may require use of content provider software with this functionality built in.

connections with each other[6], (ii) there are sufficiently many sub-pieces per piece that parallel download is always possible when multiple peers have a desired piece, (iii) a peer $i$ (or the server) uses at most $n_i$ concurrent upload connections, (iv) connections are not choked in the middle of an upload, and (v) new downloads are initiated only when the download bandwidth capacity $D_i$ is not being fully utilized.

The set of peers that a peer $i$ (or the server) is uploading to may change when (i) the peer completes the upload of a piece, or (ii) some other peer becomes interested and peer $i$ has fewer than $n_i$ active upload connections. The new set of upload targets includes (i) any peer currently being uploaded to, and (ii) additional peers up to the limit $n_i$. Using the peer selection policy, additional peers are selected from the set of interested peers that are not yet fully utilizing their download capacity. With a probability $1/n_i$ the optimistic unchoke component is used to choose a peer, and with a probability of $(n_i - 1)/n_i$ the peer that is uploading to peer $i$ at the highest rate is chosen.

The start-up delay, that is the time since arrival until a peer begins playback, is determined using a modified version of the LTA start-up rule [14]. Playback does not commence until the following two conditions are satisfied. First, the initial two segments of the video must be fully received. Second, the measured long-term average rate at which the peer has received in-order pieces must be sufficiently high such that all of the remaining pieces would be received by their playback time, should this rate be maintained. The long-term average rate is calculated as the ratio of the amount of in-order data received thus far, divided by the time since the peer first began download of a piece that was not selected using "rarest-first with ties broken randomly", or in the case no such piece download has begun, the time since the peer's arrival to the system.[7]

For simulating the transmission rates of piece transfers, it is assumed that connection bottlenecks are located at the end points and the network operates using max-min fair bandwidth sharing (using TCP, for example). Under these assumptions, each piece transfer operates at the highest possible rate that ensures that (i) no bottleneck operates above its capacity, and (ii) the rate of no transfer can be increased without decreasing the rate of some other transfer operating at the same or lower rate.

Unless stated otherwise, it is assumed that peers have three times higher download bandwidth than upload bandwidth, and that each peer concurrently uploads to at most four peers. The maximum number of server upload connections is chosen as the total server bandwidth available for the video file divided by the video playback bit rate (an integer value owing to the parameters chosen

[6] Note that the default parameters in recent versions of the mainline BitTorrent client allow peers to be connected to up to 80 other peers, which is often achieved in practice [19]. Furthermore, peers not satisfied with their performance are able to request additional peers from the tracker.

[7] Alternative start-up rules were tried, but did not impact the relative performance of the considered policies. More aggressive rules, of course, result in shorter start-up delays but increased likelihood that a piece is not received by its playback point.

in our experiments). By ensuring that each server connection can transfer data at the playback bit rate, the server is better able to assist "low buffer" peers.

A variety of workload scenarios are considered. For scenarios with a constant-rate request (peer) arrival process, the system is simulated for 6000 requests, with the initial 1000 and the last 500 requests removed from the measurements. For flash crowd scenarios (in which the arrival rate starts high and decays to zero) no warmup period was used and simulations were run until the system emptied. Except in two scenarios considered in Section 6.6, it is assumed that all peers leave the system as soon as they have received the entire file (i.e., act only as leechers).

## 6 Performance Comparisons

In this section, we compare the performance of the policies defined in Sections 3 and 4. Section 6.1 describes the metrics that will be considered in the policy evaluations. Sections 6.2-6.5 present our principal comparisons for four different workload scenarios. Section 6.6 explores the impact of differing assumptions regarding the available upload resources.

### 6.1 Performance Metrics

We use two quality of service metrics: (i) the *percentage of late pieces*, defined as the percentage of pieces that are not received by their playback point, and (ii) the *average start-up delay*, as determined by the start-up rule of Section 5. Without loss of generality, data volume is measured in units of the file size, and time in units of the total video playback duration. Hence, all data rates are expressed relative to the playback bit rate, and start-up delay is expressed relative to the time it takes to play the entire video. For example, an upload bandwidth of 1.25 means that when the peer is fully utilizing its upload bandwidth, it can upload data at 1.25 times the playback bit rate. Similarly, a start-up delay of 5% means that the delay until playback begins is equal to 5% of the total playback duration, and an arrival rate of 100 means that on average 100 peers arrive during the time it takes to entirely play back the video once.

### 6.2 Steady State Scenario

In the "steady state" scenario, peers (i) do not leave the system until having fully downloaded the file, (ii) arrive according to a Poisson process at rate $\lambda$, and (iii) are homogeneous (i.e., all peers have the same upload bandwidth $U$ and download bandwidth $D$). The server upload bandwidth is denoted by $B$.

Figure 1 shows results for the *baseline* protocol. Figures 2 and 3 show results using RPNP and UP/RPNP, respectively. When interpreting these results, it should be noted that the total bandwidth requirement (request arrival rate $\times$ file size) ranges from two to forty times the server upload bandwidth, as the request rate varies from 10 to 200. (Naturally, at least the difference between the total bandwidth requirement and the server upload bandwidth must be contributed

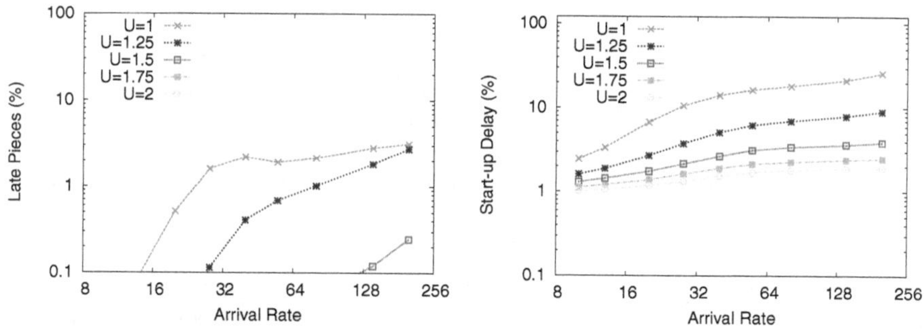


(a) Late pieces (b) Start-up delay

**Fig. 1.** Baseline; steady state scenario ($B = 5, D/U = 3$, 100 segments with 5 pieces each)

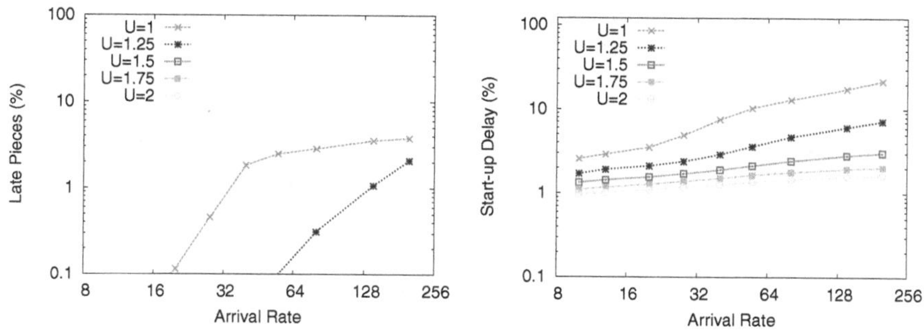


(a) Late pieces (b) Start-up delay

**Fig. 2.** Rare piece delivery to new peers (RPNP); steady state scenario ($B = 5, D/U = 3$, 100 segments with 5 pieces each)

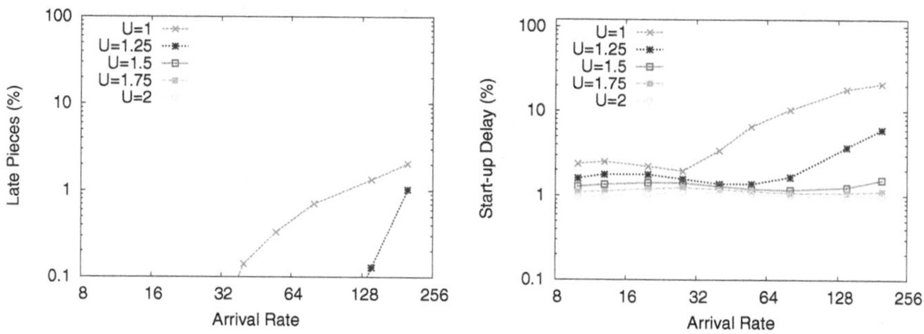


(a) Late pieces (b) Start-up delay

**Fig. 3.** Urgent piece prioritization with rare piece delivery to new peers (UP/RPNP); steady state scenario ($B = 5, D/U = 3$, 100 segments with 5 pieces each)

by peers.) To capture a wide range of workloads, start-up delays, and percentage of late pieces, these figures (as well as subsequent figures) use log scales. Note that late piece percentages under 0.1% are not shown.

Comparing Figures 1 through 3, we note that RPNP substantially improves over the baseline protocol, with respect to both start-up delay (owing to "new" peers being able to compete more effectively for the upload bandwidth of other peers) and the percentage of late pieces (owing to improved piece diversity in the system). UP/RPNP provides additional improvements in both of these metrics. The improvement in start-up delay with UP/RPNP, in comparison to RPNP, arises from a subtle side-effect of using in-order rather than Zipf-based piece selection for "low buffer" peers. Generally, a more in-order piece delivery will tend to degrade the system's piece diversity somewhat, leading to a greater fraction of uploads being initiated to a random interested peer (including new peers) rather than to a peer from which data is currently being downloaded.

### 6.3 Flash Crowd Scenario

Our second scenario is motivated by measurements of operational file sharing torrents [20]. In this scenario, peers are assumed to arrive at an exponentially decaying rate $\lambda(t) = \lambda_0 e^{-\gamma t}$, where $\lambda_0$ is the initial arrival rate at time zero and $\gamma$ is a decay factor. By varying $\gamma$ between 0 and $\infty$, a variety of arrival processes can be simulated, from constant-rate arrivals at one extreme, to a flash crowd in which all peers arrive instantaneously (to an empty system) at the other extreme.

Figures 4 and 5 show the results for the second scenario using the baseline protocol and UP/RPNP, respectively. For this workload scenario, as well as for subsequent workload scenarios, results for RPNP are omitted owing to space limitations. Here, $\lambda_0$ and $\gamma$ are selected such that the expected total number of arrivals is equal to 500. By varying $\gamma$ between $\frac{1}{8}$ and 1, we cover a wide range of intensities of flash crowds. With $\gamma = \frac{1}{8}$, 11.8% of all arrivals occur within one playback duration of the first peer arrival; with $\gamma = 1$, the corresponding value is 63.2%.

While the potentially very high initial arrival rate makes this scenario much different than the steady state scenario, UP/RPNP still achieves significant improvements in the percentage of late pieces. The cases with a high percentage of late pieces for both protocols, which occur for intense flash crowds, are due to pieces not being disseminated from the server to all peers quickly enough. To further reduce the percentage of late pieces in these cases (given the same server resources), a more conservative start-up rule would be needed.

### 6.4 Heterogeneous Scenario

The third scenario is of a heterogeneous workload with two types of peers: low bandwidth peers and high bandwidth peers. For both types of peers, the download bandwidth is three times the upload bandwidth, as assumed previously. As in the first scenario peers arrive at a constant rate $\lambda$.

Figure 6 shows the percentage of late pieces and the average start-up delay for this workload scenario. The percentage of high bandwidth peers is varied such that the system ranges from bandwidth-constrained (most peers are low bandwidth) to bandwidth-rich (most peers are high bandwidth). Results are shown for a workload in which the low and high bandwidth peers have upload

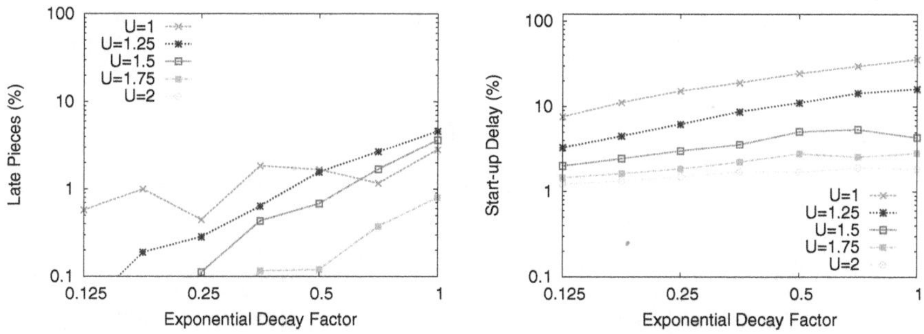


(a) Late pieces (b) Start-up delay

**Fig. 4.** Baseline; flash crowd scenario ($B = 10, D/U = 3$, 100 segments with 5 pieces each)

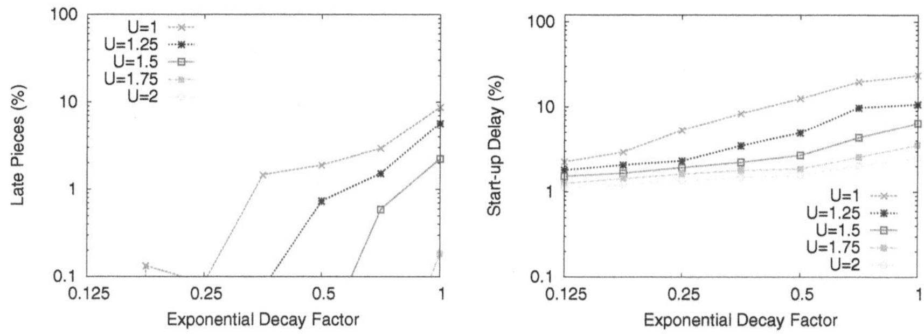


(a) Late pieces (b) Start-up delay

**Fig. 5.** Urgent piece prioritization with rare piece delivery to new peers (UP/RPNP); flash crowd scenario ($B = 10, D/U = 3$, 100 segments with 5 pieces each)

bandwidths of one and two times the playback bit rate, respectively. As in the previous scenarios, significant improvements in both the percentage of late pieces and the start-up delays are observed with UP/RPNP.

Note that for both the baseline and UP/RPNP, the high bandwidth peers generally incur both a smaller percentage of late pieces and lower start-up delays. This is a consequence of both the higher download bandwidth of these peers, and of the use of tit-for-tat coupled with their higher upload bandwidth.

## 6.5 Freeloader Scenario

Our fourth scenario is of a workload with two types of peers: contributing peers and freeloaders. We assume that both types of peers have identical download bandwidth, equal to three times their upload bandwidth, as assumed previously. However, only the contributing peers upload file pieces to other peers.

Figure 7 shows the percentage of late pieces and the average start-up delay for this workload scenario. Results are shown for a workload in which peers have an upload bandwidth of 1.25 times the playback bit rate. The percentage of

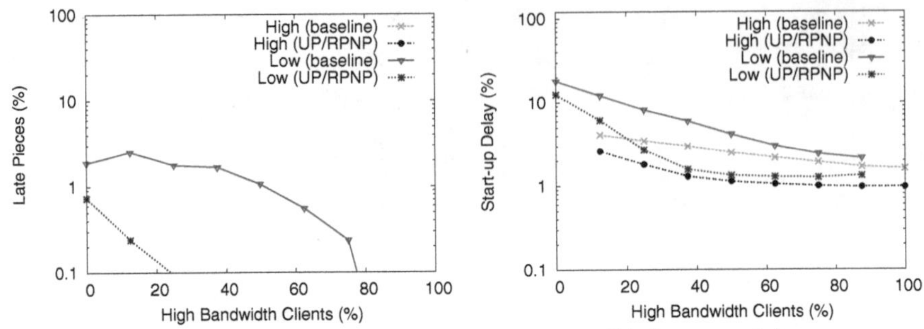


(a) Late pieces (b) Start-up delay

**Fig. 6.** Impact of heterogeneity; steady state ($B = 5, \lambda = 100, D/U = 3$, $U_{high} = 2$, $U_{low} = 1$, 100 segments with 5 pieces each)

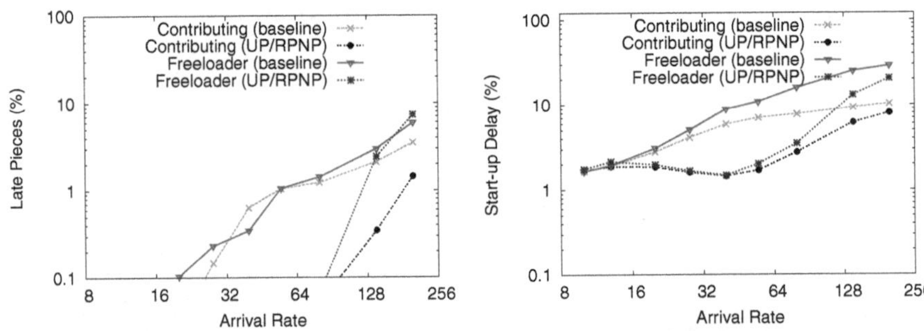


(a) Late pieces (b) Start-up delay

**Fig. 7.** Performance with freeloaders; steady state ($B = 5, D = 3.75$, $U = 1.25$, 5% freeloaders, 100 segments with 5 pieces each)

peers that are freeloaders is fixed at 5%, and the arrival rate is varied between 10 and 200. As in the previous scenarios, significant improvements in both the percentage of late pieces and the start-up delays are observed with UP/RPNP.

With both the baseline protocol and UP/RPNP, the rate-based tit-for-tat mechanism ensures that contributing peers receive substantially better performance than the freeloaders. For example, with UP/RPNP and $\lambda = 200$, freeloaders have an average start-up delay roughly three times that of contributing peers, and observe more than five times as many late pieces. While freeloaders are regularly unchoked (due to the use of optimistic unchoke), and therefore, are not completely starved, this performance advantage illustrates that the tit-for-tat policy provides peers with a strong incentive to contribute their upload resources.

### 6.6 Impact of Total Upload Capacity

Figures 8 through 10 show the percentage of late pieces and the average start-up delay as functions of the server upload bandwidth, the average time peers stay in the system after having completed download as "seeders", and the peer arrival

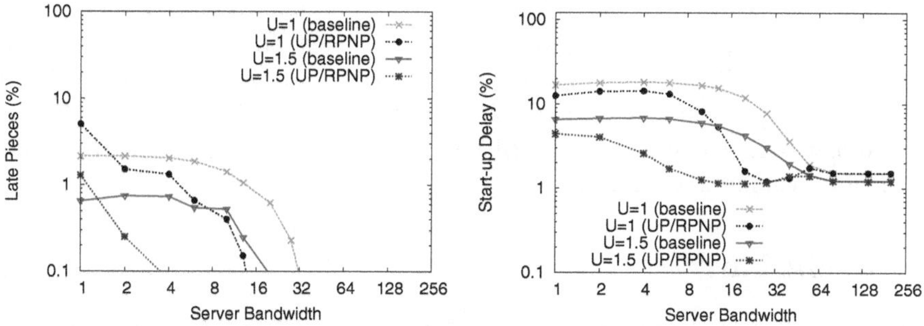


(a) Late pieces (b) Start-up delay

**Fig. 8.** Impact of server bandwidth; steady state scenario ($\lambda = 100, D/U = 3$, 100 segments with 5 pieces each)

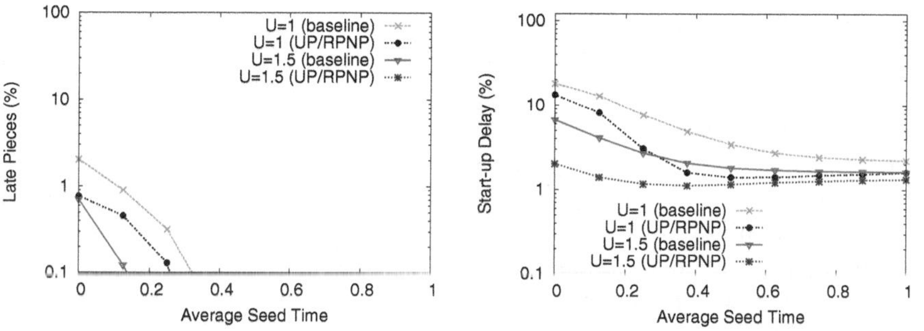


(a) Late pieces (b) Start-up delay

**Fig. 9.** Impact of the average seed time; steady state scenario ($B = 5, \lambda = 100, D/U = 3$, 100 segments with 5 pieces each)

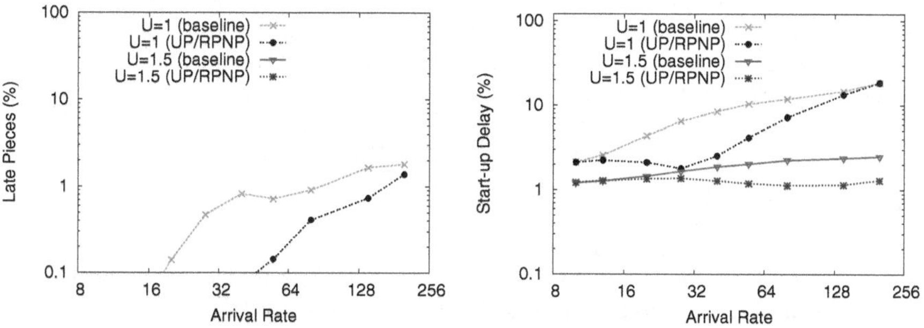


(a) Late pieces (b) Start-up delay

**Fig. 10.** Steady state scenario in which peers stay until playback completion ($B = 5, D/U = 3$, 100 segments with 5 pieces each)

rate for an example scenario in which peers stay in the system until having played back the entire file (rather than only until download completion), respectively. For simplicity, we assume that seed times are exponentially distributed.

As expected, increases in the server bandwidth (cf. Figure 8) and the available peer upload bandwidth (cf. Figures 9 and 10) have positive impacts on the quality of service. Comparing Figure 10 with Figures 1 and 3, we note that some performance improvements are possible if peers stay in the system contributing their upload bandwidth at least until having completed playback (rather than just until download is complete).

## 7 Conclusions

This paper has considered the problem of devising BitTorrent-like peer-assisted protocols for on-demand video streaming systems, in which peers are motivated to upload data to others owing to the likely beneficial impact on their own achieved performance. The challenge in this context is that of mediating the conflict between the goals of low start-up delay and consistently on-time piece delivery (which motivates piece delivery that is more "in-order"), and the requirements of effective tit-for-tat (which motivates piece delivery that is more "rarest first").

We devised new tit-for-tat compatible policies where the server, for which tit-for-tat is not an issue, gives preference to peers at imminent risk of receiving data too late for playback, and secondly to upload of rare pieces to newly arrived peers. Evaluations of the proposed policies for a variety of workload scenarios suggests that our policies are able to provide substantial improvements in quality of service while ensuring that the piece diversity is sufficient for peers to effectively employ tit-for-tat.

## References

1. Cohen, B.: Incentives Build Robustness in BitTorrent. In: Proc. Workshop on Economics of Peer-to-Peer Systems 2003, Berkeley, CA (June 2003)
2. Huang, C., Li, J., Ross, K.W.: Can Internet Video-on-Demand be Profitable? In: Proc. ACM SIGCOMM 2007, Kyoto, Japan, pp. 133–144 (August 2007)
3. Janardhan, V., Schulzrinne, H.: Peer Assisted VoD for Set-top Box Based IP Network. In: Proc. ACM SIGCOMM Workshops (Peer-to-Peer Streaming and IP-TV) 2007, Kyoto, Japan (August 2007)
4. Cui, Y., Li, B., Nahrstedt, K.: ostream: Asynchronous streaming multicast in application-layer overlay networks. IEEE JSAC 22(1), 91–106 (2004)
5. Sharma, A., Bestavros, A., Matta, I.: dPAM: A Distributed Prefetching Protocol for Scalable Asynchronous Multicast in P2P Systems. In: Proc. IEEE INFOCOM 2005, Miami, FL, pp. 1139–1150 (March 2005)
6. Hefeeda, M., Habib, A., Botev, B., Xu, D., Bhargava, B.: PROMISE: Peer-to-Peer Media Streaming using CollectCast. In: Proc. ACM MM 2003, Berkeley, CA, pp. 45–54 (November 2003)
7. Rejaie, R., Ortega, A.: PALS: Peer-to-Peer Adaptive Layered Streaming. In: Proc. NOSSDAV 2003, Monterey, CA, pp. 153–161 (June 2003)
8. Vratonjic, N., Gupta, P., Knezevic, N., Kostic, D., Rowstron, A.: Enabling DVD-like Features in P2P Video-on-demand Systems. In: Proc. ACM SIGCOMM Workshops (Peer-to-Peer Streaming and IP-TV) 2007, Kyoto, Japan (August 2007)

9. Parvez, N., Williamson, C., Mahanti, A., Carlsson, N.: Analysis of BitTorrent-like Protocols for On-demand Stored Media Streaming. In: Proc. ACM SIGMETRICS 2008, Annapolis, MD, pp. 301–312 (June 2008)
10. Cheng, B., Stein, L., Jin, H., Zhang, Z.: Towards Cinematic Internet Video-on-Demand. In: Proc. EuroSys 2008, Glasgow, Scotland, pp. 109–122 (March 2008)
11. Mol, J.J.D., Pouwelse, J.A., Meulpolder, M., Epema, D.H.J., Sips, H.J.: Give-to-Get: Free-riding Resilient Video-on-Demand in P2P Systems. In: Proc. MMCN 2008, San Jose, CA (January 2008)
12. Annapureddy, S., Guha, S., Gkantsidis, C., Gunawardena, D., Rodriguez, P.R.: Is High-Quality VoD Feasible using P2P Swarming? In: Proc. WWW 2007, Banff, Canada, pp. 903–912 (May 2007)
13. Vlavianos, A., Iliofotou, M., Faloutsos, M.: BiToS: Enhancing BitTorrent for Supporting Streaming Applications. In: Proc. Global Internet Workshop 2006, Barcelona, Spain (April 2006)
14. Carlsson, N., Eager, D.L.: Peer-assisted On-demand Streaming of Stored Media using BitTorrent-like Protocols. In: Akyildiz, I.F., Sivakumar, R., Ekici, E., de Oliveira, J.C., McNair, J. (eds.) NETWORKING 2007. LNCS, vol. 4479, pp. 570–581. Springer, Heidelberg (2007)
15. Choe, Y.R., Schuff, D.L., Dyaberi, J.M., Pai, V.S.: Improving VoD Server Efficiency with BitTorrent. In: ACM MM 2007, Augsburg, Germany, pp. 117–126 (September 2007)
16. Garbacki, P., Epema, D.H.J., Pouwelse, J., van Steen, M.: Offloading Servers with Collaborative Video on Demand. In: Proc. IPTPS 2008, Tampa Bay, FL (February 2008)
17. Gkantsidis, C., Rodriguez, P.R.: Network Coding for Large Scale Content Distribution. In: Proc. IEEE INFOCOM 2005, Miami, FL, pp. 2235–2245 (March 2005)
18. Zhang, X., Liu, J., Li, B., Yum, T.-S.P.: CoolStreaming/DONet: A Data-driven Overlay Network for Peer-to-Peer Live Media Streaming. In: Proc. IEEE INFOCOM 2005, Miami, FL, pp. 2102–2111 (March 2005)
19. Legout, A., Urvoy-Keller, G., Michiardi, P.: Rarest First and Choke Algorithms Are Enough. In: Proc. ACM IMC 2006, Rio de Janeiro, Brazil, pp. 203–216 (October 2006)
20. Guo, L., Chen, S., Xiao, Z., Tan, E., Ding, X., Zhang, X.: Measurement, Analysis, and Modeling of BitTorrent-like Systems. In: Proc. ACM IMC 2005, Berkley, CA, pp. 35–48 (October 2005)

# Video Broadcasting to Heterogeneous Mobile Devices

Cheng-Hsin Hsu and Mohamed Hefeeda

School of Computing Science, Simon Fraser University
250-13450 102nd Ave, Surrey, BC, Canada
{cha16,mhefeeda}@cs.sfu.ca

**Abstract.** We study the problem of broadcasting multiple scalable video streams to heterogeneous mobile devices, which have limited energy budgets. We show that scalable video streams should be broadcast in a different manner than nonscalable streams; otherwise energy of mobile devices could be wasted. We propose an efficient broadcast scheme for mobile TV networks that explicitly supports heterogeneous mobile devices, and we show its correctness as well as performance in terms of energy saving. We implement the proposed scheme in a real mobile TV testbed to evaluate its performance. Our results indicate that, with the proposed broadcast scheme, significant energy savings can be achieved by different heterogeneous devices. For example, using the proposed broadcast scheme allows mobile devices to achieve energy saving between 62% to 92%, while using the current broadcast scheme only allows them to achieve energy saving 62% despite how many layers they can (or opt to) receive and decode.

**Keywords:** TV Broadcast Networks, Energy Saving, Time Slicing.

## 1 Introduction

Modern mobile devices are lightweight to be carried all the time, and provide communication and computational powers that were only available to stationary computers a few years ago. These mobile devices can run many multimedia applications including mobile TV, which allows users to watch TV programs anywhere, anytime. Mobile TV is expected to be a huge market: up to 20 billion Euros with 500 million subscribers worldwide by 2011 [1]. Mobile TV can be sent over cellular networks or dedicated broadcast networks. In this paper, we only consider dedicated broadcast networks, because they can support a very large number of subscribers using a single broadcast tower.

Mobile devices are heterogeneous from several aspects, including screen resolutions, decoder features (coding standards, spatial dimensions, and frame rates), and battery capacities. For example, GSmart t600 PDA phone is equipped with a VGA (640x480) display [2], while Nokia N96 cellular phone only has a QVGA (320x240) display [3]. Mobile TV operators who wish to simultaneously support these two mobile devices will face a dilemma: broadcasting TV channels in QVGA resolution leads to lower perceived quality on GSmart t600, while broadcasting in VGA resolution results in higher communication and computational

L. Fratta et al. (Eds.): NETWORKING 2009, LNCS 5550, pp. 600–613, 2009.

overhead on Nokia N96 (and thus shorter watch times) with no visible quality improvement. One way to cope with this dilemma is to encode and broadcast every TV channel into two versions: one for each device. This multi-version approach, however, does not scale because it incurs huge bandwidth overhead, thus reduces the number of TV channels that can be concurrently broadcast. Moreover, mobile devices can be categorized into classes by not only different mobile devices but also different working conditions of the same mobile device, e.g., mobile devices with low battery levels or in poor wireless channel conditions may prefer to receive lower bit rate streams to save energy and/or reduce bit error rate. Therefore, the number of classes can be quite large, which renders the multi-version approach less practical.

To eliminate the bandwidth overhead of multi-version approach, operators can adopt *scalable video coders (SVCs)* to encode each TV channel into a single stream with multiple layers, where each layer is broadcast exactly *once*. Mobile devices can then selectively receive and decode a few (or all) layers for perceived quality that are the most suitable to them. Broadcasting scalable video streams, however, poses a challenge for the base station. This is because the base station broadcasts each TV channel in bursts with a bit rate much higher than the encoding rate of that TV channel. Mobile devices can then receive a burst of traffic and turn off their radio frequency (RF) circuits until the next burst in order to save energy. This is called *time slicing*, and it is dictated in major broadcast standards such as DVB-H (Digital Video Broadcast-Handheld) [4, 5] and MediaFLO (Forward Link Only) [6]. Preparing bursts of TV channels encoded in scalable manner is much more complex than preparing these bursts for nonscalable TV channels, because of the dependency among various layers.

In this paper, we study the burst transmission problem in mobile TV networks, where several TV channels are concurrently broadcast as scalable streams over a shared air medium to many mobile devices with *heterogeneous* resources. To the best of our knowledge, there is no existing solution in the literature to efficiently broadcast scalable video streams in mobile TV networks. We formulate and solve the burst transmission problem in mobile TV networks with scalable video streams. We show the correctness of our solution. We implement the proposed solution in a mobile TV testbed and we demonstrate its practicality and efficiency. We also empirically show that the proposed solution allows mobile devices to save energy and receive only the appropriate layers of the video streams. Solving this problem enables mobile devices to obtain the most suitable resolution and frame rate without increasing energy consumption of the devices. Moreover, solving the problem allows mobile devices to trade perceived quality for energy consumption, as they can opt to receive fewer layers to prolong battery lifetime.

The rest of this paper is organized as follows. We review the previous works in Sec. 2. In Sec.3, we formally state the considered problem and discuss methods for broadcasting scalable streams in mobile TV networks. We propose the solution in Sec.4 and evaluate the solution in Sec.5 using a real mobile TV testbed. We conclude the paper in Sec. 6.

## 2 Related Works

Multicast of scalable video streams over the Internet has been studied in the literature and many protocols and algorithms have been proposed to support multicast routing, resource reservation, robustness, and flow and congestion controls [7,8]. None of these proposals is applicable to mobile TV networks, because they are single-hop broadcast networks, rather than the multi-hop Internet. For example, in RLM (Receiver-driven Layered Multicast) [9], different layers of a video stream are sent to different multicast groups, and receivers periodically join the next higher layer's group until experiencing excessive packet loss. RLM is not useful in mobile TV networks because of their broadcast nature: *more* receivers do not incur *higher* network loads, and packet loss ratio on a mobile device is independent of how much data it receives. Most importantly, previous works in multicasting scalable video streams do not consider energy consumption on clients, as we do in this paper.

Unequal error protection (UEP) methods were proposed to improve video quality for mobile devices with bad radio receptions in mobile TV networks [10, 11]. Ghandi and Ghanbari [10] proposed to transmit the base layer and the enhancement layers with different modulation and coding schemes. This is called hierarchical modulation and channel coding, and is supported by broadcast networks like DVB-T [12]. Hellge et al. [11] proposed to use FEC bits of higher layers to protect data bits in the lower layers. The rational is that lower layers are more critical to successful decoding, and they need to be more resilient to errors. None of these works considers construction of bursts, and they are orthogonal to our work.

Previous works have studied the energy saving in mobile TV networks that broadcast TV channels in nonscalable manner. For example, it has been shown that time slicing enables mobile devices to turn off their RF circuits for a significant fraction of the time [13, 14]. The works in [13, 14] did not solve the burst transmission problem. Balaguer et al. [15] proposed an energy saving strategy by not receiving more FEC bytes once the data can be successfully reconstructed. Zhang et al. [16] considered mobile devices with an auxiliary short range wireless interface and constructed a cooperative network over this short range network to share the IP packets received from the broadcast network. The proposals in [15, 16] did not consider the burst transmission problem, and are complementary to our work.

Finally, our previous works studied the burst transmission problems and proposed time slicing schedules for mobile TV networks that broadcast nonscalable video streams to a single class of mobile devices [17–19]. That is, our previous works assumed that all mobile devices receive the entire video streams. In this paper, we solve the burst transmission problem for scalable video streams and we explicitly consider heterogeneous mobile devices that can only (or opt to) receive parts of video streams. We show that naive transmission of scalable streams could lead to wasting the energy of mobile devices.

# 3 Problem Statement

In this section, we formally describe the considered problem. We then show the limitations of the current mobile broadcast networks.

## 3.1 Burst Transmission to Heterogeneous Mobile Devices

We consider a mobile TV network in which a base station concurrently broadcasts multiple TV channels over a shared air medium with bandwidth $R$ kbps to many mobile devices with heterogeneous capability. Each TV channel is allocated a bit rate of $r$ kbps, and is divided into $C$ layers using scalable video coders. A TV channel is encapsulated and broadcast as a series of bursts, where each burst is in size $b$ kb. Each mobile device receives a burst of data and turns off its RF circuit till the next burst of the same TV channel to save energy. This is called time slicing. The energy saved by a mobile device because of time slicing is denoted by $\gamma$, and it is calculated as the ratio of time the RF circuit is in off mode to the total time [13, 14]. When computing $\gamma$, we need to consider the overhead of waking up the RF circuits on mobile devices to receive the next burst [14]. This is because it takes RF circuits some time to power up and resynchronize before data can be demodulated. This period is called overhead duration $T_o$, which can be as high as 250 msec [4]. The problem considered in this paper can be stated as follows.

**Problem 1 (Burst Transmission in Multi-Layer Broadcast Networks)** *Consider a mobile TV broadcast network with air medium bandwidth $R$ kbps shared among $S$ TV channels, where every TV channel has a bit rate of $r$ kbps. Each TV channel is encoded into $C$ layers, where each layer has a bit rate of $r_s = r/C$ kbps. Mobile devices are classified into $C$ classes so that devices in class $c$ ($c = 1, 2, \ldots, C$) receive and render all layers $\bar{c}$, where $\bar{c} \leq c$. Video streams are put into IP packets and then encapsulated into bursts of size $b$ kb. Design a burst transmission scheme to maximize energy saving of mobile devices in all classes. The burst transmission scheme assigns IP packets to individual bursts, and specifies the start time of each burst.*

Solving the above problem is critical to the quality of service in mobile TV networks, because it increases battery lifetimes for heterogeneous mobile devices. Longer battery lifetimes enable subscribers to watch more TV and provide network operators more opportunities for higher revenues due to subscription fees and advertisements.

## 3.2 Encapsulating and Broadcasting Scalable Video Streams

Video streams coded by traditional, nonscalable coders must be transmitted and decoded in their entirety, and thus may not be suitable to be broadcast to heterogeneous mobile devices. This is because all mobile devices will have to receive complete video streams even though some of them may not have resources to decode and render those streams. Scalable video coders (SVC), on the

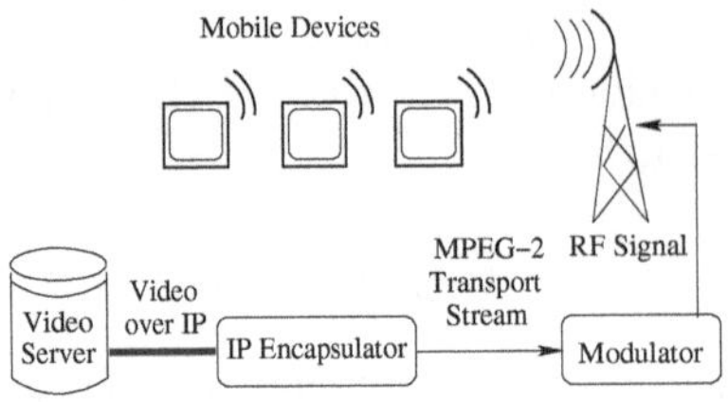


**Fig. 1.** The main components of mobile TV broadcast networks

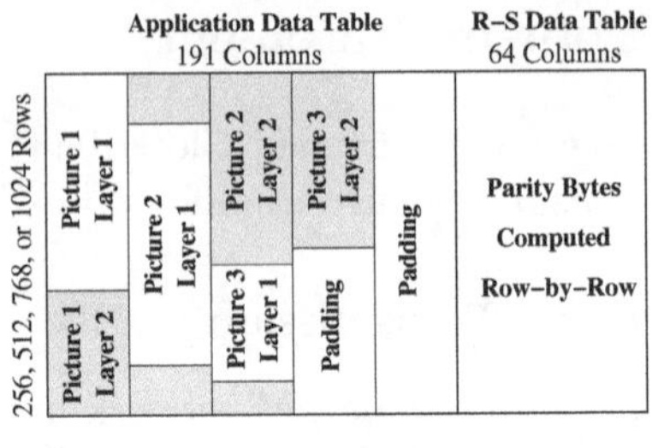


**Fig. 2.** The structure of MPE-FEC frame

other hand, can encode each TV channel into a single video stream that can be sent and decoded at various bit rates. This is achieved by simple manipulations, which extract substreams from the original stream, where each substream can be decoded and displayed at a lower perceived quality than the original (complete) stream. Modern scalable coders support several scalability modes, including spatial, temporal, and SNR (signal-to-noise ratio) scalability. With spatial scalability, several picture resolutions can be supported. With temporal scalability, different frame rates can be supported. Finally, SNR scalability supports various picture fidelities. These scalability modes are not mutually exclusive: the combined scalability is supported by recent coders such as H.264/SVC [20, Annex. G]. Interested readers are referred [21] for more details on SVC.

While SVC is promising to many applications with heterogeneous networks and clients, it is quite challenging to broadcast SVC streams over mobile TV networks in order to maximize energy saving for heterogeneous mobile devices. This is because unlike other network applications, where SVC substreams can be extracted by MANEs (Media Aware Network Elements) based on the requests from downstream subnets, there is only one air medium in each mobile TV network. Since there exists no network devices between the base station and mobile devices, substream extractions *must* be done at mobile devices, which can result in high overhead. In the following illustrative example, we show that broadcasting SVC streams in current mobile TV networks leads to no energy saving for mobile devices that cannot render the complete video stream!

As illustrated in Fig. 1, a mobile TV base station consists of several components, including a video server, an IP encapsulator, and a modulator. The video server puts the video data in RTP packets and sends these packets to the IP encapsulator. The IP encapsulator receives and encapsulates the IP packets in MPE (multiprotocol encapsulation) frames. The IP packets can be FEC-protected using Reed-Solomon (R-S) codes, which result in MPE-FEC frames. FEC is important because it provides better error resilience to bad channel conditions that are common in mobile systems. Fig. 2 shows the structure of an MPE-FEC frame, which can be divided into two parts: an application data table (ADT) carries IP packets and an R-S data table (RDT) carries the parity bytes. To compute the parity bytes, IP packets received by the IP encapsulator are *sequentially* stored column-by-column, from left to right. If there are not enough IP packets to fill ADT, zeros are padded in the remaining space. Once the ADT is full, the parity bytes are computed row-by-row, and store in the RDT. The whole MPE-FEC frame is then sent as a burst. We note that the zeros are padded in ADT to facilitate the generation of parity bytes. The padded zeros are *not* broadcast over the air, thus do not incur any communication overhead [22]. Existing IP encapsulators are not media-aware: they just encapsulate IP packets one after another. We refer to this type of burst transmission as *sequential transmission.*

To support heterogeneous mobile devices, network operators may upgrade the video server to support scalable video coding. IP encapsulator and the modulator can still encapsulate and send SVC streams by treating them as ordinary IP streams. Let us consider a small time window of 3 pictures, where each picture is encoded into 2 layers. Without loss of generality, we assume that each layer of each picture is put in a single IP packet, and these IP packets are sent by the video server in the following order: (picture 1, layer 1), (picture 1, layer 2), (picture 2, layer 1), (picture 2, layer 2), (picture 3, layer 1), (picture 3, layer 2). We further assume that the IP packets are not reordered in the network, so that they are stored in the same order within an MPE-FEC frame as illustrated in Fig. 2. This MPE-FEC frame is then broadcast. Consider a mobile device that can only display the base layer (layer 1), this mobile device, unfortunately, still has to receive and process the *complete* burst for two reasons. First, IP packets belonging to the base layer are scattered all over the frame, and a deep inspection (at RTP or video coding layer) is required to identify them. Second, each parity byte is computed over IP packets from various layers, thus it is useless if some IP packets are not received.

This illustrative example shows that simply upgrading the video server to support SVC streams results in *no energy saving* for mobile devices, and calls for new burst transmission schemes that treat IP packets of various SVC layers differently so that mobile devices can extract SVC substreams *without* receiving and processing complete bursts. We call such burst transmission schemes as *layer-aware* schemes.

## 4 Solutions: Layer-Aware Burst Transmission

We propose and analyze layer-aware burst transmission schemes in this section.

### 4.1 Parallel Services: PS

One way to achieve layer-aware burst transmission is to send each layer of a TV channel as a parallel service (PS), which can be implemented using several IP streams sent to different multicast IP addresses, or using multiple parallel elementary streams [14, Sec. 8.6]. Fig. 3 shows an example of broadcasting two TV channels with three layers, where each block inside bursts is a parallel service and carries IP packets of a specific layer only. Compared to sequential transmission, parallel service approach supports efficient demultiplexing of IP packets to

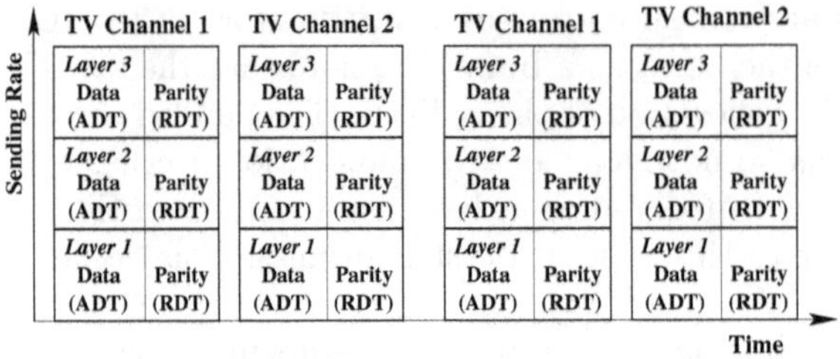


Fig. 3. Parallel Services

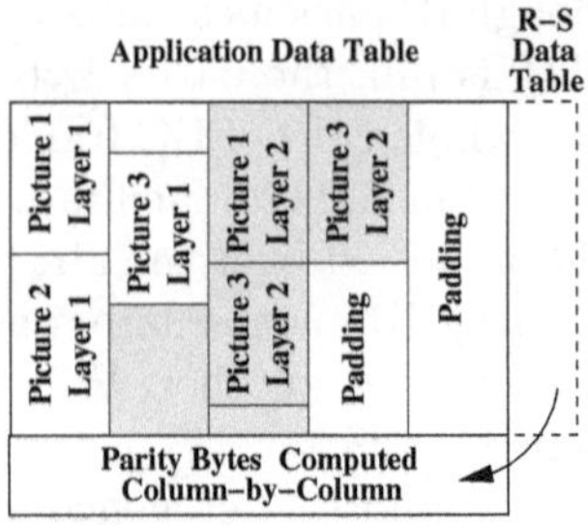


Fig. 4. Layer-Aware FEC frame

individual SVC layers based on either IP addresses or MPEG-2 PIDs (packet IDs). This frees mobile devices from inspecting IP packets and reduces their processing overhead on extracting substreams. However, as all services are sent in parallel, mobile devices still have to open their RF circuits for the complete burst duration. Therefore, all mobile devices achieve the same energy saving despite how many layers they receive and decode.

### 4.2 Layer-Aware FEC: LAF

In order to allow mobile devices that only receive a few layers to close their RF circuits earlier than each burst ends, the IP encapsulator must rearrange the received packets so that packets belonging to layer $l$ are sent before packets belonging to layer $l+1$. If we reuse the illustrative example given in Sec. 3.2, the IP packets should be sent in the following order: (picture 1, layer 1), (picture 2, layer 1), (picture 3, layer 1), (picture 1, layer 2), (picture 2, layer 2), (picture 3, layer 2), as illustrated in Fig. 4. In order to allow mobile devices to efficiently determine the boundaries between layers (in this example, between (picture 3, layer 1) and (picture 1, layer 2)), we propose to prepend the SVC layer number as a one-byte extension header before the MPE section header. Mobile devices can then demultiplex the IP packets based on this extension header.

However, even after reordering IP packets, mobile devices still have to receive complete bursts in order to perform error corrections, which again prevents them from getting higher energy saving. To address this issue, we propose to compute parity bytes *column-by-column* as illustrated in Fig. 4. Furthermore, the parity bytes of each column are sent immediately after each column of the data bytes. This allows mobile devices to perform error corrections without receiving complete bursts. That is, mobile devices can receive partial bursts and turn off the RF circuits to save energy. We call this new frame format as Layer-Aware FEC (LAF) frame. Although LAF frame allows mobile devices to receive and extract substreams while achieving proportional energy saving, it has disadvantages. First, LAF does not comply to mobile TV standards, which will cause compatibility issues between the base station and mobile devices. Second, implementing LAF requires significant changes as error corrections are usually done

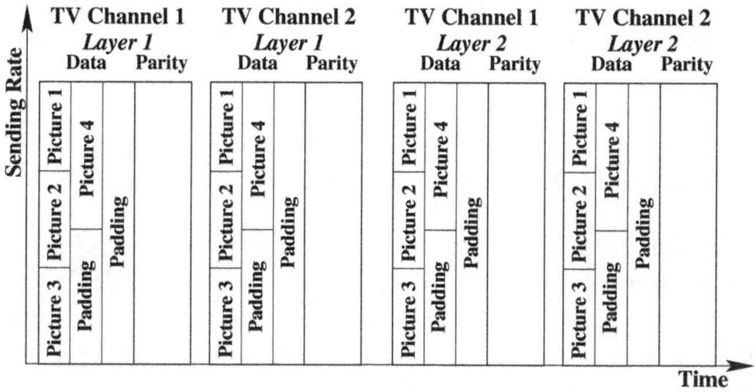


**Fig. 5.** Layer-Aware Time Slicing

in hardware/firmware for the sake of performance. Third, and more importantly, computing parity bytes column-by-column makes the FEC decoder vulnerable to bursty channel errors because it does not provide virtual time interleaving as by MPE-FEC frames [22].

### 4.3 Layer-Aware Time Slicing: LATS

Layer-aware burst transmission can also be implemented using time slicing schemes. Fig. 5 presents an illustrative example of such a time slicing scheme, which we call Layer-Aware Time Slicing (LATS). As shown in this figure, the IP encapsulator prepares a different MPE-FEC frame for each SVC layer of every TV channel: e.g., all the IP packets in the left-most burst belong to the base layer of TV channel 1, while the IP packets in the third burst belong to layer 2 of TV channel 1. Note that, as we mentioned in Sec. 3.2, the IP encapsulator adds zero padding in MPE-FEC frames only to compute parity bytes: these padded zeros are not transmitted [22]. Since each burst consists of IP packets from the same layer of the same TV channel, mobile devices know which layer those IP packets are in, even before receiving the burst. This frees mobile devices from opening the RF circuits and inspecting IP packets for substream extractions. LATS naturally works with existing MPE-FEC frame, because whenever a mobile device decides to decode layer $l$, it has to receive all IP packets in layer $l$ for successful video reconstruction. Therefore, all IP packets in ADT will be received before error corrections, and the FEC decoder (implemented in hardware/firmware) can work as-is. Last, we note that no additional signaling from the base station to mobile devices is required: to determine which bursts (layers) to receive, mobile devices only need to know the total number of layer ($C$), which is already sent to them for decoding SVC streams.

We develop the LATS scheme in the following by giving the burst start time to each layer of individual TV channels. We first compute the number of TV channels that can be concurrently broadcast as $S = \lfloor R/r \rfloor$. LATS works in a recurring window, where each window consists of $CS$ bursts of size $b$ kb. Given that the radio channel bandwidth is $R$, the window size can be computed by

$(b/R)CS$ sec. For a layer $c$, where $c = 1, 2, \ldots, C$, the starting time is given as $(b/R)(c-1)S$ because there are $S$ bursts in each layer. Finally, LATS schedules a burst start at:

$$(b/R)[(c-1)S + (s-1)] \tag{1}$$

to layer $c$ ($c = 1, 2, \ldots, C$) of TV channel $s$ ($s = 1, 2, \ldots, S$).

**Lemma 1.** *The Layer-Aware Time Slicing (LATS) scheme (Eq. (1)) specifies a feasible time slicing scheme for a recurring window of $(b/R)CS$ sec, where (i) no two bursts overlap with each other, and (ii) bursts are long enough to send data for all mobile devices to playout till the next burst. Furthermore, the energy saving achieved by mobile devices in class $c$ is given by:*

$$\gamma_c = 1 - \frac{c}{CS} - \frac{RT_o c}{bCS} \quad \text{where } c = 1, 2, \ldots, C. \tag{2}$$

*Proof.* First, since sending a burst of $b$ kb takes $b/R$ sec to transmit, by definition of Eq. (1) the resulting time slicing scheme leads to no overlapping bursts. Second, because the recurring window size is $(b/R)CS$ and the bit rate of any layer is $r_s = r/C$, the required amount of data in any layer for smooth playout is $(b/R)CSr_s = (b/R)Sr \leq (b/R)R = b$, where the inequality comes from the definition of $S$. This inequality shows that the allocated time period for each burst is long enough to carry the playout data for a layer till the next burst of the same layer.

For energy saving, since mobile devices in class $c$ receive $c$ bursts of size $b$ in every recurring window, the energy saving can be computed by $\gamma_c = 1 - \frac{(bc/R)+T_o c}{(b/R)CS}$. Manipulating this equation yields Eq. (2).

This lemma shows that LATS scheme is correct and allows mobile devices in different classes to receive and render at different perceived quality, while achieving proportional energy saving.

In the next lemma, we show that LAF scheme leads to lower energy savings than LATS scheme.

**Lemma 2.** *The Layer-Aware Time Slicing (LATS) scheme achieves higher energy saving than the Layer-Aware FEC (LAF) scheme for class $c$ mobile devices if $c \neq C$. These two schemes lead to the same energy saving for class $C$ mobile devices.*

*Proof.* LAF works in a recurring window of $S$ bursts of size $b$ kb. The window time is $(b/R)S$ sec. In every window, class $c$ devices receive a burst prefix of length $bc/C$ and turn off their RF circuits. Hence, we write the energy saving achieved of mobile devices in class $c$ as:

$$\hat{\gamma}_c = 1 - \frac{(bc)/(CR) + T_o}{(b/R)S} = 1 - \frac{c}{CS} - \frac{RT_o}{bS} \quad \text{, where } c = 1, 2, \ldots, C. \tag{3}$$

Comparing Eq. (3) against Eq. (2) yields the lemma.

**Summary:** Compared to sequential scheme, Parallel Service (PS) scheme only saves processing overhead, and does not lead to energy savings for heterogeneous devices. While Layer-Aware FEC (LAF) scheme achieves proportional energy savings, implementing it requires modifying broadcast protocols and could make broadcast networks more sensitive to bursty channel errors [22]. Moreover, LAF scheme results in lower energy savings than LATS as we proved in Lemma 2. Since LATS enables us to achieve the highest energy savings among all proposed schemes, we recommend LATS scheme and do not consider the other two in the rest of this paper. Last, we mention that although LATS scheme allocates each TV channel multiple ($C$) bursts in a recurring window ($bCS/R$ sec), these bursts are placed apart enough for the base station to fill up them. Hence, LATS scheme does not result in under-utilized bursts.

## 5 Evaluation

We first briefly describe the testbed and the experimental setup. We then present the results.

### 5.1 Mobile TV Testbed

We have implemented a testbed in our Lab for one of the most popular mobile TV standard: DVB-H [4, 5]. The testbed provides a realistic platform for analyzing the performance of the proposed burst transmission scheme. The testbed has two parts: base station and receivers. We use a commodity Linux box as the base station, and runs video server, IP encapsulator, and modulator software on it. We installed a PCI modulator [23] in the base station, which implements the physical layer of the DVB-H protocol and transmits DVB-H standard compliant signals via a low-power amplifier and an indoor antenna. We use Nokia N96 cellular phones [3] as receivers to assess the visual quality of videos. For detailed information on the signals, we add a DVB-H analyzer [24] to the testbed. This analyzer is attached to a PC via a USB port and comes with a visualization software for analysis. The analyzer records traffic streams as well as provides a very detailed information on the RF signal, the MPEs, jitter, time slicing, and so on. More details on the testbed are given in [25].

### 5.2 Setup

We have implemented the Layer-Aware Time Slicing (LATS) scheme in the testbed. For comparison, we have also implemented the current, sequential burst transmission scheme, which is denoted as CUR in the figures. We encode several video sequences at 768 kbps, which are then partitioned into four layers, where each layer has a bit rate of 192 kbps. We then configure the modulation card to use 8 MHz bandwidth, QPSK (quadrature phase-shift keying) modulation, 3/4 code ratio, 1/8 guard interval. This leads to channel bandwidth of 8.289

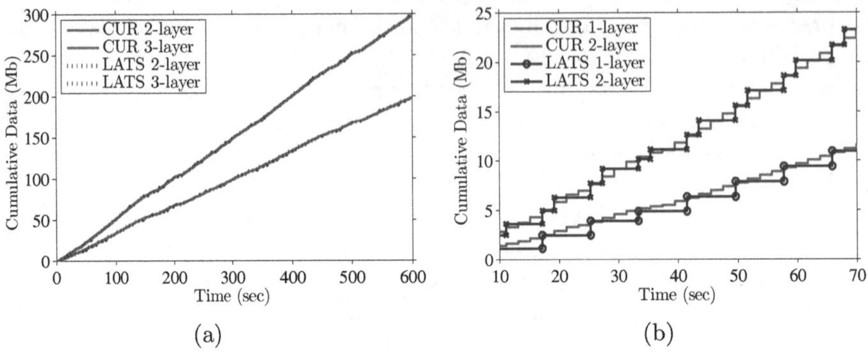


**Fig. 6.** Cumulative received data: (a) 10-min broadcast, and (b) zoom-in 60-sec period

Mbps [26]. We varied the burst size $b$ from 200 to 1600 kb to study the implications of $b$ on the energy saving $\gamma$. We broadcast 8 TV channels for 10-min using LATS scheme, and we repeat the same test using the CUR scheme.

To gather statistically meaningful results, we have instrumented the testbed to save log files for offline analysis. The log files contain start and end times of each burst as well as its size. Moreover, the log files indicate the distribution of burst data among SVC layers. For example, a burst produced by CUR scheme contains IP packets for all layers, while a burst produced by LATS only contains IP packets for a specific layer. We have developed a script to emulate mobile devices in various classes based on the log files. This script computes the cumulative size of IP packets received by each mobile device and the achieved energy saving.

### 5.3 Results

While we concurrently broadcast 8 TV channels, we only present sample results for TV channel 1. Results for other TV channels are similar and are not shown for brevity.

*Cumulative Data Dynamics:* We plot the cumulative received data in Fig. 6 for two classes of mobile devices. Results for other classes are similar. In Fig. 6(a), we plot the cumulative received data for the complete experiment. This figure shows that CUR and LATS are both feasible, and transmit the same amount of data for mobile devices in the same class. In Fig. 6(b), we zoom into a short time period. In this figure, every staircase step represents a received burst. This figure reveals that mobile devices receive many more bursts when CUR scheme is used, which leads to higher processing overhead.

*Proportional Energy Saving:* We plot the energy saving achieved by various mobile device classes. We present a sample result with $b$ = 400 kb in Fig. 7. Fig. 7(a) illustrates that LATS enables mobile devices to receive a subset of layers, and achieve *proportional energy saving*. For example, mobile devices which receive all four layers achieve 65%, while mobile devices which receive the base layer achieve more than 90% energy saving. Meanwhile, Fig. 7(b) depicts that no

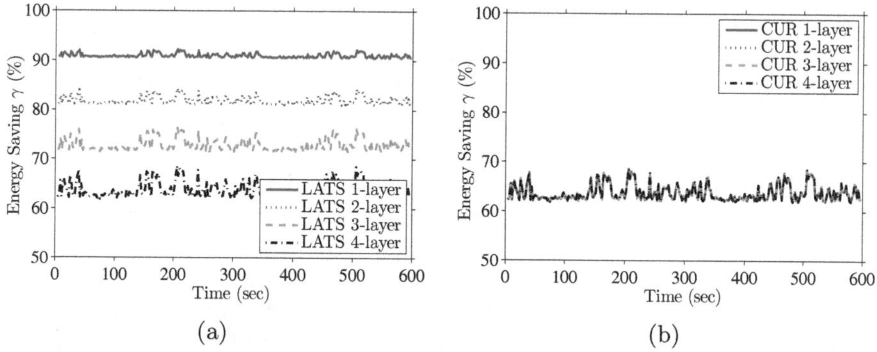


**Fig. 7.** Energy saving achieved by mobile devices in different classes: (a) LATS and (b) CUR

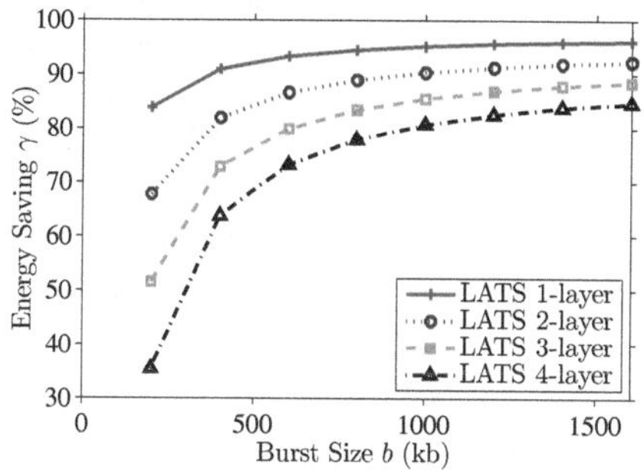


**Fig. 8.** Range of average energy saving achieved by different classes with various burst sizes

matter how many layers they receive, mobile devices in CUR scheme achieve the same energy saving. This shows that LATS is required to support proportional energy saving for heterogeneous classes.

*Implication of Burst Size on Energy Saving:* We compute the average energy saving of each mobile device class under different burst sizes. Fig. 8 shows the results. This figure reveals that larger burst sizes lead to higher energy saving. This is because larger burst sizes means fewer number of bursts, where each burst incurs an overhead duration $T_o$. However, mobile devices have to reserve more memory to receive and process bursts when the burst size is large. This figure shows that $b = 1000$ kb is a sweet spot: larger burst sizes only leads to marginal increases on energy saving.

## 6 Conclusions

We have studied the problem of efficiently broadcasting multiple scalable video streams to heterogeneous mobile devices. We showed that network operators should broadcast scalable video streams in a different manner than nonscalable streams; otherwise the energy of mobile devices could be wasted. We have presented three broadcast schemes: PS, LAF, and LATS. They explicitly support heterogeneous mobile devices. In the PS scheme, layers of the same TV channel are multiplexed by either IP addresses or MPEG-2 PIDs into a series of bursts. In the LAF scheme, layers of the same TV channel are sequentially placed in Layer-Aware FEC frames. This allows mobile devices to receive partial bursts for desired layers and turn off their RF circuits earlier to save energy. In the LATS scheme, every layer of a TV channel forms a series of bursts. This enables mobile devices to locate layers in video streams without opening their RF circuits, such that they can receive desired layers efficiently in terms of energy saving. We proved that the LATS scheme is the most efficient broadcast scheme among the three proposed scheme. Hence, we recommend the LATS scheme.

We implemented the proposed broadcast scheme in a real testbed in our Lab for DVB-H networks. We also implemented the current, sequential broadcast scheme for comparison. Our experimental results show that, with the proposed broadcast scheme, significant energy savings can be achieved by different heterogeneous devices. For example, energy saving between 62% and 92% can be achieved by receiving different number of layers. In contrast, with the current broadcast scheme, all mobile devices achieve energy saving 62% despite how many layers they can (or opt to) receive and decode.

## References


1. EU: Mobile TV across Europe: Commission endorses addition of DVB-H to EU list of official standards (2008), `http://europa.eu/rapid/pressReleasesAction.do?reference=IP/08/451&format=PDF`
2. GSmart: GSmart website (2008), `http://www.gigabytecm.com/`
3. Nokia: Nokia Nseries website (2008), `http://www.nseries.com/`
4. Kornfeld, M., May, G.: DVB-H and IP Datacast – broadcast to handheld devices. IEEE Transactions on Broadcasting 53(1), 161–170 (2007)
5. ETSI: Digital Video Broadcasting (DVB); transmission system for handheld terminals (DVB-H). EN 302 304 Ver. 1.1.1 (November 2004)
6. Qualcomm: FLO technology overview (2008), `http://www.qualcomm.com/common/documents/brochures/tech_overview.pdf`
7. Li, B., Liu, J.: Multirate video multicast over the Internet: An overview. IEEE Network Magazine 17(1), 24–29 (2003)
8. Ganjam, A., Zhang, H.: Internet multicast video delivery. Proc. of the IEEE 93(1), 159–170 (2005)
9. McCanne, S., Jacobson, V., Vetterli, M.: Receiver-driven layered multicast. In: Proc. of ACM SIGCOMM 1996, Palo Alto, CA, August 1996, pp. 117–130 (1996)
10. Haddadi, H., Rio, M., Iannaccone, G., Moore, A., Mortier, R.: Layered H. 264 video transmission with hierarchical QAM. Journal of Visual Communication and Image Representation 17(2), 451–466 (2006)
11. Hellge, C., Schierl, T., Wiegand, T.: Mobile TV using scalable video coding and layer-aware forward error correction. In: Proc. of IEEE International Conference on Multimedia and Expo. (ICME 2008), Hannover, Germany, April 2008, pp. 1177–1180 (2008)

12. Ladebusch, U., Liss, C.: Terrestrial DVB (DVB-T): A broadcast technology for stationary portable and mobile use. Proceedings of the IEEE 94(1), 183–193 (2006)
13. Yang, X., Song, Y., Owens, T., Cosmas, J., Itagaki, T.: Performance analysis of time slicing in DVB-H. In: Proc. of Joint IST Workshop on Mobile Future and Symposium on Trends in Communications (SympoTIC 2004), Bratislava, Slovakia, October 2004, pp. 183–186 (2004)
14. ETSI: Digital Video Broadcasting (DVB); DVB-H implementation guidelines. EN 102 377 Ver. 1.3.1 (May 2007)
15. Balaguer, E., Fitzek, F., Olsen, O., Gade, M.: Performance evaluation of power saving strategies for DVB-H services using adaptive MPE-FEC decoding. In: Proc. of IEEE International Symposium on Personal, Indoor and Mobile Radio Communications (PIMRC 2005), Berlin, Germany, September 2005, pp. 2221–2226 (2005)
16. Zhang, Q., Fitzek, F., Katz, M.: Cooperative power saving strategies for IP-services supported over DVB-H networks. In: Proc. of IEEE Wireless Communications and Networking Conference (WCNC 2007), Hong Kong, China, March 2007, pp. 4107–4111 (2007)
17. Hsu, C., Hefeeda, M.: Bounding switching delay in mobile TV broadcast networks. In: Proc. of ACM/SPIE Multimedia Computing and Networking (MMCN 2009), San Jose, CA (January 2009)
18. Hefeeda, M., Hsu, C.: Energy optimization in mobile TV broadcast networks. In: Proc. of IEEE Innovations in Information Technology (Innovations 2008), Al Ain, United Arab Emirates, December 2008, pp. 430–434 (2008)
19. Hsu, C., Hefeeda, M.: Time slicing in mobile TV broadcast networks with arbitrary channel bit rates. In: Proc. of IEEE INFOCOM 2009, Rio de Janeiro, Brazil (April 2009)
20. Joint Video Team: Advanced video coding for generic audiovisual services. ITU-T Rec. H.264 & ISO/IEC 14496-10 AVC (November 2007)
21. Schwarz, H., Marpe, D., Wiegand, T.: Overview of the scalable video coding extension of the H.264/AVC standard. IEEE Transactions on Circuits and Systems for Video Technology 17(9), 1103–1120 (2007)
22. Faria, G., Henriksson, J., Stare, E., Talmola, P.: DVB-H: Digital broadcast services to handheld devices. Proceedings of the IEEE 94(1), 194–209 (2006)
23. Dektec: DTA-110T PCI modulator (2008), `http://www.dektec.com/Products/DTA-110T/`
24. Enensys: Divi Catch RF-T/H analyzer (2008), `http://www.enensys.com/`
25. Hefeeda, M., Hsu, C.: Design and evaluation of a testbed for mobile TV networks. Technical Report TR 2009-03, Simon Fraser University (February 2009)
26. ETSI: Digital Video Broadcasting (DVB); framing structure, channel coding and modulation for digital terrestrial television. EN 300 744 Ver. 1.5.1 (June 2004)

# Scan Surveillance in Internet Networks (Work in Progress)

Khadija Ramah Houerbi[1], Kavé Salamatian[2], and Farouk Kamoun[1]

[1] National School of Computer Science, University of Manouba, Tunisia
khadija.houerbi@cristal.rnu.tn, farouk.kamoun@ensi.rnu.tn
[2] Lancaster University, Lancaster, UK
kave.salamatian@comp.lancs.ac.uk

**Abstract.** In recent years, many measurement studies have shown the ubiquity of scanning activities in the Internet and the growing sophistication of probing techniques that became more stealthy by stretching slowly over time or using spoofed source IP addresses. Scans are mainly generated by attackers trying to map the configuration of a target network and by computer worms trying to spread over the Internet. Although, the problem of scan detection has been given a lot of attention by network security researchers, current state-of-the-art methods still suffer from high percentage of false alarms or low ratio of scan detection. In this paper, we propose to detect changes in scanning patterns, by monitor variation of the distribution of scan features in a space spanned by IP source address, IP destination address, source port number, and destination port number. This gives insight on characteristics of scanning activities and exposes the presence of emerging scanning attacks and worms. For that, we propose to use an information theoretic-based approach to detect changes in distributions.

**Keywords:** scan monitoring, anomaly detection, Information Theory, Networks.

## 1 Introduction

A major challenge for security managers is to develop network intrusion detection systems that can timely and accurately detect network attacks. A precursor to many attacks on networks is often a reconnaissance operation more commonly referred to port scans. Scans have many legitimate uses [1], including, for system administrators, to verify the security of a network, to find web servers to index by some search engines and some application (e.g. SSH, some peer to peer as edonkey and windows applications ). However, intruders frequently perform port scanning as a mean to gather valuable information on target victims to achieve destructive attacks. Furthermore, computer worms also use port scanning to propagate; infected hosts usually perform scanning to search for new vulnerable hosts. This scanning activity takes the form of probe packets targeted at specific TCP or UDP ports on multiple target hosts. If a target host responds, then the scanning host will initiate the process of uploading malware onto that host.

L. Fratta et al. (Eds.): NETWORKING 2009, LNCS 5550, pp. 614–625, 2009.

Many measurement studies [2–4], have shown that port scans represent a sizable portion of today's Internet traffic, that exhibits steady growth, mainly since 2001, and a changing nature over time and between sites. In [2], the authors examined the history of scanning activities over 12.5 years (between 1994 and 2006) as observed at the border of the Lawrence Berkeley National Laboratory (LBNL). They showed that since 2001 scanning became an ubiquitous phenomenon (mainly due to the wide spread of Internet worms like CodeRed I and II, Nimda, Slammer, and Blaster). Following a different approach, the authors in [3], measured and characterized traffic amassed by network telescopes in two academic networks and termed it "Background radiation". This traffic, sent to unused IP addresses, is anomalous by nature, and reflects the presence of many unwanted activities ranging from scans, backscatter from a flooding attack victimizing someone else, exploit attempts and mis-configurations. The authors of [4] analyzed a set of logs from the global "DShield" repository that spanned a 4 month-long period during 2001-2002. They showed that while common scanning types (eg. vertical and horizontal) are indeed prevalent, other strategies such as coordinated and stealthy scans are also widely used.

In the next section, we will survey state-of the art of scan techniques and scan detection approaches. We will show that most of scan detection methods need connection reassembly and are limited to one scan type. Moreover, nearly all proposed methods have difficulties catching low rate scanners and often suffer from significant levels of false positives making them inadequate for automatic scan suppression.

Subsequently, we will present in section three, the approach proposed in this paper. Although, the proposed approach is not a scan detection method it aims to expose the presence of emerging scanning attacks and worms. It consists of collecting and analyzing scan traffic to track significant changes in its distributional aspects. Detecting such changes is important because it can be relevant of serious security problems such as rapid scanning worms or reconnaissance scan preceding destructive attacks. The section four exposes our experimental results. For that we used both real measurement datasets collected on Tunisian National University Network and some artificial traces obtained by mixing a real trace with experimental Nmap scan traffic. Finally, we conclude and present some perspectives in section five.

## 2 Background and Related Work

### 2.1 Scan Techniques

Port scanning can be defined as a technique for discovering open "doors" or ports on a host or a set of hosts. It aims to discover host's vulnerabilities by sending a probe to a port and listening for an answer. As most of Internet services are on TCP, scanners are mainly interested to probing TCP ports to determine their states. There are at least a dozen scan techniques implemented in Nmap utility which we can group in three major types: SYN scans, non-SYN scans, and idle scans. SYN scan techniques work against any compliant TCP stack and allow clear, reliable differentiation between the open, closed, and filtered states of a TCP port. They can be performed quickly, scanning thousands of ports per second on a fast network. The Half-open scanning technique (an example of SYN scan technique) is the most popular one. It consists of sending a single SYN packet to a specific port on a target host and then waiting for a response. A SYN/ACK response indicates that the probed port is listening (open), while a RST (reset) response signifies a closed port. If no response is received (or an ICMP unreachable error is received), the port is marked as filtered. Non-SYN scans are more subtle techniques that exploit loophole in the RFC 793 to differentiate between open and closed ports . However, many systems (Microsoft Windows, many Cisco devices, BSDI, and IBM OS/400) do not follow RFC 793 to the letter that makes these techniques less effective. Finally, Idle scan is an advanced scan method that allows for blind TCP port scanning of target hosts (meaning no packets are sent to the target host from the real attacker's IP address). It exploits predictable IP fragmentation ID sequence generation on a zombie intermediate host to glean information about the open ports on the target. IDS systems will display the scan as coming from the zombie machine used. Besides being extraordinarily stealthy (due to its blind nature), it permits mapping out IP-based trust relationships between machines.

### 2.2 Scan Detection Methods

As a result of importance of scan traffic, various scan detection methods have been proposed to detect and prevent scan activities. These methods can be classified into counting methods [1, 5, 6], and non counting ones [7–9].

Snort [6], a popular intrusion detection system, implements a simple counting method for scan detection. It marks a source IP address as a scanner if it tries to connect to more than a given number of distinct IP addresses or a fixed number of TCP/UDP ports within a time window. However the counting algorithm can erroneously mark legitimate hosts as scanners in particular when proxies or address translation mechanisms are used. For example in [1], the authors established that over a Lawrence Berkley Laboratory (LBL) traces, with Snort's default settings, 38.5% of host detected as scanner are in fact false positives. Other counting methods have built upon the observation that, unlike benign remote users, scanners are more likely to initiate failed connections to local IP addresses. Bro [5] scan detection algorithm uses this and counts only failed connection for services specified in a configurable list. For the others, it counts all connections.However the issue with Bro is similar to Snort, even if misclassification is less likely to happen because of counting only failed attempts. Threshold Random Walk (TRW) algorithm, proposed in [1], propose to use the observed disparity between the proportion of successfully established connections between legitimate remote hosts and malicious ones.TRW uses sequential hypothesis testing to distinguish between benign remote hosts who occasionally send misaddressed traffic and remote scanners who are often likely to probe non active IP addresses or ports. However, despite its advantages, one can still easily evade TRW. In [10], the authors proposed a distributed scan method, named z-scan that is using a limited number of zombies. This method is extremely effective against TRW.

Non-counting approaches have been proposed [7, 8, 11], to address problems with counting methods. The authors of [8] proposed a probabilistic approach to scan detection. It computes for each address an access probability distribution, calculated across all remote source IP addresses that are contacted by this host. Then, it compares the probability with that of scanners that are assumed as accessing each destination address with an equal probability. The authors of [11] proposed an entropy-based approach to fast scanning worms detection. The proposed approach works on flow-level and computes entropy contents of traffic parameters such as IP addresses. Significant changes in entropy indicate worm propagation. In [7], the authors formalize the problem of scan detection as a classification problem. A data mining classifier algorithm labels each pair (source IP, destination port) as scanner or non-scanner. Despite its performance, this data mining method heavily depends on an accurately labeled training trace and it is not suitable for real-time scan detection; it was used with time windows of 20 minutes.

## 3 Scan Surveillance

We have seen in the previous section that scan detection poses two tricky challenges. First, probing methods are increasingly varied and refined both for greater efficiency and to operate at lower profile to evade intrusion detection systems. Second, complete session re-assembly of traffic data, needed by many proposed scan detection methods, is impractical for on-line deployment at high-speed vantage points. Maintaining real-time assembly of many connections for many hosts requires much computing resources making the cost of such methods unaffordable for many ISPs. Moreover, as stated in [3], scan activity can be seen as an unavoidable background radiation that hits permanently our network. This means that the presence of scanners is a nuisance that is not *per se* problematic, but a change in the pattern of activity of scanners is an important sign that something is happening, *e.g.*,a new worm propagating or a cibled attack to a network. In place of trying to detect each individual scanner, we should rather monitor the scanning activity to detect change in it. When we detect a change, we can there after apply sophisticated and complex forensic techniques to detect the scanner(s) that have led to a change in scanning activity. In this paper, we will concentrate on the detection of changes in scan activity.

Our approach of monitoring scan activity has analogous foundation to what proposed in [8, 11], in the sense that we that we characterize the scanning activity by its distribution in the space spanned by IP source address (@IPsrc), IP destination address (@IPdst), source port number (# src) and destination port number (# dst) (@IPsrc,@IPdst,# src,# dst). The authors of [8] use the distribution of @IPdst or # dst as a discriminant feature between a scanner and a benign node. The authors of [11] derive the entropy from the distribution of @IPsrc is used as the feature to monitor to detect worm propagation. In this paper we are going further, and we see the joint distribution over the quadruple (@IPsrc,@IPdst,# src,# dst) as the main feature to use to detect changes in scanning activity. Our scanning monitoring algorithm therefore consists of deriving an aggregated reference distribution on an observation window of $w$ packets. This reference is thereafter compared with the distribution inferred over a running window of the same size. We detect a change in scanning pattern if the deviation from the reference window is larger than a detection threshold. In summary we are following the below three steps;

1. Aggregating suspect scan traffic collected on a chosen vantage point in separate windows of $w$ packets.
2. Computing, for each window, an empirical joint distributions of the features of interest in the scanning traffic.
3. Analyzing these empirical distributions and tracking deviation from a reference distribution to expose the presence of serious security problems that need more investigation.

We saw in previous section that detecting individual scans is a challenging task. Fortunately our objective is to detect change in scanning pattern at any observation point, *i.e.*,we should be able to monitor scanning traffic using only locally available information. So we will not use any scan detection algorithm to collect scan traffic. We propose to collect on every monitored point all SYN packets that are not followed by SYN/ACK responses within a 60 seconds interval. Collecting such traffic, although it may include benign traffic as well as the real scan traffic, has three basic benefits: first, its collection is more ressource-friendly as we are ignoring all non-SYN packets and we do not need complete session re-assembly; second, such traffic is mainly composed by scans as usually, we can neglect mis-configuration traffic and transitory network congestion or failures. Finally, isolated SYN packets form most of scan traffic. Non-SYN scans suffer from TCP/IP implementation idiosyncrasies, which limits their efficiency and therefore their use.

### 3.1 Feature Selection

Our aim here is to track changes in the distribution of scan activity in the space spanned by (@IPsrc,@IPdst,# src,# dst). This choice comes from the observation that different port scanning strategies affects feature distributions in diverse manners. A horizontal scan affects the # dst distribution of the scan traffic, so that it becomes skewed and concentrated on the vulnerable port being scanned. Moreover, if the distribution of the source IP address field is also skewed than means that the a single or a few numbers of IP addresses conduct the attack (it is probably a botnet master, looking for vulnerable machines to send them a bot), otherwise it is most likely a worm scan. In the same way, vertical scans might be detectable as a change in the distributional aspect of source and destination IP addresses, which becomes more concentrated: on the attacker IP address for the @IPsrc distribution and on the victim for @IPdst. However, coordinated scans are less obvious to detect since the attacker can probe many ports on different target IP addresses using many source IP addresses; therefore making all feature distributions appears dispersed. For this case we expect to have to use the full set (@IPsrc,@IPdst,# src,# dst) to be able to detect the changes. Thus, we will use the following traffic features to detect scanning attacks and worms:

- (@IPsrc,# dst): to detect emerging horizontal scans from botnet masters and individual attackers.
- (# dst): to detect new worm scans.
- (@IPsrc, @IPdst): to detect new or repetitive vertical scans.
- (@IPsrc,@IPdst,# src,# dst): to detect coordinated scans.

### 3.2 Distribution Inference

In most general settings, computing the traffic features joint distributions for each time window consists of having a vector of up to $2^{96}$ entries [1], where each entry represent a possible assignment of the features values. Every time the feature value assigned to an entry is observed it is incremented. Hopefully, the number of distinct TCP port pairs is near 25000 where number of distinct IP address pairs in a 24 hours scan trace (see Table 1) is about 2 millions. Although these numbers are lower than the $2^{64}$ and $2^{32}$ possible ones, they remain important, and it is still unfeasible to estimate precisely the distribution. We have therefore, to resort to estimate an aggregated histogram. One solution to build it is to apply a mask that aggregates into a single bin all features values sharing the same value of mask. Even if this solution is easy to implement it suffers from a lack of flexibility, aggregation level is not easy to control and most important it is too deterministic, *i.e.*,an attacker can guess the mask applied and make use of it to hide its scan traffic. To address these two problems we propose to apply a random hash function to each feature and to aggregate value based on the resulting hash values. This approach is easy to implement, leads to flexible aggregation, and last, it is immune to the attacker guess. To have more security one can even change frequently the random hash.

One might use the aggregated histogram to estimate precisely a parametric distribution form [12]. However in this work, we want to stay non-parametric so we will directly use the histogram without refining it into an estimated distribution.

### 3.3 Change Detection

We have postulated earlier in section 3 that we can infer change in scanning pattern by observing changes in the joint distribution of features. We need to define a way to measure the discrepancy between two distributions. This problem is indeed central in statistics. Two main approach classes can be defined to deal with this problem. A first class of approach is parametric. They consist of measuring the distance of two distributions through the change in parameters of a parametric distribution. Parametric inference and statistical tests associated with it have widely used these approaches.

[1] (@IPsrc,@IPdst,# src,# dst) is represented with 96 bits.

A second class is non-parametric and does not use any particular type of distribution. Entropy based approach [9, 11] belongs to this category. Entropy is the measure of missing information when we do not know the value of a random variable realization. The intuition beyond using entropy in anomaly detection is that we expect that a change in the distribution lead to proportional change in entropy. Unfortunately, a closer analysis of entropy shows that it is not a good indicator of distribution variation. To see this let us assume a two valued random variable with a distribution $(p, 1-p)$ and let us assume that this distribution varies to $(p+\Delta, 1-p-\Delta)$. A sensitivity analysis of the entropy $H(p)$ to variation of $p$ can be done through the derivative:

$$\frac{\partial H}{\partial p} = \log\left(\frac{p}{1-p}\right)$$

One can therefore expect that the effect of a small variation $\Delta$ on the distribution on the entropy would be approximatively $\Delta \log\left(\frac{p}{1-p}\right)$, *i.e.*,the sensitivity of the entropy variation depends strongly on the value of $p$ and not on the variation of the distribution; for small $p$ the entropy will vary largely with small variation of $p$ and for larger values the entropy will not vary as much. The sensitivity analysis shows that even very small variations of a bin probability (that could eventually result from small random fluctuations) can lead to large variation of the entropy. This elucidates why from a theoretical standpoint entropy is not a good metric to use for tracking change in distributions as small variation in distribution might lead to large variations in entropy and large variation in the distribution might not lead to important entropy variation.

In place of entropy, we propose to use the relative entropy, also named Kullback-Leibler divergence (KLD) [13] , to track changes in distributions obtained over hashed distributions. The Kullback-Leibler divergence is an information theoretic measure of the difference between two probability distributions defined on the same finite set $\mathcal{H}$: a distribution $P$ and a reference probability distribution $Q$. It is given by equation (1).

$$D(P\|Q) = \sum_{x\in\mathcal{H}} P(x)\log\left(\frac{P(x)}{Q(x)}\right) \tag{1}$$

The KLD is not defined when $\exists x \in \mathcal{H}, Q(x) = 0, P(x) \neq 0$. By applying a sensitivity analysis similar to what done for entropy to a two valued discrete distribution we obtain:

$$\frac{\partial D}{\partial p} = \log\left(\frac{p}{q}\right) + \log\left(\frac{1-q}{1-p}\right)$$

This shows that in contrast with entropy the sensitivity of KLD does not depend on the value of $p$ alone but on the ratio between $p$ and the reference distribution $Q$, *i.e.*,KLD is highly sensitive if we have a large variation compared to the reference distribution, and become less sensitive if this variation is smaller. This is perfectly what we expect from a distribution variation metric. This validates the use of KLD in place of entropy, and it explains also some of the poor results obtained when using entropy to monitor distributional changes.

Furthermore, we can formulate anomaly detection using feature distribution of the random variable $X$ as follows. Let us assume that the scan traffic follows a stable distribution $P$ and we are observing $n$ observations of scan packets. Let us $Q_n$ be the empirical feature distribution. Anomaly detection reduces to decide the following composite hypothesis test:

- $H_1$ : The observed scan traffic features follow the stable distribution $P$, *i.e.*,there is no change.
- $H_2$ : The observed scan traffic features follow an alternative distribution $P'$ *i.e.*,there is a change.

It can be shown that the classical likelihood ratio test can be replaced by the below test on KLD:

$$|D(Q_n\|P) - D(Q_n\|P')| > \frac{1}{n}\log T$$

where $T$ is the decision threshold. The Stein lemma [13] states that if one fixes the probability of false alarm (the probability that no change happened but we still decided that a change occurs) to a value and want to minimize the probability of misdetection (the probability that there are changes and we cannot detect them), this minimal misdetection probability will depend on $D(P\|P')$. However in practice we do not know the type of changes that will happen *a priori*, so the alternative distribution $P'$ is unknown. We have therefore, to replace the change detection test by the below one:

$$|D(Q_n\|P)| > \frac{1}{n}\log T$$

However, this test is not optimal anymore and we cannot derive the misdetection errors analytically, and we have to resort to Receiver Operation Characteristic (ROC) curves. These curves show the tradeoff achieved between false alarm and mis-detection rate for a given change detection scheme.

# 4 Experimental Validation

The trace used in this paper consists of all packet headers that have passed during April 4th to 5th , 2006 through the 100 Mb/s link connecting Tunisian National University Network (TNUN) to the Internet. Table 1 summarize that trace and reports the total number of packets and flows. It also shows the share of transport protocols above IP protocol.

We have assumed that all SYN packets that are not followed by SYN/ACK responses within a 60 seconds interval are scan traffic. This leads to 10 millions suspicious packets that have 18388 sources and are destinated to 56585 destinations, leading to 2 millions address pairs. The traffic targeted up to 25 000 different port numbers. The suspicious packets account for 3% of total number of packets and 42% of the total number of connections.

**Table 1.** Summary of the packet trace used for validation

| Period | Packets number in millions | Flow number in millions | Protocol share(%) TCP | UDP | ICMP |
|---|---|---|---|---|---|
| April 4-5 , 2006 | 356 | 10.5 | 97.5 | 1.8 | 0.7 |

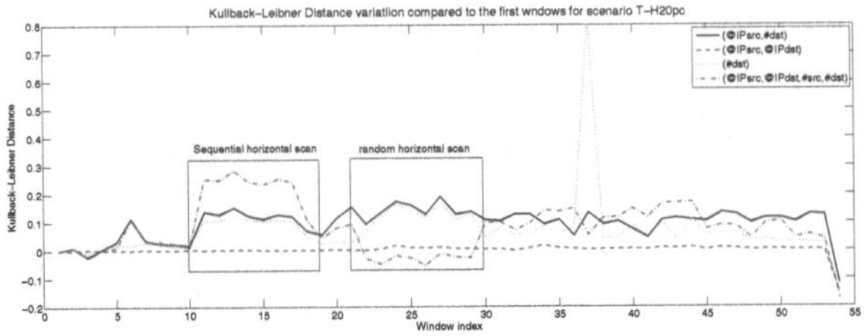


**Fig. 1.** Evolution of Kullback distance computed for four sets of features with a horizontal scan inserted

## 4.1 Validation on Artificially Changed Scan Traffic Traces

To validate the method we first create artificial changes in scanning profile and see whether we can detect these changes using the proposed methods. For this purpose, we have used Nmap tool to generate 5 scan traces: 2 horizontal scan targeted to port 80 on randomly chosen IPs (a sequential scan on all IPs of an address mask, and a random scan), and 3 vertical scans (a sequential, a random, and a targeted) to many ports on a single IP. These scan packets are injected in a real trace by mixing according to a certain percentage (referred as scan intensity) them with the originally observed real trace (T). We provide in table 2 the details of the resulting traces. We used for each set of features a hash function on 5 bits, *i.e.*,the feature distribution is obtained over 32 bins. We obtain over each fixed window of 1000 packets an empirical histogram. We used the histogram over the first window as the reference histogram for detecting changes in scanning pattern.

In Fig. 1, we present the KLD variation obtained for the scenario T-H20pc with two horizontal scans injected. The figure shows that (@IPsrc, # dst), (# dst) and (@IPsrc,@IPdst, # src, # dst), shows an noticiable change of the KLD. However unexpectedly that the distribution on (@IPsrc, @IPdst) do not show a major change. This could be explained because of the use of random hash function that distributes the value in the hash space, so that the horizontal scan

**Table 2.** Artificial scan traces summary

| Trace | Description | Intensity |
|---|---|---|
| T | Original scan trace | 0% |
| T-V20pc | Three vertical scans injected in window 100, 260, and 430 | 20% |
| T-H20pc | Two horizontal scans injected in window 100 and 260 | 20% |

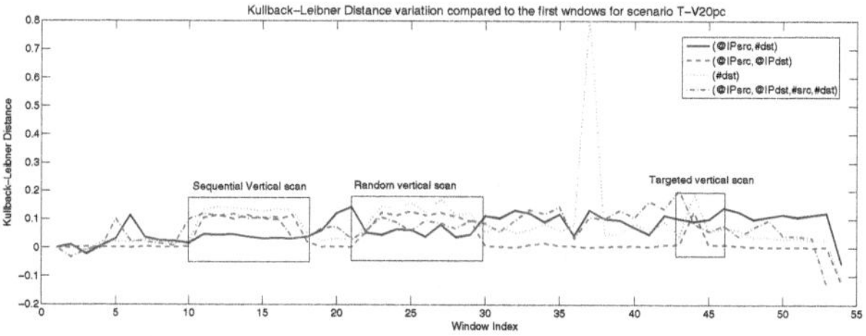


**Fig. 2.** Evolution of Kullback distance computed for four features compared to the first window for a one hour scan trace

is not affecting too much the distribution. Interestingly the KLD for (@IPsrc, # dst) and (# dst) show almost the same values during the horizontal scan injection. This could be explained by the fact that the variation in feature space is essentially related to port variations that are happening during horizontal scans. Moreover KLD variations are more emphasized on the (@IPsrc,@IPdst, # src, # dst) than on other features. This validates the use the approach proposed in the paper for detecting changes in horizontal scanning pattern. In Fig. 2, we present the KLD variation obtained for the scenario T-V20pc with three vertical scan injected. The figure shows that (@IPsrc, # dst) is not well reacting to the introduction of vertical scans, when (@IPsrc, @IPdst), as well as (@IPsrc,@IPdst, # src, # dst), are well reacting to the injected scan level changes. This validates the use the above three set of features for detecting vertical scans. In vertical scan this is mainly destination address and destination port that are changed, one can expect to not see it on the feature (@IPsrc, # dst). It is noteworthy that the feature (@IPsrc,@IPdst, # src, # dst) seems to be reactive to the two types of scans: horizontal and vertical. However, using each individual feature enable also to know if the change is resulting from vertical or horizontal scanning changes, where using (@IPsrc,@IPdst, # src, # dst) does not give this information.

## 4.2 Mining Scan Traffic Changes in Real Data

We present in the Fig. 3, the KLD variations for selected features over the whole scan traffic trace. The figure first validates the assumption that there is a stable value for KLD. This can be seen by observing that most of time the KLD distance is close to zero and has a flat structure. However some clear peaks are visible. The KLD computed over (@IPsrc, # dst) pair presents many peaks that are relative to horizontal scans. The figure also shows clearly some vertical scan that are visible by sharp peaks on the (@IPsrc, # dst) graph. However, the intensity variation is larger for horizontal scan than vertical one. This is in accordance with other studies [2–4] findings that affirmed the predominance of horizontal scans compared to other scanning strategie. Interestingly one can observe a increase and a deacrease in the KLD obtained over (@IPsrc,# dst) feature in windows 700 to 800. This increase represents a significant change in horizontal scanning behaviour. It has happened from 10 PM to 6 AM the next

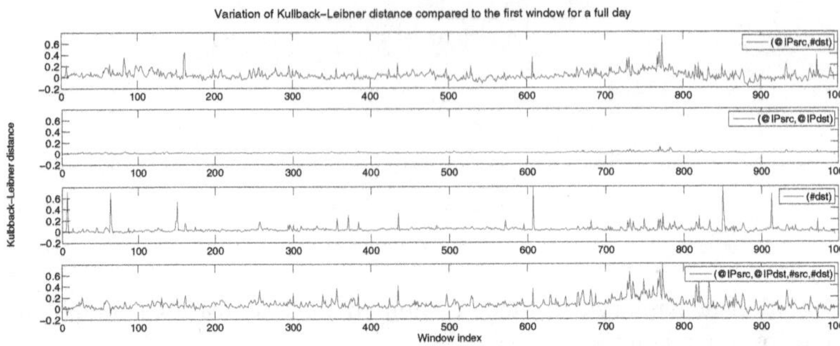


**Fig. 3.** Evolution of Kullback distance computed for four features to the first window for a one hour scan trace

day. This could be related to a significant scan activity targeting TNUN. Here also (@IPsrc,@IPdst, # src, # dst) seems to be a good tradeoff between the all the feature. It regroups peaks in (@IPsrc, @IPdst) as well as (@IPsrc, # dst) curves and it can also detect the change happening between time 700 and 800.

## 5 Conclusion

We presented in this paper a scan surveillance approach based on Kulback-distance-based approach. The proposed approach is completely non-parametric and does not need any hypothesis on the nature of scan traffic. We provided preliminary evidence that the method is effective. However, a complete analysis will need a complete study with a larger set of traces and particularly the observation of a real change in scanning pattern (for example in the advent of a new worm propagating). We are actually instrumenting equipment in network to do this real time analysis.

## References

1. Jung, J., Paxson, V., Berger, A., Balakrishnan, H.: Fast portscan detection using sequential hypothesis testing. In: IEEE Symposium on Security and Privacy, pp. 211–225 (2004)
2. Allman, M., Paxson, V., Terrell, J.: A Brief History of Scanning. In: ACM SIGCOMM/USENIX Internet Measurement Conference (October 2007)
3. Pang, R., Yegneswaran, V., Barford, P., Paxson, V., Peterson, L.: Characteristics of internet background radiation. In: 4th ACM SIGCOMM IMC conference, pp. 27–40. ACM, New York (2004)
4. Yegneswaran, V., Barford, P., Ullrich, J.: Internet intrusions: global characteristics and prevalence. In: ACM SIGMETRICS, pp. 138–147. ACM, New York (2003)
5. Paxson, V.: Bro: a system for detecting network intruders in real-time. In: 7th conference on USENIX Security Symposium, pp. 2435–2463. USENIX Association, Berkeley (1998)

6. Roesch, M.: Snort–Lightweight Intrusion Detection for Networks. In: 13th Conference on Systems Administration LISA 1999, pp. 229–238 (1999)
7. Simon, G., Xiong, H., Eilertson, E., Kumar, V.: Scan Detection: A Data Mining Approach. In: Proceedings of the Sixth SIAM International Conference on Data Mining, pp. 118–129 (2006)
8. Leckie, C., Kotagiri, R.: A probabilistic approach to detecting network scans. In: Network Operations and Management Symposium, NOMS 2002. 2002 IEEE/IFIP, pp. 359–372 (2002)
9. Lakhina, A., Crovella, M., Diot, C.: Mining anomalies using traffic feature distributions. In: SIGCOMM conference, pp. 217–228. ACM, New York (2005)
10. Kang, M., Caballero, J., Song, D.: Distributed Evasive Scan Techniques and Countermeasures. In: Hämmerli, B.M., Sommer, R. (eds.) DIMVA 2007. LNCS, vol. 4579, pp. 157–174. Springer, Heidelberg (2007)
11. Wagner, A., Plattner, B.: Entropy based worm and anomaly detection in fast IP networks. In: 14th IEEE International Workshops on Enabling Technologies: Infrastructure for Collaborative Enterprise, pp. 172–177 (2005)
12. Akodjenou-Jeannin, M., Salamatian, K., Gallinari, P.: Flexible Grid-Based Clustering. In: Kok, J.N., Koronacki, J., Lopez de Mantaras, R., Matwin, S., Mladenič, D., Skowron, A. (eds.) PKDD 2007. LNCS, vol. 4702, pp. 350–357. Springer, Heidelberg (2007)
13. Cover, T.M., Thomas, J.A.: Elements of information theory. John Wiley and Sons, Inc., Chichester (1991)

# Optimizing the Association Procedure for Low-Power 802.15.4 Nonbeacon Sensor Networks

Barbara Staehle

University of Würzburg, Institute of Computer Science, Würzburg Germany
bstaehle@informatik.uni-wuerzburg.de

**Abstract.** Building wireless sensor networks based on the IEEE 802.15.4 standard is an interesting option, as the standard enables low-power, low data rate wireless communication. Many authors analyzed and optimized the operational phase of such networks. In contrast, the initial phase, containing the 802.15.4 association procedure, has mostly been neglected. In this paper, we therefore propose four optimization possibilities for the association procedure of a nonbeacon-enabled sensor network. Our results show that significant performance improvements in terms of association probability, speed, and energy consumptions can be achieved.

**Keywords:** Wireless Sensor Networks, Nonbeacon 802.15.4, Low-Power.

## 1 Introduction

The IEEE 802.15.4 Low-Rate Wireless Personal Area Networks standard specifies PHY and MAC layer protocols to enable wireless connectivity while guaranteeing "ease of installation, reliable data transfer, short-range operation, extremely low cost, and a reasonable battery life" [1]. Together with application layer protocols specified by ZigBee [2] and WirelessHART [3], it is seen as a promising option for wireless sensor networking and has become increasingly popular for industrial purposes, monitoring and control applications.

We recall shortly the most important 802.15.4 features: each Personal Area Network (PAN) has exactly one *PAN coordinator* which should be mains powered and may function as the data sink in a 802.15.4 Wireless Sensor Network (WSN). It is able to send out *beacons*, small command messages which are used for synchronization and organization purposes. In the *beacon mode*, beacons are regularly broadcasted in order to maintain a superframe structure allowing to meet low latency or bandwidth requirements. This overhead is avoided in the *nonbeacon mode* which is suitable for WSNs with less tight synchronization, bandwidth or delay requirements like environmental monitoring or scientific data collection applications [1]. As network topologies, the 802.15.4 standard allows either the star or the peer-to-peer topology. As the latter enables larger, more complex, robust and flexible network formations, it is explicitly proposed for WSNs. In this study, we focus on a WSN operating in the 802.15.4 nonbeacon mode where all sensor nodes are Full Functional Devices (FFDs), i.e. are able to

L. Fratta et al. (Eds.): NETWORKING 2009, LNCS 5550, pp. 626–638, 2009.

relay packets on behalf of other nodes. This setting offers the highest degree of flexibility and resilience.

In low-power WSN deployments, sensor nodes are battery powered or energy harvesting, saving energy is thus the main issue. In 802.15.4 power saving mechanisms are only existing for the beacon mode, for a nonbeacon-enabled network, no possibilities for energy savings are forseen [1]. Together with a sensor MAC protocol or a higher level sleep scheduling solution, nonbeacon-enabled 802.15.4 peer-to-peer networking is nevertheless an interesting option for low power-WSNs. Very few of the proposed 802.15.4 enhancements cover the initial stages of the network life cycle, but a 802.15.4 network can not become functional before the *association procedure* (reviewed in Section 3.1) has been completed successfully. This procedure is optimized for star topologies and is thus less effective in large nonbeacon-enabled topologies.

How a nonbeaconed 802.15.4 network can start up efficiently has never been considered in depth, this study therefore focuses on the startup phase of a nonbeacon-enabled low-power 802.15.4 WSN. Related work is reviewed in Section 2. The association procedure is described in Section 3, where we reveal issues and propose optimizations for large low-power networks. In Section 4, the methodology used for evaluating different optimization possibilities is introduced. Section 5 contains insights on the individual and combined effects of the considered parameters and demonstrates that any low-power 802.15.4 WSN can operate power efficiently from the beginning. We conclude our work and point out future research directions in Section 6.

## 2 Related Work

Many performance evaluations of 802.15.4 have been published. As the 802.15.4 nonbeacon MAC protocol is barely more than plain CSMA-CA, most studies analyze and optimize aspects of the 802.15.4 beacon mode. An exemplary study of the beacon mode is the work of Kohvakka et al. [4], who analyzed the performance of 802.15.4 for large-scale WSN applications. The authors established models for the device energy consumptions and goodput in beaconed cluster tree topologies. The results allow to choose two key 802.15.4 superframe parameters for increasing the network performance and energy efficiency.

An analysis of the nonbeacon-enabled channel access was presented by Latré et al. [5] who determined maximum throughput and minimum delay for a single connection. The authors showed that the small 802.15.4 packet sizes together with the large protocol overhead significantly decreases the bandwidth efficiency. The nonbeaconed channel access has also been studied by Kim et al. [6]. Using a Markov chain model, the authors proved that the number of devices in a star topology can be optimized in order to guarantee specific QoS conditions.

As the nonbeacon mode offers only few adaptable parameters, energy efficient optimizations for the nonbeacon mode are in general done by extending or modifying the 802.15.4 protocol stack. SCP [7], an example of a WSN-MAC protocol, runs on 802.15.4 compliant radios, but replaces the CSMA-CA channel access by a synchronized adaptive channel polling scheme. Ye et al. showed that this enables sensor node duty cycles below 0.1% and enables multi-hop streaming with only small end-to-end delay. Recently, we proposed a self-organizing sleep scheduling solution operating above the 802.15.4 MAC [8]. By loosely synchronizing with their neighbors, low duty cycled sensor nodes achieve good packet delivery ratios under a distributed routing scheme.

None of the above mentioned studies considers the early stages of the network life, but all works concentrate on the situation where the network is already fully operational. One of the first performance evaluations of 802.15.4 by Zheng and Lee [9] is one of the rare works covering the initial network stage. Among many other aspects, the authors studied the association procedure for the case of beaconed networks and proposed an optimization. They used ns-2 for their simulation studies and their code is still the base of the actual WPAN ns-2.33 simulation framework which we adapted for our purposes. Recently, Zhang et al. [10] showed how to speed up the association procedure in beaconed networks. For this purpose, the number of sent primitives and thereby the collision probability of command messages were reduced. This approach is only suitable for beaconed star topologies, as sending less primitives causes conflicts to occur in networks with several FFDs.

According to Zheng and Lee, an efficient association procedure is the basis for an efficient routing tree establishment [9]. This idea was extended by Cuomo et al. [11] who exploit the parent-child relationship resulting from the association procedure for establishing a routing tree rooted in the PAN coordinator. As no additional overhead for initial route establishment is required, HERA, as this algorithm is called, outperforms AODV in simulations in terms of packet loss, delay and energy consumptions. The association procedure was not considered in this study, but of course heavily influences HERA's performance.

## 3 The Nonbeacon 802.15.4 Association Procedure

In this section, we review the key facts of the nonbeacon association mechanism as defined in [1], before we introduce our extensions for low-power WSNs.

### 3.1 Basic Mechanism

We consider 802.15.4 devices working in the unlicensed 2.45 GHz frequency band in the following. In this band, a starting PAN coordinator chooses one out of 16 available channels to operate the PAN it is going to coordinate. A node $x$ willing to *associate* to this PAN executes an active channel scan, i.e. it broadcasts a beacon request to all available channels. If the PAN coordinator or an already associated FFD receive such a request, they answer by sending a beacon. After having sent the beacon request, $x$ waits $[0.015 \cdot (2^{ScanDuration}+1)$ sec for a response [1]. As no default value for $0 \leq ScanDuration \leq 14$ is given, we adopt Zheng and Lee's proposal of $ScanDuration = 4$, resulting in 0.26 sec per channel [9]. We further apply the low-power optimization proposed by [9] to scan only 3 channels. Together with processing times and state transitions, a node spends slightly more than $t_{scan} = 0.78$ sec for a channel scan.

After the scan, $x$ sends an association request command message to one of the devices from which it received a beacon. [1] does not specify which out of several beacon senders to choose, we therefore follow again [9] and let $x$ send the association request to the quickest respondent $y$. After having received the association request, $y$ immediately sends back an acknowledgment, which $x$ answers by a data request command after $aResponseWaitTime = 0.49$ sec. Meanwhile, $y$ has to check, whether $x$ may associate. If yes, $y$ sends an acknowledgement and an association response command frame back to $x$ which in turn is acknowledged. Conditions for rejection are not mentioned in the standard [1], in our study, nodes are therefore always allowed to associate.

### 3.2 Extension for Low-Power WSNs

The 802.15.4 standard is optimized for star topologies, and proposes to realize larger deployments by using several PAN coordinators. Thus, the case where $x$ receives no beacon in response to its channel scan is not supposed to happen and thus not handled. For large low-power WSN deployments with only one PAN coordinator, it may in contrast occur quite frequently that a node receives no answer to its channel scan and fails to associate at first attempt. Possible reasons are that the PAN coordinator is out of reach, already associated FFDs are in sleep mode or packets are lost. Zheng and Lee [9] identified this problem and proposed that each node failing to associate should retry to associate $a = 1$ sec later. They proved the effectiveness of this mechanism for large beacon-enabled topologies where nodes are always active.

Our studies showed, that if nodes are periodically active and inactive, this solution may cause sensor nodes to try and retry to associate during their neighbors inactivity phases. We investigate whether for this scenario a randomization of the *association retry interval* $a$ is beneficial and examine the impact of $a$'s length on the association procedure performance. In addition, we propose two extensions for the node behavior: The *greedy* behavior causes a sensor node to immediately retry to associate after an unsuccessful channel scan. The *altruistic* behavior causes a successfully associated node to listen to the channel some time before starting its regular sleep-wake cycle. If the node receives any beacon requests, it starts to sleep as soon as it has completed all associations it participated in, otherwise the node starts to sleep after a period $t_{alt}$.

## 4 Methodology

Many factors influence the performance of a WSN and especially the association procedure in nonbeacon-enabled 802.15.4 WSNs. To investigate the performance of the association procedure under varying environmental conditions and design choices are abstracted by a set of factors which is discussed in Section 4.1. To evaluate the performance of the association process, we use the metrics introduced in Section 4.2. Details on the used energy model are given in Section 4.3.

### 4.1 Considered Factors

We aim at improving the performance of a given 802.15.4 WSN deployment by adapting the association procedure. We abstract this problem by introducing a set of *hard* factors describing the application requirements and environmental conditions which can not or only hardly be changed. The degree of freedoms for tuning the association procedure are represented by a set of *soft* factors.

#### Hard Factors

- *Node deployment (dep)*: Describes the strategy by which the positions of the sensor nodes are chosen.
- *Average node degree (d)*: The number of other nodes each sensor node may on average communicate with is determined by the transceiver transmission output power and the density of the node deployment.
- *Coefficient of variation of node activation time (c)*: Each sensor node is activated a randomly distributed time $t_0$ after the PAN coordinator has been switched on. Different values of $c$ model different starting behaviors.
- *Network clock (T)*: Determines the responsiveness of the WSN.
- *Activity factor (α)*: An energy efficient operation is abstracted by assuming that all sensor nodes are active $\alpha T$ and sleep for $(1-\alpha)T$, where $0 < \alpha \leq 1$.
- *Initial tolerance time (Δ)*: Determines when the WSN has to be functional.

#### Soft Factors

- *Association retry interval (a)*
- *Randomization of a (rand)*
- *Greedy association (gr)*
- *Altruistic association (alt)*

### 4.2 Performance Metrics

For their study of the association process, Zheng and Lee proposed two metrics: the *successful association rate* giving the percentage of devices which succeeded to associate and the *association efficiency*, indicating how much attempts the devices had to make for associating [9]. We extend this framework and use the following three network level metrics to evaluate the association process performance for a certain factor value combination:

**Association Success.** $s_A = 1$ if all nodes of the PAN could associate during $\Delta$ and 0 otherwise.

**Association Time.** $t_A$ is the time until the last node of the PAN has associated. If not all nodes could associate, $t_A = \Delta$.

**Association Power Consumptions.** $E_A$ denotes the energy consumptions of the node of the PAN which required the largest amount of energy for associating. $E_A$ does not include the energy consumptions required for sending out association responses, but focuses on the nodes trying to associate.

### 4.3 Energy Model

An exact calculation of $E_A$ has to include characteristics on all mote subsystems. As our study focuses on reducing the communication related energy consumptions, we aim only at estimating the energy consumptions of the transceiver. For this purpose, we use the state machine model proposed by Wang and Yang [12], who extracted values for current consumptions and transition times from transceiver data sheets and assumed a typical voltage of $U = 1.8\,\text{V}$. Experiments with the 802.15.4-compliant CC2420 [13] showed that this approach together with an exact model of the communication behavior closely estimates the current consumptions. Transition power consumptions are calculated as the average of the initial and final state power consumptions [12].

One component of the energy node $x$ requires for associating, $E_A(x)$, is $E_{start}$ which is consumed when the node initially activates its transceiver, i.e. for the transition from "Power Off" to "Receive" (RX) state. Next, $E_{scan}$ is consumed for each of the $n \geq 1$ channel scans. If the association procedure is not greedy, during each of the $n-1$ periods where $x$ waits for retrying to associate, $E_R$ is consumed. $E_R$ is obtained by adding the energy consumptions for transitions from and to "Power Save" state to the energy consumptions in "Power Save" for the remaining time of $a$. For the association command message exchange, $E_a$ and the optional altruistic phase, $E_{alt}$ are required. $E_{scan}$, $E_a$ and $E_{alt}$ are estimated by multiplying the energy current consumptions in RX state by the time required for these actions and the voltage $U$. As CC2420 consumes slightly more energy for receiving than for transmitting at the highest output power [13], (which we assume to be the case in our study), this estimation is close to the real consumptions and sufficient for the purposes of this study.

## 5 Simulation Results

To examine the effect of our association procedure extensions, we use the ns-2.33 802.15.4 stack and extend the association mechanism as discussed in Section 3.2. The impact of all factors besides the node deployment is examined using a $2^k$ factorial design study in Section 5.1. The obtained results allow to identify beneficial influence of the greedy and altruistic mechanism which we analyze more closely for different topologies in Section 5.2.

### 5.1 A $2^7$ Factorial Design Study

Our simulations confirmed the evident fact that the topology has the strongest influence on the performance of the association procedure. To examine the other factors' influences, we consider a simple topology where 9 sensor nodes and one PAN coordinator are arranged on a line, with the PAN coordinator at the left edge. $\Delta = 5$ min and $T = 1$ sec are used for this experiment, but the results are similar for other parameter choices. For each of the 7 remaining free factors, a high (+) and low (-) level as summarized in Table 1 are chosen.

**Table 1.** Considered experimental factors and their levels

| Factor | $d$ [nodes] | $c$ | $\alpha$ [%] | alt | gr | $a$ | rand |
|---|---|---|---|---|---|---|---|
| Level (-) | 1.8 | 0 | 1 | no | no | $0.1T$ | no |
| Level (+) | 6 | 1 | 25 | yes | yes | $5.1T$ | yes |

We model the node transceivers to correspond CC2420 transmitting at the highest (0 dBm) possible output power. The radio propagation is abstracted as a disc model, and the two-ray ground propagation model proposed by [9] is used. Together with CC2420's typical receiver sensitivity of -95 dBm [13] this causes each transmission within 20 m to succeed. We consider inter-node spacings of 20 and 5 m, each node can thus communicate with 1 or 4 neighbors per direction. Averaged over 10 nodes, this results thus in a low value of $d = 1.8$ and a high value of $d = 6$. Startup times distributed with $c = 0$ stands for all nodes starting at exactly the same time, $c = 1$ corresponds to an exponential distribution with the same mean. $\alpha$ reflects a good lower and an absolute upper bound for a low-power network duty cycle. Low and high values for the factors rand, gr and alt correspond to using these features or not. If the altruistic mechanism is used, we parameterize $t_{alt}$ to be slightly longer than the time required for a channel scan, $t_{alt} = t_{scan} \cdot 1.1 = 0.86$ sec. $a$ is chosen in dependence of the system clock and varies between a very small and a very large value. To randomize $a$, we multiply it by a random variable $X$ which is uniformly distributed in the interval (0,1]. The effects of using other distributions for $X$ have not been considered in this study. Other choices for the levels are imaginable, but our experiments showed that using other level values just slightly changes the results.

To get an overview on the factors' influence, we examine the performance of all $2^7$ factor combinations or *design points*. For each design point, the performance of the association mechanism is evaluated by the following simulation run: at time 0, the PAN coordinator is activated and begins starting a PAN. The PAN coordinator is always active during the simulation run [1]. Each of the sensor nodes is activated at a randomly distributed time $t_0$ with mean 30 seconds and coefficient of variation as specified by $c$. After the activation, each node tries to associate as described in Section 3.1 using the extensions introduced in Section 3.2 with the parameters set for this design point. After having successfully associated and after an optional altruistic period, the node starts to sleep regularly for $(1 - \alpha)T$. At the end of each simulation run, the *system responses*, i.e. the metrics $s_A, t_A$ and $E_A$ defined in Section 4.2 are collected. To obtain statistically significant results, the responses for each design point are averaged over 100 simulation runs.

**Main Effects.** The influence of factor $x$ on the system performance in terms of metric $y$ is characterized by its *main effect*, $e_x(y)$. It is obtained as the average change in $y$ due to moving $x$ from (-) to (+) while keeping all other factors fixed:

$$e_x(y) = \frac{\bar{y}_{x^+} - \bar{y}_{x^-}}{2}, \tag{1}$$

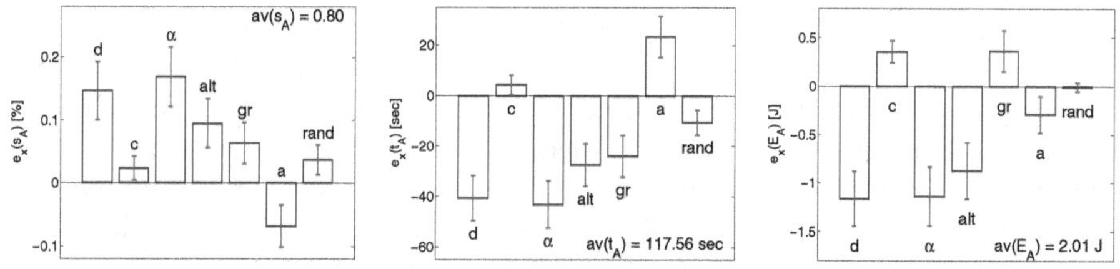


(a) $s_A$ (b) $t_A$ (c) $E_A$

**Fig. 1.** Main effects of the considered factors on the different metrics

where $\bar{y}_{x^+}$ and $\bar{y}_{x^-}$ denote $y$ averaged over all design points where x is at its high level and low level respectively [14].

In Fig. 1 we visualize the main effects $e_x$ of the 7 different considered factors. From left to right, the subfigures contain representations for the association success, the association time and the association energy consumptions of the network. Together with the main effects, their 95%-confidence intervals are shown to verify, that only some of the effects are *not* statistically significant, as their confidence intervals include 0. Additionally, we show the corresponding system response averaged over all $2^7$ design points in each figure.

Let's discuss the average system responses first. The association success $s_A$ is the probability that all nodes associate within $\Delta$. As $s_A = 0$, if all but one node were able to associate, $s_A$ averaged over all design points and 100 runs is only 0.8. The probability that an individual node associates, is slightly higher and averages to 0.88. Obviously both probabilities converge to 1 for larger values of $\Delta$. Note that the association energy consumptions for this topology averaged over all design points and runs are equal to 2 J. This corresponds to less than 0.01% of the theoretical capacity of two AA batteries [15] and seems to be neglectable when compared to the energy consumptions during the WSN lifetime. However, already in this simple topology the minimal and maximal $E_A$ of 0.2 J and 10.4 J differ strongly, the inherent energy saving potential of the association phase must thus not be underestimated.

The main effects of the factors shown in Fig. 1 illustrate that some factors have a stronger influence than others and that the factor impact differs between the metrics. A closer look reveals, that the hard parameters $d$ and $\alpha$ have a larger influence then all other parameters: the association procedure will succeed with high probability, rather quick and at low energy costs in dense and high duty cycled topologies. While this is quite evident, the influence of a randomized startup point characterized by $c$'s main effect is ambiguous. $c > 0$ is suitable for increasing the association success as it decreases the probability of packet collisions especially in dense networks, but results in slightly increased time and energy consumptions.

As the hard parameters can in general not be influenced, the analysis of the soft parameters is more profitable for optimization purposes. The randomization of the association retry interval and to a stronger degree the altruistic mechanism are the only soft parameters which have a positive influence on all three considered metrics. The energy saving potential of the altruistic mechanism is quite large as the energy savings of nodes which associate earlier outweigh the energy consumptions of nodes altruistically delaying their transition to power save mode. The greedy mechanism and a small association retry interval increase the association success and the association speed but lead to increased energy consumptions by triggering too many unnecessary channel scans.

**Interactions.** Considering just the main effects of the factors for deciding the parametrization of the association mechanism would lead to wrong decisions, if interactions between factors are existing, i.e. if the influence of one factor varies in dependence of the behavior of other factors. To examine this, we consider the *two-way interaction effect* of factors $x$ and $z$ on systems response $y$, $e_{xy}(y)$, which is computed as the average difference of $x$'s effect when $z$ is at its (+) and the effect of $x$ when $z$ is at its (-) level [14]:

$$e_{xz}(y) = \frac{1}{2}((\bar{y}_{x^+z^+} - \bar{y}_{x^-z^+}) - (\bar{y}_{x^+z^-} - \bar{y}_{x^-z^-})), \tag{2}$$

where analogous to Eq. (1), $\bar{y}_{x^+z^+}$ denotes $y$ averaged over all design points where both $x$ and $z$ are at their (+) levels.

The interactions of all considered responses are similar, we therefore only illustrate the interactions on the energy consumptions in Fig. 2. In the subfigure which is at *row* $z$ and *column* $x$ of the figure matrix, a solid line is drawn between the average system response when $x$ is at its (-) and (+) level, while $z$ is at is (-) level, $\bar{y}_{x^-z^-}$ and $\bar{y}_{x^+z^-}$, while a dashed line connects the average system response when $x$ is at its (-) and (+) level, while $z$ is at is (+) level, $\bar{y}_{x^-z^+}$ and $\bar{y}_{x^+z^+}$.

Non-parallel lines represent interactions between two factors, it gets thus quickly clear, that nearly all factors are interacting. Fig. 2 illustrates that factors with strong main effects also have strong interactions and that factor levels have to be adapted under the consideration of other factors levels. Often a factor's effect is amplified or reduced by another factor's level, as it is e.g. the case for all soft parameters which have a less stronger effect, if the activity is high and the network is dense. All in all, the analysis of interactions demonstrates, that any deeper analysis of the association mechanism has to be done considering different network configurations and that sparse and low-duty cycled network designs need to be optimized more carefully.

### 5.2 Combining the Greedy and the Altruistic Mechanism

In the last section we demonstrated, that out of the soft factors, the greedy and the altruistic mechanisms have the strongest effects on the association success and speed. We saw also, that the altruistic mechanism reduces the increased energy consumptions of the greedy mechanism (c.f Fig. 2). In this section we verify if this holds also for larger topologies and investigate to what degree the association procedure in a given sparse low-power WSN deployment can be improved using both mechanisms.

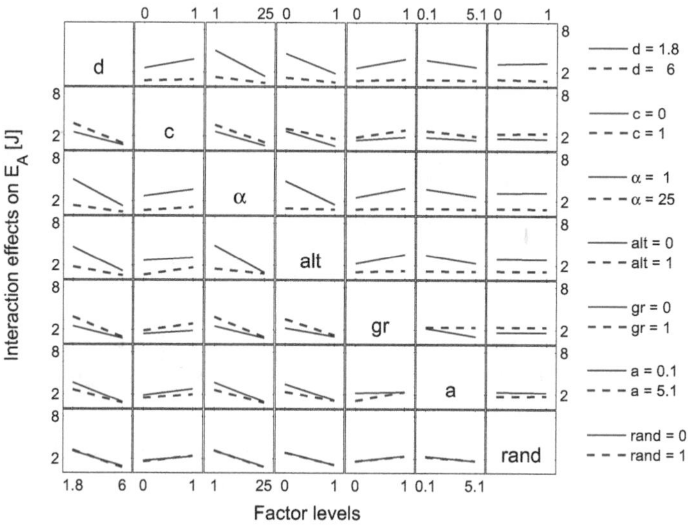


**Fig. 2.** Interaction effects on the association energy consumptions

We consider an exemplary monitoring application, where an area of $60 \times 60$ m is covered with 24 sensor nodes and one sink node acting as PAN coordinator. Small energy consumptions together with acceptable responsiveness are abstracted by $\alpha = 1\%$ and $T = 1$ sec. To examine the influence of the node deployment and the sink position, we use randomly generated topologies: A connected graph with 25 nodes results either from chosing positions on a grid with 12 m in between or uniformly distributing nodes over the square to cover. In both cases, one random position is chosen to be the position of the sink. Averaging the mean node degree over 100 different random and grid topologies results in $d = 6$ and $d = 3.2$ respectively. To analyze the benefits of the proposed mechanisms, we do not randomize the associatione retry interval and set $a = 0.5T$. Different possibilities for activating the sensor nodes are abstracted by choosing $t_0$ as randomly distributed with coefficient of variation $c \in \{0, 1e^{-3}, 1e^{-2}, 1e^{-1}, 0.577, 1, 2\}$ and mean 30 sec. For a closer examination of the altruistic mechanism, values for $t_{alt} = \{0, t_{scan}, t_{scan} + T, t_{scan} + 2T, t_{scan} + 5T, t_{scan} + 10T\}$ are used.

In Fig. 3 we depict the association success and energy consumptions resulting from $\Delta = 5$ min and a strongly varying node activation time, i.e. $c = 2$. Results averaged over 100 runs with their 95% confidence intervals are shown which are quite large due to the high variance of $t_0$. Results obtained with the greedy mechanism are shown by shaded bars. The six bars of each of the four groups depict the considered values of $t_{alt}$ in increasing order. Furthermore, we show results for the random (the two left bar groups in each subfigure) and the grid topologies (the two right bar groups). This illustrates, that due to the smaller average degree most effects are stronger in the grid than in the random topologies and demonstrates again that sparser topologies need more optimization effort.

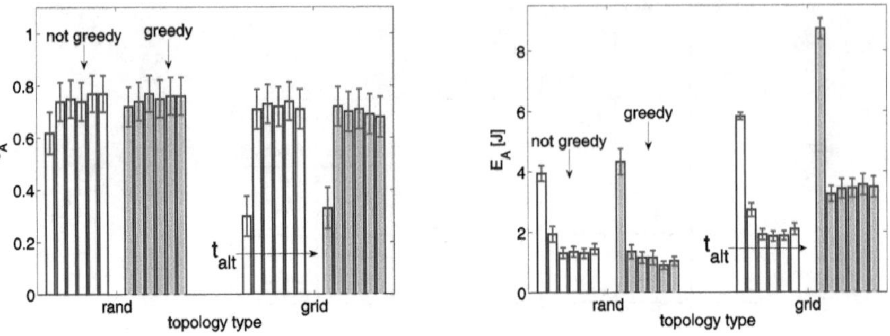


(a) $s_A$ (b) $E_A$

**Fig. 3.** Effects of the altruistic and greedy mechanism for a high startup variation

As the node activation time is varying very strongly, many nodes are starting close to $\Delta$, causing the association success to be small, especially in the sparser grid topologies. Both mechanisms are able to increase the association success and decrease the association energy consumptions but to a different degree: The clear difference between the size of the first bar in each group ($t_{alt} = 0$) and all other bars demonstrates, that the altruistic mechanisms has a stronger positive influence on both metrics than the greedy mechanism (non shaded vs. shaded bars). Combining the greedy and the altruistic mechanisms positively influence the system performance, as the energy penalties resulting from the use of the greedy mechanism are alleviated while the time and energy reductions of both mechanisms also result from their combined use.

According to the results illustrated in Fig. 3, the exact parametrization of $t_{alt}$ is affecting the system performance differently if the greedy mechanism is used. The association success is directly increasing with $t_{alt}$ we thus analyze the more interesting trade-off between association speed and energy consumptions in dependence of the startup variation. For this purpose, $\Delta$ is set to 10 minutes which allows all nodes in the considered topologies to associate in nearly all cases. Results from this experiment for both topology types were similar, but stronger for the

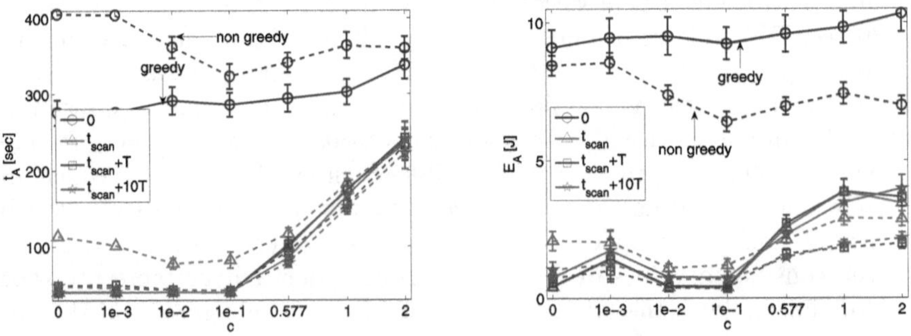


(a) $t_A$ (b) $E_A$

**Fig. 4.** Altruistic and greedy mechanism under varying starting time variations grid topologies. The association time and energy consumptions averaged over 100 simulation runs together with their 95% confidence intervals are plotted against $c$ in Fig. 4. Different values of $t_{alt}$ are distinguished by different markers. The greedy and the non greedy behavior are shown with solid and dashed lines respectively.

At first glance it gets clear, that $t_{alt} = 0$, i.e. a non altruistic behavior, leads to the slowest association and highest energy consumptions. The altruistic mechanism is improving the performance more than the greedy mechanism. The benefits of the latter are moreover strongly depending on the startup variation: While a greedy behavior together with a small $t_{alt} > 0$ increases the association speed and decreases the association energy consumptions in networks with small startup variation, this effect is inversed for larger values of $c$. This demonstrates, that for the exact parametrization of $t_{alt}$ a balanced solution between association speed and energy consumptions has to be found: the association speed always increases with $t_{alt}$, but if $t_{alt} > t_{scan}$ is chosen *too* large, too much energy is wasted by unnecessarily listening to the channel. All in all, Fig. 4 demonstrates that in the scenario we investigated, adapting the usage of the greedy and altruistic mechanism to the WSN deployment enables accelerations over 300 sec and energy savings over 5 J. Comparable numbers will result from topologies of the same size and will be larger for larger deployments.

## 6 Conclusion and Outlook

In this paper we examined possibilities for optimizing the association process in a low-power 802.15.4 nonbeacon sensor network. For this purpose, we proposed four enhancements for the procedure as described in the standard and carried out an extensive simulation study to identify the effects of our improvements. Our results showed, that the influence of factors given by the application requirements is always larger than the improvements achieved by tuning the association mechanism. However, significant performance gains in terms of association probability, speed and energy consumptions may be achieved, if the additional mechanisms we proposed are well parameterized. The altruistic and the greedy behavior are especially promising, as they allow to increase the performance of the association procedure by adapting it to the network characteristics.

All in all, our results illustrate the inherent optimization potentials of the often neglected 802.15.4 association procedure. A successful starting phase is the foundation of each successful WSN deployment and especially of 802.15.4 WSNs, our future works will therefore be dedicated to further optimizing the association procedure. As the startup behavior of the nodes is strongly influencing the network performance, we will also study optimization possibilities for the node activation strategy.

## Acknowledgements

The authors would like to thank Phuoc Tran-Gia and Dirk Staehle for valuable comments and fruitful discussions. Special thanks go to Sebastian Deschner for his help during the simulation process.

## References

1. IEEE Computer Society: IEEE Standard for Information technology – Part 15.4: Wireless MAC and PHY Specifications for WPANs (September 2006)
2. ZigBee Alliance: ZigBee Specification (January 2008)
3. HART Communication Foundation: WirelessHART Technical Data Sheet (May 2007)
4. Kohvakka, M., Kuorilehto, M., Hännikäinen, M., Hämälinen, T.D.: Performance analysis of IEEE 802.15.4 and ZigBee for large-scale wireless sensor network applications. In: PE-WASUN 2006 (2006)
5. Latré, B., Mil, P.D., Moerman, I., Dhoedt, B., Demeester, P.: Throughput and Delay Analysis of Unslotted IEEE 802.15.4. Journal of Networks 1(1) (May 2006)
6. Kim, T.O., Kim, H., Lee, J., Park, J.S., Choi, B.D.: Performance Analysis of IEEE 802.15.4 with Non-beacon-enabled CSMA/CA in Non-saturated Condition. In: Sha, E., Han, S.-K., Xu, C.-Z., Kim, M.-H., Yang, L.T., Xiao, B. (eds.) EUC 2006. LNCS, vol. 4096, pp. 884–893. Springer, Heidelberg (2006)
7. Ye, W., Silva, F., Heidemann, J.: Ultra-Low Duty Cycle MAC with Scheduled Channel Polling. In: SenSys 2006 (2006)
8. Staehle, B., Hoßfeld, T., Vicari, N., Kuhnert, M.: A Cross-Layer Approach for Enabling Low Duty Cycled ZigBee Mesh Sensor Networks. In: International Symposium on Wireless Pervasive Computing 2008, Santorini, Greece (May 2008)
9. Zheng, J., Lee, M.J.: A Comprehensive Performance Study of IEEE 802.15.4. In: Sensor Network Operations, pp. 218–237. IEEE Press, Los Alamitos (2004)
10. Zhang, F., Wang, F., Dai, B., Li, Y.: Performance Evaluation of IEEE 802.15.4 Beacon-Enabled Association Process. In: AINAW 2008 (2008)
11. Cuomo, F., Luna, S.D., Monaco, U., Melodia, T.: Routing in ZigBee: Benefits from Exploiting the IEEE 802.15.4 Association Tree. In: IEEE ICC 2007 (2007)
12. Wang, Q., Yang, W.: Energy Consumption Model for Power Management in Wireless Sensor Networks. In: SECON 2007 (2007)
13. Texas Instruments: 2.4 GHz IEEE 802.15.4 / ZigBee-ready RF Transceiver. Texas Instruments (2006)
14. Montgomery, D.C.: Design and Analysis of Experiments, 6th edn. John Wiley and Sons, Chichester (2005)
15. Energizer: Alkaline Technical Information (2008), `http://data.energizer.com`

# Impact of Misbehaviour on QoS in Wireless Mesh Networks

Szymon Szott[1], Marek Natkaniec[1], and Albert Banchs[2]

[1] AGH University of Science and Technology, Krakow, Poland
{szott,natkanie}@kt.agh.edu.pl
[2] Universidad Carlos III de Madrid, Madrid, Spain
banchs@it.uc3m.es

**Abstract.** This paper analyzes the impact of misbehaviour on QoS provisioning in wireless mesh networks. Misbehaviour occurs when a network participant decides not to cooperate. Since cooperation is fundamental for distributed environments such as mesh networks, misbehaviour can be a serious threat to them. In this work, the authors focus on the IEEE 802.11 EDCA medium access function which provides QoS in mesh networks. Simulation studies have been performed to determine what realistic forms of misbehaviour can occur and what their impact is. From these results the most beneficial forms of MAC layer misbehaviour in multihop mesh networks are derived.

**Keywords:** Mesh networks, QoS, IEEE 802.11, EDCA, misbehaviour.

## 1 Introduction

Wireless mesh networks are steadily becoming a popular approach for providing network access to people's homes, especially in suburban and rural environments. Mesh networks allow a neighbourhood to share a single Internet connection, thus solving the last mile problem. They can also bring a community together by enabling easy and reliable data exchange within the network. By utilizing the latest technology, multimedia content can be exchanged over these networks.

Fig. 1 presents an aerial view of a mesh network. Each house in this neighbourhood has a wireless router, also called a Mesh Node (MN). These MNs form a backbone mesh network to provide robust connectivity. A mesh network can therefore be thought of as an immobile ad-hoc network. One of the MNs in the figure has a connection to the Internet and serves as a gateway for the other MNs. The MNs provide network access in each home. Wireless Access Points (APs) can be attached to the MNs to provide wireless access to household devices such as laptops, PDAs, tablet PCs, etc. The MN together with the AP is called the Mesh Point (MP). Stationary PCs can be directly connected to the MNs through Ethernet links.

The IEEE 802.11 standard [1] can provide wireless connectivity throughout the mesh network. It is currently the best choice when building a mesh network, because 802.11 equipment has become popular, cheap, reliable, and secure. The MNs in the network can communicate with each other using the 5 GHz frequency band and the

L. Fratta et al. (Eds.): NETWORKING 2009, LNCS 5550, pp. 639–650, 2009.

user devices can connect with the APs using the 2.4 GHz frequency band. This makes the community-wide mesh part of the network separate from the wireless network in each household. The Enhanced Distributed Channel Access (EDCA) function ensures Quality of Service (QoS) at the Medium Access Control (MAC) layer and facilitates the exchange of multimedia content over the network. It provides traffic prioritization with four Access Categories (ACs) to provide appropriate QoS. These categories are, from the highest priority: Voice (Vo), Video (Vi), Best effort (BE), and Background (BK). In the upcoming 802.11 standard for mesh topologies – 802.11s [2] – EDCA is included as a mechanism for providing QoS. Therefore, EDCA is the main focus of the research presented in this paper.

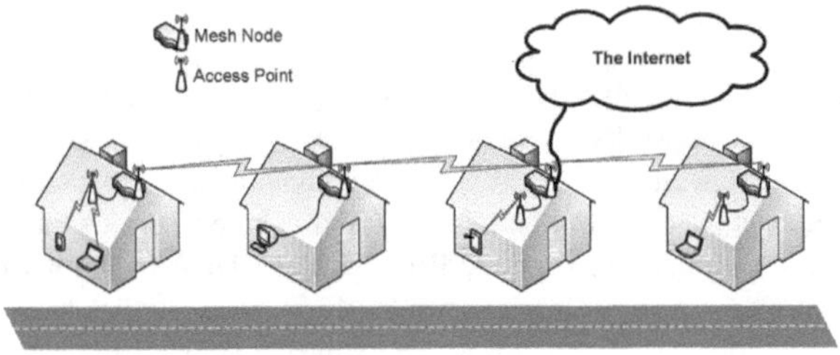


**Fig. 1.** Mesh network

Mesh networks rely on the cooperation of all participants. A problem arises if one of the participants misbehaves (i.e., decides not to cooperate with others). A mesh node may decide to misbehave in order to gain certain measurable profits (such as higher throughput). Misbehaviour is always done at the cost of the well-behaved nodes in the network. Therefore, it would be favourable if such actions were at least discouraged, if not made impossible.

Misbehaviour is a threat to networks built with the 802.11 standard because it provides no incentives to cooperate. Medium access in 802.11 is based on CSMA/CA (Carrier Sense Multiple Access with Collision Avoidance) and a set of pre-defined parameters. In EDCA, each AC has its own set of parameters: AIFS (Arbitration InterFrame Space), CWmin and CWmax, and TXOP (Transmission Opportunity) (Table 1).

**Table 1.** Values of EDCA Parameters

| AC | AIFS | CWmin | CWmax | TXOP [μs] |
|---|---|---|---|---|
| Voice | 2 | 7 | 15 | 3264 |
| Video | 2 | 15 | 31 | 6016 |
| Best effort | 3 | 31 | 1023 | 0 |
| Background | 7 | 31 | 1023 | 0 |

Any user can change these parameters to his/her own advantage. This can be done very easily with the use of the latest wireless drivers [3]. With these modifications, users can, for example, achieve better network access than their neighbours. Likewise, a vendor of wireless cards might decide on using non-standard parameters to achieve better performance. This makes misbehaviour a real threat to mesh networks. This problem has already been the subject of recent studies regarding cooperative environments such as mobile ad-hoc networks (Section 2). However, no research has been performed on the topic of providing QoS in misbehaviour-prone mesh networks.

Section 3 provides simulation results which determine the impact of misbehaviour on QoS provisioning in a multi-hop mesh environment. The focus of this work is on realistic misbehaviour, i.e., actions which are easy to perform and beneficial to the malicious user. The simulations consider modifying MAC layer parameters to either upgrade one's own traffic or to downgrade the traffic of others. These simulations show how beneficial different types of misbehaviour actually are. Finally, Section 4 concludes the paper and describes future work.

## 2 State of the Art

The problem of misbehaviour, especially in the context of mobile ad-hoc networks, has been the subject of study for the last several years. The first approaches to detecting misbehaviour were focused on the problem of not forwarding packets. Such actions are done at the IP layer and can be performed with the use of a firewall. The first benefit is that the misbehaving node has more bandwidth for its own traffic. Secondly, in the case of mobile nodes, it can extend its battery life.

The first solution to not forwarding packets was presented in [4] and later independently developed into CONFIDANT [5] and CORE [6]. This family of solutions is based on promiscuous observation of events in the network. Many types of misbehaviour can be detected, not only packets which are not forwarded, but also packet manipulation. Statistical algorithms are used to calculate a level of reputation for each node, which in turn determines cooperation. Misbehaving nodes (those with a low reputation) are gradually isolated from the network and thus such actions are discouraged.

The authors of [7] deal with the problem of MAC layer misbehaviour. They take into account several misbehaviour strategies, all dealing with manipulating the parameters of the contention window mechanism of 802.11. In their solution, it is the receiver, not the sender, which chooses the random backoff value. This value is transferred to the sender in either a CTS or ACK frame. Misbehaviour occurs when the sender deviates from that backoff.

Paper [8] presents DOMINO, an advanced software application designed to protect hotspots from greedy users. It monitors traffic, collects traces and analyzes them to find anomalies. DOMINO can detect many types of malicious and greedy behaviour, including backoff manipulation techniques. Anomaly detection is based on throughput (instead of observed backoff), which the authors acknowledge is not an optimal detection metric. The application can be seamlessly integrated with APs and it complies with standards. Additionally, a misbehaviour detection analysis in infrastructure-mode 802.11 EDCA WLANs can be found in [9]. However, both DOMINO and [9] cannot be used in distributed environments such as ad-hoc and mesh networks.

The authors of [13] present a simulation-based technique for detecting faults in wireless mesh networks. They utilize traces from a network monitor to perform simulations. The cause of the network behaviour can be detected, whether it is MAC layer misbehaviour, link congestion, or packet dropping. This is an interesting approach, however, it is not real-time and it depends on inaccurate simulations.

To summarize, there are several problems with the research efforts presented in this section. First of all, most research has been focused on WLANs operating in infrastructure mode. This is quite different from ad-hoc and mesh scenarios most notably because of the central access point. Secondly, the state of the art in misbehaviour detection is often focused on unrealistic misbehaviour. Examples include packet manipulation, selective jamming and other techniques which require expert skills. Also "adaptive" misbehaviour is considered, which is quite difficult to implement in real life. Furthermore, EDCA, with its four distinct sets of parameters, has not been taken into account in mesh network scenarios. Finally, the detection solutions are most often limited to only one layer of the OSI model (either Data Link or Network).

## 3 Analysis and Evaluation of Misbehaving Nodes

This section presents the results of an extensive simulation study of misbehaviour in mesh networks. The purpose of this analysis is to determine the impact that misbehaving users can have on QoS provisioning in such networks. All simulations were performed using the ns-2.28 simulator with a modified version of the TKN EDCA extension [10]. All the figures in this section present curves, where the error of each simulation point for a 95% confidence interval does not exceed 2% (this is too small for graphical representation).

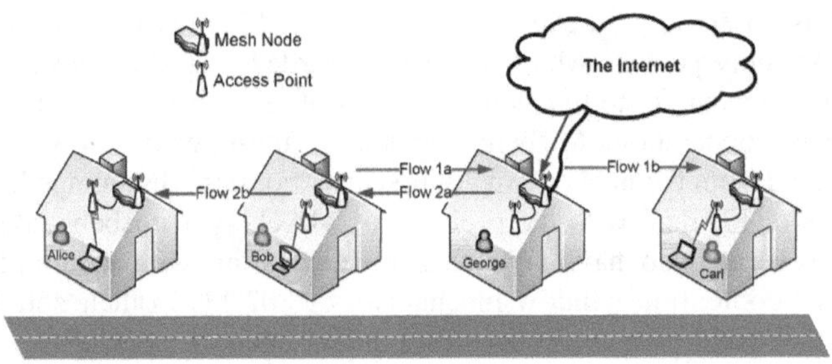


**Fig. 2.** Mesh network scenario

The simulated network topology is presented in Fig. 2. Each MN uses the EDCA function and is within range of its closest neighbour only. George's MN is a gateway to the Internet, Bob is sending a file to his friend Carl (Flow 1), and Alice is watching a video stream from the Internet (Flow 2). We can assume that UDP is used if Alice's transmission is real-time and TCP is used otherwise. Her traffic uses the highest priority (Vo) to ensure high quality of the video stream. If Bob uses a lower priority (BE) for his file transfer, the EDCA function will ensure that Alice's video stream is uninterrupted by Bob's file transfer. This is shown in the reference case (case A) in section 3.2. However, since Bob is in the path of Alice's traffic, he can misbehave by altering his medium access parameters. He can either simply degrade Alice's traffic (section 3.3) or combine this with promoting his own traffic (section 3.4). The question is: can such actions be beneficial for Bob? The answer is provided in section 3.5 which gives conclusions derived from the results of the simulations.

Since there is no impact of (and therefore no gain from) misbehaviour in non-saturated networks [11], we ensure that the simulated network is saturated. We evaluate the saturation throughput for the given topology in section 3.1. In saturation, the traffic source may not be relevant, so CBR was chosen. The packet size was 1000 B. In fact, the size of the packet is not that important because we are analyzing the behaviour of traffic priorities (and not absolute network performance). The RTS/CTS mechanism was not used since only Bob's and George's MNs generate traffic and they are neither hidden from, nor exposed to each other. The data rate of the simulated network was 11 Mbit/s and AODV was used as the routing protocol. The size of the network is small, but for one misbehaving node it is enough to show how its actions will influence network performance.

### 3.1 Saturation Throughput

In order to determine the saturation throughput of the network, the following simulation study was performed. The offered load of Flow 1 (Bob's file transfer) and Flow 2 (Alice's video stream) increased simultaneously from 64 kb/s to 12 Mb/s. The default priority (BE) was used for both flows. Both UDP and TCP were considered as the transport protocols. The results are presented in Fig. 3, which shows the average flow throughput achieved as a function of offered load.

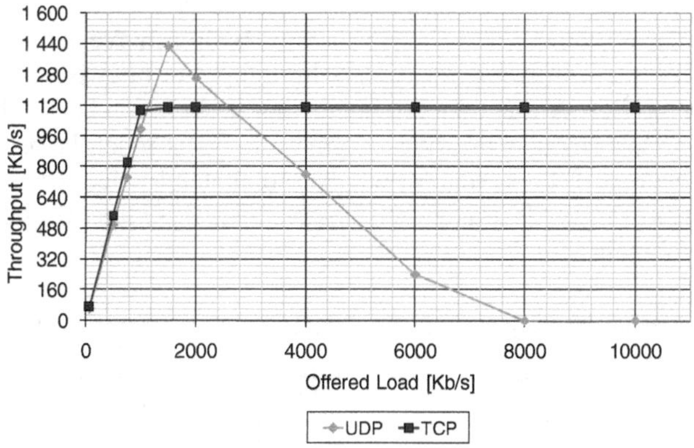


**Fig. 3.** Average flow throughput

For TCP the situation is clear – the saturation throughput is reached at approximately 1 Mb/s. This is the average end-to-end throughput of each flow. However, for UDP traffic, once a peak is reached, the throughput decreases to zero and congestion collapse occurs. This is because the interface queue present in the MAC layer of ns-2 uses the drop tail queue management algorithm. Bob's interface queue becomes completely filled with locally generated frames, leaving no room for frames that are to be forwarded. In real-life wireless cards such behaviour depends on the implementation. This does not occur for TCP traffic because this protocol adjusts its transmission speed using the additive increase/multiplicative-decrease algorithm. With respect to these results, an offered load of just over 2 Mb/s was chosen as the saturation throughput for this network scenario. In the following subsections, several different simulations were performed. Table 2 contains a brief description of all the considered cases.

**Table 2.** Descriptions of all cases

| Case | Description |
|---|---|
| A | Reference situation, no misbehaviour |
| B | Bob changes the Vo parameters in his router to resemble BK priority (simple misbehaviour) |
| C | Case B + CWmin of forwarded traffic is set to maximal value (1023) |
| D | Bob uses CWmin = CWmax = 1 and TXOP = 8160 μs for his traffic |
| E | Case D + Case B (simple misbehaviour, change of forwarded traffic priority) |
| F | Case E + CWmin of forwarded traffic set to maximal value (1023) |
| G | Case F + Bob uses AIFS = 1 |

## 3.2 Reference Case

Case A is the reference situation. Alice uses Vo priority, whereas Bob consecutively uses each of the four EDCA priorities for his file transfer. Table 3 shows the throughput results that both flows achieved in the first (Flow 1a, 2a) and the second (Flow 1b, 2b) hop. Fig. 4 presents the end-to-end throughput values for both flows. If Bob is using the same priority as Alice (i.e., Vo) they both achieve similar throughput. Otherwise, if Bob uses a lower priority, his throughput is likewise lower. This is in accordance with the EDCA function. An interesting observation is that the decrease in throughput when Bob changes priorities from Vo to Vi is much larger for TCP than UDP. The explanation of this is that Flow 1 had to contend twice for the medium and twice with a lower priority. TCP is more sensitive than UDP to congestion, especially in wireless environments.

**Table 3.** Per-hop throughput results for case A (in Kb/s)

| Flow 1 priority | UDP | | | | TCP | | | |
|---|---|---|---|---|---|---|---|---|
| | F1a | F1b | F2a | F2b | F1a | F1b | F2a | F2b |
| Vo | 1771 | 1364 | 1775 | 1363 | 1158 | 1096 | 1115 | 1055 |
| Vi | 1199 | 923 | 2111 | 2111 | 198 | 187 | 2039 | 1929 |
| BE | 1131 | 870 | 2111 | 2111 | 137 | 131 | 2095 | 1982 |
| BK | 1054 | 775 | 2111 | 2111 | 25 | 23 | 2207 | 2089 |

## 3.3 Downgrading Forwarded Traffic

In case B we assume that Bob runs a simple yet malicious script (perhaps found on the Internet) on his wireless router. This script changes the Vo parameters in his router to resemble BK priority. The priority of Alice's traffic is lowered but the frames

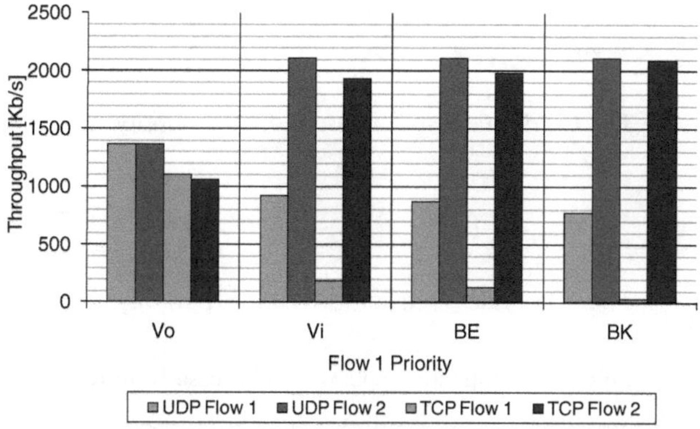


**Fig. 4.** End-to-end throughput results for case A

are not manipulated. Again, Alice uses Vo priority, whereas Bob consecutively uses each of the four EDCA priorities for his file transfer. The throughput results (Fig. 5) again reveal interesting observations. When Bob is using Vo priority he sends his traffic using his modified EDCA parameters. This means that on the first hop, his traffic is sent at BK priority, and then forwarded as Vo priority (Fig. 6). For Alice's traffic, the priorities are reversed (first hop with Vo, second with BK). Why is Bob's end-to-end throughput higher? If we look at the hop-by-hop UDP throughput for Vo priority (Table 4) we see a similar situation as before: 100% of Bob's traffic and only 33% of Alice's traffic is forwarded. Again, locally generated traffic wins with traffic that is to be forwarded. When Bob uses Vi or BE priority he achieves the throughput gain that he was expecting. This gain is obviously higher for Vi than for BE. When Bob's file transfer is using BK priority, another interesting situation occurs. The per-hop use of priorities is shown in Fig. 7. When UDP is used, Alice's flow has more throughput (because it first has Vo and then BK whereas Bob's flow always has BK). However for TCP this is not the case, even though both flows have about 95% of traffic forwarded. This seems to be a similar case to the one described in [12],

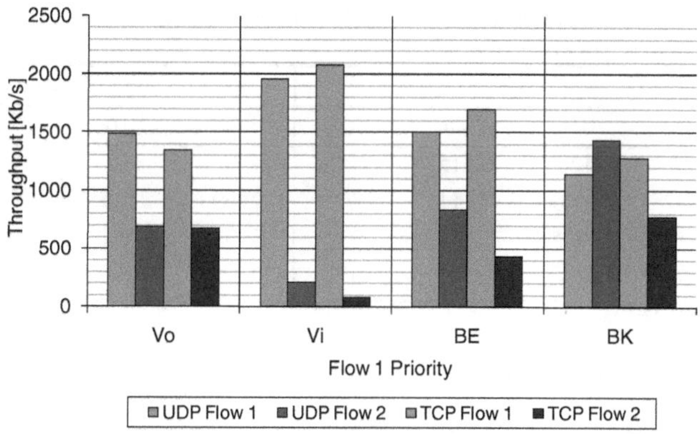


**Fig. 5.** End-to-end throughput results for case B

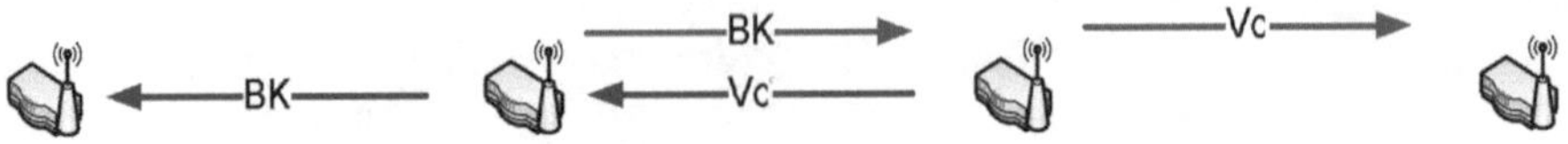


**Fig. 6.** Priorities used in Case B, Flow 1 priority: Vo

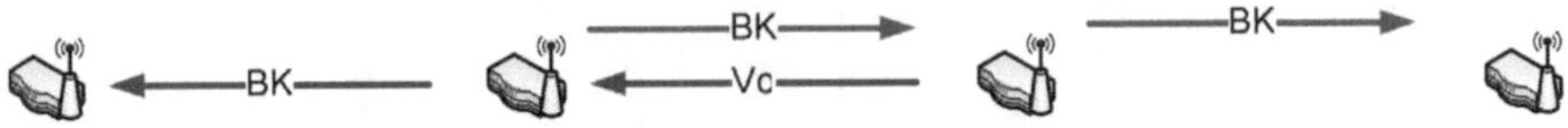


**Fig. 7.** Priorities used in Case B, Flow 1 priority: BK

**Table 4.** Per-hop throughput results for case B (in Kb/s)

| Flow 1 priority | UDP | | | | TCP | | | |
|---|---|---|---|---|---|---|---|---|
| | F1a | F1b | F2a | F2b | F1a | F1b | F2a | F2b |
| Vo | 1482 | 1482 | 2111 | 686 | 1420 | 1343 | 710 | 672 |
| Vi | 1962 | 1958 | 2111 | 201 | 2192 | 2074 | 82 | 78 |
| BE | 1616 | 1503 | 2111 | 828 | 1787 | 1691 | 450 | 425 |
| BK | 1180 | 1136 | 2111 | 1430 | 1352 | 1279 | 815 | 771 |

where it was shown that TCP may completely change throughput allocation independently of the EDCA configuration.

Case C is similar to the previous one: Bob again modifies the Vo parameters in his router. This time he increases the CWmin parameter to its maximum value (1023). Bob has now degraded the Vo priority almost as severely as possible using EDCA parameter modification. The results are presented in Fig. 8 and Table 5. When Bob uses the Vo priority for his traffic, the situation is similar to that in case B. However, in this case the throughput values are significantly lower because of the high CW parameters. For all other priorities (Vi, BE, and BK) it can be seen that misbehaviour brings meaningful gains. The fact that Bob's throughput is high even if he uses BK signifies the importance of the CW parameters on throughput.

**Table 5.** Per-hop throughput results for case C (in Kb/s)

| Flow 1 priority | UDP | | | | TCP | | | |
|---|---|---|---|---|---|---|---|---|
| | F1a | F1b | F2a | F2b | F1a | F1b | F2a | F2b |
| Vo | 428 | 428 | 2115 | 120 | 336 | 318 | 168 | 159 |
| Vi | 2052 | 2049 | 2111 | 7 | 2269 | 2147 | 3 | 2 |
| BE | 1917 | 1905 | 2111 | 34 | 2199 | 2081 | 19 | 18 |
| BK | 1775 | 1774 | 2111 | 57 | 2092 | 1980 | 34 | 32 |

### 3.4 Promoting Local Traffic

In section 3.3 (cases B and C) Bob was gaining throughput by degrading the traffic parameters of forwarded traffic. In the following cases (D to G) we assume that Bob

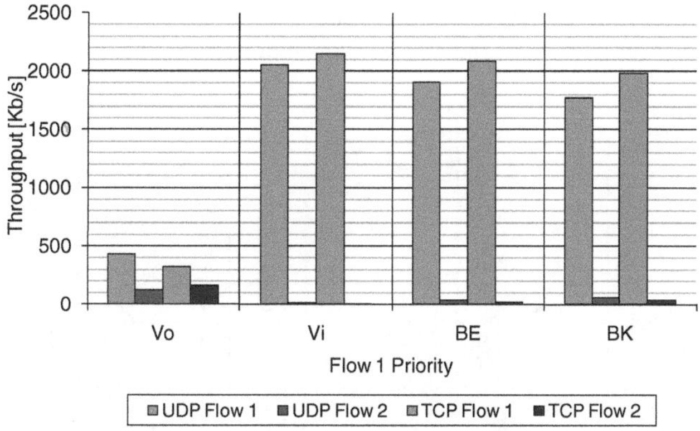


**Fig. 8.** End-to-end throughput results for case C

further manipulates EDCA parameters, this time in order to increase the medium access probability for his own traffic. In these cases Bob always uses the Vi priority for his file transfer. The results are presented in Table 6 and Fig. 9. In case D Bob uses the lowest possible CW parameters (CWmin = CWmax = 1) and the highest possible TXOP value (8160 μs). It might seem surprising that these parameters do not allow Bob to have a higher throughput than Alice. With UDP, he is able to achieve maximum throughput, but only on the first hop (Table 6). On the second hop this throughput decreases because Bob is using Vi priority, and Alice's traffic is using Vo priority. The results for TCP are similar, taking into account congestion control. In case E, Bob not only uses the most optimal EDCA parameters for Vi (like in case D) but also uses the simple misbehaviour that was presented in case B. This time, misbehaviour is advantageous for Bob in terms of achieved throughput. Case F differs from the previous one in that the CWmin parameter of Vo is increased to its maximal value (1023). The result is an even higher throughput for Bob. Finally, case G was modified from the previous one by also cheating on the AIFS value and changing it from 2 to 1. This brought a further, though minor increase in throughput.

**Table 6.** Per-hop throughput results for cases D, E, F, and G (in Kb/s)

| Case | UDP | | | | TCP | | | |
|---|---|---|---|---|---|---|---|---|
| | F1a | F1b | F2a | F2b | F1a | F1b | F2a | F2b |
| D | 2111 | 662 | 1754 | 1755 | 260 | 246 | 1979 | 1873 |
| E | 2111 | 1878 | 2111 | 229 | 2191 | 2074 | 113 | 107 |
| F | 2111 | 2060 | 2111 | 32 | 2279 | 2157 | 20 | 19 |
| G | 2111 | 2111 | 2111 | 35 | 2318 | 2194 | 29 | 27 |

### 3.5 Lessons Learned

The results from the simulations have been gathered in Fig. 10, which presents the throughput gain that a misbehaving user can achieve. The gain was calculated as the ratio of the highest throughput in each case to the throughput achieved in case A (for

Vo priority). Since the network was in saturation, it can be assumed that the gain of misbehaving Bob was equal to the loss of well-behaving Alice.

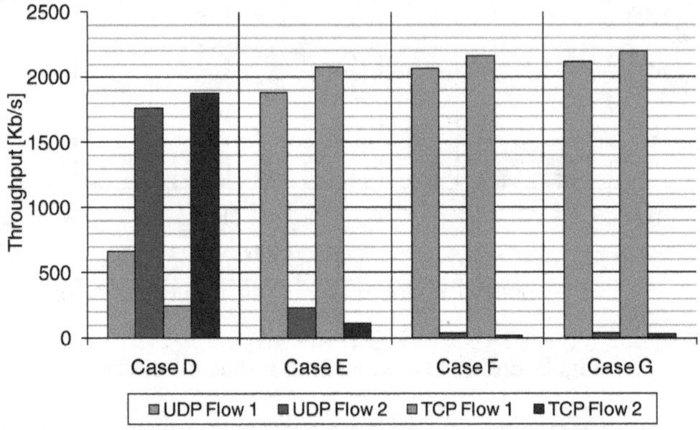


**Fig. 9.** End-to-end throughput results for cases D, E, F, and G

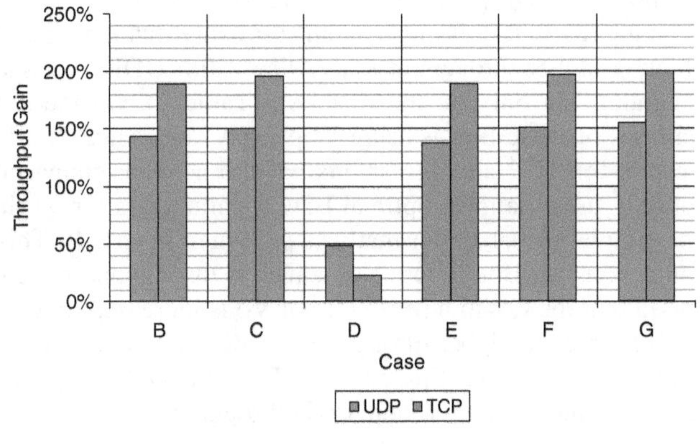


**Fig. 10.** Maximum throughput gain for misbehaving user

With the exception of case D, all the combinations of misbehaviour turned out to be very beneficial. For UDP there was a 40-50% increase, and for TCP – a 90-100% increase in throughput. The conclusion is that in all cases when Bob degraded the EDCA parameters of Alice's traffic he was able to achieve substantially higher throughput. He achieved best performance in case G, in which he both downgraded Alice's Vo traffic and promoted his Vi traffic. He changed his Vo priority parameters to resemble BK and additionally changed the CWmin of Vo to its maximum possible value. At the same time he changed the parameters of his Vi traffic to be optimal (i.e., lowest possible CWmin, highest possible TXOP, and lowest possible AIFS).

The unexpected result from these simulations is that, to achieve higher throughput in a multihop environment, it is significantly more important to degrade forwarded traffic than promote one's own. This problem has not been noticed before in literature and will influence future misbehaviour detection schemes. In multihop, EDCA-based networks, it is important to check for anomalies in the EDCA parameters used by neighbouring nodes. However, previous detection schemes focused only on detecting lowered parameters. The above results show that it is also necessary to monitor increased parameters, as this may lead to the downgrading of forwarded traffic.

## 4 Summary and Future Work

Misbehaviour occurs when a malicious user changes the settings of his/her MN in order to gain better medium access. This paper has presented the impact that realistic MAC layer misbehaviour has on QoS provisioning in mesh networks. Two forms of EDCA parameter modification were considered: downgrading forwarded traffic and promoting local traffic. It has been shown that this is a real threat to wireless mesh networks because it allows easy access to higher throughput and also degrades QoS provisioning. The main conclusion is that, in multihop scenarios, degrading forwarded traffic yields a greater advantage than cheating on medium access parameters.

Countermeasures to prevent misbehaviour are, therefore, required for mesh networks. Along this line, we envisage as future work the development of an architecture able to provide reliable multimedia content delivery, as well as, to deal with the problem of stations not adhering to standards. Based on the results presented in this paper, we will focus on detecting priority degradation of forwarded traffic. To this aim, an analytical model for detecting contention window manipulation in 802.11 EDCA mesh networks needs to be derived and some procedures to mitigate the influence of misbehaviour need to be proposed. These countermeasures should provide an incentive for the malicious users to cease their illegitimate actions.

## Acknowledgement

The research leading to these results has received funding from the European Community's Sixth Framework Programme under grant agreement n° 0384239 (NoE CONTENT). The research leading to these results has received funding from the European Community's Seventh Framework Programme (FP7/2007-2013) under grant agreement n° 214994.

## References

1. IEEE Standard for Information technology-Telecommunications and information exchange between systems-Local and metropolitan area networks-Specific requirements - Part 11: Wireless LAN Medium Access Control (MAC) and Physical Layer (PHY) Specifications, IEEE Std 802.11-2007 (Revision of IEEE Std 802.11-1999), pp. C1–1184 (2007)
2. IEEE, 802.11 TGs Mesh Networking IEEE, Protocol Proposal IEEE P802.11s/D1.07 (2007)
3. MADWiFi – Multiband Atheros Driver for WiFi, `http://madwifi-project.org`

4. Kong, J., Zerfos, P., Luo, H., Lu, S., Zhang, L.: Providing Robust and Ubiquitous Security Support for Mobile Ad-Hoc Networks. In: IEEE ICNP (International Conference on Network Protocols) 2001, Riverside (2001)
5. Buchegger, S., Le Boudec, J.Y.: Performance Analysis of the CONFIDANT Protocol: Co-operation of Nodes – Fairness In Dynamic Ad-Hoc Networks. In: Proc. IEEE/ACM Symp. Mobile Ad Hoc Net. and Comp., Lausanne, Switzerland (2002)
6. Michiardi, P., Molva, R.: CORE: a collaborative reputation mechanism to enforce node cooperation in mobile ad hoc networks. In: Communication and Multimedia Security 2002, Portoroz, Slovenia (2002)
7. Kyasanur, P., Vaidya, N.H.: Detection and Handling of MAC Layer Misbehavior in Wireless Networks. In: International Conference on Dependable Systems and Networks (DSN 2003), p. 173 (2003)
8. Raya, M., Hubaux, J., Aad, I.: DOMINO: a system to detect greedy behavior in IEEE 802.11 hotspots. In: Proceedings of the 2nd international Conference on Mobile Systems, Applications, and Services (MobiSys 2004), Boston, MA, USA (2004)
9. Serrano, P., Banchs, A., Kukielka, J.F.: Detection of malicious parameter configurations in 802.11e EDCA. In: Global Telecommunications Conference 2005 (2005)
10. Wiethölter, S., Emmelmann, M., Hoene, C., Wolisz, A.: TKN EDCA Model for ns-2, Technical Report TKN-06-003, Telecommunication Networks Group, Technische Universität Berlin (2006)
11. Szott, S., Natkaniec, M., Canonico, R., Pach, A.R.: Impact of Contention Window Cheating on Single-hop IEEE 802.11e MANETs. In: IEEE Wireless Communications and Networking Conference (WCNC 2008), Las Vegas (2008)
12. Banchs, A., Azcorra, A., García, C., Cuevas, R.: Applications and Challenges of the 802.11e EDCA Mechanism: An Experimental Study. IEEE Network 19(4), 52–58 (2005)
13. Qiu, L., Bahl, P., Rao, A., Zhou, L.: Troubleshooting wireless mesh networks. SIGCOMM Comput. Commun. 36(5) (October 2006)

# An Uplink Bandwidth Management Framework for IEEE 802.16 with QoS Guarantees

Mohamad El Masri, Slim Abdellatif, and Guy Juanole

CNRS, LAAS, Université de Toulouse, 7, Avenue du Colonel Roche, F-31077, Toulouse, France
Université de Toulouse, UPS, INSA, INP, LAAS, F-31077 Toulouse, France
{masri,slim,juanole}@laas.fr

**Abstract.** This paper presents a novel uplink bandwidth management framework for IEEE 802.16 WiMAX. This framework is destined to be used in a scenario where a large number of connections are simultaneously active, as may be the case when WiMAX is used as a backhaul for WiFi hotspots. In such a scenario we propose a simple and efficient scheduling discipline for the WiMAX uplink traffic. We also propose an admission control algorithm and a modification of the complex bandwidth request mechanisms. The framework we propose in this paper was analyzed: simulation results show a significant performance improvement in terms of overall throughput and delay when compared to recently published work.

**Keywords:** WiMAX, Quality of Service, Bandwidth Management, Scheduling.

## 1 Introduction

The IEEE 802.16 standard [1] (on which is based the Worldwide Interoperability for Microwave Access - WiMAX) defines the physical layer (PHY) and Medium Access Control (MAC) layer for Broadband Wireless Point to Multipoint and Mesh Access. WiMAX is seen by many as the solution for broadband wireless access requiring Quality of Service. A point-to-multipoint (PMP) WiMAX network is structured in a centralized Base Station (BS) and several Subscriber Stations (SS). The BS's role is to manage its Downlink access (DL) and the different SSs Uplink access (UL). Access to the medium for uplink data transfer is done in a connection oriented contention-less polling fashion. In early WiMAX forum documents such as [2], a suggested business case scenario was the use of the IEEE 802.16 access as a WiFi Hot Spot Backhaul. In such a scenario, a scalability issue should be analyzed. The lack of scalability will induce a downgrade in the Quality of Service offered to the different time sensitive flows using the network. It is clear that in such a scenario, leaving to the WiMAX BS the responsibility of managing the bandwidth provision of each connection individually becomes a serious scalability issue. This is also the case in other scenarios presented in the business case analysis [2]: for example serving high speed Internet access in rural areas where DSL services are not available.

L. Fratta et al. (Eds.): NETWORKING 2009, LNCS 5550, pp. 651–663, 2009.

Several work proposed scheduling algorithms for WiMAX [3,4,5]. Others proposed QoS architectures [6] giving scheduling and admission algorithms. But few concentrated on the scalability issue, which is essential for WiMAX to be used in the scenarios we discussed above.

We propose in this paper an uplink bandwidth management framework composed of scheduling algorithms for both the SS and the BS and an admission control algorithm. These algorithms are coupled to a modification of the complex bandwidth request mechanisms of the WiMAX standard we discussed in an earlier paper [7]. The aim of the proposed framework is to have a simpler and more flexible bandwidth management in order to solve the scalability issue while still insuring guarantees on both rate and access delay metrics for time sensitive flows. The paper is organized as follows: we first present different aspects of the WiMAX standard that are of interest for a good understanding of our work and an overview of the related work in the literature. The second section details our framework. The third section presents an analysis of the proposed framework. The paper is then concluded and the perspectives of this work are given.

## 2 Bandwidth Provision in WiMAX

WiMAX access is frame based. It manages separately an Uplink and a Downlink subframe. The subframes are duplexed either in time (TDD) or in frequency (FDD). The BS will send, at the beginning of each downlink subframe, two messages managing the scheduling. The UL-MAP will specify the multiplexing among the SSs (by TDMA or OFDMA) of the uplink subframe. The DL-MAP will specify how the downlink subframe will be organized.

### 2.1 Detailing the Services

WiMAX specifies four scheduling services to which the uplink connections are mapped. An uplink connection, depending on the service it is mapped to, is bound to use a set of rules specifying the way it requests bandwidth and will be served accordingly. The services are: The Unsolicited Grant Service (UGS), The Real-Time Polling Service (rtPS), the non-Real-Time Polling Service (nrtPS) and the Best Effort service (BE). We detail in the following paragraph the request-grant policies that each of the services can use.

### 2.2 Bandwidth Requests: The WiMAX Way

Bandwidth requests are done in WiMAX on a per connection basis. Several ways are available to allow a connection to request bandwidth or to specify needs. Upon establishing a connection, a specification of the flow using the connection is communicated to the BS. This specification can be considered as the initial bandwidth request made by the connection. Other methods to request bandwidth (or to specify the need to be polled) while the connection is active are:

- a unicast request opportunity is a period of airtime where only the destined connection can express its needs,
- a contention request opportunity is a period where several connections may express their needs in a CSMA/CA fashion contention based access, this kind of opportunities are programmed by the BS if, due to lack of space, it can not program enough unicast request opportunities,
- piggyback requests can be included by some connections in a specific type of headers: the Grant Management subheader,
- this same subheader contains the Poll Me bit which, if set, specifies the need of the SS to be polled for a bandwidth request by the BS.

*UGS :* A UGS connection is periodically granted air time without having to specifically request it. The amount of the grant is fixed upon set up of the connection, based on the Maximum Sustained traffic of the flow. A UGS connection is not allowed to use any contention based request period and is not allocated unicast request opportunities. If a UGS connection's transmit depth queue is exceeded (due to a lost UL-MAP or due to clock mismatch) it sets in outbound packets the SI bit (Slip Indicator bit) informing the BS of the situation. The PM bit (Poll Me bit) in outbound UGS packets can be used to request polls for other non-UGS connections.

*rtPS :* An rtPS connection is provided with periodic unicast request opportunities. Those opportunities will be used by the connection to express its needs depending on its queue situation. An rtPS connection is not allowed to use contention request opportunities.

*nrtPS :* An nrtPS connection is provided with regular unicast request opportunities (the standard specifies an interval on the order of one second or less). An nrtPS connection can also use contention request opportunities.

*BE :* A BE connection may be granted unicast request opportunities by the BS. It may also use contention request opportunities in order to express its needs.

## 2.3 The Grants

An SS's medium access for Uplink data transmission is done in a contentionless, polling based fashion. Within the BS, a scheduling algorithm, which is not specified in the standard, will build the UL-MAP (map of the transmission opportunities granted for the uplink direction). The UL-MAP is built based on the requests the BS received and on the initial per connection information it possesses. Two grant modes were initially defined by the 802.16 workgroup: GPC (Grant Per Connection) and GPSS (Grant Per Subscriber Station). In GPC mode, the UL-MAP specifies the time range each connection in each SS should individually use. This mode is obsolete. In GPSS mode, transmission opportunity is granted to the SS; an uplink scheduler within the SS will grant each of its connections a range in the granted time.

### 2.4 Overview of the Literature

The IEEE 802.16 standard does not give any specifications of the different QoS mechanisms to use. Recent work concentrated on these aspects. Architectures focusing on the different uplink scheduling algorithms to be used for the different classes of service were given in [3,4,5]. A QoS architecture was defined in [6] which instantiated the different blocs of a QoS architecture (admission control, classifiers, schedulers, traffic shaper and different queuing mechanisms). However few concentrated on the scalability issues. The first step in this direction was putting away from the standard the GPC mechanism which was a serious flaw in terms of scalability. Other work [8] concentrated on the optimal duration of the contention period in order to reduce collision probabilities. Our work focuses on the bandwidth management mechanisms. We propose a redesign of the request mechanisms based on an aggregated management. We also provide flexible and simplified scheduling procedures. Our design will offer rate guarantees and latency bounds to sensitive flows by adopting a Latency Rate [9] server behavior.

## 3 A Novel Bandwidth Management Framework

### 3.1 Overview of the Framework

We propose a novel bandwidth management framework for IEEE 802.16 uplink communication. The framework we propose is based on the aggregation of the bandwidth management. The aggregation here concerns the bandwidth management at the BS: instead of having to perform the scheduling on a per connection basis (which would be problematic in our case where a high number of connections will be simultaneously active), we simplify the scheduling at the BS by decentralizing this feature to the SS level, the BS will have to perform the scheduling on a per SS basis. We think that aggregating the bandwidth management will allow us to have a far simpler, more flexible bandwidth management. Our proposal will further be able to give guarantees on the rate and on the access delay of time sensitive flows. The aggregation will also cover the bandwidth request mechanisms. It will use the GPSS grant mode as specified by WiMAX. This is coupled to scheduling algorithms both on the SS side and on the BS side and an admission control algorithm allowing the guarantees to be given. The framework will act towards the flows differently whether they are time sensitive (UGS and rtPS flows), throughput requiring (nrtPS flows), or best effort flows.

The framework works as follows: all uplink flows are subject to admission control at the BS, the sum of the requested rates of all time sensitive flows of an SS (those using UGS or rtPS types of service) defines the contract of the said SS. Instead of having each connection request its own needs in terms of bandwidth, an aggregated request, covering all the needs of an SS is sent out to the BS periodically. An important feature of our proposal is to have the BS answer positively and immediately the amount of requested bandwidth falling within the SS's contract, the grant should come in the frame following the one bearing the request. This is done in order to insure the guarantees on rate and delay that the framework offers. The admission control algorithm along with a resource reservation process will be responsible of guaranteeing the whole framework will work correctly. We assume the presence of policing and shaping mechanisms that we will not detail herein. We detail in the following paragraphs the different aspects of the proposed framework.

### 3.2 Admission Control Algorithm

**Overview.** One feature of our proposal is the guarantees on delays and rate it can give to time sensitive flows. It is thus necessary to have an efficient admission control algorithm working at the BS. The admission control algorithm should guarantee that any time sensitive request falling within an SS's contract will be granted in the frame following the one that carried the request (the contract is defined as the sum of the mean rates of all admitted UGS and rtPS flows). Each new flow of an SS will be individually submitted to admission control (algorithm 1). We separate in the algorithm the procedure of admission of time sensitive flows from the admission procedure of nrtPS flows and Best Effort flows:

- UGS and rtPS flows are to be considered, for the admission control algorithm, individually and with a Frame Size granularity: i.e. the admission control algorithm will reserve for each admitted flow of these types, within all the uplink subframes covering the time of activity of the flow, enough slots (as calculated algo. 1 line 2) to serve the said flow. This is necessary if guarantees on the delay are to be given to UGS and rtPS flows. It should be noted that, since the uplink allocation unit in WiMAX is physical layer dependent, the framework should be adapted to the physical layer used. We call *slot* the generic uplink allocation unit. This can be a WiMAX minislot for a single carrier physical implementation or a combination of symbols and subchannels for OFDM and OFDMA based physical implementations.
- The cost in terms of slots of an nrtPS flow requesting admission will not be considered individually but rather as a fraction of all nrtPS flows in the network (see algo. 1 line 10-11). nrtPS flows are in no need for delay guarantees; they are to be served with the rate they ask for, however no constraint on the delay is to be enforced. Thus, the slot reservation for nrtPS flows is done on a larger scale than that of time sensitive flows.
- Best Effort flows will always be accepted (see algo. 1 line 19-20): since no guarantees are to be given to such flows and since the medium access in WiMAX is controlled and contention-less, new Best Effort flows will not affect other flows' performance. In order to avoid BE flows starvation, a fixed percentage of uplink bandwidth will be reserved (i.e. the admission control will only consider the remaining bandwidth when admitting other types of connections).

**The Algorithm.** The admission control algorithm described in 1 will need a number of parameters that we define herein:

**Algorithm 1.** Admission($Flow_i, SS_j$)

```
if Flow_i.TypeOfService == UGS OR rtPS then
   N = CalculateSlotCost(Flow_i.Rate);
   if N < TotalAvailableSlots then
      Admit(Flow_i);
      TotalAvailableSlots− = N;   SS_j.Rate+ = Flow_i.Rate;
   else
      Reject(Flow_i);
   end if
else if Flow_i.TypeOfService == nrtPS then
   AUX = nrtPSRate + Flow_i.Rate;
   N = CalculateSlotCost(AUX);
   if N < TotalAvailableSlots + nrtPSSlots then
      Admit(Flow_i);
      TotalAvailableSlots = TotalAvailableSlots + nrtPSSlots − N;
      nrtPSSlots = N;
   else
      Reject(Flow_i);
   end if
else if Flow_i.TypeOfService == BE then
   Admit(Flow_i);
end if
```

- *TotalAvailableSlots* is the number of unreserved uplink allocation slots. The variable is initialized to a percentage of the total number of uplink slots in a frame, the rest being reserved to prevent Best Effort flows from starving.
- $SS_i.Rate$ represents the contract of a given $SS_i$, it should be initialized to 0, it will be the sum of the mean rates of time sensitive flows of $SS_i$.
- *nrtPSRate* represents the total rate of all the admitted nrtPS connections, it should be initialized to 0.
- *nrtPSSlots* are the total number of allocation slots reserved by all the nrtPS connections, it should be initialized to 0.

Function $CalculateSlotCost(R)$ is used to calculate the number of slots of an uplink frame that must be reserved to serve a generated rate of $R$. This value can be calculated as:

$$SlotCost(R) = \frac{R * FS}{StationTxRate * SlotSize}$$

where $FS$ is the size of a WiMAX Frame, *StationTxRate* is a station's transmission rate and $SlotSize$ is the size of the slot as earlier defined.

### 3.3 Aggregating the Requests

The complex procedure that WiMAX describes for the bandwidth request mechanisms can find its explanation in a context with few connections per SS. Each connection will have to specifically request its needs and require them to be answered (when it is using a high priority type of service). In such a context, the SS will not have enough margin to be flexible. However in a context where each SS is serving a high number of various connections, we think it is possible to simplify the request mechanism giving the SS the flexibility to manage an across-connection aggregated request and handle the scheduling of the granted airtime without any loss in the quality of service given to the time sensitive flows. In our proposal, the complex request mechanism presented earlier (unicast polls, multicast polls, PM bits, piggybacking) will be replaced by a single per SS aggregated request, expressing the needs of the SS, and sent in a contention less fashion (as a specific UGS connection for example). The request is sent each frame, at the end of the uplink period allocated to the SS, so that the request contains an up-to-date view of the queuing situation within the SS. The aggregated request represents a photography of the situation of the queues of the SS. It contains two fields giving the BS the needed information for the uplink scheduling. The first field called the *Contracted Bytes* will contain the number of bytes requested by the SS that are within the admission control contract. The second field, called the *Additional Bytes* will contain the requested bytes that fall beyond the contract.

**Building the request.** We define here the exact content of each field of the aggregated request. As defined earlier, $SS_i.Rate$ is the sum of the mean rates of all time sensitive flows (using UGS and rtPS services). We define $W(\tau, t)$ as being the service received by the time sensitive flows during the current backlogged period (which began at $\tau$, $t$ being the current time). We also define $tr_k$ as the time of transmission of the SS's aggregate request in the current frame (which will usually be at the end of the uplink transmission time allocated to the SS as specified by the UL-MAP). Let $Q_{UGS}$ be the amount of UGS bytes enqueued within the SS at the moment the request packet is being built, the same applies to $Q_{rtPS}$, $Q_{nrtPS}$ and $Q_{BE}$. $\varepsilon_{UGS}$ is the slip amount for the UGS queue: it is the amount of enqueued bytes that do not go within the contract: $\varepsilon_{UGS} = max(Q_{UGS} - R_{UGS}, 0)$, $R_{UGS}$ being the total amount of UGS bytes falling within the contracted request. The same applies to $\varepsilon_{rtPS}$.

The aggregated request's fields are built as follows: the CB field will contain the amount of UGS and rtPS bytes enqueued within the SS that respect the SS's contract in addition to a possible slip of the UGS queue (a slip in the UGS queues is incidental); the AB field will contain the amounts of bytes enqueued within the SS that did not make it to the CB field (i.e. nrtPS and BE bytes in addition to rtPS bytes that exceed the contract due to the possible variation in an rtPS flow). Note that the sum of CB and AB at the time of build of the request is equal to the total number of bytes enqueued within the SS. Figure 1 shows a view of what the CB field represents.

$$CB = (SS_i.Rate(tr_k - \tau) - W(\tau, tr_k)) + \varepsilon_{UGS}$$

$$\text{and } AB = \varepsilon_{rtPS} + Q_{nrtPS} + Q_{BE}$$

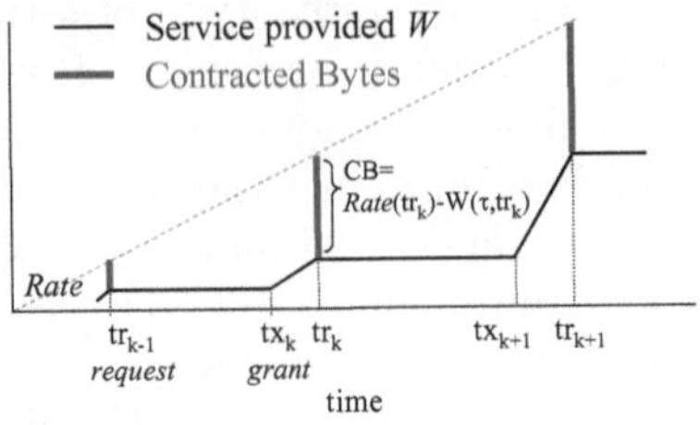


**Fig. 1.** The Contracted Bytes field

### 3.4 Granting the Requests

As it has been said earlier, the requests that fall under the terms of the contract of an SS should be granted for the frame that follows the one where the request was received. The possibility of this is insured by the admission control algorithm. The BS's uplink scheduling will first give each SS the number of slots necessary to transmit the Bytes indicated in the CB field of the aggregated request. The remaining slots are then to be distributed among the SSs. If there is enough slots to grant all of the additional requests, then the slots are to be distributed according to the requests, if not, the slots are distributed proportionally to the AB field of each SS's aggregated request. Algorithm 2 describes the grant procedure. We call $AllocatedSlots(i)$ the slots allocated to SS $i$, the array is initialized to 0. $CB_i$ and $AB_i$ are the translation in terms of slots of the values of the CB and AB fields of the aggregated request of SS $i$. $N$ is the number of SSs in the network. $AvailableSlots$ is a variable representing the number of unallocated slots and will be initialized to the total number of slots in the uplink frame.

**Algorithm 2.** Grant Requests and Build UL_MAP

**for** $i$ in 1 to $N$ **do**
  **if** $CB_i > AvailableSlots$ **then**
    ERROR;
  **else**
    $AllocatedSlots(i) += CB_i$;   $AvailableSlots -= CB_i$;
  **end if**
**end for**
**if** $\sum_{i=1}^{N} AB_i < AvailableSlots$ **then**
  **for** $i$ in 1 to $N$ **do**
    $AllocatedSlots(i) += AB_i$;
  **end for**
**else**
  **for** $i$ in 1 to $N$ **do**
    $AllocatedSlots(i) += (\frac{AB_i}{\sum_{i=1}^{N} AB_i}) * AvailableSlots$;
  **end for**
**end if**

### 3.5 Scheduling Algorithms

Scheduling algorithms can be implemented at different stages of the framework, however only one of those is crucial for the correct functioning of the system which is the SS based algorithm distributing the granted slots to its different connections. This scheduling algorithm will work as follows: it will first give to each time sensitive connection the amount of slots necessary to transmit the contracted transmissions they requested. The remaining slots will then be distributed, in a round robin fashion, first to the UGS and rtPS flows needing additional airtime, then to the nrtPS connections and finally to the BE connections. This will allow us to serve as a high priority, the delay sensitive flows. nrtPS flows will be correctly served, on the long term, due to the slot reservation made by the admission control algorithm. BE flows will avoid starvation due to the percentage of bandwidth which is initially reserved. We think that once this scheduling decided, other scheduling algorithms (like scheduling the connections of one SS in the total interval allocated to the SS) will add complexity with little effects on the delays. In fact, since each SS will be granted a part of the uplink subframe (which in our case will never go beyond 2 ms), scheduling the SS's connections inside this small range will have little effect on the delays.

### 3.6 Characteristics of the Framework

*The service with respect to time sensitive flows conforms to an LR server model.* We have proven that the designed system will act as an LR($SS.Rate$,$2FS + Max\_UL\_Size$) defined in [9] giving the framework interesting properties on the rate and bounds on the delays. Due to lack of space, the proof will not be detailed herein. In our case, time sensitive flows (UGS and rtPS) will be served following an LR model which will thus give them a guaranteed rate and a higher bound on the mean access delays. We can intuitively see in figure 1 what a worst case scenario may be: a worst case being a peak arrival within contract right after the request has been sent: in a no anticipation policy a service conforming to the contracted rate starts no later than $3F\_S$ (which is a higher bound for $2FS + Max\_UL\_Size$).

*Increased determinism.* In our proposal the single aggregated request of the SS is sent in a contention-less fashion, unlike standard WiMAX architecture where some bandwidth requests can be sent in contention zones. This property of the new design increases the determinism of the WiMAX bandwidth management.

*Flexibility and simplicity.* The aggregation of the requests and of the grants will allow a better flexibility of the bandwidth management. It will also render the whole request mechanisms simpler. The SS having a better knowledge of the state of its queues than the BS, this will allow it to adapt the request to its needs and to organize the scheduling accordingly.

*Scalability and efficiency.* Our proposal will allow an efficient usage of the network resources: the BS distributing the resources will always have an updated view of the state of the SSs. Since the request and scheduling mechanisms are simple, this will allow the BS to manage a large number of connections simultaneously.

# 4 Analysis

We analyzed our proposal by mean of simulation, using the ns-2 simulator [10]. We used the module for WiMAX proposed by Juliana Freitag Borin from the Universidade Estadual de Campinas [11]. We implemented our proposal and tested it towards a standard WiMAX implementation using a scheduler presented by Freitag et. al.[5]. Freitag et. al. adopt a WiMAX classical request procedure and a per connection management of the bandwidth. Freitag et. al.'s scheduler is designed to respect throughput requirements for the time sensitive flows. The goal of the analysis is to confirm the different properties of our framework : the possibility to have a large number of simultaneously active connections, having these connections served correctly and achieving the guarantees on the rate and on the access delays of the time sensitive flows. We detail in the following paragraphs the simulation scenarios along with the results and an analysis of the results.

## 4.1 Simulation Scenarios and Results

Different scenarios were tested with variation on the number of SSs, the number of flows per SS and the rate of the different flows. The results of those different scenarios drive to the same conclusions. We chose to present one set of parameters varying only the size of the WiMAX frame. This is to show the effect of the frame size on the access delay. Time Division Duplexing is used to duplex the DL and UL subframes. The physical rate is of 40 Mbps. The network is made of a BS and 10 SSs with the same flow profile. Within each SS, 10 flows are activated at the beginning of simulation: 3 flows using UGS (24.6 kbps CBR), 3 flows using rtPS (64 kbps ON-OFF), 3 flows using nrtPS (64 kbps CBR) and 1 flow using BE (1 Mbps CBR). Simulations are 20 seconds long. Figures 2 thru 7 present different results with frame size varying from 2 to 10 ms (we varried FS in our simulations from 2 to 20 ms). The results are one of the following:

- a cumulative distribution function (CDF) of the access delays of time sensitive flows (UGS and rtPS),
- a CDF of the access delays of nrtPS and BE flows,
- throughput of either nrtPS or BE flows.

We chose not to show graphs of the throughput of UGS and rtPS flows as they bear little information: UGS and rtPS flows achieve their generation rate.

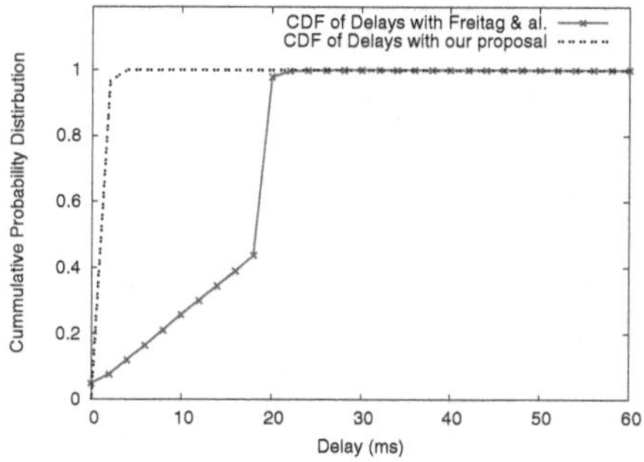


**Fig. 2.** Delays of UGS and rtPS FS=2ms

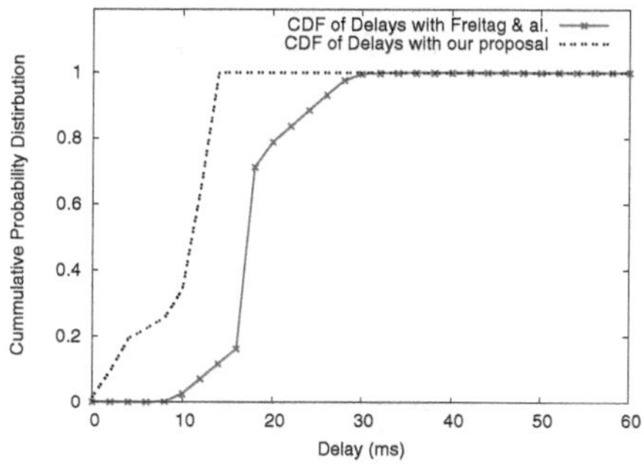


**Fig. 3.** Delays of UGS and rtPS FS=10ms

### 4.2 Interpreting the Results

**The Access Delays.** Our proposal achieves better access delays in all cases, for all kinds of flows (figures 2 to 5 allow us to compare the CDF of access delays of different flows in different scenarios using our framework or Freitag et. al.'s), we can clearly see that in all cases, our framework achieve a better overall access delay for all kinds of flows. The huge difference in delays of NRTPS and BE is due to the way WiMAX functions: WiMAX will seldom allow these kinds of flow to express their needs which is not the case in our scheme thanks to the aggregated bandwidth request. Moreover, the guarantee on the access delay of time sensitive flows can be clearly seen. In figures 2 and 3, the time sensitive flows achieve access delays that are at most equal to the double of the frame size. This is due to our scheduling policy, forcing a time sensitive request to be granted at the frame following the one bearing the request.

**The Throughput.** Our proposal achieves a better overall throughput going from 130% to 267% better if compared to the results of Freitag et. al. We see that our proposal reduces the waste in uplink resource since the BS is permanently

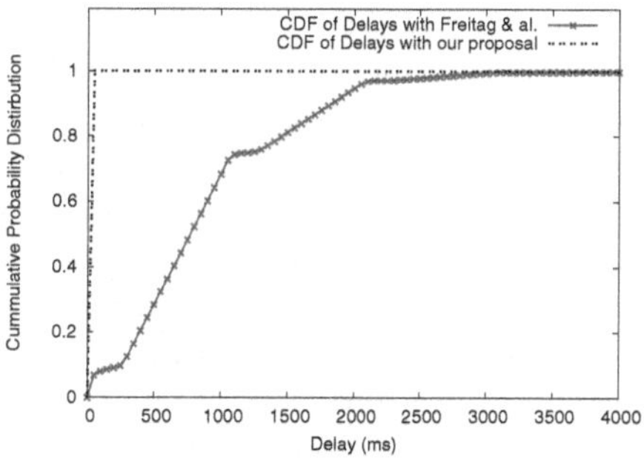


**Fig. 4.** Delays of NRTPS FS=10ms

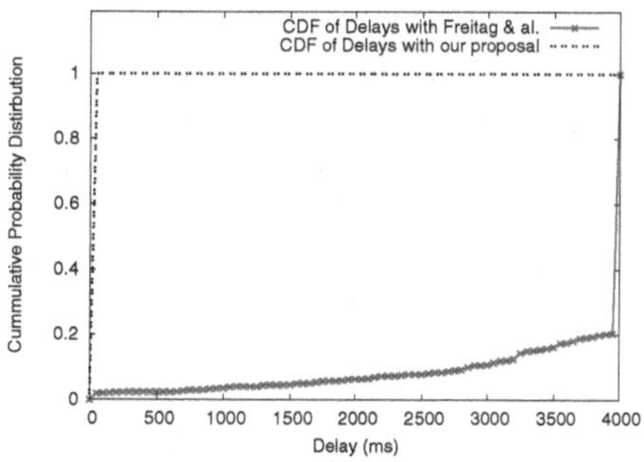


**Fig. 5.** Delays of BE FS=10ms

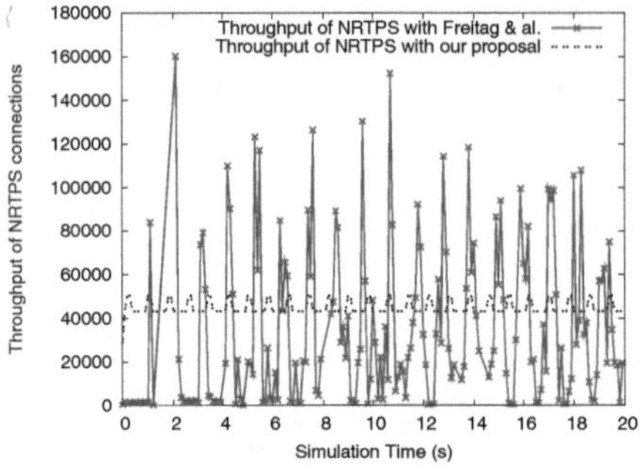


**Fig. 6.** Throughput of NRTPS FS=10ms

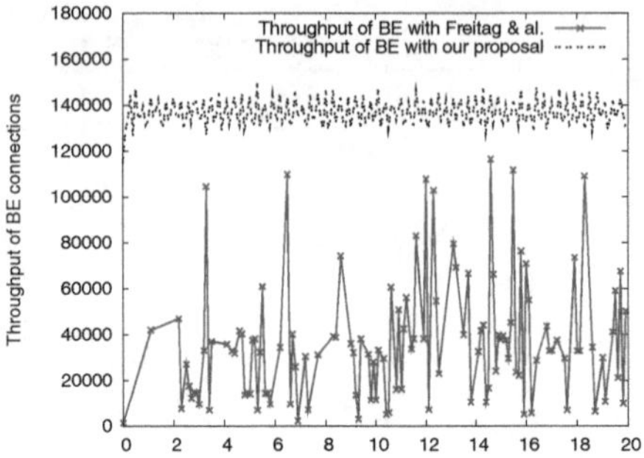


**Fig. 7.** Throughput of BE FS=10ms

updated with the situation of the queues within the SSs, giving it more flexibility to manage the bandwidth as needed. Classical WiMAX request mechanisms will give the BS scarce information about the situation of nrtPS and BE queues, making it waste precious resources when it lacks this information especially with a larger frame size. Another interesting aspect of our proposal is the stability of the service given to these flows as can be seen in figures 6 and 7. For example, in figure 6, the service given to nrtPS flows by our framework is stable around 40 KBps whereas the service offered by the Freitag et. al. is more varying. Allowing these flows to express their needs in the SS's aggregated request is the main cause of this behavior.

**Conclusions.** Our proposal proved to perform better than the standard WiMAX along with the scheduler at several levels: time sensitive flows were given better delays. The framework insured better delays for nrtPS and BE flows as well with a better throughput and much more stable service, even though a high number of connections are simultaneously active in the network.

## 5 Conclusion

We propose in this paper a bandwidth management framework for IEEE 802.16 uplink access combining a modification of the WiMAX bandwidth request mechanisms, an admission control algorithm and an uplink scheduling algorithm. This paper details the different algorithms used in our framework. The framework was analyzed using simulation and compared to standard WiMAX procedure and a scheduling algorithm proposed in [5]. Our framework allows giving guarantees on the rate and the access delays of time sensitive flows. Our framework proved to achieve an overall better usage of the network resources and achieved better access delays for the different types of service while adopting a simple procedure at the BS level allowing it to manage a large number of connections. This work will further be developed and integrated in a heterogeneous WiFi-WiMAX network in order to give a global QoS solution for heterogeneous wireless networks.

## References


1. 802.16: IEEE Standard for Local and Metropolitan area Networks Part16: Air Interface for fixed broadband Wireless Access Systems (2004)
2. Forum, W.: Business Case Models for Fixed Broadband Wireless Access based on WiMAX Technology and the 802.16 Standard (2004)
3. Liu, Q., Wang, X., Giannakis, G.: A cross-layer scheduling algorithm with qos support in wireless networks. IEEE Transactions on Vehicular Technology 55(3), 839–847 (2006)
4. Niyato, D., Hossain, E.: Queue-aware uplink bandwidth allocation and rate control for polling service in ieee 802.16 broadband wireless networks. IEEE Transactions on Mobile Computing 5(6), 668–679 (2006)
5. Freitag, J., da Fonseca, N.: Uplink scheduling with quality of service in ieee 802.16 networks. In: Global Telecommunications Conference, 2007. GLOBECOM 2007, pp. 2503–2508. IEEE, Los Alamitos (2007)
6. Delicado, J., Orozco-Barbosa, L., Delicado, F., Cuenca, P.: A qos-aware protocol architecture for wimax. In: Canadian Conference on Electrical and Computer Engineering, 2006. CCECE 2006, May 2006, pp. 1779–1782 (2006)
7. El Masri, M., Abdellatif, S., Juanole, G.: Proposal of a novel bandwidth management framework for ieee 802.16 based on aggregation. In: New Technologies, Mobility and Security, 2008. NTMS 2008, pp. 1–5 (2008)
8. Oh, S.M., Kim, J.H.: The analysis of the optimal contention period for broadband wireless access network. In: Third IEEE International Conference on Pervasive Computing and Communications Workshops, 2005. PerCom 2005 Workshops, pp. 215–219 (2005)
9. Stiliadis, D., Varma, A.: Latency-rate servers: a general model for analysis of traffic scheduling algorithms. IEEE/ACM Trans. Netw. 6(5), 611–624 (1998)
10. ns 2: The network simulator ns-2, `http://www.isi.edu/nsnam/ns/`
11. Freitag, J., da Fonseca, N.: Simulator for wimax networks. Simulation Modelling Practice and Theory 16(7), 817–833 (2008)

# A Coalitional Game Model for Heat Diffusion Based Incentive Routing and Forwarding Scheme
## (Work in Progress)

Xiaoqi Li, Wujie Zheng, and Michael R. Lyu

Department of Computer Science and Engineering
The Chinese University of Hong Kong, Shatin, N.T., Hong Kong
{xqli,wjzheng,lyu}@cse.cuhk.edu.hk

**Abstract.** We propose an incentive routing and forwarding scheme that integrates a reputation system into a monetary payment mechanism to encourage nodes cooperation in wireless ad hoc networks. For the first time in the literature, we build our reputation system based on a heat diffusion model. The heat diffusion model provides us a way of combining the direct and indirect reputation together and propagating the reputation from locally to globally. Further, we model and analyze our incentive scheme using a coalitional game, which is not the usual non-cooperative game like others. We further prove that under a proper condition this game has a non-empty stable core. From the evaluation we can see that the cumulative utility of nodes increases when nodes stay in the core.

**Keywords:** Coalitional Game, Incentive Routing, Heat Diffusion.

## 1 Introduction

The nature of wireless ad hoc networks is to let nodes cooperative together thus improve the connectivity of the whole network or execute some specific functions inside the network. However, nodes in this kind of networks may belong to different individuals or authorities and have their own interests. They may not want to help others forward routing and data packets, since that will cost their own energy and bandwidth. Consequently, it is necessary to provide incentive mechanisms to encourage cooperations among the nodes.

Incentive routing schemes for enforcing selfish agents to cooperate in wireless networks have been studied for years. One category of solution is using monetary incentives, either virtually or practically. Payment schemes need to be designed and usually are analyzed by game theoretic methods. In these schemes, the intermediate nodes declare their costs for forwarding packages. Then the routing protocol selects the lowest cost path (LCP) based on the declared costs. Afterwards the payments are rewarded to nodes on and sometimes off the LCP with the amount no less than their declared costs. However, a problem arises when nodes may purposely declare a higher cost to take advantage of the payment algorithms. So currently more research is focused on how to avoid cheating and achieve effective and also economic payments.

L. Fratta et al. (Eds.): NETWORKING 2009, LNCS 5550, pp. 664–675, 2009.

Another category is employing reputation systems to stimulate the nodes to cooperative. The common idea of these systems is that each node in the network will monitor the behaviors of its neighbors. If the neighbors are observed for not executing some functions properly, their reputations will be decreased and they will be under the threat of being blocked from the network. The key challenges are how to combine the direct neighborhood reputations together and propagate them from locally to globally. In these systems, game theoretic methods can also be used to analyze the effectiveness of the threatening mechanism. In fact we consider that the method of using monetary incentives and reputation systems are not mutually exclusive and they can be combined together to design a more flexible incentive scheme.

On the other hand, the above schemes are usually modelled as non-cooperative games. However, in wireless networks nodes cannot perform routing and forwarding behaviors individually. They must cooperate together to complete one task, so it is natural to think about modelling the wireless network behaviors as a cooperative game.

In this paper, we are going to model the routing and forwarding procedures in wireless networks as a cooperative coalitional game with transferable payoff, and propose an incentive routing and forwarding scheme that combines the idea of payment mechanism and reputation system together. Then we analyze that the game has a non-empty core, which is a stable status in cooperative game just like the Nash Equilibrium in a non-cooperative game.

Regarding the combination of reputation and payment schemes, we first need to obtain a combined and globalized reputation, and then smoothly map this reputation value to a certain amount of payment. On the basis of it, we also need to design an incentive payment scheme integrating reputation and cost together. In the formulation of a coalitional game the key challenges are: 1) how to write the value function of the coalition which represents the collective payoff of the coalition; 2) how to find the solution of this game where every node has a satisfying payoff share, so that it will not deviate from some stable status. Some games may not have such a stable solution. The objective of our paper is to solve the above questions.

We list our major contributions as follows: First, we design an incentive routing and forwarding scheme that integrates reputation information into a payment mechanism. Second, we introduce a heat diffusion model to combine the direct and indirect reputations together and propagate them from locally to globally in the way how heat diffuses. Third, unlike others, we model this incentive scheme using a coalitional game method. A characteristic value function of the coalition is designed, and we prove that this game has a core solution.

The rest of this paper is organized as follows. We first give the background of the heat diffusion model in Section 2. After describing some technical preliminaries in Section 3 we will propose our incentive routing and forwarding scheme in Section 4. The scheme is analyzed in Section 5, then we show some evaluation results in Section 6. In the end, some related work and conclusions are given in Section 7 and Section 8 respectively.

# 2 Background of Heat Diffusion Model on Weighted Directed Graph

## 2.1 Motivation

A reputation system usually needs to address two problems: 1) how to combine subjective direct reputations with indirect reputations from neighbors to make them become more objective; and 2) how to propagate the reputation from locally to globally. Previously there are different solutions to these problems such as [1] and [2]. In this work, we will employ the heat diffusion model to fulfill the requirements.

In nature, heat always flows from high temperature positions to low temperature positions via conductive media. A heat diffusion model describes this phenomenon that heat can diffuse from one point to another through an underlying manifold structure in a given time period. The higher the thermal conductivity of the medium, the easier the heat flows, which implies that the two end points have some cohesive relations. Diffusion behaviors are also affected by the underlying geometric structures. Some achievements have been made based on the heat diffusion model such as classification in machine learning field, page ranking in information retrieval [3] and marketing candidates selection in social computing [4], but to our best knowledge, there is no previous work that has been performed on the incentive routing in wireless networks.

We see that in the process of heat diffusion, each node's heat comes from all of its incoming links and diffuses out to its successors as long as it can. If we diffuse heat on a weighted directed graph, the amount of heat a node can get depends not only on the heat of its neighbors but also on the weights of the links connecting them. The higher the weight, the more thoroughly the heat can be diffused. Therefore, if we let the weight be the direct reputation value of the link, then the amount of heat will be the overall reflection of the underlying reputation information. The course of heat diffusion through all possible links can also be deemed as a propagation of the reputations.

## 2.2 Heat Diffusion on Weighted Directed Reputation Graph

We construct a heat diffusion model on the reputation graph $G = (M, E, R)$, where $M = \{1, 2, ..., m\}$ is the node set. $E = \{(i, j) | \ i$ and $j$ are in communication range and the transmission direction is from $i$ to $j\}$. The heat only flows from $i$ to $j$ if $(i, j) \in E$. $R$ is the reputation set $\{r_{ij} | \ r_{ij}$ is the direct reputation of edge $(i, j)\}$. We use $f_i(t)$ to describe the heat value of node $i$ at time $t$, beginning from an initial distribution of heat $f_i(0)$ at time zero. $\boldsymbol{f}(t)$ denotes the vector consisting of $f_i(t)$.

The heat diffusion modelling is as follows. Suppose, at time $t$ node $i$ diffuses $H_D(i, t, \Delta t)$ amount of heat to its subsequent nodes. We assume that: a) the heat $H_D$ is proportional to the time period $\Delta t$; b) $H_D$ is proportional to the heat of node $i$; c) each node has the same ability to diffuse heat; and d) node $i$ intends to distribute $H_D$ uniformly to each of its subsequent nodes, but the actual heat it can diffuse is proportional to the corresponding reputation weight of the edge. On the basis of the above considerations, we state that node $i$ will diffuse $\lambda p_{ik} f_i(t)\Delta t/l_i$ amount of heat to each of its subsequent node $k$, where $l_i$ is the outdegree of node $i$ and $\lambda_j$ is the thermal conductivity, which is the heat diffusion coefficient representing the heat diffusion ability. In the case that the outdegree of node $i$ is zero, we assume that this node will not diffuse heat to others. Then the total amount of heat node $i$ will diffuse is $\sum_{k:(i,k)\in E} \lambda p_{ik} f_i(t)\Delta t/l_i$.

On the other hand, each node $i$ receives $H_R(i,j,t,\Delta t)$ amount of heat from $j$ during a period of $\Delta t$. We also have the following assumptions: a) $H_R$ is proportional to the time period $\Delta t$; b) $H_R$ is proportional to the heat of node $j$; c) $H_R$ is zero if there is no link from node $j$ to $i$. Based on the above considerations, we obtain $H_R(i,j,t,\Delta t) = \lambda_j f_j(t)\Delta t$. As a result, the heat that node $i$ receives between time $t$ and $t+\Delta t$ will be equal to the sum of the heat flowing from all its neighbors pointing to it, which is $\sum_{j:(j,i)\in E} \lambda_j f_j(t)\Delta t$. Since the amount of heat that $j$ diffuses to $i$ should be equal to the amount $i$ receives from $j$, we have $\lambda p_{ji} f_j(t)\Delta t/l_j = \lambda_j f_j(t)\Delta t$. So we get $\lambda_j = \lambda p_{ji}/l_j$. To sum up, the heat difference at node $i$ between time $t$ and $t+\Delta t$ will be the amount of heat it receives deduced by what it diffuses. The formulation is therefore:

$$f_i(t+\Delta t) - f_i(t) = \lambda \left( \sum_{j:(j,i)\in E} \frac{p_{ji}}{l_j} f_j(t) - \mu_i \sum_{k:(i,k)\in E} \frac{p_{ik}}{l_i} f_i(t) \right) \Delta t, \quad (1)$$

where $\mu_i$ is a flag to identify whether node $i$ has any outlinks. If node $i$ does not have any outlinks, $\mu_i = 0$; otherwise, $\mu_i = 1$. To find a closed form solution to Eq.(1), we then express it in a matrix form:

$$\frac{\boldsymbol{f}(t+\Delta t) - \boldsymbol{f}(t)}{\Delta t} = \lambda \boldsymbol{H} \boldsymbol{f}(t), \text{where} \quad (2)$$

$$H_{ij} = \begin{cases} p_{ji}/l_j, & (j,i) \in E, \\ -(\mu_i/l_i)\sum_{k:(i,k)\in E} p_{ik}, & i = j, \\ 0, & otherwise. \end{cases} \quad (3)$$

Solving the above equation, we get

$$\boldsymbol{f}(t) = e^{\lambda t \boldsymbol{H}} \boldsymbol{f}(0) \quad (4)$$

The matrix $e^{\lambda t \boldsymbol{H}}$ is called the diffusion kernel, showing that the heat diffusion process continues infinite times from the initial heat diffusion step.

## 3 Technical Descriptions

Before presenting our incentive scheme and coalitional game we first give some technical notations. Our game is based on the bi-directional weighted graph $G = (M, E, P)$ described in Section 2. Suppose that $s$ is the source node and

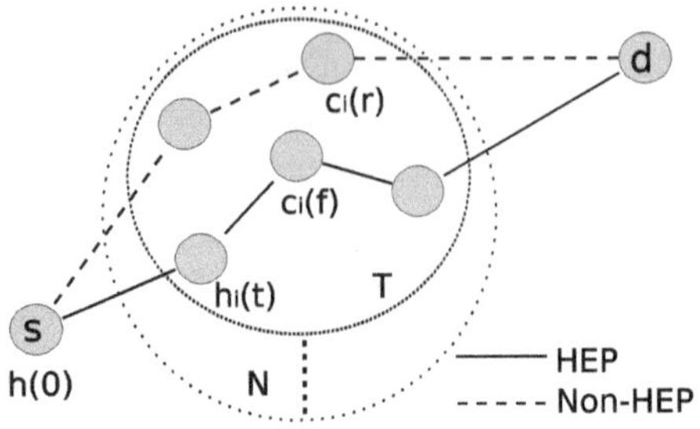


**Fig. 1.** Illustration of Notations of the Coalitional Game

$d$ is the destination node, then the player set of this coalitional game is $N = M \setminus \{s, d\}$. Coalition is denoted by any non-empty subset $T \subseteq N$, and the overall payoff of the coalition is denoted by $v(T) \in \mathbb{R}$. Then the game is expressed by $\Gamma =< N, v >$. Players will form into coalitions to help establishing the highest effective path between $s$ and $d$ with the lowest cost under the constraint that each intermediate node's heat is higher than a threshold $\theta$. If there's a tie in the total cost, $s$ will break the tie by choosing the path with the highest heat. The source can freely choose the value of $\theta$ to meet its requirement on reputation. The larger the value $\theta$ is, which means the source has a higher demand on reputations, the higher payments it will expend. All the paths established inside the coalition $T$ connecting $s$ and $d$ compose the path set $P_{sd}(T)$.

Initially, $s$ will load a certain amount of initial heat $f(0)$ and diffuse it on the reputation graph, then at time $t$, each node will be diffused $f_i(t)$ amount of heat. Correspondingly each source $s$ has an initial balance of $h(0)$, and the payment to each node $h_i(t)$ is paid by it according to $f_i(t)$. Every node evolving in the routing or forwarding procedure will cost its energy. Since the cost for sending/receiving routing and data packets are different [5], and the cost for data transmission is usually larger than that of routing packets transmission, we denote the routing and forwarding cost respectively by $c_i(r)$ and $c_i(f) \in \mathbb{R}^+$, and $c_i(r) < c_i(f)$ for all $i \in N$. Please see Fig. 1 for the illustration of notations.

## 4 Incentive Routing and Forwarding Scheme

The basic idea of achieving incentives is that nodes will be paid when they help others forwarding data or routing packets. Unlike other payment schemes that reward the nodes according to their claimed cost, our incentive routing and forwarding scheme pays the nodes by their reputations. The higher a node's reputation is, the higher payment it can get. The payment is given by the source node. The payment may be in the form of virtual currency like [6] or any other practical form. In our paper we assume that there is such a payment form and a payment operation daemon in the network.

In the scheme, the source node $s$ will originate the heat diffusion process starting from itself. Then after collecting the forwarding cost of all the

**Algorithm 1.** Incentive Routing and Forwarding Scheme

**Input**: Source $s$, destination $d$, reputation graph $G$, and heat threshold $\theta$
**Output**: $HEP_N$
**foreach** $i \in N$ **do**
  $i$ claim its forwarding cost $c_i(f)$ to $s$;
  $f_i(0) = 0$;
$f_s(0) = f(0)$;
Execute the heat diffusion process $\boldsymbol{f}(t) = e^{\lambda t \boldsymbol{H}} \boldsymbol{f}(0)$;
$s$ chooses the highest effective path to $d$ with the lowest cost, subject to $f_i(t) \geq \theta$. If there is a tie, $s$ selects the one with the highest heat;
... Data transmission process; ...
$s$ pays $h_i(t)$ to each $i$ according to $f_i(t)$;
$s$ adjusts its heat threshold $\theta$;
Updates reputation graph G;

players, $s$ will compute the lowest cost path under the constraint that the diffused heat of each intermediate node is higher than its assigned threshold $\theta$. We call this path as the highest effective path (HEP). Selfish or unreliable nodes will be degraded with respect to their direct reputations by their neighbors while enthusiastic or reliable nodes will be upgraded in their reputations. The extent of increasing or decreasing a node's reputation depends on the functions it takes. Forwarding data packets will get higher reputation increments than forwarding routing packets. Correspondingly, not forwarding data packets will get heavier punishment on reputation than not forwarding routing packets. The utility a node gets in one session is the amount of payment it receives from the source node, subtracted by the cost it expends for forwarding data or routing packets. The scheme is summarized in Algorithm 1.

Under the effect of the algorithm, we can see that by behaving cooperatively a node can get higher and higher reputations, thus the payment to it will also be increased, so as to the individual utility. To earn more utility, the node will then try to improve its reputation by actively forwarding for others.

Sometimes for one session a node's utility obtained for forwarding routing packets may be higher than that of forwarding data packets. But the increasing acceleration of the latter is larger than that of the former because of the different updating way of reputation. So in the long run the cumulative utility of the node in the latter will exceed that in the former. If a node declare a higher cost than its actual forwarding cost to avoid being selected in the HEP, it will suffer the same situation. The above are some intuitive thoughts behind the scheme; for precise analysis we give it in Section 5.

## 5 Our Coalitional Game

In this section we will analyze the proposed incentive scheme by modelling the routing and forwarding procedure as a cooperative coalitional game with transferable payoff. Furthermore, we show that the game has a non-empty core.

### 5.1 Value Function of the Coalition

The value or characteristic function is the key component of a coalitional game. For each coalition $T$, the value $v(T)$ is the total payoff that is available for division among the members of $T$. It can also be interpreted to be the most payoff that the coalition $T$ can guarantee independent of the behavior of the coalition $N \setminus T$ [7]. Now we will define the value function of the coalition in our game. As described in Section 3, $s$ is the source node and $d$ is the destination node. The player set is $N = M \setminus \{s, d\}$. When nodes join together into one coalition, they will establish one or more paths between $s$ and $d$, each of which gives the coalition a collective payoff $w_P(T)$, where $P \subseteq T$ represents a path. The collective payoff comes from each member's contributions of their reputation-based payments $h_i$, then subtracted by the costs they bear for performing routing or forwarding behaviors. For those who are only involved in the routing discovery process the cost is $c_i(r)$, and for those who have been selected in a path the cost should be $c_i(f)$. So for each path $P \subseteq T$, the corresponding payoff function for the coalition is $w_P(T) = \sum_{i \in T} h_i - \sum_{i \in P} c_i(f) - \sum_{i \notin P} c_i(r)$. Among all these payoffs, we say that the characteristic worth or the value of the coalition $v(T)$ should be the maximal collective payoff it can guarantee, which is:

$$v(T) = \max_{P \subseteq T} \left( \sum_{i \in T} h_i - \sum_{i \in P} c_i(f) - \sum_{i \notin P} c_i(r) \right) . \tag{5}$$

We call the path that has the maximal $w_P(T)$ as the highest effective path $HEP_T$, so alternatively we can write $v(T) = \sum_{i \in T} h_i - \sum_{i \in HEP_T} c_i(f) - \sum_{i \notin HEP_T} c_i(r)$. But if there is no such path inside the coalition, the coalition is inessential and worths nothing, and the value of it is 0. Then formally, we have the definition of $v(T)$ as follows.

**Definition 1 (Value Function of A Coalition).** *The value of any coalition $T \subseteq N$ is 0 when there is no path between $s$ and $d$ inside $T$. That is: $v(T) = 0$, if $P_{sd}(T) = \phi$. Otherwise, $v(T)$ is:*

$$v(T) = \sum_{i \in T} h_i - \sum_{i \in HEP_T} c_i(f) - \sum_{i \notin HEP_T} c_i(r), \quad \text{if } P_{sd}(T) \neq \phi \tag{6}$$

### 5.2 Nonemptiness of the Core

The key issue of a cooperative game is regarding how to divide earnings inside the coalition in some effective and fair way. The adequate allocation profile is then called a solution, which is a vector $\boldsymbol{x} \in \mathbb{R}^N$ representing the allocation to each player when a grand coalition is formed. The grand coalition means all the players form into one coalition. The core is one of the solution concepts for cooperative games. If a coalitional game's core is non-empty, it means that no coalition can obtain a payoff that exceeds the sum of its members' current payoffs, which means no deviation is profitable for all of its members [7]. Theoretically the core is the set of imputation vectors which satisfies the following three conditions:

$$\begin{array}{l} 1.\ x(i) \geq v(i) \\ 2.\ x(T) \geq v(T), \quad \forall T \in 2^N \\ 3.\ x(N) = v(N),\ N \text{ is the player set} \end{array} \tag{7}$$

where $x(i)$ is the payoff share of node $i$ in this game, $x(T) = \sum_{i \in T} x(i)$, and $x(N) = \sum_{i \in N} x(i)$.

The core of a coalitional game is possibly empty. Next we will analyze in which condition our game has a non-empty core, and what the possible core is. We derive the following theorem.

**Theorem 1.** *Under the condition of $h_i \geq c_i(f)$ for each player $i$, the payoff profile $x$ is in the core of the coalitional game where*

$$x(i) = \begin{cases} h_i - c_i(f), & i \in HEP_N \\ h_i - c_i(r), & i \notin HEP_N \end{cases} \tag{8}$$

*Proof.* Firstly, check the first requirement of Eq.7. Under the condition of $h_i \geq c_i(f)$, we have $x(i) = h_i - c_i(f) \geq 0$. When the coalition has only one member $i$, the value of it would be 0 if $i$ cannot establish a path between $s$ and $d$. That is $v(i) = 0$. Thus $x(i) \geq v(i)$ holds. If $i$ can connect $s$ and $d$, then $v(i) = h_i - c_i(f)$ as said by Def.1. In that case $x(i) = v(i)$ which also meets the first requirement.

Secondly, from Eq.8 and Def.1, we have $x(N) = \sum_{i \in N} x(i) = \sum_{i \in HEP_N}(h_i - c_i(f)) + \sum_{i \notin HEP_N}(h_i - c_i(r)) = \sum_{i \in N} h_i - \sum_{i \in HEP_N} c_i(f) - \sum_{i \notin HEP_N} c_i(r) = v(N)$. So the third requirement of Eq.7 also holds.

Thirdly, to prove $\boldsymbol{x}$ satisfies the second requirement $x(T) \geq v(T)$, we will list and analyze all of the different HEP situations in the grand coalition $N$ and an arbitrary coalition $T$. For those coalitions without paths inside, the values of them are 0, so we easily get $x(T) \geq v(T) = 0$. For other coalitions, there are totally four kinds of situations as illustrated in Fig.2.

The proof for these four situations are similar. Because of the space limit, we only prove the most complicated situation in Fig.2(d) here. In this case, when the grand coalition $N$ is formed, the new $HEP_N$ is different from $HEP_T$ and part of $HEP_N$ is inside $T$. For clarity we first give the following notations for Fig.2(d). We let $A = HEP_N \cap T$, $B = HEP_N \cap (N \setminus T)$, $C = HEP_T \cup A$,

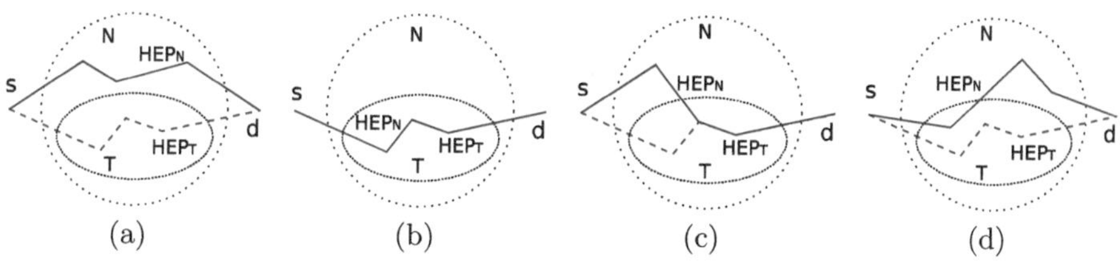


**Fig. 2.** Examples of Different HEP Situations

$D = T \setminus C$, and $E = N \setminus (HEP_N \cup HEP_T)$. According to Def.1 and Eq.8, we have:

$$x(T) - v(T) = \left[\sum_{i \in HEP_T} c_i(f) - \sum_{i \in HEP_T} c_i(r)\right] - \left[\sum_{i \in A} c_i(f) - \sum_{i \in A} c_i(r)\right] \quad (9)$$

$HEP_N$ and $HEP_T$ are two paths connecting $s$ and $d$ inside the grand coalition $N$, and $HEP_N$ dominates $HEP_T$. So based on Eq.5, we have $v_{HEP_N}(N) \geq v_{HEP_T}(N)$. Through deduction we get:

$$\sum_{i \in HEP_T} c_i(f) - \sum_{i \in HEP_T} c_i(r) \geq \left[\sum_{i \in A} c_i(f) - \sum_{i \in A} c_i(r)\right] + \left[\sum_{i \in B} c_i(f) - \sum_{i \in B} c_i(r)\right]$$

$$> \sum_{i \in A} c_i(f) - \sum_{i \in A} c_i(r)$$

Then substitute in Eq.9, we get $x(T) \geq v(T)$. So the second requirement is satisfied. In summary, under the condition of $h_i \geq c_i(f)$ for each player $i$, the proposed payoff profile $\boldsymbol{x}$ is in the core of this coalitional game. □

We can see that if only the payment a node gets based on its reputation is larger than the cost it needs to forward data packets, the core of this coalitional game exists. Nodes who want to get more payoff share $x_i$ must try to improve its reputation by helping others forwarding or increasing its link reliability, so that it can get more diffusion of the heat-based payment. In this way a virtuous cycle can be created.

## 6 Evaluations

We have theoretically proved that our incentive scheme guarantees the existence of the core when modelled as the coalitional game. Now we will evaluate the scheme in two aspects through experiments: 1) how is the general overview of all the nodes' utility and how does the network topology affect the distribution of it; and 2) how the nodes' cumulative utilities and balances evolve over time.

We conduct the evaluation on a randomly generated wireless topology with 100 nodes scattering in an area of 3000 by 3000 meters. The radio range is set to 422.757 meters. The topology is shown in Fig.3(a). There is a line connecting two nodes when they are in the communication range of each other. We label some representative nodes for further illustration. Each node has an initial balance of 100 and each directed link has a local reputation value as the weight. At each round we randomly select a source-destination pair and the source $s$ perform the incentive routing and forwarding algorithm. We assign the parameter $\lambda$ in

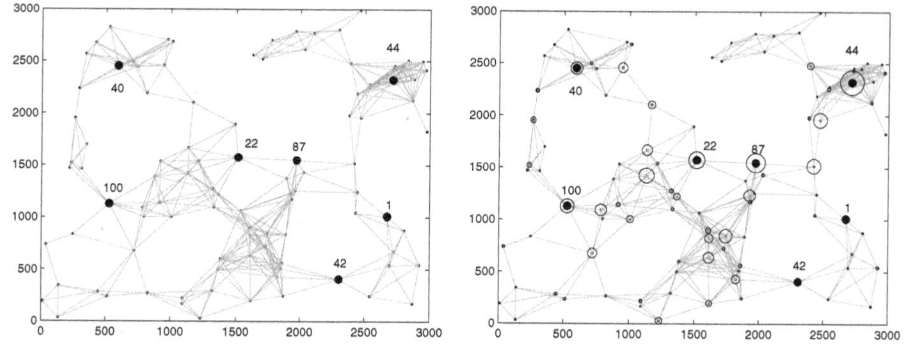


(a) (b)

**Fig. 3.** Network Topology and Overview of Nodes' Utilities

the heat diffusion equation as 1. The evaluation runs for 1000 seconds and we observe the utility and balance of each node every second.

Our first evaluation shows the overview utility at the end of the experiment in Fig.3(b). The circles around the nodes represent the cumulative utility of that node. The diameter of the circle is proportional to the amount of the utility.

We observe that in general nodes in the high density area also have large circles around them (like node 44), and on the contrary, nodes in the sparse area usually have indistinctive circles.

Our second evaluation starts from the core of the coalitional game. Fig.4(a) and 4(b) show the cumulative utilities and balances of several typical nodes respectively over the simulation time. The balance of a node may fall below the initial balance (like node 42) because the nodes have their own data transmission requests and what they earn cannot compensate what they pay.

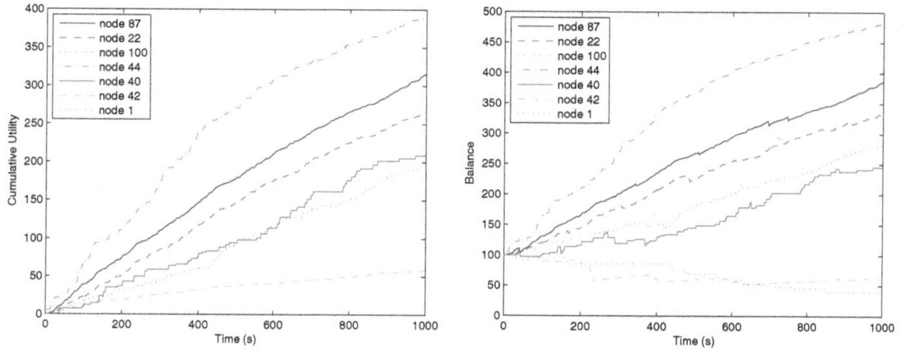


(a) (b)

**Fig. 4.** Cumulative Utility and Balance of Nodes as a Function of Simulation Time

## 7 Related Work

Many incentive routing schemes have been proposed in the past few years. Most approaches fall into one of two main categories. In the first category, nodes forwarding packets get monetary incentives for their service. In Ad Hoc-VCG [8], payments are paid to nodes consisting the actual costs incurred by forwarding data and the extra premiums. The implemented reactive routing protocol is a variation of the well-known VCG mechanism. It achieves the design objectives of truthfulness and cost-efficiency in a game-theoretic sense. Another work [6] introduces a virtual currency called *nuglets*. The source of the packet must load it with enough nuglets to pay for the trip to the destination. Cooperation is enforced in this scheme because nodes must forward packets for others in order to build up enough nuglets to get their own packets forwarded. [9] designs an incentive-compatible routing and forwarding protocol integrating VCG mechanism and cryptographic technique. Payments are implemented based on VCG protocol and the application of cryptographic techniques in the design of forwarding protocol enforces the routing decision. [10] designs a collusion-resistant routing scheme for non-cooperative wireless networks. Payments are given to nodes not only on the LCP paths but also off the paths.

In the second category, non-cooperative nodes are identified based on a reputation system and circumvented in the routing process. In CORE [1], node cooperation is stimulated by a collaborative monitoring technique and a reputation mechanism. Each node of the network monitors the behavior of its neighbors with respect to a requested function and collects observations about the execution of that function. CONFIDANT [11] differs from CORE only in that it sends reputation values to other nodes in the network, which exposes the scheme to malicious spreading of false reputation values. Liu and Issarny employ a Bayesian approach to design an incentive compatible reputation system to facilitate the trustworthiness evaluation of nodes [12]. Some also use subjective logic to calculate uncertain trust so as to design secure routing protocols [2] or incentive reputation mechanisms [13].

## 8 Conclusion and Future Work

In this paper, we present a novel incentive routing and forwarding scheme which combines reputation system and payment mechanism together to encourage nodes to cooperate in wireless ad hoc networks. Besides, we design our reputation system based on a heat diffusion model for the first time in the literature. The heat diffusion model provides us a way of combining the direct and indirect reputations together and propagating the reputation from locally to globally. Further, instead of using the non-cooperative game method, we model and analyze our incentive scheme using a coalitional game. We further prove that under a certain condition this game has a non-empty core. Through the evaluation we can see that the cumulative utility of nodes increases when the nodes stay in the core. In the future we will consider to apply other underlying reputation systems to our incentive scheme.

## Acknowledgement

The work described in this paper was fully supported by a grant from the Research Grants Council of the Hong Kong Special Administrative Region, China (Project No. CUHK4158/08E).


## References


1. Michiardi, P., Molva, R.: Core: a collaborative reputation mechanism to enforce node cooperation in mobile ad hoc networks. In: IFIP CMS, Deventer, The Netherlands, pp. 107–121. Kluwer, B.V, Dordrecht (2002)
2. Li, X., Lyu, M.R., Liu, J.: A trust model based routing protocol for secure ad hoc networks. In: Proceedings of IEEE Aerospace Conference, pp. 1286–1295 (March 2004)
3. Yang, H., King, I., Lyu, M.R.: Diffusionrank: a possible penicillin for web spamming. In: SIGIR, pp. 431–438. ACM, New York (2007)
4. Ma, H., Yang, H., Lyu, M.R., King, I.: Mining social networks using heat diffusion processes for marketing candidates selection. In: CIKM, October 2008, pp. 233–242 (2008)
5. Feeney, L.M., Nilsson, M.: Investigating the energy consumption of a wireless network interface in an ad hoc networking environment. In: INFOCOM, vol. 3, pp. 1548–1557 (2001)
6. Buttyan, L., Hubaux, J.P.: Nuglets: a virtual currency to stimulate cooperation in self-organized mobile ad hoc networks. Technical Report DSC/2001/001, Dept. of Communication Systems, Swiss Federal Institute of Technology - Lausanne (2001)
7. Osborne, M.J., Rubinstein, A.: A Course in Game Theory. The MIT Press, Cambridge (1994)
8. Anderegg, L., Eidenbenz, S.: Ad hoc-vcg: A truthful and cost-efficient routing protocol for mobile ad hoc networks with selfish agents. In: MobiCom, pp. 245–259. ACM, New York (2003)
9. Zhong, S., Li, L.E., Liu, Y.G., Yang, Y.R.: On designing incentive-compatible routing and forwarding protocols in wireless ad-hoc networks - an integrated approach using game theoretical and cryptographic techniques. In: MobiCom, pp. 117–131. ACM, New York (2005)
10. Zhong, S., Wu, F.: On designing collusion-resistant routing schemes for non-cooperative wireless ad hoc networks. In: MobiCom, pp. 278–289. ACM, New York (2007)
11. Buchegger, S., Boudec, J.Y.L.: Performance analysis of the confidant protocol: Cooperation of nodes - fairness in dynamic ad-hoc networks. In: MobiHoc, Lausanne, CH (June 2002)
12. Liu, J., Issarny, V.: An incentive compatible reputation mechanism for ubiquitous computing environments. IJIS 6(5), 297–311 (2007)
13. Kane, K., Browne, J.C.: Using uncertainty in reputation methods to enforce cooperation in ad-hoc networks. In: WiSe 2006: Proceedings of the 5th ACM workshop on Wireless security, pp. 105–113. ACM, New York (2006)

# Performance Analysis of Centralized versus Distributed Recovery Schemes in P2P Storage Systems

Abdulhalim Dandoush, Sara Alouf, and Philippe Nain

INRIA Sophia Antipolis – B.P. 93 – 06902 Sophia Antipolis, France
{adandous,salouf,nain}@sophia.inria.fr

**Abstract.** This paper studies the performance of Peer-to-Peer Storage Systems (P2PSS) in terms of data lifetime and availability. Two schemes for recovering lost data are modeled through absorbing Markov chains and their performance are evaluated and compared. The first scheme relies on a centralized controller that can recover multiple losses at once, whereas the second scheme is distributed and recovers one loss at a time. The impact of each system parameter on the performance is evaluated, and guidelines are derived on how to engineer the system and tune its key parameters in order to provide desired lifetime and/or availability of data. We find that, in stable environments such as local area or research laboratory networks where machines are usually highly available, the distributed-repair scheme offers a reliable, scalable and cheap storage/backup solution. This is in contrast with the case of highly dynamic environments, where the distributed-repair scheme is inefficient as long as the storage overhead is kept reasonable. P2PSS with centralized-repair scheme are efficient in any environment but have the disadvantage of relying on a centralized authority. Our analysis also suggests that the use of large size fragments reduces the efficiency of the recovery mechanism.

**Keywords:** Performance evaluation, peer-to-peer storage systems, recovery process, absorbing continuous-time Markov chain.

## 1 Introduction

The growth of storage volume, bandwidth, and computational resources for PCs has fundamentally changed the way applications are constructed, and has inspired a new class of storage systems that use distributed peer-to-peer (P2P) infrastructures. Although scalable and economically attractive compared to traditional systems, these storage systems pose many problems such as reliability, confidentiality and availability. To ensure data reliability and availability in such dynamic systems, redundant data is inserted in the system. However, using redundancy mechanisms without repairing lost data is not efficient, as the level of redundancy decreases when peers leave the system. Consequently, P2P storage systems need to compensate the loss of data by continuously storing additional redundant data onto new hosts. Systems may rely on a central authority

L. Fratta et al. (Eds.): NETWORKING 2009, LNCS 5550, pp. 676–689, 2009.

that reconstructs fragments when necessary; these systems will be referred to as *centralized-recovery systems*. Alternatively, secure agents running on new hosts can reconstruct by themselves the data to be stored on the hosts disks. Such systems will be referred to as *distributed-recovery systems*. Regardless of the recovery mechanism used, two repair policies can be adopted *eager* and *lazy* policy. The description of these recovery schemes and their policies are presented in Section 3.

The aim of this paper is to develop and evaluate mathematical models to characterize fundamental performance metrics (data lifetime and availability) of P2P storage systems. Our contributions are as follows

- Analysis of centralized and distributed recovery mechanisms.
- Proposition of a general model that captures the behavior of both eager and lazy repair policies and both replication-based and erasure code-based systems, and accommodates both temporary and permanent disconnections of peers.
- Numerical investigation using realistic parameters values.
- Guidelines on how to engineer the P2PSS in order to satisfy given requirements.

In the following, Section 2 briefly reviews related work and Section 3 introduces the notation and assumptions used throughout the paper. Sections 4 and 5 are dedicated to the modeling of the centralized- and distributed-recovery mechanism, respectively. In Section 6, we provide some numerical results showing the performance of the centralized and decentralized schemes. Section 7 concludes the paper.

## 2 Related Work and Background

The literature on the architecture and file system of distributed storage systems is abundant (see [1,2]; non-exhaustive list), but to the best of our knowledge, there has been no prior work on the modeling of the recovery process in P2PSS. A few studies have developed analytical models with the goal of understanding the trade-offs between the availability and lifetime of the files and the redundancy involved in storing the data. From these we cite [3] whose main purpose of [3] is the analysis of a storage system using replication for data reliability. The work [3] is the closest to ours even though the model and analysis developed therein do not apply for erasure-coded systems (we will see later that our models apply to either replicated or erasure-coded systems).

The authors of [3] develop a Markov chain analysis, then derive an expression for the lifetime of the replicated state and study the impact of bandwidth and storage limits on the system. However – and these are major differences with the work presented here, transient disconnections are not considered in their model, the recovery process is considered to be exponentially distributed for the aim of simplification, and only the distributed-repair scheme is considered. Another contribution of [3] is the analysis of the *All-pairs-ping* data set [4] that reports measures of both uptime and downtime for PlanetLab [5] nodes. The authors found that an exponential distribution is a reasonable fit for both uptime and downtime. This conjecture comes to support one of the key assumptions of the model presented in our paper, namely that "node participation can be modeled by an exponential distribution".

In our previous work [6], simplifying assumptions have been considered while modeling the P2PSS, mainly that the recovery process is exponentially distributed. However, the implications of this assumption differ between replicated and erasure-coded systems. We have found in [7], through a simulation analysis, that the exponential assumption made on the recovery process is usually not met in erasure-coded systems. This assumption will therefore be relaxed in this paper and it will be shown later on that a more precise, more realistic modeling is possible.

## 3 System Description, Assumptions and Notation

We consider a distributed *storage* system which peers randomly join and leave. The following assumptions on the P2PSS design will be enforced throughout the paper:

- A block of data $D$ is partitioned into $s$ equally sized fragments to which, using erasure codes (e.g. [8]), $r$ redundant fragments are added. The case of replication-based redundancy is equally captured by this notation, after setting $s = 1$ and letting the $r$ redundant fragments be simple replicas of the unique fragment of the block. This notation – and hence our modeling – is general enough to study both replication-based and erasure code-based storage systems.
- Mainly for privacy issues, a peer can store at most one fragment of any data $D$.
- We assume the system has perfect knowledge of the location of fragments at any given time, e.g. by using a Distributed Hash Table (DHT) or a central authority.
- The system keeps track of only the latest known location of each fragment.
- Over time, a peer can be either *connected* to or *disconnected* from the storage system. At reconnection, a peer may or may not still store its fragments. We denote by $p$ the probability that a peer that reconnects still stores its fragments.
- The number of connected peers at any time is typically much larger than the number of fragments associated with $D$, i.e., $s+r$. Therefore, we assume that there are always at least $s+r$ connected peers – hereafter referred to as *new* peers – which are ready to receive and store fragments of $D$.

We refer to as *on-time* (resp. *off-time*) a time-interval during which a peer is always connected (resp. disconnected). During a peer's off-time, the fragments stored on this peer are momentarily unavailable to the users of the storage system. At reconnection, and according to the assumptions above, the fragments stored on this peer will be available only with probability $p$ (and with probability $1-p$ it is lost). In order to improve data availability and increase the reliability of the storage system, it is therefore crucial to recover from losses by continuously monitoring the system and adding redundancy whenever needed.

We will investigate the performance of two different repair policies: the *eager* and the *lazy* repair policies. In the eager policy, a fragment of $D$ is reconstructed as soon as one fragment has become unavailable due to a peer disconnection. In the lazy policy, the repair is delayed until the number of unavailable fragments reaches a given threshold, denoted $k$. In the latter case, we must have $k \leq r$ since $D$ is lost if more than $r$ fragments are missing from the storage system. Both repair policies can be represented by the threshold parameter $k \in \{1, 2, \ldots, r\}$, where $k$ can take any value in the set $\{2, \ldots, r\}$ in the lazy policy and $k = 1$ in the eager policy. Any repair policy can be implemented either in a centralized or a distributed way. In the following description, we assume that the system misses $k$ fragments so that lost fragments have to be restored.

In the centralized implementation, a central authority will: (1) download *in parallel* $s$ fragments from the peers which are connected, (2) reconstruct at once all the unavailable fragments, and (3) upload them all *in parallel* onto as many new peers for storage. Step 2 executes in a negligible time compared to the execution time of Steps 1 and 3 and will henceforth be ignored in the modeling. Step 1 (resp. Step 3) execution completes when the last fragment completes being downloaded (resp. uploaded).

In the distributed implementation, a secure agent on one new peer is notified of the identity of *one* out of the $k$ unavailable fragments for it to reconstruct it. Upon notification, the secure agent (1) downloads $s$ fragments of $D$ from the peers which are connected to the storage system, (2) reconstructs the specified fragment and stores it on the peer's disk; (3) the secure agent then discards the $s$ downloaded fragments so as to meet the privacy constraint that only one fragment of a block of data is held by a peer. This operation iterates until less than $k$ fragments are sensed unavailable and stops if the number of missing fragments reaches $k - 1$. The recovery of one fragment lasts mainly for the execution time of Step 1; the recovery is completed then as soon as the last fragment (out of $s$) completes being downloaded.

In both implementations, once a fragment is reconstructed, any other copy of it that "reappears" in the system due to a peer reconnection is simply ignored, as only one location (the newest) of the fragment is recorded in the system. Similarly, if a fragment is unavailable, the system knows of only one disconnected peer that stores the unavailable fragment.

Given the system description, data $D$ can be either *available*, *unavailable* or *lost*. Data $D$ is said to be available if any $s$ fragments out of the $s + r$ fragments can be downloaded by the users of the P2PSS. Data $D$ is said to be unavailable if *less than* $s$ fragments are available for download, however the missing fragments to complete $D$ are located at a peer or a central authority on which a recovery process is ongoing. Data $D$ is said to be lost if there are *less than* $s$ fragments in the system including the fragments involved in a recovery process. We assume that, at time $t = 0$, at least $s$ fragments are available so that the document is initially available.

We now introduce the assumptions considered in our models.

**Assumption 1:** We assume that the successive durations of on-times (resp. off-times) of a peer are independent and identically distributed (iid) random variables (rvs) with a common exponential distribution function with parameter $\mu > 0$ (resp. $\lambda > 0$); this assumption is in agreement with the analysis in [3]. We further assume that peers behave independently of each other.

**Assumption 2:** We assume that the successive download (resp. upload) durations of a fragment are iid rvs with a common exponential distribution function with parameter $\alpha$ (resp. $\beta$); this assumption is in agreement with the analysis in [7]. Fragments downloads/uploads are not correlated.

A consequence of Assumption 2 is that each of the durations of the centralized and the distributed recovery processes is a rv following a hypo-exponential distribution [9]. Indeed, each of these durations is the summation of independently distributed exponential rvs ($s+k$ in the centralized scheme if $k$ fragments are to be reconstructed, and $s$ in the distributed scheme) having each its own rate. This is a fundamental difference with [3,6] where the recovery process is assumed to follow an exponential distribution. It is worth mentioning that the simulation analysis of [7] has concluded that the recovery time follows roughly a hypo-exponential distribution. As a consequence, the models presented in this paper are more realistic than those in [3,6].

We conclude this section by a word on the notation: a subscript "$c$" (resp. "$d$") will indicate that we are considering the centralized (resp. distributed) scheme. The notation $\mathbf{e}_j^i$ refers to a *row* vector of dimension $j$ whose entries are null except the $i$-th entry that is equal to 1; the notation $\mathbf{1}_j$ refers to a *column* vector of dimension $j$ whose each entry is equal to 1. Last, $\mathbb{1}\{A\}$ is the characteristic function of event $A$.

## 4 Centralized Repair Systems

We will focus on a single block of data $D$, and pay only attention to peers storing fragments of this block.

Let $X_c(t)$ and $Y_c(t)$ be two rvs denoting respectively the number of fragments in the system that are available for download and the state of the recovery process. Recall that, when $k$ fragments are to be reconstructed, the recovery process consists of a series of $s+k$ exponential distributions that can be seen as $s+k$ stages. We denote $Y_c(t) = j$ ($j = 0, 1, \ldots, k-1$) to express that $j$ exponential rvs have been realized at time $t$, so that $s+k-j$ are still to go. When the last stage is completed, the recovery process is completed and $Y_c(t) = 0$. Given that there could be as much as $s+r$ fragments to be reconstructed, the process $Y_c(t)$ takes value in the set $\{0, 1, \ldots, 2s+r-1\}$. As for $X_c(t)$, it takes value in the set $\{0, 1, \ldots, s+r\}$.

Consider now the joint process $(X_c(t), Y_c(t))$. When $X_c(t) \geq s$, data $D$ is *available*, regardless of $Y_c(t)$. When $X_c(t) < s$ but $X_c(t) + Y_c(t) \geq s$, $D$ is *unavailable*. When $X_c(t) + Y_c(t) < s$, $D$ is *lost*. The latter situation will be modeled by a single state $a$. Introduce the set

$$\begin{aligned}
\mathcal{T}_c := \{\ & \left.\begin{array}{l}(0,s),(0,s+1),\ldots,(0,2s+r-1),\\ (1,s-1),(1,s),\ldots,(1,2s+r-2),\\ \ldots,\end{array}\right\} \quad D \text{ is unavailable}\\
& \left.\begin{array}{l}(s,0),(s,1),\ldots,(s,s+r-1),\\ (s+1,0),(s+1,1),\ldots,(s+1,s+r-2),\\ \ldots,(s+r-1,0),(s+r-1,1),\ldots,(s+r-1,s),\\ (s+r,0)\ \}\end{array}\right\} \quad D \text{ is available}
\end{aligned}$$

$$|\mathcal{T}_c| = (s+r)^2 - r(r-1)/2 + 1.$$

Thanks to the assumptions made in Section 3, it is easily seen that the two-dimensional process $\{(X_c(t), Y_c(t)),\ t \geq 0\}$ is an absorbing homogeneous Continuous-Time Markov Chain (CTMC) with transient states the elements of $\mathcal{T}_c$ and with a single absorbing state $a$ representing the situation when $D$ is lost. Without loss of generality, we assume that $X_c(0) \geq s$. The infinitesimal generator has the following canonical form

$$\begin{array}{c c} & \begin{array}{cc}\mathcal{T}_c & a\end{array}\\ \begin{array}{c}\mathcal{T}_c\\ a\end{array} & \left(\begin{array}{c|c}\boldsymbol{Q}_c & \boldsymbol{R}_c\\ \hline \boldsymbol{0} & 0\end{array}\right)\end{array}$$

where $\boldsymbol{R}_c$ is a non-zero column vector of size $|\mathcal{T}_c|$, and $\boldsymbol{Q}_c$ is $|\mathcal{T}_c|$-by-$|\mathcal{T}_c|$ matrix. The elements of $\boldsymbol{R}_c$ are the transition rates between the transient states $(i,j) \in \mathcal{T}_c$ and the absorbing state $a$, namely, $r_c(i,j) = (s-j)\mu$, for $i = 1,\ldots,s$, and $j = s-i,\ldots,s-1$. The elements of $\boldsymbol{R}_c$ are lexicographically ordered alike the order in $\mathcal{T}_c$. The diagonal elements of $\boldsymbol{Q}_c$ are each the total transition rate out of the corresponding transient state. The other elements of $\boldsymbol{Q}_c$ are the transition rates between each pair of transient states. The non-zero elements of $\boldsymbol{Q}_c$ are:

$$q_c((i,j),(i-1,j)) = \begin{cases} i\mu, & \text{for } i = 1,\ldots,s, j = s,\ldots,2s+r-1-i;\\ & \text{or } i = s+1,\ldots,s+r-1,\\ & \quad\ j = 0,\ldots,2s+r-1-i;\\ & \text{or } i = s+r, j = 0;\\ (i+j-s)\mu, & \text{for } i = 2,\ldots,s, j = s+1-i,\ldots,s-1.\end{cases}$$

$$q_c((i,j),(i,j+1)) = \begin{cases} (s-j)\alpha, & \text{for } i = s,\ldots,s+r-k, j = 0;\\ & \text{or } i = 1,\ldots,s-1, j = s-i,\ldots,s-1;\\ & \text{or } i = s,\ldots,s+r-1, j = 1,\ldots,s-1;\\ (2s+r-i-j)\beta, & \text{for } i = 0,\ldots,s+r-2,\\ & \quad\ j = s,\ldots,2s+r-2-i.\end{cases}$$

$q_c((i,2s+r-1-i),(s+r,0)) = \beta$, for $i = 0,\ldots,s+r-1$.

$q_c((i,j),(i+1,j)) = (s+r-i)\lambda p$, for $i = 1,\ldots,s, j = s-i,\ldots,s-1$; or $i = s+1,\ldots,s+r-2, j = 0,\ldots,s-1$.

$q_c((s+r-1,j),(s+r,0)) = \lambda p$, for $j = 0,\ldots,s-1$.

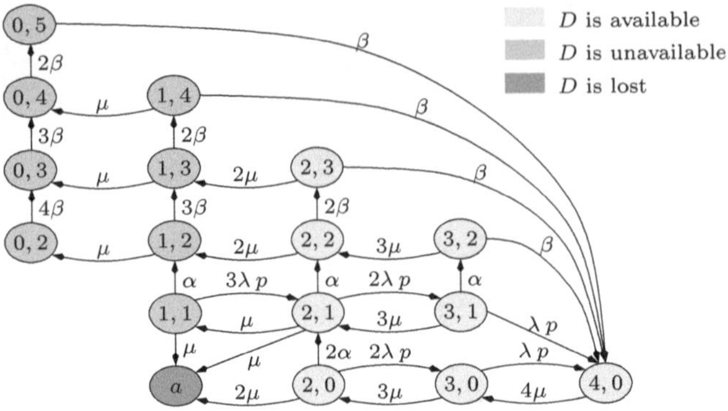


**Fig. 1.** The Markov chain $\{(X_c(t), Y_c(t)), t \geq 0\}$ when $s = 2$, $r = 2$, $k = 2$

$q_c((i,j),(i,j)) = -r_c(i,j) - \sum_{(i',j') \in \mathcal{T}_c - \{(i,j)\}} q_c((i,j),(i',j')),.$ for $(i,j) \in \mathcal{T}_c$. Note that $\boldsymbol{Q}_c$ is not an infinitesimal generator since entries in some rows do not sum up to 0. For illustration purposes, we depict in Fig. 1 an example of the absorbing CTMC with its non-zero transition rates when $s = 2$, $r = 2$, and $k = 2$.

### 4.1 Data Lifetime

This section is devoted to the analysis of the lifetime of $D$. Let $T_c(i,j) := \inf\{t > 0 : (X_c(t), Y_c(t)) = a | (X_c(0), Y_c(0)) = (i,j)\}$ be the time until absorption in state $a$, or equivalently the time until $D$ is lost, given that the initial state of $D$ is $(i,j)$. In the following, $T_c(i,j)$ will be referred to as the *conditional block lifetime*. We are interested in $P(T_c(i,j) \leq x)$ and $\mathrm{E}[T_c(i,j)]$, respectively the probability distribution of the conditional block lifetime and its expectation, given that $(X_c(0), Y_c(0)) = (i,j) \in \mathcal{T}_c$. From the theory of absorbing Markov chains, we know that (e.g. [10, Lemma 2.2])

$$P(T_c(i,j) \leq x) = 1 - \boldsymbol{e}_{|\mathcal{T}_c|}^{\mathrm{ind}(i,j)} \cdot \exp(x\boldsymbol{Q}_c) \cdot \mathbf{1}_{|\mathcal{T}_c|}, \quad x > 0,\ (i,j) \in \mathcal{T}_c \tag{1}$$

where $\mathrm{ind}(i,j)$ refers to the index of the state $(i,j) \in \mathcal{T}_c$ in the matrix $\boldsymbol{Q}_c$. Recall that the elements of $\boldsymbol{Q}_c$ are numbered according to the lexicographic order. Definitions of vectors $\boldsymbol{e}_i^j$ and $\mathbf{1}_i$ are given at the end of Section 3. Observe that the term $\boldsymbol{e}_{|\mathcal{T}_c|}^{\mathrm{ind}(i,j)} \cdot \exp(x\boldsymbol{Q}_c) \cdot \mathbf{1}_{|\mathcal{T}_c|}$ in the r.h.s. of (1) is nothing but the summation of all $|\mathcal{T}_c|$ elements in row $\mathrm{ind}(i,j)$ of matrix $\exp(x\boldsymbol{Q}_c)$.

We know from [10, p. 46] that the expected time until absorption can be written as

$$\mathrm{E}[T_c(i,j)] = -\boldsymbol{e}_{|\mathcal{T}_c|}^{\mathrm{ind}(i,j)} \cdot (\boldsymbol{Q}_c)^{-1} \cdot \mathbf{1}_{|\mathcal{T}_c|}, \quad (i,j) \in \mathcal{T}_c, \tag{2}$$

where the existence of $(\boldsymbol{Q}_c)^{-1}$ is a consequence of the fact that all states in $\mathcal{T}_c$ are transient [10, p. 45]. Inverting $\boldsymbol{Q}_c$ analytically can rapidly become cumbersome as $s$ or $r$ increases. We will instead perform numerical computations as reported in Section 6. Consider now

$$T_c((i,j),(i',j')) := \int_0^{T_c(i,j)} \mathbb{1}\{(X_c(t),Y_c(t)) = (i',j')\}\, dt$$

that is the total time spent by the CTMC in transient state $(i',j')$ given that $\{X_c(0),Y_c(0)\} = (i,j)$. It can also be shown that [11, p. 419]

$$\mathrm{E}\left[T_c((i,j),(i',j'))\right] = -\boldsymbol{e}_{|\mathcal{T}_c|}^{\mathrm{ind}(i,j)} \cdot (\boldsymbol{Q}_c)^{-1} \cdot {}^t\boldsymbol{e}_{|\mathcal{T}_c|}^{\mathrm{ind}(i',j')}, \quad (i,j),(i',j') \in \mathcal{T}_c, \quad (3)$$

where ${}^t\boldsymbol{y}$ denotes the transpose of a given vector $\boldsymbol{y}$. In other words, the expectation $\mathrm{E}\left[T_c((i,j),(i',j'))\right]$ is the entry of matrix $(-\boldsymbol{Q}_c)^{-1}$ at row $\mathrm{ind}(i,j)$ and column $\mathrm{ind}(i',j')$.

### 4.2 Data Availability

In this section we introduce different metrics to quantify the availability of $D$. We are interested in the fraction of time spent by the CTMC in any given state $(i',j')$ before absorption. However, this quantity is difficult to find in closed-form. Therefore, we resort to using the following approximation

$$\mathrm{E}\left[\frac{T_c((i,j),(i',j'))}{T_c(i,j)}\right] \approx \frac{\mathrm{E}[T_c((i,j),(i',j'))]}{\mathrm{E}[T_c(i,j)]}. \quad (4)$$

Here, $(i,j)$ is the state of $D$ at $t=0$. This approximation have been validated through simulations, as shown later in Section 6. With this approximation in mind, we introduce two availability metrics: the first can be interpreted as the expected number of fragments of $D$ that are in the system during the lifetime of $D$; the second can be interpreted as the fraction of time when at least $m$ fragments are in the system during the lifetime of $D$. More formally, given that $(X_c(0),Y_c(0)) = (i,j) \in \mathcal{T}_c$, we define

$$M_{c,1}(i,j) := \sum_{(i',j')\in\mathcal{T}_c} i' \frac{\mathrm{E}[T_c((i,j),(i',j'))]}{\mathrm{E}[T_c(i,j)]}, \quad (5)$$

$$M_{c,2}((i,j),m) := \sum_{(i',j')\in\mathcal{T}_c,\, i'\geq m} \frac{\mathrm{E}[T_c((i,j),(i',j'))]}{\mathrm{E}[T_c(i,j)]}. \quad (6)$$

## 5 Distributed Repair Systems

In this section, we model P2P storage systems that implement a distributed recovery mechanism. According to the description and assumptions listed in Section 3, the state of data $D$ can be modeled by an absorbing Markov chain $\{(X_d(t),Y_d(t)) : t \geq 0\}$, where $X_d(t)$ and $Y_d(t)$ denote respectively the number of fragments in the system that are available for download and the state of the recovery process. Unlike the centralized scheme, the distributed scheme repairs fragments only one at a time. Therefore, $X_d(t)$ takes value in the set $\{s-1, s, \ldots, s+r\}$. There are only $s$ stages in the recovery process that correspond each to an exponential with it own rate (i.e., $Y_d(t) \in \{0, 1, \ldots, s-1\}$). As in Section 4, $D$ is available when $X_d(t) \geq s$, unavailable when $X_d(t) < s$ but $X_d(t) + Y_d(t) \geq s$, and lost otherwise (situation modeled by a single absorbing state $a$). The set of transient states $\mathcal{T}_d$ is

$$
\begin{aligned}
\mathcal{T}_d := \{\ & (s-1,1), (s-1,2), \ldots, (s-1,s-1), && \} \text{ D is unavailable} \\
& (s,0), (s,1), \ldots, (s,s-1), && \\
& (s+1,0), (s+1,1), \ldots, (s+1,s-1), \ldots, && \text{D is available} \\
& (s+r-1,0), (s+r-1,1), \ldots, (s+r-1,s-1), && \\
& (s+r,0)\ \} && \\
|\mathcal{T}_d| = s&(r+1).
\end{aligned}
$$

The analysis of the absorbing Markov chain $\{(X_d(t), Y_d(t)) : t \geq 0\}$ that takes value in $\mathcal{T}_d \cup \{a\}$ is very similar to the analysis in Section 4, we will then only sketch it. In particular, $\boldsymbol{R}_d$ and $\boldsymbol{Q}_d$ have similar definitions as $\boldsymbol{R}_c$ and $\boldsymbol{Q}_c$ after replacing the subscript "$c$" with the subscript "$d$" whenever needed. The non-zero elements of $\boldsymbol{R}_d$ and $\boldsymbol{Q}_d$ are as follows

$$
r_d(s-1,j) = (s-1)\mu, \text{ for } j=1,\ldots,s-1; \quad r_d(s,j) = (s-j)\mu, \text{ for } j=0,\ldots,s-1.
$$

$$
q_d((i,j),(i-1,j)) = \begin{cases} j\mu, & \text{for } i=s, j=1,\ldots,s-1; \\ i\mu, & \text{for } i=s+1,\ldots,s+r-1, j=0,\ldots,s-1; \\ & \text{or } i=s+r, j=0. \end{cases}
$$

$$
\begin{aligned}
q_d((i,j),(i,j+1)) &= (s-j)\alpha, && \text{for } i=s,\ldots,s+r-k, j=0; \\
& && \text{or } i=s-1,\ldots,s+r-1, j=1,\ldots,s-2. \\
q_d((i,s-1),(i+1,0)) &= \alpha, && \text{for } i=s-1,\ldots,s+r-2.
\end{aligned}
$$

$$
\begin{aligned}
q_d((i,j),(i+1,j)) = (s+r-i)\lambda p, \quad & \text{for } i=s-1, j=1,\ldots,s-1; \\
& \text{or } i=s,\ldots,s+r-2, j=0,\ldots,s-1.
\end{aligned}
$$

$$
\begin{aligned}
& q_d((s+r-1,j),(s+r,0)) = \lambda p + \mathbb{1}\{j=s-1\}\alpha, \quad \text{for } j=0,\ldots,s-1. \\
& q_d((i,j),(i,j)) = -r_d(i,j) - \textstyle\sum_{(i',j') \in \mathcal{T}_d - \{(i,j)\}} q_d((i,j),(i',j')), \quad \text{for } (i,j) \in \mathcal{T}_d.
\end{aligned}
$$

For illustration purposes, we depict in Fig. 2 an example of the absorbing CTMC with its non-zero transition rates when $s=3$, $r=2$, and $k=2$.

We can now derive closed-form expressions for the distribution of the conditional block lifetime, its expectation, and the two availability metrics, as was done in Section 4, by simply replacing in (1), (2), (3), (5) and (6) the subscript "$c$" with the subscript "$d$". Alike for the centralized case, we will perform numerical computations as it is not tractable to explicitly invert $\boldsymbol{Q}_d$.

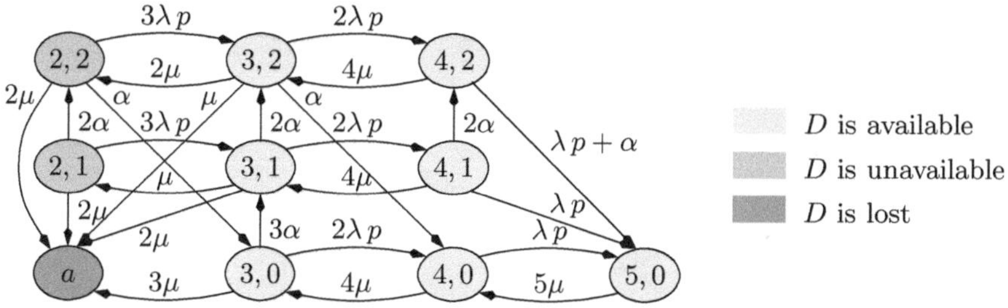


**Fig. 2.** The Markov chain $\{(X_d(t), Y_d(t)),\, t \geq 0\}$ when $s = 3$, $r = 2$, and $k = 2$

# 6 Numerical Results

In this section, we first validate the approximation made in (4) which has been made to compute the two availability metrics, then proceed with the presentations of the numerical computations. Throughout this section, we consider two sets of parameters that correspond each to a particular context. In the "PlanetLab" context, we set $1/\lambda = 61$ hours and $1/\mu = 181$ hours according to [3], $p = 0.3$ to reflect that disconnections are most likely due to software or hardware problems, and we vary the redundancy $r$ from 1 until $s$. In the "Internet" context, peers churn rate is much higher [12] (namely, hosts join and leave the system 6.4 times a day on average), which led us to set $1/\lambda = 1$ hours, $1/\mu = 3$ hours, $p = 0.7$, and to vary $r$ from 1 to $2s$. In both contexts, we let $s = 8$, $k = 1, \ldots, r$. We assume the peers' upload capacity that is dedicated to a single connection to be 10kbps (cf. [13]), and the total upload capacity of the central authority to be 400kbps. Hence, considering fragments of size 1MB, we obtain $1/\alpha = 838.8608$ seconds, $1/\beta = 20.97152 \times s$ seconds.

**Validation of** (4)**.** To this end, we have simulated the CTMC $\{X_d(t), Y_d(t) : t \geq 0\}$ in the "Internet" context but have varied $r$ from 1 to 4, yielding a total of 10 different simulation scenarios. We estimate the left-hand side of (4) by averaging the corresponding simulated results. In order to obtain a maximum estimation error of about 1% with 97% confidence interval, we need to average over 150 sampled values. Hence, each simulation is repeated 150 times. In each simulation, we have a total of $|\mathcal{T}_d| = 8(r+1)$ instances of (4), according to the possible choices of the initial state, yielding a total of $\sum_{r=1}^{4} 8(r+1)r = 320$ different instances. For each instance, we compute the relative error between the estimation of the left-hand side of (4) (the "correct" value) and the right-hand side of (4) (the approximate value, computed analytically using (2)-(3)).

We have found only 10% of the values that are larger than $0.9 \times 10^{-3}$ and, most importantly, the maximum value of the relative error is 0.0028. *We conclude that the approximation* (4) *is very good and will definitely not imperil the correctness of any computation based on it.*

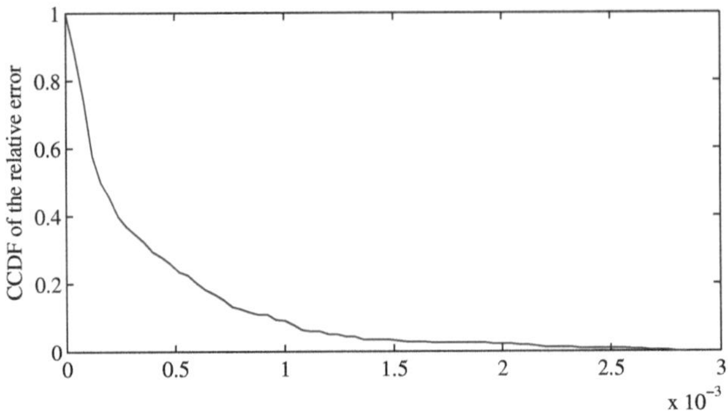


**Fig. 3.** The CCDF of the relative error induced by the approximation (4)

The empirical complementary cumulative distribution function of the relative error is displayed in Fig. 3.

**Performance Analysis.** We have solved numerically (1), (2), (3), (5) and (6) given that all $s+r$ fragments of $D$ are initially available, considering either Internet or PlanetLab context, and either the centralized or distributed recovery scheme. Results are reported partially in Table 1. It appears that, whichever the scenario or the recovery mechanism considered, the expected data lifetime increases roughly exponentially with $r$ and decreases with an increasing $k$. Regardless of the context considered, the distributed scheme yields a significantly smaller expected data lifetime than the centralized scheme, especially when the storage overhead, $r/s$, is high; cf. columns 3-4 in Table 1. The difference in performance is more pronounced in the Internet context. Regarding the expected number of available fragments, we again observe that the distributed scheme is less efficient than the centralized one. Observe how the performance deteriorates as peer churn becomes more important: compare for instance in Table 1 rows 4 vs. 16, and 7 vs. 20 (these correspond to the same storage overhead and the same value of $k$). This is particularly true for the distributed recovery mechanism. We conclude that *when peers churn rate is high, only the centralized repair scheme can be efficient should the storage overhead be kept within a reasonable value (that is $r/s \leq 2$)*. As the distributed repair scheme is more scalable than the centralized one, it will be a good implementation choice in large networks where hosts have a good availability.

**Setting the System's Key Parameters.** We illustrate now how our models can be used to set the system parameters $r$ and $k$ such that predefined requirements on data lifetime and availability are fulfilled. We assume the recovery mechanism is centralized and the context is similar to PlanetLab. We have picked one to two contour lines of each of the performance metrics studied in this paper and report them in Fig. 4. Consider point A which corresponds to $r=6$ and $k=1$ (recall $s=8$). Selecting this point as the operating point of the

**Table 1.** Expected lifetime and first availability metric

| **Internet context** $s = 8$ | | $E[T(s+r,0)]$ (in days) cent. repair | dist. repair | $M_1(s+r,0)$ cent. repair | dist. repair |
|---|---|---|---|---|---|
| $k = 1$ | $r/s = 1/2$ | 0.14 | 0.08 | 9.91 | 8.99 |
| | $r/s = 1$ | 0.85 | 0.35 | 11.72 | 10.58 |
| | $r/s = 3/2$ | 14.49 | 2.25 | 14.46 | 12.40 |
| | $r/s = 2$ | 105.80 | 23.61 | 16.55 | 14.73 |
| $k = 4$ | $r/s = 1$ | 0.71 | 0.33 | 11.66 | 9.67 |
| | $r/s = 3/2$ | 12.38 | 2.23 | 14.45 | 11.50 |
| | $r/s = 2$ | 91.59 | 23.58 | 16.48 | 13.83 |
| $k = 8$ | $r/s = 2$ | 64.16 | 22.31 | 16.30 | 12.61 |
| **PlanetLab context** $s = 8$ | | $E[T(s+r,0)]$ (in months) cent. repair | dist. repair | $M_1(s+r,0)$ cent. repair | dist. repair |
| $k = 1$ | $r/s = 1/4$ | 0.32 | 0.11 | 7.81 | 8.04 |
| | $r/s = 1/2$ | 2.15 | 1.05 | 11.01 | 8.68 |
| | $r/s = 3/4$ | 17.12 | 7.61 | 13.18 | 9.80 |
| | $r/s = 1$ | 262.16 | 46.24 | 15.11 | 12.12 |
| $k = 2$ | $r/s = 1/2$ | 0.81 | 0.37 | 10.34 | 8.19 |
| | $r/s = 3/4$ | 6.95 | 3.20 | 12.76 | 9.25 |
| | $r/s = 1$ | 110.03 | 23.34 | 14.72 | 11.37 |
| $k = 4$ | $r/s = 1$ | 13.33 | 4.34 | 13.77 | 9.81 |

P2PSS ensures (roughly) the following: given that each data is initiated with $s + r$ available fragments, then (i) the expected data lifetime is 18 months; (ii) only 11% of the stored data would be lost after 3 months; (iii) as long as $D$ is not lost, 13 fragments of $D$ are expected to be in the system; (iv) during 99.7% of its lifetime, $D$ is available for download; and (v) during 80% of the lifetime of $D$, at least $s + r - k = 13$ fragments of $D$ are available for download in the system. Observe that the storage overhead, $r/s$, is equal to 0.75.

**Impact of the Size of Fragments.** Given the size of data $D$, a larger size of fragments translates into a smaller $s$. We have computed all pairs $(r, k)$ with $s = 8$ and $s = 16$ that ensure $P(T_c(s+r,0) > 3 \text{ months}) = 0.89$ in the PlanetLab context, i.e., only 11% of the total data would be lost after 3 months. In particular, operating points $r = 6$ and $k = 1$ with $s = 8$, and $r = 12$ and $k = 7$ with $s = 16$ satisfy the above requirement, and additionally yield the same storage overhead (namely, 0.75). But, and this is important, the former point invokes the recovery process much more often (and potentially unnecessarily) than the latter point, suggesting that *large fragments size reduces the efficiency of the recovery mechanism*. This observation should be moderated by the fact that fragments size when $s = 8$ is twice their size when $s = 16$, yielding a different bandwidth usage per recovery. We currently cannot say how does the bandwidth usage per recovery vary with the size of fragments. However, we know for sure that its effect will not be the same in both centralized and distributed schemes because of the

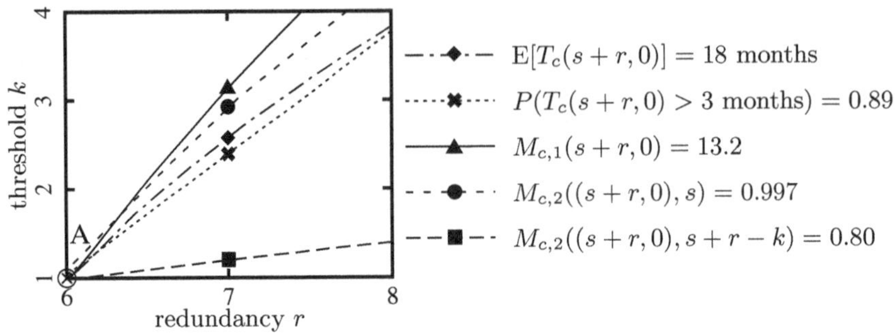


**Fig. 4.** Contour lines of performance metrics (PlanetLab context,centralized repair)

additional upload stages in the centralized implementation. A careful analysis of this issue is the objective of ongoing research.

## 7 Conclusion

We have proposed analytical models for evaluating the performance of two approaches for recovering lost data in distributed storage systems. We have analyzed the lifetime and the availability of data achieved by both centralized- and distributed-repair systems through Markovian analysis considering realistic assumptions. Numerical computations have been undertaken to illustrate several issues on the performance. We conclude that, using our theoretical framework, it is easy to tune and optimize the system parameters for fulfilling predefined requirements.

## References

1. Rowstron, A., Druschel, P.: Storage management and caching in PAST, a large-scale, persistent peer-to-peer storage utility. In: Proc. of ACM SOSP 2001, Banff, Canada, October 2001, pp. 188–201 (2001)
2. Bhagwan, R., Tati, K., Cheng, Y., Savage, S., Voelker, G.: Total Recall: System support for automated availability management. In: Proc. of ACM/USENIX NSDI 2004, San Francisco, California, March 2004, pp. 337–350 (2004)
3. Ramabhadran, S., Pasquale, J.: Analysis of long-running replicated systems. In: Proc. of IEEE Infocom 2006, Barcelona, Spain (April 2006)
4. Stribling, J.: PlanetLab - All Pairs Pings (2005), `http://pdos.csail.mit.edu/~strib/pl_app`
5. PlanetLab: An open platform for developing, deploying, and accessing planetary-scale services (2007), `http://www.planet-lab.org/`
6. Alouf, S., Dandoush, A., Nain, P.: Performance analysis of peer-to-peer storage systems. In: Mason, L.G., Drwiega, T., Yan, J. (eds.) ITC 2007. LNCS, vol. 4516, pp. 642–653. Springer, Heidelberg (2007)

7. Dandoush, A., Alouf, S., Nain, P.: Simulation analysis of download and recovery processes in P2P storage systems. Technical Report RR-6858, INRIA Sophia Antipolis (February 2009)
8. Reed, I., Solomon, G.: Polynomial codes over certain finite fields. Journal of SIAM 8(2), 300–304 (1960)
9. Harrison, P., Zertal, S.: Queueing models of RAID systems with maxima of waiting times. Performance Evaluation Journal 64(7-8), 664–689 (2007)
10. Neuts, M.: Matrix Geometric Solutions in Stochastic Models. In: An Algorithmic Approach. John Hopkins University Press, Baltimore (1981)
11. Grinstead, C., Laurie Snell, J.: Introduction to Probability. American Mathematical Society, Providence (1997)
12. Bhagwan, R., Savage, S., Voelker, G.: Understanding availability. In: Proc. of 2nd IPTPS, Berkeley, California (February 2003)
13. Guha, A., Daswani, N., Jain, R.: An experimental study of the skype peer-to-peer VoIP system. In: Proc. of 5th IPTPS, Santa Barbara, California (February 2006)

# Analyzing Network Coverage in Unstructured Peer-to-Peer Networks: A Complex Network Approach

Joydeep Chandra, Santosh Shaw, and Niloy Ganguly

Department of Computer Science & Engineering,
Indian Institute of Technology, Kharagpur-721302, India
{joydeep,santosh,niloy}@cse.iitkgp.ernet.in

**Abstract.** In this paper, we apply the theory predicting neighbor distribution of arbitrary random graphs to analyze the network coverage of the peers in unstructured peer-to-peer(p2p) networks that use $TTL$-based flooding mechanism for search and query. However, we find that for many cases, the theory cannot be directly applied to obtain correct estimate of network coverage due to the presence of certain types of edges that we refer as cross and back edges. It is also observed that the presence of cross and back edges in the p2p networks reduce the coverage of the peers and also generates large number of redundant messages, thus wasting precious bandwidth. We refine the theory and develop a model to estimate the network coverage of the peers in the presence of cross and back edges. We simulate our model for different networks with various degree distribution properties. The results indicate that our models provide good estimates of second neighbor and network coverage distribution. We perform a case study of the Gnutella networks to analyze the effects cross and back edges on network coverage and message complexity in these networks. Based on our study, we propose a new bootstrapping algorithm for Gnutella networks named HPC5 that substantially improves the network coverage and message complexity. The results have been validated using simulations.

**Keywords:** Peer-to-Peer Networks, Network Coverage Models, Overlay Networks, Gnutella.

## 1 Introduction

The unstructured peer-to-peer (p2p) networks like Gnutella [1][2], Kazaa[3] and FreeHaven[4] use broadcasting as their query and search mechanism. Thus the query and search performance of these p2p networks are directly proportional to the network coverage of the peers achieved through broadcasting. A high network coverage of the peers implies that queries reach a large subset of peers in the network, and thus yields better search performance. However, as of now, in most unstructured p2p networks like Gnutella and Kazaa, the fundamental strategy to improve coverage is to introduce more overlay links, thereby,

L. Fratta et al. (Eds.): NETWORKING 2009, LNCS 5550, pp. 690–702, 2009.

leading to huge Internet traffic. With the unbridled growth of the p2p networks in the past few years, the ISP's are facing a huge problem of network congestion and bandwidth consumption [5][6]. These problems are expected to be a big research challenge in the forthcoming days, and several recent works have started addressing these problems [5][7][8]. The scale of the problem increases as broadcast leads to redundant message generation and consequently wastage of precious bandwidth. However, properly analyzing the topological behavior of the networks and the impact of topology on coverage and redundancy can provide new insights in alleviating traffic and redundancy problems.

In this paper, we initially build up a basic analytical model to assess the network coverage of p2p networks that uses $TTL(2)$ based search and query mechanisms. We limit our study to $TTL(2)$ based networks, as search and query in popular unstructured networks like Gnutella uses $TTL(2)$ for most search cases. $TTL(3)$ is used only for rare searches; however, our models and results can easily be generalized for $TTL(3)$ searches as well. The basic model has been developed using the theory applied to derive the distribution of first and second neighbors of randomly selected node in large networks [9][10]. To the best of our knowledge, this work is a pioneering work that applies the theories used to estimate neighbor distribution in analyzing network coverage of p2p networks. Further we propose a refinement of our basic model to perfect the estimation of the neighbor distributions for networks that contain certain type of edges, which we refer as cross and back edges. The effect of these edges is to reduce the coverage of the peers and increase message redundancy. Thus for finite-sized networks with high cross and back edges, the results of the basic models tend to deviate from the simulation results. We study the impact of these edges on the network coverage of the peers and derive suitable models for the same. We compare the results of the refined model with the simulation results; the comparison reveals that the refined model produces accurate results of network coverage. Finally, we apply our derivations on Gnutella networks and estimate its coverage based on certain key statistics. We found that the existing Gnutella protocol generates a lot of redundant traffic and has low network coverage. Hence, we propose a bootstrapping mechanism named HPC5 to improve the network coverage of the Gnutella protocol. The superior performance of the proposed mechanism is validated using simulations.

The rest of the paper is organized as follows: The theoretical concepts of complex networks[1] related to our model are discussed next. In section 3 we discuss our derived model for network coverage in finite sized networks. In section 4, we analyze the Gnutella protocol and propose a new bootstrapping mechanism for Gnutella. The simulation results are stated and discussed in section 6. Finally we present our conclusion in section 7.

[1] The theories developed to explain behaviors of large dynamic networks are loosely termed as Complex Network Theory [11].

## 2 Basic Model for First and Second Neighbor Distributions

Most unstructured p2p networks use flooding as a means for search and querying. Since flooding causes huge number of query packets to flow through the network, thus consuming huge bandwidth, most p2p networks use a $TTL$-based flooding scheme. The commonly used $TTL$ value for ordinary searches in most networks is 2. Thus when a peer broadcasts a message with $TTL(2)$, the message reaches its immediate neighboring peers as well as their neighbors. Thus the coverage of a peer for a $TTL(2)$ broadcast is the sum of its first and second neighbors. Newman [10] derived models for the distribution of the number of first and second neighbors of a node in a large graph. Suppose, in a large network with $N$ nodes ($N$ is large), if $p_k$ denotes the probability of any random node in a network having $k$ first neighbors, then the first neighbor distribution — also referred as degree distribution — of the nodes can be represented using a generating function as,

$$G_0(x) = p_0 + p_1(x) + p_2(x^2) + p_3(x^3)\dots. \tag{1}$$

Thus the coefficient of $x^i$ in $G_0(x)$ gives the probability that any random node in the network will have degree $i$. The average number of neighbors of a node is given by,

$$\langle z \rangle = 1 \cdot p_1 + 2 \cdot p_2 + 3 \cdot p_3 + \dots = G_0'(1). \tag{2}$$

Another important quantity is the distribution of the outgoing edges of a node reached by following a randomly chosen edge. If $N_k$ denotes the number of nodes with degree $k$, then $p_k = \frac{N_k}{N}$, and the number of edges that leads to a node with degree $k$ equals $kN_k$. Thus, the probability, $p_k^{(o)}$, of reaching a node with degree $k$ by following a randomly chosen edge is,

$$p_k^{(o)} = \frac{kN_k/N}{\left(1 \cdot N_1 + 2 \cdot N_2 + 3 \cdot N_3 + \dots + (N-1)N_{N-1}\right)/N} = \frac{kp_k}{\langle z \rangle}. \tag{3}$$

The generating function for the distribution of the outgoing edges of a node reached by following a random edge can be represented as,

$$G_1(x) = \frac{1}{\langle z \rangle} \cdot \left(\sum kp_k x^k\right) = \frac{G_0'(x)}{G_0'(1)}. \tag{4}$$

The coefficient of $x^i$ in $G_1(x)$ gives the probability that any randomly chosen edge leads to a node with degree $i$. Suppose, we want to find the number of second neighbors of a node, $P$. Let $\hat{p}$ denote the connection probability between any two random nodes in the network. When $N$ is large and $\hat{p} \to 0$, then the probability that an outgoing edge from a neighbor of $P$ connects to another immediate neighbor of $P$, or to $P$ itself is negligible. Moreover, under these conditions, the probability that two neighbors of $P$ will have another common

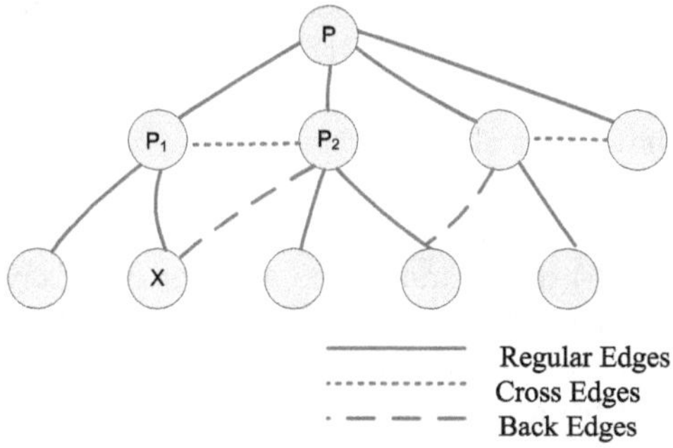


**Fig. 1.** A portion of a p2p topology. The solid lines indicate the regular edges that connect two peers. A fine broken line indicates a cross edge between two peers. A cross edge is an edge that connects directly two immediate neighbors of a peer. A heavy broken line indicates a back edge between two peers. Back edges are formed when a neighbor $P_2$ of peer $P$, connects to another peer, $X$, that is already connected to another neighbor $P_1$ of $P$.

node as neighbor is also negligible. According to power property of generating functions, $[G_1(x)]^k$ gives the distribution of the number of outgoing edges of $k$ independent nodes. Thus, the distribution of the number of second neighbors of a node $P$, is given by,

$$S(x) = \sum_k p_k \left[G_1(x)\right]^k = G_0(G_1(x)). \tag{5}$$

The total network coverage of a peer in p2p networks that use $TTL(2)$ flooding scheme is the sum of its number of first and second neighbors. Using the above stated results, we can derive the generating function for the probability distribution of the total network coverage of any peer in the network. Thus, the distribution of the total node coverage of a peer $P$ that deploys a $TTL(2)$ flooding mechanism is represented by the generating function $C(x)$ as,

$$C(x) = G_0(x) \cdot S(x) \tag{6}$$

Using these expressions, we can obtain the expected $TTL(2)$ coverage, $\langle c \rangle$ of a peer which is given as,

$$\langle c \rangle = C'(1). \tag{7}$$

*Limitations:* The above stated derivations can be used to model the expected first neighbor, second neighbor and total network coverage of any random node in a network. In unstructured p2p networks that use $TTL$-based flooding for search and query, the query messages reach the adjacent neighbors upto a number of hops, specified by the $TTL$ value. Thus these derivations can be used to model the reachability of the queries in these networks. However, these expressions provide correct reachability distributions only when the peers reached from a source node through a $TTL$-based message does not form any cycles among themselves. But, for many real cases, this condition fails to hold. Since p2p systems behave like social networks, the peers inherently form many short length cycles. The cycles that affect the coverage of the peers are referred to as *cross* and *back* edges as shown in fig. 1. A cross edge is formed when two adjacent nodes of a peer, say $P$ gets connected by an edge, whereas a back edge is formed when a neighbor (say $P_2$) of $P$ connects to another peer, say $X$, that is already connected to some other neighbor (say $P_1$) of $P$. The presence of back and cross edges reduces the coverage of a given source peer that use $TTL(2)$ flood. This can be directly interpreted from the figures 3(a) and 3(b), where the mean number of second neighbors of a peer is actually much less than predicted by our basic model. Thus in these cases, the basic model does not produce correct results for coverage of peer in a network that uses broadcast mechanism. In order to understand the actual coverage of the peers with a given degree distribution and given probability of back and cross edges, we need to develop a model that captures the effect of back and cross edges in the networks.

## 3 Network Coverage in Finite-sized Networks: Refined Model

We derive models for coverage of a peer that uses $TTL(2)$ flooding mechanism, when the degree distribution of the network is known. We also assume that the probability of a random edge being a back edge with respect to any source peer is fixed and given as $b$. We derive the second neighbor and the coverage distribution of any random peer in the network, while analytically deriving the cross-edge probability and eliminating its effects. We assume the probability that a random peer is of degree $k$ be given as $p_k$ for all possible values of $k$. We derive the peer coverage for these graphs, that deploys a $TTL(2)$ broadcast mechanism for query and search.

Let us assume that a network has $N$ peers and a random peer $P$ has $k$ first neighbors. Initially, we intend to find the distribution of the number of outgoing edges, which are not cross edges, of a first neighbor of $P$. If a peer has $j$ outgoing edges, then $i$ unique neighbors has to be chosen from $N-(k+1)$ peers — since there are $k$ first neighbors of $P$ and $P$ itself, so the total unique peers present in the network from which $P$ will have to choose is $N-(k+1)$. This can be done in $\binom{N-k-1}{i}$ ways. The rest of the $j-i$ peers has to be chosen from $k$ peers, and this can be done in $\binom{k}{j-i}$ ways. Thus, for any first neighbor of $P$ having a total of $j$ outgoing edges, the probability of having $i$ edges that are not cross edges, is given as

$$R_{k,i,j} = \frac{\binom{N-k-1}{i}\binom{k}{j-i}}{\binom{N-1}{j}}.$$

Hence, for any random neighbor of $P$ (having $k$ first neighbors), the distribution for having $i$ non-cross edges, is given by

$$R_{k,i} = \sum_{j=i}^{k+i} q_j \left[ \frac{\binom{N-k-1}{i}\binom{k}{j-i}}{\binom{N-1}{j}} \right], \tag{8}$$

where $q_j$ is the probability of having $j$ outgoing edges of a peer, reached by selecting a random edge — as obtained from the coefficent of $x^j$ of $G_1(x)$ (Eq. 4). One must note that the values of $j$ range from $i$ to $k+i$. For any value of $j < i$, the probability $R_{k,i,j}$ becomes equal to zero. Similarly, when $j > k+i$, the number of non-cross outgoing edges from $j$ must be greater than $i$ and hence $R_{k,i,j}$ is again equal to zero.

Thus, the distribution of the total number of non-cross outgoing edges from any random neighbor (out of $k$ first neighbors) of $P$, can be represented using generating function as,

$$\hat{R}_k(x) = \sum_{i'} R_{k,i'} x^{i'}. \tag{9}$$

Thus, the distribution of the total number of non-cross outgoing edges from all the $k$ first neighbors of $P$ can be found from the power property of generating functions and is given as,

$$\Gamma_k(x) = \left[\hat{R}_k(x)\right]^k = \Gamma_{k,0} + \Gamma_{k,1}x^1 + \Gamma_{k,2}x^2 + \ldots (say), \tag{10}$$

$$= \sum_m \Gamma_{k,m} x^m. \tag{11}$$

Now, suppose the probability that any random edge is a back edge with respect to the source node $P$ is known and is denoted as $b$, then for a neighbor $X$ with $t$ non-cross edges, the distribution of the number of non-back edges with respect to source node $P$ can be represented by the generating function as,

$$Q_t(x) = \sum_{\gamma \le t} \binom{t}{\gamma} (1-b)^\gamma (b)^{t-\gamma} x^\gamma, \tag{12}$$

$$= 0 \text{ for } \gamma > t. \tag{13}$$

Thus, $Q_t(x)$ gives the distribution of the number of edges, out of a total of $t$ edges, from neighbor $X$ of $P$ that connects to distinct nodes. The distribution of the actual number of unique peers to which $k$ first neighbors of $P$ connect is given by,

$$A_k(x) = \sum_{t'} \Gamma_{k,t'} Q_{t'}(x), \tag{14}$$

and the distribution of the number of unique second neighbors for any random peer in a network is,

$$\hat{S}(x) = \sum_{k'} p_{k'} A_{k'}(x). \tag{15}$$

The total coverage of the network will be given by

$$\hat{C}(x) = G_0(x) \cdot \hat{S}(x) \tag{16}$$

The model is experimentally validated for networks with Poisson and power-law degree distributions. The validation results are presented cohesively in section 6 after we explain the Gnutella model and the improvement algorithm in 4. We also verified the model for Gnutella networks based on its certain key statistics.

## 4 Analysis of the Gnutella Protocol

One of the aims of the paper is to verify the correctness of the refined model in Gnutella networks. Moreover, we propose techniques to modify the existing Gnutella protocol to eliminate back and cross edges thus significantly enhancing its coverage. We have build up a Gnutella prototype in order to carry out the experiments. The prototype is built by studying the basic model of Gnutella, its bootstrapping protocol, and its basic search technique. Certain key statistics of the Gnutella network, based on studies conducted by several researchers[1][12][13][14] is also used to develop a prototype that we discuss next. Our prototype reveals that Gnutella forms large number of back and cross edges. Since Gnutella uses $TTL(2)$ flooding for most ordinary searches, these back and cross edges generate large redundant queries at the peers. Based on these observations, we propose certain changes in the bootstrapping protocol of Gnutella that reduces query redundancy and improves network coverage (results presented in section 6).

*Basic Model:* Gnutella 0.6 is a two-tier overlay network, consisting of two types of nodes : *ultra-peer* and *leaf-peer* (the term peer represents both ultra and leaf peer). An ultra-peer is connected with a limited number of other ultra-peers and leaf-peers. A leaf-peer is connected with some ultra-peers. However, there is no direct connection between any two leaf-peers in the overlay network.

*Bootstrapping and Handshaking Protocol:* Many software clients are used to access the Gnutella network (like *Limewire, Bearshare, Gtk-gnutella*). The most popular client software, Limewire's handshake protocol is used in our prototype. Through handshaking, a peer establishes connection with any other ultra-peer. To start handshake protocol a peer first collects the address of an online ultra-peer from a pool of online ultra-peers. A peer can collect the list of online peers from *hardcoded* address/es and/or from *GwebCache* systems [15] and/or through *pong-caching* and/or from its own hard-disk which has obtained a list of online ultra-peers in the previous run [12]. A handshake protocol is used to make new connections [1,2].

*Basic Search Technique:* The network follows limited flood based query search. A query of an ultra-peer is forwarded to its leaf-peers with $TTL(0)$ and to all its ultra-neighbors with one less $TTL$ only when ($TTL > 0$). A leaf-peer does not forward query received from an ultra-peer. On the other hand ultra-peers perform query searching on behalf of their leaf peers. The query of a leaf-peer is initially sent to its connected ultra-peers. All the connected ultra-peers simultaneously forward the query to their neighbor ultra-peers up to a limited number of hops. While $TTL(2)$ is used for most searches, $TTL(3)$ is used for rare searches.

*Key Statistics:* Certain key statistics of the current Gnutella network are as follows [1][14]: Currently the number of peers in Gnutella Network is around 2000k, out of which 100k are live at any point of time[12]; the number of ultra-peers is around 15–16% of the total number of live peers. An ultra-peer connects to a maximum of 32 other ultra-peers, and to a maximum of 30 leaf peers, where as a leaf-peer connects to a maximum of 3 ultrapeers. The average number of neighboring ultra-peer of an ultra-peer is 25, whereas the average number of neighboring leaves is 22.

The key statistics points to some interesting analysis. If 15% of the live peers are ultra, then with 100k live peers, there are around $N$=15k ultra-peers. Since each ultra-peer connects to an average of 25 ultra-peers, so the average connection probability of an ultra-peer with another ultra is approximately $p^{uu} = .0017$, thus $p^{uu} \approx N^{-.67}$. As discussed in [11], when the connection probability between two random peers in a network with $N$ peers increases beyond $N^{-\frac{2}{3}}$, a large number of short cycles of length 3 and 4 are created. Hence, in a Gnutella network, a high number of cross and back edges will exist, leading to huge traffic redundancy. To handle the problem of redundancy, we propose a mechanism named HPC5 for topology generation in Gnutella networks that eliminates cycles of length shorter than five in the network. We define a topology containing cycles not less than length $r$ as a *Cycle-r* topology. The underlying rationale behind this proposal is that with a TTL(2) flooding, a cycle-5 topology will not generate any redundant messages at any node. We state our proposed HPC5 mechanism next. We simulated HPC5 and compared the network coverage and message complexity of peers with the Gnutella 0.6; the results are presented in section 6.

## 5 HPC5: Handshake Protocol for Cycle-5 Networks

As stated earlier, the major objective of HPC5 protocol is to eliminate the possibility of forming short length cycles (cycles of length 3 or 4). Figure 2 illustrates the proposed HPC5 graphically. It shows the various possibilities when peer-1 requests other online ultra-peers to be its neighbor, given that, peer-2 is already a neighbor of peer-1. In figures 2(a) and 2(b), the possibility of the formation of triangle and quadrilateral arises if a $1^{st}$ or $2^{nd}$ neighbor of peer-2 is selected. However, this possibility is discarded in fig. 2(c) and a cycle of length 5 is formed. HPC5 exactly ensures that.

Each peer maintains a list of its $1^{st}$ and $2^{nd}$ neighbors, which contains only ultra-peers (because a peer only sends request to an ultra-peer to make neighbor). The $2^{nd}$ ultra-neighbors of a leaf-peer represent the collection of $1^{st}$

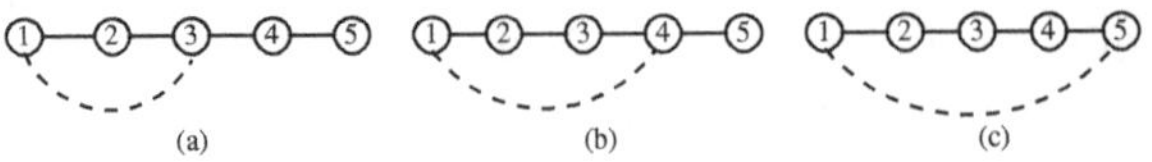

**Fig. 2.** Selection of neighbor by peer-1 after making peer-2 as a neighbor

ultra-neighbors of the connected ultra-peers. To keep updated knowledge, each ultra-peer exchanges its list of $1^{st}$ neighbors periodically with its neighbor ultra-peers and sends the list of $1^{st}$ neighbors to its leaf-peers. To do this with minimal overhead, piggyback technique can be used in which an ultra-peer can append its neighbor list to the messages passing through it.

The three steps of modified handshake protocol (HPC5) is described below.

1. The initiator peer first sends a request to a remote ultra-peer which is not in its $1^{st}$ or $2^{nd}$ neighbor set. The request header contains the type of the initiator peer. The presence of remote peer in $2^{nd}$ neighbor set implies the possibility of 3-length cycle. In fig. 2, peer-1 cannot send request to peer 2 or 3, on the other hand peer 4 & 5 are eligible remote ultra-peers.
2. The recipient replies back with its list of $1^{st}$ neighbors and the neighbor-hood acceptance/rejection message. If the remote peer discards the connection in this step, the initiator closes the connection and keeps the record of neighbors of the remote peer for future handshaking process; on acceptance of the invitation by the remote-peer, the initiator peer checks the presence of at least one common peer between its $2^{nd}$ neighbor set (say A) and the $1^{st}$ neighbor set of the remote peer (say, B). A common ultra-peer between sets A and B indicates the possibility of 4-length cycle. In fig. 2, peer 3 is in the second neighbor set of 1, and in the first neighbor set of 4. Thus 1 and 4 cannot form neighbors.
   **If** no common peer is present between sets A and B then the initiator sends *accept connection* to remote peer.
   **Otherwise** the initiator sends *reject connection* to remote peer.

Thus HPC5 prevents the possibility of forming a cycle of length 3 or 4 and generates a cycle-5 network. We simulated the network coverage of the peers as shown in fig. 5; the results indicate that our proposed protocol has much improved network coverage as compared to Gnutella 0.6 that allows formation of a Cycle-3 topology.

## 6 Simulation Results

In this section, we present simulation results generated to validate theoretical correctnes for second neighbor distribution of various networks with different degree distributions, including an arbitrary distribution generated by the Gnutella prototype, and compared the results with our derived models. Moreover, we present results of the impact of HPC5 on the network coverage and message complexity of the Gnutella network.

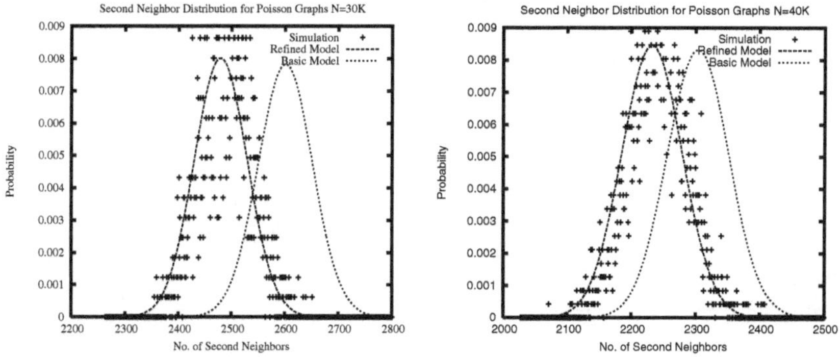


(a) N=30K (b) N=40K

**Fig. 3.** Second neighbor distribution of a random peer with $k = 51$ and 48 first neighbors for Erdos-Renyi networks (poisson degree distribution) with total peers $N = 30$K and 40K respectively. The value of $\hat{p}$ is .0017 for $N = 30$K and .0012 for $N = 40$K, and the back edge probability $b$ is .04 and .03 respectively. The points show the simulation results, the heavy broken lines show the results of refined model, compared to the results of basic model (fine broken lines).

### 6.1 Second Neighbor Distribution

We considered networks like Erdos-Renyi networks[16][17], power law networks[18], and also arbitrary networks generated by the Gnutella prototype, for our simulations. For each of these cases, we simulated the second neighbor distribution of the peers for a given first degree $k$, and estimated back edge probability, $b$. We discuss the simulation details and results for each of these networks.

*Case: Erdos-Renyi Graphs:* Erdos-Renyi graphs[16][17] are random graphs, in which any two peers in the network are connected with a fixed probability $\hat{p}$. We simulated the second neighbor distribution in Erdos-Renyi networks with $N$ =30K and 40K, for a back edge probability of .04 and .03 respectively, and for the first neighbor value $k$ =51 and 48 respectively. The connection probability $\hat{p}$ was taken as .0017 and .0012 respectively. As seen in fig. 3, the results of the simulation for $N = 30K$ and $40K$ matches well with our refined model, where as the basic model considerably deviates from the simulation results. However, it can be seen that for the basic model, the closeness of fit is more for $N = 40$K as compared to $N = 30$K. When the size of the network, $N$, is increased considerably from 30K (fig. 3(a)) to 40K (fig. 3(b)), or if the connection probability between two random nodes, $\hat{p}$ is further reduced, then the simulation results matches well with the basic model as well as our refined model. This is because, when $N$ increases or the connection probability $\hat{p}$ is considerably reduced, then the chances of forming cross or back edges by the peers become almost negligible, and thus in these limiting conditions, the basic model matches well with the simulation results. However, as $\hat{p}$ increases, the number of neighbors of the peers increases; thus the number of unique peers that has not been selected by

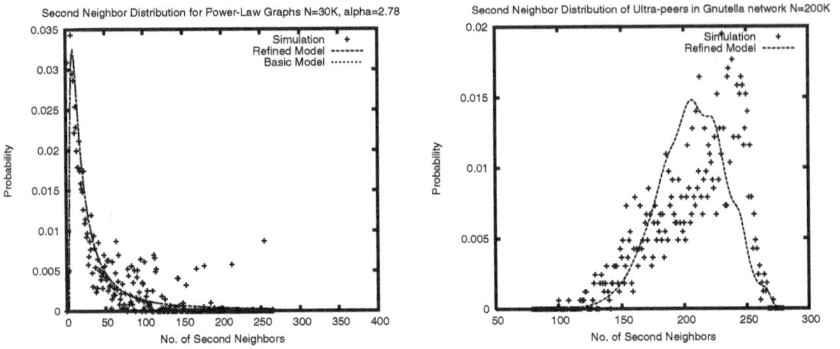


(a) Power-Law (b) Gnutella Network

**Fig. 4.** Second neighbor distribution of a peer with $k = 4$ first neighbors for power law network with $N = 30$K peers, $\alpha = 2.78$ and back probability $b = 0$, and a Gnutella network with $N \approx 27$K ultra peers and $b = .05$. The points show the simulation results, the heavy broken lines indicate the results of our refined model. For the case of power-law, the results of basic model matches exactly with the refined model.

other peer reduces. Hence cross and back edges are formed and the basic model fails to model the second neighbor of the peers precisely.

*Case: Power-Law Graphs:* We simulated the power-law graphs using the degree distribution given as $p_k \sim k^{-\alpha}$, where $\alpha$ is a constant that varies between 2 and 3 for all real networks that follows power law distribution[11][18]. The power-law network topology was generated using the well known configuration model[10]. We simulated the second neighbor distribution for varying number of peers from $N = 10K$ to $30K$, and for $\alpha$ varying from 2.31(high) to 3.0(low). Figure 4(a) shows the second neighbor distribution for peers with $k = 4$ first neighbors in a network with $N = 30$K and for $\alpha = 2.78$. Here, interestingly the results match well for both refined model, as well as the basic model. This is because, the power-law networks hold an important property; a majority of the peers in these networks have very low degree and only a very few peers have very high degree. As, most of the peers have very low connectivity, the chances of forming cross and back edges are inherently very less in power law networks, hence the simulation results match well with the basic as well as the refined model.

*Case: Gnutella Network:* Here we used the degree distribution of ultra-peers (considering only ultra-peer to ultra-peer connectivity) generated by the Gnutella prototype that we have implemented. In our Gnutella implementation we considered $N = 200$K total peers that have around 26842 ultra-peers; the ultra-peers connects to a maximum of 32 other ultra-peers, the average connectivity being 25. Figure 4(b) plots the second neighbor distribution of the peers that have $k = 9$ first degree neighbors with a back edge probability $b = .045$. The estimate of the back edge probability was obtained from the simulation of the Gnutella

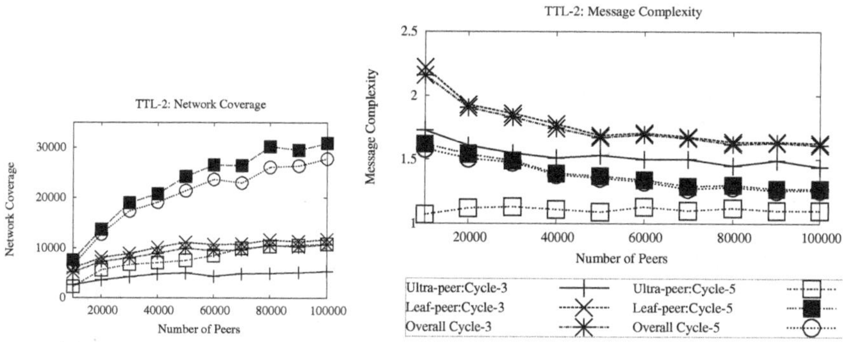


(a) Network Coverage (b) Message Complexity

**Fig. 5.** Network Coverage and message complexity with TTL 2 for cycle-3 and cycle-5 networks

prototype. We observe that the simulated results fit well with that of our model. The minor difference is due to the method in which the back edge estimation is made. Unlike our consideration in the refined model, the back edge probability in this case is not same for all the peers and hence minor differences in the simulated results can be observed.

### 6.2 Impact of HPC5 on Gnutella Networks

We simulated the effect of existing Gnutella topology and our proposed HPC5 mechanism on the *network coverage* and *message complexity* of the network. We define message complexity as the average number of messages required to discover a peer in the overlay network whereas network coverage implies the number of unique peers explored during query propagation in limited flooding. The simulation results are shown in fig. 5. The network coverage of the peers improves by a maximum amount of 10%, whereas, the message complexity of the overall networks almost reaches 1, when HPC5 is used. Thus using HPC5 leads to significant improvement in network coverage and message complexity as compared to the cycle-3 networks in traditional Gnutella.

## 7 Conclusion

In this paper, we developed suitable models that quantify the coverage of the peers in networks that perform $TTL(2)$ searches. The models based on generating function formalism provides a strong theoretical foundation needed to understand the relation between the topology of a network and the achievable performance through $TTL$-based searches. Using the derived model, we provided an insight of the topological impact on network coverage and message complexity of the peers in Gnutella. The model revealed low network coverage and high message complexity in existing Gnutella, and helped us to propose a modified bootstrap mechanism named HPC5 that showed improvement of almost 10% in network coverage and 30% in message complexity. The models can be extended further for higher values of $TTL$, and also for obtaining coverages in networks with high clustering coefficients. However, a more elegant methodology to calculate back edges needs to be developed in future.

## References

1. Gnutella and Limewire, http://www.limewire.org
2. Gnutella Protocol Specification 0.6, http://rfc-gnutella.sourceforge.net
3. Liang, J., Kumar, R., Ross, K.: The KaZaA Overlay: A Measurement Study. In: Proceedings of the 19th IEEE Annual Computer Communications Workshop, Bonita Springs, Florida (2004)
4. The FreeHaven Project, http://www.freehaven.net
5. Aggarwal, V., Feldmann, A., Scheideler, C.: Can ISP'S and P2P Users Cooperate for Improved Performance?. SIGCOMM Comput. Commun. Rev. 37, 29–40 (2007)
6. Sen, S., Wang, J.: Analyzing Peer-to-Peer Traffic Across Large Networks. In: IMW 2002: Proceedings of the 2nd ACM SIGCOMM Workshop on Internet measurment, pp. 137–150. ACM, New York (2002)
7. Bindal, R., Cao, P., Chan, W., Medved, J., Suwala, G., Bates, T., Zhang, A.: Improving Traffic Locality in BitTorrent via Biased Neighbor Selection. In: 26th IEEE International Conference on Distributed Computing Systems, 2006. ICDCS 2006, pp. 66–77 (2006)
8. Choffnes, D.R., Bustamante, F.: Taming the Torrent: A Practical Approach to Reducing Cross-ISP Traffic in Peer-to-Peer Systems. In: SIGCOMM 2008, Seattle, Washington, USA, pp. 363–374. ACM, New York (2008)
9. Dorogovtsev, S.N., Goltsev, A.V., Mendes, J.F.F.: Critical Phenomena in Complex Networks. Reviews of Modern Physics 80 (2008)
10. Newman, M.E., Strogatz, S.H., Watts, D.J.: Random Graphs with Abitrary Degree Distributions and Their Applications. Phys. Rev. E Stat. Nonlin. Soft. Matter. Phys. 64 (2001)
11. Albert, R., Barabasi, A.-L.: Statistical Mechanics of Complex Networks. Reviews of Modern Physics 74, 47 (2002)
12. Karbhari, P., Ammar, M.H., Dhamdhere, A., Raj, H., Riley, G.F., Zegura, E.W.: Bootstrapping in Gnutella: A Measurement Study. In: Barakat, C., Pratt, I. (eds.) PAM 2004. LNCS, vol. 3015, pp. 22–32. Springer, Heidelberg (2004)
13. Stutzbach, D., Rejaie, R.: Capturing Accurate Snapshots of the Gnutella Networks. In: IEEE INFOCOM, pp. 2825–2830 (2005)
14. Stutzbach, D., Rejaie, R., Sen, S.: Characterizing Unstructured Overlay Topologies in Modern P2P File-Sharing Systems. In: Internet Measurment Conference, USENIX Association, pp. 49–62 (2005)
15. GwebCache System, http://www.gnucleus.com
16. Erdős, P., Renyi, A.: On Random Graphs I. Publ. Math. Debrecen 6, 290–297 (1959)
17. Erdos, P., Renyi, A.: On the Evolution of Random Graphs. Publ. Math. Inst. Hungar. Acad. Sci. 5, 17–61 (1960)
18. Barabasi, A.L., Albert, R.: Emergence of Scaling in Random Networks. Science 286, 509–512 (1999)

# Decentralized Bootstrapping of P2P Systems: A Practical View

Jochen Dinger and Oliver P. Waldhorst

Institute of Telematics, Universität Karlsruhe (TH)
76128 Karlsruhe, Germany
{dinger,waldhorst}@tm.uni-karlsruhe.de

**Abstract.** So far, bootstrapping constitutes the only centralized task in otherwise decentralized peer-to-peer (P2P) systems. As a contribution to the development of generally applicable decentralized bootstrap mechanisms, in this paper we analyze two decentralized approaches from a practical point of view. We consider *local host caches* and *random address probing* for bootstrapping into the BitTorrent DHT as an example for a widely deployed P2P system. Based on the results of an extensive measurement study we show that local host caches allow rejoining the P2P system quickly after short times of disconnection, but are impracticable for infrequent or first-time users. Furthermore, random address probing is feasible using a direct Internet connection with high bandwidth, but is subject to practical issues raised by typical NAT routers and the distribution of ports used by BitTorrent clients. We propose two mechanisms for increasing the performance of random address probing: (1) probing multiple ports per host and (2) hash-based filter-resistant port selection, making distributed bootstrapping feasible even from a practical point of view.

**Keywords:** Internet measurement, peer-to-peer, decentralized bootstrapping.

## 1 Introduction

Since the deployment of the first peer-to-peer (P2P) systems at the end of the 1990s, P2P has become a mature technology for building distributed applications. Many examples for currently widely-deployed P2P systems exist, such as the file sharing systems KaZaA [1], Gnutella [2], BitTorrent [3], and eMule [4], the IP-telephony system Skype [5], or the video distribution system Joost [6]. In all these systems, one issue is still solved in a centralized way: the bootstrap process, i.e., the task of finding an entry point to the P2P system in order to integrate a new peer. It typically relies on dedicated bootstrap servers that are reachable under well-known addresses. This aspect is crucial, e.g., if the bootstrap server has become unavailable due to a system failure.

In this paper, we analyze approaches for decentralized bootstrapping of P2P systems from a practical point of view. Using a widely-deployed P2P system, the

L. Fratta et al. (Eds.): NETWORKING 2009, LNCS 5550, pp. 703–715, 2009.

BitTorrent DHT, as an example, we analyze two mechanisms for decentralized bootstrapping: First, we consider a *local host cache* to store addresses of several recently active BitTorrent hosts as entry points for re-entering the BitTorrent DHT. Second, we analyze *random address probing*, i.e., scanning of randomly generated IP addresses, for discovering active BitTorrent peers as entry points.

We analyze the performance of both mechanisms by performing an extensive measurement study that has lasted several weeks and discovered several million of BitTorrent peers. We found that the lifetimes of the peers, i.e., the time they are connected to the Internet using the same IP address, are sufficient to make local host caches an effective tool for re-entering the BitTorrent DHT after short periods of disconnection. However, reconnecting infrequent users may take several minutes or even fail at all, as it does for first-time users of the P2P system. Furthermore, we show that finding an active peer by random address probing, again, takes a few minutes, even in a laboratory setting with a powerful Internet connection. Furthermore, we show that this time is significantly increased by practical issues like the performance of typical NAT[1] routers as well as by the number of BitTorrent peers using non-standard ports.

To significantly speed up decentralized bootstrapping, we propose two mechanisms for increasing the performance of random address probing: (1) probing multiple ports per peer, increasing the probability to discover a client by exploiting knowledge on the port distribution, and (2) selecting a port by hashing the IP address of the client, making it predicable but difficult to filter by an Internet Service Provider (ISP). These mechanisms make successful random address probing feasible in reasonable time.

The remainder of this paper is organized as follows. Section 2 provides an overview of P2P bootstrap mechanisms as well as related measurement studies. In Section 3 we discuss how local host caches and random address probing can be used for distributed bootstrapping into the BitTorrent DHT. In Sections 4 and 5, respectively, we show by extensive measurements the shortcomings of both mechanisms. Subsequently, we propose mechanisms for speeding up bootstrapping in Section 6. Finally, concluding remarks are given.

## 2 Background and Related Work

### 2.1 Bootstrapping P2P Systems

*Bootstrapping* basically describes the process of integrating a new node into a P2P system. As an anchor point for the new peer at least the address (more precisely the IP address and port) of one active peer, i.e., a peer that currently participates at the desired P2P system, has to be discovered [8]. Several approaches for the discovery of active peers in P2P systems have been proposed (cp. [9]):

*Out-of-band mechanisms.* When Gnutella was launched in 2000, the address of active peers was exchanged through IRC [10]. Moreover, websites became popular as out-of-band mechanism to discover active peers [8, 11].

[1] Network address translation (NAT), cp. [7].

*Dedicated bootstrap servers.* Many recent P2P systems like BitTorrent use one or more (central) bootstrap server(s) with well-known DNS names or IP addresses. A node willing to join contacts a bootstrap server, which provides addresses of active peers. Even if multiple bootstrap servers exist, they may become subject to Denial-of-Service (DoS) attacks and hence single points of failure, e.g., as became obvious when the login nodes of Skype [12] have been overloaded [13]. Additionally, the number of bootstrap servers must be adjusted to the number of users (and the churn rate) of a P2P system.

*Local host cache.* To speed up reconnection, e.g., after a typical forced 24h-disconnection of a DSL-line by the ISP, nodes may maintain a list with addresses of other peers [14]. This list is denoted as *(local) host cache* and is used, e.g., by Skype [12]. It can be easily maintained while keeping the P2P routing tables up-to-date. When a node wants to re-join, it can simply try to connect to nodes from the host cache without contacting a bootstrap server. However, at least one node in the host cache must be active. Thus, the average *lifetime* of the addresses, i.e., the time the peers are connected to the P2P system using the same IP address and port, is an important factor for the performance of the host cache (see Section 4).

*Random Address Probing.* Another mechanism for decentralized bootstrapping is based on randomly probing for the addresses of active peers. This is done by sending join-request messages to randomly selected IP addresses and default ports, assuming that with sufficient high probability a peer is located at one of the IP addresses and will send a reply. This is based on the observation that the P2P systems that are currently deployed have several millions of users, as shown, e.g., in [15, 16]. This holds in particular for the DHT-extension of BitTorrent (see Section 3) and KAD, the DHT-extension of eMule [17, 18]. However, the time until the first peer is discovered heavily depends on the number of currently active peers. [19] shows that the number of eMule-peers in dial-up-networks is particularly high. Thus, to this end the probability of success can be increased by restricting the probing processes to IP addresses from such networks (*local random address probing*). By extensive measurement studies we determine the probability of success that can be expected currently in the BitTorrent DHT in Section 5.1. Furthermore, the time until a success depends on the probing rate, i.e., the number of join-requests sent per second, which may be limited by the available bandwidth and by network hardware, as we show in Section 5.2. Such factors are not considered in [19]. Last, the time depends on the number of peers using the default port, since probing multiple ports per peer obviously increases discovery time. We will elaborate on this factor in Section 5.3

*Network layer mechanisms and standard protocols.* Apart from application layer approaches mentioned, it might also be possible to use network layer mechanisms like multicast, anycast [20] or the service location protocol (SLP) [21]. These protocols could facilitate the bootstrap process. However, the information stored at the multicast or anycast routers or central SLP directory services raises scalability and robustness questions. Besides this theoretical issues, there is little support for these protocols on a global scale.

### 2.2 Related Measurement Studies

Since we perform an extensive measurement study to analyze the performance of distributed bootstrapping, we briefly recall related work in this area. Current studies of the KAD system have shown that between 3 and 4.3 millions of peers are connected at the same time and have examined the lifetimes of those peers [17][18]. However, since these studies do not focus on the bootstrap problem, lifetime is measured per P2P-specific node identifier assigned to each peer by KAD. That is, the overall lifetime of a peer is measured, not the lifetime of the addresses, so that no conclusion on the performance of local host caches can be drawn.

The lifetime of peers in the P2P systems Gnutella, KAD, and BitTorrent (in Tracker-mode) is measured in [16]. Though that work does not consider the DHT-extension of BitTorrent, results confirm some results obtained for the extension presented in this paper. The BitTorrent client Azureus uses an alternative to the DHT-extension of mainline BitTorrent client. It is examined by [15], which argues that the lifetime of peers is several hours, but does not provide detailed measurements.

In [22] a measurement study is presented that analyzes the lifetime of peers in Overnet, which is also a Kademlia-based P2P network like BitTorrent. The analysis is based on IP address and port and shows among other things that 50 % of peers in Overnet have a lifetime of 4,300 seconds. We measured 4,500 seconds for BitTorrent peers (cp. Section 4). Furthermore, we observed 2.1 % of the peers running more than one day under the same address while they measured a fraction of 2.7 %. Opposed to our work, no conclusion on the performance of distributed bootstrapping is drawn from the measurement results presented in [22].

Opposed to most related work mentioned so far we present a detailed study of connection lifetime based on the IP address and port and use it to draw conclusions on the performance of local host caches. Furthermore, we explore practical issues like randomly chosen ports or limits imposed by firewalls that are relevant for the performance of decentralized bootstrapping.

## 3 Decentralized Bootstrapping into the BitTorrent DHT

To analyze the performance of distributed bootstrap mechanisms, we use the DHT-extension of BitTorrent [23], since it is better documented than KAD, easing the experiments described in Sections 4 and 5. However, we state that most results of this paper likely hold for other systems such as KAD/eMule.

Bootstrapping into the BitTorrent DHT is based on centralized bootstrap servers with a well-known DNS name like `router.bittorrent.com`. We assume that this centralized mechanism is replaced by a combination of the two distributed bootstrap mechanisms described in Section 2.1: First, we use local host caches for fast re-entering the BitTorrent system after short periods of disconnection, ranging from, e.g., forced disconnections of DSL lines in the order of seconds to shutting down the PC when leaving for work in the order of hours. Since bootstrapping by local host caches may fail, e.g., after long periods of disconnection due to infrequent BitTorrent usage or even for first-time users, we secondly use random address probing as a fallback.

We analyze both mechanism by performing Internet measurements in the BitTorrent DHT in Sections 4 and 5. Furthermore, we consider that many typical peers use *dial-up networks* with DSL or cable modems. Typically, connections to dial-up networks use NAT routers and/or firewalls that may significantly limit the performance of random address probing as we shown in Section 5.2. Furthermore, many ISPs block the default BitTorrent UDP port 6881, forcing peers to switch to other ports with significant impact on random address probing as we show in Section 5.3.

## 4 Performance of Local Host Caches

To analyze the performance of local host caches we conduct a study of peer lifetimes in terms of addresses rather than P2P-identifiers as used in related work. We have measured peer lifetime by periodically scanning a certain number of IP addresses for BitTorrent peers using the ping mechanism of Kademlia. Therefore, we successively send Kademlia-`PING` packets to each selected IP address on the default port 6881 and wait for replies by BitTorrent DHT peers. Sending `PING` packets and listening for potential replies is performed asynchronously. That is, after sending one packet we continue to send `PING` packets to the next address immediately, without explicitly waiting for a reply. To all peers discovered during the periodical scan we send Kademlia-`PING` packets in intervals of three minutes. Peers that do not respond are timed out after 5 retries, i.e., 15 minutes conforming to the BitTorrent protocol specification [3]. Note that this methodology implies that the obtained results constitute lower bounds for the lifetimes, since peers may have already been connected to the BitTorrent system when they are discovered by the periodical scan. Thus, we refer to the measured lifetimes as *minimal lifetimes* as a lower bound on the performance of host caches.

Figure 1 plots the complementary cumulative distribution function (CCDF) of minimal lifetimes, i.e., the probability of a peer being still connected after a given period, as a function of the time. The figure shows two curves. The curve labeled "peers in scanned IP range" plots only peers within the IP range 84.128.x.x - 84.144.x.x. This range was scanned periodically with a scan rate of 300 PING packets per second (pkt/s) over a time of 27 days in February 2008. A scan of the complete IP range took about 58 minutes, i.e., the minimal lifetime for peers discovered in the second or later scans is under-estimated by at most 58 minutes. Peers that are online for less than 58 minutes may remain undiscovered by this approach. However, the curve shows that more than 8 % of the discovered peers have minimal lifetimes longer than 8 hours with a significant drop at about 24 hours. This is due to the fact that most ISP force a disconnection after this time span. The log-scale plot shown in Figure 2 illustrates this fact more clearly.

The second curve labeled "all peers" additionally considers nodes from outside the scanned IP range. Although we did not actively send Kademlia-`PING` packets to these nodes, Kademlia nodes exchange information about known peers, propagating the IP address and port of our measurement peer in the network. Within

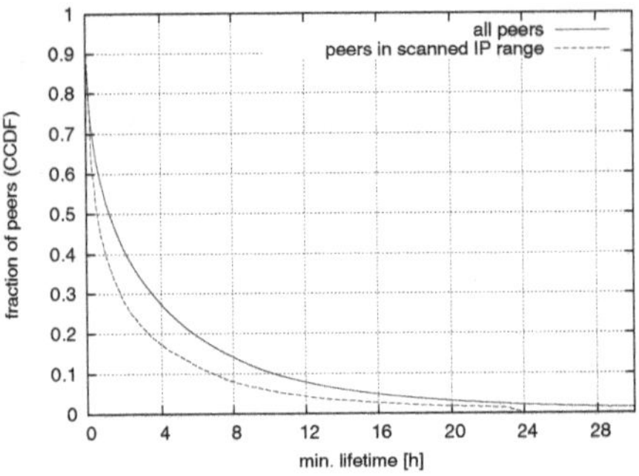


**Fig. 1.** CCDF of the (minimal) lifetime of the peers

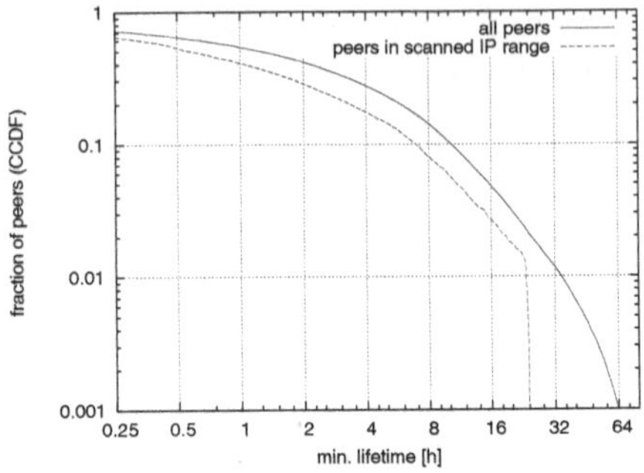


**Fig. 2.** CCDF of the (minimal) lifetime on log-scale

the 27 days we discovered 6,449 nodes within the scanned IP address range, compared to a total number of 238,628 nodes. Note that we likely underestimate the lifetime of nodes outside the scanned IP address range by more than 58 minutes. Nevertheless, Figure 1 shows that minimal lifetime for all peers exceeds minimal lifetime of peers in the scanned IP address range. In particular, more than 2% of the peers are reachable using the same address for more than 24 hours, some even for more than 7 days, implying that they are not located in dial-up networks.

In summary, the experiments show that 1.3 % of the peers in the scanned IP range had a lifetime of at least 23 hours. 0,1 % of all nodes had a lifetime of at least 64 h. Assuming a disconnection time of 64 h and a host cache size of 10,000 peers, the probability for successfully bootstrapping is almost 1. Due to the Geometric Distribution the expected number of peers that have to be probed until finding the first active peer is 1,000. We conclude that successful bootstrapping by local host caches is still feasible after a disconnection of several days. Nevertheless, bootstrapping is infeasible for low-frequency or first-time users. Thus, the next section explores whether local host caches can be complemented by random address probing.

# 5 Performance of Random Address Probing

To analyze the feasibility of random access probing approach, we investigate the expected latency using a direct Internet connection in Section 5.1. The limitations of typical DSL hardware that result in a significantly increased latency are discussed in Section 5.2. Furthermore, the distribution of the ports used by the BitTorrent clients further increases latency as shown in Section 5.3

## 5.1 Latency of Random Address Probing

In a first experiment, we apply local random address probing by scanning a list of 29.014 Class-C dial-up networks (cp. [19]) using the Kademlia-PING mechanism as described above. We use a direct Internet connection from the network of our

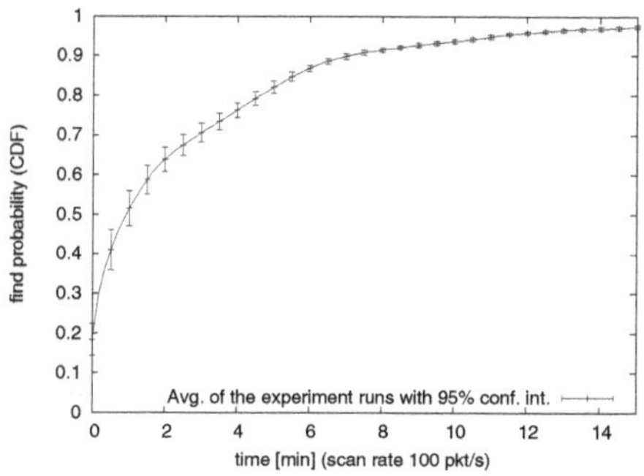


**Fig. 3.** Latency for finding a BitTorrent DHT peer by local random address probing

university. In the experiments we have completely scanned all 29.014 Class-C networks, i.e., more than 7 millions IP addresses. The experiment has been conducted in September 2007. We used the default UDP port 6881 of the BitTorrent DHT. The addresses have been scanned sequentially and the time elapsed between receiving two response messages has been logged. The scan rate has been adjusted to 100 pkt/s.

Recall that it is sufficient to discover a single BitTorrent DHT peer to join the system. Figure 3 shows the results as cumulative distribution function (CDF), i.e., it plots the probability of finding at least one peer in a particular time interval. We conducted 12 experiment runs that led to more than 5,500 values and plot the mean values together with the 95% confidence intervals. The figure shows that a peer can be located within 10 minutes with a probability of $\geq 94\%$. However, we expect that in the future the results will change to the worse to some extend, since one of the most popular BitTorrent clients $\mu Torrent$ does not use the default UDP port 6881 anymore, but a randomly selected port. Thus, the success probability of finding a peer at the default port will decrease. The current port distribution is outlined in Section 5.3.

## 5.2 Limits Imposed by Standard DSL Hardware

The measurements in Section 5.1 have been conducted using a scan rate of 100 pkt/s. With a packet size of 107 Bytes, this requires an upstream bandwidth of about 86 kbit/s. Thus, a typical DSL-line with an upstream bandwidth of about 200 kbit/s should be able to handle the scan traffic easily. However, when we repeated the experiment using DSL-lines, we found that the upstream bandwidth is not the limiting factor. In fact, scanning performance is dramatically limited by standard NAT-boxes with integrated firewalls. These boxes discard packets from hosts that have not previously been contacted from inside the home network by using *connection tables*. These tables may overflow even at low scan rates.

We found that the performance of some NAT boxes is heavily degraded, while others offer constant performance by overwriting connection table entries and,

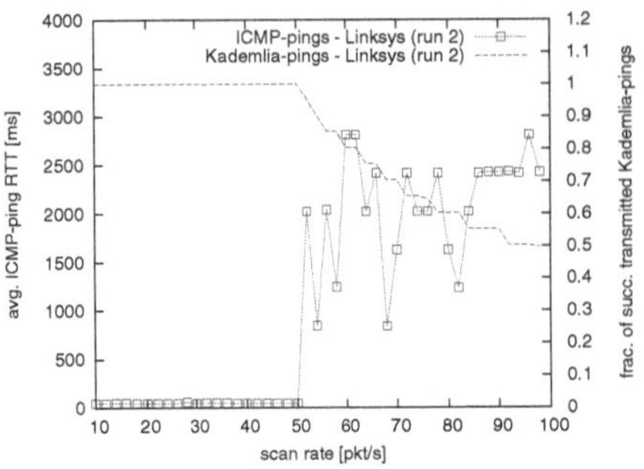


**Fig. 4.** Relation of ICMP ECHO delays and losses of Kademlia-packets

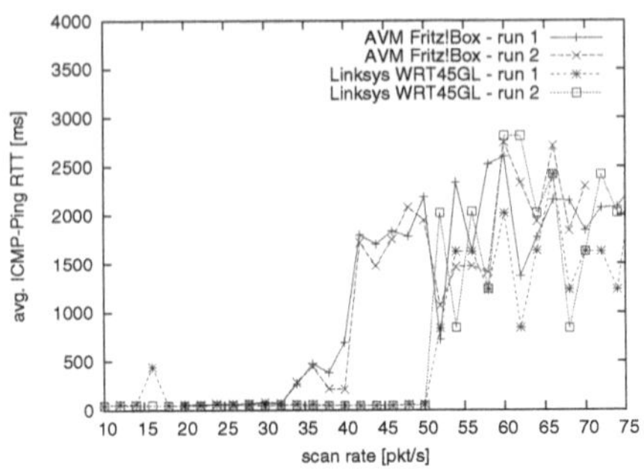


**Fig. 5.** Relation of ICMP ECHO delays and scan rate

thus, discard valid responses from BitTorrent peers (Kademlia-PINGs) that do not arrive in time. The Netgear TA612V is an example of the second class. Our experiments have shown that it uses about 4,000 entries in the connection table. With a rate of 100 pkt/s, an Kademlia-`PING` reply has to arrive within 40 seconds to be accepted. Thus, such implementation does not impose a practical limit on the scan rate.

In contrast, overloading the router due to a connection table overflow does clearly limit the scan rate up to a complete loss of functionality as we observed for the AVM FRITZ!Box 7050, which is very popular in Germany, and the Linksys WRTL45GL, which is popular on a global scale.

To gain insight into this behavior, we traced the Kademlia-`PING` packets send on the WAN port of the Linksys NAT box. We found that discarding of Kademlia-`PING` packets is strongly correlated to the discarding of ICMP ECHO packets. Thus, we send ICMP ECHO REQUEST packets in parallel to the random address probing. Measuring the average ICMP-based round trip time (RTT) is a much more convenient indicator to detect the overload than tracing all Kademlia-`PING` packets, which might even not be possible for all deployment scenarios. To illustrate this effect, for the experiments shown in Figures 4 and 5, we scanned with different scan rates for 300 seconds. Subsequently, we stopped scanning for another 300 seconds to let the NAT box recover its normal functionality, before scanning with the next rate. The available upstream bandwidth was more than 200 kbit/s. The timeout for ICMP ECHO REPLY messages was set to 4 seconds. As shown in Figure 4, the fraction of successfully transmitted Kademlia-`PING` packets decreases dramatically with increasing scan rate. For a scan rate of 100 pkt/s only 50 % of the packets are successfully transmitted. At the same time the average ICMP-based RTT increases significantly. Figure 5 plots the average ICMP-based RTT as a function of the scan rate for both NAT routers. It shows that packet losses with the AVM FRITZ!Box 7050 occur at a scan rate of about 35 to 40 pkt/s. With a Linksys WRTL45GL this effect occurred at about 45 to 50 pkt/s.

The observed behavior of NAT boxes clearly puts a limit on the scan rate, increasing the time required to successfully detect a BitTorrent DHT peer

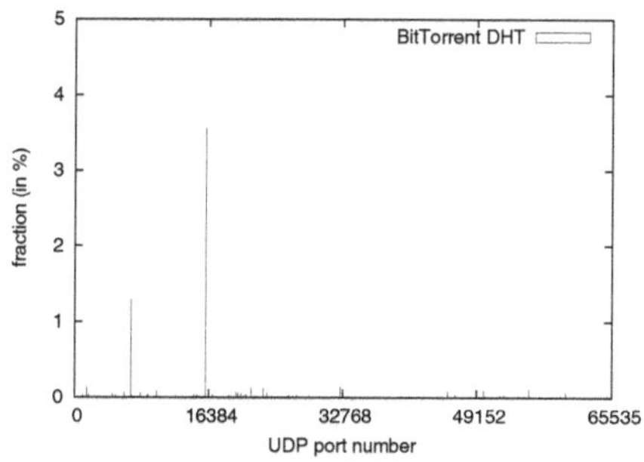


**Fig. 6.** Fraction of BitTorrent DHT peers per UDP port

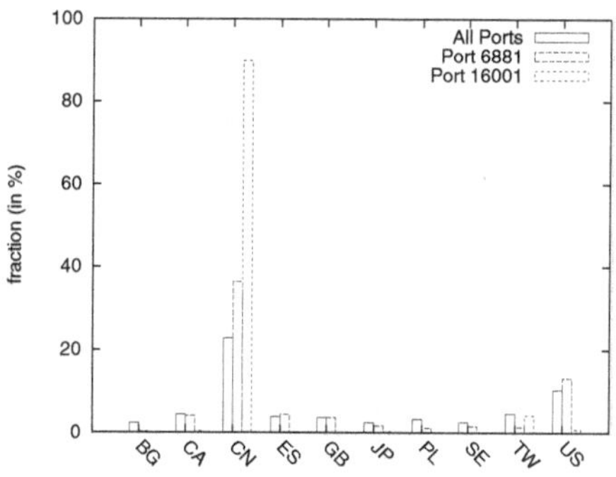


**Fig. 7.** Geographic distribution of BitTorrent DHT peers (top 15)

reported in Section 5.1. Unfortunately, the time will further increase due to the changing usage of ports as we show in the next section.

### 5.3 Port Distribution of BitTorrent DHT Peers

The mainline client of BitTorrent used the UDP port 6881 as default port for the BitTorrent DHT until version 5.2.0. With the shift to version 6.0 the basis of the mainline client changed to *µTorrent* and also the default port is not used anymore. In fact the port is chosen randomly to avoid port filtering by ISPs. To quantify the impact of the port distribution, we conducted an additional experiment: experiment setup was as follows: First, we bootstrapped in the BitTorrent DHT with our test client. Second, we sent `FIND_NODE` packets for randomly chosen keys with an rate of 1 keys/s. Over a period of 8 days in April 2008 we recorded about 5.2 million addresses (IP address and port) from peers that sent a response to our `FIND_NODE` packets.

The analysis of the captured data shows that each of the 65,536 possible ports is occupied by some peers of the BitTorrent DHT with an average of 78 peers per port and a standard deviation of 9. Figure 6 shows the distribution as histogram. Besides the small peaks that are distributed across all port numbers, we see two major peaks at 6881 and 16001. The first peak corresponds to the former default port, but the reason for the second peak is not obvious. Our assumption is that the second peak is caused by a client that uses the port 16001 as default port. We were not able to identify the client exactly, since it does not transmit the client type in the exchanged messages. However, we were able to trace its geographical origin, as Figure 7 shows by the geographic distribution of peers from the top 15 countries for all ports, port 16001 and port 6881 respectively. The figure shows that 23 % of all peers are located in China and about 10 % in the US. Moreover, we can conclude that the questionable port 16001 is used by a client that is popular in Chinese speaking countries, because about 90 % of the peers are located in China about 4 % in Taiwan, and about 1 % in Hong Kong.

To summarize the findings on the port distribution, we calculate that its entropy is $H_Q = 15.17$. Due to the $2^{16}$ possible Ports, the maximum entropy is $H_{QMax} = 16$. Assuming there is only *one* default port, the resulting entropy would be $H_Q = 0$. Thus, the search space w.r.t. random address probing is drastically expanded by the port distribution. Hence, we try to find optimizing strategies and analyze therefore the interrelation between the used ports and IP addresses in more detail in the next section.

# 6 Improving Distributed Bootstrapping

In this section, we propose two mechanisms that can improve the performance of distributed bootstrapping building upon the observations on the distribution. The first mechanism improves performance of random address probing by exploiting the fact that peers use more ports than just one "standard port". This optimizing mechanism works without changing the port selection mechanism, i.e., is optimal for the current situation. With the same rational, we proposes a second mechanism for selecting "quasi"-random ports that can be applied to future peer implementations.

## 6.1 Using More Than One Port

If the peers use more than one port as observed in 5.3, the strategy for random address probing can be potentially optimized. To illustrate that, assume the port distribution is known. Let $P_A(a)$ be the probability that a peer is run using a specified IP address $a$ and $P_P(i)$ the probability that a Peer uses Port $i$. W.l.o.g. we assume $P_P(i) \geq P_P(j)$ for $0 \leq i < j \leq 65535$.

We assume that the peers will change during the process of random address probing, because the number of potential IP addresses is huge compared to the possible scan rate. Thus, the probability for finding a peer using an IP address $a$, $P_A(a)$, is independent of the history of probed IP addresses. In contrast we argue that the ports can be checked in a short period of time and therefore $\sum_{i=0}^{65535} P_P(i) = 1$. Using these assumptions we formulate an optimization problem that is given in Formula 1. The goal is to maximize the probability to find a peer after $i$ tries and therefore determining the optimal number of ports $m$ that have to be probed per IP address. The key idea behind this is to look for the probability that a peer was not found until the $i$-th try. The quotient in the exponent results from the fact that if multiple ports are probed, less IP addresses can be probed.

$$max \left\{ 1 - \left( 1 - \left( P_a(a) \cdot \sum_{i=1}^{m} P_P(i) \right) \right)^{\frac{1}{m}} \right\} \tag{1}$$

The optimal value for $m$ can be easily determined by iterating over all possible values. To see that the solution of the optimization problem is not always trivial, consider the following exemplary probabilities $P_P(1) = 0.4$, $P_P(2) = 0.4$,

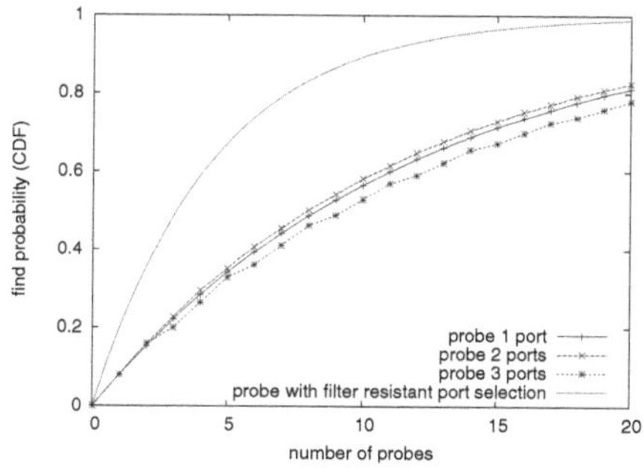


**Fig. 8.** Exemplary probability distribution for probing peers with multiple ports (see Section 6.1) and the probability distribution that results from applying the filter resistant port selection via hashing (see Section 6.2)

$P_P(3) = 0.2$, $P(i) = 0$ for $i \geq 4$. and $P(A) = 0.2$. Figure 8 illustrates the resulting probability distributions for probing one, two, and three ports per IP address (For comparison reasons the filter resistant mechanism of Section 6.2 is also included.). Here, the optimal strategy is to probe two ports, i.e., $m = 2$.

Assuming that the ports are (exactly) uniformly distributed, we were able to prove that it is more efficient to probe all ports before using a different IP address. Due to space limitations the proof is omitted here, it can be found in [24].

The optimization problem can be used to determine the best strategy in currently deployed P2P networks with a given port distribution. In future networks random address probing can be assisted by an filter resistant port selection mechanism that is presented in the next section.

### 6.2 Filter Resistant Port Selection via Hashing

To reduce the entropy of the port distribution we develop a filter resistant port selection mechanism that is optimized for random address probing and, nevertheless, cannot be trivially filtered by an ISP. The mechanism is based on the fact that the ternary content addressable memory (TCAM) in routers is used by the ISPs for establishing access control lists. TCAM is a limited resource [25]. Thus, the ISPs filter ports on the basis of IP address ranges rather than individual IP addresses.

Based on this observation, we propose the following mechanism: The IP address of a peer and a so called virtual port ($port_{virt}$) are used to calculate the actual port ($port_{real}$) that is used by the peer:

$$port_{real} := h(\ ipAddr \ \otimes \ port_{virt}\ )\ mod\ 2^{16} \tag{2}$$

$h(x)$ is a consistent hash function like SHA-1. $port_{virt}$ can be arbitrarily selected, but has to be fixed for all peers.

The characteristics of this approach are on the one hand that the ISPs can not filter peers on large scale, because each peer uses a quasi random port that depends on the IP address. On the other hand w.r.t. to random address probing the IP address is known in advance and therefore the entropy of the port distribution is $H_Q = 0$ for random address probing. Compared to the optimum that can be achieved for randomly chosen ports using the Formula 1 from above, the impact is huge as it is shown in Figure 8.

Furthermore, this mechanism can be seamlessly integrated in existing P2P networks, because current implementations can already handle arbitrary ports. Thus, the existing P2P protocol does not have to be changed and new peers using filter resistant port selection are fully compatible with existing ones.

## 7 Conclusion

In this paper, we showed the difficulties of decentralized bootstrapping into peer-to-peer systems by extensive Internet measurements. We identified two critical issues, i.e., limitations by NAT routers and the port distribution used by the clients. Hence, we proposed two mechanisms to increase the performance of decentralized bootstrapping.

Gathering real-world data by extensive measurement over several weeks, we analyze a combination of two mechanisms, local host caches and random address probing. For local host caches, we found that they enable successful bootstrapping after short periods of disconnection. After longer disconnections, however, one must resort to other bootstrapping mechanisms. Unfortunately, random address probing does not help to this end, since practical problems imposed by typical NAT hardware and the usage of non-standard ports by the peers limit its performance. To cope with these problems, we proposed two mechanisms, probing multiple peers per host and filter resistant port selection.

## References

1. Sharman Networks Ltd.: Website of KaZaA (2008), http://www.kazaa.com/
2. Gnutella Developer Forum: Gnutella: A Protocol for a Revolution, WWW-Publication (2003), http://rfc-gnutella.sourceforge.net
3. BitTorrent, Inc.: BitTorrent Website (2008), http://www.bittorrent.com/
4. n.a.: eMule Website (2008), http://www.emule-project.net
5. Skype, Ltd.: Website of Skype (2008), http://www.skype.com/
6. Joost, N.V.: Website of Joost (2008), http://www.joost.com/
7. Srisuresh, P., Holdrege, M.: IP Network Address Translator (NAT) Terminology and Considerations. RFC 2663, Informational (1999)
8. Karbhari, P., Ammar, M., Dhamdhere, A., Raj, H., Riley, G., Zegura, E.: Bootstrapping in gnutella: A measurement study. In: Barakat, C., Pratt, I. (eds.) PAM 2004. LNCS, vol. 3015, pp. 22–32. Springer, Heidelberg (2004)
9. Cramer, C., Kutzner, K., Fuhrmann, T.: Bootstrapping locality-aware p2p networks. In: Proc. 12th IEEE Int. Conf. on Networks (ICON 2004), pp. 357–361 (2004)
10. Kan, G.: Gnutella. In: Peer-to-Peer: Harnessing the Benefits of a Disruptive Technology, ch. 8, O'Reilly, Sebastopol (2001)
11. Dämpfling, H.: Website of the Gnutella Web Caching System, GWebCache (2008), http://www.gnucleus.com/gwebcache/

12. Baset, S.A., Schulzrinne, H.G.: An Analysis of the Skype Peer-to-Peer Internet Telephony Protocol. In: Proc. IEEE INFOCOM 2006, pp. 2695–2706 (2006)
13. Arak, V.: What happened on August 16, WWW-Publication (2007), http://heartbeat.skype.com/2007/08/what_happened_on_august_16.html
14. Chawathe, Y., Ratnasamy, S., Breslau, L., Lanham, N., Shenker, S.: Making gnutella-like P2P systems scalable. In: Proc. ACM SIGCOMM 2003, pp. 407–418 (2003)
15. Falkner, J., Piatek, M., John, J., Krishnamurthy, A., Anderson, T.: Profiling a million user DHT. In: Proc. 7th ACM SIGCOMM Internet Measurement Conference (IMC 2007), pp. 129–134 (2007)
16. Stutzbach, D., Rejaie, R.: Understanding churn in peer-to-peer networks. In: Proc. 6th ACM SIGCOMM Internet Measurement Conference (IMC 2006), pp. 189–202 (2006)
17. Stutzbach, D., Rejaie, R.: Improving Lookup Performance Over a Widely-Deployed DHT. In: Proc. IEEE INFOCOM 2006, pp. 2884–2895. IEEE, Los Alamitos (2006)
18. Steiner, M., En-Najjary, T., Biersack, E.: A global view of kad. In: Proc. 7th ACM SIGCOMM Internet Measurement Conf. (IMC 2007), pp. 117–122 (2007)
19. Conrad, M., Hof, H.J.: A Generic, Self-Organizing, and Distributed Bootstrap Service for Peer-to-Peer Networks. In: Hutchison, D., Katz, R.H. (eds.) IWSOS 2007. LNCS, vol. 4725, pp. 59–72. Springer, Heidelberg (2007)
20. Hinden, R., Deering, S.: IP Version 6 Addressing Architecture. RFC 4291, Draft Standard (2006)
21. Guttman, E.: Service location protocol: Automatic discovery of ip network services. IEEE Internet Computing 3(4), 71–80 (1999)
22. Qiao, Y., Bustamante, F.: Structured and unstructured overlays under the microscope: a measurement-based view of two p2p systems that people use. In: Proc. USENIX Annual Technical Conference, ATEC 2006 (2006)
23. Loewenstern, A.: BitTorrent Protocol Specification: BitTorrent Enhancement Proposal DHT Protocol, WWW-Publication (2008), http://www.bittorrent.org/beps/bep_0005.html
24. Dinger, J.: Das Potential von Peer-to-Peer-Netzen und -Systemen: Architekturen, Robustheit und rechtliche Verortung. Ph.D thesis, Universität Karlsruhe (TH) (2009) ISBN: 978-3-86644-327-3
25. Lakshminarayanan, K., Rangarajan, A., Venkatachary, S.: Algorithms for advanced packet classification with ternary cams. In: Proc. ACM SIGCOMM 2005, pp. 193–204. ACM, New York (2005)

# Performance Evaluation of Fast Startup Congestion Control Schemes (Work in Progress)

Michael Scharf

Institute of Communication Networks and Computer Engineering (IKR)
University of Stuttgart, Germany
michael.scharf@ikr.uni-stuttgart.de

**Abstract.** The Transmission Control Protocol (TCP) uses the Slow-Start mechanism at the beginning of a connection and after idle times. The Slow-Start delays the transport of data in particular if the round-trip time is large, which is undesirable for interactive applications. In order to speed up transfers, several alternatives have been proposed recently. This paper evaluates the performance and robustness of new fast startup congestion control schemes. We compare both end-to-end approaches as well as protocols that rely on additional feedback from the routers, using implementations in the Linux stack. Both testbed measurements and simulation studies quantify the potential performance improvement, the risk of packet loss, and the benefits of additional router support. Our results, which are also partly verified analytically, reveal that end-to-end fast startup mechanisms would not cause too much performance degradation if they are selectively used and carefully tuned. Additional router support would improve the fairness at the cost of a higher complexity.

**Keywords:** Congestion Control, TCP, Slow-Start, Quick-Start.

## 1 Introduction

Most Internet applications use the Transmission Control Protocol (TCP) for reliable, best effort transport. TCP uses congestion control [1] to adapt to the available bandwidth on the path. Still, after the connection setup or after idle periods it is difficult to determine an appropriate sending rate. TCP's congestion control uses the Slow-Start heuristic in these cases. This mechanism works well in the Internet, but it may require many round-trip times (RTTs) until an appropriate sending rate is reached and thus significantly delay data delivery.

This raises the question whether a faster startup is possible. The design of startup procedures for new flows is one of the remaining open issues of the Internet congestion control [2]. Recently, several new fast startup mechanisms have been developed, which all modify the Slow-Start. One solution is the Quick-Start TCP extension [3,4], which allows higher initial sending rates if the routers along the path approve a corresponding request. Since this requires modifications in the network entities, Quick-Start cannot be used in today's Internet. As a

L. Fratta et al. (Eds.): NETWORKING 2009, LNCS 5550, pp. 716–727, 2009.

potential alternative, Jump-Start [5] has been proposed. Jump-Start is a pure end-to-end mechanism. It uses rate pacing with a high initial sending rate, but reacts conservatively if the available bandwidth is exceeded. Further possibilities include e. g. increasing TCP's initial window beyond the value allowed by [6].

This paper compares several fast startup congestion control mechanisms. We study both end-to-end solutions, which only require modifications in the sender, as well as Quick-Start TCP, which is an example for a router-assisted approach. Unlike simulation studies such as [4,5], we implement the new fast startup mechanisms in the Linux kernel in order to obtain realistic results. In addition to testbed measurements, we also run simulations using the Network Simulation Cradle [7] for the user-space execution of real network stack code. To the best of our knowledge, this is the first comparative study of the new fast startup schemes and thereby complements our work on Quick-Start TCP [8,9].

The rest of this paper is structured as follows: Section 2 discusses the design space of fast startup congestion control and reviews selected proposals. In Section 3 we present an analytical model that is used to verify our experiments. Section 4 introduces our Linux implementations and our evaluation methodology. In Section 5 we compare benefits and risks of different fast startup schemes using analysis, simulations, and measurements. Section 6 concludes the paper.

## 2 Fast Startup Congestion Control

### 2.1 The TCP Slow-Start and Enhancements

The Slow-Start is one part of TCP's congestion control strategy. The basic idea is to start a transfer with a small congestion window and increase the window by one segment whenever an acknowledgment (ACK) arrives. This results in an exponential increase of the window, until the *Slow-Start Threshold* is reached and the *Congestion Avoidance* phase is entered. Originally, the initial window was one segment [1], but today a value of three segments is permitted by [6].

Several extension of the TCP congestion control, such as the Cubic algorithm [10], have been developed to improve the Congestion Avoidance phase in high-speed networks. Still, most TCP stacks use the original Slow-Start [1], which causes two problems: First, it can take a long time until the source can

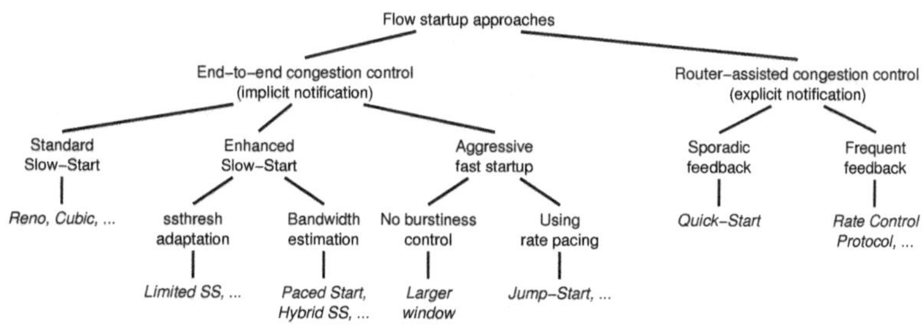


**Fig. 1.** Classification of flow startup principles

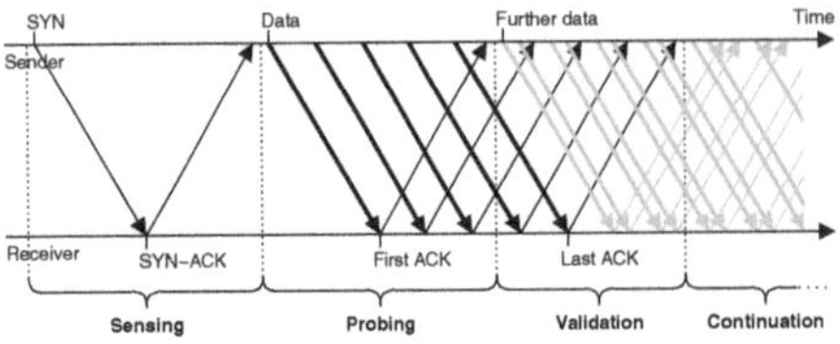


**Fig. 2.** Four phases of a fast startup

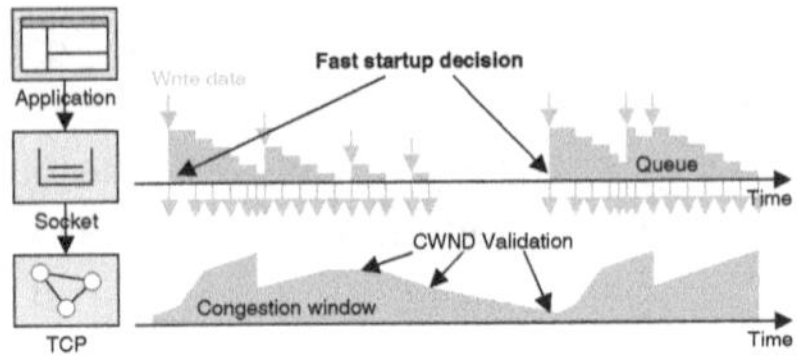


**Fig. 3.** Fast startup usage scenarios

fully utilize the available bandwidth on a path, in particular if the RTT is large. As a result, data transfers of interactive applications are unnecessarily delayed. Second, the exponential increase may be too aggressive ("Slow-Start overshoot") and cause multiple packet losses. An ideal congestion control should actually ensure that new flows converge quickly to their fair share of resources [2].

These issues are addressed by several improved startup approaches that are classified in Fig. 1. In order to overcome the overshooting problem, several enhancements have been developed, such as *Limited Slow-Start* [11]. Also, bandwidth estimation techniques have been proposed in order to determine the path capacity during the Slow-Start, e. g., with a *Paced Start* [12] or a *Hybrid Slow-Start* [13]. These enhancements still start with a small initial window.

## 2.2 Fast Startup Design Space

There are also several proposals how to speed up the Slow-Start and mitigate its performance limitations, which are surveyed e. g. in [3]. In general, a more intelligent end-to-end flow start startup mechanism could use some of the following information that is often available in end-systems: The round-trip time, cached state variables for this destination (e. g., a congestion manager), observable application communication characteristics (such as the amount of queued data in the socket), the local interface capacity, or application requirements.

Another option is to use explicit feedback from network entities along the path. Recently, several router-assisted congestion control mechanisms have been proposed. Potential solutions range from TCP enhancements, such as Quick-Start TCP [3], to completely new congestion control frameworks for a Future Internet. For instance, the Rate Control Protocol [14] suggests to uses some additional per-packet processing to speed up data transfers.

These fast startup mechanisms can be characterized by four phases that are shown in Fig. 2: During the *sensing* phase, some path characteristics are determined, potentially using explicit network feedback. Then the sender starts to send data (*probing*). The *validation* phase starts when corresponding ACKs arrive. The sender can then determine whether the initial choice was reasonable. Finally, the sender switches to the continuous congestion control (*continuation*), typically after the last ACK for the initially sent data has been received.

As depicted in Fig. 3, Slow-Starts do not only occur at the beginning of a connection, but also after longer idle periods, if the congestion window validation [15] is triggered. In this case fast *restart* mechanisms can be applied.

**Table 1.** Comparison of the considered fast startup schemes

| | **Quick-Start** | **Jump-Start** | **More-Start** | **Initial-Start** |
|---|---|---|---|---|
| **Sensing** | Router feedback | — | — | — |
| **Probing** | Approved rate (rate pacing) | Play out app. data (rate pacing) | Use given rate (rate pacing) | Large window (no pacing) |
| **Validation** | Observe loss | Count retransm. | Count retransm. | unmodified |
| **Continuation** | Revert ofter loss | Adapt after loss | Adapt after loss | unmodified |

### 2.3 Overview of Selected Fast Startup Approaches

In the following, we consider four different fast startup mechanisms:

Quick-Start TCP [3] is an experimental TCP extension that uses explicit router feedback. With Quick-Start, end-systems can ask for a high initial sending rate, e. g., during the three-way handshake. If all routers along the path approve the request, data transfers can start with this high rate. Further details about the Quick-Start mechanism and its implications can be found in [4,8,9].

Recently, Jump-Start [5] has been proposed as an end-to-end fast startup mechanism. The basic idea is to play out the queued application data during the first RTT using rate pacing. Thus, the more data is available, the higher is the initial rate. Jump-Start risks to start with a too large data rate. In order to cope with overshooting, the TCP retransmissions are counted during a validation phase. At the end of the loss recovery, the congestion window is adjusted to roughly half of the successfully transmitted segments. The authors of [5] argue that Jump-Start is not overly aggressive since most of today's TCP connections only transmit few data and would thus start with a rather small rate.

The principles of Quick-Start and Jump-Start can also be combined to a new end-to-end startup variant: Similar to Quick-Start, the applications could explicitly choose a "reasonable" initial rate, e. g., by considering application requirements or a congestion manager. Unlike Quick-Start, this rate is just used without asking routers for approval, in combination with a careful reaction to packet loss like Jump-Start. We label this combination "More-Start".

As a reference, we also consider the trivial mechanism of just increasing the initial congestion window to a larger value, without any further modifications of the TCP congestion control (Initial-Start). Table 1 highlights the differences between the four schemes and their functions during the four startup phases.

## 3 Speedup Analysis

### 3.1 Analytical Slow-Start Model

In order to verify our simulation and measurement results we use an analytical Slow-Start model developed by Bodamer (see [8]). As shown in Fig. 4, we model an end-to-end path by its TCP path capacity $R = \frac{L}{MTU} \cdot r$ and its minimum round-trip time $\tau$. Therein, $r$ is the data rate at IP layer, $MTU = 1500\,\text{byte}$ is the maximum transmission unit, and $L$ is the maximum segment size (MSS).

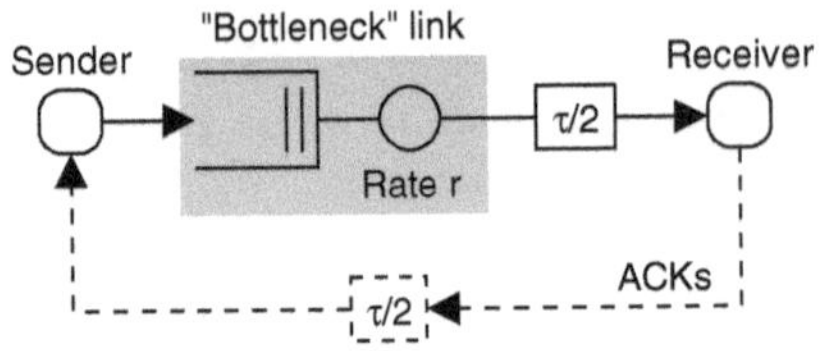


**Fig. 4.** Simplified path model

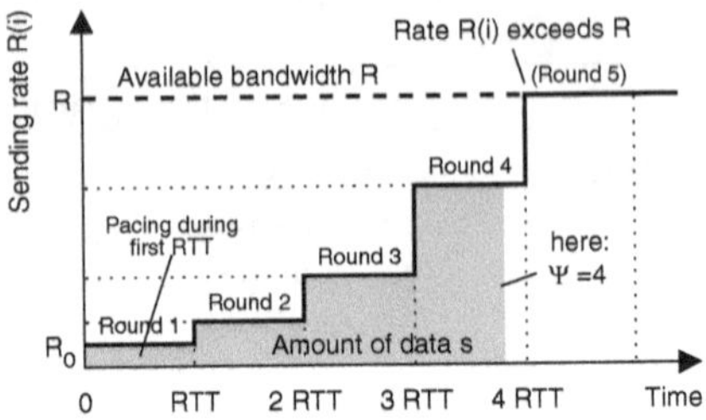


**Fig. 5.** Fast startup with rate pacing

Two further important parameters are the initial congestion window $w$ and the number of segments $b$ that a receiver acknowledges in one segment. As detailed in [8], the transfer time $\Gamma_{\mathrm{SS}}(s)$ of a certain amount of data $s$ in Slow-Start is

$$\Gamma_{\mathrm{SS}}(s) = \frac{s}{R} + \left(\tau + \frac{L}{R}\right) \cdot \psi - \frac{L \cdot w}{R} \frac{\gamma^{\psi} - 1}{\gamma - 1}. \tag{1}$$

Eq. (1) neglects packet loss and limitations by the receive window or a small initial Slow-Start Threshold. It also does not include unidirectional latencies or connection setup delays. $\psi$ is the index of the last Slow-Start round that is completely used, and $\gamma = 1 + \frac{1}{b}$. In [8] the index $\psi$ is calculated as

$$\psi = \max(\min(\underbrace{\left\lceil \log_{\gamma}\left(\tfrac{1}{w}\left(\tfrac{R\cdot\tau}{L}+1\right)\right)\right\rceil}_{\text{Window exceeds BDP}}, \underbrace{\left\lceil \log_{\gamma}\left(\left\lceil\tfrac{s}{L}\right\rceil \tfrac{\gamma-1}{w}+1\right)-1\right\rceil}_{\text{Transfer completed early}}), 0). \tag{2}$$

From this model follows $T_{\mathrm{SS}}(s) = \Gamma_{\mathrm{SS}}(s) + \tau$ as minimum response time of a client-server application. The Linux network stack can be characterized by $L = 1448$ and $b = 1$, if the *Quick-ACK* mechanism [16] is active.

### 3.2 New Model for Rate Paced Fast Startup Schemes

The fast startup mechanisms considered in this paper use rate pacing during the first RTT. This results in a different behavior that can be modeled as follows: If the initial sending rate $R_0$ is smaller than the end-to-end available bandwidth $R$, and if the sender continues in Slow-Start, the data rate is increased by a factor $\gamma$ every RTT, resulting in *rounds* as depicted in Fig. 5. During round $i$, data is continuously sent with rate $R(i) = R_0 \cdot \gamma^{i-1}$. The maximum amount of data that can be sent when round $i$ is completed is $M(i) = R_0 \cdot \tau \cdot \frac{\gamma^i - 1}{\gamma - 1}$. The transfer time is then

$$\Gamma_{\mathrm{Paced}}(s) = \tau \cdot \Psi + \frac{s - M(\Psi)}{\min\left(R, R_0 \cdot \gamma^{\Psi}\right)}. \tag{3}$$

Similar to Eq. (2), the index $\Psi$ of the last round with a rate $R(i) < R$ is

$$\Psi = \max(\min(\underbrace{\left\lceil \log_{\gamma}\left(\frac{R}{R_0}\right)\right\rceil}_{\text{Rate exceeds } R}, \underbrace{\left\lceil \log_{\gamma}\left(\frac{s\cdot(\gamma-1)}{R_0\cdot\tau}+1\right)-1\right\rceil}_{\text{Transfer completed early}}), 0). \tag{4}$$

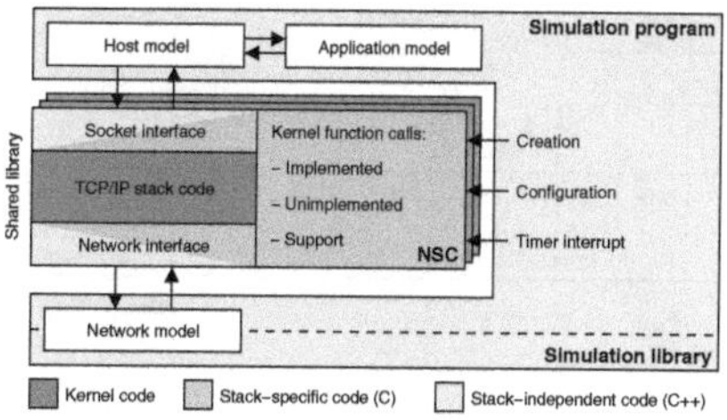


**Fig. 6.** Real code simulations with NSC

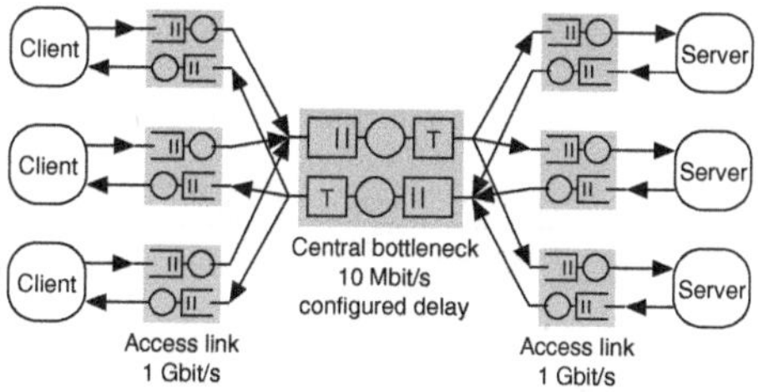


**Fig. 7.** Dumb-bell simulation topology

The minimum server response time for an approved Quick-Start rate $q$ can be derived from Eq. (3) as $T_{\mathrm{QS}}(s) = \Gamma_{\mathrm{Paced}}(s) + \tau$ with $R_0 = \frac{L}{MTU} \cdot q$. As described in [3], there is only a limited number of valid values for $q$. It must be emphasized that Eq. (3) does not apply if the Slow-Start Threshold is adapted after the validation phase (cf. [9]). In case of Jump-Start, the initial rate $R_0$ depends on the amount of queued application data. If a threshold $u$ is used, $T_{\mathrm{JS}}(s)$ can be obtained from (3) with $R_0 = \frac{\min(s,u)}{\tau}$. In the following we use $u = 64\,\mathrm{KiB}$. For Mean-Start, $T_{\mathrm{MS}}(s)$ can be determined similar to $T_{\mathrm{QS}}$, but without limitations for the granularity for $q$. The minimum server response time $T_{\mathrm{IS}}(s)$ of Initial-Start follows directly from Eq. (1) with a larger value for $w$.

# 4 Implementation and Evaluation Methodology

## 4.1 Linux Network Stack Implementations

In order to compare the different fast startup schemes, we have implemented Quick-Start, Jump-Start, More-Start and Initial-Start in the Linux network stack. Our Quick-Start TCP implementation is described in [9]. Compared to Quick-Start, our implementations of end-to-end fast startup schemes are comparatively simple, since only TCP code is affected. Still, the realization of rate pacing in Jump-Start and More-Start causes some complexity, as it requires timers and an additional state machine. These aspects, as well as potential interactions with flow control, are addressed in [9]. Our Quick-Start patch requires more than 2000 additional lines of kernel code, while e. g. Jump-Start only needs several hundred lines. But it is not sufficient to modify only the congestion control implementation. For Jump-Start, the socket processing workflow must also be adapted in order to determine the amount of queued application data.

## 4.2 Simulation and Measurement Methodology

In our measurements we use computers with Ubuntu/Fedora systems and Linux kernel 2.6.24. We always activate the non-standard-compliant *Quick-ACK* mechanism [16] to make our results reproducible and independent of the activation heuristics of the kernel. This implies $b = 1$, i. e., in our studies the Slow-Start may be significantly faster than a standard-compliant TCP stack with $b = 2$.

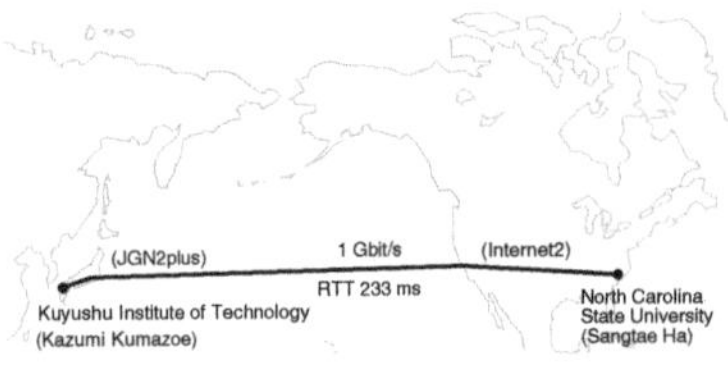


**Fig. 8.** Network path in experiments

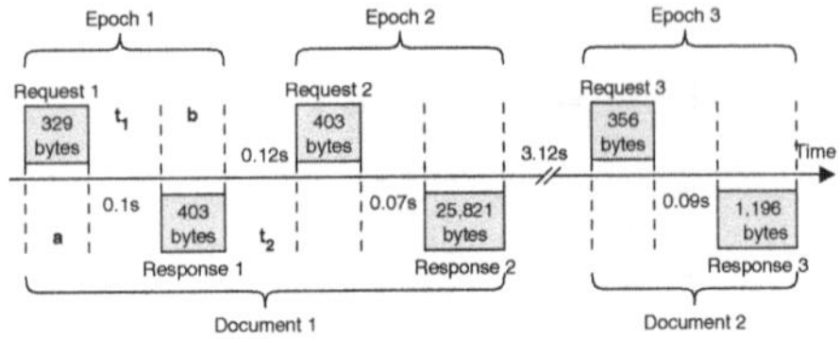


**Fig. 9.** Illustration of the $a$-$t_1$-$b$-$t_2$ model

We use the default stack configuration and the "Cubic" high-speed congestion control [10], which is Linux's default choice. The socket buffer sizes have been increased to 8 Mbyte in order to avoid limitations by the TCP flow control.

We also simulate with the Linux network stack in order to perform realistic, controlled studies with a larger number of entities. Our simulation tool uses the Network Simulation Cradle (NSC) version 3.0 [7] for the user-space execution of kernel code. The NSC architecture and its integration into our simulation tool is illustrated in Fig. 6. The corresponding wrappers and tools are documented in [17,18]. Our simulations are performed with kernel version 2.6.18, both without and with our fast startup patches.

In the simulations we use the classic dumb-bell topology [19] with $N$ client-server pairs, a central bottleneck with $r = 10\,\text{Mbit/s}$, a drop-tail buffer, and configurable delays (see Fig. 7). Unless stated otherwise, the buffer length is $B = 50$ packets and the RTT is $\tau = 200\,\text{ms}$. In our local testbed, the client and server applications run on computers that are interconnected by 10 Mbit/s Ethernet segments. The Linux network emulation "NetEm" enforces latencies. Furthermore, we also perform some real-world experiments over a long-distance high-speed path, using the experimental infrastructure shown in Fig. 8.

For workload modeling we follow the approach of [20], i. e., the characteristics of client-server applications are described by traces of requests and responses as shown in Fig. 9. As simplest model we use fixed-sized small requests and responses. Furthermore, we study scenarios with a given downlink load, which is realized by scheduling request-response vectors over persistent TCP connections. The response length is Pareto distributed (mean $m = 125$ or 10 kbyte, shape factor $\alpha = 1.1$, cutoff at 10 MB), and the inter-arrival time is assumed to be exponential. We also replay Internet traffic traces [21] in a dumb-bell topology with nine different RTTs, as recommended in [19]. Our main performance metric is the server response time, which corresponds to the duration of epochs in Fig. 9.

## 5 Performance Comparison

### 5.1 Fundamental Startup Behavior

As a first step, we compare the Slow-Start (SS) to Quick-Start (QS), Jump-Start (JS), More-Start (MS), and Initial-Start (IS) in a very simple simulation setup: Figure 10 shows downlink traces for one client and one server ($N = 1$), with a

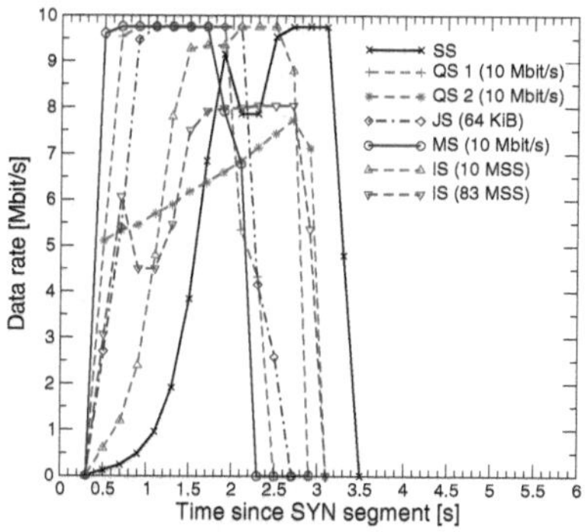


**Fig. 10.** Traces of a single startup (sampling time granularity 200 ms)

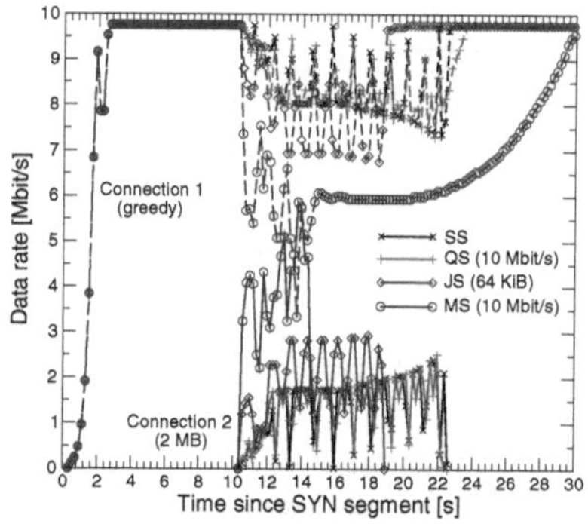


**Fig. 11.** Startup of a new flow against a long-lived competing TCP connection

request of 100 B and a response of 2 MB. Here, the Slow-Start needs over 1 s until the path is fully utilized. All fast startup mechanism are better in this scenario.

The largest approvable Quick-Start rate is $q = 5.12\,\mathrm{Mbit/s}$. If such a request is performed during the connection setup, and if it is approved, the data transfer can start immediately with this rate. After finishing the probing phase, there are different options [9]: The sender can either continue in Slow-Start ("QS 1") or adapt the Slow-Start Threshold to the Quick-Start window (about 83 segments) and continue in Congestion Avoidance ("QS 2"). The latter variant is more careful, but it may result in longer delays and is not further considered here.

Our Jump-Start implementation plays out up to $u = 64\,\mathrm{KiB}$ during the first RTT and is thus starts slightly slower. In Fig. 10 the best performance would be achieved if the sender knew approximately the available bandwidth of $r = 10\,\mathrm{Mbit/s}$ and used it (More-Start). Not shown is that a higher initial rate would, of course, cause packet loss, and the resulting completion time would be similar to Slow-Start. Finally, increasing the initial window to $w = 10$ would also work in this scenario. However, any initial value larger than the bottleneck buffer size will cause severe packet loss, resulting in no significant improvement.

A bottleneck link might already be occupied by existing flows. In order to study the convergence behavior in this case, a similar simulation scenario has been set up with $N = 2$ clients and servers and a start offset of 10 s. Figure 11 shows selected examples of the resulting flow startup. Three observations can be made: First, with Slow-Start, a short flow is unable to reach a rate of 5 Mbit/s, which would be its fair share. This long convergence is an inherent aspect of Slow-Start. Second, Quick-Start does not improve the convergence, if the routers use an admission control strategy that prevents over-commitment (cf. [8]). Due to lack of free capacity, the Quick-Start request is denied at the bottleneck, and the second connection therefore falls back to Slow-Start. Third, a fast startup e. g. with Jump-Start or More-Start improves the performance of the new flow at cost of the already established connection. It is an open and controversial question whether such an aggressive behavior of short flows is fair, nor not [2].

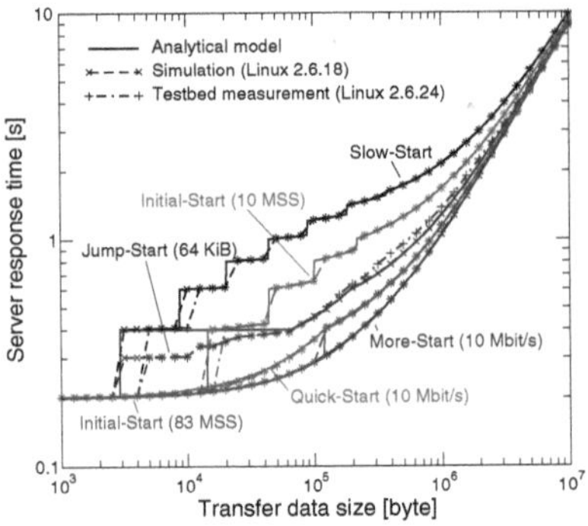


Fig. 12. Maximum possible speedup for path capacity 10 Mbit/s and RTT 200 ms

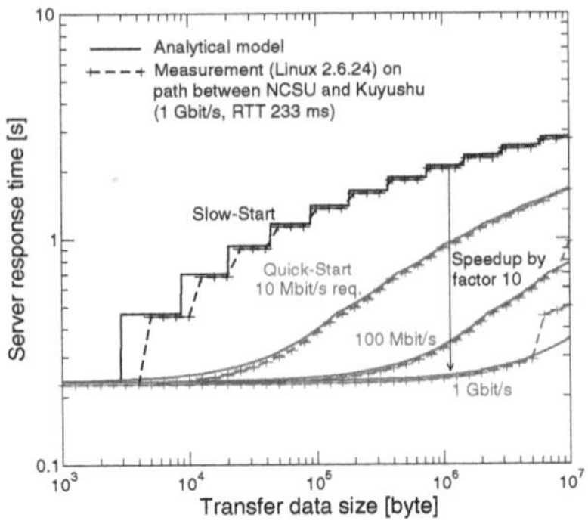


**Fig. 13.** Speedup on the high-speed path with different Quick-Start request rates

## 5.2 Speedup Compared to Slow-Start

The performance benefit of fast startup schemes depends on the amount of data $s$, the available bandwidth $r$, and the RTT $\tau$, as determined in Section 3. Figure 12 compares the analytical values for the response times $T_{\mathrm{SS}}$, $T_{\mathrm{QS}}$, $T_{\mathrm{JS}}$, and $T_{\mathrm{IS}}$ with the simulation and testbed measurement results for a large buffer ($B = 1000$ packets). The graph for standard TCP reveals the typical steps of the Slow-Start. As to be expected [4,8], a fast startup can improve the response time in particular for mid-sized response sizes. Figure 12 also reveals that model and experiments slightly differ in case of Jump-Start: If a perfect rate pacing was used, the response time for transfers up to $s = u = 64\,\mathrm{KiB}$ would be $T_{\mathrm{JS}} = \tau$. However, our real implementation only uses a limited number of timers (cf. [9]) and therefore sends faster if $s \ll u$. Figure 13 presents the results of a similar experiment with Quick-Start on a long-distance path ($r \approx 1\,\mathrm{Gbit/s}$). Due to lack of full Quick-Start router support, this study assumes that requests up to that rate are indeed approved. In such a high-speed environment, a fast startup mechanism can achieve substantial transfer time speedups up to a factor of 10.

Figure 14 studies the scenario of several connections sharing the bottleneck. The results have been obtained by simulating the exchange of requests and responses over $N = 50$ persistent TCP connections between 50 client-server pairs. In this case, Slow-Start is also used if the congestion window validation [15] is triggered. Obviously, when the load $\rho$ at the bottleneck increases, the response times get larger. A lower bound can be determined by integrating (1) and (3) over the response size distribution. Figure 14 reveals some important differences between the different schemes: First, as the load increases, the Quick-Start and Slow-Start performance is similar because of the Quick-Start admission control. Quick-Start also needs one RTT for the signaling, which reduces the speedup. An Initial-Start with $w = 83\,\mathrm{MSS}$ again turns out to be detrimental in combination with a bottleneck buffer of $B = 50$ packets. Significant overshooting also occurs when one just starts with full link speed (More-Start). An interesting result is that Jump-Start can keep the response time at a low level even for high load.

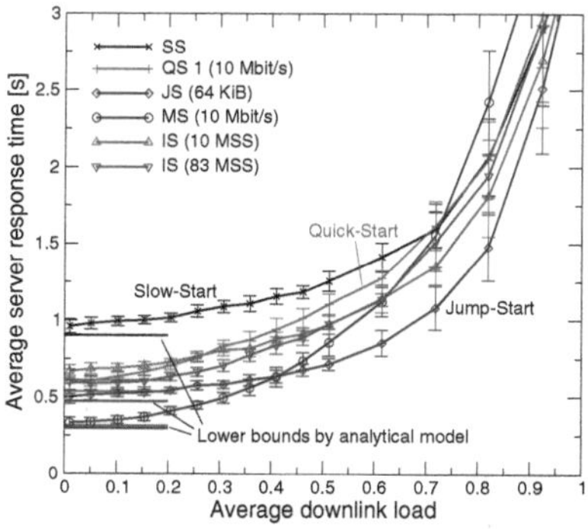


**Fig. 14.** Impact of the bottleneck load (Pareto distr. responses $m = 125\,\text{kB}$)

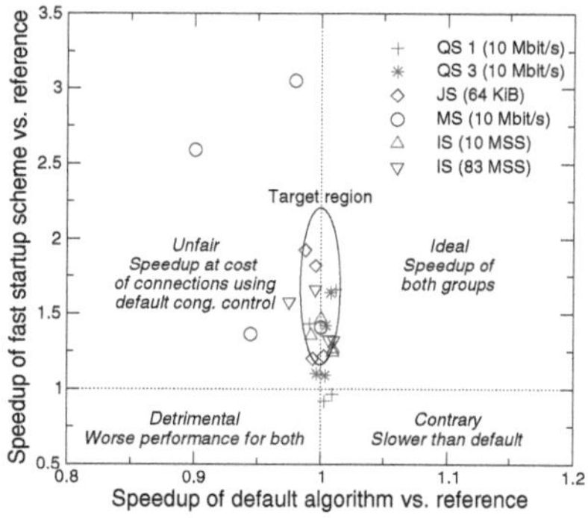


**Fig. 15.** Unfairness between competing default and fast startup connections

### 5.3 Fairness and Risk of Packet Loss

Any fast startup mechanism can result in fairness problems and risks to cause packet loss. Still, our experiments show that this may not be a severe problem if a sophisticated fast startup is used. Fig. 15 shows the result of an experiment similar to Fig. 14, except that there are now $N/2$ unmodified network stacks ("default") and $N/2$ that use a fast startup. Simulations are performed with two different traffic characteristics ($m_1 = 125\,\text{kB}$, $N_1 = 50$ and $m_2 = 10\,\text{kB}$, $N_2 = 400$) and two different loads ($\rho_1 \approx 0.05$, $\rho_1 \approx 0.3$). The x- and y-axis in Fig. 15 represent the resulting response times of the two groups. The reference value is the case that all end-systems use Slow-Start. In this representation, the interaction between the default and the fast startup mechanisms can be classified into four different cases. In the best case ("target region"), the fast startup schemes speed up their own transfers without significantly slowing down connections that use the default TCP congestion control. According to our results, both Quick-Start and Jump-Start are rather fair. The other variants can significantly affect the performance of connections that use an unmodified network stack.

### 5.4 Performance Impact for Typical Internet Workloads

The overall performance benefit of fast startup congestion control is difficult to quantify. Many TCP-based applications often transfer small objects that fit in today's initial congestion window, and most Internet RTTs are small, too. In these cases the existing Slow-Start performs reasonably well. This can also be observed in Fig. 16, which were obtained by replaying traffic traces [21] among $N = 450$ client-server pairs using realistic RTT distributions between 4 ms and 200 ms (cf. [19]). The traces are scheduled so that the downlink load of bottleneck is $\rho \approx 0.3$. If Jump-Start were used by all entities, epoch completion times that are of the order of several hundred milliseconds can be improved, but the overall benefit is not large. Using Quick-Start by default ("QS 1") even performs worse on such a moderately loaded link, since many requests get denied.

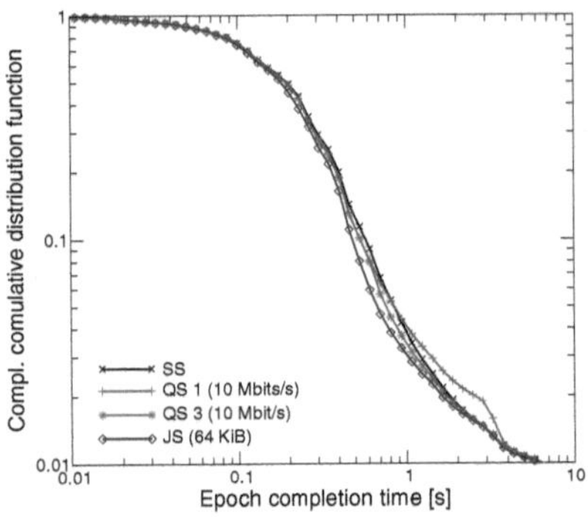


**Fig. 16.** Performance with traffic traces from [21] for a downlink load of 0.3

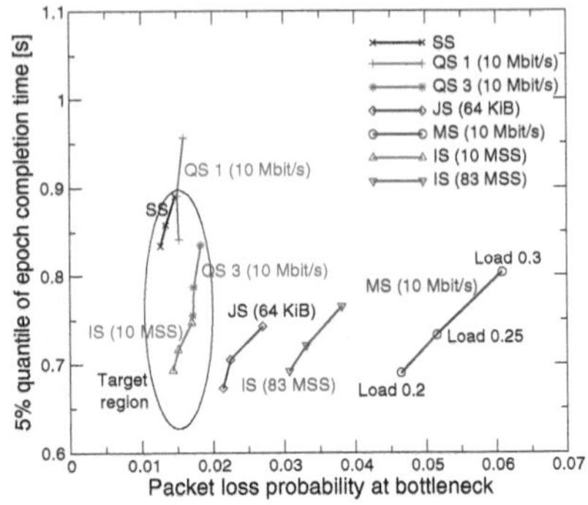


**Fig. 17.** Tradeoff between speedup and packet loss in the same setup

An improvement is possible if Quick-Start is only selectively enabled, e. g., for data transfers > 10 kB ("QS 3"). The results for the other fast startup mechanisms, which are omitted, are similar to the Jump-Start graph.

Fig. 17 reports packet loss statistics in the same setup and relates them to the 5 %-quantile of the epoch completion time. The quantile is one possibility to quantify the speedup of selected communication patterns. With default stacks, the packet loss probability is of the order of 1 % in the given scenario. Quick-Start and Jump-Start, as well as a limited increased initial window, only moderately increase packet loss, while the other variants are more aggressive. These results again indicate that both Quick-Start and Jump-Start are not overly harmful, with Jump-Start being much simpler to implement and to deploy.

## 6 Conclusion and Future Work

New fast startup congestion control approaches aim at replacing the TCP Slow-Start. This papers studies the performance of four different mechanisms by analytical models, by simulation, and by measurements. We consider the Quick-Start TCP extension, the Jump-Start proposal, a new combination of both, and the simplest approach, i. e., just to increase TCP's initial congestion window. All schemes are implemented in the Linux stack. According to our results, Jump-Start performs well, even though it causes some unfairness to competing flows using Slow-Start. Of all schemes, Quick-Start is the most conservative one, but the required router support raises several unsolved issues. In contrast, the other more aggressive alternatives could cause harm. Specifically, it is not an option just to significantly increase the initial window without rate pacing. Also, a promising solution is to enable fast startup only in selected cases and for applications that can indeed benefit, but this would require additional intelligence in the network stack. Future studies could address some algorithmic details, in particular the response to packet loss. Further work is also needed to understand the overall implications of using fast startup congestion control on Internet scale.

## Acknowledgments

This work is partially funded by the German Research Foundation (DFG) within the SFB 627. The author thanks the Kuyushu Institute of Technology and the North Carolina State University for providing access to their infrastructure.

## References

1. Jacobson, V.: Congestion avoidance and control. In: Proc. ACM SIGCOMM 1988, pp. 314–329 (August 1988)
2. Welzl, M., Papadimitriou, D., Scharf, M., Briscoe, B.: Open research issues in Internet congestion control. IRTF Internet Draft, work in progress (August 2008)
3. Floyd, S., Allman, M., Jain, A., Sarolahti, P.: Quick-Start for TCP and IP. IETF RFC 4782 (experimental) (January 2007)
4. Sarolahti, P., Allman, M., Floyd, S.: Determining an appropriate sending rate over an underutilized network path. Computer Networks 51(7), 1815–1832 (2007)
5. Liu, D., Allman, M., Jin, S., Wang, L.: Congestion control without a startup phase. In: Proc. PFLDnet 2007 (February 2007)
6. Allman, M., Floyd, S., Partridge, C.: Increasing TCP's initial window. IETF RFC 3390 (proposed standard) (October 2002)
7. Jansen, S.: Simulation with real world network stacks. In: Proc. Winter Simulation Conference (2005)
8. Scharf, M.: Performance analysis of the quick-start TCP extension. In: Proc. IEEE Broadnets (September 2007)
9. Scharf, M., Strotbek, H.: Performance evaluation of quick-start TCP with a Linux kernel implementation. In: Das, A., Pung, H.K., Lee, F.B.S., Wong, L.W.C. (eds.) NETWORKING 2008. LNCS, vol. 4982, pp. 703–714. Springer, Heidelberg (2008)
10. Rhee, I., Xu, L.: Cubic: A new TCP-friendly high-speed TCP variant. In: Proc. PFLDnet 2005 (February 2005)
11. Floyd, S.: Limited slow-start for TCP with large congestion windows. IETF RFC 3742 (experimental) (March 2004)
12. Hu, N., Steenkiste, P.: Improving TCP startup performance using active measurements: algorithm and evaluation. In: Proc. IEEE ICNP, pp. 107–118 (November 2003)
13. Ha, S., Rhee, I.: Hybrid slow start for high-bandwidth and long-distance networks. In: Proc. PFLDnet 2008 (March 2008)
14. Dukkipati, N.: Rate Control Protocol (RCP): Congestion Control to Make Flows Complete Quickly. Ph.D thesis, Stanford University (October 2007)
15. Handley, M., Padhye, J., Floyd, S.: TCP congestion window validation. IETF RFC 2861 (experimental) (June 2000)
16. Sarolahti, P., Kuznetsov, A.: Congestion control in Linux TCP. In: Proc. USENIX Annual Technical Conference (June 2002)
17. Zeeh, C.: Integration of the Linux-TCP/IP Protocol Stack into an Event-Driven Simulation Environment. Diploma thesis, University of Stuttgart, IKR (2006)
18. Proebster, M.: Performance Evaluation of Congestion Control with Explicit Router Signaling. Diploma thesis (in German), University of Stuttgart, IKR (2008)
19. Andrew, L., et al.: Towards a common TCP evaluation suite. In: Proc. PFLDnet 2008 (March 2008)
20. Weigle, M.C., Adurthi, P., Hernández-Campos, F., Jeffay, K., Smith, F.D.: Tmix: A tool for generating realistic TCP application workloads in ns-2. ACM SIGCOMM Computer Communication Review 36(3), 65–76 (2006)
21. Web site: WAN in Lab. (2008), `http://wil.cs.caltech.edu/suite/TrafficTraces.php`

# A Unified Framework for Sub-stream Scheduling in P2P Hybrid Streaming Systems and How to Do Better?*

Zhenjiang Li[1], Yao Yu[2], Xiaojun Hei[3], and Danny Hin-Kwok Tsang[1]

[1] Department of Electronic and Computer Engineering
Hong Kong University of Science and Technology
[2] Department of Electronics Science and Engineering
Nanjing University
[3] Department of Electronics and Information Engineering
Huazhong University of Science and Technology
lzjiang@ust.hk, allanyu@ese.nju.edu.cn, heixj@hust.edu.cn,
eetsang@ece.ust.hk

**Abstract.** The pull-push hybrid P2P streaming, as an emerging and promising approach, has achieved some success in delivering live video traffic. The sub-stream scheduling problem is a key design issue in a hybrid system. In this paper, we propose a max-flow model for unifying this sub-stream scheduling problem. We find that the sub-stream scheduling problem in GridMedia, CoolStreaming+ and LStreaming can be formulated into a special case of the proposed max-flow model. We further propose a min-cost flow model to combat peer heterogeneity in scheduling sub-streams. This min-cost flow model is implemented in a prototype system, LStreaming+. The accuracy of the max-flow model and the outstanding performance of LStreaming+ are demonstrated by extensive simulations. We also show that LStreaming+ achieves excellent performance in prototype experiments.

**Keywords:** P2P streaming, pull-push, hybrid, max-flow, min-cost flow.

## 1 Introduction

P2P streaming architectures have advanced in two major approaches: pull-based (mesh-pull) approach versus push-based (tree-push) approach. The pull-based systems apply a simple design principle and combat peer churn and peer heterogeneity in dynamic P2P environment. However, these pull-based systems often suffer from high traffic overhead and long start-up delay [1]. In contrast, push-based systems may achieve high throughput, low overhead and small delay if the tree structure does not break down due to peer churn. Recently, researchers are exploring a new class of pull-push hybrid architectures. Some hybrid systems, including GridMedia [2], CoolStreaming+ [3] and LStreaming [4,5], have already

* This work is supported by RGC Earmarked Research Grant 620306.

L. Fratta et al. (Eds.): NETWORKING 2009, LNCS 5550, pp. 728–741, 2009.

achieved performance improvement compared with pull-based systems in terms of throughput, signaling overhead and video viewing quality.

The general idea of pull-push streaming works as follows. The original video is divided into data chunks and each chunk is assigned with a unique chunk number (e.g. $0, 1, 2, \ldots$). Chunks are further organized into sub-streams logically. Suppose there are total $S$ sub-streams, sub-stream 1 contains chunks $1, 1+S, 1+2S, \ldots$. Video chunks may be delivered using the chunk-pulling module and the sub-stream pushing module. Each peer starts up with pulling chunks. After the initial time, the sub-stream pushing module starts to work. Then peers need to decide which sub-stream should be obtained from which neighbors. This decision process is known as **sub-stream scheduling**. After the sub-stream pushing module is enabled, the chunk-pulling module is still working to serve as backup to download those missing chunks when their playback deadline is approaching. In a hybrid system, the sub-stream pushing module sets up a local tree structure between peers based on the topology formed by the chunk-pulling module. In sub-stream scheduling, peers actively select one neighbor among all candidate neighbors to obtain a sub-stream. However, there exists no explicit "scheduling" module in conventional tree-based systems. The neighbor-relationship is formed passively after peers successfully join single or multiple trees in conventional tree-based systems. The previously proposed sub-stream scheduling schemes [2,3,4,5] are mainly designed experimentally. We find that these heuristic designs have inherent limitations. Liu *et al.* proposed to allocate sub-streams via bipartite graphs in [6]. The protocol in [6] is tailored against free-riders without an in-depth study on the sub-stream scheduling problem.

In this paper, we first highlight the importance of the sub-stream scheduling in hybrid systems on reducing overhead, improving throughput and maximizing video viewing quality. To the best of our knowledge, this is the first work thoroughly studying the sub-stream scheduling problem in P2P hybrid systems. Then we propose a *max-flow* model for unifying the sub-stream scheduling problem in the existing hybrid systems including GridMedia, CoolStreaming+ and LStreaming. We demonstrate that the sub-stream scheduling module in these three hybrid systems actually solves one special case of the proposed max-flow model, respectively. Our proposed max-flow model is solvable in polynomial time and provides useful insights for a better sub-stream scheduling design, in which we propose a *min-cost flow* model to better utilize heterogenous peers. We implement the proposed min-cost flow model in a prototype system, LStreaming+. We show that this min-cost flow scheduling scheme outperforms the existing sub-stream scheduling schemes. Compared with GridMedia and LStreaming, the total overhead reduction of LStreaming+ reaches as high as 43.9% and 20.0%, respectively. In addition, LStreaming+ achieves the highest throughput and the smoothest video playback.

The rest of this paper is organized as follows. In Section 2, we present the max-flow model and the corresponding special cases for three existing systems, GridMedia, CoolStreaming+ and LStreaming. Then we propose the min-cost flow model in Section 3 and evaluate the performance using simulations and prototype experiments in Section 4. Finally, conclusions are made in Section 5.

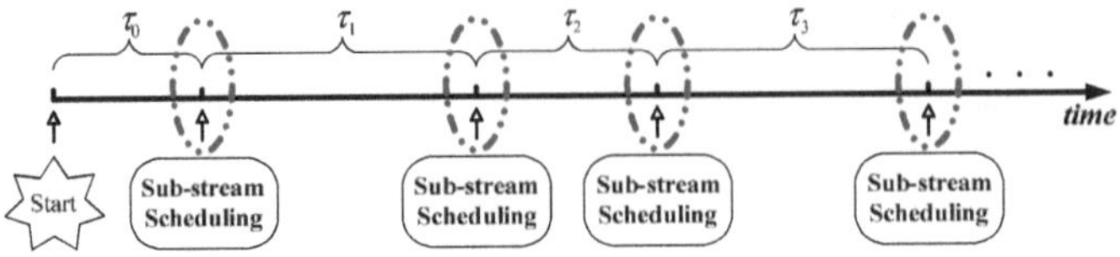


**Fig. 1.** The general scheduling pattern of peers in hybrid systems

## 2 Problem Statement and Formulation

The general scheduling pattern of peers in a hybrid system is shown in Fig. 1. During the initial $\tau_0$ time, only the chunk-pulling module is enabled so as to start up the streaming process as quickly as possible. After $\tau_0$, the sub-stream pushing module starts to work concurrently with the chunk-pulling module. Peers will conduct sub-stream scheduling for the sub-stream pushing module after $\tau_t$ time. Whether the duration of each $\tau_t$ $(t = 1, 2, ...)$ is set to be constant depends on a particular system design. If sub-stream scheduling is triggered periodically (e.g. the time-driven type), each time interval is identical; otherwise, the time intervals can vary (e.g. the event-driven type).

### 2.1 Sub-stream Scheduling Problem

In pull-push hybrid systems, the sub-stream scheduling decisions are generally controlled by the *peer qualification rule* and the *token number*. Based on the *peer qualification rule*, peers need to determine qualified neighbors in terms of the sub-stream availability to obtain sub-streams; then, each sub-stream is scheduled to be obtained from only one qualified neighbor. These selected qualified neighbors, called **push-neighbors**, push chunks of sub-streams to the requested peer. The *peer qualification rule* may be set differently in different systems. After selecting qualified neighbors, on each sub-stream scheduling event, every peer (e.g. peer $i$ in Fig. 2) needs to calculate the maximum number of sub-streams that can be obtained from each neighbor. In existing systems, this number is usually computed based on the end-to-end bandwidth between two peers. In this paper, we call this number *token number*, denoted as $T_{ih}$. It represents neighbor $h$ can be scheduled to upload at most $T_{ih}$ sub-streams to peer $i$.

We now examine peer $i$ in Fig. 2 for instance to explain the design issues in the *sub-stream scheduling*. Peer $i$ has four neighbors and five sub-streams needed to be scheduled. For sub-stream 0, peer $i$ schedules it to be obtained from neighbor 1 which is the only peer qualified for this sub-stream. On scheduling sub-streams 1 and 2, if random selection or sequential selection is used, both sub-streams 1 and 2 can be downloaded from neighbor 2. Although neighbor 3 has 2 tokens left, peer $i$ cannot download any sub-streams from peer 3. Therefore, the chunks in sub-streams 3 and 4 can only be downloaded by the chunk-pulling module later. As a result, more signaling overhead will incur. Furthermore, note that the portion of the buffer, in which the missing chunks are requested by the chunk-pulling module, is usually close to the playback pointer. If too many sub-streams

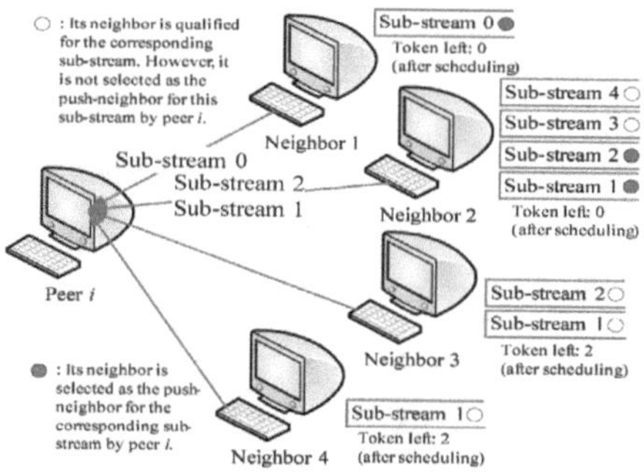


**Fig. 2.** Inefficient scheduling

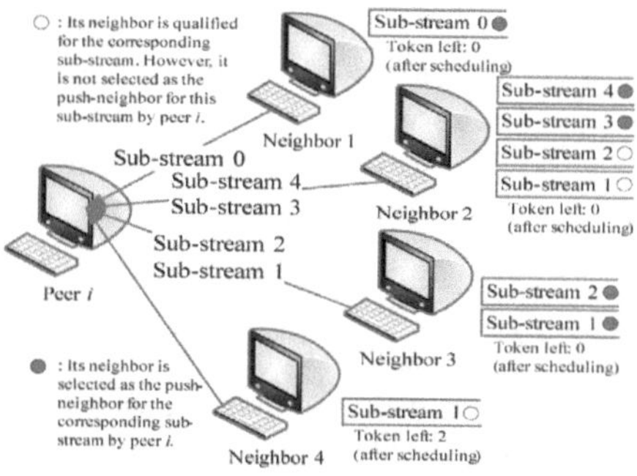


**Fig. 3.** Efficient scheduling

are not successfully scheduled (without finding a qualified push-neighbor), then a large amount of chunks can only be downloaded by the chunk-pulling module; therefore, the chance that chunks miss their playback deadlines increases and the viewing quality deteriorates accordingly. If we change to the strategy in Fig. 3, all sub-streams can be successfully scheduled. Most of the chunks will be delivered through multiple trees and only a few of them may be left to the chunk-pulling module. Hence, the viewing quality is improved with reduced signaling overhead. Although scheduling in Fig. 3 is more efficient than that in Fig. 2, a problem still remains. In Fig. 3, sub-stream 1 can be scheduled to either qualified neighbor 3 or 4, it is not clear which peer is more appropriate. This issue will be discussed in Section 3.

We have presented the importance of sub-stream scheduling on reducing signaling overhead and improving viewing quality. Since the total traffic overhead is due to signaling messages and redundant video chunks, we now briefly discuss the impact of sub-stream scheduling on video redundancy overhead. Note that a push-neighbor continues to push chunks if it does not receive any canceling messages. If the sub-stream scheduling is not conducted appropriately, peers in the system will suffer from frequent switching of push-neighbors (we call this phenomena **push-neighbor switching**). Redundant video download occurs as follows. Chunks have been pushed out from previous push-neighbors before push-neighbor switching. Due to the propagation delay and the network buffering effect, these chunks may be still in transmit. The same chunks may be disseminated via new push-neighbors later, then redundant download occurs. Frequent push-neighbor switching can incur significant video redundancy overhead, which will be further examined in Section 4. In summary, the sub-stream scheduling highly impacts viewing quality, signaling and video redundancy overhead. In the next subsection, we propose a unified model on sub-stream scheduling based on our study of existing hybrid systems.

## 2.2 Modeling

To facilitate the discussions, we tabulate the notations in Table 1. The default values are used in our simulation and prototype unless specified explicitly.

**Table 1.** Notations

| | |
|---|---|
| $e^i_{hj}$ | Decision variable to indicate whether peer $i$ obtains sub-stream $j$ from neighbor $h$ |
| $I^i_{hj}$ | Decision variable to indicate whether neighbor $h$ is qualified for sub-stream $j$ according to the *peer qualification rule* of peer $i$ |
| $\Gamma_i$ | Set of all sub-streams for peer $i$ to schedule |
| $\Omega_i$ | Set of neighbor peers of peer $i$ |
| $R$ | Playback rate of the streaming (e.g. 300kbps) |
| $\tau$ | Period of sub-stream scheduling for LStreaming+ (e.g. 10s) |
| $\chi$ | A large integer (e.g. $10^4$) |
| $L_{hi}$ | End-to-end bandwidth from neighbor $h$ of peer $i$ to peer $i$ (kbps) |
| $P^i_{hj}$ | Largest chunk number already available for sub-stream $j$ at neighbor $h$ of peer $i$ |
| $Q_{ij}$ | Starting chunk number needed by peer $i$ for sub-stream $j$ |
| $T_{ih}$ | Token number for neighbor $h$ of peer $i$ in sub-stream scheduling |
| $B$ | Number of chunks in 1 second video (e.g. 30) |
| $S$ | Total number of sub-streams in the system (e.g. 15) |

For each sub-stream scheduling, peer $i$ needs to schedule $|\Gamma_i|$ sub-streams in total, where $\Gamma_i$ represents the set of all sub-streams for peer $i$ to schedule. For any sub-stream $j$ in $\Gamma_i$, the set $\{h|I^i_{hj} = 1, \forall h \in \Omega_i\}$ includes all the qualified neighbors, from which peer $i$ can obtain sub-stream $j$. $\Omega_i$ is the set of all neighbors of peer $i$. If multiple neighbors are qualified for one sub-stream, peers in existing systems will randomly select one neighbor to obtain this sub-stream. Thus, we assume there is no difference among all these qualified neighbors in scheduling. When peer $i$ conducts *sub-stream scheduling*, it basically aims to maximize the number of sub-streams to be scheduled successfully; hence, we propose the following max-flow model for peer $i$ to unify the sub-stream scheduling operations in existing hybrid systems:

$$\text{Max} \sum\nolimits_{h\in\Omega_i} \sum\nolimits_{j\in\Gamma_i} I^i_{hj} e^i_{hj}, \tag{1}$$

$$\text{s.t.} \sum\nolimits_{h\in\Omega_i} I^i_{hj} e^i_{hj} \le 1, j \in \Gamma_i, \tag{2}$$

$$\sum\nolimits_{j\in\Gamma_i} I^i_{hj} e^i_{hj} \le T_{ih}, h \in \Omega_i, \tag{3}$$

$$e^i_{hj} \in \{0, 1\}, h \in \Omega_i, j \in \Gamma_i. \tag{4}$$

Constraint (2) ensures that each sub-stream can be scheduled to one neighbor at most. In constraint (3), the token number for neighbor $h$ of peer $i$, $T_{ih}$, is calculated by $T_{ih} = \left\lceil \frac{L_{hi}}{R/|\Gamma_i|} \right\rceil$ (we choose rounding up instead of rounding down so that peers' upload capacities can be fully utilized). The maximum number of sub-streams can be scheduled to neighbor $h$ is limited by their end-to-end bandwidth. Constraint (4) ensures that this optimization problem is a binary integer programming problem. We will show that this model can be transformed into an equivalent *max-flow* problem, which is solvable in polynomial time [7].

**Proposition 1.** *Implementations of sub-stream scheduling in GridMedia, CoolStreaming+ and LStreaming are special cases of the proposed max-flow model.*

*Proof.* Due to the page limitation, we only demonstrate the proof for GridMedia. Other two systems can be similarly proved. Note that both constraints (2) and (4) are satisfied in all these systems obviously; therefore, the proof only focuses on constraint (3) and the objective function of the max-flow model. Without loss of generality, we focus on peer $i$ in the proof. GridMedia does not consider token limit on sub-stream scheduling (i.e. $\forall h \in \Omega_i$, $T_{ih} = \chi$). This indicates that constraint (3) can be satisfied all the time. Peers in GridMedia group a fixed number of consecutive packets as a *packet group*. Moreover, peers cluster every $g$ consecutive packet groups into a *packet party* with group number from 0 to $g-1$. In GridMedia, whenever missing chunks in *packet group* 0 are requested by the chunk-pulling module, sub-stream scheduling is actually conducted at this moment. There can be multiple missing chunks in packet group 0 and each missing chunk belongs to one particular sub-stream. Suppose a packet-group-0 chunk in sub-stream $j$ is missing, if neighbor $h$ has this particular chunk (known from the buffer map), then $I_{hj}^{i} = 1$. This indicates that peer $h$ is qualified for sub-stream $j$. Since there is no bandwidth limitation, peer $i$ can always schedule sub-streams successfully. In other words, the number of sub-streams successfully scheduled is always maximized. Thus the objective function of the max-flow model holds for GridMedia. Therefore, the sub-stream scheduling scheme in GridMeida is a special case of the max-flow model.

### 2.3 Problem Transformation

In this subsection, we show that the proposed *sub-stream scheduling* model can be transformed into an equivalent classical *max-flow* problem, which can be solved in polynomial time. We briefly outline the max-flow problem here. Let a network $G = (V, E)$ be a directed graph in which each arc $(m, n) \in E$ has a nonnegative capacity $u(m, n) \geq 0$. $V$ contains two special elements $s$ and $t$, which represent *source* and *sink*, respectively. The decision variable, $f(m, n)$, denotes the amount of flow along $(m, n)$. If $(m, n) \notin E$, both $u(m, n) = 0$ and $f(m, n) = 0$. The max-flow problem can be written as follows:

$$\text{Max} \sum\nolimits_{n \in V} f(s, n), \tag{5}$$

$$\text{s.t.} \sum\nolimits_{n:(m,n) \in E} f(m, n) - \sum\nolimits_{n:(n,m) \in E} f(n, m) = 0, \forall m \in V - \{s, t\}, \tag{6}$$

$$0 \leq f(m, n) \leq u(m, n), \forall (m, n) \in E. \tag{7}$$

The time complexity of solving the max-flow problem is bounded by $O(VE^2)$ by adopting the *Edmonds-Karp algorithm* [7]. Due to the page limitation, the detail of the transformation algorithm is omitted here; nevertheless, we provide

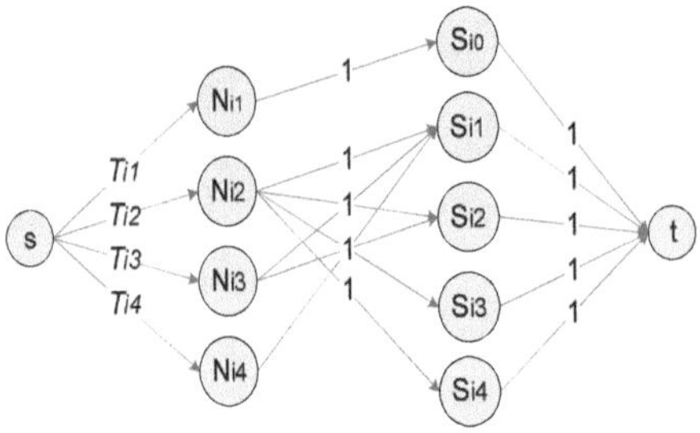


**Fig. 4.** Transformation to the max-flow problem at peer $i$

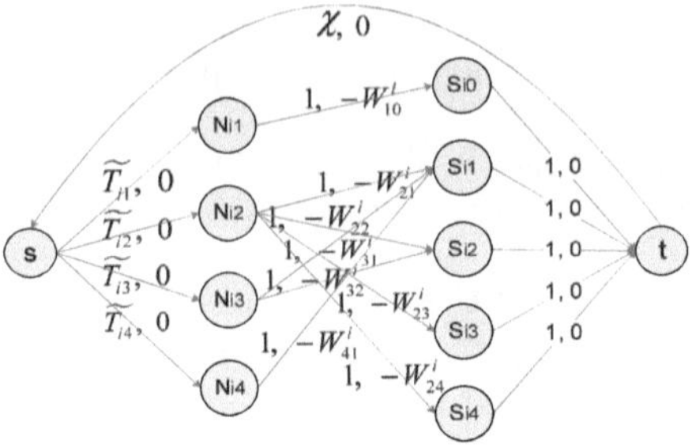


**Fig. 5.** Transformation to the min-cost flow problem at peer $i$

an illustration of the transformation shown in Fig. 4 for the example in Fig. 3. In Fig. 4, $\forall h$, $N_{ih}$ indicates that peer $h$ is a neighbor of peer $i$ and $\forall j$, $S_{ij}$ denotes that sub-stream $j$ needs to be scheduled by peer $i$. The arc between each pair of $N_{ih}$ and $S_{ij}$ ensures that neighbor $h$ is qualified for sub-stream $j$ based on the *peer qualification rule.* If $I^i_{hj} = 1$, an arc exists from $N_{ih}$ to $S_{ij}$. Nodes $s$ and $t$ in Fig. 4 are artificially introduced in order to set up the max-flow graph. The arc from $s$ to each $N_{ih}$ with capacity $T_{ih}$ satisfies the token number limitation in our model (constrain (3)), and the arc from each $S_{ij}$ to $t$ with capacity 1 guarantees that each sub-stream can be scheduled to one neighbor at most.

**Theorem 1.** *The unified sub-stream scheduling model at peer $i$ is equivalent to a max-flow problem.*

*Proof.* Applying the *flow balance equations* to all internal nodes $N_{ih}$, $\forall h \in \Omega_i$, and $S_{ij}$, $\forall j \in \Gamma_i$, in Fig. 4, we have

$$f(s, N_{ih}) = \sum\nolimits_{j\in\Gamma_i} f(N_{ih}, S_{ij}), \forall h \in \Omega_i, \tag{8}$$

$$f(S_{ij}, t) = \sum\nolimits_{h\in\Omega_i} f(N_{ih}, S_{ij}), \forall j \in \Gamma_i. \tag{9}$$

Because each sub-stream can be scheduled to at most one neighbor, we have

$$f(N_{ih}, S_{ij}) = I^i_{hj} e^i_{hj}, \forall h \in \Omega_i, \forall j \in \Gamma_i. \tag{10}$$

Plugging Eq. (10) into Eq. (8) and Eq. (9) and applying capacity constraint on each arc, we have

$$f(s, N_{ih}) = \sum\nolimits_{j\in\Gamma_i} f(N_{ih}, S_{ij}) = \sum\nolimits_{j\in\Gamma_i} I^i_{hj} e^i_{hj} \le T_{ih}, \forall h \in \Omega_i, \tag{11}$$

$$f(S_{ij}, t) = \sum\nolimits_{h\in\Omega_i} f(N_{ih}, S_{ij}) = \sum\nolimits_{h\in\Omega_i} I^i_{hj} e^i_{hj} \le 1, \forall j \in \Gamma_i. \tag{12}$$

Eqs. (11) and (12) are the first two constraints of our *sub-stream scheduling* model. Since the capacities in Fig. 4 are all binary integers, the optimal flow on each arc is either 0 or 1 based on the *Integer Theorem* [7]. Therefore $f(N_{ih}, S_{ij}) = \{0 \text{ or } 1\}$, which implies the constraint (4) of our original model is always satisfied.

The equivalence of the two objective functions is shown as follows:

$$\text{Max}\sum\nolimits_{h\in\Omega_i} f(s, N_{ih}) = \text{Max}\sum\nolimits_{h\in\Omega_i}\sum\nolimits_{j\in\Gamma_i} f(N_{ih}, S_{ij}) \quad (13)$$

$$= \text{Max}\sum\nolimits_{h\in\Omega_i}\sum\nolimits_{j\in\Gamma_i} I^i_{hj} e^i_{hj}. \quad (14)$$

The left-hand side of Eq. (13) is the objective function of the max-flow problem. Eq. (13) holds due to Eq. (8). If we apply Eq. (10) into the right-hand side of Eq. (13), we obtain Eq. (14), which is the objective function of our model. Thus, these two objective functions are equivalent. **Based on proposition 1 and theorem 1, the sub-stream scheduling in existing systems basically solves one particular max-flow problem.**

## 3 Min-cost Flow Scheduling

As shown in the proof of Proposition 1, there is no token limit imposed on GridMedia and CoolStreaming+. Without any consideration of load balancing, push-neighbor switching may incur as a direct consequence, which leads to increased redundant video download. The corresponding max-flow problems of all three existing systems usually have multiple optimal solutions. Since qualified neighbors are not differentiated in all three systems, they randomly select any optimal solution to schedule sub-streams. In practice, these qualified neighbors do not have the same network condition. In this section, we propose the *min-cost flow* scheduling scheme with the consideration of peer heterogeneity.

The tree-based streaming can achieve high throughput and low overhead when the system is stable. However, it is not able to cope well with a dynamic network. Since the throughput and overhead of the hybrid system highly depends on the sub-streaming scheduling component, a good hybrid system should adapt to peer churn quickly. LStreaming+ achieves this adaptability via the *min-cost flow* scheduling module. This module works periodically and has the ability to respond to network dynamics and schedule sub-streams to combat the network fluctuation. Since sub-stream scheduling is triggered periodically in LStreaming+, each $\tau_i$, $i = 1, 2, \ldots$, in Fig. 1 is set to be a constant $\tau$. We assign a *weight* for each neighbor on sub-stream scheduling in LStreaming+. The weight is determined as follows:

- Each LStreaming+ peer examines the download performance from push-neighbors every $\tau$ time and adjusts the previous scheduling based on neighbors' download performance in the previous $\tau$ time. For example, let $S = 15$, $B = 30$ and $\tau = 10$. Ideally, $\tau \times B/S = 10 \times 30/15 = 20$ chunks should be downloaded for one sub-stream over $\tau$ time. We define a threshold to be half of this ideal value. If the number of the downloaded chunks in sub-stream $j$ from neighbor $h$ in the previous $\tau$ time is larger than this threshold, peer $i$ sets the weight for this neighbor to be $W^i_{hj} = \chi$. This implies that peer

$i$ allows neighbor $h$ continuing to push chunks in sub-stream $j$ during the next $\tau$ time.

- For some sub-streams, if the chunk download threshold is not exceeded, they need to be scheduled to another neighbor in the next $\tau$ time. In the max-flow model, if one sub-stream of peer $i$ can be downloaded from multiple neighbors, we do not differentiate among the qualified neighbors. However, in LStreaming+, if neighbor $h$ has more fresh chunks in sub-stream $j$ compared with another qualified neighbor, peer $i$ assigns a larger weight for neighbor $h$. We denote the largest chunk number already available for sub-stream $j$ at peer $h$ (known from the exchanged information) as $P_{hj}^i$ and the starting chunk number needed by peer $i$ for sub-stream $j$ as $Q_{ij}$, respectively. Since $P_{hj}^i$ represents the chunk availability of sub-stream $j$ from neighbor $h$, we set the weight as $W_{hj}^i = P_{hj}^i - Q_{ij}$ to differentiate the qualified neighbors. A larger $P_{hj}^i - Q_{ij}$ value shows that more available chunks for sub-stream $j$ can be obtained from neighbor $h$.
- In summary, the weight ($W_{hj}^i$, $\forall h \in \Omega_i$ and $\forall j \in \Gamma_i$) for each neighbor of peer $i$ is determined as follows:

$$W_{hj}^i = \begin{cases} \chi, & \text{if the chunk download threshold is exceeded,} \\ P_{hj}^i - Q_{ij}, & \text{otherwise.} \end{cases} \tag{15}$$

We aim to maximize the total weights on sub-stream scheduling with the objective function $\text{Max} \sum_{h \in \Omega_i} \sum_{j \in \Gamma_i} W_{hj}^i I_{hj}^i e_{hj}^i$.

In practice, it is difficult to obtain the exact value of $L_{hi}$ between two peers $h$ and $i$, where $L_{hi}$ is the end-to-end bandwidth from peer $h$ to peer $i$. In LStreaming+, each peer estimates the token number for each neighbor and then conducts the sub-stream scheduling. Suppose peer $i$ has $|\Omega_i|$ neighbors. Their achievable uploading bandwidth in the previous $\tau$ time (by counting the number of received chunks) to peer $i$ are denoted as $v_1, \ldots, v_{|\Omega_i|}$. Since the total number of sub-streams to be scheduled is $|\Gamma_i|$, in LStreaming+ the token number for each neighbor is proportionally assigned based on its individual performance in the previous $\tau$ time. Let $\widetilde{T}_{ih}$ denote the estimated token number for neighbor $h$ by peer $i$, and it is calculated as follows:

$$\widetilde{T}_{ih} = \left\lceil \frac{v_h}{\sum_{k=1}^{|\Omega_i|} v_k} \times |\Gamma_i| \right\rceil, \forall h \in \Omega_i. \tag{16}$$

Then the sub-stream scheduling problem in LStreaming+ is formulated as follows:

$$\text{Max} \sum_{h \in \Omega_i} \sum_{j \in \Gamma_i} W_{hj}^i I_{hj}^i e_{hj}^i, \tag{17}$$

$$\text{s.t.} \sum_{h \in \Omega_i} I_{hj}^i e_{hj}^i \le 1, j \in \Gamma_i, \tag{18}$$

$$\sum_{j \in \Gamma_i} I_{hj}^i e_{hj}^i \le \widetilde{T}_{ih}, h \in \Omega_i, \tag{19}$$

$$e_{hj}^i \in \{0, 1\}, h \in \Omega_i, j \in \Gamma_i. \tag{20}$$

This model has similar constraints as the max-flow model in Section 2. Nevertheless, the sub-stream scheduling in LStreaming+ needs to be transformed into an equivalent *min-cost flow* problem, which can be solved in polynomial time. Let a flow network $G = (V, E)$ be a directed graph in which each arc $(m, n) \in E$ is associated with a cost $c(m, n)$. The lower bound $l(m, n)$ and the upper bound $u(m, n)$ of each arc capacity denote the minimum and maximum amount of flows that can pass through the arc. The min-cost flow problem can be written as follows:

$$\text{Min} \sum\nolimits_{\forall(m,n)\in V} c(m,n)f(m,n), \tag{21}$$

$$\text{s.t.} \sum\nolimits_{n:(m,n)\in E} f(m,n) - \sum\nolimits_{n:(n,m)\in E} f(n,m) = b(m), \forall m \in V, \tag{22}$$

$$\sum\nolimits_{k=1}^{|V|} b(k) = 0, \tag{23}$$

$$l(m,n) \le f(m,n) \le u(m,n), \forall(m,n) \in E. \tag{24}$$

In our problem, we set $b(m) = 0$, $\forall m \in V$, and $l(m, n) = 0$, $\forall(m, n) \in E$. The time complexity for solving this classic min-cost flow problem is bounded by $O(|V||E|(\log(\log U)) \log(|V|C))$ [8], where $U$ and $C$ are the largest values of the arc capacity and the arc cost, respectively.

**Theorem 2.** *The scheduling in LStreaming+ is equivalent to a min-cost flow problem.*

Due to the page limitation, the proof and the transformation algorithm are omitted here. We use Fig. 5 to illustrate this transformation for the example in Fig. 3 intuitively. Each arc in Fig. 5 is characterized by two parameters: arc capacity and cost. We generate internal nodes $N_{ih}$ and $S_{ij}$, arcs and arc capacities similar to the max-flow model. The major difference between this min-cost flow problem and the max-flow problem is on the arc cost along each arc. In the problem transformation, only the arc between $N_{ih}$ and $S_{ij}$ may have nonzero cost, because only these arcs are related to each decision variable $e^i_{hj}$ in the objective function Eq. (17). Minimizing the total negative costs is equivalent to maximizing the total positive weights.

## 4 Performance Evaluation

In this section, we evaluate the accuracy of the proposed max-flow model and the streaming performance of LStreaming+ in comparison to GridMedia (GM), LStreaming (LS) and the generic random mesh-pull scheme (MP) using simulations. We developed a discrete-event simulator coded in C++ to simulate the system behavior at the chunk level. The simulator implements the *max-flow scheduling* module and LStreaming+. In LStreaming+, we utilize the "CS2" library [9] to solve min-cost flow problems. To evaluate the accuracy of our max-flow model for characterizing the sub-stream scheduling of existing systems, we replace the original sub-stream scheduling modules of GridMedia and

LStreaming by using the max-flow scheduling module instead, and compare the performance of each modified system with its original system, respectively.

To achieve a realistic latency setup, the end-to-end latency between peers is randomly selected from the real-world node-to-node latency matrix (2500×2500) [10]. The playback rate of the stream is 300kbps and the default neighbor number is 15. All peers are DSL users with three types of upload capacities of 1Mbps, 512kbps and 128kbps, and with the corresponding download capacities of 3Mbps, 1.5Mbps and 768kbps. The fractions of these three types of peers are 10%, 50% and 40%, respectively. The upload capacity of the server is 900kbps. We simulate a 15-minute streaming session. There are 10000 peers joining the channel in total. During 5 time intervals of [0, 50]sec, [200, 250]sec, [400, 450]sec, [600, 650]sec and [800, 850]sec, 20% of total 10000 peers join the channel, respectively. The joining time of peers is uniformly distributed within each time interval. Among all peers, 5% of them do not leave after joining. The viewing durations for the remaining 95% peers are uniformly distributed from 190 seconds to 210 seconds. With the setting above, we want to simulate a dynamic network in which peer churn occurs during the entire simulation time.

### 4.1 Simulation Results

**Push-Neighbor Switching.** We characterize push-neighbor switching using the average number of switching push-neighbors when peers conduct sub-stream scheduling. In our previous analysis, duplicate chunks are downloaded mainly due to push-neighbor switching. With the same simulation settings including networking topology, link latency and peer churn, we observe that the modified system with the max-flow scheduling module (GM-MF or LS-MF) performs almost the same as its original system (GM or LS). This shows that the *max-flow* model captures the behavior of sub-stream scheduling schemes in existing systems very well. In addition, Fig. 6 shows that LStreaming+ has the least push-neighbor switching, which implies that LStreaming+ incurs the smallest amount of video redundancy overhead among all systems.

**Download Performance.** We define two traffic ratios to characterize the download performance. The *total download ratio* is the ratio between the total

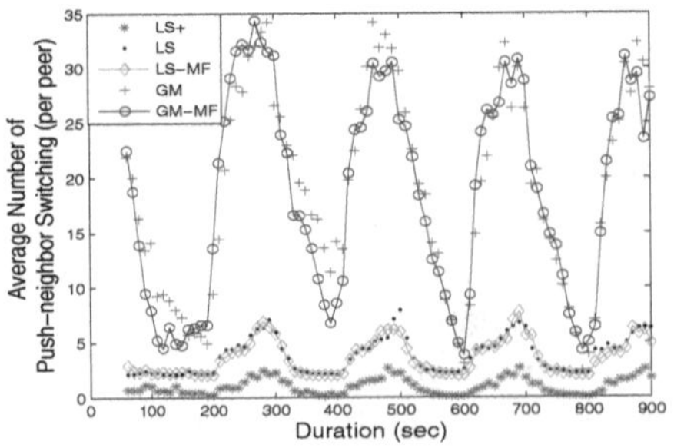


**Fig. 6.** Push-neighbor switching

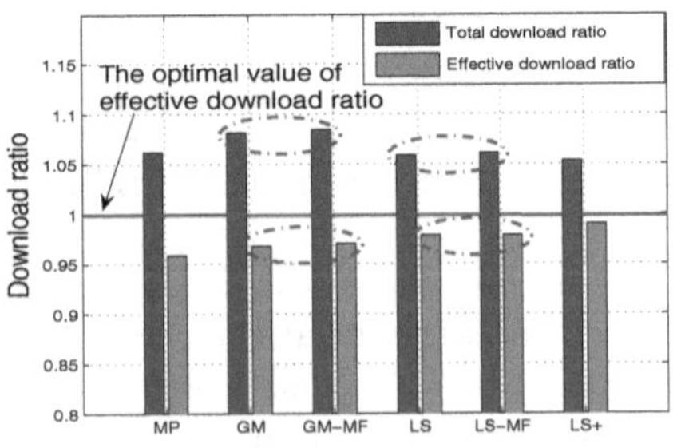


**Fig. 7.** Download performance

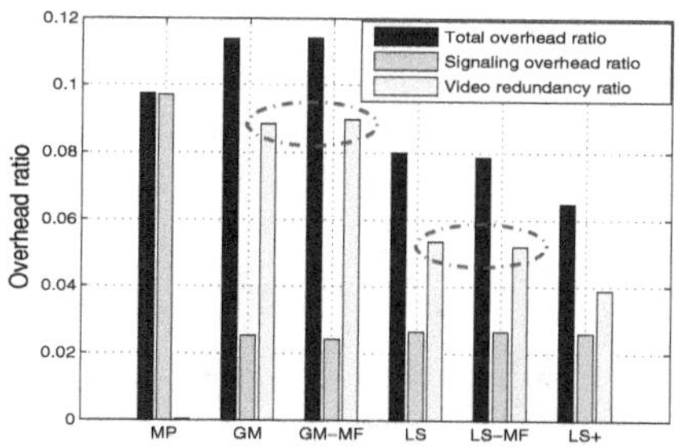


**Fig. 8.** Overhead breakdown

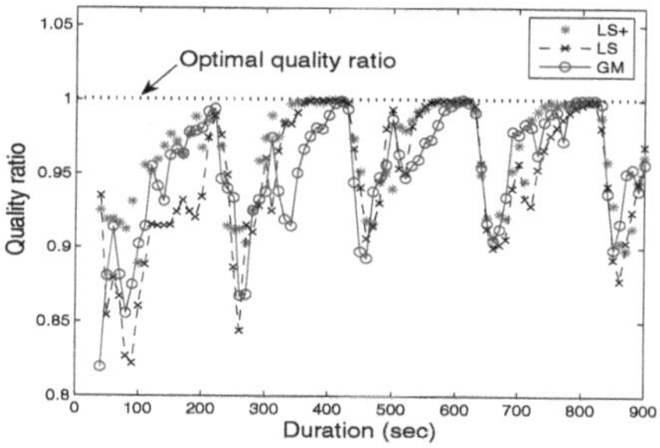


**Fig. 9.** Streaming quality

download rate and the video playback rate. The *effective download ratio* is the ratio between non-duplicate video download rate and the video playback rate. The *effective download ratio* characterizes the useful video download of the system. In Fig. 7, we observe that the modified streaming systems and the original systems experience almost the same total download ratio and effective download ratio. In Fig. 7, the effective download ratios of all hybrid systems are higher than that of a generic random pull-based system (denoted as **m**esh-**p**ull (MP) in Fig. 7). In addition, the effective download ratio of LStreaming+ is closer to the optimal value 1 compared with GridMedia and LStreaming. LStreaming+ outperforms all other schemes in terms of the quality ratio.

**Overhead Breakdown.** The total overhead traffic consists of signaling messages and redundant video chunks. In pull-push systems, the redundant video download is a potential problem. Video chunks usually have much larger size than signaling packets. If the number of duplicate video chunks is large, redundant video traffic becomes the major contributor in the total overhead traffic. In Fig. 8, each pair of the modified system and its original system suffers from almost the same amount of video redundancy ratio, which provides another good indication that the max-flow model for the existing systems is very accurate. From Fig. 8, we observe that the total overhead ratios of mesh-pull, GridMedia and LStreaming reach 9.7%, 11.4% and 8.0%, respectively. Nevertheless, the total overhead of LStreaming+ is only 6.4%. Compared with mesh-pull, GridMedia and LStreaming, the overhead reduction of LStreaming+ is 34.0%, 43.9% and 20.0%, respectively. Through Fig. 8, it is clear that the video redundancy is the major contributor to the total traffic overhead in hybrid systems. GridMedia and LStreaming suffer from video redundancy as high as 8.9% out of the 11.4% total overhead and 5.4% out of the 8.0% total overhead, respectively. In contrast, LStreaming+ achieves a much smaller video redundancy ratio, only 3.8% out of the 6.4% total overhead.

**Quality Ratio.** We define the *quality ratio* as the ratio between the number of chunks, which has been played back, and the number of chunks which should be played back up to the current time. Fig. 9 shows the quality ratio of LStreaming+

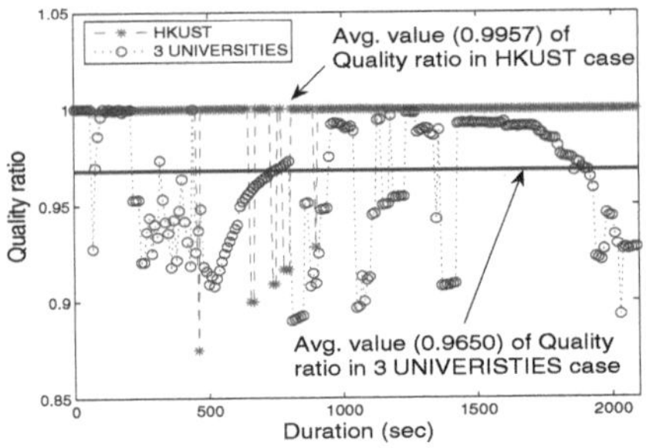


**Fig. 10.** Quality of the prototype

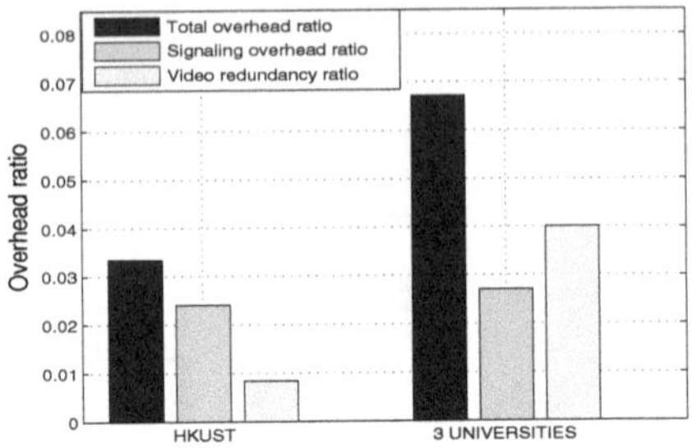


**Fig. 11.** Overhead of the prototype

outperforms the other two hybrid systems on average during the whole watching session. In particular, when peer churn occurs, LStreaming+ suffers the least and recovers the most quickly, compared with the other two systems.

### 4.2 Prototype Experiments

Our experiments are first conducted on the HKUST campus network. Then, the experiments are conducted among three universities in Hong Kong, including HKUST, HKU and CUHK. In each experiment, 100 peers join a video channel with a 300kbps playback rate. At the application level, we manually limit the upload capacity of each peer with the same upload capacity distribution used in the simulation. As shown in Fig. 10, the *quality ratio* almost maintains at the optimal value 1 during the whole watching session on the campus network. The *quality ratio* fluctuates in the inter-university experiments especially at the initial time. However, throughout the period of the experiments, there is only slight visual impact for a small number of peers. We also classify the traffic overhead in the prototype experiments in Fig. 11. Note that the total overhead of the campus experiments is as small as 3.35%. Even if in large-scale networks, the overhead ratio is merely 6.71%. Fig. 11 shows LStreaming+ effectively reduces the overhead and at the same time achieves a very good streaming performance.

## 5 Conclusion

In this paper, we propose a max-flow model for unifying the sub-stream scheduling problem in pull-push hybrid P2P streaming systems. We show that the sub-stream scheduling scheme in existing hybrid streaming systems including GridMedia, CoolStreaming+ and LStreaming, can be formulated into a special case of our proposed max-flow model. This max-flow model leads to useful insights for developing a min-cost flow model for scheduling sub-streams to better utilize heterogenous peers. We implement this min-cost flow model in a prototype system, LStreaming+. The accuracy of the max-flow model and the system performance of LStreaming+ are evaluated using extensive simulations. The prototype experiments also demonstrate that LStreaming+ achieves excellent streaming performance with significantly reduced traffic overhead.

## References

1. Hei, X., Liang, C., Liang, J., Liu, Y., Ross, K.W.: A measurement study of a large-scale P2P IPTV system. IEEE Trans. on Multimedia 9(8), 1672–1687 (2007)
2. Zhang, M., Zhang, Q., Sun, L., Yang, S.: Understanding the power of pull-based streaming protocol: Can we do better? IEEE JSAC 25(10), 1640–1654 (2007)
3. Li, B., Xie, S., Keung, G., Liu, J., Stoica, I., Zhang, H., Zhang, X.: An empirical study of the Coolstreaming+ system. IEEE JSAC 25(10), 1627–1639 (2007)
4. Li, Z., Yu, Y., Hei, X., Tsang, D.H.K.: Towards low-redundancy push-pull P2P live streaming. In: Proc. ICST QShine, Hong Kong (July 2008)
5. Li, Z., Yu, Y., Hei, X., Tsang, D.H.K.: Towards low-redundancy push-pull P2P live streaming. In: Proc. ACM SIGCOMM Demo, Seattle, USA (August 2008)
6. Liu, Z., Shen, Y., Ross, K.W., Panwar, S.S., Wang, Y.: Substream trading: Towards an open P2P live streaming system. In: Proc. IEEE ICNP, Orlando, USA (October 2008)
7. Cormen, T.H., Leiserson, C.E., Rivest, R.L., Stein, C.: Introduction to Algorithm, 2nd edn. The MIT Press, Cambridge (2001)
8. Hillier, F.S., Lieberman, G.J.: Introduction to Operations Research. McGraw-Hill, New York (1995)
9. Goldberg, A.: Network optimization library, `http://www.avglab.com/andrew/soft.html`
10. Wong, B., Slivkins, A., Sirer, E.G.: Meridian: a lightweight network location service without virtual coordinates. In: Proc. ACM SIGCOMM, pp. 85–96 (August 2005)

# A Novel Content Distribution Mechanism in DHT Networks*

Quanqing Xu[1], Heng Tao Shen[2], Bin Cui[1], Xiaoxiao Hou[3], and Yafei Dai[1]

[1] State Key Lab. for Adv. Opt. Commun. Syst. & Networks, Peking University, 100871 Beijing, China
{xqq,dyf}@net.pku.edu.cn, bin.cui@pku.edu.cn

[2] School of ITEE, The University of Queensland, Brisbane QLD 4072, Australia
shenht@itee.uq.edu.au

[3] Department of Computer Science and Technology, Stony Brook University, U.S.A.
xhou@cs.sunysb.edu

**Abstract.** DHT (Distributed Hash Table) is a structured overlay network that is widely utilized in P2P systems. Existing content distribution approaches do not completely exploit features of DHT and incur heavy network bandwidth consumption. This paper analyzes existing content distribution approaches including synchronous content distribution method in eMule based on DHT overlay networks and points out that their network loads are too heavy. We propose a novel content distribution algorithm: asynchronous distribution, in DHT networks. Compared with traditional distribution approaches, it is more effective and scalable with lower network load. We apply the techniques of vector space model and search frequency to the asynchronous distribution algorithm, which effectively improves search hit ratio and reduces network load. Simulation results based on real data from Maze system show that this approach has low network overhead and publishing cost, high search and download hit ratio.

**Keywords:** DHT, Content Distribution, Vector Space Model, Search Frequency.

## 1 Introduction

DHT, e.g., Chord [1] and Pastry [2], is a structured overlay network that is widely utilized in P2P systems, e.g., file-sharing systems like eMule[1], BitTorrent[2], and AmazingStore[3]. These file-sharing systems employ Kademlia [3] as their DHT

* This research has been supported by the National Grand Fundamental Research 973 program of China under Grant No.2004CB318204, Australian research grant (ARC DP0773483) and National Natural Science foundation of China under Grant No.60873051.

[1] http://www.emule-project.net

[2] http://www.bittorrent.com/

[3] http://amazingstore.grids.cn

L. Fratta et al. (Eds.): NETWORKING 2009, LNCS 5550, pp. 742–755, 2009.

infrastructure to improve performance. However, existing content distribution approaches, e.g., synchronous distribution algorithm in eMule, do not adequately exploit features of DHT and wastes precious network bandwidth. DHT maps object keys to overlay nodes and offers a lookup primitive to send a message to the node[4] responsible for a key. Overlay nodes maintain routing states to route messages towards the nodes responsible for their destination keys, which brings good locating performance. However, the retrieval ability of DHT is very limited and DHT does not have the requisite ability to rank search results because DHT only provides the exact match function.

We call *content distribution (publishing)*: it is a procedure that contents are provided to other people by given ways, which aims to implement resource-sharing or resource-distributing. Distributed objects are sharing-files of various formats, documents, etc. In this paper, we propose a novel Asynchronous Content Distribution approach (ACD). We analyze existing distribution approaches including synchronous publishing method in DHT networks and points out that their network load balances are heavy. We propose a novel asynchronous content distribution algorithm in DHT networks. Compared with traditional distribution approaches, it is more effective and scalable with lower network load. We apply the techniques of vector space model and search frequency to the asynchronous distribution algorithm, which improves search hit ratio and reduces network load. The main contributions of this paper are threefold:

- We present an asynchronous content distribution algorithm under DHT environments, which solves existing issues in the synchronous distribution algorithm;
- We utilize vector space model and user relevance feedback from traditional information retrieval domain to improve search hit ratio and decrease publishing cost in DHT networks;
- We confirm the effectiveness and efficiency of our methods by conducting an extensive performance measured by network overhead, publishing cost, search and download successful ratio.

The remainder of this paper is organized as follows: Section 2 presents the problem formulation and reviews related work. We design an asynchronous distribution algorithm in DHT networks in Section 3 and propose optimization strategy in Section 4. Section 5 reports the experimental results. Section 6 concludes this paper and discusses future work.

## 2 Preliminaries

### 2.1 Semantic Operations and Applications of DHTs

After DHTs appeared, there are increasing applications based on structured P2P, and many mature DHTs are utilized in real systems. Thus, it is necessary to analyze functions of a DHT and give formal definitions. From the standpoint

[4] In this paper we use the terms "node" and "peer" interchangeably.

of function, a DHT provides a hash function, publishes a content based on its key and locates the content through its key in DHT networks. From the standpoint of topology, a DHT provides a distributed routing algorithm that disperses the routing function of overlay to all the peers: each peer can lookup other peers and forward routing messages. A general DHT provides the following semantic operations:

1) **Put**($Key$, $Value$) Publishes a pair ($Key$, $Value$) into DHT networks;
2) **Get**($Key$) Returns the $Value$ of the $Key$;
3) **Lookup**($Key$) Provides general access to the node maintaining the $Key$;
4) **Routing**($Key$) Provides general access to the node responsible for the $Key$, and to each node along the routing path.

A DHT provides two basic operations [4]: Put and Get. In fact, Put and Get are functions of DHT application layer, while Lookup and Routing are functions of DHT overlay layer. Put/Get and more complicated operations can be implemented through Lookup and Routing operations. Therefore, Lookup/Routing are basic operations of a DHT. OpenDHT [5] provides interfaces including Lookup, Routing and Put/Get for upper applications.

## 2.2 Traditional Content Distribution Approaches in DHT Networks

A content distribution (publishing) and retrieval algorithm under DHT environments includes two phases: 1) publishing phase, it is a procedure that a peer publishes its local information to DHT networks by a given policy; 2) retrieving phase, it is a procedure that a peer employs a keyword to retrieve relevant results. After the publishing phase, there are a large number of contents in DHT networks. Peers can retrieve relevant contents in the retrieving phase. Therefore, different publishing policies determine index mechanisms and retrieval approaches used in DHT networks. Basically, every content distribution method in DHT networks includes these two phrases. Note that they are independent in logic but parallel in real systems. Generally, there are three interface functions:

1) **Hash(Object)** Calculates the hash value of an object;
2) **Publish**($\langle Key, Value\rangle$) Publishes a pair $\langle Key, Value\rangle$;
3) **Segment(Doc)** Segments a document.

In traditional information retrieval (IR) field, a retrieval procedure is that a relevant document list is achieved and ranked for a given query. Here we utilize concepts in IR fields. A retrieval procedure in DHT networks is that a document list is retrieved through a query that includes a term or several terms. In general, a content publishing and retrieving algorithm is classified into three categories according to publishing mode.

**Term-based Content Publishing Approach.** We present a schematic diagram of term-based content publishing approach reads as shown in Figure 1. $Node_1$ segments its documents into terms and publishes the terms into DHT networks. $Term_1$ and $Term_2$ are maintained respectively by $Node_2$ and $Node_3$ in Figure 1. The merits of this algorithm are: 1) each term of each document is

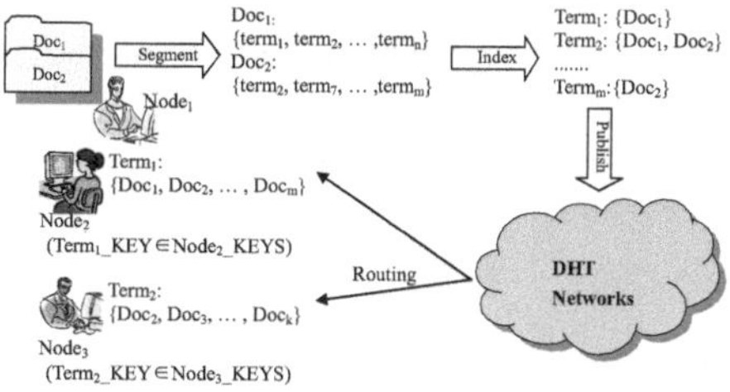


**Fig. 1.** Term-based Publishing Approach

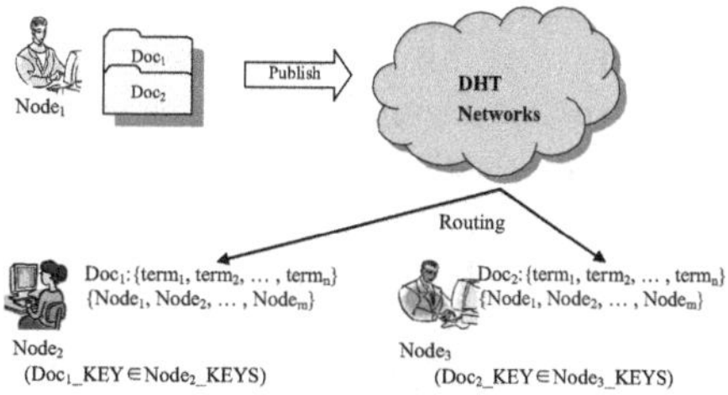


**Fig. 2.** Doc-based Publishing Approach

published; 2) a fully distributed inverted index is created; 3) a peer can easily achieve the global information of a term that it is responsible for and calculates the IDF (Inverse Document Frequency) value of this term; 4) each peer can quickly retrieve and rank search results for a given query term in DHT networks.

**Document-based Content Publishing Approach.** We give a schematic diagram of this approach is shown in Figure 2. $Node_1$ publishes its documents into DHT networks. $Doc_1$ and $Doc_2$ are maintained respectively by $Node_2$ and $Node_3$ in Figure 2. The advantage of this algorithm: each peer maintaining a document can know all the terms of the document so that the similarity between a term and a document can be calculated when a search is conducted. However, its disadvantage is that it can not use DHT to perform a search for a term. In fact, this approach is utilized in the DHT of BitTorrent: peers downloading the same file resource publish their addresses so that a downloading peer can find the peers including the file, which is helpful to improve multi-peers download efficiency.

**Hybrid Content Publishing Approach.** Analyzing merits and demerits of the above two approaches, Tang et al. [6] in eSearch presented a hybrid content publishing algorithm: executing a two-phase protocol to publish a content (document). In the first phase, it utilizes the publishing algorithm based on term objects and employs DHT routing to locate the peers responsible for top terms in the document. In the second phase, it uses the publishing algorithm based on document objects. The initiator can build the term list for the document and multicast the term list or the document itself. This publishing algorithm has two advantages: 1) The returned results can be ranked based on the similarity between a query and a document; 2) It can support complex queries. With document in hand, eSearch can search exact matches for quoted text, provide sentence context for matching terms, and support a "cached documents" features to decrease network load. However, this publishing algorithm is proposed as a full text retrieval approach in DHT networks. Each term is published with lots of documents, which causes very huge network load. Therefore, this algorithm is not suitable for fully opening P2P systems.

The first two publishing approaches are commonly used under DHT environments. The third approach aims to utilize DHT as an infrastructure for distributed full text retrieval where the published object is a document or web page, which often includes hundreds of terms. The third one is executed among relatively stable peers, e.g., all the machines in a lab or PlanetLab[5], and very large bandwidth consumption is generated. These algorithms do not consider an opening P2P system, which needs to face high node churn and limited bandwidth.

### 2.3 Other Related Work

Coral [7] is a free P2P content distribution network designed to mirror web content. One of Coral's key goals is to avoid ever creating hot spots that might prevent peers from running the Coral because of concerning about overload. Coral uses an indexing abstraction called a distributed sloppy hash table (DSHT) to achieve this goal. Self-organizing clusters of nodes are created by DSHTs to fetch information from each other, which can avoid communicating with more distant or heavily-loaded servers. Yang et al. [8] proposed a Chord-based distributed architecture for content-based publish/subscribe services. This infrastructure is a scalable platform and can simultaneously support many publish/subscribe schemas. Steiner et al. [9] studied the content management process implemented of KAD as implemented in aMule[6]. CoBlitz [10] is a scalable Web-based distribution system for large files, which distributes large files without requiring any modifications to standard Web servers and clients, since all the necessary support is located on the CoDeeN content distribution network.

## 3 Asynchronous Content Distribution Algorithm

### 3.1 Requirement Analysis of Search in P2P File-Sharing Systems

In this subsection, we briefly analyze existing problems about information retrieval that need to be considered in P2P file-sharing systems. The published and indexed objects are files shared by users in P2P file-sharing systems. These file objects have particular characteristics.

Each file has a unique file name used by a specific user. However, the same file often has many different file names applied by different users due to:

1) Different users have different habits of naming sharing files: some users like making file names normal, while others like making file names arbitrary;
2) Some users often insert many hot query terms into an ordinary file for attracting other users to download it.

Thus, each file often has different file names, which increases the retrieval difficulty in P2P file-sharing systems.

Generally speaking, there are two different requirements when initializing a query:

1) A relevant file list is achieved through a keyword used by a user. The user wants to retrieve as many files as possible and ranks these files so that the user can select the desired files and download these files quickly;

[5] http://www.planet-lab.org
[6] http://www.amule.org/

2) The user uses the key of a file object to retrieve as many users that own this file object as possible so that he/she can download it from many users.

It is easy to meet requirement 2), while it is hard to satisfy requirement 1) in DHT systems. The above-mentioned three traditional index structures can not satisfy the two requirements. Only the synchronous publishing algorithm in eMule can meet them.

### 3.2 Asynchronous Publishing

There are two kinds of objects that need to be published: file object and file name (term) object in P2P file-sharing systems. The owner of a file object is responsible for publishing the file object and its file name object in eMule. We define *synchronous publishing*, as the method where these two objects are published by the same peer. The advantage of synchronous publishing is that both a file object and its file name object are published by the same logic role. Therefore, its system architecture is relatively easy to be understood and implemented. Moreover, synchronous publishing is relatively good in performance, which is seen from that eMule is widely popular.

However, synchronous publishing does not consider the following problems:

1) Although the same file has different file names applied by different users, most of important terms are the same. However, these terms are published repeatedly by different users. Therefore, this method brings too many redundant messages;
2) Some file names have many terms. However, some of these terms are not relevant, e.g., stop words, which are not necessary to be published. A user can not publish terms according to the importance of terms because the user do not know the file names of the same file applied by other users and can not determine which terms are important to this file;
3) There is no global information for a file object. A peer responsible for a term object can not determine the relevance between this term and its corresponding files.

We comprehensively analyze these three defects and draw a conclusion that the owner of a file object published the terms of this file and results in short of useful information when publishing these terms. If the file names of a file object can be obtained, and the terms of the file object are published, this problem can be solved. Therefore, we propose an asynchronous publishing algorithm to solve the before-mentioned issues. The basic idea is that the owner of a file object publishes this file, and the maintainer of this file integrates all the file names of this file. The maintainer segments these file names to get the relevant terms, and publishes these terms in DHT networks. We introduce the details of this algorithm in the next subsection.

### 3.3 Specific Algorithms

The basic idea of the asynchronous publishing algorithm is that the owner of a file only publishes this file itself and the maintainer of this file publishes those

**Algorithm 1.** Content Asynchronous Publishing Algorithm

// Publishing *File* objects
**for** $File \in Node_1$ **do**
  File_Key=Hash(File);
  File_Value={Node_Addr,File_Name};
  Publish($\langle File\_Key, File\_Value \rangle$);
// Merging *File* objects
$Node_2$ receives the published *File* ($File\_Key \in Node_2\_Keys$) object from nodes;
/* A file list includes:(a) *Node_Addrs*, which is used to find much more download links; (b) *File_Names*, which records file names used by different peers */
$Node_2$ merges the values of *File* and maintains them;
// Publishing *Term* objects
$Node_2$ segments the *File_Name* list of the files maintained by it and get the *Terms*;
**for** $Term \in Terms$ **do**
  Key=Hash(*Term*);
  Make *File_Key* and *File_Name* be the value of *Term*;
  Publish($\langle Key, Value \rangle$);
// Merging *Term* objects
$Node_3$ receives the published *Term* object from nodes in DHT networks;
$Node_3$ maintains a file object list for each received term object;
// for all the files in DHT networks
All the nodes repeat the above steps;

terms in the names of this file. We present the asynchronous publishing algorithm as shown in Algorithm 1, the search and download algorithm as shown in Algorithm 2.

The asynchronous publishing algorithm's schematic diagram is shown in Figure 3. The algorithm integrates all the different file names of the same file, segments these file names into terms and publishes these terms. This algorithm can decrease redundant messages to some extent and integrate the file names

**Algorithm 2.** Content Search and Download Algorithm

**Input**: Term
// $Node_1$ uses a term to perform a search
*Term_Key*=Hash(Term);// $Term\_Key \in Node_2\_Keys$
// routes to $Node_2$ through DHT
$Node_2$ searches its local information based on *Term_Key*;
$Node_2$ returns the *File* object list of *Term_Key* to $Node_1$;
$Node_1$ selects $File_1$ from the *File* object list to download;
$Node_3$=Lookup($File_1\_Key$) by $Node_1$;
$Node_3$ returns all the nodes including $File_1$;
$Node_1$ downloads $File_1$ from those nodes;

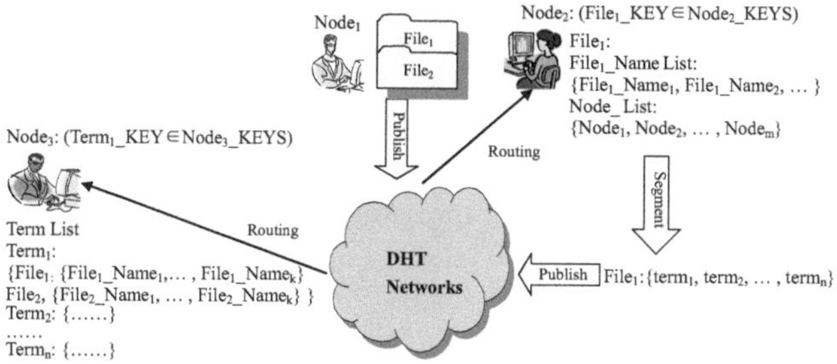


**Fig. 3.** The Schematic Diagram of Asynchronous Publishing Algorithm

of files to distinguish fake files because users may name them with fake names. Most of all, this algorithm can integrate the global information of a specific file in the same peer, it can also integrate the global information of a specific term in the same peer. We can use this global information to optimize the publishing algorithm. Although a term is published with full file name of a file in this algorithm, the full file name is much shorter than a document. Therefore, compared with the full text retrieval in DHT networks, this algorithm will not bring too much network load.

## 4 Optimizations for Asynchronous Publishing

Generally, different term in a file name has different importance to the same file object. Therefore, we can improve the asynchronous publishing algorithm based on the importance of terms in a file object when publishing these terms. Important terms are published with high frequency, while those unimportant terms are published with low frequency or are not even published, which can dramatically reduce bandwidth consumption and guarantee search performance. Moreover, the same term has different weights in different files. Thus, we can utilize some weight schemes to calculate a term's weight in a file name, which can rank search results and improve users' experience. In this section, we present two optimizations for asynchronous publishing based on weight schemes of vector space model (VSM) [11], which is widely used in information retrieval.

### 4.1 Optimization I: Term Frequency

TF*IDF (Term Frequency - Inverse Document Frequency) is a weight rule often used in information retrieval. A TF*IDF weight is a statistical measure used to evaluate how important a term is to a document in a corpus. The importance increases proportionally to the number of times (TF) that a term appears in a given document but is offset by the document frequency (DF) of the term in the corpus. In unstructured P2P networks, however, it is impossible for us to calculate the global IDF values since the calculation involves counting the document frequency of terms. On the contrary, it is easy to calculate the global IDF values since the same term is maintained by the same peer in DHT networks. Therefore, an effectively calculating formula reads as follows: $IDF_{term} = \log(N/DF_{term})$, where $N$ is a maximum unsigned integer.

Each file may have different file names for different users in P2P systems. If these file names are integrated into a "document": using the weight of a term in this document to indicate the importance of this term in this file object. The key of this approach is to get the TF and IDF values of this term. In the synchronous publishing method in eMule and the publishing methods in Section 2.2, the two values can not be achieved simultaneously. However, both of them can be obtained in the asynchronous publishing method. The concrete approach reads as follows:

In line 7 of the asynchronous publishing algorithm, when $Node_2$ segments the $File\_Name$ list of all the files maintained by it, it can count terms' frequency in file names. These terms with their TFs are published. $Node_3$ receives the published $Term$ object from nodes in DHT networks, inserts it into the local $Term$ list and counts *file (document) frequency* (DF) of this term. $Node_3$ returns this term's IDF to $Node_2$. Therefore, both $Node_2$ maintaining the file object and $Node_3$ maintaining the term object have a TF for this term in a given file and the IDF of this term. These two nodes can utilize the two values to do different things.

$Node_3$ can use TF*IDF to measure the relevance of each term in a file object so that search results can be ranked according to this relevance when a user performs a search for a term. Moreover, if a user wants to search a file, he/she can use a term considered as the most relevant term to that file. In other words, the more relevant to a file a term is, the more possible it is utilized to find this file by a user. In addition, a file has many terms, the weights of some of them are small and are not used to find this file by a user. Thus, $Node_2$ can decide the publishing frequency of a term maintained by it according to its TF*IDF. If a term has a high relevance, it will be published with normal publishing frequency. If not, it will be published with a lower publishing frequency, which can decrease network load.

This approach seems to increase a little network load: 1) The length of a message is increased because of adding TF values in the phase of publishing terms; 2) $Node_3$ needs to respond with IDF values to $Node_2$, which increases message numbers. However, both TF and IDF are so short that they are negligible and an IDF can be inserted into a response message of $Node_3$. Therefore, they have marginal influence on network load. In addition, network load can be reduced because publishing frequencies of irrelevant terms are decreased.

### 4.2 Optimization II: Search Frequency

To some extent, the TFIDF rule can distinguish the importance of each term, but it can not totally reflect this importance. Therefore, we may better distinguish the importance of each term if we consider users' search behaviors. It is easy to perform this in P2P file-sharing systems. If a user wants to search a file, he/she uses a keyword, which is considered by him/her to be the most relevant term to this file. In other words, a term is published to make users easily find corresponding files. A term often used by users for retrieval should be important. Thus, we can use search frequency to measure the relevance between a term and a file.

In line 5 of Algorithm 2, $Node_1$ may send the used keyword in line 1 to $Node_3$ when performing a lookup to find all the nodes publishing $File\_Key$ via $Node_3$ for downloading files. Therefore, $Node_3$ can know keywords used by users when these users download the files maintained by $Node_3$. $Node_3$ can locally save these keywords' frequencies for being used to retrieve. This frequency is called as *search frequency* (SF). $Node_3$ can utilize *search frequency* to optimize the proposed asynchronous publishing approach.

Note that there are no users' feedbacks at the beginning of file-sharing. Thus, this approach may be employed with TF*IDF. TF and SF are similar, which are used to measure the relevance between a term and a file. However, SF is a result of users' feedbacks and is much more important than TF. Therefore, the weight of SF is higher than that of TF when they are both utilized. This approach only adds a keyword before users find much more files and download them. This keyword is so short that the increment of network load may be negligible.

# 5 Experiments

## 5.1 Data Sets

The data sets come from shared files from 28/10/2007 to 09/11/2007 in a real P2P system: Maze[7]. There are 7,565 active users and 30,001,293 files totally. First of all, we preprocess the shared files, e.g., removing shared files in a system disk (e.g., C:\), WINDOWS installation directory (e.g., C:\WINDOWS) and programs installation directory (e.g., C:\Program Files), which are shared by free riders. Then, we choose 6,144 active users, and the files they shared are 5,543,293. To simulate users' feedbacks, we utilize the search log in 10/2007 to extract the data with high search frequency. We extract real users' queries from a real log from 28/10/2007 to 09/11/2007 in Maze system to be employed as queries in our simulations. In addition, We utilize online, offline and enter time data of active users from logs of Maze system to simulate peer churns of P2P systems, where time interval $\Delta t$ is 3~5 minutes. Kademlia [3] is utilized as our DHT infrastructure. Table 1 summarizes experimental parameters. Notations in the experiments are shown in Table 2.

## 5.2 Experimental Results

**Experiment 1: Comparison of Synchronous and Asynchronous Publishing Algorithms.** In this experiment, we compare existing synchronous publishing algorithm in eMule with the proposed asynchronous one. File objects need to be published in both algorithms. Therefore, we only concern the publishing

[7] A P2P file sharing system with millions of registered users, http://maze.pku.edu.cn

**Table 1.** Parameter Setting

| Parameter | Setting |
|---|---|
| # of nodes | 64,128,...,1024,2048,3072,...,6144 |
| Publishing ratio | 0.1,0.2,0.3,0.4,0.5,0.6,0.7,0.8,0.9,1.0 |
| # of queries | 18,846 |
| Updating ($h$) | 0.5,1.0,1.5,2.0,2.5,3.0,3.5,4.0,4.5,5.0 |
| # of copies | 1,2,3,4,5,6 |

**Table 2.** Notations

| Symbol | Description |
|---|---|
| $N_s$ | Network Size |
| $P_r$ | Publishing Ratio |
| $U_t$ | Index Update Time |
| $U_c$ | Index Update Copies |
| $N_m$ | # of Messages |

procedure about file name objects. The procedure is that term objects are published as file names are segmented. Synchronous and asynchronous publishing algorithms are only different in publishing mode. They do not affect final publishing and retrieval results because the same file and term objects are maintained by the same peer according to DHT's principle.

Compared with the synchronous publishing approach in eMule, the asynchronous one can reduce more messages as there are increasing peers in DHT networks, which is shown in Figure 4(a). The reason for this is that the asynchronous publishing approach concerns that the same file has many copies in P2P file-sharing systems and most of them have the same important keywords. For the synchronous publishing, these keywords are repeatedly published by different users so that a lot of redundant messages are produced.

In addition, we also explore changes of messages' quantity produced by these two algorithms as the publishing ratio varies with the popular degree of keywords. The smaller the publishing ratio of keywords is, the bigger the percentage of messages deceased by the asynchronous approach is, as shown in Figure 4(b). There is the same reason as explained about Figure 4(a). Moreover, the more popular keywords are, the more files including these keywords are. Therefore, the asynchronous publishing algorithm produces less messages than the synchronous one. Compared with the synchronous method, the asynchronous method can decrease bandwidth consumption, achieve better publishing performance and scalability.

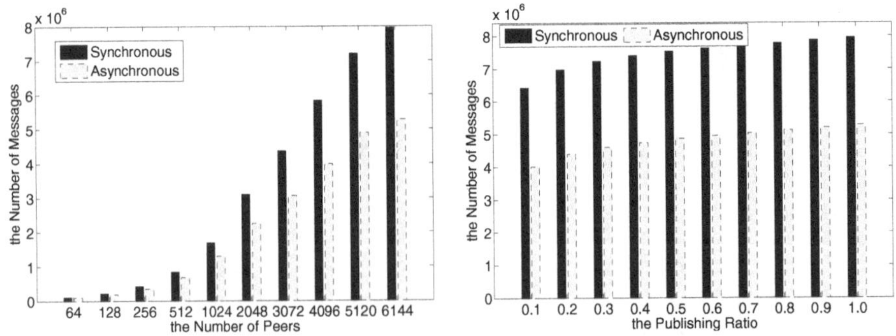


(a) Performance with Different $N_s$ ($P_r$=1.0) (b) Performance with Different $P_r$ ($N_s$=6144)

**Fig. 4.** Comparison of Synchronous and Asynchronous Publishing Algorithms

**Experiment 2: Comparison of Asynchronous Publishing and Optimized Counterparts.** This experiment mainly explores three kinds of algorithms:

1) *Random*: it is the randomly asynchronous publishing algorithm, where the published keywords are randomly selected in the asynchronous publishing algorithm;
2) *TF*IDF*: it is the asynchronous publishing algorithm with optimization I: Term Frequency, where the published keywords are selected according to the TFIDF rule;
3) *SF*IDF*: it is the asynchronous publishing algorithm with optimization II: Search Frequency, where the published keywords are selected according to the TFIDF rule with search frequency.

We utilize search hit (successful) ratio and publishing cost to evaluate these three algorithms. Search hit ratio is a percentage that indicates the proportional amount of the number of successful queries to the number of all the queries. Publishing cost ($PC$) is the bandwidth consumption when publishing file name objects in the procedure of asynchronous publishing, which is defined by:

$$PC = \sum_{peer} \sum_{term} (\|term\| + 16 + 20 \times PeerNum_{term}) \quad (1)$$

where $\|term\|$ represents the length of $term$, a file object is represented by MD5 (16 bytes), each peer's ID is represented by 160 bits (20 bytes) according to Kademlia.

For the test data sets, *SF*IDF* achieves the best performance among the three algorithms as shown in Figure 5(a) and Figure 5(b). *TF*IDF* is better than *Random* in search hit ratio, but it increases the publishing cost, which leads to heavier load in each peer. It is because when a proportion of popular keywords are published in DHT networks, their publishing frequencies are higher than those of common keywords. Based on the TFIDF rank rule, *TF*IDF* selects keywords from file names. The published keywords' records include more peers and files than those in *Random* and *SF*IDF*. *SF*IDF* is directed by search frequency, so it is better

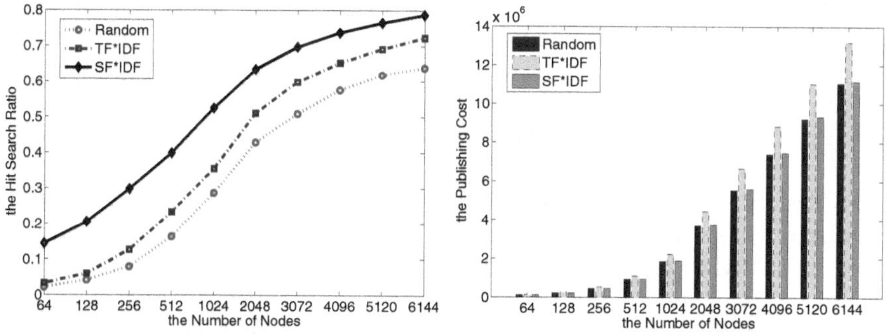


(a) Search Hit Ratio (b) Publishing Cost

**Fig. 5.** Comparison of Publishing Performance with Different $N_s$ ($P_r$=0.1)

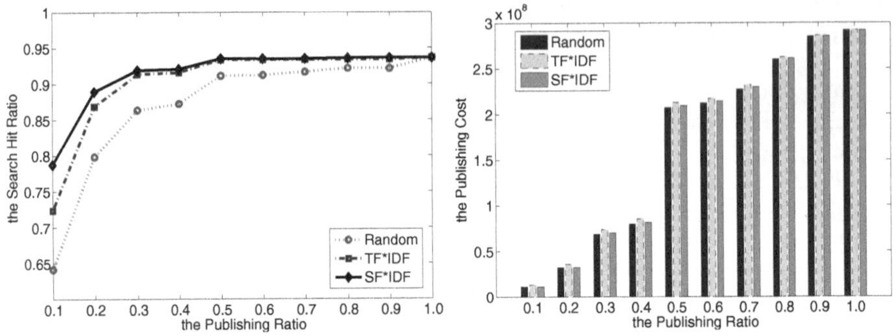


(a) Search Hit Ratio (b) Publishing Cost

**Fig. 6.** Comparison of Publishing Performance with Different $P_r$ ($N_s$=6144)

than *TF*IDF*, and is similar to *Random* in publishing cost. However, *SF*IDF* is better than *TF*IDF* and much better than *Random* in search hit ratio. From the results in Figure 6(a) and Figure 6(b), these algorithms based on the rank rule is much better than *Random* in search hit ratio as the publishing ratio is smaller. Moreover, *SF*IDF* is better than the other algorithms in the overall performance as the publishing ratio is smaller. The reasons for this are similar to those of Figure 5(a) and Figure 5(b), where they are not explained in detail.

## 6 Conclusions and Future Work

In this paper, we firstly analyzed existing publishing algorithms including synchronous publishing method in eMule and points out that their network loads are too heavy. The proposed method is more effective and scalable with lower network load than traditional publishing algorithms. We applied two optimization techniques to the proposed algorithm, which can effectively improve search hit ratio and reduce network load. Simulations based on real data from Maze system show that this approach has low network overhead and publishing cost, high search and download hit ratio.

Future work includes: Integrating a real P2P system with the proposed approach in this paper and applying this approach to full text retrieval under DHT environments, etc. Lunar[8] is an ongoing project in our research team, which aims to provide a general middleware for P2P systems, such as P2P file-sharing systems, P2P storage systems. There are several problems that need urgent solutions: how to publish contents to decrease bandwidth under DHT environments? how to design a better retrieval algorithm to rank returned results? how to design a better index mechanism to improve the availability of systems. Based on Lunar, a new Maze system: LMaze is developed. Although the proposed approach in this paper is designed for information retrieval in P2P file-sharing systems, it is also suitable for full text retrieval under DHT environments, such as Overcite [12].

[8] http://maze.pku.edu.cn/lunar.htm

## References


1. Stoica, I., Morris, R., Karger, D.R., Kaashoek, M.F., Balakrishnan, H.: Chord: A scalable peer-to-peer lookup service for internet applications. In: SIGCOMM, pp. 149–160 (2001)
2. Rowstron, A.I.T., Druschel, P.: Pastry: Scalable, decentralized object location, and routing for large-scale peer-to-peer systems. In: Guerraoui, R. (ed.) Middleware 2001. LNCS, vol. 2218, pp. 329–350. Springer, Heidelberg (2001)
3. Maymounkov, P., Mazières, D.: Kademlia: A peer-to-peer information system based on the xor metric. In: Druschel, P., Kaashoek, M.F., Rowstron, A. (eds.) IPTPS 2002. LNCS, vol. 2429, pp. 53–65. Springer, Heidelberg (2002)
4. Dabek, F., Zhao, B.Y., Druschel, P., Kubiatowicz, J., Stoica, I.: Towards a common api for structured peer-to-peer overlays. In: Kaashoek, M.F., Stoica, I. (eds.) IPTPS 2003. LNCS, vol. 2735, pp. 33–44. Springer, Heidelberg (2003)
5. Rhea, S.C., Godfrey, B., Karp, B., Kubiatowicz, J., Ratnasamy, S., Shenker, S., Stoica, I., Yu, H.: Opendht: a public dht service and its uses. In: Guérin, R., Govindan, R., Minshall, G. (eds.) SIGCOMM, pp. 73–84. ACM, New York (2005)
6. Tang, C., Dwarkadas, S.: Hybrid global-local indexing for efficient peer-to-peer information retrieval. In: NSDI, pp. 211–224 (2004)
7. Freedman, M.J., Freudenthal, E., Mazières, D.: Democratizing content publication with coral. In: NSDI, pp. 239–252 (2004)
8. Yang, X., Hu, Y.: A dht-based infrastructure for content-based publish/subscribe services. In: Hauswirth, M., Wierzbicki, A., Wehrle, K., Montresor, A., Shahmehri, N. (eds.) Peer-to-Peer Computing, pp. 185–192. IEEE Computer Society, Los Alamitos (2007)
9. Steiner, M., Carra, D., Biersack, E.W.: Faster content access in kad. In: Wehrle, K., Kellerer, W., Singhal, S.K., Steinmetz, R. (eds.) Peer-to-Peer Computing, pp. 195–204. IEEE Computer Society, Los Alamitos (2008)
10. Park, K., Pai, V.S.: Scale and performance in the coblitz large-file distribution service. In: NSDI (2006)
11. Salton, G., Wong, A., Yang, C.S.: A vector space model for automatic indexing. Commun. ACM 18(11), 613–620 (1975)
12. Stribling, J., Li, J., Councill, I.G., Kaashoek, M.F., Morris, R.: Overcite: A distributed, cooperative citeseer. In: NSDI (2006)

# CHAP: Enabling Efficient Hardware-Based Multiple Hash Schemes for IP Lookup

Michel Hanna, Socrates Demetriades, Sangyeun Cho, and Rami Melhem

Dept. of Computer Science, Univ. of Pittsburgh,
Pittsburgh, PA 15260, USA
{mhanna,socrates,cho,melhem}@cs.pitt.edu

**Abstract.** Building a high performance IP lookup engine remains a challenge due to increasingly stringent throughput requirements and the growing size of IP tables. An emerging approach for IP lookup is the use of set associative memory architecture, which is basically a hardware implementation of an open addressing hash table with the property that each row of the hash table can be searched in one memory cycle. While open addressing hash tables, in general, provide good average-case search performance, their memory utilization and worst-case performance can degrade quickly due to bucket overflows. This paper presents a new simple hash probing scheme called CHAP (Content-based HAsh Probing) that tackles the hash overflow problem. In CHAP, the probing is based on the content of the hash table, thus avoiding the classical side effects of probing. We show through experimenting with real IP tables how CHAP can effectively deal with the overflow.

**Keywords:** IP lookup, hardware multiple hashing, content-based probing.

## 1 Introduction

High speed routers require wire speed packet forwarding while the sizes of the IP tables across core routers are increasing at a very high rate [1]. IP address lookup has been a significant bottleneck for core routers. The advancement of optical networks made the situation even worse with link rates already beyond 40 Gbps. Some predict that in the near future, "Terabit" link rates will be available with affordable prices [2,3].

IP lookup proceeds as follows: the destination address of every incoming packet is matched against a large forwarding database (*i.e.*, routing table) to determine the packet's next hop on its way to the final destination. An entry in the forwarding table (called a *prefix*) is a binary string of a certain length (*prefix length*), followed by don't care bits. The adoption of *Classless Inter-Domain Routing* resulted in the need for *longest prefix match* (LPM) [4].

Existing IP forwarding engines are categorized into two main groups: hardware based and software based. The hardware based schemes are generally constrained by the size and power consumption of the engine. The software based schemes are

L. Fratta et al. (Eds.): NETWORKING 2009, LNCS 5550, pp. 756–769, 2009.

mainly constrained by the throughput, measured as the number of lookups per second. Recently, using hash techniques for IP lookup gained a lot of momentum. Hash tables come in two flavors: open addressing hash and closed addressing hash (or *chaining*). The hash table in closed addressing hash has a fixed height (number of buckets), and each bucket is an infinite size linked list. During the lookup process, a specific row index is generated by the hash function and the row is searched to find the target key. An important design goal in this case is to minimize the worst-case length of the linked lists and to balance the bucket population by using Bloom filters-like data structures [5, 6, 7].

In open addressing, the hash table has a fixed height and a fixed bucket width (number of elements per bucket). Open addressing hash has a simpler table structure than closed addressing hash and is amenable to hardware implementations. However, the issues of overflow and overflow handling have to be dealt with. Normally, the overflow is handled by means of probing [8].

The hardware schemes use special hardware such as Ternary Content Addressable Memory (TCAM) to increase the lookup throughput. Unfortunately, the TCAM approach has its own set of limitations: high power consumption, poor scalability, and low bit density. Moreover, most commodity TCAMs run at low speed compared to SRAM memory [3]. Hence many researchers proposed optimizations to the TCAM architecture [9, 10, 11, 12].

In this paper, we assume open addressing hash schemes for which a number of efficient hardware prototype implementations have been proposed recently [13, 14]. In these implementations, the hash table is stored in a set associative memory where each set stores all the elements in a bucket and the buckets are indexed through the hash function. Our goal is to fit an entire IP lookup table in a single fixed size hash table with no/acceptable overflow and with good space utilization. In addition, we want to keep both insertion/deletion into/from the table simple and straightforward. This paper makes the following contributions to the area of open addressing hash in general:

- The introduction of the new concept of content-based hash probing which more effectively tackles the overflow than other existing probing techniques.
- The application of content-based probing to multiple hash function schemes.
- The use of content-based probing and multiple hashing, together, to implement an efficient hardware-based IP lookup engine.

The rest of the paper is organized as follows. In Section 2 we briefly summarize the state-of-the-art hardware-based hash techniques. In Section 3 we describe CHAP, our main scheme. Section 4 discusses the setup procedure of CHAP and how search is done. We will then discuss the incremental updates in Section 5. Section 6 shows experimental results. Finally, we give conclusions and future work in Section 7.

## 2 Background

### 2.1 Open Addressing Hash

Searchable data items, or records, contain two fields: key and data. Given a search key, $k$, the goal of searching is to find a record associated with $k$ in the database. Hash achieves fast searching by providing a simple arithmetic function $h(\cdot)$ (hash function) on $k$ so that the location of the associated record is directly determined. The memory containing the database can be viewed as a two-dimensional memory array of $N$ rows with $L$ records per row.

It is possible that two distinct keys $k_i \neq k_j$ hash to the same value: $h(k_i) = h(k_j)$. Such an occurrence is called *collision.* When there are too many ($\geq L$) colliding records, some of those records must be placed elsewhere in the table by finding, or *probing*, an empty space in a bucket. For example, in *linear probing*, the probing sequence used to insert an element into a hash table is given as follows:

$$h(k), h(k) + \beta_0, h(k) + \beta_1, \cdots, h(k) + \beta_{m-1} \tag{1}$$

where each $\beta_i$ is a constant, and $m$ is the maximum number of probes. Linear probing is simple, but often suffers from *primary key clustering* [8].

Instead of probing, one can apply a second hash function to find an empty bucket, which is known as *double hashing* [8]. In general, the use of $H \geq 2$ hash functions is shown to be better in eliminating hash overflow than probing [15]. In this case (which we will refer to as *multiple hashing* in the rest of this paper) the probing sequence of inserting a key into the hash table is given as follows:

$$h_0(k), h_1(k), \cdots, h_{H-1}(k) \tag{2}$$

where $H$ is the maximum number of hash functions. Most work that is done in the multiple hashing domain is for closed addressing hash as in [15, 16].

Given a database of $M$ records and an $N$-bucket hash table, the average number of hash table accesses to find a record is heavily affected by the choice of $h(\cdot)$, $L$ (the number of slots per bucket), and $\alpha$, or the *load factor*, defined as $M/(N \times L)$. With a smaller $\alpha$, the average number of hash table accesses can be made smaller, however at the expense of more unused memory space.

## 2.2 Set Associative Memory Architecture Overview

We will use the CA-RAM (Content Addressable-Random Access Memory) as a representative of a number of set associative memory architectures proposed for IP lookup [13, 14]. A CA-RAM takes as an input a search key and outputs

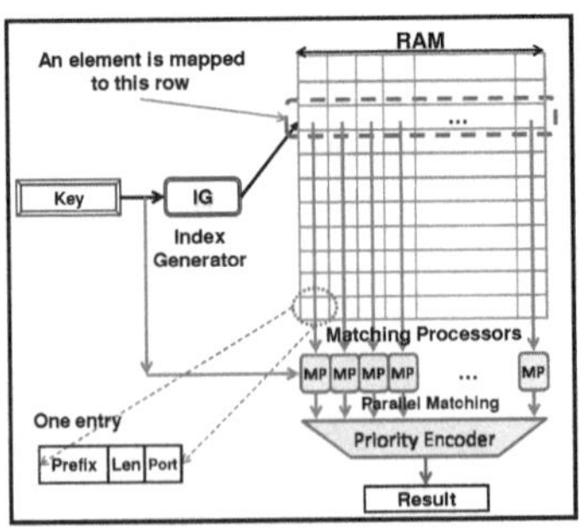


**Fig. 1.** The basic CA-RAM Architecture

the result of a lookup. Its main components are: an index generator, a memory array (SRAM or DRAM), and match processors, as shown in Figure 1.

Given a key, the index generator uses a hash function to create an index which is used to access a row of the memory array. All the keys stored in that row are fetched simultaneously and the match processors compare the row of keys with the search key in parallel, resulting in constant-time matching. Note that the format of each memory row is flexible and the matching processors are programmable.

## 3 Content-Based Hash Probing

As we mentioned in the last section, a CA-RAM row stores the elements of a bucket and is accessed in one memory cycle. Because the architecture is very flexible, we may keep some bits at the end of each row for auxiliary data; this allows for more efficient probing schemes with multiple hash functions. In this section we first present the basic content-based hash probing scheme, **CHAP(1,m)**, which is a natural evolution of the linear probing scheme described by Equation (1). We then extend this scheme to $H$ hash functions, which we call **CHAP(H,m)**.

In open addressing hash, some rows may incur overflow while others have space. While linear probing uses predetermined offsets to solve that problem as specified by Equation (1), CHAP uses the same probing sequence, but with the constants $\beta_0, \beta_1, \cdots, \beta_m$ determined dynamically for each value of $h(k)$, depending on the distribution of the data stored in a particular hash table. Specifically, the probing sequence to insert a key "$k$" is:

$$h(k), \beta_0[h(k)], \beta_1[h(k)], \cdots, \beta_{m-1}[h(k)] \tag{3}$$

This means that for each row we associate a group of $m$ pointers to be used if overflow occurs to point to other rows that have empty spaces. We call those pointers "probing pointers" and the overall scheme is called **CHAP(1,m)** since it has only one hash function and $m$ probing pointers per row.

Figure 2 shows the basic idea of CHAP when $m = 2$. In order to match the overflow excess keys to specific rows, we need to collect all the overflow elements across all the rows. We achieve this by counting the excess elements per row and finding for each row $i$ two rows in which these overflow elements can fit. These two rows indices' are recorded in $\beta_0[i]$ and $\beta_1[i]$.

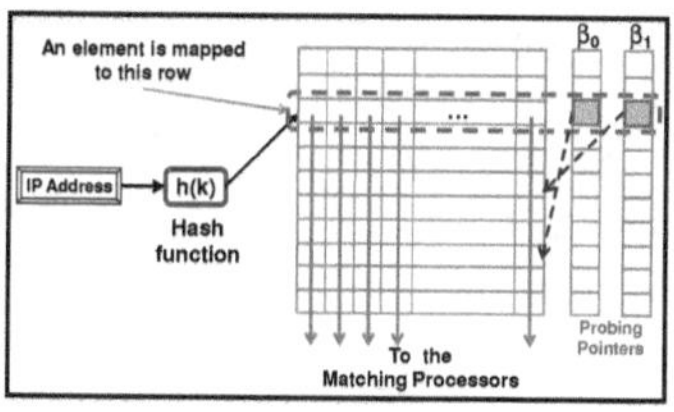


**Fig. 2.** The CHAP basic idea

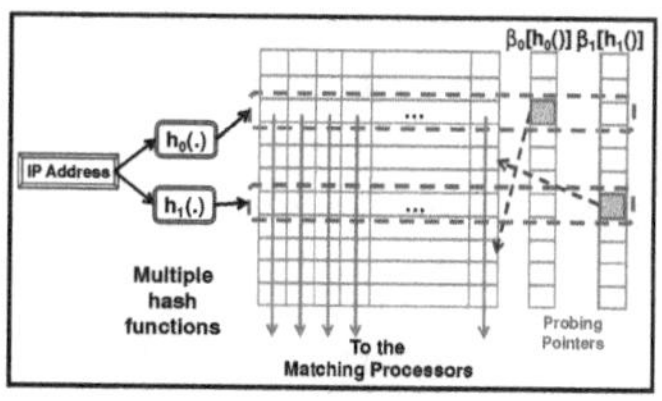


**Fig. 3.** The CHAP(2,2)

Assume that we are searching for a key $k$. If the hash function points to row $i = h(k)$ and it turns out that the input key $k$ is not in this row, we check to see if the probing pointers at row $i$ are defined or not. If defined, this means that there are other elements that belong to row $i$ but reside in either row $\beta_0[i]$ or in row $\beta_1[i]$ and might contain $k$. Consequently, rows $\beta_0[i]$ and $\beta_1[i]$ are accessed in subsequent memory cycles to find the matching key.

The content-based probing can also be applied to the multiple hashing scheme. Specifically, we refer to CHAP with $H$ hash functions and $m$ probing pointers by **CHAP(H,m)**. For example, in **CHAP(H,H)** we have $H$ hash functions and $m = H$ probing pointers. In this case, the probing sequence for inserting a key, $k$, can be defined by:

$$h_0(k), h_1(k), \cdots, h_{H-1}(k), \beta_0[h_0(k)], \beta_1[h_1(k)], \cdots, \beta_{m-1}[h_{H-1}(k)] \qquad (4)$$

In essence, we dedicate to each hash function a pointer per row. An example is shown in Figure 3 for a two hash functions CHAP scheme where a key is mapped to two different buckets. In the example, this key will have four different buckets to which it can be allocated: $h_0(k)$, $h_1(k)$, $\beta_0[h_0(k)]$ and $\beta_1[h_1(k)]$ in the given order, where $\beta_i[h_i(\cdot)]$ is the probing pointer of hash function $h_i(\cdot)$.

There are different ways to organize CHAP(H,m) when $m \neq H$ depending on whether or not the probing pointers are shared among the hash functions in a given row. In the example described above for CHAP(H,H), we assume that one probing pointer is associated with each hash function. Another organization is to share probing pointers among hash functions. Yet a third organization is to assign multiple pointers for each hash function, which is the only possible organization for CHAP(1,m), when $m > 1$. In the rest of this paper, we will limit our discussion to CHAP(1,m) and CHAP(H,H) with one pointer for each hash function and with the row order given by Equation (3). We defer the investigation of the other organization to future work.

## 4 The CHAP(H,H) Scheme

In this section we describe how to establish an IP lookup engine using CHAP(H,H). We present the setup algorithm that sets the probing pointers and maps actual IP prefixes into the CHAP hash table. With minor modifications, this algorithm can apply to the case of CHAP(1,m).

Before we describe the CHAP setup algorithm, we note that on average 98% of IP prefixes are 16 bits or longer [1]. In this work, we will use only the most significant 16 bits in our hash functions. Prefixes shorter than 16 bits (short prefixes) are not included in the hash table. A separate small TCAM can be used to store those prefixes. This small TCAM is to be searched in parallel with the main hash table on every lookup, which is a common practice [12, 17].

### 4.1 The Setup Algorithm

Algorithm 1 lays out the setup phase of CHAP. In that algorithm, $j = 0, \cdots, M-1$ is used to index the prefixes, where $M$ is the total number of prefixes in an IP routing

**Algorithm 1.** CHAP(H,H) Setup Algorithm

```
1:  Sort the IP prefixes from long to short and initialize the arrays HC[N] & OC[N][H] to zeros
2:  table_overflow = 0
3:  for(j = 0; j < M; j++)
4:       finished = false
5:       for(i = 0; i < H; i++)
6:           r_i = h_i(k_j)
7:       for (i = 0; i < H AND finished == false; i++)
8:           if(HC[r_i] < L), then
9:               HC[r_i]++
10:              finished = true
11:      for (i = 0; i < H AND finished == false; i++)
12:          if(OC[r_i][i] < λ), then
13:              OC[r_i][i]++
14:              finished = true
15:      if(finished == false), then table_overflow++
```

table. The goal is to map this table into a hash table with $2^R = N$ rows, where $R$ is the number of bits used to index the hash table. We use $i$ as an index for hash functions and $H$ as the maximum number of hash functions. An array of counters, $HC$[row index], is used to count the number of elements that will be mapped to each row of the hash table. We define a two dimensional array of counters $OC$[row index][hash function index] to count the overflow elements for each hash function per row. The maximum value of a single counter in this array is equal to $\lambda$, where $\lambda \leq L$, and $L$ is the number of prefixes per row. This bound comes from the fact that a hole, or an empty space in any row of the hash table, can never exceed $L$.

CHAP setup phase determines if the configuration parameters of the hash table is valid or not. In other words, will the parameters $L, H, \lambda$ and $N$ result in a mapping of the $M$ prefixes into a single hash table without/with acceptable overflow?

Algorithm 1 calculates the number of prefixes to be assigned to each row. By "assigned" we mean not only the prefixes that are hashed to this row, but also the overflow prefixes that are supposed to be in this row but will reside in other rows that are pointed to by this row's probing pointers. It starts by sorting prefixes from long to short, then initializing the two arrays $HC$, $OC$ and the table_overflow counter to zeros (lines 1–2). The set of hash values $\{r_0, \cdots r_{H-1}\}$ for each prefix is calculated (lines 5–6). Then, the algorithm updates the counter $HC$ by using the $H$ hash values of each prefix. If there is a spot for the current prefix in $HC$ then the algorithm will move on to the next prefix (lines 8–10). If not, it will increment the $OC$ counter (lines 11–14).

When Algorithm 1 exits, table_overflow contains the number of prefixes that could not fit in either $HC$ or $OC$. If that number is not acceptable, then the algorithm can be repeated with more hash functions. That is with $H = H + 1$. If a separate TCAM is used to store the short prefixes as described earlier, then this same TCAM can be used to store the overflow prefixes. Hence the acceptable overflow in Algorithm 1 will depend on the capacity of the added TCAM.

## 4.2 The Mapping of IP Prefixes in CHAP

The last step in CHAP is to allocate the elements into the hash table using the probing pointers. Before moving to the actual mapping of the prefixes, however, we need to assign values to the probing pointer's array. This is done by running the best fit algorithm [18] to set the values of the probing pointer. The algorithm starts by finding the largest counter value from the $OC$ array, say $OC[t][i]$, and the smallest counter value from $HC$, say $HC[J]$, (we call this a *hole*). Then the i$^{th}$ probing pointer of row $t$ is assigned the value of $J$, the row having the largest hole provided that the hole size is larger than the largest counter.

Clearly, the best fit algorithm may not find a hole for each overflow counter, which means that some keys will not be able to fit in the hash table. The number of these keys are added to table_overflow, and again, if the resulting overflow is not acceptable, then Algorithm 1 has to be re-executed with a larger value of $H$. After setting the probing pointers, the prefixes are mapped to the hash table.

### 4.3 Search in CHAP

As discussed in Section 2.2, hardware implementation of hash tables reads a full row (bucket) of the table into a buffer and uses a set of comparators to determine, in parallel, the longest prefix match among the elements in that bucket. Hence, a metric that will be used to measure the efficiency of the search in CHAP is the Average Successful Search Time, **ASST**, the average number of rows accessed for successful search.

The CHAP search algorithm itself is straightforward. Given a key $k_x$ as an input IP address, we calculate the row address $h_0(k_x)$ and then match the prefix against all the elements in this row (in parallel). If we find a hit at that row, we stop searching. If not, we try the next hash functions (*i.e.*, $h_1(k_x), \cdots, h_{H-1}(k_x)$). After we are done with all the $H$ hash functions we start looking at the probing pointers. First the probing pointer $\beta_0[h_0(k_x)]$, then $\beta_1[h_1(k_x)], \cdots, \beta_{H-1}[h_{H-1}(k_x)]$. In other words, the order of accessing the pointers used in searching is the same order used in inserting the prefixes. This constraint has to be satisfied to guarantee the LPM feature in searching. Specifically, if we do not use the insertion order when searching, we might search in a row that contains shorter prefixes than the longest prefix that should match the address $k_x$. This order is maintained through the dedication of one probing pointer per hash function. Without this dedication, we cannot preserve LPM as the probing pointers will be shared between multiple hash functions.

## 5 The Incremental Updates

An important issue in the IP forwarding engine is the incremental updates of the prefix database. The number of prefixes included in a routing table grows with time [1,2]. The updates consist of two basic operations, *Insert/Update* and *Delete* a prefix. In CHAP the delete operation is straightforward. For any prefix deletion operation we find the prefix first, then we delete it and decrement the row counter $RC$, where a counter $RC[i]$ is associated with each row $i$ to record the number of prefixes stored in that row. This counter will be used during the insert/update operation to keep track of full rows.

**Algorithm 2.** CHAP Insert Update Algorithm

1: subroutine **CHAP_Insert_Update** (prefix $k_n$)<br>
2: **for**($i = 0;\ i < H;\ i++$)<br>
3: $\quad T[i] = h_i(k_n)$<br>
4: $\quad T[i+H] = \beta_i[h_i(k_n)]$<br>
5: By searching the rows $T[0], \cdots T[2 \times H - 1]$, find:<br>
6: $\quad k_l$ = longest prefix that matches $k_n$ and $r_l$ = row that contains $k_l$<br>
7: $\quad k_s$ = shortest prefix that matches $k_n$ and $r_s$ = row that contains $k_s$<br>
8: **if**($k_l$ is not defined **AND** $k_s$ is not defined ), **then**<br>
9: $\quad$ **return**(**Insert_in_Rows**($k_n$, $T[0]$, $T[2 \times H - 1]$)) /* if no matching: insert $k_n$ in any row */<br>
10: **else if** (($|k_n| == |k_l|$) **OR** ($|k_n| == |k_s|$)), **then**<br>
11: $\quad$ Replace $k_l$ or $k_s$ with $k_n$ /* an update operation */<br>
12: $\quad$ **return** (*true*)<br>
13: **else if** ($|k_n| > |k_l|$), **then return** (**Insert_in_Rows**($k_n$, $T[0]$, $r_l$))<br>
14: **else if**($|k_n| < |k_s|$), **then return**(**Insert_in_Rows**($k_n$, $r_s$, $T[2 \times H - 1]$))<br>
15: **else, return**(**Insert_in_Rows**($k_n$, $r_l$, $r_s$))<br>
16:<br>
17: subroutine **Insert_in_Rows** (prefix $k_x$, row $r_a$, row $r_b$)<br>
18: **for**($i = r_a;\ i <= r_b;\ i++$)<br>
19: $\quad$ **if**($RC[i] < L$), **then**<br>
20: $\qquad$ insert $k_x$ in row $i$ and $RC[i]++$<br>
21: $\qquad$ **return** (*true*)<br>
22: **return** (*false*)

The basic idea of the insert/update operation, which is detailed in Algorithm 2, is to find the appropriate row $r$ that the new prefix should fit in, taking into account the LPM feature. In other words, we need to find where the new prefix should be stored according to its length to achieve LPM. If it is found that the prefix already exists in the CHAP table, the existing entry will be updated.

Algorithm 2 consists of two "Boolean" subroutines, **CHAP_Insert_Updtae()** and **Insert_in_Rows()**. The second subroutine is where the actual insertion is made, as it take a prefix $k_x$ and tries to insert it in a series of rows starting from row $r_a$ all the way to $r_b$. The first subroutine, CHAP_Insert_Updtae(), will determine the appropriate rows to insert the new prefix, $k_n$.

In the first routine a single dimension array $T[\cdot]$ of size $(2 \times H)$ is used to store the computed values of the hash functions of $k_n$ and the corresponding probing pointers (lines 2–4). For each row in $T[i]$ we match $k_n$ against all the prefixes in this row and extract both the longest prefix, $k_l$, and the shortest prefix, $k_s$, that match $k_n$ (lines 5–7). We record the rows $r_l$ and $r_s$, that include $k_l$ and $k_s$, if a matching is found.

Depending on the length of $k_n$ relative to the length of both $k_l$ and $k_s$, we will try to insert $k_n$ in one of the $2 \times H$ rows. This is done through an $if - else$ construct (lines 8–15). The first case is when neither $k_l$ nor $k_s$ are defined (i.e., no matching), thus we can insert $k_n$ into any row (lines 8–9). The second case, which is route update [1], is when $k_n$ is equal either $k_l$ or $k_s$. In this case we replace either $k_l$ or $k_s$ with the new prefix $k_n$ (lines 10–12). The third case is if the length $|k_n|$ of $k_n$ is larger than $|k_l|$, the length of $k_l$. We will try to insert $k_n$ into one of the buckets $T[0], \cdots, r_l$ if they have a space (line 13). In the next case we check to see if $|k_n| < |k_s|$ and if it is true, then we try to insert $k_n$ in a row among $r_s, \cdots, T[2 \times H - 1]$ (line 14). The final case is when $|k_s| < |k_n| < |k_l|$ and in this case we will call Insert_in_Rows() to try to put $k_n$ in any row between

$r_l$ and $r_s$ (line 15). In any case, the subroutines terminate successfully if we were able to insert $k_n$.

# 6 Evaluation

We used C++ to build our own simulation environment. This environment allows us to choose and arrange different types of hash functions. The BGP (Border Gateway Protocol) routing tables of Internet core routers were obtained from the routing information service project [19]. The statistics for the routing tables used are listed later in Table 9. When measuring the average lookup time, we used synthetic traces (uniform distribution) generated from these tables.

The hash functions used in our experiments are from three different hashing families: bit-selecting, CRC-based, and $\mathbf{H_3}$ [20] hashing families. Those families have the advantage of being simple and fast enough to be easily realized in hardware.

For a given hardware implementation, the number of rows, $N$, and the number of entries per row, $L$, are fixed and the performance of the CHAP scheme depends on two important parameters, namely the maximum overflow value of the $OC$ counters, $\lambda$, and the number of hash functions used, $H$, which is also the number of probing pointers per row in CHAP(H,H). Intuitively, if $\lambda$ is small, then the setup algorithm (Algorithm 1) may not be able to eliminate the overflow. On the other hand, if $\lambda$ is large, then Algorithm 1 may terminate with every $OC$ having smaller value than $\lambda$, but the best fit algorithm may not find holes that are large enough in the table to accommodate the values of the $OC$, thus increasing the overall overflow of the hash table.

In addition to $H$ and $\lambda$, the performance of CHAP depends on the load factor $\alpha$. Clearly, $\alpha$ depends on the size of the actual IP routing table and the size of the physical memory used to store the hash table. For a given $\alpha$, the hashing overflow depends on the aspect ratio of the memory $N/L$. In what follows, we will define a "configuration" by specifying both $N$ and $L$.

## 6.1 The Advantages of Content-Based Hash Probing

In order to show the advantage of content-based probing over linear probing, we compare the overflow generated by both CHAP(1,m) and linear probing (that has the same number of probing steps) when mapping routing tables to hash tables with specific configurations (that is with specific $L$ and $N$). We will use the table "rrc07-AS21202" obtained from [19] and use two different configurations $C_1$: $L = 200$, $N = 1024$ and $C_2$: $L = 100$, $N = 2048$. We tried many different configurations and they all led to results similar to those shown in Figure 4. In addition, these two configurations have a high average load factor $\alpha = 91.27\%$ for the "rrc07-AS21202" table, which articulates the strength of CHAP.

Figure 4 shows that for the same number of probing steps, overflow in CHAP(1,m) is less than that in linear probing. In fact, CHAP achieves 71.61% more overflow reduction than linear probing on average. Moreover, we can see

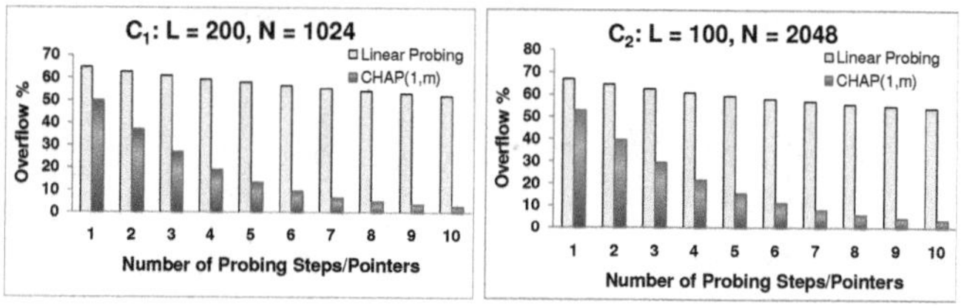


**Fig. 4.** Overflow of CHAP(1, m) vs. linear probing for table rrc07-AS13645

that the longer the probing sequence, the more effective is CHAP in eliminating overflow compared to linear probing. The main reason behind this is that CHAP is addressing the overflow reduction directly by choosing empty (or partially empty) buckets to reallocate the overflow elements. This is in contrast to linear probing which blindly tries to put the overflow elements in the nearest available bucket which may not be found within $m$ probes.

## 6.2 Sensitivity Analysis of CHAP (H,H)

In this section we study the effect of varying $\lambda$ or the maximum value of the overflow counter, $OC$, in the CHAP setup algorithm (Section 4.1). We report the results for table "rrc07-AS13645" since all other tables have similar results. We will also show the results for slight variations of the configurations $C_1$ and $C_2$. In both case we use $H = 3$.

Figure 5 shows the values of overflow versus $\lambda$ for the reported configurations. For the Figure 5(a), we set $N = 1024$ rows and $L = 180, 200, 220$ and $240$ entry per row, which results in $\alpha = 96.03\%, 91.27\%, 82.98\%$ and $76.10\%$ respectively. As for Figure 5(b), we set $N = 2048$ rows and $L = 90, 100, 110$ and $120$ entry per row which will result in the same loading factors. Note that $\lambda \in [0, L]$.

From the figures we can see that the overflow starts at some non zero value and then decreases in the range $0 < \lambda < \frac{L}{2}$. At $\lambda = \frac{L}{2}$ the overflow becomes almost zero in Figure 5(b) while it is actually zero in Figure 5(a). The low overflow determined at $\lambda = \frac{L}{2}$ means that the average hole size in the hash table is equal to $\frac{L}{2}$. For larger values of $\lambda$, the maximum hole size becomes smaller than $\lambda$ and thus we are unable to insert all the elements that were counted by $OC$ into the hash table. This increases the overflow. In the following section we will use $\lambda = \frac{L}{2}$.

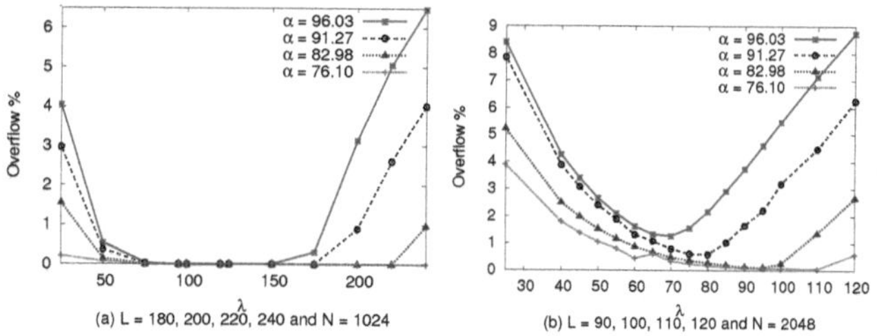


**Fig. 5.** The overflow vs. $\lambda$

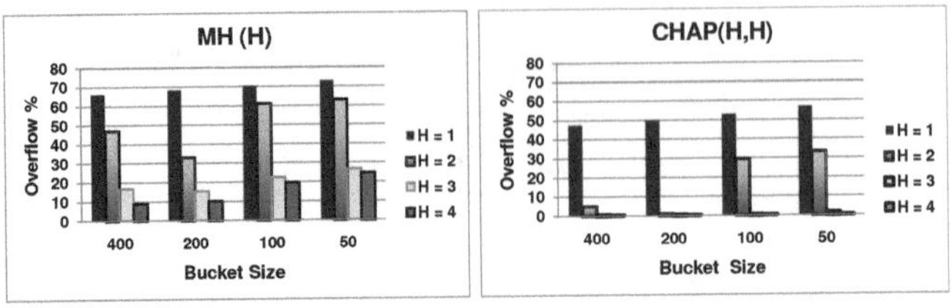


**Fig. 6.** Average overflow of 4 bucket widths for MH(H) and CHAP(H,H)

### 6.3 CHAP(H,H) Versus Multiple Hashing(H)

In this section we compare the **CHAP(H,H)** scheme against the regular multiple hash function scheme (which we call **MH(H)**) [15], where $H$ is the number of hash functions used. We compare the two schemes in terms of the ASST (Average Successful Search Time) and the overflow. Again, we use the routing table rrc07-AS13645.

In Figure 6 we show the average values of overflow for different number of hash functions (between 1 and 4) and for four different bucket sizes. To keep $\alpha$ constant at 91.27%, we set $N = 512, 1024, 2048$ and 4096 where $L = 400, 200, 100$ and 50. It's obvious from this figure that CHAP(H,H) has much less overflow than multiple hashing for the same number of hash functions.

The results shown in Figure 7 indicates that the average ASST over the four bucket sizes for MH(H) is 1.7, while it's 2.24 for CHAP(H,H). Although the difference between the two schemes seems large, we have to take into consideration that at $H = 3$ the overflow of CHAP is already zero for the bucket sizes of $L = 400, 200$ and is less than 2% when $L = 100, 50$. Thus adding more hash functions in this case makes the average memory access time worse. If we recalculate the average ASST over the four bucket sizes for CHAP (without taking $H = 4$ into account) we find it to be 1.87 which is only 10% higher than MH. What looks like the classical tradeoff between the overflow and the average memory access time can be seen in Figures 6 and 7. However, a better understanding of the tradeoff that CHAP and MH present can be obtained by comparing CHAP(H,H) with MH($2H$) since both has $2 \times H$ as the maximum number of table accesses. For example at $H = 2$ with a bucket size $L = 200$, the overflow for CHAP(2,2) is 1.01% with an average access time of 2.2 memory cycles, while for the MH(4) is 10.12% with an average access time of 2.1 memory cycles.

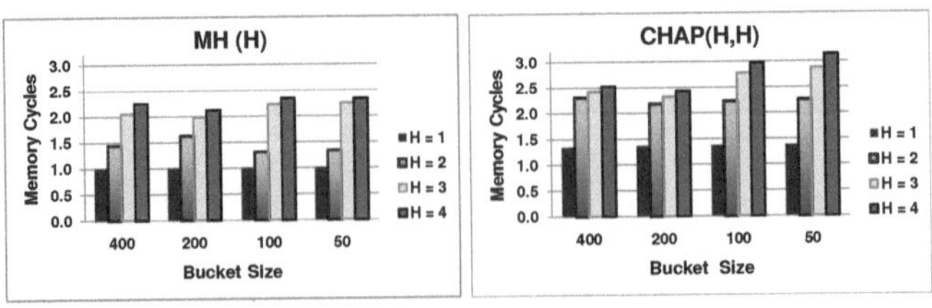


**Fig. 7.** Average ASST of 4 bucket widths for MH(H) and CHAP(H,H)

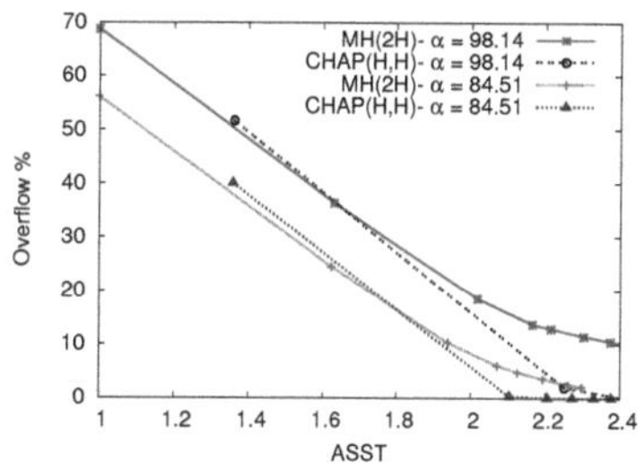


**Fig. 8.** Overflow vs. the ASST for MH(H) v.s. CHAP(H,H) for different load factors

| | No# Tables | Avg. size | Avg. $\alpha$ | A.short prefixs |
|---|---|---|---|---|
| rrc04 | 3 | 185 | 93 | 0.77 |
| rrc05 | 4 | 179 | 89.5 | 0.76 |
| rrc07 | 3 | 133 | 66.5 | 0.96 |
| rrc11 | 4 | 198 | 99 | 0.78 |

**Fig. 9.** The Statistics of the IP lookup tables used in Figure 10

In order to show that CHAP can achieve both lower overflow and average access time than MH, we plot in Figure 8 the average access time versus the overflow of both schemes for two load factors (98.14% and 84.51%) for $L = 200 \pm 25$ elements and $N = 1024$ rows. We used the same table as before and we monitored the variation of ASST and the overflow for the same load factor $\alpha$. We varied the number of hash functions from 1 to 10. Each curve on the graph corresponds to 10 values of $H$ ($H = 1, \cdots, 10$), where the left most point corresponds to $H = 1$ and the rightmost point corresponds to $H = 10$. CHAP(3,3) has zero overflow, thus we do not need to use more than three hash functions. Figure 8 shows that for the region of interest (low overflow), for any $H > 1$, the CHAP(H,H) curve has both smaller ASST and overflow than the MH(H) curve. In other words, CHAP(H,H), for $H > 1$, is more efficient (smaller ASST and fewer overflows) than the MH(H) under the same system configurations and under the same load factor.

In a different experiment we compare CHAP(3,3) with MH(6). For this experiment we map each of the 14 IP tables of [19] into a fixed hash table of 200 K entries. A summary of the properties of the 14 tables is given in Table 9, where the third column in this table indicates the average load factor (percentage) when the IP tables are mapped to a 200 K entries. All the 14 tables are from the date January 19th, 2007. The last column in the table indicates the average percentage of short prefixes in each IP routing tables group. We have indicated that these short prefixes are excluded from the main CHAP hash table and inserted in a separate TCAM table. We repeated the experiment for 4 different configurations, namely ($L = 400$, $N = 512$), ($L = 200$, $N = 1024$), ($L = 100$,

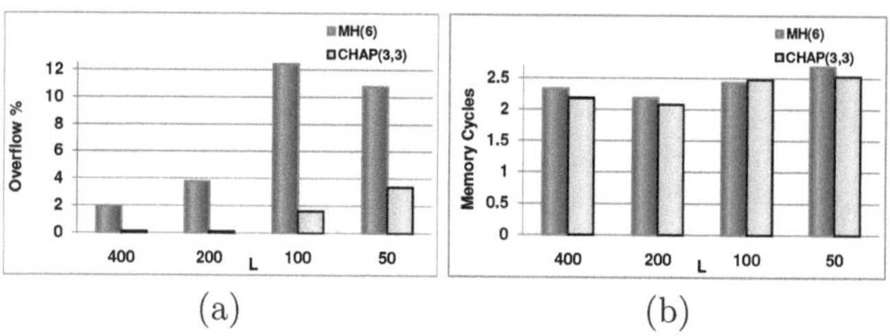


**Fig. 10.** Average (a) overflow and (b) ASST for CHAP(3,3) vs. MH(6)

$N = 2048$) and ($L = 50$, $N = 4096$), and we measured the average overflow and ASST for the 14 tables (Figure 10).

As we can see, CHAP(H,H) is better than MH(2H) in all cases in terms of both the ASST and the overflow. However there is only one case, $L = 100$ and $N = 2048$, where CHAP(H,H) has a slightly higher ASST (2.52 cycles) than the MH(2H) (2.51 cycles). However, for this configuration, MH(2H) has its worst-case overflow (12.3 %), while CHAP(H,H) incurs only 1.98% overflow. This means that CHAP(H,H) table contains around 20 K entries more than MH(2H) table at this configuration.

## 7 Conclusions and Future Work

In this paper we have described and studied CHAP, a new hash-based IP lookup scheme that eliminates the overflow problem by utilizing content-based probing and multiple hash functions. We showed that CHAP is very effective in eliminating the overflow, and at the same time, achieves a low average memory access time. We also illustrated that CHAP can be realized in hardware by taking advantage of state-of-the-art search memory architectures.

Unlike other hash-based schemes, the CHAP scheme does not require complex preprocessing of the IP tables. Simply sorting the prefixes from long to short while setting up is the only preprocessing required. CHAP uses simple functions that can be easily realized in hardware. Moreover, CHAP has simple setup and incremental update algorithms. Simulation results show that content-based hash probing is superior compared to linear probing in terms of overflow elimination. CHAP achieves 71.61% more overflow reduction than linear probing on average. The results also show that CHAP achieves lower average memory access time than the multiple hash function scheme while also reducing the overflow.

In this paper we did not study the general CHAP(H,m) scheme when $m \neq H$ and we intend to study this further in the future. Future work also includes the possibility of using CHAP in other IP protocol processing tasks (e.g., packet filtering and inspection).

## References


1. Huston, G.: Analyzing the internet's bgp routing table. The Internet Pro. J. (2001)
2. Varghese, G.: Network Algorithmics: An Interdisciplinary Approach to Designing Fast Networked Devices. Morgan Kaufmann, San Francisco (2005)
3. Jiang, W., Prasanna, V.K.: A memory-balanced linear pipeline architecture for trie-based ip lookup. In: HOTi 2007, August 2007, pp. 83–90 (2007)
4. Rekhter, Y., Li, T.: An architecure for ip address allocation with cidr. RFC (1993)
5. Bloom, B.: Space/time trade-offs in hash coding with allowable errors. Comm. of the ACM 13(7), 422–426 (1970)
6. Kirsch, A., Mitzenmacher, M.: Simple summaries for hashing with multiple choices. IEEE/ACM Trans. on Net (2007)
7. Song, H., et al.: Fast hash table lookup using extended bloom filter: An aid to network processing. In: Sigcomm 2005, August 2005, pp. 181–192 (2005)

8. Cormen, T., Leiserson, C., Rivest, R., Stien, C.: Introdcution to Algorithms. McGraw Hill, New York (2003)
9. Lakshminarayanan, K., Rangarajan, A., Venkatachary, S.: Algorithms for advanced packet classification with ternary cams. In: Sigcomm 2005, May 2005, pp. 193–204 (2005)
10. Panigrahy, R., Sharma, S.: Reducing tcam power consumption and increasing throughput. In: HOTi 2002, August 2002, pp. 107–112 (2002)
11. Shah, D., Gupta, P.: Fast updating algorithms for tcams. IEEE Micro. Mag. 21(1), 36–47 (2001)
12. Zane, F., Narlikar, G., Basu, A.: Coolcams: Power-efficient tcams for forwarding engines. In: Infocom 2003, April 2003, pp. 42–52 (2003)
13. Cho, S., Martin, J., Melhem, R.: Ca-ram: A high-performance memory substrate for search-intensive applications. In: Ispass 2007, April 2007, pp. 230–241 (2007)
14. Kaxiras, S., Keramidas, G.: Ipstash: A power-efficient memory architecture for ip-lookup. In: Micro 2003, November 2003, pp. 361–373 (2003)
15. Azar, Y., Broder, A.Z., Karlin, A.R., Upfal, E.: Balanced allocations. In: ACM SOC, pp. 593–602 (1994)
16. Vocking, B.: How asymmetry helps load balancing. ACM J., 568–589 (July 2003)
17. Basu, A., Narlikar, G.: Fast incremental updates for pipelined forwarding engines. In: Infocom 2003, pp. 64–74 (July 2003)
18. Randell, B.: A note on storage fragmentation and program segmentation. Comm. of the ACM 12, 365–372 (1969)
19. RIS: Routing information service (2006), `http://www.ripe.net/ris/`
20. Ramakrishna, M., et al.: Efficient hardware hashing functions for high performance computers. IEEE Trans. on Comp. 46(12), 1378–1381 (1997)

# PUB-2-SUB: A Content-Based Publish/Subscribe Framework for Cooperative P2P Networks
## (Work in Progress)

Duc A. Tran and Cuong Pham

Department of Computer Science
University of Massachusetts, Boston
100 Morrissey Blvd, Boston, MA 02125, USA
{duc,cpham}@cs.umb.edu

**Abstract.** This paper is focused on the content-based publish/subscribe service and our problem is to devise an efficient mechanism that enables this service in any given P2P network of cooperative nodes. Most techniques require some overlay structuralization added on top of the network. We propose a solution called PUB-2-SUB which works with any unstructured network topology. In addition, multiple independent publish/subscribe applications can run simultaneously on a single instance of PUB-2-SUB. We show that this mechanism is efficient in terms of both costs and time. Our theoretical findings are complemented by a simulation-based evaluation.

**Keywords:** P2P, publish/subscribe, prefix tree, search.

## 1 Introduction

Unlike traditional search, a query in the publish/subscribe model is submitted and stored in advance, for which the results may not yet exist but the query subscriber expects to be notified when they later become available. This model is thus suitable for search applications where queries await future information, as opposed to traditional applications where the information to be searched must pre-exist. Focusing on the publish/subscribe model, our goal is to devise a mechanism that can be integrated into a given P2P network to enable applications of this model. In particular, we are interested in distributed networks where the participating nodes are cooperative, reliable, and rather static. In these networks, such as grid computing networks and institutional communication networks, P2P can be adopted as an effective way to share resources, minimize server costs, and promote boundary-crossing collaborations [1, 2, 3, 4]. A publish/subscribe functionality should be useful to these networks.

To enable publish/subscribe applications, a simple way is to broadcast a query to all the nodes in the network or to employ a centralized index of all the queries subscribed and information published [5, 6, 7]. This mechanism is neither

L. Fratta et al. (Eds.): NETWORKING 2009, LNCS 5550, pp. 770–781, 2009.

efficient nor scalable if applied to a large-scale network. Consequently, a number of distributed publish/subscribe mechanisms have been proposed. They follow two main approaches: structuralization-based or gossip-based. The first approach [8,9,10,11,12,13] requires the nodes to be organized into some overlay structure (e.g., DHTs [14,15,16,17] and Skip Lists [18]) and develops publish/subscribe methods on top of it. The other approach [19,20,21] is for unstructured networks, in which the subscriber nodes and publisher nodes find each other via exchanges of information using the existing peer links, typically based on some form of randomization.

The structuralization-based approach is favored for its efficiency over the the gossip-based approach, but when applied to a given network, the former introduces an additional overhead to construct and maintain the required overlay structure. There may also be practical cases where the new links, that are part of the new structure, are not allowed due to the policy or technicality restrictions of the given network. On the other hand, although the gossip-based approach does not require an additional structuralization, due to the nature of gossiping, a query or a publication of new information must populate a sufficiently large portion of the network to be able to find each other at some rendezvous node with a high probability. The costs can be expensive as a result, including the communication cost to disseminate the query or publish the new information, the storage cost to replicate the query in the network, and the computation cost to evaluate the query matching condition.

We propose a publish/subscribe mechanism called PUB-2-SUB which, like the gossip-based approach, does not change the connectivity of the given network, but is aimed at a much better performance. PUB-2-SUB allows any number of independent publish/subscribe applications to run simultaneously. It is based on two key design components: the virtualization component and the indexing component. The virtualization component assigns to each node a unique binary string called a virtual address so that the virtual addresses of all the nodes form a prefix tree. Based on this tree, each node is assigned a unique zone partitioned from the universe of binary strings. The indexing component hashes queries and publications to binary strings and, based on their overlapping with the node zones, chooses subscription and notification paths appropriately and deterministically.

Because PUB-2-SUB is based on directed routing, it has the potential to be more efficient than the gossip-based approach. Our evaluation study shows that PUB-2-SUB results in lower storage and communication costs than BubbleStorm [19] – a recent gossip-based search technique. In terms of computation cost, PUB-2-SUB requires only a node to evaluate its local queries to find those matching a given information publication. The proposed technique also incurs small notification delay and is robust under network failures.

The remainder of the paper is organized as follows. We introduce the concept of PUB-2-SUB in more detail in Section 2, followed by a discussion on the evaluation results in Section 3. The paper is concluded in Section 4 with pointers to our future work.

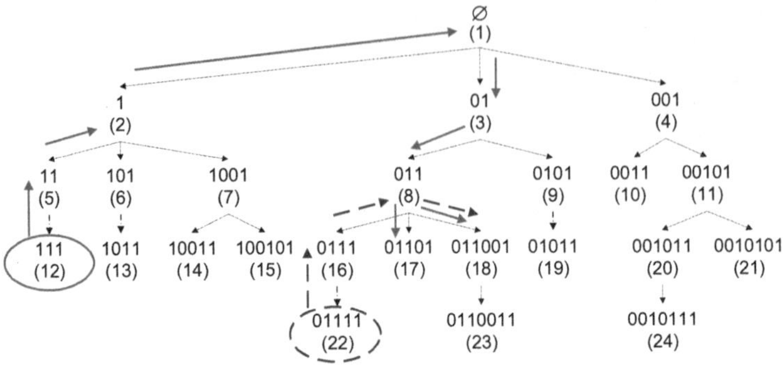


**Fig. 1.** Virtual address instance inititated by node 1 (the VAs of all the nodes form a prefix tree): solid-bold path represents the subscription path of query ['0110001', '0110101']; dashed-bold path represents the notification path of event <'0110010'>

## 2 PUB-2-SUB: The Proposed Solution

Consider a cooperative P2P network $\{S_1, S_2, ..., S_n\}$ that is constructed and maintained according to its built-in underlying protocols. The nodes should remain functional as much as possible although there may be failures that cannot be avoided, and whenever a new node joins or a failure occurs we assume that this network can re-organize itself. We are required not to modify the existing connectivity of the network; all communication must be via the provided links. PUB-2-SUB is based on two key design components, the virtualization component and the indexing component, which are described below.

### 2.1 Virtualization

A virtualization procedure can be initiated by any node to result in a "virtual address instance" (VA-instance), where each node is assigned a virtual address (VA) being a binary string chosen from $\{0,1\}^*$. Suppose that the initiating node is $S^*$. In the corresponding VA-instance, denoted by INSTANCE($S^*$), we denote the VA of each node $S_i$ by $VA(S_i : S^*)$. To start the virtualization, node $S^*$ assigns itself $VA(S^* : S^*) = \varnothing$ and sends a message inviting its neighbor nodes to join INSTANCE($S^*$). A neighbor $S_i$ ignores this invitation if already part of the instance. Else, by joining, $S_i$ is called a "child" of $S^*$ and receives from $S^*$ a VA that is the shortest string of the form $VA(S^* : S^*)$ + '0*1' unused by any other child node of $S^*$. Once assigned a VA, node $S_i$ forwards the invitation to its neighbor nodes and the same VA assignment procedure repeats. In generalization, the rule to compute the VA for a node $S_j$ that accepts an invitation from node $S_i$ is: $VA(S_j : S^*)$ is the shortest string of the form $VA(S_i : S^*)$ + '0*1' unused by any current child node of $S_i$.

Eventually, every node is assigned a VA and the VAs altogether form a prefix-tree rooted at node $S^*$. We call this tree a VA-tree and denote it by $TREE(S^*)$.

For example, Figure 1 shows the VA-tree with VAs assigned to the nodes as a result of the virtualization procedure initiated by node 1. It is noted that the links of this spanning tree already exist in the original network (we do not create any new links). In this figure, the nodes' labels (1, 2, ..., 24) represent the order they join the VA-tree. Each time a node joins, its VA is assigned by its parent according to the VA assignment rule above. Thus, node 2 is the first child of node 1 and given $VA(2:1) = VA(1:1) +$ '1' = '1', node 3 is the next child and given $VA(3:1) = VA(1:1) +$ '01' = '01', and node 4 last and given $VA(4:1) = VA(1:1) +$ '001' = '001'. Other nodes' VAs are assigned similarly. For example, consider node 18 which is the third child of node 8 (VA '011'). The VA of node 18 is the shortest binary string that is unused by any other child node of node 8 and of the form $VA(8:1) +$ '0*1'. Because the other children 16 and 17 already occupy '0111' and '01101', node 18's VA is '011001'.

A VA-tree resembles the shortest-delay spanning tree rooted at the initiating node; i.e., the path from the root to a node should be the quickest path among those paths connecting them. It can be built quickly because only a single broadcast of the VA invitation is needed to assign VAs to all the nodes. To help with indexing, in $INSTANCE(S^*)$, each node $S_i$ is associated with a unique "zone", denoted by $ZONE(S_i : S^*)$, consisting of all the binary strings *str* such that: (i) $VA(S_i : S^*)$ is a prefix of *str*, and (ii) no child of $S_i$ has VA a prefix of *str*. In other words, among all the nodes in the network, node $S_i$ is the one whose VA is the maximal prefix of *str*. We call $S_i$ the "designated node" of *str* and use $NODE(str : S^*)$ to denote this node. For example, using the virtual instance $TREE(1)$ in Figure 1, the zone of node 11 (VA '00101') is the set of binary strings '00101', '001010', and all the strings of the form '0010100...', for which node 11 is the designated node. The following properties can be proved, which are important to designing our indexing component:

1. $ZONE(S_i : S^*) \cap ZONE(S_j : S^*) \neq \emptyset$, for every $i \neq j$
2. $\bigcup_{i=1}^{n} ZONE(S_i : S^*) = \{0,1\}^*$
3. $\bigcup \{ZONE(S' : S^*) \mid S'$ is $S_i$ or a descendant of $S_i\} = \{str \in \{0,1\}^* \mid VA(S_i : S^*)$ is a prefix of $str\}$, for every $i$

## 2.2 Indexing

For each publish/subscribe application under deployment, the information of interest is assumed to have a fixed number of attributes called the dimension of this application. PUB-2-SUB supports any dimension and allows multiple applications to run on the network simultaneously, whose dimension can be different from one another's. We use the term "event" to refer to some new information that a node wants to publish. The queries of interest are those that specify a lower-bound and an upper-bound on each event attribute. For ease of presentation, we assume that events are unidimensional. The idea can easily be extended for the case of multidimensionality (see our extended work [22]).

Without loss of generality, we represent an event $x$ as a $k$-bit binary string (the parameter $k$ should be chosen to be larger than the longest VA length in the network). A query $Q$ is represented as an interval $Q = [q_l, q_h]$, where $q_l$, $q_h \in \{0,1\}^k$, subscribing to all events $x$ belonging to this interval (events are "ordered" lexicographically). As an example, if $k = 3$, the events matching a query [‘001’, ‘101’] are {‘001’, ‘010’, ‘011’,‘100’, ‘101’}.

Supposing that every node has been assigned a VA as a result of a virtualization procedure initiated by a node $S^*$, we propose that

**Query subscription:** Each query $Q$ is stored at every node $S_i$ such that the zone of this node $ZONE(S_i : S^*)$ intersects with $Q$.

**Event notification:** Each event $x$ is sent to $NODE(x : S^*)$ – the designated node of string $x$. It is guaranteed that if $x$ satisfies $Q$ then $Q$ can always be found at node $NODE(x : S^*)$ (because this node's zone must intersect $Q$).

Figure 1 shows an example with $k = 7$. Suppose that node 12 wants to subscribe a query $Q =$ [‘0110001’, ‘0110101’], thus looking to be notified upon any of the following events {‘0110001’, ‘0110010’, ‘0110011’, ‘0110100’, ‘0110101’}. Therefore, this query will be stored at nodes {8, 17, 18}, whose zone intersects with $Q$. For example, node 8's zone intersects $Q$ because they both contain ‘0110001’. The path to disseminate this query is $12 \rightarrow 5 \rightarrow 2 \rightarrow 1 \rightarrow 3 \rightarrow 8 \rightarrow \{17, 18\}$ (represented by the solid arrow lines in Figure 1). Now, suppose that node 22 wants to publish an event $x = <$‘0110010’$>$. Firstly, this event will be routed upstream to node 8 – the *first* node that is a prefix with ‘0110010’ (path $22 \rightarrow 16 \rightarrow 8$). Afterwards, it is routed downstream to the designated node $NODE$(‘0110010’:1), which is node 18 (path $8 \rightarrow 18$). Node 18 searches its local queries to find the matching queries. Because query $Q =$ [‘01011111’, ‘01100011’] is stored at node 18, this query will also be found.

The storage and communication costs for a query's subscription depend on its range; the wider the range, the larger costs. For an event, the communication cost measured as the number of hops traveled to publish an event is $O(h)$ where $h$ is the tree height ($h = O(\sqrt{n})$ in most cases). The delay to notify a matching subscriber is the time to travel this path; hence, also $O(h)$. The computation cost should be small because only one node – the designated node $NODE(x : S^*)$ – needs to search its stored queries to find those matching $x$. Our evaluation study in Section 3 indeed finds these costs reasonably small.

## 2.3 Update Methods

There may be changes in the network such as when a new node is added or an existing node fails. Supposing that the network is virtualized according to the VA-instance $INSTANCE(S^*)$, PUB-2-SUB addresses these changes as follows.

**Node Addition:** Consider a new node $S_{new}$ that has just joined the network according to the network's underlying join protocol. As a result, it is connected to a number of neighbors. We need to add this node to the VA-instance. First, this node communicates with its neighbors and asks the neighbor $S_{neighbor}$ with the minimum tree depth to be its parent; tie is broken by choosing the one with fewest children. This strategy helps keep the tree as balanced as possible so its height can be short and workload fairly distributed among the nodes. The neighbor will then assign $S_{new}$ a VA that is the shortest unused binary string of the form $VA(S_{neighbor} : S^*)$ + '0*1'. Because $ZONE(S_{neighbor} : S^*)$ is changed, the next task is for the parent node $S_{neighbor}$ to delete those queries that do not intersect $ZONE(S_{neighbor} : S^*)$ anymore. Also, this parent node needs to forward to $S_{new}$ the queries that intersect $ZONE(S_{new} : S^*)$.

**Node Removal:** When a node fails to function, it is removed from the network according to the underlying maintenance protocol. This removal however affects the connectedness of the VA-instance in place. Because the VAs of the child nodes are computed based on that of the parent node, the child nodes of the departing node need to find a new parent so the VA-instance remains valid. Consider such a child node $S_{child}$. This node selects a new parent among its neighbors. The new parent, say node $S_{parent}$, computes a new VA for $S_{child}$ (similar to node addition). Then, $S_{child}$ re-computes the VAs for its children and informs them of the changes. Each child node follows the same procedure recursively to inform all its descendant nodes downstream. The query transfer/forwarding from $S_{parent}$ to $S_{child}$ and, if necessary, from $S_{child}$ to the descendant nodes of $S_{child}$ is similar to the case of adding a new node.

In addition, because each descendant node $S_i$ of the removed node is now assigned a new VA and thus a new zone, the queries that are stored at $S_i$ before the VA adjustment may no longer intersect its new zone. These queries can be either deleted or re-subscribed to the network depending on the priority we can set at the first time they are subscribed to the network. If a query is marked as "high-priority", it is stored in the network permanently until the subscriber determines to unsubscribe it (the unsubscription procedure is similar to the subscription procedure). On the other hand, if a query is marked as "low-priority", it is associated with a lease time after which the query will expire and be deleted. How to implement these two types of priority is determined by the application developer.

The worst case with node removal is when the root node fails in which we have no way to recover other than rebuilding the entire VA instance. To avoid this unfortunate case, we propose that the root of a VA-instance be a dedicated node deployed by the network administrator and thus we can assume that this node never fails. This requirement can easily be realized in practice.

## 3 Evaluation Study

We conducted a preliminary study to evaluate the performance of PUB-2-SUB. This study was based on a simulation on a 1000-node network with 2766 links, whose topology was a Waxman uniform random graph generated with the BRITE generator [23]. A random node was chosen to be the root for the VA instance. Ten random choices for this root node were used with the simulation and the results averaged over these ten choices are discussed in this section.

An event was represented as a $k$-bit string and a query an arbitrary interval of $k$-bit strings. A query or event was initiated by a random node chosen uniformly. To cover a large domain of possible events, we set $k$ to 50 bits, thus able to specify up to $2^{50}$ different events. From this domain, 10,000 events were chosen uniformly in random. The subscription load consisted of 10,000 queries, each having a range chosen according to the following Zipf's law: in the set of possible ranges $\{2^0, 2^1, ..., 2^{49}\}$, range $2^i$ is picked with probability $\frac{1/i^{\alpha}}{\sum_{j=1}^{50}(1/j^{\alpha})}$ . We considered two models: $\alpha = 0$ (uniform distribution) and $\alpha = 0.8$ (heavy-tail distribution where a vast majority of the queries are specific).

We evaluated PUB-2-SUB in the following aspects: subscription efficiency, notification efficiency, notification delay, and failure effect. We also compared PUB-2-SUB to two versions of BubbleStorm [19] – a recent search technique designed for unstructured P2P networks: (1) BubbleStorm-64%: each query or event is sent to $\sqrt{n}$ nodes, resulting in a 64% query/event matching success rate (when there is no failure), and (2) BubbleStorm-98%: each query or event is sent to $2\sqrt{n}$ nodes, resulting in a 98% success rate (when there is no failure).

### 3.1 Subscription Efficiency

This efficiency is measured in terms of the storage cost and the communication cost. The storage cost is computed as the number of nodes that store a given query, and the communication cost as the number of hops (nodes) that have to forward this query during its subscription procedure. Figure 2(a) and Figure 2(b) show these costs respectively for every query, which are sorted in the non-decreasing order. It is observed for either cost that all queries result in a small cost except for a very few with a high cost. These high-cost queries are those with long ranges. As such they intersect with the zones of many nodes and thus have to travel more to be stored at these nodes. Despite so, on average, a query is replicated at only 15 nodes (uniform case) and 4.3 nodes (heavy-tail case), resulting in a communication cost of 25.6 hops (uniform case) and 15 hops (heavy-tail case).

These costs are much lower than that incurred by BubbleStorm. Figure 2(c) shows that BubbleStorm-64% stores an average query at 33 nodes, more than twice the storage cost of PUB-2-SUB (uniform) and eight times the cost of PUB-2-SUB (heavy-tail). The storage cost of BubbleStorm-98% is even higher. In comparison on the communication cost, as seen in Figure 2(d), a query in BubbleStorm-64% and BubbleStorm-98% has to travel 33 hops and 66 hops, respectively, which are also higher than the communication cost of PUB-2-SUB.

### 3.2 Notification Efficiency

This efficiency is measured in terms of the communication cost and the computation cost. The communication cost is computed as the number of hops (nodes) that have to forward a given event during its publication procedure, and the computation cost as the number of queries evaluated to match this event.

Because an event is routed based on the nodes' VAs, its communication cost is independent of the query model used, uniform or heavy-tail. Figure 3(a) shows

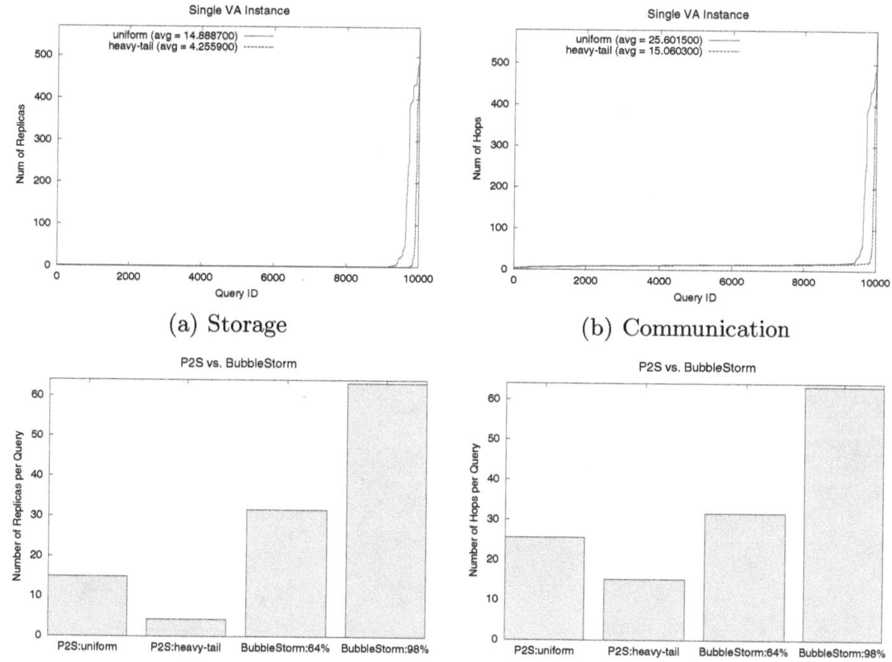


(a) Storage

(b) Communication

(c) Storage: comparison

(d) Communication: comparison

**Fig. 2.** Query subscription costs

that this cost is distributed normally from zero hop (best-case) to 25 hops (worst-case), having an average of 12 hops. The communication cost is also much lower (by approximately three times at least) when compared to BubbleStorm (Figure 3(c)). Together with the study on the subscription efficiency it is evident that PUB-2-SUB clearly outperforms BubbleStorm in both storage cost and communication cost. In terms of computation cost, Figure 3(b) shows that in the worst case about 1400 queries are evaluated to find all those matching a given event; i.e., only 14% of the entire query population. On average, the computation cost is only 563 query evaluations (uniform case) and 458 query evaluations (heavy-tail case), corresponding to 5.63% and 4.58% of the query population, respectively.

### 3.3 Notification Delay

When an event is published, there may be more than one queries subscribing to this event. To represent the notification delay for each (event, query) matching pair, we compute the ratio $a/b$ where $a$ is the hopcount-based distance the event has to travel from the publisher node to the subscriber node and $b$ is the hopcount-based distance directly between these two nodes. This ratio is at least 1.0 because even if the publisher knows the subscriber, it must already take $b$ hops to send the event to the subscriber. In practice, because the publisher

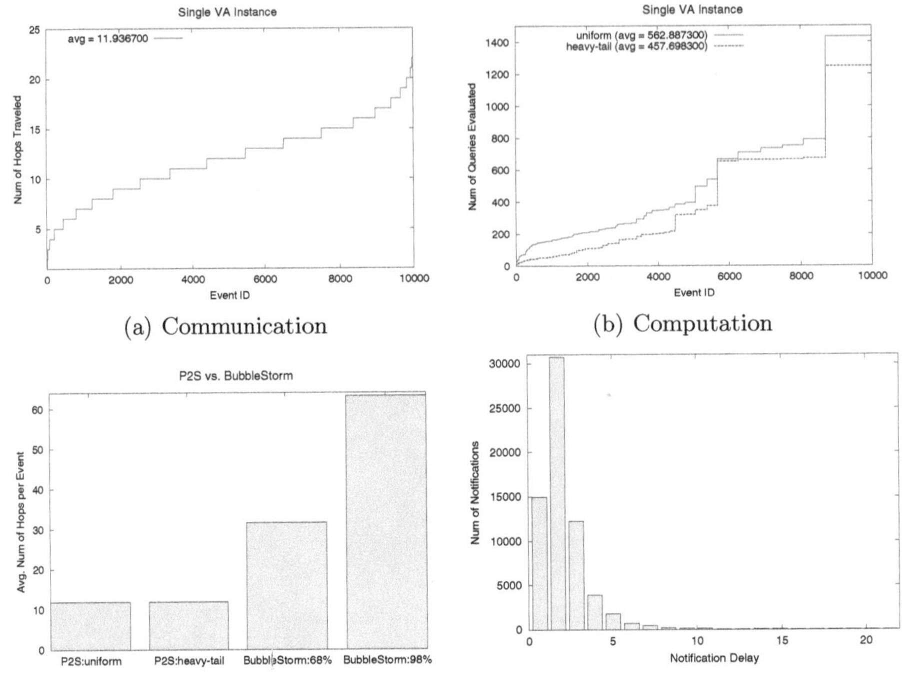


(a) Communication

(b) Computation

(c) Communication: Comparison

(d) Histogram of notification delay

**Fig. 3.** Event notification costs

and the subscriber initially do not know each other, it is impossible to obtain a perfect 1.0 ratio.

Figure 3(d) plots the histogram of notification delay incurred by PUB-2-SUB. Approximately, 70% of the notifications have a delay not exceeding 2.0 (i.e., twice the perfect delay) and 90% have a delay not exceeding 3.0 (i.e., three times the perfect delay). Thus, despite a few (event, query) pairs with high notification delay, a vast majority of events can notify their matching queries reasonably quickly.

### 3.4 Failure Effect

When a node stops functioning, an event may fail to notify its subscribers. To evaluate the failure effect, we compute "recall" – the percentage of the returned events that match a given query out of all the matching events. We consider the case where 10% of the nodes fail simultaneously and the case where 30% fail.

Figure 4(a) shows the results for the uniform-query-model case, where it is observed that 75% of the queries are successfully notified by all the matching events (i.e., recall = 100%) even when 30% of the nodes fail. The difference between the 10%-fail case and the 30%-fail case is that in the latter case most of the remaining queries (the remaining 25%) fail to receive any matching event while in the former case about half of the queries do not receive any matching

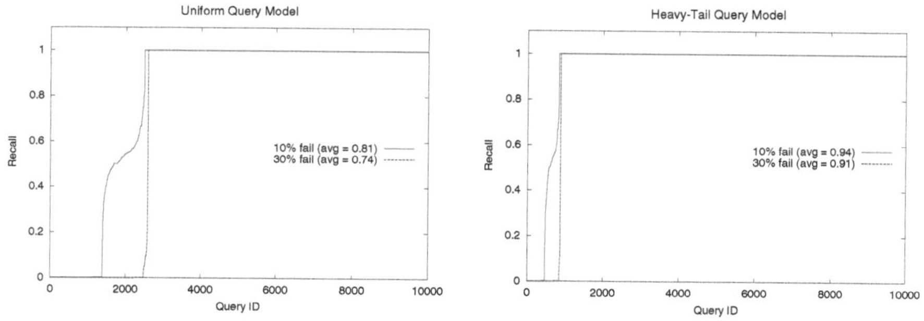


(a) Uniform query model

(b) Heavy-tail query model

**Fig. 4.** Effect of Failures: 10% of nodes fail and 30% of nodes fail

event and the other half receiving at least some portion of the matching events. On average, the recall for the 10%-fail case is 81%, and for the 30%-fail case, 74%. Higher recall is obtained when the query model is heavy-tail (see Figure 4(b)). The average recall is 91% when 30% of the nodes fail and 94% when 10% fail. The results are encouraging because in practice the query range should follow the heavy-tail model more often than the uniform model. This study is demonstrative of PUB-2-SUB's sustainable effectiveness when a large portion of the network fails to operate.

## 4 Conclusions

We have proposed a publish/subscribe mechanism, called PUB-2-SUB, which can be integrated into any unstructured P2P network. Using PUB-2-SUB, any number of content-based publish/subscribe applications can be deployed simultaneously. Unlike the gossip-based approach previously recommended for unstructured networks, the proposed technique is based on directed routing and incurs less storage and communication costs. This is evident in an evaluation study in which PUB-2-SUB is compared to a representative technique of the other approach. It is also found that our technique results in low computation cost and low notification delay and remains highly effective in cases when many nodes in the network stop to function.

We do not recommend PUB-2-SUB for use in highly dynamic networks with the nodes being on and off frequently. Instead, PUB-2-SUB works best for P2P-based cooperative networks in which the nodes are supposed to be functional most of the time and failures should not happen too often. Thus, data grid networks and institutional collaborative networks can take full advantage of the proposed technique. The work described in this paper remains preliminary. More evaluation is needed for our future work in which we will also include investigation into better methods addressing failures and load balancing.

## Acknowledgment

This work was funded in part by the US National Science Foundation under Grants CNS-0615055 and CNS-0753066 and by the University of Massachusetts under the Proposal Development Grant. The authors are thankful for these supports.

## References

1. Sun, X., Liu, J., Yao, E., Chen, X.: A scalable p2p platform for the knowledge grid. IEEE Trans. on Knowl. and Data Eng. 17(12), 1721–1736 (2005)
2. Teranishi, Y., Tanaka, H., Ishi, Y., Yoshida, M.: A geographical observation system based on p2p agents. In: PERCOM 2008: Proceedings of the 2008 Sixth Annual IEEE International Conference on Pervasive Computing and Communications, Washington, DC, USA, pp. 615–620. IEEE Computer Society, Los Alamitos (2008)
3. Shalaby, N., Zinky, J.: Towards an architecture for extreme p2p applications. In: Parallel and Distributed Computing and Systems Conference (PDCS), Cambridge, MA (November 2007)
4. Cai, M., Frank, M., Chen, J., Szekely, P.: Maan: A multi-attribute addressable network for grid information services. In: GRID 2003: Proceedings of the 4th International Workshop on Grid Computing, Washington, DC, USA, p. 184. IEEE Computer Society, Los Alamitos (2003)
5. Hanson, E.N., Carnes, C., Huang, L., Konyala, M., Noronha, L., Parthasarathy, S., Park, J.B., Vernon, A.: Scalable trigger processing. In: Proceedings of the 15th International Conference on Data Engineering, Sydney, Australia, pp. 266–275. IEEE Computer Society, Los Alamitos (1999)
6. Chen, J., DeWitt, D.J., Tian, F., Wang, Y.: Niagaracq: a scalable continuous query system for internet databases. SIGMOD Rec. 29(2), 379–390 (2000)
7. Fabret, F., Jacobsen, H.A., Llirbat, F., Pereira, J., Ross, K.A., Shasha, D.: Filtering algorithms and implementation for very fast publish/subscribe systems. In: SIGMOD 2001: Proceedings of the 2001 ACM SIGMOD international conference on Management of data, pp. 115–126. ACM Press, New York (2001)
8. Castro, M., Druschel, P., Kermarrec, A., Rowstron, A.: SCRIBE: A large-scale and decentralized application-level multicast infrastructure. IEEE Journal on Selected Areas in communications (JSAC) 20(8), 1489–1499 (2002)
9. Gupta, A., Sahin, O.D., Agrawal, D., Abbadi, A.E.: Meghdoot: content-based publish/subscribe over p2p networks. In: Jacobsen, H.-A. (ed.) Middleware 2004. LNCS, vol. 3231, pp. 254–273. Springer, Heidelberg (2004)
10. Terpstra, W.W., Behnel, S., Fiege, L., Zeidler, A., Buchmann, A.P.: A peer-to-peer approach to content-based publish/subscribe. In: DEBS 2003: Proceedings of the 2nd international workshop on Distributed event-based systems, pp. 1–8. ACM Press, New York (2003)
11. Aekaterinidis, I., Triantafillou, P.: Internet scale string attribute publish/subscribe data networks. In: CIKM 2005: Proceedings of the 14th ACM international conference on Information and knowledge management, pp. 44–51. ACM Press, New York (2005)
12. Tran, D.A., Nguyen, T.: Publish/subscribe service in can-based p2p networks: Dimension mismatch and the random projection approach. In: IEEE Conference on Computer Communications and Networks (ICCCN 2008), Virgin Island, USA. IEEE Press, Los Alamitos (2008)

13. Bianchi, S., Felber, P., Gradinariu, M.: Content-based publish/subscribe using distributed r-trees. In: Kermarrec, A.-M., Bougé, L., Priol, T. (eds.) Euro-Par 2007. LNCS, vol. 4641, pp. 537–548. Springer, Heidelberg (2007)
14. Ratnasamy, S., Francis, P., Handley, M., Karp, R., Shenker, S.: A scalable content addressable network. In: ACM SIGCOMM, San Diego, CA, August 2001, pp. 161–172 (2001)
15. Stoica, I., Morris, R., Karger, D., Kaashock, M., Balakrishman, H.: Chord: A scalable peer-to-peer lookup protocol for internet applications. In: ACM SIGCOMM, San Diego, CA, August 2001, pp. 149–160 (2001)
16. Rowstron, A., Druschel, P.: Pastry: Scalable, distributed object location and routing for large-scale peer-to-peer systems. In: Guerraoui, R. (ed.) Middleware 2001. LNCS, vol. 2218, pp. 329–350. Springer, Heidelberg (2001)
17. Zhao, B.Y., Huang, L., Stribling, J., Rhea, S.C., Joseph, A.D., Kubiatowicz, J.: Tapestry: A resilient global-scale overlay for service deployment. IEEE Journal on Selected Areas in Communications 22(1) (January 2004)
18. Pugh, W.: Skip lists: A probabilistic alternative to balanced trees. Communications of the ACM 33, 668–676 (1990)
19. Terpstra, W.W., Kangasharju, J., Leng, C., Buchmann, A.P.: Bubblestorm: resilient, probabilistic, and exhaustive peer-to-peer search. In: SIGCOMM 2007: Proceedings of the 2007 conference on Applications, technologies, architectures, and protocols for computer communications, pp. 49–60. ACM, New York (2007)
20. Wong, B., Guha, S.: Quasar: A Probabilistic Publish-Subscribe System for Social Networks. In: Proceedings of The 7th International Workshop on Peer-to-Peer Systems (IPTPS 2008), Tampa Bay, FL (February 2008)
21. Gkantsidis, C., Mihail, M., Saberi, A.: Random walks in peer-to-peer networks: algorithms and evaluation. Perform. Eval. 63(3), 241–263 (2006)
22. Tran, D.A., Pham, C.: Enabling publish/subscribe services in cooperative p2p networks. Technical Report, University of Massachusetts Boston (February 2009)
23. Medina, A., Lakhina, A., Matta, I., Byers, J.: Brite: An approach to universal topology generation. In: MASCOTS 2001: Proceedings of the 9th International Symposium in Modeling, Analysis and Simulation of Computer and Telecommunication Systems, Washington, DC, USA, p. 346. IEEE Computer Society, Los Alamitos (2001)

# Minimum-Delay Load-Balancing through Non-parametric Regression

Federico Larroca and Jean-Louis Rougier

TELECOM ParisTech, Paris, France
46 rue Barrault F-75634 Paris Cedex 13
firstname.lastname@telecom-paristech.fr

**Abstract.** Network convergence and new applications running on end-hosts result in increasingly variable and unpredictable traffic patterns. By providing origin-destination pairs with several possible paths, load-balancing has proved itself an excellent tool to face this uncertainty. Formally, load-balancing is defined in terms of a convex link cost function of its load, where the objective is to minimize the total cost. Typically, the link queueing delay is used as this cost since it measures its congestion. Over-simplistic models are used to calculate it, which have been observed to result in suboptimal resource usage and total delay. In this paper we investigate the possibility of learning the delay function from measurements, thus converging to the actual minimum. A novel regression method is used to make the estimation, restricting the assumptions to the minimum (e.g. delay should increase with load). The framework is relatively simple to implement, and we discuss some possible variants.

**Keywords:** Traffic Engineering, Wardrop Equilibrium, Convex Non-parametric Least Squares, Next Generation Internet.

## 1 Introduction

As network services and Internet applications evolve, the traffic is becoming increasingly complex and dynamic. The convergence of data, telephony and television services on an all-IP network as well as user-mobility (which implies service-mobility) directly translates into a much higher variability and complexity of the traffic injected into the network. Moreover, new architectures with relatively low link capacities (such as Wireless Mesh Networks) cannot foresee overprovisioning as a viable solution. To cope with both the traffic increasing dynamism and the need for cost-effective solutions, a self-managing network architecture is required.

*Dynamic load-balancing* has proved itself a very efficient solution to the above issues [1, 2, 3]. If an origin-destination (OD) pair is connected by several paths, the problem is simply how to distribute its traffic among these paths in order to achieve a certain objective. In these dynamic schemes, paths are configured a priori and the portion of traffic routed through each of them depends on the current demand and network condition. Since the considered time-scale is in the

L. Fratta et al. (Eds.): NETWORKING 2009, LNCS 5550, pp. 782–794, 2009.

order of minutes, a distributed algorithm is a requirement in this kind of schemes. Mathematically, the problem is generally defined in terms of a certain convex link-cost function of the link load ($f_l(\rho_l)$), and the objective is to minimize the total network cost. For instance, if the cost function is the link utilization, the objective could be to minimize the maximum utilization on the network. Due to its simplicity, this particular objective has been considered in several routing [4] and load-balancing [2, 3] mechanisms. However, it may result in inefficient resource usage [5]. Another possible approach is to define $f_l(\rho_l)$ as a measure of the congestion in the link and minimize its sum over all links. A typical link-cost function is the queueing delay function of an *M/M/1* queue [1]. However, such simplistic model may result in bigger delays [6] and a greater maximum utilization [5] with respect to the actual minimum.

In this paper we study the possibility of designing a load-balancing mechanism that makes *no* assumptions on the delay function (except for some natural hypothesis on its shape). We will assume that $f_l(\rho_l)$ exists, but is not known, and we will estimate it from measurements. The proposed framework allows to converge to the actual minimum-delay configuration (or a very good approximation of it), thus maximizing performance. Besides, its implementation is relatively easy. The required measurements (mean incoming rate and queue size) are readily available in most routers, and the greatest upgrade required of routers is to enable load-balancing itself. Moreover, adaptations to the estimation technique are made that assure convergence of the load-balancing algorithm. The parametrization of this algorithm is also discussed.

The next section discusses the network model, our particular objective, and the distributed optimization algorithm we used. Section 3 presents the non-parametric regression algorithm we used to estimate $f_l(\rho_l)$ and some necessary adaptations to jointly use it with the distributed optimization algorithm. Implementation issues and some packet-level simulations that verify the performance of the framework are discussed in Sec. 4. The paper is concluded in Sec. 5.

## 2 Greedy Load-Balancing

### 2.1 Network Model

The network is defined as a graph $G = (V, E)$. In it there are $S$ so-called *commodities* (or OD pairs), indexed by $s$ and specified in terms of the triplet $o_s$, $q_s$ and $d_s$; i.e. origin and destination nodes, and a fixed demand of traffic from the former to the latter. Each commodity $s$ can use $n_s$ paths connecting $o_s$ to $q_s$ (each noted as $P_{si}$), and can distribute its total demand arbitrarily among them. Commodity $s$ sends an amount $d_{si}$ of its traffic through path $P_{si}$, where $d_{si} \geq 0$ and $\sum d_{si} = d_s$. This traffic distribution induces the demand vector $d = (d_{si})$.

Given the demand vector, the total load on link $l$ is then $\rho_l = \sum_s \sum_{i:l \in P_{si}} d_{si}$. The presence of this traffic on the link induces a certain mean queueing delay given by the non-decreasing function $D_l(\rho_l)$. The total delay of path $P$ is defined as $D_P = \sum_{l:l \in P} D_l(\rho_l)$. As a measure of the congestion in the network, we shall use the *mean total delay* $D(d)$ defined as:

$$D(d) = \sum_{s=1}^{S}\sum_{i=1}^{n_s} d_{si} D_{P_{si}} = \sum_{l=1}^{L} D_l(\rho_l)\rho_l := \sum_{l=1}^{L} f_l(\rho_l)$$

That is to say, a weighted mean delay, where the weight for each path is how much traffic is sent through it, or in terms of the links, the weight of each link is how much traffic is traversing it. Note that, by Little's law, $f_l(\rho_l)$ is proportional to the average number of bytes in the queue of link $l$. We will then use this last value as $f_l(\rho_l)$ which is easier to measure than the queueing delay.

We are now in conditions to write the problem explicitly:

$$\underset{d}{\text{minimize}} \sum_{l=1}^{L} f_l(\rho_l) \qquad \text{s.t.} \;\; d_{si} \geq 0 \;\; \sum_{i=1}^{n_s} d_{si} = d_s \tag{1}$$

Note that no explicit constraint on $\rho_l$ was made. This is assumed to be implicitly included in the link-cost function. For instance, $f_l(\rho_l)$ goes to infinity (or a relatively high value) as $\rho_l$ reaches $c_l$ (the link capacity) and remains at infinity after this point. It should also be noted that in the framework described above the destination for a commodity is not necessarily a single node (e.g. two gateways to the internet may be seen as a single destination).

### 2.2 Wardrop Equilibrium

In this section we present and discuss how to solve problem (1) in a distributed fashion. In particular, we will consider mechanisms where each commodity greedily minimizes a certain cost function of its paths ($\phi_P$), which require minimum coordination. This context constitutes an ideal case study for game theory, and is known as *Routing Game* in its lingo [7]. The case in which the path cost is the sum over its links of a positive non-decreasing link-cost function of the load ($\phi_P = \sum_{l:l \in P} \phi_l(\rho_l)$) is known as *Congestion Routing Game* and has several important properties such as uniqueness of the equilibrium.

In a routing game, commodities are assumed to be constituted by infinitely many agents, each controlling through which path an infinitesimal amount of traffic is sent. In this context the division $d_{si}/d_s$ represents the portion of agents of commodity $s$ that have $P_{si}$ as their choice. If every agent acts selfishly, then the system will be at equilibrium when no agent can decrease its cost by changing its path choice. This defines what is known as a *Wardrop Equilibrium* (WE) [8]. Formally, a demand vector is a WE if for each commodity $s = 1 \ldots S$ and for each path $P_{si}$ with $d_{si} > 0$ it holds that $\phi_{P_{si}} \leq \phi_{P_{sj}}$ for all $P_{sj}$ with $j = 1, .., n_s$.

It can be proved that a WE results in a local minimum of the so-called potential function $\Phi(d) = \sum_{l=1}^{L} \int_0^{\rho_l} \phi_l(x)dx$ [7]. This means that the WE of a congestion routing game where the link cost is the derivative of $f_l(\rho_l)$ ($\phi_l(\rho_l) = f_l'(\rho_l)$) is the solution of (1) (if $f_l(\rho_l)$ is convex, a local minimum is actually the global minimum). A distributed algorithm that converges towards such equilibrium is described in the following subsection.

### 2.3 REPLEX: Exploration-Replication Policy

The concept of Wardrop Equilibrium was first proposed in the context of transportation to characterize the equilibrium of users who greedily want to minimize their travel time. In this context, users are assumed rational and their behavior is the mechanism through which the equilibrium is attained. In our case however, routers make the choice for every user (i.e. packets). It is then necessary to specify an algorithm that when independently ran in every router, the equilibrium is achieved as fast as possible and without oscillations. In [9] the authors present such mechanism and use it to design a load-balancing scheme in [3]. Below, we can see the algorithm that each commodity executes in turn (where $p_{si}$ is the portion of demand $d_s$ routed through path $P_{si}$).

$p'_s \leftarrow p_s$
**for** every pair of paths $P_{si}, P_{sj}$ of commodity $s$ **do**
  **if** $\phi_{P_{si}} > \phi_{P_{sj}}$ **then**
    $\delta \leftarrow \lambda \left( (1-\beta)\, p_{sj} + \frac{\beta}{n_s} \right) \frac{\phi_{P_{si}} - \phi_{P_{sj}}}{\phi_{P_{si}} + \alpha}$
    $p'_{si} \leftarrow p'_{si} - \delta$
    $p'_{sj} \leftarrow p'_{sj} + \delta$
  **end if**
**end for**
$p_s \leftarrow p'_s$

It can be seen that the portion of traffic that changes its path in a turn is proportional to the relative gain in path delay, and in a weighted mean between the portion of traffic using the new path (called *proportional sampling* in the algorithm) and $1/n_s$ (*uniform sampling*). The algorithm converges to the WE as long as $\lambda \leq k/r$, where $k > 0$ is a suitable constant and $r$ is an upper-bound to the relative slope of all $\phi_l(\rho_l)$, which is defined as follows [9]:

**Definition 1.** *A differentiable cost function* $\phi_l(x)$ *has* relative slope $r$ *at* $x$ *if* $\phi'_l(x) \leq r\phi_l(x)/x$. *A cost function has relative slope* $r$ *if it has relative slope* $r$ *over the entire range* $[0, 1]$.

Intuitively, migration from one path to the other should be slow if the cost function has abrupt changes. On the other hand, if the cost function is relatively "soft", changes may be faster. As discussed in [3], the values of $\alpha$ and $\beta$ are not very influential, and $\beta = 0.1$, $\alpha = 0$ turned out to be good choices.

## 3 Non-parametric Regression With Shape Restrictions

### 3.1 Convex Non-parametric Least Squares

The problem we address now is how to learn $f_l(\rho_l)$ from measurements (we are interested in its derivative, $\phi_l(\rho_l)$, but the queue size $f_l(\rho_l)$ is the observable quantity). For the sake of clarity we will concentrate on the problem for a single link, so we shall omit the sub-index $l$. We are given $n$ pairs of observations

$(\rho_1, Y_1), (\rho_2, Y_2), \ldots, (\rho_n, Y_n)$ (also called *training set*), where the *response variable* $Y$ (the measured mean queue size) is related to the *covariate* $\rho$ (the link load) by the equation $Y_i = f(\rho_i) + \epsilon_i$ for $i = 1, \ldots, n$.

The function $f(\rho)$ is now called the *regression function* and the measurement errors $\epsilon = (\epsilon_1, \ldots, \epsilon_n)'$ are assumed to be uncorrelated random variables with $\mathbb{E}(\epsilon) = 0$ and $\text{Var}(\epsilon) = \sigma^2 < \infty$. The problem is to "learn" $f(\rho)$ from the observations in the training set and obtain an estimation $\widehat{f}(\rho)$. The idea is to restrict the assumptions on its functional form to the minimum. So far, we have only three necessary requirements: (i) $f(\rho)$ should clearly be increasing (ii) $\phi(\rho)$ should be non-decreasing, so $f(\rho)$ should be convex (iii) $\phi(\rho)$ should have a finite relative slope in order to make REPLEX work correctly (and probably all distributed optimization algorithms).

We will now consider the first two requirements, which are by far the most restrictive. For this, we turn our attention to the recent work of Kuosmanen [10]. Let $\mathcal{F}$ be the set of continuous, monotonic increasing and globally convex functions. The *Convex Nonparametric Least Squares* (CNLS) problem is to find $\widehat{f} \in \mathcal{F}$ that minimizes the sum of squares of the residuals:

$$\min_f \sum_{i=1}^{n} (Y_i - f(\rho_i))^2 \quad \text{s.t. } f \in \mathcal{F} \tag{2}$$

Problem (2) is very difficult to solve due to the size of $\mathcal{F}$. Consider instead the following family of piecewise linear functions (where $I = \{1, \ldots, n\}$):

$$\mathcal{G}(P) = \left\{ g(\rho) = \max_{i \in I} \{\alpha_i + \beta_i \rho\} \ : \ \beta_i \geq 0;\ \alpha_i + \beta_i \rho_i \geq \alpha_j + \beta_j \rho_i \ \ \forall j, i \in I \right\}$$

It is clear that $\mathcal{G}(P)$ belongs to $\mathcal{F}$ for any arbitrary set of observations $P = \{\rho_i\}_i$. In [10] the author proves that $\mathcal{G}(P)$ may be substituted in (2) and the same optimal solution is obtained. This result allows us to transform the infinite dimensional problem (2) into the following standard finite dimensional Quadratic Programming (QP) problem:

$$\min_{\epsilon, \alpha, \beta} \sum_{i=1}^{n} \epsilon_i^2 \tag{3}$$

$$\begin{aligned} \text{subject to} \quad & Y_i = \alpha_i + \beta_i \rho_i + \epsilon_i \ \forall i = 1, \ldots, n \\ & \alpha_i + \beta_i \rho_i \geq \alpha_j + \beta_j \rho_i \ \forall j, i = 1, \ldots, n \\ & \beta_i \geq 0 \ \forall i = 1, \ldots, n \end{aligned}$$

Regarding the set of representor functions $\mathcal{G}(P)$, it may seem that a nonparametric problem was transformed into a parametric one. However, it should be noted that although we look for a piecewise linear function, the partition of the linear segments is not fixed a priori. That is to say, the number and location of the segments are endogenously determined to minimize the squared residual. Moreover, although each observation $(\rho_i, Y_i)$ has an associated $(\alpha_i, \beta_i)$, the actual number of different values is generally a very small fraction of $n$. This means that, in contrast to kernel-type regressors [11], $\widehat{f}(\rho)$ is completely represented by a number of parameters that will generally be much smaller than $n$, and that once these parameters are estimated, evaluating $\widehat{f}(\rho)$ and its derivative $\widehat{\phi}(\rho)$ is computationally very cheap. Moreover, this explicit regression function allows one to intra/extrapolate with relative confidence.

Regarding (3), although it is a standard QP problem for which mature methods to solve it exist (e.g. interior point algorithms) and that several solver software are available (for instance, we used MOSEK [12]), its size is considerable. It has a total of $3n$ variables and $n(n+1)$ restrictions. The second set of constraints, which are the key to modeling convexity, are quadratic in the number of observations. The size of the problem is clearly the major drawback of the method. However, as we will discuss in Sec. 4.1, these calculations need not be performed very frequently, and they may even be delegated to a central entity.

### 3.2 An Example

To illustrate the method, we will apply it to a training set obtained by injecting a 4 hours long packet trace (obtained from [13]) to a simple queue emulator we developed (in the absence of information on the buffer size, we assumed it infinite). The link has a capacity of 150 Mbps. Measurements are the mean queue size in kB and the mean load in kB/s over a one minute period.

Figure 1(a) shows the measurements (240 in total), $\widehat{f}(\rho)$ (as a reference, the MATLAB version of MOSEK solved (3) in less than 10 sec. in this case) and the estimation the *M/M/1* model yields ($\rho/(c-\rho)$). First of all, it should be noted that the *M/M/1* model has little to do with the real mean queue size. It consistently underestimates it, and its shape is almost a line when measurements clearly indicate a more convex curve. Regarding the estimation by CNLS, we can see that it is remarkably good, both in value and shape.

In Fig. 1(b) we show the derivative estimation through CNLS and the *M/M/1* model ($c/(c-\rho)^2$). Since the CNLS estimation is piecewise linear, its derivative is a piecewise constant function, and after no more observations are available it becomes constant. As a consequence, CNLS will produce a good estimation of $\phi(\rho)$ in the support of the observations, after which it will systematically underestimate it. The *M/M/1* model again underestimates the derivative, except at light loads where they are both small. Finally, Fig. 1(c) shows the pairs $(\alpha_i, \beta_i)$ in the plane. As we mentioned, although there are 240 different values, they are clustered around relatively few centers. Using only these cluster centers represent an insignificant lose in precision.

### 3.3 How to Use CNLS for REPLEX

The final purpose of the previously described regression was to use the estimated derivative $\widehat{\phi}(\rho)$ as a cost function on REPLEX in order to obtain a good approximation to the optimal traffic distribution. Although CNLS yields a non-decreasing cost function that approximates very well $\phi(\rho)$, it presents discontinuities. This means that $\widehat{\phi}(\rho)$ has an infinite relative slope, thus making the

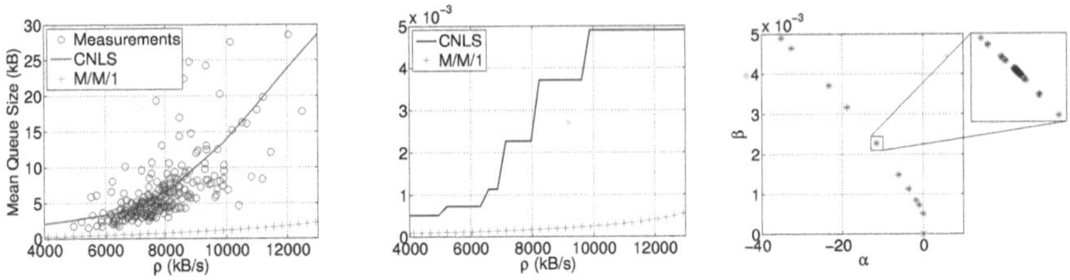


(a) Mean queue size measurements (b) The derivative function (c) The pairs $(\beta_i, \alpha_i)$ in the plane

**Fig. 1.** An example of a CNLS regression

regression method inappropriate for our purposes at first sight (cf. point (iii) in Sec. 3.1). In this subsection we discuss a possible way to approximate $\widehat{\phi}(\rho)$ through a smooth function.

Assume the regression function $\widehat{f}(\rho)$ is defined by $n'$ $(\alpha_i, \beta_i)$ parameters so that $\widehat{f}(\rho) = \max_{i=1,\ldots,n'} \alpha_i + \beta_i\rho$. A good approximation of this function is the so-called *log-sum-exp* function:

$$\widehat{f}^*(\rho) = \frac{1}{\gamma}\log\left(\sum_{i=1}^{n'} e^{\gamma(\alpha_i+\beta_i\rho)}\right)$$

This non-decreasing and convex function is clearly smooth. Moreover, the precision of the approximation can be controlled through the parameter $\gamma$ since $\widehat{f}(\rho) \le \widehat{f}^*(\rho) \le \widehat{f}(\rho) + \log(n')/\gamma$. This means that clustering the values of $(\alpha_i, \beta_i)$ not only decreases the size of the representation of $\widehat{f}(\rho)$, but also improves the precision of $\widehat{f}^*(\rho)$. Finally, its derivative is the following:

$$\widehat{\phi}^*(\rho) = \frac{1}{\sum_{i=1}^{n'} e^{\gamma(\alpha_i+\beta_i\rho)}} \sum_{i=1}^{n'} \beta_i e^{\gamma(\alpha_i+\beta_i\rho)} \tag{4}$$

We will use (4) as the link-cost function. A reasonable approximation to its relative slope, whose demonstration we omit for the sake of space, is $r \approx \gamma \max_{i=1,\ldots,n'} \beta_i$. This formula makes explicit the intuitive fact that the bigger $\gamma$ is (and better the approximation) the less soft the resulting $\widehat{\phi}^*(\rho)$ is.

The problem now is how to assign $\gamma$. As a rule of thumb, we recommend using a value such that the error in the soft approximation is approximately 30%. That is to say, $\gamma = log(n')/(0.3\bar{Y})$ where $\bar{Y}$ is the mean of $Y$. This value, as we shall now illustrate with an example, results in a good tradeoff between precision and convergence speed. We will consider the same $\widehat{f}(\rho)$ as in Fig. 1, and use three different values of $\gamma$: our recommended 30% error value ($\gamma^*$), $10\gamma^*$ and $0.1\gamma^*$. In Fig. 2 we can see the resulting $\widehat{f}^*(\rho)$ and $\widehat{\phi}^*(\rho)$ for the three considered values. First of all, a value as small as $0.1\gamma^*$ results in too much error for all

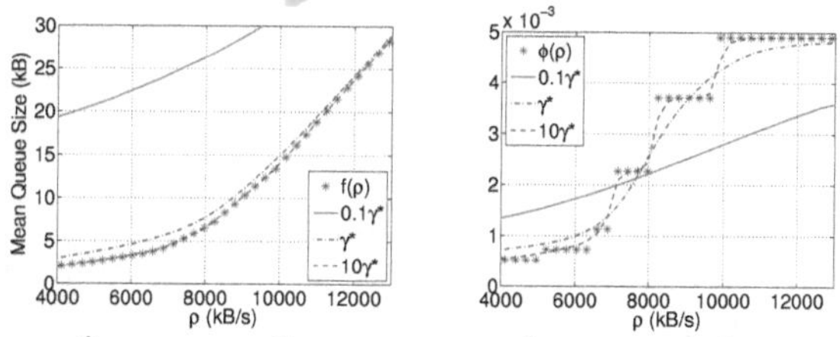


(a) $\widehat{f}(\rho)$ and $\widehat{f}^*(\rho)$ for some values of $\gamma$

(b) $\widehat{\phi}(\rho)$ and $\widehat{\phi}^*(\rho)$ for some values of $\gamma$

**Fig. 2.** An example of approximating a piecewise linear function with the log-sum-exp

practical purposes. On the other hand, the difference between $\gamma^*$ and $10\gamma^*$ is almost insignificant for $\widehat{f}^*(\rho)$, and reasonably small for $\widehat{\phi}^*(\rho)$. Moreover, note that $\gamma^*$ results in a ten times smaller relative slope than $10\gamma^*$, thus allowing a convergence speed ten times faster (cf. 2.3).

# 4 Simulations

## 4.1 Implementation Discussion

The application of our framework in a real-world network is relatively simple. Once all links have been characterized (i.e. we have the parameters $(\alpha_i, \beta_i)_l$ for all $l$), each OD pair receives $\rho_l$ from the links it uses[1], calculates its paths cost with (4), and applies REPLEX to update its traffic distribution. This process is repeated indefinitely every some seconds (in particular, we used 60 sec). This update period should be long enough so that the obtained measurements' quality is reasonable, but not too long to avoid unresponsiveness.

Regarding the learning phase (i.e. gathering the training set and performing the regression) we envisage several possibilities, differing in the degree of distribution of the resulting architecture and on what data is used at each moment.

With respect to who does which calculations, one possibility is that a central entity gathers the measurements, performs the regression and communicates the obtained parameters to all ingress routers (we assume that these routers, through which commodities inject traffic to the network, distribute this traffic). This has the advantage that the required new functionalities on the router are minimal. However, as all centralized schemes, it may not be possible to implement it in some network scenarios, and handling the failure of this central entity could be very complicated. An alternative is that ingress routers perform the regression. Links measure their load and mean queue size, communicate periodically these measurements to all ingress routers (instead of the central entity), which in turn perform the regression. However, the regression for any given link would be performed by several routers, constituting a waste of resources. A more reasonable alternative is that links (or better said, the router at the origin of the link)

[1] For this purpose, a TE-enabled routing protocol such as OSPF-TE may be used.

perform the regression. Links keep the mean queue size measurements for themselves, perform the regression and communicate the result to ingress routers. The regression could be done once a day, in the periods of low intensity (i.e. the night) so that normal operation is not affected by it.

A second aspect that has different possibilities is what characterization (i.e. $(\alpha_i, \beta_i)_l$) use at each moment. For instance, measurements could be gathered every day, the regression performed, and its result used the next day. Another possibility is to use the result of the measurements of the same day the previous week. More granularity could be added, and we could use different characterizations for different moments of the day. What granularity is needed and if it is actually necessary is an analysis that we let for future work.

### 4.2 Examples

In this subsection we will consider two relatively simple examples we implemented in ns-2 [14] that will help us gain further insight into the framework and verify its correctness in the presence of delayed and noisy measurements. We will assume that the training set has been obtained and the regression performed by a central entity. Routers only have the pairs $(\alpha_i, \beta_i)_l$, their associated $\gamma_l$ (which was calculated with the formula discussed in Sec. 3.3) and already know $\lambda$ (cf. Sec. 2.3). As a final remark, in all the simulations load balancing is performed at the granularity of flows (i.e. once a flow is routed through a path, it rests there throughout its lifetime) and is random (i.e. new incoming flows are routed through path $P_{si}$ with probability $p_{si}$).

We will begin with the simplest example: one commodity has two one-hop paths (see Fig. 3(a)). Traffic is a mixture of elastic and streaming flows. The elastic ones (whose size is exponentially distributed with mean 20 kB) are generated as a Poisson process. The streaming part of traffic is constituted of CBR flows (at a bitrate of 10 kbps and an exponentially distributed duration with mean 20 sec.) also arriving as a Poisson process, and it represents 10% of the total traffic. The measurements (467 for each link) were obtained by averaging the link load and the queue size over a minute period. In Fig. 3(b) we can see the measurements together with the resulting regression, and in Fig. 3(c) we show the corresponding cost function $\phi_l(\rho_l)$ and its soft approximation $\phi_l^*(\rho_l)$.

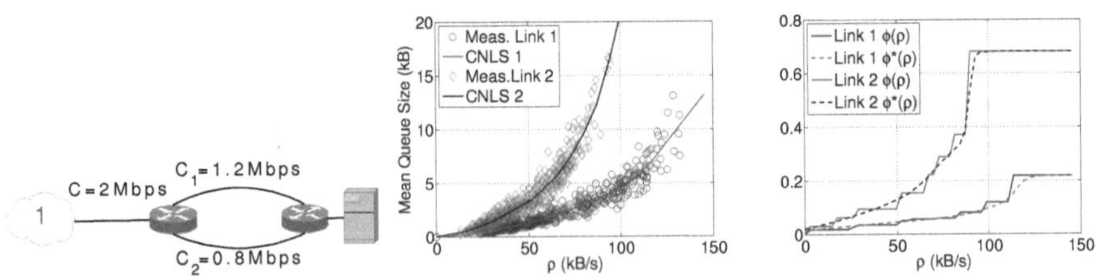


(a) The Network (b) The training set and the regression (c) $\phi(\rho)$ and its soft approximation $\phi^*(\rho)$

**Fig. 3.** The single-source case example topology and regression

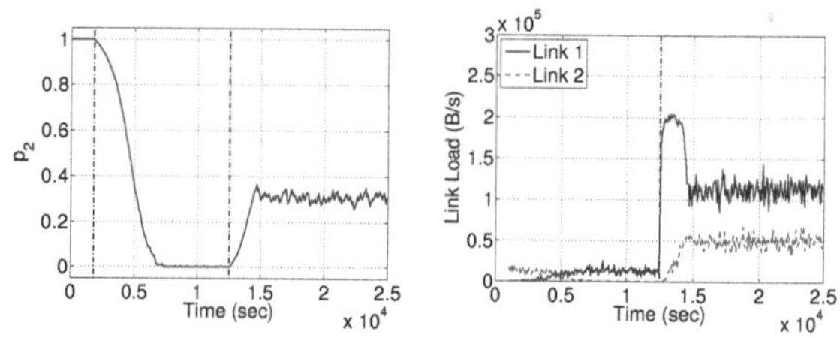


(a) $p_2$ as a function of time (b) The links load

**Fig. 4.** Simulation results in the single-source case example

Using this topology we will run a two-parts simulation. In its first half the total demand is approximately 100 kbps, after which it abruptly increases to 1200 kbps (this moment is marked in Fig. 4 by a vertical line). The traffic distribution is updated after second 1800 (marked by the first vertical line in Fig. 4(a)). Notice in Fig. 4(a) how at first $p_2$ (the portion of traffic routed through the lower path) changes relatively slow, but as it decreases the change grows faster. This is a consequence of the sampling step of REPLEX, in particular of the proportional one. The small (but inevitable) oscillations around the optimum traffic distribution should also be noted. They are inevitable since $\rho_l$ measurements are noisy (see Fig. 4(b)), but the effect of this noise on the convergence is minimized by the algorithm. Finally, notice how at the beginning of the second half of the simulation load on link 1 momentarily goes to values outside the support of the training set. As discussed in Sec. 3.2, for all such values of $\rho$, $\phi(\rho)$ is constant when it should actually increase. Although this does not prevent the algorithm from reaching the optimum, the convergence speed is slower than it should, resulting in an overloaded link for almost 30 iterations (or 30 minutes in our case). Although such an abrupt increase in traffic should be rare to say the least, this highlights the importance of a training set that encompasses as much operational points as possible.

We will now consider a somewhat more complex case scenario. The network (see Fig. 5(a)) consists of six links, all with a capacity of 1 Mbps. There are a total of 4 commodities whose destination is the same node $q$, but only commodity

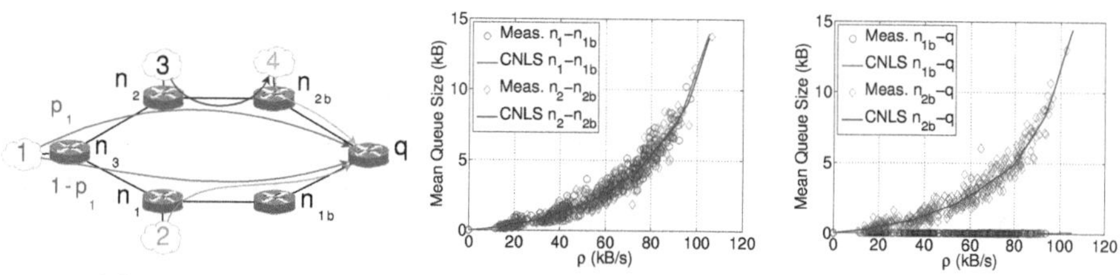


(a) The Network (b) The training set and the regression for link $n_1$-$n_{1b}$ and $n_2$-$n_{2b}$ (c) The training set and the regression for link $n_{1b}$-$q$ and $n_{2b}$-$q$

**Fig. 5.** The second example: two paths and four commodities

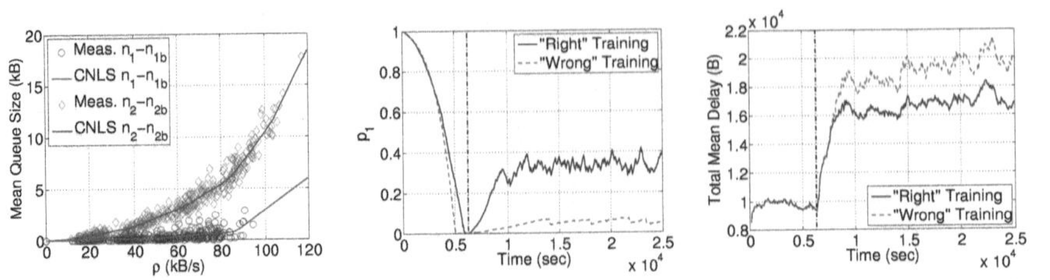


(a) Alternative training set and regression for link $n_1$-$n_{1b}$ and $n_2$-$n_{2b}$
(b) $p_1$ as a function of time
(c) The total mean delay $D(d)$ as a function of time

**Fig. 6.** An alternative training set for the second example

1 can use more than one path. The traffic generated by each commodity has the same characteristics as in the previous example.

In Fig. 5(b) and Fig. 5(c) we can see the training set and the corresponding regression for four of the links. The training set was obtained by changing the traffic intensity of the four commodities at the same time (with $p_1$ fixed at 0.5). Notice how the mean queue size of link $n_{1b}$-$q$ is near zero regardless of its load, and how its "symmetric" link ($n_{2b}$-$q$) is quiet the opposite and may be considered identical to $n_2$-$n_{2b}$. Traffic through link $n_{1b}$-$q$ does not generate significant queue because link $n_1$-$n_{1b}$ already shaped the traffic.

This example introduces the problem of links that have an insignificant queue size independently of its load. This may be due to traffic characteristics (as before), or because the link buffer is small. A link of such characteristics clearly presents a problem for our framework. For instance, let us consider an alternative training set, obtained in a situation where commodity 2 sends little or no traffic and commodity 1 sends most of its traffic through the lower path. In Fig. 6(a) we can see that although the small amount of traffic generated by commodity 2 results in a little bit of queue in link $n_1$-$n_{1b}$, the mean queue size $f_l(\rho_l)$ (and thus $\phi_l(\rho_l)$) for this link is almost zero except at big loads.

Consider now the following situation. Commodities 2, 3 and 4 all generate the same demand (approximately 450 kbps). Commodity 1 generates approximately 100 kbps during the first fourth of the simulation, and then abruptly increases its demand to the same value as the rest (this moment is marked with a vertical line in Fig. 6). Figure 6(b) shows the evolution of $p_1$ over time when using the results of both training sets. In the first part they both converge to $p_1 = 0$, but when $d_1$ increases we can appreciate the difference between the training sets. While the "right" training set (i.e. the one of Fig. 5(b)) moves towards a reasonable $p_1 = 0.4$, the other training set gets stuck in $p_1 = 0.05$ which results in a higher total delay (see Fig. 6(c)) and two almost overloaded links ($n_1$-$n_{1b}$ and $n_{1b}$-$q$).

## 5 Concluding Remarks

In this paper we presented a dynamic load-balancing mechanism that converges to an excellent approximation of the minimum-delay traffic distribution without assuming any given delay model. This was achieved on the one hand by "learning" the delay function from measurements, and on the other hand by applying a greedy load-balancing algorithm with provable convergence (and verified by our packet-level simulations). The chosen regression method is a piecewise linear fitting method, where the number and position of the lines are endogenously determined to minimize the squared residual. The cost function, which is the derivative of this regressor, was then not continuous, a fact that poses a problem to the distributed algorithm. This forced us to make a soft approximation of the regressor function, controlled by a single parameter $\gamma$, for which we gave hints on how to assign it. The few parameters the distributed algorithm requires were also discussed in the paper.

In Sec. 4 we highlighted two shortcomings of our framework which may result in overloaded links: the regression outside the support of the observations is not reliable (since the cost function does not increase any further) and links that present little or no queueing delay always have a negligible cost. A possible solution to both problems is adding to the link cost a known parametric function that is negligible with respect to $\phi_l(\rho_l)$ except at very high loads. This will increase the cost of loaded links both when no observations are available or when their actual mean queueing delay is small. Moreover, if the optimum does not load considerably any link, it will be attained with a very small error. Further development of such correction is the subject of future work.

It would also be very interesting to perform a deeper statistical analysis of the behavior of the mean queue size with respect to load. A possible analysis would be to study how often does the regression function change over time (i.e. answer the question of whether the mean queue size function changes over time, and how often it does).

Regarding the queueing model, we considered that the mean queue size is a function of the mean incoming rate only. This is naturally not true, as it actually depends on the complete packet arrival process. Methods that estimate $\phi_l$ considering the whole process exist, such as the one used in [6]. Apart from being more complicated (they require to measure the arrival and departure time of every packet), the problem with such methods is that $\phi_l$ now depends on a number of unknown and uncontrollable variables. This results in the impossibility of guaranteeing convergence to the optimum by changing the portions of traffic only, and it does result in oscillations as presented (but not explained) in [6]. A possible improvement to our model is to consider that $\phi_l$ depends on the mean load of its incoming links. For instance, in Fig. 5(a), that the mean queueing delay in link $n_1$-$n_{1b}$ depends on the load on links $n_3$-$n_1$ and the one connecting commodity 2 and $n_1$. For this now multi-dimensional regression problem the same method may be used. A deeper analysis of this alternative model represents also interesting future work.

## References

1. Elwalid, A., Jin, C., Low, S., Widjaja, I.: MATE: MPLS adaptive traffic engineering. In: INFOCOM 2001, vol. 3, pp. 1300–1309 (2001)
2. Kandula, S., Katabi, D., Davie, B., Charny, A.: Walking the tightrope: responsive yet stable traffic engineering. In: ACM SIGCOMM 2005, pp. 253–264 (2005)
3. Fischer, S., Kammenhuber, N., Feldmann, A.: Replex: dynamic traffic engineering based on wardrop routing policies. In: CoNEXT 2006, pp. 1–12 (2006)
4. Ben-Ameur, W., Kerivin, H.: Routing of uncertain traffic demands. Optimization and Engineering 6(3), 283–313 (2005)
5. Larroca, F., Rougier, J.L.: A fair and dynamic Load-Balancing mechanism. In: International Workshop on Traffic Management and Traffic Engineering for the Future Internet (FITRAMEN) 2008, Porto, Portugal (December 2008)
6. Cassandras, C., Abidi, M., Towsley, D.: Distributed routing with on-line marginal delay estimation. IEEE Trans. Comm. 38(3), 348–359 (1990)
7. Altman, E., Boulogne, T., El-Azouzi, R., Jiménez, T., Wynter, L.: A survey on networking games in telecommunications. Comput. Oper. Res. 33(2), 286–311 (2006)
8. Wardrop, J.: Some theoretical aspects of road traffic research. Proceedings of the Institution of Civil Engineers, Part II 1(36), 352–362 (1952)
9. Fischer, S., Räcke, H., Vöcking, B.: Fast convergence to wardrop equilibria by adaptive sampling methods. In: STOC 2006: Proceedings of the thirty-eighth annual ACM symposium on Theory of computing, pp. 653–662 (2006)
10. Kuosmanen, T.: Representation theorem for convex nonparametric least squares. Econometrics Journal 11(2), 308–325 (2008)
11. Wasserman, L.: All of Nonparametric Statistics: A Concise Course in Nonparametric Statistical Inference. Springer, Heidelberg (2006)
12. The MOSEK Optimization Software, http://www.mosek.com/
13. Cho, K.: WIDE-TRANSIT 150 Megabit Ethernet Trace 2008-03-18, http://mawi.wide.ad.jp/mawi/samplepoint-F/20080318/
14. The Network Simulator - ns, http://nsnam.isi.edu/nsnam/index.php/Main_Page

# A Power Benchmarking Framework for Network Devices

Priya Mahadevan, Puneet Sharma, Sujata Banerjee, and Parthasarathy Ranganathan

HP Labs
1501 Page Mill Road, Palo Alto, CA 94306, USA
{priya.mahadevan,puneet.sharma,sujata.banerjee,partha.ranganathan}@hp.com

**Abstract.** Energy efficiency is becoming increasingly important in the operation of networking infrastructure, especially in enterprise and data center networks. Researchers have proposed several strategies for energy management of networking devices. However, we need a comprehensive characterization of power consumption by a variety of switches and routers to accurately quantify the savings from the various power savings schemes. In this paper, we first describe the hurdles in network power instrumentation and present a power measurement study of a variety of networking gear such as hubs, edge switches, core switches, routers and wireless access points in both stand-alone mode and a production data center. We build and describe a benchmarking suite that will allow users to *measure and compare* the power consumed for a large set of common configurations at any switch or router of their choice. We also propose a network energy proportionality index, which is an easily measurable metric, to compare power consumption behaviors of multiple devices.

**Keywords:** Network energy management, Benchmarking.

## 1 Introduction

Energy efficiency has become crucial for all industries, including the information technology (IT) industry, as there is a strong motivation to lower capital and recurring costs. According to recent literature, the annual electricity consumed by networking devices in the U.S. is 6.06 Terra Watt hours, which translates to around 1 billion US dollars per year [1], thereby presenting a strong case for reducing the energy consumed by networking devices such as hubs, access points, switches and routers.

Unlike wireless networks, energy management[1] for networking devices such as hubs, switches and routers in wired networks has not received much attention until very recently. Researchers have proposed several strategies to make routers

[1] We use power and energy management interchangeably in this paper. There is a distinction between power management for heat density versus electricity costs; however in this paper, we do not distinguish between these two issues.

L. Fratta et al. (Eds.): NETWORKING 2009, LNCS 5550, pp. 795–808, 2009.

and switches more energy-aware such as link rate adaptation during periods of low traffic and sleeping during no traffic [1–4]. Currently, however, there are no hardware implementations of the same. More importantly, the exact energy savings by adopting these techniques are poorly understood. One impediment to innovation in this area is the lack of power measurements from live networks and a good understanding of how the energy consumed varies under different traffic loads and switch/router configuration settings. There are many vendors manufacturing a wide variety of network devices and as of yet, there has been no focus on standardized benchmarks to measure the energy consumption of these devices. We also need benchmarks to compare the effectiveness of various energy efficient improvements.

In this paper, we attempt to fill this gap by quantifying the energy consumed by a wide variety of networking gear ranging from core switches to wireless Access Points (APs) from three different networking vendors, and under different traffic and network configurations. Our goal is to build a publicly available benchmarking suite that can be deployed on any switch; our suite sets various configurations in the switch under test and generates different traffic patterns, while a power-meter connected to the switch measures the average power for the duration of the experiment.

Non-uniformity across switches from different manufacturers, and different functionality available in the switches, make the task of identifying the right set of configurations to include in the benchmark suite challenging. Additionally, we would like our benchmarking suite to be modular, so that new components for power instrumentation can be easily incorporated. One additional challenge for device manufacturers is to ensure that networking devices such as hubs, switches and routers are energy proportional, *i.e* they consume energy proportional to their load, similar to energy proportional laptops and servers [5, 6]. In this paper, we also discuss how energy proportionality can be measured for networking devices and their components. From our initial benchmarking study, we find that:

- There is great variability amongst switches with respect to their maximum rated power. The ratio of the actual power consumed by the device on an average, to its maximum rated power varies widely across different device families. Thus, relying merely on the maximum rated power can grossly overestimate the total energy consumed by these networking equipment.
- Energy consumed by a switch increases linearly with the number of linecards plugged into the switch as well as the number of active ports on each card.
- Energy consumed by a switch is largely independent of the packet size for a fixed traffic throughput.
- Ideally, devices should consume energy proportional to their load [5, 6]. To quantify this behavior, we define an energy proportionality index (EPI) for network devices (Section 3). The EPI values for the evaluated devices clearly demonstrate non-energy proportional behavior.

The rest of this paper is organized as follows: We begin with describing challenges in obtaining large-scale power measurements in Section 2. Our methodology to characterize power consumption in switches and routers is described in Section 3. In Section 4, we present results from our suite. We discuss related work in Section 5. We conclude with the implications of the results of our benchmarking study in Section 6.

## 2 Challenges

Current switches and routers don't include comprehensive energy consumption values. Device specification data sheets only report maximum rated power. This value by itself is insufficient to understand the actual energy consumption of the networking device. As we show in the rest of this paper, the actual energy consumed by switches and routers depends on various factors such as device configurations and traffic workload; thus relying only the maximum rated power will grossly overestimate the actual energy consumption, by as much as by 70%, in some of our observations.

Power instrumentation of a device, in a live network deployment, requires unplugging the device from its power supply and plugging it into a power meter. Since the device will not be functional for several minutes, this task needs to coincide with regular network downtimes. Along with the power measurements, one also needs information on exact device configurations as energy varies significantly with different configuration settings. One also needs to record actual traffic traversing through the device and measure the energy consumed at regular intervals.

Most existing networks are not equipped with internal power instrumentation to provide information on how much energy is being consumed by each component at any given time. Adding external power instrumentation for each network device in operational networks is quite cumbersome. This task involves working with network operators to find the appropriate opportunities such as maintenance downtime to install power meters in-line with the network equipment power cables. Eventually, we expect more support for internal power measurements in the next generation of network equipment (much like the battery meters on laptops). Ideally, such power measurements need to be incorporated into standardized SNMP MIBs so that they can be queried by management applications. However, the large base of legacy network equipment deployed today does not have such instrumentation built in. Motivated by these facts, we present a model-based power inference mechanism that can be scaled up to large networks.

The idea is to generate power consumption models of networking equipment using standard benchmarking. Such models have already been defined for servers (for instance see [7]). We describe such a benchmarking suite in the next section. Using the measurements obtained from our benchmarking suite, we build a model to predict the power consumed by any networking switch or router; the device's configuration and traffic flowing through it are specified as input to the model. Once the models are available, using standard SNMP MIBs or command line interface (CLI) query mechanisms, the network operator can query the configuration and traffic information from the switches/routers and plug these values into our models to derive the power consumed by a single box, multiple boxes and the whole network. These models can be also used in simulation/emulation testbeds. While models can perhaps not take the place of actual in-line power measurement, based on our small study, we believe that it is fairly effective, and certainly much better than using the rated (plate) power ratings provided by network vendors.

## 3 Benchmarking Framework

We begin by describing factors that are commonly used and most likely to affect a switch or router's power consumption. Next, we describe our efforts in building the benchmark suite, along with results from a few selected switches/routers that span from low end devices to ones that are deployed in high traffic data centers and network backbones.

### 3.1 Benchmarking Suite

Currently, there are no standard benchmarks used in power measurement studies of network components. One of our contributions is to develop such a benchmark that can be used to compare the power characteristics of a variety of network devices. Figure 1 shows the architecture of our benchmarking suite. The network device to be benchmarked is connected to the power outlet via a *power meter*. The *device configurator* modifies the various configuration states of the device according the benchmarking requirements. The benchmarking process also loads the device with varying traffic patterns using the *traffic generator*. The *benchmark orchestrator* coordinates the various components in order to synchronize the configuration, workload and measurements from the power-meter. The collected information is then processed by an *analyzer* to generate various energy-proportionality indices and other power-related metrics.

A network switch or a router consists of many different components, such as the main chassis, linecards, TCAM, RAM, processor, fans etc. A complete instrumentation of various components is difficult to perform. First, we list various factors that are likely to affect power consumption for switches and routers; further, we describe the important ones that we have incorporated in the *device configurator*. While additional metrics and tests may have to be added to our suite in the future due to technological advances in switches and routers, using our current benchmarking suite, we are able to predict within a 2% error margin, the power consumed by a variety of switches deployed today in a real operational data center.

- *Base chassis power:* Higher-end switches (typically deployed at the edge of a network, and in data centers, etc.) come with a base chassis and a fixed number of slots. In each slot, a linecard can be plugged in. In lower-end switches (typically having 24 or fewer ports), the slots and linecards are fixed and cannot be changed. In both cases, the chassis power typically includes the power consumed by components in the switch such as processor, fans, memory, *etc.*

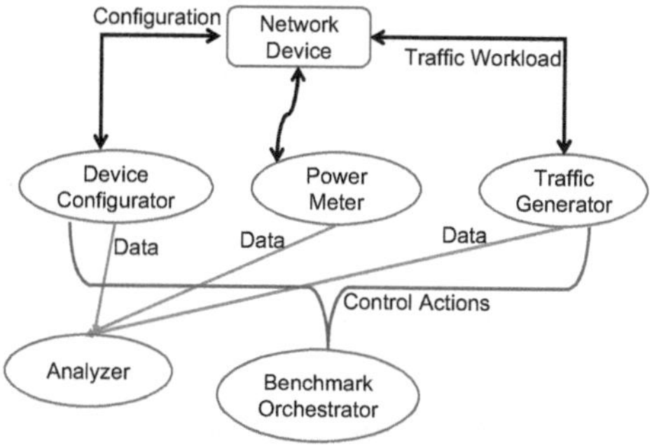


**Fig. 1.** Benchmarking suite architecture

- *Number of linecards:* In switches that support plugging in linecards, there is a limit on the number of ports per linecard as well as the aggregate bandwidth that each linecard can accommodate. This mechanism allows network operators the flexibility to only plug in as many linecards as they need. Further, it also offers choices such as the ability to plug in 24-port 1 Gbps linecard for an aggregate 24 Gbps capacity versus a 4-port 10 Gbps linecard for an aggregate of 40 Gbps capacity.
- *Number of active ports:* This term refers to the total number of ports on the switch (across all the linecards) that are active (with cables plugged in). The remaining ports on the switch are explicitly disabled using the switch's command line interface.
- *Port capacity:* Setting this parameter limits the line rate forwarding capacity of individual ports. Typically, the capacity of a full-duplex 1 Gbps port can be also set to 10 Mbps and 100 Mbps.
- *Port utilization:* This term describes the actual throughput flowing through a port relative to its specified capacity. Thus, a port whose capacity has been set to 100 Mbps and having a 10 Mbps flowing through it has a port utilization of 10%.
- *Ternary Content Addressable Memory (TCAM) :* Switch vendors typically implement packet classification in hardware. TCAMs are supported by most vendors as they have very fast look-up times. However, they are are notoriously power-hungry. The size of the TCAM in a switch is widely variable.
- *Firmware:* Vendors periodically release upgraded version of their switch/ router firmware. Different versions of firmware may also impact the device power consumption.

The traffic characteristics at each port might also affect the power consumption; we list a few traffic factors that we have incorporated in the *traffic generator component* of our benchmark suite:

- *Packet size:* Varying from 48 bytes to 1500 bytes.
- *Inter-packet delay:* Time between successive packets at a port.

– *IP options set in the packet:* While this factor might not affect power consumption at switches performing MAC forwarding, processing packets that have IP options might impact the power consumption of a router.

Given these range of factors in the *traffic generator* and *device configurator* components of our benchmarking suite that can impact switch power consumption, our goal is to define the ones that will impact the switch power consumption the most. Capturing this short list of factors correctly, will allow us to be accurate in predicting switch power consumption.

### 3.2 Network Energy Proportionality Index

As noted previously, manufacturers need to ensure that devices consume energy proportional to their load. In the rest of this section, we describe how we can calculate energy proportionality for network devices.

In Figure 2, the power consumed by a device is plotted against the load on the device (in Gbps or active number of ports) with a maximum load of G Gbps. Ideally, the power consumed should be proportional to the load, with the maximum power consumed, M watts, being as low as possible. The curve marked *ideal* represents this desired energy proportionality behavior. In practice however, the behavior of network devices follow the line marked *measured*, with the device consuming I watts even under *idle* (no load) conditions. The difference between the *ideal* and *measured* lines forms the basis of the following *energy proportionality index (EPI)* for networking components, which we define as $EPI = \frac{M-I}{M} * 100$. If $\theta$ and $\phi$ are the angles at the origin for the ideal and measured power, EPI is simply $(tan(\phi)/tan(\theta)) * 100$. We express EPI in percentage, with 100 implying that the device has perfect energy proportionality and 0 implying that the energy consumed by the device is completely agnostic to offered load. Note that EPI is independent of the maximum load that can be carried by the device and thus is most useful in comparing the energy proportionality of devices in the same class with the same maximum load. *Normalized power* [8], another metric,

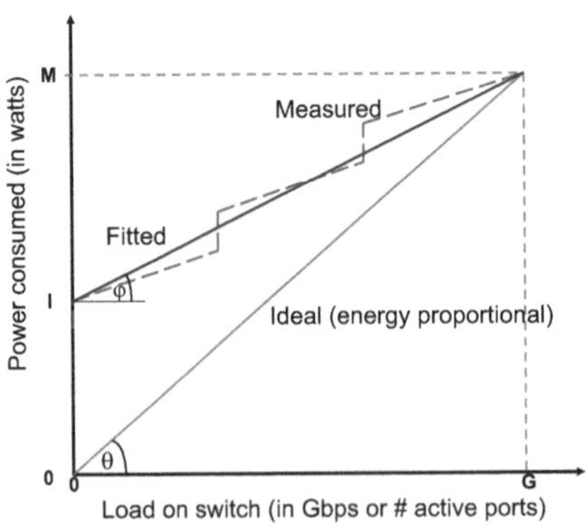


**Fig. 2.** Ideal (energy proportional) and measured power characteristics of network devices

is the maximum power consumed by the device divided by its aggregate bandwidth and is calculated as $NormalizedPower = \frac{M}{G}$. Since our devices span a wide range, from wireless access points to high end switches in data centers, we compute normalized power in milliWatts/Mbps. We use the above metrics to quantify the energy proportionality of a wide range of networking devices in the following section.

## 4 Experimental Details and Results

We build our benchmarking suite incorporating all the factors described in Section 3.1. To measure power consumed by a network device, we use a digital power-meter from Brand Electronics. In all of our experiments, we plug the power meter into an electrical outlet and the switch under test into the receptacle of the power meter. Our measurements include any power losses due to inefficiencies. We do not consider power factor in this paper. We use *expect* [9] scripts to build our *Orchestrator* and *Device Configurator* shown in Figure 1. Our configurator is modular so that we can easily add support for any other networking device as well. Our *Traffic Generator* consists of well-known programs such as iperf [10] as well as our own in-house packet generator program that sends CBR traffic with parameters to vary the packet size, inter-packet gap and total traffic throughput. As part of our benchmarking process, we run each experiment for 300 seconds and report the average power over the entire duration. Further, we run each experiment three times to minimize the effect of noise in these measurements. We observe insignificant differences between peak power and average power over the experiment duration.

As the first steps in our benchmarking process, for each network device, we measure the power consumed in the idle state (all ports inactive and no cables connected to them). We also measure the time taken for each switch to reach a steady state after booting. Table 1 lists the device categories we use in our study along with their maximum power ratings (plate power) as described in their published specifications. The measured maximum power is the power consumed when the offered load is equal to the total aggregate bandwidth the switch (or router) can support. We report all power measurements in Watts (except the last column). We anonymize the device models, but all of our experiments were performed on devices from well-known vendors such as Cisco, ProCurve, and Brocade.

Device A is a low-end network hub. While the maximum measured power for this switch is a relatively low 12.8 W, the fact that the switch has only 12 ports, with each port capable of forwarding traffic at a maximum of 100 Mbps, accounts for the rather high milliWatts / Mbps value when compared to devices B, C, D and E. Devices C, E and F can be built to different specifications with respect to the model and number of line cards that can be added to the chassis, while no such options exist for switches B and D. Switches B, C and E have the ability to support PoE (Power Over Ethernet), while D does not. Switch C is available as a modular chassis with 6-slots, with each slot capable of supporting

**Table 1.** Power consumption summary for network devices

| Label | Type | Rated Max Power | Measured Max Power (M) | Measured Idle Power (I) | Time to reach steady state | EPI (in %) | Aggregate bandwidth in Mbps | mWatts / Mbps [a] |
|---|---|---|---|---|---|---|---|---|
| A | 10/100 Hub | 35 | 12.8 | 11.7 | 22 secs | 8.59 | 1200 | 10.7 |
| B | edge LAN switch | 759[b] | 198 | 150 | 125 secs | 24.2 | 48000 | 4.1 |
| C | edge LAN switch | 875[c] | 175 | 133.5 | 119 secs | 23.7 | 48000 | 3.7 |
| D | edge LAN switch | 300 | 102 | 76.4 | 99 secs | 25.1 | 48000 | 2.1 |
| E | core switch | 3000 | 656 | 555 | 212 secs | 15.4 | 48000 | 13.7 |
| F | edge router | 300 | 210 | 168.5 | 195 secs | 19.8 | 24000 | 8.75 |
| G | wireless access point | 12.5 | 8.3 | 5.2 | 50 secs | 37.3 | 54 | 153.7 |

[a] Measured max power in milliWatts / Aggregate bandwidth in Mbps. This term is equivalent to Joules per bit.
[b] including 400W for PoE
[c] including 400W for PoE

a 24-port linecard. Each port can be set to a maximum of 1 Gbps capacity. In our benchmarking experiments, we choose 2 linecards for a total of 48 ports. Switch E is a core switch, typically used as a root switch in data centers. This particular model comes with 9 slots. We plug in a single 48-port 1 Gbps linecard in one slot and the management blade in another slot for all our benchmarking tests. Device F is a router, designed to sit at or just before the network edge. It has 4 card slots, allowing us to plug in a 12-port linecard in each slot. Each port can operate at a line speed of 1 Gbps. We plug in 2 linecards for this device, for a total of 24 ports. Device G is a wireless access point with a capacity of 54 Mbps. While the actual power consumed by this device is the lowest compared to all other models, its efficiency, as described by milliWatts/ Mbps, is the lowest (153.7 W) due to the fact that the access point is only capable of supporting traffic at 54 Mbps.

Devices B, C and D have an EPI value around 25%, indicating that their energy consumption varies slightly with traffic, though not by very much. The core switch (E) and router (F) exhibit almost no energy proportionality. The wireless access point (device G) on the other hand has the highest EPI value. It is important to note that EPI does not translate into energy efficiency. In the case of G, the milliWatts that have to be used to forward 1 Mbps of traffic is the highest amongst all the devices; thus it is the least efficient device.

For a switch having a line card with 48 full-duplex 1 Gbps ports, one way to fully load the switch is to attach servers to each port and ensure 1 Gbps of traffic going in and coming out of each port or use expensive measurement equipment from vendors such as Ixia (www.ixiacom.com). We settle on a relatively inexpensive strategy to fully load the switch without using 48 sources. We loop-back 23 port pairs causing 46 ports to be active. For the remaining two ports in the switch, we attach a commodity linux server to each port.

We run our traffic generator program from one server; every packet from this server is sent to the broadcast IP address. This technique ensures that the traffic received by the switch port to which the server is connected is now forwarded to all ports on the switch, thus ensuring that all ports get the offered load. Since we have loops in our topology, enabling spanning tree and disabling spanning tree have separate effects.[2] While our technique of creating loops in a switch is unconventional[3], it allows us to benchmark the switch power consumption with just a couple of servers instead of connecting a server to each switch port, which can be unscalable especially for larger switches.

For each active port on the switch (except model A and G), we measure the switch power consumption for 3 separate port capacities - 10 Mbps, 100 Mbps and 1 Gbps. For each capacity value, we vary the utilization of each port by changing the traffic flowing through it as well as by enabling and disabling spanning tree. As noted previously, we repeat our measurements for each configuration thrice, with each experiment lasting 5 minutes, for a total of 15 minutes.

To verify the validity of our looping technique for power measurements, we first benchmark switch D using the technique we describe above and note its power consumption for different switch and traffic configurations. Next, we validate our results by instrumenting the same switch with 48 different servers connected to each port, and for the same traffic and switch configurations. In each case, we find the results from our looping technique to be within 2 Watts of the 'attaching one server to each port' technique (the error margin was under 2%). While we could not validate all the other switches with this technique due to infrastructure constraints, based on conversations with switch vendors, we believe that our technique will hold for other switches as well.

We now describe results from our power benchmarking study with the edge LAN switch B. We obtain similar results for switches C, D and E, but don't report the results in detail here due to space constraints.

**Edge LAN switch B:** This 48-port switch, with a maximum per port capacity of 1 Gbps, is typically used in LANs as well as in data centers to connect servers in a rack. We find the power consumption to be very stable and the standard deviation and variance to be negligible. We plot average power consumed for each port capacity as a function of the number of active ports on the switch

[2] When spanning tree is enabled, the switch internally detects all the loops and ensures that for each port pair that form a loop, exactly one port is forwarding all the broadcast packets and the other port is receiving all the packets. For example, if all ports in the switch are enabled through the command line interface, and if one of the connected servers is generating 1 Gbps of traffic, each port in the switch either receives or forwards 1 Gbps of traffic. In the case when spanning tree is disabled, the loops in the topology cause the switch to flood every port with control packets. Every port in this case, both transmits and receives packets equal to its full line speed. In other words, if a port's capacity is configured to 1 Gbps, the throughput (due to the control packets) on this port is 1 Gbps in each transmitted and received state.

[3] Communications with a switch vendor confirmed that our power measurement numbers using broadcast packets would match the observed power if we had connected a single server to each port and resorted to transmitting unicast packets.

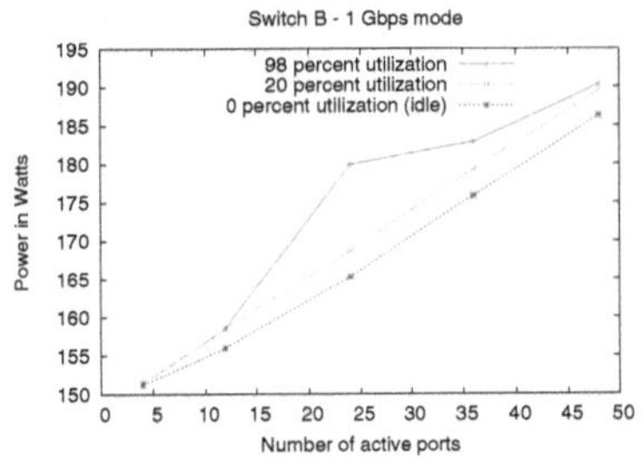


Fig. 3. Switch B - 1 Gbps port capacity

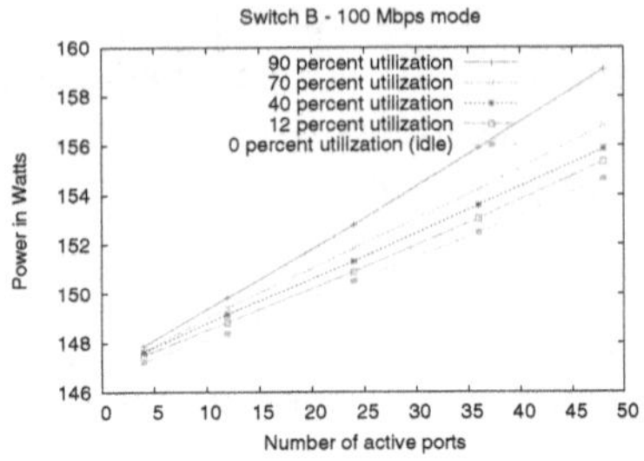


Fig. 4. Switch B - 100 Mbps port capacity

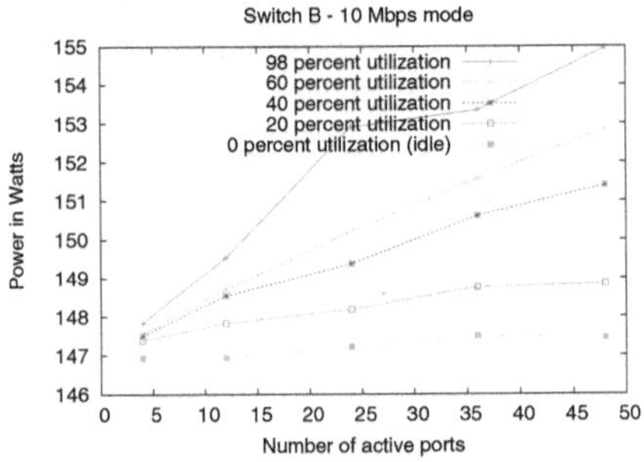


Fig. 5. Switch B - 10 Mbps port capacity

in Figures 4, 4, and 5. Port capacity influences the power consumption significantly, especially for higher number of active ports. Further, as we increase the number of active ports, the impact of port utilization (whether no load or fully loaded) on power consumption is under 5%.

Moving onto the effects of traffic, we find that for a fixed traffic throughput at any port, packet size does not impact power consumption at all. In Table 2, we show the power consumed by the switch with 48 active ports as a function of 4 different packet sizes.

Next, we determine the effect of the number of entries in the TCAM on the power consumption. For this purpose, we add rules into the TCAM to accept and deny packets from certain IP addresses. We vary the number of TCAM entries from 0 to 3044, the maximum number available for this switch. With the TCAM

**Table 2.** Impact of packet size on power consumption

| Number of active ports | Packet size in bytes | Average power in Watts |
|---|---|---|
| 48 | 10 | 151.2 |
| 48 | 500 | 151.6 |
| 48 | 1000 | 151.4 |
| 48 | 1500 | 152.3 |

appropriately filled, we again repeat all our experiments for different switch and traffic configurations described above and find that the number of the TCAM entries does not impact power consumption at all. One interesting aspect that we observe is the effect of switch firmware on energy consumption. When we upgrade to the latest version of firmware available on the manufacturer's website, and run our complete benchmarking suite again, we find that the power consumed is 8-12 W lower in every case. The idle power (I) is now 141 W, while the measured max power (M) is 178 W, for a new milliWatts/Mbps value of 3.7 and an EPI of 20.8%. We believe that concerns about energy usage is starting to motivate device manufacturers to make improvements in their firmware such as turning off unused components. While these improvements may be a contributing factor towards the lower power consumption, the EPI of the switch is still low.

We observe similar results for switches C, D, E and F. For devices where we can plug in linecards, we note that each linecard consumes 35-40 Watts. The base switch power now includes the chassis power plus the cost of each linecard. Other factors such as the number of active ports, effect of traffic, packet sizes, TCAMs, etc are similar to what we describe for switch B.

From the above results, we obtain a model to compute the total power consumed by the switch. From our measurements, we find that a linear model is able to accurately capture the total power consumption of switches/routers currently in use. As new architectural and design changes are implemented in these devices, a linear model might not be the best fit; we might have to use other models in that case.

Total power consumed consists of a fixed component (chassis power and power for each linecard) and a variable component that depends on the number of active ports, capacity of each port and utilization of each port.[4] We experiment with a combination of port line rates; for example, we set the line speed of 16 ports in the switch to 10 Mbps, 16 other ports to 100 Mbps and the remaining 16 ports to 1 Gbps. We find that the switch power consumption follows our linear model even when individual ports have different line rates set. The power model is given by $Power_{switch} = Power_{chassis} + num_{linecards} * Power_{linecard} + \sum_{i=0}^{configs} numports_{configs_i} * Power_{configs_i} * utilizationFactor$. $Power_{linecard}$ is the power consumed by the linecard with no ports turned on, and $num_{linecards}$ is the actual number of cards plugged into the switch. Variable *configs* in the summation is the number of configurations for the port line rate. $Power_{configs_i}$ is the power for a port running at line rate $i$, where $i$ can be 10 Mbps, 100 Mbps, or 1 Gbps and $utilizationFactor$ is the scaling factor to account for the utilization of each port.

**Validation:** We use our linear power model to validate the power consumed by switches within a live data center in our labs. Most of the racks in our data center had two model D switches attached to all the servers in a rack. With help from network administrators, we were successful in connecting several model D switches to power meters to record the power as real traffic was flowing through

[4] Our model is slightly different from the one proposed by [11] for routers; we also include the cost for each active port at its specified line rate and utilization.

them. Using SNMP queries, we obtain traffic flowing through each port for each switch, the number of active ports on the switch and the capacity of each port. We substitute the values from these live switches in our linear power model for switch D and find that our predicted power matches the real power measured by the power meter with an error margin of under 2%.

## 5 Related Work

Chabarek [11] *et al.* enumerate the power demands of two widely used Cisco routers; further the authors use mixed integer optimization techniques to determine the optimal configuration at each router in a sample network for a given traffic matrix. In our study, we consider a broader classification of switches and routers. Additionally, we also discuss energy proportionality and energy efficiency of these devices.

Some device manufacturers such as Cisco provide an online tool to calculate total power consumed by some of their switch and router models. However, these models are coarse-grained, and do not account for finer switch configurations such as the line rate at which each port has been configured as well as effects of traffic through each port.

At an individual switch or router level, researchers have been proposing techniques such as putting idle subcomponents (line cards, ports *etc.*) to sleep [1–4], as well as adapting link rates depending on the traffic [2–4, 12]. Allman *et al.* [13] suggest incorporating a power-aware proxy that relays keep-alive messages such as ARP replies to the switch on behalf of end devices such as desktops and laptops. This would allow the end devices to be put to sleep, while the proxy keeps the network connection alive.

Most of these proposals are in their early stages, and the underlying hardware support for implementing these schemes is not yet available, though equipment manufacturers are actively working on these proposals. While all these efforts are commendable and a step in the right direction, what has been lacking in the network community is understanding the energy savings by implementing these schemes. For example, how many Joules can we save by allowing a switch in the network edge to sleep for 30 secs? To the best of our knowledge, no in-depth measurement study exists that quantifies the actual energy consumed by a wide range of switches under widely varying traffic conditions as well the actual power savings that can be obtained by performing techniques such as link rate adaptation.

## 6 Implications and Conclusions

Benchmarking the power consumed for a variety of switch and router configurations is a challenging problem, that will continue to grow in importance. We build a benchmarking suite that incorporates several device configurations and traffic factors that can impact energy consumption and identify the main factors that impact power consumption the most. We summarize the main observations from our study:

- The power consumed depends on the number of active ports. Explicitly disabling unused ports on a line card reduces the device power consumption. As the number of active ports increases, power consumed increases linearly for all three port line speeds - 10 Mbps, 100 Mbps and 1 Gbps.
- The power consumed depends on the line speed each port is configured at, with 1 Gbps consuming the most power and 10 Mbps the least. This effect is due to the extra energy required to operate the physical layer (PHY) at higher line speeds. Thus, for ports that have low utilization, it is beneficial to configure the port line speed itself to 10 or 100 Mbps than imposing rate-limiting on traffic through the port without lowering the port speed.
- Traffic through the device (port utilization) does not have a significant effect on power consumed, as can be observed from the low EPI values of the devices. For 10 Mbps port line speed, the impact of traffic on power consumed is slightly higher than that of 1 Gbps line speed. Further, power consumption is independent of the packet size.
- Power consumed depends on other factors such as firmware version on the switch/router. In current models, number of TCAM slots used does not impact power. Making components such as TCAMs power-proportional will also lower the energy consumption of devices.

Our study has two implications for network operators and device manufacturers. Our measurements serve as motivation for device manufacturers to adopt more aggressive techniques such as turning off unused components in the devices, to make them more energy efficient. At the same time, the low EPI values of current devices suggest that techniques such as traffic consolidation might provide significant energy savings. For instance, network operators in an enterprise or data center networks might be able to consolidate traffic from multiple switches onto a single switch and turn off the used switches.

We have built a linear model for power consumed by network device; using this model, users can calculate the power consumed by their switch (or router) by specifying the configuration and traffic values for their device. Further, we define and compute the energy proportionality index for each switch, similar to what has been proposed for servers. We note that energy proportionality does not imply energy efficiency. These are two separate terms and energy management techniques should consider making devices both energy efficient as well as energy proportional. In the future, we plan to make the benchmarking suite available to build a comprehensive public repository of power characterization of all the network devices.

## References

1. Gupta, M., Singh, S.: Greening of the Internet. In: Proceedings of ACM SIGCOMM (August 2003)
2. Gupta, M., Grover, S., Singh, S.: A feasibility study for power management in LAN switches. In: Proceedings of IEEE ICNP (October 2004)

3. Gupta, M., Singh, S.: Energy conservation with low power modes in Ethernet LAN environments. In: Proceedings of IEEE INFOCOM (MiniSymposium) (May 2007)
4. Nedevschi, S., Popa, L., Iannaccone, G., Ratnasamy, S., Wetherall, D.: Reducing Network Energy Consumption via Rate-Adaptation and Sleeping. In: Proceedings of NSDI (April 2008)
5. Barroso, L.A., Holzle, U.: The case for energy-proportional computing. In: IEEE Computer (December 2007)
6. Mayo, R., Ranganathan, P.: Energy Consumption in Mobile Devices: Why Future Systems Need Requirements-Aware Energy Scale-Down. In: Falsafi, B., VijayKumar, T.N. (eds.) PACS 2003. LNCS, vol. 3164, pp. 26–40. Springer, Heidelberg (2005)
7. Rivoire, S., Shah, M., Ranganathan, P., Kozyrakis, C., Meza, J.: Modeling and Metrology Challenges for Enterprise Power Management. IEEE Computer (December 2007)
8. Al-Fares, M., Loukissas, A., Vahdat, A.: A Scalable, Commodity Data Center Architecture. In: Proceedings of SIGCOMM (August 2008)
9. Expect, `http://expect.nist.gov/`
10. iperf, `http://dast.nlanr.net/Projects/Iperf/`
11. Chabarek, J., Sommers, J., Barford, P., Estan, C., Tsiang, D., Wright, S.: Power awareness in network design and routing. In: Proceedings of INFOCOM (April 2008)
12. Gunaratne, C., Christensen, K., Nordman, B., Suen, S.: Reducing the Energy Consumption of Ethernet with Adaptive Link Rate (ALR). IEEE Transactions on Computers 57, 448–461 (2008)
13. Allman, M., Christensen, K., Nordman, B., Paxson, V.: Enabling an Energy-Efficient Future Internet Through Selectively Connected End Systems. In: Proceedings of HotNets (November 2007)

# MPLS Label Stacking on the Line Network*

Jean-Claude Bermond[1], David Coudert[1], Joanna Moulierac[1], Stéphane Perennes[1], Hervé Rivano[1], Ignasi Sau[1,2], and Fernando Solano Donado[3]

[1] Joint project MASCOTTE, I3S(CNRS-UNS) INRIA, Sophia-Antipolis, France
[2] Graph Theory and Combinatorics Group at Applied Mathematics IV Department of UPC, Barcelona, Spain
[3] Institute of Telecommunications, Warsaw University of Technology, Poland

**Abstract.** All-Optical Label Switching (AOLS) is a new technology that performs forwarding without any Optical-Electrical-Optical conversions. The most promising scheme to manage the control plane of these optical networks is Generic MultiProtocol Label Switching (GMPLS). In this paper, we study the problem of routing a set of requests in GMPLS networks with the aim of minimizing the number of labels required to ensure the forwarding. In order to spare the label space, we consider label stacking, allowing the configuration of GMPLS tunnels. We study particularly this network design problem when the network is a line. We provide an exact algorithm for the case in which all the requests have a common source and present some approximation algorithms and heuristics when an arbitrary number of sources are distributed over the line. We analyze by simulations the performance of our proposed algorithms and compare them with previous ones.

**Keywords:** Label space reduction, label stacking, GMPLS, AOLS.

## 1 Introduction

All-Optical Label Switching (AOLS) [1] is an approach to transparently route packets all-optically, allowing a speed-up of the forwarding. This very promising technology for the future Internet applications also brings new constraints and, consequently, new problems have to be addressed. Indeed, as the forwarding functions are implemented directly at the optical domain, a specific correlator is needed for each optical label processed in the node. Therefore, it is of major importance to reduce the number of employed correlators in every node, hence reducing the number of labels (as referred in the rest of the paper). The most promising scheme to manage the control plane of these optical networks is Generic MultiProtocol Label Switching (GMPLS). Therefore, for reducing the total number of labels in routers, solutions deployed by GMPLS for reducing the number of labels, such as label merging or label stacking, have to be studied.

* This work has been partly funded by the European project FET AEOLUS, the COLOR INRIA LARECO and the project "Optimization Models for NGI Core Network" (Polish Ministry of Science and Higher Education, grant N517 397334).

L. Fratta et al. (Eds.): NETWORKING 2009, LNCS 5550, pp. 809–820, 2009.

In this paper we consider the problem of routing a set of given requests with the aim of minimizing the total number of labels. We study this problem when the network is a line and when label stacking allowing to configure GMPLS tunnels is considered. Restricting the problem to the case when the network is a line will provide efficient algorithms that are necessary to better apprehend the general problem.

The first studies related to label space reduction in GMPLS networks are based on a technique called label merging (not discussed here). Saito *et al.* were the first considering this problem and they propose in [2] a linear programming mathematical model to find the most efficient routing solution in terms of labels using label merging. It is worth mentioning the heuristic proposed by Bhatnagar *et al.* in [3] with the same aim. The contributions using label merging were further extended in [4].

In [5] the authors deal with the problem of minimizing the number of used labels, when routes are given and the stack depth is limited to two. In [6], the authors extend this problem by assuming that routes should be found as well, considering that links have capacities. In these two contributions, the authors have as objective the minimization of the usage of the label space while keeping the stack depth to a maximum of two, which can be seen as a network design problem since the goal is to find the minimum capacities in the nodes to satisfy a traffic matrix.

This paper is organized as follows. In Section 2, we recall the basic concepts of GMPLS label forwarding mechanism. In Section 3, we formally state the problem addressed in this paper. In Section 4, we present a optimal polynomial-time algorithm when one source is considered in the line. In Section 5, we propose an approximation algorithm and heuristics when multiple sources are considered. Simulation results concerning these algorithms are reported in Section 6. Finally, Section 7 gives conclusion and perspectives of the work.

## 2 Label Switching Mechanism in GMPLS

In GMPLS, requests are established by the configuration of Label Switched Paths (LSPs). Packets are associated to LSPs by means of a label, or tag, placed in the header of the packet. In this way, routers - called Label Switched Routers (LSR) - can distinguish and forward packets. In addition, in GMPLS, it is allowed to carry a set of labels in packets header, conforming a stack of labels. Even though a packet may contain more than one label, LSRs must only read the first (or top) label in the stack in order to take forwarding decisions. Stacking labels and label processing, in general, is standardized by the following set of operations that an LSR can perform over a given stack of labels:

- SWAP: replace the label at the top by a new one,
- PUSH: replace the label at the top by a new one and then push one or more onto the stack, and
- POP: remove the label at top in the label stack.

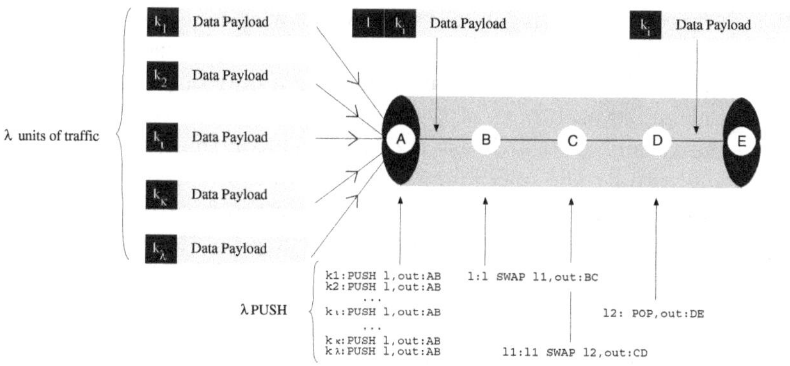


**Fig. 1.** GMPLS Operations performed at the entrance and at the exit of a tunnel

The labels stored in the forwarding table are significant only locally at the node and swapped all along the LSP.

### Label Stacking

When two or more LSPs follow the same set of links, they can be routed together 'inside' a higher-level LSP, henceforth a *tunnel*. In order to setup a tunnel, multiple labels are placed in the packet's header: a method known in the literature as *label stacking*.

As mentioned before, the LSRs in the core of the network route data solely on the basis of the topmost label in the stack. This helps to reduce both the number of labels that need to be maintained on the core LSRs and the complexity of managing data forwarding across the backbone.

Figure 1 represents the general operations needed to configure a tunnel with the use of label stacking. At the entrance of the tunnel, $\lambda$ PUSH are performed in order to route the $\lambda$ units of traffic through the tunnel. Then, only one type of operation (either a SWAP or a POP at the end of the tunnel) is performed in all the nodes along the tunnel, regardless of $\lambda$. In this figure, a stack of size 2 is used to route the $\lambda$ LSPs in one tunnel from node $A$ to node $E$. The top label $l$ is swapped and replaced at each hop: by $l_1$ at node $B$, by $l_2$ at node $C$, and is finally popped at node $D$. The $\lambda$ units of traffic, at the exit of the tunnel at node $E$ can end or follow different paths according to their bottom label $k_i$, for all $i \in \{1, 2, ..., w\}$ in the stack.

Therefore, the total cost $c(T)$ of this tunnel $T = (A, E)$ in terms of number of labels is: $c(T) = \lambda + l(T) - 1$, where $\lambda$ is the number of units of traffic forwarded through this tunnel and $l(T)$ is its length in terms of number of hops (which is 4 on this example).

The traffic can enter in any node of a tunnel but can exit in only one point, the last node of the tunnel. In other words, when some traffic is carried by a tunnel, it follows the tunnel until the end.

This cost function $c(T)$ still holds for some degenerated cases. For example, in the case of an arc (*i.e.*, a path of length 1, $l(T) = 1$), or when one unit of traffic is routed in a path (*i.e.*, a single LSP with $\lambda = 1$ whose cost is only its length). In the following, we consider as a tunnel, without loss of generality: (1) an arc routing several units of traffic, (2) a path routing a only one unit of traffic, and (3) a path routing several units of traffic (*i.e.*, $\lambda > 1$ and $l(t) > 1$). Note that strictly speaking, only the third case is considered as a tunnel.

In this paper, we fix the maximum stack size to 2. Increasing the stack size, increases also the total bandwidth consumption in the network. When the size of the stack is not limited, *label stripping* [7,8] encoding the whole path in the stack provides a feasible solution.

## 3 Modelling the LSPR Problem

This section describes the problem of routing a set of requests in GMPLS network with the aim of minimizing the number of labels. The problem is formally defined as follows:

**Label Space Reduction in a GMPLS Network: LSPR**

INPUT: a network (digraph) $G = (V, E)$ and a set of requests $\mathcal{R}$, where in the request $r \in \mathcal{R}$, $r = (s_i, u_j)$, $s_i \in V$ sends $w_r$ units of traffic to $u_j \in V$.

OUTPUT: A set $\mathcal{T}$ of tunnels enabling to route the traffic and a dipath composed of tunnels in $\mathcal{T}$ for each request $(s_i, u_j)$.

OBJECTIVE: minimize the total cost of $\mathcal{T}$, that is $c(\mathcal{T}) = \sum_{T_k \in \mathcal{T}} c(T_k)$ where the cost $c(T_k)$ of a tunnel $T_k$ which contains $\lambda_k$ units of traffic and is of length $l(T_k)$ (number of arcs in $G$ associated to the path joining the end-vertices of $T_k$) is $c(T_k) = \lambda_k + l(T_k) - 1$.

*Computation of a solution to the example of Figure 2.* Consider the line network with one source $s$, $w_1$ units of traffic destined to $u_1$ at distance $l_1$ from $s$ ($l_1 - 1$ nodes between $s$ and $u_1$) and $w_2$ units of traffic destined to $u_2$ at distance $l_1 + l_2$ from $s$. See Figure 2 for an illustration. The optimal solution depends on the values $l_i$ and $w_i$. Indeed, two solutions have to be examined.

In the first solution, a specific tunnel $(s, u_i)$ is configured for each destination $u_i$, giving two tunnels $(s, u_1)$ and $(s, u_2)$ with a total cost: $(w_1 + l_1 - 1) + (w_2 + l_1 + l_2 - 1) = w_1 + w_2 + 2l_1 + l_2 - 2$.

The second solution is composed of the two tunnels $(s, u_1)$ and $(u_1, u_2)$. The requests destined to $u_2$ will first use the tunnel $(s, u_1)$ and then the tunnel $(u_1, u_2)$. The traffic carried by $(s, u_1)$ is $\lambda_1 = w_1 + w_2$ and the traffic carried by $(u_1, u_2)$ is $\lambda_2 = w_2$. Therefore, the total cost is $(w_1 + w_2 + l_1 - 1) + (w_2 + l_2 - 1) = w_1 + 2w_2 + l_1 + l_2 - 2$.

The optimal solution is either the first one if $l_1 \leq w_2$ or the second one if $l_1 \geq w_2$.

**Lemma 1.** *In any network* $G = (V, E)$, *there exists an optimal solution* $\mathcal{T}$ *for the problem* LSPR *such that all the units of traffic of the request* $(s_i, u_j)$ *are routed in* $\mathcal{T}$ *via a unique dipath (set of consecutive tunnels) from* $s_i$ *to* $u_j$.

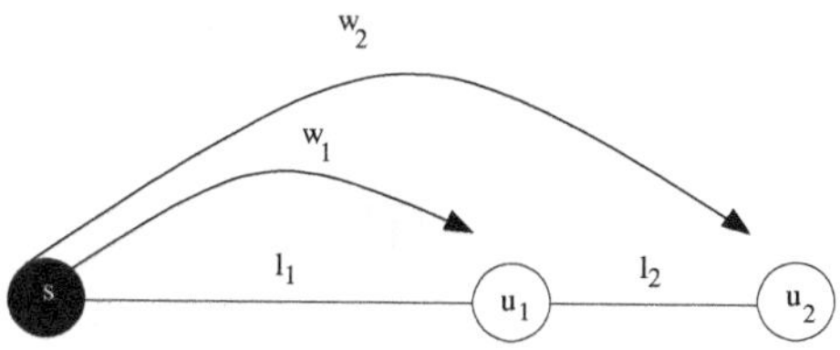


**Fig. 2.** Depending on the values $l_1$ and $w_2$, the optimal solution may be composed either of tunnels $(s, u_1)$ and $(s, u_2)$, or of tunnels $(s, u_1)$ and $(u_1, u_2)$

*Proof.* Let $\mathcal{T}$ be an optimal solution and suppose that the requests arriving at $u_j$are routed via $p > 1$ different paths $P_1, \ldots, P_m$. Let $\lambda_m$, $1 \leq m \leq p$, be the number of traffic units forwarded by $P_m$. Let $h_m$ ($h$ like hops), $1 \leq m \leq p$, be the number of consecutive tunnels in the path $P_m$. Let the order of the paths be such that $P_1$ is a path with the minimum number of consecutive tunnels $h_1$.

Then, for any other path $P_m$ $(m > 1)$ reroute the $\lambda_m$ requests routed via $P_m$ via $P_1$. We obtain a new feasible solution $\mathcal{T}'$ whose cost is

$$c(\mathcal{T}') \leq c(\mathcal{T}) + \lambda_m h_1 - \lambda_m h_m.$$

Indeed, the cost of each tunnel used in $P_m$ is decreased by $\lambda_m$, plus possibly, if some tunnel $T$ of $P_m$ becomes empty, by $l(T) - 1 \geq 0$. On the other hand, the cost of each tunnel of $P_1$ is increased only by $\lambda_m$ as the tunnel already exists. Therefore, as $h_1 \leq h_m$, we get $c(\mathcal{T}') \leq c(\mathcal{T})$ with strict inequality if $h_1 < h_m$ (the path $P_m$ is strictly longer than $P_1$) or if, in the rerouting, some tunnels of length more than 1 become empty. So $\mathcal{T}'$ is also an optimal solution.

Repeating the operation for each $P_m$ we obtain an optimal solution $\mathcal{T}^*$, where all the requests arriving at a node $u_i$ are routed in $\mathcal{T}$ via a unique dipath. □

The cost of an optimal solution $\mathcal{T}$ for problem LSPR with $|\mathcal{T}|$ tunnels and $|\mathcal{R}|$ requests is:

$$c(\mathcal{T}) = \sum_{k=1}^{|\mathcal{T}|}(l(T_k) - 1) + \sum_{r=1}^{|\mathcal{R}|} h_r w_r.$$

where $h_r$ is the number of consecutive tunnels for the request $r$ in $\mathcal{T}$, $w_r$ is the number of units of traffic of the request $r$ and $l(T_k)$ is the length of the tunnel $T_k$ in terms of number of hops. The cost $c(\mathcal{T})$ is the sum of the cost for the configuration of the tunnels $(\sum_{k=1}^{|\mathcal{T}|}(l(T_k) - 1))$ and the cost for the requests to enter the tunnels $(\sum_{r=1}^{|\mathcal{R}|} h_r w_r)$.

## 4 LSPR-L1 Problem: The Line Network, One Source

In this section, we focus on the specific case when the network $G = (V, E)$ is a directed line (a dipath) and when the number of sources is equal to 1. Focusing on the same problem with simplest constraints will provide algorithms that will be

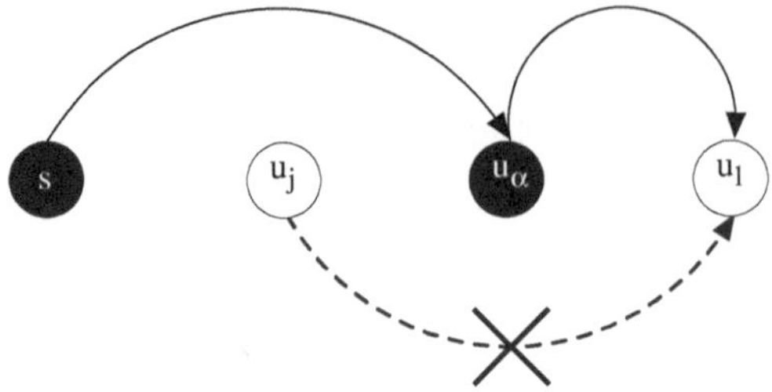


**Fig. 3.** The tunnel in dotted points is not present in an optimal solution

useful to find efficient solutions for the general problem. Let us denote by $P_{s \to u_n}$ the line where $s$ is the source and where there are $n$ requests $(s, u_i)$ with the $u_i$ indexed in the increasing order of their distance from $s$. This problem is referred as LSPR-L1 in the sequel (standing for Label Space reduction in a GMPLS Line Network with 1 source). The main result of this section is an algorithm based on dynamic programming techniques that finds an optimal solution in time $\mathcal{O}(n^3)$, as stated in Proposition 1. First, we need two technical lemmas.

**Lemma 2.** *When the network is a directed line, with source $s$, an optimal solution $\mathcal{T}$ for* LSPR-L1 *problem is such that, if $(s, u_\alpha)$ is the longest tunnel from $s$, then there is no tunnel $(u_j, u_l)$ in $\mathcal{T}$ with $j < \alpha < l$.*

*Proof.* Suppose there exists such a tunnel $(u_j, u_l)$ (see Figure 3). As $\alpha$ is the maximum index, then $u_j \neq s$, otherwise $(s, u_l)$ would have been longer than $(s, u_\alpha)$. Therefore, the number of consecutive tunnels towards $u_l$, $h_{r=(s,u_l)}$ denoted simply $h_l$, satisfies: $h_l \geq 2$. Consider the solution $\mathcal{T}'$ obtained from $\mathcal{T}$ by deleting the tunnel $(u_j, u_l)$ and adding, if it does not exist, the tunnel $(u_\alpha, u_l)$. The request $(s, u_l)$ is then routed through two consecutive tunnels $(s, u_\alpha)$ and $(u_\alpha, u_l)$. It is an admissible solution whose cost satisfies:

$$c(\mathcal{T}') \leq c(\mathcal{T}) - \lambda_l h_l - (l(u_j, u_l) - 1) + 2\lambda_l + l(u_\alpha, u_l) - 1,$$

where $\lambda_l$ is the number of requests arriving at $u_l$. As $h_l \geq 2$ and $l(u_\alpha, u_l) < l(u_j, u_l)$, $c(\mathcal{T}') < c(\mathcal{T})$. □

**Lemma 3.** *For a line $P_{s \to u_n}$ with $w_i$ units of traffic for the request $(s, u_i)$, the cost of an optimal solution is:*

$$c^*(P_{s \to u_n}) = \min_{\alpha} \left[ \sum_{i=\alpha}^{n} w_i + l(s, u_\alpha) - 1 + c^*(P_{s \to u_{\alpha-1}}) + c^*(P_{u_\alpha \to u_n}) \right],$$

*where $u_\alpha \in P_{u_1 \to u_n}$ is a splitting point that decomposes the problem into two sub-problems.*

*Proof.* By Lemma 2 an optimal solution contains a tunnel $(s, u_\alpha)$ of cost $(\sum_{i=\alpha}^{n} w_i + l(s, u_\alpha) - 1)$ plus an optimal solution on the sub-line $P_{s \to u_{\alpha-1}}$ and an optimal solution on the sub-line $P_{u_\alpha \to u_n}$. □

**Algorithm 1.** Polynomial-time algorithm computing an optimal solution for the LSPR-L1 problem

**Input**: Line $P_{s \to u_n}$ from $s$ to $u_n$, where $s$ is the source (referred also as $u_0$) and $(s, u_i)$ are the set of requests ($i \geq 1$), each of them having $w_i$ units of traffic
**Output**: Set of tunnels enabling the routing from $s$ of all the requests $(s, u_i)$
**begin**
  $C$ is a table of size $n^2$ indicating the costs all the sub-solutions;
  $S$ is a table of size $n^2$ indicating the splitting points $u_\alpha$ associated to the optimal sub-solutions;
  $W$ is a table of size $n$ storing partial sums of weigths, $W[0] = 0$, $W[j] = \sum_{i=1}^{j} w_i = W[j-1] + w_j$, and so $\sum_{i=\alpha}^{\beta} w_i = W[\beta] - W[\alpha - 1]$;
  **for** $i \in [0, n]$ **do**
    $C[u_i, u_i] = 0$;
    $C[u_i, u_{i+1}] = w_{i+1} + l(u_i, u_{i+1}) - 1$;
    $S[u_i, u_{i+1}] = u_{i+1}$;
  **for** $i \in [2, n]$ **do**
    **for** $\forall k \in [0, n-i]$ **do**
      $min = +\infty$;
      **for** $\forall \alpha \in [k+1, k+i]$ **do**
        $value =$
        $(W[k+i] - W[\alpha - 1]) + l(u_k, u_\alpha) - 1 + C[u_k, u_{\alpha-1}] + C[u_\alpha, u_{k+i}]$;
        **if** $value < min$ **then**
          $min = value$;
          $C[u_k, u_{k+i}] = c(P_{u_k \to u_{k+i}}) = value$;
          $S[u_k, u_{k+i}] = u_\alpha$;
  Compute the optimal set of tunnels from the table $S$;
**end**

**Proposition 1.** *When the network is a directed line $P_{s \to u_n}$, and all requests are issued from $s$, then an optimal solution of the* LSPR-L1 *problem can be computed in time $\mathcal{O}(n^3)$ by Algorithm 1.*

*Proof.* According to Lemma 3, to compute an optimal solution for $P_{s \to u_n}$, we need first to compute optimal sub-solutions for $P_{s \to u_{\alpha-1}}$ and for $P_{u_\alpha \to u_n}$, $u_\alpha \in \{u_1, \ldots, u_n\}$, and recursively. The algorithm computes first solutions for $P_{u_i \to u_{i+1}}$, and for computing solutions for $P_{u_i \to u_{i+2}}$, the already computed values for sub-lines $P_{u_i \to u_{i+1}}$ (say $C[u_i, u_{i+1}]$) and $P_{u_{i+1} \to u_{i+2}}$ (say $C[u_{i+1}, u_{i+2}]$) are used without any recomputation.

For example, to compute the solution on $P_{s \to u_2}$, we need the values $C[s, u_1]$ and $C[u_1, u_2]$ since we have $C[s, u_2] = \min\{(w_1 + w_2 + l(s, u_1) - 1 + C[u_1, u_2]), (w_2 + l(s, u_2) - 1 + C[s, u_1])\}$. Now, if we want to compute the solution on $P_{s \to u_3}$, we need to compute first $C[u_1, u_3]$ and $C[u_2, u_3]$, but not $C[s, u_1]$ and $C[s, u_2]$ that are already known from previous computations and stored in table $C$.

Finally, we can compute the optimal solution using dynamic programming (Algorithm 1), with time complexity $\mathcal{O}(n^3)$ and space complexity $\mathcal{O}(n^2)$. □

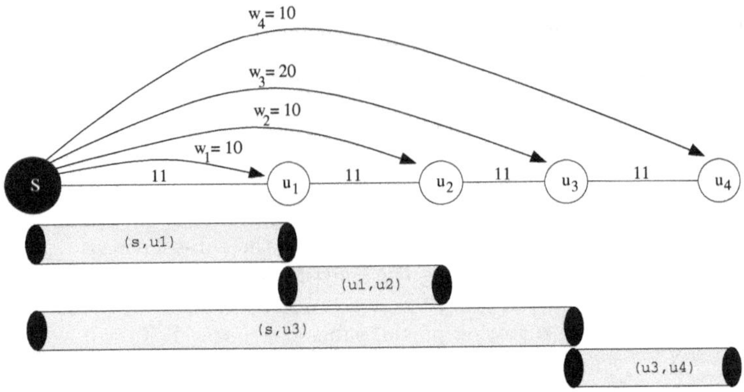


**Fig. 4.** An example with its optimal solution

*The optimal algorithm in the example of Figure 4.* Let us compute an optimal solution to the example in Figure 4 using Algorithm 1. We first have to compute the table $C$ containing the costs of the sub-optimal solutions for each sub-line.

First, the sub-paths of length 1, $P_{s\to u_1}$, $P_{u_1\to u_2}$, $P_{u_2\to u_3}$, and $P_{u_3\to u_4}$ are straightforward computed in $C[u_0,u_1]$, $C[u_1,u_2]$, $C[u_2,u_3]$, and $C[u_3,u_4]$.

Then, for the sub-paths of length 2, $P_{s\to u_2}$, $P_{u_1\to u_3}$, and $P_{u_2\to u_4}$, two splitting points are considered by the algorithm. For example, for $P_{s\to u_2}$, the optimal solution implies a splitting point $u_1$ with cost $w_1+w_2+l(s,u_1)-1+C[u_0,u_0]+C[u_1,u_2]=50$ (the splitting point $u_2$ implying a greater cost $w_2+l(s,u_2)-1+C[u_0,u_1]+C[u_2,u_2]=51$). The already computed costs $C[u_0,u_0]$, $C[u_1,u_2]$, and $C[u_2,u_2]$ have been used by the algorithm and are not computed again.

For the computation of the optimal solution on the whole line $P_{s\to u_4}$, four splitting points, $u_1$, $u_2$, $u_3$, and $u_4$ should be considered.

When the table $C$ showing the optimal costs for all the subpaths has been computed as presented in Table 1, the set of tunnels composing the optimal solution can be deduced from the splitting points. The optimal solution for line $P_{s\to u_4}$ has cost 132 and a splitting point $u_3$. Thus, the optimal solution is composed of a tunnel $(s,u_3)$ and of optimal solutions for the sub-paths $P_{s\to u_2}$ and $P_{u_3\to u_4}$. The first sub-solution has a splitting point $u_1$ which gives tunnels $(s,u_1)$, $(u_1,u_2)$. The optimal solution for the sub-path $P_{u_3\to u_4}$ is obviously the tunnel $(u_3,u_4)$.

Finally, the optimal solution is composed of tunnels $(s,u_1)$, $(u_1,u_2)$, $(s,u_3)$, and $(u_3,u_4)$.

In the special case where the requests are uniform, we are able to give a closed formula of the cost of an optimal solution, as stated as follows.

**Proposition 2.** *For a line network $P_{s\to u_n}$, with $n=2^q-1+r$, where $0\le r\le 2^q-1$, and an uniform distribution: $\forall i, w_i=1$, the cost of an optimal solution is $2^q(q-1)+1+(q+1)r$.*

The proof of this proposition is technical and is not given in this paper due to lack of space. In this specific case we can prove that $c^*(P_{s\to u_n})=c^*(P_{s\to u_{n-1}})+\lfloor\log(n)\rfloor+1$.

**Table 1.** Computation of the table $C$ and $S$ for the optimal solution of the example on Figure 4, the nodes in brackets representing the splitting points of table $S$

| | $s = u_0$ | $u_1$ | $u_2$ | $u_3$ | $u_4$ |
|---|---|---|---|---|---|
| $s = u_0$ | 0 | 20 | 50 ($u_1$) | 101 ($u_2$) | 132 ($u_3$) |
| $u_1$ | - | 0 | 20 | 61 ($u_3$) | 91 ($u_3$) |
| $u_2$ | - | - | 0 | 30 | 60 ($u_3$) |
| $u_3$ | - | - | - | 0 | 20 |
| $u_4$ | - | - | - | - | 0 |

## 5 LSPR-LM Problem: The Line Network, Multiple Sources

In this section, we study the problem of routing a set of requests on the line network when multiple sources are distributed along the line. Since sources may inject traffic anywhere in the network, Lemma 2 is not valid anymore, hence the problem seems to be inherently more complicated. As the problem cannot be decomposed as easily as previously, we present in this section a $\log(n)-$approximation algorithm and an heuristic that will be compared to the optimal solution and to previous known heuristics in Section 6. The problem is referred in the following as LSPR-LM (standing for Label Space reduction in a GMPLS Line Network with Multiple sources).

### 5.1 log($n$)-Approximation Algorithm for LSPR-LM

Consider the nodes $\{u_0, u_1, \dots, u_n\}$, that can be source or destination or both, sorted according to their position on the line from the left to the right ($u_{i+1}$ after $u_i$ on the line, $u_{i+1}$ being at distance $l_{i+1}$ from $u_i$). Suppose that the line is of length $L$, meaning that $L = \sum_{i=1}^{n} l_i$.

The algorithm consists of configuring all the consecutive tunnels $\{(u_0, u_1), (u_1, u_2), \dots, (u_{n-1}, u_n)\}$, $\{(u_0, u_2), (u_2, u_4), \dots, (u_{n-2}, u_n)\}$, $\{(u_0, u_4), (u_4, u_8), \dots, (u_{n-4}, u_n)\}$, and more generally, those of length a power of 2. See Figure 5 for an illustration of the configuration of the tunnels. Consequently, there exists a path of at most $\log(n)$ tunnels from any source to any destination, ensuring a valid routing for all the requests. When the solution has been computed, then some tunnels that are not used by any destination may be removed.

**Theorem 1.** *For a problem with $n$ sources and/or destinations, there exists a $\log(n)$-approximation algorithm for the* LSPR-LM *problem.*

*Proof.* The cost of a solution computed by the algorithm is (1) the cost of the configuration of the tunnels plus (2) the cost for entering the tunnels.

To configure each level of consecutive tunnels, at most $L-1$ labels are needed. There are at most $\log(n)$ different levels of tunnels. So, the overall number of labels needed for the configuration of tunnels is at most $(1) \leq (L-1)\log(n)$.

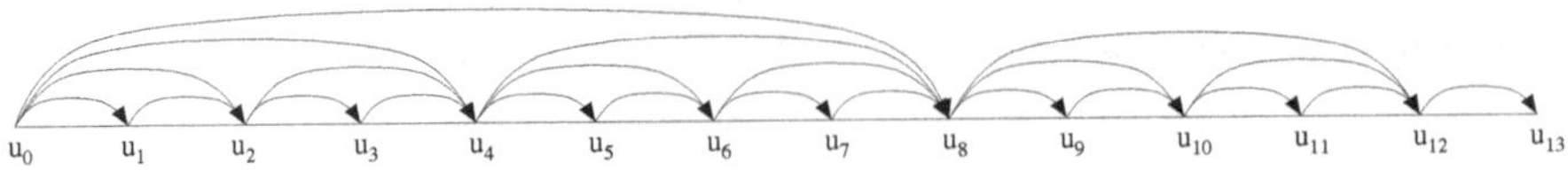


**Fig. 5.** Computing this set of tunnels gives a $\log(n)-$approximation algorithm for LSPR-LM problem

When that set of tunnels has been configured, any source can join any destination in at most $\log(n)$ hops. Therefore, the total cost needed to enter the tunnels is at most $(2) \leq \sum_{r=1}^{|\mathcal{R}|} w_r \log(n)$.

Then, the cost of this solution is at most: $(1) + (2) \leq \sum_{r=1}^{|\mathcal{R}|} w_r \log(n) + (L - 1)\log(n) = \log(n)(\sum_{r=1}^{|\mathcal{R}|} w_r + L - 1)$.

In the best case, an optimal solution will be of cost $\sum_{r=1}^{|\mathcal{R}|} w_r + (L-1)$, giving a $\log(n)-$approximation. □

### 5.2 Proposed Heuristic: Extended Dynamic Programming

This subsection presents a simple heuristic to find a solution of the problem on the line with multiple sources. Suppose that, when constructing the solution, there is a set of tunnels leading from a source $u_0$ to a destination $u_i$. Then, if another source, say $u_x$ with $x > 1$, has to transmit traffic to $u_i$, then $u_x$ may insert traffic directly in the tunnels going to $u_i$ without additional cost.

Therefore, the heuristic consists of considering only the source $u_0$, then, to affect the whole set of requests to $u_0$ and to use the polynomial algorithm just described previously for only one source. In the solution, there would be tunnels from $u_0$ to all the destinations, and the other sources will insert their traffic in the tunnels passing through them.

## 6 Simulations

In this section we analyze the performance of the proposed heuristics using simulations. The analysis consists of the comparison of the total number of labels used by the heuristics.

In our simulations, we use a line network consisting of 500 nodes. Each experiment consists of a different number of sources and destinations. The number of sources equals the number of destinations in each experiment. Between a pair of source and destination nodes, a demand is generated (with a probability of 80%) with a random capacity between one and 500 (uniform).

Figure 6 (top) shows the behavior the heuristics proposed in this article together with the Longest Segment First (LSF) heuristic [9]. The number of sources (or destinations) varies from 5 to 113 with increments of three nodes in each experiment. Each experiment was run 100 times. The results show that, even though the $\log(n)$-approximation runs in $\mathcal{O}(p \log n)$ and guarantees a

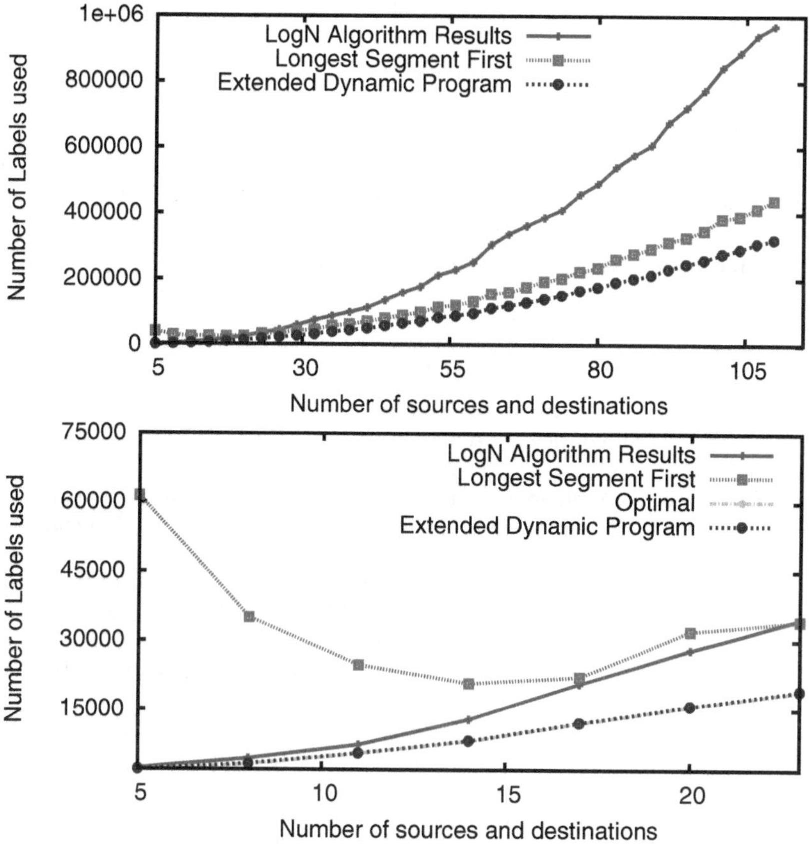


**Fig. 6.** Comparison on the number of labels used by different heuristics and magnification of the first 20 experiments including the optimal solution

bound in terms of sub-optimality, in practice the results are not as good as the proposed Extended Dynamic Programming heuristic or the LSF heuristic running in $\mathcal{O}(n^3)$ and $\mathcal{O}(np^2)$, respectively. We also observed that the requirements in memory for LSF are lower than those of the Extended Dynamic Programming heuristic; however the quality of the solution of the later always outperforms the former's. Some other previously proposed heuristics (see [6] and [10]) were tested as well with worse results, hence not considered in this analysis.

At the bottom of Figure 6, a magnification of the results in the first 20 experiments is shown. The plot showing the optimal value is also added. In these experiments, the numerical solution computed by the heuristic based in dynamic programming is within 1% (in most of the case) of the optimal value. We conjecture that this is because the demands share the same set of destinations. The proposed heuristics in this paper show a better convergence than that of LSF when the number of sources is low.

## 7 Conclusion and Perspectives

We presented in this paper the problem of routing a set of requests with the aim of reducing the total number of labels in the network. For the line, we exhibit a polynomial-time algorithm when there is a single source and a $\log(n)$-approximation algorithm, and one heuristic for multiple sources. We show the good performance of these algorithms through simulations. In future work, we plan to extend these proposed algorithms to general networks and to study the computational complexity of the LSPR problem.

## References

1. Ramos, F., et al.: IST-LASAGNE: Towards all-optical label swapping employing optical logic gates and optical flip-flops. IEEE J. Sel. Areas Commun. 23(10), 2993–3011 (2005)
2. Saito, H., Miyao, Y., Yoshida, M.: Traffic engineering using multiple MultiPoint-to-Point LSPs. In: IEEE Conference on Computer Communications (Infocom 2000), pp. 894–901 (2000)
3. Bhatnagar, S., Ganguly, S., Nath, B.: Creating Multipoint-to-Point LSPs for traffic engineering. IEEE Commun. Mag. 43(1), 95–100 (2005)
4. Solano, F., Fabregat, R., Marzo, J.: On optimal computation of MPLS label binding for MultiPoint-to-Point connections. IEEE Trans. Commun. 56(7), 1056–1059 (2007)
5. Solano, F., Stidsen, T., Fabregat, R., Marzo, J.: Label space reduction in MPLS networks: How much can a single label do? IEEE/ACM Trans. Netw. (December 2008)
6. Solano, F., Caenegem, R.V., Colle, D., Marzo, J.L., Pickavet, M., Fabregat, R., Demeester, P.: All-optical label stacking: Easing the trade-offs between routing and architecture cost in all-optical packet switching. In: IEEE Conference on Computer Communications (Infocom 2008), Phoenix, AZ, USA, April 2008, pp. 655–663 (2008)
7. Van Caenegem, R., et al.: Benefits of label stripping compared to label swapping from the point of node dimensioning. Photonic Network Communications Journal 12(3), 227–244 (2006)
8. Van Caenegem, R., et al.: From IP over WDM to all-optical packet switching: Economical overview. J. Lightw. Technol. 24(4), 1638–1645 (2006)
9. Solano, F., Fabregat, R., Donoso, Y., Marzo, J.: Asymmetric tunnels in P2MP LSPs as a label space reduction method. In: Proc. IEEE International Conference on Communications (ICC 2005), May 2005, pp. 43–47 (2005)
10. Solano, F., Fabregat, R., Marzo, J.: A fast algorithm based on the MPLS label stack for the label space reduction problem. In: Proc. IEEE IP Operations and Management, IPOM 2005 (October 2005)

# Modelling and Performance Evaluation of Improved Access Mechanisms in a Novel Multiservice OPS Architecture*

Thaere Eido[1], Ferhan Pekergin[2], and Tülin Atmaca[1]

[1] TELECOM & Management SudParis, RST Dept., 9 rue Charles Fourier, Evry, 91011 France
[2] LIPN – CNRS UMR 7030, Av. J.B. Clément, Villetaneuse 93430 France
{Thaere.Eido,Tulin.Atmaca}@it-sudParis.eu,
Ferhan.Pekergin@lipn.univ-paris13.fr

**Abstract.** Optical Packet Switching (OPS) technologies are among the most promising solutions for Next Generation Networks. In OPS networks, several mechanisms for Quality of Service (QoS) management have been developed. In this paper, we address QoS and fairness issues in a novel OPS ring architecture with a slotted synchronous transmission. The MAC protocol in the studied architecture is similar to CSMA/CA. Each node may use the available bandwidth in an opportunist manner. Therefore, the chances for each node to find available transmission resources are tightly correlated to the matrix of traffic at the other nodes. In order to remedy this problem, we elaborate two improved access mechanisms with preemptive approaches: Packet Erasing Mechanism (PEM) and Extraction and Reemission Mechanism (ERM). We evaluate the performance of the proposed mechanisms using a discrete-time analytical model. Finally, the results obtained by our model are validated and analyzed through several simulation scenarios.

**Keywords:** Medium Access Control (MAC), Metro Access Network Architectures, Modeling and Performance Evaluation, Quality of Service, Slotted Optical Packet Switched (OPS) Ring Networks.

## 1 Introduction

Packet ring-based networks were pioneered by the Cambridge Ring, followed by other important network architectures such as FDDI and RPR. Those networks are mainly based on electronic routers which provide satisfying performance but with relatively high cost. However, the French research project ECOFRAME is targeting a new optical Metro Access Network (MAN) [1], [2]. This network architecture aims to develop a new technological approach with lower cost compared to a pure electronic solution. It has also some added values in terms of reliability and performance. The studied network

* This work was partially supported by the French ANR / ECOFRAME and by the European EuroNF research projects.

L. Fratta et al. (Eds.): NETWORKING 2009, LNCS 5550, pp. 821–834, 2009.

is synchronous. Moreover, the transmission medium is subdivided into slots of equal duration. Each slot can accommodate one fixed-size optical packet.

In the scope of this paper, we address modeling and performance evaluation of optimized access mechanisms that we propose for this novel OPS architecture. We use an absolute priority scheme inside each ring node. However, it is well known that the absolute priority scheme improves the performance of the higher priority classes, but can cause excessive delays for the low-priority classes especially when the network is highly loaded [3]. Discrete time communication systems with constant slot size, including several priority management schemes, have been studied with many details in [4]. Besides, a dynamic packet transmission discipline, with multiple classes of delay-sensitive traffic and the Head of the Line (HOL) priority scheme, was proposed and analyzed in [5]. Moreover, a dynamic priority scheme with dynamic jumps has been recently investigated by analytical modeling [6]. However, while these studies have provided interesting results concerning the priority management inside a single node, none of them have taken into account the interactions through the link between several nodes at the network scale. In fact, even the absolute priority that performs a large discrimination between different classes of traffic may lack efficiency when it is applied to a ring or bus topology. Depending on the traffic matrix, some stations can have a very privileged position which allows them to monopolize a large amount of bandwidth. Hence, the priority scheme must be elaborated at the scale of the network and not only at the node scale.

Therefore, we elaborate two preemptive access mechanisms: Packet Erasing Mechanism (PEM) and Extraction and Reemission Mechanism (ERM). These mechanisms prevent the transmission of packets with high priority from being blocked by packets of lower priority sent by other ring nodes. The preemption of BE packets at intermediate nodes is controlled by a Queue-Length Threshold (QLT) policy. Thus, this mechanism implements a sort of dynamic priority. In the studied architecture the interaction between classes is performed through the ring. In addition to the state of local buffers, the use of the slots (link state) should also be considered in the analysis of such system. This original feature of this study extends the QLT priority scheme to the network level.

From the modeling point of view, the synchronous slotted mode that characterizes the studied architecture imposes the resort to a common service policy which is based on a cyclic admittance scheme [7]. However, in the studied ring network architecture, optical packets bypass intermediate nodes transparently. Moreover, the transmission medium is shared by several nodes. Hence, the transmission of new packets arriving to the system undergoes a possibility of blocking. After resolution of the proposed model, we provide expressions for delay and loss performance parameters. The results obtained are validated by simulation works.

This paper is organized as follows. The studied network architecture is described in section 2. In section 3 we present the ring node function and the main ideas that inspire the proposed model. Expressions for delay measures and packet loss ratio are provided in section 4. Novel QoS mechanism (ERM) and the extension of our model for its support are presented in section 5. Section 6 gives model validation by simulation works and performance analysis of the studied network and the proposed mechanisms. Finally, section 7 concludes our work.

## 2 Studied Network Architecture

We consider a ring network with synchronous transmission, which allows the transport of fixed-size optical packets in a slotted mode. Client traffic consists of several classes of Service (CoS). Ring nodes communicate directly without resorting to any central station (Hub).

Each ring node (station) can receive, drop or insert packets to be transmitted on the ring. The MAC protocol in the studied network is similar to the Carrier Sense Multiple Access with Collision Avoidance (CSMA/CA) scheme [8]. The node constantly watches on the transit line (link), seeking for free slots on the ring. Network nodes read the header of all incoming packets in order to decide whether to extract or not each of these optical packets from the ring. When a node is the destination of a packet, it removes the packet completely from the ring.

In the studied architecture, the traffic coming from client networks consists of small electronic packets of variable-size. These electronic packets are collected and put together into large fixed-size optical payloads. In the following sections, we refer to this procedure as the *fixed-size packet creation mechanism.* Note here that we do not deal in this study with this procedure since it has been well studied in [9]. The resulted optical packets are stored in electronic queues called *local buffers* (one buffer for each class of service). When the current slot is free, *local buffers* are served according to a head of line (HOL) policy. Since we consider synchronous slotted transmission, the node can transmit an optical packet (if any) only at the beginning of a free slot.

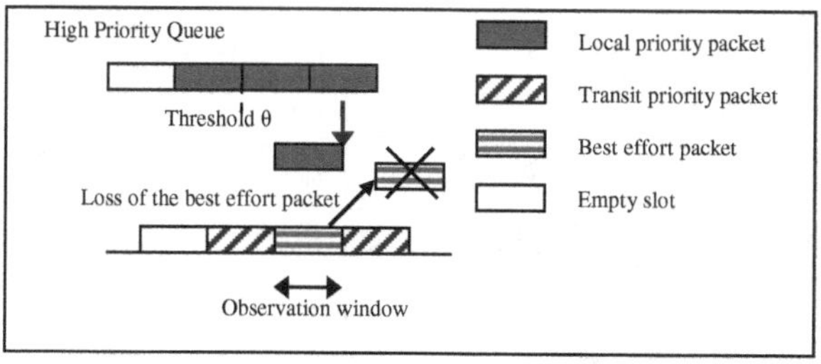


**Fig. 1.** Packet erasing mechanism

As for the QoS management, we propose two improved preemptive access mechanisms: the Packet Erasing Mechanism (PEM) and the Extraction and Reemission Mechanism (ERM). As shown in Figure 1, when the PEM mechanism is enabled, Best Effort (BE) packets can be dropped (preempted) at intermediate nodes in order to be replaced by packets of higher priority. We denote by $C$ the number of classes of service in the network and by $Nq$ the capacity of the buffer of class $q$ ($q$=1,..,$C$). Note that $q$=$C$ corresponds to the *Best Effort* class. In order to limit the preemption of BE packets, the activation of the erasing mechanism is controlled. Packets of class $q$ ($q$=1,..,$C$-1) can preempt BE packets only when the number of packets in the buffer of class $q$ is superior or equal to a threshold $\theta_q$ ($\theta_q$ =1,.., $N_q$).

# 3 Modeling of the OPS Ring

## 3.1 Network Model

We consider a ring constituted by *S* stations connected by a link. In this structure each station behaves in the same way, therefore they are characterized by the same node model but possibly with different parameters. However, the nodes in the network interacts through the link, so to describe the network behavior, the link state indicating if the slots are free or occupied, must be also determined.

Each node encloses *C* electronic buffers in order to accommodate a number of *C* Classes of Service (*CoS*). *CoS*1 is the class with the highest priority. We may also refer to these buffers as *add buffers* or *local buffers*. The capacity of a *local buffer q* ($q=1,..,C$) is the number of packets $N_q$ which can be stored inside.

We suppose that packets of class *q* arrive to the station following a Poisson process with a rate $\lambda_q$ ($q=1,..,C$). We denote by s, the state of a slot on the link. When a slot is free, its state is 0; otherwise it corresponds to the CoS ID (*1,..,C*) of the packet transmitted in this slot.

## 3.2 Computation of the State Probabilities

Hereafter we compute the state probabilities for local buffers at the stationary state of the system. We are interested only in the state of the transmission buffers. The state of the reception buffer doesn't impact the network performance.

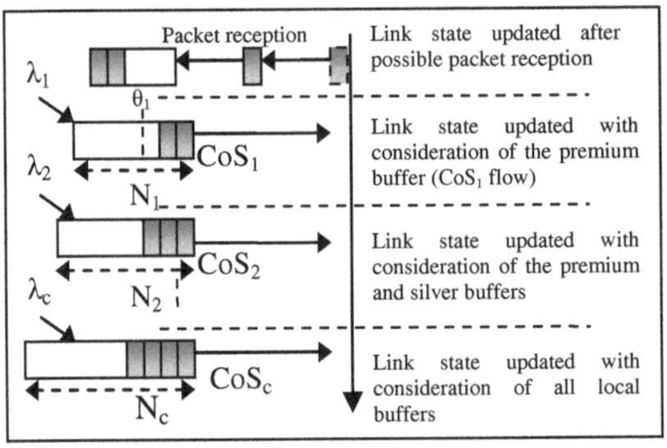


**Fig. 2.** Steps of slot state update (decomposition method)

The queueing policy at each node is based on HOL absolute priority. The service of clients of class *q* depends only on the service of potential clients in the previous queues (classes 1 to *q*-1) and on the state of the current slot seen by the first buffer in the considered node. Hence, we decompose our system to as many parts as the number of local buffers in all ring nodes. The effect of each node is reported on the link state in a circular sequence. For a timeslot observed by a single node, the state of the link is modified by taking into account the state of the local transmission buffers, considered in a sequential way according to their priority. Moreover, at the entrance of each node, the state of the current slot is updated in order to take into account the traffic matrix, i.e. the probability for the potential packet carried on this slot to be destined to the current node. Figure 2 illustrates the above described decomposition.

Due to the constant slot duration, the steady state distribution of this system depends on the observation moments that we choose. For the resolution of the studied model, we observe the system state right after a possible modification of the link state, taking into account the effect of all queues in the node under consideration. So, the steady state of the system is observed at the end of each timeslot, right after the slot passes the current node and before it goes to the next node on the ring. The model obtained after decomposition using the priority is a typical discrete time queueing model with buffers of finite capacity. It can be easily shown that the associated steady state distribution exists and it is unique.

At each node, the state of the link is modified consecutively by taking into account the effect of each local buffer $q$ ($q=0,..,C$), where $q=0$ denotes the reception buffer and $q=1,..,C$ denote local transmission buffers. Let $P_{l,q}(s)$ be the probability for the current slot on the link, seen by the buffer $q$, to be in the state $s$ ($s=0,..,C$). Hence, $P_{l,q+1}(s)$ will represent the state probability of the link after taking into account the effect of the buffer $q$.

Let $P_q(n)$ be the probability for the buffer of class $q$ to have $n$ packets inside at the end of the current timeslot (taking into account new arrivals and after one of its packets has been possibly moved to the server for transmission).

We denote by $\Omega_{q,n_q}(k)$ the conditional probability that $k$ ($k \in \{0,1\}$) packet exits from the queue of class $q$ during the current slot when $n_q>0$.

$$\Omega_{q,n_q}(k=1) = P_{l,q}(0) + \mathbb{1}(n_q \rangle \theta_q) \cdot P_{l,q}(C) \tag{1}$$

where $\mathbb{1}(.)$ is the indicator function which takes the value 1 if the associated condition is satisfied and 0 otherwise. Clearly, the second term expresses the case where a *Best Effort* packet (called *BE* or *class C*) is erased to the benefit of the class $q$.

We denote by $\alpha_q(j)$ the probability that $j$ packets of class $q$ arrive in timeslot duration $\Delta$. When the packet arrivals is characterized by a Poisson process, $\alpha_q(j) = (\lambda_q \times \Delta)^j \cdot e^{-\lambda_q \times \Delta} / j!$

During the computation, first, the state of the local buffer $q$ is updated taking into account the state of the link viewed by this queue. Then the state of the slot is updated taking into account the effect of $q$. The new state of the slot will be seen by the next queue (of class $q+1$) of the node. The last update in a node gives $P_{l,0}(s)$ for the next node on the ring. Note here that since we are interested in the stationary state of the system, the initial values of the state probabilities (of the link and the local buffers) have no impact on the results but only on the computation time.

For the node under consideration, the state of the buffer of class $q$ changes upon the arrival of new packets or the departure (transmission) of the packet at the head of it. Therefore, this state is characterized by the following transition equations:

$$P_q^+(n) = \begin{cases} \sum\limits_{i+j-k=n} P_q(i) \cdot \alpha_q(j) \cdot \Omega_{q,i}(k) & \text{for } 0 \le n < N_q - 1 \\ \sum\limits_{i+j \ge N_q} P_q(i) \cdot \alpha_q(j) \cdot \Omega_{q,i}(1) + \sum\limits_{i+j=N_q-1} P_q(i) \cdot \alpha_q(j) \cdot \Omega_{q,i}(0) & \text{for } n = N_q - 1 \\ \sum\limits_{i+j \ge N_q} P_q(i) \cdot \alpha_q(j) \cdot \Omega_{q,i}(0) & \text{for } n = N_q \end{cases} \tag{2}$$

Recall that, the probabilities $P_{l,0}(s)$ should be updated at the entry of each node depending on whether the packet at the current slot is destined or not to the current node. This update will provide the probabilities $P_{l,1}(s)$, which characterize the link state seen by the first buffer in the node. Transition equations for this update are:

$$\begin{cases} P_{l,1}(0) = P_{l,0}(0) + \sum_{i=1}^{C} P_{l,0}(i) \cdot P_{R,i} \\ P_{l,1}(i) = P_{l,0}(i) \cdot (1 - P_{R,i}) \quad for \quad 1 \le i \le C \end{cases} \tag{3}$$

where $P_{R,i}$ ($i$=1,..,C) is the probability for a packet of class $i$ to be destined to the considered node. For a uniform traffic in a ring with $S$ stations, $P_{R,i}=1/(S-1)$, ($i$=1,..,C)

Finally, the state transitions for the slot, considering the buffer of class $q$, are:

Case 1: The slot is free and no packets in queue are ready for insertion.
Case 2: The queue is not empty and the slot is free.
Case 3: The slot carries a *BE* packet, which is preempted by the PEM mechanism.
Case 4: The slot is carrying a transiting packet and no insertion event occurs.

The state transitions for the slot under the effect of a queue $q$ ($q$=1,..,$C$-1), taking into account the state of this queue at the end of the timeslot, are described as follows:

$$\begin{cases} P_{l,q+1}(0) = P_{l,q}(0) \cdot P_q(0) \cdot \alpha_q(0) \\ P_{l,q+1}(q) = P_{l,q}(q) + P_{l,q}(0) \cdot \left(1 - P_q(0) \cdot \alpha_q(0)\right) + P_{l,q}(C) \cdot \left[ \sum_{i+j \ge \theta_q} P_q(i) \cdot \alpha_q(j) \right] \\ P_{l,q+1}(C) = P_{l,q}(C) \cdot \left[ 1 - \sum_{i+j \ge \theta_q} P_q(i) \cdot \alpha_q(j) \right] \\ P_{l,q+1}(i) = P_{l,q}(i) \quad if \ 0 < i < C \text{ and } i \ne q \end{cases} \tag{4}$$

Note here that a packet that arrives to the node during a slot is stored, at least till the end of the current slot, in the local buffer that corresponds to its class of service.

## 4 Delay and Loss Expressions

### 4.1 Computation of Packet Loss Ratios

In the studied architecture, packets may be lost at local buffers because of buffer overflow or, for *BE* packets, dropped at intermediate nodes by the PEM mechanism. The *Packet Loss Ratio* $\rho_q$ is the ratio of packets which are lost because of overflow of the buffer $q$, to total generated packets of class $q$.

When there are $n$ packets in a local buffer $q$ of capacity $N_q$, the arrival of $k$ new packets ($k > [N_q - n]$) in a timeslot of duration $\Delta$ causes the loss of $k - [N_q - n]$ packets. Therefore, $\rho_q$ can be expressed as:

$$\rho_q = \sum_{n=0}^{N_q} \sum_{k=N_q-n+1}^{\infty} \alpha_q(k) \cdot P_q(n) \cdot \left[k - (N_q - n)\right] \Big/ \lambda_q \cdot \Delta \tag{5}$$

Thereafter, the performance measures concerning different stations will be written with a superscript $k$. For example, the loss probability at the add buffer of class $q$ in the station of rank $k$ ($k$=1,..,$S$) will be denoted as $\rho_q^{(k)}$. This superscript will be removed whenever there will be no risk of confusion.

The *total loss ratio* of packets of class $q$ ($q$=1,..,$C$-1) is equal to $\rho_q^{(k)}$. As for the class $C$, packets which are dropped at intermediate nodes should also be considered.

We denote by $P_{d|C}^{(k)}$ the conditional probability for a packet of class $C$, being transmitted on the slot, to be dropped at the intermediate station of rank $k$.

$$P_{d|C}^{(k)} = 1 - \prod_{q=1}^{C-1} \sum_{0 \le i+j < \theta_q^{(k)}} \left( P_q^{(k)}(i) \cdot \alpha_q^{(k)}(j) \right) \tag{6}$$

A station of rank $v$ may send packets to the stations situated between *next(v)*=[($v$+1) *mod S*] and *prev(v)*=[($v$-1) *mod S*], where *mod* denotes the *modulus* function. A *Best Effort* packet enters to the buffer of class $C$ in the station $v$ with the probability $\left(1 - \rho_C^{(v)}\right)$. Afterwards, the probability for this packet to reach its destination (that we denote by $k$) can be expressed as: $\prod_{h=next(v)}^{prev(k)} \cdot \left[1 - P_{d|C}^{(h)}\right]$

We denote by $r_{k|C}^{(v)}$ the probability for a packet of class $C$ arriving to the station $v$ to be destined to the station $k$. Therefore, the *total loss ratio* of *Best Effort* packets arriving to a station $v$ can be expressed as follows:

$$\rho_{C,PEM}^{(v)} = 1 - \left(1 - \rho_C^{(v)}\right) \cdot \left[ \sum_{k=(v+2).\mathrm{mod}.S}^{prev(v)} r_{k|C}^{(v)} \cdot \prod_{h=next(v)}^{prev(k)} \cdot \left[1 - P_{d|C}^{(h)}\right] \right] \tag{7}$$

Recall that the packets sent from $v$ to $k$ = [($v$+1) *mod S*] are not be affected by the erasing mechanism PEM, hence [($v$+1) *mod S*] is omitted in the above summation.

### 4.2 Computation of the Moments of the Queueing Delay

Each packet arriving into a local buffer should wait a time $R$ up to the beginning of the next timeslot. Afterwards, it should wait a time $W$ until it gets to the head of the queue. Because of the non-continuous availability of the link, a packet which reaches the head of the queue should wait an additional delay (called *access time* or $A$) until it finds a free slot where it is then transmitted.

The queueing delay $Q$ inside each local buffer consists of the sum of $R$, $W$ and $A$. Moreover, the waiting time $W$ is equal to the sum of access times of packets seen by the arriving one while entering to the system. Recall that for a queue having stationary distribution and state changes of one-step type (which is the case in our model), the distribution found at the incoming moments and left at outgoing moments are the same [10]. Hence, each packet entering (but not all arriving packets) to a local buffer sees the same distribution, as a packet which has just been retrieved from the queue.

Therefore, if we characterize the state of the system right after packet transmission moments, then the queueing delay $Q$ is expressed as: $Q = R+(1+X_e)\cdot A$, where $X_e$ is the queue length seen by a packet just before entering to the queue or just after leaving it. $R$, $X_e$ and $A$ are independent random variables, therefore:

$$E[Q] = E[R] + (1 + E[X_e]) \cdot E[A] \tag{8}$$

$$Var[Q] = Var[R] + (1 + E[X_e]) \cdot Var[A] + Var[X_e] \cdot E^2[A] \tag{9}$$

Let us now compute the above mentioned terms. First, the distribution of $X_e$ is to be obtained. The distribution of the queue, seen by an admitted arriving packet, is derived from the distribution $P_q(i)$ already obtained for local buffers at each timeslot right after a possible transmission. Note that an output is not possible when the queue is empty and no packet can leave behind it $N_q$ packets. Hence, we have:

$$P_q(X_e = k) = 1(0 \le k < N_q) \cdot \frac{P_q(X_- = k+1) \cdot \Omega_{q,k+1}(1)}{\sum_{i=1}^{Nq} P_q(X_- = i) \cdot \Omega_{q,\ i}(1)} \tag{10}$$

where $X\_$ is the number of clients in the queue at the end of each timeslot, right before any possible transmission occurs. We have computed in section 3.2 the queue distribution at the end of each timeslot, right after a possible service of the client at the head of the queue. This distribution is related to the one of $X\_$ by the equations:

$$P_q(X_- = n) = \begin{cases} \sum_{i+j=n} P_q(i) \cdot \alpha_q(j) & \text{for } 0 \le n \le N_q - 1 \\ \sum_{i+j \ge N_q} P_q(i) \cdot \alpha_q(j) & \text{for } n = N_q \end{cases} \tag{11}$$

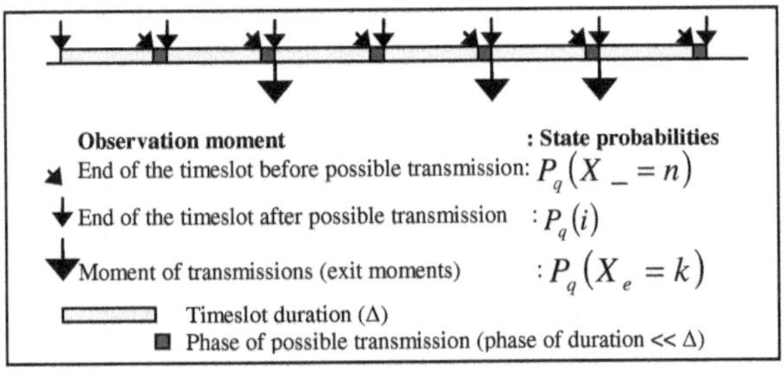


**Fig. 3.** Observation moments for the studied system

To obtain the expression of the moments of the access time we define first $\overline{\Omega_q}$ as:

$$\overline{\Omega_q} = \sum_{i=1}^{N_q} \left[ \Omega_{q,\ i}(1) \cdot P(X_- = i) \Big/ \sum_{i=1}^{N_q} P(X_- = i) \right] \tag{12}$$

Clearly, $\overline{\Omega_q}$ gives the availability of a slot on the average. Therefore, the access time for the packet at the head of each queue follows a geometric distribution:

$$E[A] = \Delta \cdot \overline{\Omega_q} \sum_{i=1}^{\infty} i \cdot (1 - \overline{\Omega_q})^{i-1} = \Delta \Big/ \overline{\Omega_q} \tag{13}$$

The previous formula includes additionally the transmission time $\Delta$ needed to deposit this packet on the ring. The variance of A is equal to:

$$Var[A] = \Delta^2 \cdot [1 - \overline{\Omega_q} / \overline{\Omega_q}^2] \tag{14}$$

Finally, we calculate the residual time $R$ to obtain the queueing delay. Optical packets arrivals are independent of the synchronized timeslots on the network ring. Since we consider Poisson arrivals, the packet inter-arrival times have therefore an exponential distribution, on each timeslot, truncated by the duration $\Delta$. The distribution function of $R$ is: $f_R(\tau) = \lambda_q \cdot e^{-\lambda_q \cdot (\Delta - \tau)} / \left(1 - e^{-\lambda_q \cdot \Delta}\right)$, for $0 \le \tau \le \Delta$.

The expectation and the variance of $R$ can be easily obtained:

$$E[R] = \Delta \cdot \left(1 - e^{-\lambda_q \cdot \Delta}\right)^{-1} - \left(1/\lambda_q\right),\ Var[R] = 1/\lambda_q^{\ 2} - [\Delta^2 \cdot e^{-\lambda_q \cdot \Delta} / \left(1 - e^{-\lambda_q \cdot \Delta}\right)^2] \tag{15}$$

# 5 Extraction and Reemission Mechanism

The PEM mechanism presents the weakness of dropping some best effort packets which may have to be retransmitted later on the network (TCP, etc.), yielding to an extra load in the network. Therefore, we have proposed the following mechanism:

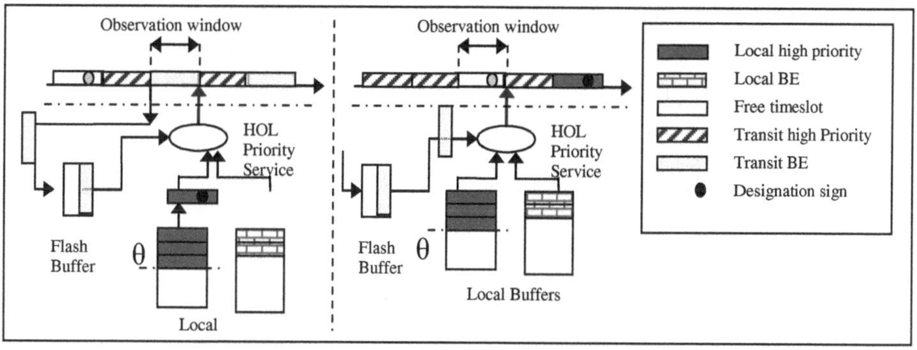


**Fig. 4.** Example of the ERM mechanism

- A buffer called *flash buffer* of a very limited capacity is added in order to store few optical Best Effort (BE) packets which can be extracted from the transit line.
- When the current slot at the transit line is occupied by a BE packet, we observe the filling levels of high priority buffers. If one of them reaches a certain threshold $\theta$, the BE packet at the transit line is preempted and it is stored in the *flash buffer*. One packet of the high priority overloaded buffer is transmitted instead.
- If the *flash buffer* is saturated, the extracted BE packet is lost.
- Afterwards, packets waiting in the *flash buffer* will be retransmitted again, prior to other packets inside the same node, on the first next available timeslots.

## 5.1 Modeling of the ERM Mechanism

In order to model the ERM mechanism, we introduce minor modifications to our model. The flash buffer should be taken into account at three different levels:

1. The effect of this buffer is reported on the link state at the node entry right after a possible reception
2. The state of this buffer is updated in order to consider possible transmission of one of its packets on the slot
3. When all local buffers are scanned, the state of the *flash buffer* is updated again in order to consider a possible new arrival (extracted BE packet)
4. Delays and loss expressions for local buffers remain valid. However, some intermediate variables should be added to compute the new state of the system.

The number of packets inside the *flash buffer* is associated to probabilities denoted $P_f$ $(n)$, where $n$ varies from 0 to the *flash buffer* capacity $N_f$. The state of the current slot on the link taking into account the *flash buffer* is expressed as:

$$\begin{cases} P_{l,1}(0) = P_{l,f}(0) \cdot P_f(0) \\ P_{l,1}(q) = P_{l,f}(q) \quad if \quad 0<q<C \\ P_{l,1}(C) = P_{l,f}(C) + P_{l,f}(0) \cdot (1 - P_f(0)) \end{cases} \tag{16}$$

In this case, $P_{l,f}(s)$ denotes the state of the slot seen by the *flash buffer* right after a possible packet reception at the entry of the considered node. The state of the *flash buffer* is updated at two levels. The first one, at the node entrance, is as follows:

$$\begin{cases} P_f^{+}(n) = P_f(n) \cdot (1 - P_{l,f}(0)) + P_f(n+1) \cdot P_{l,f}(0) \quad for\ 0 \le n < N_f \\ P_f^{+}(N_f) = P_f(N_f) \cdot (1 - P_{l,f}(0)) \end{cases} \tag{17}$$

According to the thresholds utilization rules, the probability that a Best Effort (BE) packet is extracted is as follows:

$$P_E = P_{l,f}(C) \cdot \left[ 1 - \prod_{q=1}^{C-1} \sum_{0 \le i+j < \theta_q} \left( P_q(i) \cdot \alpha_q(j) \right) \right] \tag{18}$$

Thus, the equations for the second update of the flash buffer will be expressed as follows, using the probabilities obtained in eq. 17:

$$\begin{cases} P_f^{+}(0) = P_f(0) \cdot (1 - P_E) \\ P_f^{+}(n) = P_f(n) \cdot (1 - P_E) + P_f(n-1) \cdot P_E \quad for \quad 0 < n < N_f \\ P_f^{+}(N_f) = P_f(N_f) + P_f(N_f - 1) \cdot P_E \end{cases} \tag{19}$$

The Packet Loss Ratio (PLR) at the *flash buffer* is equal to the probability of extraction while the buffer is full. Then: $\rho_f = P_f(N_f) \cdot P_E$

When the ERM mechanism is enabled, the *total loss ratio* of *Best Effort* packets arriving to a station $v$ can be expressed using the same formula obtained for the PEM mechanism in section 4.1. However, some changes are introduced to the expression of the probability $P_{d|C}^{(k)}$ for a *Best Effort* packet to be dropped (lost) at an intermediate station $k$. It is computed as:

$$P_{d|C}^{(k)} = P_f^{(k)}\left(N_f^{(k)}\right) \cdot P_E^{(k)} / P_{l,f}^{(k)}(C) \tag{20}$$

Finally, the mean Queueing delay inside the *flash buffer*:

$$E[Q_f] = E[n_f] \times E[A_f] \tag{21}$$

where $E\ [n_f]$ is the mean number of packets inside the *flash buffer* and $E\ [A_f]$ is the mean access time computed as in section 4.2.

# 6 Evaluation and Analysis of the Network Performance

## 6.1 Traffic Assumptions and Node Design Parameters

Using the ns Network Simulator, we evaluate the performance of a ring optical network with 10 nodes. Given the data provided by our industrial partners, each metro access node can connect about 25 Distant Subscriber Connection Units (DSCU), each of which supports over 2000 subscribers. Therefore, we obtain a total number of 500000 subscribers. As for our model scalability, since we used an iterative approach, we believe that the computation time is a linear function of the number of queues in the network. The same model can be applied to larger networks (e.g. 30 nodes).

The ring consists in one single wavelength of 10Gbits/s. Unless otherwise mentioned, we consider that the client traffic is uniformly offered to all ring nodes. The load offered to the network (denoted $\rho$) is defined as the ratio of the total generated client traffic with regard to the optical link capacity.

Client traffic is classified into four classes of service. To each class corresponds a relative load ratio (with respect to the total load offered to the network) and absolute ratio (percentage of the wavelength capacity required for this traffic).

**Table 1.** Parameters Related to each Class of Service

| Class | Premium (1) | Silver (2) | Bronze (3) | BE (4) |
|---|---|---|---|---|
| Rel. ratio | R1 = 0.25 | R2 = 0.25 | R3 = 0.25 | R4 = 0.25 |
| Abs. ratio | A1= R1 * ρ | A2= R2 * ρ | A3= R3 * ρ | A4= R4 * ρ |
| Buffer size | N1 = 20 | N2 = 20 | N3 = 20 | N4 = 50 |
| $\theta_q$ | $\theta_1$ | Disabled | Disabled | Undefined |

In the simulator of the network, we assume electronic client packets of variable sizes given in [11]. The traffic is constituted as follows: 40% of packets are of 50 bytes, while the rest consists equally of packets of 500 bytes and 1500 bytes.

Client traffic in the studied model consists of fixed-size optical packets arriving at a Poisson rate. However, in our network simulator, we use the *fixed-size packet creation mechanism*, which is described in section 2. In this case, optical packets are created from small variable-size packets, with IPP (Interrupted Poisson Process) arrivals. We do not deal in this study with the optical packet creation delay. This topic has been studied in [9]. The timeslot duration Δ is equal to 10μs.

As for performance metrics, we consider the total packet loss ratio (*total PLR*), the average queueing delay and the squared coefficient of variation for the delay. This last metric is expressed as: SCV = Var[Q] / $E^2$[Q].

### 6.2 Model Validation

We analyze the proposed model with a network load of 0.85. Each node is supplied with the PEM mechanism ($\theta_1 = 5$). Three cases are considered:

**Case 1 "Analytical Model":** Performance results are obtained using the resolution method proposed in the previous sections.

**Case 2 "Model Simulator":** We simulate our model with Poisson arrival distributions. This allows validation of the formulas and the resolution of our model.

**Case 3 "Network Simulator":** This case provides validation of the Poisson arrival assumption. We simulate the original network where electronic client packets arrive according to an IPP distribution. Optical packets are created as described in section 2.

Simulation results are obtained with 5% confidence intervals. Figure 5 illustrates average queueing delays and SCV results per class of service. As we can see, average delay and delay SCV are decreasing function of the class priority. Delay SCV is inferior to 1 for all classes of service. Hence we observe that the studied network is a low delay variation system. This is mainly due to the assumed Poisson arrival process and the slotted transmission in the studied model. As for the network simulator, it has been shown in [12] that the *fixed-size packet creation mechanism* contributes in reducing the traffic burstiness. As for average queueing delays, we observe in Figure5 that these delays vary from 40μs for the premium CoS to about 70μs for the best effort traffic. Therefore, with a balanced traffic matrix (which is the case here), the network can assure good performance even for the best effort traffic.

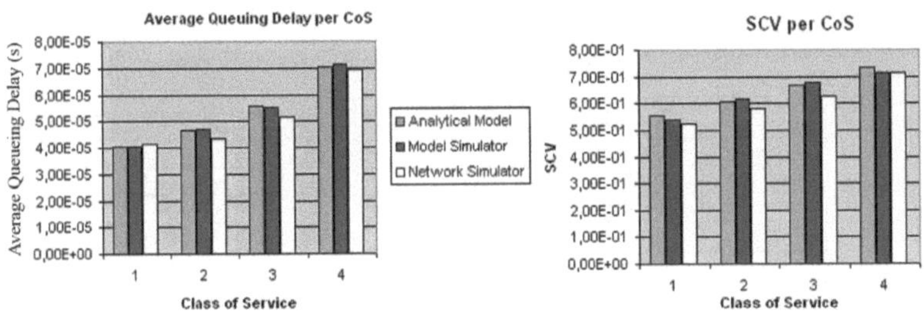


**Fig. 5.** Model Validation – Average Queueing Delay and SCV

Figure 5 confirms the validity of our model and the Poisson arrival process assumption. The results obtained are quite similar in all studied cases (see Table 2).

**Table 2.** Concordance of the three studied models (differences in % with respect to the analytical model)

| | Average Queueing Delays (%) | | | | SCV (%) | | | |
|---|---|---|---|---|---|---|---|---|
| Class of Service | CoS1 | CoS2 | CoS3 | CoS4 | CoS1 | CoS2 | CoS3 | CoS4 |
| Model Simulator | 0,24 | 1,27 | 1,45 | 1,40 | 2,03 | 1,62 | 1,48 | 2,81 |
| Network Simulator | 3,12 | 6,87 | 7,71 | 1,44 | 5,73 | 5,19 | 6,37 | 2,81 |

### 6.3 Investigation of the PEM and ERM Mechanisms

In order to investigate the performance of the PEM and ERM mechanisms, we start by a normal function scenario where client traffic is generated according to absolute ratios respectively equal to: 30% load for premium traffic, 10% for each of silver and bronze traffic, and 20% for best effort traffic. The resulting load offered to the network is therefore equal to 70%. Afterwards, we increase the absolute ratio of the premium traffic up to 60%. We compare the following study cases:

- *High Load Premium (HLP)*: In this case, we observe the priority-based Head of Line (HOL) service policy, without intervention of PEM or ERM mechanisms.
- *High Load Premium HLP: PEM $\theta_l$=n*: The network is still overloaded by additional premium traffic. The PEM mechanism is enabled with $\theta_l=n$ packets.
- *High Load Premium HLP: ERM $\theta_l$=3*: In this case, the ERM mechanism is activated with $\theta_l$ set to 3 packets and the capacity of the flash buffer to 2 packets.

The average queueing delay and the PLR per class of service, for each study case, are presented in Figure 6. Because of the Head of Line (HOL) priority scheme, the increase of the load of premium traffic at a given node is translated automatically by simultaneous increase of the average queueing delay at all other buffers in the same node. These delays are reduced by the PEM and ERM mechanisms. PLR results show the counterpart of such delay improvement. For the sake of clarity, loss ratios that are smaller than $10^{-9}$ (obtained by the model) are not shown in this Figure.

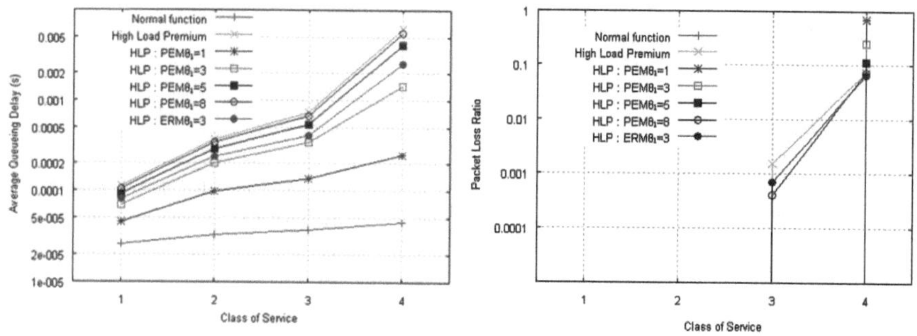


**Fig. 6.** Average Queueing Delay and PLR per Class of Service (PEM vs ERM)

We observe in the right side of Figure 6 a large loss ratio of BE packets in the case of PEM mechanism with systematic erasing ($\theta_l$=1). This PLR is reduced when the erasing threshold becomes higher or when we use the ERM mechanism, which preserves the performance of the best effort traffic. Note here that, using the ERM mechanism, Best Effort packets that enter to the flash buffer wait an average delay which is in the order of 4 timeslots (about 40μs).

## 7 Conclusion

In this work, we provided a performance evaluation of two improved access mechanisms which are applied to a novel Multiservice Optical Packet Switched ring architecture.

Based on preemptive approaches, these mechanisms implement a sort of dynamic priority. We evaluate the performance of these mechanisms by analytical modeling. Our model captures the impact of the client traffic matrix. Furthermore, it provides expression for the queueing delay and the packet loss measures.

We have used our model in order to evaluate the performance of the PEM and ERM mechanisms. Results obtained have shown that depending on the traffic matrix, the absolute priority scheme is not sufficient to guarantee the performance of the high-priority traffic. On the other hand, we have observed that both PEM and ERM mechanisms contribute in reducing the delay and PLR for the premium traffic. Compared to PEM, the ERM mechanism provides significantly lower Packet Loss Ratios for the Best Effort class (in the local buffer and in transit). Hence, the ERM mechanism provides a good compromise for the priority management in the access nodes of the studied optical ring network.

## References


1. Mathieu, C.: Toward Packet OADM. WDM product group, Alcatel-Lucent (December 2006)
2. Chiaroni, D.: Optical Packet Add/Drop Multiplexers: Opportunities and Perspectives. Alcatel-Lucent R&I, Alcatel-Lucent presentation (October 2006)
3. Lee, J.Y., Kim, Y.H.: Performance analysis of a hybrid priority control scheme for input and output queueing ATM switches. In: Proceedings INFOCOM 1998 (1998)
4. Bruneel, H., Kim, B.G.: Discrete-time models for communication systems including ATM. Kluwer Academic Publishers, Boston (1993)
5. Lim, Y., Kobza, J.E.: Analysis of a Delay-Dependent Priority Discipline in an Integrated Multiclass Traffic Fast Packet Switch. IEEE Trans. on Com. 38(5) (1990)
6. Maertens, T., Walraevens, J., Bruneel, H.: On priority queues with priority jumps. Performance Evaluation 63, 1235–1252 (2006)
7. Hébuterne, G.: A Gate with Periodic Openings and Bulk Service. In: Teletraffic Science for New Cost-Effective Systems. Elsevier Science Publishers B.V, Amsterdam (1989)
8. Carrier sense multiple access with collision avoidance. American National Standard T1.523-2001, Telecom Glossary (2000), `http://www.atis.org/glossary/definition.aspx?id=6101`
9. Chaitou, M., Hebuterne, G., Castel, H.: On aggregation in almost all optical networks. In: Wireless and Optical Communications Networks, WOCN (2005)
10. Kulkarni, V.G.: Modeling and analysis of stochastic Systems. Chapman and Hall, Boca Raton (1995)
11. IP packet length distribution (2002), `http://www.caida.org/research/traffic-analysis/AIX/`
12. Haciomeroglu, F., Atmaca, T.: Impacts of packet filling in an optical packet switching architecture. In: Advanced Industrial Conference on Telecommunications, AICT 2005 (July 2005)

# On the Impact of Clustering on Measurement Reduction*

## (Work in Progress)

Damien Saucez, Benoit Donnet, and Olivier Bonaventure

Université catholique de Louvain - CSE Deparment - Belgium

**Abstract.** Measuring a path performance according to one or several metrics, such as delay or bandwidth, is becoming more and more popular for applications. However, constantly probing the network is not suitable. To make measurements more scalable, the notion of clustering has emerged. In this paper, we demonstrate that clustering can limit the measurement overhead in such a context without loosing too much accuracy. We first explain that measurement reduction can be observed when vantage points collaborate and use clustering to estimate path performance. We then show, with real traces, how effective is the overhead reduction and what is the impact in term of measurement accuracy.

**Keywords:** clustering, BGP, reduction, measurement.

## 1 Introduction

During the past decade, we have seen the emergence of a set of applications requiring more and more quality of service (QoS). For instance, IPTV needs large bandwidth and delays as low as possible. Further, while previously a content was located in a single place, it is, nowadays, frequent that this content is replicated among a set of servers located anywhere on the five continents or even among the end-users themselves. Perfect examples of this are peer-to-peer (P2P) applications and FTP mirrors.

Using measurements collected at network vantage points to infer the Internet conditions is an important feature in such a context. Indeed, the path performance metrics, such as delay, bandwidth, or packet loss, collected by applications to a potentially large set of destinations might be a good indicator on which destination to select.

However, constantly probing the network leads to scalability issues. Indeed, probes injected in the network might burden the traffic. Further, if those probes come from multiple vantage points, they might appear as being a distributed denial-of-service attack. As previously mentioned by Cheswick et al., any networking measurement system must be engineered very carefully to avoid abuse [1].

* This work is supported by the European Commission-funded 223936 ECODE project. Mr. Donnet is funded by the Fonds National de la Recherche Scientifique (FNRS – Rue d'Egmont 5, 1000 Brussels). The authors would like to gratefully acknowledge Pierre Francois and Pascal Merindol for their comments.

L. Fratta et al. (Eds.): NETWORKING 2009, LNCS 5550, pp. 835–846, 2009.

There exist several ways for reducing the amount of required measurements. A first possibility is to modify the probing technique so that it consumes less resources. Another way is to allow collaboration between probing monitors and to cluster probe targets. *Clustering* means aggregating a subset of targets into the same hat and considering a measurements towards one of the target as being representative of the whole cluster [2,3,4].

In this paper, we investigate the second solution (i.e., collaboration and clustering). We deeply explain how measurement reduction can be achieved through collaboration between vantage points and destination clustering. We also discuss several metrics that can be used to evaluate the performance of a cluster based measurements campaign. We further explain that a greater measurement reduction can be achieved if collaboration between measurement sources is added to clustering.

In addition, we discuss five clustering techniques that can be used to reduce the measurements impact. These clustering techniques are based on available network information.

Based on real data collected, we evaluate these clustering techniques using the metrics we propose. We show that they are reasonably accurate while allowing a strong reduction in the amount of required probes.

The remainder of this paper is organized as follows: Sec. 2 provides a theorical background for measurement reduction through collaboration and clustering; Sec. 3 discusses five clustering techniques and positions them regarding the state of the art; Sec. 4 evaluates these clustering techniques and shows how clustering can reduce the measurement overhead; Finally, Sec. 5 concludes this paper and discusses future directions.

## 2 Theoretical Background

### 2.1 Measurement Reduction

Active Internet measurements reduction is a strategic issue when large-scale measurements are required. Up to now, several solutions have been proposed for delay measurements [5,6,7], Internet topology discovery [8,9], and bandwidth estimation [10,11]. However, these techniques, despite their strong advantages in term of measurement reduction, are complex to deploy and can require to modify the measurement technique itself.

If keeping the measurement mechanisms intact is a requirement when trying to reduce the probing impact, two solutions are imaginable: reducing the number of measurement vantage points (i.e., the *sources*) and reducing the number of measurement targets (i.e., the *destinations*). If both lead to a measurement reduction, they can also lead to an accuracy loss. A balance must thus be found between measurement reduction and accuracy.

These two solutions might be implemented through *collaboration* and *clustering*. Collaboration can help to reduce the number of sources involved in measurements while clustering is useful to reduce the number of destinations to probe.

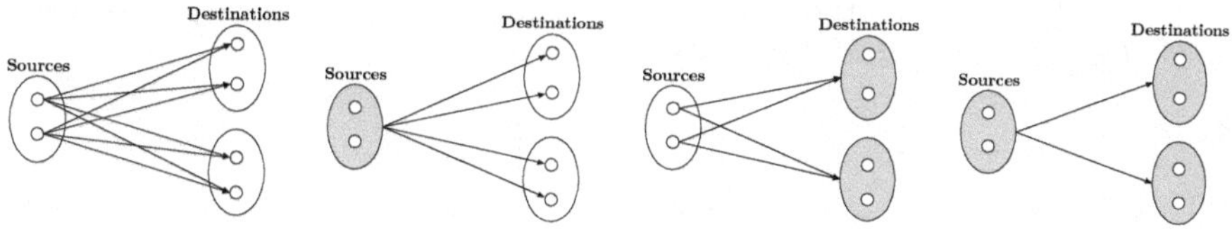


(a) Duplication (b) Collaboration (c) Clustering (d) Collaboration and clustering

**Fig. 1.** Illustration of measurements duplication and reduction

Any two nodes $a$ and $b$ could collaborate if they are topologically close. For instance, if they both belong to the same campus network. Indeed, if they share the same first hops, when measuring a destination $d$, it is very likely that probe packets will follow, for both $a$ and $b$, the same path. Or at least, they will share large path segments. Consequently, the resulting measurement will be very close and if both nodes act in isolation from each other, they duplicate unnecessarily their efforts, as illustrated in Fig. 1(a). On the contrary, if $a$ is aware of the measurements already performed by $b$ to $d$, then, $a$ no longer needs to measure $d$. In other words, collaboration avoids duplication of measurements and thus reduce the measurement overhead. Collaboration is illustrated in Fig. 1(b). How measurement sources can collaborate is still an open issue (however, efforts have been made in the context of Internet topology discovery [8]) we let for further works.

On the other hand, the key idea behind clustering is to aggregate a set of nodes under the same hat and consider that all nodes within a given hat share the same arbitrary properties. A *Cluster* $\mathcal{C}$ is defined as a set of nodes sharing the same properties. Clustering is illustrated in Fig. 1(c).

The basic assumption behind clustering is that all nodes belonging to a given cluster share the same path performances (i.e., delay, bandwidth, etc). Consequently, measuring a single point within a cluster would be sufficient. The node that is measured to estimate the cluster path performances is called the *Reference Point*. It is thus worth to notice that clustering allows to reduce the number of measurements as only the reference point has to be measured.

In the remainder of this paper, we only consider one reference point per cluster.

A given cluster is said *popular* if any destination within this cluster is often measured. We evaluate the popularity $\pi_\mathcal{C}$ of a given cluster $\mathcal{C}$ by counting the number of sources sending probes towards $\mathcal{C}$.

We can bring together collaboration and clustering by defining $\mathcal{S}_\mathcal{C}$, the *Source Cluster* of $\mathcal{C}$. $\mathcal{S}_\mathcal{C}$ represents the set of sources measuring a given cluster $\mathcal{C}$. Remind that this makes sense only if all nodes in $\mathcal{S}_\mathcal{C}$ are topologically close. Collaboration and clustering together allows a greater reduction in probing effort, as depicted in Fig. 1(d). It is worth to notice that the set of collaborating nodes is a cluster itself.

We propose a metric for evaluating the measurement reduction: the *Measurement Reduction Factor*.

**Definition 1 (Measurement Reduction Factor).** *The Measurement Reduction Factor $\rho$ for clustering technique $t$ on $\mathcal{P}$ is:*

$$\rho = \frac{|\mathcal{P}| - \sum_{C_i \in \mathcal{C}_\mathcal{P}} |\mathcal{R}_{C_i}|}{|\mathcal{P}|} \tag{1}$$

*where $\mathcal{P}$ is the set of $< s, d >$ nodes pairs such that $s$, the source, has to measure $d$, the destination. $\mathcal{C}_\mathcal{P}$ is the set of all the clusters on $\mathcal{P}$ assuming clustering technique $t$. $\mathcal{R}_\mathcal{C}$ is the set of all the reference points of $\mathcal{C}$.*

Positive values for $\rho$ means that the measurement reduction technique used effectively reduces the number of measurements. On the contrary, a negative value means that more measurements have to be performed than without reduction. For instance, if $\rho = 0.5$, the number of measurements is reduced by 50%.

Finally, note that measurement reduction is achieved on $\mathcal{P}$ if it exists a cluster $\mathcal{C}$ such that $|\mathcal{S}_\mathcal{C}| > 1$ or $|\mathcal{C}| > 1$ or both.

### 2.2 Clustering Accuracy

In a cluster $\mathcal{C}$, only the reference point is used to predict the destinations performance within $\mathcal{C}$. It would be a matter of concern if measuring a single point (or a few points) within a cluster leads to a strong measurement accuracy loss. This section defines how to estimate the accuracy of clustering techniques.

**Definition 2 (Prediction Error).** *For a path performance metric $m$ (delay, bandwidth, etc), the prediction error between node $i$ and node $j$ is:*

$$e_{ij} = \frac{|\mathtt{m}_{ij} - \widehat{\mathtt{m}}_{ij}|}{\mathtt{m}_{ij}} \tag{2}$$

*where $\mathtt{m}_{ij}$ is the measured value for $m$ between $i$ and $j$ and $\widehat{\mathtt{m}}_{ij}$ its predicted value.*

The prediction error gives the error proportion if the performance of a node is based on the predicted value instead of the directly measured value. The predicted error is expressed in percentage. Closer to zero the predicted error, more accurate the measurement prediction. Thus, a value of zero means that the prediction is perfect, while, for instance, a value of 0.5 means that there is a difference of 50% between the predicted and the actual value.

For any clustering technique $t$ (see Sec. 3), the predicted metric $\widehat{\mathtt{m}}_{ij}$ from a node $i$ to a node $j$ in a cluster $\mathcal{C}$ is the metric associated to all nodes within the cluster.

The measurement error shows how the prediction fits with the reality for a given node in a cluster. However, it can be interesting to characterize the error for the whole cluster and not only a particular node within this cluster. For such an information, statistical tools like mean, percentiles or standard deviation can be applied on the set of all cluster prediction errors.

## 3 Clustering Techniques

In this section, we discuss five clustering techniques. These techniques offer the strong advantage of being very easy to setup as they only require simple information already, or easily, available in the network. With these techniques, clusters are a set of IP prefixes such that a simple longest prefix matching is enough to determine to which cluster a given IP address belongs.

**AS Clustering:** clusters are defined based on the Autonomous System (AS) membership of Internet hosts (e.g., all the nodes of AS 2611 are put in the same cluster). This is somewhat equivalent to the notion of *super-cluster* introduced by Krishnamurthy and Wang [12] for modeling the Internet topology.

**Geographic Clustering:** clusters are defined based on the geographic localization of Internet hosts (e.g., all the nodes near Paris are within the same cluster).

**$n$-agnostic Clustering:** clusters are defined as fixed-length IP prefixes. All the nodes sharing the same $n$ bits prefix are put in the same cluster. The `/24` division proposed by Szymaniak et al. [3] is a particular case of such a technique (i.e., 24-agnostic clustering).

**BGP Clustering:** clusters are built according to BGP. Every advertised BGP prefix refers to a cluster. Each IP address belongs to the cluster with the longest matched BGP prefix. The clustering in BGP prefixes was firstly proposed by Krishnamurthy and Wang [4].

**$n$-hybrid Clustering:** clusters are built according to BGP prefixes but the minimum length of the prefixes is fixed to $n$. When the BGP prefix length is lower than $n$ bits, it is divided into as many `/n` as needed instead of the single prefix length advertised in BGP. $n$-hybrid Clustering is thus a mix between BGP and $n$-agnostic Clustering. To the best of our knowledge, we are the first to propose this technique.

Except for $n$-agnostic Clustering, all these clustering techniques rely on a dataset defining to which cluster an IP address belongs. For Geographical Clustering, the dataset must contain a correspondence between IP addresses and a geographical position. We believe that the MaxMind database [13] is accurate enough for most of the applications, even if it has been shown that it is less accurate (in term of geolocation) than active probing [14]. In the case of AS Clustering, the dataset consists of a mapping between an Autonomous System Number (ASN) and an IP prefix. BGP feed, iPlane [15] or Cymru [16] provide this information. In both BGP and $n$-hybrid Clustering, IPs are mapped to their longest matched BGP prefix. To know the advertised BGP prefixes, a BGP feed is required. It can be obtained directly through a BGP session or a looking glass (e.g., RIPE). It is worth to notice that BGP Clustering is equivalent to 0-hybrid Clustering and that, in 32-agnostic, clusters have a cardinality of one.

We propose $n$-hybrid Clustering to avoid very large prefixes announced by BGP (e.g., some `/8`). Indeed, some of them may concern hosts spread all around the world and, thus, lead to a very low measurement accuracy.

Clustering techniques for reducing the amount of required measurements have already been extensively studied by the research community. Krishnamurthy and Wang introduce the BGP Clustering in the context of web caching [4]. In addition, they propose an adaptive clustering for addresses not classified with BGP. Unfortunately, their technique is based on reverse DNS and traceroute which is not suitable as traceroute is intrusive. Instead, we map undefined IP addresses to an agnostic `/n` prefix. Szymaniak et al. show that latencies are globally equivalent within the same /24 prefix on the Internet [3]. Based on that, Szymaniak et al. suggest to use a /24 cluster division (equivalent to what we call 24-agnostic Clustering in this paper) to reduce the amount of required measurement. Finally, Brown proposes to use clustering for optimizing traffic from a local network to a set of IP addresses put into a given cluster [2]. Based on measurements to a set of *scanpoints* (equivalent to our Reference Points), an entity might decide to choose a particular Internet path for traffic towards the cluster.

In addition, others proposed techniques for monitoring networks and, possibly, reducing the amount of required measurements while keeping accurate the measured metrics [17,18,19,7,9]. In particular, Donnet et al. [9] impose a limit on the number of traceroute monitor that can probe a given destination. This is done by dividing the traceroute monitors into clusters, each cluster focusing on a particular portion of the traceroute destinations.

# 4 Evaluation

## 4.1 Methodology

Our evaluation relies on two datasets. The first dataset is taken from the CAIDA's *Archipelago* measurement infrastructure [20], the *skitter* evolution. Archipelago collects traceroute and RTT information towards all routed /24. For our study, we consider data collected in August 2008 by two Archipelago monitors: `bcn-es` (Barcelona, Spain) and `san-us` (San Diego, USA). From these dataset, we extract only complete traceroutes, i.e., traceroutes terminating at the destination. For the rest of this paper, this dataset is named *Archipelago*. For the second dataset, we collected full NetFlow traffic traces on our campus network. The traces were collected at the single 1Gbps Internet connection of the campus. A total of 45.4 TB of outgoing traffic has been monitored. However, in this paper, we do not consider 0 bytes or 0 packets flows and ignore our VPN-like traffic. Thus, after filtering, the outgoing traffic represents 7.45 TB (i.e., 22.27 Mbps on average). The filtered dataset contains 10,084 different source IP addresses and 36,263,710 destination IP addresses for a total of 60,638,413 different layer-3 flows (i.e., the number of different <src, dst> IP address pairs). In the following of the paper this dataset is named *NetFlow*.

The NetFlow dataset is used to estimate the measurements reduction and the popularity while the Archipelago dataset permits to estimate the impact of clustering on performance estimation accuracy. Unfortunately, we have not found significant datasets that allowed us to estimate measurement accuracy and

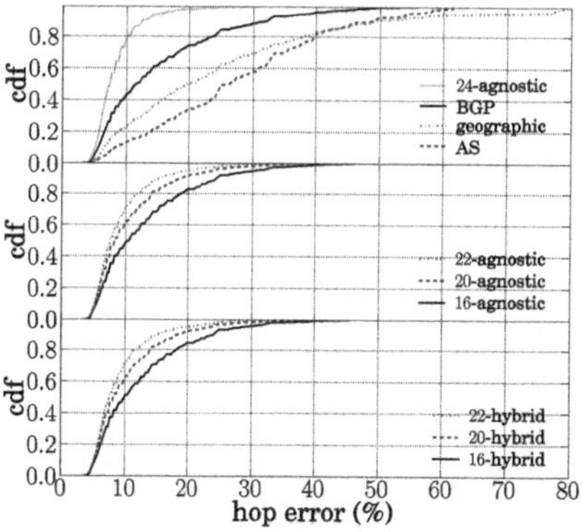


(a) `bcn-es`

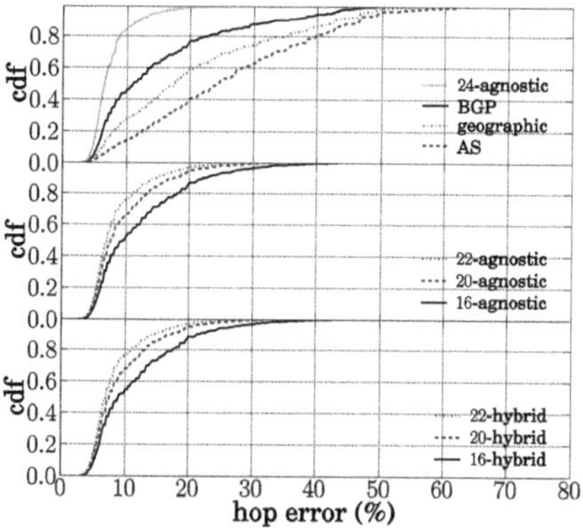


(b) `san-us`

**Fig. 2.** Prediction error applied to RTT

reduction at the same time. However, measurements from the `bcn-es` monitor is representative of the RTT we see on our campus network. We can thus consider the two dataset as not completely independent (`bcn-es` monitor is connected through the European research network like our campus).

The mapping between IP addresses and announced prefixes was done using the first BGP table dump of August 2008 retrieved from the University of Oregon RouteViews project [21]. ASN and IP addresses mapping was done using the iPlane dataset [15]. Finally, the geographic mapping was achieved using the MaxMind database [13].

In the following, we limit the $n$ value to $\{16, 20, 22\}$ for $n$-agnostic and $n$-hybrid.

### 4.2 Measurement Accuracy

In this section, we evaluate the measurement accuracy of clustering techniques. We focus on two networking metrics: the round-trip time (RTT) and the number of hops. We let jitter and bandwidth for further work.

Fig. 2 shows the prediction error (see Def. 2) applied to the RTT for the five clustering techniques introduced in Sec. 3. Fig. 2(a) focuses on the `bcn-es` monitor and Fig. 2(b) on `san-us`. For each subfigure, there are three plots, the upper one comparing BGP, Geographic, AS, and 24-agnostic Clustering, the middle one $n$-agnostic, and, finally, the below one $n$-hybrid. The horizontal axis gives the RTT Error (in %) and the vertical axis the cumulative distribution form (CDF).

A first observation is that $n$-hybrid and $n$-agnostic Clustering perform better ($n$-hybrid being the best) than other clustering techniques, most of the measurements having an error less or equal to 50%. More precisely, in 80% of the cases, the RTT Error is less or equal to roughly 25%. The AS Clustering offers the worst performance. This is somewhat expected as some ASes are too large to asses the assumption that any measurement towards the cluster reference point

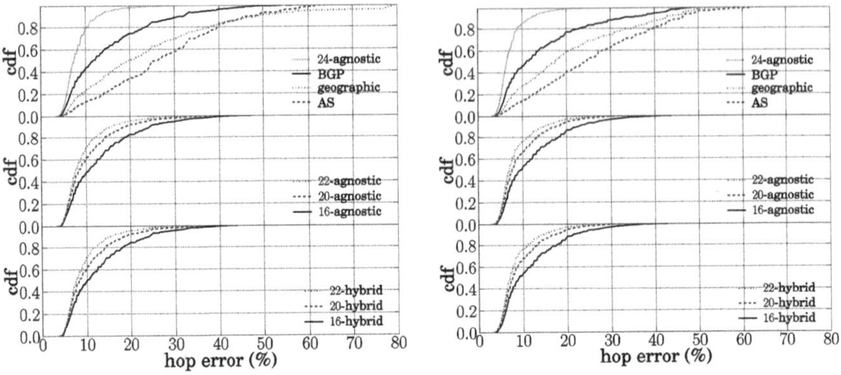


(a) bcn-es (b) san-us

**Fig. 3.** Prediction error applied to hop

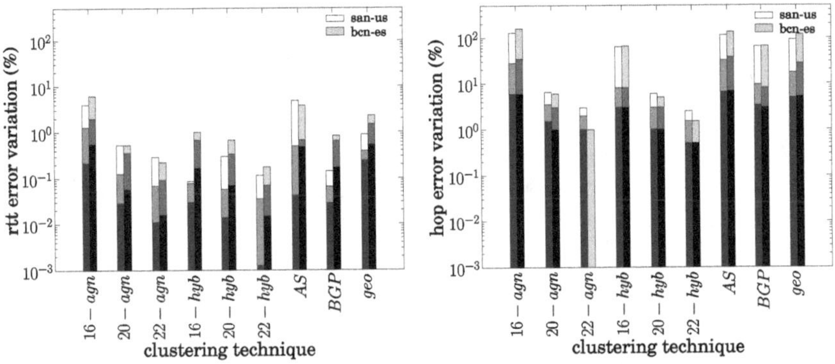


(a) RTT (b) hop

**Fig. 4.** $25^{th}$, $50^{th}$ and $75^{th}$ percentile of prediction error

is representative of the entire cluster. Indeed, an AS can be very large and may contain smaller entities that are separately administrated. From Fig. 2, we notice that the RTT Error might be very large, i.e., up to 1,000% for AS Clustering (not shown on Fig. 2). This is due to inherent limitations of our dataset. Some traceroute might take a long time to reach the destination and the RTT measured at this time might not be really representative of the actual "distance" separating the traceroute monitor and the destination.

Fig. 3 shows the prediction error applied to hop. Again, AS and Geographic Clustering provides the largest error. This is expected as they span larger areas. Further, for $n$-hybrid and $n$-agnostic, larger the cluster (i.e., smaller the $n$), larger the hop error. As for RTT error, $n$-hybrid provides the best performance.

Fig. 4 shows the *prediction variation* (the percentiles of all the prediction errors for a given cluster) applied to RTT (Fig. 4(a)) and to the number of

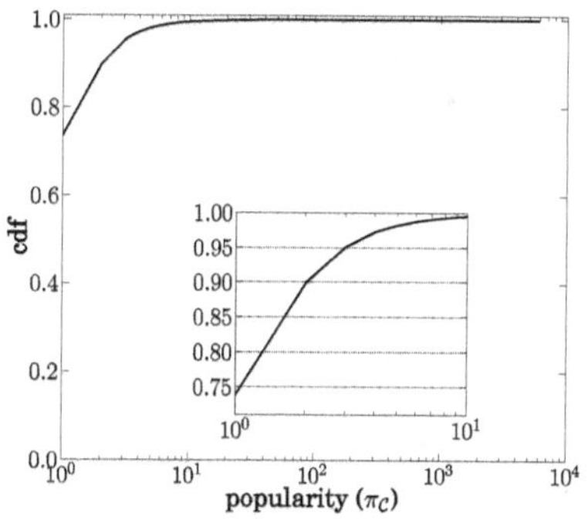


**Fig. 5.** Popularity in the NetFlow dataset

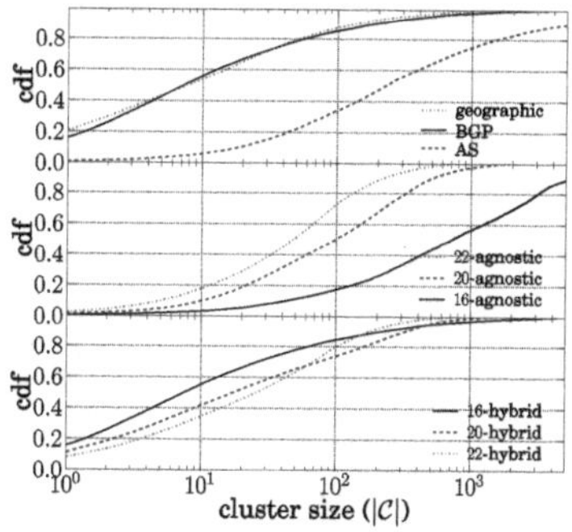


**Fig. 6.** Cluster size distribution

hops (Fig. 4(b)). The horizontal axis gives the various clustering techniques while the horizontal axis, in log-scale, gives the prediction variation. Fig. 4 plots stacked bars. The lowest bar (i.e., the darkest) refers to the 25$^{\text{th}}$ percentile of the prediction variation. The middle bar shows the 50$^{\text{th}}$ percentile (i.e., the median), while the highest bar (i.e., the lighter) gives the 75$^{\text{th}}$ percentile.

The main lesson learned from Fig. 4(a) is that, whatever the clustering technique, the RTT variation is low: the median has a maximum of 1.5% for 16-agnostic (for `san-us`). However, the situation is a little bit different for hop variation (Fig. 4(b)). The 75$^{\text{th}}$ percentile provides large hop errors, in particular for the largest clusters, i.e., 16-agnostic, 16-hybrid, BGP, AS, and Geographic Clustering.

To summarize, except in a few extreme cases, clustering provides thus pretty good measurement accuracy. In particular, this section suggests that $n$-hybrid Clustering is the most reasonable choice for a clustering technique deployment.

## 4.3 Measurement Reduction

In this section, we evaluate the measurement reduction observed on the NetFlow dataset for the clustering techniques introduced in Sec. 3.

We first determine if measurement reduction can be achieved, on the NetFlow dataset thanks to the collaboration. On Fig. 5, the horizontal axis, in log-scale, shows the popularity while the vertical axis gives the CDF.

Fig. 5 shows that 26.1% of the destinations are reached by, at least, 2 different sources. Collaboration will thus reduce measurements for this dataset. In addition, the top 250 destinations are reached by more than 2,000 different sources. Theses destinations are the major CDNs and web search engines.

Fig. 6 shows the CDF of the cluster sizes. The plot first shows that, on our dataset, clusters cover several nodes. Measurement reduction will thus be observed on our dataset.

Fig. 6 also suggests that the clustering technique influences the cluster size. For instance, 50% of the clusters cover more than 200 addresses in AS Clustering while it is the case for less than 10% in BGP or Geographic Clustering. Moreover,

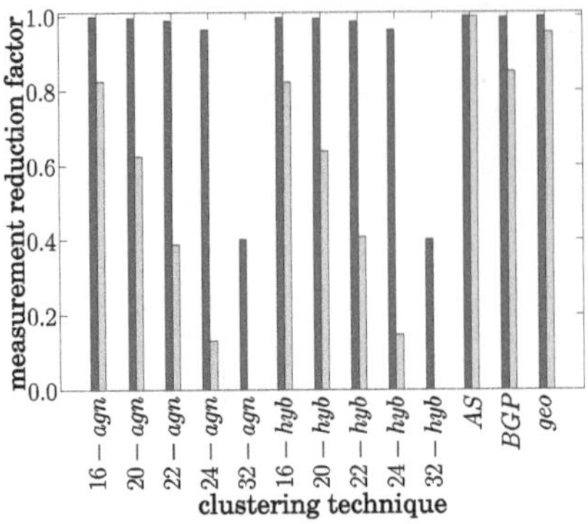


**Fig. 7.** Measurement reduction in the NetFlow dataset

the cluster size is also influenced by the $n$ parameter when applicable. With small $n$, prefixes are large and have the possibility to absorb many destinations, the cluster size being therefore important. As expected, $n$-hybrid behaves partially like BGP for small $n$ and like $n$-agnostic for large $n$.

Fig. 7 shows the effective reduction factor with respect to the different clustering techniques, with or without collaboration. The vertical axis is the reduction factor while the horizontal axis indicates the clustering techniques. On Fig. 7, light gray bars show the reduction factor when the nodes are not collaborating. On the other hand, dark gray bars show the reduction factor when nodes are collaborating.

32-agnostic and 32-hybrid Clustering (one cluster per destination) from Fig. 7 confirm that collaboration can reduce measurements (by 40% in our dataset). Light gray bars confirm that clustering can reduce measurements.

Moreover, Fig. 7 shows that, for the same clustering technique, cooperation always offers better reduction factor than no cooperation or collaboration without clustering. Confirming so that combining collaboration and clustering gives even better measurements reduction than collaboration or clustering alone.

We also see on Fig. 7 that BGP, AS, and Geographic Clustering offer an important reduction factor because the number of BGP prefixes, ASes, or locations is very small compared to the number of different addresses in the NetFlow dataset.

When considering $n$-hybrid and $n$-agnostic Clustering, the reduction factor is more sensitive to $n$ than the technique itself which is particularly true without collaboration.

Regarding to what we discussed above, we would suggest to use 20-hybrid clustering as this technique remains accurate and presents an interesting measurement reduction, even without collaboration. We would suggest not to use $n$-agnostic for small values of $n$ as it will not reflect the topology. AS and geo clustering should be avoided.

In this section, we demonstrated that they effectively reduce the number of measurements while remaining quite accurate.

## 5 Conclusion

Measuring the quality of a set of paths, in terms of delay or bandwidth, is becoming more and more important for applications and services. Indeed, the resulting measurements could allow the application to select the best location for the required service or for getting a particular content. However, constantly monitoring the network through active measurements is not desirable due to scalability issues.

A solution for rendering measurements more scalable is to consider clustering and collaboration between measurement sources. Clustering aims at aggregating a subset of destinations into the same hat and consider that a single (or a few) measurement towards the cluster is representative of the whole cluster.

In this paper, we first discussed how collaboration and clustering might lead to a reduction in probing effort. We defined metrics for evaluating the performance of any clustering technique. Those metrics evaluates the accuracy of a prediction based on clustering and the measurement reduction.

Secondly, we explained five different clustering techniques. Those techniques have the advantage of being very easy to setup, i.e., they do not required strong calculations. Based on real data collected, we evaluated those techniques with metrics we introduced. We demonstrated that clustering techniques offer quite accurate measurement predictions as well as measurements scalability.

In this paper, we only took into account two networking metrics: delay and the number of hops. Future work should reveal how other network metrics, such as bandwidth and jitter, are impacted by cluster based measurements.

## References

1. Cheswick, B., Burch, H., Branigan, S.: Mapping and visualizing the internet. In: Proc. USENIX Annual Technical Conference (June 2000)
2. Brown, G.: Internet address space clustering for intelligent route control (2004), see: `http://www.cs.indiana.edu/~geobrown/journal-new.pdf`
3. Szymaniak, M., Presotto, D., Pierre, G., Van Steen, M.: Practical large-scale latency estimation. Computer Networks 52(7), 1343–1364 (2008)
4. Krishnamurthy, B., Wang, J.: On network-aware clustering of web clients. In: Proc. ACM SIGCOMM (August 2000)
5. Ng, T., Zhang, H.: Predicting Internet network distance with coordinates-based approaches. In: Proc. IEEE INFOCOM (June 2002)
6. Ng, T.S.E., Zhang, H.: A network positioning system for the Internet. In: Proc. USENIX Annual Technical Conference (June 2004)
7. Dabek, F., Cox, R., Kaashoek, K., Morris, R.: Vivaldi, a decentralized network coordinated system. In: Proc. ACM SIGCOMM (August 2004)
8. Donnet, B., Raoult, P., Friedman, T., Crovella, M.: Efficient algorithms for large-scale topology discovery. In: Proc. ACM SIGMETRICS (June 2005)
9. Donnet, B., Friedman, T., Crovella, M.: Improved algorithms for network topology discovery. In: Dovrolis, C. (ed.) PAM 2005. LNCS, vol. 3431, pp. 149–162. Springer, Heidelberg (2005)

10. Hu, N., Steenkiste, P.: Exploiting Internet route sharing for large-scale available bandwidth estimation. In: Proc. ACM/Usenix Internet Measurement Conference, IMC (October 2005)
11. Ramasubramanian, V., Malhki, D., Kuhn, F., Abraham, I., Balakrishnan, M., Gupta, A., Akella, A.: A unified network coordinate system for bandwidth and latency. Technical Report MSR-TR-2008-124, Microsoft Research (September 2008)
12. Krishnamurthy, B., Wang, J.: Topology modeling via cluster graphs. In: Proc. ACM SIGCOMM Workshop on Internet Measurement, IMW (November 2001)
13. MaxMind: Geolocation and online fraud prevention from maxmind (2002), See: `http://www.maxmind.com/`
14. Siwpersad, S., Gueye, B., Uhlig, S.: Assessing the geographic resolution of exhaustive tabulation for geolocating Internet hosts. In: Claypool, M., Uhlig, S. (eds.) PAM 2008. LNCS, vol. 4979, pp. 11–20. Springer, Heidelberg (2008)
15. Madhyastha, H.V., Isdal, T., Piatek, M., Dixon, C., Anderson, T., Krishnamurthy, A., Venkataramani, A.: iPlane: An information plane for distributed services. In: Proc. USENIX Symposium on Operating Systems Design and Implementation, OSDI (November 2006), `http://iplane.cs.washington.edu`
16. Team Cymru: Internet security research and insight (2004), See: `http://www.team-cymru.org/`
17. Yalagandula, P., Sharma, P., Banerjee, S., Lee, S.J., Basu, S.: S3: A scalable sensing service for monitoring large networked systems. In: Proc. ACM SIGCOMM Workshop on Internet Network Measurement, INM (September 2006)
18. Leonard, D., Loguinov, D.: Turbo King: Framework for large-scale Internet delay measurements. In: Proc. IEEE INFOCOM (April 2008)
19. Mahajan, R., Zhang, M., Poole, L., Pai, V.: Uncovering performance differences among backbone ISPs with NetDiff. In: Proc. Symposium on Network Systems Design and Implementation (NSDI) (April 2008)
20. Claffy, K., Hyun, Y., Keys, K., Fomenkov, M.: Internet mapping: from art to science. In: Proc. IEEE Cybersecurity Applications and Technologies Conference for Homeland Security, CATCH (March 2009)
21. University of Oregon: Route views, University of Oregon Route Views project, `http://www.routeviews.org/`

# Topology Design for Service Overlay Networks with Economic and QoS Constraints
## (Work in Progress)

Davide Adami, Christian Callegari, Stefano Giordano, Michele Pagano, and Teresa Pepe

Dept. of Information Engineering, University of Pisa, Italy
{d.adami,christian.callegari,s.giordano,m.pagano,teresa.pepe}@iet.unipi.it

**Abstract.** Service Overlay Networks (SON) have emerged as an efficient way to provide end-to-end *Quality of Service* (QoS) support in the Internet, as required by real-time services.

The deployment of a SON is a capital-intensive investment. Thus, minimizing the economic cost is one of the key objectives for the SON operator. Moreover, end-to-end performance must be enhanced, so as to provide innovative value-added services. This paper, addresses the problem of cost and performance optimization of a SON. More specifically, we propose a set of algorithms for the selection of a SON topology, which minimizes the economic cost while taking also into account performance constraints.

**Keywords:** Service Overlay Networks, Topology Optimization, QoS support.

## 1 Introduction

Originally conceived as a network infrastructure for military applications and academic research, the Internet has rapidly turned into a business platform, where private enterprises seek to offer new value-added network services. At the same time, the wide spread of the Internet has brought heterogeneous applications with different QoS requirements, to converge over a single IP-based network infrastructure. However, today's Internet mainly provides a best effort packet delivery service and lacks in the end-to-end QoS support.

In this context, Service Overlay Networks (SONs) have recently emerged as one of the most promising solutions to provide end-to-end QoS [1] [2] [3] [4] [5]. Unlike other network architectures, SONs do not require any change in the underlying network layer. However, from an economic point of view, the deployment of SONs requires a valuable investment.

As clearly shown in several research works [6] [7], the choice of the optimal SON's topology plays a critical role to reduce the overall network costs. Nevertheless, when SONs are used to provide QoS, the topology design problem can not be only faced from an economic point of view, since performance metrics, such as bandwidth and delay, must be also taken into account.

L. Fratta et al. (Eds.): NETWORKING 2009, LNCS 5550, pp. 847–858, 2009.

For this reason, in this paper we propose a new approach for the topology design of a SON which aims at minimizing the SON deployment costs, while at the same time meeting multiple QoS performance constraints.

The rest of the paper is organized as follows. Section 2 discusses some related works, whereas Section 3 introduces the problem of the SON topology design. Next, section 4 describes a new approach for the design of a SON by means of algorithms which takes into account economic costs as well as performance metrics. The performance of such algorithms are compared in Section 5 through a large set of simulations. Section 6 concludes the paper with some final remarks.

## 2 Related Works

The use of a SON as a means to provide QoS in the Internet was first introduced in [1]. The authors studied the bandwidth provisioning problem without neglecting economic aspects. Indeed, the key challenge of this research was to provide the optimal amount of bandwidth to support end-to-end QoS-sensitive services while minimizing the bandwidth costs.

In [8] the authors addressed this issue in terms of choosing the number and the location of the overlay nodes to be deployed as well as the reserved bandwidth on each overlay link, while in [2] the authors proposed a framework for the adaptive design of a SON. In more detail, they developed a set of heuristic methods, and compared their performance. Another set of heuristics was proposed in [9] to address the same problem in large-size networks.

## 3 SON Topological Design

Let us we consider an application scenario where a SON, built on top of the traditional IP network, provides end-to-end QoS support to a number of users, who pay the SON operator by means of a service contract. The overlay network consists of overlay nodes (ONs) and a set of end-systems (ESs). The ONs are specialized nodes strategically placed across the ISPs networks by the Overlay Service Provider (OSP). Usually, one or more ONs can be deployed in each domain. Moreover, the number of ONs in an ISP may be increased, so as to improve scalability and fault tolerance. In our scenario, we suppose that:

- the number of ONs to be deployed is set in advance (how many ONs are needed and where to deploy them in certain scenarios also represent interesting research topics, but they are out of the scope of this paper);
- the connections between the ONs are provided by overlay links, usually composed of one or more physical links;
- ESs can access the network using an ON called *access ON*.

In this work, we assume that the connection between the ESs and the access ONs is possible only if they both belong to the same ISP. The OSP buys network services, such as guaranteed bandwidth, from different ISPs and, according to pre-established Service Level Agreements (SLAs), provides bandwidth guarantees connections to the ESs.

## 4 New Approach for the Design of SONs

For the overlay topology selection, we have to solve two distinct problems. First, we have to select the links that connect the ONs (transport links). Second, for each ES, we must select an access ON that allows the ES to access the ON. The total cost of the overlay network is given by the sum of the access costs and the transport costs of the resulting topology. In more detail, given a network composed of $M$ ESs and $N$ ONs, our proposed method consists of the following four steps.

**Step 1.** We only consider the ONs. Since the connection between the ONs is provided by the underlaying network layer, we can assume that it is always possible to have an overlay link $(i, j)$ between two distinct ONs. Thus, we start by considering a full-mesh topology for the overlay network, where each link is described by a metric, corresponding to the potential bandwidth that the ISP can make available on it.

**Step 2.** We assume that the ISPs can not always provide all the requested bandwidth. If the bandwidth associated to the link ($\omega_{ij}$) is less than the bandwith necessary to provide the required QoS ($\Omega$), we prune the link from the graph corresponding to the network topology. After that, we assign two distinct metrics to each remaining link: economic cost and link delay ($l_{i,j}$ and $r_{i,j}$, respectively).

**Step 3.** We determine a path $p$ (with cost $c_{ij}$) between each pair $(i, j)$ of ONs so that the total transport cost and the link delay are minimized. This is a multi-objective optimization problem. It is worth highlighting that solving this problem does not correspond to find an optimal solution for each objective function, separately. Since finding an optimal solution to this multi-objective optimization problem is too complex, in the following we use two distinct approximated methods: **Weight Method** and **$\varepsilon$-constraint technique**.

1. **Weight Method.** It is the simplest approach, since it sums the quantities to be minimized (metrics) into a single objective function, pre-multiplying them by an appropriate weight. The weight of a metric is chosen in proportion to the importance of the metric in the problem.
2. **$\varepsilon$-constraint Method.** This method corresponds to the classical method of constraint optimization.
   In our case we aim to minimize the economic cost while taking into account the constrained delay. Formally, we can write:

$$\text{minimize} \sum_{i=1}^{N}\sum_{j=1}^{N} l_{ij}\omega_{ij} \quad \text{subject to} \sum_{(ij)\in p} r_{ij} < \rho \qquad (1)$$

**Step 4.** Finally, we have to determine how to connect the ESs to the ONs. In this phase, we assume that the access delays are negligible with respect to the transport delays. The main idea is to select the access ONs, so as to minimize the total cost. The problem can be formalized as follows:

$$\text{minimize} \sum_{i=1}^{M}\sum_{k=1}^{N} w_i \alpha_{ik} y_{ik} + \sum_{i=1}^{M}\sum_{j=1}^{N}\sum_{k=1}^{M}\sum_{l=1}^{N} y_{ij} y_{kl} \omega_{ik} c_{jl} + \sum_{j=1}^{M}\sum_{l=1}^{N} w_j \alpha_{jl} y_{jl} \quad (2)$$

subject to

$$\sum_{j=1}^{N} y_{ij} = 1 \quad for \quad i = 1, .., M \quad (3)$$

where $\alpha_{ij}$ is the access cost for bandwidth unit, $w_i$ is the requested access bandwith and $y_{ij}$ is an indicator function.

$$y_{ij} = \begin{cases} 1 & \text{if } ES_i \text{ is assigned to } ON_j \\ 0 & \text{otherwise} \end{cases} \quad (4)$$

The optimization problem (2) is a variant of the well-known quadratic assignment problem (QAP). Since the QAP is a NP-hard problem, we can not find an exact solution in a polynomial time. Heuristic methods, therefore, are preferable options to solve this problem. There is a large set of heuristic algorithms for solving the QAP. In this work, we take into account two heuristics: the first one consists in the random selection of the access ONs, while the second one performs the selection on the basis of the access costs only, ignoring the transport costs.

In the following, we present six different algorithms we have developed for solving the presented problem. To be noted that step 1 and 2 (pruning phase for satisfying the banwidth constraints) are common to all these algorithms and are performed as previously described, while the approach used for the step 3 and 4 (transport links selection and access links selection, respectively) differs from algorithm to algorithm. Thus, we detail step 3 and 4 and we present the pseudo-code for each of the algorithms we have proposed. For all the algorithms, the input parameters are:

- $N$ number of ONs
- $\{\omega_{ij}\}$ bandwidth matrix
- $\Omega$ bandwidth constraint
- $\rho$ delay constraint
- $\{l_{ij}\}$ transport cost (per unit of bandwidth)
- $\{r_{ij}\}$ link delay
- $\{R_p\}$ delay of path $p$, defined as $\sum_{(i,j)\in p} r_{ij}$

Moreover, the algorithm output, which is common to all the methods, is given by the overlay network topology with his respective overall cost (given by the sum of transport costs and access costs).

### 4.1 Modified Dijkstra Random Mode - MDRM

In step 3, for the selection of the transport links this algorithm uses the *Modified Dijkstra algorithm*, which allows to find the k-shortest (least cost) paths between each pair of ONs. Among these, we select the least cost path, which satisfies the delay constraint.
In step 4, the access nodes are selected in a random mode.
Formally the algorithm can be described as follows:

**for all** $i$ and $j \in \{1, ..., N\}$ **do**
  **if** $\omega_{ij} < \Omega$ **then**
    prune link $(i,j) \Rightarrow l_{i,j} = \infty$
  **end if**
  **for all** pair of ONs, $i$ and $j \in$ {1, ..., N} **do**
    use Modified Dijkstra algorithm to find $k$ least-cost paths;
    **for** $p = 1 : k$ **do**
      evaluate path delay for the path $p$
      **if** $R_p < \rho$ **then**
        select the path
        break;
      **end if**
    **end for**
  **end for**
**end for**
select access nodes in a random mode;
evaluate access cost.

### 4.2 Modified Dijkstra Minimum Access Cost - MDAC

This algorithm is similar to the previous one. Indeed, only step 4 is different: in this case the access nodes are selected, so that the access costs are minimized. Formally, the algorithm can be described as follows:

**for all** $i$ and $j \in \{1, ..., N\}$ **do**
  **if** $\omega_{ij} < \Omega$ **then**
    prune link $(i,j) \Rightarrow l_{i,j} = \infty$
  **end if**
  **for all** pair of ONs, $i$ and $j \in \{1, ..., N\}$ **do**
    use Modified Dijkstra algorithm to find $k$ least-cost paths;
    **for** $p = 1 : k$ **do**
      evaluate path delay for the path $p$
      **if** $R_p < \rho$ **then**
        select the path
        break;
      **end if**
    **end for**
  **end for**

**end for**
select the access node which gives minimum access cost;
evaluate access cost.

### 4.3 Martins Santos Minimum Access Cost - MSMAC

As in the previous algorithm, the selection of the transport links is based on a k-shortest path algorithm, but in this case step 3 is carried out by means of *Martins Santos algorithm* [10], instead of the modified Dijkstra algorithm. Step 4 is the same as in the first case.
Formally the algorithm can be described as follows:

**for all** $i$ and $j \in \{1, ..., \text{N}\}$ **do**
  **if** $\omega_{ij} < \Omega$ **then**
    prune link (i,j) $\Rightarrow l_{i,j} = \infty$
  **end if**
  **for all** pair of ONs, $i$ and $j \in \{1, ..., N\}$ **do**
    use Martins Santos algorithm to find $k$ least-cost paths;
    **for** $p = 1 : k$ **do**
      evaluate path delay for the path $p$
      **if** $R_p < \rho$ **then**
        select the path
        break;
      **end if**
    **end for**
  **end for**
**end for**
select the access node which gives minimum access cost;
evaluate access cost.

### 4.4 Elimination

In this algorithm, step 3 is based on the use of the "standard" Dijkstra algorithm for the selection of the least cost path. But, in this case we evaluate the delay related to this path. If the path delay exceeds the imposed bound, the path does not guarantee the required level of QoS, so the path link with the maximum link delay is pruned, from the graph associated to the network topology. Then, if necessary, we iterate this procedure. Step 4 is based on the random selection of the access nodes. Formally, the algorithm can be described as follows:

**for all** $i$ and $j \in \{1, ..., \text{N}\}$ **do**
  **if** $\omega_{ij} < \Omega$ **then**
    prune link $(i, j) \Rightarrow l_{i,j} = \infty$
  **end if**
  **for all** pair of ONs, $i$ and $j \in \{1, ..., N\}$ **do**
    use Dijkstra algorithm to find least-cost path $p$;
    evaluate path delay for the path $p$

**if** $R_p > \rho$ **then**
prune the link with the maximum link delay and restart from the application of the Dijkstra algorithm
**else**
select the path
**end if**
**end for**
**end for**
select access nodes in a random mode;
evaluate access cost.

### 4.5 Additive Metric - AM

In this algorithm, step 3 is based on the Weight Method. The idea is to construct an unique metric which takes into account both cost and delay. This metric is obtained with a linear combination of the two metrics. As in the previous algorithm, step 4 is based on a random selection of the ONs. Formally the algorithm can be described as follows:

**for all** $i$ and $j \in \{1, ..., N\}$ **do**
**if** $\omega_{ij} < \Omega$ **then**
prune link $(i, j) \Rightarrow l_{i,j} = \infty$
**end if**
compute the metric $\varphi = \lambda_1 \cdot \textit{economic cost} + \lambda_2 \cdot \textit{delay}$
**for all** pair of ONs, $i$ and $j \in \{1, ..., N\}$ **do**
use Dijkstra algorithm to find least-cost $(\varphi)$ path $p$,
evaluate path delay for the path $p$
**if** $R_p > \rho$ **then**
impossible connection,
**else**
evaluate path cost;
**end if**
**end for**
**end for**
select access nodes in a random mode;
evaluate access cost.

### 4.6 Multiplicative Metric - MM

As in the previous case, step 3 is based on the computation of an unique metric, but in this case we consider the product between cost and delay. Once again step 4 is based on a random selection of the ONs. Formally, the algorithm can be described as follows:

**for all** $i$ and $j \in \{1, ..., N\}$ **do**
**if** $\omega_{ij} < \Omega$ **then**

prune link $(i,j) \Rightarrow l_{i,j} = \infty$
**end if**
compute the metric $\varphi = economic\ cost \cdot delay$
**for all** pair of ONs, $i$ and $j \in \{1, ..., N\}$ **do**
use Dijkstra algorithm to find least-cost ($\varphi$) path $p$,
evaluate path delay for the path $p$
**if** $R_p > \rho$ **then**
impossible connection,
**else**
evaluate path cost;
**end if**
**end for**
**end for**
select access nodes in a random mode;
evaluate access cost.

## 5 Simulation Results

In this section we present the results of the simulation carried out to compare the performance of the proposed algorithms, focusing at first on the description of the network scenario and then on the discussion of the achieved results.

In our simulation, we considered a network composed of $M$ ESs, where $M$ can assume the uniformly distributed values $\{10, 20, 30, 40, 50, 60, 70, 80, 90, 100\}$ and $N = 200$ ONs. Moreover, we assumed that the underlay network is formed by 38 ASs and each ES is located in a different AS. We also supposed that an ES can be connected to a given access ON, only if they belong to the same AS and that each ES is only connected to a single access ON.

From an economic point of view, we assumed that the transport costs between the ONs, as well as the access costs, are drawn from an uniform distribution in the range [5,50] monetary unit.
Instead, the reserved bandwidth is uniformly distributed in the range [10,20] Mbps, while the link delays are uniformly distributed in the range [3,15] ms.

The simulations have been conducted varying the service requests, namely varying the number of ESs, as well as the delay and bandwidth constraints. In more detail, we have considered the following settings:

- scenario 1: the number of ESs is fixed to 100, the delay constraint can assume the uniformly distributed values $\{27, 30, 33, 36, 39, 42, 45, 48, 51, 54\}$ ms, and the bandwidth constraint is equal to 13 Mbps;
- scenario 2: the number of ESs is variable, the delay constraint is fixed to 36ms, and the bandwidth constraint is equal to 13Mbps.

It is worth noticing that in all the simulations all the ESs are guaranteed to be connected to an access ON. Moreover, we have decided not to vary the number of ONs, since we consider the situation where the overlay backbone is given by the OSP.

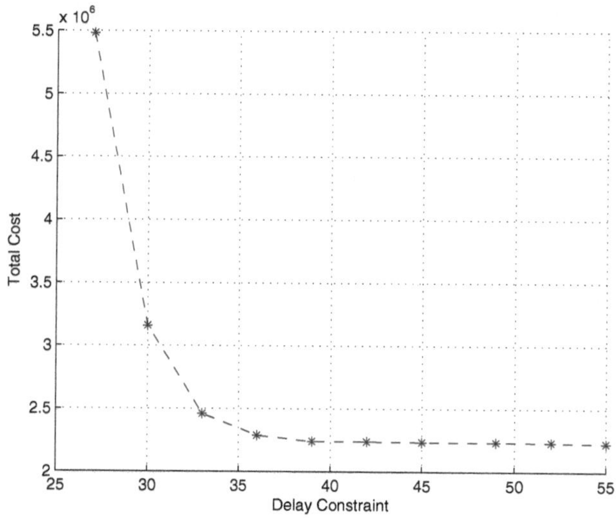


**Fig. 1.** Topology cost and delay threshold

The performance parameter taken into account to compare the proposed algorithm is given by the overall network cost, given by the sum of the access costs and transport costs. The obtained results represent the average value, computed over 100 independent simulations.

At first, we have evaluated (scenario 1) the dependence of the network costs on the given delay constraints. For sake of brevity, we only show the results related to one algorithm (i.e., MDRM), since all the algorithms present a similar trend.

In more detail, figure 1 plots the network costs versus the delay constraint. As expected, the overall costs significantly depend on the last constraint. Indeed, we have an increase of the network costs when the delay constraint becomes tighter. This result is not unexpected, since a tighter bound could imply the rejection of low cost paths.

Regarding the second scenario, we have evaluated the network costs achieved by the different algorithms, when varying the number of ESs. In figure 2, we propose a comparison among the different solutions. Since all the algorithms behave almost equally, in figure 3 we show a zoomed version of the previous figure, from which we can better analyze the algorithms behavior. It emerges that MM and AM algorithms have the highest costs demonstrating they are less suitable to simultaneously take into account both economic cost and delay in the minimization problem.

It is important to highlight that due to the possibility to assign different values to the weights $(\lambda_1, \lambda_2)$, in the AM algorithm, it is possible to achieve lower costs by giving more importance to the economic costs, as shown in figure 4. This solution corresponds, in fact, to only take into consideration the economic

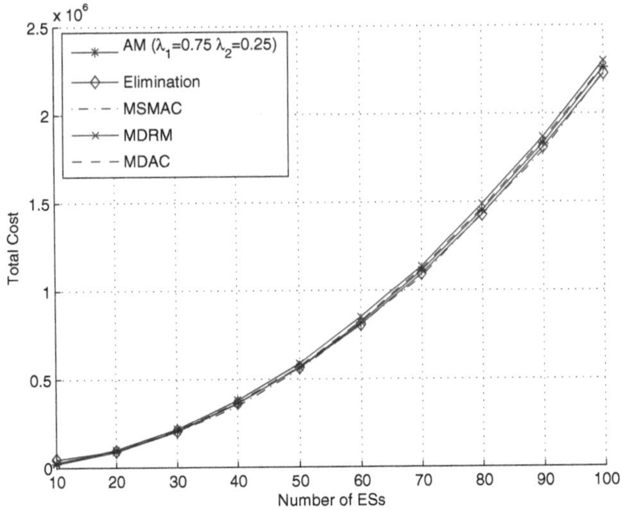


**Fig. 2.** Algorithms Comparison

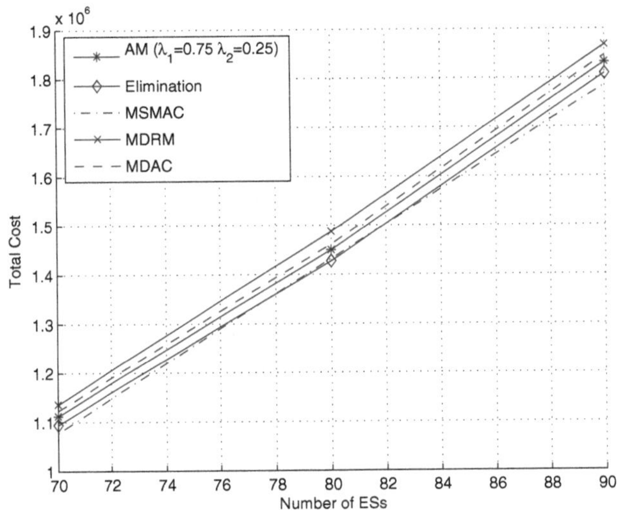


**Fig. 3.** Algorithms Comparison (zoomed version)

metric (case $\lambda_1 = 1, \lambda_2 = 0$) and so the metric could not be able to guarantee the requested QoS.

On the contrary, K-shortest path based algorithms and the Elimination algorithm allow the creation of low cost topology, while simultaneously taking into account the delay constraint.

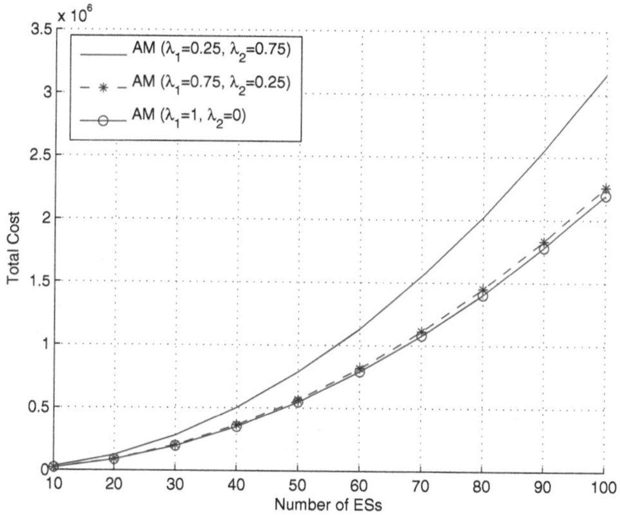


**Fig. 4.** AM algorithm performance when varying the weights

**Table 1.** Normalized Execution Time

| | **Normalized Execution Time** |
|---|---|
| **MDRM** | 4 |
| **MDAC** | 4 |
| **MSMAC** | 23 |
| **Elimination** | 8 |
| **AM** | 1 |
| **MM** | 2 |

Among these algorithms the best solution is given by the use of the MSMAC. Indeed, it achieves the best performance, in terms of network costs, and at the same time satisfies QoS constraints.

Nevertheless, it is important to also consider that this algorithm presents the highest execution time, as shown in table 1. This is due to the fact that it requires an iterated use of the Dijkstra algorithm up to finding the path that satisfies the delay constraint (Martin Santos algorithm).

Thus, also taking into account the computational costs, it emerges that the best algorithms are those based on the use of the Dijkstra algorithm, since they achieve good performance with low execution time.

## 6 Conclusion

In this paper we have proposed several algorithms to solve the topology design problem for SONs. The presented methods take into account both the economic aspects as well as several QoS constraints.

In more detail our aim has been to choose the access ONs and transport links which optimize costs and performance. To be noted that this optimization problem does not have an optimal solution, since it is a NP-hard problem. Thus, our solutions are based on several heuristics.

The performance analysis has shown that the best performance are achieved by the MSMAC algorithm, even though it has a very high computational cost.

## References


1. Duan, Z., Zhang, Z.L., Hou, Y.T.: Service overlay networks: Slas, qos, and bandwidth provisioning. IEEE/ACM Trans. Netw. 11(6), 870–883 (2003)
2. Tran, H.T., Ziegler, T.: A design framework towards the profitable operation of service overlay networks. Comput. Netw. 51(1), 94–113 (2007)
3. Lakshminarayanan, K., Stoica, I., Balakrishnan, H., Katz, R.: OverQoS: Offering Internet QoS Using Overlays. In: 1st HotNets Workshop, Princeton, NJ (October 2002)
4. Li, Z., Mohapatra, P.: QRON: QoS-aware Routing in Overlay Networks. IEEE Journal on Selected Areas in Communications 22(1), 29–40 (2004)
5. Cui, J.H., Gokhale, S.: Towards Integrated Provisioning of QoS Overlay Network. University of Connecticut, storrs, CT 06296
6. Li, Z., Mohaparta, P.: The impact of topology on overlay routing service. In: Proceedings of IEEE INFOCOM (2004)
7. Li, Z., Mohaparta, P.: On investigating overlay service topologies. Computer Network 51, 54–68 (2007)
8. Capone, A., Elias, J., Martignon, F.: Optimal design of service overlay networks. In: IEEE Telecommunication Networking Workshop on QoS in Multiservice IP Networks, Feburary 2008, pp. 46–52 (2008)
9. Vieira, S.L., Liebeherr, J.: Topology design for service overlay networks with bandwidth guarantee. In: IEEE International Workshop on Quality of Service, IWQoS, pp. 211–220 (2004)
10. Martins, E.Q., Pascoal, M.M., Santos, J.L.: Deviation algorithms for ranking shortest paths. International Journal of Foundations of Computer Science 10, 247–261 (1999)

# Bandwidth Optimization for Multicast Transmissions in Virtual Circuit Networks
## (Work in Progress)

Vincent Reinhard[1], Joanna Tomasik[1], Dominique Barth[2], and Marc-Antoine Weisser[1]

[1] Computer Science Department, SUPELEC, 91192 Gif sur Yvette, France
{vincent.reinhard,joanna.tomasik,marc-antoine.weisser}@supelec.fr
[2] PRiSM, Université de Versailles-St-Quentin. Versailles, France
dominique.barth@prism.uvsq.fr

**Abstract.** The CARRIOCAS project aims to guarantee QoS connectivity services to distributed applications in a Telecom carrier network. A large number of these applications (for example video applications) use a multicast service packet delivery. Multicast which minimizes the total used bandwidth in the MPLS network has become an important subject. We study multicast routing in the network where only some routers can duplicate packets. We prove that the construction of a multicast tree minimizing the bandwidth used in such a network is a *NP*-complete problem and we propose an heuristic algorithm to solve it. We evaluate the performance of the heuristic in terms of total bandwidth used by the multicast for different network sizes.

**Keywords:** Multicast, bandwidth optimization, MPLS networks.

## 1 Introduction

As distributed applications become more and more popular, Internet providers have become interested in furnishing to their customers not only the connectivity service but other services (storage, computation) as well. Supplying at the same time connectivity and other services was one of the motivations of the CARRIOCAS project [1][2][3][4] The latter is one of the projects of the "SYSTEM@TIC PARIS-REGION [1]competitiveness cluster". It aims to provide connectivity services and usual grid services to enable the executions of distributed application workflows in a Telecom carrier network. In the CARRIOCAS service architecture, the network service management is centralized per a network operator domain in order to enable querying for explicit bandwidth reservation. This network service management disposes of the network topology. It computes a virtual connection for each external connectivity service demand and, whenever possible, reserves a bandwidth on the connection. If no virtual connection

[1] SYSTEM@TIC PARIS-REGION and the CARRIOCAS project are supported by the French Ministry of Industry, Essonnes, Haut-de-Seines and Paris General Council.

L. Fratta et al. (Eds.): NETWORKING 2009, LNCS 5550, pp. 859–870, 2009.

with the requested bandwidth can be reserved, it rejects the reservation query. This mechanism ensures that the bandwidth needed for the CARRIOCAS applications never exceeds the bandwidth available and guarantee the QoS for applications. As GMPLS [14] is to be deployed in the CARRIOCAS network, it has to ensure both unicast and multicast communications.

There are several schemes for multicasting data in networks [6][7]. A first one is to construct virtual circuits from the multicast source to each destination. Such scheme is equivalent to multiple unicasts and the network bandwidth used by a large multicast group may become unacceptable [8]. In another scheme the multicast source sends data to the first destination and each destination acts as a source for the next destination until all destinations receive the data flow. In yet another scheme intermediate routers make copies of data packets and dispatch them to their successors in the multicast tree. This solution allows the multicast transmission to share bandwidth on the common links. Many multicast tree algorithms have been proposed and can roughly be classified into two categories [6]. The first category contains the algorithms based on the shortest path while minimizing the cost of the path from the multicast source to each destination. The second category contains algorithms based on the Steiner tree problem [9][10][11][12]. Such algorithms derived from the Steiner tree problem minimize the total cost of the multicast tree. This minimization is a *NP*-complete problem [10].

In this paper we study a solution based on the last mentioned scheme used in a MPLS [15] optical network which introduces an additional strong constraint on the construction of a multicast tree. From the technological point of view, routers able to duplicate packets introduce a supplementary delay due to O/E/O conversions and are more expensive. For these reasons network operators want to limit the number of such routers which we call "diffusing nodes". The constraint on the number of diffusing nodes invalidates the existing algorithms of multicast tree construction based on the Steiner tree [9][10] because they were proposed for networks the routers of which can all duplicate data. Such a constraint has already been considered for wavelength-routed networks [5] under the condition that all routers support the "drop and continue" functionality. Moreover, no complexity study had been made. In our work we consider that the routers do not support "drop and continue". We prove that minimizing the total cost of the multicast tree under this condition and with the constraint on the number of diffusing nodes is an *NP*-complete problem and propose an heuristic algorithm to solve it.

The rest of this paper is organized as follows. In Section II and III we propose a network model and define the bandwidth optimization problem. In Section IV, we present a bipartite graph used to study the bandwidth optimization problem. In Section V we prove that the problem is *NP*-complete and Section VI contains a pseudo-polynomial algorithm to solve it. We also propose an heuristic algorithm to solve the bandwidth optimization problem in Section VII. Section VIII contains our results concerning the performances of the heuristic. Finally, we conclude and outline future work.

## 2 Network Modeling

We model a network using a directed symmetric graph $G = (V, E)$. We define $cap(u, v)$ as the bandwidth available on the link $(u, v)$, $u, v \in V$. A multicast request is a triplet $\epsilon = (e, R, c)$, with $e \in V$ the source of the multicast, $R \subset V$ the set of destinations and $c$ the size of the data flow. In this work, we consider only multicast requests of the size $c = 1$ and note them $\epsilon = (e, R)$. We define $D_G \subset V$ as the set of diffusing nodes of $G$. An example of a network graph with a multicast request is given in Fig. 1.

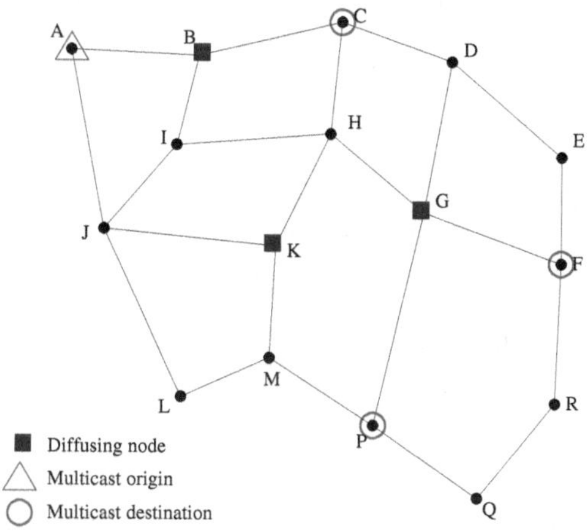


**Fig. 1.** A network graph $G$, $D_G = \{B, K, G\}$ and a request $\epsilon = (A, \{C, F, P\})$

## 3 Definition of the Bandwidth Optimization Problem

Our goal is to optimize the bandwidth used by multicasts in the network represented by graph $G$. We define a diffusion tree for $\epsilon = (e, R)$ as an arborescence $A_\epsilon$ in $G$ rooted in $e$, spanning $R$ and with all leaves included in $R$ (Fig. 2). We define $V(A_\epsilon)$ as the set of nodes of $A_\epsilon$ and $E(A_\epsilon)$ as the set of edges of $A_\epsilon$. A request $\epsilon$ is satisfied by the set of paths $C_{A_\epsilon}$ in $A_\epsilon$, defined as follows:

- every node of R is the final extremity of exactly one path in $C_{A_\epsilon}$,
- every node of $D_G$ is the final extremity of at the most one path in $C_{A_\epsilon}$,
- the origin of a path in $C_{A_\epsilon}$ is either $e$ or a node of $D_G$, In the latter case, the node of $D_G$ is also the final extremity of a path in $C_{A_\epsilon}$.
- a node of $D_G$ is in a path $p \in C_{A_\epsilon}$ only if it is the final extremity or the origin of $p$.

The load $ch_{A_\epsilon}(u, v)$ of a link $(u, v) \in E(A_\epsilon)$ is the number of paths $p \in C_{A_\epsilon}$ such as $(u, v)$ is a link of $p$. The load $ch_{A_\epsilon}$ of the arborescence $A_\epsilon$ is $\sum_{(u,v) \in E(A_\epsilon)} ch_{A_\epsilon}(u, v)$.

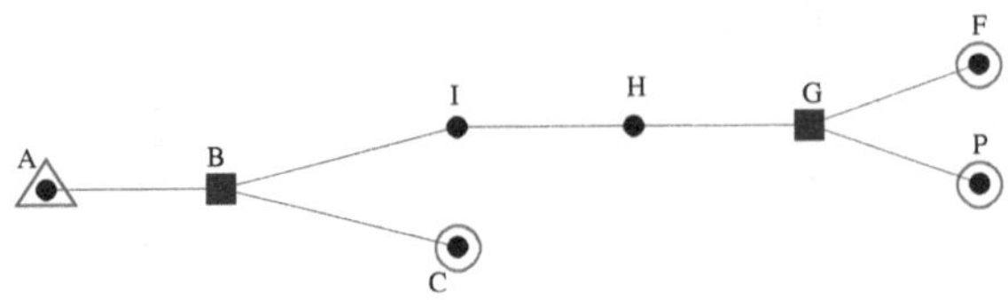


**Fig. 2.** A possible arborescence $A_\epsilon$ for $G$ and $\epsilon$ shown in Fig. 1

**Definition of the Problem *diff_tree*.** *Given a network $G$, its set of diffusing nodes $D_G$, and a multicast request $\epsilon = (e, R)$, find an arborescence $A_\epsilon$ for $\epsilon$, such as $ch_{A_\epsilon}$ is minimal.*

## 4 Complexity of the Problem

We study the complexity of *diff_tree* and we prove that its decision problem *diff_tree_dec* is *NP*-complete. First, we prove that the certificate of the decision problem is in class *P*. We then reduce the problem of the weighted set cover which is *NP*-complete [17] to the problem *diff_tree_dec.*

First, we show that the certificate of the problem *diff_tree_dec* is in class *P*. Given an instance of the problem *diff_tree_dec* (a graph of the network, a set of diffusing nodes and a multicast request) and an associated arborescence $A_\epsilon$, we can compute $ch_{A_\epsilon}$ in a polynomial time (we must compute a finite set of shortest paths). The certificate of the problem *diff_tree_dec* is therefore in class *P*. We now reduce the problem of the weighted set cover which is *NP*-complete [17] to the problem *diff_tree_dec* and we show that the solution to the problem of the set cover is equivalent to the solution to *diff_tree_dec.*

**Definition of the Weighted Set Cover Problem.** *Given a collection $C$ of subsets of a finite set $S$, with a cost $h$ associated to each subset and an integer $K$, is it possible to find a cover $S$ of size less than $K$, namely, a collection $C' \subseteq C$ such as every element of $S$ is in at least one subset $C'$ and whose cost is less than $K$.* Let $I$ be an instance of the set cover problem, namely a collection $C = \{c_1, ..., c_m\}$ of subsets of a finite set $S = \{s_1, ..., s_n\}$, a cost $h$ associated to each subset $c_i$ and an integer $K$. From this instance $I$, we can construct in polynomial time an instance $I'$ of the problem *diff_tree_dec* (Fig. 3): Let $G = (V, E)$ be a graph. We define $V$ as the set of its nodes followingly: there is a node for each subset $c_i$ from collection $C$; a node for each element $s_i$ from finite set $S$ and a node $e$. We define $E$ as the set of its edges created as follows: an edge of weight $l > h/2$ between any node representing $c_i$ from collection $C$ and every node representing an element $s_i$ of the $c_i$; an edge of weight $h$ between the node $e$ and each node representing a subset $c_i$. We define $D_G = \{c_1, ..., c_m\}$ and $K'$ an integer which adequate value will be defined later. Let $\epsilon = (e, R)$ be a multicast request in $G$, with $R = S$.

In a solution to $I'$ in $G$, we know that a node of $D_G$ is never reached from another node of $D_G$ but always from node $e$. Since $l > h/2$, the weight of a

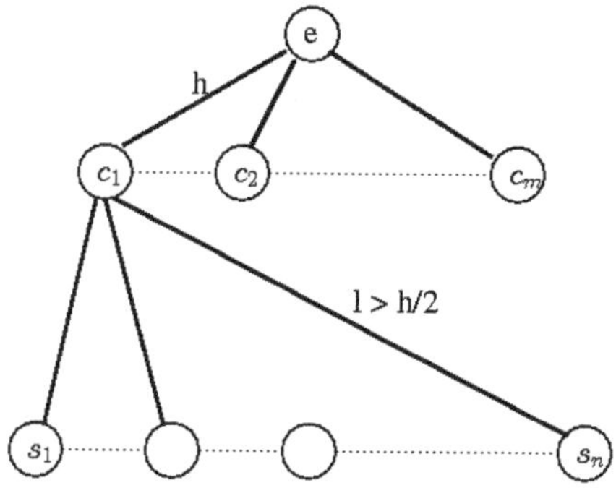


**Fig. 3.** Construction of an instance $I'$ of the problem *diff_tree_dec* from an instance $I$ of the problem of the set cover

path between two nodes of $D_G$ is always greater than $2l$, whereas the weight of any arc between $e$ and a node of $D_G$ is $h < 2l$. A solution to $I'$ in $G$ is an arborescence $A_\epsilon$ rooted in $e$. $A_\epsilon$ contains all nodes $s_i$ and $j \leq m$ nodes $c_i$ such as $ch_{A_\epsilon} < K'$. Futhermore, $ch_{A_\epsilon} = j * h + l * n$ (the weight of the $j$ edges between node $e$ and nodes $c_i$ and the weight of the $n$ edges between nodes $c_i$ and $s_i$). In instance $I$, selecting the subsets $c_i$ corresponding to the nodes $c_i$ which are also nodes of $A_\epsilon$ in $I'$ gives a cover $S$ of size $K = j * h = K' - l * n$.

Therefore, if there is a solution of weight less than $K'$ to the instance $I'$, then we can construct a solution of weight less than $K = K' - n * l$ to the instance $I$. Similarly, if there is a solution of weight less than $K$ to the instance $I$, then we can construct a solution of weight less than $K' = K + n * l$ to the instance $I'$. We can reduce any instance of the problem of the set cover to an instance of the problem *diff_tree_dec* in a polynomial time. Furthermore, we have demonstrated that the certificate of this problem is in class *P*. Therefore, the problem *diff_tree_dec* is *NP*-complete.

## 5 The Diffusion Graph

To treat a multicast request $\epsilon$ in graph $G$, we construct a new directed and weighted graph containing the three sets of nodes: $e$, $D_G$ and $R$. We call this graph a "diffusion graph" and define it as $B_{G,(e,R)}$. The weights of its arcs correspond to the weights of shortest paths in $G$. The graph $B_{G,(e,R)} = (V_B, E_B)$ is defined as follows:

- $V_B$ the set of nodes of $B_{G,(e,R)}$ such as $V_B = \{e\} \cup R \cup D_G$.
- $E_B$ the set of arcs constructed as follows:
  - $\forall u \in V_B$, $u \neq e$, the arc $(e, u) \in E_B$.
  - $\forall (u, v) \in D_G$, the arcs $(u, v)$ and $(v, u) \in E_B$.
  - $\forall u \in D_G$, $\forall v \in R$, the arc $(u, v) \in E_B$.

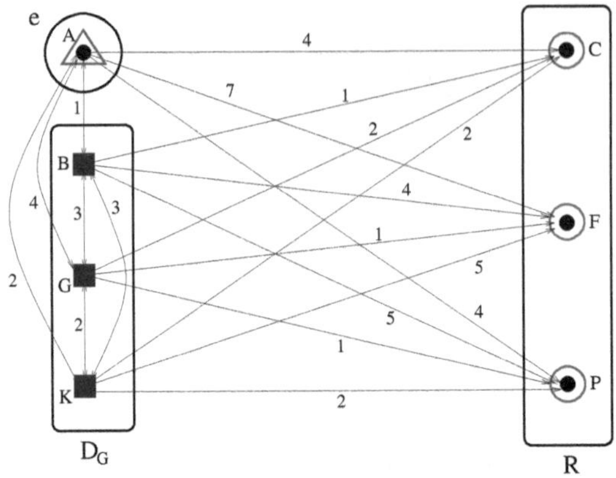


**Fig. 4.** The diffusing graph $B_{G,(e,R)}$ for $G$ and the multicast request of Fig. 1

– The weight of the arc $(u, v)$ of $E_B$ is the weight of a shortest path from $u$ to $v$ in $G \setminus \{D_G \setminus \{D_G \cap \{u, v\}\}\}$. If there is no path from $u$ to $v$ in $G \setminus \{D_G \setminus \{D_G \cap \{u, v\}\}\}$, the arc does not exist.

Fig. 4 represents the diffusion graph for the network graph from Fig. 1. The complexity of the diffusion graph construction is $O(n^3)$, where $n$ is the number of vertices in $G$.

We prove that finding a solution to the *diff_tree* problem in graph $G$ is equivalent to finding a solution to the *diff_tree* in the diffusion graph.

To find a solution to the *diff_tree* problem in graph $G$, we search an arborescence $A_\epsilon$ such as $ch_{A_\epsilon}$ is minimal. In the diffusion graph $B_{G,(e,R)}$, we search an arborescence $B_\epsilon$ such as $ch_{B_\epsilon}$ is minimal. Any arc $(u, v)$ of $B_{G,(e,R)}$ represents a shortest path from $u$ to $v$ in $G$. If we know an arborescence $B_\epsilon$ in $B_{G,(e,R)}$ for which $ch_{B_\epsilon}$ is minimal, we can determine a set of paths $C_{B_\epsilon}$ in $B_\epsilon$. We can construct a set of paths $C_{A_\epsilon}$ in $G$ such as $ch_{A_\epsilon} = ch_{B_\epsilon}$ using the corresponding shortest paths of $G$ represented by the arcs of $C_{A_{B_\epsilon}}$. From the set of paths $C_{A_\epsilon}$, we can construct the corresponding arborescence $A_\epsilon$ in $G$. Since $ch_{B_\epsilon}$ is minimal, $ch_{A_\epsilon}$ is minimal as well. Similarly, if we know $A_\epsilon$ in $G$ such as $ch_{A_\epsilon}$ is minimal, we can create $B_\epsilon$ in $B_{G,(e,R)}$ such as $ch_{B_\epsilon}$ is minimal.

## 6 A Pseudo-Polynomial Exact Algorithm

We propose an exact algorithm to determine an arborescence $A_\epsilon$ minimizing $ch_{A_\epsilon}$ in $B_{G,(e,R)}$. Let us define $S \subseteq D_G$ as a set of diffusing nodes which are also nodes of $A_\epsilon$. Knowing $S$ allows us to construct $A_\epsilon$ in polynomial time as follows. Let $\{e\} \cup R \cup S$ be the nodes of $A_\epsilon$. We chose for each node in $R$ an arc of minimum weight which has as its origin a node of $S$ and we add this arc to $A_\epsilon$. We then compute a spanning tree in the clique formed by $e$ and the nodes of $S$ and we add it to $A_\epsilon$. Any part of this algorithm is polynomial. To determine

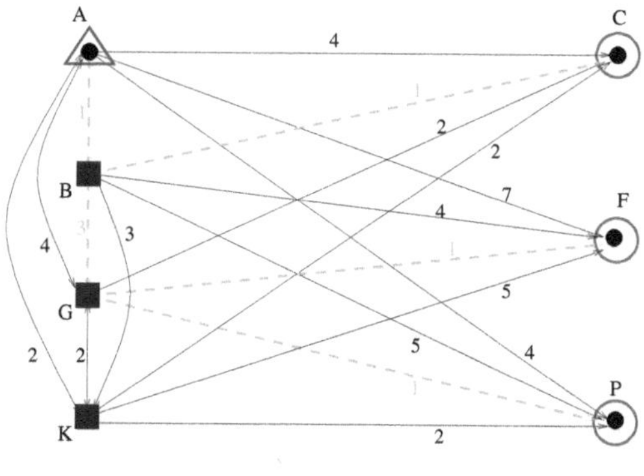


**Fig. 5.** Solution for the graph of Fig. 1 computed by the exact algorithm. $ch_{A_\epsilon} = 7$.

which nodes are in $S$, we successively generate all subsets of $D_G$. We construct for each subset a minimal arborescence (as described above). Finally, we chose as solution to the problem the arborescence with the minimal $ch_{A_\epsilon}$.

The complexity of this exact algorithm depends on the cardinality of the power set of $D_G$ and it is $O(2^{\#D_G})$. Fig. 5 shows the solution computed by this algorithm for the graph and request shown in Fig. 1.

## 7 Heuristic Algorithm

Since the pseudo-polynomial exact algorithm can not be used to solve the *diff_tree* problem when the number of diffusing nodes is large, we propose a polynomial heuristic algorithm.

This heuristic is based on an algorithm of maximal flow of minimal cost and its description is given below. In the experiments that we performed, we used the Busacker and Gowen maximal flow of minimal cost algorithm (Busacker and Gowen, 1961). First, we construct a directed graph $F_{G,(e,R)} = (V_F, E_F)$ (Fig. 6) based on $B_{G,(e,R)}$ which will be used for a flow algorithm.

- $V_F$ is the set of nodes of $F_{G,(e,R)}$ with $V_F = V_B \cup \{p\}$ where $p$ is an additional node used as a sink.
- $E_F$ is the set of arcs with capacities and costs constructed as follows:
  - $\forall (u, v) \in E_B$, $(u, v) \in E_F$, its cost is the weight of $(u, v) \in E_B$ and the capacity on $(u, v)$ is infinite.
  - $\forall u \in D_G$, $E_F$ contains $(u, p)$. The cost of these arcs is always 0 and their capacity is 1.

In $F_{G,(e,R)}$ we compute a maximal flow of minimal cost from $e$ to $p$. Since the capacity on each arc $(u, p)$ is 1, the maximal flow value is equal to the number of nodes in $R$ and since a maximal flow has to pass by all $r \in R$ every node of $R$ is in a flow. The computation of a maximal flow of minimal cost gives an arborescence in $F_{G,(e,R)}$ which contains $e$ and all nodes of $R$. This arborescence deprived of the sink $p$ is a possible $A_\epsilon$ solving the *diff_tree* problem in $B_{G,(e,R)}$. In any algorithm

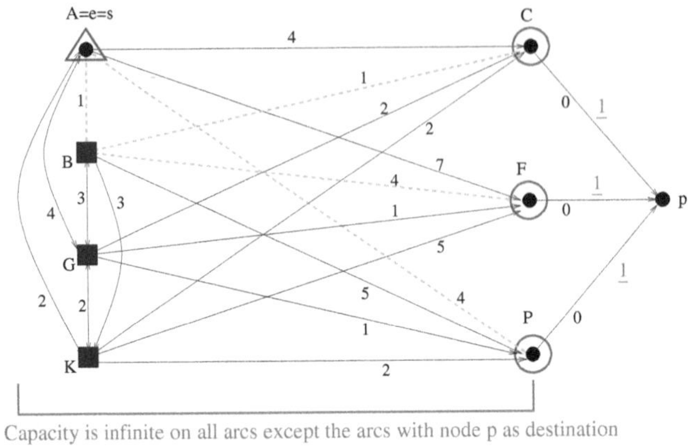


**Fig. 6.** $F_{G,(e,R)}$ constructed from $B_{G,(e,R)}$ of Fig. 4. Capacities are highlighted. The solution constructed by the heuristic is shown in dotted lines and $ch_{A_\epsilon} = 10$.

of maximal flow of minimal cost, the cost of each arc is counted as many times as the flow value on it. To compute $ch_{A_\epsilon}$ (Section III), we count the cost of each arc used in the solution only once. For this reason, we modify the maximal flow of minimal cost algorithm by setting to zero the cost of an arc which has already been used by a flow. Setting the cost of the arcs already used to zero impacts the computation of the solution. The arborescence $A_\epsilon$ corresponding to the maximal flow is not necessarily of minimal cost. A solution found by this heuristic for the graph of Fig. 1 is illustrated by Fig. 6.

The complexity of this heuristic is the same as the algorithm of maximal flow of minimal cost (for the Busacker and Gowen algorithm, $O(n^4)$).

## 8 Results

We performed experiments to study the performance of our heuristic in terms of the number of links used by the multicast tree which in our case corresponds to the bandwidth used by the multicast. We also studied the influence of the number of diffusing nodes on the performances of our algorithm.

We generate graphs $G = (V, E)$ representing a network with the BRITE generator [13] using the Waxman model (whith parameters: $\alpha = 0.15$, $\beta = 0.2$, $m = 2$, $MaxBW = 1024$, $MinBW = 10$). We define $T \subset V$ as the set of nodes of degree smaller than three.

In all experiments, to determine the location of the diffusing nodes in $G$ we use a k-center heuristic algorithm based on a dominating set algorithm [16]. This approach was chosen because the k-center algorithm distributes the centers "equally" in the graph. In our opinion this approach is adapted to deal with multicasts whose origins are not a *priori* known efficiently.

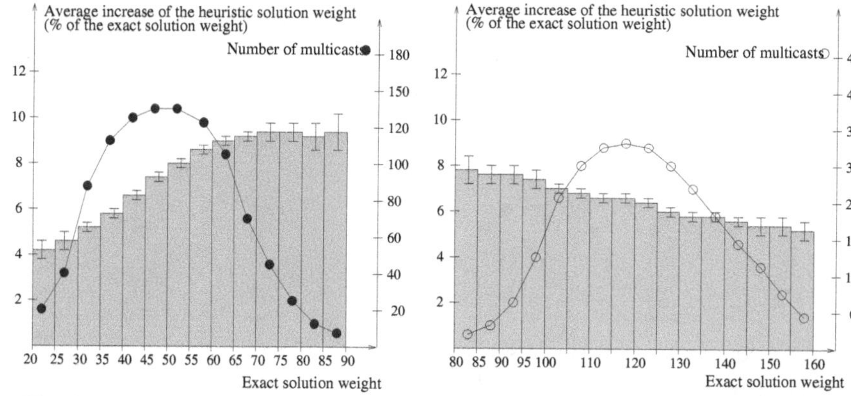


Fig. 7a Fig. 7b

**Fig. 7.** Distribution of the multicast trees depending on their weight and the performance of the heuristic compared to the exact algorithm for a network with 200 nodes. The number of destinations is generated with $N_1$ (Fig. 7a) and $N_2$ (Fig. 7b). Confidence intervals computed with the significance coefficient $\alpha = 0.05$.

We used the Busacker and Gowen algorithm as basis for our heuristic algorithm. The results obtained could be slightly different with another maximal flow minimal cost algorithm.

We generate multicast requests (multicast groups) in the network as follows. We chose an origin for multicasts uniformly in $V$. For each chosen origin, we generate different destination sets. For each destination set, we use a normal distribution $N_1 = \mathcal{N}(10\%\#T, 2\%\#T)$ to fix the number of destinations. The destinations are then chosen uniformly in $T$.

We started with a topology of 200 nodes. We arbitrarily fixed $\#D_G = 6$ and placed the diffusing nodes in the network according to the k-center algorithm. We chosed multicast sources and for each source generated 30 destination sets. We computed the average number of links used in the multicast trees constructed by both the exact algorithm and our heuristic. We constructed an histogram containing the results as follows. We created weight intervals and placed every generated multicast request in such an interval depending on the weight of its best multicast tree (constructed by the exact algorithm). We arbitrarily set the weight intervals length to 5. For each interval we computed the average weight of the multicast tree constructed by the exact algorithm and by the heuristic. We then showed in the histogram the relative average increase of the weight of the solution computed by the heuristic compared to the weight of the exact solution. We also showed for each weight interval the number of multicast requests in it. The results are shown in Fig. 7a.

What we first observe is that our heuristic always finds solutions of average weight at the most 10.5% greater than the best solution. We also observe that the performance of our heuristic decreases when the average weight of the multicast trees increases. This degradation can be explained as follows. An average greater

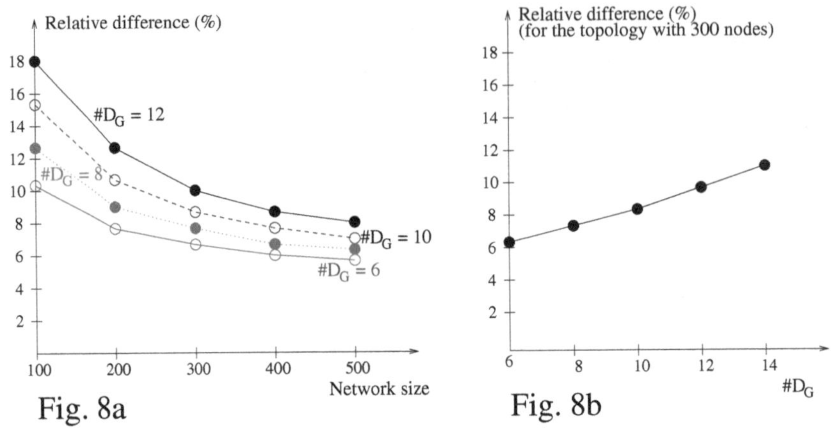


**Fig. 8.** Relative difference between average sizes of multicast trees obtained on one hand with the heuristic and on the other with the exact algorithm depending on the network size (Fig. 8a). Fig. 8b shows the same relative difference, depending on the number of diffusing nodes in the network with 300 nodes. Confidence intervals for both Fig. 8a and Fig. 8b are computed with precision 5% and significance coefficient $\alpha = 0.05$.

weight of the multicast tree is due to either more destinations or destinations farther from the multicast source (or both). When the heuristic constructs the multicast tree, it tries in priority to add destinations to the diffusing nodes already selected in the partial solution because the weight of the arcs already used is set to zero (Section VII). When the number of destinations is not large enough to select most of the diffusing nodes in the solution, the set of diffusing nodes selected by the heuristic is most often different than the set selected by the exact algorithm and the heuristic solution diverges more from the best solution when the average weight increases. It is only after all or most of the diffusing nodes have been selected in the partial solution, does the heuristic add each new destination in the best way. To observe the performance of our heuristic when the number of destinations is greater, we generated more multicast requests by using another distribution $N_2 = \mathcal{N}(25\%\#T, 2\%\#T)$ to fix the number of destinations. The results are shown in Fig. 7b. We observe that the heuristic constructs solutions whose weight is closer to the best one. In cases in which the number of destinations is relatively large, both the exact solution and the heuristic one are constructed with all or most diffusing nodes. After all diffusing nodes have been selected by the heuristic, the heuristic adds each new destination in the best way and the weight of its solution becomes closer to the best one.

In order to observe the influence of the network size on the performance of our heuristic, we generated topologies of $i * 100$, $i = 1, 2, ..., 5$ nodes preserving the same average degree of nodes. We placed a various number of diffusing nodes ($\#D_G = 6, 8, 10, 12$) in each network using the k-center algorithm. As above, for each multicast source we generated 30 different destination sets whose cardinality is given by the distribution $N_1$. We computed the number of links used by the multicast trees with both our heuristic and the exact algorithm (Fig. 8a).

We observe that for a fixed number of diffusing nodes our heuristic performs better for greater networks. We have shown in Section VI that the non-polynomial part of the exact algorithm corresponds to finding the set of diffusing nodes used in the minimal multicast tree. When the set of diffusing nodes of the multicast tree constructed by the heuristic is almost the same as the set of diffusing nodes of the best multicast tree, the heuristic finds a multicast tree whose weight is close to the weight of the best multicast tree. In our model, since we always preserve the average degree of nodes and use the distribution $N_1$ based on the number of nodes of low degree, the average number of destinations depends on the network size. When the network size becomes larger, the average number of destinations for the generated multicast groups increases and more of the diffusing nodes must be used to construct the best multicast tree. All or most of the diffusing nodes are chosen by both the exact algorithm and the heuristic. The heuristic then constructs a multicast tree whose weight is close to that of the best one.

The results depending on the number of diffusing nodes in a given network (Fig. 8b) tends to confirm that for this given network, increasing the number of diffusing nodes degrades the performance of our heuristic. For the reasons explained above, the heuristic constructs better solutions when the ratio between the number of diffusing nodes and the number of destinations is small. A greater ratio means that a smaller portion of the diffusing nodes appears in the best multicast tree and the heuristic choses the same set of diffusing nodes less often than in the previous experiments.

## 9 Conclusion and Perspectives

We studied a construction of a multicast tree optimizing bandwidth in an optical MPLS network. We proved that the construction of a multicast tree optimizing the number of links used is *NP*-complete when only a subset of the routers are diffusing nodes. We proposed a heuristic algorithm to construct the multicast tree optimizing the number of links used and we evaluated the performance of our heuristic with a pseudo polynomial algorithm. We showed that the performance of the heuristic depends on the ratio between the number of diffusing nodes and the number of destinations. We observed that our heuristic performs very well when this ratio is small and that it constructs multicast trees whose weight is close to the weight of the best ones. We are currently working on new heuristics to construct a multicast tree and we will compare them with the heuristic we proposed here. We are also examining how we can improve the placement of the diffusing nodes in the network when we are given the set of the multicast sources.

## References

1. The CARRIOCAS website, `http://www.carriocas.org`
2. Verchere, D.: Orchestrating optimally IT and network resource allocations for stringent distributed applications over ultrahigh bit rate transmission networks. In: ECOC 2007 (2007)

3. Audouin, O.: CARRIOCAS description and how it will require changes in the network to support Grids. OGF (2007)
4. Reinhard, V., Tomasik, J.: A mechanism of application-network interactions in a high-speed optical network. In: CISIS 2008 (2008)
5. Zhang, X., Wei, J., Qiao, C.: Constrained Multicast Routing in WDM Networks with Sparse Light Splitting. In: INFOCOM 2000 (2000)
6. Salama, H.F., Reeves, D.S., Viniotis, Y.: Evaluation of Multicast Routing Algorithms forReal-Time Communication on High-Speed Networks. IEEE J. on Sel. Areas in Comm. 15(3) (April 1997)
7. Myoungki, J., Yijun, X., Cankaya, H.C., Vandenhoute, M., Qiao, C.: Efficient multicast schemes for optical burst-switched WDM networks. In: IEEE ICC 2000 (2000)
8. Malli, R., Zhang, X., Qiao, C.: Benefits of multicasting in all-optical networks. In: SPIE Proceedings, All Optical Networking, November 1998, pp. 209–220 (1998)
9. Beasly, J.: An SST-based algorithm for the Steiner problem in graphs. Networks 19, 1–16 (1989)
10. Hwang, F., Richards, D.: Steiner tree problems. Networks 22, 55–89 (1992)
11. Kompella, V., Pasquale, J., Polyzos, G.: Multicasting for multimedia applications. In: INFOCOM 1992 (1992)
12. Bharath-Kumar, K., Jaffe: Routing to multiple destinations in computer networks. IEEE Transactions on Communications 31(3), 343–351 (1983)
13. Medina, A., Lakhina, A., Matta, I., Byers, J.: BRITE: An Approach to Universal Topology Generation. In: MASCOTS 2001 (2001)
14. Mannie, E. (ed.): Generalized Multi-Protocol Label Switching (GMPLS) Architecture, IETF RFC 3945 (Ocotber 2004)
15. Rosen, E., Viswanathan, A., Callon, R.: Multiprotocol label switching architecture, IETF RFC 3031 (January 2001)
16. Mihelic, J., Robic, B.: Solving the k-center Problem Efficiently with a Dominating Set Algorithm. J. of Comp. and Info. Tech. - CIT 13, 3, 225–233 (2005)
17. Karp, R.M.: Reducibility Among Combinatorial Problems. In: Complexity of Computer Computations, pp. 85–103. Plenum, New York (1972)

# Harmony - Advance Reservations in Heterogeneous Multi-domain Environments*
## (Work in Progress)

Alexander Willner[1], Christoph Barz[1], Joan Antoni Garcia Espin[2], Jordi Ferrer Riera[2], Sergi Figuerola[2], and Peter Martini[1]

[1] University of Bonn, Institute of Computer Science IV, Römerstr. 164, 53111 Bonn, Germany
{willner,barz,martini}@cs.uni-bonn.de

[2] i2CAT Foundation, c/ Gran Capità 2-4, Edifici Nexus I, 2ª planta, desp. 203, 08034 Barcelona, Spain
{joan.antoni.garcia,jordi.ferrer,sergi.figuerola}@i2cat.net

**Abstract.** Grid computing aims at offering standardized access to heterogeneous and distributed resources for scientific communities. In order to ensure certain quality of service requirements, the interconnecting networks have also be considered as Grid resources and must be taken into account for the co-scheduling process. However, most current systems do not support co-allocation of heterogeneous network resource provisioning systems and malleable advance reservations in multi-domain, multi-technology, and multi-vendor environments. Our approach, called Harmony, provides a functional service plane to unify the underlying network management systems and supports advanced reservation capabilities to utilize the available capacity and network resources in an efficient manner. The developed prototype has been demonstrated on numerous conferences and a preliminary performance evaluation of the current implementation is given.

**Keywords:** Advance Reservations, BoD, GMPLS, MPLS, QoS.

## 1 Introduction

Since its first demonstration in the context of the I-WAY [2] project more than ten years ago, the concepts of Grid computing [3] have become increasingly popular in the field of high performance computing. The vision is to virtualize any kind of resource and to combine them to a single abstract entity that is as simple to use as the power grid. One important research area in this context is the co-allocation of these different resources. In particular, the underlying network has to be considered as a first-class Grid resource that must be managed to ensure certain quality of service ($QoS$) requirements.

* Research for this paper was partially financed by the EU in the IST-034115 project PHOSPHORUS [1]. We thank our project partners for their contributions and their collaboration to this research work.

L. Fratta et al. (Eds.): NETWORKING 2009, LNCS 5550, pp. 871–882, 2009.

As shown by Foster et al. [4] resource co-allocation in Grid computing environments should be done automatically and transparently. In order to integrate the network into this scheduling process, mainly two issues have to be faced. Meanwhile the infrastructure has to support different QoS features, such as a minimum bandwidth or maximum delay on a specified path, an agreement based resource management is needed to reserve resources immediately or in advance.

In 2006, Travostino et al. [5] gave an overview of the relevance of network research in Grid environments. In a nutshell, there have been two different approaches. Either low-level provisioning mechanisms for IP networks were used such as Integrated Services (*IntServ/RSVP*) [6] and Differentiated Services (*Diff-Serv*) [7]. Or, in order to meet the very high bandwidth demands of some Grid applications, recent projects have started to use circuit-switched optical networks. However, shortcomings of the existing approaches are mostly their limited applicability in real environments or missing support of advanced scheduling algorithms.

In this paper, we assume a set of independent administrative domains that run their local network resource provisioning systems (*NRPSs*). These domains might use different network technologies and are interconnected by static predefined links. In our approach, we use an abstract service plane to provide an end-to-end service and allow to co-allocate these networks in conjunction with other Grid resources. The primary goal of this work is to provide a system that reuses existing developments already running as NRPSs, and to allow independent administrative domains to be integrated into the Grid resource scheduling process, and therefore support advanced mechanisms to utilize the transport network efficiently.

The key contribution of this paper is to provide a proof-of-concept and to deliver measurements of a real implementation that builds an abstraction layer for various existing signalling protocols. Overall, we believe this paper helps to construct a foundation for further network research and developments in the field of resource co-allocation in multi-domain, multi-technology, and multi-vendor Grid environments.

The remainder of the paper is structured as follows. We give an overview of related work in the context of dynamic circuit provisioning in Section 2. In Section 3, we state our assumptions and the proposed architecture and we introduce the challenges that are given in co-allocated multi-domain advance reservation situations. Section 4 contains implementation details of our solution. In the subsequent Section 5 a preliminary performance evaluation of our implementation is given. Finally, we close with some conclusions and describe future work in Section 6.

## 2 Related Work

Grid applications typically need to allocate and reserve multiple types of resources, such as computation, data, instrumentation, and networks. The co-allocation problem for computational Grids has been defined by Czajkowski [8]

1999. In the same year, based on the techniques and concepts of the Globus Resource Management Architecture (*GRMA*) [9], a Distributed Resource Management Architecture (*GARA*) [10, 4] that used a Globus Resource Allocation Manager (*GRAM*) job scheduler to co-allocate the network with other resources in advance was proposed. By then, the mechanisms used for network provisioning were IntServ [6] and DiffServ [7].

While GARA is popular among the Grid community as a general purpose platform allowing reservations of numerous resources it is not specialized for networks. Its API and Resource Specification Language (*RSL*) do not take network specific attributes into account. Therefore, the Network Resource Scheduling Entity (*NRSE*) [11] was introduced in the Grid Resource Scheduling (*GRS*) project in 2002. Still IntServ and DiffServ mechanisms were used to deliver guaranteed throughput over packet-based networks.

Unfortunately, bulk data transfer oriented Grid computing often requires guaranteed minimum bandwidth and minimized packet loss, which are not easily achievable in packet switching networks. Moving Terabytes or Petabytes of data among multiple sites require dedicated optical networks, e.g., based on wavelength switching, that can provide guaranteed bandwidth and performance in terms of low bit error rates.

In 2005 the Exploitation of Switched Light paths for eScience Applications (*ESLEA*) project had started. It demonstrated the usefulness of circuit-switches networks for different application areas. The ESLEA Control Plane Software (*CPS*) was implemented as a modification of the afore mentioned NRSE and was integrated into the EGEE Bandwidth Allocation and Reservation (*BAR*) architecture. At the same time the DARPA DWDM-RAM project addressed similar issues and a Network Resource Scheduling (*NRS*) service was developed to enable the efficient use of optical networks as a primary Grid resource.

***Associated projects.*** Harmony is based on existing solutions for intra-domain path provisioning and scheduling, developed within associated projects. In total, four different NRPSs are used to administer the underlying network: *ARGIA* [12] (former *UCLP* – User Controlled Light path Provisioning); Nortels proof-of-concept middleware called Dynamic Resource Allocation Controller (*DRAC*) [13]; the MPLS [14] based Allocation and Reservation of Grid-enabled Optical Networks (*ARGON*) [15] system that was developed within the German research project VIOLA[1] and also a thin adaption layer for the GMPLS control plane.

***Related Projects.*** In fact, other related projects aim at similar challenges, although from a different perspective, while the Phosphorus view is from the user to the network: Originating from the GÉANT2[2] project, the Automated Bandwidth Allocation across Heterogeneous Networks (*AutoBAHN*) system; as an achievement of the DANTE-Internet2-CANARIE-ESnet collaboration (*DICE*), an inter-domain control plane based on OSCARS was developed where an

[1] `http://www.viola-testbed.de`
[2] `http://www.geant2.net/`

inter-domain controller (*IDC*) communicate in a decentralized way to provision end-to-end multi-domain network paths; the G-lambda project that develops an interface between Grid resource management systems and network resource management systems that also support advance reservations; the GMPLS based EnLIGHTened Computing project that focuses on dynamic optical light-paths between supercomputing sites which are created and torn down in advance or on demand based upon application needs; the Dynamic Resource Allocation in GMPLS Optical Networks (DRAGON) project that aims at both the dynamic intra- and inter-domain provisioning of packet and circuit switched network resources in response to user requests for high-performance e-Science applications; and finally, the Grid-enabled GMPLS ($G^2$MPLS) Network Control plane, as a enhancement of the ASON/GMPLS Control Plane architecture that implements the concept of Grid Network Services (*GNS*).

## 3 Assumptions, Limitations and Architecture

Consider a network consisting of interconnected, independent domains that are used as transit systems or that contain different kinds of resources. A domain in this context is itself again a high performance (optical) network, with a control plane above and a network resource management system (*NRPS*) on top. Each level of this architecture performs different roles and tasks in order to allow creating on demand data plane paths for several users along a single domain. Interconnection with third party domains has to be proposed, agreed and set up by all involved parties. This leads to high delay in inter-domain path provisioning, bureaucratic overhead and hard scheduling of resource utilisation (considering both intra- and inter-domain resources) which most of the times discourage end users.

The problem we consider here in detail is how to find and reserve an-end-to end path between two or more resources on demand with certain QoS requirements, while supporting heterogeneous network technologies.As a consideration, each domain must not reveal its internal topology and the network resources should be utilized to capacity.

***General Architecture.*** The proposed architecture extends the afore mentioned hierarchy by one additional layer. The network service plane (*NSP*) consists of at least one inter-domain broker (*IDB*) and one NRPS Adapter (*HNA*) for each subjacent domain. They both share the same Harmony Service Interface (*HSI*) and allow to exchange abstract topology information and to administer network paths. The architecture has started with a centralized design (Fig. 1a) and evolved to an hierarchical (Fig. 1b) and distributed (Fig. 1c) system.

***Topology Knowledge.*** In particular, for reasons of confidentiality the internal topology information of a domain might not be opened to the service layer. This is the reason why the knowledge of the global topology is restricted to a set of basic information based on three main elements: the endpoint, the link, the domain itself, and the border points.

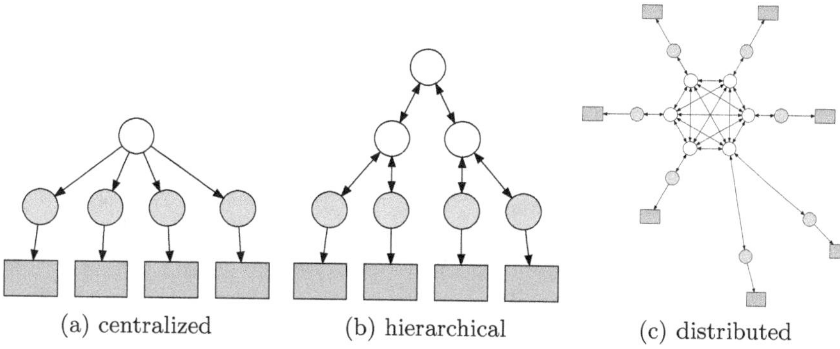


**Fig. 1.** The service plane architectures and the operating modes implemented. *White circles* are IDBs, *grey circles* are adapters, and *grey boxes* are administrative domains with their corresponding NRPS.

Based on the Transport Network assigned address (*TNA*) [16] endpoints in the NSP are identified by strings with IPv4 syntax. Each network port receives a unique TNA in the domain it is attached to and the domain itself serves one or more TNA address ranges. Two border endpoints identify uniquely an inter-domain link between two given domains.

Every domain exports its border endpoints connected to inter-domain links and the inter-domain links themselves. Hence, a transport network controlled by a single HSI capable system is seen as a cloud with a set of border endpoints. If two or more domains are controlled by a single IDB, this new super-domain will be seen as one cloud with a subset of the original border endpoints. Border endpoints connected to the other domains under control of the same IDB are kept as intra-domain endpoints and are not passed up.

***Path Computation.*** End-to-end path provisioning requires dynamic intra- and inter-domain path computation. Since only an abstracted topology knowledge is distributed, the intra-domain path computation is performed by each NRPS. The inter-domain path computation is performed inside each IDB by using Dijkstra's shortest path algorithm. Only border endpoints and inter-domain links are considered.

A reservation itself contains one or more services and each service contains one or more connections. Whereby each connection is a requested path between one source TNA and one or more destination TNAs. Different fixed or malleable QoS demands can be defined on each level.

The actual reservation is following a non-blocking tree two-phase commit protocol scheme. After the reception of a reservation request, the path computer starts looping over all known network resources in order to find feasible paths. Upon the completion of this task, the availability of the resources within all involved domains in the calculated path is requested. In case the resources are not available, they are pruned within the path computer session and the loop will start again. Otherwise the resource will be reserved.

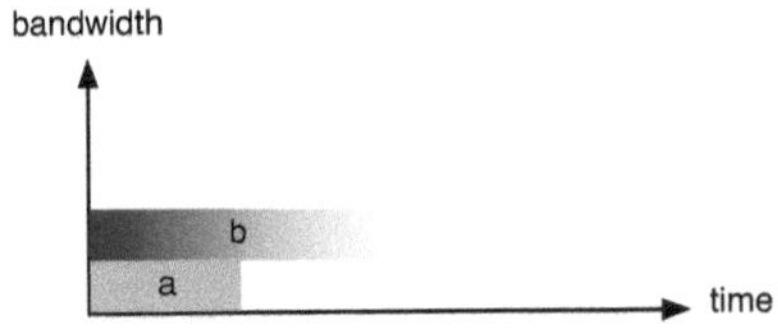


(a) immediate reservations

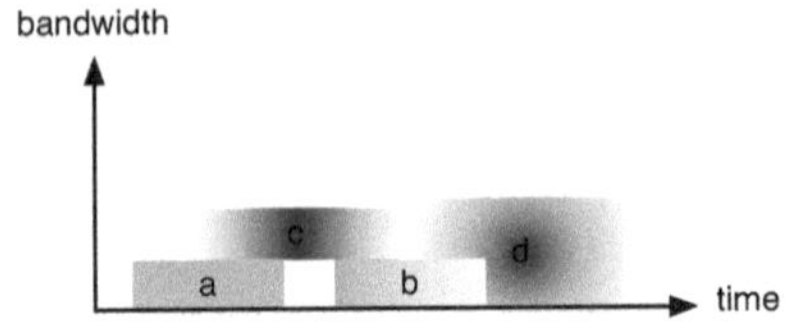


(b) advance reservations

**Fig. 2.** The different types of reservations. Fading areas represent unspecified bandwidth or time constraints.

***Malleable Multi-domain Network Resource Allocation.*** As stated in [17] and depicted in Fig. 2 reservations can be divided into two different types. First, reservations can start straightaway upon receipt of the reservation (Fig. 2a). Second (Fig. 2b), they can be scheduled with a specified starting time and a specified duration (fixed, FAR, STSD), a specified starting time and a unspecified duration (STUD), an unspecified starting time and a specified duration (deferrable, DAR, UTSD), and finally with an unspecified starting time and an unspecified duration (malleable, MAR, UTUD).

Of particular interest are the advance reservation requests with elastic starting times and an unspecified duration. These malleable reservations can be used by Grid middleware applications for file transfers. They allow the largest flexibility to schedule resources and to utilize the full network capacity.

When a malleable reservation request is received, the IDB starts looping over distinct inter-domain paths, with feasible start-times and bandwidths in order to find a set of available resources to fulfil the request. First of all, the possible paths are calculated. After that, the path computation algorithm starts looping over all the obtained paths. Inside the loop, the feasible bandwidths of the endpoints involved in the connection are retrieved and the maximum bandwidth in the range of the bandwidth provided in the request is chosen. This bandwidth is adjusted in all endpoints to the given path according to the technology-induced granularity. In a final step, the algorithm checks for the availability of the network resources. In case any NRPS returns non-available result, the IDB begins to adapt the start time and bandwidth.

***Formal Definitions.*** We define the data plane as digraph $G_d = (D, L)$. The elements of set $D$ are called *domain*, and the elements of $L$ are ordered pairs of vertices, called *inter-domain link* with $(u_d, v_d, id_d) \in L \subseteq D \times D \times ID$. An *inter domain path* in $G_d$ between two domains $d_{source}$ and $d_{dest}$ is an adjacent sequence of inter domain links $p_d(d_{source}, d_{dest}) = (u_{d_0}, v_{d_0}, id_{d_0}), \ldots, (u_{d_{m-1}}, v_{d_{m-1}}, id_{d_{m-1}})$, with $u_{d_0} = d_{source}$, $v_{d_{m-1}} = d_{dest}$, $m \in \mathbb{N}_0$ and $len(p_d) = m$. The number of parallel inter-domain links between two domains can be defined as $n(u_d, v_d) = |\{(u_{d_i}, v_{d_i}, id_{d_i}) \in L | u_d = u_{d_i} \wedge v_d = v_{d_i}\}|$. The successor domains of a domain referring to the data plane are defined as a function $s(u_d) = \{w_d | \exists (u_d, w_d, id_d) \in$

$L\}$. Let $p_d^i(d_{source}, d_{dest}) = (u_{d_i}, v_{d_i}, id_{d_i})$ be the $i$th link of a path. Then the adjacent domains of a path $p_d$ are defined as $a(p_d) = \{d_{source}\} \cup_{0 \leq i < len(p_d)} \{v_{d_i} | p_d^i\}$.

The service plane is described as the digraph $G_s = (H, C)$ with $H$ a set whose elements are called *Harmony Service Interface entity* (HSI entity), and $C$ a set of ordered pairs of vertices, called *HSI interconnection* with $(u_s, v_s) \in C$. In the centralized or hierarchical model these HSI interconnections also reflect the hierarchy. A path in $G_s$ in denoted as $p_s = (u_{s0}, v_{s0}), \ldots, (u_{s_{m-1}}, v_{s_{m-1}})$, with $m \in \mathbb{N}_0$ and $len(p_s) = m$. The child of an HSI adapter is defined as $s_{u_s} = \{w_s | (u_s, w_s) \in C\}$. This means, that the HSI entity $u_s$ delegates the reservation task to the set $s(u_s)$. Thus, we have a tree structure of HSI entities. In the distributed model we have a mesh of HSI entities which may be themselves the root of a HSI entity tree.

Let the set of Domains for which an HSI entity $h$ is responsible be $r(h \in H) \in D$ and $R(H_i \subseteq H) = \{r(h) | h \in H_i\}$. Thus, the set of responsible HSI entities of a data path is $R(a(p_d))$.

## 4 Implementation

The communication between the entities is done by using SOAP messages and the Apache Muse framework. To support workflows and resource co-allocation, the API was derived from existing interfaces used to abstract and map network resources with most limited knowledge in intra-domain Grid environments and was modified to cover aspects of multi-domain conditions [18].

The underlying testbed (Fig. 3a and Fig. 3b) has a high similarity to real network environments in the National Research and Education Networks (*NRENs*). Due to its heterogeneity and variety of administration actors, testing of new developments in the service plane is more controllable when done in a virtual testbed (Fig. 4a and Fig. 4b).

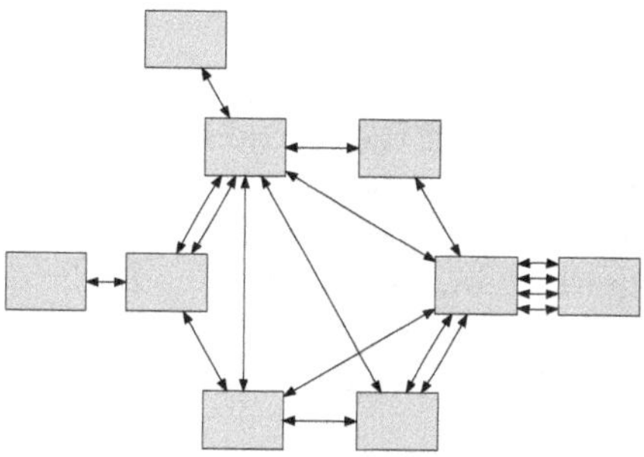

(a) real data plane

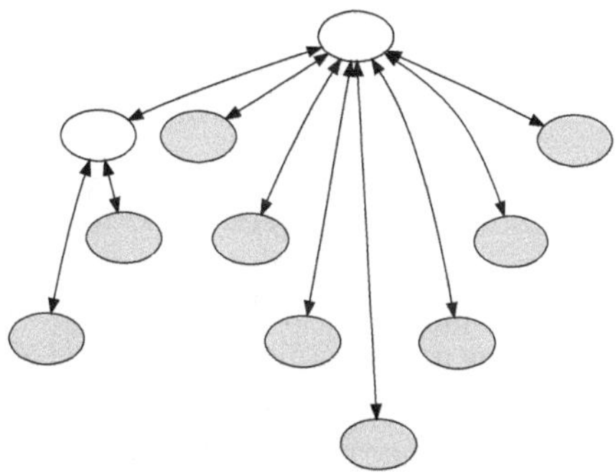

(b) real service plane

**Fig. 3.** The data and service plane testbed architecture. Whereby *grey boxes* are administrative domains with their corresponding NRPS and *arrows* are preconfigured VLANs between them. *Grey circles* are domains controlled by a single NRPS and *white circles* represent IDBs.

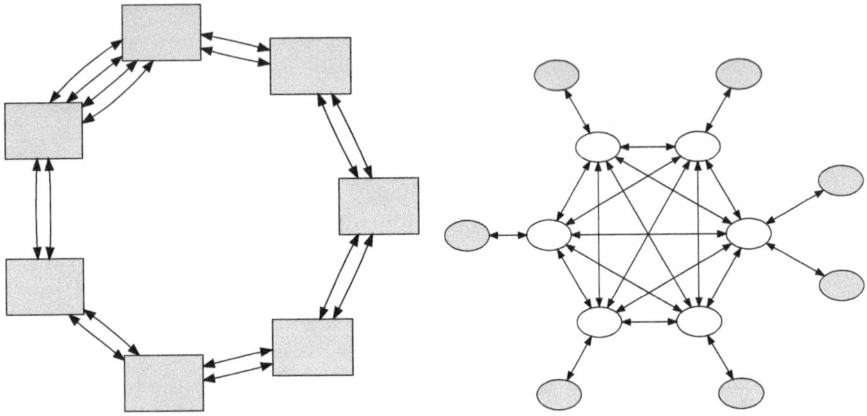

(a) virtual data plane (b) virtual service plane

**Fig. 4.** The service and the emulated data plane architecture within the virtual testbed. Whereby *grey boxes* are administrative domains with their corresponding NRPS and *arrows* represent emulated VLANs between them. *Grey circles* are domains controlled by a single NRPS and *grey circles* represent IDBs.

The data plane inter-connections within the testbed use dedicated light-paths from either GÉANT2 or GLIF[3] infrastructure. The switching equipment in the local testbeds is composed among others by the following systems: Alcatel-Lucent 1678, 1850, and 7750; Calient DiamondWave FiberConnect; Cisco Catalyst 6509, 3750; Nortel Optera 5200, OME 6500, and HDXc; and Riverstone 15008. Furthermore, hardware used to run the NSP was compiled by standard personal computer components and virtual machines. The main IDB was running on an Intel Quad Core Xeon with 2.80 GHz, 1 GByte RAM, and Fedora Core release 6.

## 5 Evaluation

***Formal Performance Analysis.*** First, we consider the centralised and hierarchical model for the formal performance analysis. We show, how the processing times of the different domains can be combined for the overall service time of a single path request. Based on this definition, we extend the approach to a distributed model. Subsequently, we show a worst case calculation of the necessary configuration steps if there are several alternative data links between the domains and reservation attempts fails.

The time a single HSI entity $h \in R(a(p_d))$ needs to process a request internally is denoted as $t_h$ and delegates the reservation to its child HSI entities. If we assume that all children of $h$ are requested consecutively, we can define $t_h^{total}$ of an HSI entity recursively as:

[3] `http://www.glif.is/`

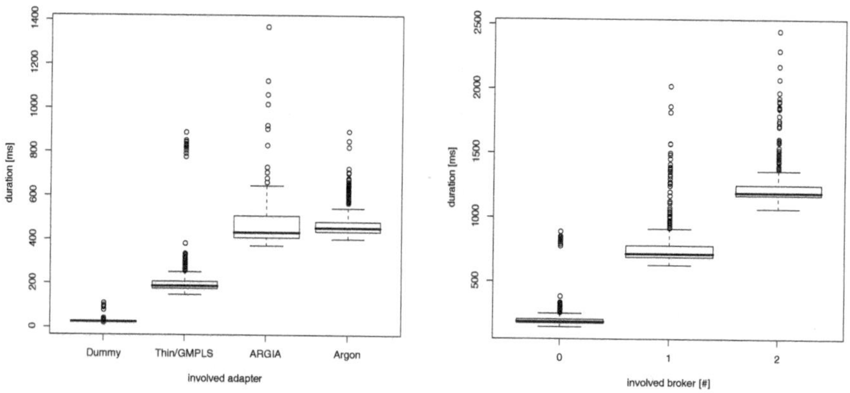


(a) Direct calls (b) Two levels of hierarchy

**Fig. 5.** Processing of a reservation request within a single IDB (500 repetitions). (a) depicts the response time for each Adapter and (b) shows the delay that is added by every IDB hierarchy level.

$$t_h^{total} = \begin{cases} t_h & \text{if } |s(h)| = 0, \\ t_h + \sum_{h_i \in s(h) \cap R(a(p_d))} t_{h_i}^{total} & \text{if } |s(h)|) \geq 1. \end{cases} \quad (1)$$

If the communication with the children is parallelized, we can define $t_{h_{total}}$ as:

$$t_h^{total} = \begin{cases} t_h & \text{if } |s(h)| = 0, \\ t_h + \max_{h_i \in s(h) \cap R(a(p_d))} t_{h_i}^{total} & \text{if } |s(h)|) \geq 1. \end{cases} \quad (2)$$

In case of using the distributed model for the service plane is used there is a mesh of HSI entities on top level $H_{top} \in H$. Each top level HSI entity $h_{top} \in H_{top}$ can itself be the root of a hierarchical HSI entity structure. Thus, we get for path request $p_d$ for a distributed service plane operating in a consecutive mode:

$$t^{total} = \sum_{h_i \in H_{top} \cap R(a(p_d))} t_{h_i}^{total} \quad (3)$$

For a path request for $p_d$ for a distributed service plane operating in a parallel mode:

$$t^{total} = \max_{h_i \in H_{top} \cap R(a(p_d))} t_{h_i}^{total} \quad (4)$$

with $t_{h_i}^{total}$ to be evaluated with Eq. 3 or Eq. 2.

If there are several links between at least two adjacent domains of a path $p_d(d_{source}, d_{dest}$ and the reservation request fails in using one of these redundant links, there might still be a chance in reserving the data path via one of the alternative links. In a consecutive model, where a local reconfiguration is possible

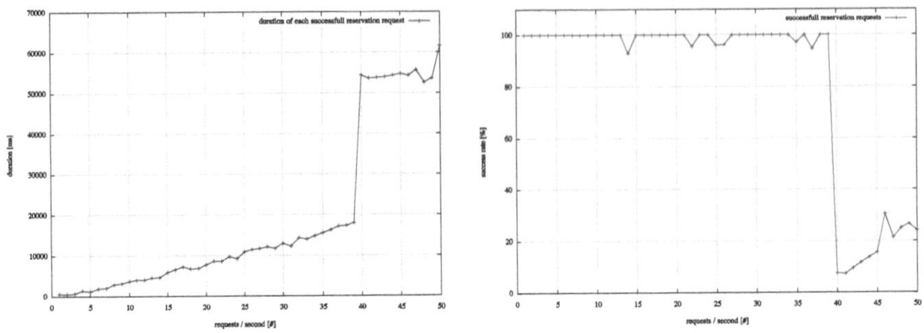

(a) Duration of each successful reservation (b) Success rate

**Fig. 6.** Processing of $n$ reservation request per second within the Service Plane. (a) depicts the duration of each successful request and (b) shows the success rate.

as a crankback mechanism, the number of configuration steps $c$ of a path $p$ is limited by:

$$O(c) = len(p) + \sum_{0 \leq i \leq len(p)-1} n(u_i, v_i) - 1 \tag{5}$$

Here, the first summand represents the successful configurations and the second summand the reconfigurations because of a failed reservation request at one of the HSI entities.

***Measurements.*** To emphasize and support the preceding analysis, measurements from the running prototype are given. As shown in Fig. 5a the mean response time of each NRPS adapter ($t_h$) varies from 200 ms to 410 ms and has a noticeably large number of outliers. The dummy adapter delay points out a 10 ms communication and processing overhead for each request. As seen in Fig. 5b each additional hierarchy level increases the total response time ($t_h^{total}$) by approximately 500 ms. Furthermore, Fig. 6a and Fig. 6b indicate that the current prototype's response time increases by arithmetic progression with reference to the amount of incoming requests. It also shows clearly a performance threshold at about 39 requests per second. After that the success rate drops dramatically and the request duration almost triples.

## 6 Conclusion and Future Work

We have shown that under the given assumptions and limitations the proposed architecture allows for multi-domain, multi-technology, multi-vendor network reservation and co-scheduling. We've tested it in a real scenario under the Phosphorus Europe wide testbed. Besides this proof-of-concept the crucial result is that even with a very limited knowledge of topology information and without an homogeneous control plane, complex malleable advance reservations are feasible and realizable.

Based on these results different network operators such as National Research and Education Networks (*NRENs*) would be able to establish bilateral agreements for end-to-end provisioning services without changing technologies or exposing confidential data.

Besides the implemented prototype that certainly is valuable to demonstrate the main functionality within a testbed with real hardware and users, simulations would provide the possibility to evaluate alternative algorithms in a more sophisticated manner. We expect to have first results using the discrete event simulation package SimJava [19] shortly. Moreover, we are currently implementing more progressive algorithms for malleable reservations in order to reduce the number of availability requests and increase the average network utilization.

Long-term objectives mainly focus on enhanced network resiliency capabilities. This includes as well the computation of disjoint paths as the delegation of responsibilities, path monitoring, and automatic re-routing and -connection.

## References


1. Figuerola, S., Ciulli, N., de Leenheer, M., Demchenko, Y., Ziegler, W., Binczewski, A., et al.: PHOSPHORUS: single-step on-demand services across multi-domain networks for e-science. In: Wang, J., Chang, G.-K., Itaya, Y., Zech, H. (eds.) Network Architectures, Management, and Applications V. Proceedings of the SPIE, vol. 6784, (2007) 67842X
2. Foster, I., Geisler, J., Nickless, B., Smith, W., Tuecke, S.: Software Infrastructure for the I-WAY High-Performance Distributed Computing Experiment. In: HPDC, vol. 96, pp. 562–571 (1996)
3. Foster, I., Kesselman, C.: The Grid: Blueprint for a New Computing Infrastructure. Morgan Kaufmann, San Francisco (2004)
4. Foster, I., Jennings, N., Kesselman, C.: Brain Meets Brawn: Why Grid and Agents Need Each Other. In: International Conference on Autonomous Agents: Proceedings of the Third International Joint Conference on Autonomous Agents and Multiagent Systems, vol. 1, pp. 8–15 (2004)
5. Travostino, F., Mambretti, J., Karmous-Edwards, G.: Grid Networks: Enabling Grids with Advanced Communication Technology. John Wiley & Sons, Chichester (2006)
6. Braden, R., Clark, D., Shenker, S., et al.: Integrated Services in the Internet Architecture: an Overview. IETF RFC 1633 (Proposed Standard) (1994)
7. Blake, S., Black, D., Carlson, M., Davies, E., Wang, Z., Weiss, W.: An Architecture for Differentiated Service. IETF RFC 2475 (Proposed Standard) (1998)
8. Czajkowski, K., Foster, I., Kesselman, C.: Resource Co-Allocation in Computational Grids. In: Proceedings of the Eighth IEEE International Symposium on High Performance Distributed Computing (HPDC-8), pp. 219–228 (1999)
9. Czajkowski, K., Foster, I., Karonis, N., Kesselman, C., Martin, S., Smith, W., Tuecke, S.: A Resource Management Architecture for Metacomputing Systems. In: Feitelson, D.G., Rudolph, L. (eds.) IPPS-WS 1998, SPDP-WS 1998, and JSSPP 1998. LNCS, vol. 1459, pp. 62–82. Springer, Heidelberg (1998)
10. Foster, I., Kesselman, C., Lee, C., Lindell, B., Nahrstedt, K., Roy, A.: A distributed resource management architecture that supportsadvance reservations and co-allocation. In: Seventh International Workshop on Quality of Service, 1999. IWQoS 1999, pp. 27–36 (1999)

11. Bhatti, S.N., Sorensen, S.A., Clark, P., Crowcroft, J.: Network QoS for Grid Systems. International Journal of High Performance Computing Applications 17, 219 (2003)
12. Grasa, E., Junyent, G., Figuerola, S., Lopez, A., Savoie, M.: UCLPv2: a network virtualization framework built on web services [web services in telecommunications, part II]. Communications Magazine, IEEE 46, 126–134 (2008)
13. Travostino, F., Keates, R., Lavian, T., Monga, I., Schofield, B.: Project DRAC: creating an applications-aware network (2005)
14. Rosen, E., Viswanathan, A., Callon, R., et al.: Multiprotocol Label Switching Architecture. IETF RFC 3031 (Proposed Standard) (2001)
15. Barz, C., Bornhauser, U., Martini, P., Pilz, M., de Waal, C., Willner, A.: ARGON: Reservation in Grid-enabled Networks. In: Proceedings of the 1. DFN-Forum on Communication Technologies. (2008)
16. Rajagopalan, B.: Documentation of IANA Assignments for Label Distribution Protocol (LDP), Resource ReSerVation Protocol (RSVP), and Resource ReSerVation Protocol-Traffic Engineering (RSVP-TE) Extensions for Optical UNI Signaling. IETF RFC 3476, Informational (2003)
17. Varvarigos, E., Sourlas, V., Christodoulopoulos, K.: Routing and scheduling connections in networks that support advance reservations. Computer Networks (2008)
18. Andree, B., Hommes, F., Peeters, B., Recio, J., Snchez, A., Schon, J., de Waal, C., Willner, A.: PHOSPHORUS Deliverable D1.1.: Requirements and specifications of interfaces architecture for interoperability between NRPS, GMPLS, Middleware. European IST project PHOSPHORUS (2007)
19. Howell, F., McNab, R.: SimJava: A Discrete Event Simulation Package For Java With Applications In Computer Systems Modelling. In: Proceedings of the First International Conference on Web-based Modelling and Simulation (1998)

# Creating Butterflies in the Core – A Network Coding Extension for MPLS/RSVP-TE*
(Work in Progress)

Thorsten Biermann, Arne Schwabe, and Holger Karl

University of Paderborn, Research Group Computer Networks,
Pohlweg 47-49, 33098 Paderborn, Germany
{thorsten.biermann,arne.schwabe,holger.karl}@upb.de

**Abstract.** Network Coding exploits network resources more efficiently than plain routing. Specifically, it reduces packet delays and packet loss rates. Such advantages are especially useful in core networks where many different users can benefit from this at once. Hence, network coding should be transparently integrated in technologies suitable for core networks. We present an extension for MPLS and RSVP-TE which can instantiate network-coded paths in core networks. The feasibility of this approach is demonstrated in a proof-of-concept simulation.

**Keywords:** Network coding, Signaling, Core, MPLS, RSVP-TE.

## 1 Introduction

The technique of Network Coding (NC), as first introduced by Ahlswede et al., has become a vivid area in current networking research. Instead of only forwarding packets, NC permits nodes on the packets' route to process the data, i.e., to modify and mix up their content. These transformations are inverted at a decoder again to deliver the original content to the destination. Depending on where which transformations happen, various benefits can be achieved.

There is a variety of applications where users can benefit from NC. In this paper, we concentrate on NC to use available network resources more efficiently than with plain routing. In particular, we look at core networks where multiple flows share physical links. In such scenarios, links are usually redundantly deployed to deal with link failures or temporary overload situations. Affected flows are switched from the broken or overloaded link to alternative links. An example for such a scenario is depicted in Fig. 1(a) and 1(b) where the link R1→R2 gets congested due to a failure or a load peak.

As soon as the congestion is detected, Flow 2 (blue/dotted) is redirected to the alternative route on the right side (S2→D1→R2). Although all links are

* The research leading to these results has received funding from the European Community's Seventh Framework Programme (FP7/2007-2013) under grant agreement n° 216041.

L. Fratta et al. (Eds.): NETWORKING 2009, LNCS 5550, pp. 883–894, 2009.

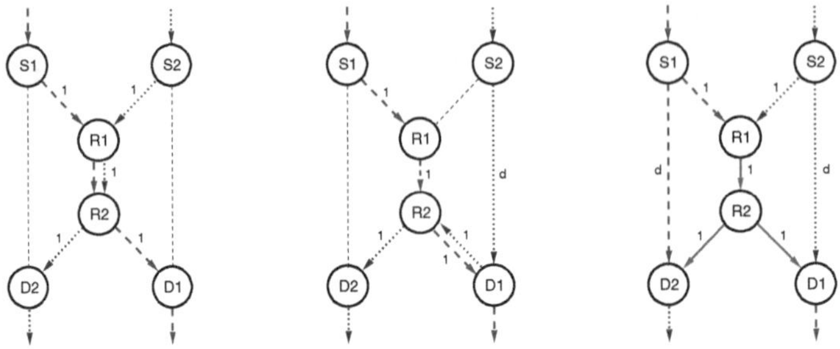


(a) Enough capacity is available. Packets of both flows are transported using classic forwarding. The overall latency is 3.

(b) Overload on R1→R2 forces the blue/dotted flow to be redirected via an alternative route. Its latency raises to $d+2$.

(c) Alternatively, NC is applied: Encoding is done at node R1; decoding at D1 and D2. Latency remains at $\max(d, 3)$.

**Fig. 1.** Comparison of classic transmission via packet forwarding and NC-enabled transmission of two data flows. The numbers next to links denote their latencies. In contrast to routing, NC permits load shifting from link R1→R2 to lateral links while retaining low latency.

now able to deal with the incoming load, depending on the alternative links' properties, the rerouted flow might suffer from noticeably higher delay.

To overcome this drawback, linear NC can be applied. This technique regards packets as vectors and allows a node to apply a linear transformation to them before they are forwarded. Linear NC achieves optimal throughput for multicast transmission, i.e., the maximum flow from the source to each receiving node. A special case of such linear transformations is calculating the XOR value of two packets. In our scenario, instead of completely redirecting one of the flows, both of them are jointly encoded at R1 by calculating the XOR of each packet pair of the flows. This reduces load on the bottleneck link and maintains low delay for both flows. Decoding is done at the destinations, as shown in Fig. 1(c).

The techniques and benefits of network coding have been widely explored [1], but the control problem – when and where to turn it on and how to signal this – has only been addressed in wireless contexts [2,3] or on the application layer [4]. Specifically, we are not aware of approaches trying to integrate NC into wired systems on the link or network layer. Especially, controlling NC for traffic flows in core networks, where many users benefit at once, is not possible so far. To achieve this, we present a signaling protocol in Sec. 3 and 4 that triggers and controls the instantiation of network-coded paths in networks using Multiprotocol Label Switching (MPLS), as a typical example of a core network technology. Our protocol is independent of the algorithm used for detecting NC possibilities, i.e., finding suitable subtopologies [5], as well as of the algorithm that decides when to actually activate coding. Both decisions are made on top of our signaling protocol. The feasibility is demonstrated in a proof-of-concept simulation in Sec. 5. Sec. 2 briefly summarizes relevant MPLS background.

## 2 Background

Data transport in core networks is often realized with *label switching*. Instead of evaluating hierarchical IP addresses, simple flat labels are prepended to packets. A switch processes incoming packets solely based on their label, i.e., all packets with the same label are treated equally.

Label switching requires two components: a data transport service which actually transports the data packets based on their labels and a label distribution protocol which sets up the Label Switched Paths (LSPs), i.e., distributes labels to be used among Label Switch Routers (LSRs). This determines the actual forwarding behavior. As MPLS [6] is a practically highly relevant transport service for label switching, we will focus on extending it by NC; in principle, the technique should also be applicable to other label switching transport services.

Resource Reservation Protocol – Traffic Engineering (RSVP-TE) [7] is one of several label distribution protocols that can be used with MPLS. It has been extended multiple times to support emerging networking technologies and consolidates many aspects of traffic engineering in a single protocol. Besides the wide adoption, we have chosen this protocol as base for our extension because there is already a Point-to-Multipoint (P2MP) extension available [8] which is necessary for applying XOR NC in a butterfly topology.

## 3 Extending MPLS to Support NC

To perform XOR NC in a butterfly, three issues have to be solved: packets must be encoded (at node R1 in Fig. 1(c)), the uncoded packets must be sent to decoding points (from S1/S2 to D2/D1), and the right encoded and uncoded packets must be matched for the decoding operation (at D1/D2). When implementing these additional functions, the following requirements have to be met:

- Using NC must not disrupt normal label switching operation.
- Modifying nodes not directly involved in NC operations, i.e., which only forward network-coded packets, is not acceptable.

To achieve this in MPLS, two modifications are required. First, LSRs which are responsible for en-/decoding need to be extended to support these operations and, second, one LSR passing the flow to be encoded must be able to add packet sequence numbers as packets must be uniquely identifiable. Hence, in contrast to traditional label switching where all LSRs are equal and perform the same operations (packet forwarding based on labels), NC requires to distinguish between different roles that LSRs may adopt:

- *Forwarding LSRs* perform traditional store-and-forward operations required for ordinary packet forwarding. This also encompasses cloning of flows for multicasting. The payload of forwarded packets is unchanged.
- *Numbering LSRs* add packet sequence numbers to MPLS flows. The packets' payload is not altered either.

- *Encoding LSRs* apply XOR coding to packets of two incoming flows. The result is one encoded output flow.
- *Decoding LSRs* receive one encoded and a corresponding plain flow. After decoding, the resulting decoded packets are forwarded.

Required modifications for implementing the different LSRs are described in the following two subsections. Setting up paths is mostly a problem for the signaling protocol and described separately in Sec. 4.

### 3.1 Encoding and Decoding

In MPLS, an LSR determines how to treat an incoming packet solely based on its label; the concrete action is determined by a Next Hop Label Forwarding Entry (NHLFE). NHLFEs are basically table entries which map an incoming label/interface combination to corresponding next-hop information (outgoing interface and next-hop address) and label manipulation instructions like label push, pop, or swap operations.

To support XOR NC, NHLFEs must be extended to not only support simple forwarding of packets but to also indicate to the LSR that two incoming flows shall be combined to one single outgoing flow. This is done by feeding packets of an incoming flow into an encoder. The partner flow for encoding is selected by a second NHLFE which feeds another flow into the same encoder instance. This encoder contains all functionality required for the actual encoding operation like buffering, XOR calculation, etc. This modification is sufficient to implement an XOR *Encoding LSR*, as well as the corresponding *Decoding LSR*.

### 3.2 Packet Sequence Numbers

To be able to successfully decode packets at destination LSRs, these nodes need to know which packets from which flow were used for encoding (e.g., trying to decode ($a$ XOR $b$) from flows A and B with a packet $b'$ from flow B obviously gives wrong results). Hence, packets forming an MPLS flow which is involved in any coding operation must carry information that allows their unique identification. The simplest way to achieve this is by adding sequence numbers to packets.

As MPLS flows usually transport an aggregate of many different flows, it is impossible to reuse sequence numbers of upper layers for this; they have to be explicitly generated for a given MPLS flow. This is done by *Numbering LSRs* with a special NHLFE. Such coding-related sequence numbers are not required over the *whole* packet lifetime. It is sufficient to add them before the flow is multicasted. E.g., in the scenario depicted in Fig. 1(c), this is done by the LSRs S1 and S2 at the latest. At the Encoding LSR (R1), both flows are combined and, hence, each packet of the encoded flow must carry two sequence numbers.

Extending the MPLS header to transport the additional packet sequence number is not feasible as this breaks compatibility to standard MPLS implementations. Normal LSRs would *not* be able to forward packets already numbered for NC later on, but this is an obvious deployment requirement. Therefore, we use

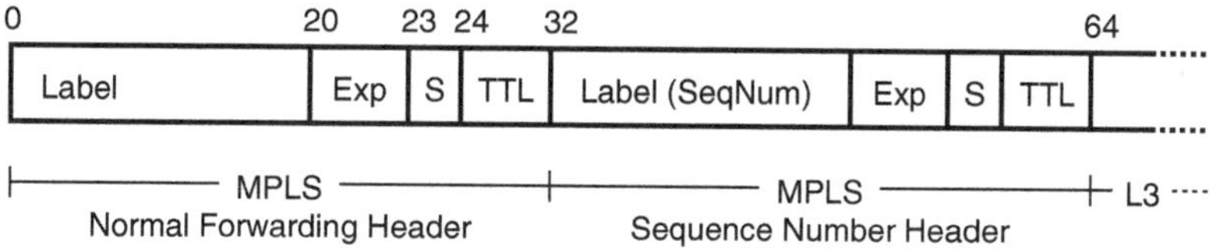


**Fig. 2.** MPLS label stacking is exploited to transport packet sequence numbers. The label field of the inner header contains the packet's sequence number.

the label stacking feature of MPLS to push an additional sequence number label on the stack, transporting the sequence numbers within the 20 bit label field. Fig. 2 shows a sample stack transporting a single packet sequence number.

Sequence numbers also allow us to treat boundary cases. For example, it could happen that at the Encoding LSR, a packet arrives and, even after waiting some time, no packet from the partner flow is available for joint coding. One option would be to stall such a packet until a partner packet for encoding arrives; an alternative is to send this packet uncoded (to avoid blocking buffers) and to indicate from which flow a packet is missing by an empty sequence number.

## 4 Extending RSVP-TE to Signal NC-Enabled Paths

The modified MPLS system can realize NC operations within the network. What is still missing is a mechanism to propagate forwarding and de-/coding configurations to involved nodes to actually turn on NC. To do so, the label distribution protocol has to be extended as well. For the design of the RSVP-TE NC protocol extension we must meet the following requirements:

- Using NC functionality must not disrupt normal RSVP-TE operation.
- Switching an LSP from normal to coded operation must be possible without disturbing the ongoing transmission.
- Assured QoS constraints must not be violated when using NC, i.e., RSVP-TE must still be able to manage reservations and guarantee them.
- Although the extension was designed with XOR NC in butterfly-like topologies in mind, it is kept as flexible as possible to easily permit other coding techniques and topologies as well.

Fulfilling these requirements guarantees a seamless integration of NC LSPs into a running network. The actual decision if and where NC is applied within the network and which pairs of flows should be coded together is orthogonal to the signaling problem addressed in this paper. Our signaling scheme supports any kind of decision algorithm on top. The decision can even be made independently from the nodes forming a coded path. The only limitation is that the decision must be communicated to a node on the path to start the signaling process.

Based on the presented assumptions, the next sections describe how a NC LSP is actually set up (Sec. 4.1) and teared down again (Sec. 4.2).

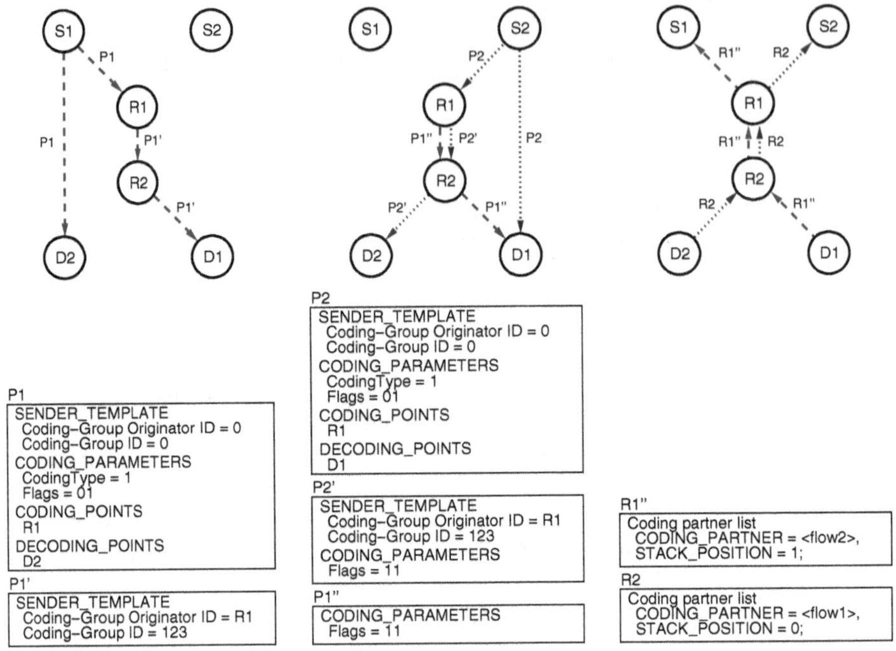


(a) Path 1 is set up first. PATH message P1 is modified at R1. The resulting message is called P1'.

(b) Path 2 is also set up. Besides P2 being modified to P2' at R1, path 1 is also refreshed using P1".

(c) The setup of path 2 and refreshing of path 1 is acknowledged to the sources via RESV messages.

**Fig. 3.** Sample for joint setup of two network-coded LSPs. S1/S2 are ingress LSRs, R1 is the Encoding LSR, and D1/D2 are the egress LSRs which do the decoding. Red/dashed arrows denote signaling messages belonging to path 1; blue/dotted arrows the signaling of path 2. The first RESV reply for path 1 is omitted. The boxes denote the signaling messages (only NC-relevant data which has changed is shown).

### 4.1 Setting Up Network-Coded Paths

RSVP-TE uses end-to-end semantics to set up paths by sending a PATH message from the ingress to the egress LSR. It is acknowledged by a RESV message. Hence, the use of NC has to be signaled by the ingress LSRs of both flows to be encoded.

There are two possible constellations how network-coded LSPs can be set up. In the first and simpler case, both ingress LSRs are aware that their LSPs will be used for NC. Hence, they are both initially set up with NC enabled. In the second case, one or both sources start without the knowledge that their LSP will be used for NC later on. Hence, the paths are set up without NC enabled. This means that the NC-unaware LSPs must be converted with RSVP-TE's rerouting mechanisms [7, Sec. 4.6.4] afterwards. This procedure is not discussed here as it is identical to the standard behavior of RSVP-TE.

**Simultaneous Setup of Network-Coded Paths.** In this scenario, both ingress LSRs S1 and S2 know at instantiation time that their LSPs will be encoded. This information can either be propagated automatically by the algorithm which decides when and where NC is applied, or can be given manually.

The setup process is clarified in Fig. 3, which illustrates important steps of our signaling scheme. The LSPs start at the nodes S1/S2 and end at D1/D2. Path 1 (starting at S1) is set up first with initially no NC active. It is refreshed after the creation of Path 2 to be jointly encoded. To transport the additionally required NC parameters, the `PATH` and `RESV` messages are extended. These modifications will be described in detail later on. Note that at points where the path is multicast, `PATH` messages are copied and sent along both subpaths, akin to the P2MP upon which we built.

The path setup starts at S1 by sending the first `PATH` message P1 to R1 and D2. After R1 added itself as `Coding-Group Originator` and assigned a new `Coding-Group ID`, the message P1' is sent to R2 and finally to D1. As coding is not yet activated, D1 responds with a standard `RESV` message (not shown). Next, S2 also starts its path setup by sending P2. R1 receives P2 and detects the coding possibility with the first LSP based on the `DECODING_POINTS`. Hence, coding is enabled in P2' and path 1 is refreshed using a new `PATH` message P1".

Signaling of label stack positions for particular sequence number labels is only done by the Decoding LSRs when coding is actually enabled. Fig. 3(c) shows this signaling within the `RESV` message after encoding has been enabled by R1.

*Modified PATH message.* The intention of setting up an NC-enabled path is signaled by modifying its `SENDER_TEMPLATE` object and by adding four additional objects to the `PATH` message. These changes are discussed in the following.

`SENDER_TEMPLATE`. Two LSPs which shall be jointly encoded must be marked such that Numbering, Encoding, and Decoding LSRs know about this intention. Therefore, two additional fields (`Coding-Group Originator ID` and `Coding-Group ID`) are added to the `SENDER_TEMPLATE` object. Both are initialized with zeros by the source LSR and will be filled by the Encoding LSR. Similar to the P2MP extension, the `Originator ID` specifies the ID of the Encoding LSR (usually its IP address), whereas the `ID` is assigned by the Encoding LSR for this particular encoding operation. In combination, both IDs identify the group of LSPs which are jointly encoded on a global scope.

`CODING_PARTNERS`. There are two ways of establishing encoded LSPs: explicit and automatic partner flow selection. For the explicit partner selection, the desired partner flows are specified in the `CODING_PARTNERS` object by a list of subobject pairs. Each pair consists of the partner flow's `SESSION` and `SENDER_TEMPLATE` object. This uniquely identifies LSPs for joint encoding. In case there is no `CODING_PARTNERS`. object in the `PATH` message, an Encoding LSR may choose any LSPs for joint encoding. The LSR's decision is signaled by adding a `CODING_PARTNERS` object referencing the selected partner flow.

`CODING_PARAMETERS`. This field is used to define the coding operation that is actually applied at the Encoding LSR (`CodingType`) and transports several status flags concerning the setup procedure of the coded path (`Flags`).

The `CodingType` field carries a simple identifier of the coding technique that has to be applied at Encoding LSRs. The meaning of these identifiers has to be unique within a NC MPLS domain.

As mentioned, the flags are used to signal important parameters during the path setup. Their meaning usually differs for different coding techniques. For the butterfly NC we implemented the following flags:

- `0x01` *Sequence numbers active.* The Numbering LSR sets this flag. This will usually be the LSR where the path is multicasted. It is required to check whether encoding and decoding is possible and to prevent multiple Numbering LSRs.
- `0x02` *Coding active.* The Encoding LSR enables this flag when the encoding operation is active. It is required to prevent multiple Encoding LSRs and to signal active encoding to potential Decoding LSRs to enable buffering. Furthermore, intermediate LSRs, e.g., R2 in Fig. 3, react on two LSPs sharing the same `Coding-Group Originator ID` and `Coding-Group ID` which both have the active coding flag set. In this case both paths must be merged to multicast the encoded packets.

`CODING_POINTS`. This object contains a list of LSRs which are permitted to take over the role of the Encoding LSR. LSRs listed in this object watch for existing and future LSPs to enable the encoding operation whenever possible. The decision which flows are coded depends on the LSPs' `CODING_PARTNERS` objects. The format is identical to the `EXPLICIT_ROUTE` object in RSVP-TE.

`DECODING_POINTS`. This object has the same format as the `CODING_POINTS` object. It defines a list of LSRs which are able to decode data flows which have been jointly encoded with this flow, i.e., which will receive a plain copy of this flow. Each LSR which is contained in this list looks for other flows that are jointly encoded to take the role of the Decoding LSR.

`PATH_TIME`. Packets that are required for decoding traverse two different paths through the network. As these two paths will usually have different delays, the packets going over the faster path have to be buffered at the decoder until the corresponding packets for decoding arrive. This difference in delay and the available buffer size limit the maximum possible data rate.

As it is required to take this into account for resource reservations, this difference in delay has to be estimated on both branches of a coding path, e.g., on path S1→D2 and S2→R1→R2→D2 in the sample topology in Fig. 1(c). This is done with the `PATH_TIME` object. It contains two fields, a *minimum time* and a *maximum time*. Every LSR which forwards the `PATH` message adds the minimum and maximum observed latency for the link over which it received the message. Hence, at the decoder, the maximum delay difference of both flows is available to calculate whether the data rate requirements of the path can be fulfilled with the available buffer space or not. If not, the decoder must deny the path setup.

The required multicasting of a data flow is realized with the standard P2MP extension for RSVP-TE. LSRs where the actual copying of the flow is desired are configured in the `S2L sub-LSP descriptor list` [8, Section 4.5].

*Modified RESV message.* When a `PATH` message arrives at the egress LSR, RSVP-TE sends back a `RESV` message to the ingress `LSR` to actually set up the LSP and distribute labels between the routers. For the NC-enabled paths, i.e., where the `coding active` flag is set in the `PATH` message, labels carrying the packet sequence numbers will be added to the packets. As Encoding LSRs will receive two flows which have been augmented with sequence numbers, the encoded output flow contains both flows' sequence numbers (cf. Sec. 3.2). Hence, the Decoding LSRs must be aware of which sequence number label belongs to which flow. To signal this, the `RESV` message is extended with a `Coding partner list`.

`Coding partner list`. To retain the approach of RSVP-TE to signal labels from the destination to the source, a Decoding LSR signals the desired label ordering to the Encoding LSR. This is done by adding a `Coding partner list` to `RESV` messages. The list contains pairs of `CODING_PARTNER` and `STACK_POSITION` objects. The `CODING_PARTNER` objects are identical to the ones used in the `PATH` message and identify the partner flow which is used for encoding. The stack position is the expected index of the partner flow's sequence number labels within the label stack. For butterfly NC, the `Coding partner list` will only contain one entry (for the partner flow).

*Signaling costs.* The number of signaling messages required to set up two NC LSPs is analyzed in this paragraph. The setup procedure can be divided into three phases: setup of the first LSP, setup of the second LSP, and refreshing the first LSP to activate coding. Eq. 1 shows the set of `PATH` messages $P_1$ and Eq. 2 the set of `RESV` messages $R_1$, sent during the first phase (cf. Fig. 3(a)). A capital $P$ stands for a `PATH`, an $R$ for an `RESV` message. Indices denote the link over which the message is sent.

$$P_1 = \{P_{\mathrm{S1}\to\mathrm{D2}}, P_{\mathrm{S1}\to\mathrm{R1}}, P_{\mathrm{R1}\to\mathrm{R2}}, P_{\mathrm{R2}\to\mathrm{D1}}\} \tag{1}$$

$$R_1 = \{R_{\mathrm{D2}\to\mathrm{S1}}, R_{\mathrm{D1}\to\mathrm{R2}}, R_{\mathrm{R2}\to\mathrm{R1}}, R_{\mathrm{R1}\to\mathrm{S1}}\} \tag{2}$$

As the first two phases are symmetric, they require the same amount of signaling messages, i.e., $C(P_2) = C(P_1) = 4$ and $C(R_2) = C(R_1) = 4$, where $C(X)$ denotes the cardinality of the set $X$. This means that there are 16 signaling messages sent in total for setting up the two initial LSPs.

For refreshing the first LSP to also activate coding, additional signaling is necessary. The `PATH` and `RESV` messages that are used for this are listed in Eq. 3.

$$P_3 = \{P_{\mathrm{R1}\to\mathrm{R2}}, P_{\mathrm{R2}\to\mathrm{D1}}\} \qquad R_3 = \{R_{\mathrm{D1}\to\mathrm{R2}}, R_{\mathrm{R2}\to\mathrm{R1}}, R_{\mathrm{R1}\to\mathrm{S1}}\} \tag{3}$$

This results in a total of 21 signaling messages. Although this is more than the required 12 messages for the uncoded case, many of them are exchanged in parallel, such that the temporal overhead is very small. Furthermore, none of the two LSPs has to wait for the partner flow, i.e., data transfer can start immediately after path initiation.

### 4.2 Tearing Down Network-Coded Paths

Tearing down one of the jointly encoded paths is straightforward. The source LSR of the LSP to be torn down sends a `PathTear` message down the path. The Encoding LSR receives this message and stops encoding both flows from then on. As packets of the partner path still have to reach their destination, they are sent uncoded and the special sequence number 0 is used to signal this to the decoder (cf. Sec. 3.2).

Using this method guarantees a disruption-free operation of paths whose partners are torn down during NC operations. Depending on the application scenario, a conversion of the multicast LSP to a unicast LSP after such an event might be reasonable to save capacity. This is the case if the multicast capability has only been set up to do NC and is not further required. Such a conversion is done by the ingress LSR by additionally setting up a normal unicast LSP which uses the already allocated network ressources. After switching the traffic to the unicast LSP, the NC multicast LSP is torn down.

## 5 Evaluation

To demonstrate the feasibility of our protocol extensions, we implemented them in a simulator. The system model and the observed results are discussed next.

### 5.1 System Model

We implemented the topology already presented in Fig. 1(c) using the OMNeT++ discrete event simulator and the MPLS and RSVP-TE models from the INET framework. The data sources and destinations are outside the butterfly. All links are equal and have a capacity of 10 Mbit/s and a latency of 2 ms. The sources, connected to S1 and S2, periodically send packets of 1435 byte to the destinations, which are connected to D1 and D2, with inter-packet times of 13 ms. This is the minimum possible inter-packet time before congestion occurs at the chosen link configuration. RSVP-TE refreshes paths every 5 s.

To demonstrate NC, the following events occur at the simulated times $t$:

$t = 0$ s: S1 creates a normal LSP without coding functionality (LSP1-1). The source at S1 starts sending packets which are transported through LSP1-1.
$t = 4$ s: S1 creates an NC-enabled LSP (LSP1-2) in parallel to LSP1-1.
$t = 4.5$ s: The data packets from S1 are switched from LSP1-1 to LSP1-2. LSP1-1 is torn down. The encoding of LSP1-2 is still inactive.
$t = 5$ s: S2 creates an NC-enabled LSP (LSP2-1). This triggers the joint coding for LSP1-2 and LSP2-1 at R1.
$t = 6$ s: Source at S2 starts sending packets.
$t = 15$ s: Source at S1 stops sending packets. LSP1-2 is torn down.

During the simulation, the end-to-end delay $d$ is measured for each packet to see changes in the packet forwarding behavior. Furthermore, packet loss is monitored to detect possible problems during path switching. Nevertheless, the point of this simulation is *not* a stochastic performance evaluation but rather serves to validate the protocol. Therefore, a variation of parameters is not necessary.

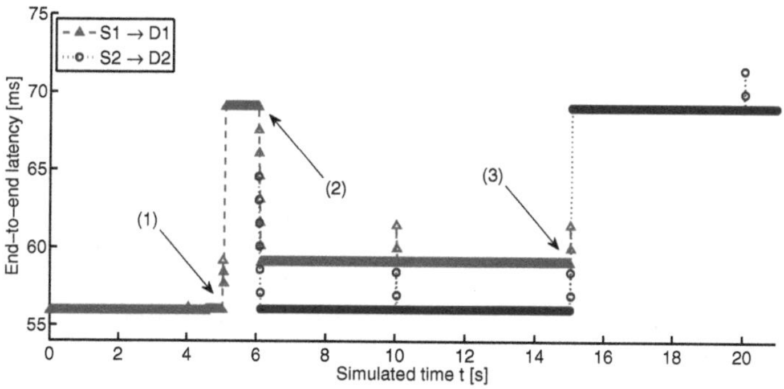


**Fig. 4.** End-to-end delay $d$ over simulated time $t$

## 5.2 Results

The end-to-end delays of both flows, measured during an example simulation run, are plotted in Fig. 4. No packet was lost during the simulated time.

When S1 sends its packets via the uncoded LSP1-1, an end-to-end delay of $d = 56\,\text{ms}$ can be observed. This corresponds to the sum of five link delays plus the corresponding packet transmission times. At event (1), enabling NC for LSP1-2 causes $d$ to raise to 69 ms. The difference is exactly the maximum buffering time of each packet at the encoder before it is sent uncoded.

At (2), data is sent via LSP2-1. LSP1-2's latency reduces again as now packets can be encoded before the maximum encoder buffering time elapses. Whereas LSP2-1's delay reduces to the minimum possible 56 ms, the delay of LSP1 stays at approximately 59 ms. This offset compared to LSP2-1 results from S2's packets arriving slightly later at the Encoding LSR than those of S1. Hence, as both sources send with identical rates, the packets of LSP1-2 have to wait until a partner packet of LSP2-1 arrives. Obviously, this changes in other simulation runs if the starting offsets of S1 and S2 are different.

The teardown of LSP1-2 happens at event (3). As from then on, packets of LSP2-1 have to wait for the maximum encoder buffering time until they are sent out uncoded, the delay of LSP2-1 raises to the already previously seen 69 ms. As LSP2-1 is not converted to disable the NC, the delay remains at this level.

Another phenomenon which can be seen every 5 s (at 5, 10, 15, and 20 s in Fig. 4) is a short increase of the end-to-end delay. It is caused by the path refresh messages (`PATH` and `RESV`) which are periodically sent and have to be queued in addition to the data packets. This temporarily increases the delay.

The absence of any packet loss shows that our protocol was able to switch between different operation modes without disrupting ongoing transfers. The observed packet delay corresponds to the expected behavior and demonstrates the operability of our approach; it behaves accordingly for different initial conditions.

## 6 Conclusion

We presented an extension to the MPLS and RSVP-TE protocols to support creating network-coded paths. The decision whether to use coding or not does not have to be made during the paths' initiation; seamless switching between non-coded and coded operation is possible. Another advantage of the protocol extension is that only those LSRs have to be modified which are actually involved in the coding operation. All others, e.g., intermediate routers which forward coded packets, can remain untouched. Compared to traditional packet forwarding, our extension enables new load distribution schemes which are especially useful for the operation of core networks.

## References

1. Fragouli, C., Boudec, J.Y.L., Widmer, J.: Network coding: An instant primer. SIGCOMM Comput. Commun. Rev. 36(1), 63–68 (2006)
2. Katti, S., Rahul, H., Hu, W., Katabi, D., Médard, M., Crowcroft, J.: XORs in the air: Practical wireless network coding. SIGCOMM Comput. Commun. Rev. 36(4), 243–254 (2006)
3. Chaporkar, P., Proutiere, A.: Adaptive network coding and scheduling for maximizing throughput in wireless networks. In: MobiCom, pp. 135–146. ACM, New York (2007)
4. Gkantsidis, C., Rodriguez, P.R.: Network coding for large scale content distribution. In: INFOCOM 2005, vol. 4, pp. 2235–2245 (2005)
5. Wang, C.C., Shroff, N.B.: Beyond the butterfly – a graph-theoretic characterization of the feasibility of network coding with two simple unicast sessions. In: Proc. IEEE Intl. Symp. on Information Theory, Nice, France (June 2007)
6. Rosen, E., Viswanathan, A., Callon, R.: Multiprotocol Label Switching Architecture. RFC 3031 (Proposed Standard) (January 2001)
7. Awduche, D., Berger, L., Gan, D., Li, T., Srinivasan, V., Swallow, G.: RSVP-TE: Extensions to RSVP for LSP Tunnels. RFC 3209 (Proposed Standard), Updated by RFCs 3936, 4420, 4874, 5151 (December 2001)
8. Aggarwal, R., Papadimitriou, D., Yasukawa, S.: Extensions to Resource Reservation Protocol - Traffic Engineering (RSVP-TE) for Point-to-Multipoint TE Label Switched Paths (LSPs). RFC 4875 (Proposed Standard) (May 2007)

# Why Is This Web Page Coming Up so Slow? Investigating the Loss of SYN Packets (Work in Progress)

Dragana Damjanovic, Philipp Gschwandtner, and Michael Welzl

Institute of Computer Science,
University of Innsbruck, Austria
{dragana.damjanovic,philipp.gschwandtner,michael.welzl}@uibk.ac.at

**Abstract.** Before the first valid calculation of the round trip time for a connection, TCP sets an initial value for the retransmission timeout to 3 seconds which, in the case of the first packets getting lost, introduces a long delay. For short transfers, like web traffic, this could have a significant influence on the performance. We performed measurements to investigate how often this happens. As our measurements show, control packets (SYN and SYN/ACK packets) do get lost and delays of 3 seconds or even more (further timeouts) occur. By means of a simple example implementation, we indicate that this problem could be solved.

**Keywords:** network protocol, measurements, connection management.

## 1 Introduction

Sometimes, when browsing the Web, some page takes just too long to appear. Usually an impatient user clicks refresh one or more times and the page loads. This already happened at least once to everybody. One possibility for this behavior could be the loss of SYN or SYN/ACK packets. Web applications like browsers mainly use TCP as a transport protocol. TCP starts a connection with three-way handshaking, sending a SYN packet, waiting for a response (SYN/ACK packet) and then sending an ACK for the SYN/ACK packet which is the first packet of a TCP connection that can contain data. As recommended in [1], before the first round trip time (RTT) measurement has been done, TCP should set the retransmission timeout (RTO) to 3 seconds. Therefore in case of SYN or SYN/ACK packets getting lost TCP waits 3 seconds (or longer for further back-offs) before retransmitting them. Especially for short-lived flows, this delay can significantly degrade the performance. Reference [2] suggests that TCP may increase its initial window from one to between two and four segments. This would make TCP more robust to losses in the starting phase when the window is 1 and any loss would force a timeout, but in the case of the connection's opening packets getting lost, TCP has no choice but entering timeout.

In [3] the authors recognize this problem of lost packets belonging to the connection opening phase and their simulations show how the response time

L. Fratta et al. (Eds.): NETWORKING 2009, LNCS 5550, pp. 895–906, 2009.

can be significantly increased by just avoiding the loss of the SYN/ACK packet. They suggested the use of ECN [4] for SYN/ACK packets, but — as studies show in 2000 only 1.1% [5] of the servers were ECN-capable and till 2004 it has increased just to 2.1% [6] — ECN is not widely used. Therefore utilizing ECN for SYN/ACKs would not improve the actual performance until the number of ECN capable servers increases.

Drawn by the idea of the impact the loss of packets belonging to the connection's opening phase can have on the performance of web browsing, we investigate on how often this happens. As our measurements show this problem does exist (in average more than 0.5% of the connections experience this). Therefore we give a simple example implementation that can overcome this drawback.

In the next section we investigate related work in the area of safe connection establishment and some work done in speeding up web servers' response times. In Section 3 we present the results of our Internet measurements and in 4 we introduce and validate an example implementation that improves the performance of web browsing, followed by the conclusion of our findings.

## 2 Related Work

Since web traffic usually consists of very short flows (request-response connections with web clients sending a request and server answering to the request), not just the loss of packets belonging to the connection opening phase but also the whole connection opening phase itself introduces a significant delay. In the past there have been proposals for protocols more suited for transaction-oriented connections that would eliminate this delay, proposals for protocols built on top of TCP as well as protocols build from scratch.

The Stream Control Transmission protocol (SCTP) [7] is a protocol well suited for web traffic. SCTP uses a concept based on associations instead of connections, each association can have multiple streams and ordered delivery on separate streams is also supported. In the case of HTTP transfers with TCP each request is sent over a separate connection, whereas SCTP enables multiple requests to be sent over the same association. For opening a connection it uses a four-way handshake, but to reduce the delay the last two packets of the connection opening phase can carry data.

TCP for Transactions (T/TCP) [8] is a protocol suggested as an extension to the TCP protocol for distributed applications like web applications (the scenario we are investigating), that would benefit from a transaction oriented transport protocol that does not always introduce additional packet exchanges for opening and closing connections. Hosts opening a connection for the first time perform the three-way handshake and host information is cached on each side. This cache is used for validating upcoming SYN packets. Data is sent even within the first packet of a connection (SYN packet) and in case a cache entry for this host exists and the sequence number is valid (i.e. it is not a duplicate) the response to this data is sent in the next packet. In all other cases the response is sent after the three-way handshake is performed. This is an easy way to speed up short flows, like web traffic, but it has many security drawbacks: it consumes more resources than normal TCP before a connection is open, therefore it can be attacked more easily by SYN flooding, it is more vulnerable to IP address spoofing, etc.

The Versatile Message Transaction Protocol (VMTP) [9] is another example of a transaction oriented protocol. It uses unique connection IDs for safe connection management and will fall back to three-way handshaking only in special cases (e.g. one of the hosts restarted). It is not built on top of TCP and provides better security and naming, statelessness of transactions etc. This protocol is designed for the area of distributed programming, for communication between servers and clients belonging to the same organizational entity, and it is better suited for those purposes than for an open system like the Web.

A transport protocol designed for wide area networks should implement a connection management mechanism for connection opening and closing that is immune to problems introduced by lost, duplicated or out-of-sequence packets. In the past there have been a couple of proposals for safe initialization and closing of a connection: three-way handshaking, unique connection IDs and timers. As already mentioned TCP uses a three-way handshake for opening but also for closing connections [10]. Protocols that use unique connection IDs usually need an extra mechanism for secure opening or closing, like three-way handshaking or a timer. VMTP is an example of this kind of protocol and as mentioned before it uses a three-way handshaking in some cases. The Delta-t transport protocol [11] uses a timer-based connection management. Connections are handled without any connection management packet exchanges. Communication is established between ports of processes and Delta-t defines a stream as the unique triple (destination port, source port, stream number), where ports are identified by 64 bit addresses unique for the whole network. The Delta-t connection management is built on the idea that there are permanent error- and flow- controlled connections between all possible streams. The time-based mechanism automatically recognizes whether a connection is in the default state and accordingly allocates or deallocates the connection state without any packet exchange. Delta-t and VMTP are just two examples of protocols that deal with connection opening without any packet exchange, but none of these protocols are used in the Internet today.

## 3 Measurements

To investigate how often the stated problem occurs we observed web traffic. Here we only consider the relevant part for our topic, which is the connection opening phase. The loss of connection opening packets can be detected on the client side and on the server side. There are 4 possible cases that can be distinguished:

**client side**

**duplicate SYN:** If a second duplicate SYN is sent by the client, either the first SYN or the corresponding SYN/ACK must have been lost.

**duplicate SYN/ACK:** If a duplicate SYN/ACK is received, we can safely deduce that the corresponding ACK has been lost.

**server side**

- **duplicate SYN:** If a second duplicate SYN packet is received by the server, the client did not receive the corresponding SYN/ACK.
- **duplicate SYN/ACK:** If a second duplicate SYN/ACK is sent, either the first SYN/ACK or the corresponding ACK was lost.

Measurements were taken on the server side as well as on the client side, using tcpdump and subsequent analysis of the trace files. For every packet, the TCP header is inspected to determine whether it is an SYN, SYN/ACK or ACK responding to the received SYN/ACK and therefore a part of the three-way handshake of TCP. Then, the analyzer checks for duplicate SYN or SYN/ACK packets (packets that hold the same source and destination IP address and port as well as the same sequence number) to detect if packets belonging to the connection opening phase got lost. We also measured the time difference between such repeated packets. If this time difference matches the standard initial RTO time (3 seconds or further back-offs with the RTO multiplied by 2 each time) we can safely assume for the majority of the cases that the duplicate packet was automatically transmitted. For all the other cases, other factors have to be taken into account.

### 3.1 Web Mail Server (Server)

The University of Innsbruck provides two separate Linux-based servers for web mail that we used for our server-side measurements, one for the employees (http://web-mail1.uibk.ac.at) and a second one for the students (http://web-mail2.uibk.ac.at). We recorded roughly 208.000.000 packets for a duration of 12 days. The following results were obtained by doing server-side measurements at the former. A total of 737.188 requests for initiating a connection have been logged, 730.523 of which were successful (99,1%). 5162 (0.7%) of those successful connections required more than one SYN packet, which shows a real occurrence of lost SYN packets. Figure 1(a) shows the results of this measurement as a histogram, illustrating the time interval between transmitting duplicate SYN packets. There are peaks of duplicate SYNs at 3 and 6 seconds, which match the first two time intervals for the RTO timer. The fact that there are many SYN packets that approximately fit into a time interval but do not do so precisely can be explained with variances in the calculation of the initial RTO [1]. Therefore we can assume that those duplicate SYN requests were automatically generated.

Furthermore there are smaller peaks at the slightly higher time values 3.4 seconds, 6.8 seconds and 13.6 seconds, which also suggests an automated generation with just a higher starting value or increasing delay in network. Another peak is visible at approximately 0-0.3 seconds. The duplicate SYN packets arriving at those small intervals (and also at any other time interval than the standard ones used by RTO) might occur due to potentially faulty TCP implementations, deliberately changed RTO settings or possibly other factors like lower-level frame collisions causing switches to forward a duplicate of a frame. However, the phenomenon of those short interval packets cannot be explained by impatient users

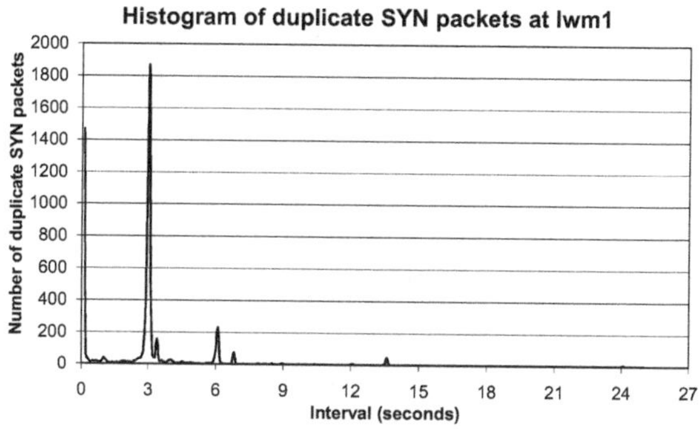


(a) Webmail Server lwm1

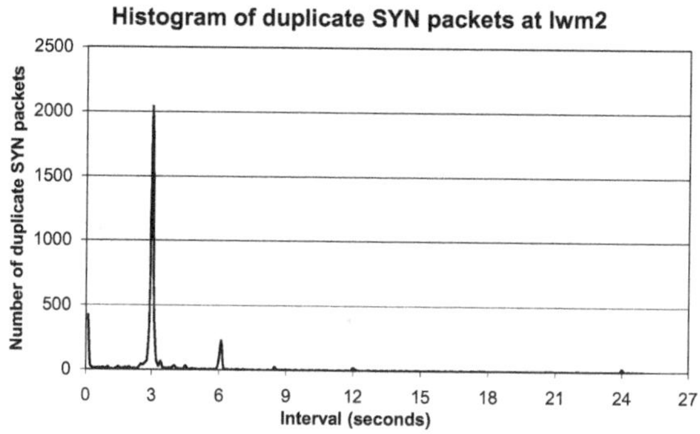


(b) Webmail Server lwm2

**Fig. 1.** Webmail Servers lwm1 and lwm2

clicking on "refresh" or accidental double-clicks, since it is not possible to force the standard TCP implementations to resend a second SYN containing the same source IP and port. We investigated further on this topic but found no way at the application layer of causing TCP to resend a duplicate SYN. The reaction of impatient users still solves the problem however, since today's web browsers simply open a new TCP connection for each refresh request.

The second measurement at the student's web mail server showed similar results. Again, the highest peaks are positioned at the standard initial RTO interval times and smaller peaks at slightly increased timings as before. Also quite a number of duplicate SYNs were received during non-automatic interval times, suggesting either miscalculation or similar phenomena as described above (mainly at 0 to 6 seconds, but continuously small occurrences up until 18 seconds). The results are illustrated in Figure 1(b).

### 3.2 Internet Proxy (Client)

Moreover we recorded traffic at the university's Internet proxy using the client-side measurement method. For a duration of two days we recorded 154.000.000

packets. A total of 1.721.382 connection attempts have been logged, 1.708.442 of which were successful (99,2%). Multiple SYN/ACKs (and SYNs) were necessary in 2.148 of those successful connections. Figure 2(a) illustrates the interval between the original and duplicate SYN/ACKs. First of all there are quite less SYN/ACKs than SYN packets compared to our previous measurements, which can be explained by the nature of the client-side measurement. The server sends a duplicate SYN/ACK only if the first SYN/ACK or its corresponding ACK got lost. If the former is the case, the client counts only one SYN/ACK - a phenomenon that also occurred in our other measurements - which results in less SYN/ACKs being registered on the client side. However this is not the case for measuring SYN packets, since we notice every transmission of SYNs on the client side, regardless of their success. Furthermore the figure illustrates a rather high number of duplicate SYN/ACKs at 0 to 4 seconds, which might be a result of high server load and therefore delayed response times. Additionally there are peaks at 1.5 seconds, 9 seconds and 16.5 seconds, which do not match any of the suggested RTO intervals. Hence we assume that these peaks are caused by miscalculation of the RTO or non-compliance of the specification.

### 3.3 Free Tracelogs of LBNL/ICSI (Server and Client)

In addition to the tests we performed at the servers of our university, we examined the logfiles of the LBNL/ICSI Enterprise Tracing Project, which contain both server- and client-side measurement data that is freely available for download.[1] The files contain three months of measurement data consisting of approximately 5.000.000 packets involving HTTP. The data that is relevant to our topic includes 232.852 connection attempts, 219.867 of which were successful (94.4%). For 4103 (1.9%) of those successful connections multiple SYNs were necessary. The results of duplicate SYNs are illustrated in Figure 2(b) and show that, compared to our own measurements, there are much less duplicate SYNs at non-automatic retransmission time intervals. This can be explained by different measuring conditions, since the data downloaded from LBNL/ICSI was retrieved in a local area network, whereas our measurements most probably include everyday technologies like ADSL, cable Internet and WLAN. Since they are likely to be more error-prone than wired Ethernet connections, it is deductible that they show more irregularities due to lost packets. Apart from the peaks at the standard RTO interval times, there are peaks at 0.5 and 20 seconds. Since they are too protruding to be random deviations it might be possible that some specific program or deliberately changed RTO setting is responsible for those peaks.

### 3.4 Summary of the Measurements

Additionally to the measurements described thoroughly above, we took measurements at Alupress AG[2], a company operating in the field of metal processing,

[1] http://www.icir.org/enterprise-tracing/download.html
[2] http://www.alupress.net

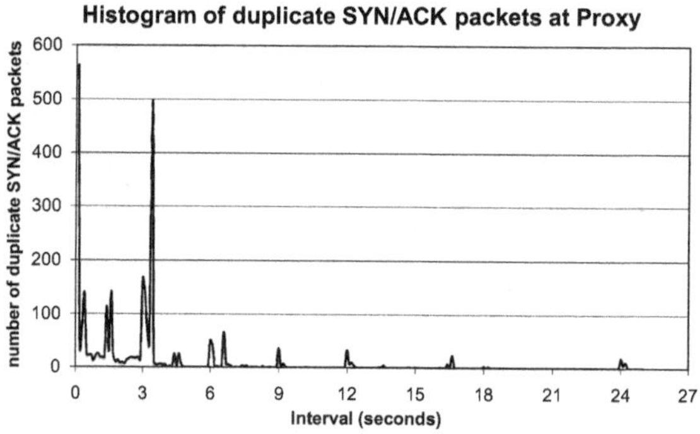


(a) Internet Proxy Server

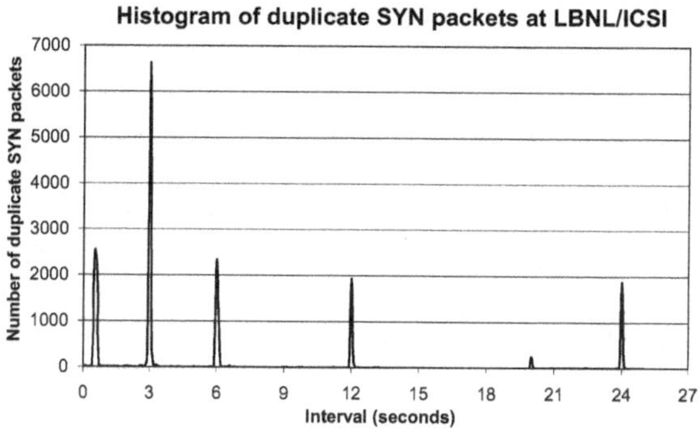


(b) LBNL/ICSI Tracefiles

**Fig. 2.** Measurements taken at the Internet Proxy Server of UIBK and LBNL/ICSI

that showed similar results and are therefore not described in detail. In all our measurements (as illustrated by Table 1) we found that lost packets belonging to the three-way handshake occur at a relatively high rate of 0.5%. Furthermore we deduce from the logfiles of our own measurements that the phenomenon of duplicate SYN packets arriving only 0 to 0.3 seconds after the original SYN

**Table 1.** Summary of all measurements

| | UIBK Webmail LWM1 | UIBK Webmail LWM2 | Alupress Proxy | UIBK Proxy | LBNL/ ICSI |
|---|---|---|---|---|---|
| Connection tries total | 737.188 | 665.829 | 75.493 | 1.721.382 | 232.852 |
| Connection tries successful | 730.523 | 659.913 | 75.049 | 1.708.442 | 219.867 |
| Connection tries not successful | 6.665 | 5.916 | 444 | 12.940 | 12.985 |
| multiple SYNs | 5.968 | 4.838 | 480 | 19.452 | 22.268 |
| multiple SYNs for successful conn. | 5.162 | 4.064 | 214 | 5.696 | 4.103 |
| multiple SYNs for unsuccessful conn. | 806 | 774 | 266 | 13.756 | 18.165 |
| multiple SYNs outside standard interval | 690 | 577 | 48 | 23 | 428 |
| SYNs per successful connection | 1,007 | 1,006 | 1,003 | 1,003 | 1,019 |
| duplicate SYN every X connection | 142 | 162 | 351 | 300 | 54 |

packet seems to be confined to a relatively small number of source hosts. They might use very small RTO settings that match those time intervals or otherwise modified TCP implementations that deviate from the suggested standard behaviour. Since one of the requirements of ZID - our university's central information technology service, who gave us permission to record and analyse traffic data - was the anonymization of host IP addresses, we cannot investigate further on the origin of those particular SYN packets.

## 4 Enhancing Performance

There are different ways the performance in case of lost control packets could be improved. A simple solution to enhance the performance is to change the TCP settings in the operating system, e.g. setting the initial retransmission value to something smaller than 3 seconds, but this will then apply for every TCP connection and possibly introduce unnecessary retransmissions and could even cause TCP to fail in certain cases of extreme delay. Therefore the Internet standard recommends a rather conservative value. For example setting the value to 100 ms would create problems for establishing connections with a longer RTT. Since two SYN packets have a higher probability to arrive at their destination, a possibility would be to send each SYN packet twice with a short time interval between. This would reduce the probability of failures at the cost of a little bit more traffic in the network. A better solution, that would reduce traffic, would be to perform some statistical evaluation on how long it normally takes to get an ACK for a SYN packet and to set the retransmission timeout accordingly. Here we present an example implementation that shows that performance can be improved. We implemented a small tool (*syn_optimizer*) that runs at the application layer. The tool keeps a copy of sent packets belonging to the connection opening phase and in case the corresponding acknowledgement does not arrive, it retransmits the packet. The tool can be applied just for a certain port (in this case port 80) and the delay before retransmission of a non-acknowledgement packet is also configurable.

The loss of packets belonging to TCP's initialization can be recognized on the client as well as on the server side and the problem can be solved on both sides. As already mentioned on the client side a loss of a SYN or SYN/ACK packet can be noticed, on the other hand, the server side cannot recognize the loss of a SYN packet. Therefore the server side implementation has some drawbacks.

### 4.1 Client Side

On the client side the tool monitors sent packets and keeps a copy of sent SYN packets. In case a corresponding SYN/ACK packet has not been received after a certain delay (SYN packet or SYN/ACK packet has been lost) the SYN packet is retransmitted. The delay and number of retransmissions can be set via an input parameter. Furthermore the tool will only retransmit packets with a certain destination port (for web traffic: port 80). The delay parameter should not be set too low because it can cause unnecessary retransmissions on a connection

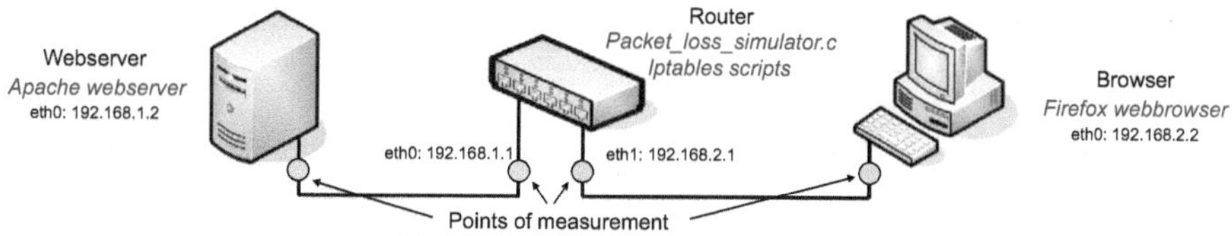


**Fig. 3.** Illustration of our testbed

with a longer RTT. In the case of this tool being installed at the net edge and then applying the changes for all traffic the choice of the delay should be even more conservative. The tool can be called like in the following example:

$syn\_optimizer - d\ 200\ \ - r\ 3\ \ - c\ \ - p\ 80$ ,

where d denotes the delay, r represents the number of retransmission, c indicates it is the client side (s for the server side) and p stands for port.

We tested the tool using a small testbed consisting of three computers, as illustrated in Figure 3. All computers run Linux (kernel version 2.6). One of the computers is used as a router and has two network cards (Eth0: Intel PRO/1000 and Eth1: Intel PRO/100) The other two computers are used as a web server (using the Apache Web Server, with Eth0: Intel PRO/100) and as a client (using Firefox, with Eth0: Broadcom Tigon 3 Gigabit).

To test *syn_optimizer* we simulated the packet loss on the router. We ran a couple of tests: without any packet loss, with one SYN packet lost, with one SYN/ACK packet lost, with both SYN and SYN/ACK packets lost and finally with 3 SYN packets lost. In each case the waiting time before *syn_optimizer* would retransmit a packet was set to 200 ms and the number of retransmissions was set to 3. Using the Wireshark[3] tool we observed packets sent on the link between the client and the router as well as packets sent between the router and the server.

**Table 2.** The test with 1 SYN packet lost; without *syn_optimizer*

| Packets sent from the client to the router: | | | | | |
|---|---|---|---|---|---|
| No. | Time | Source | Destination | Protocol | Info |
| 1 | 0.000000 | 192.168.1.2 | 192.168.2.2 | TCP | 40877 > http [SYN] Seq=0 Win= |
| 2 | 2.996952 | 192.168.1.2 | 192.168.2.2 | TCP | 40877 > http [SYN] Seq=0 Win= |
| 3 | 3.000432 | 192.168.2.2 | 192.168.1.2 | TCP | http > 40877 [SYN, ACK] Seq=0 |
| 4 | 3.000539 | 192.168.1.2 | 192.168.2.2 | TCP | 40877 > http [ACK] Seq=1 Ack= |
| 5 | 3.000689 | 192.168.1.2 | 192.168.2.2 | HTTP | GET / HTTP/1.1 |
| 6 | 3.000864 | 192.168.2.2 | 192.168.1.2 | TCP | http > 40877 [ACK] Seq=1 Ack= |
| 7 | 3.001320 | 192.168.2.2 | 192.168.1.2 | HTTP | HTTP/1.1 304 Not Modified |
| 8 | 3.001437 | 192.168.1.2 | 192.168.2.2 | TCP | 40877 > http [ACK] Seq=552 Ac |
| 9 | 8.001572 | 192.168.2.2 | 192.168.1.2 | TCP | http > 40877 [FIN, ACK] Seq=18 |
| Packets sent from the router to the server: | | | | | |
| No. | Time | Source | Destination | Protocol | Info |
| 1 | 0.000000 | 192.168.1.2 | 192.168.2.2 | TCP | 40877 > http [SYN] Seq=0 Win= |
| 2 | 0.000119 | 192.168.2.2 | 192.168.1.2 | TCP | http > 40877 [SYN, ACK] Seq=0 |
| 3 | 0.000279 | 192.168.1.2 | 192.168.2.2 | TCP | 40877 > http [ACK] Seq=1 Ack= |
| 4 | 0.000433 | 192.168.1.2 | 192.168.2.2 | HTTP | GET / HTTP/1.1 |
| 5 | 0.000587 | 192.168.2.2 | 192.168.1.2 | TCP | http > 40877 [ACK] Seq=1 Ack= |
| 6 | 0.001037 | 192.168.2.2 | 192.168.1.2 | HTTP | HTTP/1.1 304 Not Modified |
| 7 | 0.001182 | 192.168.1.2 | 192.168.2.2 | TCP | 40877 > http [ACK] Seq=552 Ac |
| 8 | 5.001267 | 192.168.2.2 | 192.168.1.2 | TCP | http > 40877 [FIN, ACK] Seq=1 |

[3] http://www.wireshark.org/

**Table 3.** The test with 1 SYN packet lost; with *syn_optimizer*

Packets sent from the client to the router:

| No. | Time | Source | Destination | Protocol | Info |
|---|---|---|---|---|---|
| 1 | 0.000000 | 192.168.1.2 | 192.168.2.2 | TCP | 57993 > http [SYN] Seq=0 Win= |
| 2 | 0.258130 | 192.168.1.2 | 192.168.2.2 | TCP | 57993 > http [SYN] Seq=0 Win= |
| 3 | 0.259776 | 192.168.2.2 | 192.168.1.2 | TCP | http > 57993 [SYN, ACK] Seq=0 |
| 4 | 0.259903 | 192.168.1.2 | 192.168.2.2 | TCP | 57993 > http [ACK] Seq=1 Ack= |
| 5 | 0.260003 | 192.168.1.2 | 192.168.2.2 | HTTP | GET / HTTP/1.1 |
| 6 | 0.260174 | 192.168.2.2 | 192.168.1.2 | TCP | http > 57993 [ACK] Seq=1 Ack= |
| 7 | 0.260630 | 192.168.2.2 | 192.168.1.2 | HTTP | HTTP/1.1 304 Not Modified |
| 8 | 0.260751 | 192.168.1.2 | 192.168.2.2 | TCP | 57993 > http [ACK] Seq=552 Ac |
| 9 | 5.258758 | 192.168.2.2 | 192.168.1.2 | TCP | http > 57993 [FIN, ACK] Seq=1 |

Packets sent from the router to the server:

| No. | Time | Source | Destination | Protocol | Info |
|---|---|---|---|---|---|
| 1 | 0.000000 | 192.168.1.2 | 192.168.2.2 | TCP | 57993 > http [SYN] Seq=0 Win= |
| 2 | 0.000135 | 192.168.2.2 | 192.168.1.2 | TCP | http > 57993 [SYN, ACK] Seq=0 |
| 3 | 0.000319 | 192.168.1.2 | 192.168.2.2 | TCP | 57993 > http [ACK] Seq=1 Ack= |
| 4 | 0.000418 | 192.168.1.2 | 192.168.2.2 | HTTP | GET / HTTP/1.1 |
| 5 | 0.000571 | 192.168.2.2 | 192.168.1.2 | TCP | http > 57993 [ACK] Seq=1 Ack= |
| 6 | 0.001008 | 192.168.2.2 | 192.168.1.2 | HTTP | HTTP/1.1 304 Not Modified |
| 7 | 0.001163 | 192.168.1.2 | 192.168.2.2 | TCP | 57993 > http [ACK] Seq=552 Ac |
| 8 | 4.999125 | 192.168.2.2 | 192.168.1.2 | TCP | http > 57993 [FIN, ACK] Seq=1 |

Table 2 shows packets seen on the link between the Web server and the router and on the link between the router and the client. In this case the first SYN packet was lost and the tool was not applied. As can be seen the first retransmission of the lost SYN packet appeared after almost 3 seconds and an HTTP GET command was sent after 3.00 seconds. The results when *syn_optimizer* was applied are shown in Table 3. We set a delay of 200 ms for resending the SYN packet and the SYN packet was retransmitted after about 258ms. A possible reason for this extra delay of about 60 ms is that our tool runs on the application level. The HTTP GET command was sent after 260 ms which was 2.74 seconds faster than without our tool. Some test showed that the response can be even 20.29 seconds faster with the tool. The test with the loss of 3 SYN packets showed that without our tool applied the HTTP GET was sent after 21.02 seconds compared to just 0.73 seconds with our tool in use. The summary of all tests is shown in Table 4.

### 4.2 Server Side

As we already mentioned a performance optimization in case of SYN/ACK packet loss can be done on the server side too. In this case our tool will monitor sent packets and capture SYN/ACK packets. If the corresponding acknowledgement for a SYN/ACK packet does not arrive after a certain delay, our tool will retransmit the SYN/ACK packet.

**Table 4.** Improvement using *syn_optimizer* on the client side

| number of lost SYN packets | number of lost SYN/ACK packets | without *syn_optimizer* | with *syn_optimizer* | difference |
|---|---|---|---|---|
| 1 | 0 | 3.00 | 0.26 | 2.74 |
| 0 | 1 | 3.00 | 0.30 | 2.70 |
| 1 | 1 | 6.50 | 0.45 | 6.05 |
| 3 | 0 | 21.02 | 0.73 | 20.29 |

We used the same test setup as in section 4.1 and observed the behavior of our tool in following cases: 1 SYN packet was lost, 1 SYN/ACK packet was lost, both SYN and SYN/ACK packets were lost and 3 SYN/ACK packets were lost.

In the case of the loss of a SYN packet our tool did not show any improvement, because the loss of a SYN packet cannot be detected on the server side. Since the results for this scenario are similar we will discuss one in detail. The client retransmitted the lost SYN packet after the TCP timeout expired (after 3 seconds). Since the next two SYN/ACK packets were lost too, the TCP timeout triggered two more times: once at the server (at 3.800 seconds) and once more at the client (at 8.99 seconds, the RTO was doubled). With the optimizer in place, at 0.0 seconds the server received a SYN packet and immediately sent a SYN/ACK packet. This packet was saved by the tool and since no acknowledgement was received in the next 200 ms, *syn_optimizer* resent the SYN/ACK packet at 0.316 seconds. Because of not receiving any acknowledgments in next two 200 ms intervals the SYN/ACK packet was retransmitted again at 0.560 ms and 0.801 ms. After the third retransmission the acknowledgment was received. The HTTP GET command was sent after 0.8 seconds which was 8.2 seconds faster than without *syn_optimizer* (in that case the time interval was 9.0 seconds). The summary of our other tests can be seen below:

| number of lost SYN packets | number of lost SYN/ACK packets | without *syn_optimizer* | with *syn_optimizer* | difference |
|---|---|---|---|---|
| 1 | 0 | 2.99 | 3.00 | -0.01 |
| 0 | 1 | 3.00 | 0.33 | 2.67 |
| 1 | 1 | 6.27 | 3.28 | 2.99 |
| 0 | 3 | 9.00 | 0.80 | 8.20 |

As is evident using this simple tool can increase the performance of web browsing. By using it on the client side, the performance can be improved in all cases. On the server side the tool cannot detect the loss of SYN packets, but in any case congestion is more likely to happen on the way from the server to clients rather than in the opposite direction. Still, installing such a tool at the server will improve performance of all users connecting to the server.

## 5 Conclusion

The results of our measurements show that the loss of SYN packets is occurring at a relatively high rate of 0.5%, forcing the host to resend duplicate SYN packets. While some of those duplicate packets are sent after time intervals that do not match the suggested standards, the majority of them are retransmitted at the standard RTO interval times, which can significantly delay short lived data transfers. To overcome this problem we investigated the possibility of reducing the RTO time interval, which has the disadvantage of possibly introducing unnecessary retransmissions for connections with higher round-trip times. Therefore we show, by means of a simple example implementation, that the performance can be improved. Our tool keeps copies of sent control packets and retransmits them in case no response has been received for a certain amount of time. It can be configured only to operate on a certain port and for a specific timeout. The results show that data transfer time (in our tests an HTTP GET command) can be improved by between 2.74 and 20.29 seconds. Similar improvements can also be observed when applying the tool on the server side for the retransmission of SYN/ACK packets.

## Acknowledgements

We would like to thank both Walter Müller and Benjamin Kaser, who assisted us in gathering the measurement data for this paper.

## References

1. Paxson, V., Allman, M.: Computing TCP's retransmission timer. RFC 2988, Internet Engineering Task Force (November 2000)
2. Allman, M., Floyd, S., Partridge, C.: Increasing TCP's Initial Window. RFC 2414 (Experimental), Obsoleted by RFC 3390 (September 1998)
3. Kuzmanovic, A.: The power of Explicit Congestion Notification. In: SIGCOMM 2005: Proceedings of the 2005 conference on Applications, technologies, architectures, and protocols for computer communications, pp. 61–72. ACM, New York (2005)
4. Ramakrishnan, K., Floyd, S., Black, D.: The Addition of Explicit Congestion Notification (ECN) to IP. RFC 3168 (Proposed Standard) (September 2001)
5. Padhye, J., Floyd, S.: Identifying the (TCP) Behavior of Web Servers. In: ACM SIGCOMM (2001)
6. Medina, A., Allman, M., Floyd, S.: Measuring the Evolution of Transport Protocols in the Internet. SIGCOMM Comput. Commun. Rev. 35, 37–52 (2005)
7. Stewart, R.: Stream Control Transmission Protocol. RFC 4960 (Proposed Standard) (September 2007)
8. Braden, R.: T/TCP – TCP Extensions for Transactions Functional Specification. RFC 1644 (Experimental) (July 1994)
9. Cheriton, D.: VMTP: a transport protocol for the next generation of communication systems. SIGCOMM Comput. Commun. Rev. 16(3), 406–415 (1986)
10. Postel, J.: Transmission Control Protocol. RFC 793 (Standard), Updated by RFC 3168 (September 1981)
11. Watson, R.W.: The Delta-t transport protocol: Features and experience. In: Proc. IEEE 14th Conf. Local Comput. Networks, Minneapolis, MN, October 1989, pp. 399–407 (1989)

# Gravity-Based Local Clock Synchronization in Wireless Sensor Networks*
## (Work in Progress)

Markus Wälchli, Reto Zurbuchen, Thomas Staub, and Torsten Braun

Institute of Computer Science and Applied Mathematics
University of Bern
Neubrückstrasse 12
3012 Bern - Switzerland
{waelchli,zurbuche,staub,braun}@iam.unibe.ch

**Abstract.** Contention-based MAC protocols follow periodic listen/sleep cycles. These protocols face the problem of virtual clustering if different unsynchronized listen/sleep schedules occur in the network, which has been shown to happen in wireless sensor networks. To interconnect these virtual clusters, border nodes maintaining all respective listen/sleep schedules are required. However, this is a waste of energy, if locally a common schedule can be determined. We propose to achieve local synchronization with a mechanism that is similar to gravitation. Clusters represent the mass, whereas synchronization messages sent by each cluster represent the gravitation force of the according cluster. Due to the mutual attraction caused by the clusters, all clusters merge finally. The exchange of synchronization messages itself is not altered by LACAS. Accordingly, LACAS introduces no overhead. Only a not yet used property of synchronization mechanisms is exploited.

**Keywords:** Wireless sensor networks, synchronization, virtual clustering, Networking 2009.

## 1 Introduction

Energy-efficient, contention-based MAC protocols maintain low duty cycles. This means the sensor nodes follow periodic listen/sleep cycles. In the listen cycle they are able to communicate with neighbor nodes and can forward pending data. In the sleep cycle they shut down their radio to preserve energy. In order to synchronize their listen/sleep cycles with neighboring nodes, SYNC messages are periodically exchanged. Nodes maintaining the same listen/sleep cycle are organized into clusters by this synchronization mechanism. This synchronization of common listen/sleep cycles is called virtual clustering. In theory it is possible that different, possibly disjoint, listen/sleep cycles occur.

* The work presented in this paper was partly supported by the National Competence Center in Research on Mobile Information and Communication Systems (NCCR-MICS), a center supported by the Swiss National Science Foundation under grant number 5005-67322.

L. Fratta et al. (Eds.): NETWORKING 2009, LNCS 5550, pp. 907–918, 2009.

In [1] it has been shown that the existence of multiple virtual clusters is also common in reality. Already in a multi-hop network consisting of 50 nodes, four virtual clusters have been encountered. Accordingly, border nodes interconnecting the clusters in [1] had to wake-up up to three times more than normal cluster nodes. This implies decreased sleep cycles and higher energy consumption for those border nodes. Therefore, it is desirable to agree on a common schedule to discharge the border nodes. On the other hand, maintaining a global schedule within the whole network is unnecessary and implies overhead. Common schedules are only locally required, because nodes can only communicate with their neighbors.

To achieve local synchronization we propose a Local Adaptive Clock Assimilation Scheme (LACAS). Its basic idea is similar to the principle of gravitation. In the context of virtual clusters this means that larger clusters attract smaller clusters more than vice versa, until the clusters finally merge. This applies to all clusters which are present in a network or will evolve later. Therefore, different present clusters always converge towards one single cluster. Of course, this requires that the clusters are connected to each other, which presumes the existence of border nodes.

The rest of the paper is organized as follows: In Section 2 relevant related work is presented. Section 3 introduces the local clock synchronization scheme. Simulation results are provided in Section 4. The paper ends with conclusions in Section 5.

## 2 Related Work

S-MAC [2], T-MAC [3] and DW-MAC [4] are energy-efficient contention-based protocols for wireless sensor networks. All three are based on low duty-cycles and require SYNC messages to synchronize the listen/sleep schedules of the nodes. T-MAC furthermore supports the adjustment of its wake-up period according to pending data traffic. In both protocols each node maintains its own listen/sleep schedule. These schedules are synchronized whenever possible in order to reduce control traffic overhead. Nodes maintaining the same listen/sleep schedule build a virtual cluster. New sensor nodes initially listen to the wireless medium for a specific amount of time to overhear and adapt an existing schedule. If no SYNC message has been received during this period, a node choses its own schedule. Having determined its schedule, any subsequently overheard unknown schedule is adapted too. Thus, virtual clusters are interconnected. The interconnecting nodes are called border nodes and follow multiple schedules, i.e., the schedule of each cluster they are a member of. Accordingly they consume much more energy than normal cluster nodes. Apart from virtual clustering, SYNC messages are also used to adjust clock drifts between network nodes.

TDMA-based MAC protocols such as [5], [6] require the exchange of periodic SYNC messages. However, unlike contention-based protocols, the operation of TDMA-based protocols is based on the concept of clusters. In general, they require a cluster leader which allocates slots to its cluster members. Thus, the problem of virtual clustering is not present as the nodes are per se organized into clusters. On the other hand, TDMA-based protocols require very precise synchronization and scale rather poorly.

Apart from protocols that require synchronization, asynchronous contention-based MAC protocols [7], [8], [9], [10] have been proposed. [7], [8] and [9] are based on preamble-sampling. These protocols send long preambles to reach neighboring nodes that are currently asleep. RI-MAC [10] avoids the transmission of preambles. In RI-MAC receiver nodes announce their availability by beacon messages. Based on the reception of such a beacon, a sender node transmits its pending data to the receiver. The approach achieves low duty cycles. Asynchronous protocols do not face virtual clustering, but require the exchange of preambles or beacons. Moreover, broadcast operations are poorly supported.

The problem of virtual clustering, i.e., of coexisting schedules, has been addressed in [1]. To solve the problem, an additional schedule age is introduced. The authors motivate that different schedules must have entered the network at different time points and thus have different ages. The schedule age is announced in the SYNC message. Over time all nodes converge toward the oldest schedule in the network. To prevent network partitions all other schedules need to be temporary stored too. The maintenance and distribution of the schedule age requires additional information. This paper will show that no schedule age is needed if local schedule consistency is sufficient.

## 3 Local Adaptive Clock Assimilation Scheme (LACAS)

Nodes implementing an energy-efficient contention-based MAC protocol such as S-MAC follow periodic listen/sleep schedules. Nodes with the same schedule are virtually organized into clusters. To support communication between different virtual clusters, border nodes interconnecting the according clusters are required. This synchronization mechanism is shown in Fig. 1.

Node A is member of a virtual cluster, whereas nodes B and C are members of another disjoint virtual cluster. This is indicated by the listen periods in Fig. 1. All nodes could be in transmission range of each other. However, in the example above, it is only required that node A can hear node B and node B can hear both nodes A and C. From time to time each node remains awake for an entire frame length $f_i$ in order to scan for present schedules. In Fig. 1 this is node B

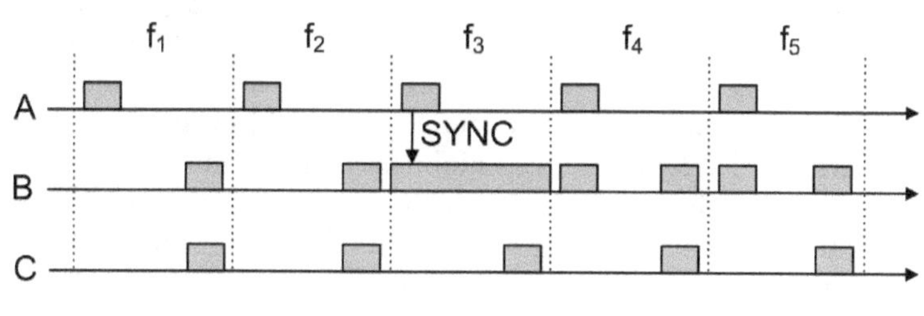


**Fig. 1.** Periodic sleeping and virtual clustering in S-MAC and T-MAC

in frame $f_3$. B learns the cluster of node A in frame $f_3$, because it overhears the SYNC message sent by node A. Only this SYNC transmission is shown in Fig. 1. Node B becomes a border node as it interconnects two clusters. This means that node B synchronizes to both known schedules after frame $f_3$.

Experiments have shown [1] that in S-MAC already in a multi-hop network consisting of 50 nodes four different virtual clusters evolved. Moreover, it has been shown that border nodes had to listen to up to three different schedules. In all four experiments more than 44% of all network nodes followed at least two schedules. In two of the four experiments 34%, respectively 47%, of all network nodes had to listen to three virtual clusters. Obviously, the border nodes thus have a higher average energy consumption than normal cluster nodes.

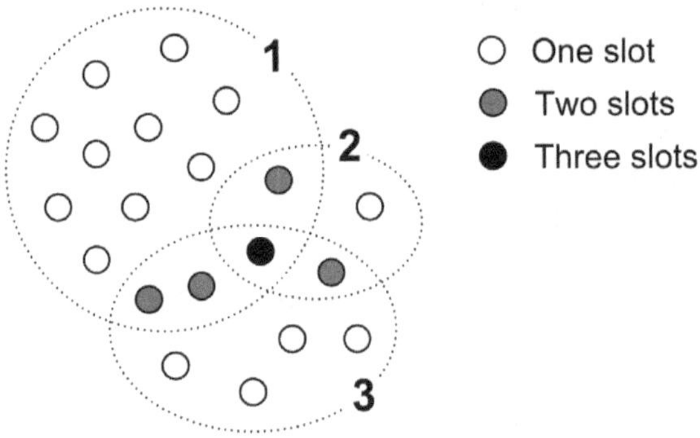


**Fig. 2.** Drawback of virtual clustering

The problem is illustrated in Fig. 2. The gray and black nodes operate as gateway nodes between the clusters and have to listen to multiple schedules. Accordingly, these nodes sleep less and their batteries deplete sooner. If this happens, network connectivity might be broken and the network might be disconnected, even though sufficiently many working network nodes could still exist. Therefore, it is desirable to avoid virtual clustering. In [1] it has been shown that surprisingly many nodes follow multiple schedules. The temporary failure of communication links and effects of strongly varying radio ranges (communication gray zones [11]), have been identified as main reasons.

While the authors of [1] use a global mechanism to solve the problem, we propose a local adaptive clock assimilation scheme (LACAS) that achieves local synchronization. A global solution requires system-wide synchronization towards one global schedule. This implies overhead in terms of signaling and requires the storage of fallback mechanisms, i.e., of temporary valid local schedules. LACAS avoids these drawbacks. Maintaining a global schedule is unnecessary, because of the locality of communication links between network nodes. LACAS avoids the drawback of virtual clustering and leads to a uniform distribution of the energy consumption that is required for synchronization.

LACAS implements a mechanism similar to gravitation. Translated into the virtual clustering problem, this means that larger clusters attract smaller ones more than vice versa, until the clusters finally merge. In LACAS, the cluster nodes represent the mass and the number of sent SYNC messages represent the gravitation force. Because all sensor nodes implement the same contention-based transmission scheme, large clusters broadcast in average more SYNC messages than small ones and accordingly cause more attraction.

LACAS only exploits the information exchanged by synchronization messages. Therefore, no additional control traffic is generated. Moreover, the loss of SYNC messages does not affect the principle of LACAS, but only temporarily decreases the gravitation force of a cluster.

Only border nodes are part of multiple clusters. Accordingly, only border nodes attract clusters. Whenever a border node evolves, it spans its listen period over all schedules it knows. Furthermore, every node adapts its own listen/sleep schedule to a given percentage $\alpha$ (e.g., $\alpha = 5\%$) towards the schedule of the cluster from which it has received a SYNC message. Accordingly, the parameter $\alpha$ controls the attraction caused by a SYNC message. In a first step of a merging process, the schedule of a border node is expanded and then it starts to contract again.

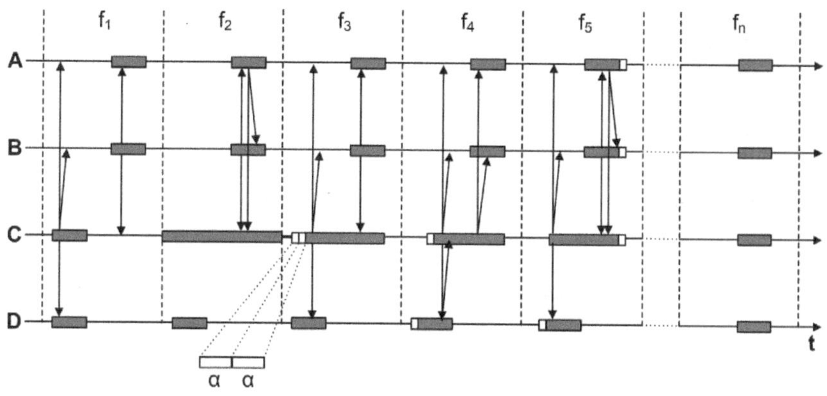


**Fig. 3.** Gravitation principle of LACAS: Neighbor clusters are detected in specific frames. Then, the clusters fuse (slowly).

A merging process is shown in Fig. 3. The dark bars indicate listen periods, while the white bars indicate schedule adaptations. Node C stays awake for a whole frame $f_i$, i.e., a whole listen/sleep period, in $f_2$. Having detected another schedule, it spans its listen period over both known schedules (see $f_3$). This listen period contracts then to the normal schedule length merging both connected clusters. The parameter $\alpha$ controls the gravitation force. High values for $\alpha$ lead to high attractions and fast convergence. On the other hand, this can temporary break connections between clusters. However, only the convergence time of the clusters is affected. The gravitation mechanism itself is not compromised. In the worst case, nodes are successively transferred from the smaller cluster to the larger. The problem is discussed in detail below. The contraction of the schedule of node C continues after period $f_5$ and ends in period $f_n$.

The operation in Fig. 3 is discussed in the following: Initially nodes A and B form cluster 1, while nodes C and D form another cluster 2 (see frame $f_1$ in Fig. 3). Because T-MAC is used, all nodes stay periodically awake for an entire frame $f_i$ in order to detect other clusters. In Fig. 3 this is node C in frame $f_2$. Having learned both clusters, C spans its own listen period over both schedules. Moreover, as it has received two SYNC messages from nodes A and B from cluster 1, node C moves its listen period for $2\alpha$ towards the listen period of cluster 1 (frames $f_2$ and $f_3$ in Fig. 3). In frame $f_3$, C is able to transmit its own SYNC message and receives another one from node B from cluster 1. Accordingly, the listen period of node C again moves for $\alpha$ towards the listen period of cluster 1. Due to SYNC from node C, node D is attracted too ($f_4$ in Fig. 3). In frame $f_4$ node C is able to transmit a SYNC message in both schedules. Accordingly, cluster 1 and node D are attracted towards C. C itself moves towards D as it has received a SYNC message from D (frame $f_5$ in Fig. 3). The merging process continues in Fig. 3 after frame $f_5$. After a while, both clusters will fuse. In the example, cluster 1 transmits in general more SYNC messages than cluster 2 (It has three members, whereas cluster 2 has only two). Accordingly, both clusters will merge closer to the original schedule of cluster 1.

The choice of the adaptation parameter $\alpha$ is crucial considering performance. A small value implies a long merging period. On the other hand, if a large value has been chosen for $\alpha$, the fast convergence towards a large cluster might disrupt the connection between a border node and its smaller cluster. This disconnection is no basic problem, because the clusters are connected again when a border node remains awake for a whole frame, but it increases merging time. In the worst case, nodes pass successively from the smaller cluster to the larger. The convergence of LACAS is however not affected. Moreover, growing clusters have in general also a growing number of border nodes and therefore their gravitation force increases too. In the current deployment we have chosen a value for $\alpha$ of 5%. Thus, all sensor nodes are able to synchronize within a few minutes. Expecting a network lifetime of at least several months, the synchronization time seems tolerable. Finally, only the listen periods of the according MAC protocol are optimized. Any subsequent data exchange period is not affected by LACAS.

## 4 Evaluation

LACAS has been implemented on top of T-MAC in the network simulator OMNeT++ [12]. All network nodes wake up randomly within the first 30 simulation seconds, and begin immediately to synchronize. T-MAC enhanced with LACAS has been used as MAC protocol for a topology control algorithm [13]. Because LACAS has been implemented using a cross-layer approach, we evaluate the performance of LACAS in a larger context. The topology control mechanism establishes a routing backbone after a synchronization and neighborhood learning period of 400 seconds. Non-backbone nodes temporarily shut-down their radios to save energy. This has some impact on the convergence time of LACAS as fewer SYNC messages are sent due to the temporary unavailability of the non-backbone nodes.

The parameters of T-MAC have been set according to [3]. All nodes follow a periodic listen/sleep frame of 610 ms, whereof they are awake for at least 13.5 ms, i.e., if no data transmission is pending. This minimum wake-up period consists of the synchronization period, which is 7 ms, and the traffic-adaptivity period TA, which is required by T-MAC and has a duration of 6.5 ms. Each node remains awake for a whole frame in every 35th frame, i.e., every 21.35 s. This is required in order to detect neighbor nodes which follow different listen/sleep cycles.

Three different network sizes of 50, 100 and 200 nodes have been simulated. Each node has in average 12 neighbors. The network topology was randomly generated taking network connectivity into account. Any experiment has been repeated 20 times. The spectrum of the schedule lengths present at a specific time point is indicated by the standard deviation.

The properties of the sensor nodes are configured according to values from the Embedded Sensor Board (ESB) platform [14]. The nodes operate in the 868 MHz frequency band, the transmission range is about 37 m, whereas the interference range is approximately 52 m. The data rate is 115.2 kbps. The energy consumption in transmission mode is 5.2 mA. Idle listening and receiving both require about 4.7 mA, whereas the radio in sleep mode only needs 5 $\mu$A.

### 4.1 Convergence Time of Schedule Length with LACAS

In this section the convergence time of LACAS is investigated. The independent initial wake up of the network nodes in the first 30 s of the simulation leads to multiple coexisting schedules in the beginning. The evolution of the schedule length of each network node has been monitored over the first hour of operation. The schedule length has been captured every 5 s. The evolution of the mean schedule length in a network consisting of 50 nodes is shown in Fig. 4.

Fig. 4(a) shows the evolution of the schedule length over the whole first operation hour. The schedule length converges to a length of approximately 10 ms within the first 200 seconds. Of course, in this convergence period the distribution of the schedule length is high in the network. There are nodes that

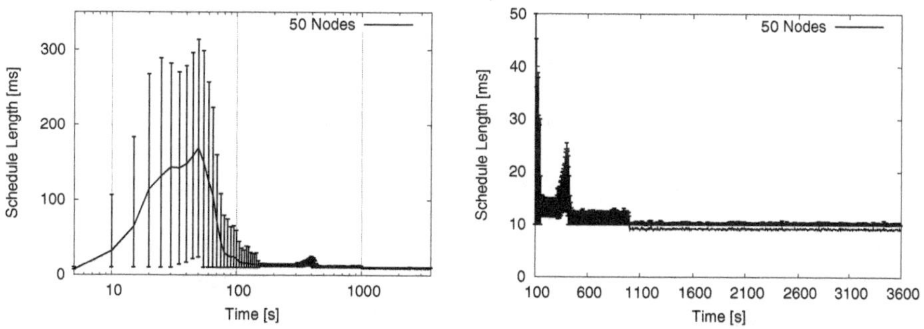


(a) Average schedule length evolution over the first hour (log-scaled).

(b) Average schedule length evolution ignoring the first 100 seconds.

**Fig. 4.** Convergence of schedule length in a network consisting of 50 nodes follow common schedules and thus have already a short schedule length. On the other hand, there are numerous nodes interconnecting different schedules, which results in a temporary large schedule length.

Fig. 4(b) shows the evolution of the mean schedule length after the first 100 seconds. The peak at second 400 is due to the backbone scenario as described above. At second 300 the parameter $\alpha$ is adapted from 0.05 to 0.5 to achieve faster convergence. This leads to the temporary peak. After the adaptation a little better performance can be achieved, though. Sleeping non-backbone nodes lead to a smaller amount of SYNC messages, which further reinforces the effect. The peak is in the order of a duplication of the schedule length. The adaptation of $\alpha$ could be implemented in LACAS without cross-layer optimization too. The mean schedule length converges to 13 ms without adaptation and to 10 ms with adaptation. The average schedule length remains stable after 200 seconds without adaptation. With adaptation stability is achieved after 450 seconds. The point in time when stability is achieved is protocol-dependent, though.

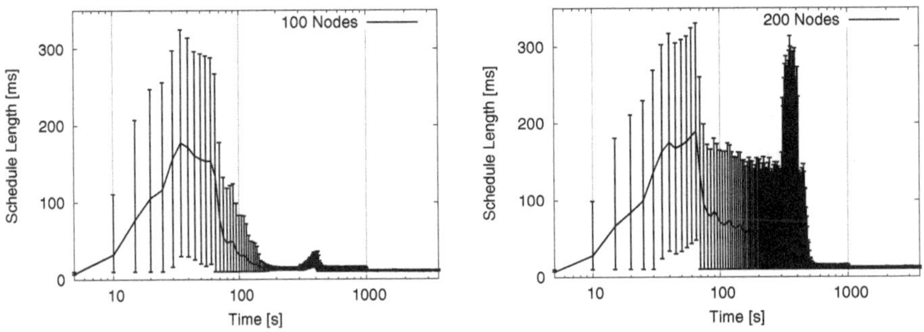


(a) Network consisting of 100 nodes. (b) Network consisting of 200 nodes.

**Fig. 5.** Average schedule length evolution over the first hour (log-scaled)

Fig. 5 shows the impact of the network size. The performance of LACAS in a network consisting of 100 nodes is depicted in Fig. 5(a). The performance is very similar to the performance in the network consisting of 50 nodes. However, the convergence needs more time in the network consisting of 200 nodes (Fig. 5(b)). Compared to an intended network lifetime of several months or more, this delay is still insignificant, though. The convergence time of LACAS increases with network size. This is due to the hop-by-hop impact of the gravitation principle. Due to gravitation, clusters show an impact similar to the movement of a ripple through water over multiple hops.

The effect is illustrated in Fig. 6. Clusters of nodes such as the nodes in $L_1$ attract nodes at the boundaries. The nodes in $L_2$ have again an impact on their border nodes, i.e, on the nodes in $L_3$. The effect causes complex mutual influences. Moreover, due to the increased time needed for dissemination, clusters located far away from each other have a longer lasting impact on each other than nearby clusters. The probability of presence of such clusters grows with network size. Thus, the convergence time increases with network size too.

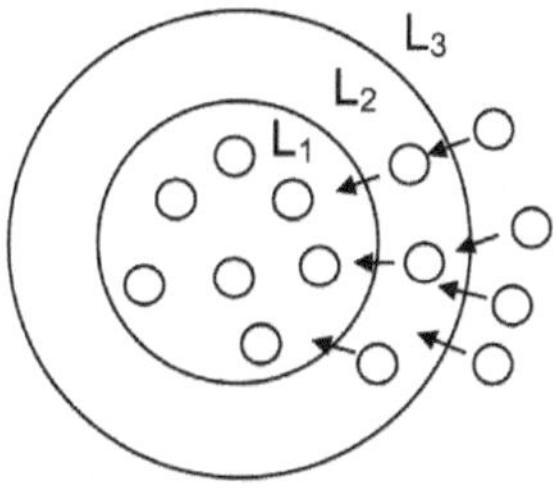


**Fig. 6.** Ripple effect of gravitation over multiple hops

Considering the network size of 200 nodes in Fig. 5(b), LACAS is not able to converge to a short schedule length before cross-layer adaptation occurs. The cross-layer impact leads again to a temporary duplication of the average schedule length, which becomes in this case much longer. However, the schedule length converges quickly to 10 ms after the adaptation. This fast convergence is again due to the cross-layer approach. Without optimization the convergence would look similar to Fig. 4(a) or 5(a). It would only require some more time.

The average schedule length has converged in all evaluated network topologies and sizes to a length of approximately 10 ms. This is not surprising, because the convergence (gravitation) is basically a local process. Mainly local communications have an impact on local convergence and any local communication is independent of the network size. On the other hand, due to the ripple effect it is not possible to achieve the schedule length of 7 ms of T-MAC. Accordingly, there is a trade-off between avoiding the border nodes and the achievable schedule length. On node-level, every border node with disjoint schedules consumes more energy than any node running LACAS. Concerning the overall energy consumption, LACAS preserves energy if the following inequation applies ($\forall i : n_i \in \mathbb{N}$):

$$n_1 \cdot 7ms + n_2 \cdot 14ms, +... + n_k \cdot k \cdot 7ms > n_t \cdot 10ms; \quad \sum_{i=1}^{k} n_i = n_t \tag{1}$$

where $n_i$ is the number of nodes maintaining a given number of schedules and $n_t$ is the total number of nodes in the network. Inequation 1 assumes that the different schedules are disjoint. Else, overlaying schedules would need to be included.

### 4.2 Analysis of Power Consumption

Unlike the convergence of LACAS, a realistic virtual clustering is difficult to simulate. Virtual clustering mainly occurs due to physical impacts such as communication gray zones [11] and temporary unavailable communication links, which are hardware- and environment-dependent and accordingly difficult to simulate properly. Approximating those impacts in simulations might falsify the simulations rather than improve them. Finally, these effects have little impact on the convergence of LACAS due to the robustness of the gravitation principle.

**Table 1.** Percentage of nodes maintaining a certain number of schedules (from [1])

| | Number of Schedules | | | |
|---|---|---|---|---|
| | **1** | **2** | **3** | **4** |
| Exp. 1 | 56% | 44% | - | - |
| Exp. 2 | 32% | 68% | - | - |
| Exp. 3 | - | 66% | 34% | - |
| Exp. 4 | 9% | 44% | 47% | - |

**Table 2.** Power consumption (in mAs) per listen/sleep cycle for T-MAC and LACAS

| | T-MAC | LACAS |
|---|---|---|
| Exp. 1 | 2.37 | 2.35 |
| Exp. 2 | 2.76 | |
| Exp. 3 | 3.85 | |
| Exp. 4 | 3.92 | |

In order to assess the power consumption needed by LACAS we adopt the results obtained in real world experiments in [1]. The costs of T-MAC and LACAS are computed according to these values and inequation 1. The long-sleep impact of non-backbone nodes has not been considered in this evaluation, because it is based on a cross-layer optimization. Accordingly, all nodes follow a normal listen/sleep schedule. The sensor network in [1] consisted of 50 nodes running S-MAC. The percentage of network nodes maintaining a certain number of schedules are listed in Table 1.

The results would be the same if T-MAC had been used due to the identical synchronization mechanism. As mentioned above, ESB nodes need 4.7 mA in idle listening state. We take this value to estimate the power consumption of LACAS. Furthermore, we assume that the different schedules, which evolved in the experiments in [1], are disjoint (see also inequation 1). SYNC messages that would have to be sent in the synchronization periods are not considered. Table 2 shows the power consumption of T-MAC and LACAS to maintain all schedules of all network nodes in one listen/sleep cycle. The results apply as soon as the networks are stable, i.e., after the convergence to the common schedule length of 10 ms in the case of LACAS, respectively after all virtual clusters have evolved in the case of T-MAC. Therefore, the values in Table 1 can be used. LACAS maintains only one schedule. Accordingly, the expected power consumption of LACAS would be the same in all four experiments.

The estimations in Table 2 show that in a network consisting of 50 sensor nodes, depending on the experiment, more or less power can be saved with LACAS, i.e., between 0.02 and 1.57 mAs in one listen/sleep cycle of 610 ms. Even though LACAS has a slightly longer minimal schedule length than T-MAC, LACAS is estimated to perform at least as well as T-MAC in all four experiments performed in [1]. Considering a network lifetime of months or more, the possible energy savings are promising. The estimations shown in Table 2 concern the average energy consumption over all nodes. Border nodes with disjoint schedules consume more energy and would therefore even sooner run out of energy.

## 5 Conclusions

In this paper a simple local clock synchronization scheme has been proposed. In general, nodes are not synchronized after deployment. Merging nodes with similar sleep/listen cycles into clusters leads to the problem of high energy consumption for the nodes connecting the clusters. Therefore, we have proposed a simple local synchronization mechanism (LACAS). LACAS provides system-wide local clock consistency and avoids the drawback of virtual clustering. It has been shown that the overhead of LACAS is marginal. Moreover, the synchronization procedure converges fast, i.e., within minutes for the simulated networks and remains stable thereafter.

In simulations the fast convergence of the algorithm has been shown. LACAS just exploits attraction information exchanged by SYNC messages. If messages are lost, attraction is decreased temporarily, but the functionality of LACAS is not affected. The according cluster just shows currently lower attraction. Due to the difficulty to realistically simulate virtual clustering, the energy consumption of LACAS has been estimated offline based on real-world results collected in related work. The estimations indicate that LACAS would save energy in a real-world implementation.

## References

1. Li, Y., Ye, W., Heidemann, J.: Energy and latency control in low duty cycle MAC protocols. In: Proc. of the IEEE Wireless Communications and Networking Conference (WCNC 2005), New Orleans, LA, March 2005, pp. 676–682 (2005)
2. Ye, W., Heidemann, J., Estrin, D.: Medium access control with coordinated adaptive sleeping for wireless sensor networks. IEEE/ACM Transactions on Networking 12(3), 493–506 (2004)
3. van Dam, T., Langendoen, K.: An adaptive energy-efficient MAC protocol for wireless sensor networks. In: Proc. of the 1st international conference on Embedded networked sensor systems (SenSys 2003), Los Angeles, CA, pp. 171–180 (2003)
4. Sun, Y., Du, S., Gurewitz, O., Johnson, D.B.: DW-MAC: a low latency, energy efficient demand-wakeup MAC protocol for wireless sensor networks. In: Proc. of the 9th ACM international symposium on Mobile ad hoc networking and computing (MobiHoc 2008), Hong Kong, China, pp. 53–62 (2008)
5. van Hoesel, L., Havinga, P.: A lightweight medium access protocol (LMAC) for wireless sensor networks: Reducing preamble transmissions and transceiver state switches. In: Proc. of the International Conference on Networked Sensing Systems (INSS 2004), Tokyo, Japan (June 2004)
6. Rajendran, V., Obraczka, K., Garcia-Luna-Aceves, J.: Energy-efficient collision-free medium access control for wireless sensor networks. In: Proc. of the 1st international conference on Embedded networked sensor systems (SenSys 2003), Los Angeles, CA, pp. 181–192 (2003)

7. El-Hoiydi, A., Decotignie, J.-D.: WiseMAC: An ultra low power MAC protocol for the downlink of infrastructure wireless sensor networks. In: Proc. of the 9th IEEE International Symposium on Computers and Communications (ISCC 2004), Alexandria, Egypt, June 2004, vol. 1, pp. 244–251 (2004)
8. Polastre, J., Hill, J., Culler, D.: Versatile low power media access for wireless sensor networks. In: Proc. of the Second ACM Conference on Embedded Networked Sensor Systems (SenSys 2004), Baltimore, MD, USA, November 2004, pp. 95–107 (2004)
9. Buettner, M., Yee, G.V., Anderson, E., Han, R.: X-MAC: a short preamble MAC protocol for duty-cycled wireless sensor networks. In: Proc. of the 4th international conference on Embedded networked sensor systems (SenSys 2006), Boulder, Colorado, USA, pp. 307–320 (2006)
10. Sun, Y., Gurewitz, O., Johnson, D.B.: RI-MAC: a receiver-initiated asynchronous duty cycle MAC protocol for dynamic traffic loads in wireless sensor networks. In: Proc. of the 6th ACM conference on Embedded network sensor systems (SenSys 2008), Raleigh, NC, USA, pp. 1–14 (2008)
11. Zhao, J., Govindan, R.: Understanding packet delivery performance in dense wireless sensor networks. In: Proc. of the First ACM Conference on Embedded Networked Sensor Systems (SenSys 2003), Los Angeles, CA, November 2003, pp. 1–13 (2003)
12. Varga, A.: Omnet++: a component-based, modular and open-architecture simulation environment (2008)
13. Wälchli, M., Zurbuchen, R., Staub, T., Braun, T.: Coordinated sleep MAC for energy-constrained wireless sensor networks. In: Proc. IEEE Global Communications Conference (IEEE GLOBECOM 2009) (2009) (submitted)
14. Scatterweb: The self-organizing wireless communication platform (October 2007)

# Two ID-Free Distributed Distance-2 Edge Coloring Algorithms for WSNs (Work in Progress)

André C. Pinho[1,2], Alexandre A. Santos[1,2], Daniel R. Figueiredo[1], and Felipe M.G. França[1]

[1] PESC/COPPE, Universidade Federal Rio de Janeiro, Rio de Janeiro, 21941-972, Brazil
{felipe,daniel}@cos.ufrj.br

[2] Information Technology Division, Centro Tecnológico do Exército, Rio de Janeiro, 23020-470, Brazil
{apinho,alexandre}@ctex.eb.br

**Abstract.** One of the most important problems for Wireless Sensor Networks (WSNs) is energy consumption since it ultimately determines the lifetime of the system. Medium Access Control (MAC) protocols based on schedules (e.g., TDMA) play an important role, since collisions and idle listening can be avoided, effectively reducing energy consumption. The problem of determining good transmission schedules for WSNs can be mapped to the *distance-2* edge coloring problem in graphs, where edge colors represent slots in a TDMA-based MAC protocol, for example. In this paper, we propose and evaluate two new probabilistic and distributed *distance-2* edge coloring algorithms that require no global node identifiers. We obtain analytical results for the worst-case convergence time. Moreover, we use simulations to evaluate the performance of the algorithms with respect to several metrics. Our findings indicate a tradeoff between convergence time and message overhead versus number of colors used.

**Keywords:** WSN, MAC, algorithms, distance-2.

## 1 Introduction

### 1.1 Wireless Sensor Network

Wireless Sensor Networks (WSNs) are composed of a large number of low cost, low power, multi function, small sensor nodes capable of sensing, processing and communicating [1]. Among the characteristics that distinguish WSNs from other wireless networks, one of most important is limited access to energy or even finite energy. Since the radio communication subsystem of sensor nodes is greatly responsible for the consumption of energy, the Medium Access Control (MAC) protocol plays an important role, as it determines when the radio transmits and listens for transmissions. In particular, energy efficiency is ranked as one of the most important attributes of MAC protocols for WSNs, leaving behind attributes such as fairness, latency and channel utilization [4].

L. Fratta et al. (Eds.): NETWORKING 2009, LNCS 5550, pp. 919–930, 2009.

It is known that MAC related energy consumption in WSN is mainly due to packet collisions caused by hidden terminals and continuous idle listening [4]. The hidden terminal problem can be illustrated using the communication graph shown in Figure 1. In this graph, vertices correspond to sensors and edges indicate that two sensors are within direct communication range of each other. The hidden terminal problem occurs among vertices $A$, $B$ and $C$, since nodes $A$ and $C$ are not aware of each other. Thus, when the transmissions of the two nodes overlap in time, a packet collision will occur in node $B$, and a retransmission will be required. The second concern is inherent to the characteristics of short-range radios commonly used WSN applications. The power consumption of listening in these radios has the same order of magnitude as power consumption of receiving or transmitting packets [3]. Therefore the energy wasted from continuous idle listening periods can become a significant burden for obtaining a prolonged network lifetime. Thus, schedule-based protocols emerge as a promising approach for developing energy efficient MAC protocols. It is worth noticing that, in the same spirit of what was presented in [12], our work addresses the issues of communication scheduling and sensing scheduling in a fully decoupled manner. In this work, we focus solely on the communication scheduling.

### 1.2 Distance-2 Coloring

The problem of developing a schedule-based MAC protocol for WSNs can be well illustrated using the communication graph, as shown in Figure 1. In particular, given the communication graph, one can determine a collision-free schedule among the nodes that avoids hidden terminals and allows for selective listening. For example, communication links can be assigned a time slot, such as in a TDMA-based MAC protocol, which would then determine when nodes should transmit and listen for transmissions.

The problem of determining a collision-free schedule for nodes of WSNs that avoids hidden terminals can be mapped to the *distance-2* edge coloring problem in graphs. In this mapping, edge colors represent time slots in a TDMA-based MAC protocol, for example. Let $G$ be an undirected graph, representing the communication graph. We say that two edges of $G$ are within distance-2 of each other if either they are adjacent or there is some other edge that is adjacent to both of them. A distance-2 edge coloring of $G$ is an assignment of colors to edges so that any two edges within distance-2 of each other have distinct colors. There are two equivalent ways to view the coloring constraint: *(i)* any two edges in any 3-edge path cannot have the same color; *(ii)* any two neighboring nodes cannot have any adjacent edge with the same color. Note that such coloring would clearly avoid the hidden terminal problem, if colors correspond to a transmission schedule.

The *distance-2* edge coloring problem can be mapped to the *distance-2* **vertex** coloring problem in graphs. We say two vertices of $G$ are within distance-2 of each other if they are adjacent or if they are adjacent to a common vertex (i.e., have a neighbor in common). A distance-2 vertex coloring of $G$ is an assignment of colors to vertices so that any two vertices within distance-2 of each other have distinct colors. Given a *distance-2* vertex coloring of $G$ we can obtain a *distance-2* edge coloring by associating with each edge the non-ordered pair of the colors of the nodes incident to the edge. Note that this mapping is clearly a *distance-2* edge coloring of the graph, since any two neighboring nodes will not have edges with repeated colors. Thus, the *distance-2* edge coloring problem can be indirectly solved via the *distance-2* vertex coloring of the graph.

### 1.3 Our Contribution

In this paper we propose and evaluate two new probabilistic distributed *distance-2* edge coloring algorithms that require no global identifiers on the nodes. The first algorithm presented, known as $Edge^3 - Sched$, performs a *distance-2* edge coloring directly. The second algorithm, known as $Node^2 - Sched$, performs a *distance-2* vertex coloring, which is then mapped to a an edge coloring. In this paper, both algorithms are considered with single purpose of defining a collision-free link scheduling. Our numerical evaluation of the two algorithms indicates a trade off in their performance, in particular, with respect to convergence time and number of colors used. Our objective is not to obtain the optimal (minimum) number of colors required to produce a *distance-2* edge coloring. Note that an optimal color assignment, despite having potentially very long convergence time, does not necessarily produce the best schedule for a given application. We thus aim at algorithms that can efficiently produce an edge coloring in the lack of global node identifiers.

The remainder of the paper is organized as follows. Some related work is presented in Section 2. Section 3 describes the two new distributed algorithms in detail. Subsequently, the simulation setup is reported in Section 4, with evaluation and results being presented in Section 5. Finally we present our conclusions and future work in Section 6.

## 2 Related Work

The problem of developing schedule-based collision-free MAC protocols for WSNs has been extensively studied in the recent years. Trigoni et al. [7], propose a new protocol that carefully schedules message transmissions to avoid packet collisions. Their protocol assumes nodes are organized in a grid structure and a collision-free scheduling is established through simultaneous link activation according to predefined distances between the nodes and transmission direction. Under this scenario, sensors' radios can work in a deterministic duty cycle saving energy by turning off and avoiding idle listening periods.

Kulkarni et al.[9] proposed a self-stabilizing, deterministic algorithm for TDMA-based MAC in WSNs where a sensor node is aware only of its neighbors. The proposed algorithm relies on the initial slot assignment that is dynamically determined by the nodes through periodically diffusion of messages sent by a base station.

Ergen et al. [10] proposed a distributed algorithm based on distributed coloring of nodes. This work consider a network comprising a single access point

(AP) and several sensor nodes that use the same AP for transferring their messages. They also obtained an upper bound for these schedules as a function of total number of packets generated in the network.

Pantazis et al. [12] proposed a TDMA-based scheduling scheme that balances energy-saving and end-to-end delay. In their approach, sensor nodes send neighboring and routing information to a gateway, which constructs the schedule. The schedule is then sent to the sensor nodes using flooding techniques.

Gandham et al. [6] proposed a distributed algorithm for obtaining a TDMA-based MAC schedule based on the edge-coloring problem. Their work aims in minimizing the number of colors produced and considers that nodes have a global identifiers.

Balakrishnan et al. [5] considers the 2-distance edge coloring problem to determine the maximum capacity of WSNs. Their algorithm assumed nodes have a unique global identifier and are running globally synchronized clocks.

Our work differs from the above in following aspects: (i) our algorithms are fully decentralized (there is no central node); (ii) no global identifiers are needed across the set of sensor nodes, a strong requirement for scalable WSNs as mentioned in [1]; (iii) our probabilistic algorithms are fully independent of network topology, including the case of disconnected networks (i.e., communication graph is composed of multiple connected components). It is worth noting that lack of global node identifiers necessarily requires a non-deterministic approach to break the symmetry (i.e., color edges), as demonstrated in [2].

## 3 Distributed Algorithms

The two distributed algorithms proposed in this paper use an election mechanism to determine which nodes, on each step, can color their adjacent edges or themselves. In summary, the algorithms work as follows. Each node generates a random *sort* number which is then sent to their $h$-neighborhood in the communication graph (the $h$-neighborhood of node $v$ is the set of nodes that are at distance at most $h$ from $v$). This message is sent through multi-hop communication. The nodes that have the highest sort number among the sort numbers it receives (known as *winners*) then color the edges adjacent to them. Once a node is a winner, it ceases to participate in future elections, performing only message retransmissions (i.e., aiding multi-hop communication) afterwards. The process continues until all edges (or nodes) have been colored. The parameter $h$ will be defined in following subsections for each algorithm proposed.

### 3.1 $Edge^3 - Sched$

As briefly discussed in Section 1.3, the $Edge^3 - Sched$ algorithm performs a distance-2 edge coloring of the communication graph directly. Observing Figure 1, and assuming edges *(C,D)* and *(C,F)* have been colored, note that when node $A$ attempts to color its edges, in particular edge *(A,B)*, it cannot use the colors already assigned to edges *(C,D)* and *(C,F)*. Moreover, note that edges

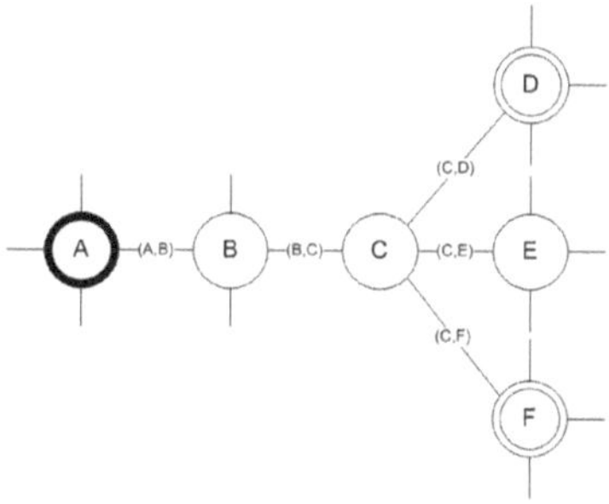


**Fig. 1.** $Edge^3 - Sched$ constraints

*(B,C)* and *(C,E)* cannot be colored at the same time because there could be a conflict. In this way the election mechanism needs to reach a 3-neighborhood (nodes $B$, $C$ and $E$) and information about colored edges needs to reach a 2-neighborhood (nodes $B$ and $C$). A node stops when its 1-neighborhood and 2-neighborhood has all its edges colored.

Before describing the details of this algorithm, it is necessary to define some terminology. *winners*: nodes elected to color edges incident to them; *node status*: active or inactive, as the nodes participate or not in the election process; *constraints*: colors already used to color edges or nodes; *sort number*: a random number chosen uniformly in the the interval from 0 to 999 used to determine the winners.

Basically the $Edge^3 - Sched$ functions as follows: all nodes in the network starts in the *active status* exchanging information about their *constraints*, *sort number* and *status*. The *constraints* and status are transmitted to the 2-hop neighborhood and the *sort number* to the 3-hop neighborhood. Once a node becomes a winner, it color its incident edges respecting the constraints it receives. It then changes status, going to the *inactive* status. An *inactive* node becomes only a relay aiding multi-hop communication. The process terminates when all nodes in the 2-hop neighborhood become *inactive*.

### 3.2 $Node^2 - Sched$

Differently from the $Edge^3 - Sched$, the $Node^2 - Sched$ algorithm achieves a distance-2 edge coloring indirectly, via a distance-2 vertex coloring of the communication graph. It is worth reminding that, although $Node^2 - Sched$ performs a vertex coloring, the main goal is to establish a distance-2 edge coloring that will lead to a link scheduling. The Figure 2 shows an example of distance-2 node coloring and its corresponding distance-2 edge coloring. The value within the nodes in the figure represents their color. An edge is colored using the non-ordered pair of the color of the nodes incident to the edge. We note that this indirect edge coloring mechanism obeys the restriction imposed by distance-2 edge coloring, namely, that any two neighbors cannot have edges with the same color.

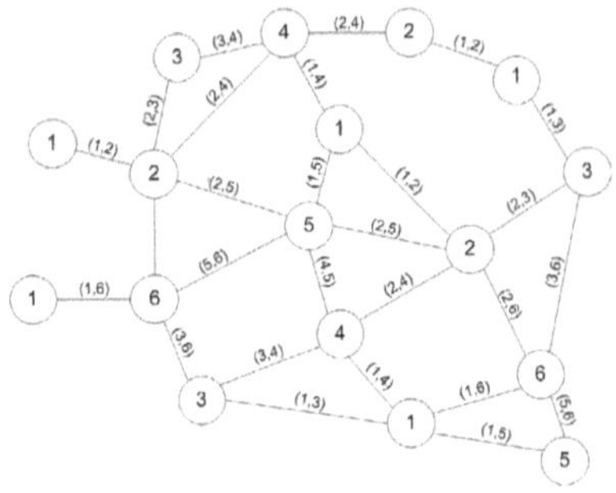

**Fig. 2.** Edge Coloring by Node Coloring

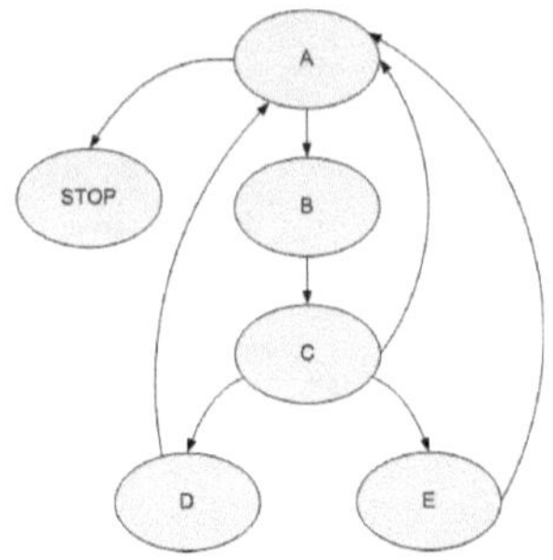


**Fig. 3.** $Node^2 - Sched$ state machine

The detailed state machine of $Node^2 - Sched$ algorithm is depicted in Figure 3 and its functionality is described below. Note that the terminology used in the $Edge^3 - Sched$ is perfectly compatible with the description of this algorithm.

### $Node^2 - Sched$ State Machine

- **state A:** all nodes send to their 1-hop neighbors the randomly chosen *sort number* and their *status*. Nodes receive these messages and store the respective sort numbers. The inactive nodes go to state F and stop. All others go to state B;
- **state B:** all active nodes select the highest sort number among the numbers received and its own. Afterwards the nodes send the maximum sort number to all active 1-hop neighbors. The active nodes receive these messages and compare the values with their own sort number. Nodes that have a sort number greater than any other received become winners. All nodes then go to state C;
- **state C:** all nodes send to their active neighbors their condition: *winner* or not. This procedure aims in establishing a distributed knowledge of nodes' conditions in the communication graph. The *winner* nodes go to state E; the *winner* neighbors go to state D; the other nodes return to state A;
- **state D:** the *winner* neighbors send to their *winners* their own color and the colors of their neighbors. These are the *constraints* that the *winners* will take into consideration when coloring themselves. After this, *winner* neighbors wait for messages containing the new color of nodes. Afterwards, nodes update their *status*. In particular, a node becomes *inactive* if the node has already colored itself and all its neighbors have also colored themselves. Nodes then return to state A;
- **state E:** Nodes wait to receive the *constraints* from their neighbors in *state E* and then color themselves. Afterwards the nodes update their status (nodes in this state become inactive if all their neighbors have been colored), send its own color to their neighbors and return to state A.

### 3.3 Complexity Analysis

We generalize the proof presented in [13] by developing the following complexity analysis. Let $G = (N, E)$ be the communication graph for which $distance - 2$ edge (or vertex) coloring will be performed using one of the algorithms presented. During the execution nodes can color incident edges (or itself) while others wait for a chance to do the same. Therefore, we define nodes that continue executing the algorithm and have incident edges not colored (or itself) as probabilistic nodes, while the nodes that have all its incident edges (or itself) colored as deterministic nodes.

Considering $h$ as the hop distance ($h$-neighborhood) to be considered for a node to become a winner and $n$ the distance to be considered so that a node finishes its execution, we have the following configuration for the two presented algorithms: $Edge^3 - Sched : h = 3, n = 2$ and $Node^2 - Sched : h = 2, n = 1$.

1. Execution
   (a) Every node sends its sort number and constraints to all its neighbors;
   (b) Every node, upon receiving a message re-sends it $(h - 1)$ hops;
   (c) Every node that has superiority on the $h$-neighborhood colors its edges (or itself) and cease to be a probabilistic node, performing only retransmissions;
   (d) Every node stops executing when there is no probabilistic nodes on its $n$-neighborhood.
2. Correctness
   The algorithms are correct if they avoid the hidden terminal problem and if they do not present any deadlock condition. Therefore, we can start the analysis as follows:
   (a) The nodes compete to be a winner in their $h$-neighborhood using the sort number (probabilistic nodes). After an specific node is declared a winner, this node will no longer compete (becoming a deterministic node). Thereby every node will be a winner at some point in the execution of the algorithm. After becoming a deterministic node, a node waits all other nodes in its $n$-neighborhood to become deterministic nodes. This stop condition is sufficient to ensure that the algorithm will never suffer deadlock or starvation;
   (b) Since the algorithm ensures the association of time slots considering the $h$-neighborhood constraints then there remains whether the synchronization mechanism will ensure correct use of the links. As the algorithm ensures that every pair of neighbors in the network has distinct link colors, we have that collisions cannot occur.
3. Notations and Analysis
   Let $N_k \subseteq N$ denote the set of probabilistic nodes after $k$ algorithm steps for $k$ integer and positive. Let $viz_{N_k}(v)$ denote the set of probabilistic neighbors

in the $h$-hop neighborhood of $v$. Let $G_k = (N_k, E_k)$ denote the induced subgraph of $G$ after $k$ steps. Let $d_i^k$ be the sort number generated by the node $v_i$ in the $k$. Nodes that have the largest $d_i^k$ in their $h$-neighborhood will be the winners.

(a) Let $S_i^k$ be the event that corresponds to node $v_i$ being a winner in step k;
(b) Let $S^k$ be the event that corresponds to some node being a winner in step k;
(c) Let $B_j^k(v_i, \alpha)$ be an event that occurs whenever $d_i^k = \alpha > d_j^k$ for some $v_j \in viz_{N_k}(v_i)$, that means that $v_i$ wins some probabilistic node in $h$-neighborhood. It is important to note that the $B_j^k(v_i, \alpha)$ represents independents events for all $v_j \in viz_{N_k}(v_i)$.
(d) Considering that the sort number was generated from a die with $f$ faces, it defines the following probability:

$$Pr(d_i^k = \alpha) = \frac{1}{f} \Rightarrow Pr(B_j^k(v_i, \alpha)) = \frac{\alpha}{f}; \tag{1}$$

(e) Thus we can continue the complexity development. Now consider the probability that node $v_i$ is a winner in step k:

$$Pr(S_i^k) = \sum_{\alpha=0}^{f-1} Pr(d_i^k = \alpha) * Pr(S_i^k / d_i^k = \alpha) \tag{2}$$

$$Pr(S_i^k / d_i^k) = Pr(\bigcap_{v_j \in viz_{N_k}(v_i)} B_j^k(v_i, \alpha)) = \prod_{v_j \in viz_{N_k}(v_i)} (\frac{\alpha}{f}) \geq (\frac{\alpha}{f})^{\Delta_{kh}} \tag{3}$$

where $\Delta_{kh}$ is the number of probabilistic nodes in $h$-neighborhood in step k.

(f) Replacing 3 in 2 and observing that $Pr(d_i^k = \alpha) = \frac{1}{f}$:

$$Pr(S_i^k) \geq \sum_{\alpha=0}^{f-1} (\frac{1}{f})(\frac{\alpha}{f})^{\Delta_{kh}} = (\frac{1}{f^{\Delta_{kh}+1}}) \sum_{\alpha=0}^{f-1} \alpha^{\Delta_{kh}}, \forall v_i \in N_k \tag{4}$$

(g) Using lower limit presented in [14]:

$$\sum_{\alpha=0}^{f-1} \alpha^{\Delta_{kh}} \geq (\frac{(f-1)^{\Delta_{kh}+1}}{\Delta_{kh}+1} + \frac{(f-1)^{\Delta_{kh}}}{2}) \;, \quad \text{where } \Delta_{kh} \geq 1 \tag{5}$$

(h) Finally replacing 5 in 4 :

$$Pr(S_i^k) \geq (\frac{1}{\Delta_{kh}+1})(1 - \frac{1}{f})^{\Delta_{kh}+1} + (\frac{1}{2f})(1 - \frac{1}{f})^{\Delta_{kh}} \tag{6}$$

(i) Considering a complete graph with $n$ nodes as the worst case, and therefore $\Delta_{kh} = n - 1$:

$$Pr(S_i^0) \geq (\frac{1}{n})(1 - \frac{1}{f})^n + (\frac{1}{2f})(1 - \frac{1}{f})^{n-1} = (\frac{2(f-1)+n}{2fn})(1 - \frac{1}{f})^{n-1} \tag{7}$$

$$Pr(S^k) = Pr(\bigcup S_i^k) \geq Pr(\bigcup_{i=1}^{n} S_i^0) = \sum_{i=1}^{n} Pr(S_i^0) \tag{8}$$

$$Pr(S^k) = (\frac{2(f-1)+n}{2f})(1 - \frac{1}{f})^{n-1} \tag{9}$$

(j) We can now determine the average attempts required for a node to be a winner in step k

$$T(n) = (\frac{2fn}{2(f-1)+n})(\frac{f}{f-1})^{n-1} \tag{10}$$

(k) Thus, the worst case convergence complexity is:

$$O(f(\frac{f}{f-1})^{n-1}) \tag{11}$$

## 4 Experimental Setup

We have developed a simulation environment to evaluate the two proposed algorithms. The underlying communication network was implemented using a multi-threaded java simulation, where each sensor node corresponds to the instance of an unique thread. Each thread independently operates the state machine described in Section 3. Threads communicate with other threads in accordance with the underlying communication network, which determines the direct neighbors of a node. We adopt an unicast communication model where all communication is fully reliable.

We consider two types of topologies to evaluate the algorithms: a regular grid and a random node placement. In the grid topology, we consider networks with 25, 49, 100, 200 and 400 nodes, regularly distributed on a square grid, such that radio coverage creates links only between the closest neighbors (maximum degree is 4). In the random node placement, nodes are randomly (uniformly) distributed in dimensionless 200x200 square area. We consider scenarios with 50, 100, 150, 200 and 400 nodes and radio coverage radius of 30. All nodes within distance 30 of a given node correspond to direct neighbors in the communication graph. Each scenario was simulated 100 times and we report sample averages of all metrics.

The following performance metrics were considered: convergence time, number of messages exchanged, number of bits transmitted (i.e., message sizes), and number of colors used by the algorithm. The convergence time is measured in steps, which correspond to transitions in the state machine, and a node is said to have converged once it enters the stop state.

# 5 Evaluation and Results

## 5.1 Grid Topology

Figures 4, 5, 6 and 7 show a direct comparison of the performance of the two algorithms on convergence time, messages exchanged, bits transmitted, and number of colors used, respectively. We first note that $Node^2 - Sched$ converges much faster than $Edge^3 - Sched$, and for larger topologies its at least 4 times faster. Consequently, the number of messages exchanged by $Node^2 - Sched$ is much smaller, reaching a factor of 6 for larger topologies. This also reflects on number of bits transmitted, which for larger topologies is around 10 times smaller. On the other hand, the $Node^2 - Sched$ algorithm uses more colors than its counterpart to color the edges. In particular, $Node^2 - Sched$ uses around 2 times more colors than $Edge^3 - Sched$ for larger topologies. The results clearly indicate a tradeoff between convergence time and message efficiency versus number of colors used when considering the two proposed algorithms.

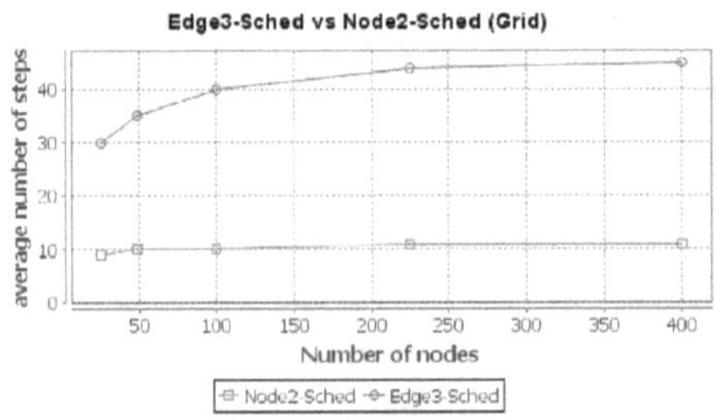


**Fig. 4.** Edge vs Node (steps)

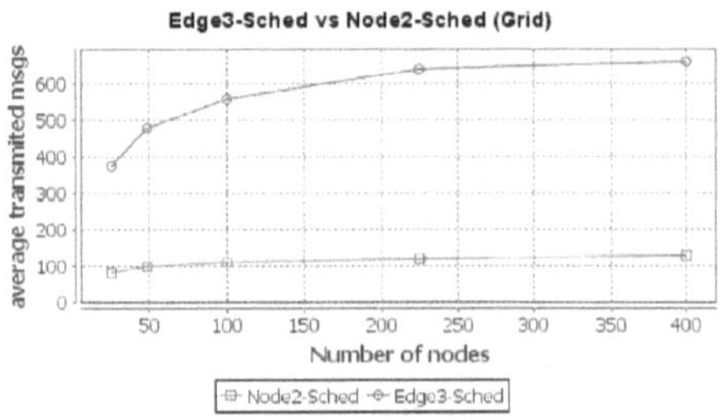


**Fig. 5.** Edge vs Node (messages)

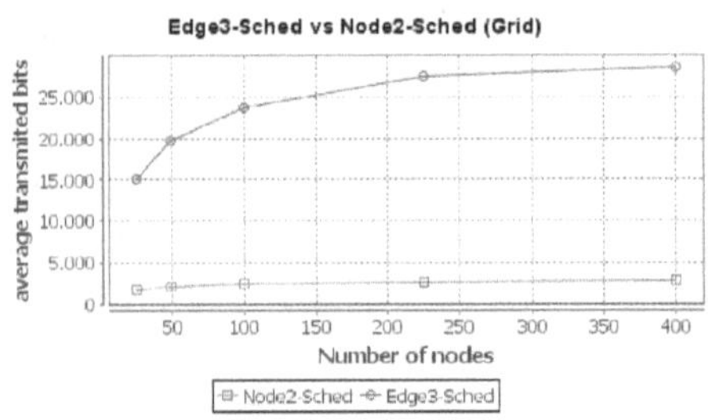


**Fig. 6.** Edge vs Node (bits)

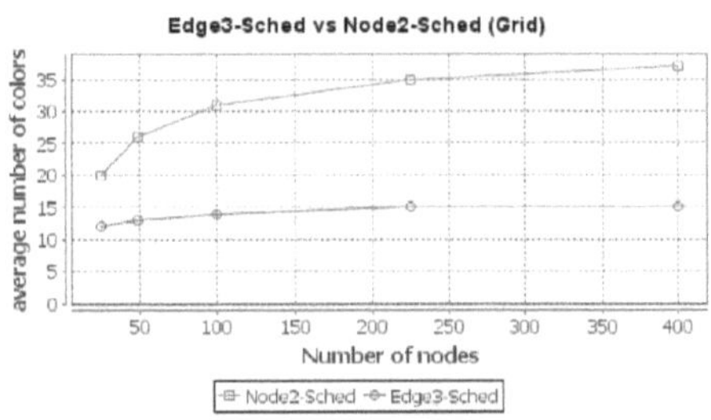


**Fig. 7.** Edge vs Node (colors)

## 5.2 Random Node Placement

We now consider the networks formed by random placement of node, as described in Section 4. Figures 8, 9, 10 and 11 present a direct comparison of the performance of the two algorithms on convergence time, messages exchanged, bits transmitted, and number of colors used, respectively. Similarly to the results obtained in the grid topology, we note that $Node^2 - Sched$ converges much faster

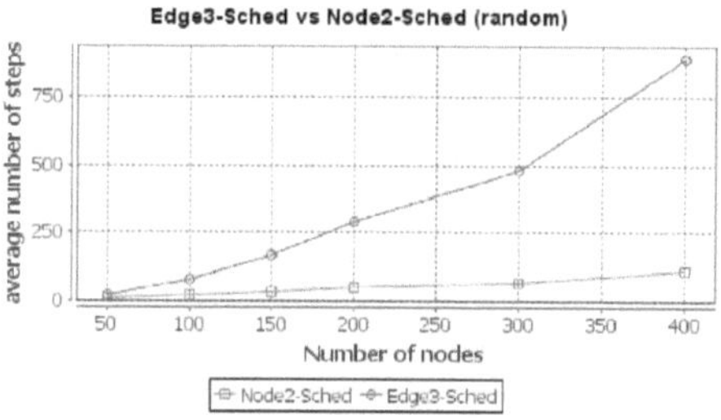


**Fig. 8.** Edge vs Node (steps)

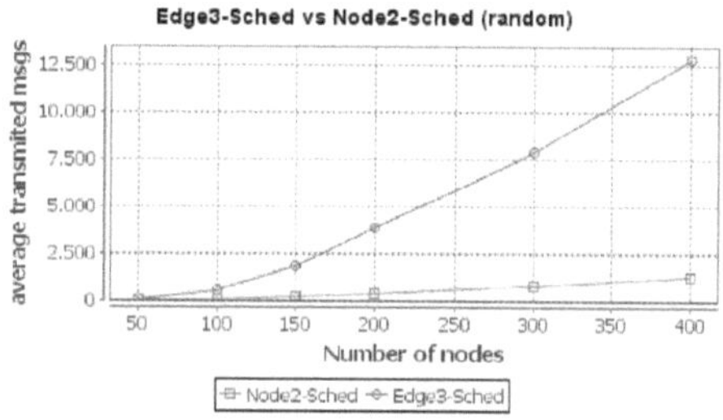


**Fig. 9.** Edge vs Node (messages)

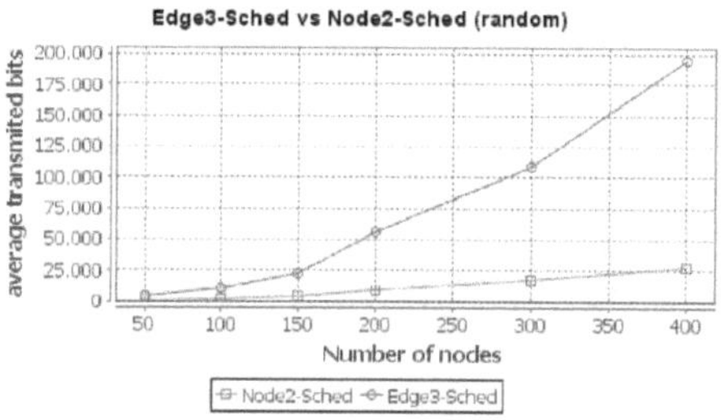


**Fig. 10.** Edge vs Node (bits)

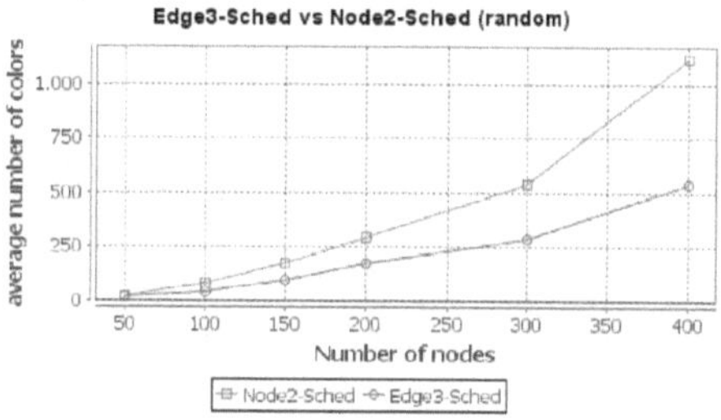


**Fig. 11.** Edge vs Node (colors)

than $Edge^3 - Sched$ and requires many fewer messages and bits transmitted. However, such performance differences have become more pronounced, specially with higher number of nodes. Interestingly, though, the number of colors used by the $Node^2 - Sched$ did not increased when compared to $Edge^3 - Sched$. Again, as with the grid topologies, we observe a trade-off between convergence time and overhead (in messages and bits) versus number of colors used by the algorithms. It is worth mentioning that, except for the number of colors used, $Node^2 - Sched$ has superior performance when compared to the link-scheduling algorithm presented in [6], which corresponds to a very similar scenario.

## 6 Conclusion

We have proposed and evaluated the performance of two distance-2 edge coloring algorithms for WSNs. To the best of our knowledge, these are the first probabilistic, distributed and global ID-free algorithms that can fully yield support to decentralized TDMA-based MAC protocols. Simulations have shown that $Edge^3 - Sched$, a direct edge coloring algorithm, has a higher overhead in terms of convergence time and message exchanged, while using less colors to color the edges. In the other hand, the $Node^2 - Sched$ algorithm has a much smaller overhead while requiring more colors in its coloring. We note that either algorithm could serve as the basis for a TDMA-based MAC protocol, depending on the application requirement. Note that minimizing the number of colors in the coloring may not necessarily lead to the optimal schedule, in terms of energy efficiency, and will depend on the application requirements.

## References


1. Akyildiz, I.F., Su, W., Sankarasubramaniam, Y., Cayirci, E.: A Survey on Sensor Networks. IEEE Communications Magazine 40(8), 102–114 (2002)
2. Angluin, D.: Local and Global Properties in Networks of Processors. In: Proc. of 12th ACM Symposium on Theory of Computing, pp. 82–93 (1980)
3. Reason, J., Rabaey, J.M.: A Study of Energy Consumption and Reliability in a Multi-hop Sensor Network. ACM Mobile Comp. and Comm. Rev. 8(1), 84–97 (2004)
4. Ye, W., Heidemann, J.: Medium Access Control in Wireless Sensor Networks, USC/ISI Tech. Rep. ISI-TR-580 (October 2003)
5. Balakrishnan, et al.: The Distance-2 Matching Problem and its Relationship to the MAC-layer Capacity of Ad hoc Wireless Networks. IEEE J. on Selected Areas in Communications 22(6), 1069–1079 (2004)
6. Gandham, S., Dawande, M., Prakash, R.: Link Scheduling in Sensor Networks: Distributed Edge Coloring Revisited. J. of Par. and Distr. Comp. 68(8), 1122–1134 (2008)
7. Trigoni, N., Yao, Y., Demers, A., Gehrke, J., Rajaraman, R.: WaveScheduling: energy-efficient data dissemination for sensor networks. In: Proc. of the 1st Inter. Work. on Data Management for Sensor Networks, pp. 48–57 (2004)
8. Arumugam, M., Kulkarni, S.: Self-Stabilizing deterministic TDMA for Sensor Networks. In: Chakraborty, G. (ed.) ICDCIT 2005. LNCS, vol. 3816, pp. 69–81. Springer, Heidelberg (2005)
9. Kulkarni, S., Arumugam, M.: SS-TDMA: A self-stabilizing MAC for sensor networks. In: Sensor Network Operations, Wiley-IEEE Press (2006)
10. Ergen, S.C., Varaja, P.: TDMA scheduling algorithms for sensor networks, Tech. Rep., University of California, Berkley (2005)
11. Cui, S., Madan, R., Goldsmith, A., Lall, S.: Joint routing, Mac, and link layer optimization in sensor networks with energy constraints. In: Proc. of IEEE ICC 2005, vol. 2, pp. 725–729 (2005)
12. Pantazis, N.A., Vergados, D.J., Vergados, D.D., Douligeris, C.: Energy Efficiency in Wireless Sensor Networks using Sleep Mode TDMA Scheduling. Ad Hoc Networks 7(2), 322–343 (2009)
13. Arantes Jr., M., França, F.M.G., Martinhon, C.A.: Randomized Generation of Acyclic Orientations upon Anonymous Distributed Systems. J. of Par. and Distr. Comp. 69(3), 239–246 (2009)
14. Spiegel, M.R.: Mathematical Handbook of Formulas and Tables. Makron, MacGraw-Hill (1992)

# A Performance Model for Maintenance Tasks in an Environment of Virtualized Servers
## (Work in Progress)

Tien Van Do[1,*] and Udo R. Krieger[2]

[1] Department of Telecommunications
Budapest University of Technology and Economics
H-1117, Magyar tudósok körútja 2., Budapest, Hungary
do@hit.bme.hu
[2] Faculty Information Systems and Applied Computer Sciences
Otto-Friedrich-Universität, D-96045 Bamberg, Germany
udo.krieger@ieee.org

**Abstract.** This paper introduces the CPP/M/c model with working vacations to describe queueing phenomena that arise in an advanced computing environment of virtualized servers operated by the infrastructure owners. In the proposed queue the inter-arrival times of jobs requesting servers follow a Generalized Exponential distribution. To model a maintenance activity, we assume that a certain number of servers simultaneously goes to a maintenance state for a random period when they complete the service of requests and find no further jobs in the waiting line. We derive an expression for the steady-state probabilities and prove a conditional stochastic decomposition property. By a relatively simple model we are able to prove a property which has a significant impact on the organization of maintenance activities of virtualized servers. It means that instead of migrating virtual servers to expensive physical backup servers during software maintenance, a wise and simple strategy based on the vacation approach can be used. Moreover, it is theoretically proved that the system is not overloaded if we organize the maintenance according to the vacation model. We believe that our model can be useful for administrators to choose an appropriate parameter set for the maintenance activities.

**Keywords:** Virtualized services, performance management, CPP/M/c model, working vacations policy.

## 1 Introduction

At present, virtualization constitutes a main trend in information systems and advanced business engineering. Recent studies have shown that a proportion of 39% among 808 of the largest companies worldwide apply server virtualization

* Corresponding author.

L. Fratta et al. (Eds.): NETWORKING 2009, LNCS 5550, pp. 931–942, 2009.

to achieve new business goals and to provide more efficient services to their customers. Disaster recovery, avoidance of service outage and dynamic load balancing represent some of the most important areas for the application of the rapidly evolving virtualization concepts. Compared to existing service technologies 25% of the cost or even more can be saved by these means.

In this context, virtualization means either to let a federation of servers appear as multiple computing entities or to let many computing entities appear as a single computer. The latter is commonly called server aggregation or grid computing. It is identified by an IDC research report (http://www.idc.com) that virtualization of system resources in severs with an x86 compatible instruction set is one disruptive technology. In the near future it may initiate a paradigm shift in IT industry providing new powerful services like enhanced server hosting.

As indicated by various studies it is the rationale behind this trend that virtualization can reduce the infrastructure and IT management cost. The reason is that it substantially improves the utilization of the physical infrastructure, i.e. servers, storage systems and network components, while it can provide the same safety and performance compared to a solution where each ASP obtains a separate physical machine/server from the owner of the infrastructure. It is another advantage that the infrastructure can provide in a flexible manner different service packages concerning specific operating systems running on top of the same hardware.

From a practical perspective, it is observed that virtualization is a well founded area. However, there are no theoretical investigations which consider contention problems arising in the virtualized environment of a server farm. To model the interaction between application service providers and an infrastructure provider this paper studies the CPP/M/c queue with a compound Poissonian arrival process (CPP) and working vacations.

Such vacation queues have been an intensively studied research topic of queueing theory, cf. [3,5,7,8,9]. However, most of those studies assume a Poissonian arrival model [3,5,7,9] or a model of single arrivals [8]. Regarding the performance evaluation of practical systems, this assumption limits the application of vacation queues.

Recently, queues with working vacations have obtained a big attention, see, e.g., the work of Servi et al. [7] and Liu et al. [5]. It is motivated by the performance evaluation of Wavelength Division Multiplexing (WDM) in optical systems. In this respect, the multi-server queue introduced here is indeed a generalization of the M/M/c system with synchronous vacations [9] regarding two different aspects, namely, the Poissonian batch arrivals and working vacations.

The rest of the paper is organized as follows. In Section 2, we first provide a description of the CPP/M/c model with working vacations (WV), develop then a matrix-analytic solution approach and prove some interesting property of this CPP/M/c WV-queue. Then some illustrative numerical results are presented in Section 3. Finally, we summarize our findings in the conclusions.

# 2 Analyzing the Maintenance Performance by a Versatile Queueing System

## 2.1 Description of the Maintenance Model

In a virtualization environment three different roles can be identified (see Figure 1):

- users/applications,
- application service providers and
- owners of the hardware infrastructure.

Applications and related services, e.g. Web servers with Web, information retrieval and business services, are provided by application service providers that require virtual machines from an infrastructure owner to run their virtualized application servers.

In this environment two interrelated categories of Service Level Agreement (SLA) can be defined:

- an SLA between users and application service providers specifying the service requirements, e.g. the response time and availability of a service, etc,
- an SLA between application service providers and an infrastructure owner.

The SLA between users and application service providers are complicated and they also depend on the nature of the hosted applications.

To operate the infrastructure efficiently, it is recognized that advanced management tools are needed. In this respect, system management activities should also include the tasks of managing both virtual servers and physical resources efficiently.

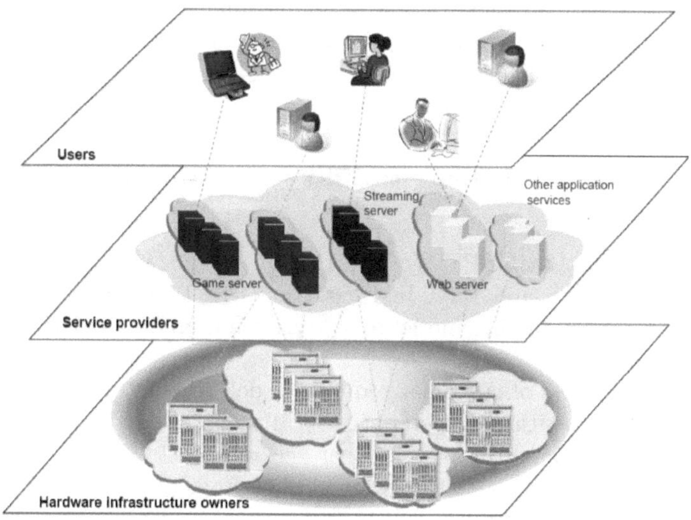


**Fig. 1.** Utility Computing Environment Based On Virtual Machines

In this paper, we consider the interaction between application service providers and an infrastructure owner. We suppose that there are $c$ virtual servers available in the server pool of the infrastructure owner. To realize a pay-as-you-go approach, application service providers can initiate requests for servers to the provider of the infrastructure and server releases after task completion.

We assume that server requests arrive in batches following the Compound Poisson Process (CPP) (cf. [4]). This means that the inter-arrival times follow a Generalized Exponential (GE) distribution. The arrival process is motivated by the fact that GE is the only distribution of least bias [4] if only the mean and variance of inter-arrival times can be reliably computed by the available measurement data. This situation typically arises in virtualized computing environments exploiting the capabilities of monitoring systems. It has been shown by recent studies [1,2] that the CPP is sufficiently accurate to model Internet traffic in a Web server environment (i.e. the relevant CPP parameters have been estimated by the captured Internet traffic) and that it can be applied to the performance evaluation of wireless telecommunication systems.

To create a reliable computing system with these $c$ servers, the provider of the infrastructure can initiate specific maintenance actions, e.g. software updates, a virtual server live migration etc., when any $d$ servers become idle after a service completion instant. This kind of maintenance activities are modeled in such a way that $d$ servers take a simultaneous vacation. During such a vacation period, the residual $c-d$ servers do not take a vacation even if they are idle. To ensure the mathematical tractability of the model, we assume that the durations of the vacation periods are independent, identical exponentially distributed random variables with parameter $\theta$. The service rate of each server which is not in a vacation state is given by an independent exponential distribution with parameter $\mu$. A server on vacation can serve customers following an independent exponential distribution with rate $\mu_v$. Note that an application service provider who receives the allocation of a server which is on vacation may pay less as a form of compensation.

### 2.2 Analysis of an Advanced Multi-server Model with Working Vacations

Here, we consider the CPP/M/c multi-server queue with working vacations, infinite waiting room and First In First Out (FIFO) service principle that we have derived as performance model to analyze maintenance tasks in a virtualized server environment.

The arrival process of customer requests is determined by a Compound Poisson Process (CPP) with parameters $(\lambda, \omega)$. It means that the probability distribution function of the inter-arrival times $\tau$ is defined by $\mathbb{P}\{\tau = 0\} = \omega \in (0,1)$ and $\mathbb{P}\{0 < \tau < t\} = (1-\omega)(1-e^{-\lambda t})$. Therefore, the arrival process can be seen as a batch Poisson process whose batches of the random size $S$ arriving at some epoch follow a geometric distribution $\mathbb{P}\{S = s\} = (1-\omega)\omega^{s-1}, s \geq 1$, with mean $\mathbb{E}(S) = 1/(1-\omega)$ and variance $\mathrm{Var}(S) = \omega/(1-\omega)^2$.

The requests are served by $c$ servers following a specific working-vacations policy with independent, identical, exponentially distributed service and vacation times with rates $\mu, \mu_v, \theta$, respectively. Let us suppose that there are no servers on vacation due to maintenance activities. Then a simultaneous vacation period of $d$ servers starts if there are $d$ idle servers after a service completion. At the end of a simultaneous vacation period of these $d$ servers, three alternatives are possible:

– if there are no waiting customers, the $d$ servers stay idle and are ready to serve any arriving new customers;
– if there are $c-d < j < c$, customers in the system, $j-c+d$ returning servers immediately start serving these customers and the other $c-j$ returning servers become idle;
– if there are $j \geq c$ customers in the system, the $d$ returning servers all start serving these customers immediately.

At any time $t$ the state of the system $Y(t) = (I(t), J(t))$ can be completely specified by two integer-valued random variables:

– $I(t) = \begin{cases} 0 & \text{if } d \text{ servers are on vacation at time } t \\ 1 & \text{if there are no servers on vacation at time } t \end{cases}$
– $J(t)$ represents the number of customers in the system at time $t$ including any in service or the waiting room.

The system is now modeled by a continuous-time discrete state Markov process $Y = \{I(t), J(t)\}$ on a rectangular lattice strip $S = \{0,1\} \times \mathbb{N}_0$ due to our Markovian assumptions. We denote its corresponding steady-state probabilities by $\pi = \{\pi_{i,j}\}_{(i,j)\in S}$, where $\pi_{i,j} = \lim_{t\to\infty} P\{I(t) = i, J(t) = j\}$, and let $\mathbf{v}_j = (\pi_{0,j}, \pi_{1,j})$ be the partitioned vector of state probabilities.

The one-step transitions of the Markov chain $Y$ have a specific tridiagonal block structure since the possible transitions are driven by following events:

(a) changing the status of $I(t)$, i.e. from the vacation to non-vacation of servers. Then $A_j(i,k)$ denotes the corresponding transition rate from state $(i,j)$ to state $(k,j)$, $i,k \in \{0,1\}, j \geq 0$. Let

$$A = A_j = \begin{bmatrix} 0 & \theta \\ 0 & 0 \end{bmatrix}, \ \forall j \geq 0; \quad \text{and} \quad A^* = \begin{bmatrix} -\theta & \theta \\ 0 & 0 \end{bmatrix}.$$

(b) the arrivals of customers. Then $B_{i,j,s}$ is the rate of the $s$–step upward transition from state $(i,j)$ to state $(i,j+s)$, $i \in \{0,1\}, j \geq 0$, caused by a batch arrival of size $s$ and

$$B_{i,j,s} = (1-\omega)\omega^{s-1}\lambda, \quad j \geq 0, i \in \{0,1\}, s \geq 1.$$

(c) the departures of customers. $C_j(i,k)$ is the transition rate from state $(i,j)$ to state $(k,j-1)$; $i,k \in \{0,1\}, j \geq 0$. Then we get:

$$C_j = \begin{cases} \begin{bmatrix} j\mu & 0 \\ 0 & j\mu \end{bmatrix}, & 1 \le j \le c-d \\ \begin{bmatrix} (c-d)\mu + \mu_v & 0 \\ (c-d+1)\mu & 0 \end{bmatrix}, & j = c-d+1 \\ \begin{bmatrix} (c-d)\mu + (j-c+d)\mu_v & 0 \\ 0 & j\mu \end{bmatrix}, & c-d+1 < j \le c \\ \begin{bmatrix} (c-d)\mu + d\mu_v & 0 \\ 0 & c\mu \end{bmatrix} = C, & j > c. \end{cases}$$

Note that by a transition from $(1, c-d+1)$ to $(0, c-d)$ after a service completion with rate $(c-d+1)\mu$ we get a simultaneous vacation of $d$ servers.

Let $\mathrm{Diag}(x)$ denote the diagonal matrix defined by a row vector $x$ and $E \in \mathbb{R}^{2\times 2}$ be the identity matrix. We introduce the following notations

$$\begin{aligned} \Lambda &= \mathrm{Diag}[\lambda, \lambda] = \lambda E; \quad \Omega = \mathrm{Diag}[\omega, \omega] = \omega E; \\ B_s &= B_{j,s} = \mathrm{Diag}[(1-\omega)\omega^{s-1}\lambda, (1-\omega)\omega^{s-1}\lambda], \quad j \ge 0, \end{aligned}$$

and obtain

$$\begin{aligned} B_s &= \Omega^{s-1}(E-\Omega)\Lambda = \omega^{s-1}(1-\omega)\lambda E, \quad j \ge 1, \\ \Lambda &= \sum_{s=1}^{\infty} B_s = \lambda E. \end{aligned}$$

**Lemma 1.** *The necessary and sufficient condition for the existence of the steady-state probabilities of the process $Y = (I, J)$ is determined by*

$$\frac{\lambda}{c\mu} + \omega < 1 \quad \Leftrightarrow \quad \rho = \frac{\lambda}{(1-\omega)\cdot c\mu} < 1 \tag{1}$$

*Remark 1.* The standard condition (1) states that the traffic intensity $\rho$ must be less than one to achieve the ergodicity of $Y$. Neither the rate $\theta$ of the vacations period nor the number $d$ of simultaneous servers on vacations have an impact on the stability of the system. In other words, the system will not be overloaded due to a maintenance activity. Indeed, this is good news for a system administrator who shall organize the maintenance tasks of idle virtual machines.

*Proof.* The steady-state balance equation of the M/G/1-like upper Hessenberg system can be written as follows:

$$\sum_{s=1}^{j} \mathbf{v}_{j-s} B_s + \mathbf{v}_j \left[A^* - \Lambda - D^{C_j}\right] + \mathbf{v}_{j+1} C_{j+1} = 0, \qquad \forall j \ge 1. \tag{2}$$

Here $D^{C_j}$ are diagonal matrices whose diagonal elements are the sum of the elements in the rows of $C_j$. Note that by construction $D^{C_j} = C_j$ holds for all $j \ne c-d+1$.

For $j \geq c+1$ we can write

$$\sum_{s=1}^{j} \mathbf{v}_{j-s} B_s + \mathbf{v}_j \left[A^* - \Lambda - C\right] + \mathbf{v}_{j+1} C = 0. \tag{3}$$

Substituting $B_s = \Omega^{s-1}(E - \Omega)\Lambda$ into this equation (3), we get

$$\sum_{s=1}^{j} \mathbf{v}_{j-s} \Omega^{s-1}(E - \Omega)\Lambda + \mathbf{v}_j \left[A^* - \Lambda - C\right] + \mathbf{v}_{j+1} C = 0 \quad \forall j \geq c+1, \tag{4}$$

and

$$\sum_{s=1}^{j-1} \mathbf{v}_{j-1-s} \Omega^{s-1}(E - \Omega)\Lambda + \mathbf{v}_{j-1} \left[A^* - \Lambda - C\right] + \mathbf{v}_j C = 0 \quad \forall j \geq c+2. \tag{5}$$

If we multiply equation (5) by $\Omega$ and then subtract the result from equation (4), we obtain the three-term recurrence equations

$$\mathbf{v}_{j-1}[\Lambda - A^*\Omega + C\Omega] + \mathbf{v}_j \left[A^* - \Lambda - C - C\Omega\right] + \mathbf{v}_{j+1} C = 0\ ,\ j \geq c+2,$$

$$\mathbf{v}_{j-1} Q_0 + \mathbf{v}_j Q_1 + \mathbf{v}_{j+1} Q_2 = 0\ ,\ j \geq c+2, \tag{6}$$

where $Q_0 = \Lambda - A^*\Omega + C\Omega,\ Q_1 = A^* - \Lambda - C - C\Omega,\ \ Q_2 = C.$

$Q(x) = Q_0 + Q_1 x + Q_2 x^2$ is defined as the characteristic matrix polynomial associated with the equations (6). It is proved in [6] that the solution of these matrix equations (6) is closely related to the eigenvalues and left-eigenvectors of the polynomial $Q(x)$. If $(x, \psi)$ is an eigenvalue-eigenvector pair of $Q(x)$, then it holds

$$\psi Q(x) = 0,\ \ \det[Q(x)] = 0.$$

Consequently, we obtain:

$$\det[Q(x)] = \det \begin{bmatrix} q_{00}(x) & \theta x - \theta\omega \\ 0 & q_{11}(x) \end{bmatrix} = q_{00}(x) q_{11}(x)$$

$$q_{00}(x) = \lambda + ((c-d)\mu + d\mu_v)\omega + \omega\theta - (\lambda + (c-d)\mu + d\mu_v + ((c-d)\mu + d\mu_v)\omega + \theta)x + ((c-d)\mu + d\mu_v)x^2$$

$$q_{11}(x) = \lambda + c\mu\omega - (\lambda + c\mu + c\mu\omega)x + c\mu x^2 = (1-x)(\lambda + c\mu\omega - c\mu x)$$

Therefore, $Q(x)$ has four eigenvalues

$$x_1 = \frac{1}{2G}\{H + G - \sqrt{(H+G)^2 - 4G(\lambda + ((c-d)\mu + d\mu_v)\omega + \omega\theta)}\}$$

$$x_2 = \frac{1}{2G}\{(H + G + \sqrt{(H+G)^2 - 4G(\lambda + ((c-d)\mu + d\mu_v)\omega + \omega\theta)}\}$$

$$x_3 = \lambda/(c\mu) + \omega,\ \ x_4 = 1,$$

where

$$G = (c-d)\mu + d\mu_v$$
$$H = \lambda + ((c-d)\mu + d\mu_v)\omega + \theta$$

holds.

Note that $\psi_1 = (1, (\theta\omega - \theta x_1)/q_{11}(x_1))$ is the left-hand-side (LHS) eigenvector of $Q(x)$ for the eigenvalue $x_1$, and $\psi_3 = \psi_4 = (0,1)$ are the LHS eigenvectors of $Q(x)$ for the eigenvalues $x_3$ and $x_4$, respectively.

Since $\omega < 1$ holds, we have

$$(\lambda + ((c-d)\mu + d\mu_v)\omega + \omega\theta) < H,$$
$$4G(\lambda + ((c-d)\mu + d\mu_v)\omega + \omega\theta) < 4GH,$$
$$(H+G)^2 - 4G(\lambda + ((c-d)\mu + d\mu_v)\omega + \omega\theta) > (H+G)^2 - 4GH,$$
$$0 < x_1 < \frac{1}{2G}(H + G - |H-G|) \leq 1,$$
$$x_2 > \frac{1}{2G}(H + G + |H-G|) \geq 1.$$

Applying results from [6], it is a necessary and sufficient condition for the ergodicity of the Markov chain $Y$ that the number of eigenvalues of $Q(x)$ inside the unit disk is given by 2. Therefore, $x_3 < 1$ is required which yields condition (1). □

The steady-state balance equations of $J(t) \in \{0, \ldots, c+1\}$ can be written in the following form:

$$\mathbf{v}_0 \left[A^* - \Lambda\right] + \mathbf{v}_1 C_1 = 0$$

$$\sum_{s=1}^{j} \mathbf{v}_{j-s} B_s + \mathbf{v}_j \left[A^* - \Lambda - D^{C_j}\right] + \mathbf{v}_{j+1} C_{j+1} = 0\,, \quad 1 \leq j \leq c+1,$$

For $j \geq c+1$ the steady-state probabilities can be expressed as follows (cf. [6]):

$$\mathbf{v}_j = \alpha\psi_1 x_1^j + \beta\psi_3 x_3^j$$
$$\pi_{0,j} = \alpha x_1^j$$
$$\pi_{1,j} = \alpha \frac{\theta\omega - \theta x_1}{q_{11}(x_1)} x_1^j + \beta x_3^j \tag{7}$$

where $\alpha$ and $\beta$ are coefficients that have to be determined by the boundary conditions.

Furthermore, we have to satisfy the normalization equation:

$$\sum_{j=0}^{\infty} \sum_{i=0}^{1} \pi_{i,j} = 1. \tag{8}$$

Consequently, we have to determine the vectors $\mathbf{v}_j$, $0 \leq j \leq c$, $\alpha$ and $\beta$. The total number of these unknowns is given by $2(c+1) + 2 = 2(c+2)$. To determine these unknowns, we have the steady-state balance equations of the levels $j = 0, \ldots, c+1$ and the normalization equation. Thus, $2(c+2)+1$ is the number of boundary equations, among those only $2(c+2)$ equations are independent.

It can be observed from the steady-state balance equations of $J(t) \in \{0, \ldots, c\}$ that $\mathbf{v}_j$, $1 \leq j \leq c$ and $j \neq c-d+1$, can be expressed as a function of $\mathbf{v}_0$, i.e $\pi_{0,0}$ and $\pi_{1,0}$, and $\mathbf{v}_{c-d+1}$. Therefore, we have only six unknowns $(\pi_{0,0}, \pi_{1,0}, \pi_{0,c-d+1}, \pi_{1,c-d+1}, \alpha, \beta)$, which can be solved efficiently using the steady-state balance equations of the states $J(t) = c$, $J(t) = c-d$, $J(t) = c+1$ and the normalization equation.

### 2.3 Conditional Stochastic Decomposition

In the following, we prove a conditional stochastic decomposition property for the CPP/M/c queue with working vacations.

**Lemma 2.** *If the ergodicity condition for the CPP/M/c queue with working vacations holds, then the conditional steady-state queue length $J_b = lim_{t\to\infty}\{J(t) - c - 1 | J(t) > c, I(t) = 1\}$ provided that the server system is not on a working vacation can be decomposed into the sum of two independent random variables*

$$J_b = J_0 + J_c.$$

*Here $J_0$ is the conditional steady-state queue length of the CPP/M/c queue without vacations and $J_c$ is the additional steady-state queue length due to vacations.*

*Proof.* The probability that the server is busy and the number of jobs is larger than $c$ is determined by:

$$P_b = P\{J(t) > c, I(t) = 1\} = \sum_{j=c+1}^{\infty} \pi_{1,j} = \sum_{j=c+1}^{\infty} \left(\alpha \frac{\theta\omega - \theta x_1}{q_{11}(x_1)} x_1^j + \beta x_3^j\right)$$
$$= \alpha \frac{\theta\omega - \theta x_1}{q_{11}(x_1)} \frac{x_1^{c+1}}{1-x_1} + \beta \frac{x_3^{c+1}}{1-x_3}$$

The probability generating function of $J_b$ can be expressed as follows:

$$G_{J_b}(z) = \sum_{j=0}^{\infty} P\{J_b = j\} z^j = \sum_{j=0}^{\infty} \frac{\pi_{1,j+c+1}}{P_b} z^j$$
$$= \frac{1}{P_b} \sum_{j=0}^{\infty} \left(\alpha \frac{\theta\omega - \theta x_1}{q_{11}(x_1)} x_1^{j+c+1} + b x_3^{j+c+1}\right) z^j$$
$$= \frac{1}{P_b} \left(\alpha \frac{\theta\omega - \theta x_1}{q_{11}(x_1)} x_1^{c+1}/(1 - x_1 z) + \beta x_3^{c+1}/(1 - x_3 z)\right)$$

The steady-state probabilities of the CPP/M/c queue without vacations can be obtained by setting $\theta = 0$, $d = 0$ and $\mu_v = \mu$. The probability that the number of customers in the CPP/M/c queue without vacations is given by $\pi_j = \beta^* x_3^j$ for $j \geq c+1$, where $\beta^*$ is an appropriate coefficient. Therefore, $G_{J_0}(z) = \beta^* \frac{x_3^{c+1}}{1-x_3 z}$ follows for the probability generating function of $J_0$. These relations yield the stated result. □

## 3 Illustrative Numerical Results

In this section we present some numerical results to illustrate the impact of the model parameters on the formulation of an effective maintenance policy, i.e. how many servers should be simultaneously on vacations. For demonstration purposes, we investigate the average number of customer requests waiting for free servers

$$\mathbb{E}(L_Q) = \sum_{j=c+1}^{\infty} (j-c) \cdot (\pi_{0,j} + \pi_{1,j}) = \sum_{j=c+1}^{\infty} (j-c) \cdot \left( \alpha[1 + \frac{\theta\omega - \theta x_1}{q_{11}(x_1)}] x_1^j + \beta x_3^j \right)$$
$$= \frac{\alpha[1 + \frac{\theta\omega - \theta x_1}{q_{11}(x_1)}] x_1^{(c+1)}}{(1-x_1)^2} + \frac{\beta x_3^{(c+1)}}{(1-x_3)^2}$$

as major performance metrics and select some illustrative parameter set. Other characteristics like the mean number of requests in the system

$$\mathbb{E}(L) = \sum_{j=1}^{\infty} j \cdot (\pi_{0,j} + \pi_{1,j}),$$

or the mean number of active servers $\mathbb{E}(N_V) = \sum_{j=1}^{\infty} \min(j,c) \cdot (\pi_{0,j} + \pi_{1,j})$ and the throughput $\eta = \sum_{j=1}^{\infty} \mathbf{v}_j \cdot C_j \cdot \begin{pmatrix} 1 \\ 1 \end{pmatrix} = \sum_{j=1}^{\infty} \sum_{i=0}^{1} \pi_{i,j} \cdot (C_j(i,0) + C_j(i,1))$ can be computed in a similar manner.

In Figure 2(a) we plot the average number of waiting requests $\mathbb{E}(L_Q)$ versus $d$ for the following parameter set of a high load regime: $c = 100$ servers, $\omega = 0.2$, $\theta = 1.0$, $\mu = 5.0$, $\mu_v = 2.5$, $\rho = \lambda/[(1-\omega)c\mu] \in [0.7, 0.9]$. It generates batch arrivals of mean size $\mathbb{E}(S) = 1.25$ and variance $\mathrm{Var}(S) = 0.3125$ for a high traffic intensity $0.7 \leq \rho \leq 0.9$ and assumes that the average service time $1/\mu$ of requests needs only 20 % of the mean maintenance time $1/\theta$ while during these maintenance periods the latter service time is extended by 100 % compared to the normal operation mode.

Considering the average number $\mathbb{E}(L_Q)$ of requests waiting in the system, it is observed that increasing the load $\rho$ from 0.7 to 0.8 or from 0.8 to 0.9 generates an increment of one order of magnitude. To show the impact of the size $S$ of arriving batches and the influence of $\omega = 1 - 1/\mathbb{E}(S)$ and $\mathbb{E}(S) \in \{1.25, 1.67, 2, 5, 10\}$, respectively, we use the set of the same parameters but fix the load at $\rho = 0.8$. Figure 2(b) illustrates the average number of waiting requests $\mathbb{E}(L_Q)$ versus $d$ and $\omega$. In Figure 3 $\mathbb{E}(L_Q)$ is plotted against $d$ and $\mu_v$ for the load $\rho = 0.9$ and a mean batch size of $\mathbb{E}(S) = 1/(1-\omega) = 1.25$.

It is observed that batch arrivals have the strongest impact on the average number of waiting customers. The impact of the offered load $\rho$ and the service rate $\mu_v$ during maintenance can be handled by choosing an appropriate number $d$ of servers under maintenance.

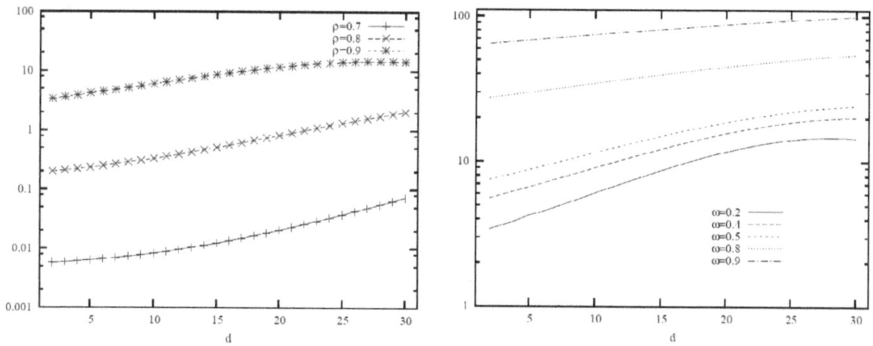


(a) (b)

**Fig. 2.** Average number $\mathbb{E}(L_Q)$ of waiting requests versus $d$ for different traffic load $\rho$ (left) and different control parameter $\omega = 1 - 1/\mathbb{E}(S)$ of the mean batch size $\mathbb{E}(S)$ (right)

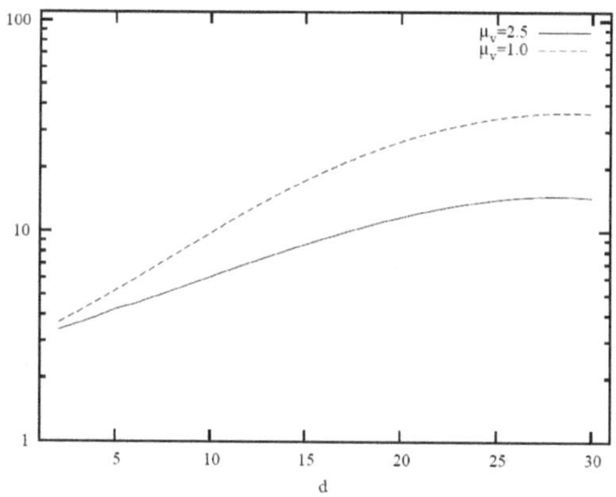


**Fig. 3.** Average number $\mathbb{E}(L_Q)$ of waiting requests versus $d$ and $\mu_v$

## 4 Conclusions

To model the queueing and congestion phenomena arising from maintenance tasks of a virtualized server environment, we have presented in this study a CPP/M/c multi-server system with Poissonian batch arrivals and working vacations.

In the proposed queueing system the inter-arrival times of jobs requesting service by a virtualized server follow a Generalized Exponential distribution. To model the maintenance activities, we have assumed that a certain number of servers goes simultaneously to a maintenance state for a random period when they have completed the service of jobs and find no further requests in the waiting line.

Analyzing the arising Markovian model by matrix-analytic methods, we have derived a new expression for the steady-state probabilities and proved a conditional stochastic decomposition property. The validation of the approach in a testbed and the estimation of the parameters by gathered data is a topic of future research.

In conclusion, we believe that the proposed Markovian multi-server system with working vacations can serve as a useful tool to define efficient maintenance policies in the virtualized environment of current server farms.

## References

1. Do, T.V., Chakka, R., Harrison, P.G.: An integrated analytical model for computation and comparison of the throughputs of the UMTS/HSDPA user equipment categories. In: MSWiM 2007 Proceedings of the 10th ACM Symposium on Modeling, Analysis, and Simulation of Wireless and Mobile Systems, pp. 45–51. ACM, New York (2007)
2. Do, T.V., Krieger, U.R., Chakka, R.: Performance modeling of an Apache Web server with a dynamic pool of service processes. Telecommunication Systems 39(2), 117–129 (2008)
3. Doshi, B.T.: Queueing systems with vacations – a survey. Queueing Syst. Theory Appl. 1(1), 29–66 (1986)
4. Kouvatsos, D.: Entropy maximisation and queueing network models. Annals of Operations Research 48, 63–126 (1994)
5. Liu, W., Xu, X., Tian, N.: Stochastic decompositions in the M/M/1 queue with working vacations. Oper. Res. Lett. 35, 595–600 (2007)
6. Mitrani, I., Chakka, R.: Spectral expansion solution for a class of Markov models: Application and comparison with the matrix-geometric method. Performance Evaluation 23, 241–260 (1995)
7. Servi, L.D., Finn, S.G.: M/M/1 queues with working vacations (M/M/1/WV). Perform. Eval. 50(1-4), 41–52 (2002)
8. Tian, N., Zhang, Z.G.: Stationary Distributions of GI/M/c Queue with PH Type Vacations. Queueing Syst. Theory Appl. 44(2), 183–202 (2003)
9. Zhang, Z.G., Tian, N.: Analysis on queueing systems with synchronous vacations of partial servers. Perform. Eval. 52(4), 269–282 (2003)

# Towards Automated Secure Web Service Execution
## (Work in Progress)

Béla Genge and Piroska Haller

"Petru Maior" University of Târgu Mureş, Department of Electrical Engineering,
N. Iorga Str., No. 1, (540088) Târgu Mureş, Romania
{bgenge,phaller}@engineering.upm.ro
http://www.upm.ro

**Abstract.** Existing solutions for authentication and authorization in Web services make use of technologies such as SAML or WS-Security. These provide a static solution by using a set of predefined protocols. We propose a semantic security protocol model from which security protocol specifications are generated and automatically executed by participants. The proposed model consists of a sequential component, implemented as a WSDL-S specification, and an ontology component, implemented as an OWL specification. The correctness of the proposed model is ensured by using a set of rules and algorithms for generating it based on a protocol model given by the user. We validate our approach by generating and implementing several specifications for existing protocols such as ISO9798 or Kerberos protocols.

**Keywords:** Security protocols, automated execution, ontology, Web services.

## 1 Introduction

Security protocols are "communication protocols dedicated to achieving security goals" (C.J.F. Cremers and S. Mauw) [1] such as confidentiality, integrity or availability.

Existing technologies, such as the Security Assertions Markup Language [16] (i.e. SAML) or WS-Security [17] provide a unifying solution for the authentication and authorization issues through the use of predefined protocols. By implementing these protocols, Web services authenticate users and provide authorized access to resources. However these approaches are rather static, meaning that in case of new security protocols, services must be reprogrammed.

In this paper we propose a semantic security protocol model (SSPM) for generating security protocol specifications that can be automatically executed by participants. The SSPM has two components: a sequential model and an ontology model. The first component is implemented as a WSDL-S [4] specification while the second component is implemented as an OWL [12] specification. The role of the WSDL-S implementation is to describe the message sequences and

L. Fratta et al. (Eds.): NETWORKING 2009, LNCS 5550, pp. 943–954, 2009.

directions that must be executed by protocol participants. The role of the OWL implementation is to provide semantic information such as the construction, processing and implementation of cryptographic operations (e.g. encryption algorithm, encryption mode, key).

The SSPM is constructed from a given SPM and it must maintain the protocol's security properties. For this we propose several generating rules and algorithms that provide a mapping for each component from SPM to SSPM. The correctness of the proposed rules and algorithms results from the one-to-one mapping of each component and from the correctness of SPM.

The paper is structured as follows. In section 2 we present the basic protocol model. In section 3 we present the semantic model with generating rules and algorithms. Example usage and experimental results are presented in section 4. We relate our work to others in section 5. We end with a conclusion and future work in section 6.

## 2 Protocol Model

Protocol participants communicate by exchanging *terms* constructed from elements belonging to the following basic sets: $\mathsf{P}$, denoting the set of role names; $\mathsf{N}$, denoting the set of random numbers or *nonces* (i.e. "number once used"); $\mathsf{K}$, denoting the set of cryptographic keys; $\mathsf{C}$, denoting the set of certificates and $\mathsf{M}$, denoting the set of user-defined message components.

In order for the protocol model to capture the message component types found in security protocol implementations [16], [17] we specialize the basic sets with the following subsets:

- $\mathsf{P}_{DN} \subseteq \mathsf{P}$, denoting the set of distinguished names; $\mathsf{P}_{UD} \subseteq \mathsf{P}$, denoting the set of user-domain names; $\mathsf{P}_{IP} \subseteq \mathsf{P}$, denoting the set of user-ip names; $\mathsf{P}_U = \{\mathsf{P} \setminus \{\mathsf{P}_{DN} \cup \mathsf{P}_{UD} \cup \mathsf{P}_{IP}\}\}$, denoting the set of names that do not belong to the previous subsets;
- $\mathsf{N}_T$, denoting the set of timestamps; $\mathsf{N}_{DH}$, denoting the set of random numbers specific to the Diffie-Hellman key exchange; $\mathsf{N}_A = \{\mathsf{N} \setminus \{\mathsf{N}_{DH} \cup \mathsf{N}_T\}\}$, denoting the set of random numbers;
- $\mathsf{K}_S \subseteq \mathsf{K}$, denoting the set of symmetric keys; $\mathsf{K}_{DH} \subseteq \mathsf{K}$, denoting the set of keys generated from a Diffie-Hellman key exchange; $\mathsf{K}_{PUB} \subseteq \mathsf{K}$, denoting the set of public keys; $\mathsf{K}_{PRV} \subseteq \mathsf{K}$, denoting the set of private keys;

To denote the encryption type used to create cryptographic terms, we define the following *function names*:

$$
\begin{aligned}
FuncName ::= &\ sk && (symmetric\, function)\\
\mid &\ pk && (asymmetric\, function)\\
\mid &\ h && (hash\, function)\\
\mid &\ hmac && (keyed\, hash\, function)
\end{aligned}
$$

The encryption and decryption process makes use of cryptographic keys. Decrypting an encrypted term is only possible if participants are in the possession of the decryption key pair. In case of symmetric cryptography, the decryption key is the same as the encryption key. In case of asymmetric cryptography, there is a public-private key pair. Determining the corresponding key pair is done using the function $\_^{-1} : \mathsf{K} \rightarrow \mathsf{K}$.

The above-defined basic sets and function names are used in the definition of *terms*, where we also introduce constructors for pairing and encryption:

$$\mathsf{T} ::= . \mid \mathsf{P} \mid \mathsf{N} \mid \mathsf{K} \mid \mathsf{C} \mid \mathsf{M} \mid (\mathsf{T},\mathsf{T}) \mid \{\mathsf{T}\}_{FuncName(\mathsf{T})},$$

where the '.' symbol is used to denote an empty term.

Having defined the terms exchanged by participants, we can proceed with the definition of a *node* and a *participant chain.* To capture the sending and receiving of terms, the definition of nodes uses *signed terms.* The occurrence of a term with a positive sign denotes transmission, while the occurrence of a term with a negative sign denotes reception.

**Definition 1.** *A* node *is any transmission or reception of a term denoted as* $\langle \sigma, t\rangle$, *with* $t \in \mathsf{T}$ *and* $\sigma$ *one of the symbols* $+, -$. *A node is written as* $-t$ *or* $+t$. *We use* $(\pm\mathsf{T})$ *to denote a set of nodes. Let* $n \in (\pm\mathsf{T})$, *then we define the function* $sign(n)$ *to map the sign and the function* $term(n)$ *to map the term corresponding to a given node.*

**Definition 2.** *A* participant chain *is a sequence of nodes. We use* $(\pm\mathsf{T})^*$ *to denote the set of finite sequences of nodes and* $\langle \pm t_1, \pm t_2, \ldots, \pm t_i\rangle$ *to denote an element of* $(\pm\mathsf{T})^*$.

In order to define a participant model we also need to define the preconditions that must be met such that a participant is able to execute a given protocol. In addition, we also need to define the effects resulting from a participant executing a protocol.

Preconditions and effects are defined using predicates applied on terms: $CON\_TERM : \mathsf{T}$, denoting a generated term; $CON\_PARTAUTH : \mathsf{T}$, denoting participant authentication; $CON\_CONF : \mathsf{T}$, denoting the confidentiality of a given term); $CON\_INTEG : \mathsf{T}$, denoting the integrity of a given term; $CON\_NONREP : \mathsf{T}$, denoting the non-repudiation property for a given term; $CON\_KEYEX : \mathsf{T}$, denoting a key exchange protocol.

The set of precondition-effect predicates is denoted by $\mathsf{PR\_CC}$ and the set of precondition-effect predicate subsets is denoted by $\mathsf{PR\_CC}^*$. The types attached to each protocol term are modeled using the following predicates: $TYPE\_DN : \mathsf{T}$ to denote distinguished names, $TYPE\_UD : \mathsf{T}$ to denote user-domain names, $TYPE\_NT : \mathsf{T}$ to denote timestamps, $TYPE\_NDH : \mathsf{T}$ to denote Diffie-Hellman random numbers, $TYPE\_NA : \mathsf{T}$ to denote other random numbers, $TYPE\_NDH : \mathsf{T} \times \mathsf{T} \times \mathsf{T} \times \mathsf{P} \times \mathsf{P}$ to denote Diffie-Hellman symmetric keys, $TYPE\_KSYM : \mathsf{T} \times \mathsf{P} \times \mathsf{P}$ to denote symmetric keys, $TYPE\_KPUB : \mathsf{T} \times \mathsf{P}$ to denote public keys, $TYPE\_KPRV : \mathsf{T} \times \mathsf{P}$ to denote private keys, and $TYPE\_CERT : \mathsf{T} \times \mathsf{P}$ do denote certificate terms.

The set of type predicates is denoted by $\mathsf{PR\_TYPE}$ and the set of type predicate subsets is denoted by $\mathsf{PR\_TYPE}^*$. Based on the defined sets and predicates we are now ready to define the participant and protocol models.

**Definition 3.** *A* participant model *is a tuple* $\langle prec, eff, type, gen, part, chain\rangle$, *where* $prec \in \mathsf{PR\_CC}^*$ *is a set of precondition predicates,* $eff \in \mathsf{PR\_CC}^*$ *is a set of effect predicates,* $type \in \mathsf{PR\_TYPE}$ *is a set of type predicates,* $gen \in \mathsf{T}^*$ *is a set of generated terms,* $part \in \mathsf{P}$ *is a participant name and* $chain \in (\pm\mathsf{T})^*$ *is a participant chain. We use the* MPART *symbol to denote the set of all participant models.*

**Definition 4.** *A* security protocol model *is a collection of participant models such that for each positive node* $n_1$ *there is exactly one negative node* $n_2$ *with* $term(n_1) = term(n_2)$. *We use the* MPROT *symbol to denote the set of all security protocol models.*

## 3 Semantic Security Protocol Model

In this section we propose a new semantic security protocol model (SSPM) based on which we construct security protocol specifications that can be automatically executed by protocol participants. The proposed model must maintain the security properties of the protocol and must provide sufficient information for participants to be able to execute the protocol.

Protocols are given using their SPM model described in the previous section. Based on this model we generate the corresponding SSPM that has two components: the sequential model (SEQM) and the ontology model (ONTM). The first component is implemented as a WSDL-S specification while the second component is implemented as an OWL specification. In the remaining of this section we provide a description of each component and we provide a set of rules to generate SSPM from a given SPM.

### 3.1 Sequential and Ontology Models

We use the symbol URI to denote the set of *Uniform Resource Identifiers*, CONC to denote the set of all concepts and $\mathsf{CONC}^*$ to denote the set of subsets with elements from CONC.

**Definition 5.** *An* annotation *is a pair* $\langle uri, c\rangle$, *where* $uri \in \mathsf{URI}$ *and* $c \in \mathsf{CONC}$. *The set corresponding to a* SSPM *is denoted by* ANNOT *and the set of subsets with elements from* ANNOT *is denoted by* $\mathsf{ANNOT}^*$.

By consulting the WSDL-S specification we define a message as a pair consisting of the message direction and an annotation.

**Definition 6.** *A* message *is a pair* $\langle d, a\rangle$, *where* $d \in \{in, out\}$ *and* $a \in \mathsf{ANNOT}$. *We define* MSG *to denote a set of messages and* $\mathsf{MSG}^*$ *to denote the set of subsets with elements from* MSG.

Next, we define the sequential model as a collection of preconditions, effects and messages, based on the previous definitions.

**Definition 7.** *A* sequential model *is a triplet* $\langle s\_prec, s\_eff, s\_msg\rangle$, *where* $s\_prec \in \mathsf{ANNOT}^*$ *is a set of preconditions,* $s\_eff \in \mathsf{ANNOT}^*$ *is a set of effects and* $s\_msg \in \mathsf{MSG}^*$ *is a set of messages.*

The ontology model follows the description of OWL.

**Definition 8.** *An* ontology model *is a triplet* $\langle conc, propr, inst\rangle$, *where* $conc \in \mathsf{CONC}$ *is a set of concepts,* $propr \in \mathsf{PROPR}$ *is a set of properties and* $inst \in \mathsf{INST}$ *is a set of instances. An element from propr is a pair* $\langle \alpha, \beta\rangle$, *where* $\alpha$ *is a unique id and* $\beta$ *is a syntactic construction denoting the property name.*

*Let* $pr_1 = \langle \alpha_1, \beta_1\rangle$ *and* $pr_2 = \langle \alpha_2, \beta_2\rangle$. *Then* $pr_1 = pr_2$ *iff* $\alpha_1 = \alpha_2$ *and* $\beta_1 = \beta_2$. *We define the function* $(\_)_{\mathrm{id}}$ *to map the* $\alpha$ *component and the function* $(\_)_{\mathrm{nm}}$ *to map the* $\beta$ *component of a given property.*

*We use* $\mathsf{PROPR}$ *to denote the set of all properties and* $\mathsf{INST}$ *to denote the set of all instances. We use* $\mathsf{PROPR}^*$ *to denote the set of all subsets with elements from* $\mathsf{PROPR}$ *and* $\mathsf{INST}^*$ *to denote the set of all subsets with elements from* $\mathsf{INST}$.

In order to handle the previously defined ontology model we define the function $(\_)_{\mathrm{d}} : \mathsf{PROPR} \rightarrow \mathsf{CONC}$ to map the domain concept of a given property, $(\_)_{\mathrm{c}} : \mathsf{PROPR} \rightarrow \mathsf{CONC}$ to map the category concept of a given property, $(\_,\_)_{\mathrm{ci}} : \mathsf{CONC} \times \mathsf{PROPR} \rightarrow \mathsf{INST}$ to map the instance corresponding to a domain concept and property, $(\_)^{\mathrm{s}}_{\mathrm{e}} : \mathsf{CONC} \rightarrow \mathsf{CONC}^*$ to map the set of concepts for which the given concept is parent, $(\_)_{\mathrm{p}} : \mathsf{CONC} \rightarrow \mathsf{PROPR}^*$ to map the set of properties for which the given concept is domain.

### 3.2 Generating the Semantic Security Protocol Model

In order to generate the SSPM for a given SPM, we start with a core ontology model (OM) (figure 1) that contains concepts found in classical security protocols. The core OM was constructed by consulting security protocols found in open libraries such as SPORE [15] or the library published by John Clark [5].

The core ontology is constructed from 7 sub-ontologies. The sub-ontologies that must be extended with new concepts for each SSPM are denoted in figure 1 by interrupted lines, while the permanent sub-ontologies are denoted by continuous lines. The new concepts are generated from SPM, however, information that is not available in the SPM must be provided by the user.

The *SecurityProperty* sub-ontology contains concepts such as *Authentication*, *Confidentiality* or *Session_key_exchange*. The *TermType* sub-ontology includes concepts related to term types used in security protocol messages such as *SymmetricKey*, *PublicKey* or *ParticipantName*. Concepts related to cryptographic specifications such as encryption algorithms or encryption modes are found in the sub-ontology *CryptoSpec*. In order to model modules needed to extract keys, names or certificates we use the *LoadingModule* sub-ontology. The *ParticipantRole* sub-ontology defines concepts modeling roles handled by protocol participants such as *Initiator*, *Respondent* and *Third_Party*.

The *Knowledge* sub-ontology contains 5 concepts: *PreviousTerm*, *AccessedModule*, *InitialTerm*, *GeneratedTerm* and *DiscoveredTerm*. Each concept defines

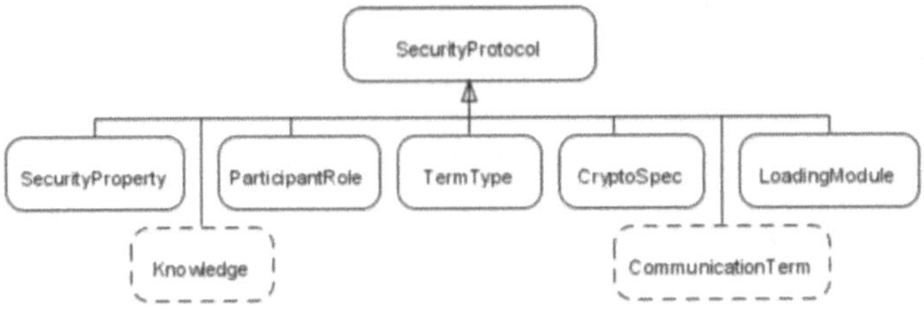


**Fig. 1.** Core ontology of SSPM

a class of terms specific to security protocols: terms from previous executions, modules, initial terms, generated terms and discovered terms.

The last sub-ontology is *CommunicationTerm*, which defines two concepts: *SentTerm* and *ReceivedTerm*. This sub-ontology is extended for each SEM-S with concepts that are sent or received. For each concept, functional properties are defined denoting the operations performed on the terms corresponding to concepts. The concepts used to extend the core ontology are specific to each protocol, however, the defined properties are applied on all constructions. From these properties we mention: *hasKey*, *isStored*, *isVerified*.

In order to generate the SSPM from a given SPM we define a set of rules and generating algorithms. The developed rules use the $\_ \leftarrow_{\mathrm{r}} \_$ operator to denote set reunion and the $\_ \leftarrow_{\mathrm{a}} \_$ operator to denote a value transfer.

The first two rules generate the predicate concepts corresponding to preconditions *prec* from a SPM, where the function $gc : \mathsf{T} \rightarrow \mathsf{CONC}$ is used to generate the concept corresponding to a given term and the function $gcc : \mathsf{PR\_CC} \rightarrow \mathsf{CONC}$ is used to generate the concept corresponding to a given precondition predicate:

$$\frac{pr \in prec \;\; pr = CON\_TERM(t)}{c \leftarrow_{\mathrm{a}} gc(t) \;\; s\_prec \leftarrow_{\mathrm{r}} \{\langle uri, c\rangle\} \;\; (InitialTerm)^{\mathrm{s}}_{\mathrm{e}} \leftarrow_{\mathrm{r}} \{c\}}\mathrm{pr\_term},$$

$$\frac{pr \in prec \;\; pr \neq CON\_TERM(t)}{s\_prec \leftarrow_{\mathrm{r}} \{\langle uri, gcc(pr)\rangle, \langle uri, gc(t)\rangle\}}\mathrm{pr\_propr}.$$

The rules generating the effects have a similar structure because of the *eff* set. For each positive or negative node there is a corresponding concept in the *SentTerm* and *ReceivedTerm* sub-ontologies, generated by the following rules:

$$\frac{n \in chain \;\; sign(n) = +}{c \leftarrow_{\mathrm{a}} gtx(term(n)) \;\; s\_msg \leftarrow_{\mathrm{r}} \{\langle out, c\rangle\} \;\; (SentTerm)^{\mathrm{s}}_{\mathrm{e}} \leftarrow_{\mathrm{r}} \{c\}}\mathrm{msg\_tx},$$

$$\frac{n \in chain \;\; sign(n) = -}{c \leftarrow_{\mathrm{a}} grx(term(n)) \;\; s\_msg \leftarrow_{\mathrm{r}} \{\langle in, c\rangle\} \;\; (ReceivedTerm)^{\mathrm{s}}_{\mathrm{e}} \leftarrow_{\mathrm{r}} \{c\}}\mathrm{msg\_rx}.$$

The concatenated terms corresponding to each transmitted or received term are modeled using similar rules. For each sent term the SSPM must provide the construction operations and for each received term the SSPM must provide processing operations. Sub-concepts of *SentTerm* are connected to sub-concepts of *Knowledge* through the *isExtracted* property, generated according to the following rule, where we used the function $\mathsf{PR\_CC}^* \to \mathsf{ID}$ to generate a new property id:

$$\frac{c \in (SentTerm)^{\mathrm{s}}_{\mathrm{e}}}{p \leftarrow_{\mathrm{a}} \langle gid(propr), isExtracted\rangle \;\; (c)_{\mathrm{p}} \leftarrow_{\mathrm{r}} \{p\} \;\; (p)_{\mathrm{c}} \in (Knowledge)^{\mathrm{s}}_{\mathrm{e}}}\text{con\_extr.}$$

Processing of received terms is done according to the type of the given term and to the knowledge available to the user. The modeled operations introduce constraints on the type and location of knowledge through the following rules, where we used the $E\_SYM : \mathsf{CONC}$ predicate to denote symmetric encryption:

$$\frac{c \in (ReceivedTerm)^{\mathrm{s}}_{\mathrm{e}} \;\; p \in (c)_{\mathrm{p}} \;\; (p)_{\mathrm{nm}} = isDecrypted}{c' \leftarrow_{\mathrm{a}} (p)_{\mathrm{c}} \;\; E\_SYM(c') \vee E\_SYM(c') \;\; (c') \in (DiscoveredTerm)^{\mathrm{s}}_{\mathrm{e}}}\text{con\_decr,}$$

$$\frac{c \in (ReceivedTerm)^{\mathrm{s}}_{\mathrm{e}} \;\; p \in (c)_{\mathrm{p}} \;\; (p)_{\mathrm{nm}} = isStored}{c' \leftarrow_{\mathrm{a}} (p)_{\mathrm{c}} \;\; (c') \in (DiscoveredTerm)^{\mathrm{s}}_{\mathrm{e}}}\text{con\_stored,}$$

$$\frac{c \in (ReceivedTerm)^{\mathrm{s}}_{\mathrm{e}} \;\; p \in (c)_{\mathrm{p}} \;\; (p)_{\mathrm{nm}} = isVerified}{c' \leftarrow_{\mathrm{a}} (p)_{\mathrm{c}} \;\; (c') \in \{(DiscoveredTerm)^{\mathrm{s}}_{\mathrm{e}} \setminus (AccessedModule)^{\mathrm{s}}_{\mathrm{e}}\}}\text{con\_verif.}$$

In the *Knowledge* sub-ontology, each concept has an *isOfType* property attached based on which participants can decide on the operations to execute. For each type, additional properties are defined such as the *hasSymmAlg* or *hasKey* properties for symmetric encrypted terms. The rules based on which these properties are generated are specific to each type. For example, the following rules define the algorithm type and key for an encrypted term that must be processed or constructed:

$$\frac{c \in (Knowledge)^{\mathrm{s}}_{\mathrm{e}} \;\; E\_SYM(c)}{p \leftarrow_{\mathrm{a}} \langle gid(propr), hasSymmAlg\rangle \;\; (c)_{\mathrm{p}} \leftarrow_{\mathrm{r}} \{p\} \;\; (c,p)_{\mathrm{ci}} \in (Symmetric)^{\mathrm{s}}_{\mathrm{e}}}\text{sim\_alg,}$$

$$\frac{c \in (Knowledge)^{\mathrm{s}}_{\mathrm{e}} \;\; E\_SYM(c)}{p \leftarrow_{\mathrm{a}} \langle gid(propr), hasKey\rangle \;\; (c)_{\mathrm{p}} \leftarrow_{\mathrm{r}} \{p\} \;\; (p)_{\mathrm{c}} \in (Knowledge)^{\mathrm{s}}_{\mathrm{e}}}\text{sim\_key.}$$

The rules presented above are executed by algorithms. For example, modeling positive nodes in SSPM is done through the use of algorithm 1. Here, the set of knowledge *KNOW* corresponding to each executing participant grows with the construction and reception of each new term. We used the function $mpart : \mathsf{T} \to \mathsf{T}^*$ to map the set of concatenated terms and the keyword "Exec" to denote the execution of sub-algorithms.

### 3.3 Correctness of SSPM

In the generation process of SSPM from a given SPM, we consider a correct SPM constructed by the user. With the large number of attacks reported in

**Algorithm 1.** Model positive and negative nodes

**Require:** $n \in (\pm\mathsf{T}), sign(n) = +$
**for all** $t \in mpart(term(n))$ **do**
  Let $c = gc(t)$
  Let $p \Leftarrow$ @con_extr(c)
  **if** $t \in KNOW$ **then**
    $(p)_{\mathrm{c}} \leftarrow_{\mathrm{a}} c$
  **else if** $t = \{t'\}_{f(k)}$ **then**
    $(GeneratedTerm)^{\mathrm{s}}_{\mathrm{e}} \leftarrow_{\mathrm{r}} \{c\}$
    Exec $ModelEncryptedGenerated(t)$
  **else if** $t \in gen$ **then**
    $(GeneratedTerm)^{\mathrm{s}}_{\mathrm{e}} \leftarrow_{\mathrm{r}} \{c\}$
    Exec $ModelPlainGenerated(t)$
  **else**
    $(DiscoveredTerm)^{\mathrm{s}}_{\mathrm{e}} \leftarrow_{\mathrm{r}} \{c\}$
    Exec $ModelDiscoveredLoaded(t)$
  **end if**
  $KNOW \leftarrow_{\mathrm{r}} t$
**end for**

the literature [6], [7], it is vital for new protocol models to maintain the security properties of protocols for which security properties have been proved to hold.

In order to prove the correctness of the generated SSPM, we consider $\Gamma$ representing the set of all information included in an SSPM. The information generated by the proposed rules can be divided into three components: mapped information, user-provided information and participant knowledge-based information.

The set of mapped information is denoted by $\gamma_{map}$ and represents information originating directly from SPM. The set of user-provided information is denoted by $\gamma_{up}$ and represents information originating from the user (e.g. cryptographic algorithms). The set of knowledge-based information originates from the knowledge available when running the protocol and is denoted by $\gamma_{know}$.

By using the above sets $\Gamma = \gamma_{map} \cup \gamma_{up} \cup \gamma_{know}$. The correctness of the information contained in $\gamma_{map}$ results from the original protocol model, while the correctness of the information contained in $\gamma_{up}$ results from the assumption that the user provides correct information.

The information contained in $\gamma_{know}$ is generated based on the design principles of *fail-stop* [11] protocols. These principles state that the correctness of each received term must be verified and the protocol execution must be stopped immediately in case of invalid terms. By using these principles, the rules we proposed generate verification properties for each received term found in the participant's knowledge set. Protocols that do not follow these rules can not be modeled with our method.

The the correctness of the generated SSPM follows from the correctness of the information generated in the $\Gamma$ set, constructed from the three sets $\gamma_{map}, \gamma_{up}, \gamma_{know}$ for which the correctness has been discussed above.

```
...
<xsd:element name="Msg1Request">
  <xsd:complexType>
    <xsd:sequence>
      <xsd:element name="Term1" type="xsd:base64Binary"
                   wssem:modelReference=".../SecProt.owl#SentTerm1">
    </xsd:sequence>
  </xsd:complexType>
</xsd:element>
...
<wsdl:operation name="Msg1">
  <wsdl:output message="tns:Msg1Request"/>
</wsdl:operation>
<wssem:effect name="SessionKeyExchange"
              wssem:modelReference=".../SecProt.owl#SessionKey"/>
...
```

**Fig. 2.** Part of the sequential model's implementation

# 4 Experimental Results

In this section we exemplify the construction of a SSPM from a given SPM and provide a few experimental result from implementing several generated SSPM.

## 4.1 Constructing the SSPM for the "BAN" Protocol

In order to provide an example for constructing an SSPM for a given SPM, we use the well-known "BAN Concrete Secure Andrew RPC" protocol [15]. This is a two-party protocol providing a session key exchange using symmetric cryptography. The protocol assumes that participants are already in the possession of a long-term key $K_{ab}$.

Because of space considerations, we only provide the construction of the SSPM for the $A$ participant. Based on this, the construction of the SSPM for the second participant is straight-forward.

The precondition set $prec_A$ for participant $A$ is $prec_A = \{CON\_TERM(A), CON\_TERM(B), CON\_TERM(K_{ab})\}$ and the effect set $e\!f\!f_A$ for the same participant is $e\!f\!f_A = \{CON\_KEYEX(K_{ab})\}$. The set $type_A = \{TYPE\_UD\ (A), TYPE\_UD(B), TYPE\_KSYM(A, B, K_{ab}), TYPE\_KSYM(A, B, K), TYPE\_NA\ (N_a), TYPE\_NA(N_b)\}$ defines the type corresponding to each term and the set $gen_A = \{N_a\}$ defines the terms generated by participant $A$. The participant name is $part_A = A$ and the participant chain is $chain_A = \langle +(A, N_a), -\{N_a, K, B\}_{sk(K_{ab})}, +\{N_a\}_{sk(K)}, -N_b\rangle$.

By applying the rules and algorithms described in the previous sections we generate the SSPM model. Due to space considerations, instead of describing the actual SSPM we describe the implementation of the model. The sequential model is implemented as a WSDL-S specification, while the ontology model is implemented as a OWL specification.

Part of the resulted WSDL-S specification is given in figure 2 and part of the graphical representation of the OWL specification is given in figure 3.

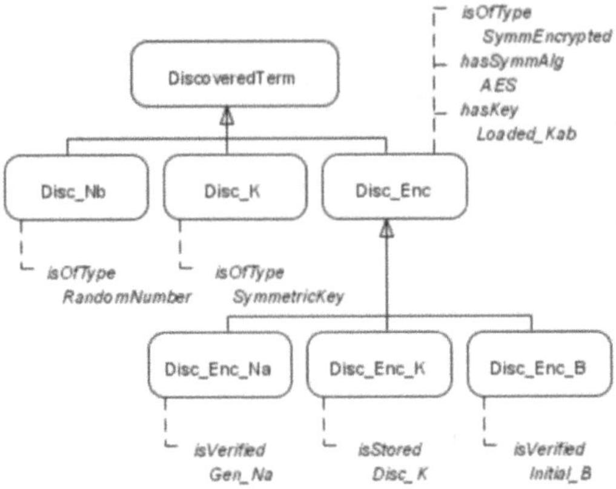


**Fig. 3.** Part of the ontology model's implementation

### 4.2 Executing the Generated Specifications

In order to prove that the SSPM model contains sufficient information for participants to execute the generated specifications, we generated several WSDL-S and OWL specifications corresponding to initiator and respondent protocol roles.

In order to execute the specifications, messages were encoded and transmitted according to the constructions provided by the WS-Security standard [17]. In the experiments we conducted, participants downloaded the specification files from a public server and they were able to execute the protocols based only on the received descriptions. The participants hardware and software configurations: Intel Dual Core CPU at 1.8GHz, 1GByte of RAM, MS Windows XP.

Part of the experimental results are given in table 1, where the values correspond to milliseconds. The "Spec. proc" column denotes the specification processing time, the "Msg. constr." column denotes the message construction time (for output messages) and the "Msg. proc." column denotes the message processing time (for input

**Table 1.** Protocol execution timings

| Protocol participant | Spec. proc. (ms) | Msg. constr. (ms) | Msg. proc. (ms) | Total (ms) |
|---|---|---|---|---|
| BAN Init. | 14.58 | 11.81 | 3.68 | 30.08 |
| BAN Resp. | 14.03 | 2.86 | 1.62 | 18.52 |
| ISO9798 Init. | 13.07 | 35.784 | 23.30 | 72.16 |
| ISO9798 Resp. | 13.51 | 6.876 | 12.24 | 32.63 |
| Kerb. Init. 1 | 22.63 | 0.83 | 0 | 23.47 |
| Kerb. Init. 2 | 12.61 | 0.55 | 1.58 | 14.76 |
| Kerb. Init. 3 | 2.23 | 3.34 | 0.94 | 6.52 |
| Kerb. Resp. 1 | 19.28 | 0 | 0.41 | 19.69 |
| Kerb. Resp. 2 | 10.81 | 3.379 | 1.67 | 15.87 |
| Kerb. Resp. 3 | 5.25 | 11.41 | 3.59 | 20.26 |

messages). The table contains two two-party protocols ("BAN Concrete Andrew Secure RPC", or more simply BAN, and ISO9798) and one three-party protocol (Kerberos). The performance differences between the BAN and ISO9798 protocols are due to the fact that ISO9798 makes use of public key cryptography, while BAN uses only symmetric cryptography.

## 5 Related Work

In this section we describe approaches we found in the literature that mostly relate to our proposal.

An approach that aims at the automatic implementation of security protocols is given in [2]. This approach uses a formal description as a specification which is executed by participants. The proposed specification does not make use of Web service technologies, because of which inter-operability and extendability of systems executing the given specifications becomes a real issue.

Abdullah and Menasc propose in [3] a specification that is constructed as an XML document from which code is generated. The resulted code is then compiled and executed by participants. Because of this aspect, our proposal is more dynamic in the sense that applications can download and execute new protocols based on the developed specifications automatically, without having to stop program execution.

The authors from [8] propose a security ontology for resource annotation. The proposed ontology defines concepts for security and authorization, for cryptographic algorithms and for credentials. This proposal was designed to be used in the process of security protocol description and selection based on several criteria. In contrast, our ontologies, have a more detailed construction. For example, the ontology from [8] defines a collection of cryptographic algorithms, however, it does not define the algorithm mode, which is an implementation-specific information.

There have been several other security ontologies proposed [9], [10] which can be used to complete our core ontology with additional concepts and properties, for generating more complex protocol models.

## 6 Conclusion and Future Work

We developed a novel method for the automated execution of security protocols. Our approach is based on a semantic security protocol model from which security protocol specifications are generated. The sequential component of the proposed model is implemented as a WSDL-S specification while the ontology component is implemented as an OWL specification.

Constructing the SSPM model is not a trivial task and can induce new flaws in correct protocols that can lead to attacks. In order to ensure a correct construction process, we developed several generating rules and algorithms that map each component from the input protocol model to a component in the SSPM model.

As future work we intend to develop a service-based middleware to support secure distribution of these specifications. The middleware will also be able to create new protocols based on already existing protocols and distribute the new specifications to Web services.

## References


1. Cremers, C.J.F., Mauw, S.: Checking secrecy by means of partial order reduction. In: Leue, S., Systa, T. (eds.) Scenarios: Models, Transformations and Tools. LNCS, vol. 3466. Springer, Heidelberg (2005)
2. Mengual, L., Barcia, N., Jimenez, E., Menasalvas, E., Setien, J., Yaguez, J.: Automatic implementation system of security protocols based on formal description techniques. In: Proceedings of the Seventh International Symposium on Computers and Communications, pp. 355–401 (2002)
3. Abdullah, I., Menascé, D.: Protocol specification and automatic implementation using XML and CBSE. In: IASTED conference on Communications, Internet and Information Technology (2003)
4. Akkiraju, R., Farrell, J., Miller, J., Nagarajan, M., Schmidt, M., Verma, K.: Web Service Semantics - WSDL-S. A joint UGA-IBM Technical Note (2005)
5. Clark, J., Jacob, J.: A Survey of Authentication Protocol Literature: Version 1.0. York University (1997)
6. Gavin, L.: Some new attacks upon security protocols. In: Proceedings of the 9th CSFW, pp. 162–169. IEEE Computer Society Press, Los Alamitos (1996)
7. Cremers, C.J.F.: Compositionality of Security Protocols: A Research Agenda. Electr. Notes Theor. Comput. Sci. 142, 99–110 (2006)
8. Kim, A., Luo, J., Kang, M.: Security ontology for annotating resources. In: Meersman, R., Tari, Z. (eds.) OTM 2005. LNCS, vol. 3761, pp. 1483–1499. Springer, Heidelberg (2005)
9. Blanco, C., Lasheras, J., Valencia-Garcia, R., Fernandez-Medina, E., Toval, A., Piattini, M.: A systematic review and comparison of security ontologies. In: Proc. of the Third International Conference on Availability, Reliability and Security, pp. 813–820 (2008)
10. Denkera, G., Kagal, L., Finin, T.: Security in the semantic web using owl. Information Security Technical Report 1(10), 51–58 (2005)
11. Gong, L.: Fail-Stop Protocols: An Approach to Designing Secure Protocols. In: Proceedings of the 5th IFIP Conference on Dependable Computing and Fault-Tolerant Systems, pp. 44–55 (1995)
12. World Wide Web Consortium, OWL Web Ontology Language Reference, W3C Recommendation (2004)
13. Gutmann, P.: Cryptlib Encryption Toolkit, http://www.cs.auckland.ac.nz/pgut001/cryptlib/index.html
14. OpenSSL Project, version 0.9.8h, http://www.openssl.org/
15. Laboratoire Specification et Verification, Security Protocol Open Repository, http://www.lsv.ens-cachan.fr/spore
16. Organization for the Advancement of Structured Information Standards, SAML V2.0 OASIS Standard Specification (2007), http://saml.xml.org/
17. Organization for the Advancement of Structured Information Standards, OASIS Web Services Security, WSS (2006), http://saml.xml.org/

# Performance Study of a Video Application over Multi Hop Wireless Networks with Statistic-Based Routing
## (Work in Progress)

Alexander Klein[1] and Jirka Klaue[2]

[1] University of Wuerzburg, Institute of Computer Science, Germany
lastname@informatik.uni-wuerzburg.de
http://www3.informatik.uni-wuerzburg.de
[2] EADS Innovation Works, Munich, Germany

**Abstract.** In recent years, technologies like Ultra Wideband (UWB) were developed that can be used to boost the data rate of wireless nodes to a new level. The higher data rate makes wireless networks capable of transporting multimedia content for real time applications like In Flight Entertainment (IFE), surveillance applications, or structural health monitoring. In this paper we measure and simulate the performance of a video application for IFE over a high data rate multi hop wireless network using a Directed Diffusion based routing protocol. Therefore, we take a look at the perceived video quality instead of focusing solely on the packet delivery ratio. Furthermore, we discuss the parameters of the Statistic-Based Routing (SBR) protocol which are relevant for this particular application and focus on their impact on the video quality during topology changes.

**Keywords:** routing, wireless, multi-hop, video, performance.

## 1 Introduction

Frequent topology changes represent a serious problem for IFE since most of the traffic has to be transmitted in real time which limits the possibility of retransmitting lost packets. Depending on the used wireless technology, carry-on luggage or clothes that are placed on the back rest might lead to topology changes. For that reason, the routing protocol has to frequently send out messages to detect topology changes and reroute traffic.

Thus, we have to deliberate over the question whether wireless solutions can be competitive to wired solutions for our particular application of IFE. First of all, the video quality that is recognized by a customer has to be approximately on the same level to allow a comparison of wired and wireless solutions which mainly depends on the routing protocol if we assume that the used physical layer is capable to achieve a high point-to-point data rate. Other characteristics that have to be taken into account are the costs for deployment and general maintenance work. The installation costs of wired solutions in already assembled planes

L. Fratta et al. (Eds.): NETWORKING 2009, LNCS 5550, pp. 955–966, 2009.

are more expensive than the installation costs of wireless solutions. Flexibility is a very important issue since the airline companies are interested in adapting the configuration of a plane, e.g. by extending the business or the economy class.

Our work is organized as follows. In Section 2 we take a look at wireless network simulation tools and the way they simplify the signal propagation to allow large-scale simulations. The problem of topology change detection is discussed in Section 3. We describe the simulation of the routing protocol which is used to evaluate the performance of the video application in Section 4 and its implementation in Section 5 respectively. The framework that is used to estimate the performance of the received video is introduced in Section 6. The results are presented and analyzed in Section 7. Finally, we summarize the results and introduce our future fields of research.

## 2 Related Work

As a consequence of rapid improvements in technology and miniturization, Wireless Multimedia Sensor Networks (WMSN)become more and more interesting for a large number of new applications [1]. However, sensor nodes with UWB or other high data rate wireless interfaces are hardly available at the moment. Thus, simulation is the most common approach to estimate the performance of WMSNs. Many simulation tools like ns-2 [2] or OPNET Modeler [3] come with simplified propagation models, e.g. free space, which neglect most of the characteristics that have great impact on the communication in a multi-hop wireless network. Often, these models are even further simplified to allow the simulation of large-scale networks within a justifiable amount of time. Kotz et al. [4] summarized the typical assumptions that are made by simulations like circular transmission area, equal transmission range, and symmetric links. Another work that is published by the same research group [5] deals with the problem of simulation validation by using measurements from an environment typical to the one of interest. Their results show that it is important to use measurements in order to find out which propagation model meets the desired requirements.

Therefore, we decided not to focus solely on the standard performance metrics like packet delivery ratio, end-to-end delay and jitter since they do not necessarily represent the perceived link quality by the user. If we take a closer look e.g. at video encoding and decoding, we recognize that the video quality over a lossy link strongly depends on the way packets are lost. It is important to know whether consecutive packets are lost as a consequence of topology changes, single packets which might be caused by interference, or a low signal to noise ratio. For that reason, we focus on the Mean Opinion Score (MOS) of a video transmission to evaluate the performance of our network.

Kuladinithi et al. [6] implemented the AODV [7] protocol using the programming language Java to allow the experimental evaluation of different AODV implementations on various systems. Inspired by the their implementation we decided to implement the SBR protocol [8] to evaluate its performance in a testbed and compare the measurements with the simulation results gathered from the OPNET Modeler simulation.

## 3 Toplogy Change Detection

Many popular routing protocols like AODV [7] and OLSR [9] use time outs to detect link breaks. In the following we refer the time interval as downtime that is required by the routing protocol to detect a topology change and to find a new valid route. The problem is to find an optimized expiry interval in order to minimize the downtime on the one hand, and the probability that a link is untruly assumed to be broken due to short temporary interference on the other hand. For that reason, we have chosen the SBR protocol since it is able to detect link breaks within a short amount of time.

SBR is based on the concept of Directed Diffusion which is a data-centric and application aware paradigm which was introduced by C. Intanagonwiwat et. al. [10]. Each node in the network broadcasts (hello) messages. These messages are similar to route requests in AODV and are forwarded by intermediate nodes. The neighbor through which a new hello message was received is rated by using a cumulative function. Thus, a higher routing value corresponds to a higher link quality as a consequence of the cumulative function. Figure 1 shows the downtime of the SBR protocol in the average case and in the worst case depending on the development of the corresponding routing entry values.The graphs represent the routing entry values of node A and B towards a destination node X from the perspective of another node C. At time $t_0$ node C receives the first routing message via node A and at time $t_{start}$ from node B respectively. The link between node A and node C breaks at $t_{loss}$. The alternative route via node B becomes the route with the highest routing entry at $t_{handover}$.

Node C forwards its data traffic for a destination node X either to node A or node B. Node C always selects the neighbor with the highest routing entry towards the destination as next hop. Therefore, the downtime represents the time until the routing value of the alternative route becomes higher than the previously best route. A topology change is detected quickly if the routing values are on an equal level from the time when a link break occurs. Thus, the worst case is represented by a scenario in which the routing entry of a single neighbor is very high whereas the routing values of all other neighbors are very low.

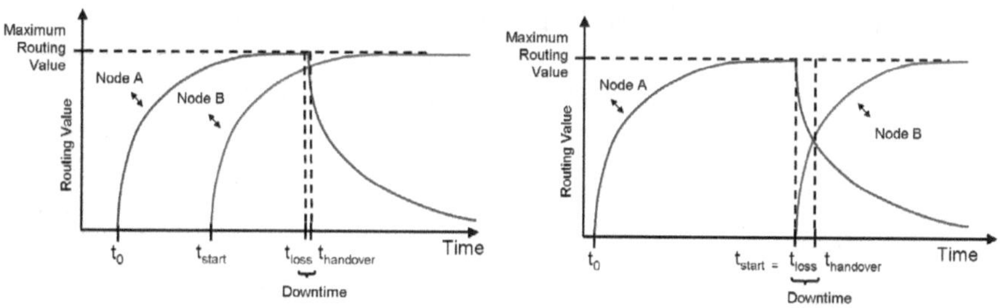


(a) Average case (b) Worst case

**Fig. 1.** Topology change detection

## 4 Simulation

We use a framework for WSNs which we have developed with the OPNET Modeler 14.0 [3] software to simulate all layers of the communication stack. For this particular simulation, we decided to use the IEEE 802.11 mac and physical layer that is provided by OPNET to allow the comparison of the simulated performance of the video application with the performance of a testbed using IEEE 802.11g interfaces. Therefore, we put our network and application layer on top of OPNETs mac and physical layer to simulate the communication stack. Trace files were recorded from video applications which are used for traffic generation within the simulation in order to generate realistic traffic patterns.

In addition, we implemented a filter process which allows us to dynamically modify the signal propagation to simulate unpredictable signal loss which leads to major topology changes in the network. Furthermore, the process is used to limit the signal propagation such that only a certain topology can be used which corresponds exactly to the topology in the testbed. The statistics of the data traffic are collected by a centralized node to allow a more flexible evaluation and visualization. Thus, we extended the data source and sink OPNET models to meet our requirements of traffic generation and evaluation.

## 5 Implementation

The programming language Java was used to implement the routing protocol to allow its usage on different Operating Systems (OS). Most common OSs, e.g. Linux and Windows, come with tools that allow the modification of their routing table without much effort. Therefore, the Java routing application has to detect which OS is used in order to know which commands are used by the OS. This enables us to manipulate routes in the table without the need of notifying other applications. The implementation consists of three major packages. The first one is represented by the network package which is used to receive and transmit data packets via the IEEE 802.11g interface. We use the Jpcap 0.7 [11] library which is based on WinPcap to grab packets from the interface. The second package covers configuration, routing table, and time management functions, e.g. timer and statistic tasks, which are then used by the routing protocol. The behavior of the routing protocol and the used messages build the third package. Incoming packets are detected and evaluated by a receiver task which sends a callback to the routing task to further evaluate the packet. The routing task then decides what actions have to be performed according to the content of the packet, e.g. modification of the routing table, changing of routing entries, forwarding or dropping of the packet. Additionally, periodic tasks like hello message transmission timer or routing entry decrease timer send callbacks to the routing application.

We added a filter class to the network package which covers the same functionality as the filter process within the OPNET simulation. Thus, we can restrict the topology of the testbed according to the topology used in the simulation. Furthermore, time triggered topology changes can be used to study the behavior of the protocol to deal with link breaks depending on its configuration.

## 6 Video Quality Evaluation

The standard method to assess the performance of video transmission systems is to calculate the Peak Signal to Noise Ratio (PSNR) between the source and the received (possibly distorted) video sequence.It is a differential metric which is calculated image-wise and very similar to the well-known SNR but correlating better with the human quality perception [12]. The PSNR calculation yields a quality indicator for each image of the video sequence in relation to the original image. Thus, this metric is only meaningful if the quality of the original image sequence is high in terms of human perception which is not necessarily the case. For instance, if the video sequence is passed through a state-of-the-art video encoder to reduce the bit-rate the compressed video will be already distorted since modern video-codecs – like MPEG-4 or H.264 – are usually lossy. Loss of packets will lead to decoding errors at the decoder/player while delay can cause buffer under-runs. Both will ultimately cause the loss of images at the player. Since modern video-codecs make extensive use of the temporal redundancy (encoding only the differences) in most videos, the loss of single images also leads to the distortion of all following images that are differentially encoded based on the lost image. Lost frames usually will cause the video player to "freeze" – to show the last successfully received and decoded image. It is important for an image-by-image metric to reproduce this behavior in case of transmission losses or delay in order to avoid alignment issues between the source and the received video. For a better illustration of the meaning of quality measures for non-experts the ITU-R developed a quality indication scale which is tied to the quality impression of human observers [14]. This scale is shown in Table 1.

ITU-R recommendation BT.500 [14] further describes a methodology to gain these quality indicators by subjective assessment series (by a group of humans). Such a scale is often called Mean Opinion Score and used in several quality assessment systems. In [13] there is a mapping of PSNR values to MOS values which can be used to roughly estimate the human quality perception for videos with relatively low motion (like for instance videos from surveillance cameras). This mapping from PSNR to MOS is shown in Table 2 and used in this paper.

A MOS value is assigned to each image according to Table 2 which is based on the PSNR values that are calculated for every single image of a received video sequence. These values are averaged over all images of a sequence to produce a single quality indicator for a video transmission as proposed by the methodology

**Table 1.** ITU-R quality and impairment

| Scale | Quality | Impairment |
|---|---|---|
| 5 | Excellent | Imperceptible |
| 4 | Good | Perceptible |
| 3 | Fair | Slightly annoying |
| 2 | Poor | Annoying |
| 1 | Bad | Very annoying |

**Table 2.** PSNR[dB] to MOS conversion

| PSNR | MOS |
|---|---|
| $> 37$ | 5 (Excellent) |
| 31 - 37 | 4 (Good) |
| 25 - 31 | 3 (Fair) |
| 20 - 25 | 2 (Poor) |
| $< 20$ | 1 (Bad) |

described in [14]. However, averaging can be problematic for long videos since short temporal distortions will not influence the average significantly. To avoid this effect another depiction of the MOS values of the single images of the video sequence is given in this paper in addition to the average values. The percentage of images with a certain MOS is displayed and compared to the original video.

In this paper we evaluate the performance of a wireless video transmission system by means of measurements and simulations taking into account the general principles of video quality evaluation. Therefore, we use a video quality evaluation tool-set – called EvalVid – which provides the necessary tools [15].

## 7 Results

We selected one of the standard video sequences which is used by a variety of video encoding and transmission studies by, e. g., the Video Quality Experts Group [16]. This video sequence is called "Hall Monitor" and consists of 300 frames in CIF resolution (352x288 pixel) with 30 Hz frame rate. It is a relatively low-motion sequence so that the PSNR to MOS mapping shown in Table 2 can be applied. Due to the fact that it is only 10s long, we concatenated the sequence six times. Since the video is recorded with static camera and there is little motion in the scene, the influence of this concatenation on the video encoder performance is low – even at the junctions. The resulting one minute long video was then encoded with the state-of-the-art H.264 video encoder x264 [17] with an average target bit-rate of 128 kbit/s. A key-frame was encoded every second in order to have a good balance between coding efficiency and error recovery capabilities. To give a better impression of the video sequence used, Figure 2 displays a sample image together with the bit-rate profile and the PSNR between the encoded and the original video.

A second video sequence with more motion was selected in order to stress the performance evaluation methodology with content appropriate for IFE. Figure 3 shows a sample image from the one minute long scene from the Movie "Star Wars III". The resolution is 360x216 pixels and the frame rate is 25 Hz. With current video encoding technology it is not possible to achieve an acceptable

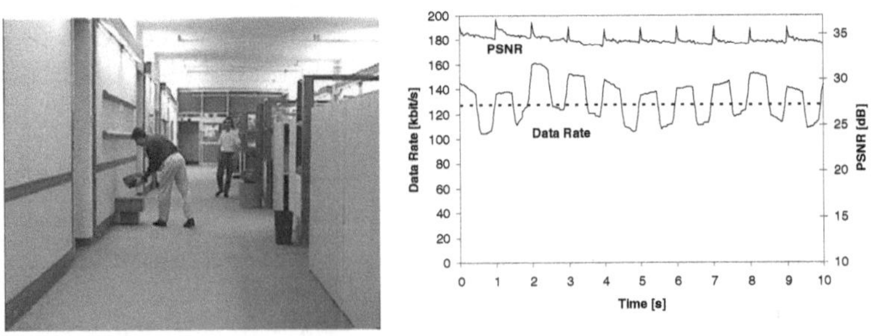


(a) Sample image (b) Video profile & quality

**Fig. 2.** Profile of "hall" video clip, (a) sample image (b) data rate & PSNR profile

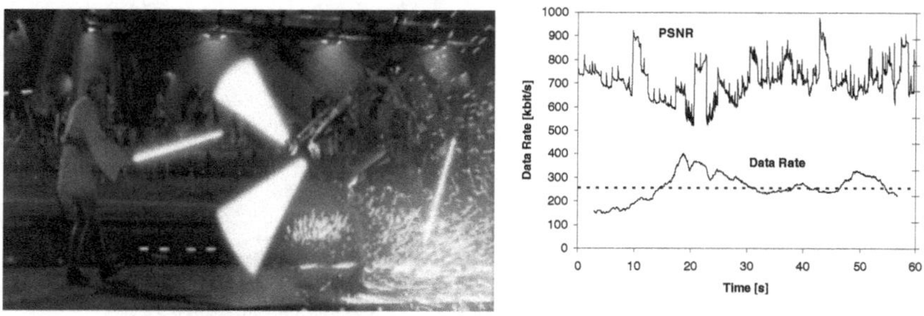


(a) Sample image (b) Video profile & quality

**Fig. 3.** Profile of “sw3” video clip, (a) sample image (b) data rate & PSNR profile

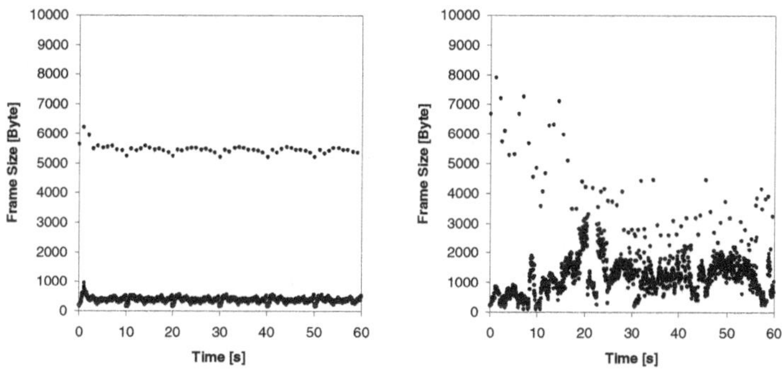


(a) hall clip (b) sw3 clip

**Fig. 4.** Frame sizes of (a) “hall” video clip and (b) “”sw3“” video clip

PSNR with an average target bit-rate of 128 kbps. Consequently, the video clip was encoded with a target bit-rate of 256 kbps. The different content of the selected clips is also reflected in the variations of the size of the encoded frames as shown in Figure 4. While the only variations in the hall clip are basically the different sizes of the I and P frames, the frame size fluctuations in the sw3 clip are much higher.

In order to calibrate the simulation with the measurements we performed a set of test runs in a specific scenario. Five wireless nodes are initially connected to each other and transmit the encoded hall video. sequence using RTP (Real-time Transport Protocol) [19] from Node 1 to Node 5. The nodes in the simulation and the testbed are configured such that they are forced to build a string topology. Thus, they are only able to receive messages from their direct neighbors. Node 5 represents an exception since it temporarily connects to the other nodes as shown in Figure 5. We have chosen this extraordinary example due to the fact that it is the worst case scenario for the routing protocol. A description of the connectivity during the simulation and the measurement is given in Figure 5. Note, the $u_1$, $u_2$, and $u_3$ are random variables which are selected at the

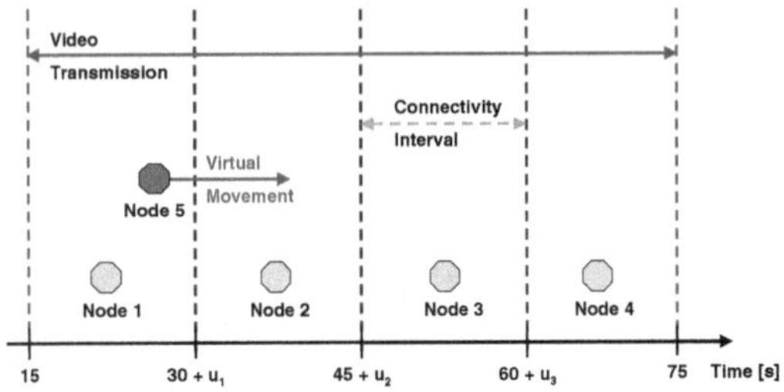


**Fig. 5.** Illustration of the connectivity during video transmission

beginning of the simulation according to a uniform distribution between -1s and 1s. The variables are required to shift the disconnection times in order to avoid the alignment with I frames.

In the 15+u seconds interval, the direct connections between Node 5 and Node 2, 3 and 4 respectively were detached which caused the system to find a new route. This represents a relatively harsh scenario since abrupt disconnections represent the worst case for real-time applications. The most relevant parameters represent the hello message interval and the routing decrease interval which are set to 1s. Due to the fact that the other parameters have no significant impact in our scenario we skip their description since a detailed description of all parameters is given in [8]. Furthermore, the implementation and the simulation will be made available to download from [21].

Using EvalVid, a trace of the video file was generated, containing the size and type of each video packet transmitted over RTP. Additionally, an IP-level packet trace was created using Wireshark [20] at the transmitting and receiving node. These traces were used by EvalVid to calculate packet and frame loss figures as well as reconstructing the received (possibly distorted) video files. The received videos were then decoded using FFmpeg [18] to be able to calculate the PSNR and MOS figures for the video quality evaluation. Figure 6 compares the frame loss of the measurements and the simulations while Figure 7 shows the corresponding MOS values for the received video. The overall frame loss is slightly higher for the measurements which is caused by single packet losses due to interferences, multi-path propagation, and moving obstacles. Moreover, the percentage of key I frames lost in the simulation was slightly higher which was quite suprising. A closer look at the trace files revealed that the starting times of the disconnections were varying more during the measurements due to the human reaction time. Against, in the simulations the disconnection interval was quite stable and accidentally always during an I frame transmission. This effect is avoided in the following parameter study by equally distributing the disconnection intervals in a certain range. The bars in Figure 7 show the percentage of frames with a certain MOS in comparison to the reference videos (rightmost) MOS distribution. The reference video reflects the coding loss and consists of 100% frames with a MOS of 4 (good). In contrast to the raw frame loss these results include the quality degradation caused by frames that could not be correctly decoded due to losses of previous frames. Though the key frame loss in

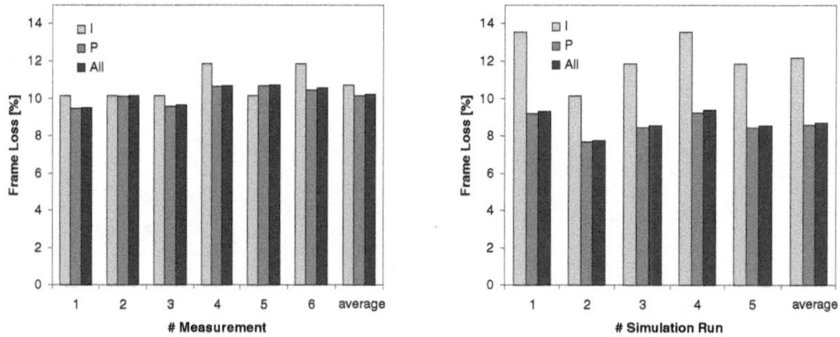


(a) Measurement (b) Simulation

**Fig. 6.** Hall - Comparison of frame loss, (a) measurements (b) simulations

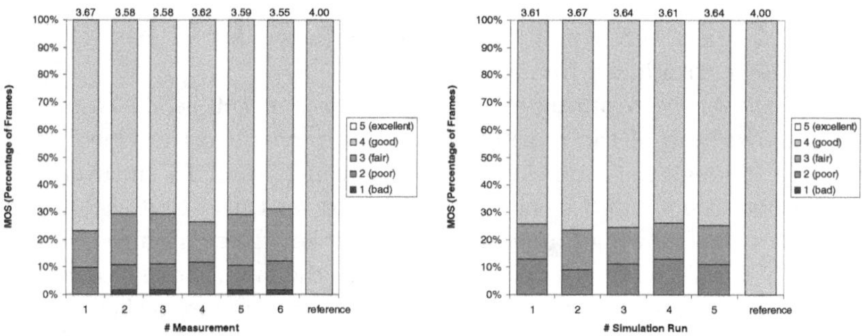


(a) Measurement (b) Simulation

**Fig. 7.** Hall - Comparison of MOS, (a) measurements (b) simulations

the simulations was higher the quality of the video was worse in the measured scenario. This is caused by the rare random single packet losses during the measurements which influence all following P frames after the packet loss until the next I frame. The impact that single packet losses have on the MOS and PSNR depends on the used encoder, its configuration e.g. the I frame rate, and the type of video which is encoded e.g. action sequence or landscape stills.

Considering the differences between the measurements and the simulation, the loss and MOS statistics are similar enough such that we can focus on the simulation in order to evaluate the performance of the testbed. In the following we want to demonstrate how to use the simulation for performance evaluation and parameter optimization of the routing protocol to achieve an acceptable video quality even in the case of abrupt disconnections. Thus, we varied the hello message interval of SBR from 1.0s down to 0.1s in steps of 0.1s and transmitted the video 100 times for each setting. The scenario simulated was again the multi-hop setup with three abrupt disconnections in a 15s interval. The exact

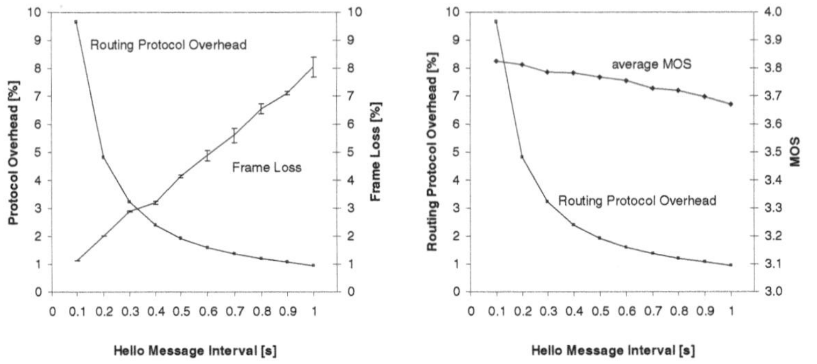


(a) Frame loss (b) MOS

**Fig. 8.** Hall - Average frame loss (a) and average MOS (b) against the routing overhead

disconnection times were equally distributed in a window of $\pm 1$s to avoid the exact alignment with a key frame.

Figure 8 shows the resulting average frame loss as well as the average MOS against the overhead of the routing protocol in percent of the video traffic for the hall clip. Though the frame loss varies between 1% and 8% the average MOS only varies between about 3.7 and 3.8. The reason for this is that each lost frame can influences the following frames up to the next key frame. Figure 10 shows the percentage of frames with a certain MOS. Due to the fact that the expressiveness of the average MOS is limited in case of longer videos. Figure 9 shows the frame loss and MOS statistics for the high-motion scene sw3. Although the frame loss rate is not higher than in the low-motion hall clip, the average MOS is suffering more from the losses. This results from the higher differences between adjacent frames which lead to a higher sensitivity to lost frames. Another factor is the appearance of frames with a very low MOS (1-2). In fact, the disturbances of the

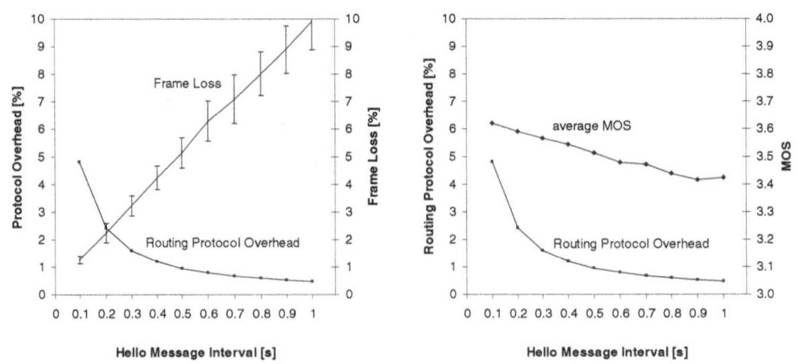


(a) Frame loss (b) MOS

**Fig. 9.** SW3 - Average frame loss (a) and average MOS (b) against the routing overhead

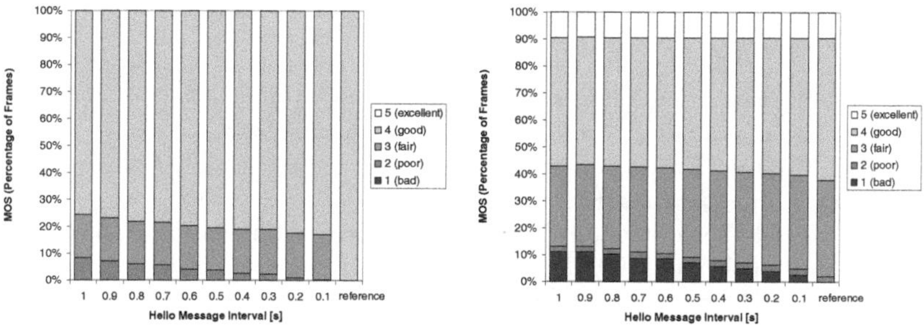


(a) hall (b) sw3

**Fig. 10.** Percentage of frames with certain MOS values depending on the hello message interval in comparison to the reference video quality

video quality are short in both investigated cases. Figure 10 shows the number of frames with a certain MOS in comparison to the undistorted reference. In contrast to the average MOS curves in Figure 8(b) and Figure 9(b), it is shown here that the quality impact on the sw3 clip is much smaller than on the hall clip, which results from the faster recovering in case of losses due to the higher number of intra-coded parts.

The overhead of the routing protocol rises exponentially with the downsizing of the hello message interval. It is acceptable up to around 2-3% of the application traffic, since this is in the range of the protocol overhead of RTP (1.7% in this scenario). The relative routing overhead is lower for the sw3 scenario since the bit-rate is higher than in the hall scenario. The MOS distribution bars in Figure 10 show that the difference in quality between the message interval of 0.3s and 0.4s is not noticeable by human observer. Considering the smaller overhead a hello message interval of 0.4s of the SBR protocol would be optimal in this scenario regarding the perceived video quality.

## 8 Conclusions and Future Work

We studied the performance of the network to deal with frequent topology changes depending on the duration of the hello message interval of the SBR protocol. The parameter hello message interval was adjusted in order to achieve an acceptable video quality which was measured by comparing the MOS value of the source and the received video.

The implemented simulation and performance evaluation framework will be used to optimize intra-aircraft applications ranging from surveillance to entertainment. Since the criticality of the applications is different, it is essential that the evaluation methodology reflects the specific requirements in terms of perceived quality. We are confident that the performance evaluation framework introduced in this study is useful for other researchers who want to assess the performance of a wide range of wireless video transmission systems.

## References


1. Akyildiz, I.F., Melodia, T., Chowdhury, K.R.: Wireless Multimedia Sensor Networks: A Survey. IEEE Wireless Communications Magazine 14(6), 32–39 (2007)
2. Network Simulator, ns-2, http://www.isi.edu/nsnam/ns/
3. OPNET Modeler, University Program, http://www.opnet.com/services/university/
4. Kotz, D., et al.: Experimental Evaluation of Wireless Simulation Assumptions. In: Proceedings of the 7th ACM International Symposium on Modeling, Analysis and Simulation of Wireless and Mobile Systems, pp. 78–82. ACM Press, New York (2004)
5. Liu, D., et al.: Simulation Validation Using Direct Execution of Wireless Ad-Hoc Routing Protocols. In: Proceedings of the 18th Workshop on Parallel and Distributed Simulation PADS, pp. 7–16. ACM Press, New York (2004)
6. Kuladinithi, K., et al.: Experimental Evaluation of AODV Implementations. In: CEWIT 2004, USA (2004)
7. Perkins, C.E., Belding-Royer, E.M., Das, S.R.: Ad Hoc On-Demand Distance Vector (AODV) Routing RFC 3561, IETF MANET Working Group (August. 2003)
8. Klein, A., Tran-Gia, P.: A Statistic-Based Approach towards Routing in Mesh Networks. In: IEEE MASS, Pisa, Italy, October, pp. 1–6 (2007)
9. Clausen, T., Jaqcquet, P.: Optimized Link State Routing (OLSR) RFC 3626, IETF Networking Group (October 2003)
10. Intanagonwiwat, C., Govindan, R., Estrin, D.: Directed Diffusion: a Scalable and Robust Communication Paradigm for Sensor Networks. In: Proceeding of ACM MobiCom 2000, Boston, MA, pp. 56–67 (2000)
11. Jpcap - Java library for capturing and sending network packets, JPCAP, http://netresearch.ics.uci.edu/kfujii/jpcap/doc/
12. Riley, M., et al.: Digital Video Communications. Artech House (1997)
13. Ohm, J.-R.: Multimedia Communication Technology. Springer, Heidelberg (2004)
14. ITU-R, Recommendation BT.500-10, Methodology for the Subjective Assessment of the Quality of Television Pictures (2000)
15. Klaue, J., et al.: EvalVid - A Framework for Video Transmission and Quality Evaluation. In: Proceedings of the 13th International Conference on Modelling Techniques and Tools for Computer Performance Evaluation, pp. 255–272 (2003)
16. Video Quality Experts Group, http://www.vqeg.org/
17. Merritt, L., et al.: x264 - a free h264/avc encoder, http://www.videolan.org/developers/x264.html
18. FFmpeg - Multimedia Framework, http://ffmpeg.mplayerhq.hu/
19. Schulzrinne, H., et al.: RTP: A Transport Protocol for Real-Time Applications, RFC 3550
20. Wireshark - Network Protocol Analyzer, http://www.wireshark.org/
21. Overview of Currently Available Routing Protocols (under construction), http://www.routingprotokolle.de/

# Author Index

FSC
www.fsc.org
MIX
Papier | Fördert gute Waldnutzung
FSC® C083411

Kolibri 360 GmbH
Ferdinand-Jühlke-Straße 7
99095 Erfurt, Deutschland
produktsicherheit@kolibri360.de